The Central European Magdalenian

Vertebrate Paleobiology and Paleoanthropology Series

Edited by

Eric Delson
Vertebrate Paleontology, American Museum of Natural History,
New York, NY 10024, USA
delson@amnh.org

Eric J. Sargis
Anthropology, Yale University
New Haven, CT 06520, USA
eric.sargis@yale.edu

Focal topics for volumes in the series will include systematic paleontology of all vertebrates (from agnathans to humans), phylogen reconstruction, functional morphology, Paleolithic archaeology, taphonomy, geochronology, historical biogeography, an biostratigraphy. Other fields (e.g., paleoclimatology, paleoecology, ancient DNA, total organismal community structure) may b considered if the volume theme emphasizes paleobiology (or archaeology). Fields such as modeling of physical processes, geneti methodology, nonvertebrates or neontology are out of our scope.

Volumes in the series may either be monographic treatments (including unpublished but fully revised dissertations) or edited co lections, especially those focusing on problem-oriented issues, with multidisciplinary coverage where possible.

More information about this series at http://www.springer.com/series/6978

The Central European Magdalenian

Regional Diversity and Internal Variability

Andreas Maier

Institute of Prehistoric Archaeology, Friedrich-Alexander-Universität Erlangen-Nürnberg, Erlangen, Germany

Springer

Andreas Maier
Institute of Prehistoric Archaeology
Friedrich-Alexander-Universität Erlangen-Nürnberg
Erlangen, Germany

ISSN 1877-9077 ISSN 1877-9085 (electronic)
Vertebrate Paleobiology and Paleoanthropology
ISBN 978-94-017-7205-1 ISBN 978-94-017-7206-8 (eBook)
DOI 10.1007/978-94-017-7206-8

Library of Congress Control Number: 2015940534

Springer Dordrecht Heidelberg New York London

Cover illustration: Artistic depiction of Magdalenian hunter-gatherers in their environment.
Image by Florian Sauer (florian.sauer@ftsauer.de).

Printed on acid-free paper

Springer Science+Business Media B.V. Dordrecht is part of Springer Science+Business Media (www.springer.com)

Foreword

On the one hand, this book presents a well-informed comprehensive overview about all aspects of Magdalenian culture in Central Europe: the site catalogue contains all essential information from about 650 assemblages. On the other hand, this book comes up with a description of the regional and chronological structure of the Magdalenian culture: the author has identified five regional groups between the Rhone valley to the west and the Vistula valley to the east, which existed roughly between 20,000 and 14,000 years ago.

The book combines two levels of information: it contains a review of essential publications on the Central European Magdalenian culture, and, at the same time, it delivers firsthand data from the author's own studies of 26 lithic assemblages from the entire investigated area.

The volume of data concerning the Magdalenian has grown enormously, and this wealth of information has made it difficult to compile a scenario of Magdalenian demographic and cultural history. The author advocates for an initial phase of Central European resettlement after the Last Glacial Maximum fed by two distinct regional sources: both southwestern Europe and southeastern Europe may have contributed to the Magdalenian population increase, leading to the reoccupation of the vast area between the Rhine and Vistula Rivers.

The reoccupation of Central Europe after the Last Glacial Maximum constitutes a major topic of the interdisciplinary Collaborative Research Center (CRC) 806 "Our Way to Europe—Culture-Environment Interaction and Human Mobility in the Late Quaternary," funded by the German Research Foundation. This dissertation had been included within the 2009–2014 CRC program. In parallel, Andreas Maier won a University of Cologne A.R.T.E.S. graduate fellowship which gave him independent funding as an associate member of the CRC. Consequently, a second CRC dissertation project has been focusing on a demographic model of Magdalenian population history.

Cologne
June 2014

Jürgen Richter

Acknowledgments

Without the enduring support and encouragement of many colleagues and friends, I never would have been able to complete this research project successfully.

First, I take this opportunity to express my sincere thanks to my supervisors, Prof. Dr. Jürgen Richter and Prof. Dr. Andreas Zimmermann.

I also owe special thanks to the A.R.T.E.S. School of Graduate Research at the University of Cologne, in particular to Prof. Dr. Andreas Speer, Prof. Dr. Michael Bollig, and Prof. Dr. Michael Quante, as well as to all members of the "Life Studies" stream for providing me with an inspiring academic atmosphere at A.R.T.E.S. and for a very fruitful time. I also thank Dr. Jörg Holzkämper for his friendly and continuous support and the CRC 806 "Our Way to Europe" for financing my research trips.

My sincere thanks likewise go to all of the colleagues who permitted me to conduct my technological analyses at their research institutions and who helped to make my stays outside Cologne very pleasant and enjoyable. I am grateful to Dr. Christophe Cuplillard, Dr. Erwin Keefer, and Dr. Karin Mansel, who granted me access to the collections of Grappin, Brillenhöhle, and Kaufertsberg, as well as to Dr. Frank Gelhausen for giving me the opportunity to study the material from Oelknitz at Monrepos. I must also thank Dr. Mario Küßner for his permission to study the assemblages of Teufelsbrücke. I am especially thankful to Dr. Marc Cartonnet for accommodating me at his guesthouse, where I could study the material of Chênelaz, and for a trip to the site. I very much enjoyed my stay at the Archaeological Institute in Prague and the interesting conversations I had there with Dr. Slavomil Vencl about various topics, including the material from Hostim, which I was allowed to analyze. I am grateful to Jaroslav Jiřík and JiříFróhlich for hosting me at the Pracheňské Museum at Písek, for granting me access to the collections of Lhota, and for showing me the site. I also want to warmly thank Dr. Petr Neruda and Dr. Zdeňka Nerudova for welcoming me at the Anthropos Institute in Brnó and for giving me the opportunity to study the collections of Balcarka Cave and Hranice; my thanks goes as well to Dr. Petr Škrdla for granting access to the collection of Mokra V. I want to express my gratitude to Halina Królik for permitting me to study the material of Wilczyce and Rydno II/59 and for providing me with delicious Polish pastries. I wish to thank Urszula Kurzątkowska for helping me organize my trip to Lublin and for mediating my contact with Tadeusz Wiśniewski, who kindly granted me the opportunity to study the material of Klementowice-Kolonia. I extend my very warm thanks to Dr. Denise Leesch for hosting me so cordially in Neuchâtel and for giving me the chance to study the assemblages of Monruz and Kastelhöhle-Nord. I had a very pleasant stay and enlightening conversations, to which both Dr. Werner Müller and Marina also contributed significantly. Likewise, I am very thankful to Julia and Edward Prescott, Judith Normann-Bausch, and Thomas Sode for hosting me so generously in Munich, Stuttgart, and Weimar.

I thank Dr. Jutta Meurers-Balke and Dr. Arie J. Kalis for initiating my involvement with the Magdalenian palynological record and for encouraging my research, as well as to Dr. Bernhard Weninger for patiently answering my many questions regarding radiocarbon dating. My most cordial thanks go to Isabell Schmidt, who so thoroughly proofread this work and whose constant support and interest, helpful criticism, and good ideas contributed substantially to its completion. For my valuable discussions with Inga Kretschmer and the exchange of ideas and information we shared, I also want to say thank you. Many thanks also go to Dr. Georg Roth for patiently answering all of my statistical questions and for providing indispensable help with R. In the same sense, I thank Dr. Andreas Bolten for helping me familiarize myself with ArcGIS. I am deeply grateful to Dr. Johanna Blokker for her proofreading of the final draft of this study, which substantially improved its readability. Thanks also go to Lee Clare for his open ear for language issues.

I also thank Eric Delson and my reviewers for the Vertebrate Paleobiology & Paleoanthropology series, whose most appreciated comments and critiques helped to improve this manuscript considerably, as well as Judith Terpos and particularly Zachary Romano for editorial work.

My best thanks go to my colleagues, friends, and family members, particularly to Julia Hecking, Helga, Marlene, Caterina, Gerda, and Karl.

My most cordial thanks to all of you!

Andreas Maier

Contents

List of Tables

List of Figures

Abbreviations

5to15	Butt size between 5.1 and 15 mm^2
A	Antler
A7	Striking angle $\leq 70°$
A7t8	Striking angle between 70° and 80°
A80	Striking angle $\geq 80°$
AC	Accuracy
Al	*Alces*
App	Appendix
Ar	Artiodactyla
ARMD	Average distance traveled during a residential move
Art	Artemisia
AS	Ability to separate
AT	Austria
Ax	*Alopex*
b	Bone
B/B	*Bos/Bison*
Bal	Balcarca Cave
BB	Backed bladelet
Bb	Bulb present
bb	Burnt bone
BBC	Rectangular backed bladelet
Bbd	Diffuse bulb
BBD	Double backed bladelet
BBDe	Denticulated backed bladelet
BBDT	Double backed truncated bladelet
BBP	Pointed backed bladelet
BBT	Truncated backed bladelet
BBTT	Bitruncated bladelet
BDR	*Baguettes demi-rondes*
BE	Belgium
Bet	*Betula*
BFr	Backed fragment
Blw	Brillenhöhle, lower layer
Bo	Borer
Boi	Bois Laiterie
BP	Before present (uncalibrated ^{14}C-date)
BSc	Bulbar scar present
Bu	Burin
Bup	Brillenhöhle, upper layer
c	Charcoal
C	Classification of the assemblage
Ca	*Capreolus*

CA	Correspondence Analysis
cal BP	Calibrated ^{14}C-date before present
Ce	*Cervus*
CEM	Central European Magdalenian
CF	Confidence
CFvec	Confidence for each (true) group
CH	Switzerland
Che	Chênelaz
Cl	*Coelodonta*
Cn	*Canis*
Co	Country
COR	Correlation values
CR	Correctness rate
Cr	*Crocuta*
Cs	Castor
CTR	Contribution
Cv	*Cervid*
CZ	Czech Republic
DE	Germany
DR	Dorsal reduction
DRh	Dorsal reduction, hinged negatives
DRhp	DR, hinged and pointed negatives
DRn	No dorsal reduction
DRp	Dorsal reduction, pointed negatives
E	East
E	Spain
E	*Equus*
EE	*En éperon* preparation present
F	Faunal Grouping (1–5)
F	France
FBU	Flat-faced burin
G	Assigned regional group
GI	Greenland Interstadial
GISP	Greenland Ice Sheet Project
Gra	Grappin (Arlay)
GS	Greenland Stadial
Gu	Gulo
h	Horn
h	Höhle
Ho	*Homo*
Hos	Hostim
Hra	Hranice
i	Ivory
Ib	Ibex
ID	Identification number
In	Indeterminate
INR	Inertia
Ipt	Impact point present
Jun	*Juniperus*
Kas	Kastelhöhle Nord
Kl8	Klementowice-Kolonia, 80ties
Kln	Klementowice-Kolonie, new
Klw	Kaufertsberg, lower layer

Kup	Kaufertsberg, upper layer
L	Local
La	*Lagopus*
Lab.-Nr.	Laboratory number
Lat	Laterally retouched piece
Lat	Latitude in decimal degrees
LBO	Long borer
LBU	Long burin
Le	*Lepus*
LGM	Last Glacial Maximum
Lho	Lhota
Lil	Lip laterally present
Lip	Lip present
Long	Longitude in decimal degrees
LU	Luxembourg
lw	Lower
M	Material
M level	Magdalenian level
Ma	*Mammuthus*
Mi	Microfauna
Mo	Mollusk
Mok	Mokra V
Mon	Neuchâtel-Monruz
n	Not determined
N	North
n Tr.	Number of Triangles
NADW	North Atlantic Deep Water
NAP	Non-arboreal pollen
NASP	North Atlantic Seabord Programme
NDP	Notched and denticulated pieces
NGRIP	North Greenland Ice Core Project
NL	The Netherlands
NW	Northwest
o	Organic
Oel	Oelknitz
OeO	Oelknitz, old
OeY	Oelknitz, young
OLUP	Older Late Upper Paleolithic
Ori. Tr.	Origin of Triangles
Ov	Ovib0s
ov15	Butt size $\geq$ 15.1 mm^2
p	Peat
Pa	*Panthera*
PB	Pointed blade
Pfs	Straight profile
Pin	*Pinus*
PL	Poland
Poa	Poaceae
QLT	Quality
R	*Rangifer*
Ris	Rislisberghöhle
RSC	Laterally retouched scraper
Ryd	Rydno II/57

S	South
Sa	Salix
Sc	End scraper
Sc	Impact scar present
Sdr	Triangular striking-platform remnant
Sfc	Facetted striking-platform remnant
Sg	*Saiga*
sh	Shell
ShP	Shouldered point
Sir	Irregular striking-platform remnant
Sln	Linear striking-platform remnant
sm	Small mammals
so	Soil
Sov	Oval striking-platform remnant
Sp	Species
SpP	Splintered pieces
SPR	Striking-platform remnant
Spt	Punctiform striking-platform remnant
Src	Rectangular striking-platform remnant
SSC	Short scraper
Ssm	Smooth striking-platform remnant
SST	Sea surface temperature
Std	Standard deviation
Sus	Striking-platform remnant "chapeau de gendarme"
t	Tooth
Te3	Teufelsbrücke, level 3
Te4	Teufelsbrücke, level 4
to5	Butt size ≤ 5 mm^2
Tri	Triangle
Tru	Truncated piece
up	Upper
Ur	*Ursus*
Vu	*Vulpes*
W	West
w	Wood
Will	Wilczyce
Zi	Zinken

Chapter 1
Introduction to the Central European Magdalenian: Area, Corpus, and Major Questions

Among the many disciplines in the humanities, Paleolithic archaeology is a comparatively young branch of research. In 1874, R. Virchow was of the opinion that "prehistory is not a discipline and will probably not become one" (Virchow 1874: VII). However, 140 years later, Paleolithic archaeology is a well-established subject at many universities worldwide and arouses interest in large parts of the general public. Nevertheless, at a time when economic efficiency holds primary importance in both society and in science and the curation of the world's cultural heritage and achievements is regarded rather a cost-intensive minority hobby than an indispensable necessity for the identity formation of the global community, Paleolithic research undoubtedly seems to many people to be the ultimate waste of public funds. To be sure, the immediate contribution of this subject to social questions currently affecting the larger global community is not always obvious. If it were true that Paleolithic investigations amounted to nothing more than a chronological succession of changing artifact quantities and inaccurately dated events, then it would generate observations which are at best interesting only to the researchers themselves, and its *raison d'être* might indeed be questionable. However, the study of past societies is much more than the establishment of a simple sequence of isolated observations; it allows each person to contextualize himself or herself against the background of the history of our species. It makes a difference, for example, if a person conceives himself as the summit of God's creation, a member of the "master race," or one element in the long and interlinked chain of human evolution. Here, the discovery of the first Neanderthal skeleton or the development of the "Out of Africa" theory are clear instances where contributions made by prehistoric archaeology influenced humankind's perception of itself in substantial ways.

It is often claimed that archaeological data, unlike the data of any other anthropological discipline, potentially holds information on all scales in the temporal, spatial, and social dimensions. Observations may stretch from individual activities at a particular time and place to the development of continent-wide populations over hundreds of millennia. It is this unique quality that yields the potential to reveal and understand the basic and long-term processes—normally unnoticed during everyday activities—that structure human ways of life. Nonetheless, despite this general potential, the attempt of gaining knowledge of the nature of human societies that exceeds the quality of commonplace statements from Paleolithic data has proven itself a difficult task. As a matter of fact, it is more often the case that knowledge from other disciplines of the social science, particularly ethnography, is used to interpret the Paleolithic record than it is the other way round (cf. Wobst 1978). Thus, the question remains as to in how far prehistoric archaeology is capable to contribute to social and cultural debates?

More than 40 years ago, Fritz and Plog already suspected that "unless archaeologists find ways to make their research increasingly relevant to the modern world, the modern world will find itself increasingly capable of getting along without archaeologists" (Fritz and Plog 1970: 412). However, many studies in Paleolithic archaeology are still purely descriptive and characterized by what Wylie described as the "paralyzing cautiousness" of "narrow empiricism" (Wylie 1985: 69). A simple positivistic gathering of empirical "facts" does not constitute in and of itself a meaningful arrangement of data. Neither can mere tallies of artifacts give deeper insight into the nature of past societies, because "percentage frequency of stone tool types is not an appropriate analytical measure for the investigation of something as complex as social organization" (Stiles et al. 1979: 3). If archaeologists wish to make a relevant contribution to current debates in the social sciences, they must go beyond pure description and find ways to produce meaningful interpretations of the excavated remains. Therefore, productive archaeological research always depends on hermeneutics. It is important to emphasize that hermeneutic reasoning does not mean to indulge in free speculation (see Seiffert 1991: 104f.); on the contrary, it is "based on extensive verification of sources and has an indirectly but markedly data-oriented component" (Bernbeck 1997: 279).

Andreas Maier, *The Central European Magdalenian*, Vertebrate Paleobiology and Paleoanthropology,
DOI 10.1007/978-94-017-7206-8_1

The necessity of working with interpretations, hypotheses, and analogies should not lead to the conclusion reached by Tilley that "the hallmark of all archaeological writing is the imaginative faculty" (Tilley 1996: 339). Even though it is true that "we are not transparently representing the real in our discourses," Tilley seems to carry his line of argument too far by saying that "our relationship to the past is a metaphorical one, and can be no other" (Ibid.). Nor does it seem justified to reinterpret archaeological research as sublime speculation, amounting to nothing more than an endlessly changing conglomerate of "rhizomes, montages and collages, Bartelbooth's jigsaw puzzles, actant-networks, and hypermedia documents," and to recommend that every written page would be best dipped "in a detergent solution, from which clean and unmarked sheets of paper will emerge" (Holtorf 2000: 175).

There can never be certainty about results in archaeological research. However, this realization should not be taken as an invitation to abandon all constraints of scientific procedure and to practice archaeology as a new form of literary action and randomized creative writing. Considering the epistemological problems discussed below, it is vital to insist on a disciplined way of generating questions and answers that operates as closely as possible to the archaeological evidence. In his work "Truth and Method," Gadamer reasons as follows:

> Thus there is undoubtedly no understanding that is free of all prejudices, however much the will of our knowledge must be directed toward escaping their thrall. Throughout our investigation it has emerged that the certainty achieved by using scientific methods does not suffice to guarantee truth. This especially applies to the human sciences, but it does not mean that they are less scientific; on the contrary, it justifies the claim of special human significance that they have always made. The fact that in such knowledge the knower's own being comes into play certainly shows the limits of method, but not of science. Rather, what the tool of method does not achieve must—and really can—be achieved by a discipline of questioning and inquiring, a discipline that guarantees truth. (Gadamer 2006: 484)

As an attempt to review the archaeological record of the Magdalenian hunter-gatherers in Central Europe, this study is essentially concerned with basic research, which reveals its benefits usually only in the long run. However, by applying a disciplined way of questioning and inquiring, this study also wishes to provide a meaningful interpretation of the collected data. The Magdalenian, characterized as it is by extensive repopulation after the Last Glacial Maximum, by the formation and maintenance of land-use patterns and networks, and finally by the gradual transformation toward the Late Paleolithic, is a potential source of new insights on topics of more universal concern, such as colonization processes, regional variability, cultural transmission, and interaction patterns. Thus, the results presented offer many points of contact for research in adjacent fields and provide a rich base for further reflections and specific case studies about the nature of human societies, as well as other currently debated topics in social and cultural sciences.

Since fieldworks began at the eponymous rock shelter of La Madeleine in 1863, 150 years of research have made the Magdalenian one of the best known Paleolithic entities. Many things have changed since these early days. Excavation techniques and methods of documenting finds have improved significantly (e.g., Erdmann 1980; Gersbach 1989; Harris 1989), and the introduction of radiometric dating methods has enhanced the understanding of past processes significantly. Moreover, there have been lively debates about archaeological models, theories of cultural transmission, and the diffusion of concepts in human societies as well as about interpretive approaches, which have led to the transformation of many stereotypes about the nature of prehistoric people and their ways of life (e.g., Sollas 1924; Childe 1951; Clarke 1972b and articles therein; Binford 1980; Hodder 1982b; Wylie 1985; Shanks and Tilley 1987; Ucko 1995 and articles therein; Bentley and Shennan 2003; Eerkens and Lipo 2005).

Thanks to detailed analyses of numerous individual sites in recent decades, ample information on many different topics is available in the archaeological literature. These high-resolution studies of discrete artifact concentrations have been complemented by a series of comparative studies analyzing the internal spatial and temporal structure of the Magdalenian. However, these studies vary widely in both their spatial scale and richness of detail. Generally speaking, investigations addressing larger regions tend to use more coarse-grained data, whereas studies with a higher-resolution deal with small regions only (cf. Valoch 1960, 2009; cf. Hanitzsch 1969: 189; Feustel 1974: 191ff.; Weniger 1982; Bahn 1985; White 1985; Rozoy 1988b; Otte and Keeley 1990; Eriksen 1991; Otte 1992; Rensink 1993; Jochim et al. 1999; Połtowicz 2006; Küßner 2009: 192ff.; Sisk 2011). The present study, in contrast, is the first attempt to analyze the Magdalenian record in the large-scale region of Central Europe using high-resolution data.

The Term "Central Europe" in This Study

Central Europe is not a distinctively defined geographical region, nor is it a clear political unit. Rather, its boundaries change over time and according to political opinions. In this study, the term Central Europe refers to the area covered by the states of the Netherlands, Belgium, Luxembourg, eastern France, Germany, Switzerland, Austria, Poland, Czech Republic, and Slovakia. The inclusion of Belgium, the Netherlands, Luxemburg, and eastern France might seem idiosyncratic at first glance. Indeed, by today's standards, it does not correspond to the most quotidian use of the term, although the Benelux states are occasionally considered a part of Central Europe (cf. Ahlheim and Preuß 1983). From a geomorphological point of view, the chosen delimitation respects the contiguity of the mid-latitude mountains of the French Jura, Vosges, and Rhenish Massif, and from a vegetational point of view, it coincides roughly with the extent of the floristic region "Central Europe" (Frey and Lösch 2010). Most important, however, is the fact that this definition has not been chosen to fit present-day political opinions or to reflect precisely environmental conditions, but to meet the terms which are given by the spatial distribution of Magdalenian sites itself.

Here, it is important to stress that the way in which the borders of an investigation area are defined has considerable influence on the outcome of research, especially when artificial boundaries are imposed on the object of study. Because of the nonuniformitarian distribution of sites, the problem of edge effects may distort the result significantly (O'Sullivan and Unwin 2003: 34; see Chap. 2, section "Spatial Data"). Edge effects arise because sites which are located at the edges of the investigated area can only be analyzed in relation to sites located toward the center of the area, whereas more central sites allow for analyses in all directions. Moreover, consistent site clusters may be divided by arbitrary boundaries. Eventually, the resulting distortions may bias the analytical results considerably and thus obscure prehistoric reality. To avoid such bias, an attempt has been made to draw the borders in closest possible accordance with the archaeological record. Therefore, the distribution pattern of the Magdalenian sites itself is chosen as a basis for the definition of the boundaries of the investigated area, although it is certainly true that the assignment of a site to an archaeological entity already implies an act of interpretation.

Taking this distribution as a starting point (Fig. 1.1), three out of four borders are already virtually self-defining. The easternmost Magdalenian sites known so far are found along the Vistula and San rivers in Poland. However, raw material analyses from several sites indicate site catchment areas extending eastward as far as the banks of the Bug river and the headwaters of the Dniester river. Consequently, these currents delimit the research area to the east. The southern border is set along the southern fringe of the Outer Western Carpathians and the northern slopes of the Alps. In the north, the margin of the distribution of sites roughly coincides with the northern edge of the Central European low mountain range.

So far, the delimitation strictly respects the distribution of Magdalenian find-spots, and no sites are excluded. For the

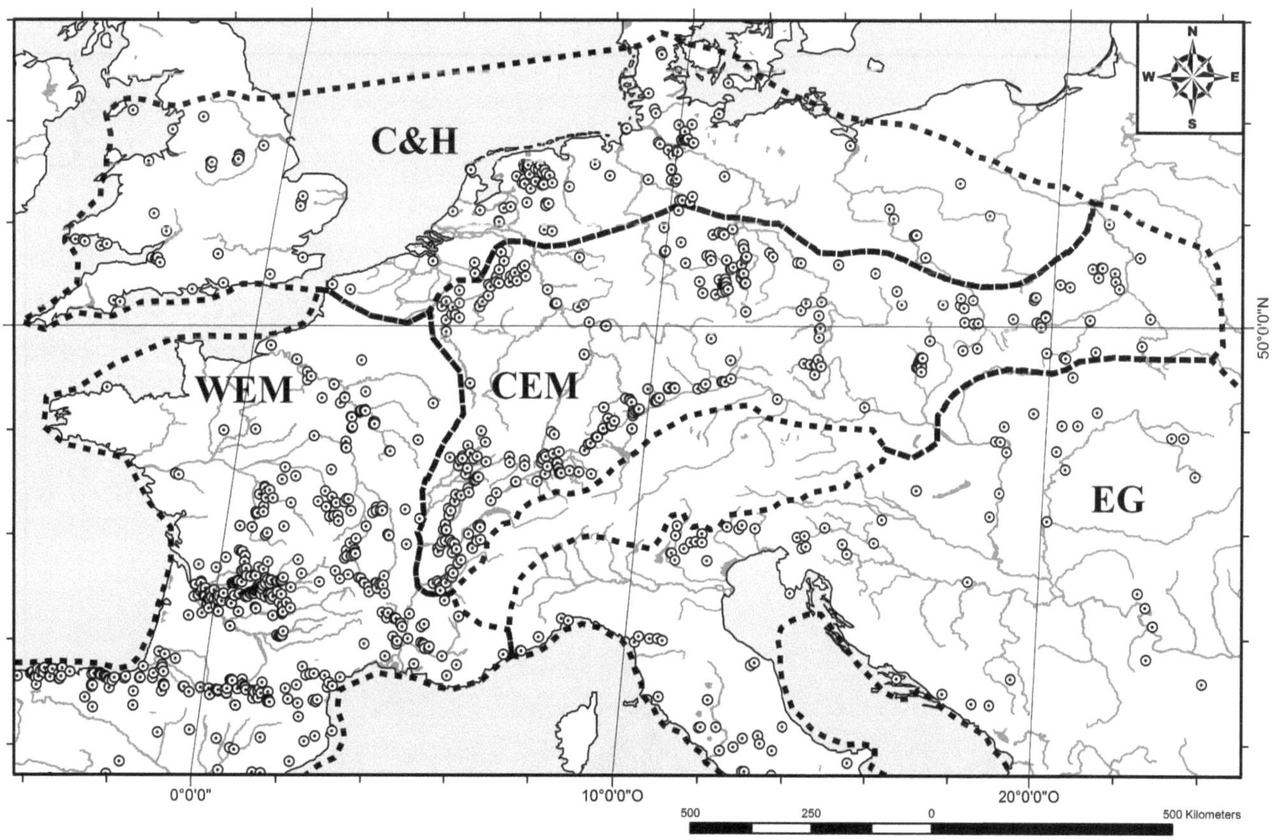

Fig. 1.1 Distribution of Late Upper Paleolithic sites in Western and Central Europe

determination of the western border, however, this practice cannot be sustained. In order not to divide any obvious site clusters, the western border is set along the currents of the Isère, Rhône, Saône, and Meuse rivers. This demarcation both respects the integrity of site clusters in the Rhône valley and in the Ardennes and accounts for the spaces without any known Magdalenian sites that separate these clusters from those in the Massif Central and the Paris Basin (see Fig. 1.1). All together, the investigated area stretches about 1600 km from east to west and 400 km from north to south, covering approximately 600,000 km^2.

Around 15,000 cal BP, Hamburgian and Creswellian hunter-gatherers settled in the northwestern European Plain and the present-day British Isles and coexisted with the later Magdalenian (Barton et al. 2003; Grimm and Weber 2008; Jacobi and Higham 2009). In contrast to Northern Europe and the area of the CEM itself, the bordering regions in the east and the south were occupied continuously by hunter-gatherers throughout the Middle and Late Upper Paleolithic. This continuity is expressed in the terms "Epigravettian" and "Epigravettiano" for the Late Upper Paleolithic entities of Eastern Europe and Italy, respectively, which were coined to emphasize the relation to the preceding Gravettian. However, especially in regions where both entities occur, it is almost impossible to decide typologically whether a site belongs to the Magdalenian or to the Epigravettian. Nevertheless, for the definition of the investigated area, the classification given in the literature is adopted to delimit the eastern border of this study.

The Sites Assigned to the Central European Magdalenian

Having set the geographical and archaeological demarcations of this study, it is possible to give a first overview of the recorded assemblage corpus. Since this study aims at a comprehensive analysis of the CEM record and since the number of sites and assemblages is much too large to be (re)analyzed by the author, this study has to relay in many parts essentially on information published in the archaeological literature (for methods and problems, see Chap. 2). Generally, every assemblage assigned to the Magdalenian or with a supposed Magdalenian component was recorded. Assemblages which are assigned to the Upper Paleolithic in general, however, were not considered. In doing so, 651 Magdalenian assemblages from 540 sites were identified (Fig. 1.2) and included in the database (Tables A.1 and A.2). Here, information was gathered with regard to their research history, geographic con-

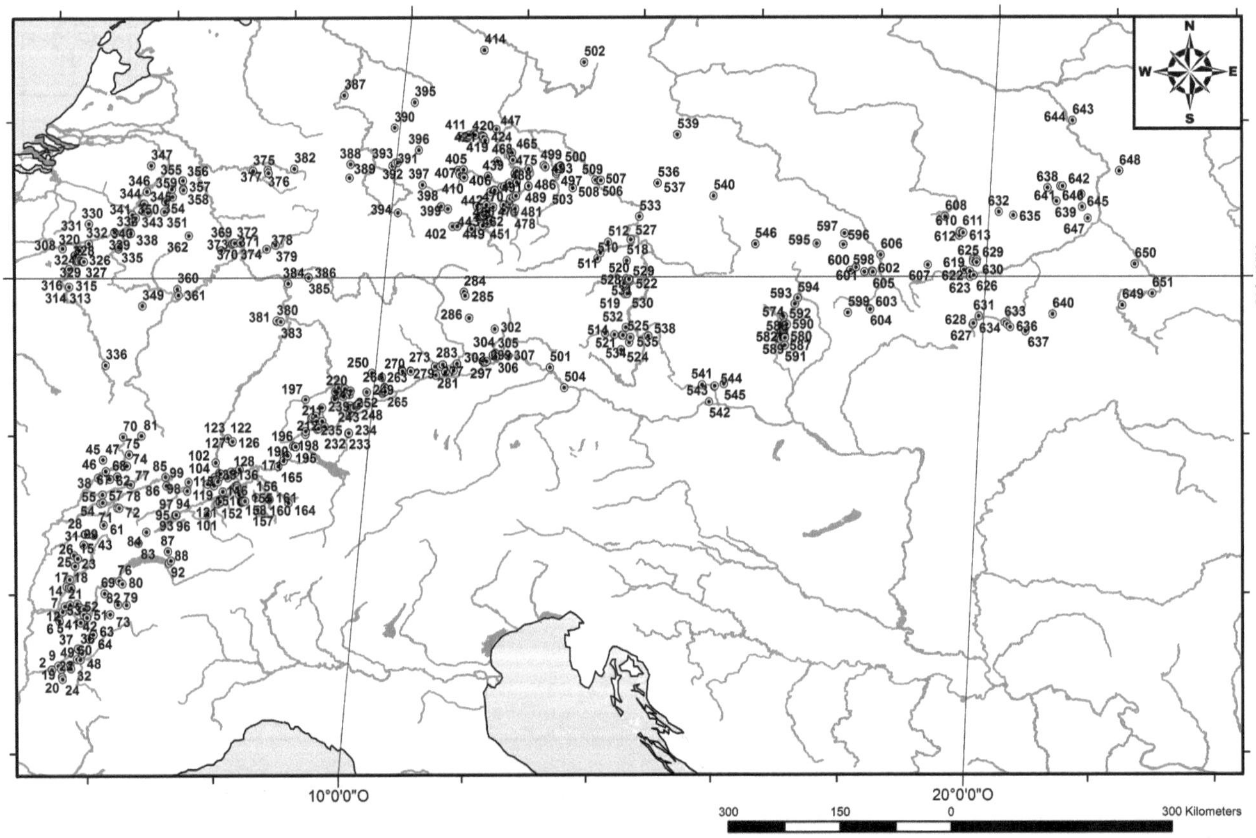

Fig. 1.2 Distribution of all recoded assemblages

text (including location and setting), raw material and mollusk shell spectrum, tool typology, blade production technology, ^{14}C-dates, faunal remains, and sculpted and engraved objects.

Additionally, 171 Hamburgian inventories from 161 sites were gathered as a reference sample for which only the geographic coordinates and the tool inventory were recorded (Table A.3).

Eventually, the database comprises thus information on 822 assemblages from 701 archaeological sites. Furthermore, data from 59 palynological sites was recorded for an analysis of spatial and temporal variability of the Central European vegetation during the Magdalenian (Tables A.4 and A.5).

Purpose and Conduct of the Investigation

Analyzing the Central European Magdalenian as a whole using a large-scale perspective together with a very fine-grained and detailed data base generally aims at gaining deeper insight in the nature and inner structure of the Central European Magdalenian in order to better understand the life-ways of Late Upper Paleolithic hunter-gatherers. The first step toward this goal is to better understand the spatial diversity and internal variability of the archaeological record. The basic question in this study therefore is whether or not regional groups of some sort can be detected in the CEM record and, if yes, how these groups interrelate. To answer this question, it is necessary to seek for spatial and temporal trends in the CEM record and to analyze how these trends are constituted. Thereby, it will be possible to identify differences and similarities between certain regions. If regional groups can be identified, a number of secondary questions arise, such as what can be said about the nature of these groups? What do the observed results imply for the history of the recolonization of Central Europe? What does the Central European Magdalenian stand for in terms of social relations and how can the obtained results be interpreted meaningfully with regard to sociocultural topics such as cultural transmission, communication, interaction, or territoriality?

In order to keep the conduct of this study as comprehensible as possible and to clearly separate the analytical from the interpretive aspects, it is divided into three parts.

Part I explains the methodological and theoretical basis of this investigation prior to analysis, discusses the methods and problems of collecting data for a large-scale, literature-based study (Chap. 2), and makes explicit the theoretical assumptions guiding the investigation (Chap. 3). Part II then contains the actual analyses and investigates the site pattern and inner structure of the CEM on different scales and from different analytical angles. It starts with Chap. 4 by reviewing past research on the CEM with a special emphasis on the temporal significance of lithic triangles. In Chap. 5, the spatial and temporal variability of climatic, vegetational, and faunal data is investigated to obtain a better understanding of the environmental setting of the CEM. Chapter 6 contains the analyses of the overall geographical pattern, typological composition, and technological features of the recorded sites and is the centerpiece of this study. It starts by critically reviewing the sites assigned to the CEM and aims at detecting possible factors determining the large-scale structure of the overall distribution pattern of CEM sites. Then, the raw material procurement pattern is examined, and regional interaction patterns are used to infer regional groups. In a second step, potential sources of fossil and subrecent marine mollusks are examined to detect supra-regional interaction patterns. Subsequently, the diversity in the spatial distribution of typological, technological, and representational concepts is investigated and counterchecked with the previously found results. In Part III, the results of all these different and independent analyses are then compared and interpreted. Chapter 7 focuses on territories and land-use patterns, and Chap. 8 investigates the implications for the recolonization of Central Europe. Here, the current theory of the recolonization of Central Europe after the Last Glacial Maximum is presented, and four major objections are raised to it. Eventually, a new hypothesis more consistent with the archaeological record is presented. Chapter 9 eventually aims to give a meaningful interpretation of the social system of the CEM.

Part I
Methodological and Theoretical Framework

Chapter 2
Collecting Data for Large-Scale, Literature-Based Studies

The available data on the Central European Magdalenian (CEM) is—just as for any other Paleolithic entity—by its nature a source of knowledge with multiple biases. This holds particularly true when the collection of data has to rely on information from the archaeological literature. These biases need to be made explicit, but they should not discourage the researcher to address those questions which demand for the investigations of ample sets of data and seek to detect and explain large-scaled spatial and temporal processes. This chapter therefore discusses several problems of data collection as they occur during the buildup of databases from site publications for large-scale investigations. It starts by highlighting actual and potential biases during the collection of data. Subsequently, the purposes and methods of the data collection is presented with regard to nine major aspects: (1) site history and documentation, (2) geographical data including location and setting of a site, (3) tool typology, (4) blade production technology, (5) site catchment with regard to raw material procurement and far-distance imports of mollusk shells, (6) sculptures and engravings, (7) faunal remains, (8) climatic and palynological data, and (9) radiometric dating. Each of the aspects mentioned is related to a set of specific problems which need to be considered for a sound and comprehensible verification of the validity and diagnostic capacity of the collected data.

Addressing the question of regional variability in a large-scale region such as Central Europe requires a database that is both comprehensive in its general scope and detailed with regard to the investigated aspects. Since such a database comprises a great amount of data, it was unavoidable in practical terms to focus on selected aspects (see below) during data collection. Thus, a deductive rather than an inductive approach was chosen for the collection of data. While inductive sampling attempts to collect an ample set of basic data which is unbiased in terms of preliminary considerations, deductive sampling emphasizes the necessity of generating specific questions and hypotheses prior to the process of collecting data (Hill 1972; Collins 1973: 54; Bernbeck 1997: 49ff.). The debate between proponents and opponents of these approaches has often been heated (Kaplan 1964: 268; Hanson 1967; Nagel 1967: 10; Binford 1972; Hill 1972; Clarke 1972a). Researchers working deductively have been characterized as ignorant of the archaeological material, whereas those working inductively have been portrayed as naïve and unaware of the theoretical background of their own interpretations (Bernbeck 1997: 61). However, "since no research is entirely inductive or entirely deductive" (Hill 1972: 62), a strict division between these two approaches is a "false dichotomy" (Ibid). That said, it is nevertheless worth calling attention to the existence of these categories because it certainly makes a difference which approach prevails in a study.

Working with archaeological data always implies accepting to work with a set of data that is incomplete and whose gaps and inconsistencies often defy evaluation. First of all, the units of potential information are almost exclusively excavated stone, bone, and antler finds and therefore reflect only a very limited range of Magdalenian daily life. Second, the transmission of these finds is itself fragmentary and depends on factors that operate randomly on the archaeological record, such as site formation processes, organic preservation, and which sites are detected in the first place (Rigaud and Simek 1987; Schiffer 1987). Thus, even before archaeological fieldwork starts, time and nature have created a sample, the quality and selection criteria of which often remain unclear to the researcher. With the onset of archaeological activities, the extent of imponderabilities usually increases. In most cases, it is very difficult to assess the unknown assortment brought about by the size and methods of the excavation or the random samples of surface collections.

Because it is not possible to review all relevant assemblages personally within the course of this triennial investigation, this study had to rely essentially on the archaeological literature in order to access the necessary information.

Andreas Maier, *The Central European Magdalenian*, Vertebrate Paleobiology and Paleoanthropology,
DOI 10.1007/978-94-017-7206-8_2

All in all, more than 1300 articles and monographs were analyzed (for a detailed list of site-specific literature, see Table A.1). These texts originated over a period of about 150 years and deal with sites situated in nine European countries. However, collecting data from the literature is a complex process. Besides the problems that arise from the reading of foreign language texts—the investigated texts were written in Spanish, French, English, Dutch, German, Czech, and Polish—the heterogeneous quality of information was a major obstacle for the collection of data. Further complicating matters is the usage of different and sometimes incommensurable classification systems for the description and analysis of the finds as well as nonempirical and paradigmatic assumptions about the nature of the archaeological record (see Chap. 3, section "Types as Analytical Units"), which guide the interpretation of the excavated objects. Other challenges arise from the strong increase in collaboration between Paleolithic research and other scientific fields that have developed in recent years. The incorporation of data from other disciplines such as ethnography, paleoclimatology, or palynology has become a common procedure and an indispensable factor for the comprehension of past societies and their development. However, the interdisciplinary use of data holds as well some pitfalls and should not be done without taking into account the specific methods and problems of the disciplines involved (see section "Environmental Data").

In order to account for the problems arising from large-scale, literature-based collection of information, specific adaptations are necessary to overcome the disparate quality of the available data and to prepare and transform them into a consistent and analyzable record set. Below, these adaptations are explained in detail for each type of gathered data. Figure 2.1 shows the call form for the recording of information on background data and most of the cultural data. Recording strategies for the remaining types of data are explained in the respective sections.

Background Data

Data on Site History and Documentation

For every site, information to its research history, such as the excavator(s) and date(s) of excavation(s), were noted. Additionally, the relevant bibliographic references were listed, and indications were given as to the availability of artifact drawings and the completeness of information on the lithic artifacts. Although these aspects were not directly an object of investigation, they are indispensable for the verification of sources and the assessment of the reliability of the sampled information.

Site

Research History	**Geographic Information**			**Dating**	**Raw material**	**Mollusk Shells**
Excavation date(s)	**Location**	**Setting**		Date BP	Name	Species
Excavator(s)	Longitude	Cave	Valley	Material	Proportion %	Origin
not excavated	Latitude	Rock shelter	Slope	Lab.-Nr.	Origin	Source distance
Level of publiction	Altitude	Open Air	Top	References	Source distance	
References						

Assemblage	**Fauna**	**Vegetation**		**Sculptures**	**Engravings**
Level/Concentration	Species	*Pinus* %	Poaceae %	Material	Material carrier
Size in m²	MNI	*Betula* %	*Artemisia* %	Representation	Representation
Classification	NISP	*Juniperus* %	NAP %		

Finds		**Tool classes**		**Tool types**		**Technological Features of Blades**	
n lithic artifacts	n organic tools	Sc	Tru	SSC	BBDT	Dimensions	Lip
n raw nodules	n points	Bu	Lat	RSC	BBTT	Preservation	Bulb
n chunks	n needles	Zi	BB	LBU	BBC	Cortex cover	Bulbar scar
n cores	n barbed points	Bo	BP	FBU	BBP	SPR Kind	Esquillement du bulbe
n flakes	n awls	NDP	Tri	LBO	BBDe	SPR Form	Impact point
n blades	n B. demi-rondes	SpP	ShP	BBD		SPR Size	En éperon preparation
n lithic tools	n bâtons percés	PB	BFr	BBT		Dorsal reduction	

Fig. 2.1 Call form for the recording of data used in this study (for abbreviations, see List of Abbreviations)

Geographical Data

For each site, a set of geographic information was recorded. It comprises its location and its setting. The location of a site is rather concerned with numerical, spatial data, whereas the data describing the setting of a site is rather qualitative. Because the nature of both kinds of data differs considerably from one another, their specific problems will be discussed separately.

Spatial Data

Because this study is primarily concerned with the spatial variation within the CEM, information on the position of a site is basic to all subsequent analyses. In order to pinpoint the geographic position of a site, its longitude and latitude (in decimal degrees) and altitude (in meters above sea level) were recorded (see Fig. 2.1 and Table A.1). If the accurate location could not be obtained from the literature, geographical description and information on the municipal assignment were used to localize the site on digital, geo-referenced maps as precise as possible. The three-dimensional coordinates of the location were subsequently used as an approximation of the true position of the site. The offset between the approximated and the true location introduces a bias which stays within the limit of a few kilometers. With regard to the scale of this study, this bias is regarded to be negligible for both the distribution maps produced in the course of the investigation and the analysis of spatial trends. The distortion of altitudes probably changes according to the overall appearance of the landscape and may be especially pronounced in areas with a marked relief where elevation varies considerably within a comparably small area. The offset for the degrees of longitude and latitude, in contrast, is supposed to be rather invariant with regard to the landscape.

Comparing the geographical distribution of archaeological remains is one of the basic methods employed in research on prehistory. Not surprisingly, distribution maps have become a widely applied instrument throughout all fields of archaeological research. A major function of such maps is the visualization and subsequent interpretation of patterns. Because this study also relies to a large extent on the interpretation of distribution maps, it seems advisable to discuss some specifics of spatial data in brief.

Two phenomena which are important for the interpretation of spatial data are the nonuniformity of space and spatial autocorrelation (O'Sullivan and Unwin 2003: 28f.). Nonuniformity describes the simple fact that geographical space is not a homogeneous unit, but differs considerably from one place to another in terms of its physical properties, such as elevation, geology, and hydrology. Spatial autocorrelation describes the fact that events occurring close to one another are more likely to exhibit similar characteristics than are widely separated events (Ibid.). This observation certainly holds true for geographic data, and it seems to be a reasonable assumption for archaeological data as well, although this may not always be the case.

Pattern-generating processes are usually expected to differ from random ones in two ways, namely, in terms of their first-order and second-order effects. First-order effects reflect "variations in the density or *intensity* of a process across space" [Ibid.: 79 (italics in original)]. If, for instance, the mapped events are animals which rely on a certain kind of plant for their subsistence, then it is highly likely that there will be a clustering of events in zones where these plants are commonly found. For this reason, the absolute location of an event is an important factor, since the event receptiveness of the investigated area varies with variations in the environment across space. Second-order effects, on the other hand, reflect interaction between locations, and thus relative position becomes more relevant (Ibid.). These interactions may result in the suppression or stimulation of nearby events. One instance is the presence of a powerful town, which is likely to suppress the development of other powerful towns in its immediate vicinity. The diffusion of Neolithic techniques and traditions, on the other hand, can serve as an example of the stimulation of nearby events. At bottom, it seems advisable to keep in mind that both first-order and second-order effects may have a strong influence on the generation of observed patterns, even though it is certainly difficult (or sometimes even impossible) to distinguish them from one another, especially when the analyzed patterns are complex.

When using a geographic information system (GIS) to analyze point patterns, the observed set of events must take account of certain complicating factors as well as fulfill certain requirements. In the following, the factors and requirements relevant to the study at hand are discussed using the terms established by O'Sullivan and Unwin (2003: 78f.).

Mapping on the Plane

When compensating for the distortions introduced by the map projection used, point pattern analysis can be inadequate if the events are scattered over a very wide geographical area. The method applied must therefore take this problem into account. For the questions with which this study is concerned, the accuracy of the geographic position of the events is considered sufficiently high, even with the distortions introduced by the projection.

Objective Determination of the Study Area and the Edge Effect

The way in which the boundaries of an investigated area are drawn can result in different analytical outcomes. Thus, their definition should be as objective as possible and should preferably respect the existing contours of the phenomena under investigation. So-called edge effects arise where an artificial boundary is imposed: events located close to the arbitrary boundary are stripped of the contextual data provided by territory which happens to lie outside this boundary. The resulting asymmetry in the data must of course be accounted for. For the study at hand, this problem was taken into consideration during the process of defining the boundaries of the investigated area (see Chap. 1).

One-to-One-Correspondence Between Objects and Events

Each event should only correspond to one point object. In this study, every site is represented by a single event on the map. If sites contain more than one Magdalenian assemblage, each assemblage is treated as an independent event.

Precise Indication of the Location of Events

The events in the pattern should display the location of the objects precisely, that is they should not be centroids of areal units or arbitrary points on line objects. Except for deviations introduced by the imprecise measurement of site coordinates, all events dealt with in this study are placed in their geographically correct position within the sites they represent. With respect to the scale of the study, the amount of deviation seems negligible.

Representative Enumeration or Census of the Objects of Interest

The pattern should not represent a sample but should embrace all relevant entities in the study region, since many methods of point pattern analysis react very sensitively to missing events. This requirement is highly problematic for archaeological research. Since we cannot know either the total number of Magdalenian sites that ever existed or the full extent of the possible biases influencing the actual pattern, it is well-nigh impossible to estimate the degree to which this requirement has or has not been fulfilled. Fortunately, the results of more than one century of research can be assumed to provide a not unrepresentative glimpse into the settlement pattern of the CEM (see Chap. 6, section "Possible Factors for the Formation of the Large-Scale Site Pattern"). Uncertainties about the reasons for gaps in the distribution of sites will probably remain one of the major problems in archaeology, however.

Landscape-Related Data

The second kind of geographical data which is important for this study concerns the setting of sites. Generally, three kinds of setting are distinguished, namely, cave sites, rock shelter sites, and open-air sites. This classification is of interest because all three kinds of sites are differently prone to recurrent occupation and consequently to the formation of palimpsests. It is assumed that locations at a prominent geographical setting, which provide a natural shelter and at the same time a rather limited amount of space, as it is often the case for rock shelters and for caves in particular, are especially prone to recurrent stays of hunter-gatherers and hence assumingly foster the archaeological visibility of these people and the formation of palimpsests. Information on this aspect was available for all sites. Statements on other landscape-related features, such as the distance to the next river or whether the site is located at a valley bottom or on a plateau, are, in contrast, much scarcer and often difficult to extract from the descriptions of the sites' geographical setting. Because these particulars are crucial for the reconstruction and interpretation of the land-use pattern of hunter-gatherer groups, the respective data often had to be obtained from digital maps subsequent to the mapping of the sites which was performed according to the recorded spatial data.

Radiocarbon Data

Without doubt, ^{14}C-dating has significantly improved the chronological framework of the Late Upper Paleolithic. The advent of AMS dating in particular has enhanced the measurement's precision in a substantial way. Today, an ample set of ^{14}C-dates for the CEM is available. For each assemblage, the existing radiocarbon dates were listed together with the respective laboratory number. Where possible, the nature of the dated material (such as bone, antler, or charcoal) as well as the species (such as horse, reindeer, or pine tree) was noted together with the specific bibliographic references. In total, 502 radiocarbon measurements from 126 sites could be recorded, about the half of which being AMS dates (see Table A.6 and the references therein).

However, the quality of these dates is quite variable. There are a number of factors influencing both the quality and the archaeological significance of an individual measurement, among them the reliability of the (stratigraphic) association between the prehistoric occupation and the dated sample and the appropriateness and accuracy of the sample's chemical and physical treatment in the laboratory (for a more detailed discussion, see e.g., Keller 1987; Housley et al. 1997).

Not all of the problems associated with the archaeological application of ^{14}C-dates are resolved simply by meeting the requirements for a reliable measurement, however. A methodological question which is still under debate is whether or not ^{14}C-dates should first be calibrated before they are used to make archaeological arguments (e.g., Housley et al. 1997, 2000; Blockley et al. 2000a, b). In this study, the view is taken that ^{14}C-measurements absolutely must be calibrated if they are to be more than vague estimations of unconnected events. The radiocarbon timescale is a nonlinear sequence which is based on fluctuating atmospheric ^{14}C-concentrations (Fig. 2.2). As a consequence, linear calendar time is deformed by compressions and expansions within this sequence. Hence, 500 ^{14}C-years usually do not correspond to 500 calendar years, but may comprise a larger or smaller interval on the calendar time scale. Thus, the relationship between two dated events can only be evaluated appropriately if their relative distance on the calendar time scale is known. The fact that the standard deviations of calibrated dates are often larger than those of uncalibrated dates does not mean that calibrated dates are less precise, since the increased standard deviation only accounts for the deformation of calendar time within the ^{14}C-time scale. Especially when it comes to the reconstruction of diachronic processes, the relative position of events on the calendar time line is the most important information. The usage of uncalibrated dates as the basis for arguments in diachronic debates obstructs the understanding of the event's true chronological position and simulates both false relations between the events and a pseudoprecision of measurements which does not exist in calendric reality. The calibration of ^{14}C-dates is therefore vital for a true understanding of diachronic processes and developments. Proponents of the use of uncalibrated dates often refer to inconsistencies in the various approaches to calibration and deficits in the current calibration curves (e.g., Housley et al. 2000).

However, when dates are not comparable with regard to their true position on the calendar time scale, the entire task of measuring ages becomes pointless. It thus seems much better to work with imperfect calibration curves, whose weak

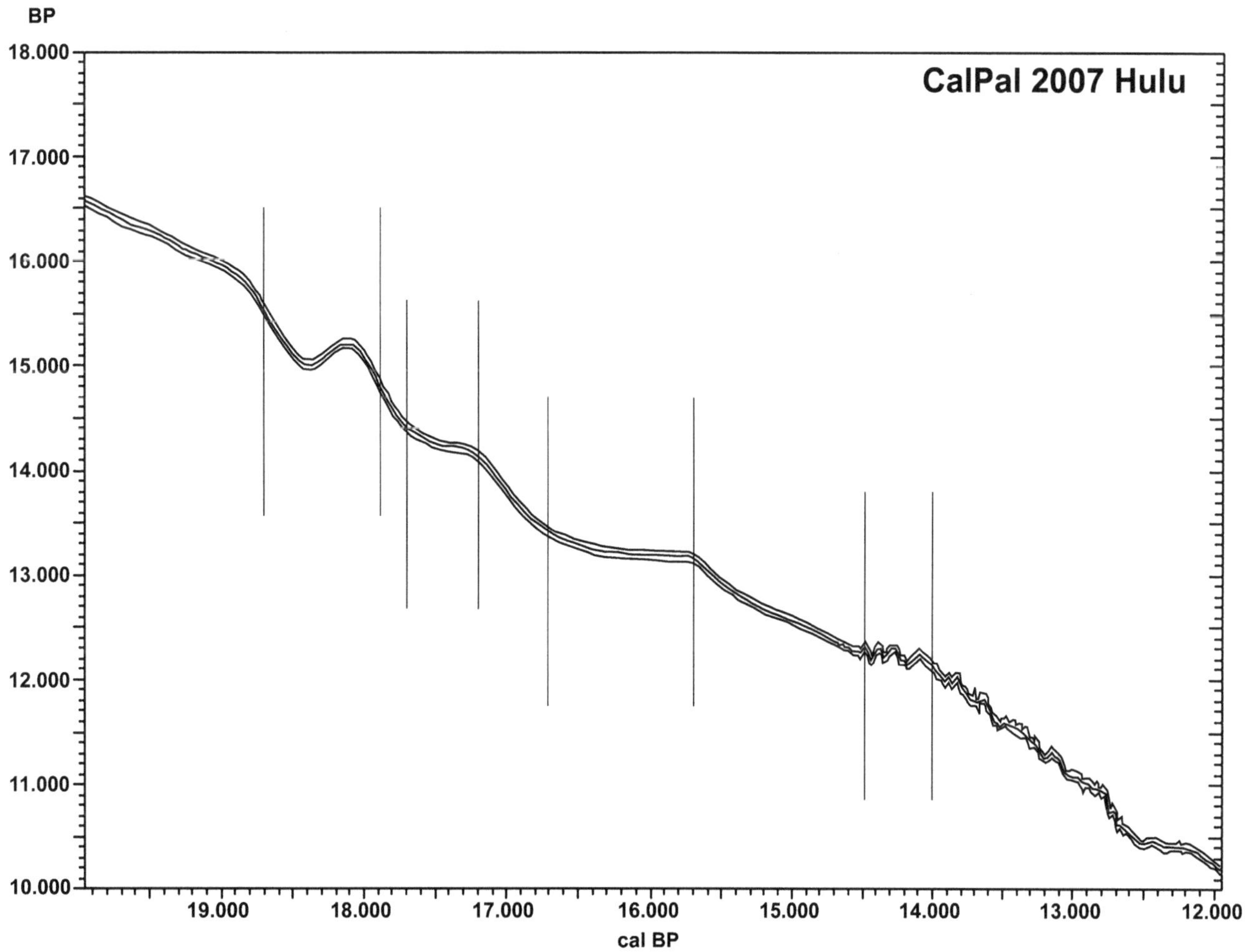

Fig. 2.2 Late glacial radiocarbon calibration curve. Pairs of lines enclose plateaus and oscillations: CalPal 2007 Hulu (Weninger et al. 2012)

spots are made explicit, than to use uncalibrated dates with vague mutual relationships. Moreover, the calibration of late Weichselian radiocarbon data has improved significantly in recent decades. Today, the Hulu calibration curve represents a very good tool for the translation of radiocarbon dates into calendric ages. In this study, dates are therefore generally given as calendric ages (cal BP) and are calibrated with the CalPal 2007 Hulu curve (Weninger and Jöris 2008; Weninger et al. 2012).

That said, plateaus and even oscillations in the calibration curve do persist and may eventually limit the possible resolution of this dating method. Relevant for the CEM is an oscillation between roughly 18,700 and 17,900 cal BP and three plateaus between about 17,700 and 17,200 cal BP, 16,700 and 15,700 cal BP, and 14,500 and 14,000 cal BP (Fig. 2.2). Within these plateaus and in the range of equivalent parts of the oscillation, all measurements will result in similar calendric ages.

Environmental Data

Climate

The recolonization of Central Europe by Late Upper Paleolithic hunter-gatherers took place toward the end of the Weichselian Glacial. Today, the chronological, biostratigraphic, and climatic structure of this period is relatively well investigated, at least in its general lines (see Chap. 5). Nevertheless, many questions and unsolved problems become apparent when environmental reconstructions, even very coarse-grained ones, are done on a regional scale and for shorter intervals. It is exactly these regional scenarios, however, that are crucial for the understanding of similarities and differences in the land-use patterns of hunter-gatherer groups, because different environmental settings are likely to result in different adaptation and subsistence strategies. Among the major factors influencing Europe's climate are the temperature and circulation of the North Atlantic and atmospheric circulation (Ruddiman and McIntyre 1981; Broecker and Denton 1989; Meyer and Kottmeier 1989; McCabe and Clark 1998; Boyle 2000; Seager et al. 2002; Renssen et al. 2007; Williams 2009). On the one hand, the coastal areas of the Western European landmass are warmed by the North Atlantic Current, which carries the warm water of the Gulf Stream; on the other hand, oceanic circulation and the evaporation of the North Atlantic affect atmospheric circulation and precipitation over the continent. Today, a number of parameters from terrestrial and marine archives are available to reconstruct the climate of the past. These parameters, such as isotope concentrations or ice-rafted sand deposits, are indirect indicators—so-called proxies—of phenomena such as wind intensity, evaporation rate, or the position of the Oceanic Polar Front. In combination, they allow for insights into climatic development in the past, even though the exact mechanisms behind the complex processes of their generation, circulation, interaction, and deposition are not yet completely understood. A major problem for the interpretation of proxies involves the extent to which locally sampled data can be seen as representative for larger areas or far-distant regions, as well as the difficulty of accounting for regional variability (Schlummer et al. 2014). A large part of the data available today, especially on isotopes, derives from ice cores that were drilled in Greenland; other cores come from marine sites in the Atlantic or Mediterranean. They provide general information on large-scale climatic trends. For small-scale, regional studies, in contrast, only a rather small amount of data from lake sediments, fenlands, or speleothems exists; in addition, such data does often not extend back to periods prior to the onset of the Late Glacial Interstadial Complex around 14,700 cal BP. Thus far, the applicability of available proxies for small-scaled climate reconstructions has therefore been limited, and attempts to use it must be done with care. If the characteristics of the regional topography are taken into account in combination with data from other palynological and faunal records, it seems possible to obtain at least a basic idea of the environmental situation in the relevant region.

Vegetation

Palynological analyses certainly play a key role in the reconstruction of regional environmental settings. This is mainly due to three factors: first, the sampled point data from a pollen site already contains information about the regional vegetation, since anemophilous plants which grow at greater distances from the sampled site are also represented in the sampled pollen spectrum due to the wide distribution of pollen grains from wind-pollinated plants. Therefore, even a relatively small number of sampled sites already make possible some statements about the character of a landscape. Second, the spectrum of pollen types and plant species positively identified in a sample provides proxies well suited to the estimation of the mean temperatures during the vegetation period. Finally, information on vegetation and temperatures can be used to draw conclusions about the faunal elements which correspond to the reconstructed situation. This is particularly the case for many herbivores, which tend to be found in specific environments (Koenigswald 2002: 140). In the course of the present study, palynological data of 59 sites from all over the investigated area could be recorded (Tables A.4 and A.5). The application of palynological data is not unproblematic, however. Three major obstacles hamper the interpretation of its results in the context of the Late Glacial:

1. First and foremost, the availability of reliable and firmly dated regional pollen profiles varies widely within the investigated area: densely sampled zones contrast sharply with large areas from which almost no data is available.
2. The terminology of the Late Glacial biostratigraphic units in northern Europe is plagued by major discrepancies (Hoek 1997: 31f.; Bos 1998: Tables 4.2 and 4.3, 115–177; de Klerk 2004). Generally speaking, there are two contrasting terminological systems: on the one hand, there is the "Bio- und Chronostratigraphie des Weichsel-Spätglazials in Nordwestdeutschland (Eifel)" (Litt and Stebich 1999; Litt et al. 2007), a system based on varve chronology that is being applied with increasing frequency in German archaeological research; on the other hand, there is the terminology of the North Atlantic Seaboard Programme (NASP) and the Integration of Ice core, Marine and Terrestrial Records (INTIMATE) group (Berglund et al. 1994; Walker et al. 1994; Björck et al. 1998; Lowe et al. 2001, 2008), which is standard in the United States and most European countries. The parallel use of these two systems by the international archaeological community often leads to confusion because their nomenclatures use homonymic terms for different periods. These terms are generally based on the biostratigraphic classification system introduced by Johannes Iversen in 1942 (Iversen 1942) for his palynological analyses in Bølling Sø (Denmark) and later adopted by both Thomas van der Hammen (1949a, b) for his pollen diagram from Hijkermeer (the Netherlands) and Rudolf Schütrumpf (1958) for the pollen diagrams from the Ahrensburg Tunnel Valley (Germany). In 1968 this classification was expanded by Burchard Menke, who introduced a Meiendorf Interval and a Grömitz-Oscillation, both thought to precede the Bølling and Allerød Interstadials. An extensive set of ^{14}C-dates shows that the Bølling Interstadial sensu Menke (1968) and the Bølling Interstadial sensu Iversen (Hoek 1997) are not contemporaneous and thus differ in their palynological compositions: whereas the Meiendorf Interval sensu Menke is characterized by a *Hippophaë* maximum, *Hippophaë* (sea buckthorn) plays only a marginal role during the Bølling Interstadial (sensu Menke). In contrast, a *Hippophaë* maximum does characterize the Bølling sensu Iversen. This is by no means a surprise, since the Bølling sensu Iversen and the Meiendorf sensu Menke are chronologically identical and both correspond to the GI-1e (sensu Weninger and Jöris 2008). With regard to current analyses of macro-remains, algae, and daphnia, as well as a series of AMS dates of the Bølling Sø sediments (Bennike et al. 2004), this type locality is gaining considerably in importance, as is the designation Bølling for the first Interstadial of the Late Glacial.
   In this study, the classification follows the international terminology used by the NASP and INTIMATE group. This choice was made for two reasons: first, the investigation comprises sites which are located in nine different European states. To avoid misunderstandings, the use of the international terminology applied in most of these countries seems most appropriate. Second, the varve-based chronology of the "Bio- und Chronostratigraphie des Weichsel-Spätglazials in Nordwestdeutschland (Eifel)" is not totally in accordance with the ^{14}C-based chronology of Weninger and Jöris (2008). For the time around the Bølling sensu Iversen (GI-1e), the chronological classification according to Litt et al. (2007) shows a shift of about 200 calendar years in comparison with the classification according to Weninger and Jöris (2008).
   Because radiometric dates are crucial for the temporal classification of CEM sites and because in this study their calibration is done with the calibration curve CalPal 2007 Hulu (Weninger and Jöris 2008), the international classification—which is based on 239 ^{14}C-dates from 102 pollen diagrams in the Netherlands alone (Hoek 1997)—sets a better frame of reference.
3. Likely the most difficult question concerning the application of palynological data in the evaluation of late Weichselian environmental conditions is the interpretation of the calculated percentage values of the different pollen types, spores, and non-pollen palynomorphs found in the analyzed samples. Three major aspects of this issue must be taken into account.

First, there is the fact that the percentage values depend on the number and the kind of types that form the basis of the calculation. The decision as to which types are taken into account may change the character of a pollen diagram significantly, but this decision is not always made explicit. For example, *Cyperaceae* (sedges) often formed abundant populations along Late Glacial lakeshores; if their pollen grains are integrated into the calculation base, the proportion of other taxa often falls considerably.

Second, similar percentage measurements of anemophilous plant pollens may nevertheless signify very different proportions of the respective plants in the regional landscape, depending on the net pollen production of the regional vegetation. This disparity is not only problematic in very detailed reconstructions and landscape models, but complicates even basic estimations, such as the ratio of woodland to open country (see below).

Third, interpretations of the percentage values for the late Weichselian Glacial depend on essentially very coarse-grained estimations about the general character of the landscape. For regions which are generally assumed to be treeless, even relatively high values of arboreous pollen are often interpreted as the result of long-distance wind-borne transport. This is the case, for instance, in a recent study of vegetational development in Poland (Ralska-Jasiewiczowa et al. 2004a): samples from northeastern Poland dated to about 17,000 cal BP (14,000 ± 100 ^{14}C BP) exhibit values of *Pinus* pollen between 27 and 43 %, but are still interpreted as long-distance transport into a treeless landscape (Latałowa et al. 2004, 167).

Another crucial question affecting the general assumptions made about the character of the vegetation in a given area concerns the amount of time needed by this vegetation, in particular by trees, to respond to climatic changes. Whereas studies from Switzerland point to an almost immediate response (e.g., Eicher and Siegenthaler 1976), analyses from northwestern Europe suggest a significant lag between a shift in temperatures and a corresponding change in the vegetation (e.g., Litt and Stebich 1999). The main difference between these two regions seems to be their relative distance to potential refugia of arboreal plant taxa. Consequently, the respective proportions of the tree pollens in the diagrams must be interpreted differently. In order to account for these differences, it is necessary to refine our general assumptions about Late Weichselian vegetation, and here the identification of the glacial refugia of arboreal plants and their potential routes of redispersal is vital. The analysis of macro-remains combined with subsequent ^{14}C-dating is able to demonstrate the presence of certain taxa at a certain place and time. Despite the considerable progress made during recent years in this field of research (Willis et al. 2000; Stewart and Lister 2001; Willis and van Andel 2004; Jankovská and Pokorný 2008; Svenning et al. 2008), there are still many unresolved questions. In order to broaden the set of available data, it would be advisable to perform a wood identification of all charcoal remains from Paleolithic sites which are selected for radiocarbon dating.

Fauna

Archaeozoological analyses make use of three major groups of remains: bones, teeth (including tusks), and antlers. Despite this well-defined set of objects, however, and despite the fact that the number of these anatomical items per animal is known, it is not always easy to compare the results of such studies. As is the case for the lithic inventory, it must be assumed that a large portion of the faunal remains were not recognized or not recorded, particularly in older excavations. Additionally, it seems that in older publications, only those pieces were considered that were comparatively easy to identify, resulting in an additional bias. Today, more precise standards for recording and presenting faunal data in the archaeological literature exist, but these sometimes render a wider comparison difficult. In particular, the fact that some authors present the assumed minimum number of individuals (MNI), while others provide only information about the weight or number of identified bone, tooth, and antler specimens (NISP), is problematic for a broader analysis (see Serangeli 2006: 3). For the study at hand, information on faunal assemblages was recorded for 129 assemblages from 89 sites. However, the quality of the information varies considerably and very detailed description alternate with only summery remarks on the present species. Data on the proportion of horse and reindeer—the most important hunting game during the Magdalenian—could be obtained for all 129 collections (Table A.7). For the remaining species, namely, mammoth, rhinoceros, bison, wisent, ibex, chamois, moose, red deer, roe deer, saiga, musk ox, and wild boar, only their presence per site was documented (Table A.8).

Still other open questions concern assumptions about the behavior and environmental adaptation of animals during the Late Glacial. In exceptional cases, these assumptions are based on mummified specimens (e.g., from permafrost areas), but for the most part, investigators must rely on observations of modern fauna. Such analogous reasoning is not without difficulties, since the behavior of animals today does not necessarily reflect that of animals in the past. An obvious problem arises in the case of extinct species, such as the mammoth or woolly rhinoceros. Here, conclusions about their behaviour are often drawn from comparisons with their closest living relatives. Also in the case of species which still endure today, however, such as reindeer, red deer, or horses, the transfer of modern observations onto the remote past remains an assumption in itself. Although the occurrence of most herbivores seems closely associated with particular types of vegetation, horses (*Equus ferus*) and red deer (*Cervus elaphus*), for instance, are found in cold-adapted steppe or tundra environments, as well as in rather temperate and forested biotopes (Koenigswald 2002: 140). Moreover, recent observations of roe deer in regions which have been cleared of trees for agricultural purposes show an adaptation of this species to the altered environmental conditions and the development of an eco-type called field-row deer (Feldreh). Field-row deer live in groups of 30–40 animals and exhibit behaviors typical of gregarious animals in steppe environments (Kurt 2002: 153f.). For theories on Magdalenian land-use patterns, estimations regarding the habitat, migration routes, and traveled distances of reindeer are often an important factor. However, observations of caribous (reindeer) in North America differ considerably from those of reindeer in northern Europe, and the reported co-occurrence of reindeer, red deer, and even wild boar in Spain may indicate the presence of reindeer in a taiga environment (Serangeli 2006: 107). These problems should be kept in mind when applying archaeozoological data in archaeological research.

Cultural Data

Mobility Patterns

During their seasonal, annual, and multi-annual migrations, hunter-gatherers carry along their belongings from one camp to the next. At each stay, objects which become obsolete or get lost are left behind, whereas new items are acquired. These new items are carried along until they are finally discarded or get lost for their part. For some objects, it is possible to determine the area of their origin. Thus, inferences can be drawn from the relative position of an object in comparison to its presumed area of origin about the directions and distances traveled by the carrier of these objects. With regard to the CEM, three classes of objects are of special interest: lithic raw material, mollusk shells, and amber.

Lithic Raw Material

Siliceous rock was an important resource during the Magdalenian and probably gathered personally by its users embedded in the procurement of other daily necessities (see Chap. 6, section "Procurement Pattern of Lithic Raw Materials"). Depending on the time and the circumstances of their formation, the appearance of these rocks varies considerably in color, matrix, and fossil inclusions. Often the composition of these attributes is so particular that it is possible to pinpoint the origin of a certain raw material found at an archaeological site, especially when microscopic analyses are applied (e.g., Floss 1994, 2012a, b, c, d); Affolter 2009b; Affolter and Bressy 2009; and therein articles 2–7), and thanks to extensive analyses during the last decades (e.g., Floss and Siegeris 2012), many sources of siliceous rock are meanwhile well known which were already exploited during the Magdalenian. Therefore, the analysis of lithic raw materials may reveal insights into the catchment area of a site as well as into the direction and distance traveled by a group of hunter-gatherers (cf. Rensink 1993: 104ff.; Floss 1994: 320ff.). However, the assignment of a lithic artifact to a specific source region is often not unambiguous and for several reasons fraught with uncertainty. A major obstacle to a precise determination of sources is the fact that there are many overlaps in the texture and matrix of siliceous rocks in different deposits, especially when they originated in similar geological contexts. Therefore, a macroscopic identification in terms of color and inclusions is frequently difficult, if not impossible. This applies especially for smaller artifacts, because the textures may vary considerably even within one piece of raw material. Small samples may therefore be unrepresentative and perhaps misleading. Further complicating matters is the redeposition of raw materials by rivers or glaciers that transport some materials over large distances and thus increase the potential source area considerably. This most extreme example is probably the distribution of Baltic flint, which was transported by the advancing Scandinavian glaciers and can be found almost ubiquitously in the Central European northern lowlands. An accurate identification of its source is therefore impossible as is the reconstruction of catchment areas.

However, raw material analyses provide today the only possibility to directly trace the movement of Paleolithic hunter-gatherers, and they are not only a permissible but very effective tool for the investigation of Magdalenian catchment areas, land-use patterns, and territories (see Chap. 3, section "Territories, Territoriality, and Land Tenure") and, thus, for the detection of regional groups. In order to access this potential information, all available data on the assemblages' raw material spectrum and the assumingly related source regions was recorded (see Fig. 2.1; Table A.9). If the source allocation remained ambiguous, the source closest to the respective find site is generally recorded as the most probable point of origin. However, if one of the ambiguous source regions is located near the source of another raw material clearly identified in the same assemblage or between the clearly identified source and the find site, then this source is taken as the most probable place of provenance.

Mollusk Shells and Amber

Fossil and subrecent mollusk shells were often used as pendants or ornaments by Magdalenian hunter-gatherers (e.g., Capitan and Peyrony 1928; Rähle 1983a, b, Hladilová 1999). In contrast to lithic raw materials, shells were generally transported over much larger distances and can thus be considered indicative of supra- and interregional relations among hunter-gatherer groups (see Chap. 6, section "Procurement Pattern of Mollusk Shells and Amber"). While it is comparatively well feasible to pinpoint the sources of specific siliceous rocks, locating the origins of shells is often more difficult. This is because of the broad distribution of many species in the aquatic sediments of the Jurassic and Tertiary (Oligocene and Miocene) seas, such as the Western and Central Paratethys which covered the Alpine Molasse basin, as well as the Vienna and Pannonian Basin and the Carpathian

Foredeep (Rögel and Steininger 1983; Walter 1995: 394; Rögel 1998). This circumstance entails two major difficulties when it comes to the applicability of the respective data for the investigation of the inner structure of the CEM.

The first problem arises when a single shell may originate in two or more distinct source regions. This is the case, for example, with fossil specimens of *Pirenella plicata* (Brug.), which can be found in the tertiary sediment of Belgium as well as in the Paris and Mainz Basins (Rähle 1983b: 156). In order to cope with this problem in a consistent and comprehensible way, the mode of recording is chosen analogously to that of lithic raw materials: the source closest to the respective find site is usually recorded as the most probable point of origin. If, on the other hand, one of the shells in an assemblage can be assigned to a specific area where the other species occur as well, then this source is taken as the most probable place of provenance (see Table A.10). This approach is selected due to two reasons. First, it seems reasonable to suggest that closer resources were exploited preferentially. Second and more important, in selecting this approach, a false attribution tends to bias the results toward the already evaluated medium-distance relations indicated by the raw material procurement pattern. The identification of long-distance relationships on the other hand, and hence the actual purpose of the analysis in this chapter, maintains its credibility and argumentative capacity.

The second problem arises when the source region of a shell covers a very large area, as is the case for the coasts of the Mediterranean Sea and Atlantic Ocean. In this case, it is not possible to specify the source region any further. Amber, which is also found in some CEM assemblages, occurs mainly in the Baltic region but is occasionally found elsewhere Central Europe as well, for instance, in the glacial moraines (Ganzelewski and Slotta 1996). Thus, it is not possible to determine its precise place of origin either. Nevertheless, even the rough determination of far-distant origins and the resulting information about the major directions of transportation are of high analytical value.

Tool Types

Since lithic artifacts are by far the most common group of Paleolithic finds and often the only source of information, the collection of typological data from the archaeological literature aimed to record the typological spectrum as detailed as possible. The major obstacle to do so was, however, the considerable variation in the quality of the analyzed assemblage descriptions. A first major difference is whether tool classes are presented in the literature or tool types. Tool classes combine a number of similar tool types and exhibit thus only a minor degree of typological information. The tool class of scrapers, for instance, consists among others of the types "end scraper with lateral retouches" and "short scraper." Additionally, not all publications list the artifact assemblage with the desirable details and sometimes tools with two or more modifications are simply subsumed as "combined tools." Moreover, many texts do not give the prices number of the items per tool class or type, but give only relative statements for their abundances. Especially in the absence of summarizing tables of the amount of excavated objects, expressions such as "some scrapers," "about 20 burins," or "a large number of organic points" should be abandoned in favor of the accurate number of the respective items.

Where possible, lithic tools were counted according to the number of their functional units or working edges; thus an artifact with two working edges, for instance, a burin at one end and a scraper at the other, was recorded as one scraper and one burin. In order to cope with the strongly varying quality of the publications, the typological data was recorded in two tables representing two different levels of typological resolution: tool classes and tool types (see Fig. 2.1). Initially, for each assemblage, the number of tools per class was recorded, i.e., the number of scrapers (SC), burins (BU), *zinken* (ZI), borers (BO), notched and denticulated pieces (NDP), splintered pieces (SP), pointed blades (PB), truncations (TRU), laterally retouched pieces (Lat), backed bladelets (BB), backed points (BP), triangles (TRI), shouldered points (ShP), and backed fragments (BFr). When drawings are presented, they were used to complete or at least improve the typological information.

If the information allowed for a more detailed collection of data, the same assemblage was recorded again in a second table according to the present tool types. It includes the number of short scrapers (SSC), laterally retouched scrapers (RSC), long burins (LBU), flat-faced burins (FBU), long borers (LBO), double backed bladelets (BBD), truncated backed bladelets (BBT), double backed truncated bladelets (BBDT), bitruncated backed bladelets (BBTT), all-round retouched backed bladelets (BBC), backed bladelets with a pointed end (BBP), and denticulated backed bladelets (BBDE) (for descriptions of these types, see Chap. 2, section "Definition of Magdalenian Tool Types").

In total, information on lithic tool classes was recorded for 550 assemblages in 389 sites, whereas information on lithic tool types could be recorded for 393 assemblages in 280 sites only. For reasons of clarity, the collected data is presented in a combined table. Here, the counts for the detailed types is, of course, subtracted from the counts for the tool classes in order to avoid double counting (see Table A.11).

For organic tools, the standards of publication appear to be even more heterogeneous. Here, a consistent collection of data was possible only for the tool classes (see Fig. 2.1). Table A.12 gives an overview of the total number of organic tools mentioned in the archaeological literature and lists the recorded data.

Definition of Magdalenian Tool Types

This section gives an overview of the definitions of the tool types relevant for the study at hand. Initially the lithic tool types are presented. A second part describes the properties of the organic tools types. Unless otherwise indicated, the terminology applied follows Hahn (1991), Brézillon (1983), and Floss ((Ed.) 2012).

End Scrapers

The defining element of end scrapers (Fig. 2.3: 1–6) is a continuous, curved retouch on the dorsal face of one end of an elongated blank (Kind 2012: 415). Additionally it was recorded, whether or not the lateral edges of the scrapers are retouched (Fig. 2.3: 1, 3). Those scrapers with a preserved proximal part which are smaller than 3 cm are counted as short end scrapers (Ibid: 416) (Fig. 2.3: 6). Some artifacts are indicated in the literature as "core scrapers." If drawings of the respective pieces were accessible, they always showed cores or core chunks. Therefore, all "core scrapers" were classified as cores or core chunks, respectively (cf. Valentin 2009: 189). Sometimes laterally retouched blades and flakes are published as "side scrapers." However, these flakes mainly show only an irregular or noncontinuous retouch. Therefore, they are recorded as laterally retouched pieces.

Burins

Burins (Fig. 2.3: 7–12) in general exhibit on one or more edges of the blank a thin and elongated negative, the burin facet (Pasda 2012a: 421). This facet is obtained by performing a so-called burin blow on a platform. The platform can be prepared by retouches or remain unprepared. Some types of burins are considered to have spatial and temporal significance for a regional or diachronic division of the Magdalenian. The first type to discuss is a burin, whose proximal end is modified into a long tip by a concave retouch. Afterward, the retouched tip serves as the striking platform for the burin blow. This burin type is regarded as typical for the middle and late Magdalenian and usually referred to as Lacan burin (Brézillon 1983: 179f.). Sometimes the spelling Lacam burin is used too (Cheynier 1965: 132; Hahn 1991: 185). However, because some authors regard rather the long, concave retouch as the discriminatory element (cf. Hanitzsch 1972: 73), and not so much the succession of retouch and burin blow, all burins with a long concavely retouched tip are counted and referred to as long burins (Fig. 2.3: 11–12). Parrot beak burins are characterized by a burin blow that is carried out on a convex retouch, producing a rather short burin facet that ends in a hinge. This type is said to be rather rare or absent in the CEM record and typical for the late Magdalenian in southwestern France. In the event that the burin facet is located on the ventral face of the blank, the resulting type is referred to as a flat-faced burin. Analogous to "core scrapers," some artifacts are classified as "core burins." These artifacts exhibit usually several negatives and correspond in their morphology with cores or core chunks, respectively, as which they are recorded (cf. Valentin 2009: 189).

Borers

Borers (Fig. 2.3: 13–16) are characterized by a tip that is modified on both sides by retouches. The tip axis is usually congruent with the blank axis but may also be oblique. Long borers (Fig. 2.3: 16) exhibit a particular long tip (Floss 2012a).

Zinken

Zinken (Fig. 2.3: 17, 18) differ from borers only in a small but crucial detail. They, as well, have a tip with bilateral retouches. The tip axis is often oblique to the blank axis but may be also congruent with it. The vital difference is that *Zinken* carry one or more small, elongated negatives on the dorsal face of the tip that are struck from the ventral face and form a very small cutting edge. They are said to be particularly characteristic for the Hamburgian but occur also in several Magdalenian assemblages (Floss 2012a).

Backed Bladelets

Backed bladelets (Fig. 2.4: 1–11) are small, thin, and elongated tools with a steep retouch on at least one edge. The longitudinal edges should be more or less parallel, and the width should not exceed 10 mm. Depending on the position of the retouches, backed bladelets are divided into several types (Bolus 2012). Simple backed bladelets show a continuous retouch on one longitudinal edge (Fig. 2.4: 1, 2). Double backed bladelets show continuous retouches on both longitudinal edges (Fig. 2.4: 3, 4). Truncated backed bladelets show a continuous retouch on one longitudinal and one transversal edge (Fig. 2.4: 5, 6). Double backed truncated bladelets show continuous retouch on both longitudinal and one transversal edge (Fig. 2.4: 7, 8). Bitruncated backed bladelets show a continuous retouch on one longitudinal and both transversal edges. Rectangular backed bladelets show continuous retouches on all edges. Pointed backed bladelets show a continuous retouch on at least one longitudinal edge and a pointed tip (Fig. 2.4: 9). Denticulated backed bladelets show a continuously and a denticulated retouched edge (Fig. 2.4: 10–11).

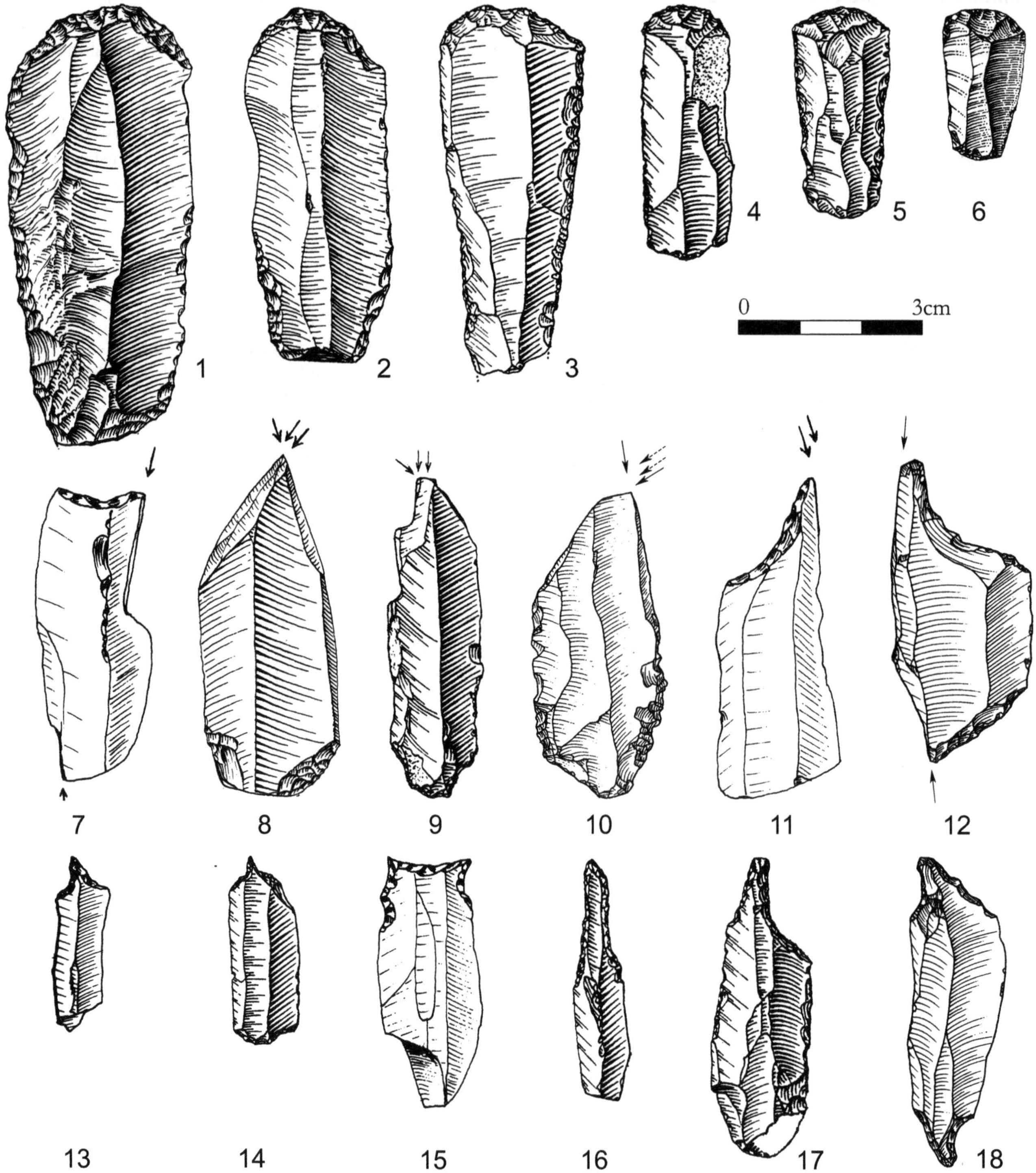

Fig. 2.3 Endscrapers, burins, borers, and *zinken* of CEM sites. 1–5: End scrapers (1 and 3 with lateral retouches); 6: short end scrapers; 7: multiple burin; 8–10: dihedral burins; 11, 12: long burins (Lacan burins); 13, 14: borers; 15: multiple borer; 16: long borer; 17–18: *zinken*. 1, 2, 8: Speckberg (Hahn 1982); 3, 4, 5, 9, 14, 16, 17: Hostim (Vencl 1995); 6: Teufelsbrücke, level 4 (Feustel 1980); 7, 11: Trou de Chaleux (Dewez 1987); 10, 12, 13, 18: Monruz: (Bullinger et al. 2006); 15: Trou du Frontal (Dewez 1987). Scale 1:1

Backed Points

Backed points (Fig. 2.4: 12–17) are elongated pieces with a steep lateral retouch on one edge forming one or two tips with the unretouched edge. The retouched part can be straight (Fig. 2.4: 12), convex (Fig. 2.4: 13–15) or angled. Angled backed points may have one angle (Creswell point, Kent point; Fig. 2.4: 17) or two angles (Cheddar point, trapezoidal point; Fig. 2.4: 16). The transition between backed and angled points is sometimes fluent (Floss and Weber 2012b: 510ff.).

Shouldered Points

Shouldered points (Fig. 2.4: 18) are named after a concave retouch at the lower part of the tool. The point itself is formed by a steeply retouched truncation. Both retouches can merge into one another to form a continuous retouch or, as it is the case for Hamburgian points, can be separated by an unretouched part (Floss 2012b; Floss and Weber 2012b: 509f.). The transition between continuously retouched shouldered points and angled backed points is sometimes fluent.

Pointed Blades

Pointed blades (Fig. 2.4: 19–21) are blades with a bilateral, rather flat retouch which forms a point that usually tapers also in the profile (Pasda 2012b).

Laterally Retouched Pieces

Blades or flakes with lateral retouches (Fig. 2.4: 22), which are often noncontinuous and of various forms, were recorded as laterally retouched pieces (Pasda 2012c).

Truncated Pieces

Truncated pieces (Fig. 2.4: 23, 24) are characterized by a retouch on a transversal edge of an elongated blank. The retouch can be concave, convex, or straight. The transition between convex truncations and end scrapers is sometimes fluent (Pasda 2012d).

Notched and Denticulated Pieces

Artifacts with a single retouched notch (Fig. 2.4: 25) were recorded as notched pieces, whereas pieces with two or more retouched notches were recorded as denticulated pieces (Hahn 1991: 202ff.).

Splintered Pieces

Splintered pieces (Fig. 2.4: 26–29) carry flat, stepped, or hinged negatives on the ventral and/or dorsal face in longitudinal and sometimes transversal direction (Le Brun-Ricalens 2012).

Magdalenian Triangles

Magdalenian triangles are small- to medium-sized thin tools with a steep retouch on at least one edge (Figs. 2.5 and 2.6). The longitudinal edges form a more or less acute angle and meet in a more or less pointed tip. A major distinction is made between isosceles and scalene triangles as well as between scalene triangles and scalene bladelets (for a detailed discussion see Chap. 4, section "The Long Debate About Lithic Triangles").

Organic Points

Organic points (Fig. 2.7) made of bone or antler are characterized by an elongated morphology and an all-around modification which removes all edges and protuberances as well as a clear terminal point. The basal part is usually single bevelled, double bevelled, or forked.

Bâtons Percés (Perforated Batons)

Bâtons percés (Fig. 2.8) are relatively large and elongated objects. They carry one or more holes which are usually situated at one end of the tool.

Needles

Needles are thin and elongated objects with a pierced eye at the basal part and a pointed distal end.

Baguettes Demi-Rondes (Half-Round Rods)

Baguettes demi-rondes are elongated, half-rounded rods with a curved topside and a flat bottom side (Fig. 2.9). They were probably used pairwise (Bosinski 1990: 180).

Magdalenian Barbed Points

Magdalenian barbed points show an elongated shaft from which barbs spring in usually one or two rows. The barbs

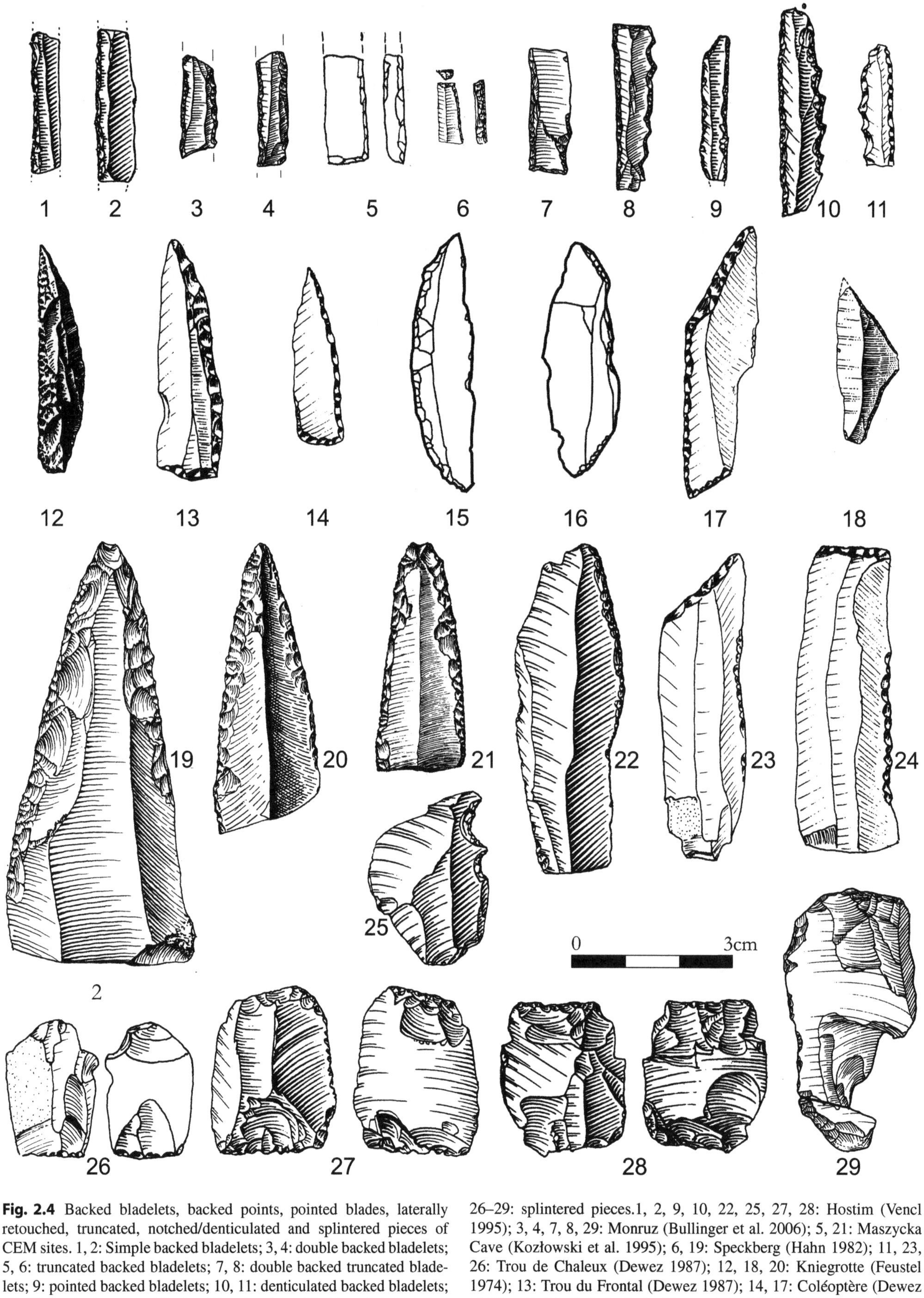

Fig. 2.4 Backed bladelets, backed points, pointed blades, laterally retouched, truncated, notched/denticulated and splintered pieces of CEM sites. 1, 2: Simple backed bladelets; 3, 4: double backed bladelets; 5, 6: truncated backed bladelets; 7, 8: double backed truncated bladelets; 9: pointed backed bladelets; 10, 11: denticulated backed bladelets; 12–17: backed points; 18: shouldered points; 19–21: pointed blades; 22: laterally retouched blade; 23, 24: truncations; 25: notched piece; 26–29: splintered pieces.1, 2, 9, 10, 22, 25, 27, 28: Hostim (Vencl 1995); 3, 4, 7, 8, 29: Monruz (Bullinger et al. 2006); 5, 21: Maszycka Cave (Kozłowski et al. 1995); 6, 19: Speckberg (Hahn 1982); 11, 23, 26: Trou de Chaleux (Dewez 1987); 12, 18, 20: Kniegrotte (Feustel 1974); 13: Trou du Frontal (Dewez 1987); 14, 17: Coléoptère (Dewez 1987); 15, 16: Bois Laiterie (Sano et al. 2011); 24: Trou de Nutons (Dewez 1987). Scale 1:1

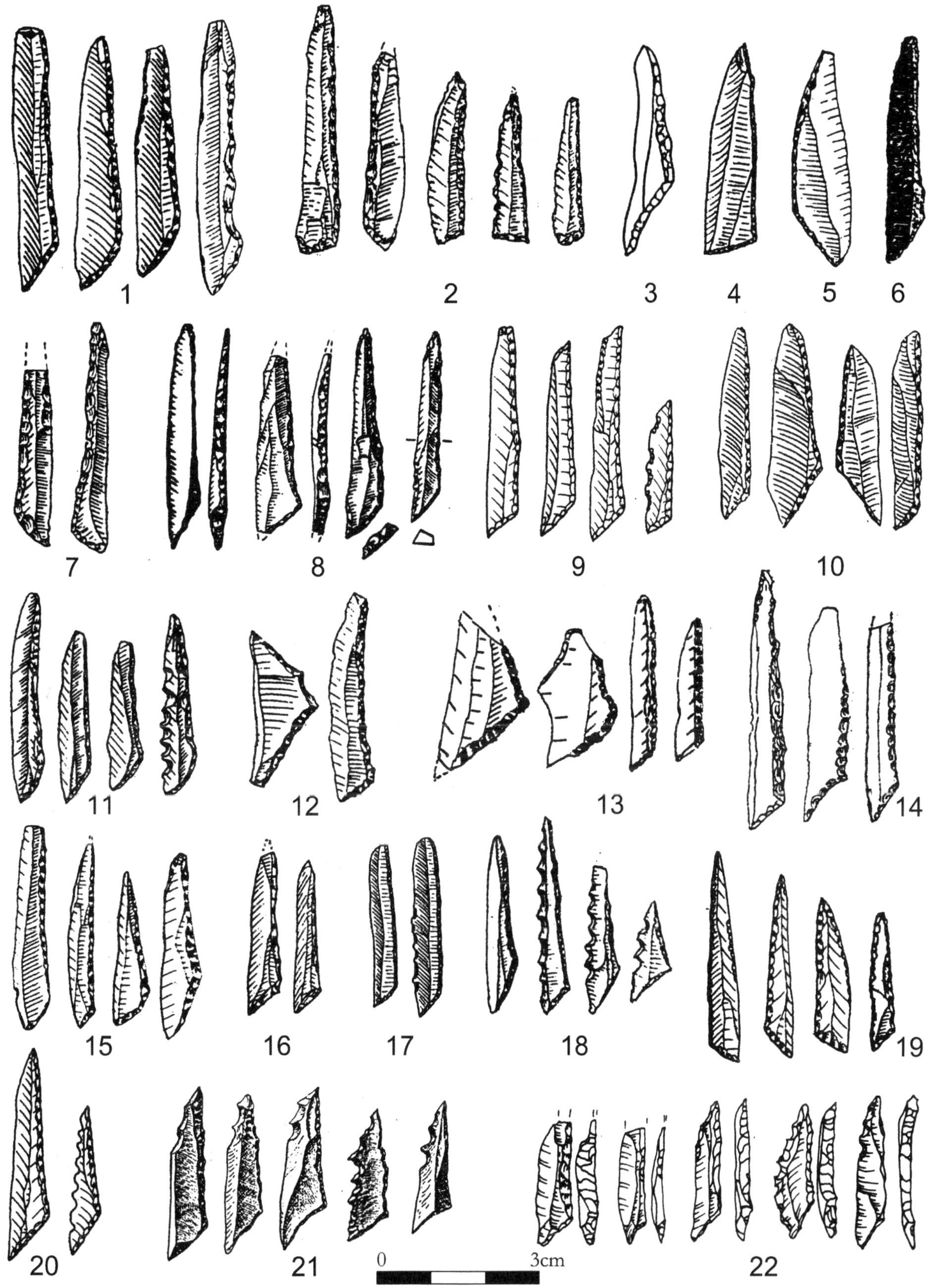

Fig. 2.5 Triangles of CEM sites. 1: Laugerie-Haute; 2: Goutte Roffat; 3: Cueva de la Paloma; 4: Grotte d'Abzac; 5: Cap-Blanc; 6: Combe Cullier; 7: Bellet; 8: Chair-à-Calvin; 9: Abri Lafaye; 10: Saint-Germain-la-Rivière; 11: Abri du Martinet; 12: Abri Reverdit; 13: Baume-Loire I; 14: La Souquette; 15: Saint-Cirq; 16: Grande Bize; 17: Jean Blanc; 18: Puy-de-Lacan; 19: Abri Fontalès; 20 Abri Plantade; 21: Kniegrotte; 22: Dzierżysław 35. 1–21: (Höck 2000); 22: (Ginter et al. 2005). Scale 1:1

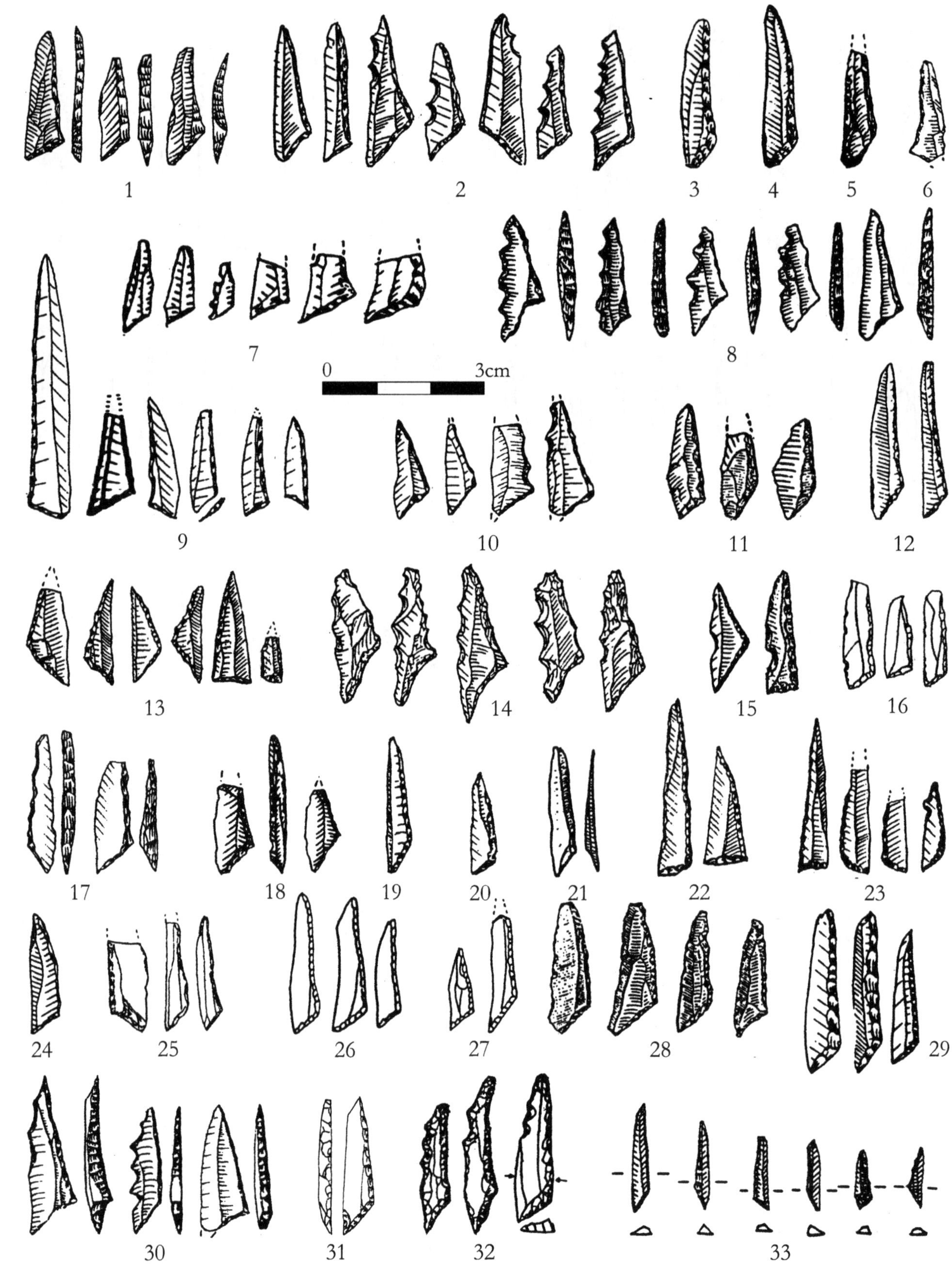

Fig. 2.6 Triangles of CEM sites. 1: Canecaude I; 2: Castelmoron; 3: Chez-Galou; 4: Gazel; 5: Granet I; 6: Huguenots; 7: Chariez-Guillotine; 8: Crabillat; 9: Duruthy; 10: Farincourt I; 11: Faustin; 12: La Forge; 13: Gare de Couze; 14: Garrigue; 15: Jolivet; 16: Lascaux; 17: Lassac; 18: Grotte de l'OEil; 19: Métairie de Belcayre; 20: Putois II; 21: Raymondon-Chancelade; 22: Rond du Barry; 23: Sâlpétrière; 24: La Teulera; 25: La Vache; 26: Aitzbitartre IV; 27: Cendres; 28: Parpallo; 29: Urtiaga; 30: Birseck-Ermitage; 31: Eitensheim; 32: Hranice; 33: Cauna de Belvis. 1–9, 11–33 (Höck 2000); 10: (David 1996). Scale 1:1

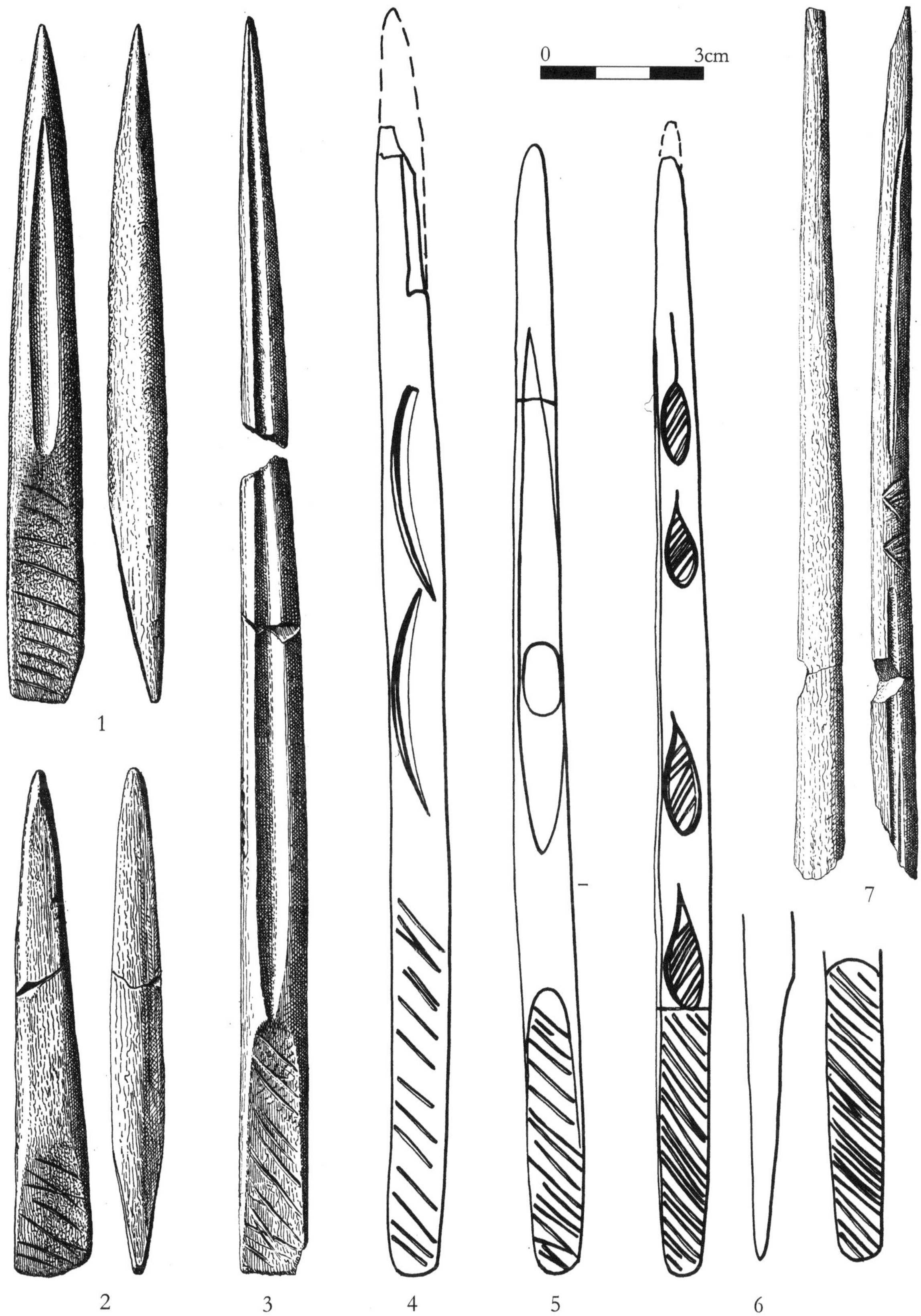

Fig. 2.7 Organic points of CEM sites. 1–3, 7: Kniegrotte (Feustel 1974); 4–6: Maszycka Cave (Kozłowski et al. 1995). Scale 1:1

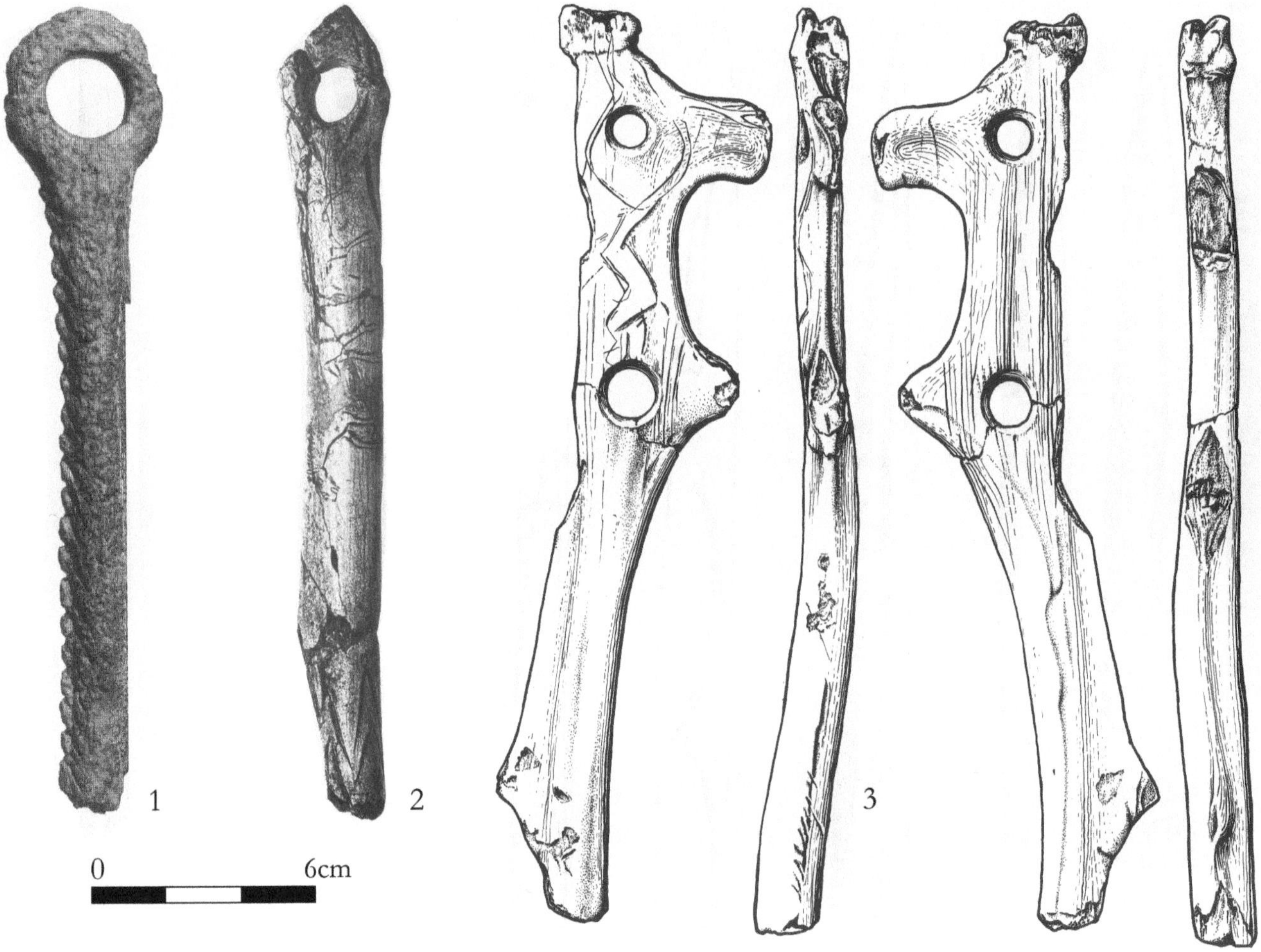

Fig. 2.8 Bâtons percés of CEM sites. 1–3: Bâtons percés (Perforated batons).1: Pekárna (Valoch 2001); 2: Schweizersbild (Bosinski 1982); 3: Kesslerloch (Höneisen 1986)

vary in number and angle, and the basal part of the point is often conical (Julien 1982).

Navettes

A *navette* is basically a shaft with two conical ends, each showing an incision of some centimeters in depth. *Navettes* were probably used for hafting lithic implements (Allain et al. 1985).

Blade Technology

The morphology of a blank and its striking-platform remnant (SPR) are the result of a complex interaction of different factors. In addition to the knapper's personal skills, these include the striking gesture, the material and size of the hammer, the kinetic percussion energy, the preparation of the core and striking platform, as well as the quality of the raw material. However, for the successful production of a blank, the number of potential combinations of these factors is not arbitrary but limited by several constraints. If, for example, a heavy mineral hammer is applied to a core with an acute-angled and comparatively fragile edge, the impact will probably only damage the core without producing the desired blank. On the other hand, a relatively light organic hammer may not generate enough kinetic energy to detach blanks from cores with a striking angle of more than 80°. As a consequence, typical combinations of feature characteristics are related to different knapping techniques (e.g., Pelegrin 2000).

The main objective of Magdalenian *débitage* concepts was certainly the production of blades and bladelets. The vast majority of cores were designed to meet the requirements for the production of these blanks, from which most Magdalenian tools were made. The *chaîne opératoire* of Magdalenian blank production is—specifically for the sites in the Paris Basin—relatively well investigated (Pigeot 1987; Audouze et al. 1988; Schmider 1992; Bodu 1993;

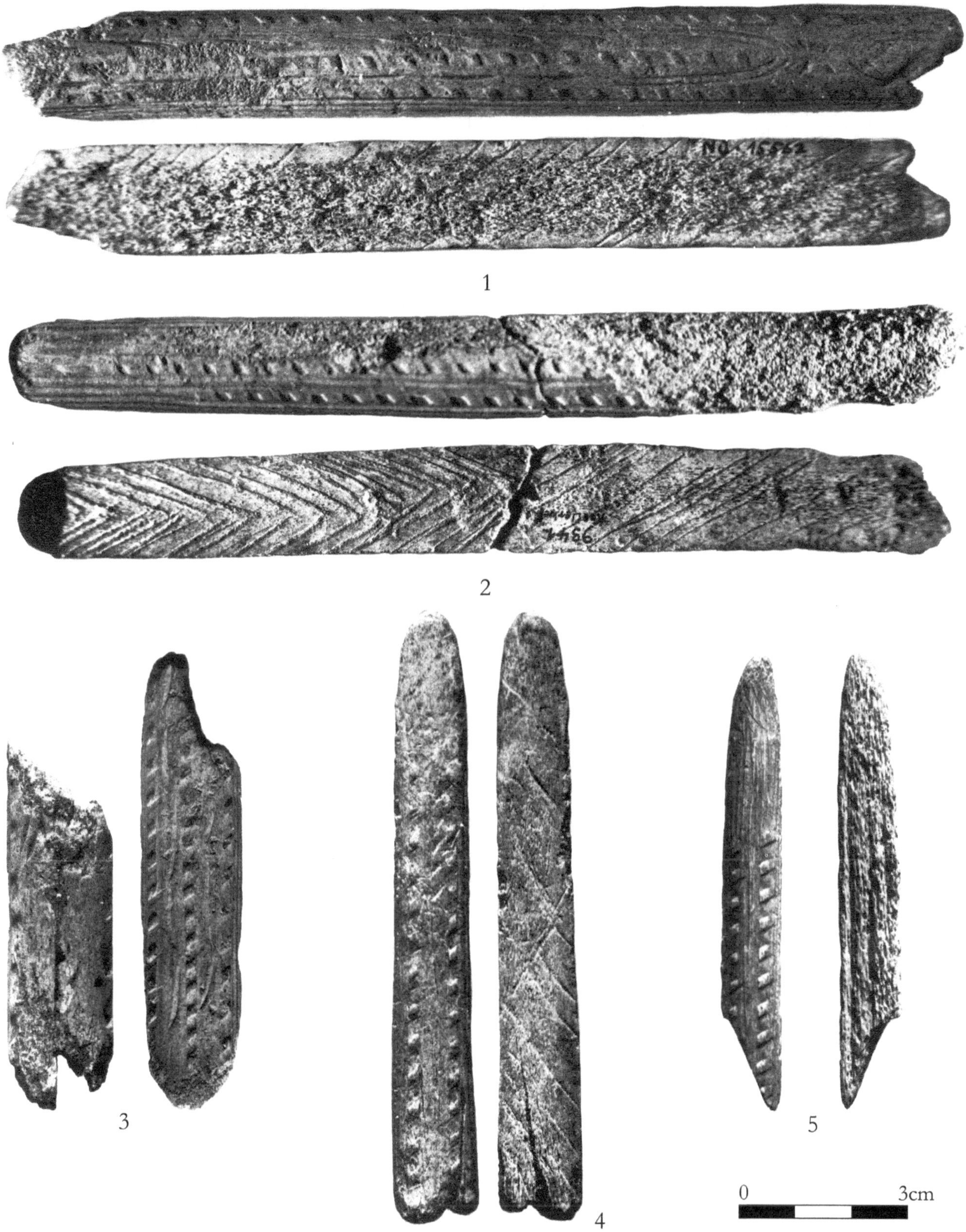

Fig. 2.9 Baguettes demi-rondes with raised humps of CEM sites. 1–4: Kesslerloch (Bosinski 1982); 5: Freudenthalhöhle (Bosinski 1982). Scale 1:1

Valentin 1995; Floss and Weber 2012a). The raw material is usually of high quality. Initially prepared core blanks ideally have a biconvex outline with two crests or a crest opposed to a back and a rhomboid cross section. The striking platform is at an acute angle to the reduction face. Despite a clear preference for unipolar reduction, sometimes a second striking platform is present. It serves for another sequence of blank production or is used for core renovation, which is undertaken extensively if necessary. Preparation and renovation of the core are usually done with a mineral hammer. The main focus of the Magdalenian *chaîne opératoire* was, however, on the serial production of long and curved blades as well as bladelets, a process which was normally performed with an organic hammer. Bladelets were obtained either independently or embedded in the process of blade production on the same core (Bodu et al. 2006: 62ff.). For better fracture control, the impact-receiving part of the striking platform was often prepared by shaping it into spur-like form, a technique which is called *en éperon* preparation (Cheynier 1956: 347, Fig. 5; Karlin 1972: 268, Fig. 180; Weiner 1989: 209). Some late Magdalenian sites in the Paris Basin, however, exhibit features which differ from the "classical" *chaîne opératoire* described above (e.g., Bodu 1993; Valentin 2000; Bodu et al. 2006; Weber 2006). Here, the use of a soft stone hammer (Madsen 1983) for the production of relatively short blades with a straight profile is technologically conspicuous. In this case, "soft" stone refers to the relatively low scratch resistance of the hammer's material, as in the case of sandstone, limestone, or flint nodules with a thick, chalky cortex (Pelegrin 2000). Generally speaking, soft stone hammers are applied in two versions: the tangential version and the *rentrante* version. In the tangential version, the hammer is directed to the edge of the core, as it is with organic hammers, and produces rather elongated blades. In the *rentrante* version, the hammer hits the striking platform closer to the center and therefore produces comparatively shorter and thicker blades (Pelegrin 2000: 77ff.; Valentin 2000: 254). The fact that soft stone hammers are commonly used in the Late Paleolithic supports the hypothesis of a continuous technological development from late Magdalenian to Azilian (Valentin 1995, 2000; Bodu and Valentin 1997; Valentin et al. 2004).

Since blades and bladelets were the most important *débitage* products, it is expected that the methods for their extraction can reveal interesting aspects of technological behavior which might be connected to an exchange of ideas and, hence, may give evidence of interaction between hunter-gatherer groups. Unfortunately, the archaeological literature on the CEM contains rather poor technological data concerning the features of the SPRs. Moreover, the standards with regard to analysis of features and taking of measurements are very inconsistent. Therefore, it is not possible to use the published data for an extensive comparison. To cope with this problem, 26 assemblages from 20 sites in Belgium, France, Switzerland, Germany, Czech Republic, and Poland are investigated (see Chap. 6, section "Correspondence Analysis: Multivariate Investigation of Spatial and Temporal Trends in Blade and Bladelet Production"). Four criteria played a major role in their selection: one aim was to choose preferentially those assemblages that came from a clear stratigraphic context. Second, the inventories should be ^{14}C-dated in order to trace a possible temporal development. Third, the spatial distribution should cover the whole of the investigated area. Last, the number of artifacts per assemblage should be large enough to ensure that a sufficient number of blades and bladelets with an intact SPR is present. Admittedly, not all criteria could be met for every assemblage; nevertheless, the selection is able to give useful insights into CEM blade production. In the case of smaller assemblages, all available platform remnants were analyzed, whereas for larger assemblages samples of 100 blades and bladelets were chosen at random. Since this selection of specimens was nearly exhaustive even for many larger assemblages, the analyzed samples are regarded as sufficiently ample to be statistically representative (see Chap. 6, section "Correspondence Analysis: Multivariate Investigation of Spatial and Temporal Trends in Blade and Bladelet Production"). In the following, the analyzed features are described and explained briefly in order to clarify their technological significance. The explanations are based on the works of J. Pelegrin (2000) and B. Valentin (2000), from which the majority of the information presented is taken. The recording followed the system of Auffermann et al. (1990), which was adjusted for those features which are not included in the original classification system (see Tables A.13, A.14, A.15, A.16, A.17, A.18, A.19, A.20, A.21, A.22, A.23, A.24, A.25, A.26, A.27, A.28, A.29, A.30, A.31, A.32, A.33, and A.34).

Recording of Technological Information

Blanks

Since the technological analysis focuses exclusively on the production of blades and bladelets, the recorded types of blanks were:

1. Blade/bladelet
2. Primary crested blade
3. Secondary crested blade

Blank Dimensions and Preservation

For each blank, the length, width, and thickness were measured (in mm), and its state of preservation was noted. To be sure, only those blanks with a preserved SPR could be considered for the analysis.

Length	Width/thickness
1. Preservation complete	1. Preservation complete
2. Only the proximal part is preserved	2. Width incomplete
3. Thickness incomplete	
4. Width and thickness incomplete	

Cortex Cover

The dorsal cortex:

1. Was absent
2. Covered less than 1/3 of the blank
3. Covered between 1/3 and 2/3 of the blank
4. Covered more than 2/3 of the blank
5. Was complete

Striking-Platform Remnant (SPR) Kind

SPRs can be:

1. Not preserved
2. Cortex covered
3. Cleft
4. Smooth
5. Facetted

Striking-Platform Remnant (SPR) Form

With regard to their morphology, SPR can be:

1. Not preserved
2. Oval/pointed oval
3. Punctiform
4. Linear
5. Triangular
6. Rectangular
7. Irregular
8. Ventral-sided splintered (Fig. 2.10: 14–17)
9. Chapeau de gendarme (Fig. 2.11: 25)

Striking-Platform Remnant (SPR) Size

SPRs of blades which were produced with an organic hammer are normally several square millimeters in size. This may also be the case for blades produced with a soft stone hammer; however, the latter can also have very small, hardly measurable SPRs. In order to access this information length and width (in mm) of the SPR were measured.

Dorsal Reduction

The diverging properties of mineral and organic hammers require different preparations of the striking platform, adapted to the respective characteristics of the materials applied. The relative softness of organic hammers allows an intrusion of the core's edge into the hammer's matrix at the moment of impact. The blank is therefore "torn off" of the core rather than chipped off, as is the case with mineral hammers. When using an organic hammer, it is therefore better to prepare the striking platform in such a way that an acute angle or a small overhang allows the edge to intrude into the hammer. In contrast, such a preparation would be completely inappropriate for a mineral hammer, even a comparatively soft one. Because of the potentially higher kinetic energy and harder matrix, a mineral hammer would crush such a fragile edge without detaching the desired blank, especially when used with a tangential gesture. Therefore, even when using a relatively soft mineral hammer, all overhangs must be removed in order to stabilize the striking platforms' edge. This trimming of the edge is usually referred to as dorsal reduction (Fig. 2.10: 1–13).

For the technological analysis, the following expressions of the feature "dorsal reduction" were recorded:

1. Proximal part not preserved
2. No dorsal reduction
3. Pointed negatives
4. Hinged negatives
5. Abrasion
6. Pointed negatives and abrasion
7. Hinged negatives and abrasion
8. Pointed and hinged negatives
9. Pointed and hinged negatives and abrasion
10. Indeterminate reduction

Lip

The formation of a lip is characteristic for organic hammers, but also occurs sometimes when a soft stone hammer is applied in a tangential way. However, in the latter case, the lip is usually only laterally present and is absent close to the impact point. When the soft stone hammer is applied in the *rentrante* version (for explanation, see Chap. 6, "Technological Analysis of Blade and Bladelet Production"), a lip is usually lacking. The recorded feature expressions were thus:

1. Not preserved
2. Removed by modification
3. Removed by other means
4. Present
5. Absent
6. Laterally present

Bulb

Blanks produced with an organic hammer usually have no bulb or only a diffuse one. The same can be observed of some blanks obtained with a soft stone hammer used in a tangential way. However, in the latter case, marked bulbs also occur, and these become increasingly frequent when the hammer is used in the *rentrante* version.

Bulbs can be:

1. Not preserved
2. Removed by modification
3. Removed by other means
4. Markedly present
5. Absent
6. Double
7. Diffuse

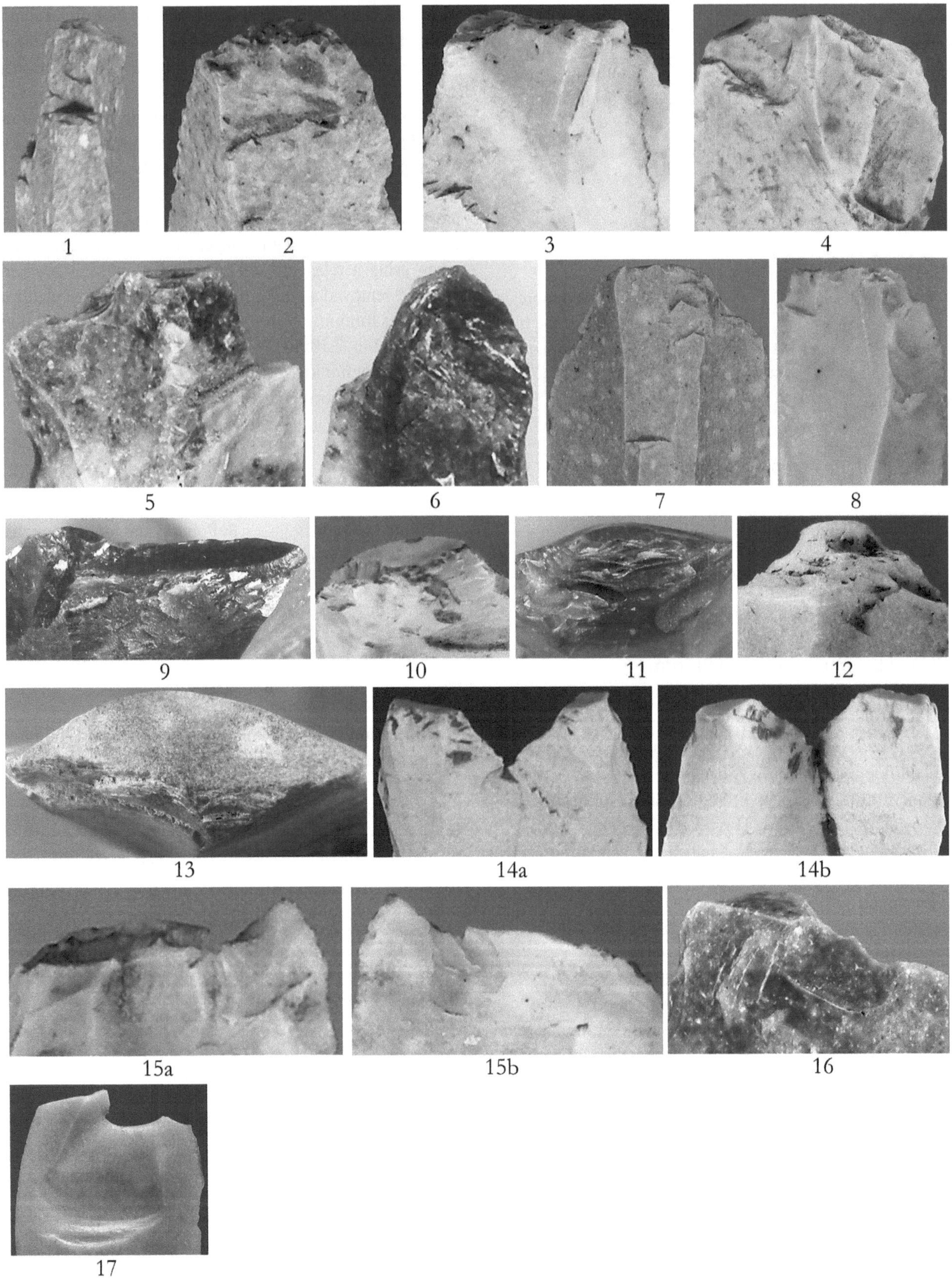

Fig. 2.10 Dorsal reduction and ventrally splintered pieces. 1–13: Dorsal reduction; 14–17: ventrally splintered pieces. 1–3: Chênelaz; 4–5: Hranice; 6: Kaufertsberg (lower level); 7: Klementowice; 8: Monruz; 9: Brillenhöhle; 10: Hranice; 11: Kaufertberg (upper level); 12–13: Kastelhöhle Nord; 14a, 14b: Chênelaz; 15a, 15b, 16: Hranice; 17: Monruz

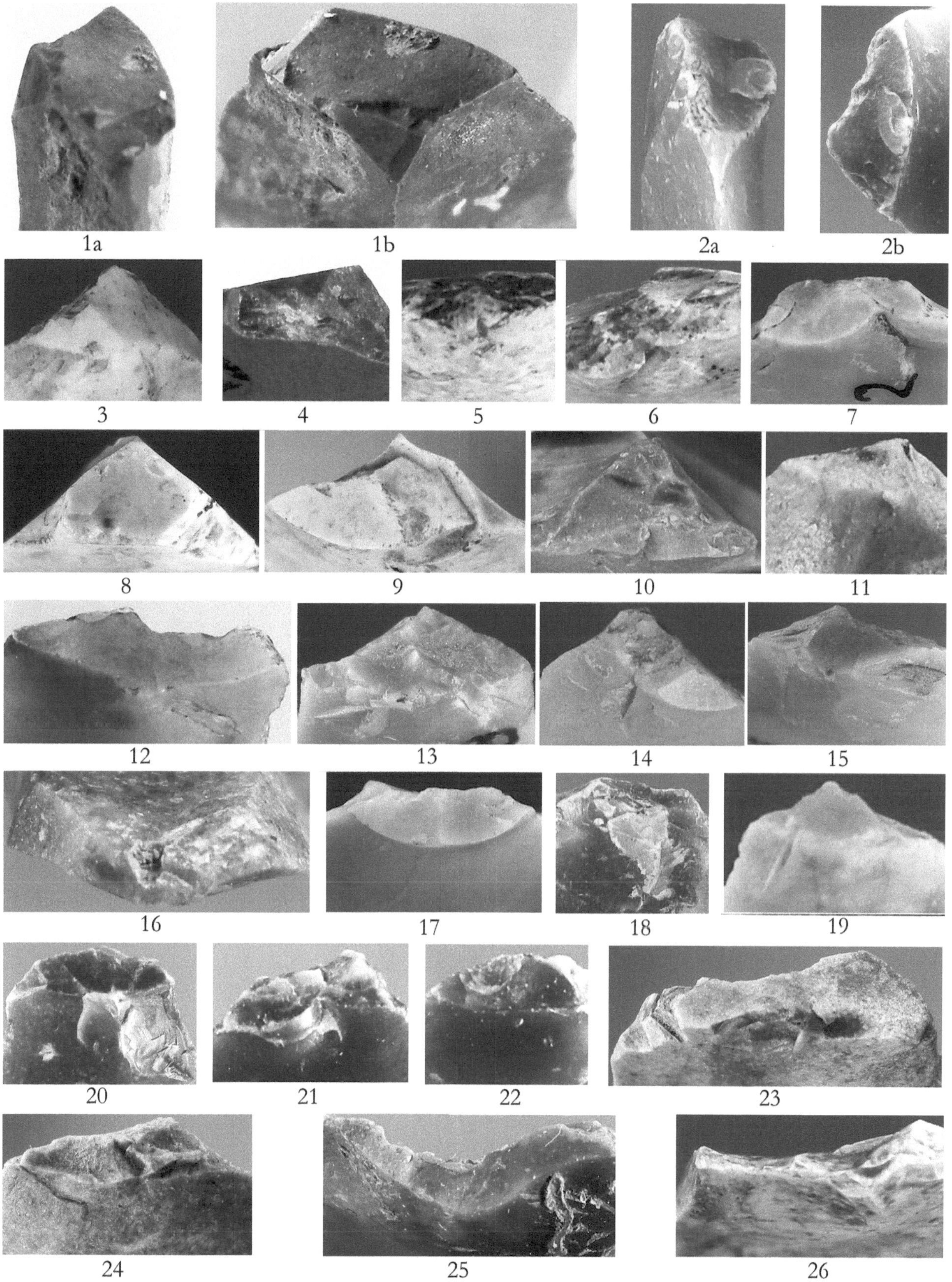

Fig. 2.11 Inherent *éperon*, impact points, SPRs. 1a and 1b: Inherent éperon; 2a and 2b: Éperon and impact point; 3–24: impact point. 25, 26: Chapeau de gendarme and linear striking platform remnant. 1a, 1b: Kaufertberg (upper level); 2a, 2b: Oelknitz; 3: Balcarova Cave; 4: Brillenhöhle; 5, 6: Chênelaz; 7: Grappin; 8: Hostim; 9: Hranice; 10: Kastelhöhle Nord; 11: Klementowice; 12: Kaufertsberg (upper level) 13: Putim; 14: Rislisberghöhle; 15: Rydno; 16, 17: Monruz; 18: Oelknitz; 19, 20: Teufelsbrücke (level 4); 21, 22: Teufelsbrücke (level 3); 23, 24: Wilczyce; 25: Oelknitz; linear; 26: Klementowice

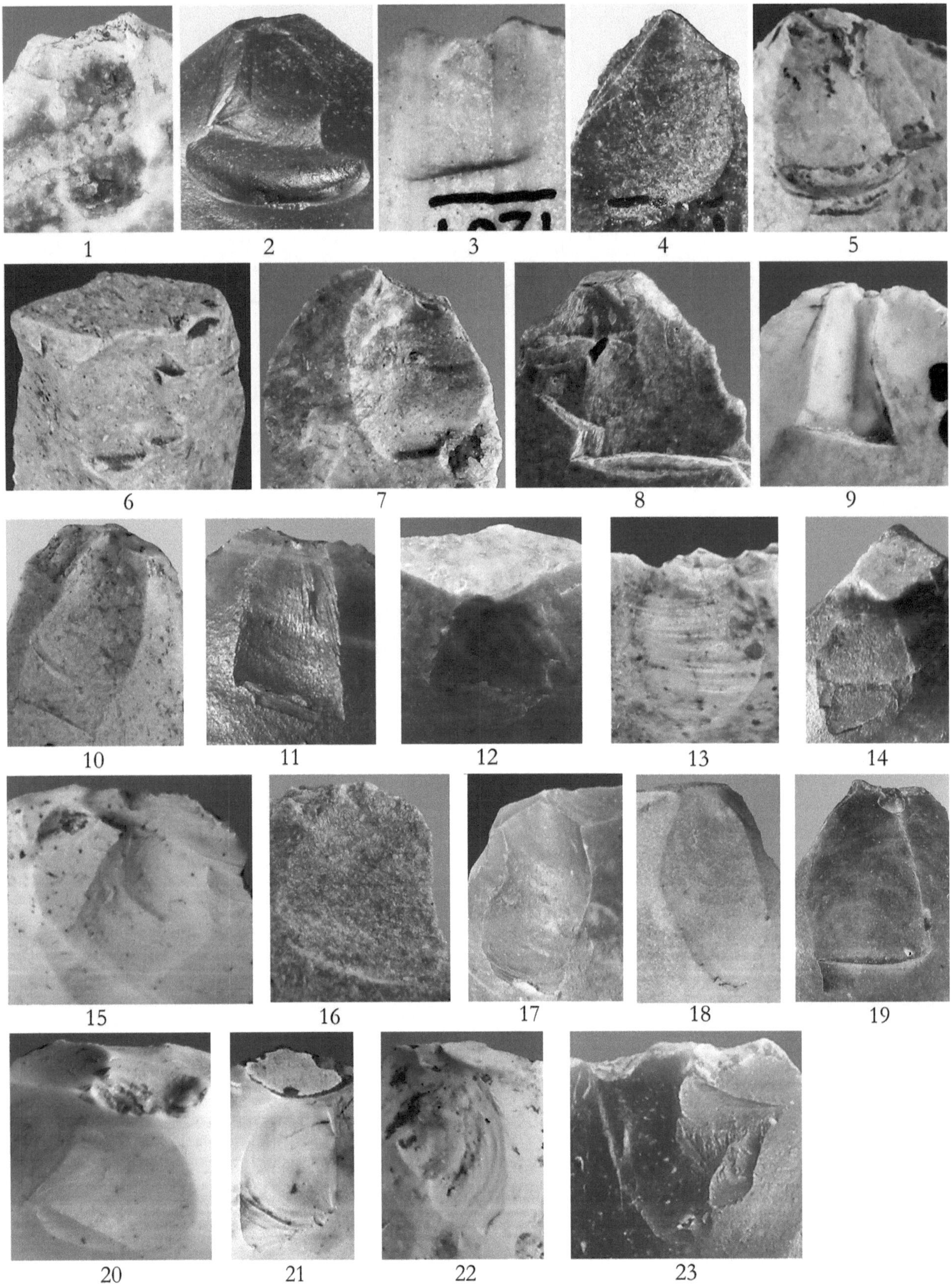

Fig. 2.12 *Esquillements du bulbe* and bulbar scars. 1–19: Esquillements du bulbe; 20–23: Bulbar scars. 1, 2: Balcarova Cave; 3, 4: Brillenhöhle; 5–7: Chênelaz; 8, 9: Grappin; 10: Kastelhöhle Nord; 11: Kaufertsberg (upper level); 12, 13: Klementowice; 14: Lhota; 15: Mokra V; 16: Monruz; 17: Putim; 18: Rislisberghöhle; 19: Rydno. 20: Chênelaz; 21: Hranice; 22: Mokra V; 23: Teufelsbrücke (level 3)

Bulbar Scar

Organic and soft stone hammers may cause bulbar scars (Fig. 2.12: 20–23). However, these are more frequently observed in the context of the latter hammer type, especially when applied in the *rentrante* version. The recorded expressions of this feature are:

1. Not preserved	5. Absent
2. Removed by modification	6. Double
3. Removed by other means	7. Multiple
4. Present	8. Splintered

Esquillements du Bulbe

According to Pelegrin (2000: 79f.), this feature is rather rare and is exclusively indicative of the usage of a soft stone hammer. It is a scar on the ventral side of the artifact, running parallel to the striking direction and starting directly at the SPR (Fig. 2.12: 1–19). It must not be confused with a bulbar scar, which is located below the SPR and usually runs perpendicular to the striking direction. With reference to the other features, the recorded expressions were:

1. Not preserved	5. Absent
2. Removed by modification	6. Double
3. Removed by other means	7. Diffuse
4. Markedly present	

Impact Point

Because of its softness, an organic hammer does not usually leave a scar at the impact point, whereas hard and soft stone hammers do (Fig. 2.11: 3–24). Therefore, this feature is considered to be diagnostic for mineral hammers. Possible expressions of this feature are:

1. Not preserved	4. Present
2. Removed by modification	5. Absent
3. Removed by other means	6. Double

En éperon Preparation

A very distinct kind of striking-platform preparation is the *en éperon* technique (Cheynier 1956; Karlin 1972). It results in the formation of a small spur or *éperon* (Fig. 2.13: 1–27). This preparation offers several advantages in combination with an organic hammer: it adjusts the flaking angle locally, allows for higher precision of the blank detachment, and helps to direct the percussive energy. Since a mineral hammer would crush this fragile preparation, it normally occurs exclusively in connection with an organic hammer. An *éperon* can be:

1. Not preserved	4. Present
2. Removed by modification	5. Absent
3. Removed by other means	6. Inherently Present

The last expression "inherently present" refers to the observation that the negatives of previously obtained blades sometimes lead to the formation of a spur-like morphology of the striking platform (Fig. 2.13: 1). In several cases, these spurs were not reduced but apparently served—without further preparation—the same purpose as intentionally prepared *éperons*.

Flaking Angle

The angle is measured between the dorsal face of the blank and the SPR (Fig. 2.14). Whereas the application of organic hammers is usually restricted to acute flaking angles rarely exceeding 75°, the situation for soft stone hammers is different: used with a tangential gesture, the flaking angle normally lies between 70° and 85°, and used in the *rentrante* version, it may be up to 90°.

Sculpted and Engraved Objects

Given the virtual absence of cave paintings in Central Europe, engraved and sculptured objects yield special scientific importance within the entirety of CEM finds. In contrast to tools, which are designed to fulfill the requirements of concrete, technical actions such as scraping hide, the purposes of these items are probably mainly related to social, spiritual, and symbolic interactions. In other words, whereas tools facilitate human-environment interactions, engravings and sculptures unfold their actual potential in human-human interactions. This peculiarity is an inherent property of those artifacts and as such still efficacious: the ostensible immediacy, with which many viewers of these objects feel confronted with the conceptual world of the manufacturer is compelling, even though the cultural context of their creation is bygone. During the act of involvement, the viewer may often have the impression to get an idea of the manufacturer's intensions and the symbolic content of the artifact. This feeling of mental relatedness is usually stronger than during the observation of technical artifacts, such as tools. However, this impression is probably misleading, because the shire inexhaustible manifoldness of contextual possibilities, symbolic conceptions, and cultural meanings, which set the frame for the production and reception of sculptures and engravings, render the chance of the understanding of their original intension from a remote distance vanishingly low.

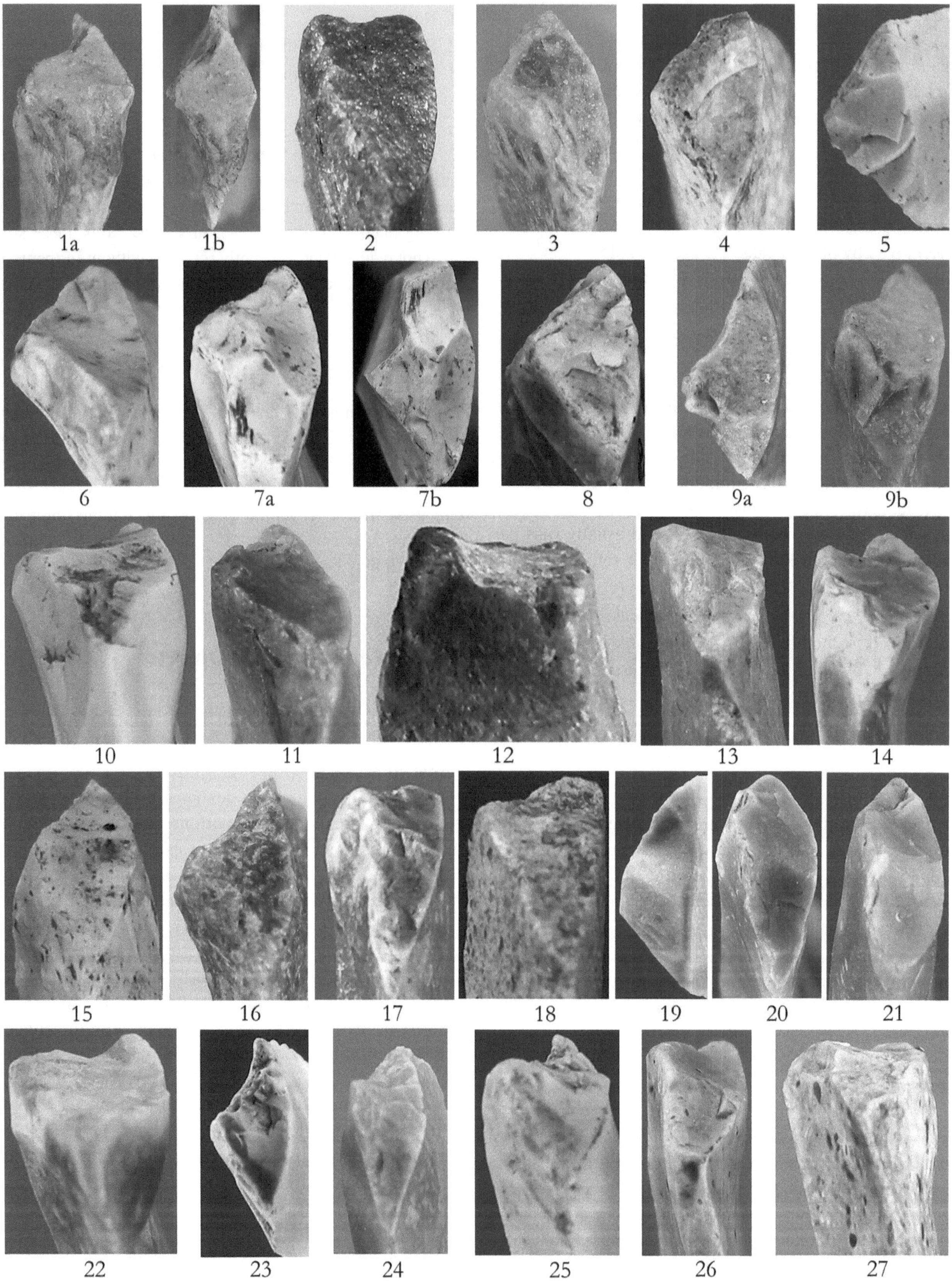

Fig. 2.13 *En éperon* preparation. 1a, 1b: Balcarova Cave; 2, 3: Brillenhöhle; 4, 5: Chênelaz; 6: Grappin; 7a, 7b: Hostim; 8: Hranice; 9a, 9b, 10: Kastelhöhle Nord; 11, 12: Kaufertsberg (lower level); 13, 14: Klementowice; 15: Mokra V; 16: Monruz; 17: Oelknitz; 18: Putim; 19: Rislisberghöhle; 20–22: Teufelsbrücke (level 3); 23–25: Teufelsbrücke (level 4); 26, 27: Wilczyce

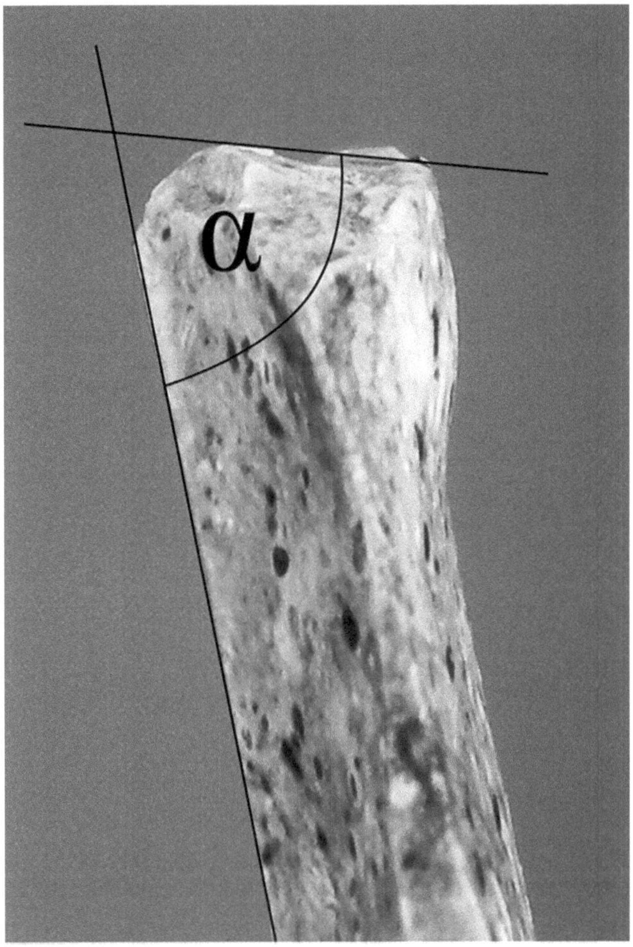

Fig. 2.14 Measuring of the flaking angle

This applies certainly for figured representations but holds especially true for abstract signs and groups of lines, the representative character of which is even sometimes doubtful. Tools, on the other hand, are bound by their functional constraints to a limited set of possible usage contexts, which makes their immediate understanding in comparison less complicated (for a detailed discussion, see Chap. 3, section "Types as Analytical Units"). Abstract symbolic meanings and creative intensions are far beyond the reach of archaeological analysis and irretrievably gone together with their bearers and cannot be reconstructed from the corpus of artifacts itself. Therefore, the questions if Paleolithic sculptures, engravings, or paintings represent artworks in the sense of fine arts or rather artisan craftworks in the sense of decorative or applied arts as well as if they are related to hunting and fertility magic or totemic ideas, if they represent clan signs or are simply "l'art pour l'art" (for more details on these subjects, see Bahn and Vertut 1988: 149ff.; Lorblanchet 1997: 85ff.), must remain unsolved. Eventually, all studies that seek to reveal the original meaning will never be more than subjective interpretations, no matter which ideas are used as explanatory concepts, and the paintings, engravings, and sculptures will keep their enigmatic character. Nonetheless, an analysis of these objects in terms of their morphology and spatial distribution may provide interesting insights into the distribution of certain concepts and consequently into the regional structure of the CEM.

During the collection of data, information on Magdalenian sculptures and engravings was recorded from 78 sites (see Fig. 2.1). Here, only the figured representations and geometrical signs are considered, and the material of each item was recorded together with the corresponding representations (e.g., geometrical signs, cervid, or musk ox). For the personal ornaments, the shape of each manufactured object and its material were noted. Groups of accidentally crossing lines and grooves on bones, antlers, or stone slabs, which appear to be rather the result of cutting or arbitrary movements than of the intention to create whatever kinds of representations, were not considered and will not be discussed.

Chapter 3
Theoretical Framework

Sites, Assemblages, and the Problem of Palimpsests

In the present study, every find-spot of Magdalenian artifacts is considered a site. This includes mixed finds as well as surface collections and single finds. An assemblage, on the other hand (also referred to as an inventory or collection), describes a group of artifacts found together in a concentration or stratigraphic layer (sometimes termed archaeological horizon, layer, or level) that cannot be subdivided any further by archaeological means (cf. Laville et al. 1980: 13f.). Ideally, an assemblage represents the artifact spectrum of a single occupation. However, especially in sites with an extensive stratigraphy, but also in open air sites, the separation of individual archaeological horizons, not to mention single occupations, is often problematic. In the common case of palimpsests, it is difficult to estimate the duration of material accumulation and often impossible to tell how many occupations took place at a given site. This has consequences for the value of assemblages as analytical units.

Many palimpsests consist exclusively of occupations by people who drew on the same typological spectrum. If a site was repeatedly visited for the same purpose by people with the same typological background, as seems to be the case, for example, at Bois Laiterie (Straus and Otte 1998; Sano et al. 2011; Sano 2012), the palimpsest-forming process increased the absolute number of artifacts but did not alter the character of the type association. The palimpsest in this case can be described as an intensification of the initial situation which retains the specialized character of the assemblage and does not obscure past behavior. On the other hand, when a palimpsest comprises occupations dedicated to different activities performed by people using the same set of types, it is very difficult (and often enough impossible) to separate the occupations. In this case the palimpsest always biases the assemblage's character toward diverse inventories. Assemblages having a highly diverse typological spectrum, very high numbers of artifacts, and long occupation continuity must thus be treated with care. Another possibility is a palimpsest whose type spectrum is sufficiently distinct to separate out some of the components which were generated by hunter-gatherers of different Paleolithic entities. Most problematic are those assemblages which exhibit features usually ascribed to different but typologically similar Paleolithic entities existing contemporaneously or in direct chronological succession. Depending on the researcher's paradigm, such inventories might be interpreted as palimpsests whose components are identifiable with reference to certain types regarded as unambiguous markers, or they might be seen as consistent assemblages reflecting contact or gradual transition from one entity to another.

It would be both naïve and wrong-headed to treat all Paleolithic assemblages as if they represented inherently consistent events. On the other hand, dismissing palimpsests as of almost no analytical importance in terms of its content is overstating the case. The cooccurrence of artifact types within an assemblage can always be considered to display a nonrandom association and to be the residue of distinct activities. The fact that an assemblage often represents several occupation events does not contradict this statement. In extreme cases, finds from different activities and periods are mixed together in one inventory. Even then, however, the assemblage is an aggregate of consistent events and by no means a product of random chance. Hence, assemblages are permissible categories of archaeological research and retain an analytical value even in the case of palimpsests.

Types as Analytical Units

Typological classifications are indispensable for the comparison of Paleolithic assemblages, because typological information is often the only kind of information available in the literature. Nevertheless, typologies bear a lot of problems

Andreas Maier, *The Central European Magdalenian*, Vertebrate Paleobiology and Paleoanthropology,
DOI 10.1007/978-94-017-7206-8_3

at the same time and are often even a major source of confusion. For large-scale investigations that are necessarily based in major parts on the analysis of archaeological publications, typologies are thus both boon and bane. This section therefore discusses the concept of types and typologies as they are used and understood in this study. It addresses the question of the ontological status of types as well as the question whether these etic categories possess emic significance and explains their value as indicators for interaction. It then presents the way typological data was recorded and which definitions of the organic and lithic tool types were used for the following analyses.

Types are the basic analytical unit of Paleolithic archaeology. They must not be confounded with tokens—in other words, with specific artifacts, which are the material realizations of a type. In principle, a type is constituted by a fixed set of properties, regardless of their nature; these may be morphological, technological, or functional. Therefore, once the definition of a type is made explicit, it is universally comprehensible and independent of individual perception. It is a well-known fact, however, that the assignment of artifacts to types may differ considerably from one researcher to another. Especially in the early days of archaeological research, types often lacked concrete definitions, and even today these definitions are far from being consistent, but can vary considerably from one research tradition to another [see, for instance, the different and indistinct definitions for *Zinken* in Brézillon (1983: 379) and Hahn (1991: 187)]. This problem does not lie with the analytical unit "type" itself, but with the underlying definitions and the ability of archaeologists to classify specific tokens within a system of types. Such inconsistencies complicate comparative studies such as the one presented here.

Many attempts have been made to develop typologies and thereby objectify the observations of morphological variability among lithic artifacts (e.g., Spaulding 1953; Sonneville-Bordes and Perrot 1954, 1955, 1956a, b; Klejn 1962; Bordes 1967; Laplace 1972; Isaac 1977). These attempts were intended to facilitate the intersubjective comparability of assemblages and enhance the quality of typological observations among archaeologists. However, it must be acknowledged that the selection and definition of the criteria considered significant for the identification of variability and, hence, the classification of artifacts, was often intuitive and arbitrary in nature (Clarke 1968: 188). Moreover, many definitions, even some within the same taxonomical system, do not follow a consistent method of criteria deduction but are rather random and arbitrary amalgamations of morphological, technological, and functional attributes and often ignore processes such as reworking and resharpening as a source of variability (Ioviţă 2009: 1447). In the end, the desired standardization of typological comparisons was not accomplished. On the contrary, definitional inconsistencies together with the growing body of typological systems have led to confusion and, in consequence, have made comparative studies even more difficult. Again, this is not a problem intrinsic to typologies in general, but is an effect of insufficient cooperation and consultation among archaeologists on a national and international level. Nevertheless, some promising attempts to improve typological systems have been made: on the one hand, there is the effort to better integrate technological properties in a consistent way in order to allow for an "analysis of choices instead of simple catalogues of shapes" (Valentin 2009: 189); on the other hand, computer programs such as SECANTO (Schmidt and Mom 2012) are increasing the reproducibility and thus the objectivity of typological investigations. The development of such programs may help to solve many of the typological problems facing archaeological research today.

The question however remains of how to generate good typologies and how best to evaluate them. Bettinger and Eerkens suggest that typologies "should be evaluated mainly in pragmatic terms: 'good' ones work—they reveal the patterns in which we are interested; 'bad' ones don't" (Bettinger and Eerkens 1999: 231). While compelling, this view seems to carry some risks. Every typological system will reveal a pattern, because the archaeological record is not homogeneous in time or in space. The fact that the observed pattern fits the question at hand or conforms to the researcher's expectations is no positive evidence of the typology's quality. Bettinger and Eerkens admit that "it is occasionally worth asking why our 'good' typologies work—and more importantly, why and where they don't" (Ibid.). Indeed, this question is worth asking constantly, because it always needs to be demonstrated that the revealed pattern is of genuine significance for the question under investigation.

Another argument states that all typologies are equally suited to measure variability and that their applicability depends on the question at hand. If a researcher defines specific criteria of whatever kind and applies them to the archaeological record in a consistent manner, then the resulting taxonomies can be understood as permissible instruments for measuring variability, and the "differences or similarities noted are 'real' and demand explanation" (Binford 1972: 114). However, there is a structural problem with this view arising from the general nature of typologies. The design of taxonomical classifications is such that they always bias morphological observations toward homogeneity. This is because the assignment of unique, handmade items to a restricted number of classes inevitably reduces complexity and leads to a loss of morphological information. Thus, observed similarities may simply be a product of the typology applied, because "variability is being perceived through the same set of typological filters" (Clark 1994). This reduction of complexity does have its advantages, however, since it allows for the comparison of a large set of inventories (Bettinger and Eerkens 1999: 231).

In sum, despite a number of significant problems, types and typologies are permissible and useful categories in Paleolithic research when they are applied in a consistent manner. Their utility, on the other hand, depends on the question under study. In any case, it is necessary to show that the observed typological patterns have a demonstrable relationship to the object of inquiry and to be aware that the design of typologies with their tendency to homogenize variability means that differences should be regarded as more important than similarities.

Facts or Fiction

In Paleolithic research, fundamental debates are still being fought concerning the intrinsic conceptions and nonempirical assumptions that guide epistemological processes and interpretations and thus influence archaeological reasoning. For example, most archaeologists would likely agree that Paleolithic entities such as the Aurignacian, Gravettian, or Magdalenian consist primarily in repeated associations of certain lithic and organic artifacts occurring in a certain area over a certain time span. However, this consensus immediately dissolves when archaeologists are asked how these repeated associations came into being and what they stand for in terms of social interactions. Here, interpretations usually oscillate between two extremes, which correspond to two diametrically opposed paradigms. Simply put, the one states that these classifications designate distinct Paleolithic social groups, while the other states that they do not (cf., e.g., Taylor 1967; Deetz 1968; Fritz and Plog 1970; Hill 1972; Casti 1989; Clark 1994). These two irreconcilable but inherently consistent ways of thinking—also characterizable as scientific realism and scientific antirealism—dominate the discussion of the nature of these Paleolithic classifications until today.

Scientific realism (or empiricism) assumes that the archaeological record consists of intrinsically meaningful and self-evident facts that can be discovered. These facts are considered to be independent of intellectual traditions and to have an objective reality apart from their defining concepts. Consequently, the smallest analytical unit of Paleolithic archaeology, the type, is thought of as a "mental template" that was present in the minds of Paleolithic people and was rediscovered by archaeologists. Some types are considered to have a very distinct temporal and spatial distribution and are therefore treated as the archaeological equivalent of a paleontologist's index fossils, unique to a certain group of people (e.g., Otte and Keeley 1990; Kozłowski 1992a; Mellars 1992; Stringer and Gamble 1993). Some sites which show a very clear stratigraphic order of well-separated archaeological horizons are assumed to be extensively generalizable in terms of their find composition and are hence treated like geological type sites (Clark 1994). The repeated association of distinct morphological features and index fossils is interpreted accordingly as the remains of people belonging to a distinct social unit, for example, an ethnicity. The temporal and spatial distributions of types, therefore, mark the respective boundaries of these groups.

The paradigm of scientific antirealism (or hypothetical realism) (Casti 1989) directly opposes the empiricist paradigm in many respects. In this view, objectivity, understood as the independence from any particular observer of the description of an issue, is problematic, because observation is always dependent on a particular mind and therefore subjective. Knowledge is thus perceived as a contingent and hypothesis-dependent approximation of reality. And because types are created in the mind of archaeologists, they cannot be seen as givens or facts, but are regarded as having no objective reality, no intrinsic meaning, and hence no existence apart from the researcher's perception of them (Binford and Binford 1966; Binford 1972, 1973; Gallay 1986: 55ff.; Trigger 1989: 13; Clark 1994). This means that they do not simply exist to be discovered, but were created by the twentieth-century archaeologists for analytical purposes, and therefore have no demonstrable relationship to concepts in the minds of Paleolithic people. Consequently, they have no relevance for the description of Paleolithic social units of any kind. Moreover, Clark argues that "retouched stone tools simply do not assort themselves in space and time according to 'key' sites and sequences, nor, upon reflection, is there any reason to expect 'key' sites and sequences to have broadly generalizable properties with respect to their retouched tool components" and that "with the widespread application of absolute dating methods, they [types] have become obsolete" (Clark 1994: 331).

A researcher's choice of one or the other, or of a position somewhere between them, has wide-ranging implications for the entire investigative process, since both paradigms affect the important question of the ontological status of types as analytical units, and consequently the estimation of their diagnostic value and argumentative capacity, and the way analytical results are finally interpreted (e.g., Bordes and Sonneville-Bordes 1970; Binford 1972, 1973; Clarke 1972a, 1994). Both views, when taken in their pure form, seem to carry their respective line of argument too far. The realists' paradigm risks adopting an overly simplistic and naïve attitude. Exponents of antirealism tend to discard a useful analytical tool too easily, not only because there are still many sites which, due to poor or no organic preservation, cannot be dated absolutely, but also because etic categories can possess a certain emic significance. In this study, a type is therefore understood as an abstract concept referring to an object with distinct morphometrical, technological, and functional properties created by researchers for analytical purposes. Being thus etic in nature, the next section explores the potential of emic insights provided by types.

Etic Concepts with Emic Significance

The basic assumption adopted in this study to bridge the gap between the empiricists' presumed objectivity and the subjective view of scientific antirealism is the concept of intersubjectivity (Husserl 1992). Intersubjectivity is two or more people sharing the same observation and perception. Knowledge is seen in this concept as something which all observers can agree on, independent of any single observer's personal opinion or perspective. Thus, this position rejects both the notion of an objective reality and the viewpoint that all observations are arbitrary singularities of isolated, subjective minds. Intersubjective experiences are usually shared between contemporaneous individuals (cf. Husserl 1992), where the shared experiential and cultural horizon sets a frame of reference whose parameters allow each individual to comprehend and mentally reenact the observations of the other. For Paleolithic research, the question is therefore whether and to what extent intersubjective experiences are possible between non-contemporaneous individuals whose mental conceptions were acquired in different cultural and experiential spheres. Or to put it more concretely, is it possible for today's archaeologists to reenact the observations of Late Upper Paleolithic people?

To begin with, Paleolithic artifacts can be thought of as "fossilized ideas" (Deetz 1967: 45) or "petrified coherences of theoretical and practical interaction between people and their environment. As stony or bony manifestations, they combine needs and thoughts with environmental conditions and constraints" (Gehlen 1986: 11). Paleolithic artifacts represent "cognitive processes which last in this form independently of their functionality at the time. Therefore, they objectify, pass on, and stabilize individual thinking, cognition, and, hence, knowledge through reproduction, usage, maintenance and storage" (Fischer 2004: 82ff.).

Assuming that the general cognitive structure of *Homo sapiens sapiens*' brain has not changed significantly, if at all, since the Last Glacial Maximum, it should be possible—at least in principle—for people living today to comprehend Late Upper Paleolithic knowledge as such. However, common points of reference that may provide orientation for cross-cultural intersubjective observations are needed. Paleolithic artifacts preserve, among other things, information on specific problem-solving strategies, i.e., the observations of specific problems and needs as well as the respective mental responses to them. The more complex these problems or needs are and the higher the potential variability of possible responses is, the more difficult it will likely be for outside observers to deduce the respective observations. In general it can be said that the fewer the constraints of how artifacts must be constituted to be used for their appointed tasks, the higher is their potential variability and the harder is an intersubjective understanding of them.

Problems related to the spiritual sphere are probably among the hardest to deduce. For members of a common culture, the execution of spiritual tasks is bound to certain rules and conventions which are known to all. Thus, the material auxiliary devices used to execute these tasks usually allow only for a restricted set of alterations (cf. Deetz 1967: 45). Between members of different cultures, in contrast, these rules may be fundamentally different, and as a consequence, the objects applied may also differ considerably. Summoning spirits, for instance, is theoretically possible with any kind of object, and only cultural rules determine whether crucifixes should be involved or rather antlers, feathers, or what have you. In this case, there is no possible way for an outside observer to tell whether the one or the other object is more suitable for the purpose. By the same token, it is almost impossible to deduce its use for summoning only by looking at an object without specific knowledge of its cultural background. The high degrees of freedom in the choice of shape, texture, and handling of objects used in spiritual tasks depend on the fact that there are no physical or logical constraints which would render them unsuitable.

Tools, on the other hand, are strongly bound to physical functional constraints, which remain stable over time. Cutting meat, for example, is easiest with a sharp cutting edge. This fact alone reduces the number of possible objects which may serve to accomplish this particular task considerably. Conversely, sharp edges are likely to serve in any culture, among other things, for cutting purposes. Functional specialization thus effectively limits the number of tasks that a specific Upper Paleolithic tool may respond to and allows for the identification of its intended purpose using morphological, technological, functional, and traceological analyses. On the basis of such analyses, the researcher can intersubjectively rethink the way a specific problem was observed and solved by Paleolithic hunter-gatherers (cf. Ricoeur 1984: 5f.). For instance, a blank with a semicircular retouch at one end which exhibits microscopic traces of working hide can be interpreted as a tool to remove fibers and tendons from a piece of animal skin. Hence, it is possible to mentally reenact the following sequence: (a) observation of the need to remove tendons from hide, (b) production of a tool which fits the task, and (c) accomplishment of the task and solution of the problem. The objection that the same tool could also have been used to serve other yet unrecognized purposes is valid, but by no means invalidates this inference, which remains true. Consequently, tools more than other artifacts have the potential to provide the basis for an intersubjective cross-cultural experience.

These observations hold a range of implications for the relationship between archaeological types and Paleolithic societies. Paleolithic research shows that tools with specific combinations of distinct morphological, technological, and functional properties occur repeatedly at different sites. Together with the finding that "in every culture, there are conventions which dictate the form of artifacts" (Deetz 1967: 45), it is reasonable to assume that the observed reoccurrences are the material expression of common concepts which guided a given artifact's production. In an act of intersubjective observation, these concepts can by "rethought" and expressed as types by archaeologists. As such they are not just arbitrary categories that are forcibly imposed on the archaeological record with which they share no genuine connection. On the contrary, they are carefully developed from the observation that specific tokens are replicated uniformly at different sites throughout a certain time span. Moreover, trace wear analysis has shown that those Upper Paleolithic tool tokens which are assigned to a certain type—such as "scraper," "borer," "bruin," or "projectile point"—were indeed mainly used to fulfill the specific tasks which are conceptually related to the respective type (Sano 2012: 215ff.).

At this point, it can be said that while Upper Paleolithic tool types do not replicate the everyday practices of Upper Paleolithic people, they likely reflect to a large extent concepts that underlay them. If this were not the case, the reoccurrence of specific combinations of morphological, technological, and functional properties would be difficult to explain. It is highly unlikely that such concepts were radically separate from all the other conventions which presumably guided the Upper Paleolithic hunter-gatherer way of life. Consequently, it may be assumed that these concepts were part of the cultural conventions. Thus, the category "type," generally thought of as etic in nature, can be also considered to possess emic significance and meaning.

Types as Indicators for Interaction

The logic of taking material cultural traits as an approximation of interaction is based on the finding that humans do not behave in the same way at all times and all over the world. Regardless of when or where it takes place, artifact production is therefore restricted to and guided by a limited set of culturally inherent concepts (Deetz 1967: 45). The process that keeps cultural traits (traditions) constant is cultural transmission, which "functions like an inheritance system, producing significant similarities between those handing down the information and those adopting them" (Boyd and Richerson 1985: 46ff.). At the same time, phenomena such as innovation, guided variation, indirect bias, copying errors, or drift cause alterations of traditions (see Neiman 1995; Bettinger and Eerkens 1999; Bentley and Shennan 2003; Eerkens and Lipo 2005; Shennan 2008: 77). Drift (or sampling error in cultural transmission) is a special process that affects the transmission of cultural traits. And the "Markovian structure of drift makes it likely that isolated groups will diverge from identical starting frequencies [of cultural traits], even in the absence of innovation that is historically unique to each group" (Neiman 1995: 31). As Buchanan and Hamilton explain, because "drift is a consequence of sampling, it is amplified in smaller populations in which the number of people to copy from, and the number of objects or traits to copy are limited" (Buchanan and Hamilton 2009: 280). This process becomes even more pronounced when parts of a population become isolated (Shennan 2000, 2001; Henrich 2004). Thus, two populations that start with the same set of traits will diverge from each other in the case of non-interaction, and strong similarities in typological, technological, or other cultural traits between both populations rather indicate intensive interaction. Here, similarities seem to hold different significance for the identification of interaction intensity than do differences. Whereas the exploitation of the same resources and the use of similar technological, typological, and representational concepts appear to be highly indicative for interaction (Binford 1972: 116), divergences in these operations do not necessarily reflect a lack of interaction; rather, they may be the result of individual choice and conscious delimitation or may even represent negative responses to interaction with other groups (Hodder 1982b). It is true that the "distribution of material culture traits, and the maintenance of group identity are not necessarily and wholly structured by patterns of interactions" (Hodder 1977: 269). Still, with the above said in mind, the intensity of interaction as mirrored in the degree of similarity of typology, technology, or subsistence strategies, or as evidenced by the joint exploitation of local resources, can be taken as a measure for the intensity of social relations, since "similarity over time is a function of the number of transmission episodes" (Neiman 1995: 31), meaning it is dependent on the degree of interaction. Tracing the traditions of types maintained by individuals with limited temporal and spatial frames of activity may thus provide access to the geographic and diachronic interaction of Paleolithic hunter-gatherers. In this way it may contribute to understanding the reasons for variation, which in turn is a crucial step in understanding the variability and diversity of the Paleolithic archaeological record.

Several schemes of cultural transmission are distinguished in the archaeological literature. A major difference is drawn between unbiased and biased transmission. Unbiased transmission (Boyd and Richerson 1985) is defined as the one-to-one passing on of knowledge from one individual to another. Here, "each individual is equally likely to be copied, while each variant is copied in proportion to its frequency" (Bentley and Shennan 2003: 460). As a result, variation among the population is transmitted and amplified over time (Eerkens and Lipo 2005: 322). Biased transmission appears in many manifestations. They differ in terms of the bias affecting the transmission process and, hence, in terms of the rules guiding the decisions of individuals to acquire or abandon cultural traits or to develop variants. The most important such schemes are conformist-biased transmission (Eerkens and Lipo 2005: 323f.), prestige-biased transmission (Eerkens and Lipo 2005: 325)—both are understood by S. Shennan (2008: 77) as expressions of a context bias—content bias (Shennan 2008: 77), indirect bias (Boyd and Richerson 1985; Bettinger and Eerkens 1999: 236), and guided variation (Ibid.; Ibid.), the latter two subsumed by S. Shennan under the hypernym results bias (Shennan 2008: 77).

Different aspects of Paleolithic artifacts are considered to be of different diagnostic value for the detection of tradition and thus interaction. A major distinction is often made between style and function (Sackett 1977; Dunell 1978). The functions of tools or other artifacts, such as scraping hide or indicating a person's social status, depend on laws and conventions and therefore allow for only minor alterations within the realizations of their material auxiliary devices (see above). Style, on the other hand, is restricted to a much lesser extent and exhibits much greater variability. Whereas function is thought to be subject to natural selection, style is seen as depending rather on stochastic processes (Dunell 1978: 199). Another view is that style and function operate in parallel most of the time (Bettinger et al. 1996; Boone and Smith 1998). In the current discourses of cultural evolution, the division of style and function is often related to the question of whether the aspect of the archaeological record being investigated is payoff correlated or not (e.g., Brantingham 2007). Here, payoff describes the "differential performance or utility of socially or individually learned behaviors, including the items of material culture that are integral to those behaviors" (Brantingham 2007: 396). In other words, a trait is payoff correlated if it contributes to an individual's fitness. In this regard, functional traits are considered as being payoff correlated, whereas stylistic traits are payoff independent. Consequently, it is assumed that stylistic traits are much more open to innovation, copying errors, drift, or individual choice (e.g., Neiman 1995). Since similarities in style are therefore less likely to occur as a result of coincidence, their variation exhibits traditions—and thus interaction—more pronouncedly than do variations in functional traits (Morrow and Morrow 1999: 227; Buchanan and Hamilton 2009). Moreover, S. Shennan argues that "if a particular cultural attribute, for example, the sharpness of a lithic cutting edge, is very strongly determined by its function, then it will contain no signal of its transmission history" (Shennan 2008: 79). Since tools are strongly determined by their function, this statement might suggest that they are a rather unsuitable class of artifacts for analyzing interaction. However, a sharp cutting edge (or any other kind of functional part) does not occur isolated. This "transformative part" is usually connected to other parts of the tool, which serve the transmission and reception of the work energy and function as a handle (Boëda 2013). The latter two parts of the tool allow for a sufficiently high variation that it is possible to trace variation and thus interaction.

Some researchers argue that typological data is simply much too coarse grained to support the detection of traditions. Three main arguments as summarized by Clark (1994: 332f.) are as follows:

(1) Basic tool types occur virtually everywhere and are consistent over very long periods of time, and variations in type frequencies must have a functional basis.

It must be conceded that some basic tool types occur almost everywhere and endure over very long time periods. However, others are relatively strongly restricted in both their spatial and temporal occurrences (e.g., *navettes*, Lacan burins, or parrot-beak burins). The claim that the reason for differences in inter- and intraregional type frequencies, and therefore assemblage composition, "can only be functional in nature" (Clark 1994: 334) remains to be demonstrated. In contrast, the fact that several contemporaneous tool types serve the same function strongly suggests that the "way of doing things" (Hodder 1990: 45) may be part of the reason as well.

(2) The high mobility of hunter-gatherers tends to obscure group patterns over the long term.

Paleolithic hunter-gatherers were highly mobile. Nevertheless, this does not mean that they moved around randomly; it is more likely that different regional groups exploited different areas, whose resource locations and environmental features were known to them. On this basis it can be supposed that these regional groups left traces in distinct patterns. Moreover, the surprising spatial integrity of catchment areas throughout the duration of the CEM and the fact that regional differences in preferences of tool concepts are observable contradict the hypothesis that the high mobility of hunter-gatherers necessarily obscures group patterns over the long term.

(3) The resolution is not sufficiently fine grained to identify identity-conscious social units as necessary bearers of traditions, and Paleolithic entities simply last too long and cover too broad an area to be treated as analogous to them.

Identity-consciousness is an individual mental state that can never be discovered through excavation, no matter how fine grained the analysis (see section "Ethnicity"). However, the identification of identity-conscious social units is not a prerequisite for the detection of traditions. Interaction and not identity-consciousness is the important factor for cultural transmission. The fact that some Paleolithic-type concepts remain unchanged over large areas and long time periods is by no means an argument against their having been guided by traditions and cultural transmission. Such constancy is much more an expression of a low rate of innovation, which may have several causes: a low population density (Shennan 2001: 15), conformist-biased transmission (Kohler et al. 2004), and/or intensive interaction which compensates for the effect of drift (Neiman 1995).

Given the low population density during the Upper Paleolithic, random meetings of usually isolated groups were likely very rare, and it is reasonable to assume that dispersed and unrelated subpopulations would have differed considerably in their typological concepts. The investigation of the CEM record reveals striking redundancies of certain concepts at different sites, a fact which demands explanation. The probability that these redundancies came into being as a result of coincidence is virtually nil. Thus, it is more reasonable to interpret the observed similarities as evidence of interaction between the individuals who applied these concepts. Thus, it is assumed in this study that despite the mentioned objections, similarities between types (and other cultural traits) can be taken as indicators of interaction and the frequency of such redundancies, as well as the strength of the similarities and differences between artifacts, can thus be taken as a measure of the degree of interaction.

The Use of Ethnographic Analogies in this Study

The use of analogy in prehistoric archaeology has generated many arguments for and against [see, e.g., Ascher (1961), Wobst (1978), Hodder (1982a), Murray and Walker (1988), Eggert (1995), Gramsch (2000) and articles therein; for extensive explications, see Wylie (1985) and Bernbeck (1997)]. Most important are analogies derived from ethnographical research. However, the applicability of data obtained through observation of modern-day hunter-gatherer groups to the interpretation of behavior in Paleolithic times is still a topic of considerable debate (Binford 1972, 1980, 1982, 1983a, 2001; Clarke 1972a; Wobst 1978; Gamble 1986; Kelly 1995; Van Reybrouck 2000; Wiermann 2000). A major objection to the use of ethnographic analogies is the uncritical borrowing of seemingly suitable examples to explain observed patterns in the archaeological record. Such a handling of ethnographic data fosters an increasing appearance of "glimpses of Nunamiut, !Kung and other groups faintly camouflaged in Paleolithic costume" (Gamble 1986: 28). It is tempting to assume the existence of a relation between recent hunter-gatherers and those of the remote past only on the strength of similarities: while it is true that there are still reindeer hunters who use rather simple technologies and that ethnographic terms are easily translatable back into the Paleolithic (Ibid.: 29), neither fact justifies an assumption of relatedness. Hence, researchers must take care that their use of analogies does not lead to circular reasoning (Clarke 1972a). The fact that Paleolithic patterns may correspond to recent ethnological examples in no way indicates that these assumptions are correct (Gamble 1986: 59; cf. Clarke 1972a: 40).

Since there is no way to observe Paleolithic behavior directly, it stands to reason to try understand the past through the present (Spaulding 1968: 37). Thus, a way is needed to establish a link between ethnographic and archaeological observations that is stronger than mere similarities. In order to "productively using ethnographic data in the service of archaeological goals" (Binford 2001: 2) and thus making justified assumptions about the past, disciplined reasoning is required. The link with the potential to connect recent and prehistoric hunter-gatherers is their subsistence. The consumption of environmental resources, rather than agricultural production, implies a mobile way of life and a relatively small group size of people living together permanently (Gamble 1986: 29ff.). These preconditions, in turn, are basic to patterns of social relation within and between hunter-gatherer groups. Since there seems to be a "necessary ecological linkage between environmental variables and the organizational properties of hunter-gatherers" (Binford 2001: 51), these properties can be inferred from environmental parameters. Thus, a disciplined selection of hunter-gatherer cases based on environmental parameters and subsistence strategies is prerequisite to meaningful comparisons. To be sure, identical environments do not necessarily result in identical behavior – they simply make similar behavior more likely. The environment imposes constraints on the distribution and availability of resources. However, within the given parameters, societal norms and decisions specify the way the environment is exploited.

Eventually, ethnographic analogies are considered not only permissible but essential epistemological tools for Paleolithic research. They are a rich source of ideas for the interpretation of the Paleolithic record and a good background against which those ideas can be tested. Hence, such information can be a useful tool for the development of concepts and reflections about past societies. The lifestyles of recent hunter-gatherer groups should not, however, be understood as a one-to-one translation of practices during the Paleolithic and thus taken as an explanatory concept in themselves. However, they provide "a means to test and strengthen our procedure of analysis from statics to dynamics" (Gamble 1986: 30)—that is, from dead artifacts to living behavior.

Territories, Territoriality, and Land Tenure

The analysis of spatial variability raises the question as to in which way the observed regional differences in the exploitation of resources might be the result of territoriality. Therefore, this special aspect deserves special attention. Generally, "territory" is not a consistently defined technical term in archaeological research and is assigned to different connotations or is simply used intuitively without precise explanation (see Stanner 1965; Heinz 1972; Higgs and Vita-Finzi 1972; Peterson 1975; Wobst 1976; Binford 1982). One widely adopted conception of territory is that given by Noble in the 1930s: he described it as "any defended area" (Noble 1939: 267). A problem for the application of this definition in Paleolithic research is that "defense" is often mentally connected to "physical aggression." This is perhaps the reason why some researchers reject the hypothesis of territoriality in Paleolithic societies. Reynolds, for instance, wrote with reference to Vallois (Vallois 1961: 229) that "the evidence indicates that early Paleolithic man was cooperative, not territorial, and had social and sexual relationships over wide areas. Societies still living in a nomadic hunter-gatherer ecology, such as the Bushmen of the Kalahari, or the Hadza of east Africa, show little territoriality or inter-group aggression" (Reynolds 1966: 449). Peterson, in contrast, came to the conclusion that "man is [...] territorial in a number of senses. He has a personal space halo, notions of property in land, a willingness to defend areas of ground, and limits to the range of his daily, yearly, and lifetime wanderings" (Peterson 1975: 63). Reviewing numerous examples of territoriality among recent hunter-gatherer societies, R. Kelly came to the conclusion that "Upon reconsideration of ethnographic evidence, we see that no society has a laissez-faire attitude toward spatial boundaries" (Kelly 1995: 185). In contrast to territoriality, which, according to Kelly, "means the exclusive use of a defended area" and thus "focuses attention on only one aspect of a range of behaviors differentially implemented to regulate resource use under different conditions" he uses the term land tenure to refer to "different ways of regulating people and land" (Ibid.: 163).

Even without ethnographic examples which may speak for or against defensive behavior among hunter-gatherer groups, there is another and more practical objection to such a definition of territory in Paleolithic research. Given the low population density at the time and the likelihood that wandering hunter-gatherers exploited large areas, it would simply not have been possible for a mobile group to "defend" its exploitation range by patrolling its borders (Peterson 1975: 60). Still, the exploited area provides the subsistence base for hunter-gatherers and is thus of utmost economic importance. Furthermore, evidence from ethnographic research indicates that hunter-gatherers generally do not move randomly across a landscape but stay within certain boundaries, which are in turn respected by neighboring groups (e.g., Peterson 1975: 60f.; Binford 1982: 8).

The analysis of CEM raw material procurement patterns shows that regional agglomerations of sites generally correspond to overlapping site catchment areas, indicating a common exploitation of the same outcrops by groups of people settling in the same region. Notably, some raw materials are found exclusively within the assemblages of one regional group, although sites of neighboring groups are located relatively nearby. This is the case, for example, for raw materials found in the High Rhine region or at the confluence of the Danube and Isar rivers (see Chap. 6, section "Procurement Pattern of Lithic Raw Materials"). This pattern, with regionally overlapping site catchment areas on the one hand and the selective usage of raw material sources by only one of two neighboring groups on the other, strongly indicates the existence of five non-overlapping exploitation territories. Given that the observed pattern was generated over a period of some 2000 years, its remarkable temporal and spatial integrity emphasizes the necessary existence of social conventions that were passed on from one generation to the next; otherwise, the clarity of the pattern would be very difficult to explain. The absence of such conventions for the procurement of raw materials or major ruptures within their transmission probably would have generated a much more chaotic picture, reflecting changes in raw material procurement. In this case, it would be highly unlikely that borders between the different regional catchment areas would still be observable today. This finding gives reason to suggest that Magdalenian hunter-gatherers were conscious of the limits of their exploitation territories and recognized their and their neighbors' respective boundaries. Since the demarcation of these areas is exclusively based on raw material procurement and is therefore purely resource related, it seems more reasonable to describe the observed pattern in terms of territoriality than to employ the socially far more complex term "land tenure."

With all this in mind, however, in this study, a territory is not understood as a tightly circumscribed and impermeable area, but is instead seen as an area exploited during several annual cycles by a specific regional hunter-gatherer group, with borders that are recognized by neighboring regional groups. These boundaries are in accordance with social mechanisms and conventions permeable to members of other groups. The precise nature of these mechanisms and conventions is nevertheless beyond the resolution of the archaeological record and can only be suggested through the application of analogical reasoning. What can be inferred directly from the record, in contrast, are the spatial limits within which the movements and hence the daily operations of a regional group took place. Individuals exploiting a certain region together while deliberately sparing the resources of another nearby region can be considered a territory-conscious unit. Thus, it is possible to identify territory-conscious interaction units which existed in the past. With regard to the CEM, at least five such units can be identified (see Chap. 7, "Territories and Land-Use Patterns of the Five Regional Groups").

Ethnicity

Another issue which is often associated with the interpretation of the Paleolithic record or with the idea of reconstructing Paleolithic territories is that of ethnicity. There is a long-standing and ongoing discussion about whether or not archaeological entities such as the Magdalenian represent a kind of ethnic group (e.g., Childe 1956; Bordes and Sonneville-Bordes 1970; Binford 1972, 1973; Bordes 1972, 1973; Hodder 1982b; Wotzka 1993; Clark 1994; Jones 2008). The term "ethnicity" itself is problematic, being underdetermined and lacking a consistent definition, not to mention burdened with political implications. Today, ethnicity or ethnic identity is rather seen as a dynamic, social product of the interdependencies of external perception and self-perception of a group (Barth 1969; Brumann 2007). Understood as an internal, mental state of human beings, as opposed to an observable characteristic measurable from the outside, ethnicity reveals itself as inappropriate for the archaeological description of Paleolithic societies, where the opportunity to directly question the individuals under study does not exist. Similar problems occur with the term "identity-conscious social unit" (cf. Clark 1994: 331). Much like ethnicity, social identity-consciousness is a mental state that may be expressed in many ways, none of which necessarily include material goods. Nor do territorial boundaries always delimit units of identity-conscious people, and several identity-conscious social units may share the same territory. Consequently, in Paleolithic research it cannot be assumed that the production and usage of similar objects implies that the manufacturers and users conceived of themselves as belonging to the same social unit (cf. Hodder 1982b; Brumann 2007). Moreover, it is doubtful whether all members of an entity which produced similar objects were identity-conscious about their social unit or if all those who did possess such consciousness shared the same perception of their belonging or membership. The same holds true for the exploitation of resources in a certain area. From a systemic point of view, it is therefore reasonable to say that the question of the identity of a system, be it an individual or a social unit, must be posed and answered within the system itself and not by an exterior observer (Luhmann 2009: 15). Therefore, although the concept of the identity-conscious social unit is less problematic than that of ethnicity, it is equally beyond the reach of archaeological analysis.

Part II
Analyzing the Central European Magdalenian

Chapter 4
Past Research on the Magdalenian and Its Current Implications

The Rock Shelter of La Madeleine

Édouard Armand Lartet (1801–1871) a French paleontologist and "the leading figure in early Paleolithic research" (Trigger 1989: 95), and his friend and patron Henry Christy (1810–1865), an English banker and ethnologist, discovered the site of La Madeleine in the Vézère Valley in 1863 (Lartet and Christy 1864). There, a chapel dedicated to Mary Magdalene overlooks a small village and gave its name to both the rock shelter where the site is located and to the settlement. Lartet and Christy excavated the central part of the 250 m long, southwardly oriented rock shelter on the northern bank of the Vézère in autumn and winter 1863–1864. They found a rich collection of bone and flint artifacts as well as objects of art, such as a thin ivory plate with engravings, which were recognized as the outlines of a Mammoth by the Scottish paleontologist Hugh Falconer (1808–1865), who was present coincidentally, when the plate was discovered by a workman in May 1864. The excavation went on until 1865, when Christy died of pneumonia. He left unfinished his part of a book entitled *Reliquiae Aquitanicae; being contributions to the Archaeology and Paleontology of Périgord and the adjoining provinces of southern France*, in which he and Lartet intended to publish their results and discoveries from the La Madeleine rock shelter and other sites. Four years later, in 1869, Louis Laurent Gabriel de Mortillet (1821–1898), a French archaeologist, named the type of bone industry excavated in La Madeleine *Magdalénien* in his article *Essai d'une classification des caverns et des stations sous abri fondée sur les produits de l'industrie humaine*. Lartet continued working on the book until his breakdown in health in the autumn of 1870, followed by an apoplectic stroke in January 1871. After Lartet's death, the English geologist and paleontologist Thomas Rupert Jones (1819–1911) completed the text and published the book in 1875.

Following the works of Lartet and Christy, the rock shelter of La Madeleine was the object of several episodic excavations, of which those conducted by Elie Massénat and Paul Girod around 1895 and by Emile Rivière about 1901 have been documented and published (Bouvier 1977). From 1910 to 1913 and after the First World War in 1926, Denis Peyrony (1869–1954), a French school teacher and archaeologist, resumed work at the rock shelter. First, Peyrony and his workmen examined the back dirt of the old excavation, which is estimated by Bouvier (1977) at 1400 tons. After removing the back dirt, Peyrony recognized differences in the color and composition of the remaining in situ sediments, which were visible in the stepped profile (Capitan and Peyrony 1928). He defined three geological horizons and attributed them to three archaeological levels. The recording of the finds was carried out eventually according to these three levels. Within this three-stage sequence, Peyrony observed a change in the bone and antler industry from the bottom to the top. In the lower level, he found so-called proto-harpoons or primitive harpoons. These are bone or antler points with sawlike edge preparation. The next level then is characterized by points with a single row of barbs (single-row harpoon), and in the upper level, these points show barbs on two sides (double-row harpoon).

The Subdivision of the Magdalenian into Six Phases and the Critique to Which It Is Subject

Already in 1912, H. Breuil subdivided the Magdalenian into six phases (I–VI, Table 4.1) in his famous contribution to the congress in Geneva: *Les subdivisions du Paléolithique supérieur et leur signification*. For the older phases I to III,

Andreas Maier, *The Central European Magdalenian*, Vertebrate Paleobiology and Paleoanthropology,
DOI 10.1007/978-94-017-7206-8_4

Table 4.1 Typological subdivision of the Magdalenian

Phases	After	Typical objects	This study
0	Bordes	OG: single-bevelled points LI: mostly on flakes, transversal burins, multiple borers, *raclettes* rare/absent; additional bladelet production	Badegoulian
I	Breuil/Peyrony	OG: single-bevelled points LI: mostly on flakes, transversal burins, multiple borers, *raclettes*; additional bladelet production	Badegoulian *à raclettes*
II	Breuil/Peyrony	OG: thick, single-bevelled points (also grooved) LI: mostly on blades/bladelets: backed bladelets, multiple burins, scalene bladelets, end scrapers	Lower Magdalenian
III	Breuil/Peyrony	OG: single (long-)bevelled, grooved, short points (type "Lussac-Angles"), double-bevelled, long, thin points, *baguettes demi-rondes* LI: scalene triangles	
à navettes	Allain et al. (1985)	OG: *navettes*, phalliform objects	
IV	Breuil/Peyrony	OG: "proto-harpoons," *baguettes demi-rondes*, *rondelles*, *contours découpées*	Middle Magdalenian
V	Breuil/Peyrony	OG: single-row barbed points, long double-bevelled points (circular cross section)	Upper Magdalenian
VIa	Breuil/Peyrony	OG: double-row barbed points (barbs close to the shaft, thick points with quadrangular base; no more *baguettes demi-rondes*, *contours découpées* or *rondelles* LI: points de Teyjat, parrot beak burins, points Laugerie basse	
VIb	Breuil/Peyrony	OG: double-row barbed points (angular barbs well separated from the shaft) LI: short scrapers, triangles, rectangles, segments, backed points	

OG Organic tools, *LI* Lithic tools

the Grotte du Placard in Charente was his most important site of reference. He noted:

> Trois vastes assises (Magd. 1 à 3) assez facilement discernables y sont observées, sans qu'aucune œuvre d'art comparable à la série classique […] Ce n'est qu'au-dessus que se développent les assises à harpons d'abord rudimentaires (Magd. 4), puis à une (Magd. 5) et à deux rangées de barbelures (Magd. 6), ce qui fait six divisions fondamentales magdaléniennes avec leur outillage suffisamment reconnaissable, et leur position précisée par des superpositions stratigraphiques bien constatées. (Breuil 1937: 46)

The crucial elements for the definition of the Magdalenian I, II, and III were thus the bone and antler points and their changes in morphology throughout the stratigraphic sequence of the site of Grotte du Placard. For the Upper Magdalenian on the other hand, Breuil used the division proposed by Peyrony in La Madeleine, as well as corresponding observations from other sites in the Périgord and the Pyrenees by Edouard Piette, to establish the three phases IV, V, and VI according to the changes in harpoon morphology mentioned above. Later D. and E. Peyrony integrated elements of the lithic industry into this division, which was by then almost exclusively build up on bone and antler tools. For the Magdalenian phases I–III, they used their stratigraphic observations at the site of Laugerie-Haute (Peyrony and Peyrony 1938) and for the phases IV–VI the rock shelter of La Madeleine and Villepin (Capitan and Peyrony 1928; Peyrony 1936).

At Laugerie-Haute, Peyrony and Peyrony (1938) describe the lithic industry of the Magdalenian I in level I′ as characterized by numerous flakes of different dimensions and forms, backed with abrupt retouches, thick scrapers, and a low abundance of backed bladelets, together with end scrapers on blades, burins, and fine borers (Ibid.: 47ff.). In contrast, the industry of the Magdalenian II in level I″ is marked by a low amount of thick scrapers and flakes with abrupt retouches, while backed bladelets are quite numerous. But the most characteristic feature is the occurrence of triangles (*triangles scalènes*) which motivated Peyrony to call this facies *Magdalénien à triangles scalènes* (Peyrony 1936). End scrapers on blades, burins, and borers are present like in the previous level (Peyrony and Peyrony 1938: 54) and are also frequent in the assemblage of the Magdalenian III in level I‴. Here, the backed bladelets are again abundant, but triangles are almost absent (Ibid.: 56).

The lowest level at *La Madeleine*, corresponding with the Magdalénien IV, contains various forms of burins, including some rare pieces which resemble a little the type *bec de perroquet* (parrot beak burin). There are long and short end scrapers, sometimes with a lateral retouch. Backed bladelets are numerous as well as backed points (Capitan and Peyrony 1928: 20ff.). The intermediate level containing the Magdalénien V is according to its lithic industry very similar to the previous one. But it is distinguished by some atypical shouldered points as well as burins *bec de perroquet*, which occur here in little higher number and are somewhat less massive than in the lower level (Ibid.: 63f.). In the upper level, finally both tool types are quite abundant and occur in various forms. Additionally there are some backed points, some with a retouched base and some double points (Ibid.: 81ff.).

At the Abri de Villepin, situated about 50 m downstream from La Madeleine, Peyrony subdivided the Magdalenian VI once more into a lower and an upper facies, of which the latter contrasts with the former by the presence of harpoons with nearly rectangular barbs that are very clearly detached from the shaft, long truncated blades, and scrapers on flakes (Peyrony 1936: 271).

In the course of subsequent research, Breuil's subdivision of the Magdalenian into six phases (Table 4.1) was often criticized and has undergone several adaptations and alterations. Most of these reclassifications unite neighboring phases of Breuil's system in changing combinations resulting in two to four phases. As this chapter does not aim to give a comprehensive overview of all of the proposed modifications—some of which yield only minor changes—it is sufficient at this point to focus on those critiques which contribute substantial arguments to the discussion and are thus most relevant for this study.

An important alteration to the six-phase classification system was the introduction of an additional phase 0, which was said to precede the Magdalenian I. This phase 0 was defined according to observations by F. Bordes at the rock shelter of Laugerie-Haute Est (Sonneville-Bordes 1967). Today, however, the Magdalenian 0 (sensu Bordes and Sonneville-Bordes) and the Magdalenian I (sensu Breuil) are referred to as Badegoulian (Cheynier 1939, 1951; Vignard 1965; Allain 1968, 1989). The eponymous rock shelter of Badegoul was excavated by André Cheynier, who ascribed the finds in 1939 to a so-called Proto-Magdalénien, which he subdivided into three phases in 1951. To avoid confusion with the Proto-Magdalénien defined by Cheynier and the pre-Solutrean Proto-Magdalénien defined by D. and E. Peyrony (1938), E. Vignard proposed the term Badegoulian for Cheynier's archaeological entity in 1965, which he considered to be independent of the Magdalenian. Typologically, the Badegoulian is characterized by the frequent use of flakes for tool production and the occurrence of transversal burins, multiple, "star-shaped" borers, as well as nosed scrapers, carinated scrapers, and side scrapers. Dihedral burins and backed bladelets are rather scarce, and lithic triangles are virtually absent. J. Allain emphasized the separation between Magdalenian and Badegoulian based on his works at the Fritsch rock shelter and subdivided the Badegoulian into two phases: an older phase without *raclettes* and a younger one with *raclettes* (Allain and Fritsch 1967; Allain 1968, 1989; Bosselin and Djindjian 1988).

In 1985, J. Allain, R. Desbrosse, J.K. Kozłowski, and A. Rigaud introduced an additional Magdalenian facies, the Magdalenian *à navettes*, which is mainly characterized by the occurrence of a specific organic artifact, the so-called *navette*. Chronologically, this tool appears to be contemporaneous with organic points of the type *Lussac-Angles* and with lithic triangles (Bosselin and Djindjian 1988: 324) and thus to be restricted to the period before the advent of barbed points. However *navettes* occasionally seem to co-occur with so-called proto-harpoons (Allain et al. 1985: 116). Generally, *navettes* are interpreted as hafting devices, the split ends of which served to receive lithic implements. According to Allain et al. (1985), the occurrence of this tool is restricted to a rather short period of time, and thus they suggest that the Magdalenian *à navettes* fills a gap in the classical system between the phases I–III from Laugerie-Haute and the phases IV–VI from La Madeleine. It is however conspicuous that the appearance of *navettes* roughly coincides with the occurrence of so-called *baguettes demi-rondes*. These half-round rods, which were used pairwise, appear for the first time in assemblages assigned to the Magdalenian III of Breuil's classification model (Bosinski 1990: 180), and they disappear during the Upper Magdalenian (Leroi-Gourhan 1988: 644). *Baguettes demi-rondes* are often interpreted as thick, large, composite projectile points, which could only with difficulty (or not at all) be manufactured as massive points from the compact tissue of reindeer antlers (Ibid.). However, the interpretation of these half-round rods as composite hunting weapons is not totally convincing. A major point of critique here is the elaborate plastic decoration of some of these tools, particularly during the later phases of their occurrence, as they are reported from Isturitz, for instance (Bosinski 1990: 180, 199). Such a design appears to accord better with a highly curated tool with a long use-life than a piece of consumption equipment such as a projectile point. As a pair of *baguettes demi-rondes* and a *navette* are often very similar in their dimensions and overall morphology, it seems reasonable to assume that they had a similar function. Consequently, *baguettes demi-rondes* and *navettes* are perhaps best understood as hafting devices, the former in a composite version, the latter a massive variant of the same concept.

One of the most substantial critiques of the six-phase classification was expressed in a 1988 study by B. Bosselin and F. Djindjian on the chronological significance of Magdalenian and Badegoulian lithic tool types. For a profound evaluation of the archaeological record, they calculated an extensive Correspondence Analysis using data on 19 tool types from 103 sites in France. They eventually came to the conclusion that "Breuil's six-phase evolutionary model of 1912, which was taken up and extended by D. and E. Peyrony (1938) and Sonneville-Bordes (1967) must be abandoned in its entirety" (Bosselin and Djindjian 1988: 322). According to Bosselin and Djindjian's analysis, only four groups of lithic tool types could be found which possess a temporal significance. The oldest group corresponds to an older Badegoulian with transversal burins and a low proportion of *raclettes*, while the second group corresponds to a younger Badegoulian and shows high proportions of *raclettes*. The two remaining groups are attributed to the Magdalenian and differ mainly in terms of the presence (older group, M0) and absence (younger group, M1) of lithic triangles (which is in accordance with results of this study, see Chap. 3, section "The Long Debate About Lithic Triangles"). M0 corresponds for the most part to Breuil's phase II, while M1 comprises Breuil's phases III–VI (Bosselin and Djindjian 1988: 312f.). A third Magdalenian group (M2), which is characterized typologically by a high number of microliths (30–70 %), is contemporaneous with both the other groups (M0 and M1) and probably represents a different site function (Ibid.).

What becomes clear from this short description is that the Magdalenian is not a historical fact that was discovered as a whole, but has been constructed over a long period of time by many researchers. It took about 70 years from the first excavation until the first complete concept about its internal chronology and organic and lithic industry was established. This concept was drawn from observations at a small set of randomly discovered sites in a rather small region and later extrapolated to an almost continental scale. While expanding the scope of application of the locally developed classification, the original concept of "Magdalenian" as developed by Breuil and Peyrony was altered and adapted several times, although its major outlines are still being used today.

The Long Debate About Lithic Triangles

There is a long and ongoing debate concerning whether or not the occurrence of lithic triangles has a certain temporal significance within the Magdalenian. This debate was initiated by Peyrony at the *Congrès des Sociétés de Paris* in 1934. There, he presented a Magdalenian facies which he called *Magdalénien à triangles scalène* (Peyrony 1936). On the basis of his observations at the rock shelter of Laugerie-Haute, he considered this facies to be contemporaneous with Breuil's Magdalénien II. The morphological definition of triangles and their respective typological designation is, however, anything but consistent (cf. Peyrony 1936; David 1947; Cheynier 1951; Couchard 1960; Allain 1979). Nevertheless, a major distinction is made between isosceles and scalene triangles. Another division is made within the category of scalene forms. Here, the crucial point is whether the two longitudinal edges do or do not form a point (for several examples, see Figs. 2.5 and 2.6). In the first case, they are called scalene triangles, whereas in the latter, they are referred to as scalene bladelets or *lamelles scalènes* (Tixier 1963; Demars and Laurent 1992). Some researchers consider scalene bladelets to be typical of the Magdalenian II and see their co-occurrence with scalene triangles as characteristic of the Magdalenian III (cf. Demars and Laurent 1992: 108). Others argue that triangles have no temporal significance whatsoever (cf. Höck 2000: 112, 2012).

For the CEM, the question of the temporal significance of triangles and their value as index fossils is important, particularly for the debate about the spatial and temporal structure of the early phase of the recolonization process. It is therefore necessary to examine the evidence in detail. First, however, it is useful to consider some points concerning the function and the hafting of triangles. It has often been suggested—and meanwhile it is widely acknowledged—that Magdalenian triangles were part of composite tools. They were used as barbs; that is, they were glued onto organic points to produce composite barbed points (cf. Bouyssonie et al. 1935; Cheynier 1951; Höck 2000, 2012). In this view, the denticulation of some triangles is understood as an additional hafting preparation that made it possible to tie them to the organic shaft using organic fibers. This interpretation is supported by a small sculpture of a barbed point from the Kniegrotte that appears to replicate this kind of hafting (Höck 2000). Moreover, many Upper Magdalenian barbed points, which are produced from a single piece of antler or bone, show ornamental lines that sometimes run in pairs from one barb to the next or are depicted on the barbs themselves (cf. Julien 1982). These decorations might be seen as a typological rudiment (Allain and Rigaud 1986). If the interpretation of triangles as parts of composite barbed points is accepted, then it is reasonable to assume that they became obsolete with the emergence of Upper Magdalenian massive barbed points that are carved in one piece.

In her dissertation on the site of Kniegrotte, Christiane Höck analyzes a comprehensive compilation of 71 triangle-bearing assemblages in terms of the chronological significance of the triangles, using information from the archaeological literature. She eventually concludes that triangles have no temporal significance whatsoever, but occur throughout the entire Magdalenian (Höck 2000: 112). However, a critical review of this compilation allows an opposite line of reasoning. Three major points need to be discussed.

First of all, Höck uses assemblages of very heterogeneous quality for her analysis. For the chronological discussion, she even considers inventories from surface collections and other contexts where the consistency of the finds, and hence their diagnostic value, are highly questionable. More than 40 assemblages derive from old excavations, and it is often impossible to tell which layer the artifacts belong to, or whether the assemblage represents a consistent unit, or how many archaeological levels are mixed up in the collections. Moreover, for 38 assemblages, the number of triangles is less than 5 specimens, and 12 collections include only 1 triangle. Three assemblages are even taken from surface collections, which may include any kind of material. For nine assemblages, no drawings of the triangles are available. Second, Höck's study uses 47 ^{14}C-measurements dating 31 assemblages from 25 sites. However, it does not include a critical examination of the diagnostic relevance of these dates in terms of the age determination of the triangle-containing assemblages, aside from some brief comments on five of the sites. Third, the possibility of an admixture of Magdalenian collections with Mesolithic triangles is not sufficiently considered.

In the following paragraphs, the compilation of 71 assemblages investigated in Höck's study is therefore reanalyzed with respect to the points mentioned above. Additionally, the ^{14}C-dated and triangle-bearing sites of Dzierżysław 35 and Schweizersbild are included in the discussion. Thus, 73 inventories are considered in the investigation (see Tables 4.2, 4.3, and 4.4 for information on location and other details).

Table 4.2 Triangle-bearing assemblages which are typologically assigned to the Lower and Middle Magdalenian

Site	Co	Excavated	M level	Ori. triangle	n Tr.	Phase	References
Abzac	F	1893–1940	Ml; Mu	Ml	1	I/II	Peyrony (1947), de Sonneville-Bordes (1960)
Bellet	F	1908–1960	M	ex. 1908	2	II	Couchard (1957), Couchard (1960)
Blanchisserie	F	1979–80	M	M	2[a]	II	Onoratini et al. (1996)
Cap-Blanc	F	1909, 30, 68/69	Ml; Mu	Ml	1	III	de Sonneville-Bordes (1960)
Chariez-Guillotine	F	1955	M	M	6	III	Thévenin (1956)
Crabillat	F	Peyrony	M	M	161	II	Peyrony (1941), de Sonneville-Bordes (1960)
Farincourt I	F	1806, 1878	M	M	7	III	Pernot (1996)
Farincourt II	F	1878; 1933–37	M	M	2	III	Pernot (1996)
Forge	F	1925, 27	M	M	6	III	Darpeix (1934), de Sonneville-Bordes (1960)
Goutte Roffat	F	1909; 30–56	M	M	29	II	Larue et al. (1955)
Granet I	F	1974ff.	M	M	1	III	Lenoir (1983)
Hranice	CZ	1953	M	M	ca. 50	III	Neruda and Kostruhn (2002)
Huguenots	F	1990	Ml; Mu	Ml	1	II	Onoratini and Joris (1995), Onoratini et al. (1996)
Jaurias	F	1944ff.	Ml; Mu	Ml	2[a]	III	Lenoir (1983)
Jean Blanc	F	1882ff.; 81–89	Ml; Mu	Ml	3	III	Peyrony and Peyrony (1934), Cleyet-Merle (1989)
Julivet	F	1927–34	Ml; Mu	?	3	III	de Sonneville-Bordes (1960)
La Paloma	E	1914/15	M=8	Ml	1	III	Utrilla (1981)
Lafaye	F	1864/5	M=2	?	6	III	Cartailhac (1903)
Métairie de Belcayre	F	1875ff.	M	M	1	III	de Sonneville-Bordes (1960)
Placard	F	1856ff.; 60–68	M	M	1[a]	III	Roche (1971)
Plantade	F	1864/5, 1960s	Ml; Mu	?	2	Mid.	Pajot (1969)
Plateau Parrain	F	1959–63	M	M	1[a]	III	Gaussen (1994)
Putoise II	F	1955	M	M	1	Mid.	Cleyet-Merle and Madelaine (1991)
Puy-de-Lacan	F	1860–1905; 1929	M=2	M	ca. 300	III	de Sonneville-Bordes (1960)
Raymondon-Chancelade	F	1876ff; 1927–29	M=4	2, 3	3/7	II/III	de Sonneville-Bordes (1960)
Reverdit, Abri	F	1878–1935	M=2	1, 2	2	III	Delage (1935)
Roc Saint-Cirq	F	1936/37; 52/53	Ml; Mu	Ml	23	III	de Sonneville-Bordes (1960)
Roche de Castelmoron	F	1976	M	M	139	II	Le Tensorer (1981)
Sâlpétrière	F	1954 and before	?	7–13	7	III	Escalon de Fonton (1962)
Vilshofen	D	1994	G/M	M	20	III	Weißmüller (1995)

For abbreviations, see legend of Table 4.3 and list of abbreviations. For drawings of the triangles, see Figs. 2.5 and 2.6

[a]No drawings available

Assemblages for Which No Radiocarbon Dates Are Available

It is important to state at the outset that for 41 sites, no radiocarbon dates are available. Thus, their assessment must be based on typological observations. In 30 cases (73 %), these observations do not contradict an assignment of the assemblage to Magdalenian II or III (Table 4.2). Only at the site of Abri Lafaye is there a representation of a single-row barbed point engraved in an antler fragment. This does not, however, represent an argument for a younger date of the triangles, first of all because the engraving could as easily represent a composite barbed point (see above), and second because the site was excavated in 1864/1865, which heightens the possibility of an admixture of older and younger finds. All of the other associated finds which may serve as temporal indicators are fluted bone and antler points—often of the type *Lussac-Angles*—and *baguettes demi-rondes*, which are in accordance with a Magdalenian II or III assignment. In one case (Grotte du Placard), a *contour decoupé* may be associated with the triangle assemblage.

On the other hand, there are 11 remaining sites where the associated finds may indicate the occurrence of triangles in Upper Magdalenian assemblages (Table 4.3). These cases will therefore be discussed in detail.

The Cueva de Aitzbitarte IV (Fig. 2.6: 26) was repeatedly excavated from 1892 onwards and was also often subject to illicit excavations. Thus when J.M. Barandiarán conducted excavations at the site between 1960 and 1964, the cave sediments were already heavily altered. The Magdalenian collection probably represents two separate Magdalenian horizons. However, because the sedimentological analysis does not allow for a separation and the faunal remains have been recorded together, the assemblage must be classified as mixed and hence must be excluded from the analysis (Barandiarán 1967; Utrilla 1981; Höck 2000).

The Abri du Chez-Galou (Fig. 2.6: 3) was discovered in 1905. Among the 235 artifacts, there are some burins but only one backed bladelet, one backed point, and one triangle. Given the early date of the excavation, it is likely that small items such as backed bladelets and triangles were not recorded and that separate layers were excavated as a unit. Together with the fact that a backed point is among the assemblage, it is highly probable that the assemblage is mixed (de Sonneville-Bordes 1960).

Extensive illicit excavations have led to serious disturbance of the cave sediments at the site of Cueva de Cendres (Fig. 2.6: 27). In a 2 m^2 test trench, directly underneath the completely disturbed layer, the excavators separated a 63 cm thick layer with Magdalenian finds, such as scalene triangles, *lamelles scalènes*, backed bladelets, and two barbed point fragments (Cacho 1989). The coherence of these finds is doubtful for two reasons. First, in a test trench only 2 m^2 in size, it is not always easy to judge the extent of a disturbance. Second, it is very likely that a layer 63 cm thick includes residues of several Magdalenian occupations. Since the coherence of the finds remains highly doubtful, this site should also be excluded from the analysis.

The assemblage of Eitensheim-Windhöhe (Fig. 2.6: 31) is a surface collection with artifacts from Middle to Upper Paleolithic (Rieder 1988, 1989). The assemblage's consistency is therefore more than doubtful and must not be considered in an investigation of chronology.

The Abri Fontalès (Fig. 2.5: 19) was excavated repeatedly between 1865 and 1960. The upper layer contained an assemblage with mixed finds from the Magdalenian through

Table 4.3 Triangle-bearing assemblages within which Lower and Middle Magdalenian artifacts are intermixed with Upper Magdalenian finds, or Upper Magdalenian finds are intermixed with Mesolithic ones

Site	Co	Excavated	M level	Ori. triangle	n Tr.	Phase	References
Lower and Middle Magdalenian intermixed with Upper Magdalenian							
Aitzbitarte IV	E	1892–1917; 60–64	M	M	3	II–IV?	Utrilla (1981)
Chez-Galou	F	1905	M	M	1	II–VI?	de Sonneville-Bordes (1960)
Cenders	E	1975/5, 81	M	M	3	II–VI?	Villaverde Bonilla (1981)
Fontalès	F	1936–1960	M	M	13	II–VI?	Pajot (1969)
Garrigue	F	1970s	M	M	101	II–VI?	Lenoir (1983)
Parpallo	E	1872–1931	M=4	Mu	Min. 17	II–IV?	Aura Tortosa (1989)
La Souquette	F	1902/3; 11	Mix	Mix	6	III–IV?	de Sonneville-Bordes (1960)
Upper Magdalenian intermixed with Mesolithic							
Baume Loire I	F	1968	M	M	2	V–VI	Crémilleux (1972)
Grotte de l'OEil	F	1967–1968	Mix	Mix	3	V–VI	Sacchi (1986)
La Teulera	F	1962ff.	Mix	Mix	3	V–VI	Sacchi (1986)
Eitensheim-Windhöhe	D	?	Mix	Mix	1	V–VI	Rieder (1989)

For abbreviations, see list of abbreviations. For drawings of the triangles, see Figs. 2.5 and 2.6

to the Gallo-Roman epoch. In the lower part of the stratigraphy, two Magdalenian layers were recorded. The excavators emphasize that the site had already been largely ransacked and that the different layers blended into each other and were hardly distinguishable (Darasse and Guffroy 1960). The associated finds from these two levels include points of the *Lussac-Angles* type and grooved organic points as well as proto-harpoons, single-row barbed points, and double-row barbed points. Together with the stratigraphic observations, these finds speak clearly in favor of a mixed assemblage. Consequently, it is not suitable for a chronological analysis.

Garrigue (Fig. 2.6: 14) is an open-air site that was excavated in the 1970s. In addition to a rich collection of Magdalenian triangles, several backed points were also found (Lenoir 1983). Since a number of points are of the *Malaurie* type, the assemblage represents at least two occupations and may have a Mesolithic component as well. Therefore, it cannot be seen as a consistent collection.

The Cueva de Parpalló (Fig. 2.6: 28) was excavated for the first time in 1872. In the following years, the cave was subjected to several illicit excavations and was repeatedly ransacked. During the later excavations conducted by L. G. Pericot from 1929 to 1931, the Magdalenian horizon was divided into four layers and assigned to the Magdalenian I–IV. The uppermost layer yielded several triangles, *baguettes demi-rondes* and a fragment of a single-row barbed point (Pericot 1942). Again, the heavy disturbance of the sediment and the early excavation date make an admixture of the triangles and the barbed point highly likely.

Abri La Souquette (Fig. 2.5: 14) was completely excavated from 1902 to 1911. In addition to triangles, the Magdalenian inventory contains a shouldered point, a double-row barbed point, and a grooved organic point and is almost certainly intermixed (de Sonneville-Bordes 1960).

For the Abri de Baume Loire I, Höck mentions 4 triangles. Judging from the drawings (Fig. 2.5: 13), 3 out of these 4 artifacts should rather be classified as a backed bladelet, a fragment with a steep, lateral retouch, and a fragment of an angled backed point. The latter interpretation is supported by the presence of other backed points in the assemblage. The remaining piece might be classified as a scalene bladelet, but can be seen as a truncated backed bladelet as well. Given that there is only one artifact which might perhaps be identified as a scalene bladelet and taking into account the occurrence of backed points, the Abri de Baume Loire I should be seen as a Late Paleolithic site, in the context of which an oblique truncated backed bladelet is neither surprising nor suspicious (Crémilleux 1972, 1977; Philibert 1982).

A Bronze Age burial was probably the objective of an illicit excavation at the Grotte de l'OEil (Fig. 2.6: 18). The ransacked sediments yielded an admixture of medieval as well as Bronze Age and Magdalenian finds (Sacchi 1986). Thus, the co-occurrence of triangles, an organic point of the *Belvis* type, and a double-row barbed point in one assemblage must be seen as the result of intermingling.

The assemblage of the open-air site of La Teulera (Fig. 2.6: 24) is a surface collection with triangles, shouldered points, and backed points (Sacchi 1986). The coherence of these finds is highly unlikely.

Assemblages for Which Radiocarbon Dates Are Available

For 27 sites, radiocarbon dates are available for the assessment of the chronological position of the triangle-bearing layers. For a critical review of the ^{14}C-date's diagnostic value, the table given in Höck's study (2000: Table 46) was extended and corrected (see Table 4.4 and the references therein). All calibrations of the dates, weighted averages, and probability calculations were performed with CalPal 2007 (Weninger et al. 2012).

The oldest dates of triangles in a Magdalenian horizon come from the Lascaux cave. The weighted average of both dates gives an age of about 20,500 cal BP. Similar dates come from Cueva de Urtiaga and Laugeri-Haute I″. From Lassac and Cueva de Rascaño, there are two dates around 20,000 cal BP and 19,000 cal BP respectively. All of these dates accord with stratigraphic and typological observations. Three measurements are available for the site of Cueva de Abauntz. The oldest date is about 19,000 cal BP, and the two younger dates give a weighted average of 16,800 cal BP, pointing to a minimum of two occupations, both of which may have brought up the triangles. A very consistent series of three dates from the Grotte de Pégourié gives a weighted average of about 18,500 cal BP, which is close to the date from Abri de Chaire-à-Calvin and Abri de Saint-Germain-la-Rivière (for drawings of the triangles, see Figs. 2.5 and 2.6).

At the cave site of Rond du Barry (Fig. 2.6: 22), three Magdalenian levels have been defined. The two upper levels have yielded triangles. In the second level, the triangles are associated with three single-row barbed points, several grooved organic points, and lithic backed points. Three radiocarbon measurements could be obtained from this level. The oldest gives a date of about 18,500 cal BP and is perfectly consistent with the dates from Grotte de Pégourié and Chair-à-Calvin. In contrast, the weighted average of the two other dates is 15,160 cal BP. The most plausible explanation is that the second layer comprises at least two occupations: one at 18,500 cal BP with triangles and another at 15,000 cal BP with barbed points. The backed points may represent a further occupation, but they may be also part of the Late Magdalenian assemblage. The few finds from the uppermost level are described as conspicuously small in size and are probably of Late Paleolithic or even partly Mesolithic age (De Bayle Des Hermens 1971, 1974, 1979).

Table 4.4 Triangle-bearing assemblages for which radiocarbon dates are available

Site	Co	Excavated	M level	Ori Tr	n Tr	Phase	Lab.-Nr.	Lev/Sam	^{14}C BP	Std	Cal BP	Std	WA	Std	WAC	Std	P%	Reference (dates)
Grotte de Lascaux	F	1940–1963	M	M	3	II	GrN-1632	/c /c	17,190	140	20,650	160						Leroi-Gourhan and Évin (1979)
							SA-102		16,000	500	19,300	570						
Cueva de Urtiaga	E	1928–1936, 54–59	Ml; Mm; Mu	Ml; Mm	4	Low.	GrN-5817	Ml/b	17,050	140	20,500	160						Utrilla (1981)
Laugerie-Haute	F	1862–1959	M I′–I‴	I″; 4–6, 8	46	II	Ly-973	I″ X8/?	17,040	440	20,430	540						Délibras and Évin (1976)
				I‴; 1–2	5	III	Ly-974	I‴ 2, 3/?	13,970	480	16,950	650						
Lassac	F	1964–1966	M	M	4	II/III	Gif-2981	M/Bn	16,750	250	20,090	310						Sacchi (1986)
Cueva de Rascaño	E	1912/12; 74/75	Ml; Mm; Mu	Ml	1[a]	III	BM-1453	Ml/?	15,988	193	19,150	240						Utrilla (1981)
Cueva de Abauntz	E	1976–1979	M	M	2	Mid.	Ly-1965	M/?	15,800[b]	350	19,110	390	13597[b]	152	16770	180	6	Utrilla (1981, 1982, 1995)
							Beta-65723		14,470[b]	480	17,690	530						
							OxA-5983		13,500[b]	160	16,530	330						
Grotte de Pégourié	F	1967ff.	Ml; Mu	Mu	1[a]	II	Ly-1394	Mu (8)	15,540[c]	520	18,720	610	15446[c]	258	18450	340	96	Séronie-Vivien et al. (1981)
							Ly-1836	Ml (9a)	15,470[c]	390	18,570	460						
							Ly-1834	Mu (8a)	15,340[c]	460	18,480	470						
Chair-à-Calvin	F	1960–1971	Ml; Mu	Ml; Mu	12	II/III	n.c.	Mu, Ml/?	15,440	440	18,570	490						Delpech (1989)
Grotte de Rond du Barry	F	1900;1965–1979	Ml; Mm; Mu	Mm; Ml	7;13	Low.	Gif-3038	Ml/?	17,100	450	20,510	550						
						Up.	Gif-2672	Mm/?	15,400	400	18,500	430						
						Up.	Gif-3492	Mm/?	12,800[d]	170	15,330	250	12687[d]	145	15160	220	20	De Bayle de Hermen (1979)
						Up.	Gif-2671	Mm/?	12,380[d]	280	14,610	500						
St-Germain-La-Rivière	F	1928–1934; 1960/1970s	M=5	Lev. 1,2,3	Min. 4	Mid.	Gif-5479	4+5/?	16,200	600	19,510	670						Lenoir (1983)
							Gif-5478	3/?	15,300	410	18,410	400						
Abri Flageolet II	F	1960s?	M	M	16[a]	III	Ly-918	M/?	15,220[e]	320	18,320	310	15024[e]	290	18220	290	15	Höck (2000)
						Up.	Ly-917	M/?	14,110[e]	690	16,970	900						
Grotte Gazel	F	1850–1950; 60–71	M	M	3	IV	Gif 2655	M/c	15,070	270	18,250	280						Sacchi (1986)
Combe Cullier	F	1900–1906, 1967	M	Ex.1906	3	Mid.	Ly 978	M/?	15,030	330	18,220	310						Délibras and Évin (1974)
Dzierżysław	P	1997/1998	M	M	122	M	Ki-8997	M/?	14,850	280	18,120	320						Ginter et al. (2002), Ginter and Połtowicz (2004)
							Ki-8951	M/?	13,700[f]	350	16,620	570	13356[f]	44	16580	100	5	
							GdA-69	M/?	13,500[f]	80	16,760	90						
							n.c.	M/?	13,370[f]	80	16,480	230						
							GdA-70	M/?	13,220[f]	70	16,120	300						
Grande Bize	F	1827–1933	Ml; Mu	Mu;	2	mid.	Ly-1675	Mu(3)/?	14,770[g]	160	18,090	280	14748[g]	53	17860	30		Sacchi (1986), Tavoso (1987)
							Ly-1232	Mu(3)/?	14,530[g]	510	17,750	570						
Canecaude I	F	1965–1976	M	M	8	mid.	Gif 2708	M(2)/b	14,230	160	17,420	200						Sacchi (1986)
Duruthy	F	1873, 1957–1975	M=4	Lev. (III)	1	III	Ly-861	5/?	14,180	220	17,390	230						Arambourou (1978)
				Lev. (IV)	3	IV	Ly-860	4l/?	13,840	220	17,070	230						
				Lev. (VI)	15	IV	Ly-859	4u/?	13,510	220	16,440	430						
						VI	Ly-858	3/?	11,150	220	13,070	210						

Martinet	F	1922–1935	M	M	52	V–VI	Ly-1605	M/b	14,098[h]	239	17,340	310	14076[h]	234	17330	310	66	BANADORA
							Ly-5069	M/b	13,600[h]	1100	16,060	1880						
Schweizersbild	CH	1891–1993	M	M	2	M	OxA-5745	M/b	13,940	100	17,090	70						Street (2000)
							OxA-5744	M/b	11,780	90	13,680	90						
La Vache	F	1866ff., 1940–1950	M=4	1, 2, 3, 4	11	VI	Gif-?	m²-71/?	13,770[i]	140	16,970	100	13622[i]	75	16880	60	31	Pailhaugue 1998
							Gif-?	m²-71/?	13,650[i]	130	16,870	110						
							Gif-?	m²269/?	13,490[i]	120	16,660	190						
							Gr-2026	4/?	12,850[j]	140	15,410	200	12691[j]	72	15210	130	14	
							Gif-7603	2/?	12,800[j]	140	15,320	200						
							Gr-2025	2/?	12,540[j]	105	14,970	190						
Kniegrotte	D	1876ff.	M	M	122	Mid.	OxA-4851	M/b	14,470	140	17,610	160						
							Bln-1564	M/b	13,585	165	16,710	230						
							OxA-4852	M/b	13,520[k]	130	16,690	190	13272[k]	57	16250	290	23	Hedges et al. (1992)
							OxA-4832	M/b	13,310[k]	110	16,270	360						
							OxA-4846	M/b	13,190[k]	130	16,100	350						
							OxA-4850	M/b	13,160[k]	140	16,070	360						
							OxA-4848	M/b	13,150[k]	130	16,060	350						
							OxA-4849	M/b	13,130	120	16,030	340						
							OxA-4845	M/b	13,120	130	16,020	350						
							OxA-4853	M/b	13,090	130	15,980	350						
Bois-Ragot	F	1969–1990	Ml; Mu	Mu(6)	1[a]	V–VI	OxA-10262	6/b\a?	12,645[l]	65	15,130	170	12,636[l]	40	15,130	130	86	Chollet and Dujardin (2005)
							OxA-12080	6/b\a?	12,630[l]	50	15,120	140						
Faustin	F	1950s, 70–77	M	M	4	VI	Ly-2700	M/b	12,370	220	14,580	430						Lenoir (1983)
C. d. Belvis	F	1964–1969	M=4	M	54	V–VI	Gif-2950	M/c	12,270	280	14,470	470						Sacchi (1986)
Birseck	CH	1910–1914	M	M	13	M	B-4261	M/b	12,040[m]	80	14,000	170	11,970[m]	62	13,860	70	16	Höneisen et al. (1993)
							B-4260	M/b	11,860[m]	100	13,750	120						
Faurélie II	F	1960s–1972	Ml; Mu	Ml	?[e]	V–VI	Gif 3649	Ml/?	11,780	180	13,690	210						Tixier (1976)
Gare de Couze	F	1884;1912; 61/62	M=9	2,3,4	Min. 9	VI	Ly 976	3/?	11,730	310	13,770	420						Délibras and Évin (1976)

For abbreviations, see List of Abbreviations. Dates not given in boldface are considered unrelated to the occupation events that led to the discard of the triangles

M level: Stratigraphic information concerning the kind and number of Magdalenian levels per site. M=only one Magdalenian level documented and generally assigned to the Magdalenian; Ml=level assigned to the Lower Magdalenian; Mm=level assigned to the Middle Magdalenian; Mu=level assigned to the Upper Magdalenian; M=n=number of Magdalenian levels documented at the site. Ori Tr: Origin of the lithic triangles n Tr= number of documented triangles. Phase: Typological attribution to the phases according to Breuil/Peyrony or the division into Lower Magdalenian (low.), Middle Magdalenian (mid.), and Upper Magdalenian (up.) (see Table 4.1). Lev/Sam=level of origin for the radiocarbon sample. WA=weighted average of uncalibrated radiocarbon dates; WAC=calibrated weighted average. P%=Probability (chi-square test) that dates considered in the weighted average are statistically identical. A percentage value of five is interpreted to allow for the calculation of a weighted average. Calibration and weighted average calculation was done using CalPal 2012 (Weninger et al. 2012). For abbreviations see legend of Table 4.3 and list of abbreviations. For drawings of the triangles, see Figs. 2.5 and 2.6

[a]b-m: dates used to calculate the weighted averages

For the rock shelter of Flageolet II, the Grotte Gazel, and the Grotte de Comb-Cuiller, radiometric dates of about 18,200 cal BP have been obtained. A similar date of about 18,100 cal BP is known from Dzierżysław 35. However, four other dates give a weighted average of about 16,600 cal BP. The triangle assemblages of the two cave sites Gand Bize and Canecaude I are dated to about 17,900 cal BP and 17,400 cal BP, respectively. The assemblage of the site of Schweizersbild, from which two triangles are reported, represents an admixture of at least two occupations: one around 17,000 cal BP, which is in accordance with the occurrence of triangles, and another around 13,700, which probably represents a post-Magdalenian occupation. Two dates from the site of Abri du Martinet gave a weighted average of 17,330 ± 310 cal BP.

At the Abri de Duruthy, 4 Magdalenian levels have been distinguished, overlaid by an Azilian and a Chalcolithic layer (Arambourou 1978). The four levels are classified as Magdalenian III, IV, V, and VI. Triangles are recorded from the Magdalenian levels III, IV, and VI, and ^{14}C-dates are available for all of them. The three dates from the levels III and IV are situated between 17,400 cal BP and 16,400 cal BP. Level VI, on the other hand, is dated to 13,000 cal BP and is separated from the lower layers by level V, which has yielded no triangles. The triangles of level VI are considerably smaller than those from levels III and IV (Fig. 2.6: 9). Additionally, in at least one case, the retouched parts form an acute rather than an obtuse angle, which does not make sense if it was meant to be used as a barb (see above). Consequently, as indicated by the radiocarbon date, level VI very probably contains a Mesolithic component. This is typologically expressed by microlithic triangles which have nothing whatsoever to do with the Magdalenian triangles of levels III and IV.

The archaeological layer of the cave site of Grotte de la Vache (Fig. 2.6: 25) ranges from 15 to 90 cm in thickness and in many places is impossible to subdivide stratigraphically. During the excavation it was partitioned into 4 artificial levels (Gailli 2003: 50). Additionally, it seems that the cave sediments were reworked several times during the Paleolithic occupation (Robert and Allard 2003: 36). Thus, the archaeological layer probably represents a palimpsest of several occupations. Two weighted averages of ^{14}C-dates point to at least two occupations at about 16,900 cal BP and 15,200 cal BP, the first of which corresponds to the other measurements for Magdalenian inventories with triangles.

The weighted average of absolute dates measured on human-modified bones from Kniegrotte (Fig. 2.5: 21) gives an age of about 16,250 cal BP, which is at the same time the youngest reliable date for a Magdalenian assemblage with triangles (see Table 4.4).

At the cave site of Cauna de Belvis (Fig. 2.6: 33), the archaeological layer is up to 40 cm thick and is subdivided geologically into four levels having no archaeological significance, according to Sacchi (1986). Bones of level 3 gave a radiocarbon date of around 14,500 cal BP. On the basis of typological observations, the whole assemblage is classified as a homogeneous ensemble. Yet the thickness of the archaeological layer and the fact that a subdivision according to the sediment's color and grain size was possible speak in favor of a palimpsest rather than a homogeneous assemblage. It therefore seems inappropriate to take an arbitrary radiocarbon date as representative for the entire archaeological layer. The presence of very small cores and small backed points as well as the extremely reduced size of the triangles themselves would rather seem to indicate an admixture with the remains of a subsequent Mesolithic occupation.

The rock shelter of Abri Faustin was already partly destroyed by quarrying activity when M. Lenoir started his investigations in 1970. The excavated assemblage includes three triangles (Fig. 2.6: 11) as well as numerous other microliths and backed and shouldered points. It is highly likely that the collection represents an admixture of Upper Magdalenian and Mesolithic artifacts. A radiocarbon date of approximately 14,500 cal BP is consistent with a possible Late Magdalenian occupation, whereas the triangles are most probably remnants of a Mesolithic occupation (Lenoir and Terraza 1971; Lenoir 1983).

At Abri de Faurelié II, a sequence of archaeological horizons was excavated that comprises Magdalenian levels as well as an Azilian occupation. The upper Magdalenian horizon (couche 4) is classified typologically as Magdalenian VI (Tixier 1974). From the lower horizon, it is reported that no characteristic tools of the Magdalenian VI assemblage are present. Despite the presence of triangles and the absence of other diagnostic artifacts, Tixier classifies the assemblage of horizon 5 as Magdalenian V (Tixier 1974: 192). However, because of a lack of diagnostic Upper Magdalenian artifacts and the presence of triangles, the assemblage of the horizon 5 should be reclassified as Magdalenian II or III. A ^{14}C-date of about 13,700 cal BP is consistent with neither a Lower nor an Upper Magdalenian occupation and points to an admixture with younger finds, which perhaps could also be a possible source of the triangles.

The Magdalenian horizon of Gare de Couze (Fig. 2.6: 13) is about 70 cm thick and is subdivided into five levels, all of which are considered Magdalenian VI (Fitte and Sonneville-Bordes 1962). The fact that backed points as well as triangles become more abundant as one moves from the bottom to the top of the stratigraphy suggests that the collection in fact represents a palimpsest of Magdalenian, Azilian, and Mesolithic occupations. The triangles are rather small in size and isosceles in form. At least two of the triangles even have a straight base. Typologically, they therefore resemble Mesolithic much more than Magdalenian triangles (Taute 1971). The single ^{14}C-date obtained for this site (14,190–13,350 cal BP) may thus date an Azilian occupation, but cannot be considered representative for the whole assemblage.

During transformation of the Birseck-Ermitage cave (Fig. 2.6: 30) into a hermitage in the middle of the nineteenth century, the Paleolithic layers were partly destroyed. Additionally, a Neolithic burial was dug into these layers. Two dates of around 14,000 cal BP do not correspond to the preserved steppe fauna (Höneisen et al. 1993). Together with the early date of excavation (1910–1914), the coherence of the assemblage and consequently the ^{14}C-date's diagnostic value for the age of the lithic assemblage is very doubtful.

The assemblage of the Magdalenian level 6 at Grotte de Bois-Ragot yields among other things 41 barbed points and 908 backed bladelets, 33 of which are truncated (Chollet and Dujardin 2005: 53ff.). One of these truncated backed bladelets has an oblique truncation and therefore resembles a scalene triangle (Le Mignot 2005: 57). Against the background of all the other finds, it is soundest to interpret this artifact as a truncated backed bladelet and not as a triangle (Chollet and Dujardin 2005: 57). Hence, the assemblage of Bois-Ragot is seen as a regular Upper Magdalenian collection. A consistent series of ^{14}C-dates for level 6—the two oldest of which are presented in Table 4.4—supports this view (Ibid.: 402).

Conclusion

In the case of the 41 undated assemblages, 30 are in perfect accord with a typological classification to the Magdalenian II or III. The remaining 11 all appear to represent admixtures of Lower and Upper Magdalenian levels or even Magdalenian and Mesolithic layers.

All 20 sites where there is a reliable connection between the triangle-bearing layer and the radiocarbon measurements obtained provide dates older than 16,000 cal BP (see Table 4.4). For the remaining seven sites, it can be shown that the respective assemblages represent mixed collections or have no demonstrable relation to the ^{14}C-dates. Moreover, there are no consistent inventories which comprise both triangles and barbed points.

In the end, it can be stated that there are no reliable arguments which would support the occurrence of lithic triangles in a Magdalenian context later than 16,000 cal BP (13,200 ^{14}C BP). Thus, Magdalenian lithic triangles have a pronounced temporal significance.

These observations are in perfect accordance with the occurrence of another chronologically very distinctive—and in this case truly complementary—tool type, namely, the organic barbed point. This kind of projectile is correlated with the Upper Magdalenian (Breuil's phases V and VI). The strict temporal division of single- and double-row barbed points that was initially posited has since been shown to be untenable. Both types appear together and are contemporaneous, although double-row barbed points seem to have reached peak distribution later than did those with only a single row (de Sonneville-Bordes 1960: 356; Weniger 1995: 175).

Apart from the so-called proto-harpoons, which occur apparently already around 13,500 ^{14}C BP, the first massive barbed points appear at about 13,200 ^{14}C BP or 16,000 cal BP (Weniger 1995: 175ff.; Álvarez Alonso 2007; Pétillon 2008: 68). Older ^{14}C-measurements, which are sometimes mentioned in the literature, are all very unreliable. Either they are in conflict with younger dates from the same level, as is the case for Rond du Barry E or Pierre-Châtel, or the stratigraphic context, and therefore the connection between the date and the barbed point is dubious, for instance, at Verlaine (Weniger 1995: 171f.).

It is quite striking that at the same time there are no reliably dated assemblages with lithic triangles younger than approximately 16,000 cal BP (13,200 ^{14}C BP). This observation is remarkable for its distinctness, but not surprising, since lithic triangles and barbed points both represent the same kind of tools and, from a functional point of view, are therefore two sides of the same coin.

Theories About the Internal Structure of the Central European Magdalenian

All the studies that aimed to analyze the internal structure of the Magdalenian (e.g., Valoch 1960, 2009; Hanitzsch 1969; Feustel 1974; Weniger 1982; Bahn, 1985; White 1985; Rozoy 1988b; Otte and Keeley 1990; Eriksen 1991; Otte 1992; Rensink 1993; Jochim et al. 1999; Połtowicz 2006; Küßner 2009; Sisk 2011) deal in one way or another with the question of regional groups and seek explanations for the uneven distribution of Magdalenian sites. However, very different approaches were used to address this question: on the one hand, typological similarities and differences were taken to be decisive for the definition of groups. On the other, the geographic position of a site and environmental factors was given more importance. As this investigation is concerned with regional variability, it is important to present and evaluate both attempts briefly.

Jean-Georges Rozoy published already in 1988 an article in which he distinguishes at least 16 Magdalenian site groups according to their geographical position. He suggests that they might correspond to 12–15 distinct groups of people with close relations that maintained the homogeneity of the Magdalenian record. Rozoy points out that typology is vital for the distinction of these groups, but that it must not be used as the only criterion (Rozoy 1988b:

144). Another study which was carried out in an exemplary manner is the investigation of Magdalenian sites in Northern Switzerland, the Danube region, the Rhineland, and Eastern Germany by G.-Ch. Weniger (1989). The extensive analysis of different factors of Magdalenian lifeways and their reciprocal evaluation provide a reliable basis for conclusions about hunter-gatherer diversity and regional variability.

Much less reliable and very problematic in terms of confirmability as well as replicability are those attempts that rely exclusively on typological reflections in order to detect groups within the Magdalenian. Moreover, the classification systems used to delimit these groups are highly arbitrary and particularly prone to circular reasoning. Nevertheless, terms and concepts that were generated on its basis can still be found even in recent contributions (see below). Therefore, it seems necessary to elaborate on this topic. In order to illustrate the problems of this approach, a discussion is presented which was initially mainly concerned with Eastern German sites but later became expanded. However, a similar discussion could be presented as well for the Magdalenian sites in Switzerland (see Bandi 1960, 1968; Feustel 1961b; de Sonneville-Bordes 1963).

Already Hermann Schwabedissen (1954: 73 f., 95) reflects on the nature and structure of regional groups during the Late Glacial. Based on typological observations he postulates a so-called *Döbritzer Gruppe*, which he says to encompass the Eastern German sites of Kniegrotte, Herdloch (Ranis), Oelknitz, Kahla-Löbschütz, Wallendorf (formally Friedensdorf/Kriegsdorf), Saaleck, Gera-Liebschwitz (Binsenacker), and perhaps Burk as well as the Moravian site Malomerice-Borky.

In 1957, Rudolf Feustel states that such a grouping is admittedly reasonable to show some major relationships, but that it neglects at the same time the differences within these assemblages (Feustel 1957: 16). Therefore, he proposes that the term *Döbritzer Gruppe* should only be used for those sites with assemblages similar to the Kniegrotte inventory. Instead he introduces the label *Ölknitzer Gruppe* for the Eastern German assemblages from Oelknitz, Kahla-Löbschütz, Gera-Liebschwitz (Binsenacker), Ahlendorf, Groitzsch A, Saaleck, the Swiss assemblage of Moosbühl, and the Moravian assemblage of Malomerice-Borky I (Feustel 1959: 226ff.). Furthermore, he assigns the finds from Etzdorf (Eastern Germany) to an independent group and writes that the diversity of tool types was probably triggered by the formation of more or less isolated and coherent social entities (Feustel 1957: 25 [Anm. 72], 26). Later he defined as well a *Lausnitzer Gruppe* (Feustel et al. 1963) which he thinks includes possibly besides Abri Theure the open-air site of Bad Frankenhausen (both Eastern Germany) (Feustel 1961a: 24).

Volker Toepfer also concludes that probably several different groups existed within the Eastern German Magdalenian (Toepfer 1958: 156), and Karl J. Narr observes three typological entities in the CEM which occur throughout the whole investigated area: a backed bladelet-borer group, an angled point-shouldered point group, and a Probstfels-Rissener group (Narr 1963: Karte 4).

In his article on the structure of the Eastern German Magdalenian, Helmut Hanitzsch (1969) criticizes the classification outlined above as insufficient and partly wrong. According to his opinion, this is because many publications on the relevant assemblages are incomplete and present badly drawn representation of the finds. Additionally, he states that important types remained unidentified, whereas unimportant ones got overemphasized (Ibid.: 181). After analyzing many assemblages in detail and defining a set of types he considers significant, Hanitzsch divides the Central German Magdalenian in four temporal levels and assigns them several groups (Ibid.: 186ff.).

The first level is represented by the sites of Abri Theure, Bad Frankenhausen, Burk, and Aschersleben, which he says are characterized by the absence of most of the types considered as significant. Level II is then marked by the *Nebraer Gruppe* with the sites of Nebra, Gera-Liebschwitz (Binsenacker), Kahla-Löbschütz, and Halle-Galgenberg. In the third level, he defines a *Saalecker Gruppe* comprising Saaleck and Ahlendorf, a *Groitzsch*-D-*Gruppe* and *Groitzsch-C-Gruppe*. The last level (IV) consists of the two one-site groups of *Groitzsch-B-Gruppe* and *Etzdorfer Gruppe*. In contrast to previous researchers, the sites of Kniegrotte and Oelknitz are seen as mixed assemblages of levels I–III and II–III, respectively.

Five years later, Feustel notes that the reason for the differences between Hanitzsch and himself—as well as himself and Kozłowski, who published a classification of four chronologically and typologically defined groups in 1971 (Kozłowski 1971)—regarding the classification of Eastern German Magdalenian sites, are mainly due to the choice and assessment and therefore anticipated weighting of certain types as well as the over- or underestimation of differences between the industries (Feustel 1974: 191). He reinforced the existence of the *Ölknitzer Gruppe*, but subdivided it into several facies as there are (Ibid.: 193):

- Facies Kniegrotte (Eastern Germany) with probably Hranice and Nová Drátenicka (Moravia), Schweizersbild and Birseck-Ermitage (Switzerland), and Farincourt (France)
- Facies Ölknitz with Saalfeld (Eastern Germany) as well as Pekárna and possibly Kůlna 6 (Moravia)
- Facies Ahlendorf (Eastern Germany) with Kesslerloch (Switzerland)

- Facies Moosseedorf (Switzerland) with Saaleck, Groitzsch A (Eastern Germany), Veyrier (France), and Ochozská Cave (Moravia)
- Facies Goyet-Chaleux with other sites in Belgium

Finally, he concludes that these facies probably represent different waves of immigration. The first two more or less contemporaneous ones running southwest-northeast and west-east, respectively, are followed by two waves coming again from southwest (Ibid.: 197).

Even in recent studies, this kind of classification is still in use and is an important part of the contextual evaluation and interpretation of sites (Mania 1999; Terberger 2002; Küßner 2009). Starting from his observations at Andernach Martinsberg, Terberger (2002: 137f.) defines an extended version of the *Nebraer Gruppe*, the so-called *erweiterte Nebraer Gruppe*. This group comprises sites from the Paris Basin, the Ardennes, the Rhineland, and of course the Saale region. The members of this group are interpreted as a group of interaction marked by social relations who conceivably formed an ethnic union (Ibid.: 138). Küßner (2009: 193) notes that he fully agrees with Terberger's definition of the *erweiterte Nebraer Gruppe*. He also supports the opinion of Feustel (1961a) that the site of Bad Frankenhausen forms together with Abri Theure the *Lausnitzer Gruppe* (Küßner 2009: 194), a group which consists so far only of these two sites. These groups are seen in a temporal succession, whereby the *erweiterte Nebraer Gruppe* is assigned to the period before and the *Lausnitzer Gruppe* after the onset of the Late Glacial Interstadial Complex (Ibid.: 195).

There are three major points of criticism concerning this view:

First, what becomes clear from the discussion between Feustel and Hanitzsch is the arbitrariness of individual impressions and preferences that govern the decisions on whether or not a certain type is relevant for the distinction of groups. It is the idea that "our problem may not be one of a paucity of evidence, but one of defining which elements of the evidence are significant" (Thomas 1991: 2f.). However, significance should never be "defined" by researchers but should result from statistical testing and analytical investigations. Otherwise the investigation risks ending up in circular reasoning. And this is exactly the case when certain types are considered significant for group detection and afterward groups are detected according to these types. This kind of group "detection" should absolutely be avoided.

Second, the *erweiterte Nebraer Gruppe* is characterized by only a few arbitrary typological criteria. Because discriminating types are very low in number, it is not surprising at all, but rather a consequence that similarities, even with distant sites, can be observed. These similarities display no large-scale relations (as suggested by Terberger 2002: 137) but a case study for the perception of variability through an insufficient typological filter which biases the observations toward homogeneity (cf. Clark 1994: 332). Therefore, if typology is used to detect groups, all available typological data should be considered initially, and significant types should be found subsequently by statistical testing.

Third, before typological differences are interpreted as significant for regional or temporal groups, other factors should be considered. The fact that a lot of sites assigned to the *erweiterte Nebraer Gruppe* are interpreted as aggregation or base camps (Nebra, Gera-Liebschwitz, Andernach, Gönnersdorf, Kanne, Orp, Etiolles) and at least one of the two-site group *Lausnitzer Gruppe* (Abri Theure) is a hunting camp raises the question whether the observed typological differences are simply an expression of complementary site function rather than the result of social relation or temporal development.

This discussion underlines the strong necessity to overcome interpretations which are guided by arbitrary decisions, circular reasoning, and disregarded alternatives. What is needed instead is a combined analysis in which geographical information are counterchecked with typology, technology, and site catchment analysis.

Chapter 5
Environmental Diversity

Climate and Vegetation

In the course of the CEM, the environment was not a stable and changeless biotope with homogeneous plant cover and a uniform fauna. On the contrary, this was a time of profound climatic alterations with far-reaching consequences for vegetation, animals, and humans alike. In the past as now, the diverse topography of Central Europe, its various landforms, and the changing gradient between the Continental and Atlantic climate zones favored regional differences in the flora and fauna of this region. The following sections will therefore provide an overview of climatic conditions from the Last Glacial Maximum up to and including the Bølling Interstadial (GI-1e) and the corresponding vegetational development. Subsequently, some aspects of faunal distribution during the CEM will be discussed on the basis of hunting remains and faunal representations in engravings and sculptures.

The Last Glacial Maximum

The discussion of the climatic and vegetational development starts with the description of the conditions during the Late Glacial Maximum (LGM) in order to provide a basis of reference that may serve as a background against which the diachronic changes can be projected. Here the LGM is understood as the time between 24/23,000 and 19/18,000 cal BP as suggested by the EPILOG group (Mix et al. 2001). During the LGM, the Oceanic Polar Front had shifted southward to about the latitude of Northern Spain (about 44°N); north of this line, the Atlantic was probably covered with sea ice during winter, but during summer, the pack ice zone retreated considerably (Ruddiman and McIntyre 1973; Florineth and Schlüchter 2000). A number of geomorphological features indicate continuous permafrost at that time down to the latitude of central France and the Alps (Washburn 1979). As a consequence of the Oceanic Polar Front's southward shift, the main track of the precipitation-bearing mid-latitude cyclones had moved southward as well (Meyer and Kottmeier 1989). This displacement of the westerlies' storm tracks to a position south of the Alps gradually cut off from moisture supply those areas of Europe that are today exposed to strong westerly precipitation, leading to dry and continental conditions in Northern and Western Central Europe and to an only insignificant advance of the glaciers in the Pyrenees, the Vosges, and the Massif Central during that time (Florineth and Schlüchter 2000: 305). Paleowind indicators such as dune orientation, the grain-size trends of coversand and loess, the provenance of sediments, or wind-polished rock surfaces can be used to deduce the dominant wind directions of the past (Meyer and Kottmeier 1989: 11; Renssen et al. 2007: 282). Together with model-based simulations, they are good tools for paleowind reconstructions. For Western and Central Europe, cold and dry westerlies seem to have been dominant throughout the whole late Weichselian Glacial (Meyer and Kottmeier 1989: 17; Renssen et al. 2007). A strong impact of northeastern winds, as may be indicated by a high amount of K^+-ions in the NGRIP ice core (see Fig. 5.1), which are interpreted as a proxy for the activity of a Siberian high-pressure cell (Mayewski et al. 1997), is - in contrast to the period between 26,000 and 24,000 cal BP - rarely detectable. The influence of the anticyclonic high-pressure cell above the Fennoscandian and British ice sheet and the resulting katabatic air current seems to be restricted to a narrow, scarcely 100 km wide zone in front of the ice margin (Meyer and Kottmeier 1989: 17; Christiansen and Svensson 1998; Renssen et al. 2007: 289).

In contrast to Northern and Western Central Europe, Eastern Central Europe, Southern Europe, and the Middle East appear to be relatively moist; this is indicated by comparatively high lake levels in Southern Europe as well as in the Middle East (Prentice et al. 1992) and the position of the

Andreas Maier, *The Central European Magdalenian*, Vertebrate Paleobiology and Paleoanthropology,
DOI 10.1007/978-94-017-7206-8_5

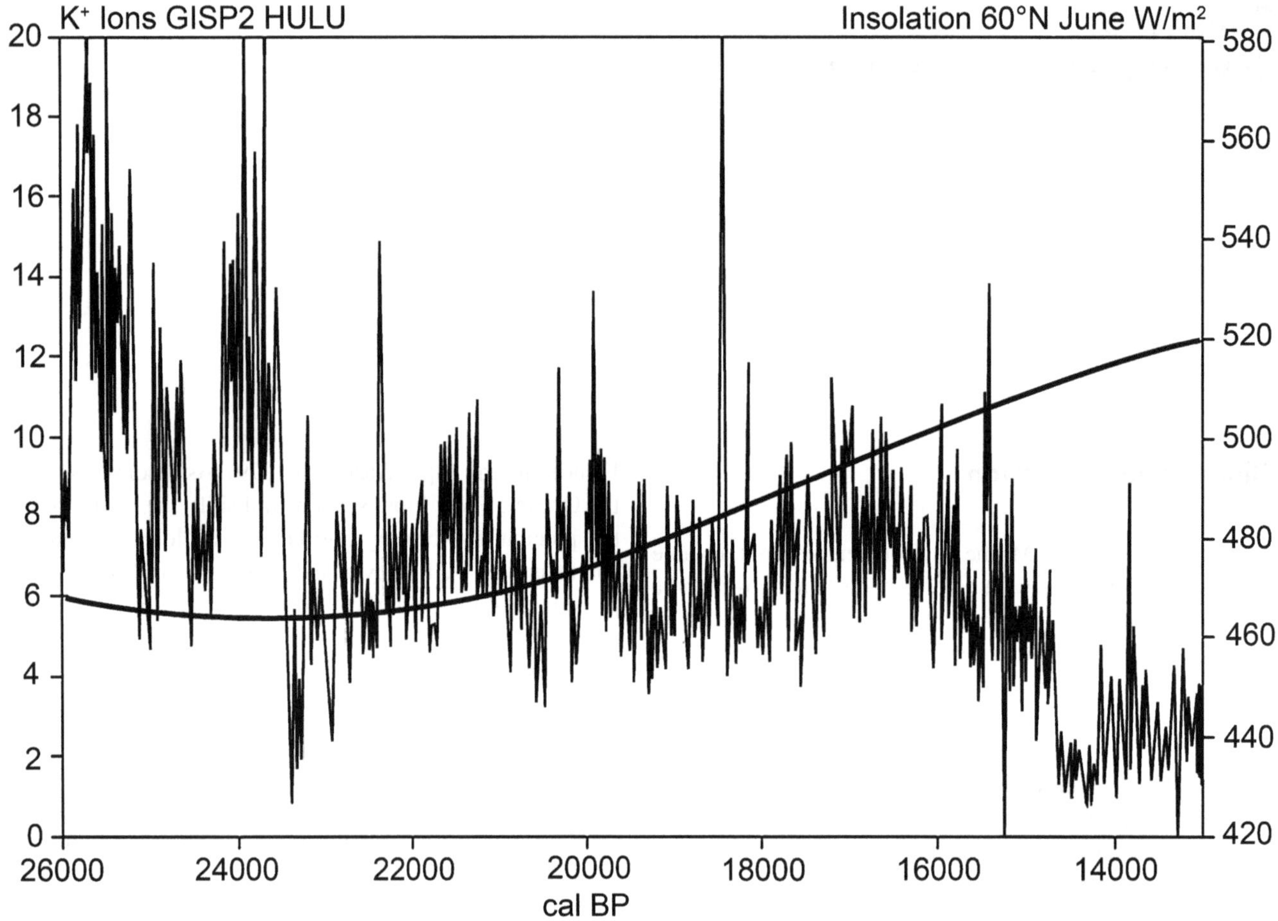

Fig. 5.1 K^+ concentration in the Hulu calibrated GISP2 ice core (Andersen et al. 2006, 2007; Svensson et al. 2006) plotted against solar insolation in June at 60°N in watt per m^2 (from CalPal 2007: Weninger et al. 2012)

main ice divide of the Fennoscandian ice sheet, which had shifted northeastward and was located over the Gulf of Bothnia (Kleman et al. 1997). At the same time, the highest ice dome of the Alps was situated in their eastern part. Together with the observations mentioned above, this fact is seen by Florineth and Schlüchter (2000) as strongly supporting the hypothesis of a southerly position of the westerlies with cyclonic circulation over the Mediterranean and a split of the storm track, with one branch continuing to the Middle East and another one heading north. This caused rather moist condition in the eastern part of the Alps, the Carpathians, and at the southern edge of the Fennoscandian ice sheet (see Florineth and Schlüchter 2000).

These climatic changes naturally affected European vegetation and led to a withdrawal of nemoral and boreal trees and shrubs from the northern latitudes to more southwardly located stands. The prevailing opinion on this topic during the last few decades found its expression in the so-called southern refugia hypothesis, which states that temperate tree species were restricted to the Iberian and Italian peninsulas and the Balkan during the LGM (Bennett et al. 1991; Taberlet and Cheddadi 2002; Svenning et al. 2008). However, the results of recent pollen and macro-remain analyses, as well as phylogeographic studies and environmental modeling, have led to different conclusions and to the proposal of a northern refugia hypothesis (Willis et al. 2000; Stewart and Lister 2001; Willis and van Andel 2004; Svenning et al. 2008). Within this hypothesis, it is acknowledged that the refugia of nemoral tree species were probably restricted to Southern Europe, whereas reliable evidence has been brought forward that some boreal tree species survived the LGM in southern Central Europe and as far north as the Carpathians. The best evidence here are macro-remains from tree species such as *Larix*, *Pinus*, *Betula*, and *Salix*, which were found in the southern and western Carpathian region and are ^{14}C-dated to the time of the LGM (Willis et al. 2000; Willis and van Andel 2004; Jankovská and Pokorný 2008). These observations accord well with the climatic assumption of comparably moist conditions in this area. The permafrost soil and very low precipitation rates probably inhibited tree growth in the remaining areas of Central and Western Europe. Especially the amount of moisture available during the vegetation period was probably the most important limiting factor for their dispersal. This assumption is supported by observations in today's continental steppe environments: the lack of trees in these biotopes is apparently not caused by very cold winter temperatures, but by an insufficient moisture supply. If trees are planted in steppes and watered regularly, they grow well during the first years. However, with increasing size, the amount of available water during the vegetation period becomes deficient, leading to a dieback of the trees (Walter 1943).

Between LGM and Oldest Dryas: The Late Pleniglacial Optimum

It can be stated that the extensive recolonization of the area between the French Jura and the Moravian Karst did not start before 18,000 cal BP (see Chap. 8), even though some ^{14}C-dates, partly worthy of discussion, may indicate a sporadic presence of hunter-gatherers already at an earlier date (see Housley et al. 1997; Terberger and Street 2002). According to the δ^{18}O-curve of the NGRIP Greenland ice core (Björck et al. 1998; Rasmussen et al. 2006), which is understood as a proxy for evaporation and consequently for temperature (Grootes et al. 1993; Mayewski et al. 2004), the climate grew continuously warmer from the end of the LGM onward, reaching a first maximum between roughly 18,000 and 17,000 cal BP. This period was therefore the warmest part of the late Pleniglacial, with temperatures higher than at any time throughout the Younger Dryas and only slightly colder than during the late Allerød (GI-1a) (Fig. 5.2). Occasionally, the term Lascaux Interstadial is used to account for environmental changes which are observable throughout this time interval (Leroi-Gourhan and Renault-Miskovsky 1977; Leroi-Gourhan 1980; Renault-Miskovsky and Leroi-Gourhan 1981; Delpech 1992). However, because this period is not an interstadial in the proper meaning of the word and the exact temporal borders are only poorly defined, this term will not be used in this study. Instead, the respective period will be referred to as Late Pleniglacial Optimum.

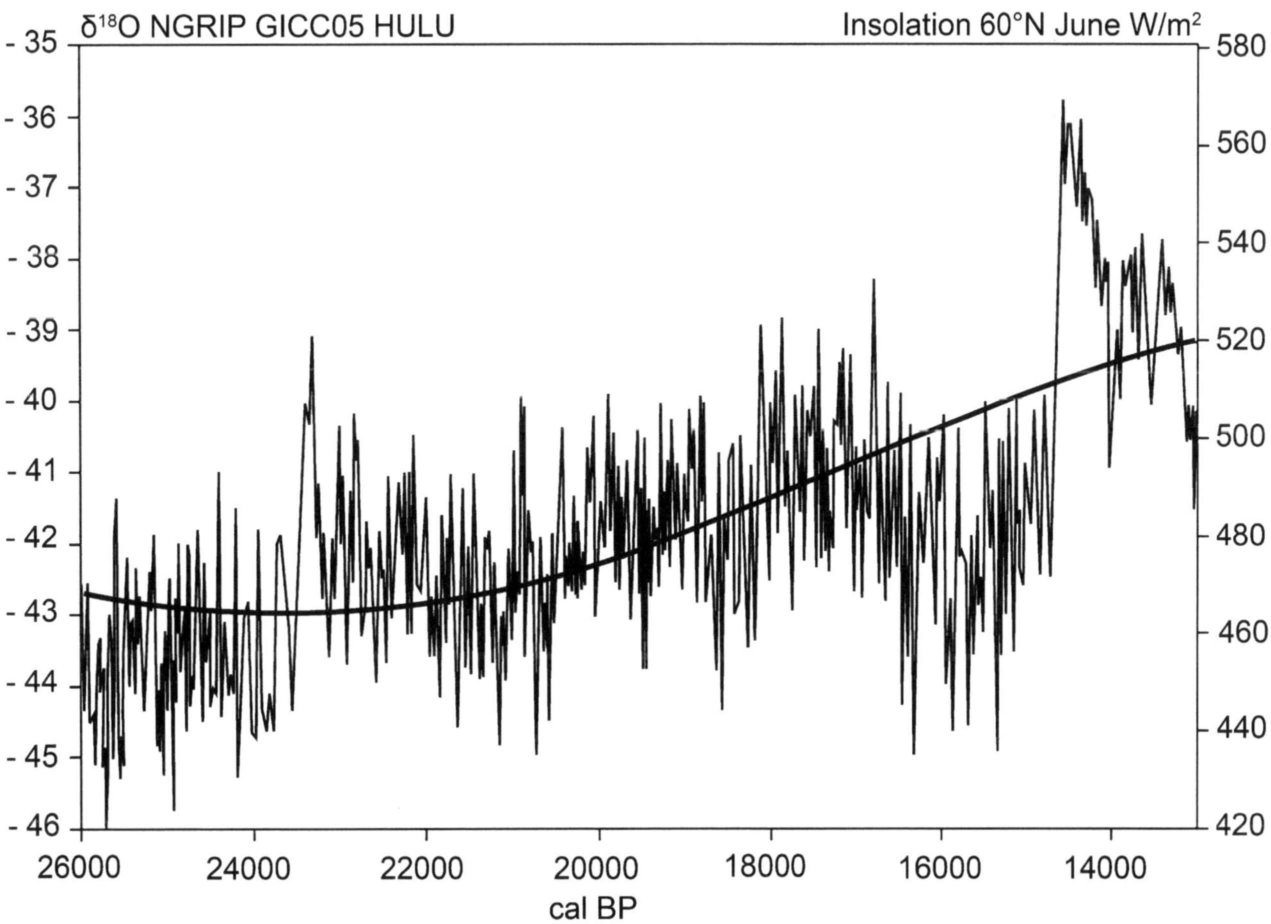

Fig. 5.2 δ^{18}O concentration in the Hulu calibrated NGRIP ice core (Andersen et al. 2006, 2007; Svensson et al. 2006) plotted against solar insolation in June at 60°N in watt per m^2 (from CalPal 2007: Weninger et al. 2012)

The climatic amelioration just described was accompanied by a constant rise in solar insolation (Fig. 5.2), which probably led to a gradual northward shift of the permafrost zone and the Oceanic Polar Front. As a consequence of the latter, the main track of the precipitation-bearing mid-latitude cyclones presumably moved northward as well, exposing Northern and Western Central Europe to westerly precipitation once again. These changes certainly did not remain without consequences for vegetation, and it is likely that trees started redispersing from their LGM refugia. However, as a lot of water was still bound in the glacial ice sheets and precipitation thus likely remained at a low level, the amount of moisture available during the vegetation period was probably still low in many parts of Western and Central Europe. Additionally, relatively high levels of K^+-ions indicate that cold and dry winds from the northeast may still have had an inhibiting influence on vegetation (Fig. 5.1).

Three boreal tree species, namely, Scots Pine (*Pinus sylvestris*) and Silver and White Birch (*Betula pendula* and *B. pubescens*), played an important role during the early period of arboreal redispersal. *Pinus sylvestris* is a species with pronounced pioneer features: it tolerates wide variations in annual temperature, a broad range of precipitation, and soil and orographic conditions and is able to expand rapidly in open country (Latałowa et al. 2004: 165f.). On the other hand, it is sensitive to low temperatures during the development of its shoots and requires a mean July temperature of about 12 °C (Iversen 1954; Szafer 1964). *Pinus* pollen grains, which are light and equipped with air sacs, disperse easily over large distances. Additionally, the amount of *Pinus* pollen produced per year is immense in comparison to that of grasses, for example. The dendriform birch species *Betula pendula* and *B. pubescens* produce large amounts of wind-dispersed pollen and are both rather frost-resistant pioneer trees which have similarly broad habitat requirements to *Pinus sylvestris* (Ralska-Jasiewiczowa et al. 2004b: 57f.). The additional requirements of the flood-tolerant *B. pubescens* are only a mean July temperature of not lower than 10 °C (Zarzycki 1979; Paus 1992).

Because both pine and birch produce large amounts of easily dispersed pollen, it is difficult to deduce their true proportions in steppe or tundra environments from the respective percentages in the pollen diagram alone, since steppes and tundras are characterized by a comparatively low inherent pollen production. Moreover, the *Pinus* (*diploxylon*) pollen type includes, besides *Pinus sylvestris*, the low-growing shrub *P. mugo* and the shrubby *P.* ^{x}*rhaetica* (Latałowa et al. 2004: 165). It has been assumed that a percentage value of 25 % *Pinus* pollen in the pollen diagram represents a threshold for the local presence of *Pinus sylvestris* in a forested landscape (Huntley and Birks 1983). The pollen of dwarf birch (*B. nana*), a low-growing shrub which today is distributed mainly in tundra environments, is generally distinguishable from *B. pubescens* and *B. pendula* pollen. However, in the literature, all birch species tend to be represented in a single pollen curve, especially in older studies.

The question of whether or not dendriform pines and birches were growing in the vicinity of a pollen analysis sample spot can therefore only be answered unambiguously if other parts of the plants, such as fruit, twigs, leaves, or needles are present as well. The analysis of macro-remains is thus the most reliable source of information on the presence or absence of pines and birches in the investigated area during the late Weichselian Glacial. For this reason, identification of the wood represented in charcoal from Paleolithic sites, especially before performing ^{14}C-dating, should become a standard procedure in archaeological research. However, as macro-remains analyses are virtually nonexistent for most of the investigated area, the following comparison must rely on palynological data. For this reason, an ample set of pollen diagrams has been analyzed. It must be stressed that this compilation is not intended to be exhaustive, but rather attempts to cover the investigated area as extensively as possible—an attempt which, admittedly, is not completely successful, especially for the oldest period.

In order to obtain comparable results, all diagrams are constructed the same way. First, a selection of pollen taxa and pollen groups was chosen which was thought to be suitable for characterizing a landscape's vegetation, namely, *Pinus*, *Betula*, *Juniperus*, Poaceae, *Artemisia*, and nonarboreal pollen (NAP). Then, all values were measured individually from the published diagrams. Smaller measurement errors, as do naturally occur, are negligible because the focus of the analysis lies on larger spatial trends and not on single observations. The measured results were added together, and the sum was set to 100 %. Afterward, percentage values for each pollen taxon and pollen group were calculated relative to the respective sum. Pollen diagrams from the Late Pleniglacial Optimum are rather scarce for the area under investigation. However, some diagrams yield pollen spectra which are either ^{14}C-dated or located directly below the characteristic rise of *Artemisia* pollen, which defines palynologically the onset of the Oldest Dryas (see below), a period of climatic cooling which followed the Late Pleniglacial Optimum (see Fig. 5.2). Therefore, those pollen spectra which are situated immediately under the samples from the Oldest Dryas and are not separated from them by a hiatus in the pollen sequence can be seen as representative for the preceding Late Pleniglacial Optimum.

Figure 5.3 shows a collection of pollen spectra from Central Europe for the Late Pleniglacial Optimum. There are three rather surprising observations that can be made by analyzing this map. First, there is a pronounced diversity in the composition of the pollen spectra. Second, this different diversity is by no means an expression of randomly varying proportions but, to the contrary, shows a clear spatial trend.

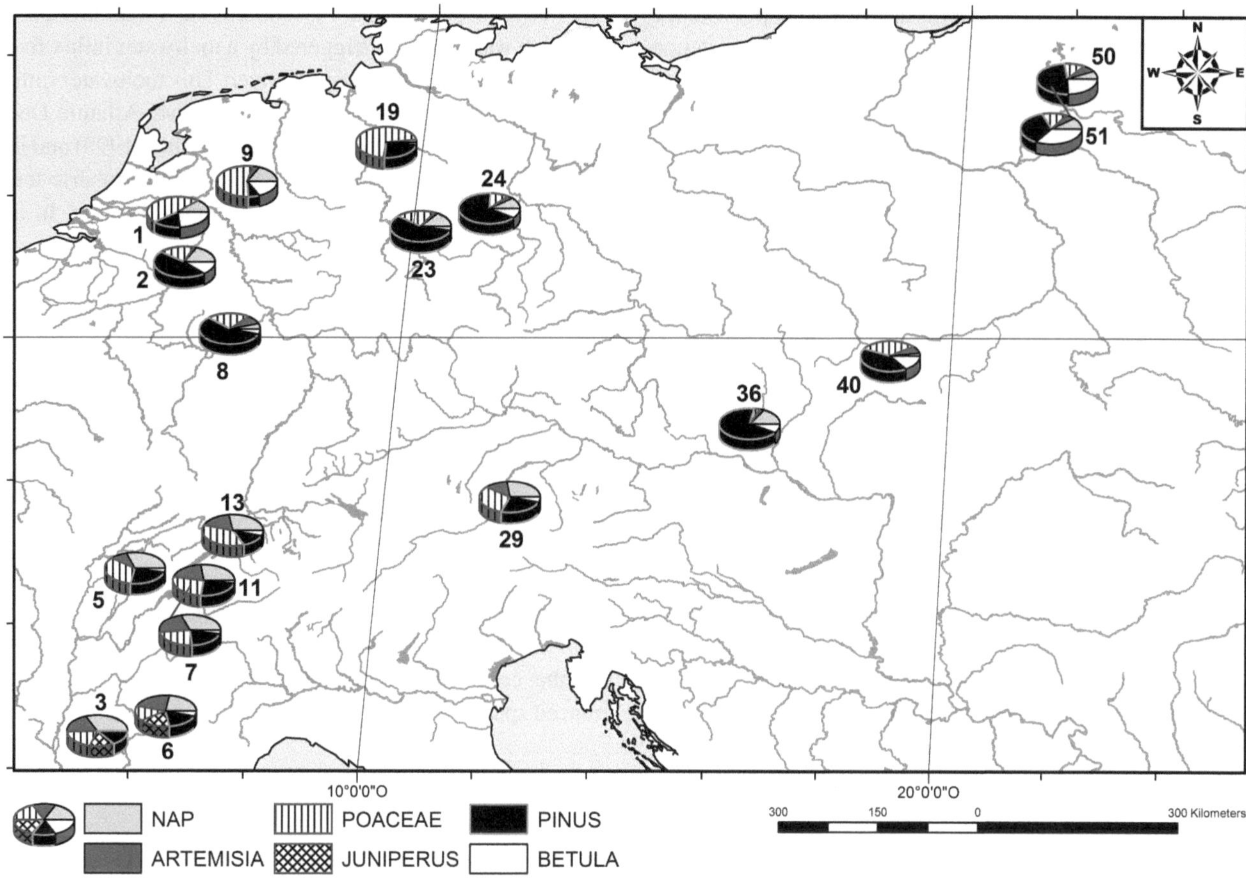

Fig. 5.3 Pollen spectra from the time of the Late Pleniglacial Optimum (see Table A.4 and references therein and Table A.5)

Third, the amount of *Pinus* pollen in some diagrams far exceeds 50 %.

A first group of pollen spectra with similar ratios of the analyzed taxa groups can be located in the southwest part of the investigated area. Here, pollens of *Pinus*, *Artemisia*, Poaceae, and NAP are more or less equally abundant, indicating an open landscape with herbs and grasses. *Juniperus* is only present in the far south. Whether the amount of *Pinus* pollen, which sometimes reaches values of 20–30 %, indicates scattered stands of pines or is exclusively the result of wind-borne far-distant transport cannot be decided without corresponding results from macro-remains analyses. *Betula* pollens are virtually absent. Despite a lack of pollen diagrams between the rivers Aar and Isar, it seems that this kind of biotope was distributed in the northern and western parts of the Alps as well as in the French and Swiss Jura.

Further to the north and east, the amount of *Pinus* pollen increases drastically, reaching values of about 50–70 % in the western Carpathian region, the Hercynian Mountains, and the Eifel. Explanations of *Pinus* pollen values of more than 50 % by long-distance wind-borne transport alone are not convincing, especially if paleowind reconstructions are taken into account (see above) which speak in favor of dominating west winds. If the glacial refugia of *Pinus* are assumed to be located in the Carpathian region, this massive transport would have taken place against the dominant wind direction. Thus, it seems more reasonable to assume a spread of pine trees along the valleys of waterways that source in the Carpathians and their surroundings, such as the Vistula, Elbe, and Oder rivers. On the one hand, seeds were transported effectively downstream by the current, and on the other, the unstable conditions of the dynamic braided river systems constantly created new potential stands (Ellenberg 1978: 368f.), as has been documented for the Pupplinger Au (Seibert 1958). It can therefore be assumed that at least scattered stands of pines must have been present in the vicinity of the locations where the pollen samples were taken.

The northwest of the investigated area, in contrast, is characterized by Poaceae values of 50–70 %. This is particularly marked in the Hämelsee spectra (Merkt and Müller 1999), where Poaceae and *Pinus* are almost exclusively present, which may point to a steppe environment with isolated pines. It is somewhat conspicuous that *Betula* pollens occur in considerable numbers only in the very northwesternmost and northeasternmost sites where they vary between 10 and 20 %. Here, in the lowlands of the North European Plain, the

Pleniglacial braided rivers systems and the likely persistence of permafrost soil possibly led to the formation of a steppe tundra with grasses, herbs, and patches of *Betula nana*.

The Oldest Dryas

The relatively warm period of the Late Pleniglacial Optimum was followed again by a period of colder conditions, the so-called Oldest Dryas or Greenland Isotope Stadial 2a (GS-2a). This period lasted roughly for 2000 years, from 16,900 to 14,700 cal BP (Björck et al. 1998; Rasmussen et al. 2006). In the Greenland ice core records, the $\delta^{18}O$ values started to decline steadily at about 16,900 cal BP, and around 16,000 cal BP reached a level comparable to the conditions of the LGM. At the same time, the level of K^+-ions decreased as well but still remained comparatively high (see Fig. 5.1). The temperature decline indicated by the $\delta^{18}O$ values took place despite a continuous increase in solar insolation (see Fig. 5.2). However, it coincided with a cooling of the sea surface temperatures (SST) in the Gulf of Mexico of about 2 °C between 16,000 and 14,7000 cal BP (Williams 2009: 18) which was probably triggered by a meltwater influx from lake Agassiz via the Mississippi River. This meltwater influx resulted in turn in the disruption of the North Atlantic Deep Water (NADW) formation (McCabe and Clark 1998) and led to the decline of North Atlantic SST and atmospheric temperatures in Western, Northern, and Central Europe. In the Antarctic record, in contrast, the warming trend that began about 21,000 cal BP continued throughout this period without interruption (Ibid.: 376).

Palynologically, the onset of the Oldest Dryas is defined by a rise in *Artemisia* pollen (Iversen 1973: 29). This period ends with the abrupt warming of the Bølling Interstadial (GI-1e), which is in turn defined by a maximum of *Hippophaë* pollen. Figure 5.4 shows a map that plots those spectra of Central European pollen diagrams that show a rise in *Artemisia* pollen before the occurrence of a *Hippophaë* maximum.

Again, it becomes clear that Central Europe was already an environmentally diverse region during the Oldest Dryas. As was the case for the Late Pleniglacial Optimum, a pronounced spatial trend is visible that separates the investigated

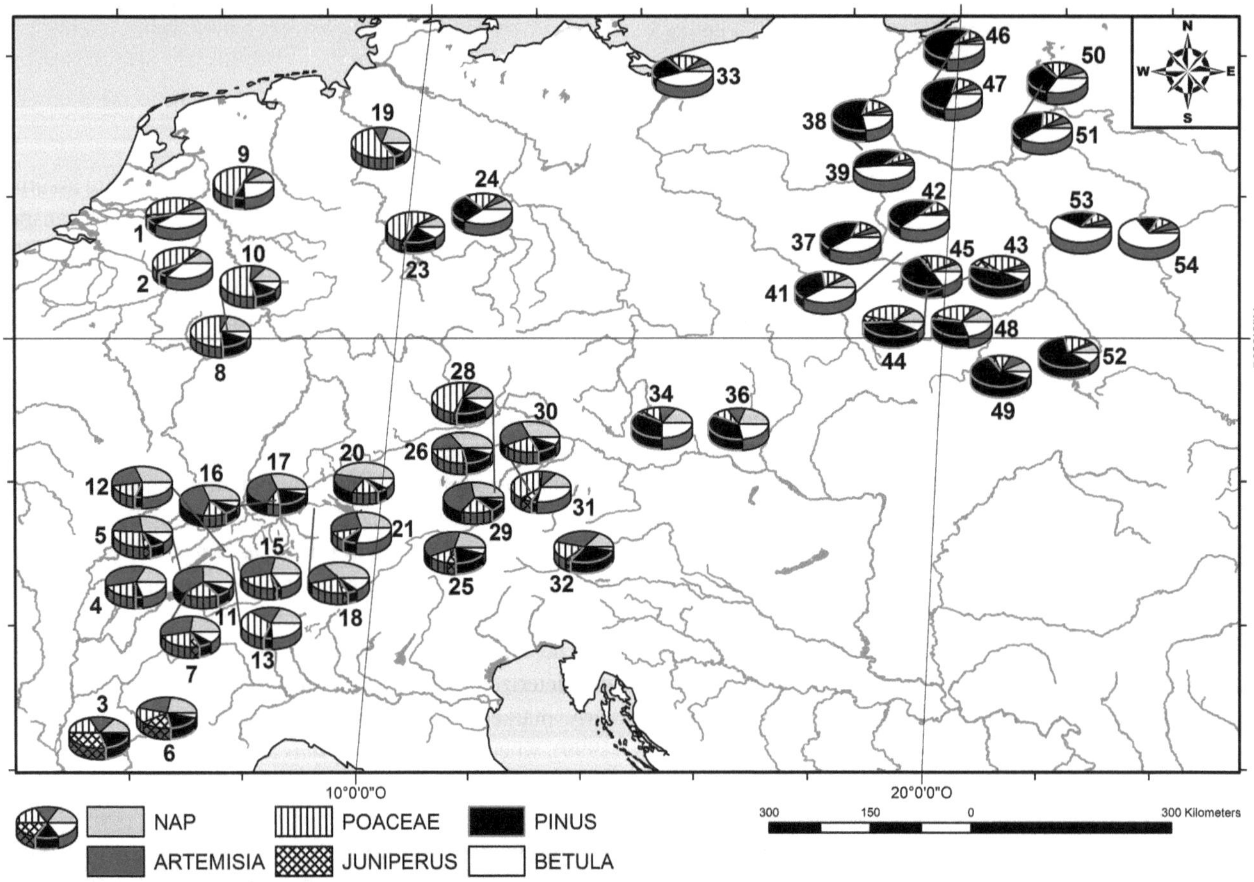

Fig. 5.4 Pollen spectra from the time of the Oldest Dryas (see Table A.4 and references therein and Table A.5)

area into three major environmental regions: the southwest, the northwest, and the east.

The southwestern region comprises the Alps, the French and Swiss Jura, and the Swabian and Franconian Alb. Generally, this region is dominated by pollens from herbs and grasses. *Pinus* pollens exceed 20 % only in the southeasternmost diagram, but in most spectra appear in proportions of less than 10 %. Values for *Betula* pollen are low as well and rise above 20 % only at the site of Moosham (Schmeidl 1971) and some sites between Lake Leman and Lake Constance. Higher percentages of *Betula* pollen are probably the result of local stands of *B. nana*, whereas the presence of *Pinus* pollen might be exclusively the result of long-distance wind-borne transport. The dominant pollen taxon in the diagrams is *Artemisia* with proportions often between 35 and 40 %; the second most abundant group is other non-arboreal pollens. Both increased considerably in comparison to the previous period. Poaceae, on the other hand, decreased slightly, and *Juniperus* was still restricted to the southern parts of the area. On the whole, the southwestern region was likely a treeless steppe tundra during the Oldest Dryas with *Artemisia* and other herbs dominating the landscape.

In the northwestern region, which comprises the Rhenish Massif and the lowlands adjacent to the north, the environmental conditions were quite different. Pollens of Poaceae are most abundant in the respective spectra with proportions between 40 and 55 %. *Artemisia* and non-arboreal pollens, which are dominant in the southwestern region, rarely exceed 10 %. In contrast to the previous period, during which *Pinus* pollen dominated the spectra in the uplands, its proportion rose only slightly above 20 % in the far east of the region. Therefore, even though the presence of scattered pines in the landscape cannot be ruled out, the presence of *Pinus* pollen might be the result of wind-borne transport. The percentage values of *Betula* pollen, on the other hand, increased significantly, reaching values of up to 40 % in the northwestern area. Thus, it seems that the northwestern region was mainly a tundra steppe. Unlike the biotope of the southwestern region, the steppic elements appear to prevail, especially in the uplands, indicating a comparatively dryer climate than in the south. At sheltered places with a sufficient water supply, scattered pines may have grown too. In the lowlands, permafrost and braided river systems probably created moister conditions. The seasonal thawing of the permafrost's active layer during the vegetation period probably led to waterlogging as the permanently frozen layer prevented the water from seeping away; this may have fostered a more tundra-like vegetation. *Betula* pollen may therefore represent mainly *B. nana*, although isolated occurrences of *B. pubescens* cannot be ruled out.

A remarkable feature is the sharp border that delimits this region to the east. Whereas the site of Luttersee (Chen 1982) on the western side of the Hercynian Forest still accords completely with the other spectra of the Rhenish Massif, the site of Gaterslebener See (Müller 1953), which is located only about 100 km away on the eastern side of the Hercynian Forest, has a completely different pollen composition and displays the characteristics of other eastern spectra.

The spectra of the eastern region, which ranges approximately from the Hercynian Forest in the northwest and the Bohemian Forest in the southwest up to the valleys of the San and Vistula as well as the lowlands of Masuria in the east, indicate a completely different environment in comparison to the ones described above. Here, the pollen spectra of the Oldest Dryas are characterized mainly by arboreal pollens from *Pinus* and *Betula*. Except for two sites, one in the area of the Szczecin Lagoon, the other east of the San river, the proportion of *Pinus* pollen at no point drops below 25 % but remains at an average level of 40 %. In the mountainous area of the Carpathians and the Polish Jura, *Betula* pollen is often present in proportions of less than 10 %. Further to the north and east, however, the level rises up to 65 %. Among the non-arboreal pollens, those of Poaceae are most important and reach values of up to 30 %. *Artemisia* and other NAP only play a negligible role.

The high proportions of pine and birch pollen cannot be convincingly explained by wind dispersal alone. Especially in the eastern area, the respective amounts stayed rather constant in comparison to the Late Pleniglacial Optimum. In the western areas, however, *Pinus* pollen values dropped considerably and were only half as high during the Oldest Dryas as they were during the previous period. Apparently, the eastern region was characterized by a kind of park steppe, with a spatial gradient in tree density ranging from sparse occurrences of scattered specimens in the west to more aggregated stands in the east. A piece of *Pinus* charcoal from the site of Wilczyce was ^{14}C-dated to 12,770 ± 120 ^{14}C BP (15,260 ± 230 cal BP) (see Table A.6), thereby confirming the presence of pine trees north of the Vistula river in the eastern parts of the Holy Cross Mountains during the Oldest Dryas. *Pinus* charcoal fragments from the open-air site of Putim in Bohemia were even dated to 13,010 ± 60 ^{14}C BP (15,580 ± 80 cal BP) indicating pine stands throughout the Oldest Dryas in the Otava river valley. The severe climatic conditions of the Oldest Dryas thus pushed back the redispersal of boreal trees which started during the period of climatic amelioration after the LGM. The cooling trend which was triggered by a change in the North Atlantic circulation system (see above) apparently affected the different parts of Central Europe in different ways. In the western area, which is more exposed to the influence of the Atlantic climate, increasing coldness and aridity led to a dieback of trees and favored the development of a steppe tundra, in which the steppe and tundra elements varied according to soil moisture. In the eastern and more continental part of Central

Europe, the decreasing precipitation that resulted from colder North Atlantic SST likewise hampered arboreal growth. Especially in the western part of Eastern Central Europe, trees could likely only survive at favorable locations and drew back into rather sheltered stands with a sufficient water supply. In the eastern part of this region, however, it seems that the climatic effects of the Oldest Dryas became less pronounced, at least during the vegetation period, since east of the Oder river, a park steppe with patches of pines and birches apparently persisted throughout this period.

The Bølling Interstadial

The Greenland ice core records show a sharp rise in $\delta^{18}O$ values around 14,700 cal BP. This is interpreted as an increase in Greenland air temperature of 9 ± 3 °C, marking the beginning of the Bølling Interstadial (GI-1e) (Björck et al. 1998; Rasmussen et al. 2006). During the transition from the Oldest Dryas to the Bølling Interstadial, the mean annual temperature in the northern hemisphere rose to today's level within a few decades or even years (Steffensen et al. 2008). At the same time, the level of K^+-ions dropped sharply (see Fig. 5.1), indicating that the influence of cold and dry Siberian winds on the European climate became insignificant. This sudden climatic amelioration coincides with a rise in SST in the Gulf of Mexico of about 3.5 °C (Williams 2009: 26) and is very probably connected with an increase in NADW formation and a resulting warming of the North Atlantic (Weaver et al. 2003).

In the palynological record, the Bølling Interstadial (GI-1e) is marked by a maximum of *Hippophaë* pollen (Iversen 1973: Fig. 29), which is often connected with a rise in *Juniperus*, *Salix*, *Betula*, or *Pinus*. Hence, it is comparatively easily recognizable in the pollen diagrams. Figure 5.5 shows Central European pollen spectra from this period. According to the displayed findings, the investigated area can be subdivided again into three environmentally distinct regions, as was the case for the Oldest Dryas. The borders of these regions stay surprisingly stable, as do the general characteristics of the individual spectra's pollen composition everywhere but in the southwestern region.

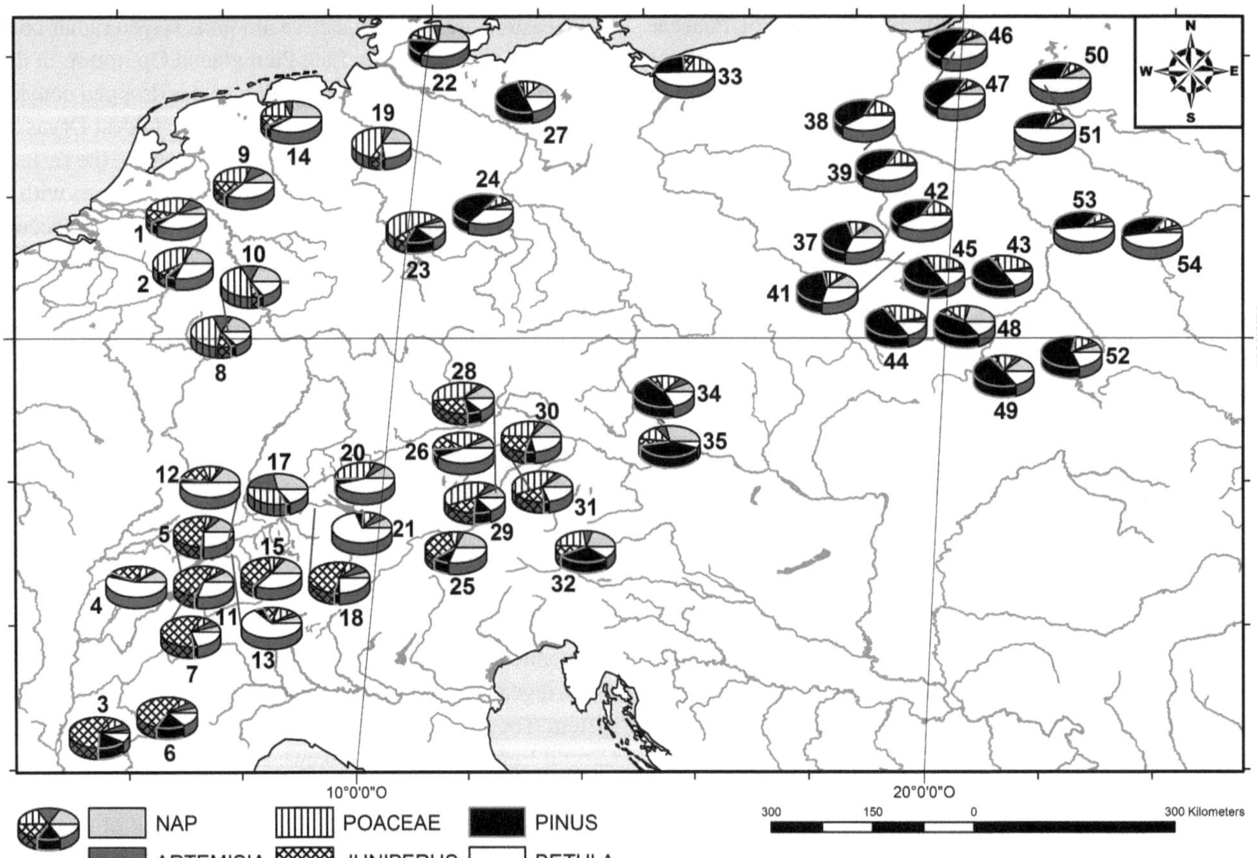

Fig. 5.5 Pollen spectra from the time of the Bølling Interstadial (see Table A.4 and references therein and Table A.5)

In this latter region, which still comprises the Alps and the Jurassic formations from the French Jura up to the Franconian Alb, the climatic amelioration of the Bølling Interstadial led to a spread of *Juniperus,* which became the dominant taxon in its southwestern area and reached values of more than 20 % in the northeast. In comparison to the previous period, the internal differentiation of this region increased perceptibly. Proportions of Poaceae pollen are high northeast but play only a marginal role in the southwest, whereas *Betula* pollens are highest in the area of Lake Neuchâtel and Lake Constance. *Pinus* pollen is virtually absent in most diagrams and only occurs in considerable amounts toward the east and the southwest. This diversification might have been the result of soil moisture. In areas where water supply was sufficiently high, *Juniperus* was highly competitive in a landscape that was otherwise treeless and mostly devoid of shrubs. In dryer areas, however, grasses were dominant. In the valley bottoms and around larger lakes, where waterlogging may have occurred and the soil was swampy for at least a part of the year, dendriform and shrubby birches had competitive advantages. Pines, so it seems, did not yet redisperse successfully in this region during the Bølling Interstadial.

The pollen spectra of the northwestern region, including the Rhenish Massif and the lowlands adjacent to the north, are, in contrast, comparably homogeneous with regard to their overall composition. Nevertheless, a spatial trend is visible here too. In the uplands, non-arboreal pollens are the most abundant, and Poaceae pollen reaches average values of 40 %. In the lowlands, the proportion of Poaceae pollen declines gradually, first in favor of *Betula* pollens, which gain considerably in significance and rise to 40 %, and further to the north in favor of other NAP. As is assumed for the southwestern region, soil moisture is probably the most important factor for this internal diversification. The amount of *Pinus* pollen does not exceed 5 % in most diagrams and only rises to about 20 % in the far east. Here, in the area of the Hercynian Forest, the sharp border between the northwestern and the eastern regions, which could be observed for the Oldest Dryas, clearly persisted during the Bølling Interstadial.

In the eastern region, the vegetation pattern changed drastically. Arboreal pollens are by far the most abundant component of the spectra with average values around 70 %. Whereas birch pollens are dominant east of the San and Vistula rivers as well as in the Szczecin Lagoon, their abundance is equal to that of pine pollen in the area of Łódź. In all other areas, however, *Pinus* is the dominant pollen type. Among the non-arboreal pollens, those from Poaceae are most abundant. Generally, the landscape in the eastern region seems to be a park steppe with extensive stands of pines and birches. Whether the former or the latter prevailed locally probably depended on soil moisture, with pines in dryer and birches in wetter places.

Fauna

Just as the vegetation during the CEM was not a uniform community of plants in time or in space, neither was the distribution of faunal elements. Following Koenigswald, the occurrence of most herbivores is usually closely connected to a particular type of vegetation; throughout the Late Pleistocene, there were mainly two types of faunal communities: the mammoth (*Mammuthus primigenius*) fauna and the straight-tusked elephant (*Elephas antiquus*) fauna (Koenigswald 2002: 140). The former community comprised, in addition to the eponymous mammoth, other species of larger mammals such as reindeer (*Rangifer tarandus*), saiga antelope (*Saiga tatarica*), musk ox (*Ovibos moschatus*), woolly rhinoceros (*Coelodonta antiquitatis*), and steppe bison (*Bison priscus*). All of these species were adapted to cold environments with a mostly treeless steppe, tundra, or steppe-tundra environment (Guthrie 1990; Koenigswald 2002: 140, Fig. 176), although reindeer appear to have existed sometimes in taiga environments (Weniger 1982: 91; Serangeli 2006: 105; see below). The straight-tusked elephant community, in contrast, preferred a completely different environmental setting, with a temperate climate and a rather forested landscape; this fauna included species such as fallow deer (*Dama dama*), roe deer (*Capreolus capreolus*), wild boar (*Sus scrofa*), and aurochs (*Bos primigenius*). Additionally, there were herbivorous species which occurred regularly in both kinds of environments, namely, horse (*Equus ferus*), red deer (*Cervus elaphus*), and megaloceros (*Megaloceros giganteus*) (Koenigswald 2002: 140, Fig. 176).

While this can serve as a general but somewhat idealized pattern, there are some findings which must be considered in order to do justice to the much higher complexity of prehistoric reality. It must be recognized that "[t]he habitats of Central Europe during the Weichsel/Würm Glaciation are not represented anywhere today and species which differ markedly in their ecological requirements today are often found together in non-analogue glacial environments" (Musil 2003: 167). At the Thuringian site of Teufelsbrücke, for instance, the co-occurrence of roe deer with reindeer and even saiga antelope during the Magdalenian is reported (Feustel 1980; Serangeli 2006: 98). Moreover, studies from Northern Europe suggest that smaller groups of temperate animals may have survived in sheltered areas where deeply incised valleys provided a favorable microclimate (Gautier et al. 1997; Stewart and Lister 2001). Finally, there are some traces from Spanish sites which seem to point to the existence of a kind of taiga-reindeer there, even though the taphonomic situation is not entirely clear; more reliable are Preboreal and Boreal reindeer finds from Lower Saxony and Baden-Wuerttemberg, which at least indicate a coexistence of reindeer and temperate fauna (Serangeli 2006: 107).

Moreover, recent observations of roe deer, a species which is usually thought to be strongly associated with forested environments, show a surprising development: in regions which have been cleared of forests for agricultural purposes, row deer have adapted to the altered environmental conditions. Here, they live in groups of 30–40 animals and exhibit behaviors which are typical of gregarious animals in steppe environments. Therefore, these row deer are seen as a distinct eco-type and are referred to as "field-row deer" (Feldreh) (Kurt 2002: 153f.).

In the next section, the occurrence of faunal remains in Magdalenian sites will be discussed for a number of species. However, the analysis will focus on the two most important Magdalenian prey species, namely, reindeer and horse, because both animals are represented in considerable quantities in a sufficient number of sites to ensure representative results (for a detailed overview see Tables A.7 and A.8).

Horse and Reindeer Remains

Already before the term "Magdalénien" was coined by Gabriel de Mortillet (1869), Édouard A. Lartet had labeled this period "l'Âge du Renne" (Age of the Reindeer) in his chronological subdivision of the Stone Age, which was based on the predominance of certain species in the respective paleontological record (Lartet 1861). This appellation is still occasionally used in archaeological research today. Indeed, reindeer was doubtless one of the major pillars of Magdalenian subsistence, and at many Magdalenian sites, reindeer remains outnumber those of other larger mammals. What reasons can be brought forward to explain the importance of this animal for Late Upper Paleolithic hunter-gatherers? If they were only a source of meat, hide, and bones acquired through an encounter strategy (see Binford 1978, 1980), then any other larger game would have been equally suitable. Thus, the question of what made reindeer so attractive remains.

A physical feature that distinguished reindeer from most other Pleistocene animals in Central Europe was certainly its production of antlers, an important raw material for the production of organic tools. On the other hand, shed antlers could be collected without actually needing to bring down a reindeer, and organic artifacts could be manufactured from bone as well, even though antler might be more durable. However, in terms of nutritional value, reindeer provide a significant advantage in comparison to other animals: even though a reindeer weighs only half as much as a horse, it provides 13 times the amount of marrow, which is furthermore less fluid than that of horses. Thus, reindeer could provide an ideal source of high caloric nutrition, which was especially important in cold environments (Enloe 2003: 28f.). Another reason for the importance of reindeer to Magdalenian subsistence may be related to their behavior. Because all available observations on this point have been obtained from Holocene reindeer, the uniformitarian assumption that these animals' habits have not changed significantly since the CEM is the basis for the following reflections. In spring, small bands of female reindeer leave their winter stands and move to their natal grounds where they bear their young, usually during the first half of June. Afterward, they move together to the summer stands. The males leave the winter stands only after the young are born. In the summer stands, the bands disperse randomly. At the end of summer, males, females, and young congregate in order to migrate together to the winter stands, where they disperse again (Bodu et al. 2006: 31f.). The fall migration lasts a few weeks at most (Weniger 1982; Serangeli 2006). Thus, reindeer form herds twice a year, namely, during fall and spring migration, a predictable behavior pattern that can be exploited specifically. The fall migration, during which body weight and fat content are at their maximum, is the best time for hunting reindeer, especially when storage plays a role for winter subsistence strategies (Enloe 2003: 24). Looking at the ethnographic record, the Nunamiut Inuit, for instance, acquire over 70 % of their yearly supply of food during the seasonal caribou migrations, with about 15 days of hunting in spring and fall, respectively (Binford 1979: 256).

Horses, in contrast, migrate only on a regional scale. The spring congregation is followed by the bearing of the young and the subsequent rut. Afterward, during the late spring and summer, horses live in dispersed bands of two types: harems consist of one stallion and several mares with their foals, while bachelor bands are formed by groups of 2–6 stallions living together. At the beginning of the warm season, horses usually remain in rather close proximity to each other. As the food supply declines, the distance between bands as well as between individual animal increases, reaching its maximum in winter. Besides seasonal migrations from winter grazing grounds in the valley to summer grazing pastures on higher land, band movement is usually determined by the availability of food resources and shelter against cold winds (Weniger 1982; Bodu et al. 2006: 32; Serangeli 2006).

Looking at the faunal remains from CEM sites, it is clear that reindeer and horse were the most important game during this period. Nevertheless, the proportions of the two could vary considerably. Thus, the question arises of what reasons Magdalenian hunters had for choosing to target the one animal or the other. Did hunting preferences play a role, or was seasonality the important factor? Or was it simply a question of abundance and availability, determined by different environmental conditions? In order to address these issues, the faunal composition of 129 CEM assemblages, for which the respective information could be obtained from the archaeological literature, was analyzed (see Table A.7). Focusing exclusively on the number of horse and reindeer, five prey classes were created according to the ratio of both species. Class 1 contains those assemblages in which the occurrence of horse remains is proven and reindeer remains are absent (n = 19; 7 %), while Class 2 comprises all assemblages which are dominated by horse but yield also some reindeer remains (n = 25; 19.4 %). A more or less equal number of horse and reindeer remains is observed in assemblages of Class 3 (n = 47; 36.4 %), whereas the reindeer are predominant in Class 4 (n = 34; 26.4 %). In Class 5, the presence of reindeer is attested, but horse remains are lacking completely (n = 14; 10.9 %). In Fig. 5.6, all 129 assemblages are mapped and labeled according to their classification. It becomes clear that the distribution of the prey classes is anything but uniform. Rather they show a clear spatial trend. In the southwest, the faunal assemblages are clearly dominated by reindeer. From there, the proportion changes in a gradual transition to the north and east, where the assemblages are dominated by horse.

To the southwest of the Danube's source region, reindeer is clearly the most important prey species. All assemblages of Class 5 and the great majority of those of Class 4 are located here. In contrast, horse remains prevail only in two assemblages, those from the sites at Hauterive-Champréveyres and Neuchâtel-Monruz (Leesch 1997; Leesch et al. 2004). These two sites are located close to each other at the northern shore of Lake Neuchâtel. At Champréveyres, two major Magdalenian levels were excavated. Whereas the faunal assemblage of the main level in sector 2 yielded a balanced ratio of two horses and two reindeer (MIN), the principal level in sector 1 is clearly dominated by horse (Leesch et al. 2004: Fig. 159). Furthermore, three additional collections of rather sparse finds show similar features (Leesch et al. 2004: 42, 114). In the Magdalenian assemblage of Monruz, horse remains clearly outnumber those of reindeer (Müller 2006: 124). In another collection from sector 2, where a Magdalenian occupation is probably intermixed with Azilian finds (Leesch et al. 2004: 182ff.), horse remains prevail as well (Müller 2004: 198). Following the current of the Danube river downstream, horse remains gain significantly in importance, especially between the Danube's tributaries Große Lauter and Brenz. Here, about 55 % of the sites have yielded assemblages with a balanced proportion of horse, and reindeer remains and about 30 % are horse dominated. East of the Brenz river, information on faunal assemblages is available for two sites only.

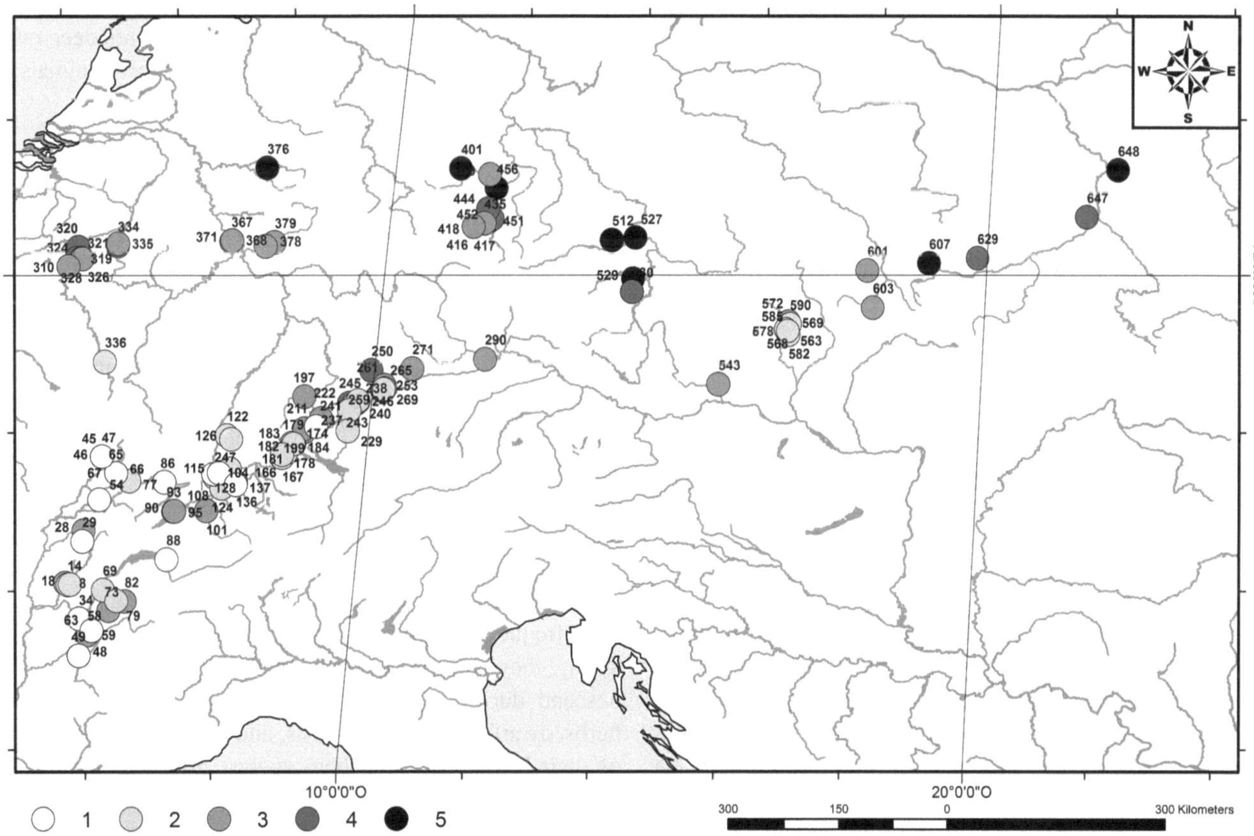

Fig. 5.6 Proportion of the skeletal remains of horse and reindeer in the CEM's faunal record. (1) Exclusively reindeer, (2) reindeer dominated, (3) equal proportion of horse and reindeer, (4) horse dominated, (5) exclusively horse (see Table A.7)

Both Kaufertsberg and Kastlhänghöhle show a balanced ratio of horse and reindeer remains. The tendency toward a predominance of horses among the prey species increases further when looking at the sites in the Rhenish Massif. In this region, there is not a single site with a predominance of reindeer reported in the literature. In contrast, 60 % of all inventories show a prevalence of horse remains, and in 40 % of the cases, an even proportion is observable. The northeasternmost assemblage from this region that has been analyzed, the assemblage found at the Feldhofhöhle, is the only one in which no reindeer remains are reported at all. From the sites along the rivers of Saale and Mulde, only nine assemblages could be analyzed in terms of their faunal remains. However, the picture is similar to the one obtained for the Rhenish Massif, even though the horse-dominated inventories gain further significance and amount to 70 %, whereas those assemblages with balanced ratios decreased to 30 %. Again, only one site (Bad Frankenhausen) in the northwestern part of the region has yielded no reindeer bones so far. Among the sites of the Vltava cluster, information about the faunal assemblage is only available for four sites. Nevertheless, it is remarkable that three of them yield no reindeer remains at all, and the fourth collection is horse dominated. In the case of the easternmost site cluster, assemblages with a balanced ratio are just as abundant as horse-dominated collections with seven cases, followed by five cases of reindeer-dominated inventories. In this region, an inner spatial segregation is visible. All five reindeer-dominated sites are clustered in the Moravian Karst. Moreover, they prevail in this area, followed by four assemblages with an even proportion and only two sites with a predominance of horse bones (see Valoch 1960: 84f.). In contrast, the four assemblages east of the Oder river are clearly horse dominated, and the assemblages of Dra Majera and Klementowice Kolonia contain no reindeer bones at all.

From the observed pattern, it is possible to suggest some answers to the questions raised above. The clear spatial trend of reindeer predominance in the southwest as opposed to the preponderance of horse remains in the north and east is difficult to explain in terms of seasonally driven game selection. If seasonality were the determining factor behind this pattern, a more homogeneous distribution of prey classes would be expected with spatial differentiation on a more regional level, displaying group migrations throughout a seasonal land-use cycle. Regional hunting preferences accord better with the observed pattern. However, there is no region in the investigated area where an exclusive specialization on horse or reindeer can be proven, especially when other game is taken into account (Gaudzinski and Street 2003). Another possibility is to assume the specific nutritional value of the species as a main argument for prey selection. In this case, one should expect a prevalence of reindeer in the north, because the large amount of marrow offered by this animal provides an ideal source of high-calorie nutrition, especially important in cold environments (Enloe 2003: 28f.). In light of this, it seems counterintuitive that reindeer-dominated assemblages are in fact found in the south of the investigated area, but are lacking so far north of 50°N. In the final analysis, it seems that the best explanation for the gradually changing spatial trend in the ratio of horse and reindeer remains in the faunal assemblages is their respective abundance in different environmental settings. Indeed, G.-Ch. Weniger (1989) observed this trend in his investigation of Magdalenian sites in Northern Switzerland, the Danube region, the Rhineland, and Eastern Germany and interpreted it as "congruent with assumed climatic and environmental changes" (Weniger 1989: 333). The wider spatial frame of the current study confirms this view. Together with the evaluation of the pollen spectra (see above), the following interpretation appears most plausible.

In the southwestern part of the investigated area, a treeless tundra with a high number of herbaceous plants characterized the landscape throughout the Late Pleniglacial Optimum and the Oldest Dryas (see above). Reindeers were probably better adapted to this environment than horses and therefore had competitive advantages. In the northern and eastern regions, however, where steppe and forest steppe vegetation prevailed, horses, which feed predominantly on grasses (Weniger 1982: 94), appeared to be the more successful competitors. Therefore, the large-scale trend of the horse-reindeer ratio probably reflects the respective abundance of these animals in biotopes that gradually merge into each other.

Other Animal Remains in the Faunal Record

The distribution of remains of other animals is admittedly scarcer than that of horse and reindeer. Nevertheless, for the most part, these remains show spatial trends in their distribution frequencies too. The following remarks are based on data collected from the archaeological literature (for an overview, see Table A.8) and are intended to give an impression of the overall distribution of the hunted fauna, while making no claim to completeness.

Ibex (*Capra ibex*) and chamois (*Rupicapra rupicapra*) can be found throughout the investigated area, but occur most frequently in the southwestern part (Figs. 5.7 and 5.8). Today, they live above the alpine tree line in summer and descend during the winter season. They feed mainly on herbs, dwarf-shrubs, and lichens, and the rocky environment of their habitat protects them against predators. However, male ibexes tend to stand on the top of rocks, making them a comparatively easy target for ranged weapons (Lenardi 1990: 106; Serangeli 2006: 130f.). Thus, the predominance of ibex and chamois remains at southwestern sites coincides

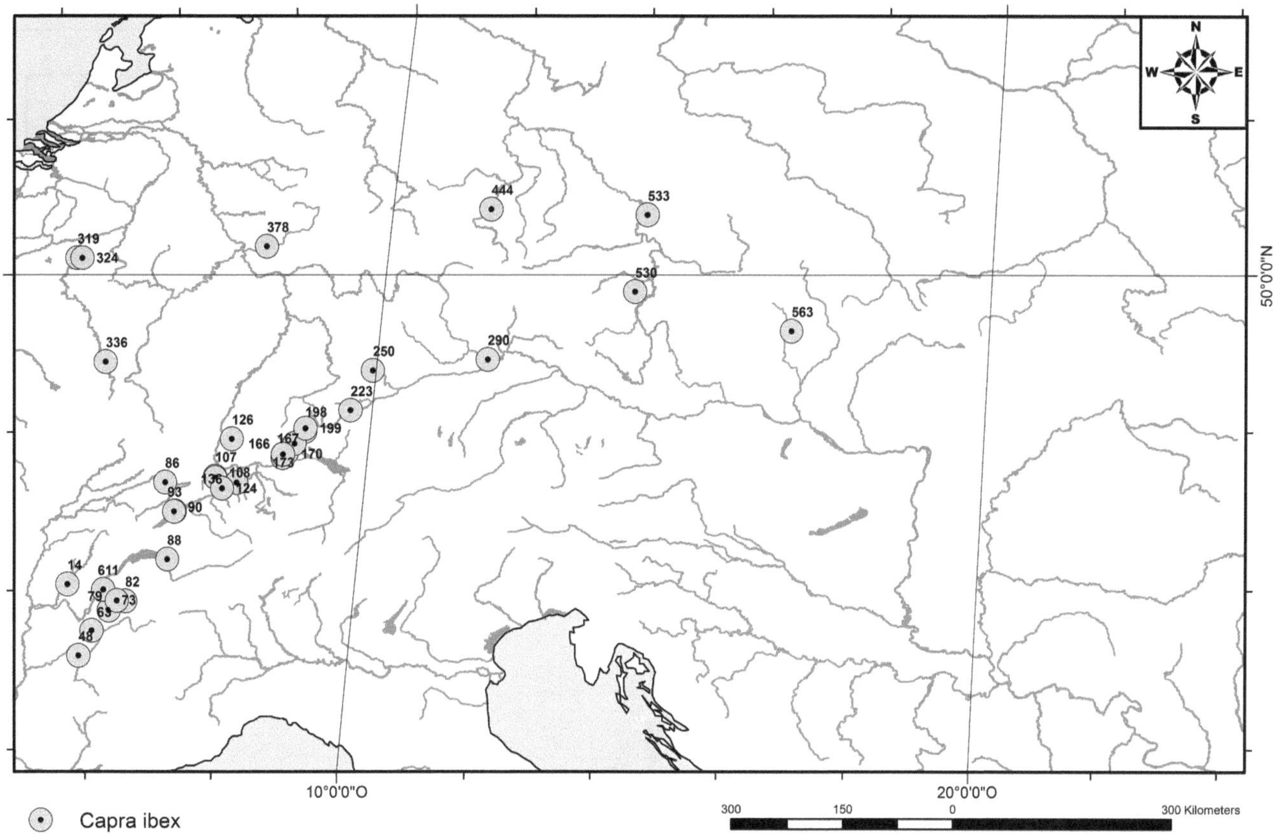

Fig. 5.7 Distribution of skeletal remains of *Capra ibex* in Central European Magdalenian sites (see Table A.8)

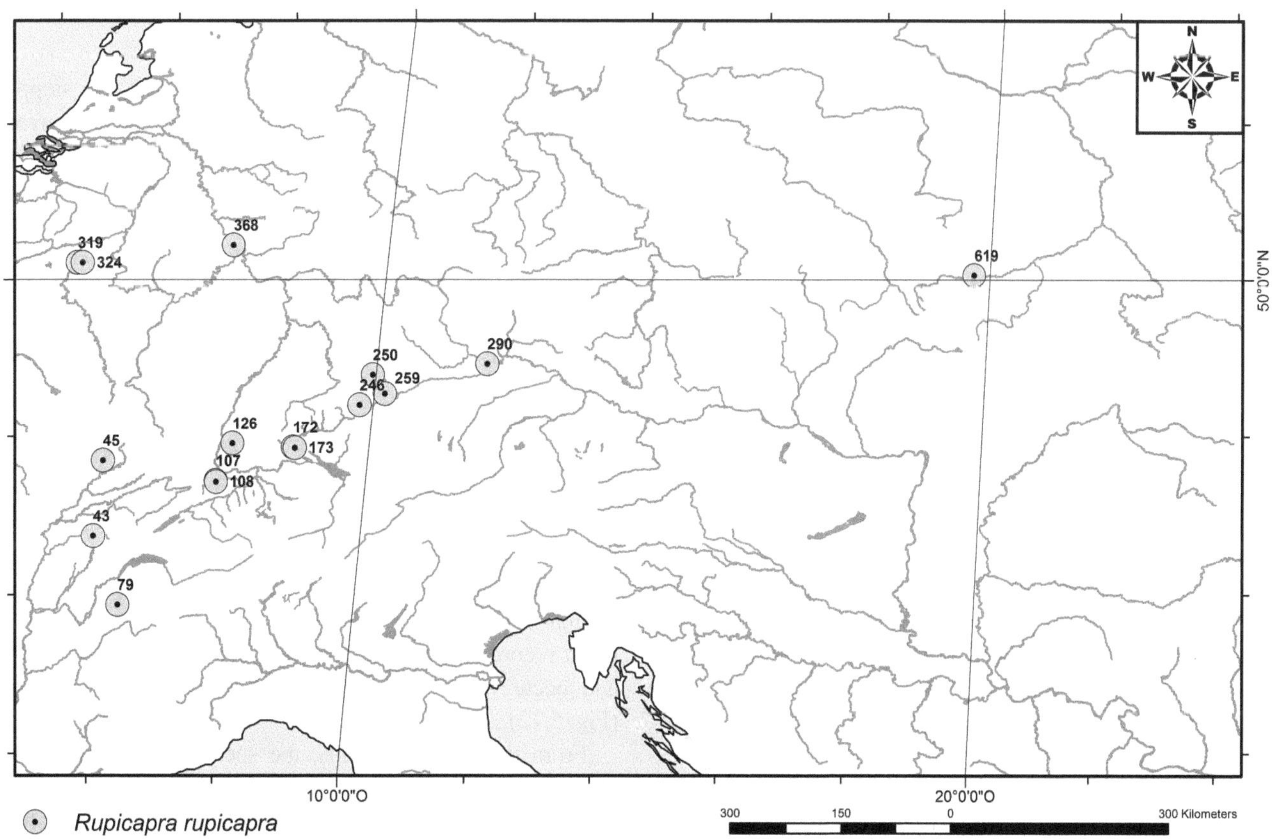

Fig. 5.8 Distribution of skeletal remains of *Rupicapra rupicapra* in Central European Magdalenian sites (see Table A.8)

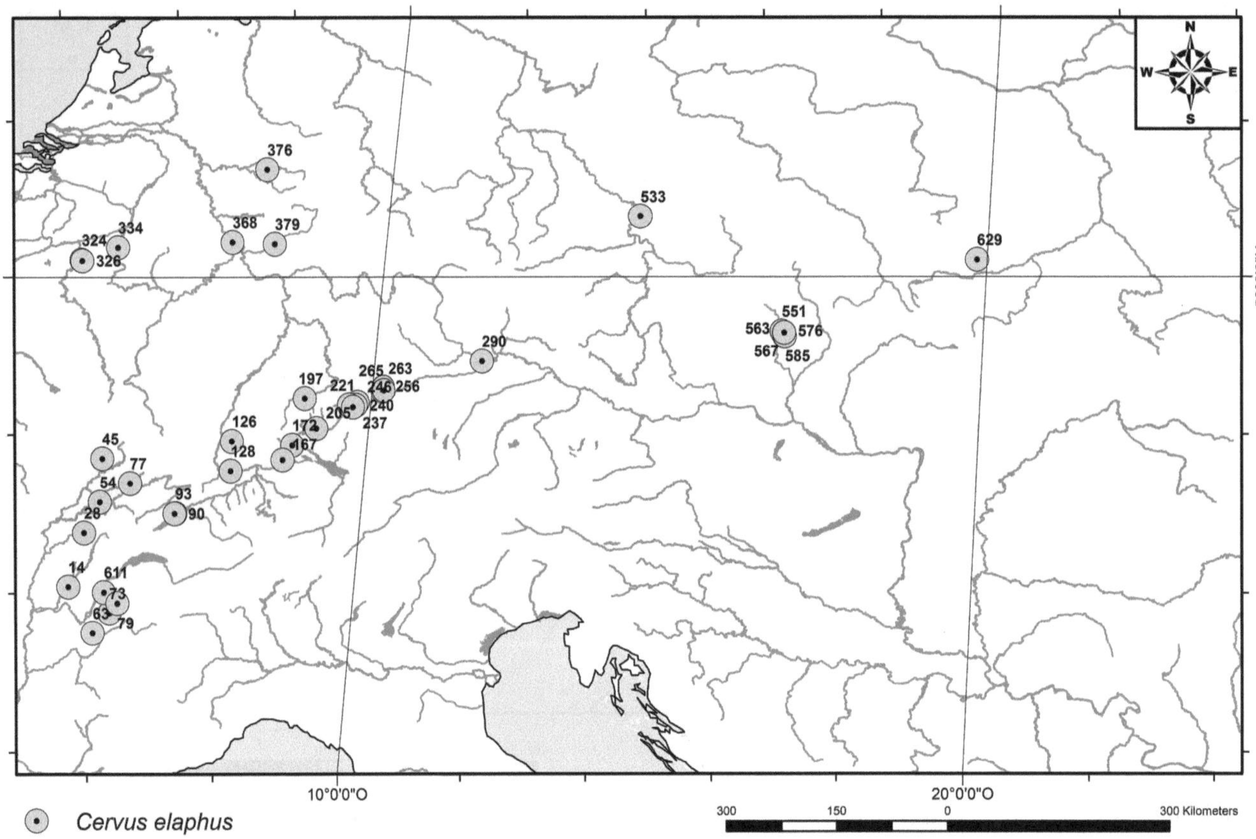

Fig. 5.9 Distribution of skeletal remains of *Cervus elaphus* in Central European Magdalenian sites (see Table A.8)

with the favorable conditions of the rocky habitat and the prevailing tundra vegetation and is probably the result of a greater abundance of these species.

A similar distribution pattern is observable for red deer (*Cervus elaphus*). This species is rather versatile in its environmental demands (Koenigswald 2002: 140; Serangeli 2006: 89) and has been found in faunal assemblages from the French up to the Polish Jura. It is conspicuous, however, that there have been no reports from eastern German sites so far (Fig. 5.9).

Elk (*Alces alces*), a generally flexible species in terms of environmental conditions and one that today lives in forest, taiga, and steppe habitats (Serangeli 2006: 101) is, except for Gönnersdorf, only reported from the southwestern part of the investigated area (Fig. 5.10).

Today, the occurrence of wild boar (*Sus scrofa*) is mainly associated with temperate and forested areas. However, it is reported that they can also be found in alpine pastures during the summer season (Heptner et al. 1966). Additionally, these animals have a very broad diet (Serangeli 2006: 84), leading to a high adaptive potential. In the CEM context, wild boar is mainly found in the western part with a special emphasis on the southern region, but is also reported at the Eastern German site of Oelknitz (Fig. 5.10).

Saigas (*Saiga tatarica*) are well adapted to dry steppes and have a very broad food spectrum, but tend to avoid regions with uneven and rutted surfaces (Serangeli 2006: 143). In the investigated area, saiga remains are reported only at sites north of the Saône-Danube-Vistula line (Fig. 5.11) and hence in environments that are characterized by dry and steppic conditions. In the rocky areas of the Alps, Jura, Swabian Alb, or the Carpathians, which were presumably moister, this species is missing so far from the faunal records.

In many environmental concepts, the appearance of roe deer (*Capreolus capreolus*) is related to the presence of dense forests, as is claimed for instance by Serangeli (2006: 97). However, recent observations of roe deer in regions which have been cleared of forests for agricultural purposes show an adaptation of this species to these altered environmental conditions (see Chap. 5, section "Fauna"). In the CEM record, roe deer remains are distributed rather evenly and occur in the western as well as in the eastern part (Fig. 5.12).

From the faunal remains, the species of steppe wisent (*Bison priscus*) and Aurochs (*Bos primigenius*) are difficult to distinguish, and both have been found in steppic as well as forested environments, the latter perhaps indicating a more

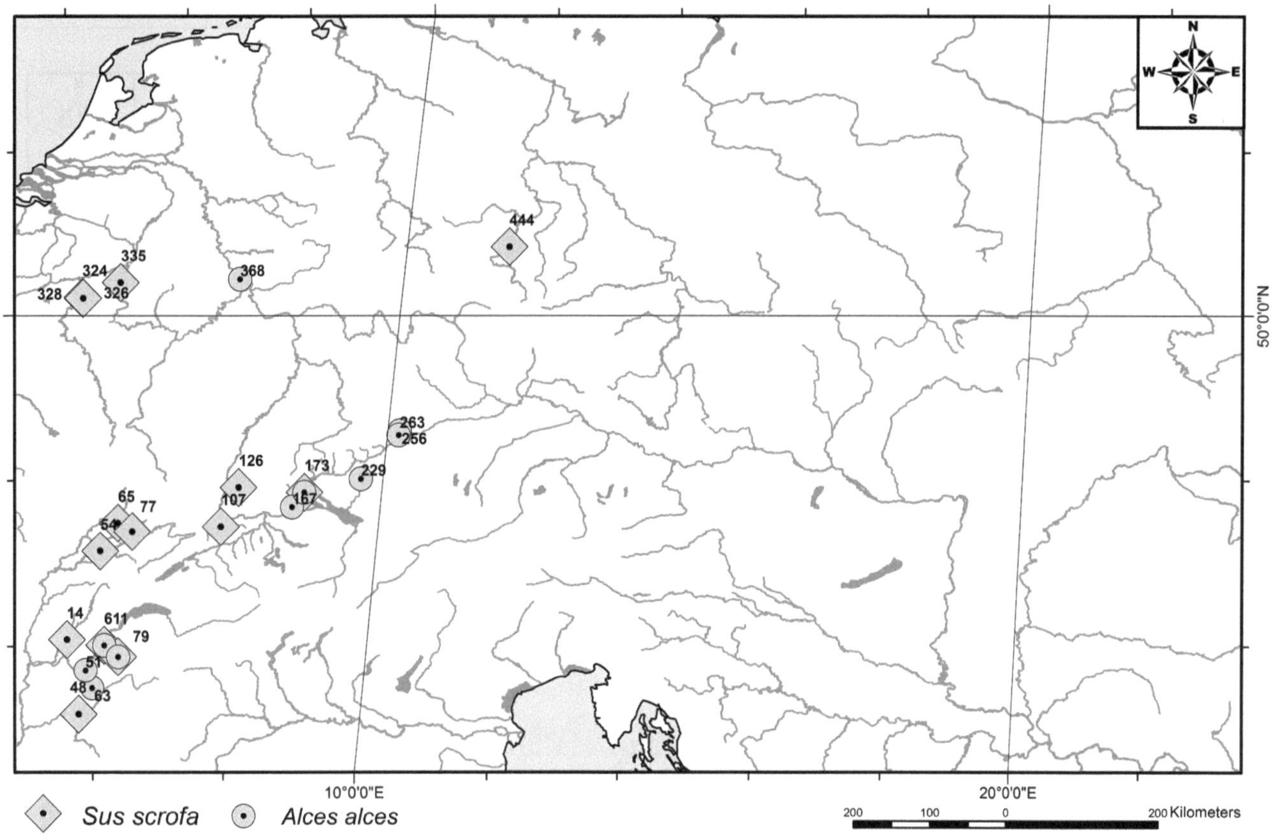

Fig. 5.10 Distribution of skeletal remains of *Alces alces* and *Sus scrofa* in Central European Magdalenian sites (see Table A.8)

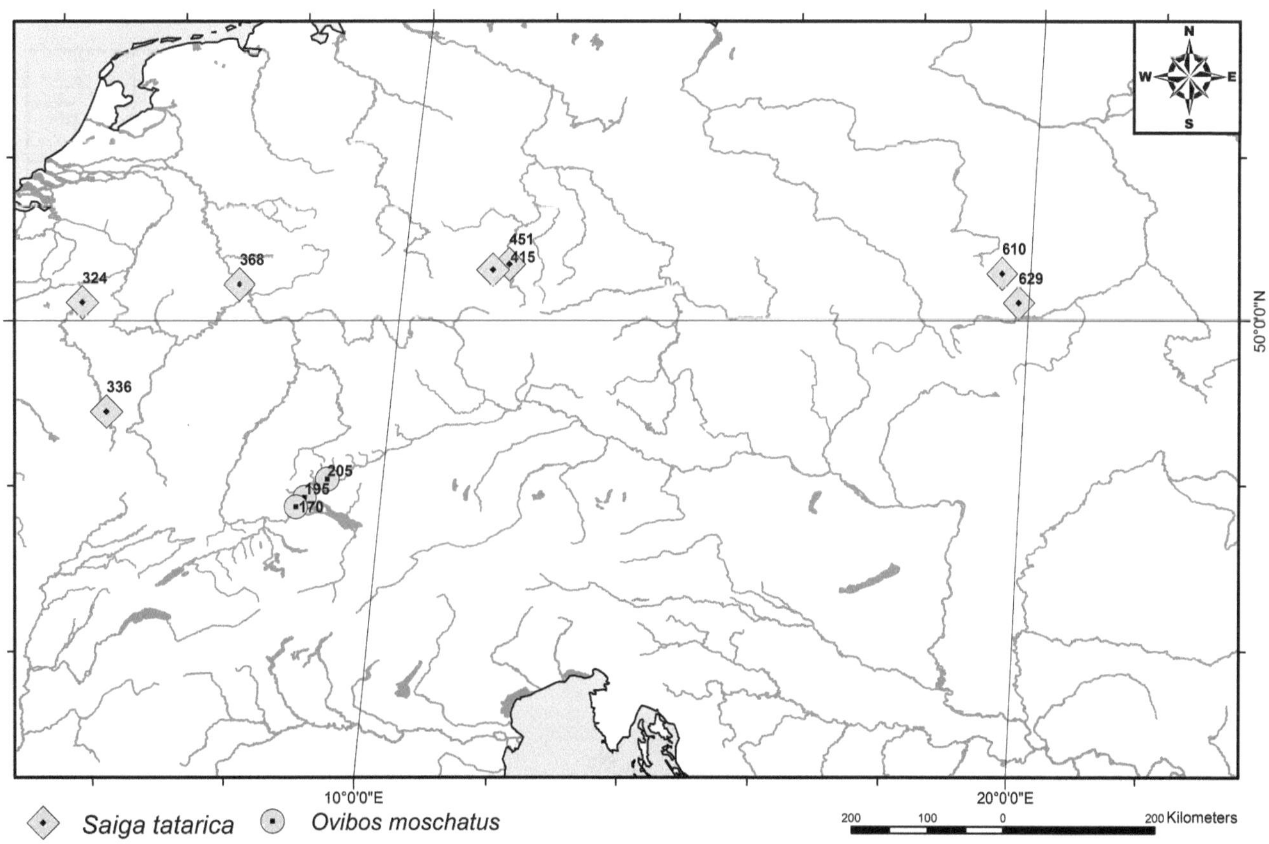

Fig. 5.11 Distribution of skeletal remains of *Saiga tatarica* and *Ovibos moschatus* in Central European Magdalenian sites (see Table A.8)

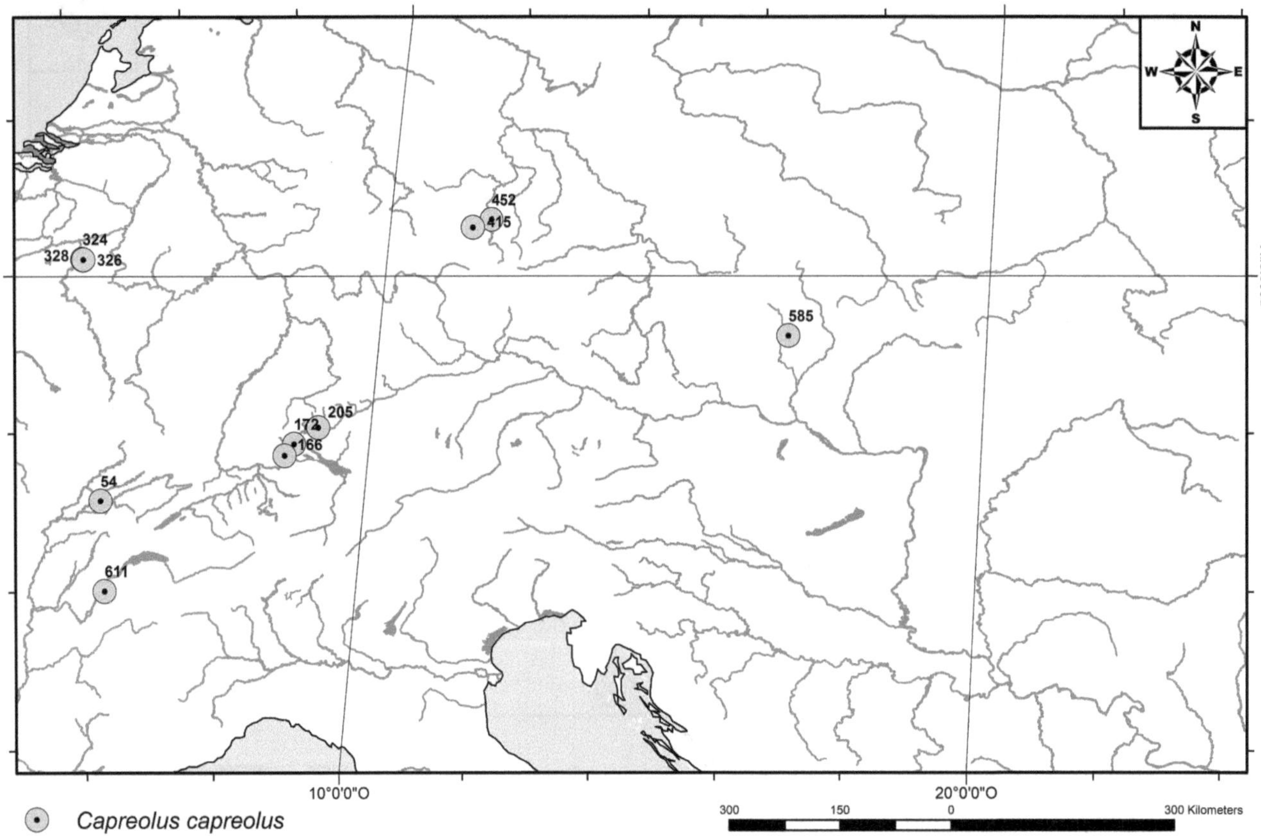

Fig. 5.12 Distribution of skeletal remains of *Capreolus capreolus* in Central European Magdalenian sites (see Table A.8)

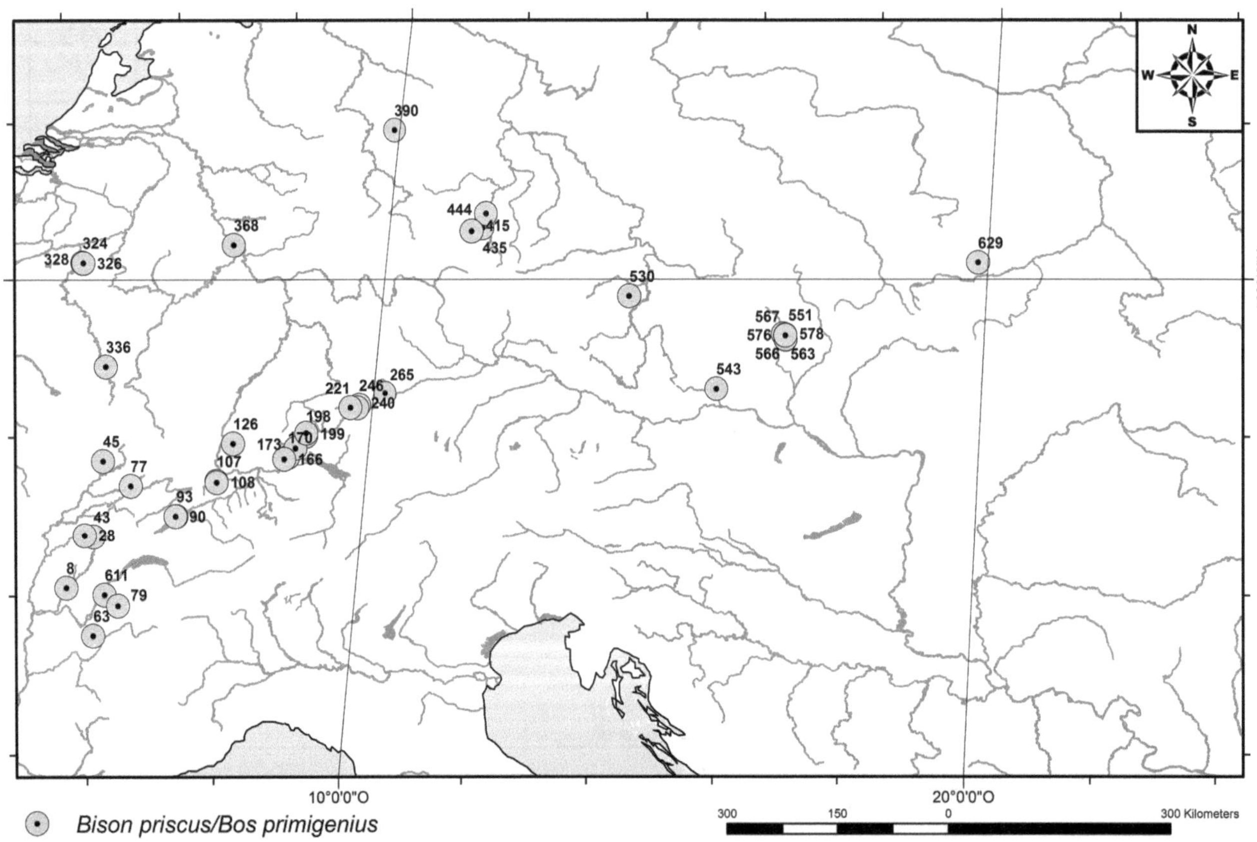

Fig. 5.13 Distribution of skeletal remains of *Bison priscus/Bos primigenius* in Central European Magdalenian sites (see Table A.8)

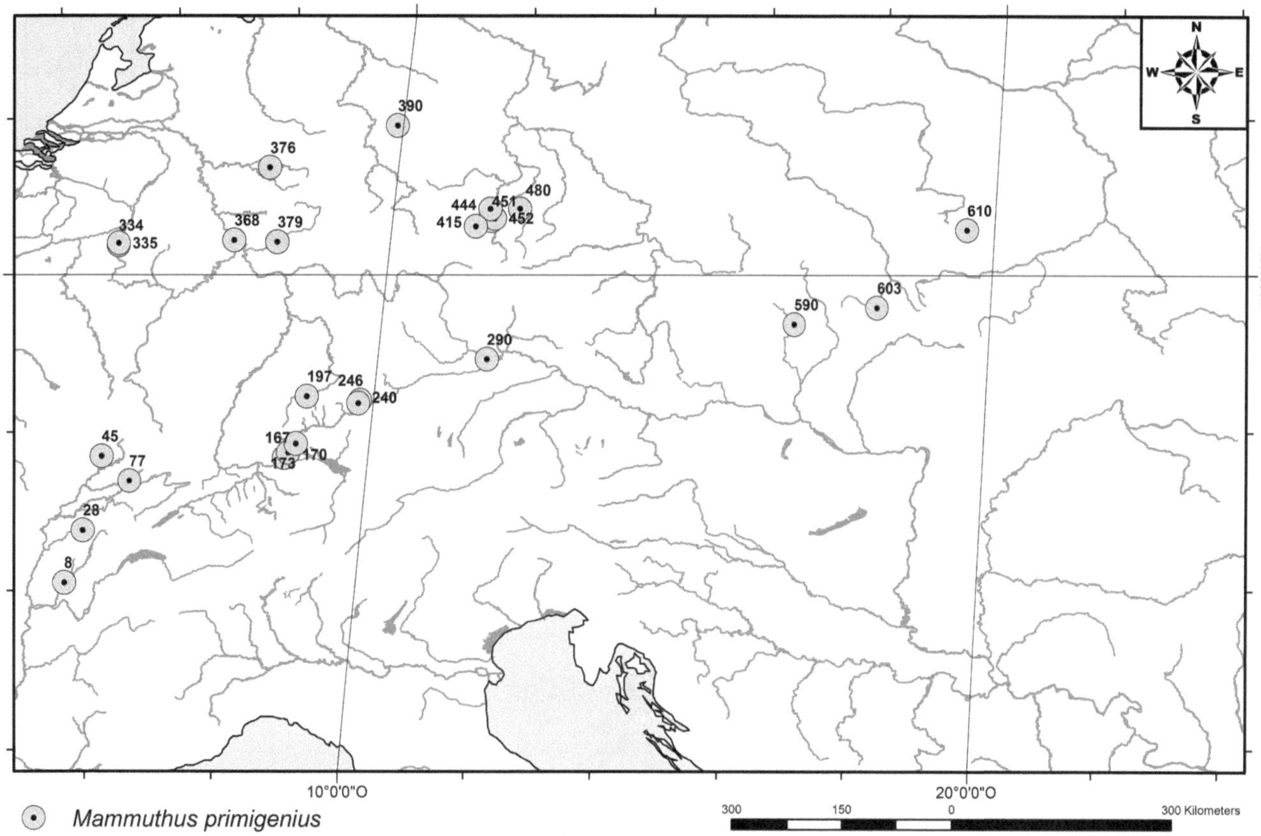

Fig. 5.14 Distribution of skeletal remains of *Mammuthus primigenius* in Central European Magdalenian sites (see Table A.8)

temperate climate (Serangeli 2006: 115f.). Bones assigned to these animals have been found throughout the whole investigated area. Nevertheless, a slight prevalence in the western part is discernable (Fig. 5.13). The musk ox (*Ovibos moschatus*) is adapted to very cold and dry climatic conditions (Serangeli 2006: 123f.). Its occurrence in the CEM context is only reported at three sites in a very small area between the High Rhine and the Danube (Fig. 5.11).

Mammoths (*Mammuthus primigenius*) lived in cold and rather dry steppe and tundra environments, where they fed mainly on grass, twigs, and mosses (Serangeli 2006: 69f.). They occur frequently in the faunal assemblages across the whole investigated area without any discernable spatial preferences (Fig. 5.14).

The distribution of woolly rhinoceros (*Coelodonta antiquitatis*), in contrast, shows a clearly recognizable spatial trend. Its remains appear occasionally in western and southwestern assemblages, but the majority of sites with bones from this animal are reported in eastern Central Europe (Fig. 5.15). Together with horses, the woolly rhinoceros is thus the only species with an eastern emphasis in its distribution. With regard to its environmental demands, the woolly rhinoceros seems to have required similar conditions to mammoths and musk oxen: it fed on steppe and tundra plants and was adapted to cold and dry environments (Serangeli 2006: 62). The markedly different distribution of these species in the faunal record of the CEM is therefore difficult to explain in terms of environmental constraints.

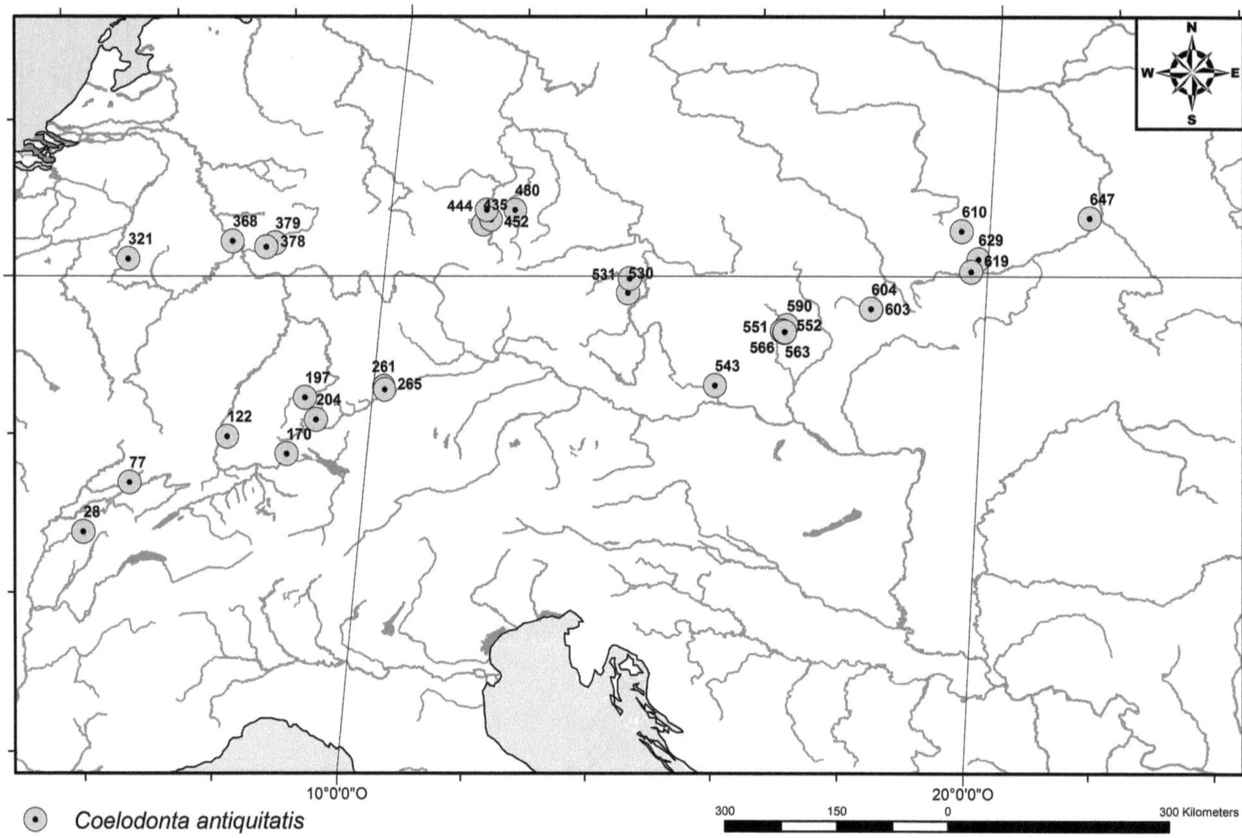

Fig. 5.15 Distribution of skeletal remains of *Coelodonta antiquitatis* in Central European Magdalenian sites (see Table A.8)

Chapter 6
Cultural Diversity and Regional Grouping

Just as its environment, the Magdalenian in Central Europe is traditionally perceived as a rather homogeneous unit (e.g., Bordes 1958; de Sonneville-Bordes 1960). Its uniform appearance is explained by a unidirectional expansion of a growing but closely interconnected population of hunter-gatherers from Franco-Cantabria to the Vistula and San rivers (e.g., Bosinski 1990). Nevertheless, several attempts have been made to distinguish different regional groups within this entity and to analyze the spatial variability of the archaeological record (e.g., Valoch 1960, 2009; Hanitzsch 1969; Feustel 1974: 191ff.; Weniger 1982, 1989; Bahn 1985; White 1985; Rozoy 1988b; Otte and Keeley 1990; Eriksen 1991; Otte 1992; Rensink 1993; Jochim et al. 1999; Połtowicz 2006; Küßner 2009: 192ff.; Sisk 2011; see also Chap. 2, section "Theories About the Internal Structure of the Central European Magdalenian"). Pursuing the same objective, this chapter analyses the distribution of sites within the investigated area as well as the internal variability of worked and used objects and their raw materials in order to shed some light on the spatial and temporal diversity of the Central European Magdalenian (CEM) record. It starts by critically reviewing the recorded assemblages and identifying major factors that determine their distribution pattern. In a second step, lithic raw material acquisition and sources of fossil and subrecent marine mollusks will be analyzed. The information deriving from the former issue will serve mainly as a proxy to estimate site catchment areas and interaction on a regional scale, whereas the latter artifact class is thought to represent supra-regional interaction. The merging of both aspects, then, will give a first impression of the inner structure of the CEM and will indicate the outlines of regional and supra-regional groups. Then, the typological spectrum and the applied knapping techniques of the CEM assemblages will be investigated statistically. Last, some aspects of CEM sculptures and engravings will be analyzed with regard to similarities and differences in different regions. The joined evaluation of the individual results will serve to sharpen and possibly correct the impressions already obtained and will finally be used to distinguish regional groups within the CEM which will then be discussed in detail in Part III.

Critical Review and Classification of the Recorded Assemblages

Figure 1.2 shows the distribution of all sites in the investigated area that, according to the literature, are assigned to the Magdalenian. Two questions arise from this overview: first, how reliable are these assignments, which are sometimes taken from publications dating to the first half of the last century? And second, what are the reasons for the distribution pattern visible on the map? Even though the latter question clearly holds much higher explanatory potential regarding the nature of the CEM, it is nevertheless crucial to answer the former question first, since falsely allocated sites may distort the pattern considerably. To keep the analytical value of assemblages as high as possible, it is important to strive for transparency with regard to the selection criteria of the analyzed assemblages. To this end, every assemblage considered in this study is classified individually according to the probability of its being assigned correctly to the Magdalenian. Absolute dating is, of course, an important factor for the classification. However, as absolute dates are only available for some assemblages, typological criteria must be considered as well. In order to keep this process transparent and comprehensible, the criteria applied are presented and explained below. In the part of the Central Europe archaeological record that has been satisfactorily verified and is well dated, some artifact types occur virtually exclusively, or at least mainly, within the period between 20,000 and 14,000 cal BP. These types are *baguettes demi-rondes* and *navettes* for the organic artifacts, as well as Magdalenian triangles, long

Andreas Maier, *The Central European Magdalenian*, Vertebrate Paleobiology and Paleoanthropology,
DOI 10.1007/978-94-017-7206-8_6

borers, and Lacan burins for the lithic artifacts (for definitions of tools, see Chap. 2, section "Definition of Magdalenian Tool Types"). The presence of these types is therefore considered sufficiently indicative to identify a Magdalenian component in an assemblage. This does not of course exclude the possibility of an admixture with material from later or earlier periods. Furthermore, a high number of backed bladelets and at the same time no or comparatively few backed points and the occurrence of Magdalenian single- as well as double-row barbed points (so-called harpoons) are also considered to be a strong indication of Magdalenian occupation. *Zinken* and shouldered points are more frequent within Hamburgian sites but also occur regularly in smaller numbers in Magdalenian assemblages. In Switzerland, shouldered points are even regarded as highly indicative for the late Magdalenian (Nielsen 2009: 95). Therefore, this study ascribes these types a certain diagnostic value.

In simple terms, the classification of the recorded assemblages results in a probability ranking with three levels. According to the points mentioned above, assemblages are considered:

Surely or very probably Magdalenian if they:

- Are dated reliably to the period between about 20,000 and 14,000 cal BP
- Have no absolute dates but yield a combination of several diagnostic artifacts

Probably Magdalenian if they:

- Have no absolute dates but show at least one diagnostic artifact feature and no contradictive features

Probably not Magdalenian if they:

- Are dated absolutely to another period and have no diagnostic artifacts
- Have no absolute dating and show contradictive features such Mesolithic microliths or a high number of backed points and at the same time comparably few or no backed bladelets

These categories can account for a majority of 433 assemblages. However, for 218 collections, a reliable allocation to one of these classes was not possible. For this purpose, two additional categories have been introduced. First, there are those inventories which probably represent an admixture of at least two archaeological entities. This is assumed to be the case if an assemblage has reliable ^{14}C-dates of Magdalenian age but a contradictive artifact ensemble, or if it yields clearly Magdalenian artifacts but has contradictive dates. Second, there are some assemblages which do not allow for any classification at all, since the available information in the literature is insufficient or the diagnostic value of the collection is inadequate. In these cases, the respective collections are considered not determinable. In the end, 382 out of 651 assemblages (59 %) are considered to belong surely/very probably (Class 1) or probably (Class 2) to the Magdalenian, and in 16 cases (2 %), the inventory is considered intermixed with material from other archaeological entities (Class 3) (Table 6.1, see Table A.1 for the classification of each assemblage). For 51 collections (8 %), however, it seems most probable that they represent post-Magdalenian occupations (Class 4), and in about 202 (31 %) of the cases, the assemblage is classified as not determinable (Class 5) (Fig. 6.1).

It is important to note that the classification listed in Tables 6.1 and A.1 and mapped in Fig. 6.1 (see Fig. 6.2 for the distribution of site-classes 1–3) is by no means the only possible way to divide the sites assigned to the Magdalenian. However, it has the virtue of being simple in its structure and therefore easy to comprehend, as well as flexible enough to accommodate the considerable differences in sample size and in the state and quality of the published information. Most important, the criteria used for classifying are independent of the sites' geographical position and raw material spectrum. Thus, these important parameters can be used to investigate the internal structure of this set more closely without risking circular reasoning.

Possible Factors for the Formation of the Large-Scale Site Pattern

In the previous section, all known sites assigned to the CEM were presented, reviewed critically, and classified on the basis of archaeological observations. This section will now go a step further and evaluate possible explanations for the

Table 6.1 Classification of all recorded assemblages

Classification	Class 1	Class 2	Class 3	Class 4	Class 5	Total
n assemblages	225	157	51	16	202	651
%	35	24	8	2	31	100

Class 1: Surely or very probably Magdalenian; Class 2: Probably Magdalenian; Class 3: Probably mixed; Class 4: Probably post-Magdalenian; Class 5: Not determinable. For a detailed list of all individual classifications, see Table A.1

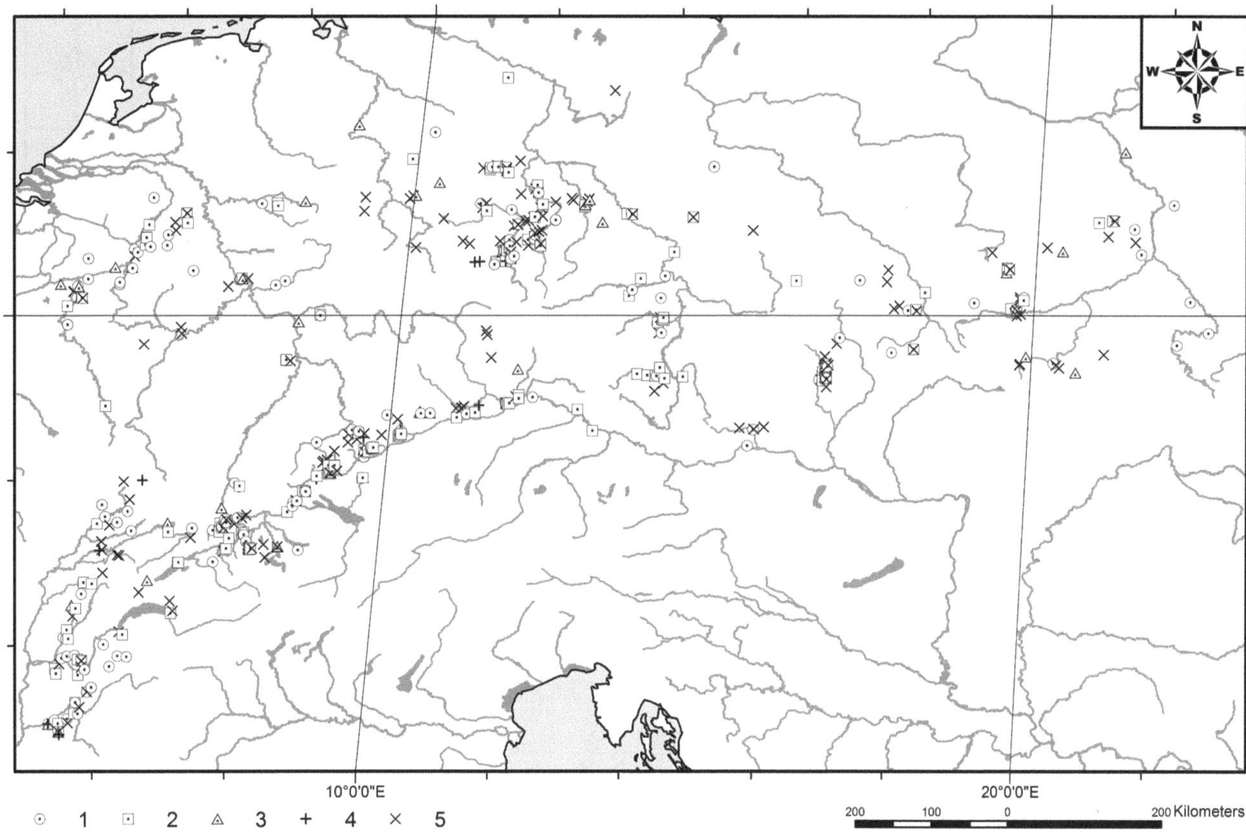

Fig. 6.1 Classification of all recorded assemblages. 1: Surely or very probably Magdalenian, 2: Probably Magdalenian; 3: Probably mixed; 4: Probably not Magdalenian; 5: Not determinable (For a detailed overview see Table A.1)

overall site pattern of the CEM. Here, the focus lies on the identification of the main factors which are responsible for the spatial distribution of the sites. Hence, the following reflections give particular consideration to the general structures on the spatial macroscale—in other words, the whole investigated area. The inner structure of the CEM and more detailed analyses of small-scale patterns are addressed in the sections that follow.

For a first analysis of the CEM site pattern, it is important to take only those sites into account for which an allocation to the Magdalenian is at least fairly certain. Consequently, only the 398 inventories classified as surely, very probably, or probably belonging to the Magdalenian (classes 1 and 2) are taken into account, plus the sites where Magdalenian finds are intermixed with remnants from other archaeological entities (class 3). The 253 assemblages that are considered post-Magdalenian or are not determinable (classes 4 and 5) are excluded and will not be considered in the following reflections (see Fig. 6.1 and Table A.1).

A comparison of Fig. 6.1, which shows all 651 assemblages, and Fig. 6.2, in which 253 inventories are excluded, shows that the site pattern does not change significantly in its overall structure. Rather, it can be stated that the exclusion of questionable or not determinable assemblages affects the whole investigated area more or less homogeneously. Nevertheless, two regions seem to be affected particularly by the exclusion process. The most affected region is the Weserbergland and surroundings, an area more or less consistent with the catchment area of the Eder, Fulda, and Werra river system. All sites located here are questionable or not determinable in their allocation to the Magdalenian. A similar picture arises for the Franconian Jura. All four sites that are located north of the Regensburg Basin are excluded. This is mainly because the collections from sites such as Rennerfels of Hohlenstein bei Happurg are very small and undiagnostic. Moreover, they were excavated during the early days of Paleolithic research.

Looking at the distribution pattern of the remaining assemblages, it is clear at first glance that the CEM sites are not evenly distributed over the investigated area. On the contrary, regions where sites are densely clustered alternate with empty areas without any sign of occupation. From this observation arises an important question: How did this pattern come into being and how should it be interpreted? To answer this question, three major possibilities have to be considered. First, it may be that the pattern mirrors archaeological

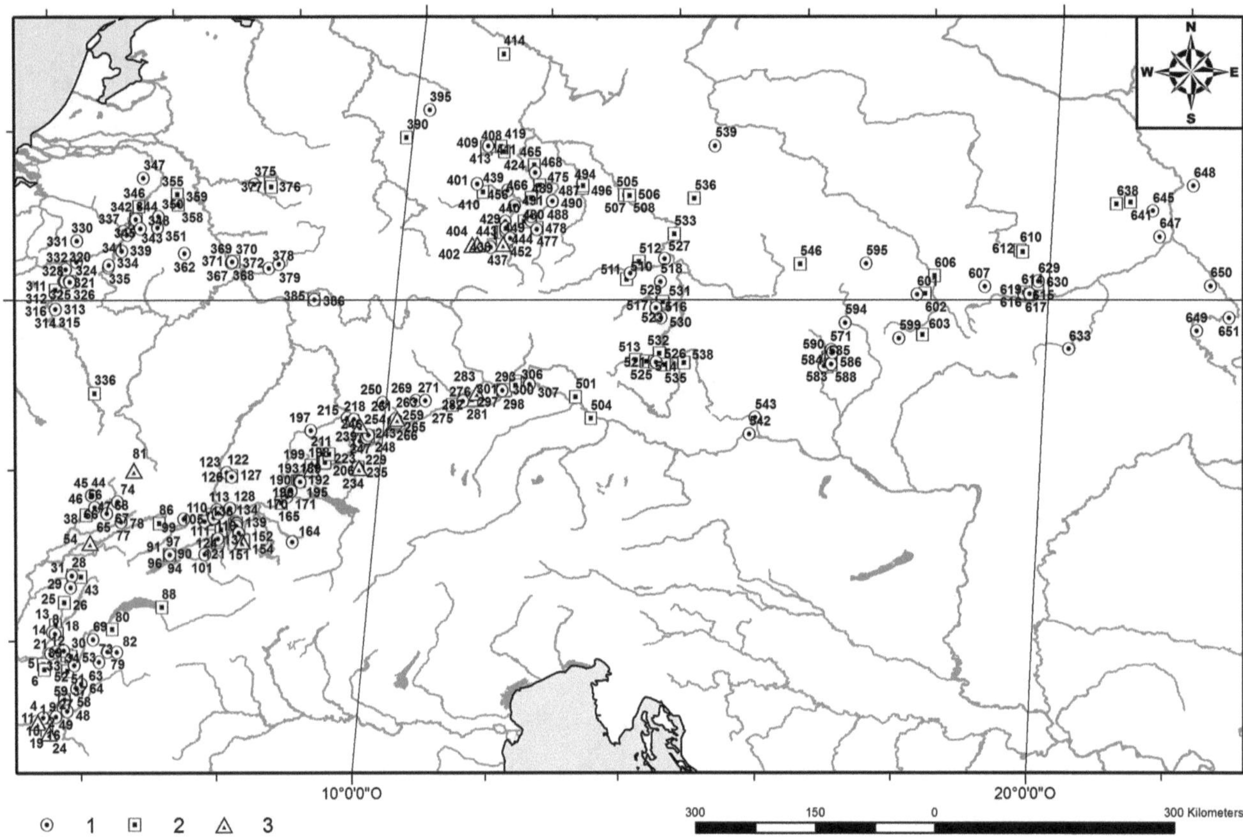

Fig. 6.2 Spatial distribution of the Central European Magdalenian assemblages considered in this study. 1: Surely or very probably Magdalenian; 2: Probably Magdalenian; 3: Probably mixed (for a detailed overview see Table A.1)

research activities and that empty spaces thus indicate areas of insufficient investigation. Second, the pattern may be the result of erosional processes that affected certain regions more strongly than others and caused the depletion of large areas. Third, the site distribution could reflect prehistoric reality and consequently; explanations must be sought for why some areas were more attractive for Magdalenian hunter-gatherers than others.

Research Activities

Archaeological research in Central Europe was quite extensive and intense throughout the last century. A significant proportion of sites from all epochs were discovered by non-professionals making voluntary surveys of the fields around their hometowns. Usually, these volunteers collect finds from all periods without a special bias. If, therefore, in an extensively surveyed region no signs of Magdalenian occupation can be traced among the collected material, this is definitely not due to a lack of research. For the Rhineland and Westphalia, for instance, a thorough review of public and private collections over 3 years brought to light only one new Magdalenian site (Eierberg), which is located in the immediate surrounding of the two already known sites of Kamphausen and Galgenberg (Holzkämper et al. 2014). This observation confirms the existing picture of a gap in the Magdalenian occupation for large areas of Westphalia. For France, where a lot of systematic research was undertaken, Rozoy (1988b) draws a similar conclusion:

> "Ces groupes [...] séparés par des zones vides ou peu fréquentes que les recherches les plus systématiques ne comblent pas, les nouveaux sites sont presque toujours découvertes dans les zones déjà garnies. Il n'est plus possible d'attribuer ces blancs à des lacunes de la recherche" (Rozoy 1988b: 143).

Of course, not all regions of the investigated area show the same standard of research. However, many of them are well surveyed, and it must be assumed that surveying activities which led to the discovery of other Stone Age sites would have recognized Magdalenian artifacts as well, if they had been present. The distribution of Late Paleolithic and Mesolithic sites therefore indicates that insufficient research activities cannot be held responsible for the formation of the Magdalenian site pattern. Looking at the respective distribution maps for the investigated area (e.g., Baales 1996: Fig. 166; Cziesla 1998: Fig. 1; Ikinger 1998: Figs. 113, 120, 130, 138, 148, 157, 158; Jochim 1998: Fig. 22; Nielsen 2009: Fig. 1365; Gehlen 2010: Fig. 233), it becomes apparent that sites from these periods occur in considerable numbers, even in those areas that are free of Magdalenian occupations. The chronological distribution of the discovery years recorded for 979 Upper Paleolithic sites (Bocquet-Appel et al. 2005: Fig. 2) indicates a decline in discoveries after 1970. According to Bocquet-Appel et al., this decline in discovery frequency against the background of a rather constant archaeological pressure suggests that "sites still to be discovered will not fundamentally modify the spacetime pattern of the sites which have already been found" (Ibid.: 1660). In summary, a lack of research cannot be maintained as an explanation for the CEM site pattern.

Erosion

Another hypothesis suggests that erosion is mainly responsible for the distribution of known CEM sites. The first counterargument is given again by the maps of Late Paleolithic and Mesolithic sites. Taking into account that the intensification of agriculture triggered heavy erosional processes, it is consistent to assume that Magdalenian, Late Paleolithic, and Mesolithic sites are affected equally by it. Applying erosion as an explanatory concept for the observed pattern would thus imply that it operated selectively by depleting Magdalenian sites while at the same time leaving Late Paleolithic and Mesolithic sites intact. Additionally, it would have to be explained why erosion depleted some regions completely, such as those along Moselle, Neckar, or Main rivers while affecting others of similar or even more pronounced relief, such as those along Isère, Rhône, Aar, or Upper Danube, much less. Hence, erosion cannot be said to offer a plausible or satisfying explanation for the overall site pattern of the CEM, although it certainly played a role on smaller scales.

Having ruled out research intensity and erosion as main agents shaping the present-day CEM record, it is possible to assume that the CEM site pattern reflects to a great extent past reality. This assumption raises the question of whether it is possible to identify some geomorphological and environmental key elements which made a certain region more attractive for Magdalenian people than others.

Rivers

Looking again at the map in Fig. 6.2, it becomes apparent that the majority of CEM sites cluster along larger rivers. Generally, these waterways are of first and second orders or are at least comparable to second-order rivers in terms of their length and water volume. In the Jura Mountains, these are the Saône, Rhône, Ain, Isère, and Aar rivers. Further to the east, the Rhine, the upper Neckar, and above all the Danube are of major importance. In the Rhine-Meuse area, these two rivers seem to be vital for the spatial distribution of sites. In addition, the Ruhr and lower courses of the Lahn and the Main contribute to the structuring of the site pattern. The river system comprised by the Elbe, Saale, White Elster, and Mulde, together with the Leine river form the base frame for the spatial distribution of the eastern German sites. In Bohemia, the Vltava and its tributaries Ohře and Otava appear to be most important for the location of sites. For the position of sites in Moravia, the Svitava and Morava rivers seem to serve as guidelines and north of the Carpathian Mountains; the site distribution is connected to the currents of the Odra and the Vistula, as well as their direct tributaries. Rivers may serve as water sources, landmarks, and guides in familiar as well as in unfamiliar landscapes (Kelly 2003: 48ff.; Nieves Zedeño and Stoffle 2003; and other articles in Rockman and Steele 2003). Therefore, it is highly likely that they played a major role in both the recolonization process of Central Europe and in the post-colonization life of CEM hunter-gatherers. This hypothesis is supported by the observation that only 12 of the 405 mapped assemblages (3 %) come from sites that are more than 25 km away from one of the above-mentioned rivers (Fig. 6.3). All other inventories were found at sites closer than this threshold value, which is easily walkable within 1 day. The observation that rivers played an important role in the settlement pattern of Magdalenian hunter-gatherers was also expressed by R. White (1985: 90ff.) in his study about land-use patterns in the Périgord.

Generally, all major rivers are suitable as water sources and landmarks. Thus, it is rather surprising that many of the larger second-order rivers, such as the Moselle, Neckar, Main, Fulda, Werra, or Weser, are almost or even completely free of Magdalenian occupation. The question immediately arises as to what other factors might have made an area attractive or unattractive for CEM hunter-gatherers and could thus explain the conspicuous gaps in the site distribution. Here, lithic row material sources hold great explanatory potential.

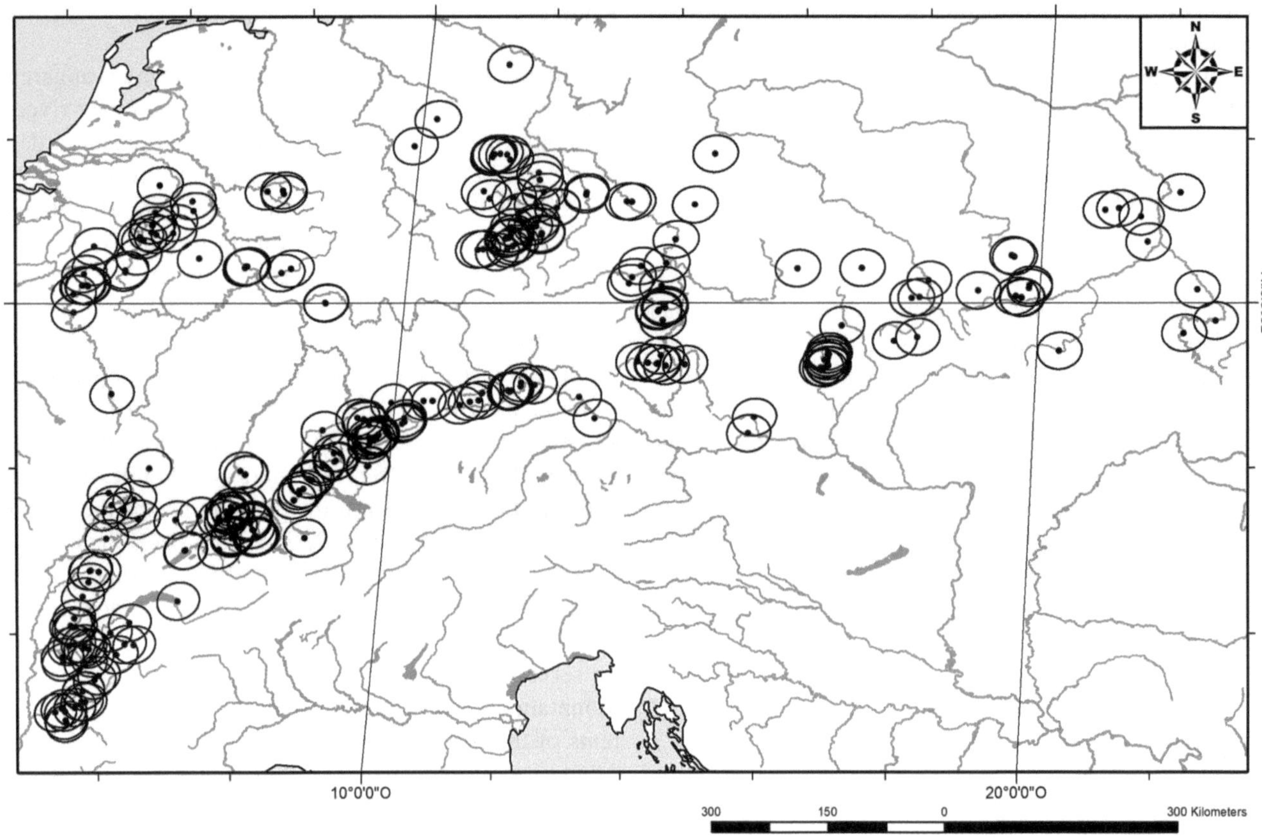

Fig. 6.3 Distribution of Central European Magdalenian sites with regard to the courses of major rivers. *Circles* indicate an area of 25 km radius around each site

Lithic Raw Material

Lithic raw material was one of the basic pillars of Magdalenian technology. Especially compared to Late Paleolithic industries, it seems that Magdalenian hunter-gatherers put emphasis on a high quality of this material. In this context, high quality means that the crystal structure of the rock is as fine grained and homogeneous as possible. In particular, rocks with a high silicon dioxide (SiO_2) content and preferably a cryptocrystalline matrix, which break with conchoidal fractures and give sharp edges, are suitable for the production of lithic tools (Floss 2012c) and were of great importance during the Magdalenian. Thus, it can be assumed that areas yielding high-quality raw materials were preferred by Magdalenian people. The type of chert that occurs extensively and is most suitable for Magdalenian needs is probably Upper Cretaceous flint which is composed of about 96 % silicon dioxide (Deecke 1933; Wetzel 1933; Floss 1994). Aside from primary sources, this raw material is distributed virtually ubiquitously in the North European lowlands in the form of glacial erratics. Together with Jurassic chert (*Jurahornstein sensu* Floss 1994: 104), it was the most important raw material during the Magdalenian. In contrast, Middle Triassic chert (*Muschelkalkhornstein*) is of lower quality, since it often has an oolitic structure and regularly shows limy, unsilicified sections or veins of Chalcedony as well as hollows from fossils (Ibid.: 107f.). A characteristic of Late Triassic chert (*Keuperhornstein*) is its very pronounced fissuring, which strongly decreases its usability for knapping, especially for the selective Magdalenian hunter-gatherers. Nevertheless, Late Triassic chert is known from Middle Paleolithic, Late Paleolithic, and Mesolithic sites (Ibid.: 108f.). Other rocks with a high silica content include radiolarite, lydite, and *Kieselschiefer*; these occur in Central Europe mainly in Ordovician, Silurian, Devonian, and Mississippian (*Unterkarbon*) contexts. In the Alps, radiolarite is also found in Triassic and Jurassic formations (Ibid.: 66).

With regard to these observations, the CEM sites are plotted against the background of those geological formations that were identified above as yielding potentially high-quality raw materials (Fig. 6.4). The analysis of the distribution shows that only 11 % of the assemblages ($n=44$) are located more than 5 km away from one of the respective

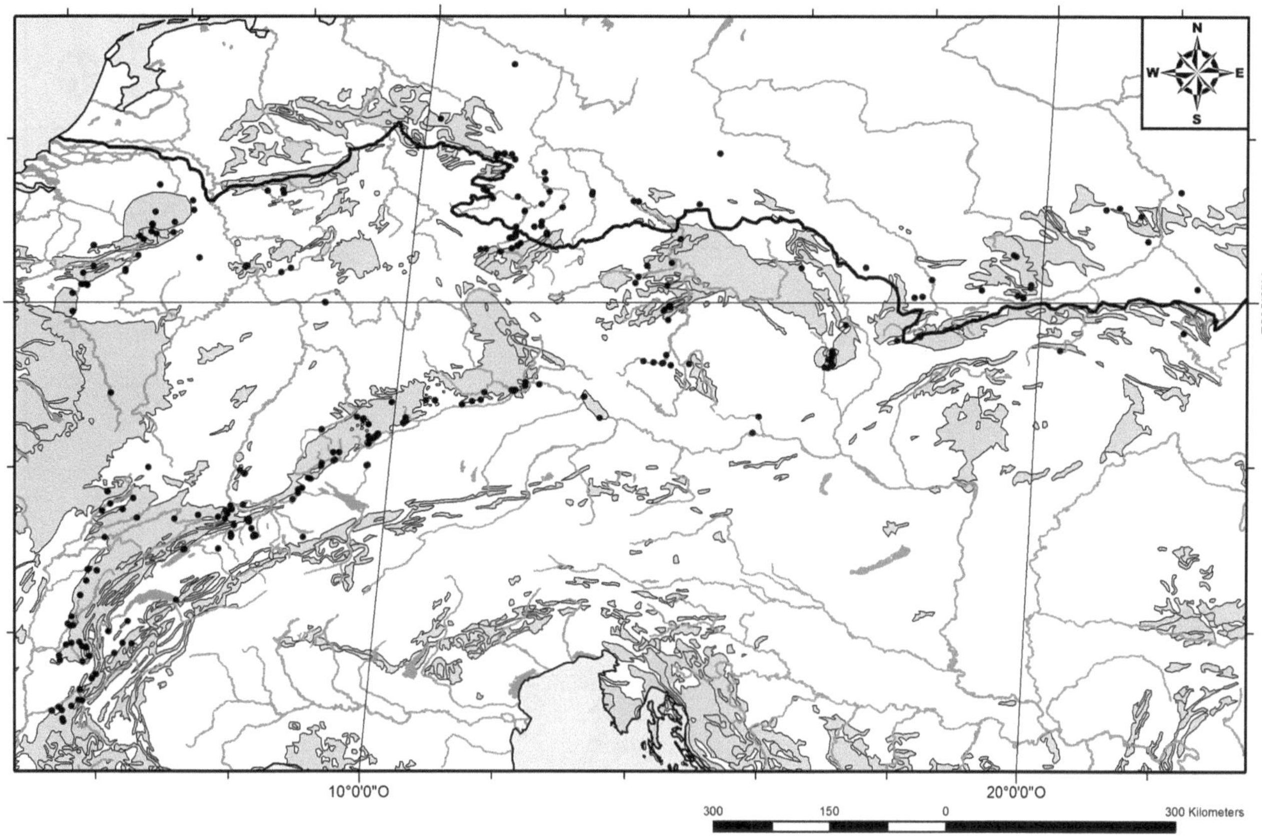

Fig. 6.4 Distribution of CEM sites in comparison to geological formations which potentially yield high quality raw materials (*grey patches*). *Black line* roughly marks the southern fringe of the maximum distribution of Baltic (erratic) flint

geological patches. A large part of these assemblages ($n=17$) can be summarized according to another geographical feature which all of them have in common: their location along the shores of paleolakes. This applies to several collections from Central Switzerland (Nielsen 2009: 199f.) as well as to the sites in the area surrounding the Federsee. Another group of seven outliers is located in the Otava valley in Bohemia. Here, Cretaceous sediments can be found in the Budweis Basin, adjacent to the west.

If the threshold between a site and the next region of high-quality raw material outcrops is set at 25 km, a distance that is easily walked within 1 day, the relationship between raw material availability and site location becomes very clear. Here, only nine assemblages (2 %) are located farther away from a potential high-quality raw material source than this boundary value. Four of these inventories come from sites in the area surrounding the Mainz Basin: the collections from Dreieich-Götzenhain East and North as well as the assemblages of Fußgönheim I and II. Whereas Dreieich-Götzenhain seems to be a regular Magdalenian site, though the excavation is still in process, the collections of Fußgönheim I and II are characterized typologically by backed and shouldered points and are rather untypical of the CEM. The outcrops of fossil mollusk shells are perhaps a reason for the occupation of this region. A similar situation can be observed for the two sites of Gudenushöhle and Kamegg, which lie rather isolated in the Austrian Waldviertel. In the adjacent Vienna Basin, rich outcrops of fossil mollusk shells can be found. Rather isolated lies the site of Kartstein in the Eifel region. The remaining two inventories from Dolni Pořiči 8 and Slaník 1 in the Otava valley are with tool assemblages of two, three, and four pieces, respectively, very small in number. Moreover, their locations exceed the threshold value of 25 km only slightly, and they all are part of a larger site cluster. In the end, only the two sites of Dreieich-Götzenhain and Kartstein belong without doubt to the Magdalenian and are located more than 25 km away from a high-quality raw material source.

In order to sharpen the image of the relationship between CEM sites and the geological context, Fig. 6.5 shows the same sites plotted against Precambrian, Carboniferous, Permian, Triassic, Paleogene, and Neogene deposits, all of which provide rather low-quality raw materials in general. It is quite striking that the respective geological deposits and the plotted sites are almost mutually exclusive in terms of

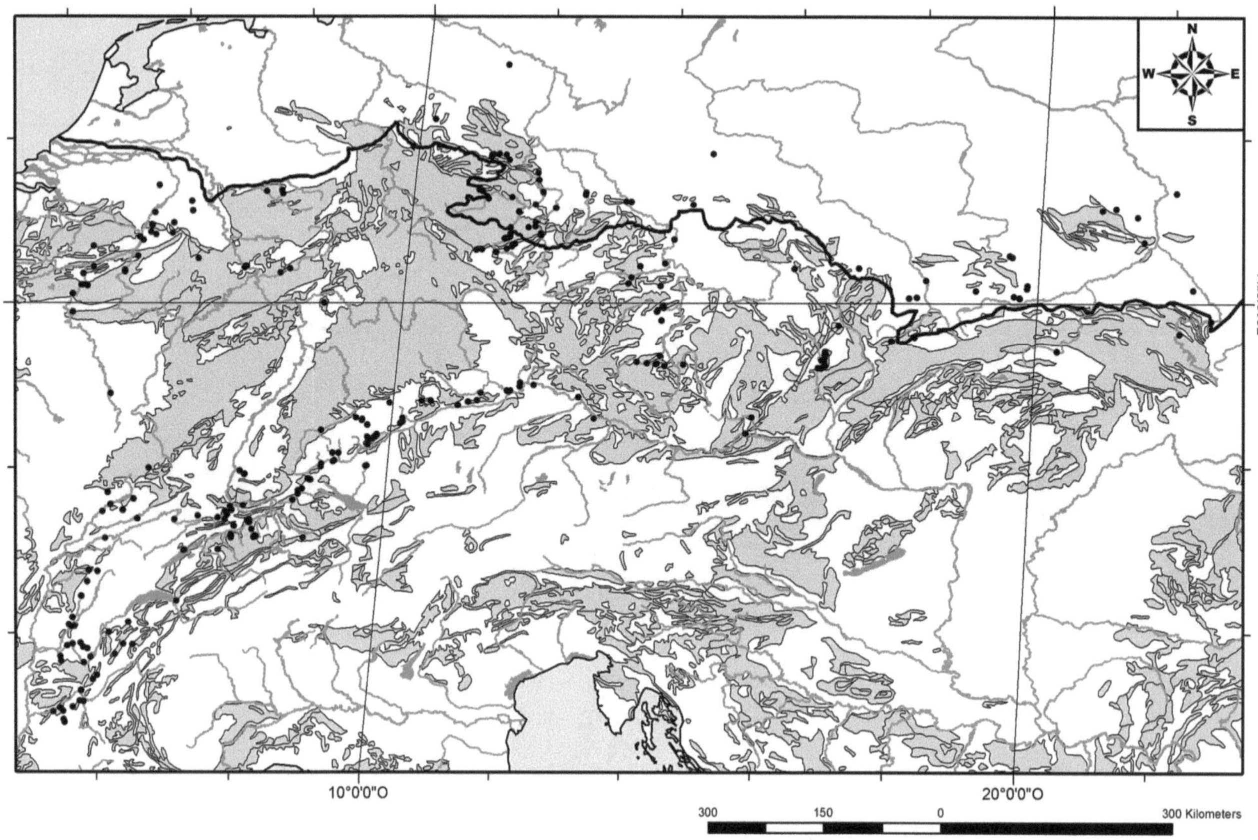

Fig. 6.5 Distribution of CEM sites in comparison to geological formations which potentially yield rather low quality raw materials (*grey patches*). *Black line* roughly marks the southern fringe of the maximum distribution of Baltic (erratic) flint

their geographical position. This is particularly evident with regard to the Triassic sediments in the Vosges and Central Germany, the Precambrian patches in Bohemia, and the Paleogene and Neogene deposits in Moravia and the Carpathians. Despite of a high density in adjacent areas, almost no Magdalenian sites are located in these regions.

In sum, it can be stated that Magdalenian hunter-gatherers did not wander randomly about the landscape but likely chose their settlement areas according to the availability of certain resources. Even though rivers and sources of high-quality lithic raw material were certainly not the only determinants guiding the decisions of Magdalenian hunter-gatherers, these two factors hold a very high explanatory potential with regard to the distribution pattern of CEM sites on the macroscale. If both features are taken into account and a day-trip distance of 25 km to each of these features is considered, these factors may explain the location of 98 % of all 398 assemblages which are considered as belonging to the CEM. Thus, it seems that the overall distribution of CEM sites is strongly influenced by first-order effects (see Chap. 2, section "Spatial Data").

Procurement Pattern of Lithic Raw Materials

Siliceous rock, such as Upper Cretaceous flint, Jurassic chert, or radiolarite, was certainly one of the major raw materials used throughout the Magdalenian and figured among the daily necessities of hunter-gatherers in this period. Thanks to extensive analyses performed during recent decades, it is often possible to pinpoint the origin of the raw materials found at an archaeological site (see Chap. 2, section "Lithic Raw Materials, Mollusk Shells, and Amber"). Intensive studies of individual assemblages show that the material processed on-site was often procured in the immediate or regional vicinity of a camp. Nevertheless, rock acquired from sources more than 50 km away from the camp site also occurs frequently in Magdalenian assemblages (see below). Lithic raw material acquisition was probably not done through direct procurement but was rather "embedded" (Binford 1979: 259) in daily micro-moves around the camp, as well as in seasonal macro-moves, when larger distances were traveled (cf. Floss 1994: 325). Based on ethnographic observation, Binford argues as follows:

> Raw materials used in the manufacture of implements are normally obtained incidentally to the execution of basic subsistence tasks. Put another way, procurement of raw materials is embedded in basic subsistence schedules. *Very rarely, and then only when things have gone wrong, does one go out into the environment for the express and exclusive purpose of obtaining raw material for tools*. For example: a fishing party moves in to camp at Natvatrauk Lake. The days are very warm and fishing is slow, so some of the men may leave the others at the lake fishing while they visit a quarry on Nassaurak Mountain, 3.75 miles to the southeast. They gather some material there and take it up on top of the mountain to reduce it to transportable cores. While making the cores they watch over a vast area of the Anaktuvuk valley for game. If no game is sighted, they return to the fishing camp with the cores. If fishing remains poor, they return to the residential camp from which the party originated, carrying the cores. Regardless of the distance of Nassaurak Mountain from the residential camp, what was the procurement cost of the cores? Essentially nothing, since the party carried home the lithics in lieu of the fish which they did not catch. They had transport potential, so they made the best use of it; the Eskimo say that only a fool comes home empty handed! (Binford 1979: 259 [italics in original])

Moreover, there are ethnographic examples of large blocks of raw material being procured in the course of seasonal moves and transported over distances of more than 150 km (Holen 1990; Floss 1994: 326f.). The exchange of lithic raw materials is only rarely documented in the ethnographic record. H. Floss (1994: 320ff.) argues convincingly that exchange is a rather unlikely scenario for the procurement of raw materials in the CEM. Rather, embedded procurement is the most probable strategy for lithic materials which occur within the annual range of a hunter-gatherer group, even though materials from distant sources are exploited preferentially at a site. In the very rare case that raw materials come from exceptionally distant source located well beyond the possible annual range of a hunter-gatherer group, exchange seems the most plausible explanation, however. With regard to the CEM, the vast majority of procurement distances (97 %) is below or equal to 300 km as the crow flies (Fig. 6.6). Larger distances are only reported from the eastern part of the investigated area and are often plagued with uncertainties (see section "The Polish-Moravian Group"). Thus, this value is taken as a threshold distance for embedded procurement.

Having excluded direct procurement and exchange as rather improbable strategies for general raw material procurement patterns during the CEM, it can be assumed that within a radius of about 300 km around a site the present raw material was acquired by embedded procurement. Thus, the hunter-gatherers who left behind remains of their activities at a certain site had themselves been present in the source region of the raw material they brought with them. The spectrum of the acquired raw material therefore provides a starting point for estimating the minimum range of the catchment area that belongs to a certain site (cf. Rensink 1993: 104ff.; Floss 1994: 320ff.). Overlapping or mutually exclusive

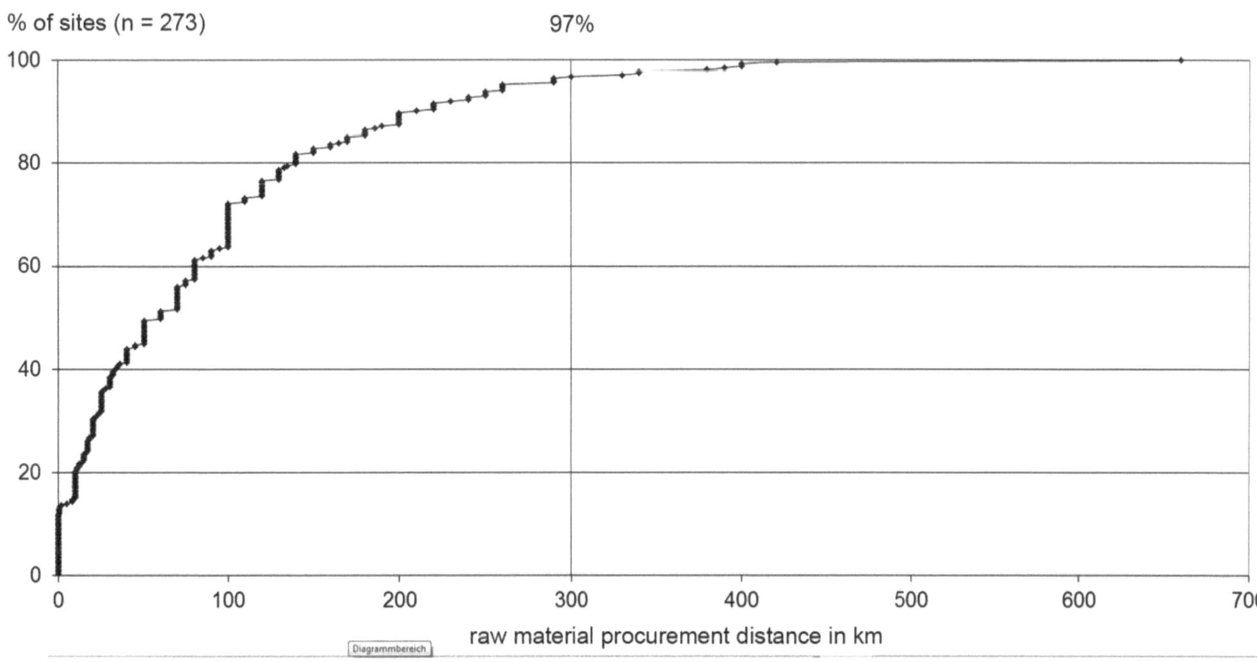

Fig. 6.6 Observed distances of raw material procurement in the CEM record

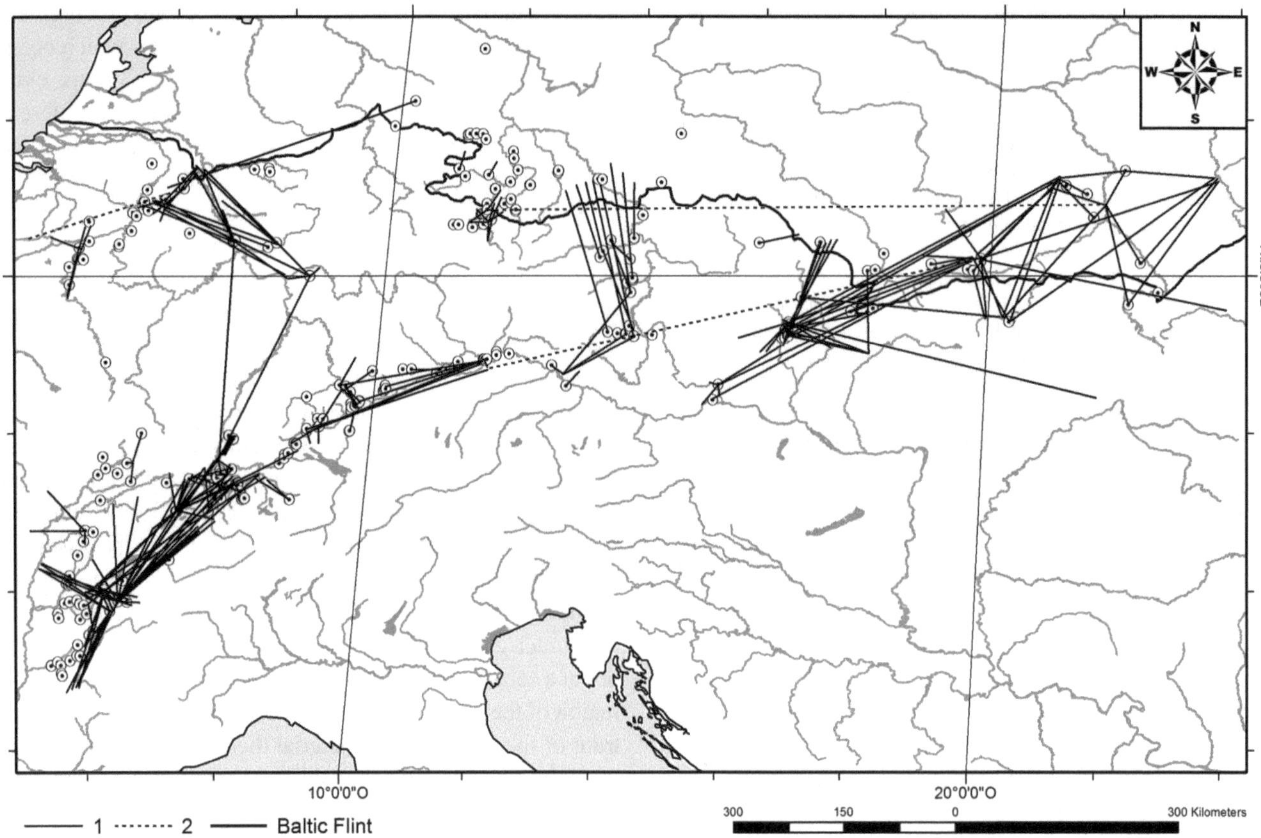

Fig. 6.7 Raw material procurement pattern of Central European Magdalenian sites. *Lines* connect the probable sources of certain raw materials and the site locations where specimens of these materials have been found (1: Reliable identification; 2: Questionable identification)

catchment areas and the collective or non-collective exploitation of raw material sources can, in turn, give information about the interaction patterns of those people who produced the assemblages. Extensive overlaps and common sources point to intensive interaction. In contrast, mutually exclusive areas and the usage of completely different raw material sources suggest a comparatively low level of interaction. Intensive interaction would indicate a strong coherence of the people involved, whereas weak or no interaction would rather point to independence and segregation.

For the investigation of the site catchment areas, information on 744 individual raw material transports from 151 CEM assemblages was analyzed (Table A.9). Hence, information on raw material procurement is available for 38 % of the 393 inventories considered here. In Fig. 6.7, lines connect the probable sources of certain raw materials and the site locations where specimens of these materials have been found, in order to illustrate the distances and directions of the respective imports. Those imports which were acquired within a distance of 15 km around a site are not indicated on the map, for the sake of preserving its readability and clarity. To be sure, it is possible that further research will alter the actual pattern. However, the sample size is considered sufficiently large to provide reliable insights into the CEM interaction pattern, insights that go beyond random observations.

With reference to Fig. 6.7, it is notable that raw materials are not procured evenly from all around a given site but that their acquisition seems primarily to be directed along certain axes. For the French and Swiss Jura, the route between the southern part of the Jura Mountains via the Molasse basin and the Aar valley up to the southern slopes of the Black Forest is certainly of major importance. In the case of the sites in the area of the Danube, Maas, Rhine, and Vltava, the direction of the raw material transports seems to be oriented along the course of these rivers.

Generally, sites located close to each other or in the same region exhibit strong overlapping in their catchment areas and show a collective exploitation of the regional raw material sources. This finding is by no means surprising but rather meets the intuitive expectation. However, this principle does not hold true for all cases. Sometimes, sites in neighboring areas regularly yield raw materials from mutually exclusive sources (see below). The following section therefore analyzes raw material acquisition with regard to breaks and linkages within the procurement pattern in order to gain some insights into regional exploitation ranges.

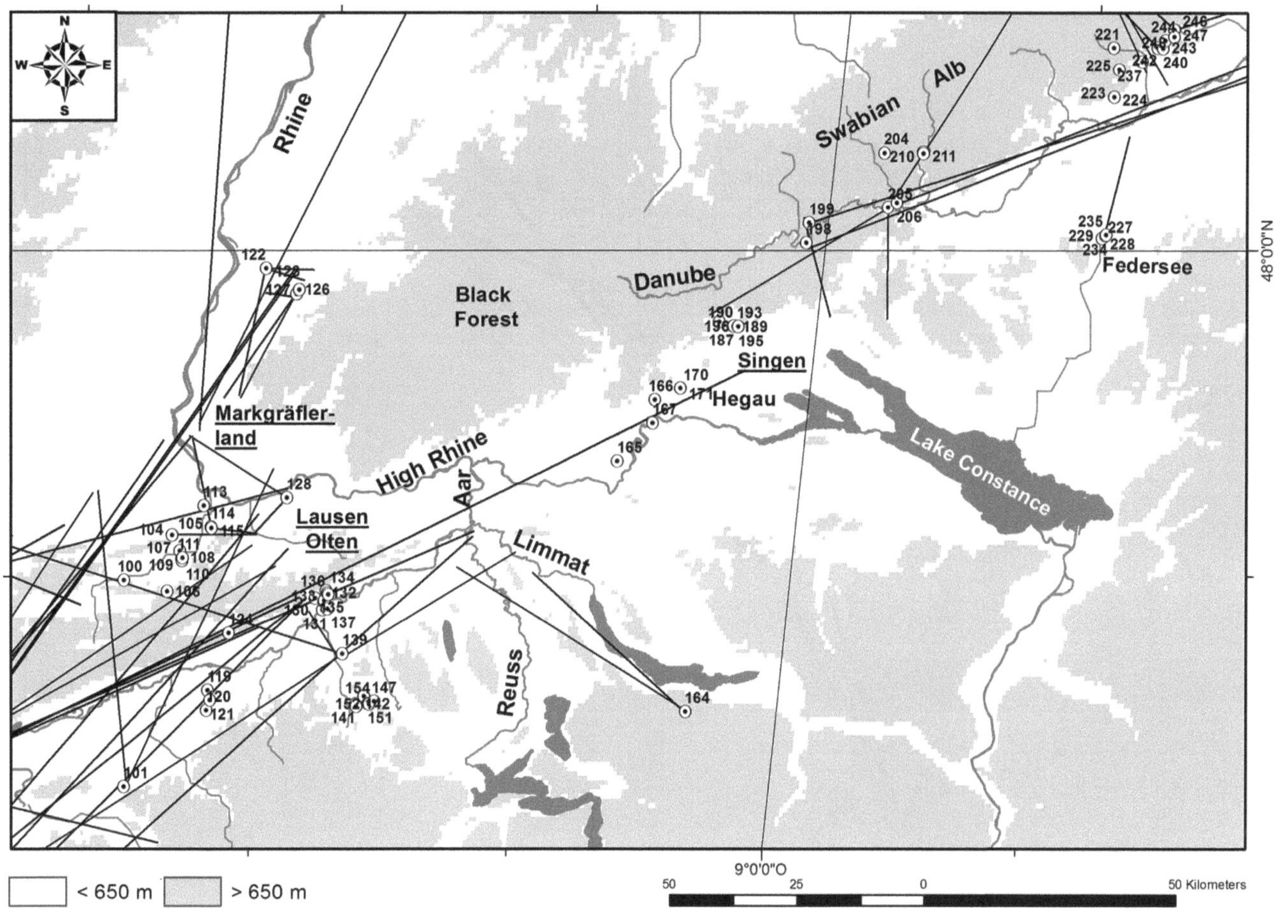

Fig. 6.8 Raw material procurement pattern in the border area between the Swiss Jura and the Swabian Alb

The most conspicuous example of CEM hunter-gatherer groups using different and almost mutually exclusive raw material sources despite the spatial proximity of their sites can be observed in the border area between the Swiss Jura and the Swabian Alb or more precisely the region where the Limmat, Reuss, and Aar rivers flow together and join the High Rhine (Fig. 6.8). Despite a high density of CEM sites east and west of this region, no site has yet been identified in this area itself. If the raw material procurement pattern of the neighboring sites on both sides of this area is taken into account, this gap takes on considerable significance. With regard to raw material acquisition, there are virtually no indications of interaction between the sites of the Hegau and Swabian Alb in the east and the sites of the Swiss Jura—including the Aar and High Rhine site cluster—in the west. In contrast, the site cluster in the Swiss Jura exhibits in itself a very tightly meshed procurement pattern and thus a high internal regional connectivity, and long-distance connections are almost exclusively oriented toward the southwest. The only exception is reported from the Magdalenian horizon of Monruz. Here, about 0.5 % of the lithic finds are identified as Phonolite, which is thought to come from the area around Singen in the Hegau region, which is located approximately 180 km to the northeast (Affolter 2006: 76, Fig. 104).

For the Swabian Alb, the inner connectivity is a bit less pronounced than it is the case in the French and Swiss Jura. However, all long-distance relations are oriented toward the east. Moreover, Swabian sites, such as Buttentalhöhle (ID 198), Probstfels (ID 199), or Burghöhle Dietfurt (ID 205), yield raw materials from the Kelheim region in confluence area of the Altmühl and Danube rivers about 240 km away to the east (see Fig. 6.10). At the same time, raw materials from the outcrops around Olten and Lausen, which were intensively exploited by Magdalenian hunter-gatherers in the Swiss and French Jura, are completely lacking in the assemblages of the Hegau and Swabian Alb, although they are located only about 60–150 km to the west. Eventually, it can be stated that the site cluster of Aar and High Rhine on one side and the one of Swabian Jura on the other show virtually no overlap in their site catchment areas, although their closest sites are only about 70 km away from each other.

Another example of the selective usage of raw material sources by only one of two neighboring groups can be found in the region where the Isar river joins the Danube. Here, in

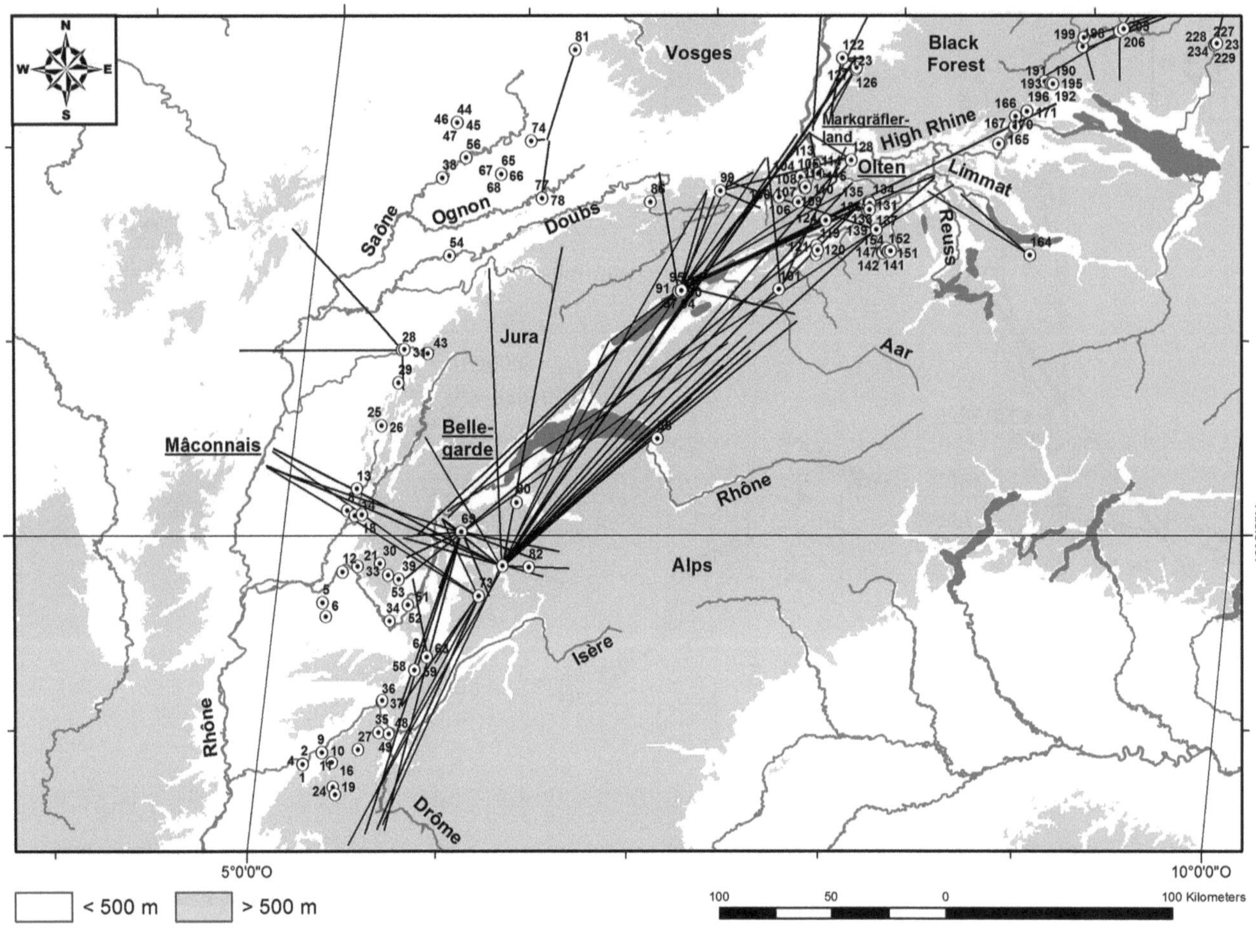

Fig. 6.9 Raw material procurement pattern of the assemblages in the Circum-Jurassic Group

the vicinity of the towns of Vilshofen and Ortenburg are several outcrops of high-quality chert. Two Magdalenian sites, Metten (ID 501) and Vilshofen-Kuffing (ID 504), are reported in this area, which is situated about 100 km away from the next two regions with Magdalenian site clusters, the Regensburg Basin and the Otava valley in southern Bohemia. However, despite being located at equal distances from the Vilshofen and Ortenburg area, chert from this source is exclusively present at Bohemian sites (see Fig. 6.12).

The observations above allow the following conclusions to be drawn: first, the acquisition of raw materials did not take place evenly in all directions around a site, but followed certain routes and respected borders. Second, these borders are the same for all sites within a certain region. Third, this internal coherence speaks strongly in favor of an intensive interaction and synchronized resource exploitation among the hunter-gatherer population in a certain region. Therefore, the assemblages within one of these discrete regions of overlapping site catchment areas can be interpreted as the remains of a territory-conscious interaction unit (see Chap. 3, section "Territories, Territoriality, and Land Tenure"). With regard to site location (Fig. 6.2) and the observable pattern in Fig. 6.7, five territory-conscious interaction units or, simply put, regional groups can be distinguished.

The Circum-Jurassic Group

The first unit comprises the sites of the French and Swiss Jura, which form a rough triangle between the Alps in the southeast, the Black Forest and the Vosges Mountains in the north, and the mountain ranges along the left banks of the Rhône and Saône rivers in the west (Fig. 6.9). The whole region is comparatively rich in outcrops of high-quality raw material, and the raw material procurement pattern of the Circum-Jurassic Group is restricted almost entirely to the

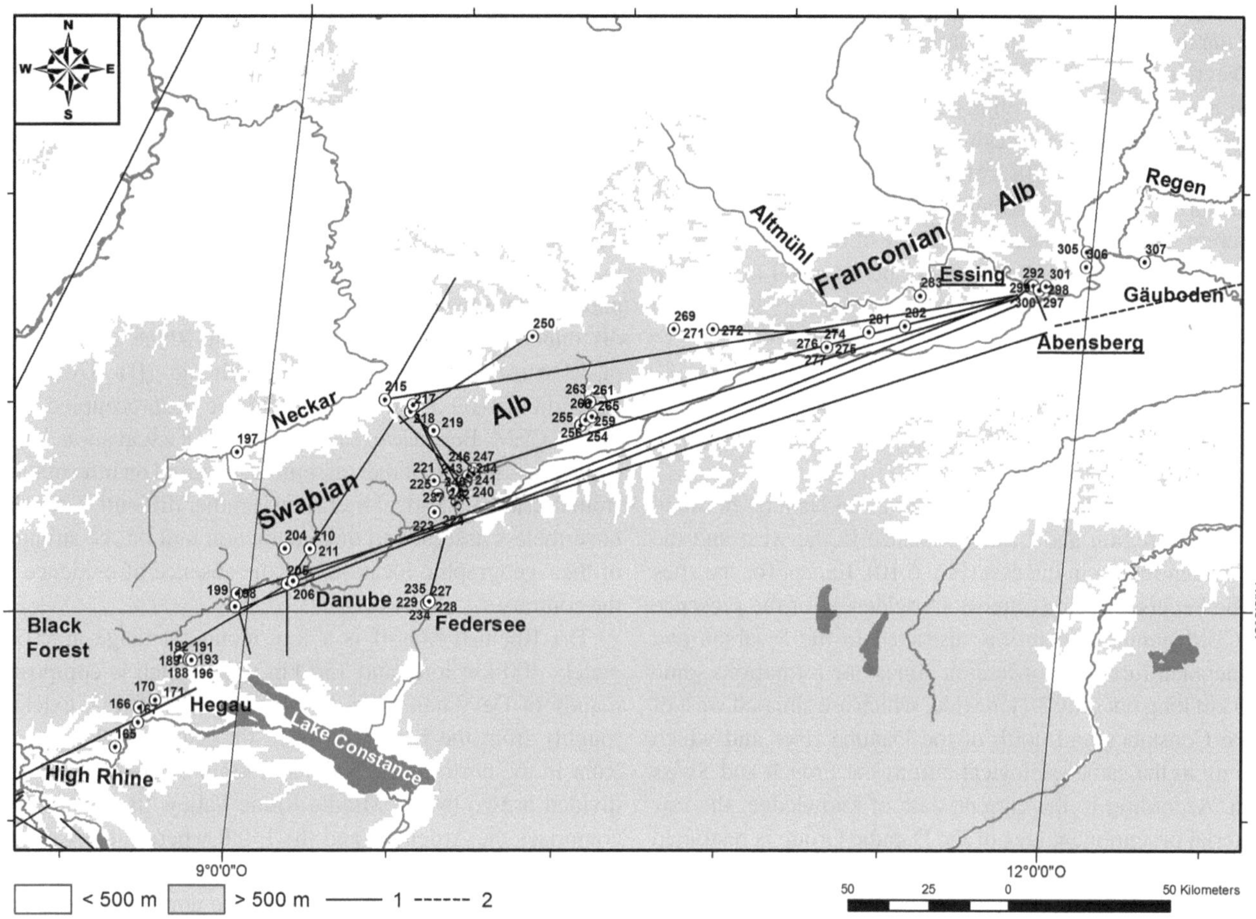

Fig. 6.10 Raw material procurement pattern of the assemblages in the Danube Group, 1: considered reliable; 2: verification needed

area covered by the cluster's site distribution itself, especially rich are the regions of Olten and Bellegarde-Seyssel (Affolter and Bressy 2009; Cattin 2009). In the Late Jurassic (Malm) deposits of the Jura, nodular or flat varieties of chert are common and extensively distributed (Floss 1994: 104).

Spatially, this site cluster is separated from the neighboring Danube cluster by a zone only about 70 km wide, where the Limmat, Reuss, Aar, and Rhine rivers meet (see above). In terms of the raw material procurement pattern, however, this separation is very clear. While the assemblages around the Jura Mountains on the one hand and along the Danube river on the other show a close-meshed pattern of overlapping site catchments, virtually no overlap can be observed between the two site clusters (see above, Fig. 6.8). Integrated into the Circum-Jurassic procurement pattern are the three sites of Munzingen (ID 122), Teufelsküche (ID 126), and Steinberg (ID 127), all situated southwest of Freiburg im Breisgau in an area where patches of Jurassic rocks are present. Less well connected, on the other hand, are the sites located near the Doubs, Ognon, and Saône rivers. Because of chert originating from outcrops northwest of the Jura that was found at the shelters of Veyrier (ID 79) (see Stahl Gretsch 2006), as well as the general proximity of these sites to those on the Aar and Rhône and the fact that no other connection contradicts an allocation, this site cluster is considered to be a part of the French and Swiss Jura unit as well. The structure of the procurement pattern indicates that the Molasse basin, which stretches between the Jura Mountains and the Alps from Lake Geneva via Lake Neuchâtel up to the Aar valley, was an important route for exchange and communication. This hypothesis is supported by assemblages from the sites of Hauterive-Champréveyres (ID 93–97) and Neuchâtel-Monruz (ID 89–91), where raw material imports from both ends of this corridor—namely, the region of Bellegarde-Seyssel in the southwest and the region of Olten in the northeast (Leesch 1997; Leesch et al. 2004)—are reported. At Monruz, about 0.5 % of the lithic finds are identified as Phonolite, which is thought to come from the area around Singen in the Hegau, located approximately 180 km to the northeast (Affolter 2006: 76, Fig. 104; Cattin 2009).

This is the only hint of an import of raw material from outside the cluster's site distribution. From the site ensemble of Veyrier, raw materials from the region of the Aveyron river are reported. This region lies on the western side of the Massif Central at a distance of about 450 km as the crow flies (Stahl Gretsch 2006: 150f.). However, the finds were discovered during quarry works and originate from at least four different locations. Thus, the integrity of the Veyrier assemblage is rather doubtful and cannot be regarded as reliable evidence.

The Danube Group

The second interaction unit lies along the Danube between the Hegau region and Lake Constance in the west and the Gäuboden region in the east (Fig. 6.10). Except for the sites in the Federsee region, almost all evidence for the presence of CEM hunter-gatherers is restricted to the Swabian and Franconian Jura, a set of Jurassic limestone formations some 350 km long but only 30 km wide, which are situated west of Lake Constance and north of the Danube river and which belong to the same geological unit as the French and Swiss Jura. According to the current state of knowledge, the raw material procurement area of the Danube Group is restricted to the extent of this site cluster, and there is no evidence for imports from other parts of the investigated area. The almost ubiquitous availability of high-quality raw materials is probably the reason for a marked trend toward local acquisition, the range of which usually does not exceed 40 km. At the same time, banded Jura chert from the region of Abensberg Arnhofen in the Regensburg Basin occurs in many collections and was found at a distance of more than 250 km away in assemblages of the western Alb. It is striking that this banded chert is distributed almost exclusively west of the Regensburg Basin. To the east, in contrast, the distribution ends with the extent of the Danube site cluster. The only exception seems to be the assemblage of Maszycka Cave in the Polish Jura, where a further occurrence of banded chert is reported.

It should also be mentioned that Albrecht et al. (1977b: 166) report three flakes from the site of Gnirshöhle 1 (ID 195) in the Upper Danube region that they say resembles Baltic flint. Since this allocation is rather uncertain, however, it is not considered in this study. The spatially distinct structure of the Danube site cluster and the coherent pattern of overlapping catchment areas make it possible to distinguish this regional group from the surrounding clusters and to interpret it as a consistent and territory-conscious interaction unit (see above, cf. Fig. 6.8).

The Meuse-Rhine Group

The sites of the Meuse-Rhine Group are located along the river valleys which traverse the Rhenish Massif and the Lower Rhine Bay (Fig. 6.11). The sites in the northwestern part of the investigated area are primarily located along the currents of the Meuse and the Rhine, as well as their tributaries. Here, the drainage system of the Meuse contains the majority of sites, the site density in the eastern and southern part of the Rhenish Massif is considerably lower. Rather disconnected from the sites west of the Rhine are the sites of Martinshöhle (ID 375), Feldhofhöhle (ID 376), and Balverhöhle (ID 377), which may have been occupied during the CEM. Because of their early date of excavation, however, an evaluation of their assemblages based on information from the archaeological literature is rather difficult. They are nevertheless assigned to this interaction unit on the strength of their geographic location and the absence of evidence to the contrary.

The Rhenish Massif is a low mountain range approximately 400 km long and 150 km wide which is composed mainly of Devonian rocks (Walter 1995: 160ff.). It extends roughly from the river valleys of the Sambre, Meuse, and Ruhr in the north to the Nahe and Main in the south and is divided in two by the Middle Rhine Valley. Its eastern part comprises the Ardennes and the Eifel, whereas the western part can be subdivided from north to south into the Sauerland, the Westerwald, and the Taunus. At the northern margins of the Ardennes, patches of Cambrian, Carboniferous, and Cretaceous sediments as well as loess can be found close to one another. The Westerwald and the Neuwied Basin, a broader part of the Rhine valley between the Westerwald and the Eifel, is characterized geologically by the occurrence of volcanic rock (Walter 1995: 182ff., Abb. 74; Zahn 1996: 74–75). In the region of Charleville-Mézières, outcrops of chert can be found, whereas flint from the Cretaceous sediments can be found north of the Ardennes. Here and in the Lower Rhine Bay, Meuse gravels and Meuse eggs also occur in a large area (Floss 1994: Fig. 29), and Baltic flint occurs north of about 51° N (see Fig. 6.4). Tertiary quartzite can be found in many parts of the Rhenish Massif, while chalcedony occurs mainly in the region of Bonn (Floss 1994: Fig. 8, 11). Baltic flint and raw materials from the Meuse region can be found in CEM assemblages as far south as the Neuwied Basin, the Lahn valley, and even the lower Main area, indicating common exploitation of these areas by a regional hunter-gatherer population. The presence of chert from the Markgräflerland—a region between the Upper Rhine valley and the Black Forest—at the sites of Gönnersdorf (ID 369–371) and Dreieich-Götzenhain (ID 385) (Floss 1994; Terberger et al. 2009a: 22) is interpreted as an indication of

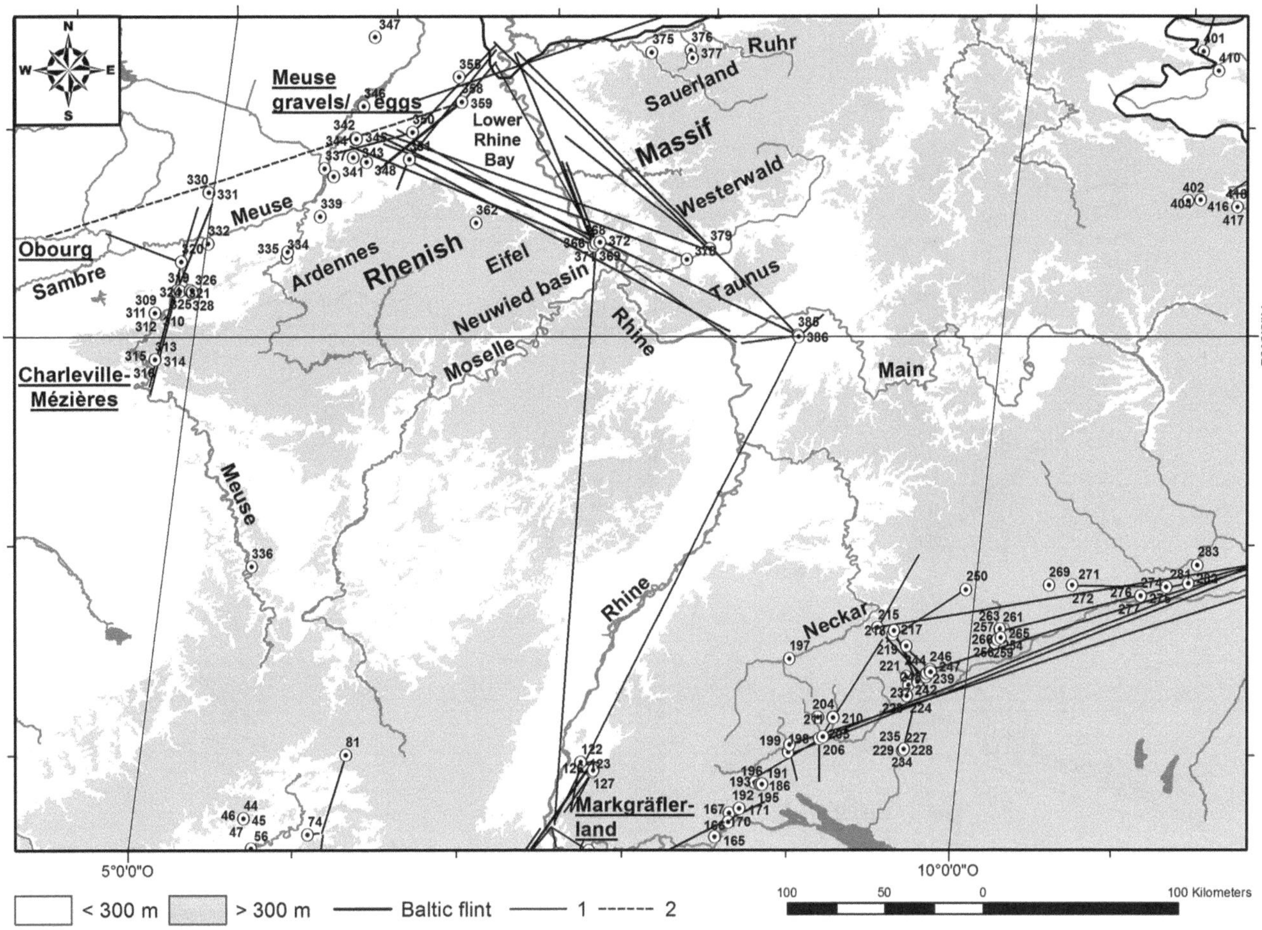

Fig. 6.11 Raw material procurement pattern of the assemblages in the Meuse-Rhine Group, 1: considered reliable; 2: verification needed

interaction between the Circum-Jurassic Group and the Meuse-Rhine Group. Three arguments can be offered to support this: first, the Markgräflerland is assigned to the interaction unit of the French and Swiss Jura, as is indicated by the spatial proximity and the overall raw material procurement pattern; second, these raw materials are the only imports in the Meuse-Rhine Group that originate from south of 50° N; and third, the transportation distance clearly exceeds that of all other imports observed for this group. Chert from the site of Dreieich-Götzenhain may have been imported from the Swabian or the Franconian Alb. However, the acquisition of these materials from secondary deposits in the vicinity of the site cannot be ruled out (Serangeli and Terberger 2006). In contrast to the numerous transports along a north-south orientation, evidence for long-distance west-east transport is largely absent. The reported occurrence of Obourg flint at the site of Kamphausen (ID 359) (Thissen 1997) could not be confirmed during a recent review of the material. A possible explanation for this conspicuous pattern is the availability of high-quality raw material throughout the central area of the Meuse-Rhine group, which would have rendered long-distance imports unnecessary. Besides these imports, a possible export of raw material from the Meuse area is mentioned at the site of Gadenstedt (ID 395, cf. Fig. 6.7), north of the Hercynian Mountains (Veil and Hiller 1998; Simoneit and Veil 1999). This is interesting inasmuch as this kind of evidence for relations between the regional groups in the eastern and western parts of the investigated area is otherwise very scarce.

The Vltava-Saale Group

The sites of the fourth regional group are located within the drainage system of the Elbe river, mainly along its most important tributaries, the Vltava and the Saale (Fig. 6.12). The Vltava-Saale Group is characterized geographically by several low mountain chains, which also subdivide it. To the west, the site distribution is delimited by the Thuringian-Franconian Highlands (including the Thuringian Forest, the Thuringian Slate Mountains, the Franconian Forest and

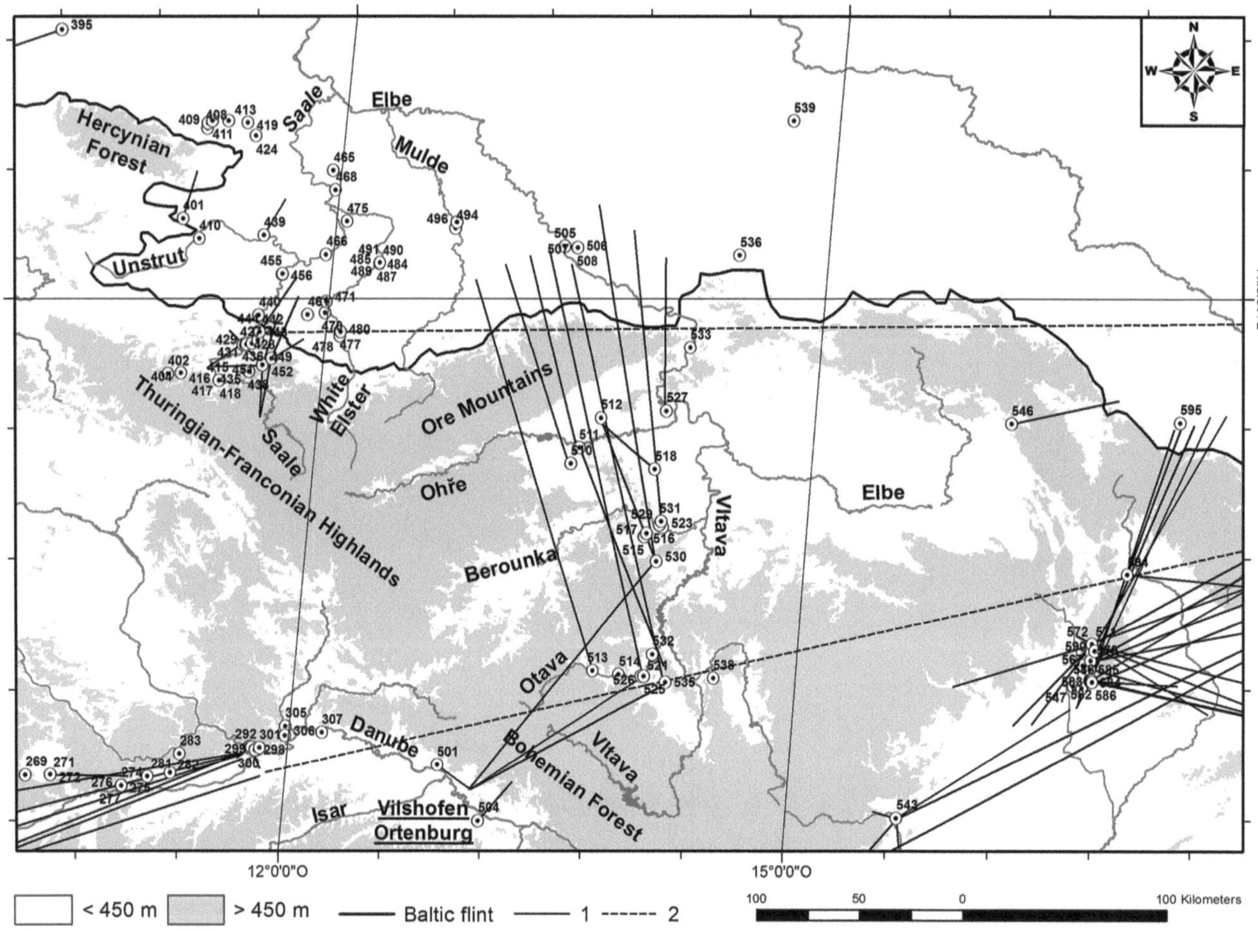

Fig. 6.12 Raw material procurement pattern of the assemblages in the Vltava-Saale Group, 1: considered reliable; 2: verification needed

the Fichtelgebirge), the Upper Palatinate Forest, and the Bohemian Forest. The Erzgebirge (Ore Mountains), which run more or less perpendicular to the Thuringian-Franconian Highlands, divide the group into a northwestern and a southeastern part.

The Vltava rises in the Bohemian Forest and subsequently flows through the South Bohemian Basin (Jihočeské pánve), which is generally characterized by granite and metamorphic rock. However, in the vicinity of Budweis (České Budějovice), Cretaceous sediments can be found (Zahn 1996: 74–75) which potentially yield raw material of good quality (see Fig. 6.4). In the vicinity of the Berounka river, there are patches of Ordovician, Silurian, Devonian as well as Carboniferous and Cretaceous sediments (Walter 1995: Abb. 102; Zahn 1996: 74–75). The North Bohemian tableland between the Berounka river, the Ore Mountains, and the Elbe (Labe) river is dominated in its eastern part by Cretaceous sediments, whereas the western part is marked by Tertiary sediments and volcanic rock (Walter 1995: Abb. 127). The Saale rises in the borderland between the Thuringian Forest and the Ore Mountains and then crosses Lower Triassic (*Buntsandstein*) and Middle Triassic (*Muschelkalk*) sediments before flowing through the Leipzig Lowland Bay, which is characterized geologically by loess deposits in its southern part and by moraine deposits in the north (Walter 1995: Abb. 114; Zahn 1996: 74–75); in the latter area, Baltic flint can also be found. West of the Saale river lies the Thuringian Basin. This region is dominated geologically by Late Triassic sediments (Keuper) and is surrounded by Middle Triassic formations. Lower Triassic sediments are characteristic of the outer part of the basin (Walter 1995: Abb. 114). Chert from Late Triassic sediments is often marked by clefts, rendering this raw material rather unsuitable for Magdalenian tool production (Floss 1994: 108).

In the northwestern part of this cluster, the lithic assemblages are mainly composed of local raw materials and thus seem primarily to reflect the local and regional activity range of the hunter-gatherers. Local chert (e.g., hornstone and radiolarite) and quartzite is frequently found, as is Baltic flint. Some authors have posited raw material transports from

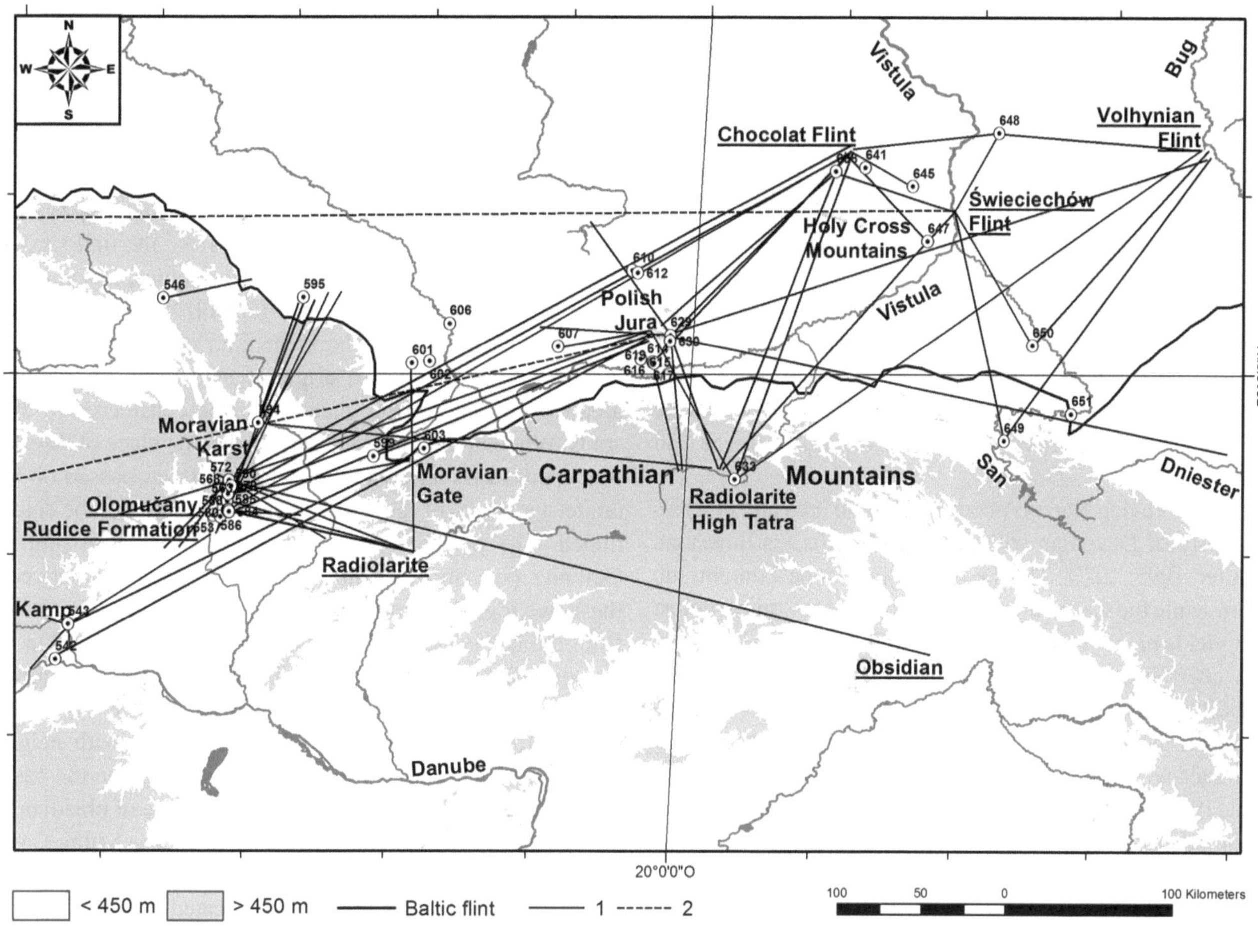

Fig. 6.13 Raw material procurement pattern of the assemblages in the Polish-Moravia Group, 1: considered reliable; 2: verification needed

northern Switzerland, southern Germany, and the region of Becov in Bohemia to sites such as Bad Frankenhausen (ID 401), Kniegrotte (ID 451), Teufelsbrücke (ID 415–418), or Groitzsch (ID 483–491) (Hanitzsch 1972: 55; Feustel 1974, 1977: 28, 1980; Malina 1974: 98). However, recent studies show the link between these sources and the specimens in the assemblages to be anything but compelling, since the respective raw materials also occur in the immediate vicinity of the sites (Höck 2000: 57f.; Küßner 2009: 88, 100f.). The only long-distance import of the Vltava-Saale Group is reported from the site of Oelknitz (ID 444). Here, Kozłowski (1987: 36) mentions the presence of one blade made of Świeciechów flint, the outcrops of which are found in the vicinity of the Holy Cross Mountains in Poland, about 800 km away from the site (see Fig. 6.13). Since macroscopic assignments of raw materials to their potential sources are always plagued by several unknowns, this attribution must be considered questionable. On the other hand, until it is proven incorrect, it should be kept in mind as a possible clue to the relations and interactions of this group with other hunter-gatherer groups in the eastern part of the investigated area. In addition to local raw materials, Baltic flint is present at many sites in the southeastern part of the cluster and was therefore transported over distances of up to 230 km. A type of dotted chert found at the Bohemian sites of Hostim (ID 530), Putim (ID 525), and Žd'ár 1 (ID 535) has been identified as Flintsbach chert (Přichystal 2006: 357). This observation extends the exploited area of the fourth interaction unit as far south as the sites of Metten (ID 501) and Vilshofen-Kuffing (ID 504) and thus to the region where the Isar joins the Danube.

The lithic assemblage found at Gadenstedt (ID 395) in Lower Saxony is partly composed of Meuse gravels that were transported about 350 km (Veil et al. 1998; Veil and Hiller 1998; Simoneit and Veil 1999). Against the background of the information obtained so far, it is not possible to attribute the Gadenstedt assemblage to a particular interaction unit. Its rather isolated geographic position at a distance of 80 km away from the closest neighboring site, located in the Hercynian Mountains to the southeast, as well as its raw material spectrum which points toward the west, would seem to suggest an intermediate position between the sites in the Meuse-Rhine area and those in Eastern Germany.

The Polish-Moravian Group

The fifth and easternmost site cluster covers the largest area of all the interaction units and at the same time shows the farthest average transportation distances (Fig. 6.13). Its sites cover the area from the Kamp valley and the Moravian Karst via the Moravian Gate and the Krakow Jura up to the Vistula, San, and even the Bug and Dniester rivers. Thus, it stretches more than 700 km from west to east; its north-south extension is considerably smaller, however, with maximum values around 200 km. The uneven distribution of sites probably reflects the structure of the landscape, with more and less favorable settlement regions (see section "The Large-Scale Site Pattern as a Reflection of Raw Material Sources and Rivers").

The Moravian Karst is about 100 km^2 in area and consists primarily of Devonian with some Carboniferous limestone (Walter 1995: 287). The Devonian limestone contains no chert, while the Carboniferous limestone in the southernmost part yields pieces of black chert up to 10 cm thick; however, this chert has not yet been found among the Moravian Paleolithic assemblages. Important local sources of raw material include the Jurassic limestone formation near Olomučany and the so-called Rudice Formation (Svoboda et al. 1995: 136). Further to the east lies the Moravian Gate, a natural pass approximately 65 km long and 15 km wide between the Eastern Sudetes and the Western Carpathians. At its northeastern part, outcrops of Cretaceous sediments can be found (Zahn 1996: 74–75). The Polish Jura or Kraków-Częstochowa Upland is an Upper Jurassic limestone formation characterized by escarpments of sometimes more than 500 m in thickness (Walter 1995: 115). In the northeast, the Jurassic chain merges into the Miechów depression, separating it from the Holy Cross Mountains. The depression itself is covered in the south with Cretaceous sediments and in the north with moraine sediments (Walter 1995: 117; Zahn 1996: 74–75). In their southwestern part, the Holy Cross Mountains (Góry Świętokrzyskie) are built up mainly by formations of the Cambrian, Ordovician/Silurian, Devonian, and Carboniferous rocks, whereas Jurassic limestone occurs frequently at the northeastern slope (Walter 1995: 110f., Fig. 52; Zahn 1996: 74–75). Here, outcrops of high-quality Chocolate and Świeciechów flint are located. Southeast of the Holy Cross Mountains, the San river cuts through Loess and moraine deposits before it joins the Vistula (Zahn 1996: 74–75). South of the Polish Jura lie the Mountains of the High Tatra. This massif is a part of the Beskids which coincide for their part with the Outer Western Carpathians. Tertiary rocks are the dominant geological formation of the Beskids, although patches of Jurassic to Cretaceous sediments appear frequently, especially in the northern part (Zahn 1996: 74–75).

Looking at the raw material spectrum of this regional group, there are three imports which clearly exceed the normal transportation distances and overstep the boundaries of the group's territory. From Maszycka Cave (ID 629), occurrences of chert from the vicinity of Abensberg in the Regensburg Basin are reported (Kozłowski et al. 1995: 191ff.). If correct, the source of this raw material is thus located at a distance of about 660 km from the find site, as the crow flies. Despite remaining uncertainties, this appears to be the most reliable evidence for Polish-Moravian relations in this region. At the same site, other raw material identifications are said to originate also from the Danube area (Ibid.), but are too uncertain to be considered in this study. Whether this raw material reached its place of discard as a result of direct contact or whether it was passed on by hunter-gatherers of the Vltava-Saale Group cannot be determined. The possible occurrence of Świeciechów flint at Oelknitz (see above) points to a closer relation between these two regional groups. Again from Maszycka Cave, flint from the headwaters of the Dniester river about 350 km to the southeast has been reported (Ibid.). Whether the acquisition involved direct procurement by hunter-gatherers of the Polish-Moravian group or reflects interaction with neighboring groups to the east cannot be determined on the basis of the currently available information. A piece of obsidian is reported from the Magdalenian layer 5 of the Kůlna Cave (ID 571) in the Moravian Karst (Valoch 1988: 25). The closest source area of this raw material is located on the southern slopes of the Carpathian Mountains, about 400 km away. Again, the current state of information is much too poor to make reasonable assumptions about the acquisition process. The presence of Volhynian flint in many assemblages of the Polish-Moravian Group shows that its territory extended in the east at least as far as the Bug river (Połtowicz 2006).

In summary it can be said that the raw material procurement pattern of the CEM indicates the existence of five territory-conscious interaction units. In addition to the generally discrete and coherent character of these units, long-distance raw material imports likewise point to the existence of contact between them. The latter aspect is analyzed more closely in the next chapter.

Procurement Pattern of Mollusk Shells and Amber

Shells, the calcareous exoskeleton of many invertebrate animals, are frequently found in CEM sites. However, while it is comparatively easy to pinpoint the sources of specific siliceous rocks, locating the origins of shells is often more difficult (see Chap. 2, section "Lithic Raw Materials, Mollusk Shells, and Amber"). Fossil or subrecent shells of marine mollusks in particular seem to have attracted the interest of Paleolithic hunter-gatherers. Many of the excavated specimens exhibit a pierced hole, indicating their usage as pendants or ornaments sewn on cloth; this is suggested by observations of Paleolithic burial sites, such as those at Sungir (Rogačev and Anikovič 1984), Dolní Věstonice (Klíma 1988), or La Madeleine (Capitan and Peyrony 1928). Because of their ornamental character, shells certainly had a completely different significance compared to raw materials for tool production and were probably more symbolic in meaning (Gamble 1980; Wiessner 1984). It can therefore be assumed that the pattern of their procurement not only differs in its spatial structure, but also reflects as well a different sphere of Magdalenian everyday life (Rensink 1993: 114ff.; Taborin 1993; Floss 1994: 337). Even though direct procurement cannot be excluded, it seems more plausible to assume that the shells of far-distant origin reached their place of discovery embedded in a longer chain of events. It may be, for instance, that shells were exchanged between members of different regional groups. Another possibility is that they were carried into distant areas as personal ornaments when people went to live with another group after getting married. In contrast to lithic raw materials, which were consumed on a daily basis and were usually procured on a regional scale, shells were generally transported over much greater distances. Because the average transport distance of shells exceeds that of most lithic raw materials by a considerable margin, it is reasonable to assume that their acquisition took place in a different way. Much like the lithic artifacts found at exceptional distances from their sources (see above), shells are considered to be indicative of supra- and interregional relations among hunter-gatherer groups. From this point of view, it is expected that the procurement pattern of shells contrasts distinctly with the one previously obtained for lithic raw materials.

An appropriate tool to analyze and compare the procurement pattern of shells and amber is a distribution map (Fig. 6.14). To generate this map, 198 indications of occurrences of different fossils and subrecent mollusk shells from a total of 62 assemblages were analyzed (Table A.10). In order to ensure readability and to give a clear depiction of the pattern, not every specimen-source relation was converted into an individual line. Instead, all items of an assemblage that have been identified as coming from the same source are bundled into one line, irrespective of the abundance or diversity of species. For the same purpose, the mouth of the

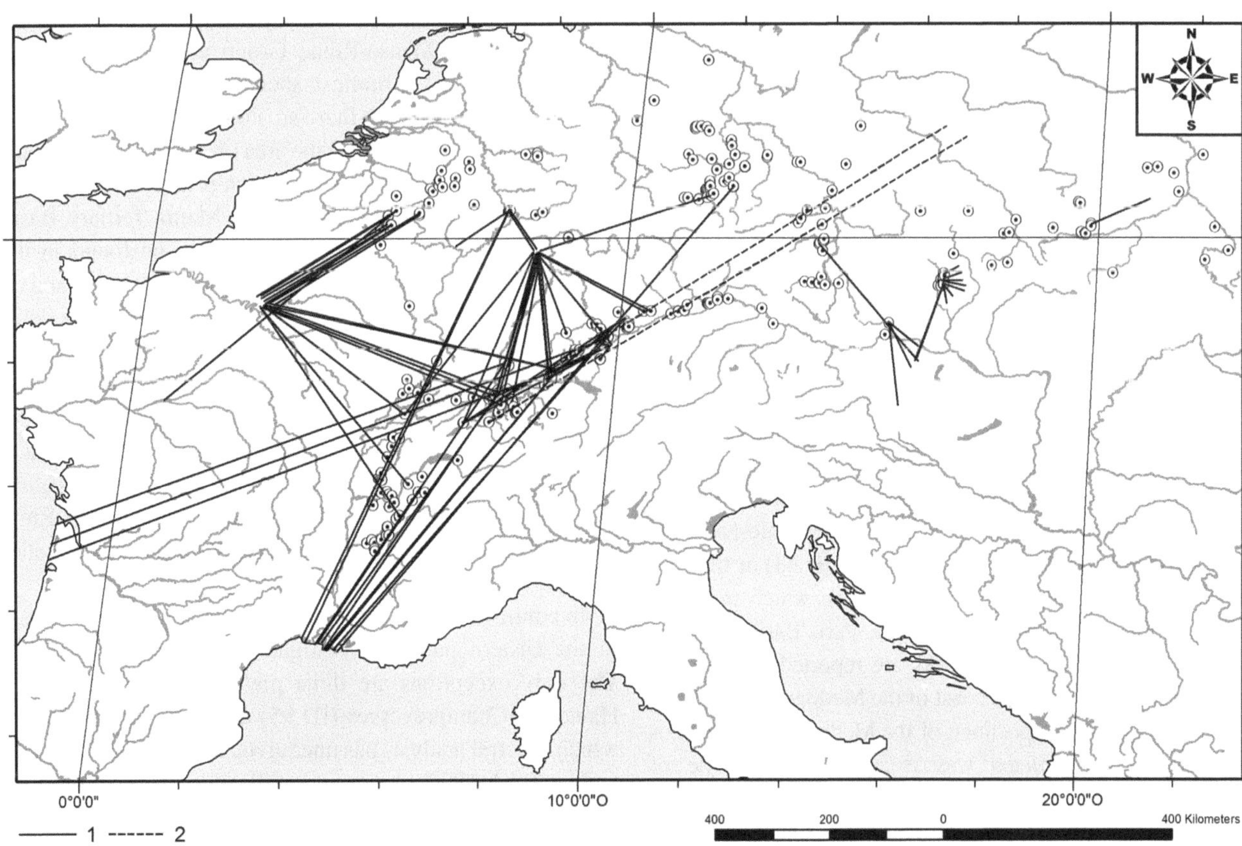

Fig. 6.14 Distribution pattern of mollusk shells (1) and amber (2). *Lines* connect the probable sources of certain shells and the site locations where specimens of these shells have been found (for detailed information see Table A.10)

Dordogne is chosen as a dummy to represent the source area of the Atlantic Ocean, the Rhône outlet serves as a dummy for the Mediterranean coast, and the eastern bank of the Odra river is chosen to represent an eastern origin of Baltic amber.

Already at first sight, it is evident that the overall pattern of Fig. 6.14 differs entirely from the one observed for lithic raw materials. It even seems that both patterns complement one another. Whereas the connection lines for the siliceous rocks tend to run within geographically distinct site clusters and hence delimit them from each other, the lines representing the procurement pattern of shells run primarily between site clusters and therefore provide an inter-cluster linkage. Looking at the overall structure of this procurement pattern more closely, a pronounced contrast between the eastern and western site clusters becomes apparent. The sites in the French and Swiss Jura, along the Danube, in the Lower Rhine area, and in the Ardennes, corresponding with the three westerly regional groups (see above), appear to be very closely interconnected. In contrast, the two regional groups in the east, namely, in Eastern Germany and Bohemia as well as in the Moravian Karst and Poland, show a less tightly meshed pattern. In order to analyze these differences more closely, the next section examines the respective regions in a more detailed manner.

The Circum-Jurassic Group

With subrecent shells from the Atlantic and Mediterranean, as well as fossils from the Kirchberg layers and the Paris, Mainz, and Steinheim basins, the site cluster in the French and Swiss Jura appears to be integrated in a long-span exchange network. In particular, the sites along the courses of the Aar and the High Rhine yield assemblages of mollusk shells which are both high in number and diverse in terms of the species represented. For the sites in the vicinity of the Isère, Rhône, Doubs, and Saône rivers on the other hand, the evidence for connections is much weaker and is restricted to the Paris and Mainz Basins, both of which seem to be important sources of fossil shells for the sites in the Jura region. Five assemblages of the Circum-Jurassic Group yield fossil shells of either the genus *Turritella* (Birseck-Ermitage ID 114, Kastelhöhle-Nord ID 110, Rislisberghöhle ID 124, Jean Pierre I ID 63) or the species *Bayana lactea* (Abri Cabônes ID 54), which probably stem from the Tertiary sediments of the Paris Basin. At three sites, shells of subrecent mollusks are reported which were presumably collected at the coast of the Mediterranean Sea or the Atlantic Ocean. A specimen of the Mediterranean species *Homalopoma sanguineum* was excavated at the site of Kohlerhöhle (ID 108) (Sedlmeier 1988). Malacological investigations of the assemblage from Munzingen (ID 122) identified one shell as belonging to this same species (Albrecht 1981b), while another shell is presumably of an Atlantic provenance (Bosinski 2008b: 284); though difficult to identify (Padtberg 1925), the latter shell was classified by W. Rähle as probably belonging to the species *Nucella lapillus*, which today can be found from the coasts of Norway down to the shores of North Africa (Pasda 1994: 71). The cave site of Birseck-Ermitage has yielded a shell which may belong to the species *Glycymeris pilosa* and may therefore be of a Mediterranean provenance (Sedlmeier 1988: 3, Table 1).

It is not possible to reconstruct the exact manner in which these shells were transported to their place of discovery. However, the Rhône valley was presumably an important axis of communication and likely also played an important role in the exchange of objects (e.g., Floss 1994: 338, 2000; Álvarez Fernández 2001). With regard to the raw material procurement pattern, the same holds true for the Alpine Molasse basin. Interaction with the Danube group is indicated by the presence at the sites of Monruz (Bullinger and Thew 2006: 155f.) and Kohlerhöhle (Sedlmeier 1988: 3, Table 1) of fossil shells of the species *Gyraulus trochiformis*, *Viviparus suevicus*, and *Brotia escheri* originating from the outcrops at Steinheim and the Upper Danube. Moreover, Mediterranean shells found at sites of the Danube Group (see below) probably reached their destination via the Circum-Jurassic Group and can be seen as indicators of exchange between these interaction units too.

Interactions between hunter-gatherers of the Circum-Jurassic and the Meuse-Rhine Group are more difficult to assess on the basis of mollusk shell distributions. This is primarily due to the fact that there are no major outcrops of fossil shells located directly within the area of one or the other interaction unit. An important source which is located between the site clusters of the two groups is the Mainz Tertiary Basin. However, many of the species which can be found in this region also occur in the Tertiary sediments of Belgium and the Paris Basin (Rähle 1983b: 156), which makes an exact determination of a given shell's origins difficult or impossible. Specimens which are thought to come from this area are found in both regional groups and thus indicate exchange or at least a common usage of the respective outcrops. Mediterranean species found at sites of the Meuse-Rhine Group (see below) probably reached their place of discovery via the Rhône-Rhine axis (Floss 1994: 338) and therefore indicate interaction between the two groups.

In contrast, evidence for interaction with regional groups in the eastern part of investigated area is almost absent. The only exceptions are three pieces of fossil resin from Hauterive-Champréveyres (ID 95) and Moosbühl (ID 101), which spectral analysis has unequivocally identified as Baltic amber (Beck 1997). An exact localization of its origin is not possible, however, because Baltic amber can be found in the deposits of numerous glacial moraines in Central Europe (Ganzelewski and Slotta 1996).

The Danube Group

In terms of frequency, the most important shell source for the Danube site cluster outside of the Swabian and Franconian Alb is the Mainz Basin. Seven sites have yielded shells which may come from this area. However, since many of these species also appear in the Tertiary sediments of Belgium and the Paris Basin, a clear allocation to a specific source area remains impossible (Rähle 1983b: 156). Shells which may likewise originate in the Mainz Basin are also reported at the sites of Kesslerloch (ID 170), Schweizersbild (ID 166), and Freudenthalhohle (ID 167) in the Randen area, at Petersfels (ID 173 ff.) in the Hegau region, at Napoleonskopf (ID 197) in the Neckar valley, at the Hohle Fels at Schelklingen (ID 241–243) in the central Alb, and at Kaufertsberg (ID 271) and Hohlenstein bei Edernheim (ID 269) in the Nördlinger Ries (for more details see Table A.10). Long-distance imports from the Atlantic and Mediterranean coasts as well as from the Paris Basin are exclusively found at sites in the Hegau region and the central Alb. At the site of Petersfels, subrecent shells of the Mediterranean species *Cyclope nerita* ($n=2$) and *Homalopoma sanguineum* ($n=1$) *as well as Pirenella plicata*, *Trochus obliquatus*, *Cyclope neritea*, *Cytherea* spp., *Glycymeris* spp., and *Oleurotoma* spp. were recovered (Rähle 1981, 1983b). At the immediately adjacent site of Gnirshöhle I (ID 195), Albrecht et al. (1977b) mention two shells of the Atlantic species *Nucella lapillus*. One specimen identified as *Astarte montagui* may be of the same origin. In the region of the central Alb, the Hohle Fels at Schelklingen is of interest. Here, subrecent shells of the Mediterranean species *Cyclope nerita* ($n=2$) and the Atlantic species *Littorina obtusata* ($n=1$) have been excavated (Rähle 1981). Found at the site of Petersfels, fossil specimens of *Bayana lactea* ($n=1$) likely come from the Tertiary sediments of the Paris Basin (Rähle 1983b) as do another two finds of fossils of *Sycum* spp. from the Gnirshöhle I assemblage. Fossils from the Steinheim Basin and the Kirchberg layers are reported as well (Albrecht et al. 1977b; Rähle 1981, 1983b; Eriksen 1991: 184, Table VI.14). Looking at the modern archaeological record, it is noticeable that no shells have been reported from the Franconian Alb so far (see Fig. 6.14). Their occurrence appears to be restricted to the sub-clusters in the western part of the Danube Group, including assemblages from the Hegau to the Nördlinger Ries. Whether this pattern reflects past reality or is simply the result of post-depositional biases and archaeological research activities is difficult to determine. However, the wide distribution of banded chert from the Abensberg region indicates that the Franconian Alb was indeed a substantial part of the Danube Group's settlement area. In summary, it can be stated that all long-distance imports found in the Danube Group come from northwestern, western, or southwestern sources, which would seem to indicate intensive relations with the hunter-gatherers of the Circum-Jurassic and the Meuse-Rhine Groups. However, for the Meuse-Rhine Group, an assessment of the intensity of interaction is problematic because there remains some question as to whether the Mainz Basin can in fact be considered a part of the territory of this group (see section "The Meuse-Rhine Group"). Exports of fossil shells from the Danube region into the territory of the Circum-Jurassic Group underline the reciprocal character of the relation between the two groups. Except for the possible occurrence of chert from the Abensberg region at the site of Maszycka Cave (see above), there is no evidence whatsoever for relations between the Danube Group and the regional groups in the eastern part of the investigated area.

The Meuse-Rhine Group

An exact localization of the sources of fossil mollusk shells in the area of the Meuse-Rhine Group and its adjacent regions is often impossible. This is because many fossil mollusk species occur in the Tertiary sediments of Belgium as well as in the Paris and Mainz Basin (Rähle 1983b: 156). Hence, a precise an unequivocal allocation to one of these possible regions of origin often cannot be achieved (see Chap. 2, section "Lithic Raw Materials, Mollusk Shells, and Amber"). It is conspicuous that no site, so it seems, has produced evidence for the usage of shells from the Tertiary sediments of Belgium. This observation raises the question of whether the identification of sources has not somehow been biased by the researchers' own expectations. However, as these analyses cannot be verified in the frame of this study, the following reflections must rely on the published results. Consultation of the archaeological literature reveals thus a procurement pattern with an interesting structure. The sites north of the Meuse and east of the Ourthe river have not yielded any shells so far. Those sites located in the Ardennes, on the other hand, seem exclusively to yield fossils from the west, specifically from the Paris Basin, which appears to be the source of the majority of imported shells in the Meuse-Rhine Group (Fig. 6.14). An analogous situation is observed for the Danube Group (see above), where shells from the Mainz Basin seem to occur exclusively in the Swabian Alb but are absent from the Franconian Alb (see above). It thus seems that the relations between the hunter-gatherers of the Meuse-Rhine Group and those of the Paris Basin were quite intense. In the Rhineland, on the contrary, the sites of Andernach-Martinsberg (ID 364–365) and Gönnersdorf (ID 368–370) exhibit only southward-oriented connections toward the Mainz Basin (Floss 1994: 136, Fig. 117). Surprisingly, no other site of the Meuse-Rhine Group seems to yield specimens from this comparatively close region, whereas such finds occur quite frequently among the assemblages of the Circum-Jurassic and Danube Group. The only specimens of subrecent Mediterranean mollusks reported from the Meuse-Rhine Group also come from the sites of Andernach-Martinsberg and Gönnersdorf. Shells of *Homalopoma sanguineum* and *Cyclope neritea* found at these sites (Bosinski 2008b: 322) strengthen the suggestion of an interaction with the Circum-Jurassic Group, as was already indicated by the chert from the Markgräflerland. The Rhône-Rhine corridor was probably an important axis along which the communication and exchange processes between the regional hunter-gatherer groups took place (e.g., Floss 1994: 338, 2000; Álvarez Fernández 2001).

The Vltava-Saale Group

According to the archaeological literature, fossil or subrecent mollusk shells occur in only three assemblages of the Vltava-Saale Group. At the open-air camp of Gera-Zwötzen (Schafgraben, ID 480), a subrecent shell of *Turritella communis* is reported that may come from the coast of the Mediterranean, the North Sea, or the Atlantic (Küßner 2009: 60). Because the distance to the Atlantic coast is greater, a Mediterranean origin is given preference in this study (see section "Procurement Pattern of Mollusk Shells and Amber"). However, for the interpretation, this difference is of minor importance since both origins suggest a western or southwestern connection. The assemblage of fossil shells from the Kniegrotte (ID 451) is quite abundant. The dominant species is *Cyrena convexa*, followed by *Glycymeris* spp.; one specimen of *Potamides plicatus galeotti* is also present. All of these shells were probably collected in the Mainz Basin, but they may also come from Southern Germany or the Paris Basin (Höck 2000: 144f.). Given the current state of knowledge, it is impossible to determine the route by which these shells might have reached their place of deposition. Different plausible scenarios can be put forward. For example, it is possible that the shells changed hands during interactions between hunter-gatherers of the Vltava-Saale Group and people from one or more regional groups in the western part of the investigated area. For the specimens from the Mainz Basin, it is also conceivable that the shells were collected by the hunter-gatherers of the Vltava-Saale Group themselves, for instance, in the context of a seasonal meeting between several groups in this area. Of importance is the fact that all of these scenarios point to an interaction with hunter-gatherer groups in the western part of the investigated area. The intensity of this interaction was probably much weaker than that among the three western regional groups (see above). A piece of amber, which has since gone missing, was possibly acquired in moraine deposits north of the site. For Bohemia, mollusk shells are reported from the Děravá Cave (ID 515). The specimens are classified as *Arca diluvii* and *Turritella bicarinata* (Fridrich and Sklenář 1976: 26). For shells of *Arca diluvii* occurring at the site of Kamegg (ID 543), the Vienna Basin was identified as the probable source region (Papp 1952). As this is the closest reported origin of this species, it is assumed that the specimens from Děravá Cave also come from this region. Hence, in contrast to the previously discussed patterns, which display more north-south or eastward oriented transportation directions, this example shows a transport vector from southeast to northwest. These shells therefore seem to reflect interactions between hunter-gatherers of the Vltava-Saale Group and the Polish-Moravian Group.

The Polish-Moravian Group

In contrast to all of the other regional groups, the assemblages found in the area of the Polish-Moravian Group yield no specimens of mollusk shells which could not have been obtained within the area itself. The majority of the shells from CEM sites in the Moravian Karst are part of the assemblages from the Pekárna Cave (ID 579). Other sites have yielded only a few specimens. Some species such as *Dentalium badense*, *Turritella* spp., *Ancilla glandiformis*, and *Andara turonica* can be found locally in the Karst region (Hladilová 1999: 131). However, other forms such as *Melanopsis impressa* ssp. or *Mitra substriatula* were probably imported from localities in the Vienna Basin and the Carpathian Foredeep (Ibid.). Local gathering of fossil mollusk shells has been documented for the Maszycka Cave (ID 629) as well. A specimen of *Arca diluvii* found at the site of Děravá Cave in the Brdy area north of the Budweis Basin may originate from the Vienna Basin. It is the only direct evidence that suggests interaction between the hunter-gatherers of the Polish-Moravian Group and the Vltava-Saale Group. The Budweis Basin can be reached from the Vienna Basin by walking upstream along the Kamp river, where the site of Kamegg is located. This site also yielded shells of *Arca diluvii*. Following the valleys of the Kamp and its tributary, the Zwettl, the Lužnice river can be reached after a walk of only about 10 km. Following its course downstream, it leads directly to the Budweis Basin. Amber (Walchowit), which is found at some Moravian sites, occurs locally in the region. Baltic amber, on the other hand, can sometimes be found in the moraine deposits north of the Sudetes and Carpathians, where they probably have been transported by glaciers (Ganzelewski and Slotta 1996). Some shells found at the Maszycka Cave probably come from an outcrop in the region of Korytnica (Kozłowski et al. 1995: 240).

Morphological Observations

The fact that many mollusk shells were probably worn as personal ornaments draws attention not only to the procurement distances, but also to their morphological appearance, which is quite variable. Within the CEM, four larger morphological groups can be distinguished. A first group is constituted by the tubular shells of scaphopod mollusks, usually referred to as *Dentalium* spp. (group 1, Fig. 6.15). Second, all bivalves can be assigned to one unit (group 2, Fig. 6.16). Mollusks with a coiled shell can be divided into those which are shaped as an elongated cone (group 3, Fig. 6.17) and those which have a rather flat, ammonite-like cone (group 4, Fig. 6.18). The diversity of the observed morphological forms raises the question of whether regional differences can be identified with regard to the spatial distribution of these four morphological groups.

The shells of groups 1–3 are distributed throughout the entire investigated area (Figs. 6.15, 6.16 and 6.17). In contrast, shells of group 4 are only found in a rather small region between Lake Neuchâtel and the Blau river in the Swabian Jura, as well as in Andernach-Martinsberg and Gönnersdorf (Fig. 6.18). However, the four distribution patterns do not reveal any reliable signs of a regional preference for certain morphological forms. The restricted occurrence of ammonite-shaped coiled shells coincides spatially with the region containing assemblages with both the highest number and the highest diversity of shells in the whole investigated area. Therefore, this observation seems simply to reflect the relative abundance of shells, rather than selection on the basis of morphological preferences. In contrast to this rather coarse-grained classification, other studies have analyzed the distribution of shells on the level of individual species (Álvarez Fernández 2005, 2009; Bullinger and Thew 2006). Here, it seems that in general, the relative frequency of the occurrence of species in CEM assemblages decreases as the distance to their potential sources increases, an observation which is supported by the presented analysis (see Table A.10). Thus, it seems that the closer a site is located to a source of mollusk shells, the higher is the probability that it contains specimens from this source. It is however remarkable that the sites along the Danube and in the Meuse-Rhine

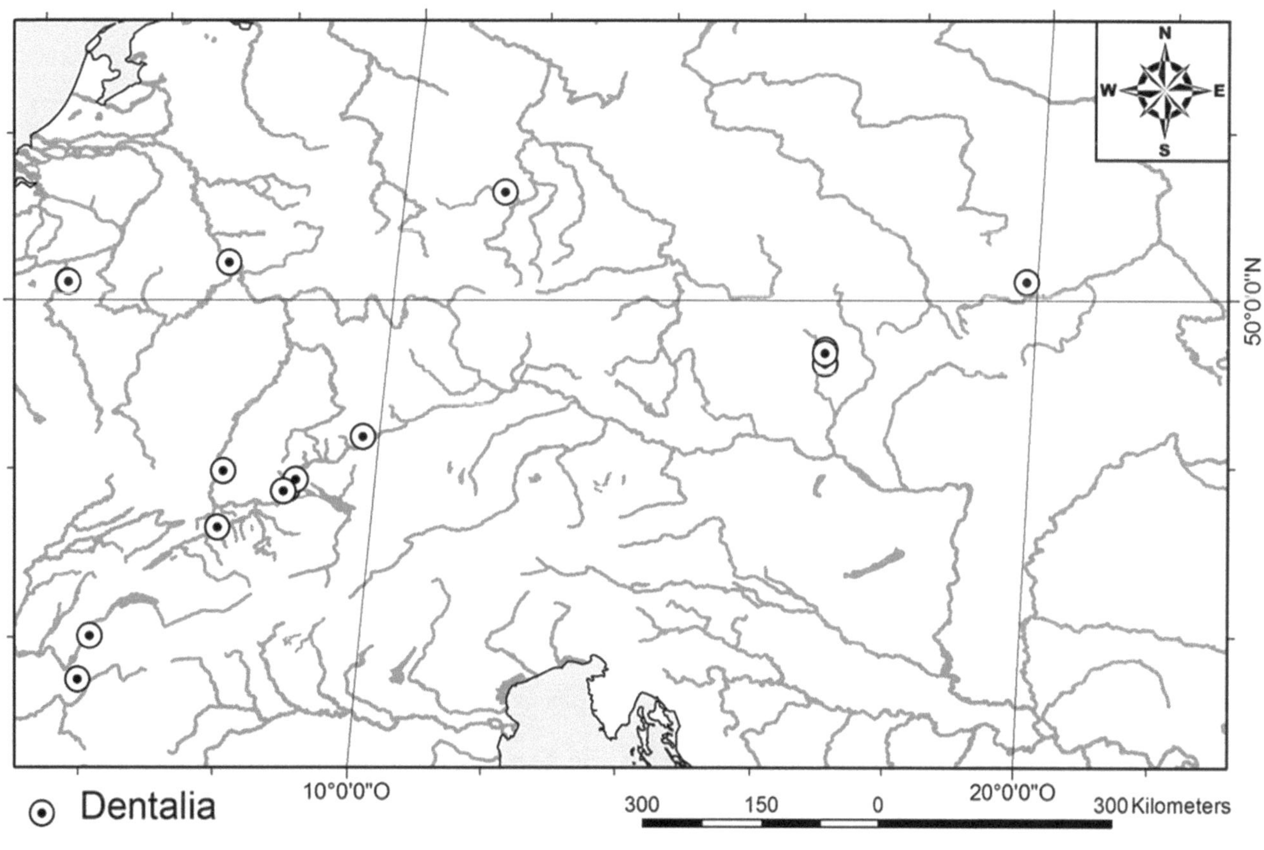

Fig. 6.15 Distribution pattern of Dentalia

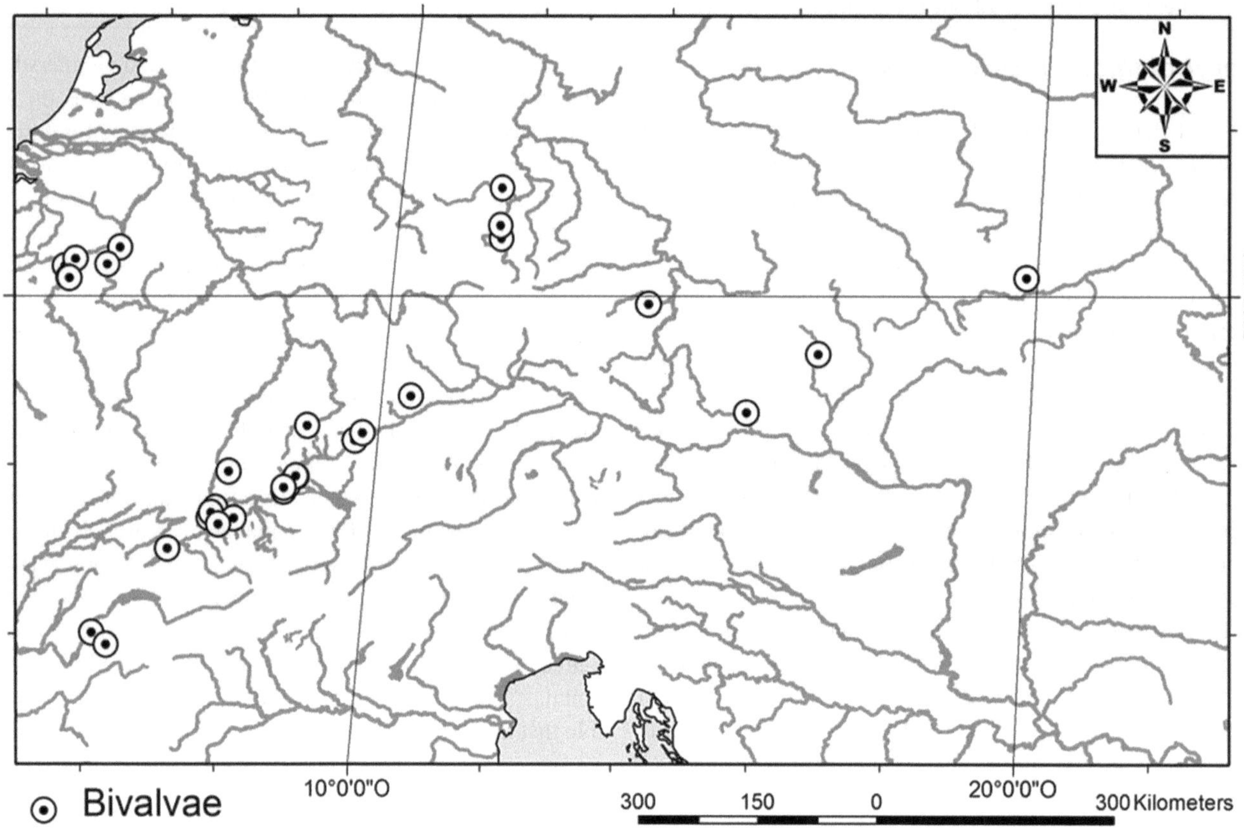

Fig. 6.16 Distribution pattern of Bivalvae

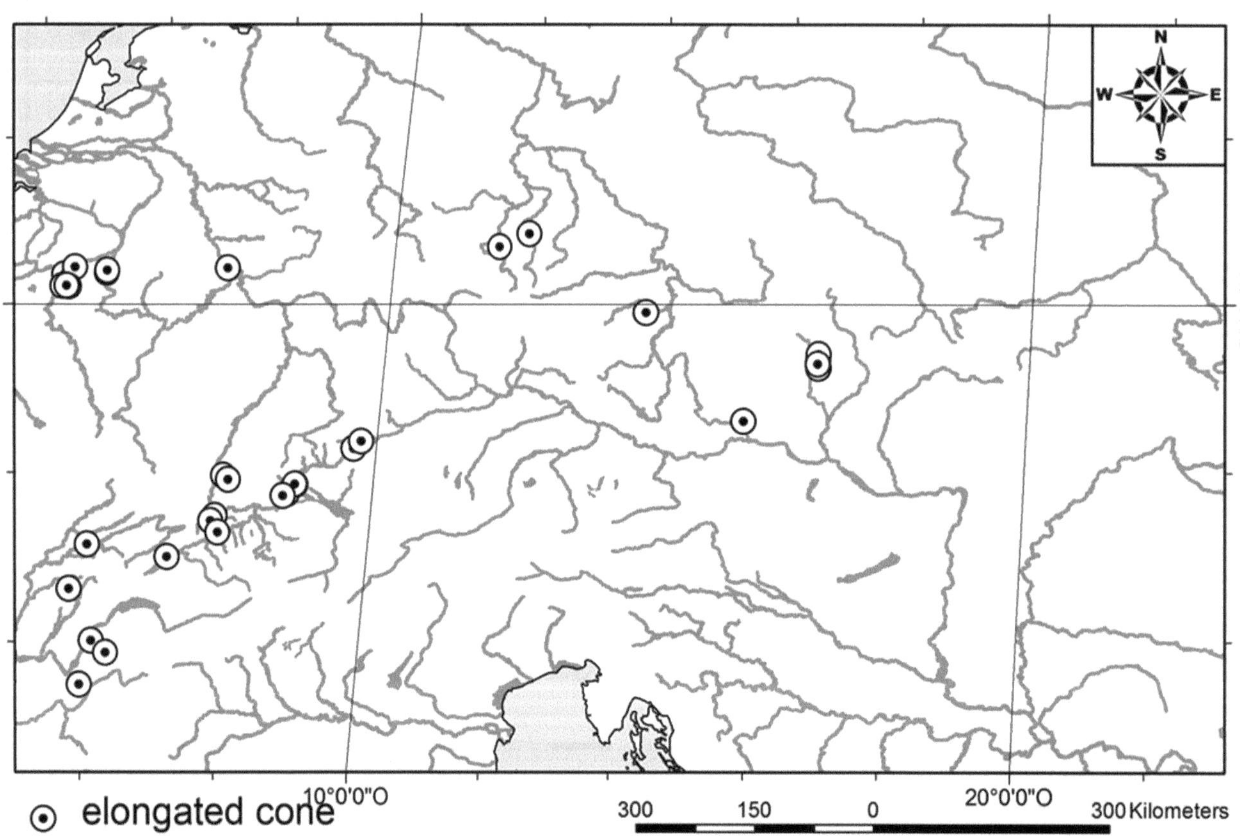

Fig. 6.17 Distribution pattern of shells in the form of elongated cones

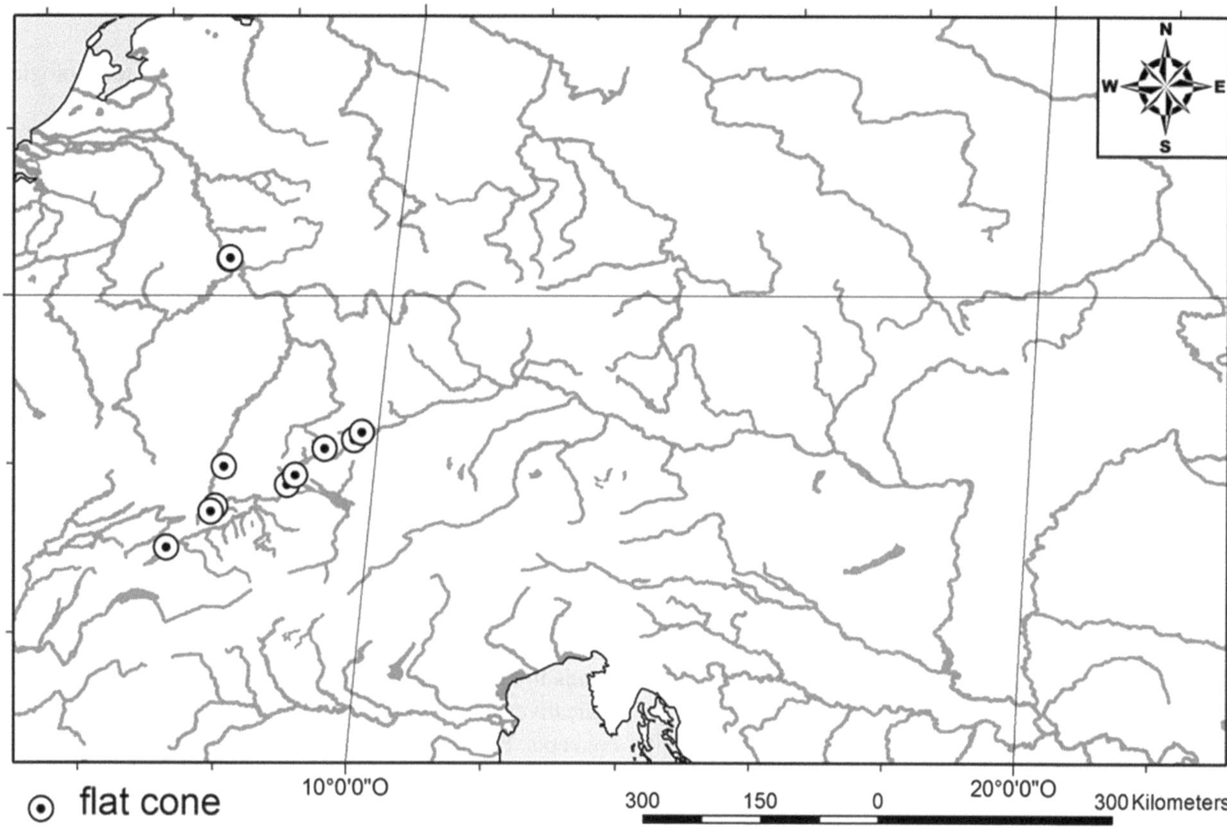

Fig. 6.18 Distribution pattern of shells in the form of flat cones

area contain shells of Mediterranean and Atlantic origin with respective distances of 650, 800, and 950 km, whereas shells from sources in the Vienna Basin located only 550 and 750 km away are completely missing (Fig. 6.14).

In conclusion, it can be stated that the distribution of subrc-cent and fossil mollusk shells complements and to a certain extent completes the impression of the large-scale Magdalenian interaction pattern, which was obtained in the previous chapter by analyzing the structures of the raw material procurement pattern. Against this background, the five regional groups identified above (see section "Procurement Pattern of Lithic Raw Materials") can be aggregated into two supra-regional groups, one in the west and another in the east. The western group comprises the sites in the Jura region, as well as those along the Danube and in the Meuse-Rhine area. Here, a spatially long-span network of intensive interaction connects the three regional groups with each other, as well as with the Paris Basin and the areas adjacent to the west, up to the coasts of the Mediterranean Sea and Atlantic Ocean. Toward the east, however, this tightly meshed web ends abruptly. Only the finds from the sites of Kniegrotte and Gera-Schafgraben indicate weak contacts to the site cluster in Eastern Germany; however, these are not at all comparable with the intensive interaction reconstructed for the western part, and further evidence for an exchange between the western and eastern groups is lacking. The eastern group, in contrast, exhibits a less tightly meshed exchange pattern. This is probably due to the fact that most of the excavated shells are likely to originate in the Moravian Karst, thc Carpathian Forcdccp, and thc Vicnna Basin and could therefore be gathered locally. Long-distance transports may therefore simply not be visible in these finds. Nevertheless, some finds from the Pekárna and Děravá Cave indicate interaction at least between hunter-gatherers in Bohemia, Moravia, and the Vienna Basin.

Typological and Technological Analysis

As demonstrated above, CEM hunter-gatherers did not constitute a homogeneous unit. Clear signals for regional variability, interaction, and boundaries within the CEM can be derived from information on the transport distances of lithic raw materials and mollusk shells indicating regionally distinct groups living in different biomes and maintaining relations of differing intensity with their neighbors. The observed site catchment pattern resulting from raw material acquisition indicates five interaction units, which are divided into two supra-regional groups according to the distribution of shells: the first group comprises the three western interaction units in the Jura region and along the rivers of Meuse, Rhine, and Danube, and the second group comprises the two eastern interaction units in Eastern Germany and Bohemia, as well as Moravia and Poland. An analysis of the typological and technological spectra of the CEM is considered a suitable tool to countercheck this twofold structure of regional and supra-regional groups. This is because they can be expected to vary independently from procurement patterns of raw materials and mollusk shells, unless all of the processes that led to the material manifestation of these different aspects of hunter-gatherer life are governed by the same structure of an underlying interaction network. A continuous pattern in the distribution of typological and technological concepts would point toward continuous flow of information and thus a more or less equally intensive interaction between hunter-gatherers in all regions (see Chap. 3, section "Types as Indicators of Interaction"). Such an observation would differ strongly from the twofold structure described above and a reassessment regarding the social implication would be necessary. If, however, a discontinuous pattern is observed within which the three western groups and the two eastern groups have more typo-technological concepts in common among themselves than with any group of the respective other supra-regional group, the hypothesis of an underlying interaction network that is itself discontinuous and organized according to the recognized regional and supra-regional groups would receive strong support (for further discussions about expected and observed patterns, see Chap. 8, section "The Current Theory and the Major Critiques to Which It Is Subject").

In order to analyze the spatial variability of the typological composition of the assemblages, two analyses are performed. First, the distribution of different lithic tool types according to their respective proportions in the assemblages (Table A.11) is mapped and compared. Second, the attribution of sites to regional groups is reviewed with reference to their typological spectra by performing a Discriminant Function Analysis. The technological variability is investigated by conducting a Correspondence Analyses which focuses on the knapping technique applied in blade and bladelet production (see Chap. 2, section "Blade Technology"). Apart from the question of spatial variability, the following analyses also address the question of chronological differences in the typological and technological spectra of the CEM. Thus, interdependencies between the chronological position of an inventory and its typo-technological composition are also investigated. With regard to the technological aspects, the hypothesis of a temporal development shifting from the usage of organic hammers in the older periods toward an increasing application of soft stone hammers in the younger stage as suggested by B. Valentin, P. Bodu, and others (Valentin 1995, 2000; Bodu and Valentin 1997; Valentin et al. 2004) will be examined.

Finally, the results of the typological and technological analyses are evaluated collectively.

Spatial Distribution of Concepts of Lithic Tools

To analyze similarities and differences in the spatial distribution of lithic tool types (for a definition of types, see Chap. 2, section "Definition of Magdalenian Tool Types"), their proportional occurrence in each assemblage is first mapped as a percentage. The analysis performed here includes all assemblages classified as belonging to the CEM (classes 1–3, see section "Critical Review and Classification of the Recorded Assemblages") and yielding ten or more lithic tools ($n = 248$; Fig. 6.19). First, the tool types and classes are mapped separately, and their distribution patterns are discussed individually. In a second step, the observations thus obtained are compared and evaluated collectively. Here, possible spatial trends in the abundance of certain tokens are understood as indicative of the distribution of the corresponding typological concepts. Finally, differences in these distributions are interpreted as reflecting the range of influence of these ideas and, hence, intensity of interaction.

End Scrapers

End scrapers (see Fig. 2.3), one of the most common tool classes of the Upper Paleolithic, occur everywhere in the investigated area and are present in almost every assemblage (Fig. 6.20). There are no noticeable differences in their abundance between the regional site clusters. Within each cluster, the variability of their occurrence in each assemblage is likely related to site function. Short scrapers, on the other hand, are rather rare among artifacts from the CEM. Their distribution (Fig. 6.21) shows that they occur primarily in the southern part of the investigated area, whereas they are virtually absent in the Meuse-Rhine area. However, where they do appear, they are especially numerous in areas where the number of end scrapers in the assemblages is also generally high. Therefore, it seems that during the CEM, short scrapers were not the expression of a distinct morphological concept that existed as an alternative to end scrapers, but were rather the result of reduction processes due to repeated reworking. Whether such an intensive reduction may reflect an attempt to deal more economically with raw material resources remains a question for further research. When the percentage values for laterally retouched scrapers alone are mapped (Fig. 6.22), the general character of their distribution pattern does not differ significantly from that of scrapers overall.

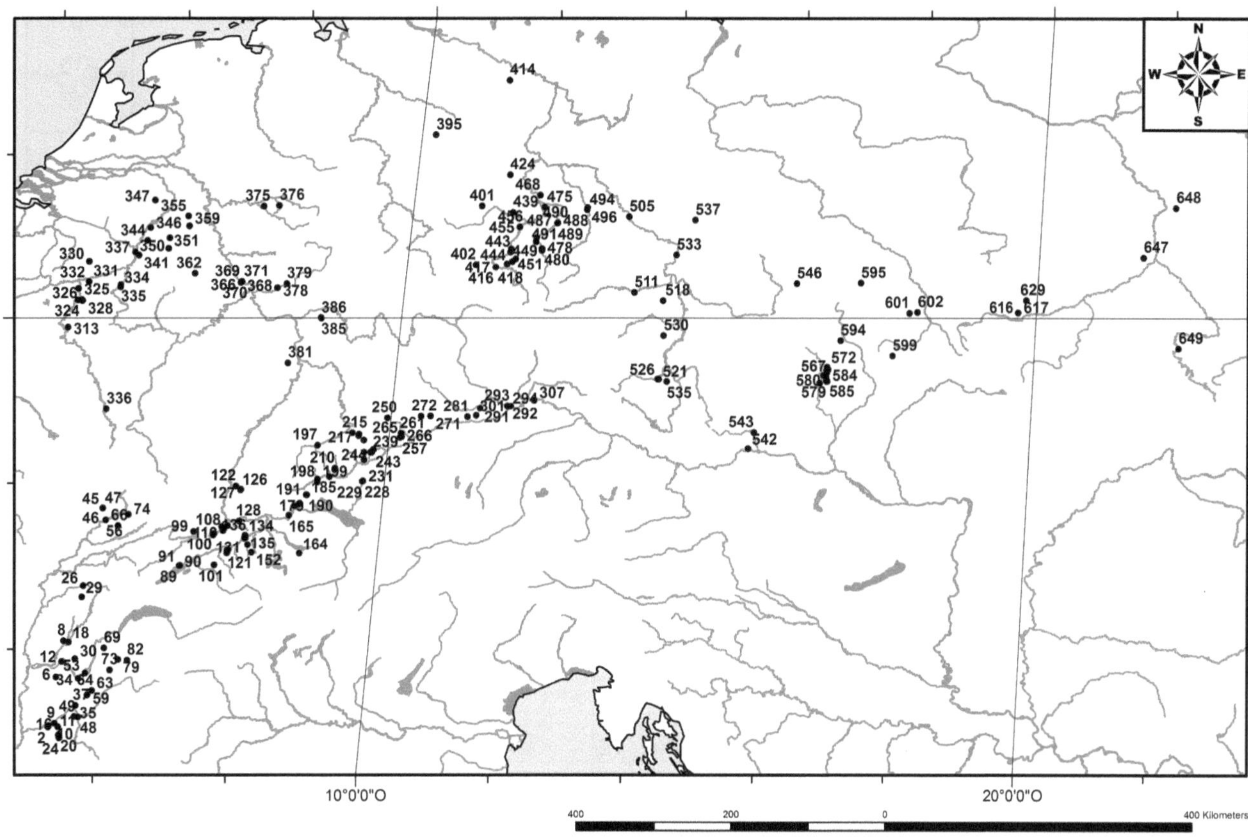

Fig. 6.19 Distribution of assemblages included in the typological analysis of lithic tools (see Table A.11)

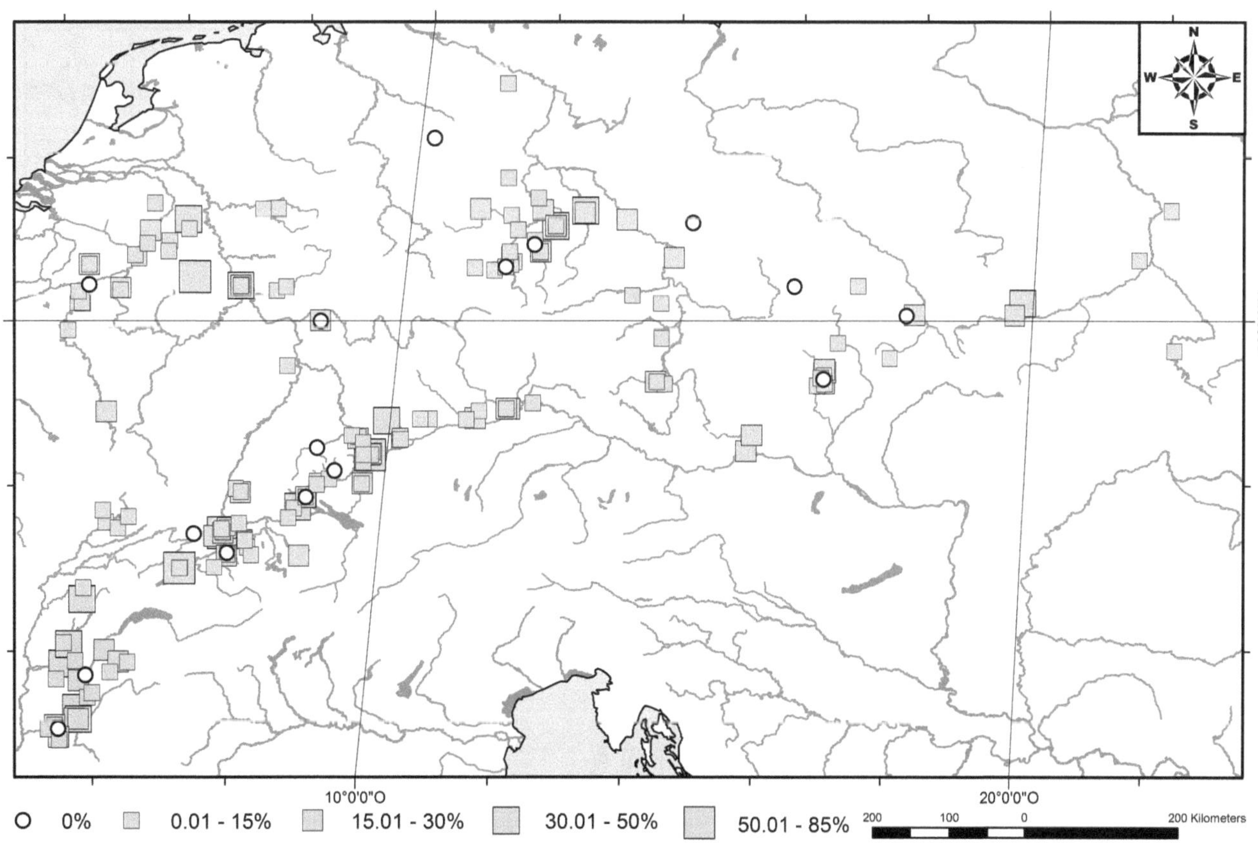

Fig. 6.20 Distribution of end scrapers in Central European Magdalenian assemblages (see Table A.11)

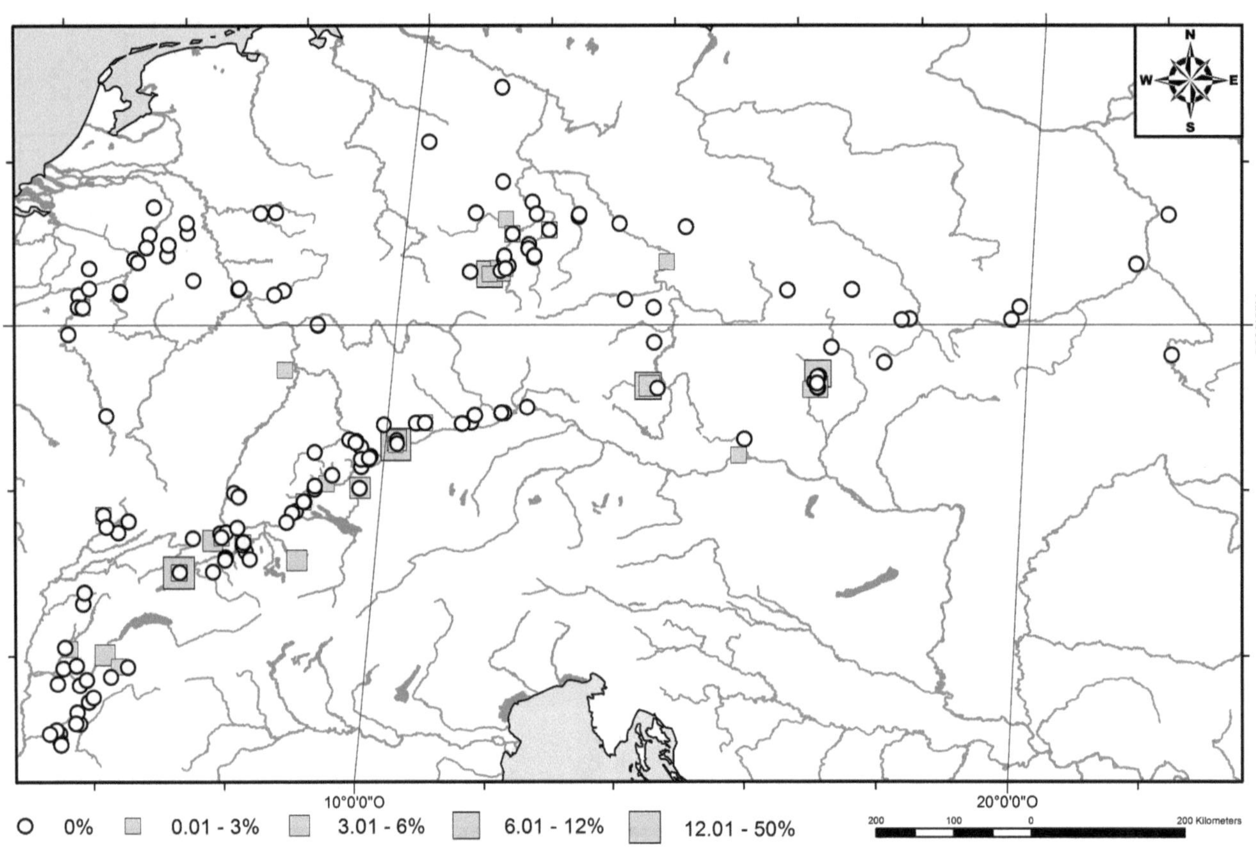

Fig. 6.21 Distribution of short scrapers in Central European Magdalenian assemblages (see Table A.11)

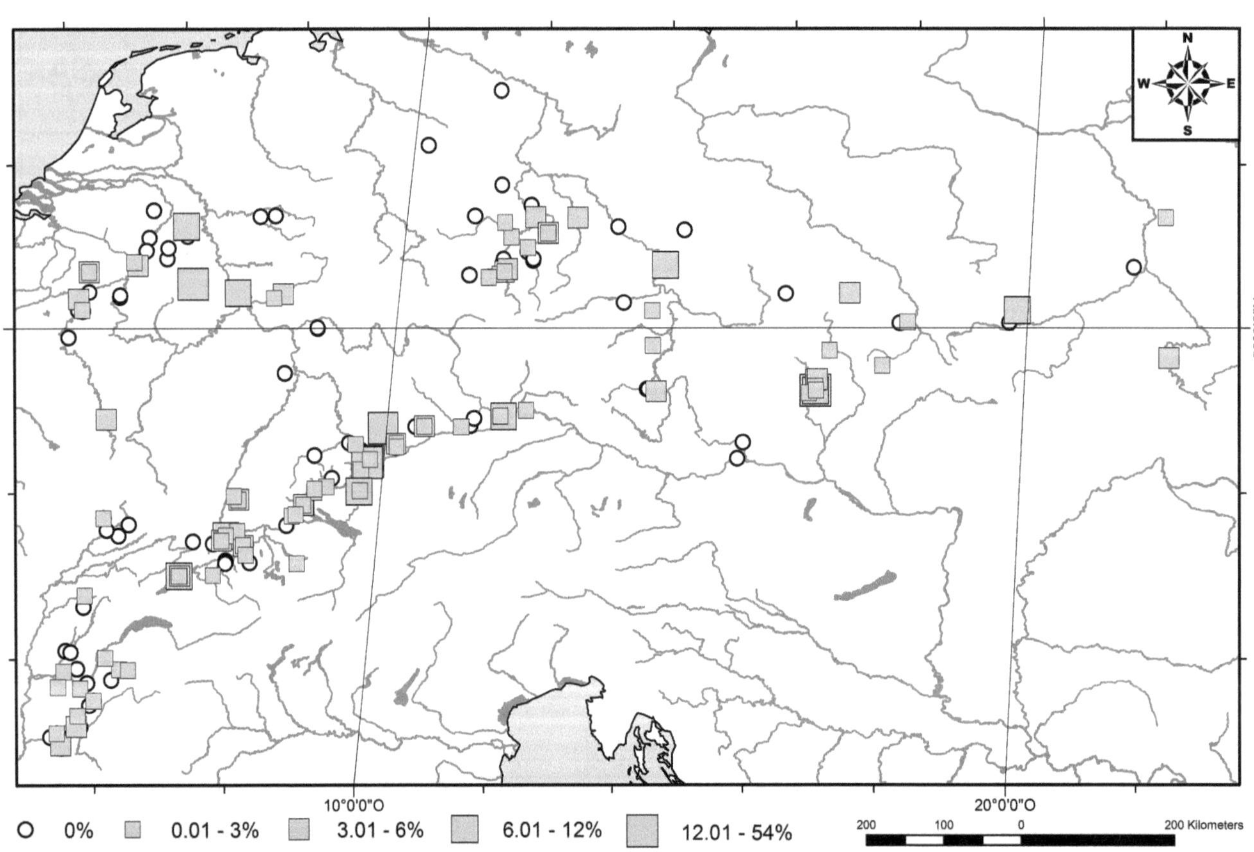

Fig. 6.22 Distribution of laterally retouched scrapers in Central European Magdalenian assemblages (see Table A.11)

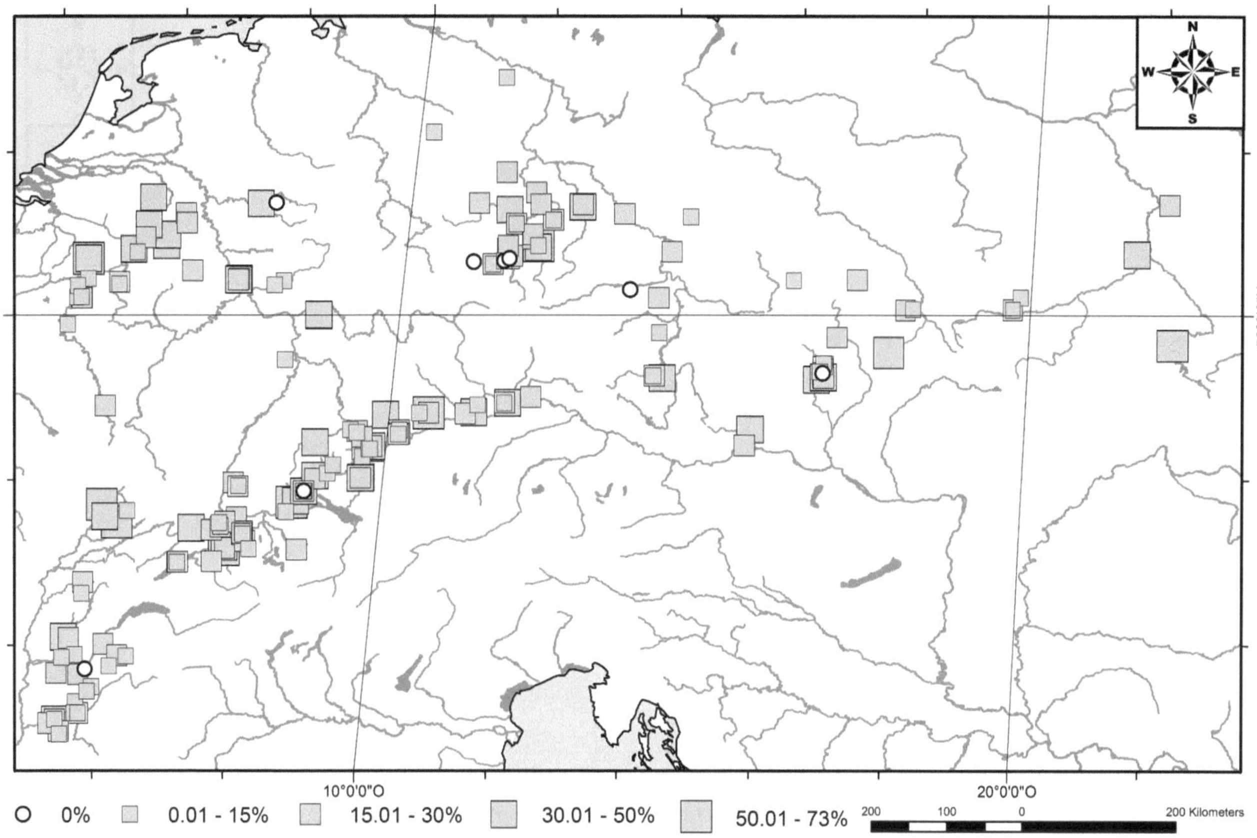

Fig. 6.23 Distribution of burins in Central European Magdalenian assemblages (see Table A.11)

Laterally retouched scrapers can be found in all site clusters, and the main difference to Fig. 6.20 seems simply to be a decreased frequency, which affects the whole investigated area rather evenly.

Burins

Burins (see Fig. 2.3) are one of the most characteristic tool classes of the Magdalenian and have a wide morphological range. When all specimens of this class are analyzed collectively (Fig. 6.23), their distribution emerges as very similar to that obtained for scrapers. Burins occur everywhere in the CEM in comparatively high numbers, and it is very rare that they are completely missing from an inventory. Differences in their numbers between sites are probably a result of site function. Long burins (or Lacan burins) are a very distinctive type of burins. With their concave retouch and elongated form, they are easy to recognize. Generally, they are agreed to be especially numerous in assemblages of the middle and late Magdalenian (Hahn 1991: 185). In contrast to the overall distribution of burins, their occurrence pattern exhibits a strong spatial trend (Fig. 6.24): whereas they are virtually absent in the southwest of the investigated area, their numbers increase toward the northeast, and they are most abundant in the sites of the Meuse-Rhine cluster. This kind of patterning cannot be explained with differences in site function, but appears to reflect the distribution of a distinct morphological concept. Another kind of burin, which is thought to have spatial and temporal significance within the Magdalenian, is the flat-faced burin (Hanitzsch 1972: 109). According to the literature, these are generally a scarce feature in CEM assemblages and exceed 3 % in an assemblage only rarely (Fig. 6.25). They occur in all interaction units but appear to be especially numerous in the Meuse-Rhine area and along the Danube. In the Jura region, in contrast, they are conspicuously underrepresented.

Borers

The overall distribution pattern of borers strongly resembles that of scrapers and burins (Fig. 6.26). Borers are a common feature in CEM assemblages and occur quite frequently throughout the whole investigated area. Differences in the percentage values within the regional site clusters are probably best explained in terms of site function. Within the class of borers, it is possible to distinguish a special type having an

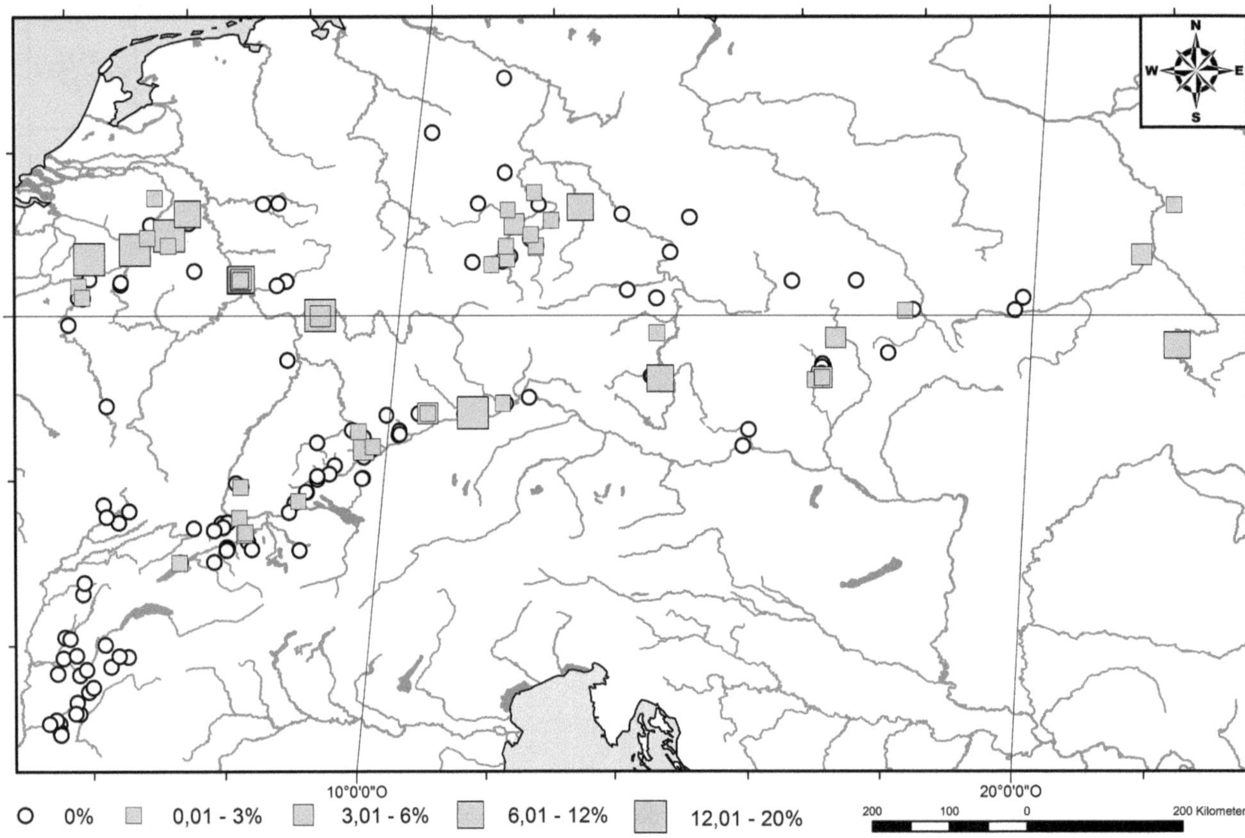

Fig. 6.24 Distribution of long burins in Central European Magdalenian assemblages (see Table A.11)

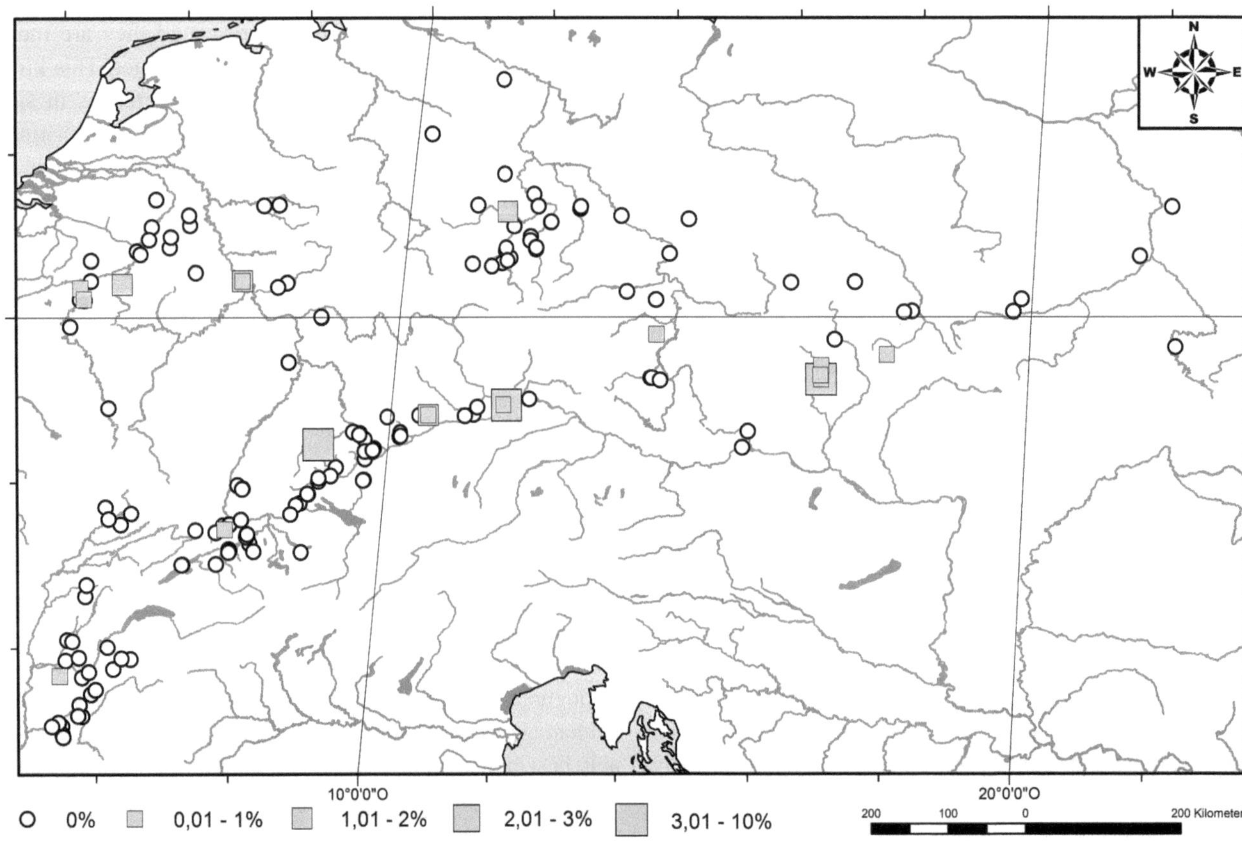

Fig. 6.25 Distribution of flat-faced burins in Central European Magdalenian assemblages (see Table A.11)

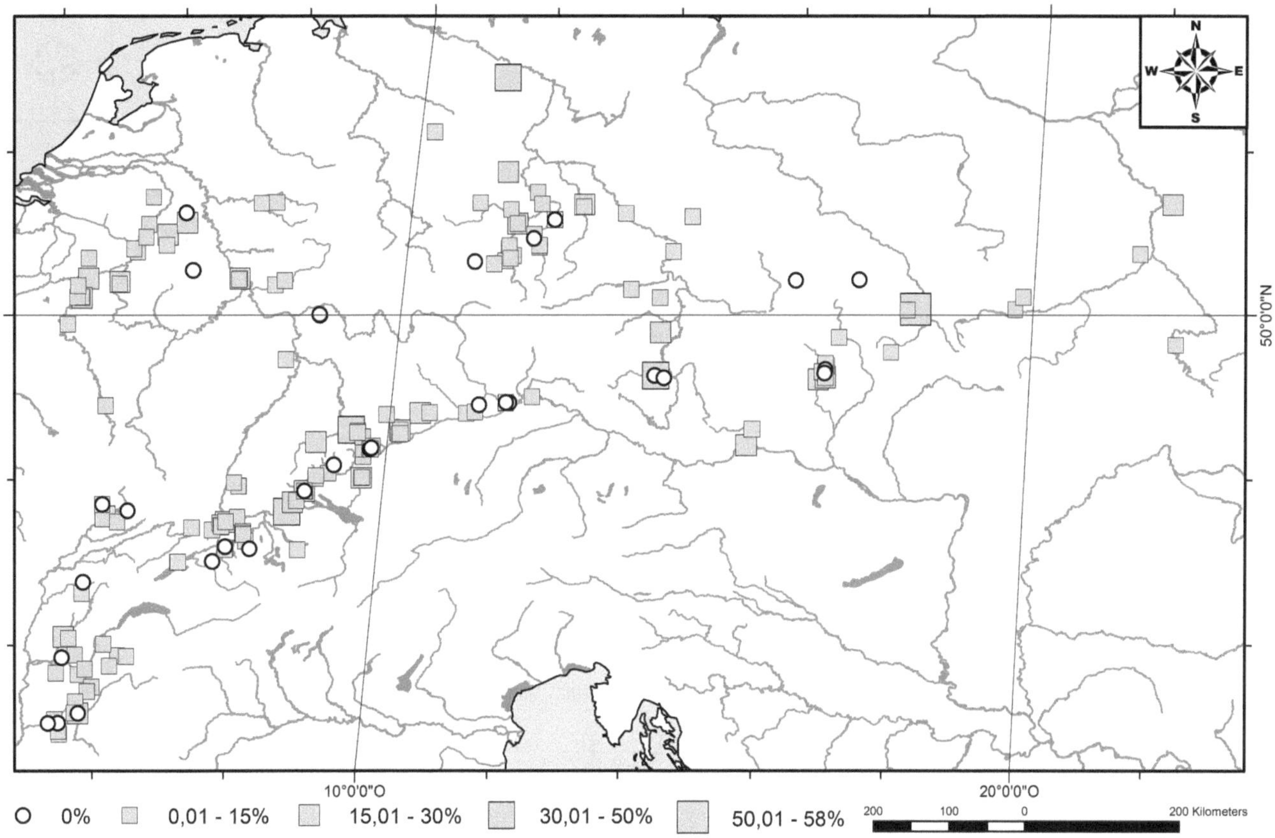

Fig. 6.26 Distribution of borers in Central European Magdalenian assemblages (see Table A.11)

elongated tip. Because these rather delicate tools are prone to breakage and probably gradually shortened by reworking during their period of usage, their true numbers may be generally underrepresented in the archaeological record. However, as this applies for the whole investigated area, it should not bias the overall picture of their spatial distribution (Fig. 6.27). In the area of France, Switzerland, and along the Danube, long borers are mostly absent or occur only in comparatively small numbers, with the exception of the site of Hohlenstein bei Ederheim (ID 269), where they make up 18 % of the assemblage (6 out of 33 tools) (Narr 1965). However, the early date of the excavation (1910–1912) may raise doubt about the completeness of the assemblage. Toward the north and east, in contrast, this proportion increases. The highest numbers of long borers can be observed among the assemblages in Eastern Germany.

Zinken (Beaked Borers)

Likely serving a similar purpose to borers, *zinken* (or beaked borers) seem to be functionally related to this tool class (see Chap. 2, section "Definition of Magdalenian Tool Types"; Fig. 2.3). They are most numerous at Hamburgian sites, but also occur in many assemblages of the CEM. The distribution pattern of *zinken* exhibits a clear spatial trend with increasing frequencies from the southwest to the northeast (Fig. 6.28). While they are virtually absent southwest of the Rhine, they can already be found at some sites between the Black Forest and the Franconian Alb. In the Meuse-Rhine area, Bohemia, Moravia, and Poland, *zinken* can be found at several sites. Their main area of distribution within the CEM, however, is located in Eastern Germany, in the drainage area of the Saale and Mulde rivers.

Backed Bladelets

These small tools (see Fig. 2.4), which are a characteristic feature of Magdalenian artifact assemblages, were probably used as implements and hafted laterally on organic projectile points in a parallel or oblique way (e.g., Allain 1979: 100ff.; Leroi-Gourhan 1983). Because of their small size, they are probably underrepresented in collections from older excavations, especially those of caves and rock shelters. As this problem holds true for the whole investigated are, it should not bias their spatial distribution toward certain regions; nevertheless, it should be kept in mind. The overall distribution

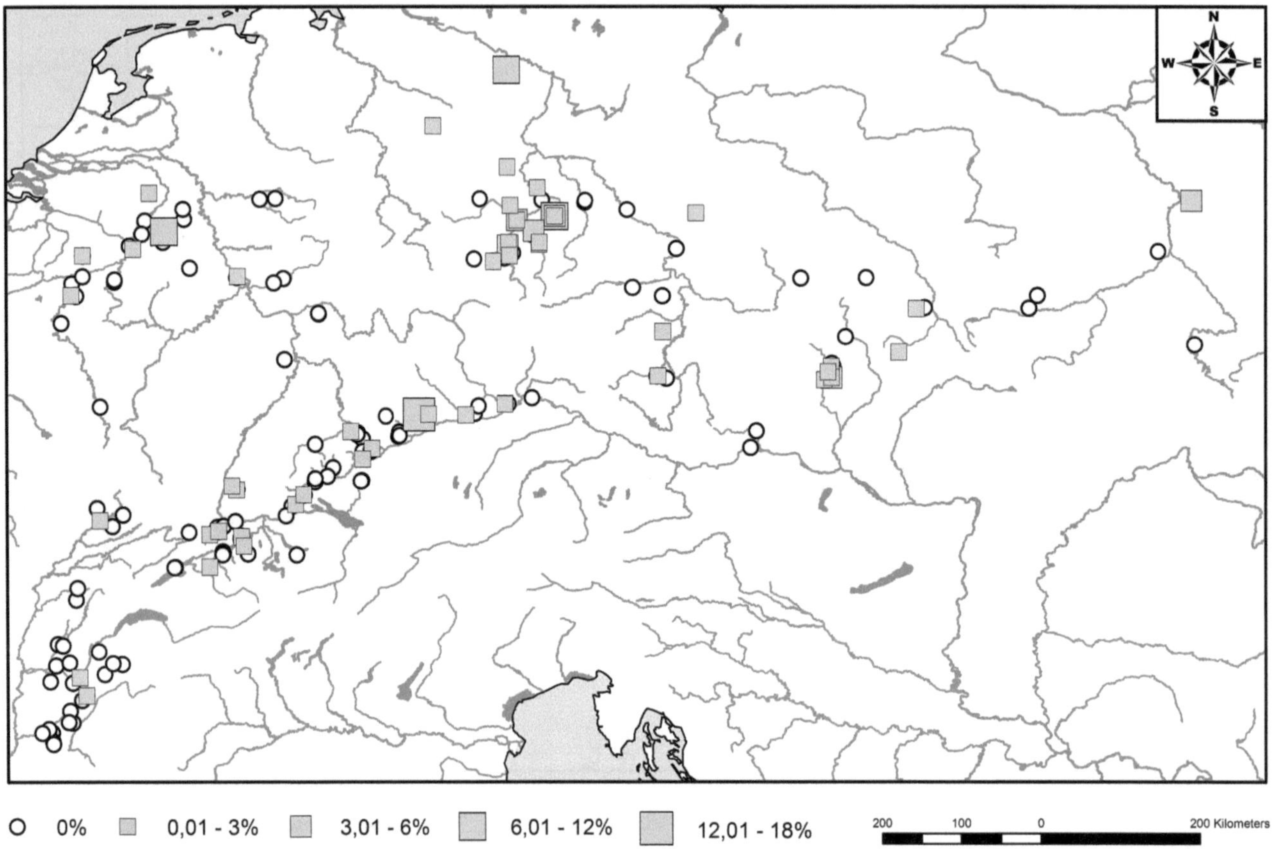

Fig. 6.27 Distribution of long borers in Central European Magdalenian assemblages (see Table A.11)

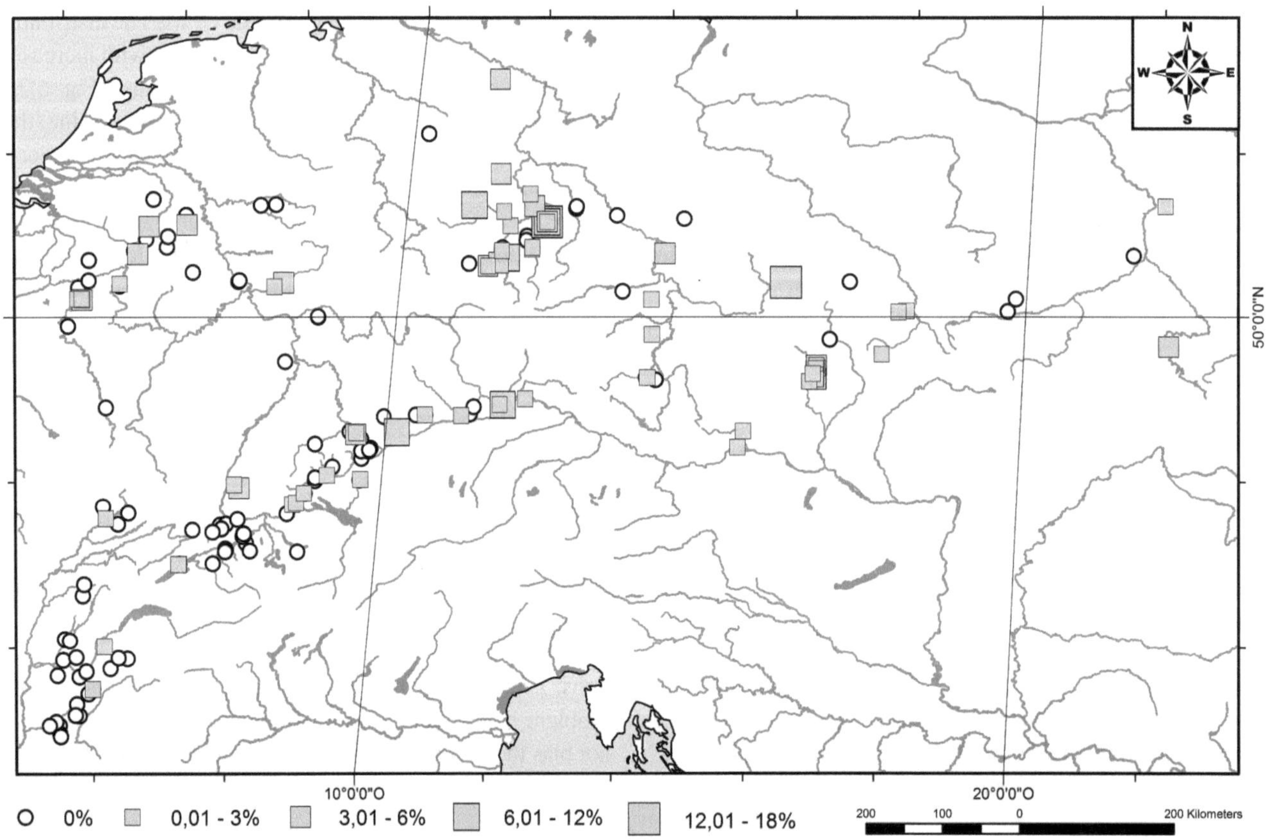

Fig. 6.28 Distribution of *zinken* in Central European Magdalenian assemblages (see Table A.11)

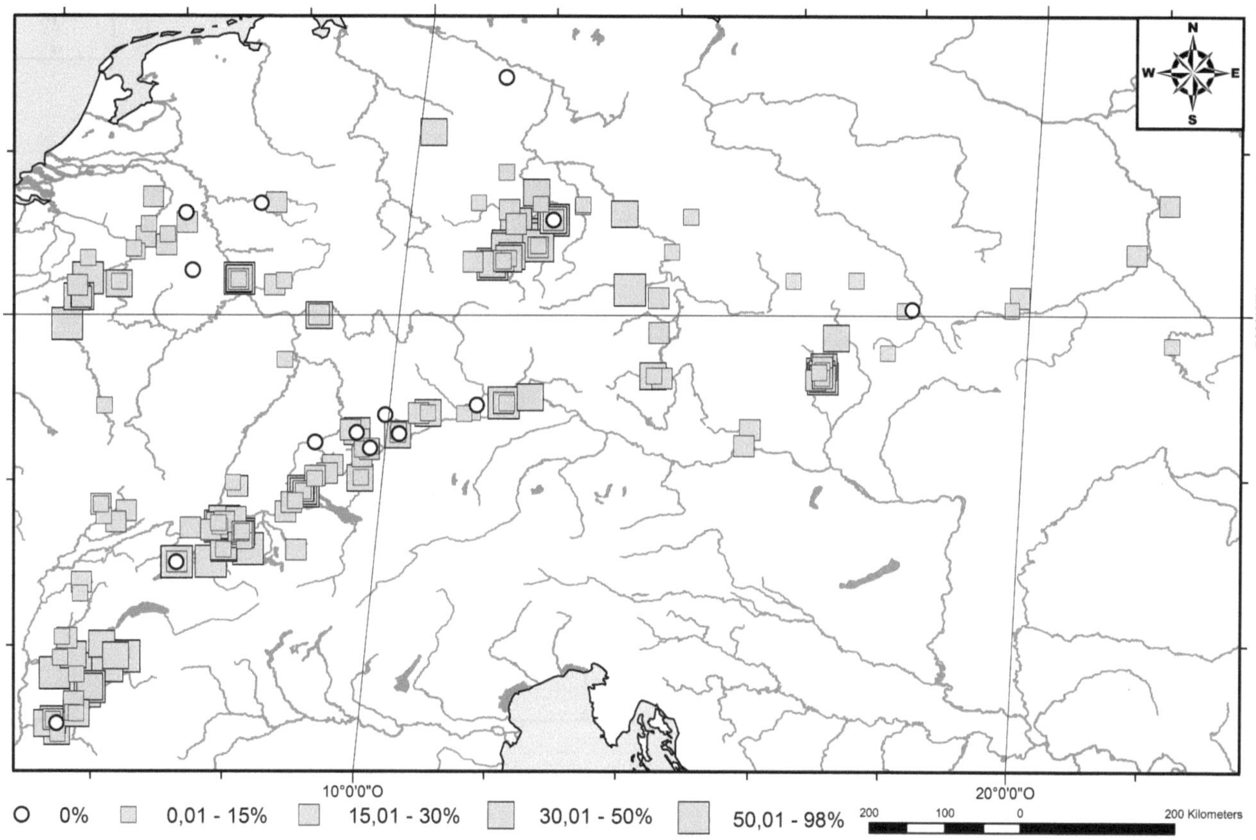

Fig. 6.29 Distribution of backed bladelets in Central European Magdalenian assemblages (see Table A.11)

of backed bladelets (Fig. 6.29) coincides more or less with those of scrapers, burins, and borers, although their abundance in the Danube area is comparatively low. Because of the strong connection between backed bladelets and hunting activities, the differences within the regional clusters are probably the result of site function. Among backed bladelets, a number of types can be distinguished according to the modification of their edges (see Chap. 2, section "Definition of Magdalenian Tool Types"). By far the most numerous type is the simple backed bladelet with a continuous retouch on one longitudinal edge. Its distribution matches the overall distribution (Fig. 6.30). Truncated backed bladelets, double backed bladelets, and double backed truncated bladelets occur in the whole investigated area (Figs. 6.31, 6.32 and 6.33). They seem to be most abundant in areas where the number of backed bladelets is generally high. Thus, it seems that their appearance is simply an expression of morphological variability, rather than an indication of a different morphological concept. The same holds true for rectangular, pointed, and bitruncated backed bladelets, which generally appear rather rarely (Figs. 6.34, 6.35 and 6.36). Denticulated backed bladelets, however, are mainly distributed in the southwestern part of the investigated area (Fig. 6.37). From there, their proportion in the assemblages decreases toward the northeast, where they occur in a small number of collections in Eastern Germany, Bohemia, and Moravia. In the Meuse-Rhine area, they are virtually absent. Thus, this morphological variety seems to be the only one with a certain significance in terms of the distribution of typological concepts.

Truncated Pieces

Truncated pieces (see Fig. 2.4) do not show any particular spatial trends (Fig. 6.38). They are ubiquitous, and their distribution pattern resembles those observed for scrapers, burins, borers, and backed bladelets.

Laterally Retouched Pieces

This tool class comprises all blades and flakes with at least one lateral retouch (see Fig. 2.4). Their distribution pattern is quite similar to that one of scrapers, burins, borers, backed bladelets, and truncated pieces and does not reveal any clear spatial trend or clustering (Fig. 6.39).

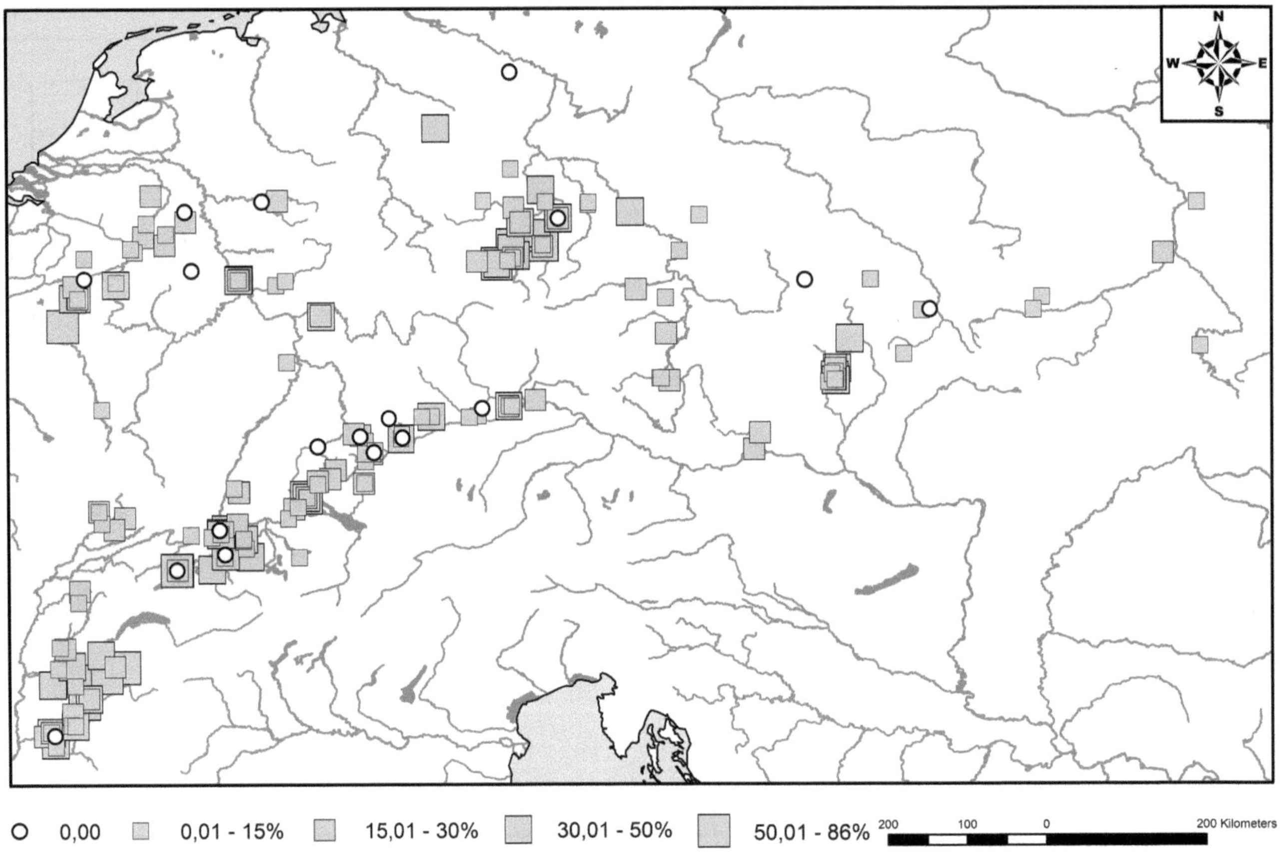

Fig. 6.30 Distribution of simple backed bladelets in Central European Magdalenian assemblages (see Table A.11)

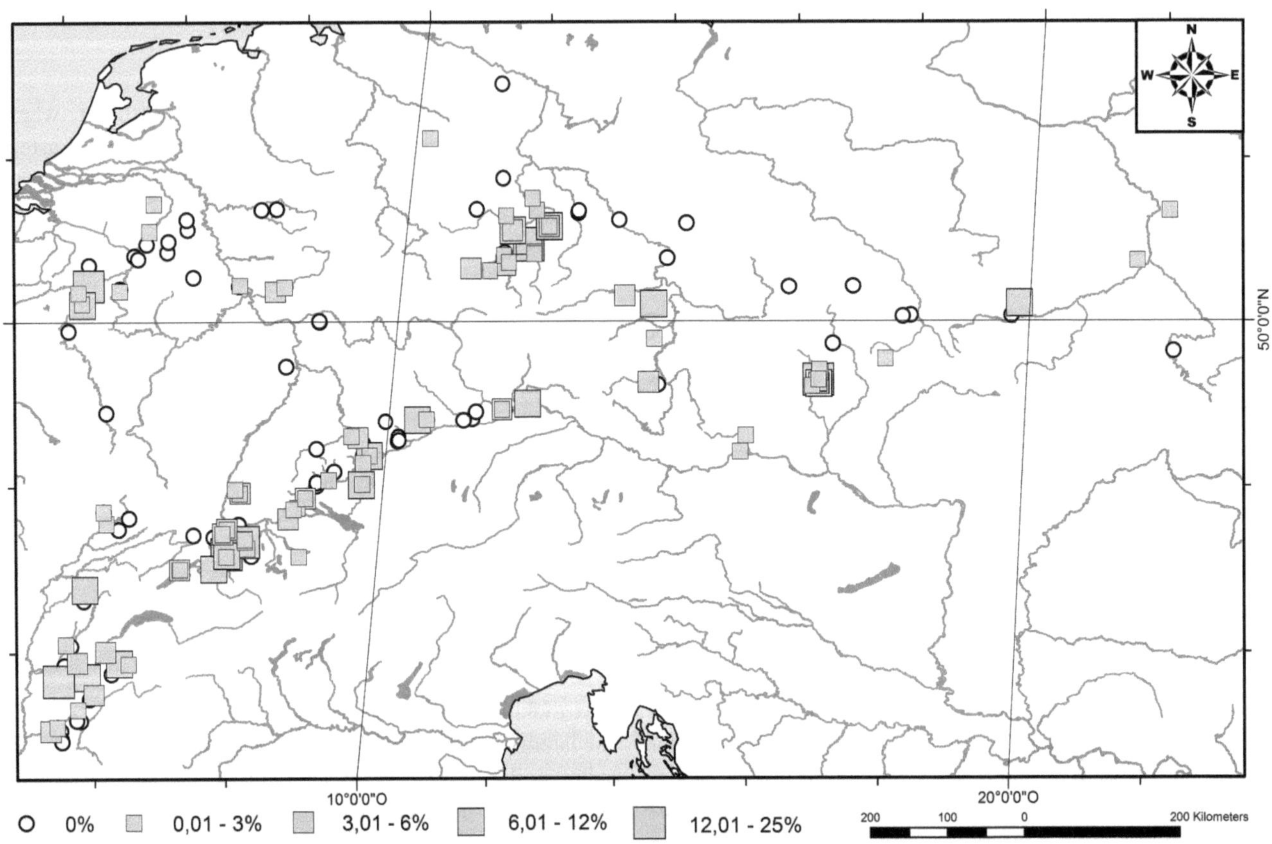

Fig. 6.31 Distribution of truncated backed bladelets in Central European Magdalenian assemblages (see Table A.11)

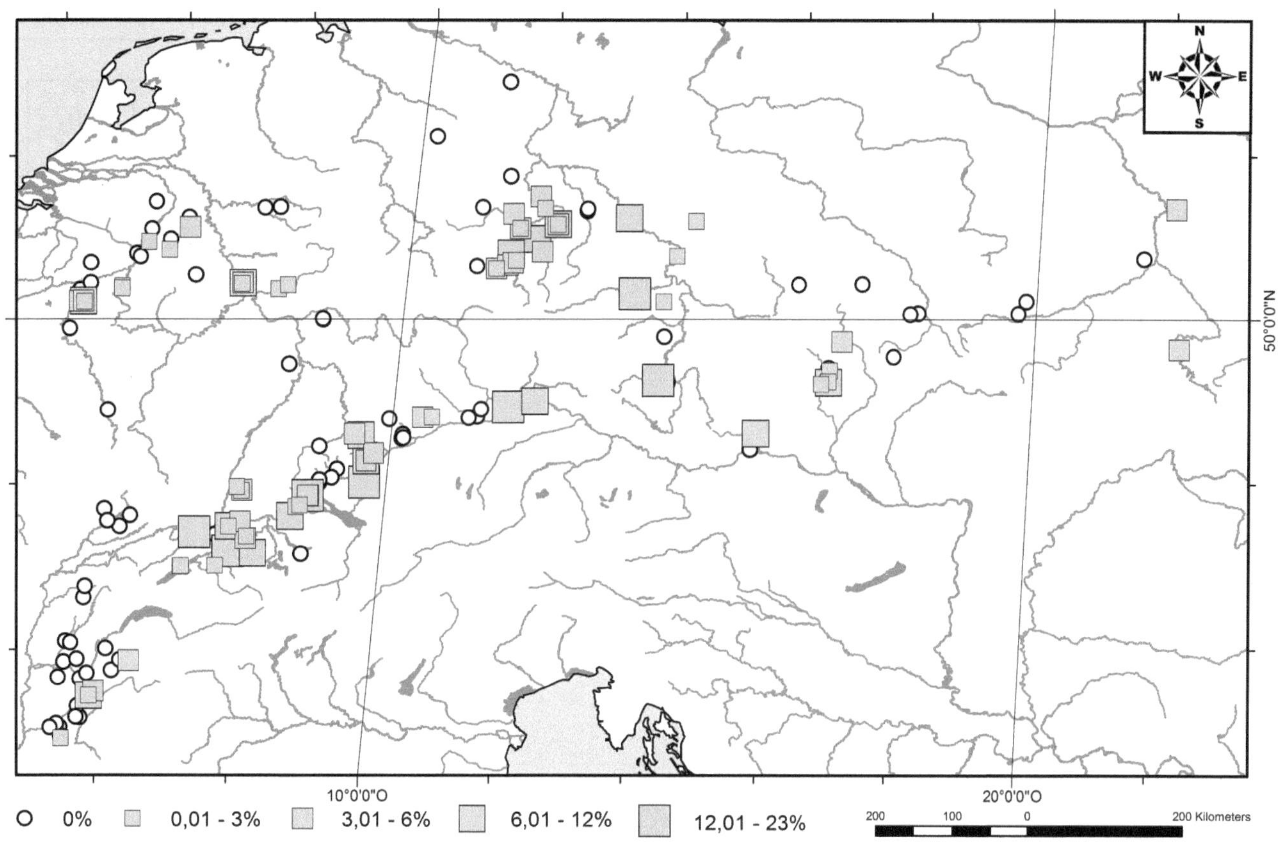

Fig. 6.32 Distribution of double backed bladelets in Central European Magdalenian assemblages (see Table A.11)

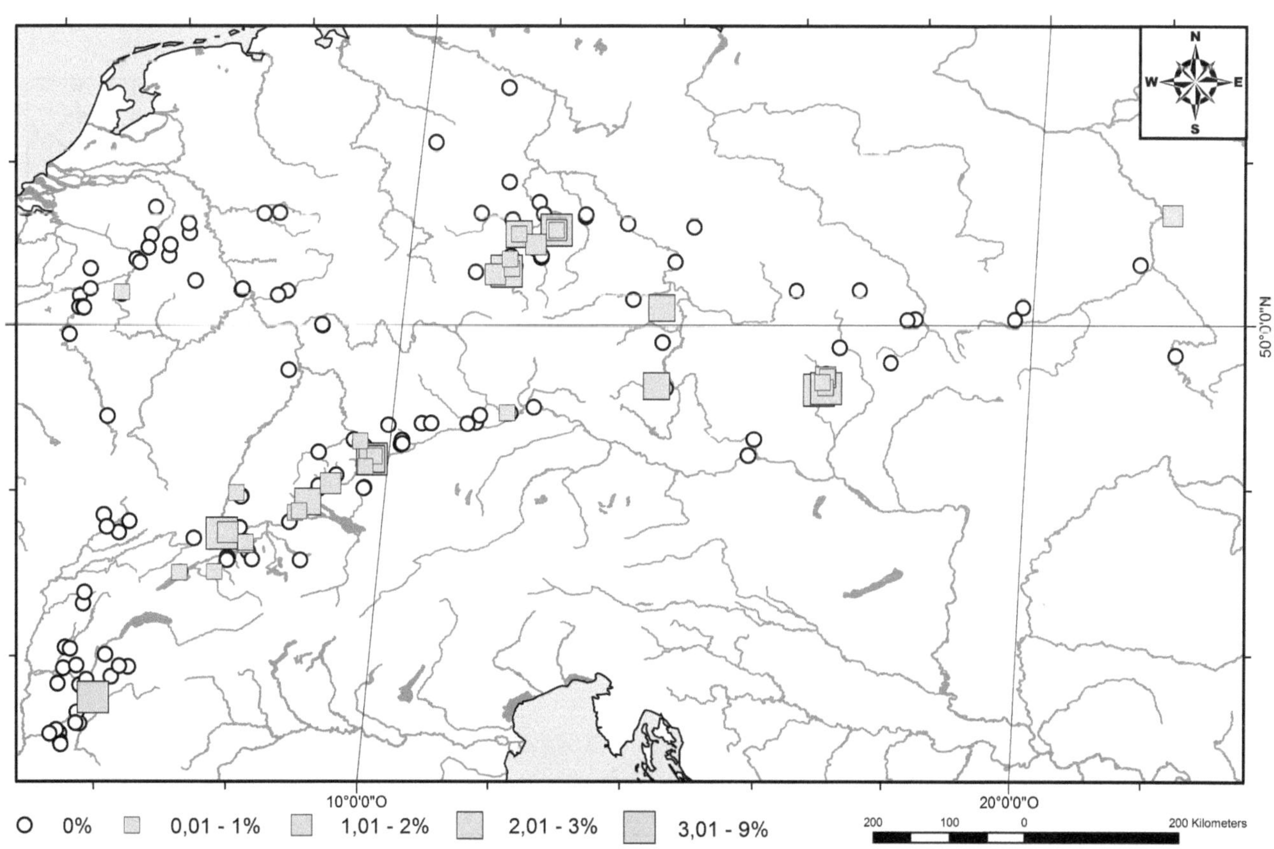

Fig. 6.33 Distribution of double backed truncated bladelets in Central European Magdalenian assemblages (see Table A.11)

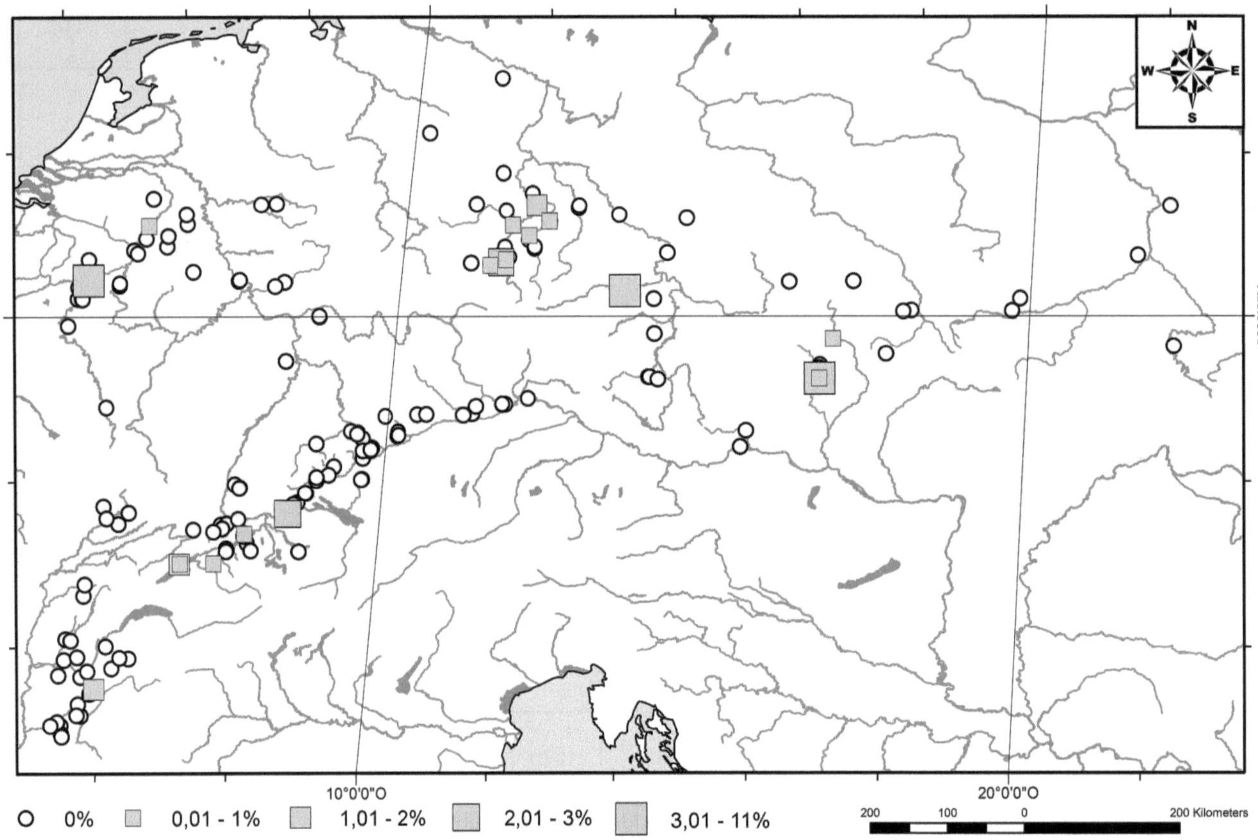

Fig. 6.34 Distribution of rectangular backed bladelets in Central European Magdalenian assemblages (see Table A.11)

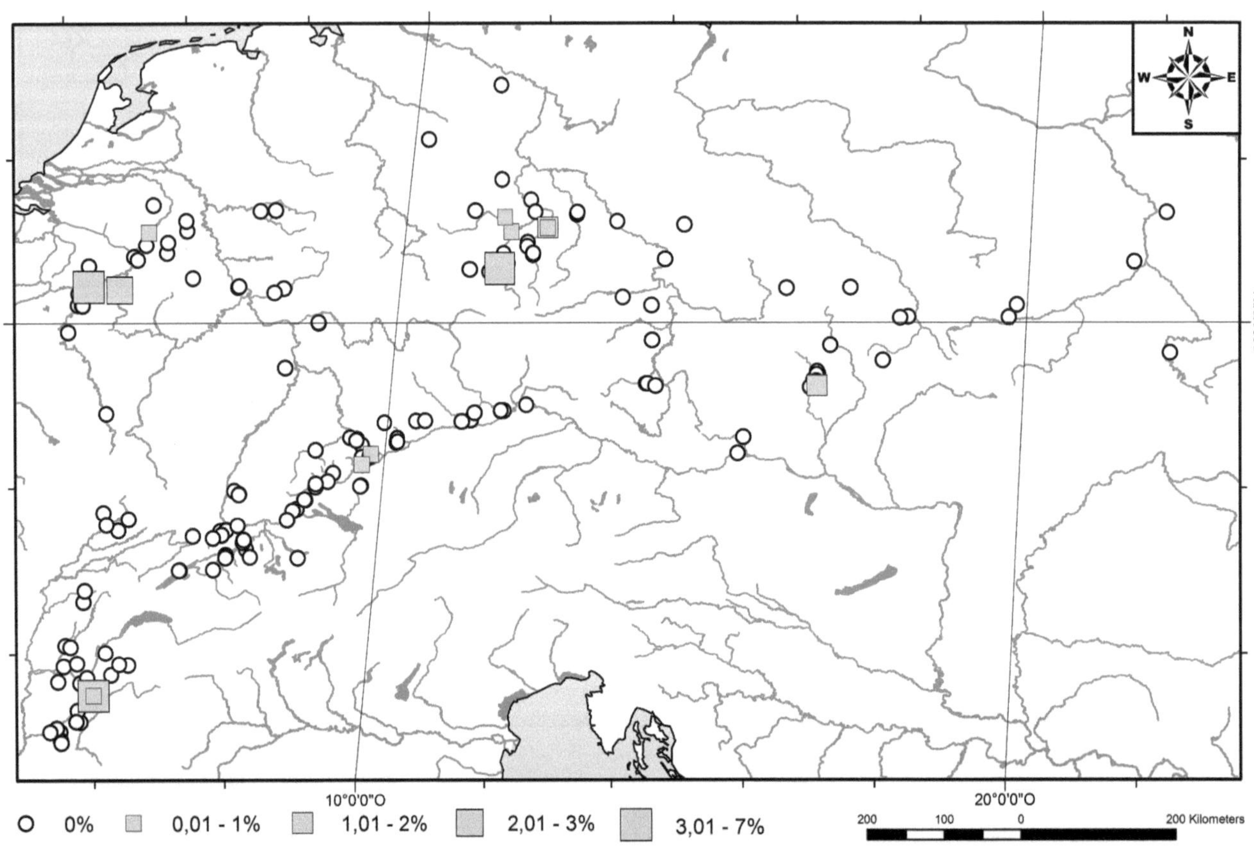

Fig. 6.35 Distribution of pointed backed bladelets in Central European Magdalenian assemblages (see Table A.11)

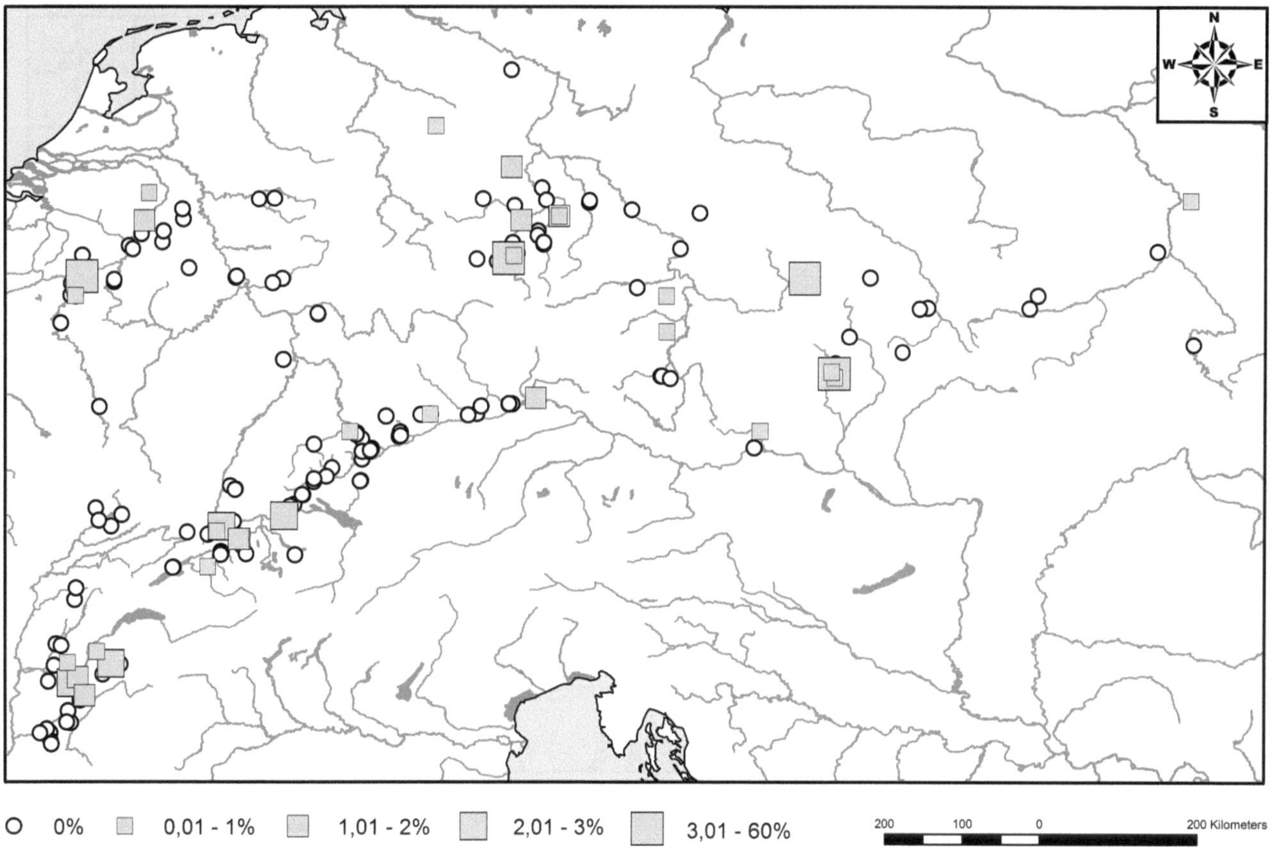

Fig. 6.36 Distribution of bitruncated backed bladelets in Central European Magdalenian assemblages (see Table A.11)

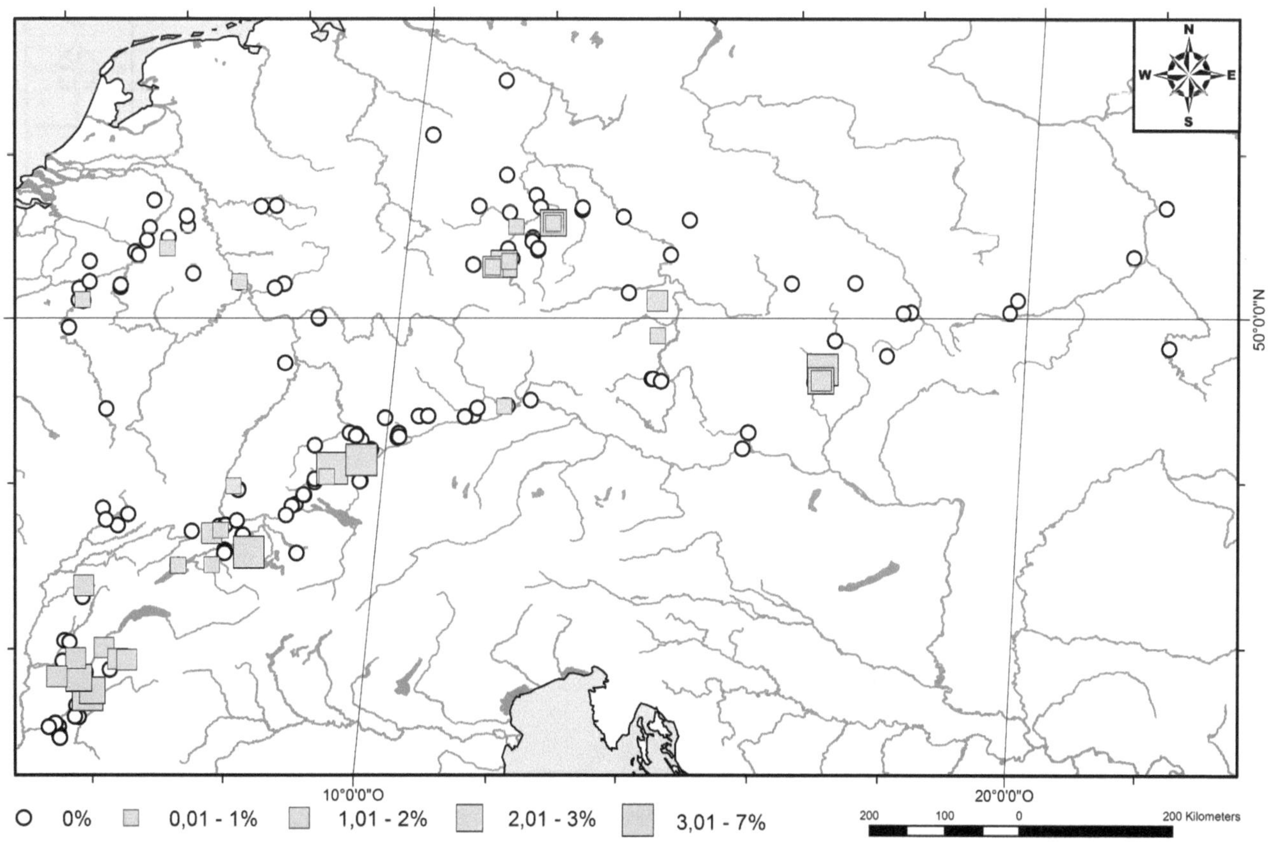

Fig. 6.37 Distribution of denticulated backed bladelets in Central European Magdalenian assemblages (see Table A.11)

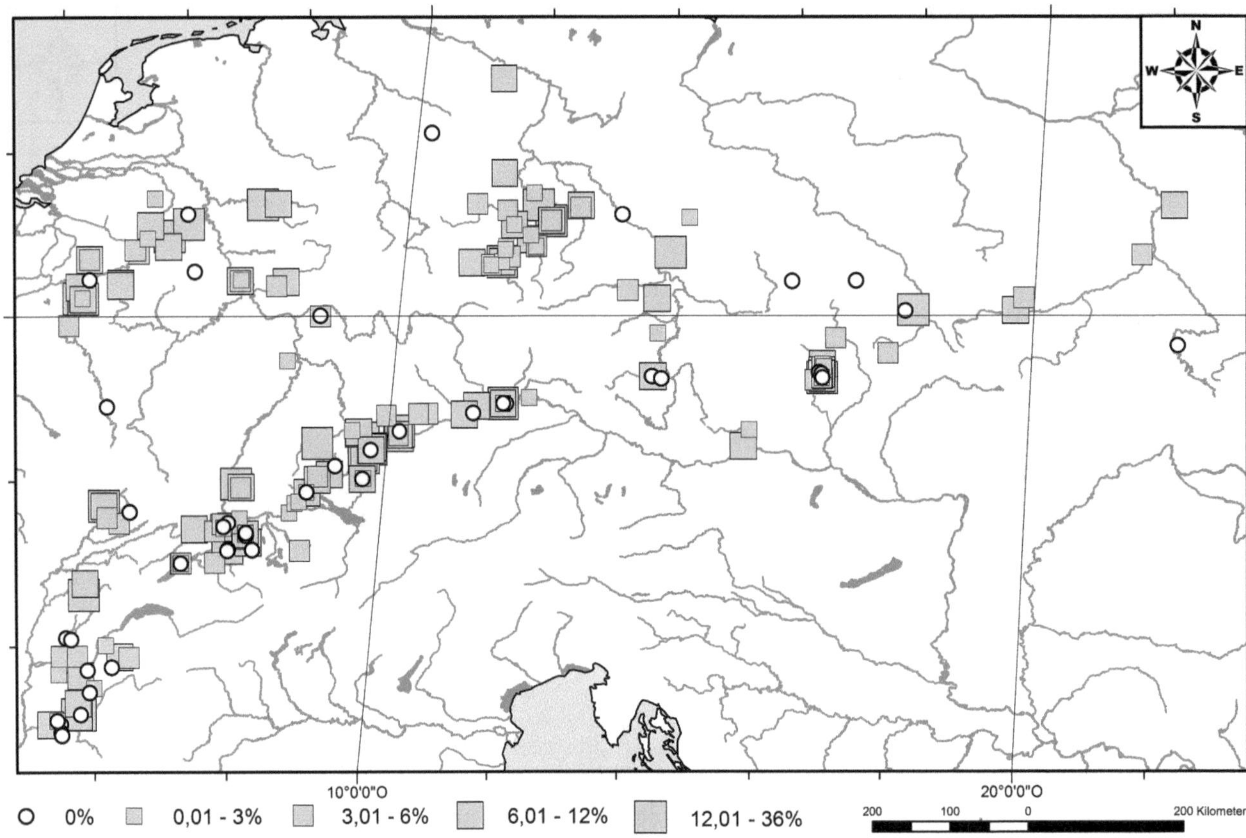

Fig. 6.38 Distribution of truncated pieces in Central European Magdalenian assemblages (see Table A.11)

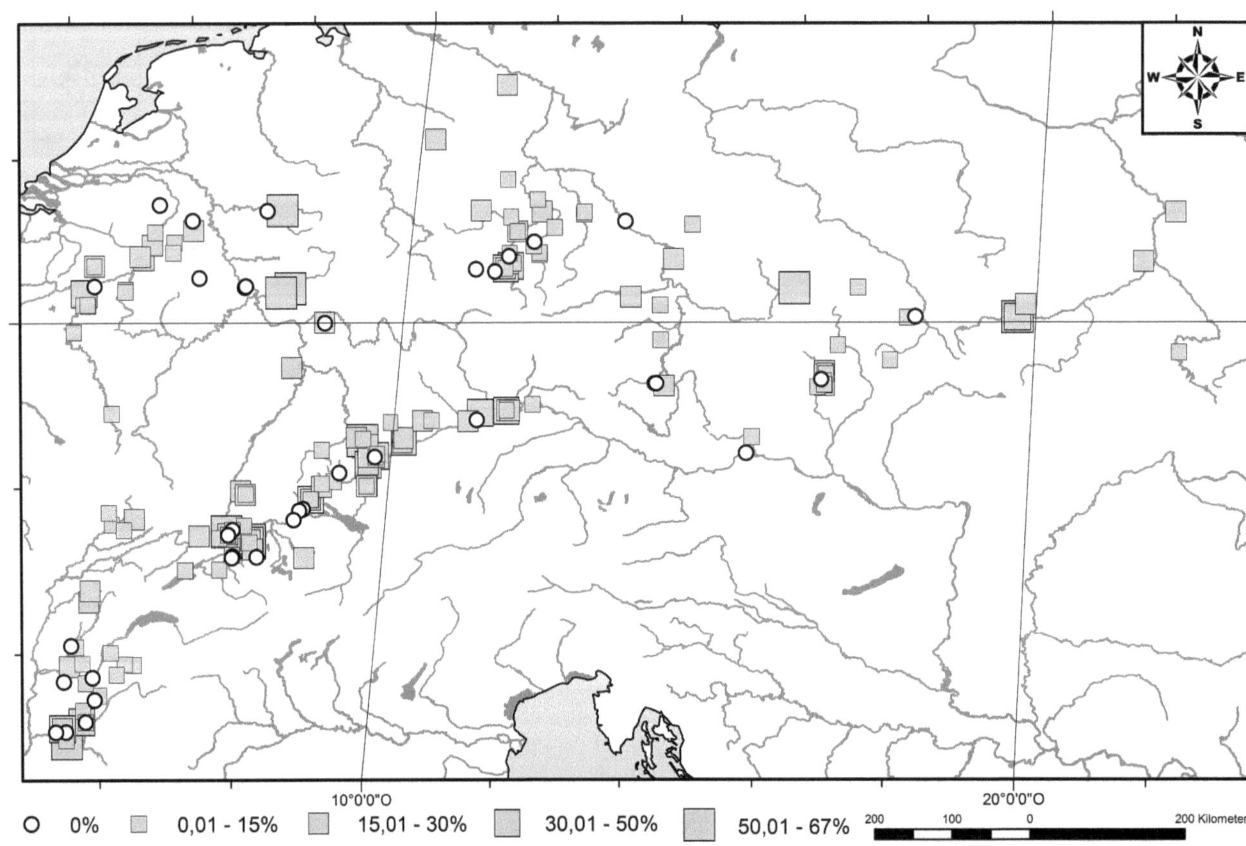

Fig. 6.39 Distribution of laterally retouched pieces in Central European Magdalenian assemblages (see Table A.11)

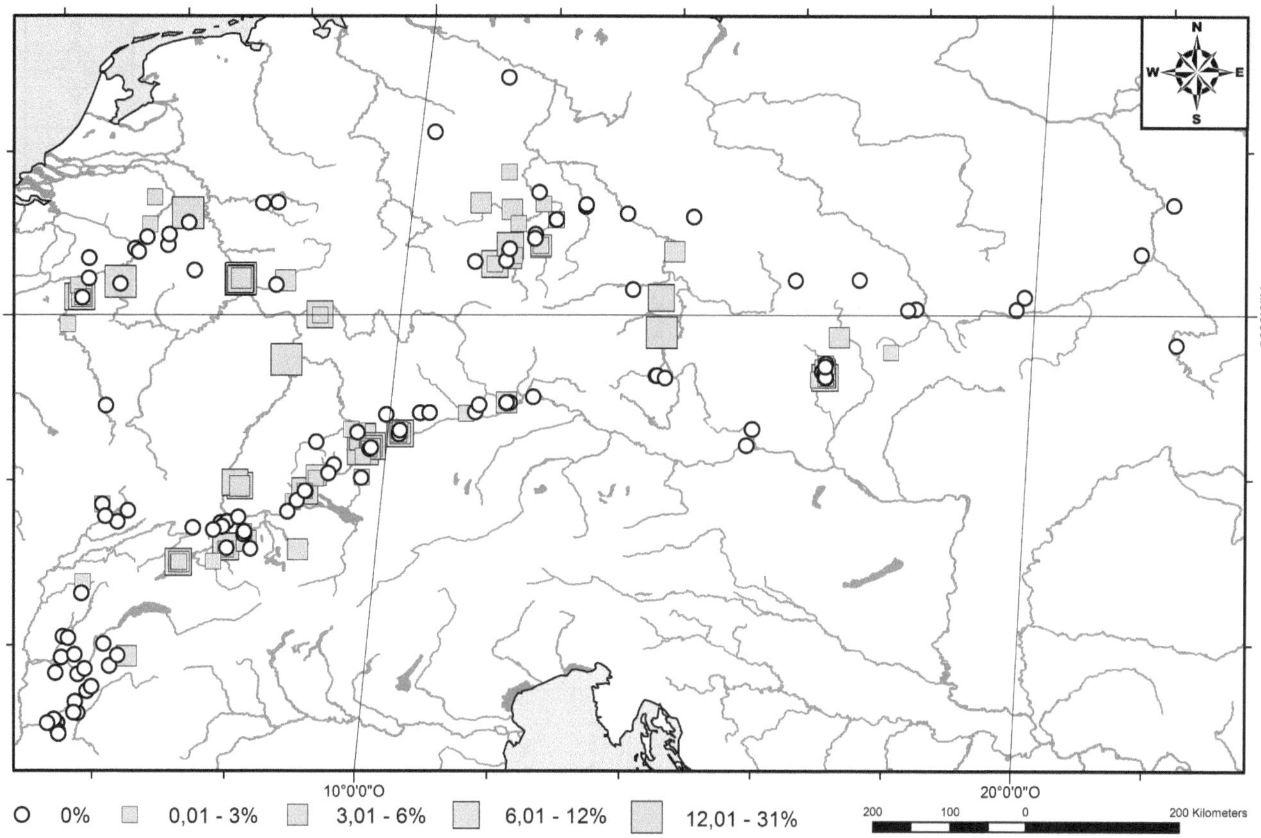

Fig. 6.40 Distribution of splintered pieces in Central European Magdalenian assemblages (see Table A.11)

Splintered Pieces

As was observed in the case of other tool types mentioned above, the spatial pattern of splintered pieces (see Fig. 2.4) shows a gradient from the southwest, where they are rare or absent, toward the northeast, where their frequency increases (Fig. 6.40). They are mainly found in sites in the Meuse-Rhine area, Eastern Germany, and Bohemia. Further to the east, their number decrease again, and they are virtually absent in Poland.

Notched and Denticulated Pieces

In comparison to splintered pieces, the distribution of notched and denticulated ones (see Fig. 2.4) exhibits a pattern with an opposing spatial trend (Fig. 6.41). While they are rather numerous in France, Switzerland, and along the Danube, they occur only rarely in the northern and eastern site clusters. It is interesting to note that the western part of the Meuse-Rhine cluster still shows moderately high values for notched and denticulated pieces, whereas they appear very rarely in the eastern part.

Pointed Blades

Pointed blades (see Fig. 2.4), which are reported occasionally at Magdalenian sites, belong to the least common types of tools among the assemblages of the CEM. In comparison to all other types, they show a unique pattern (Fig. 6.42): the highest density of assemblages with comparatively high percentage values of pointed blades can be found along the Danube and especially in the eastern part of the Franconian Alb, a region which was rather unremarkable so far. From here, the number of inventories with a high proportion of pointed blades decreases in all directions.

Backed Points

Backed points (Fig. 2.4) are a common feature of CEM assemblages. However, in some cases there is still debate as to whether they are an original part of the respective inventories or are intrusive items from underlying Gravettian horizons or later Azilian occupations. In any case, their spatial distribution is rather even with regard to regional groups (Fig. 6.43). Nevertheless, backed points seem to be slightly

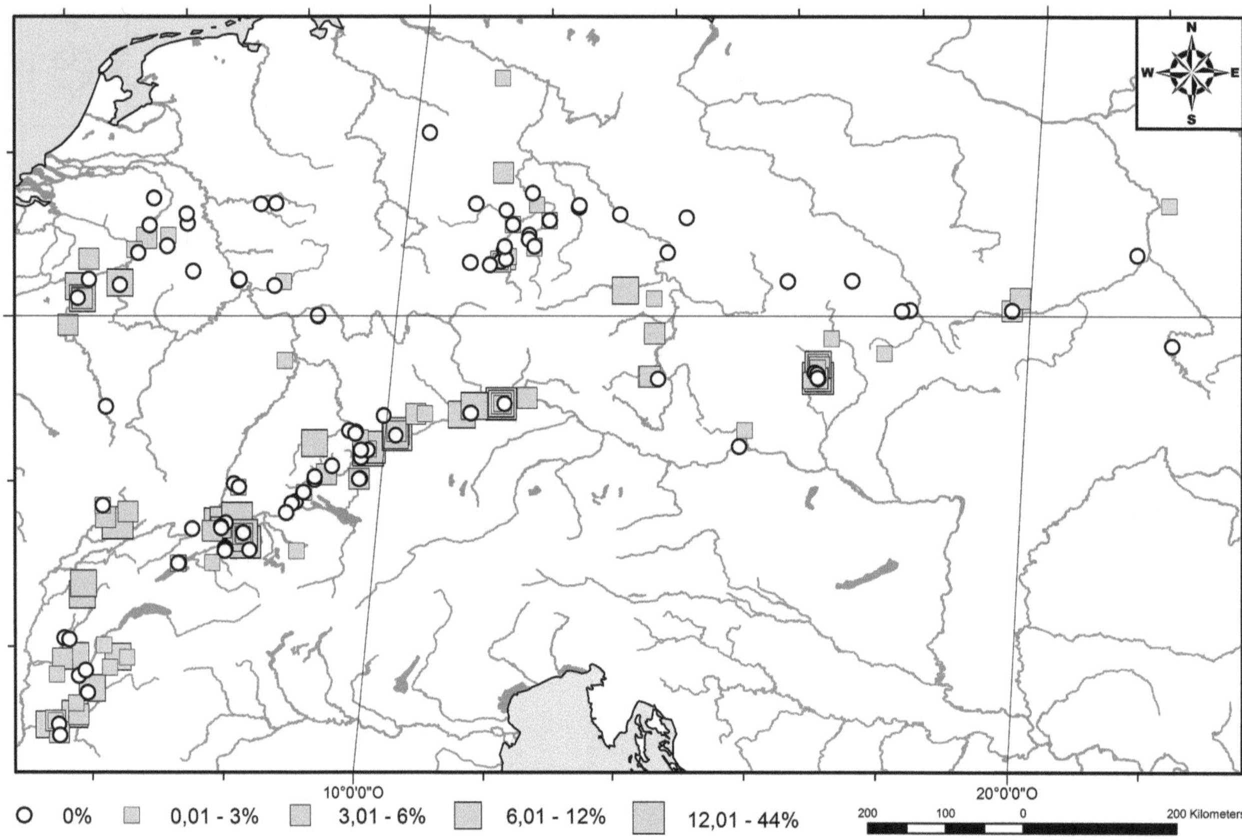

Fig. 6.41 Distribution of notched and denticulated pieces in Central European Magdalenian assemblages (see Table A.11)

more abundant in the southern and western parts of the investigated area and in Eastern Germany and are less numerous in Bohemia, Moravia, and Poland.

Shouldered Points

Shouldered points (see Fig. 2.4) are generally less common in Magdalenian inventories than backed points. However, they are without doubt a true component of CEM assemblages. In Switzerland, for instance, they are seen as a characteristic feature of this archaeological entity (Nielsen 2009: 95). Therefore, it is not surprising that they appear rather frequently in this area (Fig. 6.44). In addition, they occur in some assemblages along the Danube and are quite abundant in collections from Eastern Germany. In contrast, they are virtually absent in the French Jura and the Meuse-Rhine area as well as in Moravia and Poland.

Triangles

Triangles (see Figs. 2.5 and 2.6) figure among the rarest types in the CEM and appear to be restricted to the older assemblages (see Chap. 4, section "The Long Debate About Lithic Triangles"). They are mainly distributed in France and Switzerland, as well as in Moravia and Poland (Fig. 6.45). They also occur in the Kniegrotte assemblage in Eastern Germany. A single specimen from the small inventory of Eitensheim in the Franconian Alb is somewhat questionable. This inventory is a surface collection among which only 16 tools are of Magdalenian character. They include, besides backed bladelets, a shouldered point and two Lacan burins (Rieder 1988, 1989; Hoeck 199). However, it certainly cannot be ruled out that this triangle is of Mesolithic age (see Chap. 4, section "The Long Debate About Lithic Triangles").

Concluding Remarks

A closer examination of CEM lithic tool types reveals that there are two main groups of tools: those which are rather evenly distributed over the investigated area and those which exhibit clear spatial trends and clusters. Several post-depositional processes as well as excavation methods may have and certainly did influence and alter the original composition of the assemblages. However, these processes occurred in more or less equal measure over the whole investigated area and are therefore not considered as causing a directed spatial bias.

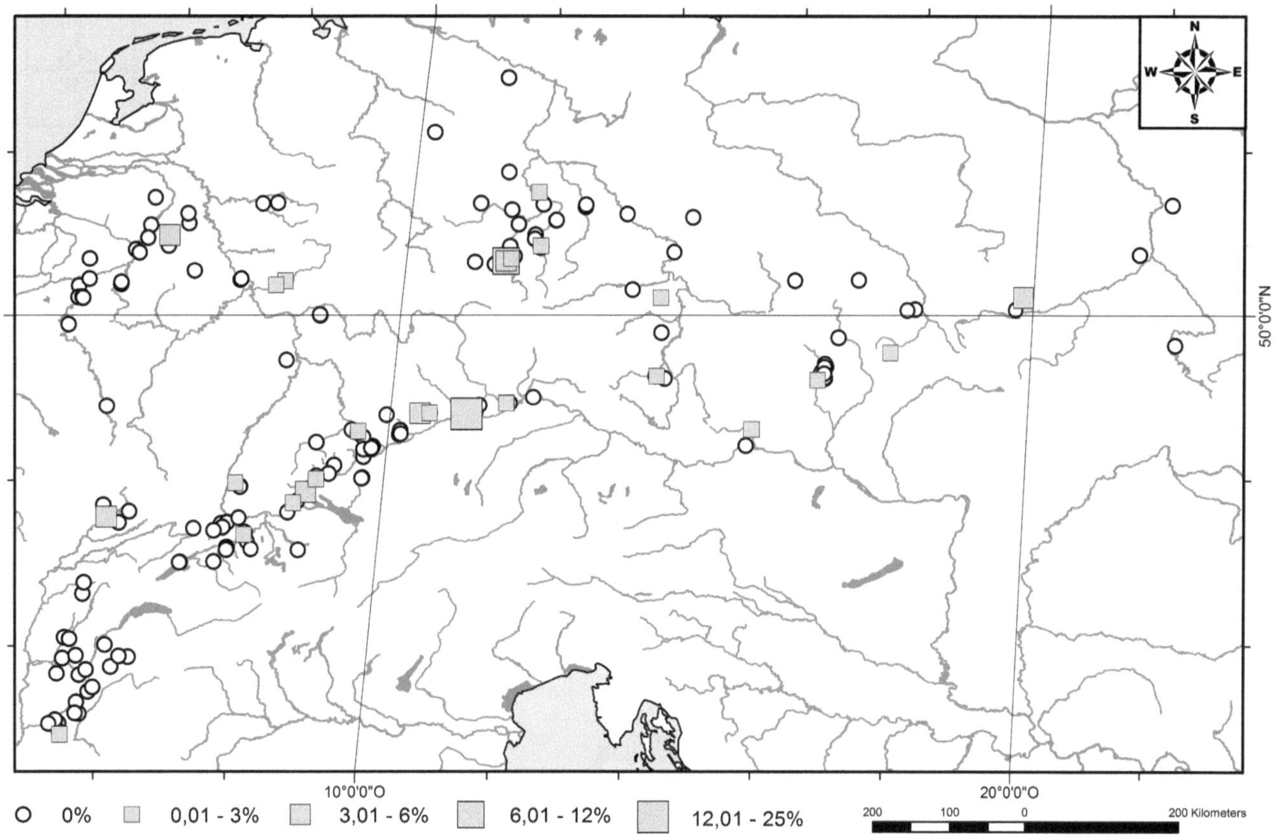

Fig. 6.42 Distribution of pointed blades in Central European Magdalenian assemblages (see Table A.11)

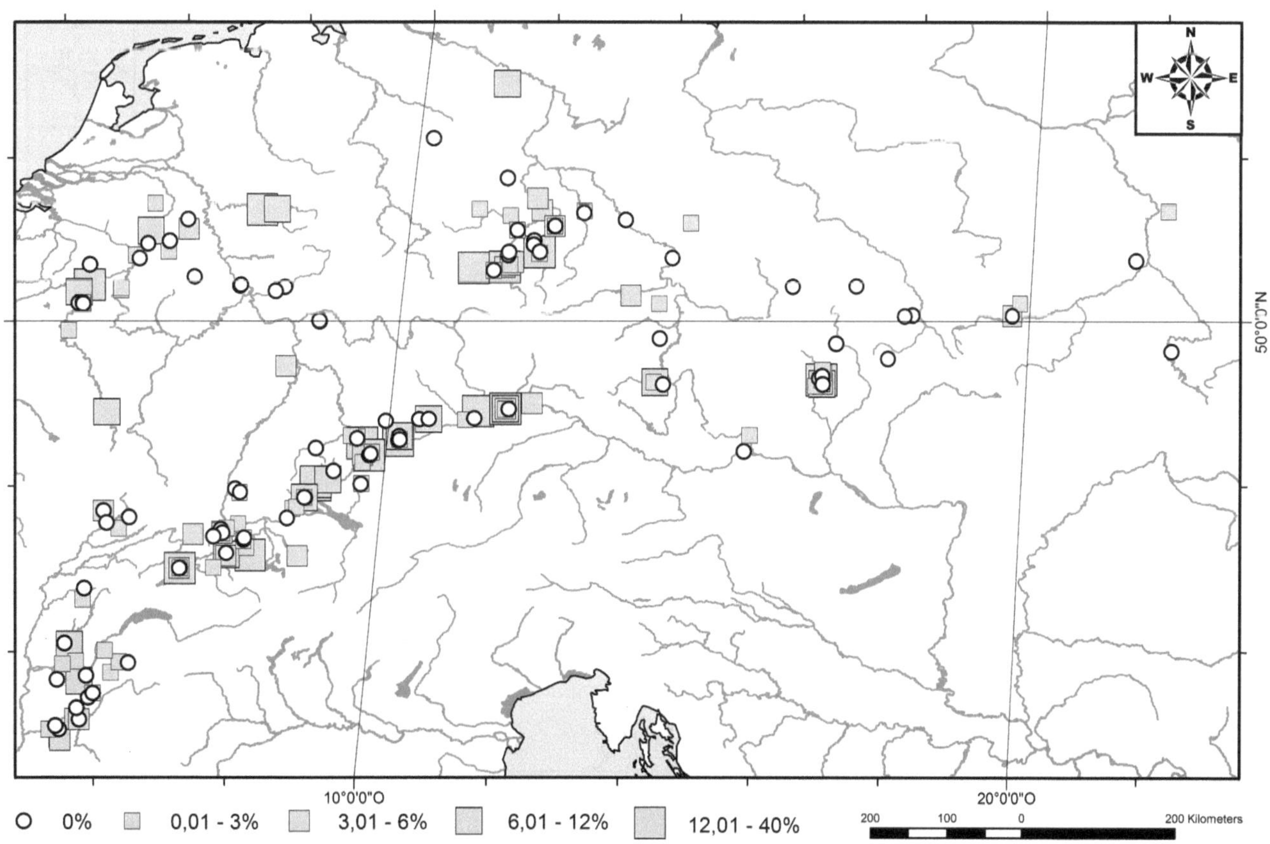

Fig. 6.43 Distribution of backed points in Central European Magdalenian assemblages (see Table A.11)

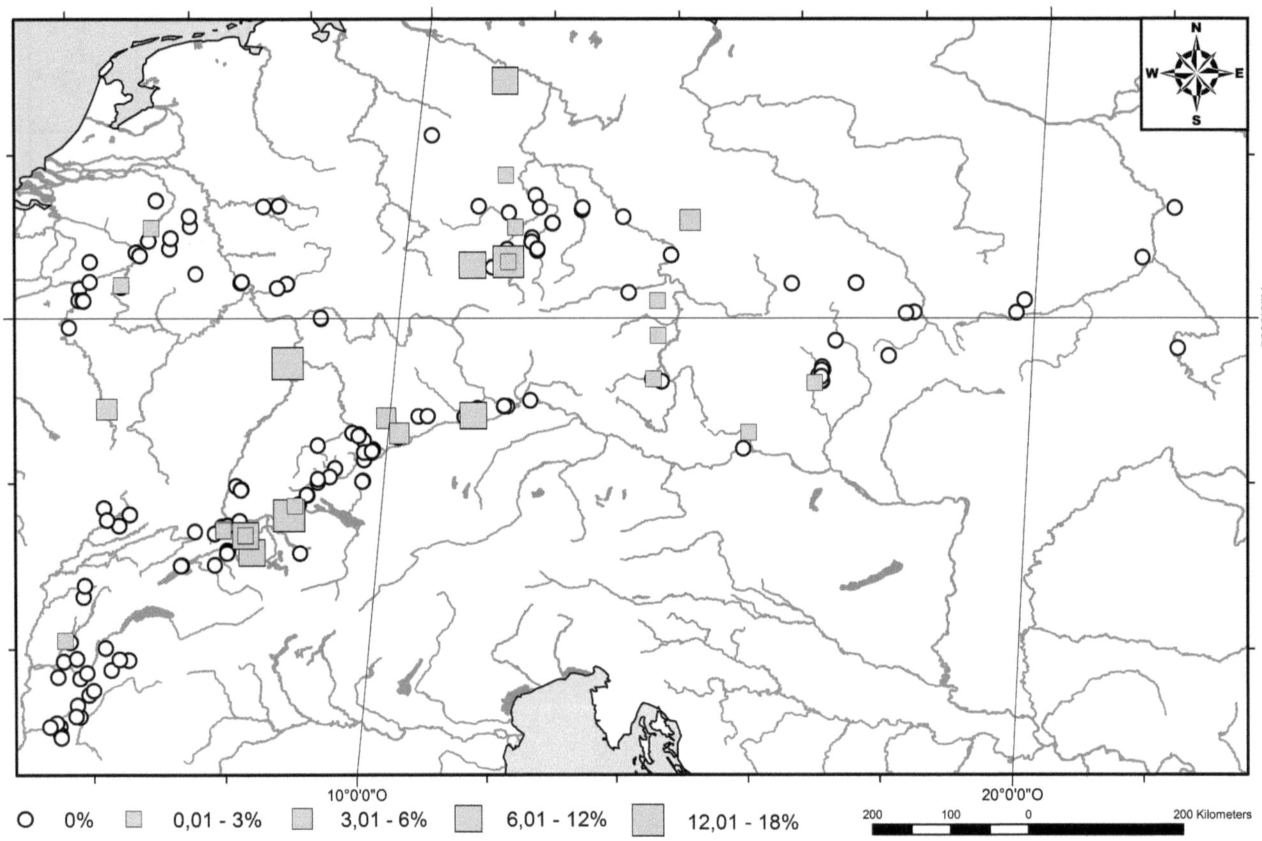

Fig. 6.44 Distribution of schouldered points in Central European Magdalenian assemblages (see Table A.11)

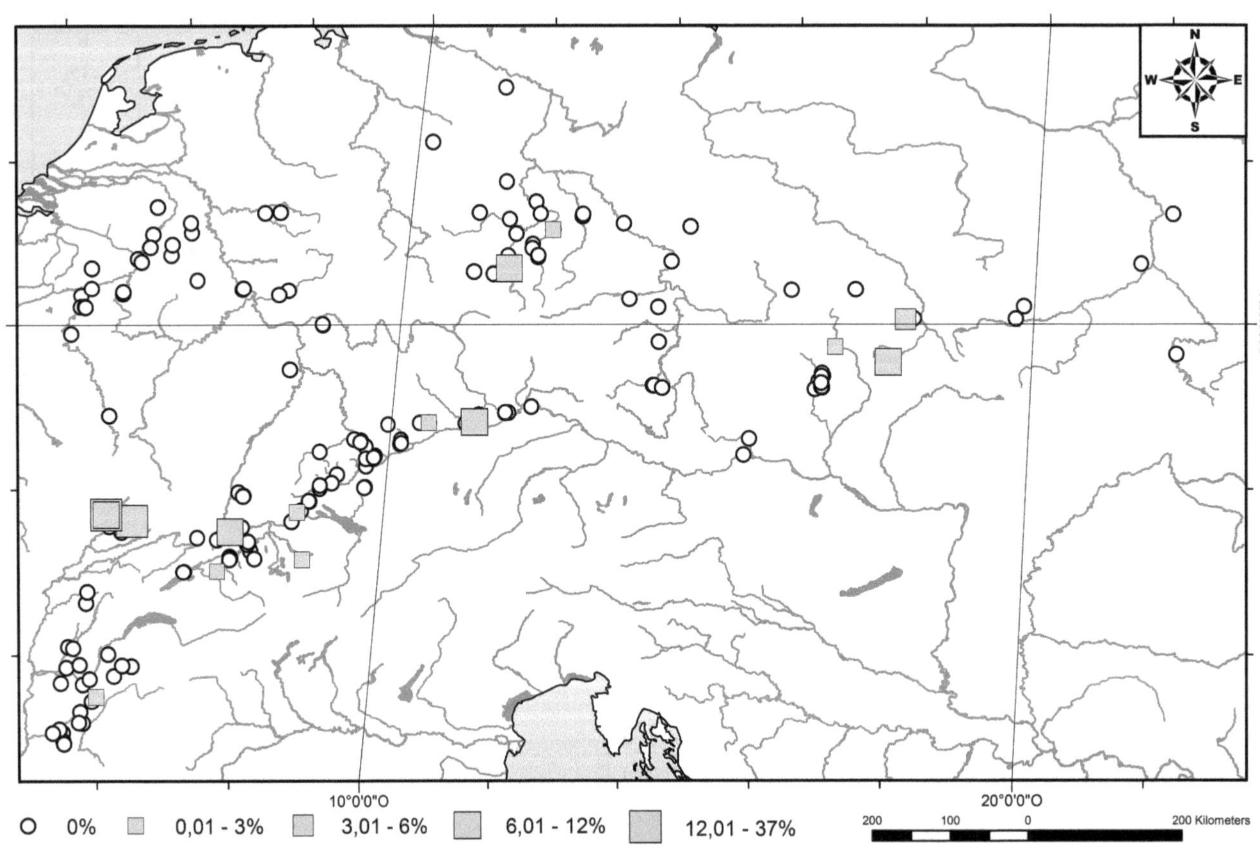

Fig. 6.45 Distribution of triangles pieces in Central European Magdalenian assemblages (see Table A.11)

If natural processes, operating randomly on the archaeological record, can be excluded as important factors in the generation of directed spatial trends, it is permissible to assume that the observed pattern reflects prehistoric reality to some extent. On this basis, the trends—which are sometimes surprisingly pronounced—can be interpreted as indicative of the distribution of morphological concepts. Hence, they gain significance as indicators of an exchange of ideas and therefore intensity of interaction between regional groups.

In summary it can be stated that the main and most common tool classes, such as scrapers, burins, borers, backed bladelets, truncated pieces, and laterally retouched pieces do not exhibit any kind of directed spatial trends or clustering whatsoever. They are rather abundant in all regional groups, and variations in their relative frequencies can be best explained in terms of differences in site function. As opposed to these rather ubiquitous classes, a group of nine morphologically distinct tool types shows a clear spatial patterning. These nine types can be assigned to four principal trends according to increasing and decreasing frequencies. The first spatial trend shows a west-east direction and is indicated by backed points (Fig. 6.43). These tools are quite abundant in the Jura region and along the Danube as well as in the Meuse-Rhine area and Eastern Germany, but are comparatively rare in Bohemia, Moravia, and Poland. The second spatial trend runs from southwest to northeast and appears among the notched and denticulated pieces (Fig. 6.41). These are fairly abundant in the Jura and Danube region, but appear to be comparatively underrepresented in the areas adjacent to the north and east. The third trend is diametrically opposed in its principal direction, exhibiting a strong gradient from northeast to southwest. Long burins, long borers, *zinken*, and splintered pieces (Figs. 6.24, 6.27, 6.28 and 6.40) are most abundant in Poland, Moravia, Bohemia, Eastern Germany, and the Meuse-Rhine area. However, their frequencies decline considerably in a southward direction, and they are only rarely found among assemblages in the region of the Jura Mountains. The fourth group comprises those types which occur mainly in the Jura and Danube area, as well as in Eastern Germany, Bohemia, and Moravia, but are virtually absent in the Meuse-Rhine area and in Poland. These are short scrapers, denticulated backed bladelets, and shouldered points (Figs. 6.21, 6.37 and 6.44).

Generally, it can be stated that with regard to the most common tool classes, the CEM appears as a homogeneous unit, within which differences in tool frequencies mirror different site functions. However, if the analytical resolution becomes more fine grained and distinguishes different types, several spatial trends, sometimes diametrically opposed in their direction, can be revealed. These trends do not necessarily correspond with the interaction pattern that could be observed on the basis of the distribution of mollusk shells (see section "Procurement Pattern of Mollusk Shells and Amber"). Nevertheless, it is interesting to note that the assemblages of the five regional groups found in section "Procurement Pattern of Lithic Raw Materials" appear to reflect preferences for certain types as a whole: except for the notched and denticulated pieces, which are more numerous in the western part of the Meuse-Rhine area than in the eastern part, the boundary lines of the spatial trends seem to respect the boundaries of the regional groups to a large extent. In section "Linear Discriminant Function Analysis: Multivariate Investigation of Spatial and Temporal Trends Among Concepts of Lithic Tools," these observations on the distribution of individual types will be counterchecked using multivariate statistics in order to see if their combination and simultaneous evaluation brings to light any emergent properties which are not visible on this analytical level. Prior to this, however, the spatial distribution of organic tools shall be analyzed more closely.

Spatial Distribution of Concepts of Organic Tools

Gathering data on Magdalenian organic tools from the archaeological literature presents similar structural difficulties to those encountered with lithic tools (see Chap. 2, section "Tool Types"). Indeed, the widely varying quality of the available publications and the often insufficient morphological information they provide makes constructing a detailed and sound data base even more difficult at times than it is for lithic tools. Because the intended morphological resolution cannot be achieved for a considerable proportion of the investigated assemblages, the analysis of the data is carried out using a more coarse-grained classification system. The categories of organic artifacts analyzed are the following: points, needles, barbed points, awls, *batons percés*, and *baguettes demi-rondes*. In total, 153 CEM assemblages from 122 (Fig. 6.46) sites are reported to yield organic tools; Fig. 6.47 shows their distribution and the number of specimens per site. In 72 % of the assemblages, the number of tools does not exceed 10 pieces, and the median value amounts to three organic tools per inventory. Only in seven cases (4 %) are 100 or more organic tools mentioned, the highest number (1255 artifacts) being reported from the site of Schweizersbild (Table A.12). With regard to the spatial distribution, it is notable that the sites in Bohemia, Moravia, and Poland, with the exception of the two cave sites of Maszycka and Pekárna, generally exhibit a relatively low number of organic tools in comparison to those sites in the western part of the investigated area.

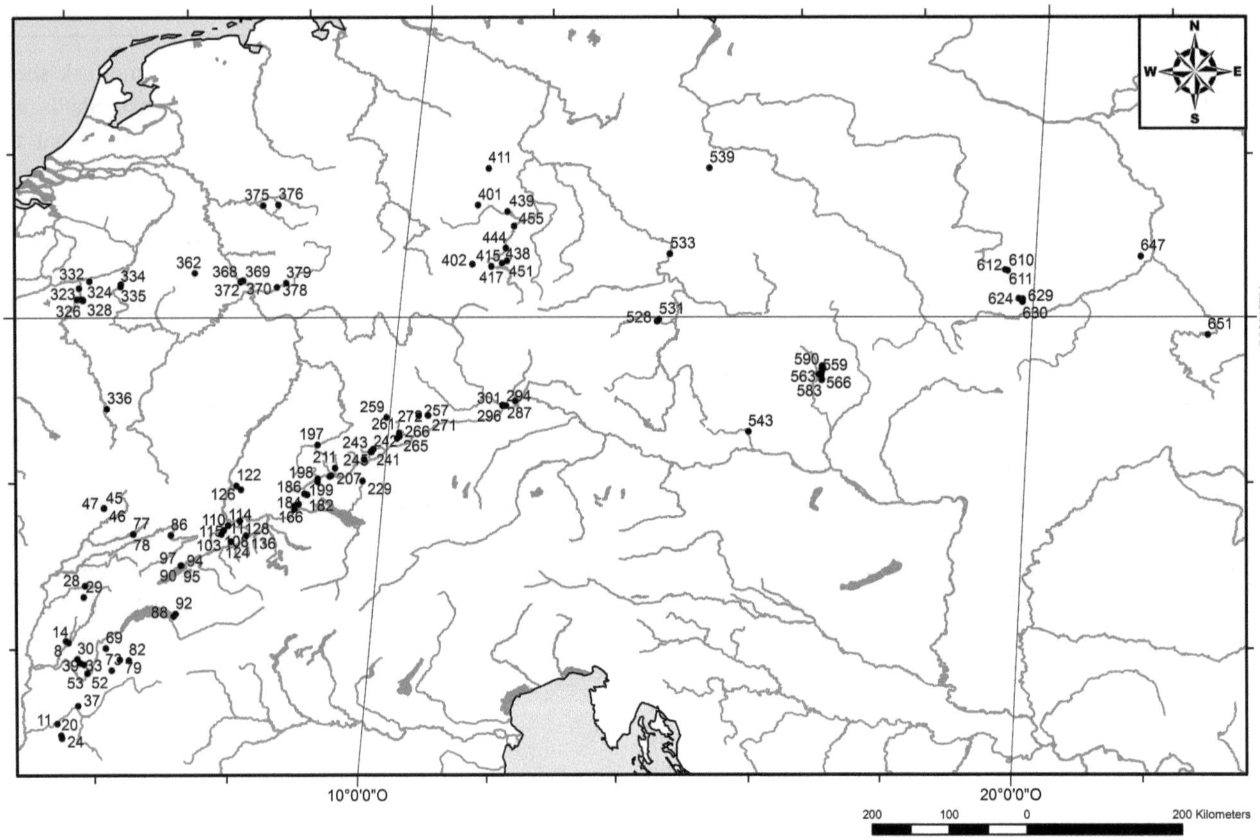

Fig. 6.46 Distribution of assemblages included in the typological analysis of organic tools (see Table A.12)

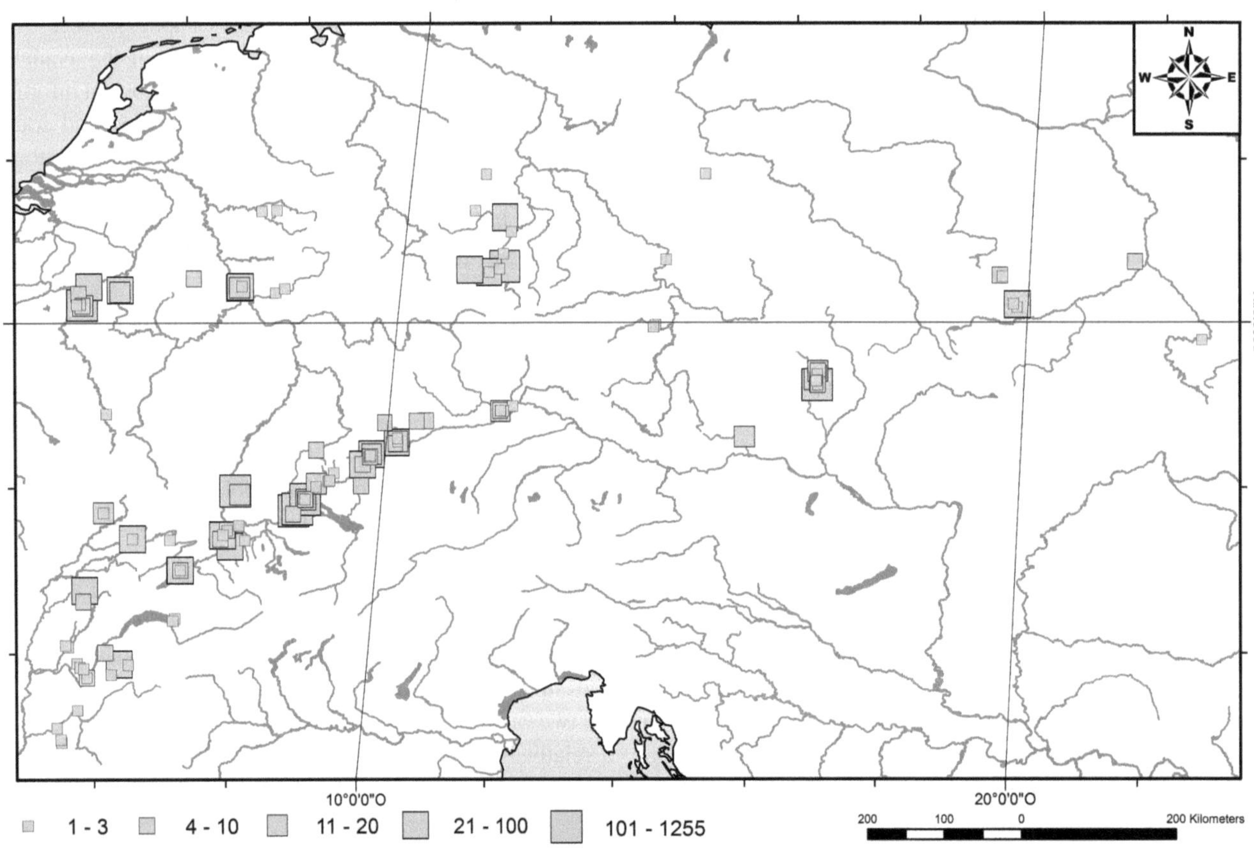

Fig. 6.47 Distribution of organic tools in Central European Magdalenian assemblages (see Table A.12)

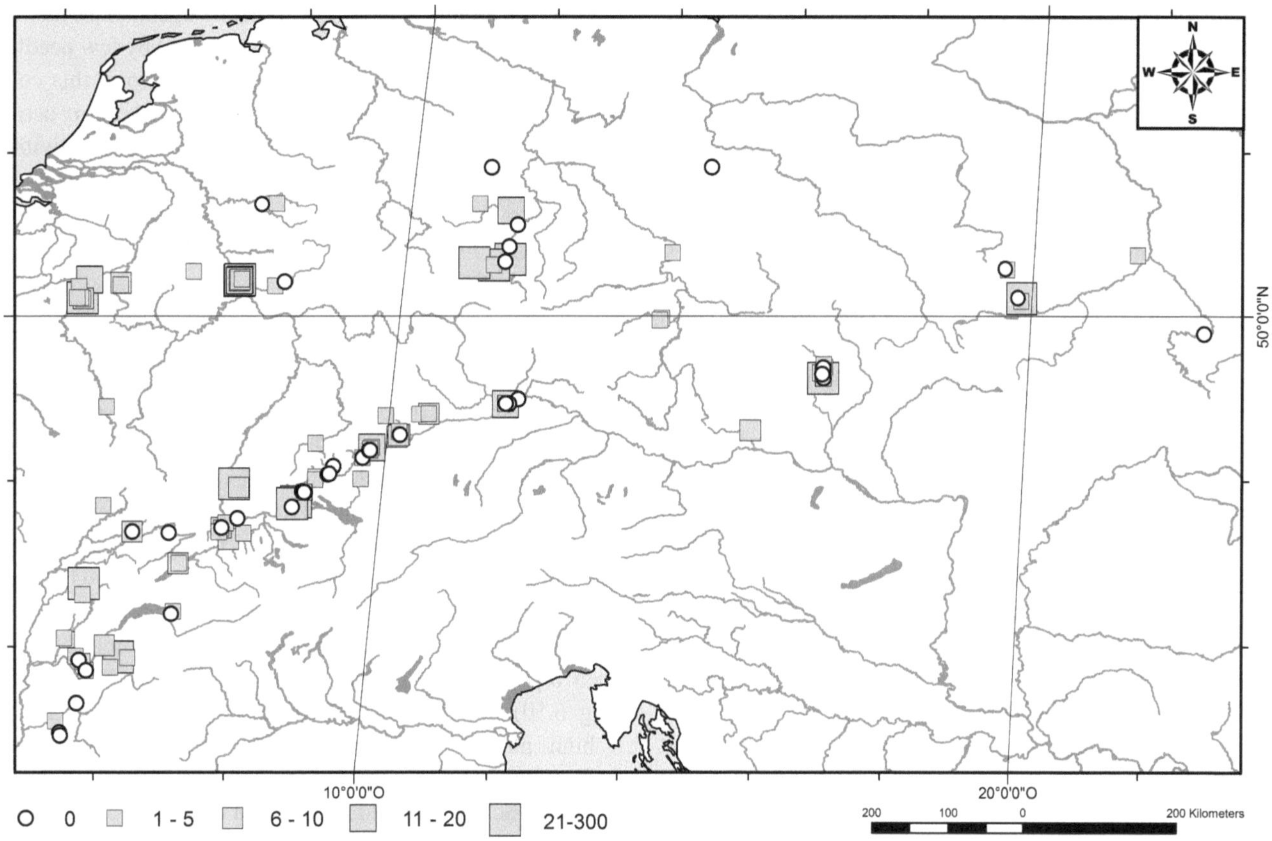

Fig. 6.48 Distribution of organic points in Central European Magdalenian assemblages (see Table A.12)

This disparity should be considered from different angles. For example, an important factor in the discovery of organic artifacts is the general state and quality of organic preservation at a site. As there is no reason to believe preservation is generally poorer in the eastern than in the western part of the investigated area, however, this factor is of low explanatory value and can thus be dismissed on the large scale. Natural processes are also unlikely to have been responsible for the formation of this distribution pattern. Instead, the pattern seems to reflect the relative abundances of organic artifacts within the tool-kit of CEM hunter-gatherers and therefore to be indicative of choice. Some clues are offered by the evaluation of pollen diagrams (see Chap. 5, section "Climate and Vegetation"). Here, the high amount of arboreal pollen in the eastern part of the investigated area indicates a continuous availability of wood in this region during the late Pleniglacial, the Oldest Dryas, and the Bølling Interstadial. Thus, there is an increased probability that this material was used for the production of tools and that it substituted in some degree for animal-sourced material such as bone, antler, or ivory. Because of the higher decomposability of wood in comparison to these animal materials, wooden artifacts tend not to be preserved, which results in an underrepresentation of organic artifacts in the eastern part of the investigated area. Against the background of the overall distribution of organic artifacts, the six tool classes mentioned above are analyzed with regard to their spatial pattern.

Points

For 105 out of 153 assemblages (69 %), organic points (see Fig. 2.7) are mentioned in the archaeological literature. In 72 of these cases (68 %), 5 or fewer specimens are reported. Only four sites, namely, Kesslerloch, Pekárna, Schweizersbild, and Kniegrotte, have yielded more than 100 organic points (see Table A.12). They are thus the most common organic tool class and show high relative as well as absolute abundances in the entire investigated area (Fig. 6.48).

Generally speaking, organic points also make up the highest proportion of any tool class within the group of organic artifacts in each assemblage. This is the case, for example, in the southwestern part of the investigated area, along the course of the river Rhône, where many collections with a low number of organic tools nevertheless show a high percentage of points. Similar observations can be made for several inventories from the Swabian and Franconian Alb and the Meuse valley, as well as for the Wildweiberlei and the Feldhofhöhle in the valleys of the Lahn and Hönne rivers and for the Kartstein in the Eifel mountain range. Further to the east, this applies equally to Bad Frankenhausen in Eastern Germany; to Rytiřská, Na Průchodě, and Tetín in Bohemia; to several sites in the Moravian Karst; and to Puchaczowa and Kostkowice in the Polish Jura. High proportions of points can likewise be observed in assemblages having an ample set of organic artifacts, such as Grappin, Munzingen, Andernach-Martinsberg, Teufelsbrücke, Pekárna, or Maszycka. Some assemblages, however, do not follow this general trend; some even display an opposite trend as is the case for the Grotte Bibi, Bobache, and Freydières in the Isère valley. In these small organic inventories, points are completely absent. Since organic points were very probably related to hunting activities, their respective numbers at different sites may be indicative of the function of those sites. Their presence in smaller assemblages such as those from Napoleonskopf or Bocksteinhöhle would therefore seem to characterize these sites as most likely the remnants of hunting camps.

Needles

Needles are reported from 54 assemblages (35 %) and thus occur about half as often as points. This is likely indicative for site function, but is probably also a result of the comparatively small size of these tools, which makes them easy to overlook and may lead to their underrepresentation in collections, especially those from older excavations, since their small width means that needle fragments may even fall through wider meshes during the sieving of excavated sediments. In the investigated area, more than half of the needle-bearing inventories contain five specimens or fewer. The highest number is reported from the backdirt of the Petersfels excavation ($n=288$) and the site of Schweizersbild ($n=240$). The former example may support the hypothesis that this tool class is structurally underrepresented in older excavations. With regard to the total number of organic tools (Fig. 6.47), needles appear to be especially abundant in the assemblages of Grotte Bibi, Romains (layer 3), Neuchâtel-Monruz, and Hauterive-Champréveyres, as well as in those from Rislisberghöhle, Felsställe, Kamegg, Žitného, Býči skála, and Wilczyce (Fig. 6.49). They appear to be underrepresented, on the other hand, in the assemblages of Grappin, Munzingen, Brillenhöhle, and Maszycka Cave, as well as in those from the eastern German sites of Bärenkeller and Nebra.

It is striking that many assemblages with a high number of needles, especially Grotte Bibi, Rislisberghöhle, Kohlerhöhle, and Felsställe, at the same time contain a low number of points. In contrast, Munzingen, Bärenkeller, Nebra, and Maszycka Cave are characterized by few needles but many points. It seems reasonable to interpret this contrasting pattern as reflecting site function, with hunting activities (points) on the one side and more domestic activities (needles) on the other.

Barbed Points

Barbed points occur in the entire investigated area but are mentioned in only 35 assemblages (23 %). They are thus much scarcer than simple points. Most of the time they occur singly ($n=26$, 74 %), and only exceptionally are they found in higher numbers, as at the site of Kesslerloch ($n=10$) or the lower layer of Brillenhöhle ($n=9$). Their distribution lacks discernable spatial differences. Using ^{14}C-measurements, their first appearance can be dated to approximately 16,000 cal BP (see Chap. 4, section "The Long Debate About Lithic Triangles"). Except for single finds of barbed points, as have been reported, for instance, from Węgliny or Przemysl 3 (Fig. 6.50), this tool class is often found in assemblages with a high number of simple points, such as Kesslerloch, Brillenhöhle (lower layer), Andernach-Martinsberg (concentration I), Schweizersbild, or Pekárna (see Table A.12). This appears to support their interpretation as hunting gear (Julien 1982: 137f.; Weniger 1995).

Baguettes Demi-rondes

The rarest class of organic tools in CEM assemblages comprises the so-called *baguettes demi-rondes* or half-round rods (Fig. 2.9), which were probably used pairwise (Bosinski 1990: 180). They are reported in only 18 collections (12 %) and in half of these cases are found singly. The highest number of specimens is reported from the sites of Farincourt ($n=10$) and Rigney 1 ($n=9$), which are located in the vicinity of the Saône and Ognon rivers. In contrast to the other classes of organic tools described so far, *baguettes demi-rondes* show a discernable trend in their spatial pattern. Despite being distributed ubiquitously, they are most abundant in southwestern part of the investigated area, for instance, at Birseck-Ermitage and Kaufertsberg, and they show decreasing proportional frequencies toward the north, where they are reported from Goyet 3, Andernach-Martinsberg, and Gönnersdorf, as well as toward the east, where they have been found at the sites of Kniegrotte, Krusza Skála (Kostkowice), Žitného, and Kůlna Cave (Fig. 6.51). Typologically remarkable are five specimens found at the sites of Kesslerloch and Freundenthalhöhle; they are the only known specimens so far from Central Europe to be decorated

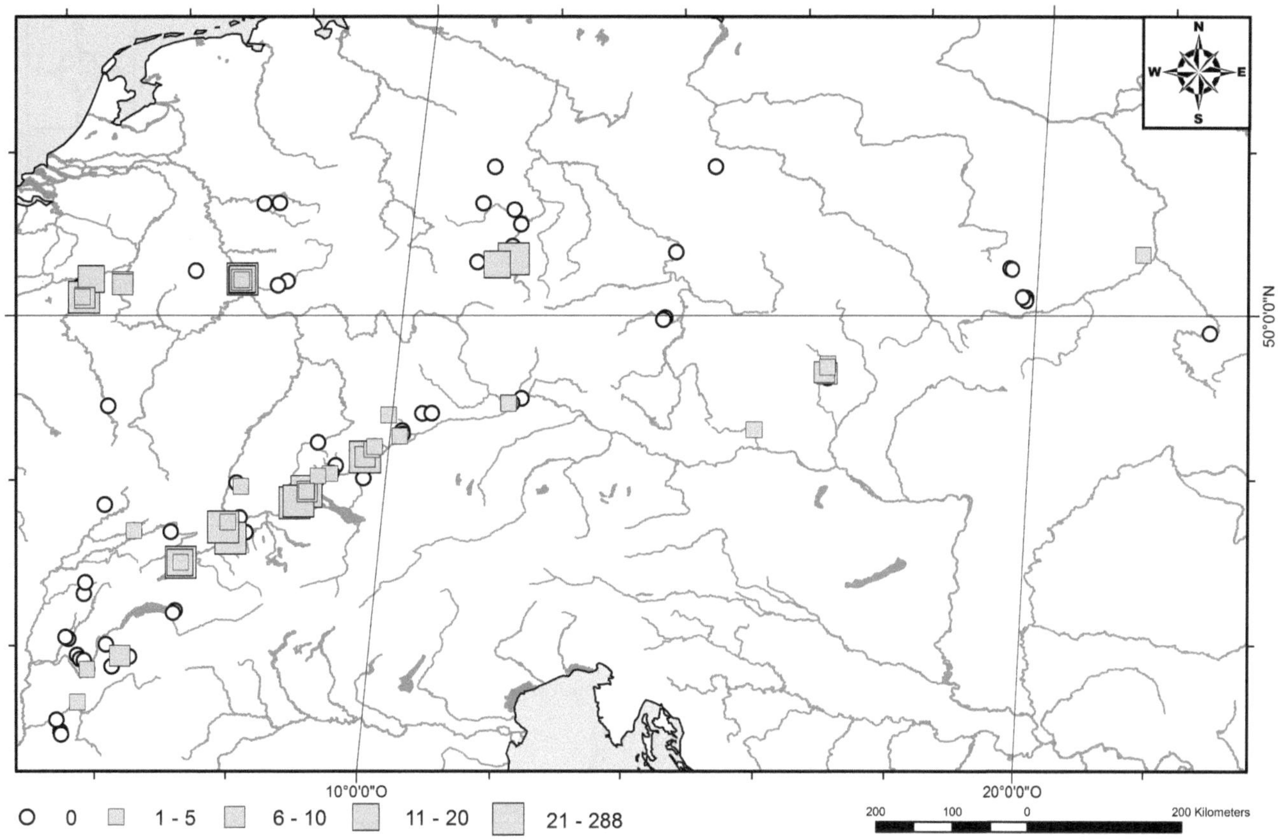

Fig. 6.49 Distribution of organic needles in Central European Magdalenian assemblages (see Table A.12)

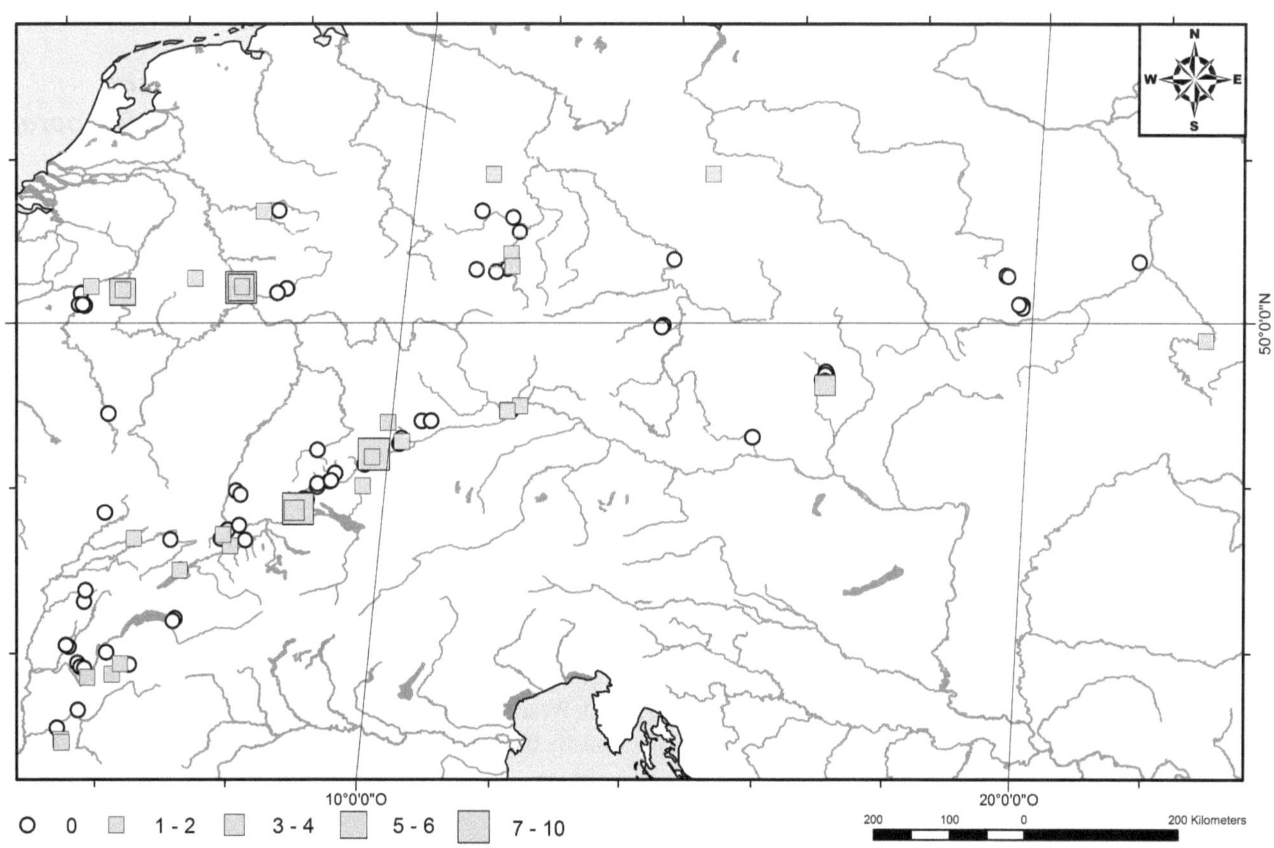

Fig. 6.50 Distribution of organic barbed points in Central European Magdalenian assemblages (see Table A.12)

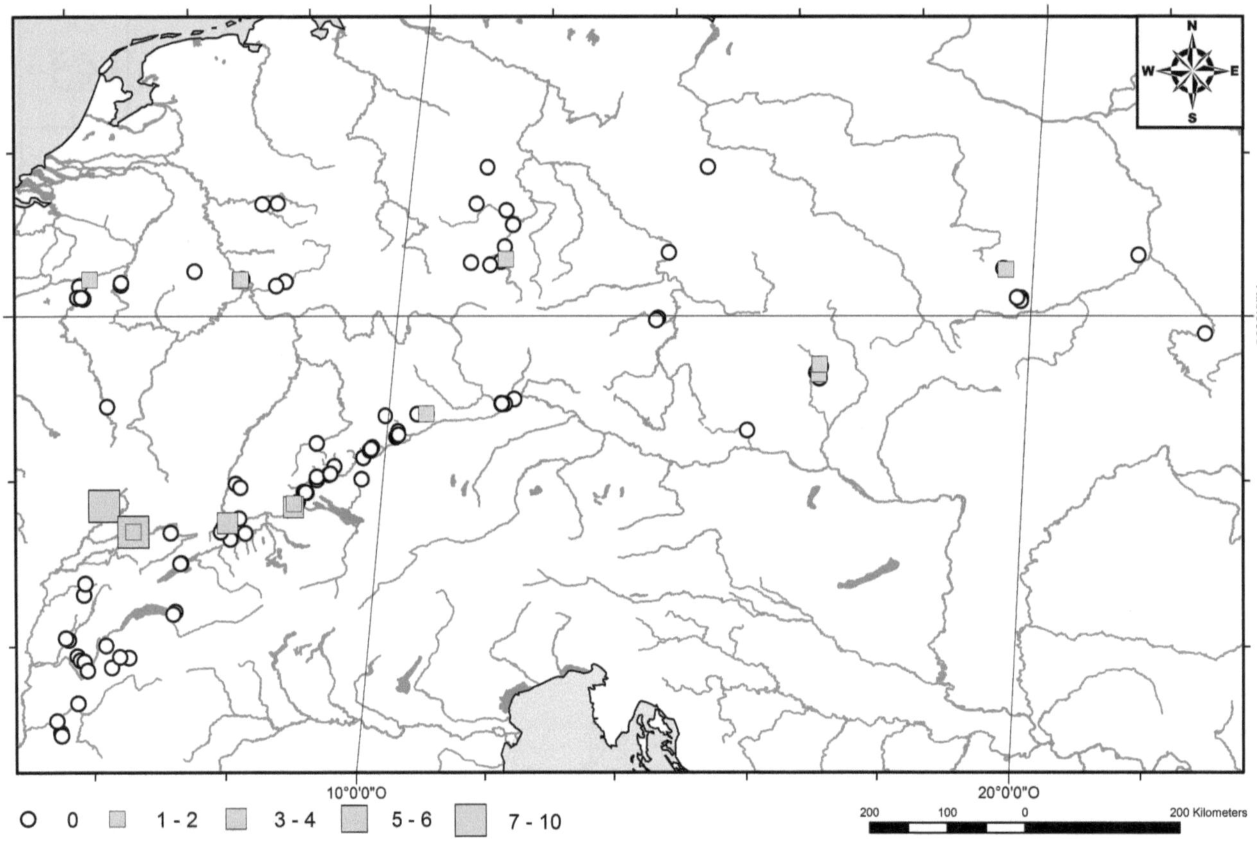

Fig. 6.51 Distribution of baguettes demi-rondes in Central European Magdalenian assemblages (see Table A.12)

with raised humps on their convex side (Fig. 2.9). Similar pieces are known, however, from the Franco-Cantabrian region (see section "Figural Representations").

Bâtons Percés

Bâtons percés (Fig. 2.8) are recorded in 30 collections (20 %) and thus occur about twice as often as *baguettes demi-rondes*. This tool class is distributed ubiquitously. They have most often been found singly ($n = 17$, 57 %), for instance, at Les Hoteaux in the French Jura, Nikolaushöhle in the Swabian Alb, or Výpustek in the Moravian Karst, but they also occur in larger numbers, as at the site of Pekárna cave, where 20 specimens were excavated, or at Schweizersbild, where 15 *bâtons percés* are reported (Fig. 6.52). In larger assemblages, they can be associated with simple points and barbed points, as is observable in the collections from Kesslerloch, Schussenquelle, Andernach-Martinsberg, and Kniegrotte.

Linear Discriminant Function Analysis: Multivariate Analysis of Spatial and Temporal Trends Among Concepts of Lithic Tools

Up to this point, the spatial distribution of each lithic tool type has been analyzed individually. In order to evaluate the typological composition of the assemblages as a whole as well as to test if these compositions differ significantly between the regional groups and to determine which types are especially important for the distinction, a Linear Discriminant Function Analysis is performed.

The Method

Linear Discriminant Function Analysis (LDA) is a multivariate statistical classification method (Fisher 1936; Handl 2002: 324–338; Legendre and Legendre 2003: 616–633; Leyer and Wesche 2007: 188–194; Claude 2008: 112–116). It is usually based on a multivariate table (Tables A.35 and A.36).

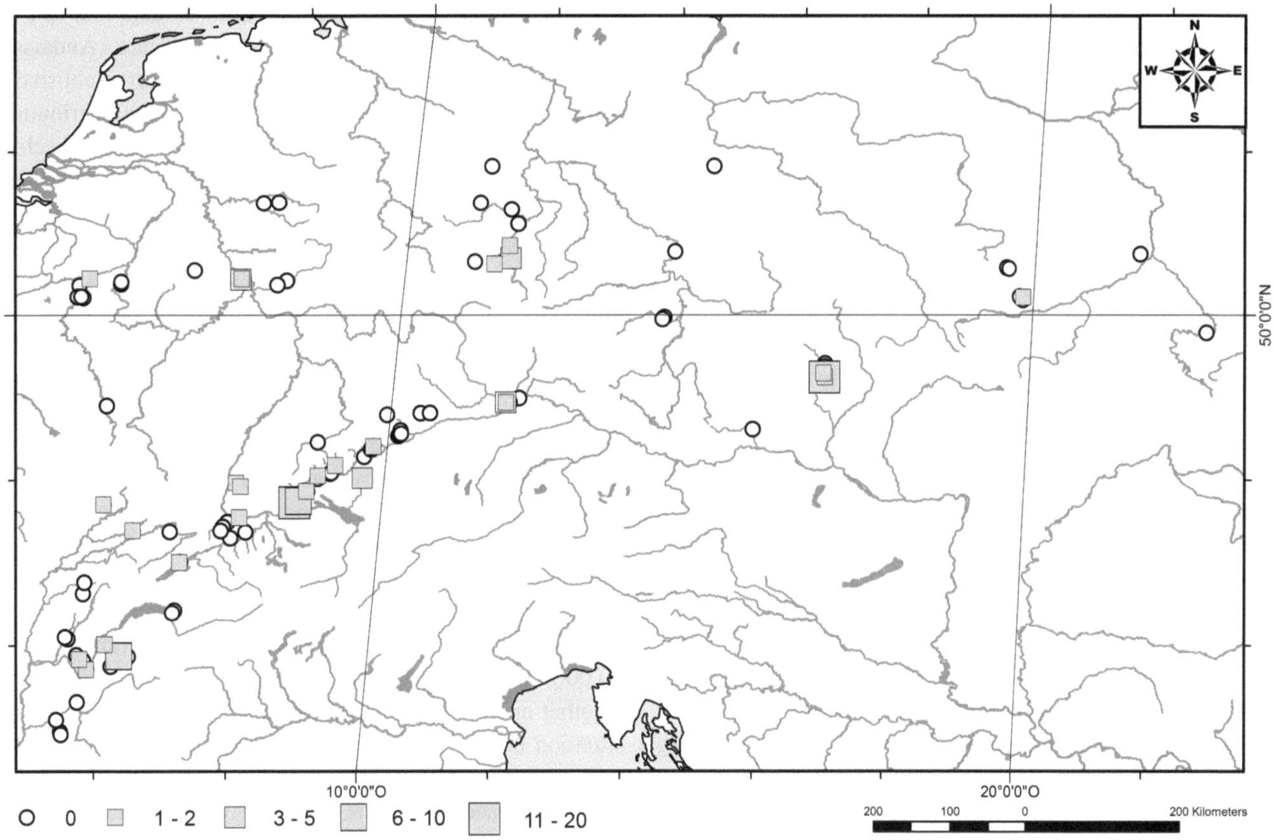

Fig. 6.52 Distribution of bâtons percés in Central European Magdalenian assemblages (see Table A.12)

The rows of the table display the cases (here: assemblages) which are assigned a priori to two or more disjunct groups (here: five regional groups in the spatial LDA and four temporal groups in the diachronic LDA), each case being assigned to only one group. The columns, on the other hand, contain the independent variables against which the a priori grouping is tested. In this study, the independent variables are the Hellinger-transformed abundances (Legendre and Gallagher 2001) of different tool types (see below). An LDA provides information as to which variables contribute to the differences between the groups and calculates the degree of probability with which each case belongs to any of one those groups. Eventually, the cases are reassigned to the group showing the highest degree of probability for membership. These probabilities are described as a posteriori, since they are the result of the LDA. A priori and a posteriori assignments may be identical, or they may differ from one another.

Initial Transformation of Raw Data

In the present study, the raw data—i.e., the abundances of artifact types in each assemblage—is transformed prior to the analysis in order to account for the fact that many rare types, such as long borers or truncated backed bladelets, occur in rather small numbers compared to common types such as end scrapers or simple backed bladelets. This is done using the Hellinger transformation, which is calculated as follows (Legendre and Gallagher 2001: 275):

$$y'_{ij} = \sqrt{\frac{y_{ij}}{y_{i+}}}$$

Here, j is the abundance of types (columns $j=(1\ \dots\ p)$) per assemblage i (rows $i=(1\ \dots\ p)$) of the abundance table $Y=\left[y_{ij}\right]$ of size $n \times p$. The transformed abundances y' are calculated by dividing the values of each cell of the abundance table *yij* by the row sums y_{i+} and extracting the square root of the result obtained. In other words, a Hellinger-transformed abundance of a tool type equals the square root of the percentage of this type. The Hellinger transformation thereby ensures that rare types are given a proportionally greater weighting than common types, which would otherwise probably obscure the observation of regional differences. This can be the case, for instance, if the number of types per assemblage is transformed into simple percentages, where the different proportions of types per assemblage must amount to 1. An analysis of simple proportional data would suffer from the "statistical noise" of the common types, since the function-related differences in their abundances would overwhelm the comparatively weaker signals of the rare types.

Nonparametric Test for Intergroup Variance

Before starting the analysis, it is advisable to test whether the predefined a priori groups differ from each other; that is, whether they capture a part of the observed dispersion. This is usually done using a multivariate analysis of variance (MANOVA) (cf. Claude 2008: 116–118). In order to examine the hypothesis that the regional groups used in this study explain—to a certain extent—the typological dispersion of the assemblages, a nonparametric test or permutational MANOVA of differences among a priori groups is performed (Anderson 2001; McArdle and Anderson 2001). This test is based on a measure of distance between pairs of observations and uses "permutations of the observations to obtain a probability associated with the null hypothesis of no differences among groups" (Anderson 2001: 33). Analogous to the transformed raw data, the underlying distance matrix is generated using the Hellinger distance, which has the following form (Legendre and Gallagher 2001: 275):

$$D_{\text{Hellinger}}(x_1; x_2) = \sqrt{\sum_{j=1}^{p}\left[\sqrt{\frac{y_{1j}}{y_{1+}}} - \frac{y_{2j}}{y_{2+}}\right]^2}$$

The Hellinger distance is thus simply the Euclidean distance—computed using the Pythagoras Theorem—of the Hellinger-transformed data. If the test produces a significant result, the null hypothesis can be rejected, and the grouping can be assumed to explain a certain amount of information (variance) embedded in the multivariate table. Put another way, a significant result proves that the groups differ from each other. The test also gives the determination coefficient R^2, which shows how much difference among the tool assemblages can be explained by the a priori groups alone.

Test for the Homogeneity of the Intragroup Variance/Covariance

An important presupposition for the calculation of an LDA is that the intragroup variances and covariances are homogeneous—i.e., that the intragroup relations between the variables are similar within all groups (Decker et al. 2010: 499). Traditional tests of intragroup homogeneity are very sensitive to departures from the assumption of multivariate normality. In archaeological studies, where the variables represent counts of abundances of types, this assumption is not always reasonable, since the abundances of types are often strongly skewed. A suitable method for examining the homogeneity of multivariate dispersion is a test suggested by M. J. Anderson which is robust to skewed data and can be based on any dissimilarity or distance measure (Anderson 2006). Since this approach uses permutations to establish the significance of its results, no assumptions on the distribution type of the variable have to be made. Anderson's approach is an elegant way to escape the problems caused by traditional statistics. In brief, it computes a principal coordinate analysis (PrCooAn) of the distance matrix of the cases—here, a Hellinger distance matrix (for mathematical details and formulas, see Anderson 2006; for the principal coordinate analysis, see Legendre and Legendre 2003: 424–444). Then, the distance between a case and its group medoid (the multivariate median) in the PrCooAn space is calculated. Since the original Hellinger distances can be understood as a "summary" of the intragroup variabilities (i.e., the variable relations), the group medoid distances in the PrCooAn space function as a representation of these relations. Eventually, a simple analysis of variance of this univariate(!) "summary" of the multivariate relations is conducted. In doing so, the group memberships are permutated several times (here: 10,000), and the corresponding results are compared to see whether another group structure might explain the observed situation better. If the test produces a significant result, the assumption of homogeneous variances cannot be maintained. Consequently, the analysis' discriminatory power might be reduced, and inferences regarding its significance may be distorted. However, as long as the LDA results in an empirically meaningful outcome, these distortions are to a certain extent negligible, since the basic relations between the elements of the analysis are apparently not obscured. In other words, conducting an LDA in the case of a significant result may not be the best choice in the ideal world of pure mathematics, but in the empirical world, it is still a suitable approach for dealing with the question at hand.

The Discriminant Functions

Once the preconditions have been outlined, the Discriminant Function Analysis can be conducted. In the first step, the group allocation for each case is set (a priori grouping). As a rule, the number of groups should not exceed the number of feature variables, each case should be assigned to only one group, and the number of cases per group should not be too small (Klecka 1980: 8f.; Decker et al. 2010: 499). In the second step, a discriminant function is estimated. Generally speaking, it is as follows (Decker et al. 2010: 511):

$$Y = b_0 + b_1 X_1 + b_2 X_2 + \cdots + b_j X_j$$

Here, *Y* is the discriminant variable, *Xj* is the feature variable *j*, *bj* is the discriminant coefficient for the feature variable *j*, and b_0 is a constant term with $j=(1, 2, \ldots, p)$, *p* being the number of variables. Thus, each original feature variable is multiplied with a certain value. Afterward, all products are summed and added to a constant, resulting in a new, synthetic variable. This variable, the discriminant function, is generated in such a way that it allows for an optimal distinction between the groups, as well as it captures the maximum of the dispersion between the groups and makes it possible to evaluate the discriminatory potential of the variables (Klecka 1980: 16; Backhaus et al. 2008: 186). For each case (assemblage), the discriminant function generates a discriminant value, and each group can be represented by its mean discriminant value, the so-called centroid. The dissimilarity of groups can then be expressed as the difference of their centroids. However, because groups with a high dispersion and a wide overlap are difficult to distinguish by their centroids alone, it is necessary to consider the differences in the groups' dispersions. This can be done by using a discriminant criterion (*Γ*), which is calculated by dividing the dispersion between the groups by the dispersion within the groups (for more details, see Backhaus et al. 2008: 188ff.; Decker et al. 2010: 508):

$$\Gamma = \frac{\text{Dispersion between the groups}}{\text{Dispersion within the groups}}$$

The dispersion between the groups is also called explained dispersion, whereas the dispersion within the groups is referred to as unexplained dispersion. Here, the term "explain" is used in a mathematical sense and can be understood to mean "can be connected with." Thus if a grouping "explains" dispersion, the researcher can be confident that some of the information contained in the multivariate table reasonably describes the differences between the groups in question. When there are more than two groups, the number of discriminant functions is $C-1$, with C being the number of groups (Klecka 1980: 16). In most cases, the entire intergroup dispersion cannot be captured by a single discriminant function alone, which is why additional functions must be computed. These functions are calculated in such a way that they are orthogonal to each other. Thus, they capture the maximum of the remaining dispersion between the groups (Klecka 1980: 16; Backhaus et al. 2008: 200), since two orthogonal functions capture two aspects of the entire dispersion, which in turn relate to two independent pieces of information.

The Standardized Canonical Discriminant Coefficient

The standardized canonical discriminant coefficient *b* shows the importance of the feature variable within each discriminant function. It is calculated as follows (Decker et al. 2010: 511):

$$b = v\left(\frac{SS_W(y)}{I-G}\right)^{\frac{1}{2}}$$

Here, *G* is the number of groups, *I* is the number of cases, and *SSW* is the matrix of the unified intragroup sum of squares (with *y* representing the discriminant function vector). *v* represents the eigenvectors of the matrix resulting from a division of the between-group variance/covariance-matrix by the within-group variance/covariance-matrix.

Wilks' Lambda

The more dispersion is captured by a function, the more important it is. One of the most common criteria to evaluate the quality of the discriminant function is the so-called Wilks' Lambda (cf. Klecka 1980: 38f.; Decker et al. 2010: 511f.):

$$\Lambda = \frac{\text{unexplained dispersion}}{\text{total dispersion}}$$

For more than two groups, this is as follows (Decker et al. 2010: 512), with the eigenvalue γk for the *k*-th discriminant functions and with *K* being the number of discriminant functions:

$$\Lambda = \prod_{k=1}^{K} \frac{1}{1+\gamma_k}$$

However, if the test for the homogeneity of the intragroup variance/covariance gives a significant result—i.e., the assumption of multivariate normality is violated—a test for Lambda's significance becomes obsolete. This is because Wilks' Lambda is based on the assumption that the observed variables follow a multivariate normal distribution, and in the case of non-normality, Wilks' Lambda does not behave like the random variable to which it is compared in the test (cf. Klecka 1980: 40f.). If nothing indicates a departure from the test's assumptions, Lambda can be taken to estimate whether the discriminant functions significantly contribute to the distinction between the groups. Usually, this is done by a stepwise omission of discriminant functions, beginning with the first one and a subsequent testing of the significance of all remaining functions together. Wilks' Lambda for this so-called residual variance Λq is calculated as follows (Ibid.):

$$\Lambda_q = \prod_{k=q+1}^{K} \frac{1}{1+\gamma_k} \qquad \forall_q = 0,1,\ldots,K-1$$

Lambda is transformed into a Chi-square distributed value *Vq* as follows (Ibid.):

$$V_q = -\left[I - \frac{J+G}{2} - 1\right] \ln \Lambda_q$$

The degrees of freedom are computed by multiplying the difference between the number of variables and the number of omitted axes with the difference between the number of groups and the number of omitted axes plus one. In the formula for *Vq*, *I* is the number of cases, *J* is the number of variables, *G* is the number of groups, and $\ln \Lambda q$ is the natural logarithm of Lambda. The resulting Chi-square value is then used to evaluate the function's significance (Bartlett 1947).

Additional Quality Measures for Discriminant Functions

Additionally (and in the case of multivariate non-normality), the result of an LDA can be evaluated by comparing the known (a priori) group allocation of a case with its (a posteriori) allocation as predicted by the LDA. U. M. Garczarek proposed the following quality measures (Garczarek 2002: 157–159):

Correctness rate (CR): Adds all cases with identical a priori (original) and a posteriori (calculated) group allocations and gives the percentage of identical or "correctly" classified cases.

Accuracy (AC): Measures the distance between the a priori and the a posteriori group allocations and gives a value between 0 (completely different) and 1 (identical), since most cases (here: assemblages) may not be classified with full confidence to one group alone.

Ability to separate (AS): Measures the distance between the a posteriori probabilities (the calculated results are not necessarily unambiguous but may be distributed over several groups) and the a posteriori allocation (the final attribution of a case to a group) and sums up the respective values. Again, this quality measure exhibits values between 0 (completely ambiguous) and 1 (clear separation).

Confidence for each (true) group (CFvec): Measures the average a posteriori probability of the cases within the groups after their classification to an a posteriori group and gives a value between 0 (ambiguous attribution) and 1 (unambiguous attribution).

Confidence (CF): Measures the overall average probability of attributions for all a posteriori cases with a result between 0 and 1.

Visualization of the Results

Commonly, a two-dimensional scatter plot is used to visualize the LDA results. The first two discriminant functions serve as axes which define the space within which the cases (here: assemblages) are displayed according to their discriminant values, which serve as the cases' coordinates. The distances between the cases correspond to the Mahalanobis distances, which represents the Hellinger distances inasmuch as they are captured by the two discriminant functions displayed. In a second plot, the significance of each variable (here: tool types) for the discriminant functions is represented by arrows. The angles between the arrows as well as between the arrows and the discriminant axes correspond to the correlation between the variables (tool types) and the correlation between the variables and the axes, respectively. Arrows and points are not displayed jointly in order to assure an optimal visual representation of the information, namely, the dissimilarity of the cases on the one hand and the correlations of the variables on the other. In a joint plot, where the cases are optimally displayed, the accordingly scaled variable arrows would be very small. When, in contrast, the correlations between the variables are displayed correctly, the topology of the cases is distorted.

The Course of Analysis

The Discriminant Function Analysis was calculated using the using function "lda" of the package "MASS" (Venables and Ripley 2002) for R (R_Development_Core_Team 2011) version 2.13.0. The nonparametric test for intergroup variance was performed with the algorithm "adonis" (Anderson 2001), and the test for the homogeneity of the intragroup variance/covariance was conducted using the R function "betadisper" (Anderson 2006). Both tests are implemented in the package "vegan" (Oksanen et al. 2012). Furthermore, additional quality measures for discriminant functions were obtained with the R function "ucpm" (Garczarek 2002) implemented in the package "klaR" (Weihs et al. 2005), and the standardized canonical discriminant coefficients as well as the variable projections into LDA space were calculated with the function "candisc" of the package "candisc" (Friendly and Fox 2011).

For the investigation of spatial variability, only those assemblages yielding at least 100 lithic tools were taken into account (see Table A.35). This selection was made in order to minimize the influence of functional variability on the typological spectrum and to reinforce stylistic aspects. Specialized short-term camps (e.g., hunting stations) usually yield only a restricted and rather specialized set of tools, which occur in the whole investigated area. Thus, they potentially bias a comparative analysis toward a homogeneous appearance. Regional preferences for stylistic variations are expected to accumulate rather in larger assemblages. Even though the decision to set the number of tools at 100 is somewhat arbitrary, it seems appropriate since it excludes inventories of specialized short-term camps while leaving a sufficiently high number of collections to perform a reliable statistical analysis.

The investigation of the typological spectrum is carried out in two steps. First, the assemblages are analyzed according to their spatial position. Here, an initial investigation includes a comparison sample of Hamburgian assemblages in order to get an impression of the discriminatory power of the method and to set a frame of reference within which the results of the subsequent analysis—calculated with Magdalenian assemblages only—can be interpreted. Second, the influence of the assemblages' temporal position is investigated. Because the total number of radiometrically dated assemblages is rather low, inventories which contain fewer tools have to be considered as well. In order to obtain a sufficiently large number of ^{14}C-dated Magdalenian inventories, tool assemblages with at least 20 tools are also included (Table A.36). Finally, the results of both analyses are compared and evaluated collectively.

Analysis of Spatial Grouping

As explained above, it is necessary to set the group allocation for each case (i.e., each assemblage) prior to the analysis. Since the spatial analysis in this study is aimed at assessing the extent to which the tool composition of an assemblage depends on its geographical location, the group allocations correspond to the five regional groups which it was possible to identify by analyzing the raw material procurement pattern (see section "Procurement Pattern of Lithic Raw Materials"). This ensures that the groups derive from the object of study itself and are not imposed arbitrarily on the archaeological record. The analysis is based on a Hellinger-transformed table of the abundances of each of the lithic tool types in each inventory (Table A.35 shows the raw data—i.e., the abundances per type—and percentages). Two major questions are of interest. The first addresses whether there are significant differences between the regional groups according to the typological composition of their assemblages. The second concerns which types are most important for the groups' distinction. To answer these questions, a two-fold analysis is performed. In the first step, the calculation includes a comparison sample of Hamburgian sites, and then the Magdalenian assemblages alone are evaluated.

LDA of Hamburgian and Magdalenian Assemblages

In the course of the climatic amelioration after the Last Glacial Maximum, Central Europe was initially resettled by Magdalenian hunter-gatherers. Later, the onset of the Late Glacial Interstadial Complex (GI 1) correlates approximately with the simultaneous emergence of the Hamburgian and Creswellian around 15,000 cal BP. The bearer of these Paleolithic entities recolonized the northwestern European Plain and the present-day British Isles and coexisted with the later Magdalenian (Barton et al. 2003; Grimm and Weber 2008; Jacobi and Higham 2009). A now-obsolete theory about the relationship between the Magdalenian and the Hamburgian interpreted the sites in the Northern Plain simply as traces of Magdalenian hunter-gatherers, who exploited the northern regions seasonally too. This view is implausible because faunal evidence from several sites proves a year-round occupation of this area, and the diversity of the typological spectra between Magdalenian, Hamburgian, and Creswellian sites cannot be explained meaningfully by different functional requirements of hunting gear during a seasonal cycle. In current research, the Creswellian and Hamburgian are considered to be independent but closely related entities which are rooted in the Magdalenian (Barton et al. 2003; Terberger 2006).

Table 6.2 LDA of spatial grouping (Magdalenian and Hamburgian)

Group	1	2	3	4	5	6	Total
n	36	19	23	22	14	22	136
%	26.5	14.0	16.9	16.1	10.3	16.2	100

Number of assemblages per Magdalenian regional group and number of Hamburgian assemblages
1: Jura region, 2: Danube region, 3: Meuse-Rhine region, 4: Eastern Germany and Bohemia, 5: Moravia and Poland, 6: Hamburgian

The consideration of Hamburgian assemblages in the analysis is useful because it allows for an assessment of the discriminatory power of the method applied. In addition, the relation between Hamburgian and Magdalenian assemblages creates a frame of reference which enhances the evaluation of intra-Magdalenian variability, since it makes possible the interpretation of the Magdalenian's inner variance against the background of differences which can be observed between two distinct Paleolithic entities. Hamburgian collections seem to be an appropriate choice to serve as a reference sample, since their typological composition is similar to those of the Magdalenian, but at the same time, their occurrence is geographically restricted to the Northern European Plain and thus does not overlap much with the distribution of Magdalenian sites (see Fig. 1.1).

The first part of the analysis comprises 136 assemblages (Table 6.2) which are assigned to regional groups as follows:

According to the results of the "adonis" test for intergroup variance, the location-based grouping into the six groups presented above explains a significant amount of multivariate dispersion (Table 6.3). The determination coefficient (R^2) shows that 29 % of the observed typological differences can be explained by the regional grouping and thus by the geographical position of the assemblages involved. Also significant is the result of the test for the homogeneity of the intragroup variance/covariance (Table 6.3), which indicates different relationships between the occurrences of tool types in the groups and consequently a departure from the assumption of multivariate normality. Since LDA presupposes these relations to be stable across the groups, further evaluation of Wilks' Lambda is not appropriate in this case and could reduce or distort the discriminatory power of the analysis.

After these preliminaries, the LDA for the spatial grouping is conducted. The confidence values for each group (Table 6.4; cf. Garczarek 2002: 157–159) show the a posteriori attribution of Hamburgian assemblages (group 6) to be unambiguous, while the values among the Magdalenian groups 1–5 (and group 2 in particular) are less clear. The fuzzier allocation

Table 6.3 LDA of spatial grouping (Magdalenian and Hamburgian)

Nonparametric test for intergroup variance (adonis)						
	df	Sum of Sqs	Mean of Sqs	F model	Probability	R^2
Groups	5	11.136	2.227	10.638	0.000	0.290
Residuals	130	27.217	0.209	–	–	0.710
Total	135	38.352	–	–	–	1
Test for the homogeneity of the intragroup variance/covariance (betadisper)						
	df	Sum of Sqs	Mean of Sqs	F value	Probability	Permutations
Groups	5	0.234	0.047	4.191	0.001	9999
Residuals	130	1.453	0.011	–	–	–

Test results for inter- and intragroup differences

Table 6.4 LDA of spatial grouping (Magdalenian and Hamburgian)

Quality measures									
Confidence values for each group								AC	AS
Group	1	2	3	4	5	6	Average	0.639	0.786
Cf value	0.850	0.721	0.835	0.793	0.822	1.000	0.842		

Quality measures for the LDA classification. 1: Jura region, 2: Danube region, 3: Meuse-Rhine region, 4: Eastern Germany and Bohemia, 5: Moravia and Poland, 6: Hamburgian/North European Plain
AC Accuracy, *AS Ability* to separate

Table 6.5 LDA of spatial grouping (Magdalenian and Hamburgian)

Proportion of trace						
	Function 1	Function 2	Function 3	Function 4	Function 5	Total
% of Variance	85.3	6.9	3.9	2.4	1.5	100

Percentages of the dispersion captured per discriminant function

of the Magdalenian assemblages is also expressed in the rather low values for the LDA's accuracy (0.639) and ability to separate the groups (0.786). The six groups result in five discriminant functions (Table 6.5). The first two functions together account for 92 % of the multivariate dispersion, of which the first function alone captures 85 %.

The two-dimensional scatter plot, which displays the position of the analyzed assemblages in terms of the first two functions, thus captures the largest proportion of the present variance (Fig. 6.53). It shows a surprisingly clear distinction between the Hamburgian and Magdalenian assemblages, which are separated from each other mainly by means of the first function. The second function, in contrast, is responsible for an inner segregation of the Hamburgian and the Magdalenian clusters. Figure 6.54 and the standardized canonical discriminant coefficients (Table A.37) reveal that simple backed bladelets (BB), shouldered points (ShP), laterally retouched pieces (LR), and *zinken* (ZI) have the highest discriminatory power with regard to the first function, whereas *zinken*, simple backed bladelets, splintered pieces (SP), long burins (LBU), and long borers (LBO) are most important for the second function.

The combined evaluation of the classification results shows that 77 % of the original (a priori) location-based assignments are in accordance with the calculated (a posteriori) typology-based classification of the Discriminant Function Analysis (Table 6.6). If the cases were classified randomly to the six groups, a "hit ratio" of only 16.7 % (1/6) would be expected. Interestingly, there seems to be a very strict typological distinction between Hamburgian and Magdalenian assemblages. In fact, no assemblage from one of these groups has been reassigned to the other. Within the Magdalenian groups, however, reclassifications occur occasionally, especially between geographically neighboring groups (Table 6.6).

In summary, it can be stated that from a typological point of view, Hamburgian and Magdalenian assemblages exhibit a pronounced segregation from one another. The typological variability within the Magdalenian and Hamburgian entities is thus considerably lower than the variability between these two entities. This observation supports the segregation of the Magdalenian and the Hamburgian into two distinct entities and contradicts models which present them as different facies of the same phenomenon. The clear distinction also underlines the discriminatory potential of the applied

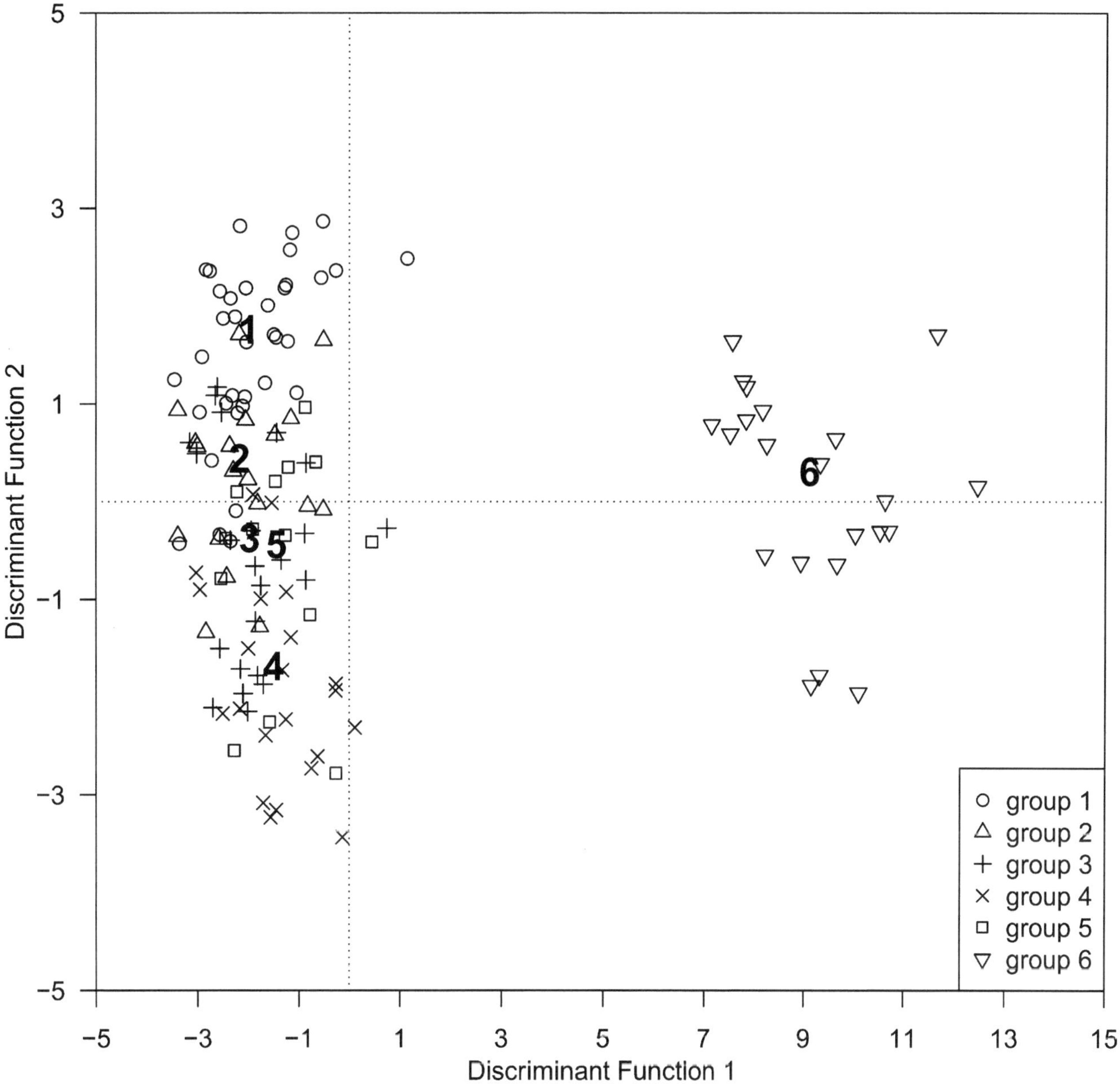

Fig. 6.53 LDA of spatial grouping (Magdalenian and Hamburgian). Positions of Magdalenian and Hamburgian assemblages in the space of the first two discriminant functions (1: Jura region; 2: Danube region; 3: Meuse-Rhine region; 4: Eastern Germany and Bohemia; 5: Moravia and Poland; 6: Hamburgian/North European plain). *Bold-face* numbers indicate the position of the groups' centroids. For centroid coordinates, see Table A.38 and for assemblage coordinates, see Table A.39

method. In order to analyze the intra-Magdalenian variance more closely, its assemblages are investigated once more in the following section, this time without the Hamburgian comparison sample.

LDA of Magdalenian Assemblages

The pronounced typological differences between the Hamburgian and Magdalenian assemblages dominated the previous analysis and thus concealed the intra-Magdalenian variance to a certain extent. In order to investigate the latter in more detail, the Hamburgian sites are now excluded. The group allocation of the remaining assemblages is maintained (Table 6.7). This results in the following distribution.

As observed in the conjoint analysis of the Magdalenian and Hamburgian assemblages, the test for the intergroup variance produces a significant result (Table 6.8) and thus strongly indicates that the a priori Magdalenian groups differ from one another in terms of their typological composition. The R^2 value, however, is considerably lower, since only 9 % of the observed dispersion can be explained by the geographic

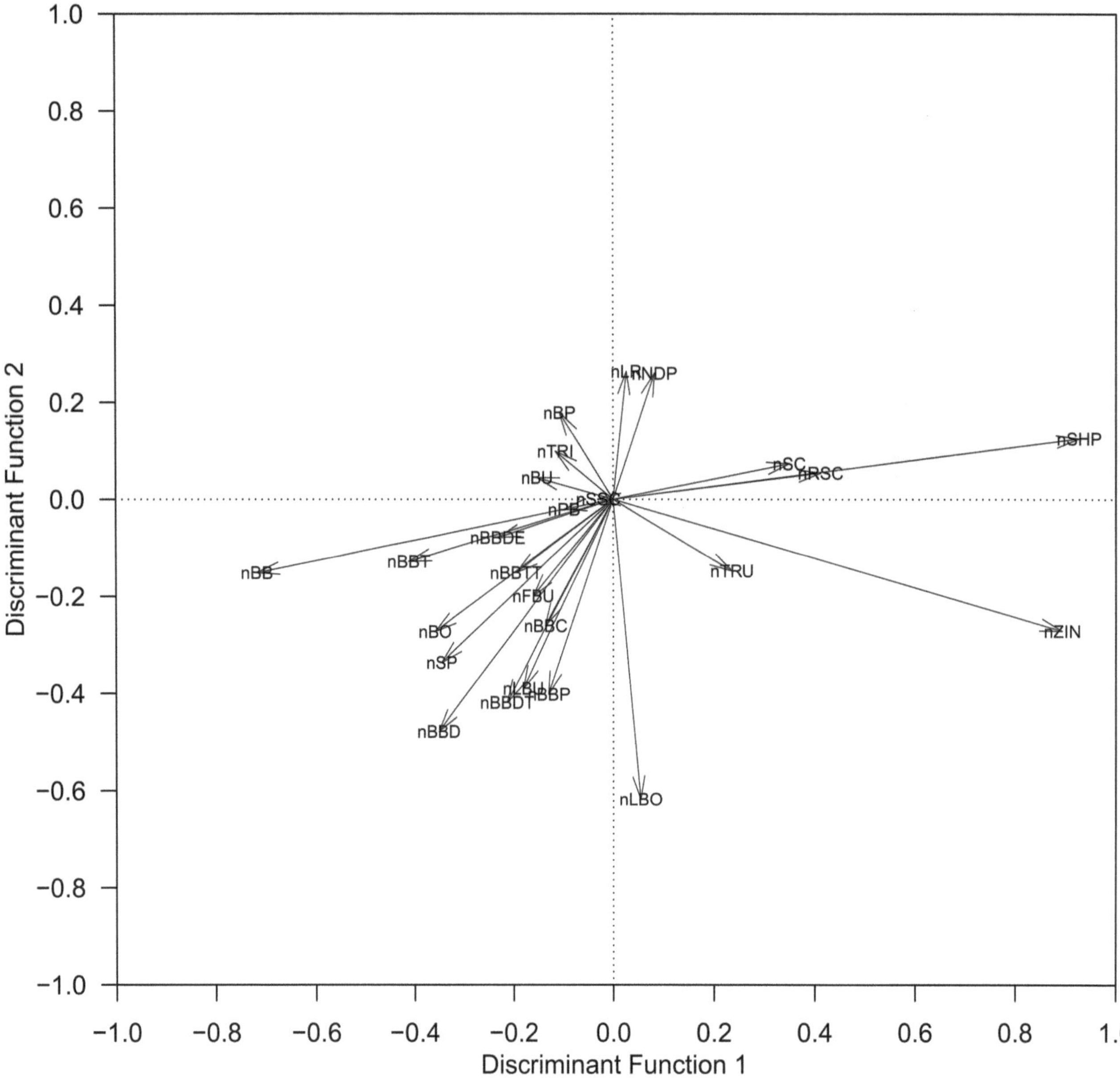

Fig. 6.54 LDA of spatial grouping (Magdalenian and Hamburgian). *Arrows* indicate the influence of tool types on the position of the assemblages in Fig. 6.53. For arrow coordinates, see Table A.40

position of the assemblages (i.e., the regional grouping of the Magdalenian sites). This result illustrates the strong internal coherence of Magdalenian assemblages, particularly against the background of the conjoint analysis. Again, the test for the homogeneity of the intragroup variance/covariance is likewise significant, and the relations among the variables (the tool frequencies) can consequently be expected to vary across the assemblages. Thus, an evaluation of Wilks' Lambda seems inappropriate, since the method can be expected to be influenced by distortions (see above).

With regard to the confidence values for each group (Table 6.9), it can be stated that with the exception of group 2 (Danube region), the results for the a posteriori attribution are comparatively high. Nevertheless, the strong internal coherence also leads to a decrease in the accuracy value (0.584) and the LDA's ability to separate (0.745) in comparison to the conjoint analysis of Magdalenian and Hamburgian assemblages described above.

The first two out of four discriminant functions together capture 76 % of the observed variance (Table 6.10). The two-dimensional scatter plot, which displays the assemblages' position according to the first two discriminant functions, therefore accounts for more than 3/4 of the variance. According to the standardized canonical discriminant coefficients (Table A.41), long burins (LBU), long borers (LBO), simple backed bladelets (BB), and splintered pieces (SP) have the highest

Table 6.6 LDA of spatial grouping (Magdalenian and Hamburgian)

Classification results													
n								%					
	1	2	3	4	5	6	Total	1	2	3	4	5	6
1	31	2	2	0	1	0	36	86.1	5.6	5.6	0	2.8	0
2	1	13	2	0	3	0	19	5.3	68.4	10.5	0	15.8	0
3	5	2	15	1	0	0	23	21.7	8.7	65.2	4.3	0	0
4	1	0	2	16	3	0	22	4.5	0	9.1	72.7	13.6	0
5	0	2	2	2	8	0	14	0	14.3	14.3	14.3	57.1	0
6	0	0	0	0	0	22	22	0	0	0	0	0	100
Total	38	19	23	19	15	22	136	CR: 77.2 % are classified correctly					

Cross-tabulation of given and predicted group memberships
Each row corresponds to an a priori group and each column to the classification (a posteriori group) of the LDA (1: Jura region, 2: Danube region, 3: Meuse-Rhine region, 4: Eastern Germany and Bohemia, 5: Moravia and Poland, 6: Hamburgian/North European Plain) with the left table giving the numbers and the right table displaying the row percentages. Thus, row 1 in the left table should be read as follows: 31 assemblages of group 1 were classified as group 1, 2 as group 2, 2 as group 3, 0 as group 4, 1 as group 5, and 0 as group 6
CR Classification results

Table 6.7 LDA of spatial grouping (Magdalenian)

Group	1	2	3	4	5	Total
n	36	19	23	22	14	114
%	31.6	16.6	20.2	19.3	12.3	100

Number of assemblages per Magdalenian regional group
1: Jura region, 2: Danube region, 3: Meuse-Rhine region, 4: Eastern Germany and Bohemia, 5: Moravia and Poland

Table 6.8 LDA of spatial grouping (Magdalenian)

Nonparametric test for intergroup variance (adonis)						
	df	Sum of Sqs	Mean of Sqs	F model	Probability	R^2
Groups	4	2.390	0.597	2.745	0.000	0.092
Residuals	109	23.727	0.218	–	–	0.908
Total	113	26.117	–	–	–	1
Test for the homogeneity of the intragroup variance/covariance (betadisper)						
	df	Sum of Sqs	Mean of Sqs	F value	Probability	Permutations
Groups	4	0.162	0.040	3.804	0.006	9999
Residuals	109	1.157	0.011	–	–	–

Test results for inter- and intragroup differences

Table 6.9 LDA of spatial grouping (Magdalenian)

Quality measures								
Confidence values for each group							AC	AS
Group	1	2	3	4	5	Average	0.58	0.75
Cf. Values	0.86	0.733	0.815	0.807	0.829	0.816		

Quality measures for the LDA classification. 1: Jura region, 2: Danube region, 3: Meuse-Rhine region, 4: Eastern Germany and Bohemia, 5: Moravia and Poland
AC Accuracy, *AS* Ability to separate

Table 6.10 LDA of spatial grouping (Magdalenian)

Proportion of trace					
	Function 1	Function 2	Function 3	Function 4	Total
% of variance	51.9	23.8	14.4	9.9	100

Percentages of the dispersion captured per discriminant function

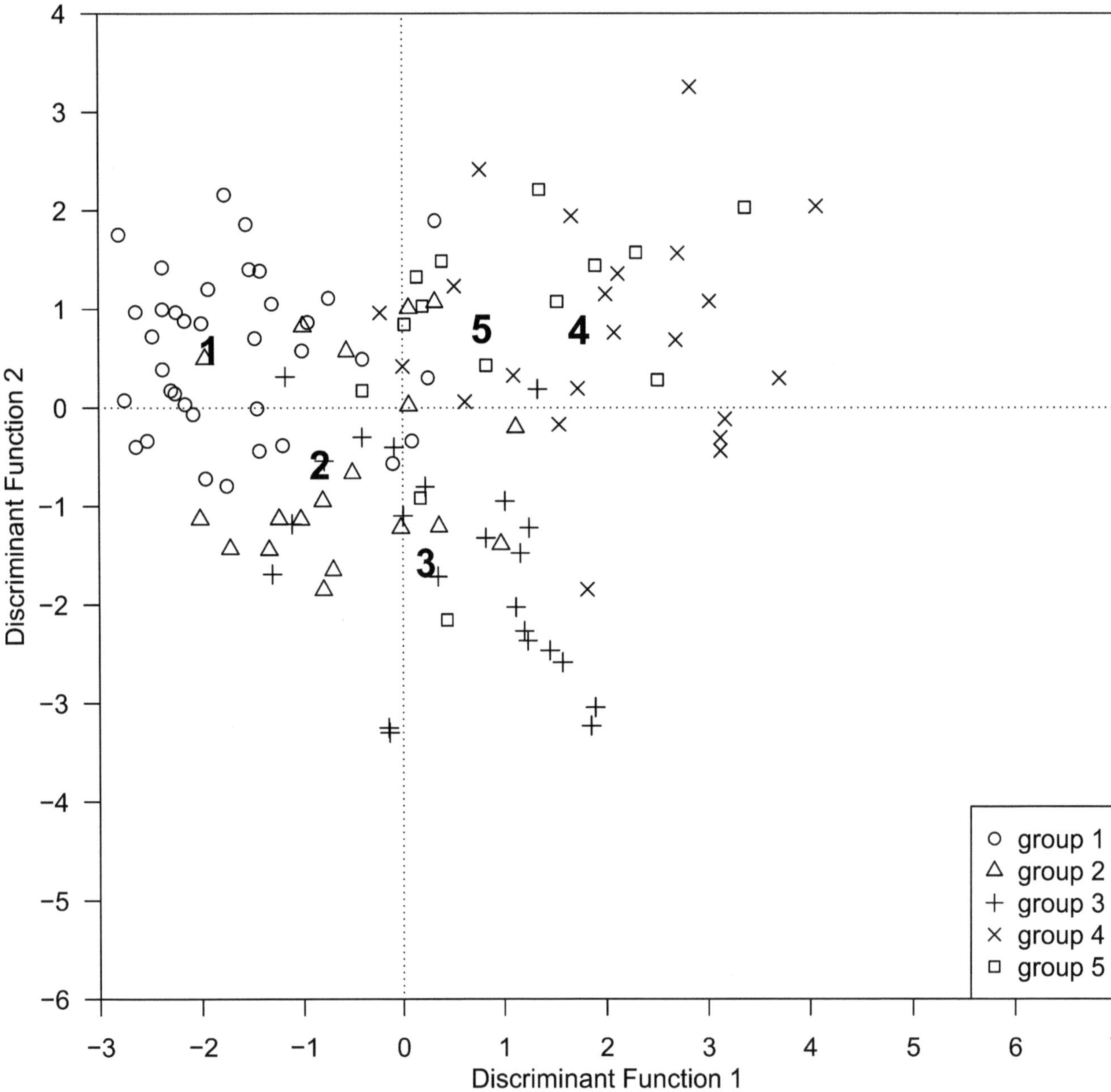

Fig. 6.55 LDA of spatial grouping (Magdalenian). Positions of assemblages in the space of the first two discriminant functions (1: Jura region; 2: Danube region; 3: Meuse-Rhine region; 4: Eastern Germany and Bohemia; 5: Moravia and Poland). *Bold-face* numbers indicate the position of the groups' centroids. For centroid coordinates, see Table A.42 and for assemblage coordinates, see Table A.43

discriminatory power for the first function, while long borers (LBO), simple backed bladelets (BB), long burins (LBU), and double backed bladelets (BBD) are most important for the second function.

Figure 6.55 displays the assemblages in the space of the first two discriminant functions. Here, the five groups are not segregated as clearly as might be expected given the quality measures of Table 6.9. However, there is a notable separation into two clusters: one comprising the groups 1–3, the other the groups 4 and 5. The dispersion within both clusters is best explained by two orthogonal axes capturing different aspects of the multivariate typological information. Within the first cluster, the cases (i.e., assemblages) of groups 1 and 3 are comparatively well separated, but show a strong overlap with the cases of group 2. The cases of groups 4 and 5, in contrast, overlap almost completely, indicating a very similar typological composition. The two clusters overlap in the center of the scatter plot, with the largest degree of overlap showing between groups 2 and 5.

For the interpretation of Fig. 6.55, it thus has to be kept in mind that the plot is only a two-dimensional representation of a four-dimensional space. The cases (assemblages) of

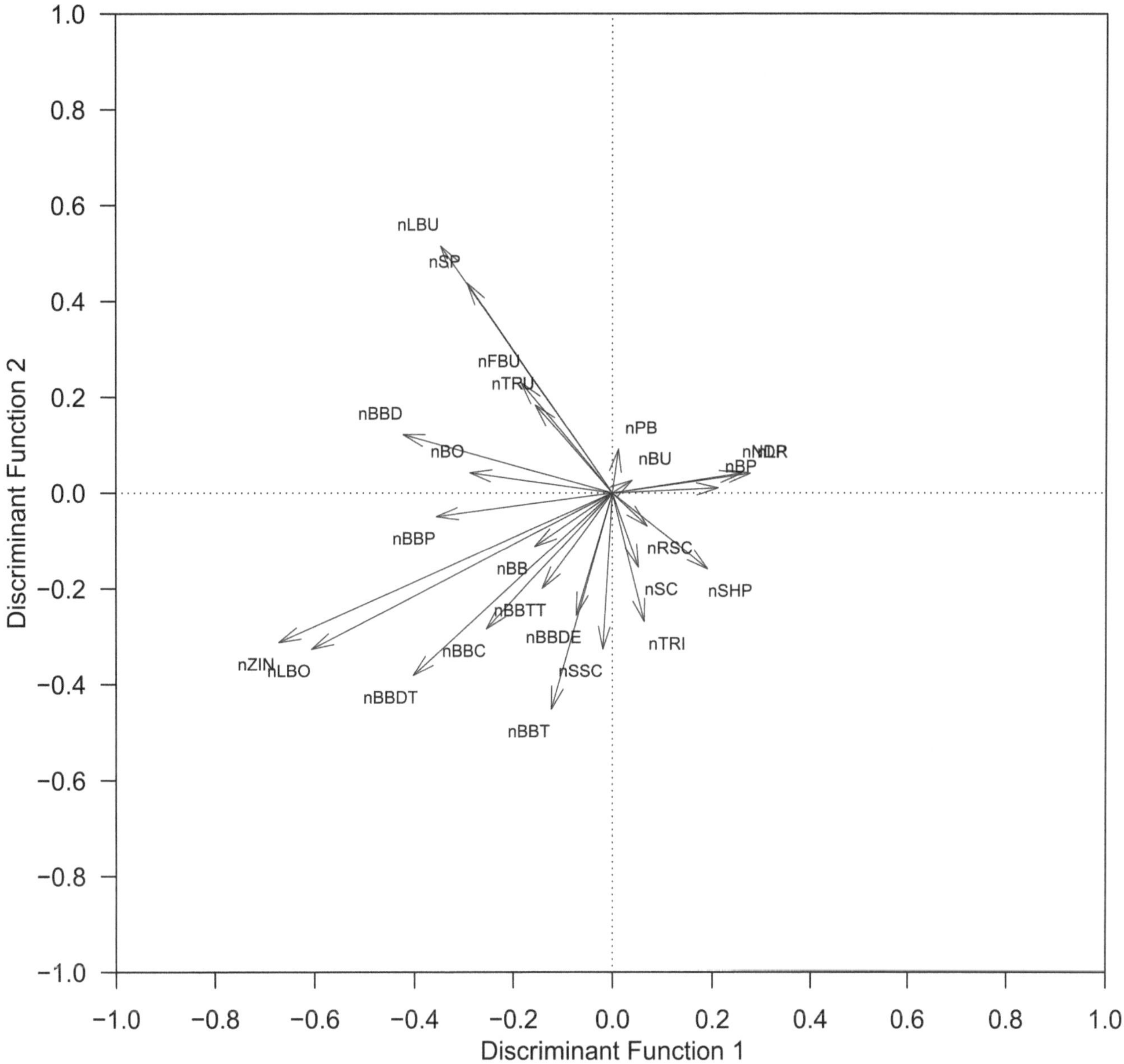

Fig. 6.56 LDA of spatial grouping (Magdalenian). *Arrows* indicate the influence of tool types on the position of the assemblages in Fig. 6.55. For arrow coordinates, see Table A.44

groups 1–3 are located on an axis which lies oblique to the first and second discriminant function. Long burins and splintered pieces on the one hand and shouldered points (ShP) on the other are the main contributors to this axis (Fig. 6.56). A visual animated inspection of the LDA result using the package "rgl" (Adler and Murdoch 2011) shows groups 4 and 5 to be best described by the third discriminant function, i.e., by an axis running perpendicular to the first two axes, which, in turn, mainly separate groups 1–3. For the separation of groups 4 and 5 from groups 1 to 3, backed points (BP), denticulated backed bladelets (BBD), and notched and denticulated pieces (NDP) are most important (see Table A.41, Function 3). A graphical illustration of this situation is omitted here, since a snapshot of the three-dimensional point cloud would appear almost unreadable in two dimensions.

Despite some overlapping between the groups, Table 6.11 shows that in 76 % of all cases, the classification predicted by the LDA (a posteriori) coincides with the original (a priori) group allocations. If this classification were carried out randomly, a "hit ratio" of 20 % (1/5) would be expected.

Regarding the outcome of the analysis, three observations are especially remarkable. First, it is striking that those cases which are attributed to groups other than the original one occur mainly in neighboring groups. In other words, they appear preferentially in those groups which represent geographically

Table 6.11 LDA of spatial grouping (Magdalenian)

Classification results											
n							%				
	1	2	3	4	5	Total	1	2	3	4	5
1	32	2	1	0	1	36	88.9	5.6	2.8	0	2.8
2	2	15	1	0	1	19	10.5	78.9	5.3	0	5.3
3	4	2	16	1	0	23	17.4	8.7	69.6	4.3	0
4	1	0	2	16	3	22	4.5	0	9.1	72.7	13.6
5	0	2	2	2	8	14	0	14.3	14.3	14.3	57.1
Total	39	21	22	19	13	114	CR: 76.3 % are classified correctly				

Cross-tabulation of given and predicted group memberships
Each row corresponds to an a priori group and each column to the classification (a posteriori group) of the LDA (1: Jura region, 2: Danube region, 3: Meuse-Rhine region, 4: Eastern Germany and Bohemia, 5: Moravia and Poland) with the left table giving the numbers and the right table displaying the row percentages. Thus, row 1 in the left table should be read as follows: 32 assemblages of group 1 were classified as group 1, 2 as group 2, 1 as group 3, 0 as group 4, and 1 as group 5
CR Classification results

adjacent regional groups. Reclassified assemblages from a cluster of groups 1, 2, and 3, for instance, are almost exclusively predicted to belong to other groups of the same cluster. Only 3 out of 78 assemblages a priori assigned to the groups 1, 2, and 3 are reclassified to the group 4 ($n=1$) or 5 ($n=2$) (Table 6.11). All other reclassifications take place within these three groups (for a detailed overview, see Table A.45).

Second, it is remarkable that the relative position of the groups in the diagram largely corresponds to the real geographic relations between the regional groups they represent. Group 1, which corresponds to the Circum-Jurassic Group, is positioned closest to group 2, the equivalent of the Danube group. The cases of group 3, which represent the northwestern Meuse-Rhine Group, are mostly located opposite to and at a greater distance from those of the first group. Groups 4 (Eastern Germany and Bohemia) and 5 (Moravia and Poland) are located close to each other, while the cases of group 4 are situated closer to group 3 and those of group 5 closer to group 2 (cf. Fig. 6.55).

Third and perhaps most astonishing, the segregation into two clusters—groups 1–3 on the one hand and groups 4 and 5 on the other—mirrors the interaction pattern that it was possible to observe by analyzing the distribution pattern of mollusk shells (see section "Procurement Pattern of Mollusk Shells and Amber"). Here, the strong interaction between the regional groups in the Jura, Danube, and Meuse-Rhine areas is reflected by the fact that the corresponding groups 1–3 are located on a single axis in the lower left part of Fig. 6.55. The clear separation of these three western groups from the two in the eastern part of the investigated area (Eastern Germany and Bohemia as well as Moravia and Poland) in terms of mollusk shell exchange finds its analogy in the fact that the cases of groups 4 and 5 are oriented along an axis which runs perpendicular to that of groups 1–3. The LDA result therefore traces the segregation of the CEM into five regional groups on the one hand and into two supra-regional groups on the other with regard to the influence of the geographical position of sites on the typological spectra of their assemblages. This is surprising inasmuch as this insight was not predictable on the basis of the spatial analysis performed in the previous chapter, when all tool types were analyzed individually. It only becomes visible in the joint evaluation of all typological information using multivariate statistics.

In order to analyze more closely the geographic position of those assemblages that are attributed to groups other than the predefined ones, the outcome of the Discriminant Function Analysis is mapped in Fig. 6.57. As is already apparent in the scatter plot, a spatial overlap is visible between the assemblages of groups 1, 2, and 3, which correspond to the assemblages in the Jura, Danube, and Meuse-Rhine regions. Here, it is interesting to note that all of the cases in group 2 which are reattributed to groups 1 and 3 are restricted to the western part of the Swabian Alb, whereas the assemblages in the Jura region that are reattributed to the Danube Group (group 2) and the Meuse-Rhine Group (group 3) are restricted to its northern part. This observation corresponds to the distribution pattern of mollusk shells (see section "Procurement Pattern of Mollusk Shells and Amber"), which shows that the sites in the western part of the Swabian Alb were much more strongly involved in exchange processes than were those in the eastern part or in the Franconian Alb. In this sense, it is conspicuous that only the northeasternmost assemblage of group 1, namely, Munzingen (ID 122), is reattributed to group 3. This peculiarity gains relevance when it is seen against the background of the raw material procurement pattern. The raw material with the southernmost origin that can be traced in the assemblages of the Meuse-Rhine Group, namely, in those from Gönnersdorf and Dreieich-Götzenhain (Floss 1994; Terberger et al. 2009a; see section "Procurement Pattern of Lithic Raw Materials"), comes from the Markgräflerland in the vicinity of Munzingen. For the Meuse-Rhine area, it can be stated that assemblages with

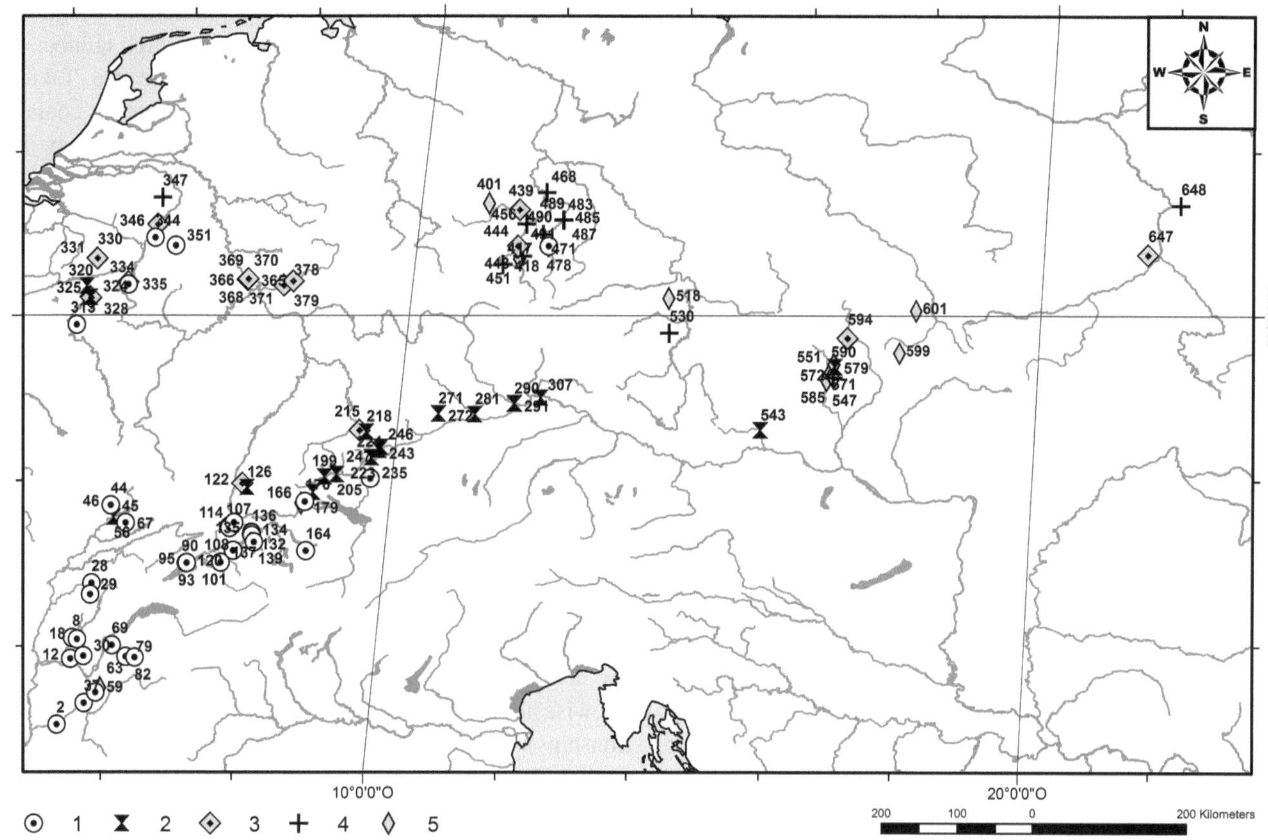

Fig. 6.57 LDA of spatial grouping (Magdalenian). Map of a posteriori classifications of Magdalenian assemblages

a changed allocation—except for the assemblage from Griendtsveen (ID 347) in the north—are exclusively reclassified to the Jura and Danube Groups and that these reclassifications occur throughout the entire area. In fact, Griendtsveen is the only inventory in groups 1, 2, and 3 which was reclassified as belonging typologically to group 4.

In the Eastern German and Bohemian site cluster, the inventories of Oelknitz (ID 444) and Nebra (ID 439) are reattributed to the Meuse-Rhine Group, whereas the site of Gera-Schafgraben (ID 480) is reclassified as belonging to the Circum-Jurassic Group. The assemblages of Kvíc (ID 518), Kniegrotte (ID 451), and Bad Frankenhausen (ID 401) are, in contrast, classified as belonging typologically to the Polish-Moravian Group. The reclassification of the Kniegrotte assemblage to the Polish-Moravian Group is remarkable inasmuch as it represents the oldest Magdalenian occupation in Eastern Germany and as such exhibits typological relations to the eastern part of the investigated area. At the same time, the Kniegrotte assemblage is—aside from Gera-Schafgraben—the only eastern German collection for which the occurrence of mollusk shells from western sources is reported (section "Procurement Pattern of Mollusk Shells and Amber"). Conspicuously, the latter inventory is also the only one in the eastern part of the investigated area which is reattributed to the Circum-Jurassic Group. These observations thus speak in favor of an exchange of ideas between the western and the eastern supra-regional groups, an exchange which was already established during the initial phase of the occupation of Eastern Germany (for further discussion, see Chap. 8, section "Towards a New Theory of the Recolonization of Central Europe"). Group 5, which represents the Polish-Moravian site cluster, comprises the lowest number of assemblages (14 cases). Moreover, 43 % are reclassified to other groups by the LDA. Here, it is noteworthy that the westernmost assemblage at Kamegg (ID 543) is reattributed to the Danube Group adjacent to the west.

Analysis of Temporal Grouping

The temporal analysis is based on 78 assemblages each containing at least 20 lithic tools (Table A.36). They are sorted into four temporal groups (Table 6.12) according to their calibrated ^{14}C-dates (Table A.6).

Just as in the two previous analyses, the investigation of temporal variability starts with the test for the intergroup variance in order to see whether the temporal grouping can explain the typological diversity contained in the multivariate table of assemblages and Hellinger-transformed abundances of tool types. Here, the resulting *p*-value of 0.056 falls only slightly short of significance (usual threshold 0.05; Table 6.13). It can therefore be concluded that defining temporal groups by intervals of 1000 years is perhaps too fine grained for the question of typological variance between 16,000 and 13,000 cal BP. With regard to the LDA results, the relatively low confidence value for the LDA classification of group 3 (Table 6.14) indicates that the cases of this group (15,000–14,000 cal BP) might be responsible for the insignificant intergroup differences.

The fact that only a small part of the typological dispersion can be explained by the temporal groups alone ($R^2=6$ %; Table 6.13) points to the same conclusion. The strong diachronic similarities of the Magdalenian assemblages—especially between 16,000 and 14,000 cal BP—is also expressed in the low values for the LDA's accuracy (0.543) and its ability to separate (0.678), which show a clear decrease in comparison to the spatial analyses. However, significance strongly depends on the sample size and an increase in firmly dated assemblages with a high number of tools may shift the test results toward significance. Taking into account the fact that three out of the four groups contain only 12–16 cases, the determination coefficient may be improved by further research, especially with regard to early and late sites. Furthermore, considering that the departure of the (arbitrary) confidence threshold is minimal (0.06 %) and that the LDA gives an empirically meaningful result against which the fuzzy structure of some groups can be explained, the slim margin by which the threshold is exceeded can be considered negligible. Nevertheless, it has to be noted that the temporal grouping based on 78 inventories performs more weakly in explaining the intergroup differences than does the regional one, which is based on 114 assemblages. The smaller number of available dated assemblages and the low threshold of 20 tools per inventory probably add an additional amount of unstructured dispersion, so that the method is no longer able to clearly discern the "signal" from the "noise." Also in contrast to the previous analyses, the test for homogeneity of the intragroup variance/covariance is likewise not significant (Table 6.13), indicating that the relationships between the tool frequencies are homogeneous or similar. Thus, the technical requirements of the LDA are met, and an evaluation of the LDA quality by Wilks' Lambda criteria may supply reliable results.

The first two discriminant functions displayed in Fig. 6.58 capture 83 % of the observed typological variance; only the first is significant according to Wilks' Lambda (Table 6.15). This is underlined by the canonical correlation, which measures the ability of a discriminant function to reliably discern between the groups.

Table 6.12 LDA of temporal grouping

Group	1	2	3	4	Total
n	14	36	16	12	78
%	17.9	46.2	20.5	15.4	100

Number of assemblages per Magdalenian temporal group
Group 1: Older than 16,000 cal BP, Group 2: 16,000–15,000 cal BP, Group 3: 15,000–14,000 cal BP, Group 4: 14,000–13,000 cal BP

Table 6.13 LDA of temporal grouping

Nonparametric test for intergroup variance (adonis)						
	df	Sum of Sqs	Mean of Sqs	F model	Probability	R^2
Groups	3	1.068	0.356	1.491	0.056	0.057
Residuals	74	17.659	0.239	–	–	0.943
Total	77	18.727	–	–	–	1
Test for the homogeneity of the intragroup variance/covariance (betadisper)						
	df	Sum of Sqs	Mean of Sqs	F value	Probability	Permutations
Groups	3	0.004	0.001	0.121	0.948	9999
Residuals	74	0.834	0.011	–	–	–

Test results for inter- and intragroup differences

Table 6.14 LDA of temporal grouping

Quality measures							
Confidence values for each group						AC	AS
Group	1	2	3	4	Average	0.543	0.678
Cf values	0.770	0.806	0.710	0.835	0.784		

Quality measures for the LDA classification. Group 1: Older than 16,000 cal BP, Group 2: 16,000–15,000 cal BP, Group 3: 15,000–14,000 cal BP, Group 4: 14,000–13,000 cal BP
AC Accuracy, *AS* Ability to separate

Table 6.15 LDA of temporal grouping

Proportion of trace and canonical correlation				
	Function 1	Function 2	Function 3	Total
% of variance	49.6	33.7	16.7	100
Canonical correlation	0.729	0.660	0.525	–
Significance of function(–s)				
Test of function(s)	Wilks' Lambda	Chi-square	df	Probability
1–3	0.191	103.335	75	0.017
2–3	0.409	55.912	48	0.202
3	0.724	20.191	23	0.630

Percentage of variance captured per discriminant function, discriminatory power (canonical correlation, above), and test of the functions' significance according to Wilks' Lambda

Table 6.16 LDA of temporal grouping

Classification results									
n						%			
	1	2	3	4	Total	1	2	3	4
1	9	2	1	2	14	64.3	14.3	7.1	14.3
2	2	31	3	0	36	5.6	86.1	8.3	0
3	0	2	14	0	16	0	12.5	87.5	0
4	2	2	0	8	12	16.7	16.7	0	66.7
Total	13	37	18	10	78	CR: 79.4 % are classified correctly			

Cross-tabulation of given and predicted group membership
Each row corresponds to an a priori group and each column to the classification (a posteriori group) by the LDA (1: Older than 16,000 cal BP, 2: 16,000–15,000 cal BP, 3: 15,000–14,000 cal BP, 4: 14,000–13,000 cal BP) with the left table giving the numbers and the right table displaying the row percentages. Thus, row 3 in the left table should be read as follows: 0 assemblages of group 3 were classified as group 1, 2 as group 2, 14 as group 3, and 0 as group 4
CR Classification results

With regard to the standardized canonical correlation coefficients, it emerges that shouldered points and triangles are most important for the constitution of the first function, whereas burins, borers, and simple backed bladelets have the highest discriminatory relevance for the second function (Table A.46). The fact that tool types which are comparatively numerous throughout the whole period of the Magdalenian, such as simple backed bladelets, burins, and borers, are mainly responsible for the second function indicates that functional differences in the assemblages have a comparatively strong impact on the typological dispersion in this analysis. However, for 80 % of the a posteriori cases, the typology-based chronological classification corresponds to the original radiocarbon-based assignments (Table 6.16). If the assemblages were assigned randomly, the expected "hit ratio" would be 25 % (1/4). Moreover, those cases which are attributed to groups other than the predefined ones are mostly reassigned to the next older or younger group. This trend is particularly strong in groups 2 and 3.

Looking at Fig. 6.58, it becomes apparent that the cases of groups 1 and 4 are comparatively well separated, whereas those of groups 2 and 3 show considerable overlap. With regard to Fig. 6.59, it can be said that the separation of the oldest group (group 1) depends particularly strongly on the presence of triangles, while the youngest group (group 4) is separated from the other groups mainly in terms of backed points. Conspicuously, the assemblages attributed to the oldest and youngest groups are scattered over a wider area within the discriminant space than are the assemblages of groups 2 and 3, which appear to be much more clustered and thus more consistent and homogeneous in terms of their typological composition. So it seems that the exchange of ideas and a corresponding homogenization of the typological spectra increased after 16,000 cal BP and remained at a comparatively high level for about 2000 years. Toward the end of the Magdalenian, around 14,000 cal BP, this homogeneity decreased again. A reason for this development at the end of the Magdalenian may be the incipient regionalization that characterized the Late Paleolithic. In the end, it seems that those assemblages which are dated between 16,000 and 14,000 cal BP are typologically very similar, whereas older and younger assemblages differ markedly from this group as well as from one another.

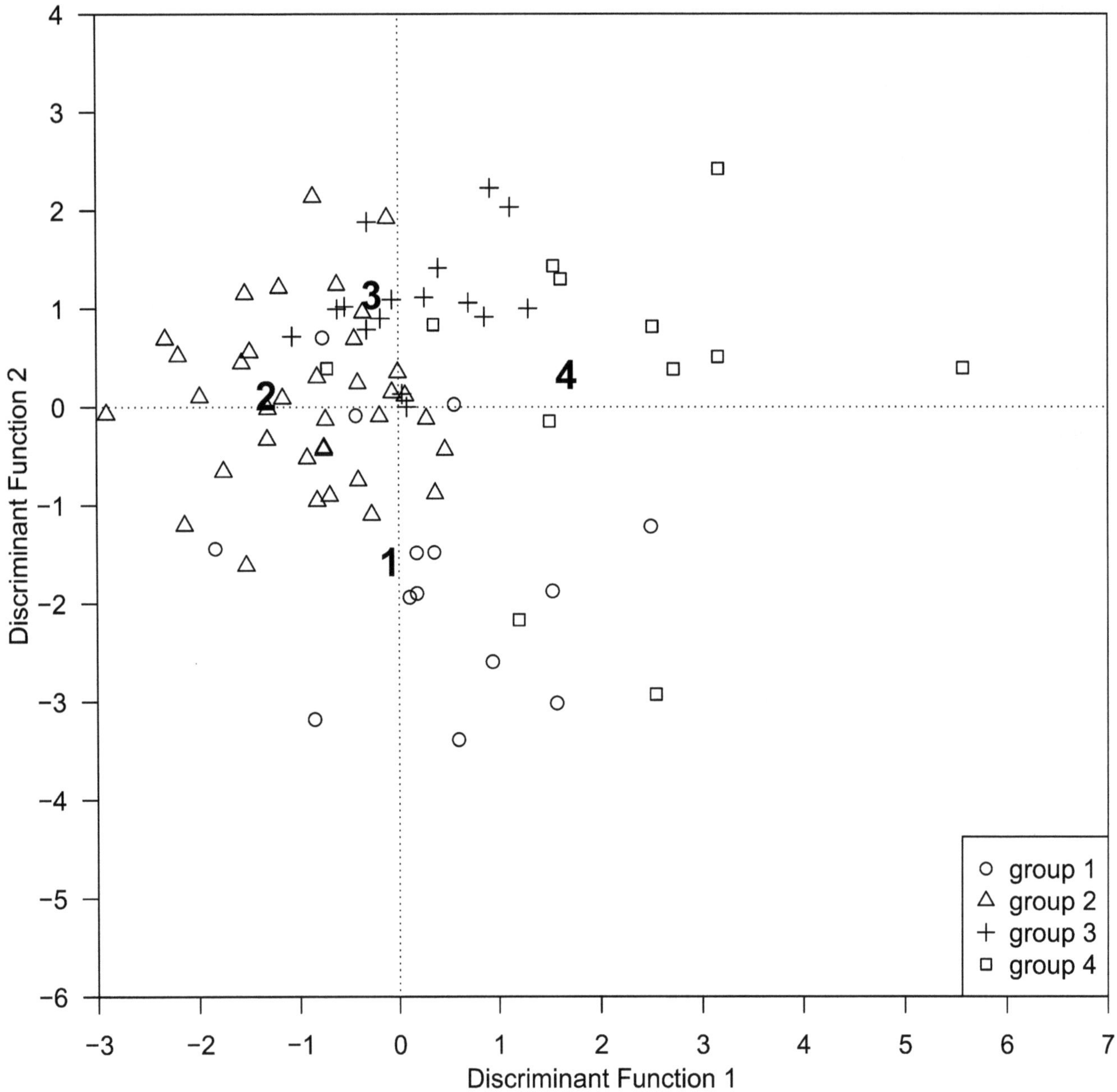

Fig. 6.58 LDA of temporal grouping. Positions of assemblages in the space of the first two discriminant functions (1: Older than 16,000 cal BP; 2: 16,000–15,000 cal BP; 3: 15,000–14,000 cal BP; 4: 14,000–13,000 cal BP). *Bold-face* numbers indicate the position of the groups' centroids. For centroid coordinates, see Table A.47, and for assemblage coordinates, see Table A.48

Correspondence Analysis: Multivariate Investigation of Spatial and Temporal Trends in Blade and Bladelet Production

The analysis of technological characteristics must account for the fact that a blank with its striking-platform remnant is the result of a complex interaction of factors (see above). Indeed, one might say that in the morphology of each blank, the specific combination of factors is preserved as they came together in the moment of the blank's detachment. Most of this information is, of course, stored in the butt itself as the impact-receiving part of the blank. A problem which arises during the analysis of this information is the dichotomy that generally exists between detailed records of many individual variables on the one hand and an overall interpretation of their combination on the other. Whereas the former is necessary to describe the conditions at the moment of detachment as precisely as possible, the latter is indispensable for a comprehensive understanding of the knapping technique as a whole. Thus, a method of analysis is required which is able

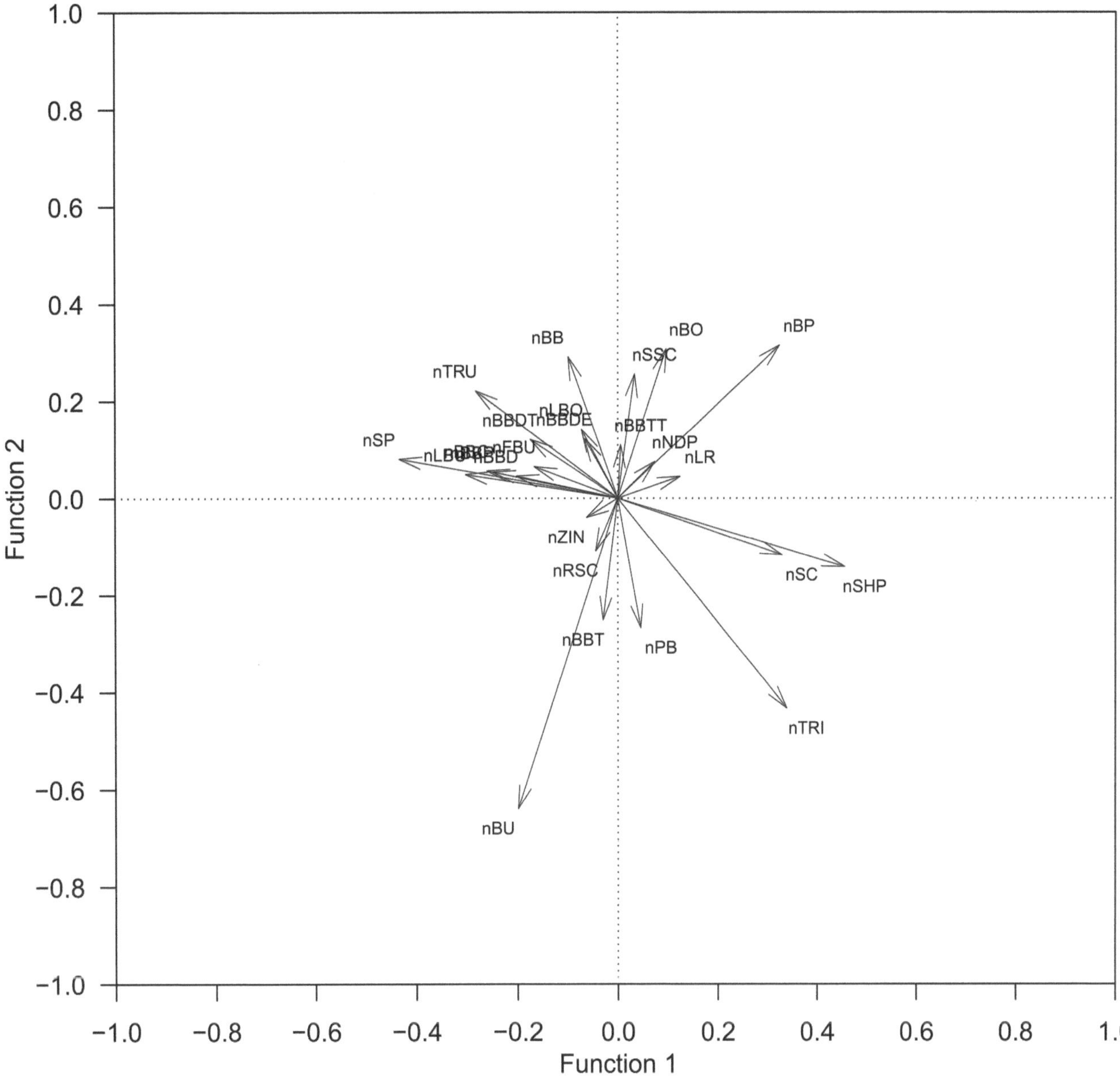

Fig. 6.59 LDA of temporal grouping. *Arrows* indicate the influence of tool types on the position of the assemblages in Fig. 6.58. For arrow coordinates, see Table A.49

to bridge the gap between individual, unconnected variable observations, and the complete multivariate information represented in the combination of all variables. Such approach would guarantee a maximum of objectivity. When dealing with artifact counts, especially counts of different attributes, a method well-suited for using absolute frequencies is the so-called Correspondence Analysis (Greenacre 2007). It offers not only an appropriate statistical evaluation of the data, but also provides a descriptive visualization of the results.

The Method

The Correspondence Analysis (CA) is a multivariate statistical method developed around 1960 by the French statistician J.-P. Benzécri, who called it *analyse des données* (Benzécri 1973). Later it became widely applied in the social sciences (see Greenacre 2007). In archaeological research, it continues to be successfully used for seriations and the formation of relative chronologies (Müller and Zimmermann 1997; Shennan 2004: 308f.), as well as for the analysis of spatial patterns (Loew 2006: 124ff.) and the comparison of assemblage compositions (Küßner 2009: 113f.; Scharl 2010). In all cases, the CA is used to sort a contingency table of objects and their features according to the main similarities and differences in their feature characteristics, as represented by their different frequencies. Usually, the rows of the table contain the objects of the analysis, while the columns contain the features. Thus, each row describes an object according to its feature characteristics, and each column describes a feature with regard to its occurrence in different objects.

In a first step, the content of each cell of the table—in other words, the abundance of the collective occurrences of each respective object and feature—is divided by the total number of all summed abundances to calculate their relative abundances. The sums of the abundances in each row and column, the so-called marginal distributions, can be used to calculate the expected abundance for each object-feature combination (Backhaus et al. 2008: 545; Roth 2011). These expected values express what the table would be like if it contained only average values. Average values are computed using the marginal distributions of the rows and columns and the total number of counts. The marginal distributions divided by the total number of counts can be understood as a representation of the average row of the average column (Greenacre 2007: 18f.; Roth 2011). Now, the table with the expected values can be used as a reference point to determine the amount of deviation from this average, which is contained in the original contingency table, also called the observed table. The deviation is calculated using the following equation (Greenacre 2007: 26f.; Backhaus et al. 2008: 546; Roth 2011):

$$\frac{\left(\text{observed table} - \text{expected}(\text{average})\,\text{table}\right)^2}{\text{expected}(\text{average})\,\text{table}}$$

This equation produces a table which contains the squared deviations from an average table composition as a fraction of the average table composition.

The way the CA solution is computed makes it possible to measure the similarity of rows and columns using a weighted Euclidian distance, the so-called chi-square (χ^2) distance (Greenacre 2007: 25f., 38f.; Roth 2011). This equation is as follows (after Leyer and Wesche 2007: 52):

$$\chi^2\,\textbf{distance} = \sqrt{\text{total sum}} \times \sqrt{\left(\Sigma \frac{1}{\text{sum of column}} \times \left(\frac{\text{cell value row 1}}{\text{total of row 1}} - \frac{\text{cell value row 2}}{\text{total of row 2}}\right)^2\right)}$$

For columns, the approach is analogous. In the end, the way a CA is computed makes it possible to combine the chi-square distance for both rows and columns into a joint result.

The sum of the abundances in a row or column is called the mass (Greenacre 2007: 24; Roth 2011). The higher the mass of a row, the more important a difference from the average row becomes, since the abundance is used to calculate the average in the first place. Therefore, rows (and columns) with a high mass influence the CA result more strongly than do those with a low mass. The dissimilarity between all of the rows (or columns) in a table and the average row (or column) is called the total inertia (Greenacre 2007: 30; Roth 2011). The more dissimilarity is present in the table, the higher the inertia becomes; rows (or columns) which show especially marked differences in comparison to the average row (or column) contribute more to the inertia than do those with minor differences. The inertia is calculated by dividing the chi-square values by the number of all summed abundances (Greenacre 2007: 32; Roth 2011).

A major purpose of a CA is to display the rows and columns of a contingency table in a joint space with a minimum of dimensionality and a maximum of preserved information (Backhaus et al. 2008: 547). The results of the calculation "(observed – expected)^2/expected" can be understood as vector coordinates for the row elements, which characterize the position of the objects and features in relation to the average object in an n-dimensional space; this space has as many dimensions as there are corresponding columns, minus one (Greenacre 2007: 48). The approach for determining the columns' elements is the same, save that the operation is repeated with a transposed table. If the position of each object and feature is represented by a point, one can think of their entirety as a point cloud in a multidimensional space. The CA seeks to draw new axes (dimensions) into this point cloud in such a way that the first axis captures the cloud's maximum extent and therefore the majority of the inertia. The second axis is chosen in such a way that it runs perpendicular to the first one and captures the cloud's second largest extent and so forth until a number of dimensions is reached which is one less than the lowest value of either the total rows or total columns (Greenacre 2007: 41f.; Roth 2011). Mathematically, the positions of the axes are calculated using a matrix algebra algorithm called singular value decomposition (Greenacre 2007: 47; Backhaus et al. 2008: 548). Because most of the information (inertia) in the n-dimensional cloud is now captured by just a few axes, a CA can be understood as a method for dimension reduction (Greenacre 2007: 41; Roth 2011). The point where all of the axes cross is the origin of the new coordinate system, which coincides with the position of the average row (or column).

Unfortunately, a single value decomposition gives a row solution, where the row and column points are located according to the so-called standard coordinates (Greenacre 2007: 62). That is, they are not located in the same space, because both row and column points are located at positions that give the maximal extension of the multivariate space for the respective other. With columns in standard coordinates, the average row should lie at the origin of the coordinate system and vice versa. One possibility for displaying both sets of coordinates in a joint space is scaling. Here, the row standard coordinates are multiplied with the singular values resulting in so-called principal coordinates. Because the singular values are always less than 1, the row principal coordinates are "shrunk" into the columns' space (Greenacre 2007: 68f.; Roth 2011). In the end, the graphical output is an asymmetrical biplot, where the columns are displayed according to their standard coordinates and the rows according to their principal coordinates. The advantage of this kind of representation is that it makes it possible to interpret not only the relative spatial position of row points *or* column points separately from one another, but also row and column points together as a whole (Greenacre 2007: 68f.).

Course and Corpus of the Analysis

The following CA was calculated using R (R_Development_Core_Team 2011) version 2.13.0, package "ca" (Greenacre and Nenadic 2010). First, the position of the assemblages with regard to the features is analyzed. In a second step, the results so obtained are juxtaposed with the available chrono- and chorological information to investigate possible spatial and temporal trends.

The Assemblages

During the process of data collection, 26 assemblages from 20 sites were recorded (Tables A.13, A.14, A.15, A.16, A.17, A.18, A.19, A.20, A.21, A.22, A.23, A.24, A.25, A.26, A.27, A.28, A.29, A.30, A.31, A.32, A.33 and A.34). In the case of smaller assemblages, all available platform remnants were analyzed, whereas for larger assemblages, samples of 100 blades and bladelets were chosen at random. Since this selection of specimens was nearly exhaustive even for many larger assemblages, the analyzed samples are regarded as sufficiently ample to be statistically representative. For the two sites of Putim and Borecnice 2, however, the available number of blades and blade fragments with an analyzable butt were considered too small to be representative and were therefore excluded from the analysis. Thus, the CA was performed with 24 assemblages from 18 sites. In the cases of Brillenhöhle, Kaufertsberg, and Teufelsbrücke, two stratigraphically separated horizons were analyzed, respectively. For Klementowice-Kolonia, the artifacts of the first excavation (1981–1982) and the current excavation were recorded separately, because the relationship between the two was still under debate at the moment of data acquisition (pers. comm. T. Wiśniewski). The special circumstances at the site of Oelknitz also required differentiated handling: here, several pits were excavated, some of which have been radiocarbon dated (see Table A.6). Because the dates show a clear division of the pits into two temporal groups, the artifacts found in those pits were recorded separately. Hence, the recorded artifacts from Oelknitz are divided into three assemblages: the first comprises the artifacts from the older pits, the second contains the specimens from the younger ones, and the third encompasses the pieces from an undated context. Unfortunately, the first two assemblages are comparatively small, having only 18 and 13 pieces, respectively.

The Technological Features

For each artifact, a set of 15 technological attributes with 57 possible manifestations was recorded (see Chap. 2, section "Blade Technology"). Besides the measurements of the length, width, and thickness of the completely preserved specimen, special emphasis was given to those attributes that characterize the preparation and morphology of the striking-platform remnant. For the CA, the individual manifestations for each artifact were counted per site (see Tables A.13, A.14, A.15, A.16, A.17, A.18, A.19, A.20, A.21, A.22, A.23, A.24, A.25, A.26, A.27, A.28, A.29, A.30, A.31, A.32, A.33 and A.34), and their abundance was transformed into a contingency table in which the rows contain the different assemblages and the columns contain the different manifestations, i.e., the features. In the second step, it was necessary to eliminate some features from the table because of overlaps in terms of their technological meaning. For instance, presence and absence of an impact point represent two sides of the same coin. If both manifestations contributed to the CA, this doubling would bias the result by overemphasizing this issue. For this reason, only the positive manifestations were taken into account. The only exception is the absence of a dorsal reduction, since in contrast to an impact point or bulbar scar, which can either be present or absent, the decision to forego the dorsal reduction is only one of several options and is not in diametric contrast to the possible positive manifestations. In the end, a selection of 28 features was chosen which seem appropriate to describe the nature of the applied striking techniques (Table A.50).

The features used in the Correspondence Analysis, displayed in Figs. 6.60 and 6.62 and listed in Table A.50, are:

Striking-platform remnant smaller than 5 mm^2	to5
Striking-platform remnant between 5 and 15 mm^2	5to15
Striking-platform remnant over 15 mm^2	ov15
Striking-platform remnant smooth	Ssm
Striking-platform remnant facetted	Sfc
Striking-platform remnant oval	Sov
Striking-platform remnant punctiform	Spt
Striking-platform remnant linear	Sln
Striking-platform remnant triangular	Sdr
Striking-platform remnant rectangular	Src
Striking-platform remnant irregular	Sir
Striking-platform remnant "chapeau de gendarme"	Sus
Dorsal reduction present	DR
Dorsal reduction absent	DRn
Dorsal reduction with pointed negatives	DRp
Dorsal reduction with hinged negatives	DRh
Dorsal reduction with pointed and hinged negatives	DRph
Lip present	Lip
Lip laterally present	Lil
Bulb diffuse	Bbd
Bulb markedly present	Bd
Impact scar present	Sc
Impact point present	Ipt
Bulbar scar present	BSc
En éperon preparation present	EE
Striking angle over 80°	A80+
Striking angle between 70 and 80°	A7t8
Striking angle below 70°	A-7
Blade profile straight	Pfs

Additionally, the total number of analyzed blades (n), the assignment to one of the regional groups (G), and temporal groups (t) is listed. For a more detailed explanation of the features, see Chap. 3, section "Blade Technology."

Results of the Technological Analysis

To visualize the results, the relation between the assemblages and the technological features is displayed in an asymmetrical biplot (Fig. 6.60), where the position of the assemblages is set according to their principal coordinates (Table A.51) and the location of the features corresponds with their standard coordinates (Table A.52). The first axis captures 40 % of the inertia and the second 26 % (Table A.53). This represents two thirds of all of the multidimensional information present in the plot. The "broken stick rule," which distributes inertia randomly over all axes, shows that the third axis already displays a lower-than-random amount of the inertia (Fig. 6.61). In order to see the precise extent to which the position of an axis is based on the dissimilarity between which features (or objects), the so-called contribution (CTR) can be consulted (Table A.52). Simply put, the contributions identify the differences which an axis measures.

The Features

In Fig. 6.60, the features are represented by triangles, the size of which displays their mass, i.e., their importance for the CA solution, and the transparency of which indicates the quality. The quality is the sum of the correlation of the first and second axes and displays the extent to which the information of the feature is represented in the two-dimensional plot (Table A.52).

In the case of the first axis, the feature *en éperon* (EE: $x=3.879$, $y=0.755$) contributes to the realization of the axis at a level of 38 %, while the feature "smooth striking-platform remnant" (Ssm: $x=-1.461$, $y=-0.922$) makes the second highest contribution at 10 %. For the second axis, the feature "impact point" (Ipt: $x=-1.229$, $y=2.231$) contributes at al level of 20 % and the feature "lip" (Lip: $x=0.330$, $y=-1.484$) contributes at 18 % (cf. CTR values in Table A.54). If the inverse question is asked—i.e., what percentage of the dissimilarity of a row or column is represented by the axis—the correlation values (COR) give the respective information (Table A.54). From these values, it becomes clear that the feature "butt size between 5.1 and 15 mm^2" (5to15: $x=-0.065$, $y=-0.007$), for instance, is poorly represented on both axes.

The position of the first axis is mainly determined by the feature *en éperon* and therefore represents the tendency of other features and assemblages to be technologically related or unrelated to this kind of striking-platform preparation, which is usually strongly associated with the usage of an organic hammer. In almost diametrical opposition lies the feature "smooth SPR," indicating that smooth butts and *en éperon* preparation are mutually exclusive. Together, they contribute at a level of 48 % to the inertia measured by the first axis. The position of the second axis is mainly defined by two contrasting pairs of features. On the one hand, the features "impact point" and "lip" show an opposed location, while on the other, the features "dorsal reduction with pointed negatives" (DRp: $x=0.951$, $y=2.196$) and "dorsal reduction with hinged negatives" (DRh: $x=-0.385$, $y=-1.285$) are situated across from each other. Whereas an impact point is thought to be indicative of the usage of a mineral hammer, a lip is usually related to an organic striking instrument. All four features together contribute to the inertia measured by the second axis at a level of 59 %. Interpreting both axes jointly, a clear spatial trend can be seen, one which divides the coordinate system into two parts. The features located on the left side correspond with the usage of a soft stone hammer, whereas those in the lower part are especially

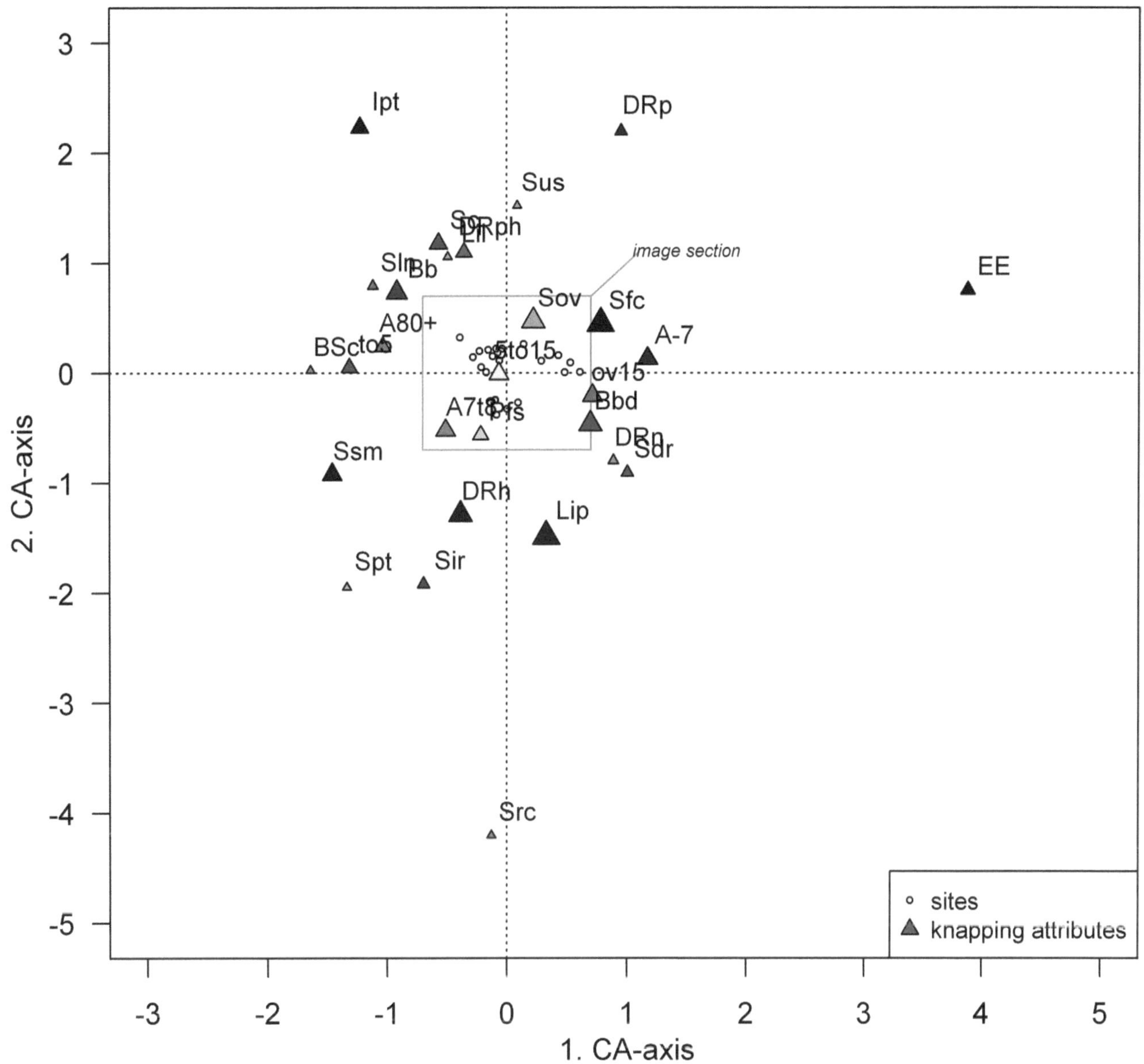

Fig. 6.60 Correspondence analysis. Asymmetrical biplot with features (knapping attributes) in standard coordinates and assemblages in principal coordinates. The size of the symbols displays the features' mass and the transparency the features' quality (Table A.54); for the feature coordinates, see Table A.55; for abbreviations, see List of Abbreviations. For a magnification of the inner sector (image section), see Fig. 6.62

opposed to an *en éperon* preparation. On the right side are situated features with a strong connection to the usage of an organic hammer, and those in the upper part occur especially frequently with an *en éperon* preparation. The figure thus mainly displays the contrast between the application of a mineral and an organic hammer, with special emphasis on the usage of *en éperon* preparation.

The upper left quadrant is exclusively characterized by features that usually occur in connection with a soft stone hammer. Here, the feature "impact point" is situated on the upper left margin and is clearly separated from the other column points. Listed in order of increasing distance from the opposed feature *en éperon*, these are an intensive dorsal reduction with pointed negatives and hinges (DRph: $x=-0.356$, $y=1.096$), a lateral-only lip (Lil: $x=-0.491$, $y=1.056$), a marked bulbar scar (Sc: $x=-0.567$, $y=1.176$) and a marked bulb (Bb: $x=-0.919$, $y=0.733$), a linear SPR (Sln: $x=-1.121$, $y=0.793$), a striking angle over 80° (A80+: $x=-1.037$, $y=0.239$), a small butt size (≤ 5 mm^2) (to5: $x=-1.318$, $y=0.047$), and at the farthest right position, an *esquillement du bulbe* (BSc: $x=-1.644$, $y=0.022$).

The four features located in the outer part of the lower left quadrant, namely, smooth SPR (Ssm: $x=-1.461$, $y=-0.922$), dorsal reduction with hinged negatives (DRh: $x=-0.385$,

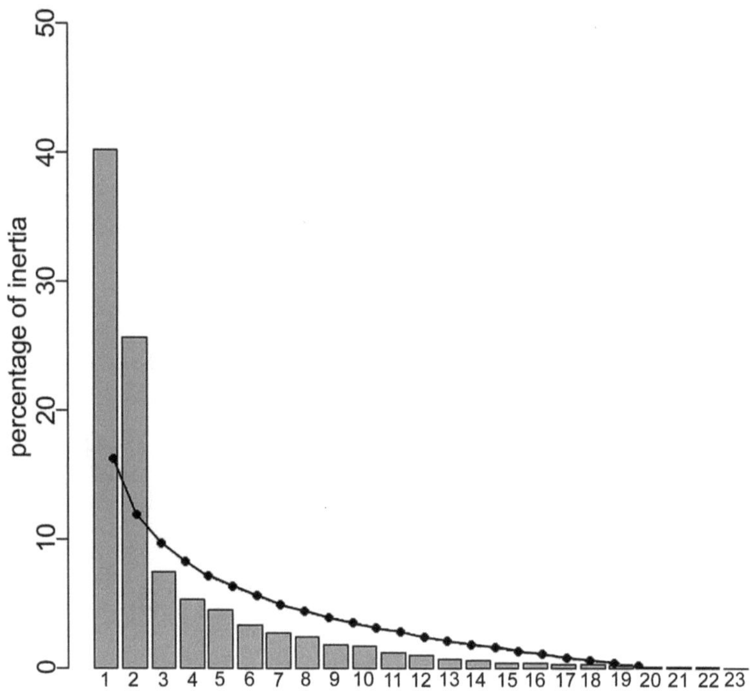

Fig. 6.61 Correspondence analysis. "Broken stick" distribution of the inertia

$y = -1.285$), point-shaped SPR (Spt: $x = -1.337$, $y = -1.948$), and irregular SPR (Sir: $x = -0.693$, $y = -1.918$), are likewise characteristic of the usage of soft stone hammers. However, in contrast to the features in the upper left quadrant, these seem to appear more in opposition to the *en éperon* preparation specifically than in opposition to the usage of an organic hammer in general.

In the center, close to the origin of the ordinates where an average-composition row or column would be located appears the medium-sized butt's class (5.1–15 mm^2) (5to15: $x = -0.065$, $y = -0.007$). This is not surprising inasmuch as these average butt sizes often occur in the context of both kinds of hammers, even though the high degree of transparency of the triangle indicates that this feature is not very well represented in the two-dimensional biplot. The surrounding features "oval-shaped SPR" (Sov: $x = 0.223$, $y = 0.469$), "straight blank profile" (Pfs: $x = -0.217$, $y = -0.564$), and "striking angles between 70° and 80°" (A7t8: $x = -0.508$, $y = -0.522$) are likewise not very well represented. However, whereas oval-shaped butts tend to coincide more with the application of an organic hammer, the two latter features seem to be more strongly related to the usage of a mineral hammer.

Features displayed to the right of the y-axis are usually associated with the usage of an organic hammer. These are a dorsal reduction with pointed negatives (DRp: $x = 0.951$, $y = 2.196$), a faceted SPR (Sfc: $x = 0.780$, $y = 0.447$), and a striking angle lower than 70° (A-7: $x = 1.173$, $y = 0.130$). All are especially close connected with the *en éperon* preparation. Furthermore, this area includes a rather large butt size (≥15 mm^2) (ov15: $x = 0.711$, $y = -0.205$), a diffuse bulb (Bbd: $x = 0.695$, $y = -0.466$), and absence of dorsal reduction (DRn: $x = 0.886$, $y = -0.798$), a triangular butt (Sdr: $x = 1.005$, $y = -0.906$), and a marked lip (Lip: $x = 0.330$, $y = -1.484$).

The Assemblages

Because the assemblage coordinates are "shrunk" into the feature's space, they cluster in the center of the coordinate system. Figure 6.62 shows a magnification of this innermost sector. Again, the size of the symbol represents the mass of the assemblage, while the transparency indicates the quality of its representation in the biplot. Looking at Fig. 6.62, it is apparent that the sites can be divided into three groups: one in the upper left quadrant, one in the upper right quadrant, and one in the lower quadrants.

The group in the upper right quadrant is characterized by technological features connected with an organic hammer. It consists of six assemblages from three sites; of these, the three small assemblages from Oelknitz (OeO: $x = 0.141$, $y = 0.258$; OeY: $x = 0.290$, $y = 0.110$; Oel: $x = 0.430$, $y = 0.157$) and the two collections from Klementowice-Kolonia (KL8: $x = 0.528$, $y = 0.093$; Kln: $x = 0.482$, $y = 0.006$) are located closer to the larger assemblages from their own respective sites than they are to other assemblages. The largest group, comprising 12 assemblages from 11 sites, is located in the upper left quadrant, which is characterized by features usually connected with the usage of a mineral hammer. The third group in the lower part of the coordinate system, however, seems to be divided. The two assemblages to the right of the

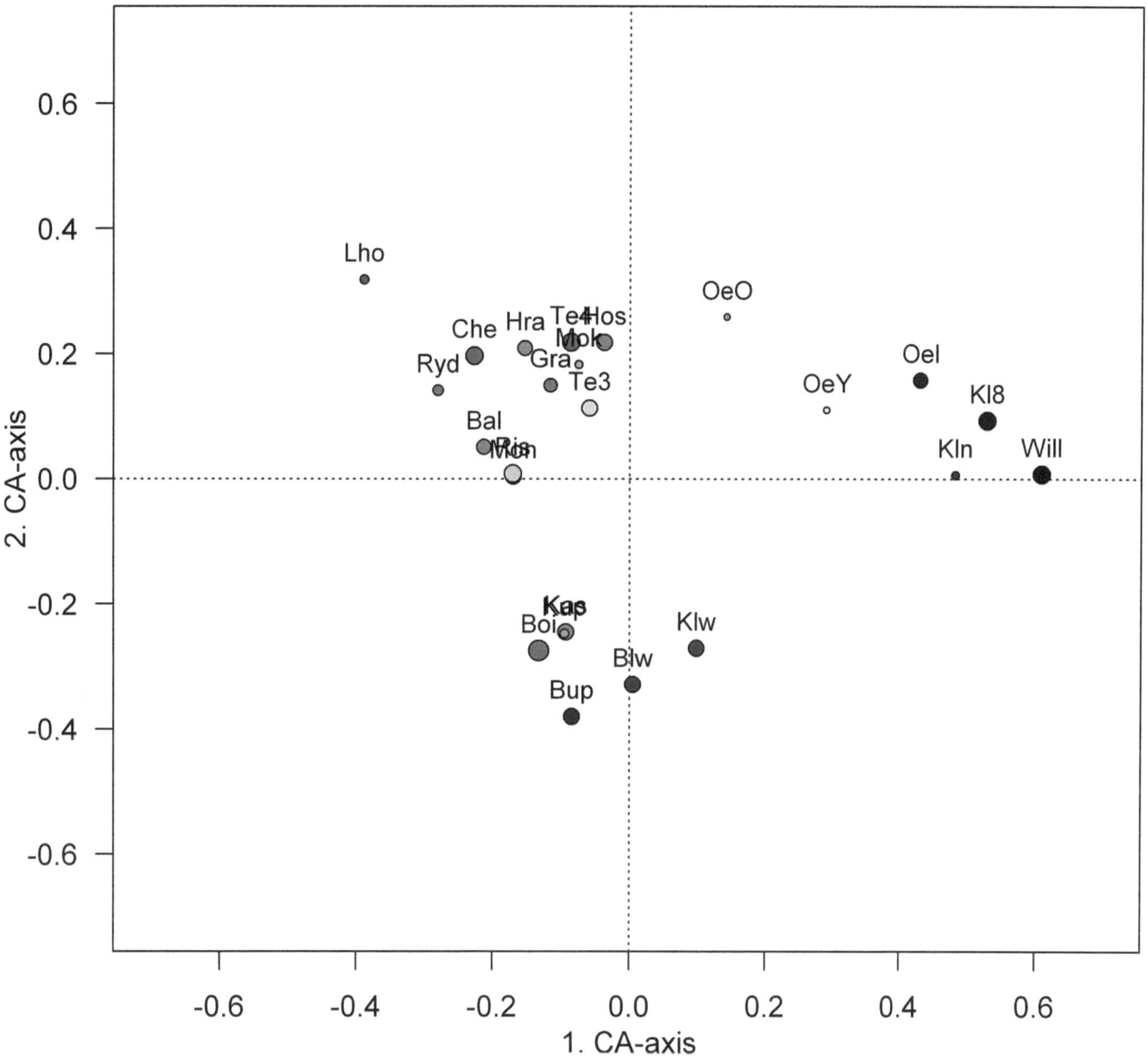

Fig. 6.62 Correspondence Analysis. Asymmetrical biplot with features (knapping attributes) in standard coordinates and assemblages in principal coordinates; magnification of the inner sector. The size of the symbols displays the features' mass and the transparency the features' quality (Table A.53); for the feature coordinates, see Table A.52; for abbreviations, see List of Abbreviations

y-axis are characterized more strongly by features of organic hammers, whereas the four collections to the left are mainly marked by features connected with a mineral hammer.

It is remarkable that the assemblages from Monruz (Mon: $x=-0.169$, $y=0.004$) and Rislisberghöhle (Ris: $x=-0.170$, $y=0.008$) as well as Kastelhöhle (Kas: $x=-0.093$, $y=-0.244$) and the upper level of Kaufertsberg (Kup: $x=-0.095$, $y=-0.246$) are situated pairwise at the same position in the two-dimensional biplot. These situations are very rare in multivariate statistics (pers. comm. G. Roth). Although their representation is not ideal (see Quality (QLT) in Table A.52), this coincidence is astonishing considering the number of features (28) that are taken into account for this analysis. This observation gains even more significance if the spatial proximity of Monruz and Rislisberghöhle (70 km) and the strong interaction pattern between the regional groups of Kastelhöhle and Kauferstberg are considered (see section "Procurement Pattern of Mollusk Shells and Amber"). The pronounced technological similarities also speak in favor of a very intense interaction between the producers of these assemblages.

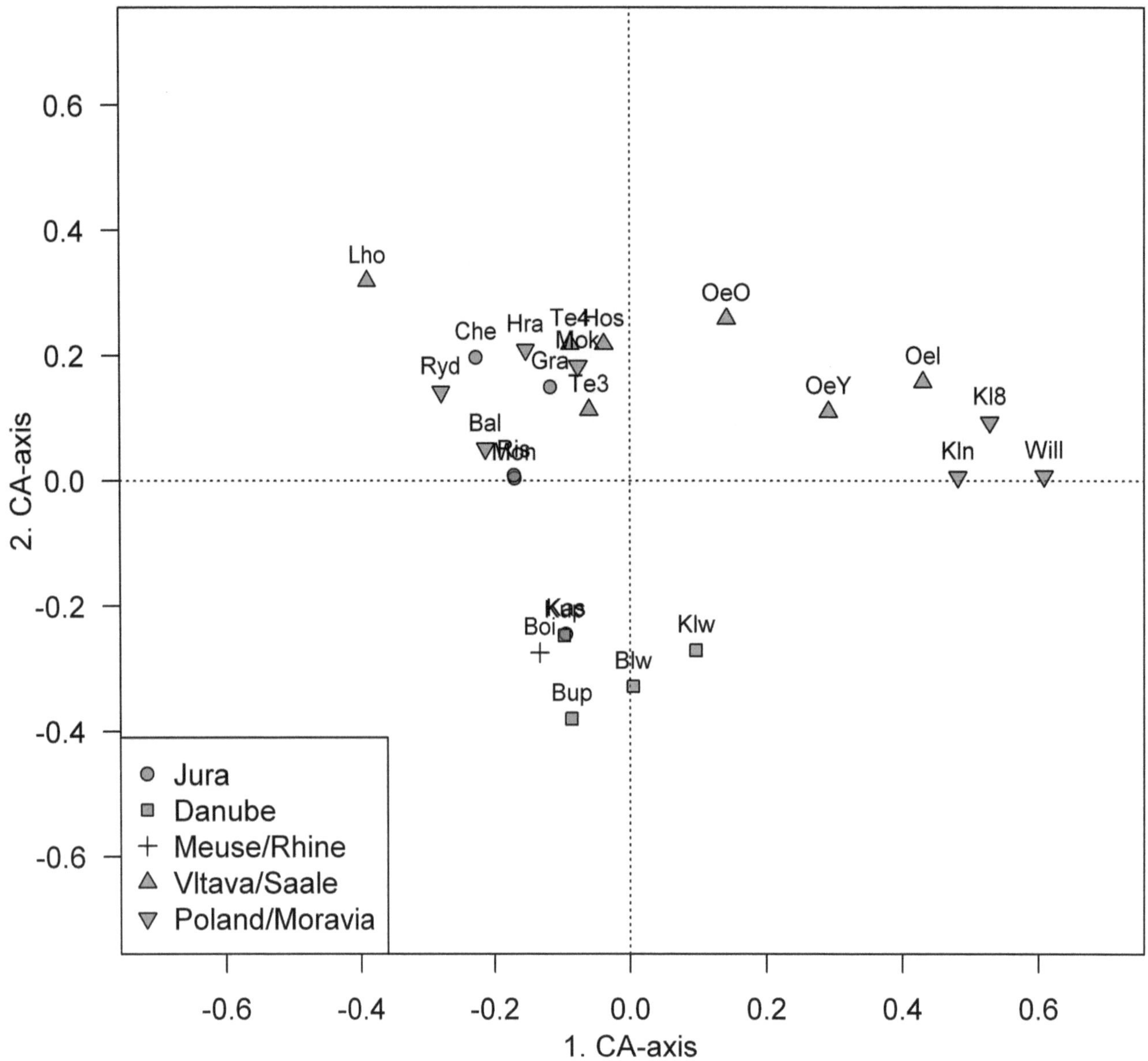

Fig. 6.63 Correspondence Analysis. Asymmetrical biplot with assemblages in principal coordinates labeled according to the five regional groups (for abbreviations, see List of Abbreviations)

Spatial and Temporal Trends Among Technological Concepts

Reviewing the observations made so far, the question arises of whether the technological diversity reflects a spatial or temporal trend. In order to address this issue, the assemblages are sorted into temporal and regional groups which correspond with the spatial and temporal classes that were used in the Discriminant Function Analysis of the typological tool spectrum in (see above).

Spatial Trends

In Fig. 6.63, the assemblages are displayed according to their allocation to five spatial groups (see Table A.50): the Circum-Jurassic Group, the Danube Group, the Meuse-Rhine Group, the Vltava-Saale Group, and the Polish-Moravian Group. It becomes apparent that the assemblages in the upper right quadrant belong either to the Vltava-Saale Group or to the Polish-Moravian Group. In the lower part of the biplot, in contrast, only assemblages from the Jura, Danube, and

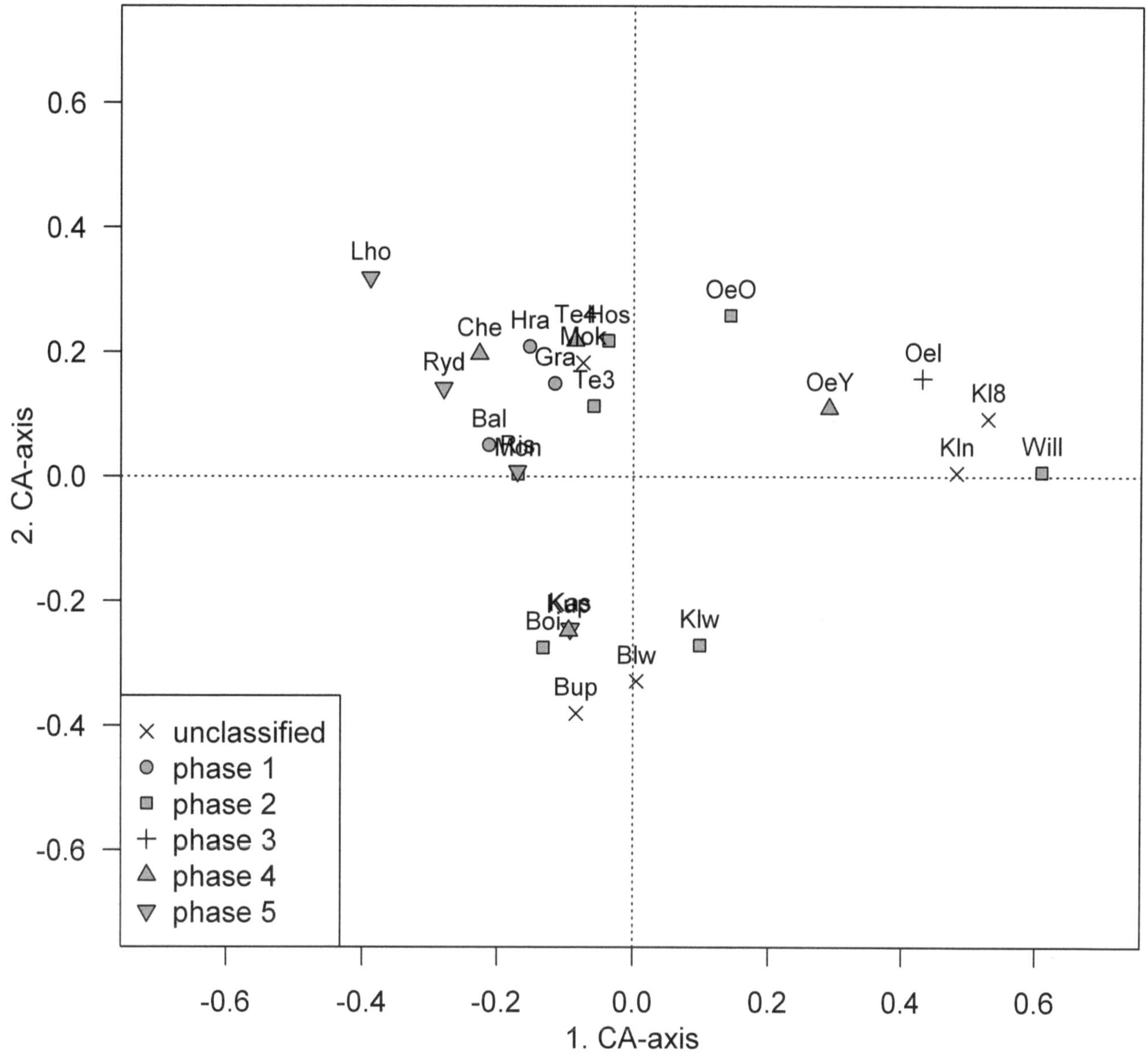

Fig. 6.64 Correspondence analysis. Asymmetrical biplot with assemblages in principal coordinates labeled according to phases (1: >16,000 cal BP; 2: 16,000–15,000 cal BP; 3: 16,000–14,000 cal BP; 4: 15,000–14,000 cal BP; 5: <14,000 cal BP; for abbreviations, see List of Abbreviations)

Meuse-Rhine Group appear, and indeed sites of the two latter groups occur exclusively in this area. The majority of the inventories of the Circum-Jurassic Group, however, are located in the upper left quadrant, where they form an enclave surrounded by assemblages of the Vltava-Saale and Polish-Moravian Group.

Thus, it can be stated that sites of the three western groups are found mainly in the lower part of the coordinate system, whereas those of the three eastern groups are restricted to the two upper quadrants. It is remarkable that assemblages from the eastern sites in particular yield especially high numbers of blades with *en éperon* preparation.

Temporal Trends

For a more detailed analysis of possible temporal trends, the magnified section of the biplot is displayed again in Fig. 6.64. Here, the different symbols represent the chronological positions of the assemblages. In correspondence with the classes of the Discriminant Function Analysis, the temporal classification is set as follows: phase 0 = unclassified; phase 1 = older than 16,000 cal BP; phase 2 = between 16,000 and 15,000 cal BP; phase 4 = between 15,000 and 14,000 cal BP; and phase 5 = younger than 14,000 cal BP. Since the assemblage from Oelknitz with an undated context (Oel) cannot be clearly attributed to one of these phases, an additional phase 3 = between 16,000 and 14,000 cal BP was introduced.

The temporal distribution of the assemblages in Fig. 6.64 shows a segregation into three groups. Collections belonging to the oldest group are exclusively found in the upper left quadrant and are located close to each other. According to this sample, the usage of soft stone hammers prevailed during this period. Typologically, the inventories of this temporal group are highly diverse in comparison to the assemblages of the second and third temporal group, which form a rather homogeneous unit from a typological point of view (see section "Analysis of Temporal Grouping"). Technologically, the assemblages assigned to the phases 2–4 also seem to be very similar and show intermingled locations in the biplot. Nevertheless, the inventories of phase 2 tend to be situated slightly more toward the bottom right, whereas those of phase 3 appear to be located rather more in the top left. Generally, they are the only collections that occur on the right side of the coordinate system and are therefore the only ones that are characterized by a dominant usage of an organic hammer and the regular application of the *en éperon* preparation. Inventories younger than 14,000 cal BP are located on the outer left margin of the distribution and are therefore mainly characterized by features associated with a soft stone hammer. Interestingly, the two youngest assemblages, those from Rydno II/59 ($x=-0.279$, $y=0.141$) and Lhota ($x=-0.389$, $y=0.319$), which are probably of Late Paleolithic age, are situated farthest left of all the collections and thus yield the highest number of blades produced with a mineral hammer.

With regard to the observations obtained so far, it is possible to argue in favor of a temporal trend which becomes visible on the CA biplot. The results of the statistical analysis of 24 assemblages from 18 sites in terms of 28 technological features can be interpreted as follows.

Blades and bladelets in assemblages which are dated older than 16,000 cal BP are often obtained with a soft mineral hammer. Between 16,000 and 14,000 cal BP, the application of an organic hammers as well as the corresponding *en éperon* preparation gained significantly in importance, and organic hammers became the dominant striking instrument. From 15,000 cal BP onward, however, a slight increase in the usage of mineral hammers is visible. In the biplot, this becomes apparent in the slight top left shift of the assemblages of phase 3 in comparison of those of phase 2. In assemblages younger than 14,000 cal BP, the usage of mineral hammers increases strongly again, and the *en éperon* preparation disappears almost completely.

Against this background, it is conspicuous that the collections from the lower horizons of Brillenhöhle (Blw: $x=0.004$, $y=-0.328$) and Kaufertsberg (Klw: $x=0.096$, $y=-0.270$) are located on the right side of the y-axis, whereas the inventories from the upper horizons at these sites (Bup: $x=-0.084$, $y=-0.380$; Kup: $x=-0.095$, $y=-0.246$) are located on the left side. This gradient indicates a shift in the usage of striking instruments from organic hammers in the lower and therefore older assemblages, to mineral hammers in the upper and consequently younger ones. A similar shift is observable for the two collections from Teufelsbrücke (Te3: $x=-0.059$, $y=0.113$; Te4: $x=-0.086$, $y=0.218$). Although both assemblages are characterized by features connected with mineral hammers, the younger one (Te4) shows an increasing importance of this striking instrument.

Typological and Technological Variability

The results of the individual and multivariate typological and technological analyses are remarkable, especially with regard to three main aspects: the pattern of presence and absence of individual tool types and technological features in contrast to their proportional abundances and evaluation in multivariate statistics; the dependence of an assemblage's typo-technological spectrum on its spatial position; and the further dependence of this spectrum on the assemblage's temporal position. In the following sections, these aspects are discussed in detail.

Spatial Aspects

When only the presence or absence of lithic tool types and technological features in the spectrum of the CEM assemblages is taken into account, very little variability is discernable with regard to the type's geographical distribution, if any differences can be observed at all. On the contrary, it seems that single tokens of all considered types can be found almost everywhere in the area of the CEM. However, if the proportional ratios of the tokens' abundance are taken into account, spatial trends and sometimes pronounced regional differences become visible. The spatial trends for the most part exhibit a southwest-northeast-oriented opposition, a phenomenon that was already observed during the analysis of the distribution of horse and reindeer remains (see Chap. 5, section "Horse and Reindeer Remains"). However, the spatial variability within the typological composition of the CEM assemblages becomes even clearer when multivariate statistics are applied. The Discriminant Function Analysis shows that the typological composition of an assemblage largely corresponds to its geographic position, or more precisely, to a regional grouping that was developed on the basis of raw material procurement patterns.

Hamburgian and Magdalenian

The strongest typological differences can be observed between Hamburgian and CEM assemblages and thus between two archaeological entities. This is not particularly surprising. Nevertheless, the clarity of the separation remains striking. With regard to the discussion whether the Hamburgian might represent a seasonal facies of the Magdalenian, the results speak clearly in favor of a separation of these two archaeological entities and thus for the independence of the Hamburgian.

Circum-Jurassic Group

With regard to typology and technology, the Circum-Jurassic Group's closest relations are observed with the Danube Group. The Discriminant Function Analysis reveals strong similarities between the typological compositions of the assemblages of the two groups. This is expressed in a relatively low abundance of both long burins (Lacan burins) and long borers in both groups and comparatively high ratios of denticulated backed bladelets as well as notched and denticulated pieces (cf. Figs. 6.36 and 6.41). Splintered pieces are especially numerous in the eastern part of the Circum-Jurassic Group and the western part of the Danube Group (cf. Fig. 6.40). *Zinken*, on the other hand, increase from southwest to northeast. Shouldered points are considered as a characteristic feature of the Magdalenian in Switzerland (Nielsen 2009: 95) and occur quite frequently within the assemblages there (cf. Figs. 6.28 and 6.44). Figure 6.57 maps the results of the Discriminant Function Analysis. It becomes apparent that the assemblages in the western part of the Danube Group exhibit especially strong similarities to those of the Circum-Jurassic Group, whereas within the latter, the assemblages similar to those of the Danube Group are more dispersed. The relation between the Circum-Jurassic Group and the Meuse-Rhine Group seem to be almost equally strong. Here, there is a conspicuous typological similarity between the northeasternmost site of the Circum-Jurassic Group, namely, Munzingen, and the sites of the Meuse-Rhine Group; in turn, four sites of the Meuse-Rhine Group are typologically similar to those of the Circum-Jurassic Group (see Fig. 6.57 and Table A.45). This relation is reflected not only in the occurrence of mollusks from the Mainz Basin in both groups, but also by raw material found in Gönnersdorf which originates from the Markgräflerland in the vicinity of Munzingen and Teufelsküche. In contrast, typological evidence for relations between the Circum-Jurassic Group and the Vltava-Saale Group or the Polish-Moravian Group is very weak. The reallocation of one assemblage of the Circum-Jurassic Group to the Polish-Moravian Group and one assemblage of the Vltava-Saale Group to the Circum-Jurassic Group (see Table 6.11) indicates rather faint relations. Nevertheless, it is interesting to note that of all the assemblages belonging to the Vltava-Saale Group, only Gera-Schafgraben, a collection which yields a mollusk shell of a presumably western origin (section "Procurement Pattern of Mollusk Shells and Amber"), is typologically similar to the Circum-Jurassic Group.

For the technological analysis, the sites of Chênelaz, Grappin, Neuchâtel-Monruz, Rislisberghöhle, and Kastelhöhle-Nord have been analyzed. Although this sample is certainly too small to be considered representative for the whole site cluster of the Circum-Jurassic Group, it is nevertheless interesting to note that all assemblages exhibit a rather intensive usage of soft stone hammers for the production of blades and bladelets. The strong technological coherence of assemblages from the Circum-Jurassic and the Danube Groups is emphasized by the remarkable coincidence that the inventories of Kastelhöhle-Nord in the Birs valley and Kaufertsberg (upper level) are located almost at the same position in the asymmetrical biplot of the Correspondence Analysis (see Fig. 6.62). Such a concurrence, which can likewise be observed for the assemblages of Neuchâtel-Monruz and Rislisberghöhle, is astonishing considering the high number of 28 features which were taken into account in the analysis. These technological similarities exclude mere coincidence as a satisfactory explanation. Rather, they point to a conformist-biased transmission as well as intensive interaction that compensated for the effect of drift and led to this surprising conformity (Neiman 1995; Kohler et al. 2004) (see Chap. 9, section "The Social System of the Central European Magdalenian").

Danube Group

With regard to the typological spectrum, an especially strong relation can be observed between hunter-gatherers of the Danube Group and the Circum-Jurassic Group (Fig. 6.57). The typological proximity of the two Groups finds its expression in the proximity of their assemblages in the plotted LDA results, which are not only located close to each other, but are also represented by the same axis. The same holds true for the assemblages of the Meuse-Rhine Group. Typological similarities to the two eastern regional groups (the Vltava-Saale and the Polish-Moravian Group) are, however, less pronounced. This separation is mirrored in the fact that the typological dispersion of the Vltava-Saale and the Polish-Moravian Group in the Discriminant Function Analysis is best explained by an axis which runs perpendicular to the one of the three western assemblages. With regard to the relative abundance of lithic tool types, it can be said that borers are underrepresented in the Franconian Alb in comparison to the western part of the Danube Group's area (see Fig. 6.26). Other noteworthy typological observations are described above in comparison with the Circum-Jurassic Group.

For the technological investigation, the lower and upper levels of the sites of Brillenhöhle on the central Alb and Kaufertsberg in the Nördlinger Ries are analyzed. The

Correspondence Analysis shows that all four assemblages are relatively similar with regard to their technological features. Interestingly, the two lower and the two upper layers of both sites are more similar to each other than they are to the respective older or younger layers from the same site. Features indicating the usage of organic hammers for blade and bladelet production prevail in the lower layers, whereas soft stone hammers seem to become more prevalent in the upper layers (see section "Spatial and Temporal Trends Among Technological Concepts"). Strong technological similarities with the Circum-Jurassic Group have been pointed out above. The assemblage of Bois Laiterie (Meuse-Rhine Group), which already shows a pronounced typological similarity to the sites of the Danube Group (Table A.45), also exhibits a clear affinity to the assemblages of this interaction unit when it comes to knapping technology. The analyzed assemblages of the regional groups in the eastern part of the investigated area are, in contrast, technologically rather distinct. Thus, the findings of the technological analysis correspond surprisingly well with the results of the typological analysis and the spatial pattern of raw material and mollusk shell distributions; strong coherences can be observed between the Danube and the Circum-Jurassic and Meuse-Rhine Groups, whereas relations between the Danube and the eastern groups appear rather weak.

Meuse-Rhine Group

The biplot of the Discriminant Function Analysis (Fig. 6.55) reveals that the assemblages of the Meuse-Rhine Group are typologically most closely related to those of the Danube Group and Circum-Jurassic Group. Nevertheless, while the two latter appear to be almost interchangeable in terms of their typological spectra, the sites of the Meuse-Rhine Group have a markedly different imprint. Denticulated backed bladelets, for example, are especially numerous in the southwestern part of the investigated area but are virtually absent in the northwest (see section "Spatial Distribution of Concepts of Lithic Tools"). The same applies in a less pronounced way to short end scrapers. Shouldered points, which are quite frequent in inventories from Switzerland, are again almost absent in the Meuse-Rhine area. Long burins (Lacan burins), in contrast, are quite abundant in the Meuse-Rhine area but are rather rare in the Circum-Jurassic and Danube Groups. A similar situation can be observed for *zinken* and, to a somewhat lesser extent, for long borers. These tool types are, on the other hand, comparatively numerous in the Vltava-Saale Group. When the results of the Discriminant Function Analysis are mapped (see Fig. 6.57), it becomes apparent that reclassifications of assemblages of the Meuse-Rhine Group result mainly in an allocation to the Circum-Jurassic ($n=4$) and Danube Groups ($n=2$). In contrast, almost no reclassifications are made to the Vltava-Saale ($n=1$) and Polish-Moravian Groups ($n=0$). In this context, it is interesting to note that only the northernmost assemblages of the Circum-Jurassic and the Danube Groups are reassigned to the Meuse-Rhine Group (see Fig. 6.57). The number of assemblages that are assigned to the two regional groups in the eastern part of the investigated area but are reclassified to the Meuse-Rhine Group ($n=4$) is slightly higher than the number of those reclassified to the Circum-Jurassic Group ($n=1$) or the Danube Group ($n=2$). To be sure, the Meuse-Rhine Group shares many typological features in common with the Circum-Jurassic and Danube Groups. On the other hand, several features of the Meuse-Rhine Group emphasize its individual character and also suggest relations to the eastern groups.

In terms of technology, the situation of the Meuse-Rhine Group is hard to assess on the basis of the investigated samples, since only the site of Bois Laiterie was analyzed. However, the assemblage is technologically comparable with the inventories of the upper layers from Brillenhöhle and Kaufertsberg in the Danube Group, as well as with the inventories of Kastelhöhle-Nord in the Circum-Jurassic Group (Fig. 6.62). The technological features speak in favor of frequent application of a soft stone hammer for the production of blades and bladelets. At the same time, an organic hammer also seems to have been used. The *en éperon* preparation technique, however, played only a marginal role.

Vltava-Saale Group

A problem for the typological comparison is the strong imbalance in the number of analyzable assemblages from the northwestern and southeastern cluster. The former clearly outnumber the latter, a fact which may bias the spatial pattern of the tool distribution. Keeping this in mind, it can be stated that with regard to the typological composition of the lithic tool assemblages, the sites of the Vltava-Saale Group differ noticeably from the three regional groups in the western part of the investigated area (Fig. 6.55). The typological variation within the assemblages of the Vltava-Saale Group is best explained by an axis running perpendicular to the axis that captures most of the typological dispersion of the three western regional groups. Pronounced typological similarities, on the other hand, can be observed between inventories of the Vltava-Saale Group and the Polish-Moravian Group, speaking in favor of an intense interaction. The typological dispersion of both groups is not only best explained by the same axis, but reassignments in the LDA results also occur more often between these two groups than between either of them and another group from the western part of the investigated area (see Table 6.11). On the same basis, it can be said that with regard to the western regional groups, the strongest typological accordance can be observed with the Meuse-Rhine Group, whereas the exchange of typological concepts between the members of the Vltava-Saale Group and those of the Danube Group and Circum-Jurassic Group appears to have been rather weak. Against this background, the occurrence of mollusk shells from (probably) the Mainz Basin at the site of Kniegrotte takes on increased significance. The spatial distribution of individual tool types results in a similar pattern, particularly with regard to long burins (Lacan burins), long borers, and *zinken* (see section "Spatial Distribution of Concepts of Lithic Tools"). These types occur quite frequently in the Meuse-Rhine Group and the Vltava-Saale Group, and they are fairly abundant in the Polish-Moravian Group. In the Danube Group, in contrast, they appear only rarely, and they are virtually absent from the Circum-Jurassic Group.

For the technological analysis, seven assemblages from four sites are analyzed. The technological spectrum covered is comparatively large. On the one hand, it includes assemblages in which the production of blades was mainly undertaken with an organic hammer and extensive use of the *en éperon* preparation technique (Oelknitz). At the sites of Hostim and Teufelsbrücke, on the other hand, the technological features of the blades and bladelets point to the primary use of soft mineral hammers. These differences seem to reflect a temporal trend (see section "Correspondence Analysis: Multivariate Investigation of Spatial and Temporal Trends Among Technological Concepts"), an assumption which is supported by the observations of the inventory of Lhota. This inventory, which is probably of post-Magdalenian age, exhibits exclusively those technological features which are usually related to the use of soft stone hammers, whereas evidence for the usage of organic hammers is absent. As is the case for typology, the assemblages of the Vltava-Saale Group show the most congruities with those of the Polish-Moravian Group also from a technological point of view (see Fig. 6.62). Similarities also exist to inventories of the Circum-Jurassic Group. The analyzed collections of the Danube Group and Meuse Rhine Group, however, show a rather divergent technological character.

Polish-Moravian Group

Typologically, the closest relations of the Polish-Moravian Group are those with the Vltava-Saale Group (see above). The typological similarity is also reflected in the number of reassigned assemblages. With regard to the distribution of individual tool types and tool classes, the occurrence of long burins (Lacan burins) and *zinken* and comparatively low numbers of notched and denticulated pieces appear to be features held in common by the Polish-Moravian Group, the Vltava-Saale Group, and the Meuse-Rhine Group.

For the technological analysis, the blade and bladelet production in seven assemblages from six sites was investigated. With regard to the frequencies of certain technological features, two technological groups could be distinguished using a Correspondence Analysis (see section "Correspondence Analysis: Multivariate Investigation of Spatial and Temporal Trends Among Technological Concepts"). On the one hand, there are the assemblages of Klementowice-Kolonia and Wilczyce, which are characterized by the predominance of organic hammer use and the most frequent occurrence of the *en éperon* preparation technique among all of the investigated assemblages. Here, the closest technological relations exist with the assemblages of Oelknitz in the Vltava-Saale Group. The remaining assemblages of Hranice, Balcarka Cave, Mokra V, and the possibly post-Magdalenian assemblage of Rydno II/59 are all characterized by the predominant use of soft stone hammers. Again, technological similarities exist with assemblages of the Vltava-Saale Group, namely, Hostim and Teufelsbrücke. The sites of Grappin and Chênelaz in the French Jura also exhibit a similar technological pattern.

Temporal Aspects

Apart from the geographic location, the temporal position of an assemblage seems to influence likewise the typo-technological appearance of an assemblage. Those assemblages which are older than 16,000 cal BP are typologically marked by the presence of lithic triangles. Organic hammers apparently did not play a major role in the production of blades. Impact scars, which are characteristic of the application of a mineral hammer (Pelegrin 2000: 77ff.), amount to 45 % and 43 % in the assemblages of Grappin and Balcarka Cave, as well as 63 % in Hranice. At the same time, these collections show only 11 %, 9 %, and 5 % *en éperon* preparation, respectively (cf. Table A.50).

Around 16,000 cal BP, composite barbed points are abandoned in favor of compact, non-composite barbed points, thus causing the disappearance of lithic triangles from the tool spectrum (see Chap. 4, section "The Long Debate About Lithic Triangles"). Interaction between the CEM regional groups must have been rather strong between 16,000 and 14,000 cal BP, since the analyzed assemblage are comparatively homogeneous in terms of their typological and technological spectra. During this phase, organic hammers became more important as striking instruments and the *en éperon* preparation reached peak implementation frequency in the technological concepts of CEM hunter-gatherers. This coincides well with the observations of many researchers working in the Paris Basin, who report high frequencies of *en éperon* preparations in assemblages from the later part of the Magdalenian (e.g., Karlin 1972; Pigeot 1987; Audouze et al. 1988). Besides the fact that *éperons* are known since the Aurignacian and also occur in the Protomagdalenian (Périgordien VII) (Surmely and Alix 2005), it seems that they were most frequent between 16,000 and 14,000 cal BP. Among the investigated samples, the highest percentages of *en éperon* preparation are found in inventories from Poland, Bohemia, and Thuringia, ranging from 12 % to 69 %. In the remaining regions, only the assemblage of Kaufertsberg (lower horizon) is situated between these margins with a proportion of 19 %. All other collections show values below 12 %.

After this phase of relative typo-technological stability between 16,000 and 14,000 cal BP, the intensity of change increased again around the transition from the late Magdalenian to the post-Magdalenian period from 14,000 to 13,000 cal BP. Typologically, the increase in lithic projectile points features prominently. Technologically, the occurrence of *éperons* decreased and mineral hammers became the dominant striking instrument once again. This observation is in accordance with findings from the Paris Basin and northwestern Europe, which show that mineral hammers were applied increasingly often toward the end of the Magdalenian and in the Late Paleolithic. Against this background, it is not surprising that the assemblage from Lhota, which is presumably of Late Paleolithic age, exhibits impact scars on 71 % of the analyzed blades, whereas *éperons* are completely absent.

Against this background, it can be said that on an intra-entity scale, the geographical location of a CEM site and its relative position in the interaction network has a discernible impact on the typological and technological composition of its tool assemblage. In the end, the results of the multivariate analyses support the twofold structure of the CEM with five regional groups aggregated to two larger units with assemblages in the Jura, Danube, and Meuse-Rhine region on the one hand and assemblages in Eastern Germany, Bohemia, Moravia, and Poland on the other. It also seems that the typological composition of CEM assemblages reflects their chronological position, with the spatial signal being slightly more clearly discernible (see R^2 values in Tables 6.8 and 6.13). This is perhaps a reflection of the larger number of inventories in the spatial analysis. Thus, also from a typo-technological point of view, the Magdalenian should not be perceived as a homogeneous unit, but rather as a heterogeneous complex with pronounced spatial and temporal diversity.

Some Aspects of Central European Magdalenian Sculptures and Engravings

The large corpus of engravings and sculptures is certainly one of the most conspicuous phenomena of the CEM and attracted the interest of many researchers such as A. Leroi-Gourhan, M. Lorblanchet, G. Bosinski, K. Valoch, and J. Svoboda to name just a few. The fact that cave paintings are very rare or virtually absent within the investigated area increases the scientific importance of sculptures and engravings and underlines their special character within the entirety of CEM finds. The original meaning of these objects is elusive, and studies that seek to reveal it must remain subjective interpretations, regardless of which ideas are used as explanatory concepts (see Chap. 2, section "Sculpted and Engraved Objects"). Nonetheless, an analysis of these objects in terms of their morphology and spatial distribution may provide interesting insights into the distribution of certain concepts and consequently into the regional structure of the CEM. Some aspects of the corpus of CEM engravings and sculptures are therefore discussed in the following sections. During the collection of data, information on Magdalenian sculptures and engravings was recorded from 78 sites. Here, only the figured representations and geometrical signs are considered. Groups of accidentally crossing lines and grooves on bones, antlers, or stone slabs, which appear to be more the result of cutting or arbitrary movements than the intention to create representations of any kind, will not be discussed.

Geometrical Representations

The first category of representations to be discussed can be subsumed under the hypernym geometrical signs. It includes all nonfigurative engravings which display groups of straight, curved, or crossing lines in which a willful order is discernable. In the CEM, these signs are usually found engraved on objects of bone and antler as well as on stone slabs. CEM assemblages yielding geometric signs and ornaments with a clearly representational character were analyzed by J. Svoboda in 1976, using a set of 92 signs and comparing their occurrence at 26 sites (Svoboda 1976; Svoboda et al. 1996: 188ff.). The focus of his analysis was mainly on the nature of the signs. Svoboda subdivided the set into two major groups: the first comprises combinations of straight lines and zigzags and is called "static," while the second consists of arches, wavy lines, and ovals and is called "dynamic" (Svoboda et al. 1996: 188). Because examples of both groups appear in all of his investigated areas and even within the same assemblages, he concludes that these groups have no detectable meaning for the temporal or geographical subdivision of the Magdalenian (Svoboda et al. 1996: 190).

One of the most problematic aspects of the analysis of signs is the classification process, which must always rely on subjective judgments. Even if the willful character of a geometric, nonfigurative engraving can be discerned—something which is not always unequivocally possible—there exist no universal intersubjective parameters which would allow for an unambiguous taxonomy. The infinite possibilities of variations and combinations which follow no rules or constraints render any attempt at classification arbitrary. Is an engraving of the form "XIX" one sign, two signs, or three signs? Is an engraving "/Λ\\V/" conceptually the same as "//Λ\\\V///"? As there are no intersubjectively unambiguous criteria which can be used to answer such questions, any classification of such engravings is a priori inadmissible. Comparisons of signs in terms of similarities and dissimilarities are equally problematic, because there are no parameters to evaluate the relative distances between them—to determine, for example, whether a zigzag line is closer to a straight line or to a wavy line. Ultimately it must be accepted that there are no intersubjective criteria that would allow for a taxonomical classification of geometric, nonfigurative signs. However, these methodological problems should not lead to the conclusion that Paleolithic research must reject this category out of hand as a source of knowledge. It is simply necessary to treat the analyses and conclusions with the appropriate care.

In the course of this study, the set of signs compiled by Svoboda was reviewed and re-analyzed with special emphasis on its possible implications for the understanding of regional variability within the CEM. For the reasons given above, it was decided not to attempt to establish a classification system. Thus, the following observations are of a rather general character. It can be stated that a group of relatively simple signs, such as parallel, zigzag, or wavy lines, arrows, circles, pointed ovals, and *x*-shaped forms, as well as herringbone or ladder patterns, appear to be rather ubiquitous, occurring throughout the investigated area (Fig. 6.65). More complex structures and composite patterns, on the other hand, appear in most cases at only a single site (Fig. 6.66). Thus, there is no indication of a regional clustering of specific signs. This observation contrasts with the findings of the previous analyses in this study, which reveal, at least on a regional scale, the application of common technological and morphological concepts for the production of lithic tools.

A possible explanation for this disparity can be derived from the observed dichotomy of cultural traits into payoff-correlated or payoff-independent traits (e.g., Brantingham 2007: 396). In this view, stylistic elements are seen as being selectively neutral or payoff independent in the process of cultural evolution (Dunell 1978: 199). The payoff independence of these traits favors individual choice and thus variability, since "selective neutrality raises the probability that, at a given level of design complexity, innovations will be historically unique" (Neiman 1995: 31). At best, it can be argued that the geometrical, nonfigurative engravings at CEM site are divisible into two groups. On the one hand, there is a common pool of basic geometrical forms which occur ubiquitously in the entire investigate area. On the other hand, an ample body of composite and more complex signs which appear exclusively within a single assemblage seems to reflect rather individual choices and preferences (e.g., Sackett 1977; Wiessner 1983; Hegmon 1995). The high morphological variability even within a small spatial region such as the Moravian Karst or northern Switzerland, as well as the substantial lack of repetitions, makes it improbable that these signs carried a communicational significance beyond their ornamental function, for instance, as group markers or notation of language, as has been argued for Magdalenian geometric signs in France (Pozzi 2004: 191; Otte 2009: 233).

Figural Representations

Among the figural representations, those which depict animals are by far the most numerous. In addition to these, engravings and statuettes of female figures are a common feature of the Magdalenian. In the following section, some aspects of this corpus are discussed and analyzed in terms of spatial distribution as well as regional similarities and differences.

Horse and Reindeer Representations

Comprising 174 representations at 21 different sites, horse is clearly the most frequently depicted animal in the CEM (Figs. 6.67, 6.68 and Table A.56). In terms of absolute numbers, mammoth is the second most commonly depicted species, with 84 individual depictions appearing at nine sites. However, at Gönnersdorf alone, 74 mammoth depictions have been found (Bosinski 2008a). Cervids, on the other hand, are evidenced in 51 specimens at 12 sites (Figs. 6.69 and 6.70). Generally, horse depictions are about 3.5 times more common than are those of cervids. When the site of Gönnersdorf, which yields by far the highest number of animal representations in the CEM (Ibid.), is excluded, the number of horse depictions decreases to 102 and cervids to 36 specimens, and the ratio between the two becomes just 3:1.

The spatial distribution of engravings showing horses and cervids is presented in Fig. 6.71. Despite the numerical imbalance between the horse and reindeer representations, it is quite apparent that engravings of cervids are most abundant in the southwest part of the investigated area. Depictions of horses, on the other hand, predominate in the northern and eastern parts. The sites of Roc-la-Tour and Trou de Chaleux in the Ardennes show a balanced proportion.

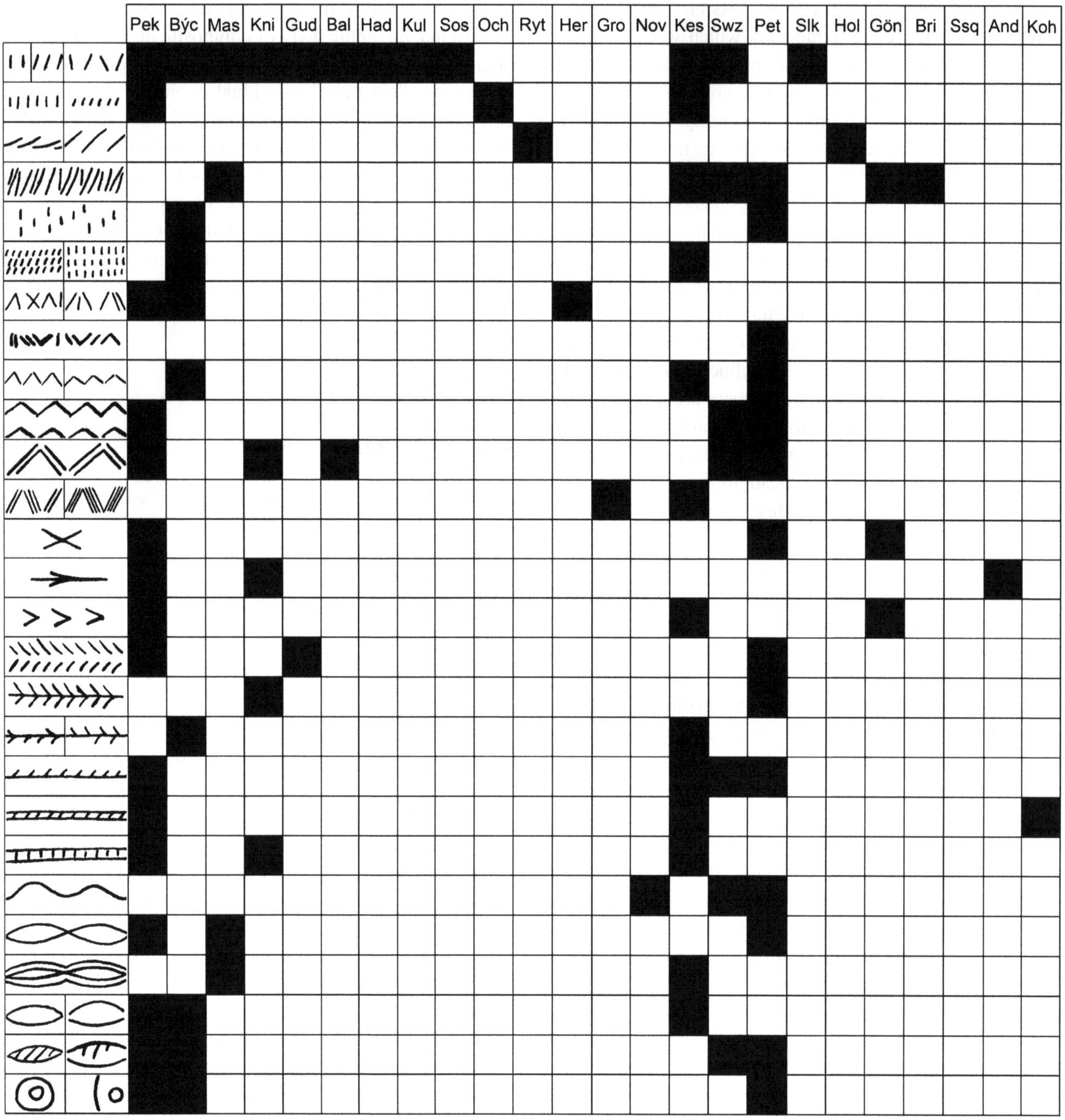

Fig. 6.65 Distribution of relatively simple geometrical signs in the CEM (after Svoboda 1976). Abbreviations see Fig. 6.66.

From Thuringia and Bohemia, no unambiguous depiction of a cervid has been reported so far. That said, 27 animal depictions at nine sites do not allow for an allocation to a distinct species: either the outlines are too vague or heavy fragmentation of the image carrier prevents a clear determination. At the open-air site of Saaleck, there are, besides a horse, two unrecognizable animal depictions (Bosinski 1982). Two other ambiguous engravings have been found at Hostim and Nebra (Vencl 1995; Mania 1999). At the Moravian cave site of Býči skála, a painting has been discovered that might represent a cervid. It was discussed as being potentially of Magdalenian age (Oliva 1995; Valoch 2001) until it was radiocarbon dated to the Middle Ages (Svoboda et al. 2005). The remaining unidentifiable depictions originate at five sites which already yield both horse and cervids engravings. Therefore, even if all of the ambiguous specimens turn out to represent cervids—a scenario which is not very probable—the overall pattern will not change significantly.

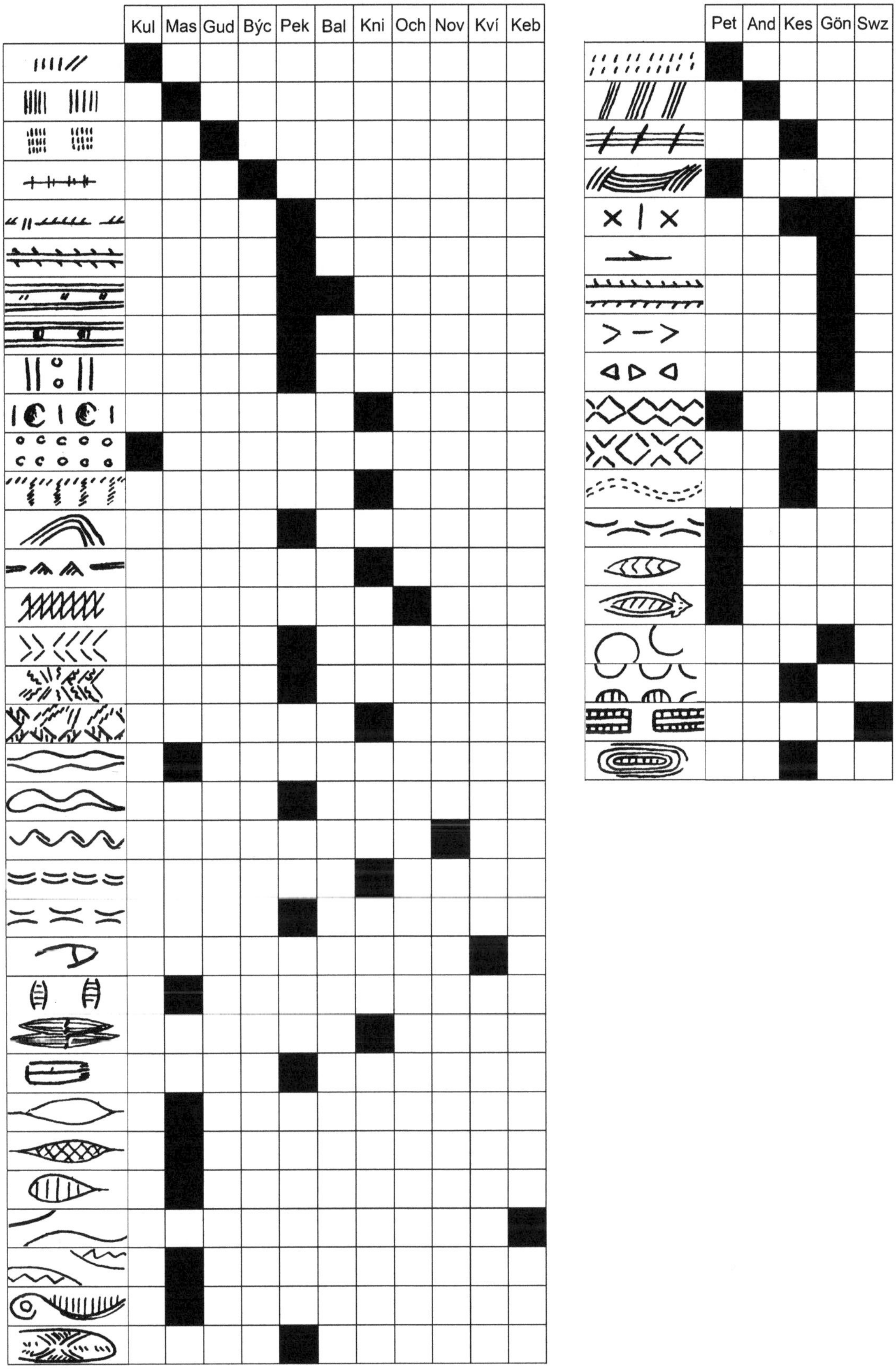

Fig. 6.66 Distribution of more complex geometrical signs in the CEM (after Svoboda 1976). Abbreviations: *And* Andernach-Matinsberg, *Bal* Balcarka Cave, *Bri* Brillenhöhle, *Býc* Býči skála, *Gön* Gönnersdorf, *Gro* Groitzsch, *Gud* Gudenushöhle, *Had* Hadí, *Her* Herdloch (Ranis), *Hol* Hohler Stein (Nördlingen), *Keb* Keblice, *Kes* Kesslerloch, *Kni* Kniegrotte, *Koh* Kohlerhöhle, *Kul* Kůlna, *Kví* Kvíc, *Mas* Maszycka, *Nov* Nová Datenická Cave, *Och* Ochozska, *Pek* Pekárna, *Pet* Petersfels, *Slk* Hohle Fels (Schelklingen), *Ryt* Rytířská, *Sos* Šošůvská Cave, *Ssq* Schussenquelle, *Swz* Schweizersbild

Fig. 6.67 Horse engravings I. 1: Hostim (Vencl 1991); 2: Mittlere Klause (Bosinski 1982); 3: Gönnersdorf (Bosinski 2008a); 4: Andernach (Bosinski 2008a)

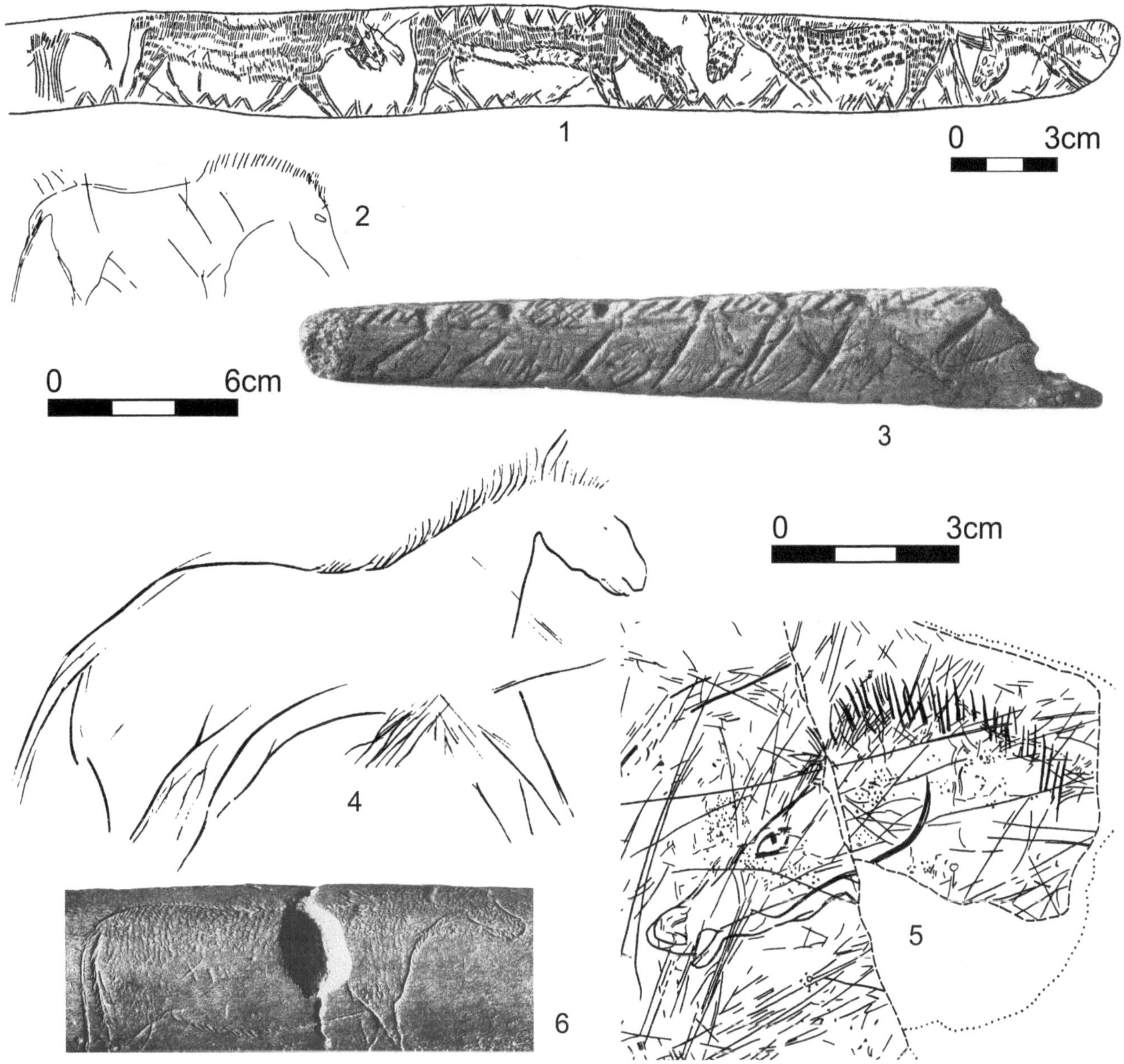

Fig. 6.68 Horse engravings II. 1: Pekárna (Klima 1965); 2: Oelknitz (Bosinski 1982); 3: Petersfels (Bosinski 1982); 4: Petersfels (Bosinski 2008a); 5: Saaleck (Bosinski 1982); 6: Kesslerloch (Bosinski 1982)

Interestingly, the pattern of the spatial distribution of horse and reindeer representations closely resembles that of the faunal remains (see Fig. 5.6). In both cases, reindeer is overrepresented in the southwestern part and underrepresented in the northeastern part of the investigated area, while the opposite is true for horse. The only difference in the distribution of representations in comparison to skeletal remnants is that the border between the areas dominated by horse and reindeer has shifted toward the southwest. Thus, it seems that those animals which were preferentially hunted, and thus were presumably more common in the environment, at the same time served more often as inspiration for representations than did less common animals.

This general spatial trend contrasts with observations of the ratio between depicted and hunted species from several sites such as La Vache, Abri Morin, Lascaux, Pair-non-Pair, Comarque, Villars, the Ardèche caves, Gargas, Altamira, Ekain, or Tito Bustillo (Bahn and Vertut 1988: 156f.). In all of these cases, there is no discernable relationship between the number of depicted and hunted specimens per species. In Gönnersdorf, on the other hand, the most-depicted species coincides with the most-hunted one: horse (Bosinski 2008a: 153). The deviation of specific situations at single sites from more general spatial trends has been observed before while analyzing the typological composition of CEM assemblages (see section "Spatial Trends in the Distribution of Tool

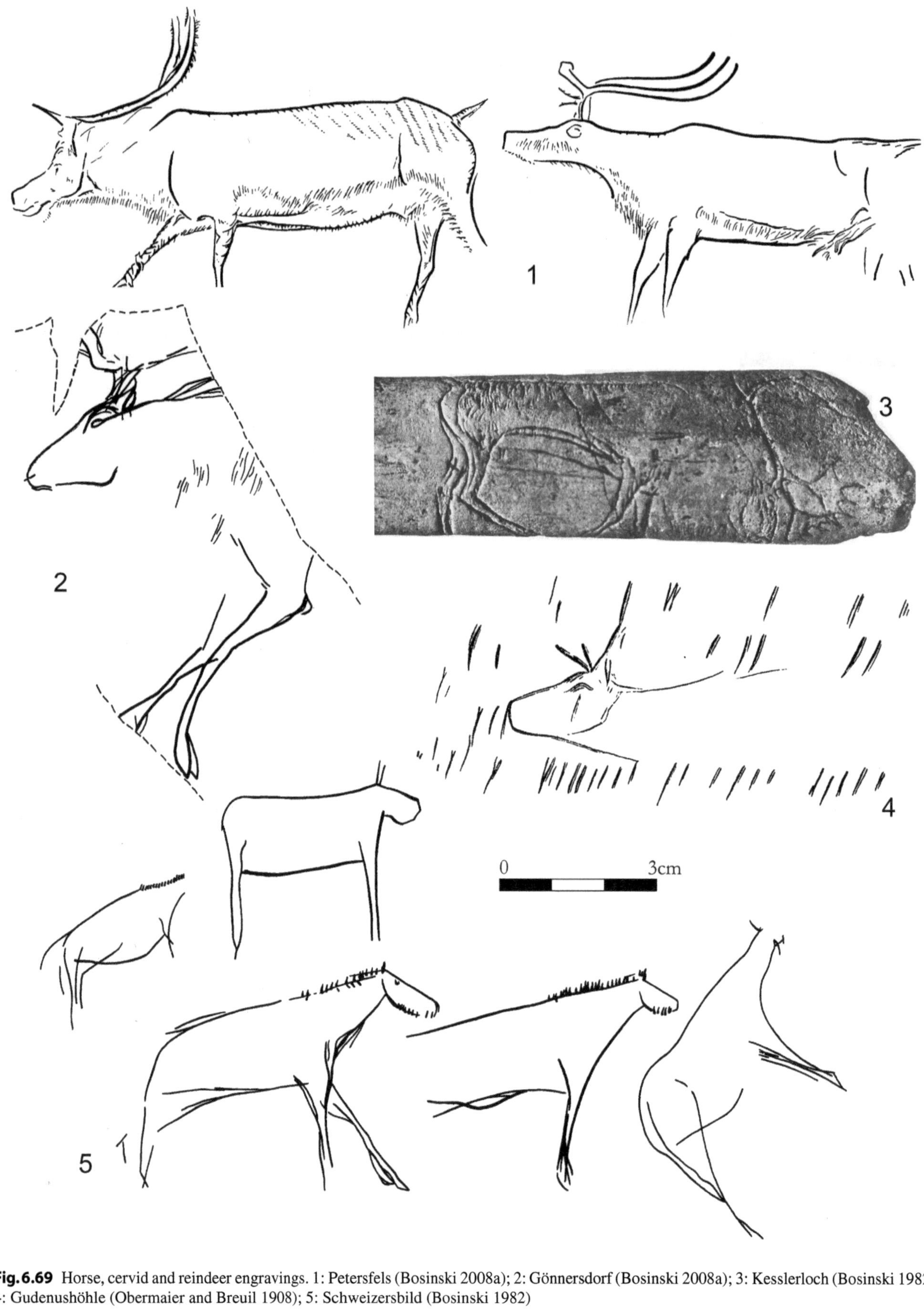

Fig. 6.69 Horse, cervid and reindeer engravings. 1: Petersfels (Bosinski 2008a); 2: Gönnersdorf (Bosinski 2008a); 3: Kesslerloch (Bosinski 1982); 4: Gudenushöhle (Obermaier and Breuil 1908); 5: Schweizersbild (Bosinski 1982)

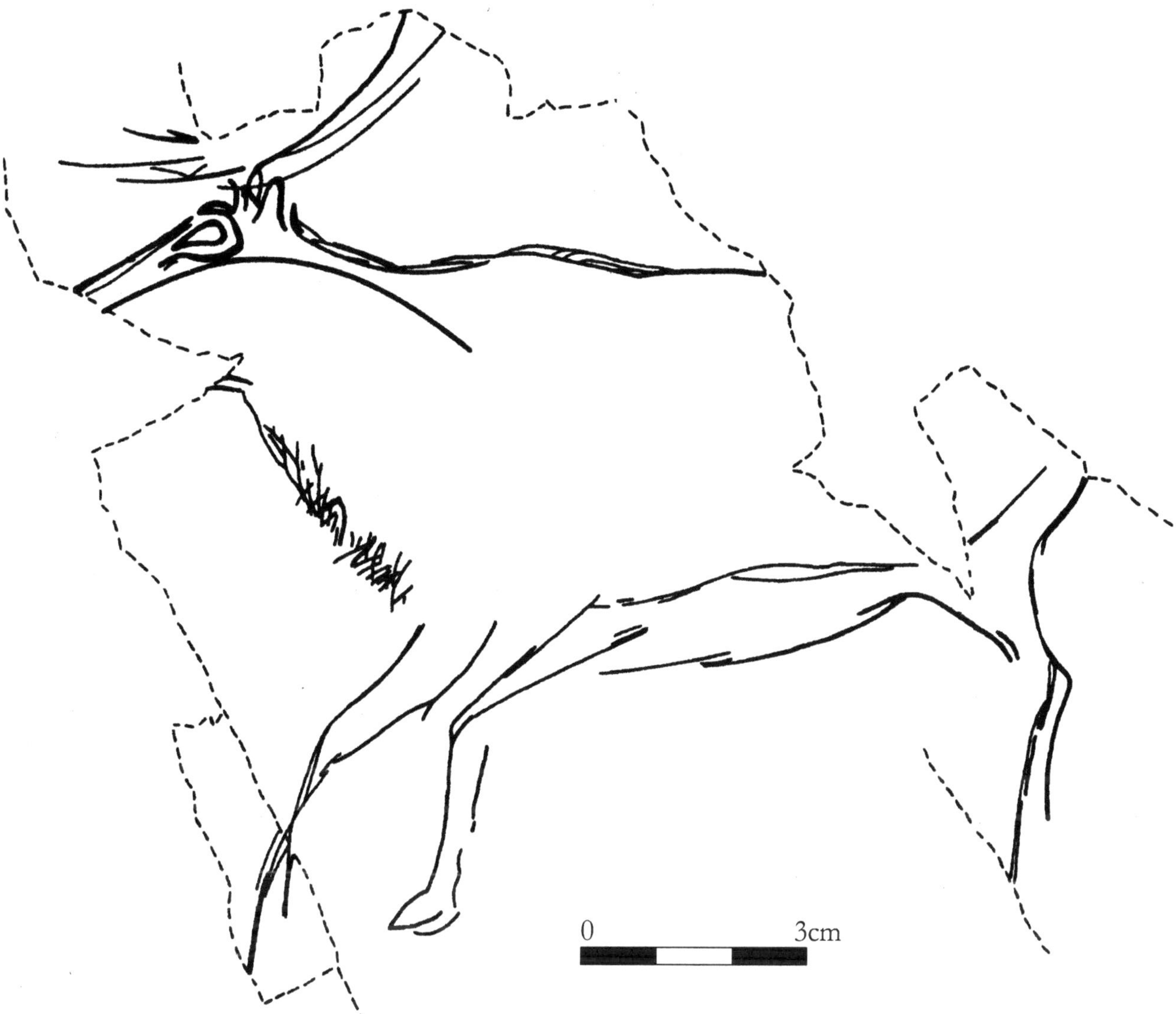

Fig. 6.70 Reindeer engraving. Andernach-Martinsberg (Bosinski 1994)

Concepts"). Thus, it seems that to determine whether those species that were depicted the most were also hunted the most, it is necessary to analyze larger samples of assemblages in order to minimize the variance that is introduced by individual observations. This approach would seem to contradict the view that "it only makes sense to compare an individual cave's figures with the bones from the same site" (Bahn and Vertut 1988: 157). However, archaeological resolution is not nearly fine grained enough to prove that bones and paintings or engravings found together at a site actually belong to the same occupational event; in fact it is quite possible that representations and faunal remains are often the result of different occupations. Hence, it is not at all necessary to restrict comparisons to single sites only—indeed, quite the opposite. Against this background, it is remarkable that C. R. Rice and L. A. Paterson, who carried out two extensive studies in southwestern France and Cantabrian Spain (Rice and Paterson 1985, 1986), came to the following conclusion:

> The analyses from both France and Spain have confirmed that the overall pattern of faunal cave art is clearly related to the sizes of the five major species portrayed (major variable) and the relative prevalence of the five species in the archeofauna of local environments (minor variable). Given the high correlation between species' size and meat yield (***r*=.98**), it is also accurate to say that the differential food yields of the five species are highly related to the proportional representation of each species in the art. (Rice and Paterson 1986: 665)

At this point in the analysis, it can be stated that the proportion of faunal remains of horse and reindeer in CEM assemblages is reflected in the frequencies with which these

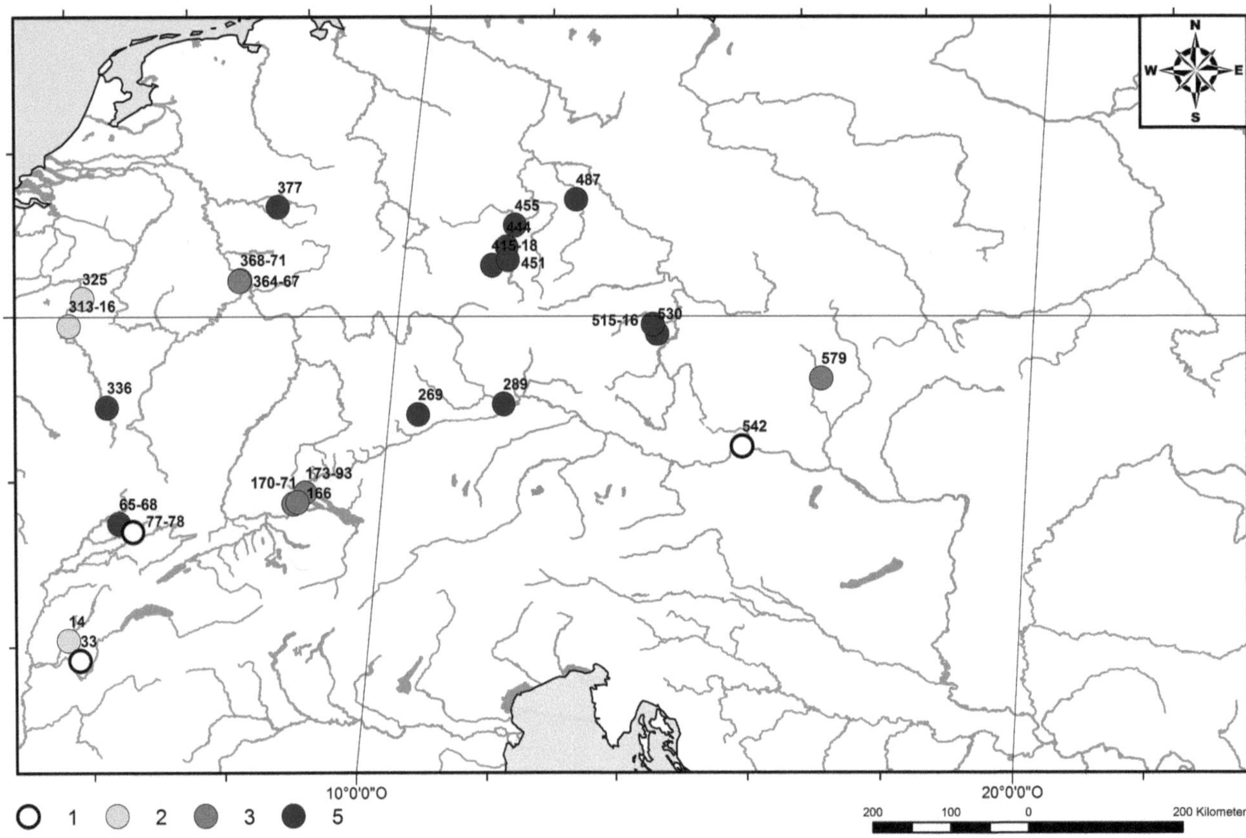

Fig. 6.71 Proportion of representations of horse and reindeer/cervids in the CEM record. 1: Exclusively reindeer; 2: Predominantly reindeer; 3: Equal proportions of horse and reindeer; 5: Exclusively horse; (see Table A.56)

species are represented. Whether this trend also holds true for other species of animals is investigated in the following section.

Other Animal Representations

Representations of animals other than reindeer and horses are comparatively rare in terms of their occurrence at different sites. Nevertheless, their distribution pattern is discussed briefly in this section. Confirmed ibex representations (Fig. 6.72) are comparatively frequent and are known from seven CEM sites from the French and Swiss Jura region, as well as from the Ardennes in the west up to Bohemia in the east (Fig. 6.73). The spatial distribution of ibex representations therefore coincides with the distribution of its remains in the faunal assemblages, although the latter show an increased frequency in the southwest (see Chap. 5, section "Other Animal Remains in the Faunal Record"), whereas the former occur everywhere at low rates. Chamois, in contrast, is only depicted at the site of La Colombière (see Table A.56).

Seven sites yield engravings of mammoths from an unequivocally CEM context (Fig. 6.74). The attribution to the Magdalenian of an engraved ivory fragment from the site of Obere Klause in the Altmühl valley (Fig. 6.74: 2) is uncertain, as it was found in an almost sterile layer clearly below the Magdalenian level, close to the underlying "Solutrean" level i (Obermaier and Frauenholz 1926: 31; Bosinski 1982: 39). As is the case for the skeletal remains (see Fig. 5.14), mammoth engravings are distributed more or less ubiquitously throughout the investigated area, although so far no specimens have been reported from the Swabian Alb (Fig. 6.73).

Depictions of woolly rhinoceros (Fig. 6.75) have been found at five CEM sites (Fig. 6.76). Against the background of the distribution of this species' skeletal remains, which occur mainly in eastern Central Europe, the fact that 3 out of these 5 representations were found in Eastern Germany and Bohemia gains some significance (see Fig. 5.15). Although the total number is much too low to permit valid statements about spatial trends, the tendencies of the two distributions are in accord.

Depending on their state of preservation and fragmentation, representations of wisent and aurochs are sometimes as difficult to distinguish, as are their skeletal remains (Bosinski

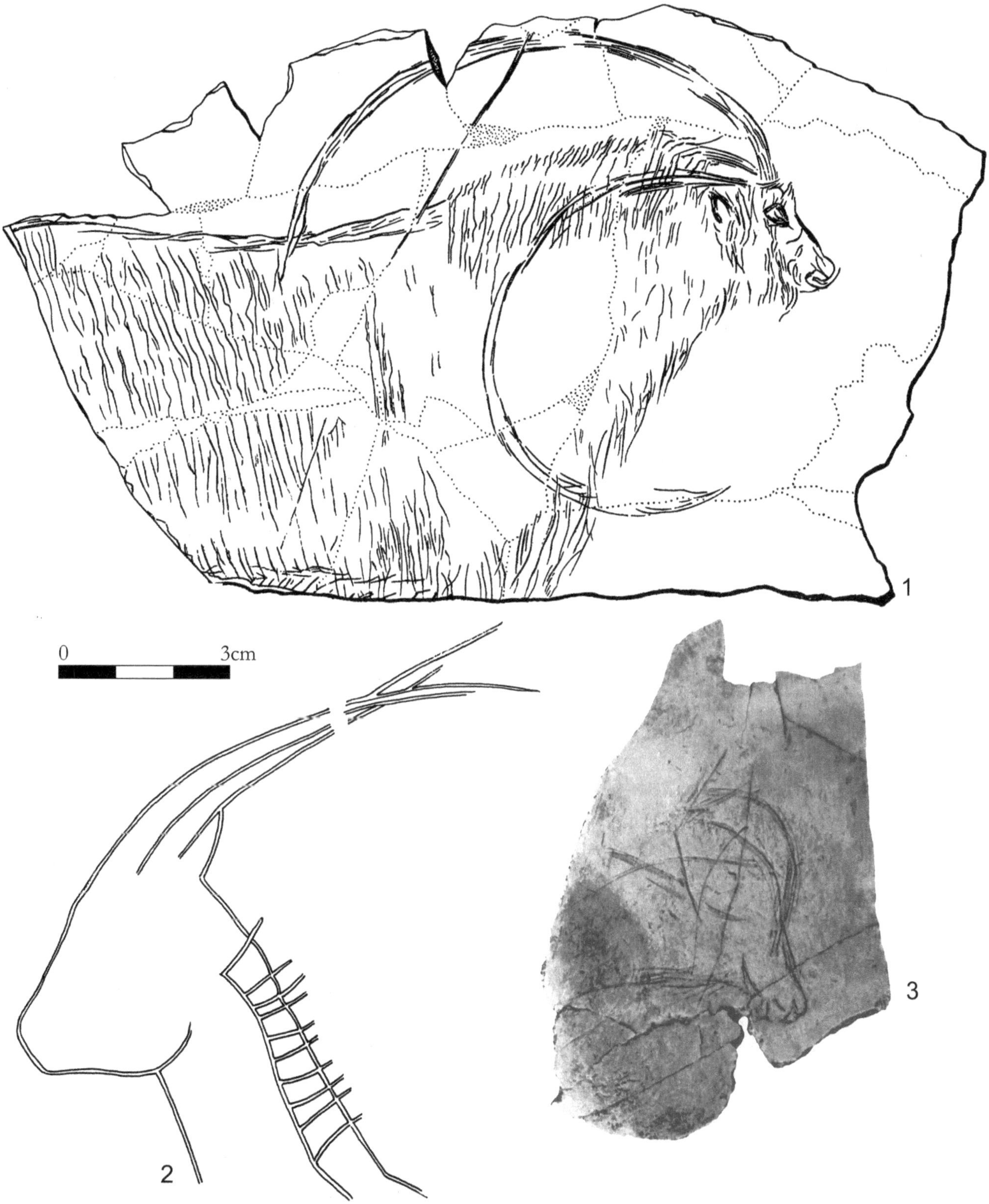

Fig. 6.72 Ibex engravings. 1: Tmaň, Děravá cave (Bosinski 2008a); 2: Grotte Goyet (Dewez 1987); 3: Rislisberghöhle (Bosinski 1982)

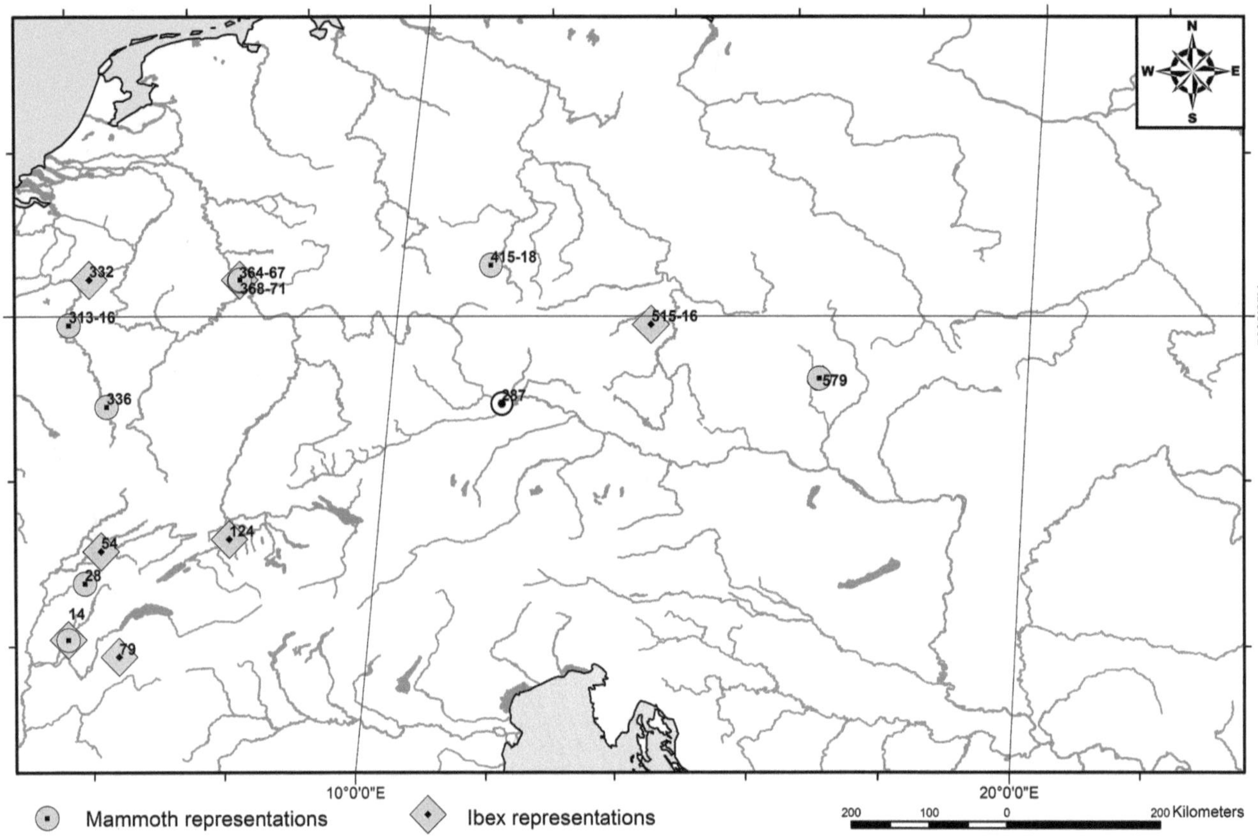

Fig. 6.73 Distribution of ibex and mammoth representations in the Central European Magdalenian (*empty symbols*: attribution to CEM questionable)

2008a: 63). Depictions of these animals are reported from seven sites (Figs. 6.76 and 6.77). With the exception of the famous engraving from Pekárna Cave (Valoch 2001), all were found in the western part of Central Europe. This coincides with the spatial distribution of their remains in the faunal assemblages, which likewise prevail in western Central Europe (see Fig. 5.13).

Relatively unambiguous engravings of red deer (Fig. 6.77) are reported from only three CEM sites, namely, Roc-la-Tour, Andernach, and Gönnersdorf. Cervid representations from Les Hoteaux and Schweizersbild may also depict red deer (see Table A.56; cf. Bosinski 2008a: 60). In the latter case, however, the rough realization of the depiction makes a clear identification rather difficult. The low quantity of representations (Fig. 6.78) contrasts with the rather frequent appearance of red deer remains in the faunal assemblages, while their restriction to the western part of Central Europe reflects the abundance of red deer bones in this region (see Figs. 5.9 and 5.12).

Saiga antelope (Fig. 6.79) is depicted only at Gönnersdorf (Bosinski 2008a: 77f.) and Pekárna Cave (Kozłowski 1992b), where bones of this animal have also been found (Fig. 6.78). Musk-ox representations (Fig. 6.80) are only known from a spear-thrower fragment discovered at the site of Kesslerloch (Bosinski 2008a: 67) and an engraving from La Colombière (Fig. 6.78). It is notable that Kesslerloch is one of the three sites from which musk-ox remains are reported and which are situated in the small area between the High Rhine and the Danube (see Fig. 5.11). An engraving from the site of Kniegrotte that is sometimes classified as a musk-ox in the archaeological literature (e.g., Feustel 1974; Höck 2000) does not show clearly distinguishable attributes of this species. In particular, the characteristics horns are missing, and thus it could equally portray a wisent.

From Gönnersdorf and Andernach-Martinsberg come the only representations of seals discovered in Central Europe to date. Seal representations are quite numerous, however, in the Charente and Pyrenees regions and have also been reported at Grotte Cosquer at the Mediterranean coast (Bosinski 2008a: Fig. 67). It is possible that the person who depicted the seals knew these animals from his or her own experience. Even if this is not the case, the seal representations point to a close relationship with hunter-gatherers living at the Mediterranean or Atlantic coasts.

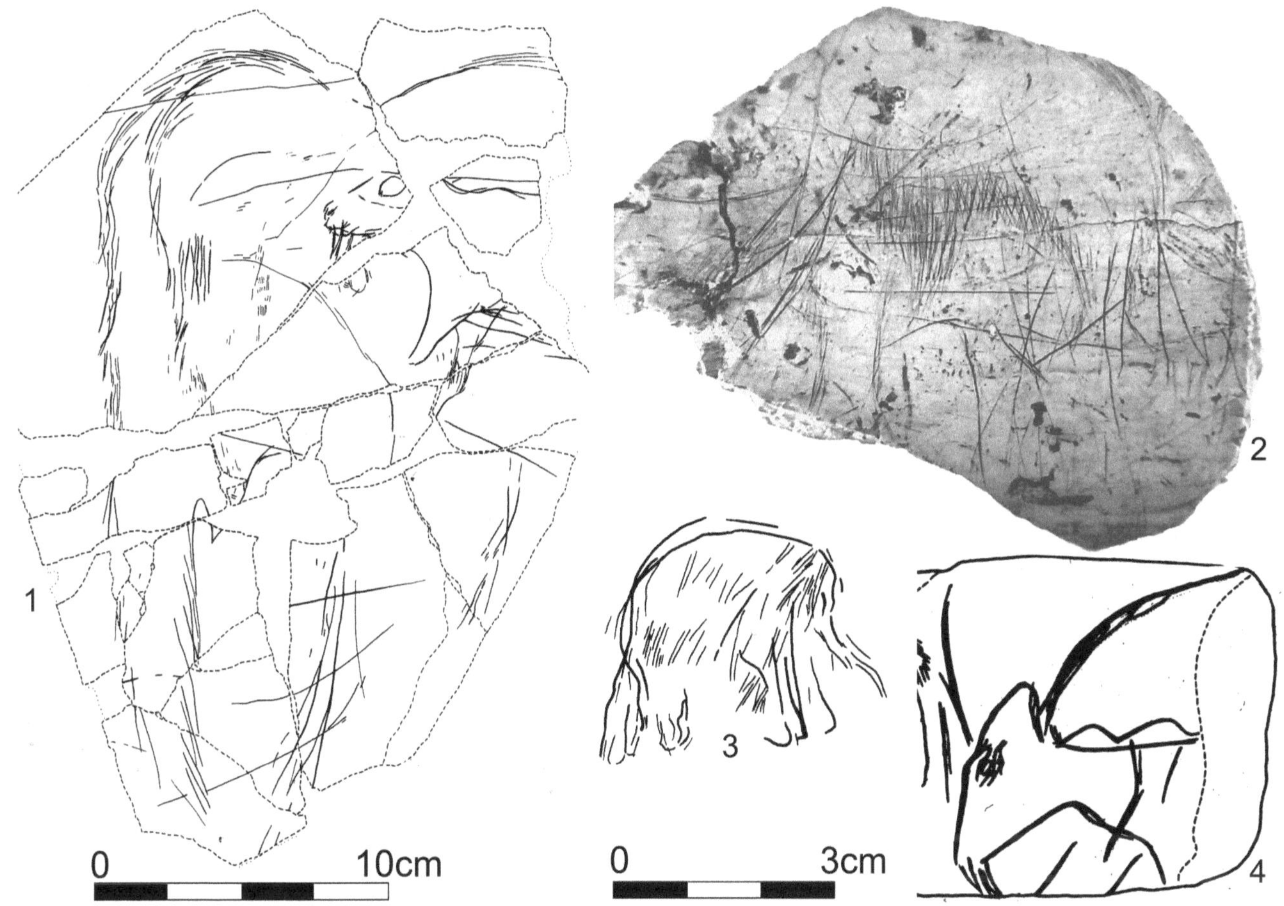

Fig. 6.74 Mammoth engravings. 1: Andernach (Bosinski 1994); 2: Obere Klause (Bosinski 1982); 3: Gönnersdorf (Bosinski 2008a); 4: Teufelsbrücke (Feustel 1980)

In the end it can be said that in the case of most species, the spatial trend in the distribution of their skeletal remains corresponds to the trend in the distribution of their representations. Given the small sample size of faunal representations, the analysis of their spatial pattern reveals a surprisingly clear tendency: in areas where certain species occur more often in the faunal assemblages, their representations are usually more frequent as well. However, it is not at all necessary to assume a dependent relationship between these two spheres or to resort to theories such as hunting magic in order to find explanations. On the contrary, it is quite possible that the numbers of faunal remains and representations both simply express of the relative abundances of a species in a certain region. This hypothesis is supported by a comparison of reconstructed vegetation and the occurrence of certain species across the investigated area (see Chap. 5). Thus, it appears likely that CEM hunter-gatherers mainly hunted the most abundant game in their regional environment. At the same time, they probably took the inspiration for their engravings and sculptures from their everyday experiences and depicted preferentially the animals they saw most often.

Female Representations

Anthropomorphic representations are rather rare in the CEM. When they do occur, their realization is mostly very reduced and standardized. Likely the most common representational scheme is that of the female representations of the so-called Gönnersdorf-Lalinde type (e.g. Bosinski et al. 2001). Here, the upper part typically takes the form of a headless, rod-shaped torso that transitions in the lower part into a prominent triangular or rounded bottom, which in turn narrows toward the often bent knees that usually terminate the figure. In some cases, breasts are also depicted, and the most complete specimens show bent arms with hands held forward (see Fig. 6.81). Realizations of this concept can be found throughout the investigated area. Figure 6.82 shows the location of sites where finds are reported. While 13 sites have yielded unambiguous representations, there are six instances in which the identification of the find as a female figure of the Gönnersdorf type remains questionable and one instance in which such an identification must be rejected.

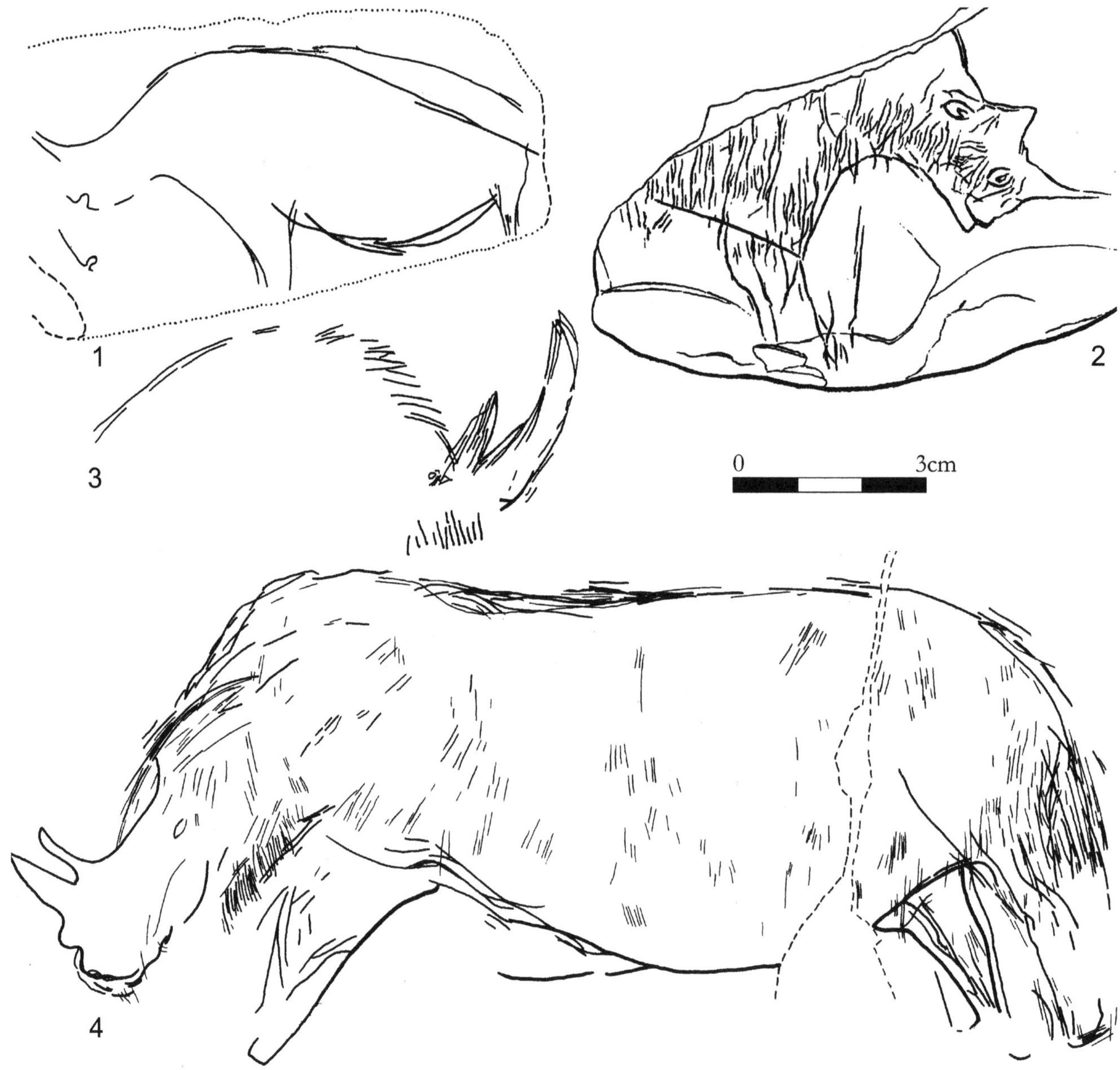

Fig. 6.75 Woolly rhinoceros engravings. 1: Teufelsbrücke (Wüst 1998); 2: Děravá Cave (Klíma 1985); 3: Kniegrotte (Höck 2000); 4: Gönnersdorf (Bosinski 2008a)

From the site of Neuchâtel-Monruz, three small pendants of jet are reported (Bullinger 2006) that are shaped like figurines of the Gönnersdorf type. A fragment of a sculpted artifact unearthed at the site of Hollenberg-Höhle-3, assigned to the Circum-Jurassic Group, is interpreted as a piece of such a figurine (Sedlmeier 1982: 49f.). Morphologically strikingly similar pendants have been excavated at the sites of Petersfels in the area of the Danube Group, underlining once again the internal coherence of these two interaction units. From the site of Farincourt, cave II, a sculpted fragment is reported that may be interpreted as a broken female figurine. On the other hand, it can also be seen as a part of a phallus, especially given that sculptures of this kind have been found there too (David 1996: 133f.). A peculiar piece has been excavated at the rock shelter of Mégarnie in Belgium. The object in question is a stalagmite fragment which is broken at both ends. While one side of the piece is rod shaped, the other has a naturally curved form. The overall morphology thus strongly resembles that of other statuettes (Fig. 6.81: 2). As dripstones do not occur naturally at the rock shelter, the piece must have been deliberately transported to the site. This argues in favor of assuming that it was selected because of its shape (Dewez 1976). A stone slab from Schweizersbild bears, in addition to three cervids, a

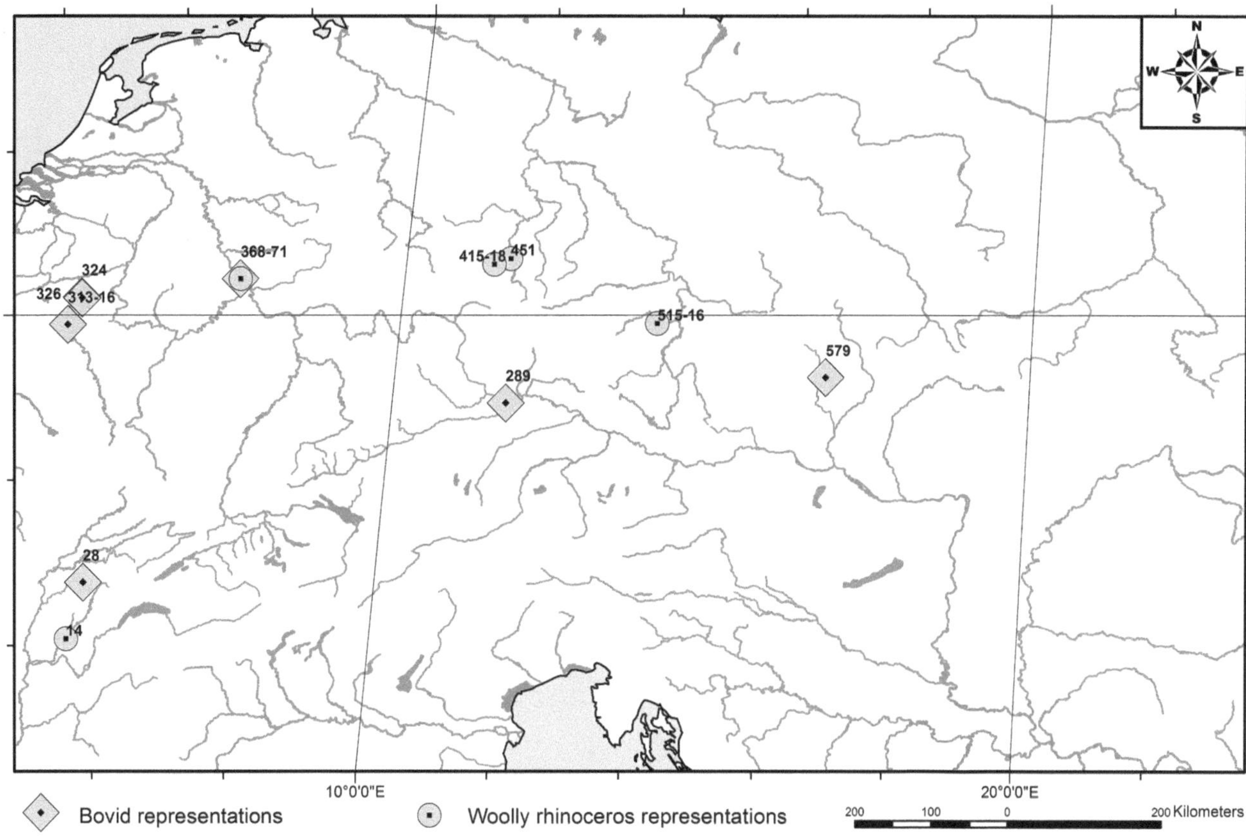

Fig. 6.76 Distribution of woolly rhinoceros and bovid representations in the Central European Magdalenian

group of engraved lines (Fig. 6.81: 6) which might represent the outline of a female figure of the Gönnersdorf type (Bosinski 1982). A small jet fragment found at Hollenberg-Höhle 3 (Sedlmeier 1982: 49f.) may also possibly belong to a statue of this type.

In a recent publication, G. Bosinski (2011) provides a description of the engravings discovered at the Mäanderhöhle in the Franconian Alb and interprets them as female figures of the Gönnersdorf type. From the figures presented in the text, however, this interpretation appears questionable. A sculpted jet fragment from the Hohler Fels at Schelklingen is discussed by J. Hahn (1977c). He points out the possibility that the fragment was once part of a female figurine of the Gönnersdorf type, but compares it as well to the triangular pendants from Petersfels (Fig. 6.81: 5). In this study, an allocation of this piece to the category of female figurines is considered rather improbable. From the site of Samborowice 2a, a "schematic statuette" has been reported (Foltyn and Foltyn 2007: 60). However, the accompanying drawings show that this artifact is a blade core, and its interpretation as a schematic statuette is not at all convincing. Its attribution to this artifact category must therefore be considered questionable, indeed inadmissible.

Regardless of whether or not the six questionable cases are accepted as reliable examples of female figures of the Gönnersdorf type, the specimens in the remaining 13 sites document the distribution of this very distinct morphological concept throughout the entire investigated area. Moreover, engravings as well as bone, ivory, or even flint sculptures of this general type are found at sites from the Atlantic coast in the west up to the Dnieper river in the east (Bosinski 1990: 235, 2011). The remarkable similarity among objects from widely separated places is certainly some of the best evidence for the distribution of certain ideas as common knowledge across the whole of Europe during the Late Upper Paleolithic and consequently for the existence of a wide-ranging interaction network at that time.

Phalliform Objects

Another pan-Magdalenian phenomenon is the occurrence of phalliform sculptures. They are found in the entire investigated area, for instance, at the sites of Rigney 1, Grappin,

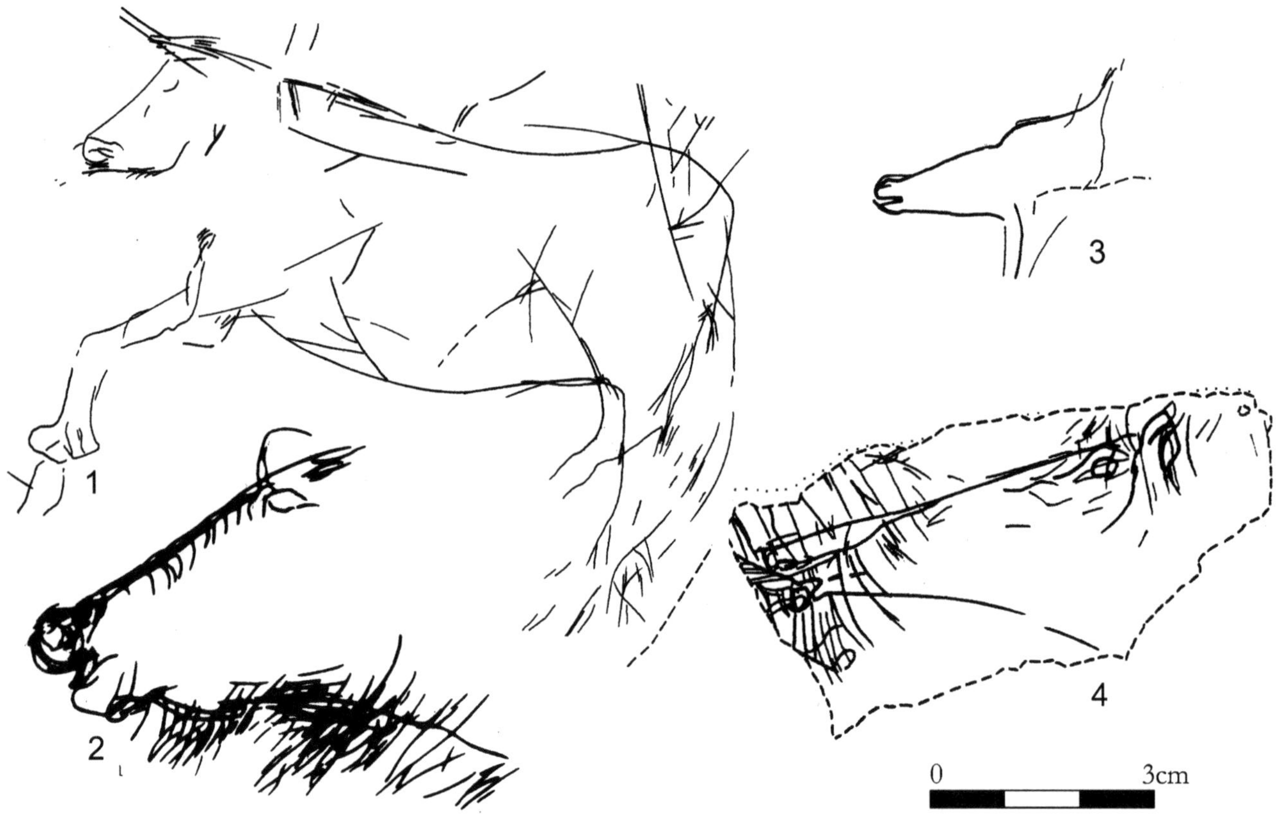

Fig. 6.77 Bovid and red deer engravings. 1–3: Gönnersdorf (Bosinski 2008a); 4: Andernach (Bosinski 1994)

Farincourt, Kniegrotte, and Maszycka Cave, as well as in southwestern France at sites such as Laugerie Basse and Grotte du Placard (Allain et al. 1985; Bosinski 1990; David 1996). These representations vary from more abstract to highly naturalistic depictions and are often realized on *bâtons percés*. Phalliform-sculpted objects are usually connected with the early phases of the Magdalenian, in particular with the *Magdalenién à navettes* (see Chap. 4, section "The Classification of the Magdalenian into Six Phases and the Critique to Which It Is Subject"). Just like the female representations, they provide evidence for the ubiquitous distribution of similar concepts throughout the Magdalenian hunter-gatherer community already during these early stages.

Other Noteworthy Objects

Finally, some engravings and sculptures which are rare or even unique in the context of the CEM are discussed in three sections: first pieces that occur in only a very restricted area of the CEM are discussed, followed by objects that have equivalents outside of the CEM area, and then by objects for which no direct analogies are known.

The first group of artifacts includes three sculptures (Fig. 6.83) found at the sites of Kesslerloch, Petersfels, and Kleine Scheuer im Rosenstein (Bosinski 1982: 30, 36). The sculptures are all made of jet and have a flat bottom side and a convex top side into which regular lines have been engraved. Their morphology is reminiscent of beetles, and they are perhaps representations of insects. Interestingly, all of these assemblages belong to the Danube site cluster and therefore to a single territory-conscious interaction unit (see Chap. 7, section "The Danube Group"). A fourth sculpture made of ivory and found at Grotte de Coléoptère is also interpreted as a kind of beetle representation, which is why this site is known as "Beetle Cave" (Fig. 6.83: 1). Although the four sculptures in the first group exhibit several differences in their realization, the typological concept seems to unite them.

At the sites of Kesslerloch and Freudenthalhöhle, which are located only 10 km away from each other, small sculpted rods sharing a similar shape have been excavated (Fig. 6.84). In terms of morphology, they recall the shape of female representations of the Gönnersdorf type. Thus, they can be seen as strongly schematized realizations of this concept and can probably be allocated to the spectrum discussed above. However, their strong internal similarities suggest that they

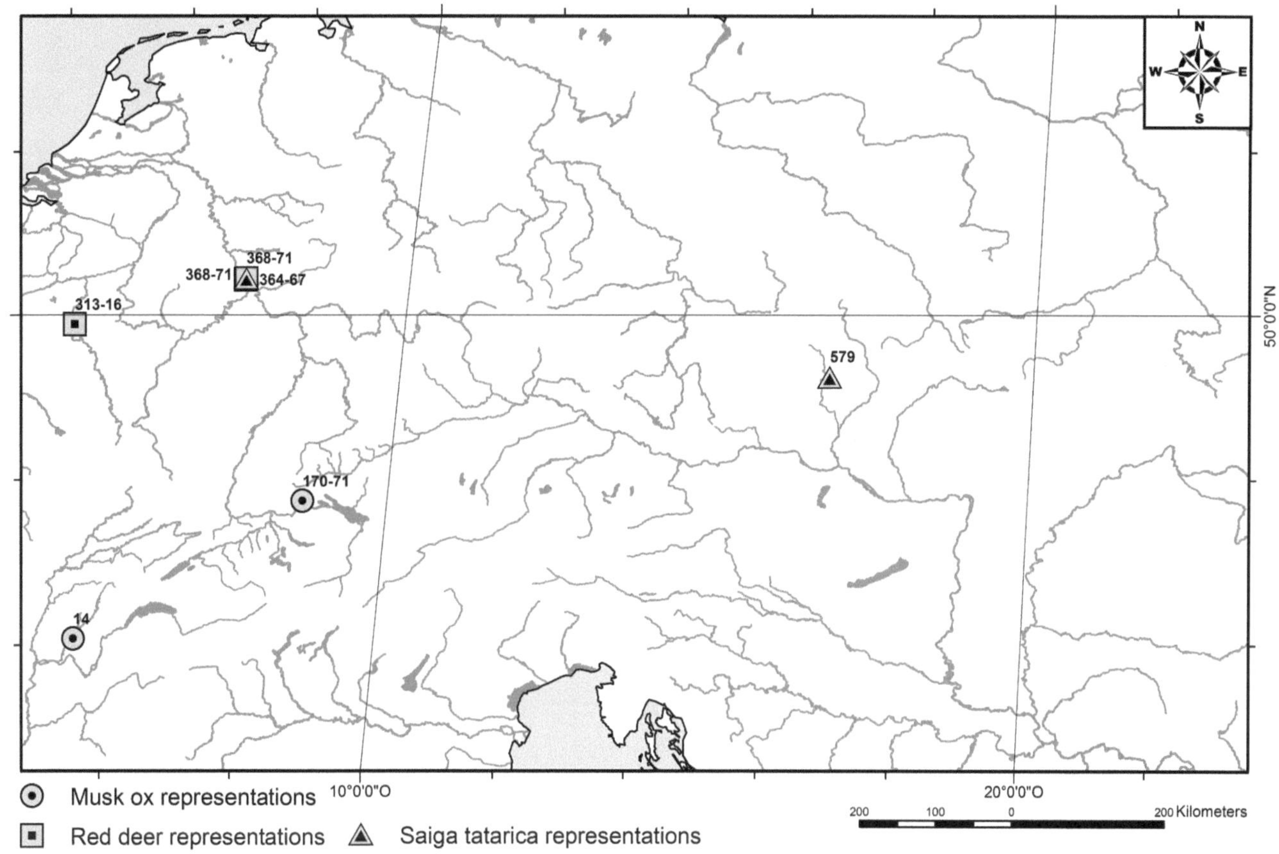

Fig. 6.78 Distribution of red deer, saiga, and musk-ox representations in the Central European Magdalenian

may have been produced by the same person or at least by persons who were in close contact with one another.

From the same two sites (Kesslerloch and Freudenthalhöhle) come fragments of *baguettes demi-rondes* with small raised humps (Fig. 2.9), the only specimens of this kind known from Central Europe. Similar artifacts, however, occur quite frequently in the Dordogne region and Cantabrian Spain (Bosinski 1982: 56, Fig. 13). In accordance with the distribution pattern of fossil and subrecent mollusk shells (see section "Procurement Patterns of Mollusk Shells and Amber"), and despite having several unique characteristics, they show that the hunter-gatherers of this interaction unit were integrated into a far-reaching exchange and communication network that extended from Cantabrian Spain up to the Danube area.

At the site of Mittlere Klause, a *bâton percé* bearing the sculpted and engraved representation of a face in frontal view (Fig. 6.85: 1) has been excavated. A comparable piece (Fig. 6.85: 2) is reported from the site of Grotte de Placard (Capitan et al. 1924: 101). The two representations have traits in common: in both the forehead, the elongated ridge of the nose, and the nostrils are represented in low relief. On both sides of the nose ridge run parallel stripes which intensify the plastic effect, and the eye area is slightly sunken. The similarities in the manner of representation point to a common conceptual background and thus further support the existence of close relations between hunter-gatherers in the Danube area and in southwestern France.

Finally, three sculptures must be discussed for which directly comparable pieces are so far lacking. Two of them were excavated at the cave site of Kniegrotte in Eastern Germany (Feustel 1974; Höck 2000). One object is a small ivory sculpture of a barbed point or harpoon. It probably represents a composed artifact, the barbs of which were attached to a pointed shaft and fixed with a twisted string running crosswise from barb to barb (Höck 2000). The second and perhaps most curious find is a foot-shaped ivory pendant that was probably attached to a string by two holes in its heel. These holes were presumably drilled after two previously-used holes situated closer to the edge of the piece had torn off. The whole pendant is covered with line ornaments of different shapes. From Andernach-Martinsberg come reports of a piece of shed reindeer antler that was given the shape of a bird's head through only minor alterations of the natural piece (Fig. 6.86). Although several other natural pieces have been found that remind of this form, the statue from Andernach is the only reliably identifiable specimen of this kind known to date (Bosinski 2008a, 125).

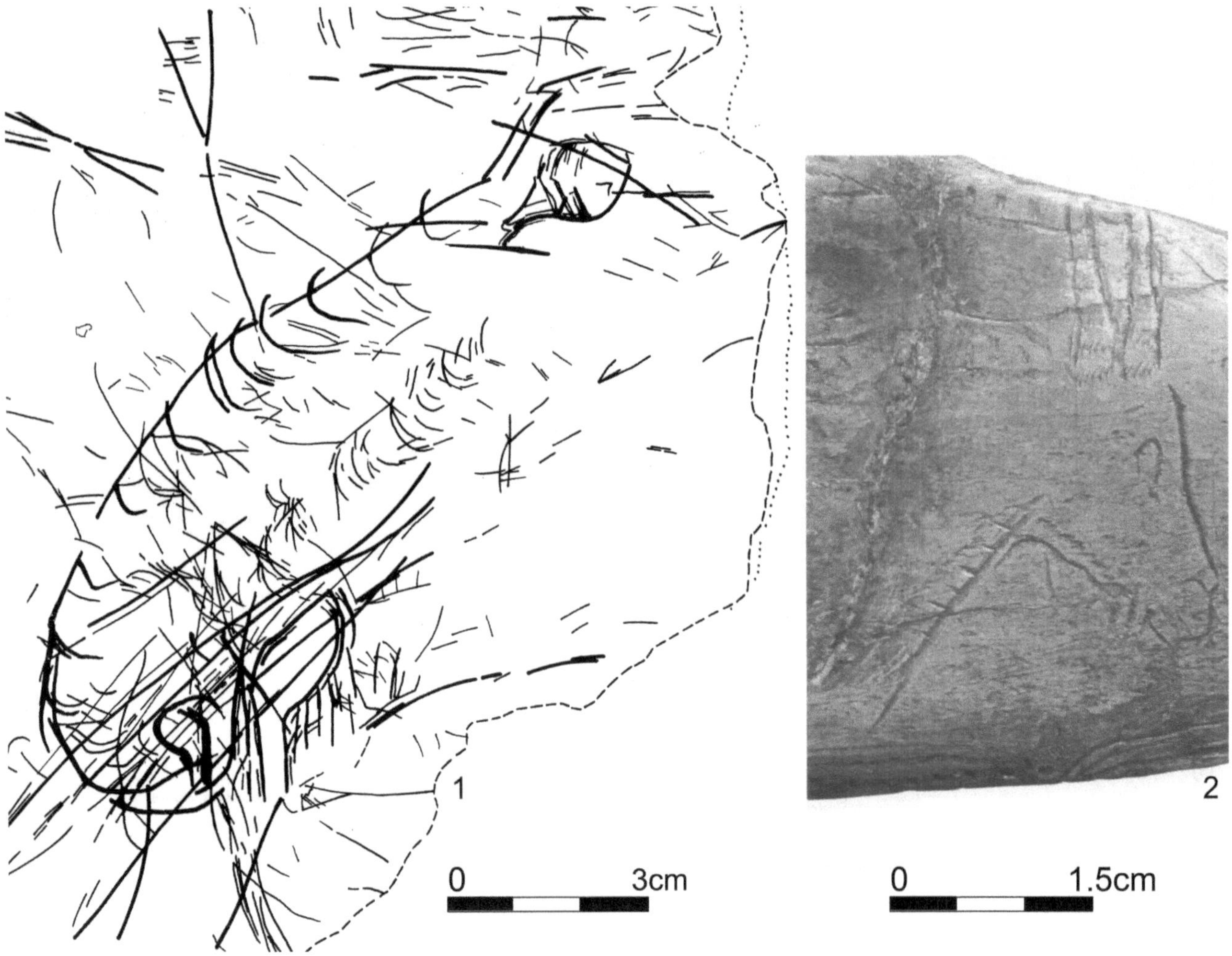

Fig. 6.79 Saiga engravings. 1: Gönnersdorf (Bosinski 2008a); 2: Pekárna (Bosinski 2008a)

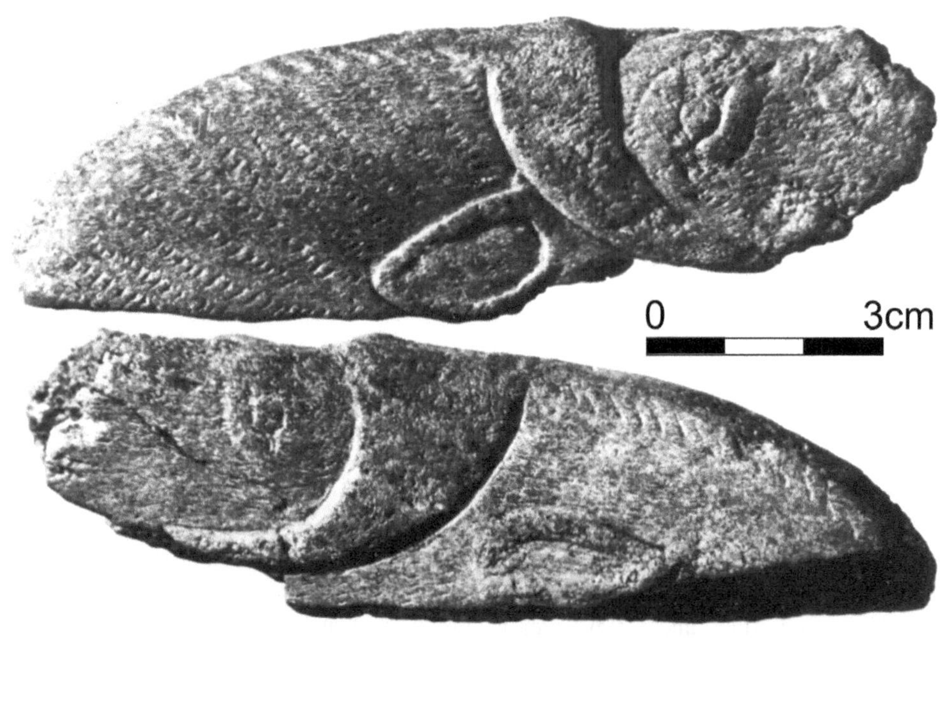

Fig. 6.80 Musk-ox sculpture. Kesslerloch (Bosinski 1982)

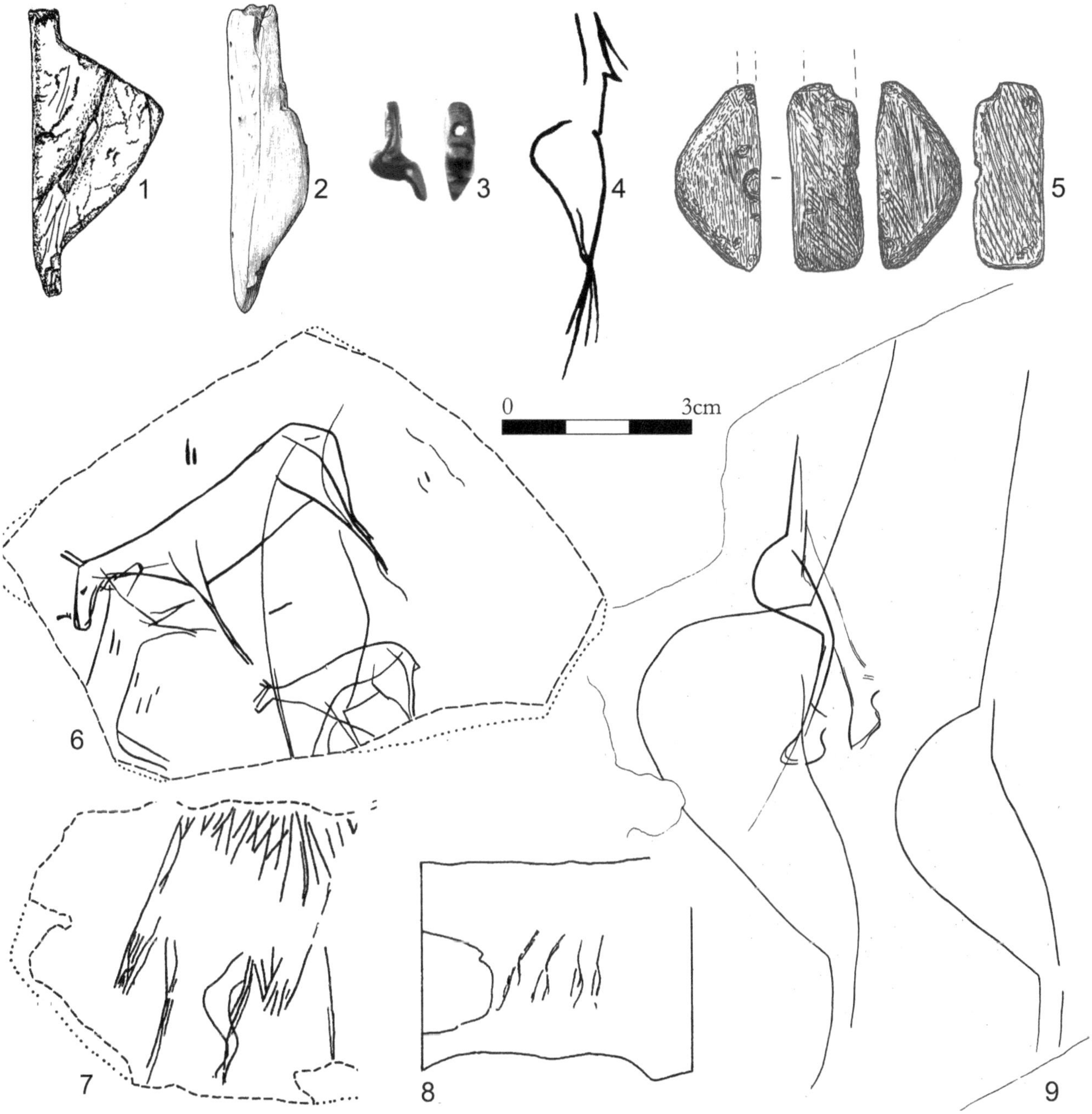

Fig. 6.81 Female representations of the Gönnersdorf type. 1: Andernach (Höck 1993); 2: Mégarnie (Dewez 1976); 3: Neuchâtel-Monruz (Bullinger 2006); 4: Andernach (Bosinski 1994); 5: Hohle Fels bei Schelklingen (Hahn 1977c); 6: Schweizersbild (Bosinski 1982); 7: Teufelsbrücke (Wüst 1998); 8: Petersfels (Albrecht and Berke 1980); 9: Hohlenstein bei Edernheim (Bosinski 1982)

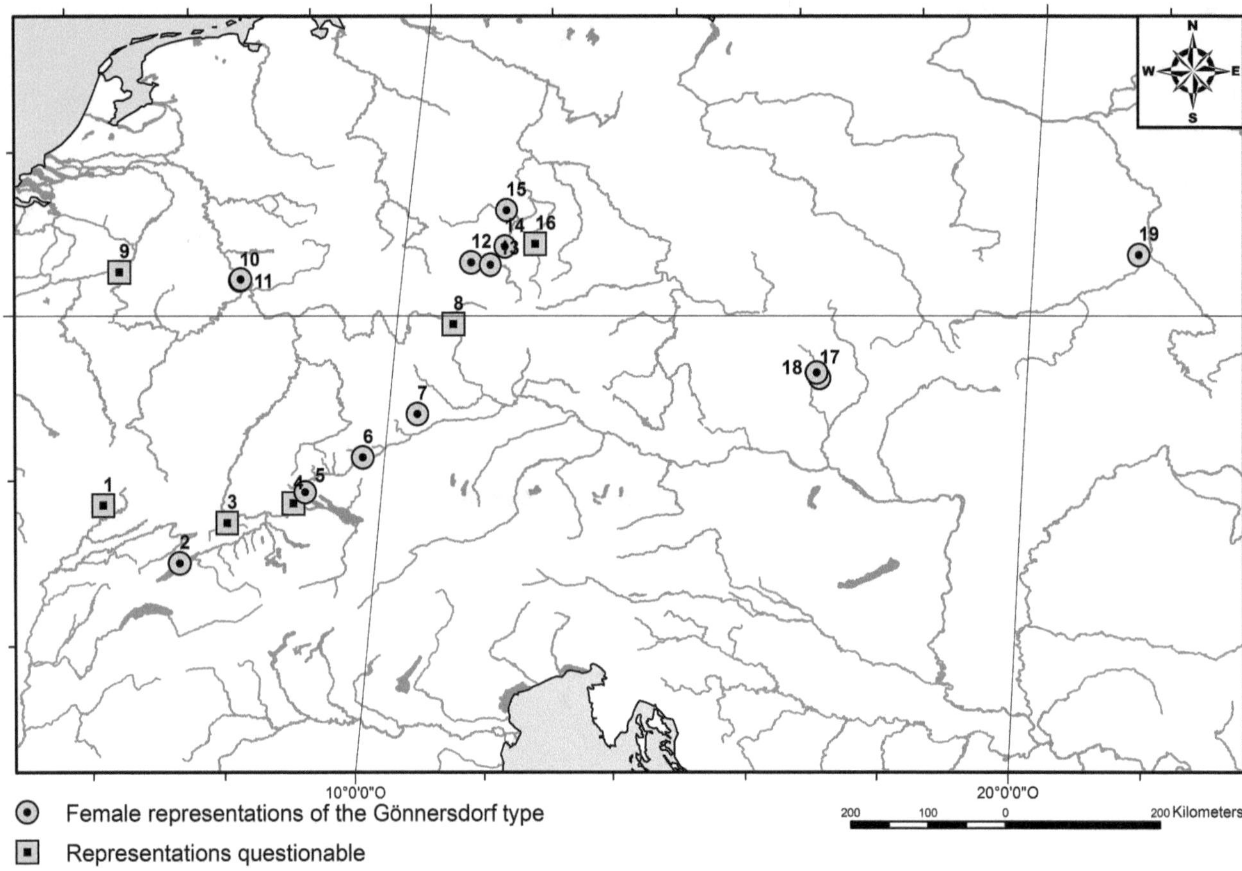

Fig. 6.82 Distribution of female representations of the Gönnersdorf type in the Central European Magdalenian (1: Farincourt; 2: Neuchâtel-Monruz; 3: Hollenberg-Höhle; 4: Schweizersbild; 5: Petersfels; 6: Felsställe; 7: Hohlenstein bei Ederheim; 8: Mäanderhöhle; 9: Abri de Mégarnie; 10: Andernach-Martinsberg; 11: Gönnersdorf; 12: Bärenkeller; 13: Teufelsbrücke; 14: Oelknitz; 15: Nebra; 16: Hyänenhöhle; 17: Pekárna cave; 18: Býči skála cave; 19: Wilczyce)

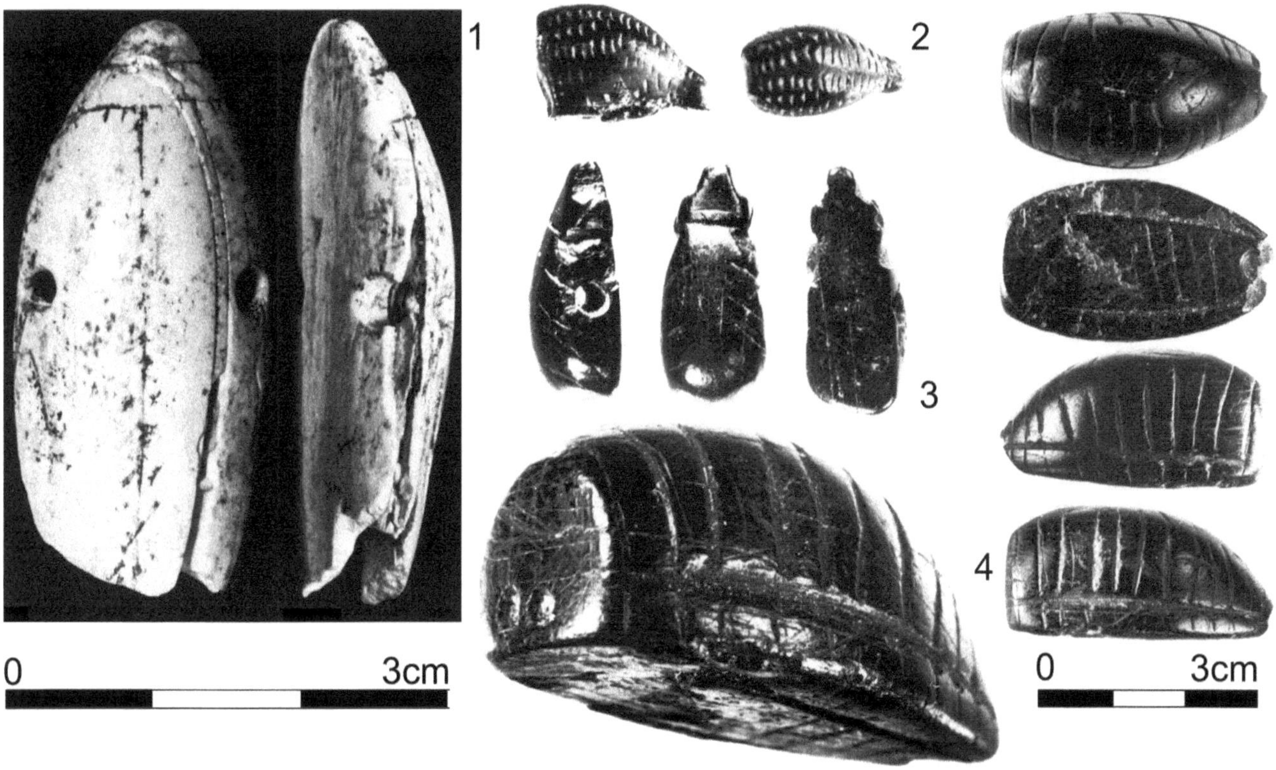

Fig. 6.83 Beetle-shaped sculptures. 1: Grotte de Coléoptère (Dewez 1987); 2: Kesslerloch (Bosinski 1982); 3: Petersfels (Bosinski 19822); 4: Kleine Scheuer im Rosenstein (Bosinski 1982)

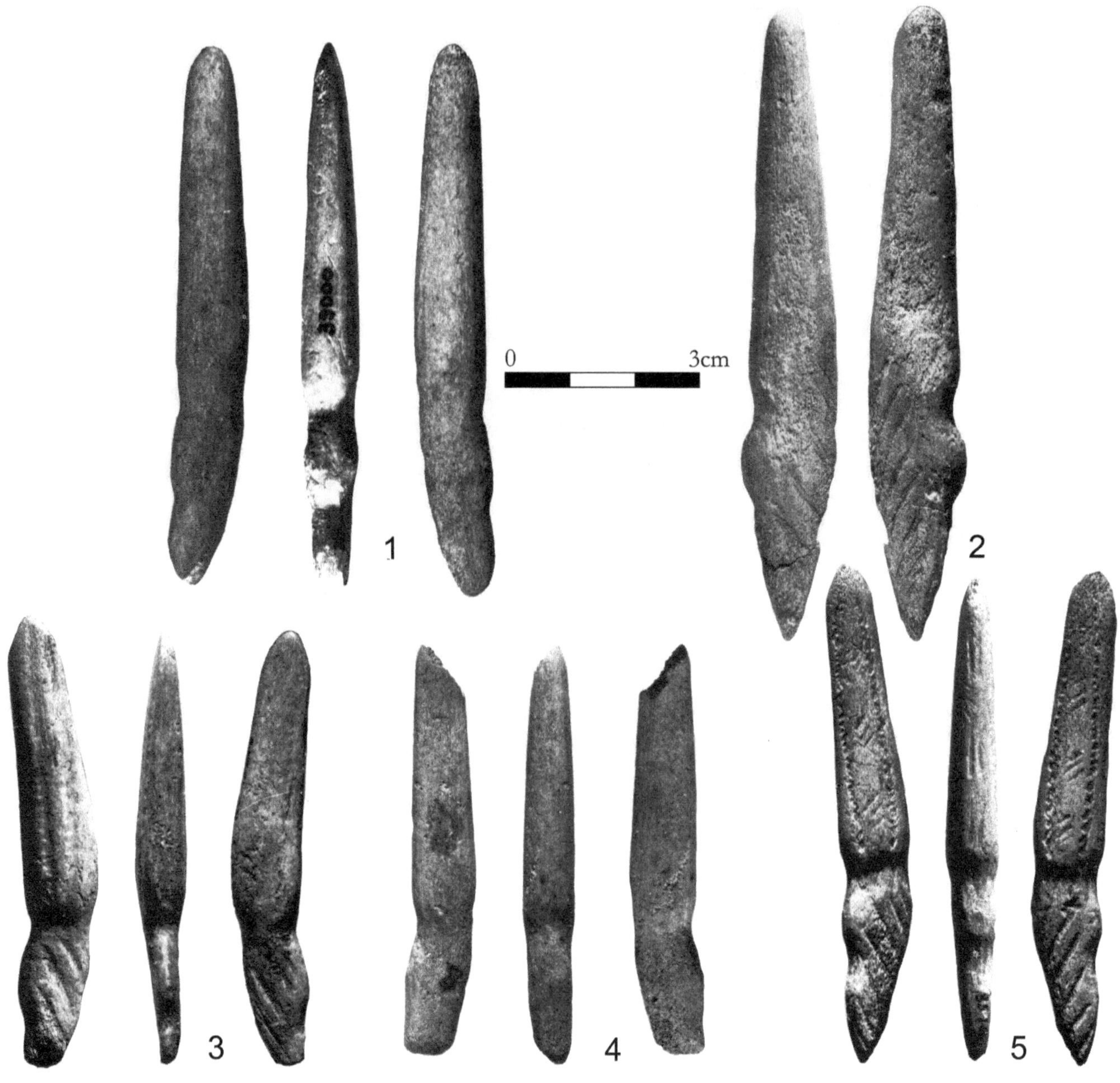

Fig. 6.84 Sculpted rods. 1: Schweizersbild (Bosinski 1982); 2–5: Kesslerloch (Bosinski 1982)

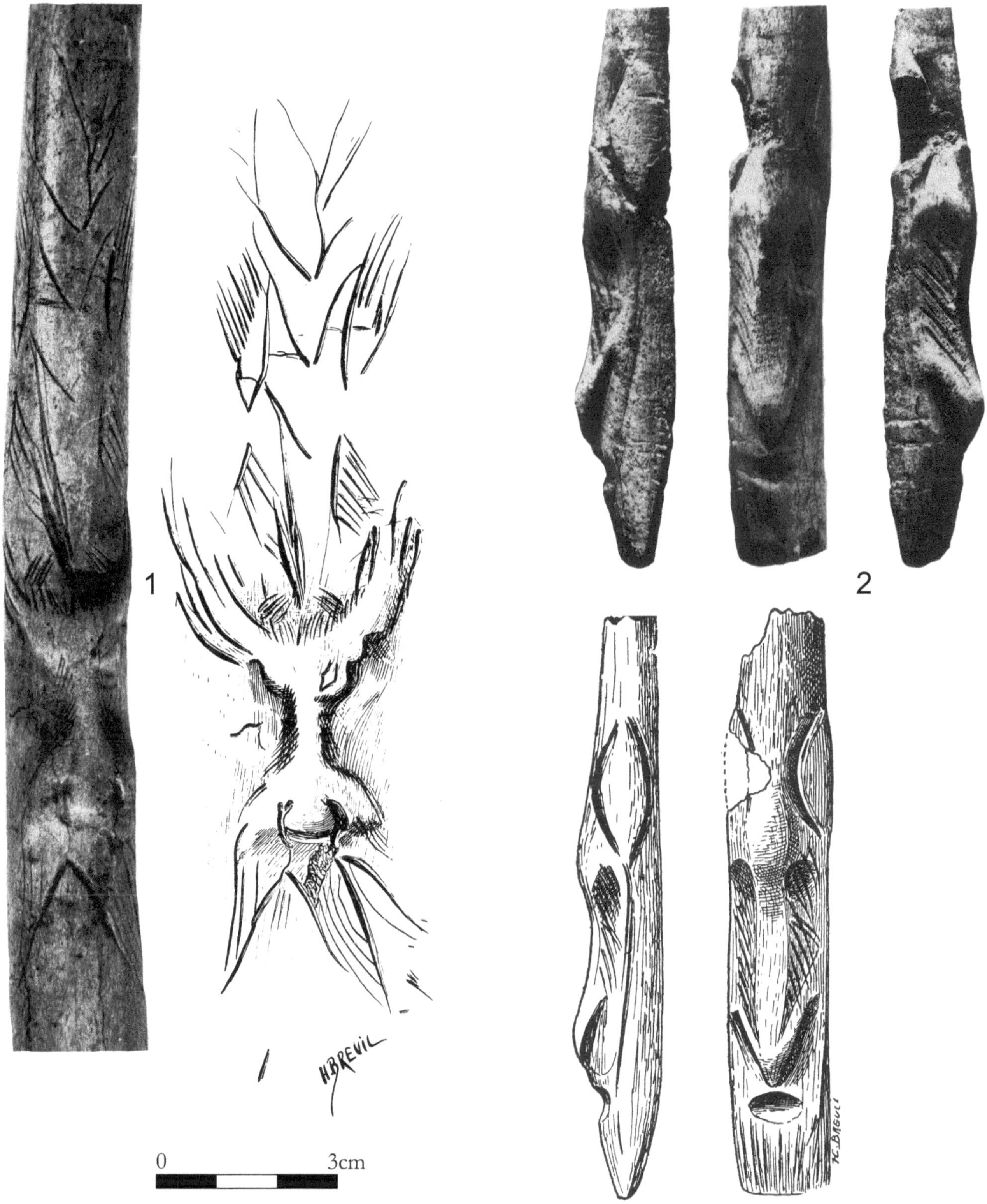

Fig. 6.85 Animal faces in front view. 1: Mittlere Klause (Bosinski 1982); 2: Placard (Bosinski 1982; Capitan et al. 1924)

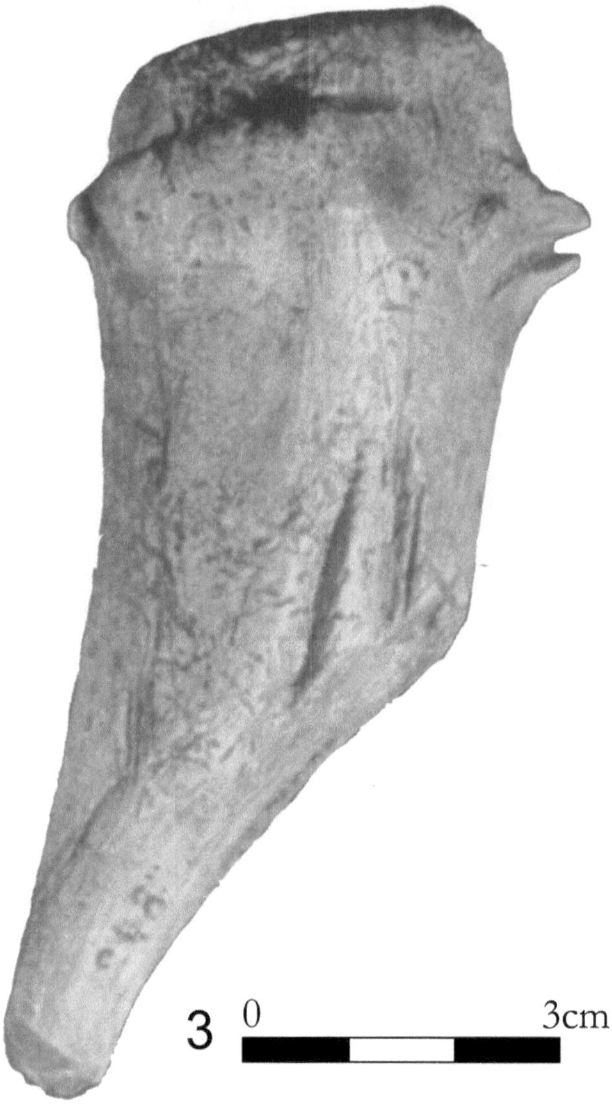

Fig. 6.86 Sculpted bird head. Andernach-Martinsberg (Bosinski 2008a)

Part III
Interpreting the Central European Magdalenian

Chapter 7
Territories and Land-Use Patterns of the Five Regional Groups

The analyses presented in the previous chapters provide various kinds of evidence from several spheres of hunter-gatherer life, all of which show that the CEM is not a homogeneous unit, but rather exhibits a pronounced regional diversity. The results speak in favor of five regional groups or territory-conscious interaction units, namely, the Circum-Jurassic Group, the Danube Group, the Meuse-Rhine Group, the Vltava-Saale Group, and the Polish-Moravian Group. In the following sections, these groups will be described in detail. Before embarking on the description, however, it seems advisable to provide some general information on the number of assemblages per group as well as to outline some general assumptions about the archaeological record and to briefly recall some organizational concepts of mobile hunter-gatherer societies.

In total, 651 assemblages from 540 sites were analyzed. During a critical review, all assemblages were classified according to their typological content and, if possible, their absolute dates. The resulting classes thus represent the probability with which the individual inventories can be assigned to the CEM (Table 7.1; for a more detailed overview, see Table A.1 and Chap. 6, "Critical Review and Classification of the Recorded Assemblages"). Of the total number of 651 assemblages, 626 (519 sites) could be assigned to one of the five regional groups. The remaining 25 assemblages (21 sites), on the other hand, could not be allocated, primarily due to their isolated geographic position.

The Circum-Jurassic Group shows the highest number of assemblages (n = 164), whereas the Meuse-Rhine Group includes only 73 assemblages. With regard to assemblages with more than 100 and more than 1000 lithic tools, respectively, this picture changes markedly. Here the Meuse-Rhine Group contains the highest proportion of the inventories (Table 7.2). The Polish-Moravian Group and Vltava-Saale Group each comprise over 100 assemblages. However, both groups show very high ratios of assemblages classified as not determinable (Table 7.1). At the same time, they yield high numbers of open-air sites and surface collections (Table 7.2).

In the following, the description of each group begins with a short statistical characterization which summarizes the number of different kinds of sites, such as caves and rock shelters, open-air camps, and surface collections, or sites with more than 100 tools. In a second section, information on the landscape of the settlement area and on the environmental setting is presented. Next, the geographic position of the sites is examined more closely and is used as the basis for estimating the minimum sizes of the groups' territories (see Chap. 3, "Territories, Territoriality, and Land Tenure"). Here, three versions are calculated. For the first and second version, buffer of 25 and 40 km, respectively, are calculated around each site, and the area covered by these buffers is used to estimate the extent of a possible territory. The third version, in contrast, takes into account the information given by the raw material procurement pattern as well as the individual geomorphological situation for each group. Finally, the observations are used to interpret a possible scenario for the settlement system and land-use pattern of each regional group, based on the results of this study and ethnographic observations.

Given the well-founded assumption that the acquisition of lithic raw materials in the CEM was embedded in everyday and seasonal subsistence activities and was not subject to specialized procurement trips or exchange (Binford 1979: 259f.; Floss 1994: 325; see Chap. 6, "Procurement Pattern of Lithic Raw Materials"), the area indicated by the procurement pattern can be taken as a minimum estimation for the area exploited by a single regional group and thus their territory (see Chap. 3, "Territories, Territoriality, and Land Tenure"). Ethnographic observations of cyclical land-use patterns among, e.g., the Nunamiut, Aborigines, and Naskapi hunter-gatherers, indicate that the annual range exploited by a regional group covers only a part of its total territory (Binford 1983a: 114f.). Hence a group exploits "a series of discrete areas, occupying each one until the environment becomes degraded" (Ibid.: 114). The exploitation to the point of degradation of such a "discrete area" can take several consecutive years and can therefore include several

Andreas Maier, *The Central European Magdalenian*, Vertebrate Paleobiology and Paleoanthropology,
DOI 10.1007/978-94-017-7206-8_7

Table 7.1 Classification of all recorded assemblages with reference to their assignment to one of the five regional groups

Regional group	1	2	3	4	5	No	Total
Class 1 (surely/very probably CEM)	67	52	35	36	32	3	225
%	41	35	47	26	30	12	35
Class 2 (probably CEM)	35	41	15	40	18	8	157
%	22	29	21	29	17	32	24
Class 3 (probably not CEM)	12	15	8	6	6	4	51
%	7	11	0	5	7	16	8
Class 4 (probably mixed)	7	5	0	4	0	0	16
%	4	4	0	3	0	0	2
Class 5 (not determinable)	43	30	15	51	53	10	202
%	26	21	21	37	46	40	31
Assemblages total	164	143	73	137	109	25	651
%	100	100	100	100	100	100	100
Assemblages total	164	143	73	137	109	25	651
%	25	22	11	21	17	4	100
Sites total	137	97	58	124	103	21	540
%	25	18	11	23	19	4	100

1: Circum-Jurassic Group; 2: Danube Group; 3: Meuse-Rhine Group; 4: Vltava-Saale Group; 5: Polish-Moravian Group. The assemblages in the "No" column could not be allocated to one of the five groups. For a detailed list of all individual classifications, see Table A.1

Table 7.2 Different types of assemblages with reference to their assignment to one of the five regional groups

Assemblages	Group 1	Group 2	Group 3	Group 4	Group 5
From cave sites	63	68	25	22	58
%	41	53	38	17	56
From rock shelter sites	29	22	0	1	5
%	19	17	0	1	5
From open-air sites and surface collections	60	38	40	108	40
%	40	30	62	82	39
With more than 100 lithic tools	48	24	27	22	15
%	32	19	42	17	14
With more than 1000 lithic tools	6	2	3	6	2
%	4	2	5	5	2

1: Circum-Jurassic Group; 2: Danube Group; 3: Meuse-Rhine Group; 4: Vltava-Saale Group; 5: Polish-Moravian Group. Not included are assemblages of Class 3 (probably not CEM)

annual land-use cycles. The Nunamiut, for instance, exploit such an area for about 10 years before moving on. Meanwhile, the resources in the remaining areas of the territory can regenerate, and it is only over the course of decades that the entire territory gets used (Ibid.). By analogy (see Chap. 3, "The Use of Ethnographic Analogies in This Study"), it can be postulated that the members of CEM interaction units probably exploited their territory in periodical cycles. Each year, only a part of the territory was used, whereas the remaining areas had the opportunity to regenerate (Fig. 7.1).

Thus, the archaeological pattern as it is known to date represents the accumulated results of daily, seasonal, annual, and multi-annual migrations. The area exploited by a group of hunter-gatherers during an annual cycle is referred to as its "annual range" (Binford 1982: 7, 1983a: 112f.) or "annual territory" (Vita-Finzi and Higgs 1970: 30; Higgs and Vita-Finzi 1972). Within this annual range, hunter-gatherers use several kinds of camps, the nature and diversity of which depends mainly on the group's subsistence strategy, which usually falls between the two poles of collecting and foraging. A common camp type in both systems is the so-called residential camp or base camp (Binford 1980: 9), which serves as a central point from which all kinds of everyday activities and day trips are undertaken. The area used for these operations is referred to as "camp range" (Binford 1982: 7) and can be subdivided into four zones. The zone immediately surrounding the camp which can be exploited during daily trips and therefore does not exceed a certain radius (the "foraging radius," Binford 1982: 7) is called the "foraging range" (Binford 1980: 15) or "site territory" (Higgs and Vita-Finzi 1972: 30). The area beyond this foraging range is referred to as the "logistical range" (Binford 1982: 8) or "site catchment"

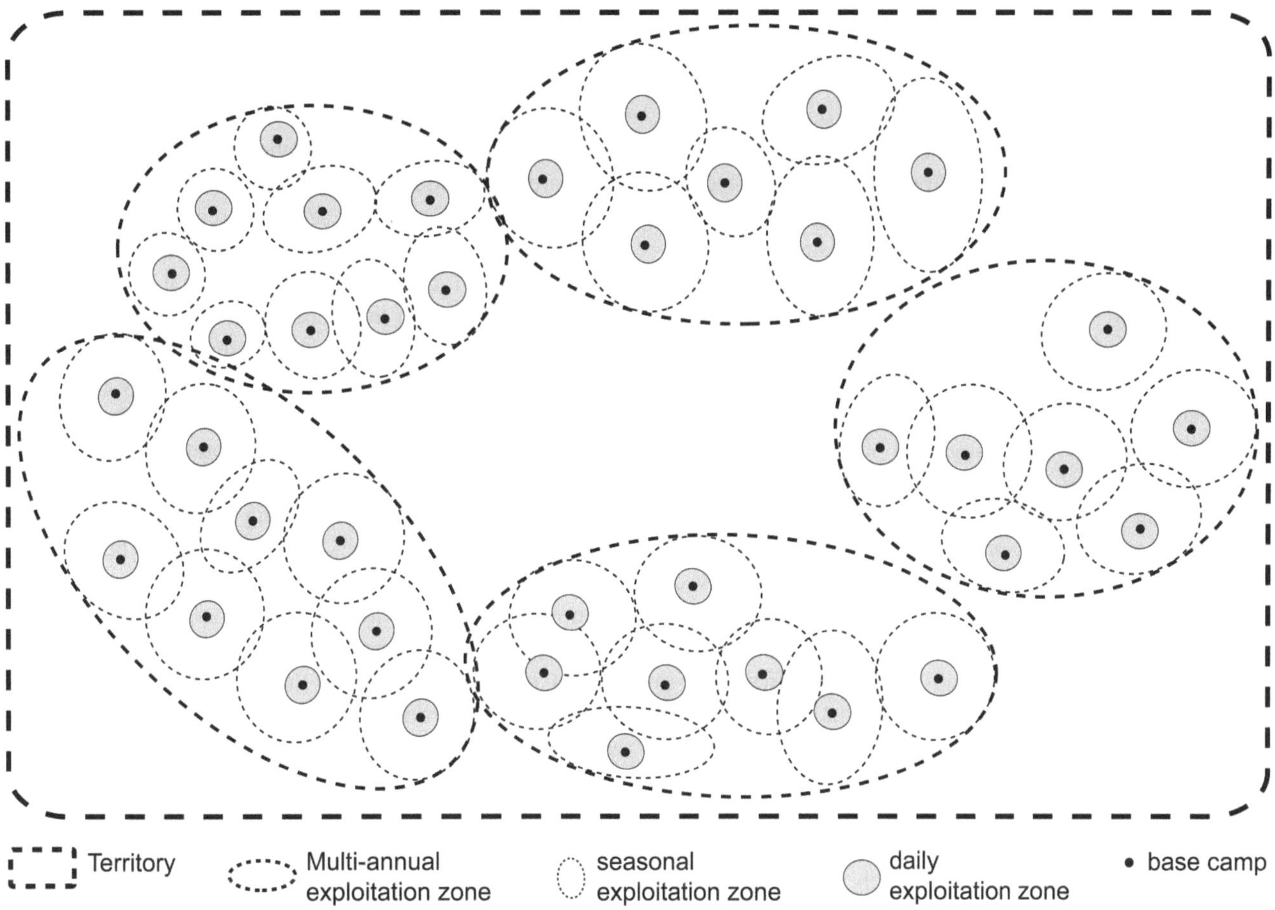

Fig. 7.1 Idealized model of hunter-gatherer land-use based on ethnographic observations

(Higgs and Vita-Finzi 1972: 30). Within this broader range, which is usually exploited by logistically organized groups with a collecting subsistence strategy (Binford 1980), special task groups undertake extended trips to distant resource patches. Beyond the logistical range lies the "extended range" (Binford 1982: 8), an area whose resource distribution is usually known to hunter-gatherers but not exploited by them at all times. The area covered by the extended range can thus be seen as the long-term range that a hunter-gatherer group exploits through many annual cycles. Outside of this range lie those areas which are simultaneously occupied by other groups, which may be relatives or partners for trading and marriage. Binford (1982: 8) refers to these areas as visiting zones. After a period of time, when the resources in the immediate vicinity of the residential camp have been exhausted and huntable game becomes rare, the group moves the camp to its next location. These "residential moves" (Binford 1980, 1982) may vary considerably in frequency and distance.

As mentioned above, Binford distinguishes two major kinds of subsistence strategies: collecting and foraging. This dichotomy provides an excellent background for reflections on Paleolithic hunter-gatherer land-use patterns, especially if the two strategies are not conceived as fixed categories but rather as end-points on a continuum, as Binford himself proposes (1980: 19). At one end of the spectrum is the foraging strategy. Here, the hunter-gatherer groups perform several seasonal residential moves among a series of resource patches. In other words, they move the consumers to the resources (Ibid.: 15). They do not store goods but rather procure them daily on an encounter basis within the foraging range and return each day to the residential camp (Ibid.: 5f.). At the opposite end of the spectrum stands the collecting strategy, where logistically organized special task groups undertake procurement operations, sometimes over several days, in order to move specific resources to the consumers in the residential camp (Ibid.: 15). Occasionally, they procure goods in very large quantities for storage (Ibid.: 10). The logistical strategy of collectors is favorable in environments with an incongruent distribution of goods, where a group is close to one critical resource but far away from another. Storage, as a response to temporal incongruity or unpredictability of resources, especially food, leads to spatial incongruity, which, in turn, favors a logistical organization (Ibid.: 15).

The subsistence strategy has direct implications for the potential character of the archaeological record. Foragers mainly use two kinds of camps, residential camps and "locations" (Ibid.: 9). The former, where many diverse activities are carried out, may vary considerably according to group size and duration of the occupation. The latter represents a short-term stay with very limited activities. Collectors, on the other hand, produce a more complex pattern with additional camp types: "field camps," which are related to special task groups and differ in size, structure, and composition according to the respective task; "stations," where information (e.g., on game) is gathered; and "caches," where goods are stored (Ibid.: 10f.). Several combinations of these camp types may occur as well. Higgs and Vita-Finzi (1972: 30) refer to these camps as "'transit', 'transitory', or occasional sites.".In contrast to residential mobility, where the residential camp is moved from one location to another, this kind of mobility around a residential camp is called "logistical mobility" (Binford 1980: 15f.). Another important factor for the archaeological visibility of hunter-gatherer activities is the degree of redundancy of land- and site-use from year to year. The more scattered and ubiquitous the distribution of resources, the less redundant their exploitation will presumably be (Ibid.: 7).

The sharp delimitation of the raw material procurement pattern in the CEM record suggests that the outer borders of the group's territory remained relatively stable throughout the CEM. The internal subdivision of this territory into daily, seasonal, annual, and multi-annual "exploitation zones" was likely subject to more pronounced fluctuations. Nevertheless, it should be possible to trace at least in part the internal subdivisions of a regional group's territory with regard to its site distribution and raw material procurement pattern. Inequalities in the distribution of sites together with geographic features may help to identify some of the areas which were subsequently used throughout multi-annual periods. A precondition for identification, of course, is that the exploitation zones stayed more or less stable over time. This is more likely to be the case in geographically differentiated areas than in regions with a more or less homogeneous landscape and an even distribution of resources.

A first step toward the identification of potential exploitation zones is to establish a measure for the relative distance traveled during a residential move; this makes it possible to decide whether two sites are located closer to or further apart from each other than the estimated average distance. In order to obtain such a measure, it is not sufficient to simply calculate the mean of all nearest-neighbor distances per regional group. Instead, one must reflect initially about the observed site pattern and the resulting implications for the possible structure of the settlement system. As can be observed on the macroscale, the sites within a regional group are not distributed evenly throughout the observed area. On the contrary, even on a regional scale of up to 15,000 or 20,000 km^2, they form agglomerations and highly inhomogeneous patterns. Furthermore, it can be assumed that the selection of a camp location is a nonrandom process which is carried out according to several criteria. In this connection, it is striking that in some areas of about 100–400 km^2, sites are densely clustered, whereas immediately adjacent areas larger in sizes are virtually empty of Magdalenian finds. Looking at the ethnographic record, it can be said that residential moves are usually undertaken when the resources in the vicinity of a site become depleted. Typically, a new camp ground is selected at some distance from the previous camp. The previously exploited area will usually be resettled during the next seasonal cycle at the earliest, after the resources there have had some time to regenerate. Among recent hunter-gatherer groups, the distances traveled during a residential move may vary considerably, but does usually not come below 5 km (cf. Binford 2001: Table 5.01). Drawing an analogy from these observations, a minimum threshold for a residential move among CEM hunter-gatherers can be estimated at 5 km. CEM sites located less than 5 km apart are thus considered to be the product of at least two different stays during two seasonal cycles. Consequently, their distance from each other is not taken into account for the estimation of the median distance traveled during a residential move. In this study, these local agglomerations of sites which are located less than 5 km apart are interpreted as belonging to the same seasonal "camp ground." These camp grounds were likely exploited once a year during an annual land-use cycle. To calculate the median distance of a residential move, a buffer with a radius of 5 km is calculated around each site. Overlapping buffer areas are aggregated into "camp ground areas." Both single sites with a 5 km buffer area and the constructed camp ground areas are subsequently treated as individual spatial units, and the nearest-neighbor distances for all these units are calculated. Then the median distance is determined, which is interpreted as representing the approximate average distance traveled during a residential move (Average Residential Move Distance, or ARMD).

As the ARMD is calculated for each regional group individually, it can be used as a group-specific measure for interpreting the distances underlying the raw material procurement pattern in terms of seasonal, annual, and multi-annual group migrations. Raw material procured at distances smaller than the ARMD were likely collected during the stay at the local camp, whereas those procured at distances twice the length of the ARMD or more can be seen as remnants of previous seasonal stays. Long-distance transportation of raw materials from areas outside the reconstructed exploitation zone but still within the limits of the group's territory can be interpreted as indications of the displacement of a hunter-gatherer group from one multi-annual exploitation zone to the next (cf. Floss 1994: 326f.). The expected result is then a coarse-grained approximation of the land-use pattern of the respective regional group and can therefore only give a rough impression of the possible settlement system of past interaction units.

The Circum-Jurassic Group

The discussion of the five regional groups begins in the southwestern part of the investigated area. Here, the first regional group which it was possible to identify through the previous analyses is located in the vicinity of the mountain range of the French and Swiss Jura. The first traces of Magdalenian occupation in this region were excavated at the site of Grotte Grappin at Arlay (Cupillard and Welte 2006) and are dated between 18,900 and 18,200 cal BP (see Chap. 8, "The Eastward Expansion"). In the following, the characteristics of this group are highlighted and aspects of its territorial structure and land-use pattern are discussed.

Group Stsatistics

With 164 assemblages at 137 sites, the inventories in the Circum-Jurassic Group clearly outnumbers those of all the other interaction units (Table 7.3). In 102 cases, the assemblage's allocation to the CEM seems certain or very probable. For 43 inventories, on the other hand, the classification is not determinable due to a substantial lack of information, and seven collections represent an apparent admixture of CEM artifacts and younger finds. The majority of the assemblages were excavated at cave sites (41 %), which together with collections from rock shelters (19 %) amount to almost two-thirds of the total number of inventories. The remaining 40 %, however, come from open-air sites and surface collections.

Despite hosting the most assemblages of all the regional groups, the Circum-Jurassic Group appears less well equipped with regard to the proportion of inventories yielding many tools. The group shows only the third most abundant percentage of assemblages comprising inventories of more than 100 lithic tools (32 %) and the second most abundant percentage of inventories of more than 1000 tools (4 %) (cf. Table 7.2).

Landscape and Environmental Setting

The sites of the Circum-Jurassic Group are located between the Alps in the southeast, the Black Forest and Vosges in the North, and the mountain ranges along the left bank of Rhône and Saône in the west. The major geographic features characterizing the landscape are the Rhône-Saône valley, the Jura Mountains, the Molasse basin and the northwestern slopes of the Alps (see Fig. 7.2). The Jura is a fold mountain range that extends approximately 370 km from the Isère river in the southwest to the High Rhine in the northeast. In its eastern part, the Jura merges into the Molasse basin, which separates it from the main chain of the Alps. In the west, it is delimited by the Rhône-Saône rift valley and in the north by the Burgundy Gate. With a maximum width of about 70 km, this limestone massif is slightly bow shaped. Its highest elevations are found in the southern part, rising slightly over 1,700 m above sea level and descending slowly in a northeasterly direction (Walter 1995: 385ff.). Because water seeps easily in the karstic ground, the Jura itself is a rather dry area in comparison to the valleys of the Rhône and Saône rivers and the Molasse Basin adjacent to the northwest and southeast. The retreat of the Alpine glaciers, for instance, in the Isère, Rhône, Aar, or Reuss valleys, formed a landscape in the Molasse basin which was mainly characterized by numerous smaller and larger melt-water lakes which filled

Table 7.3 Statistical data on the Circum-Jurassic Group

Archaeological record	n	%	Land-use	
Sites total	137		Area r=25 km	61,000 km²
Assemblages total	164	100	Area r=40 km	83,000 km²
Probably Magdalenian	102	63	Area geographically	47,000 km²
Probably mixed	7	4	ARMD	20 km
Probably younger	12	7	Altitude (Median)	408 m
Not determinable	43	26	Environment	
Sites in this study	128		Vegetation	Tundra
Assemblages in this study	152	100	Main game	Reindeer
Cave	63	41	Earliest Magdalenian settlement	
Rock shelter	29	19	Between 18,900 and 18,100 cal BP	
Open-air/Surface coll.	60	40		
>100 tools	48	32		
>1000 tools	6	4		

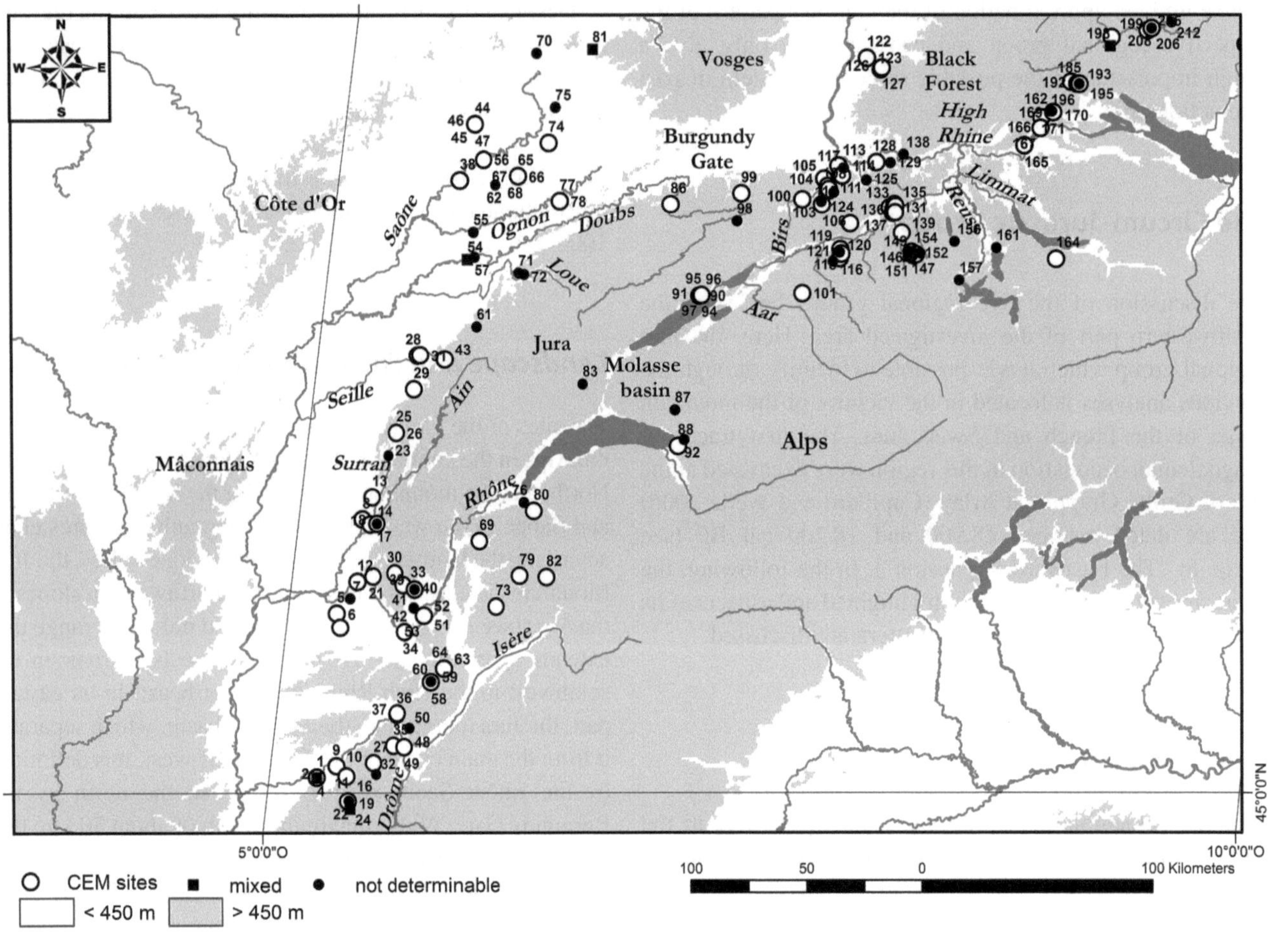

Fig. 7.2 Distribution of sites in the Circum-Jurassic Group

the valleys and depressions of the area (Évin et al. 1994; Walter 1995: 408; Nielsen 2009: 186; Mevel 2013: Fig. 1). In contrast to this lake-dominated landscape, the northwestern part of the Circum-Jurassic Group's area was marked by the braided currents of the larger rivers such as Saône, Rhône, and Doubs, which flowed as broad streams through the wide valley. Thus, orogenetic, karstic, and glacial processes led to the creation of a highly diverse landscape with pronounced elevation differences, steep valleys, broader plains, and mountainous regions, as well as to the formation of numerous caves and rock shelters. As a consequence, during the period of the CEM, a rich patchwork of different environmental conditions and biotopes could be found together with natural shelters within a relatively small spatial range. Throughout the Late Pleniglacial and the Oldest Dryas, the vegetation was mainly that of a treeless steppe-tundra characterized by herbaceous plants. In this environment, reindeer was probably an abundant species of game. Horse, on the other hand, was presumably less common (see Chap. 5, "Horse and Reindeer Remains"). According to the faunal remains, ibex, chamois, and red deer were quite numerous; rarer finds of mammoth and woolly rhinoceros show that these species were present in the landscape as well. During the Bølling Interstadial (GI-1e), juniper shrubs and dendriform birches spread and became the dominant elements of the landscape. The whole region is comparatively rich in outcrops of high-quality raw material; especially rich are the regions of Olten and Bellegarde-Seyssel (Affolter and Bressy 2009; Cattin 2009). In the Late Jurassic (Malm) deposits of the Jura Mountains, nodular or flat varieties of chert are common and extensively distributed (Floss 1994: 104).

Location of Sites and Estimation of the Group's Territory

The sites of the Circum-Jurassic Group are distributed in a roughly triangular area between the Alps in the southeast, the Black Forest and Vosges in the north, and the mountain ranges along the left bank of Rhône and Saône in the west (Fig. 7.2).

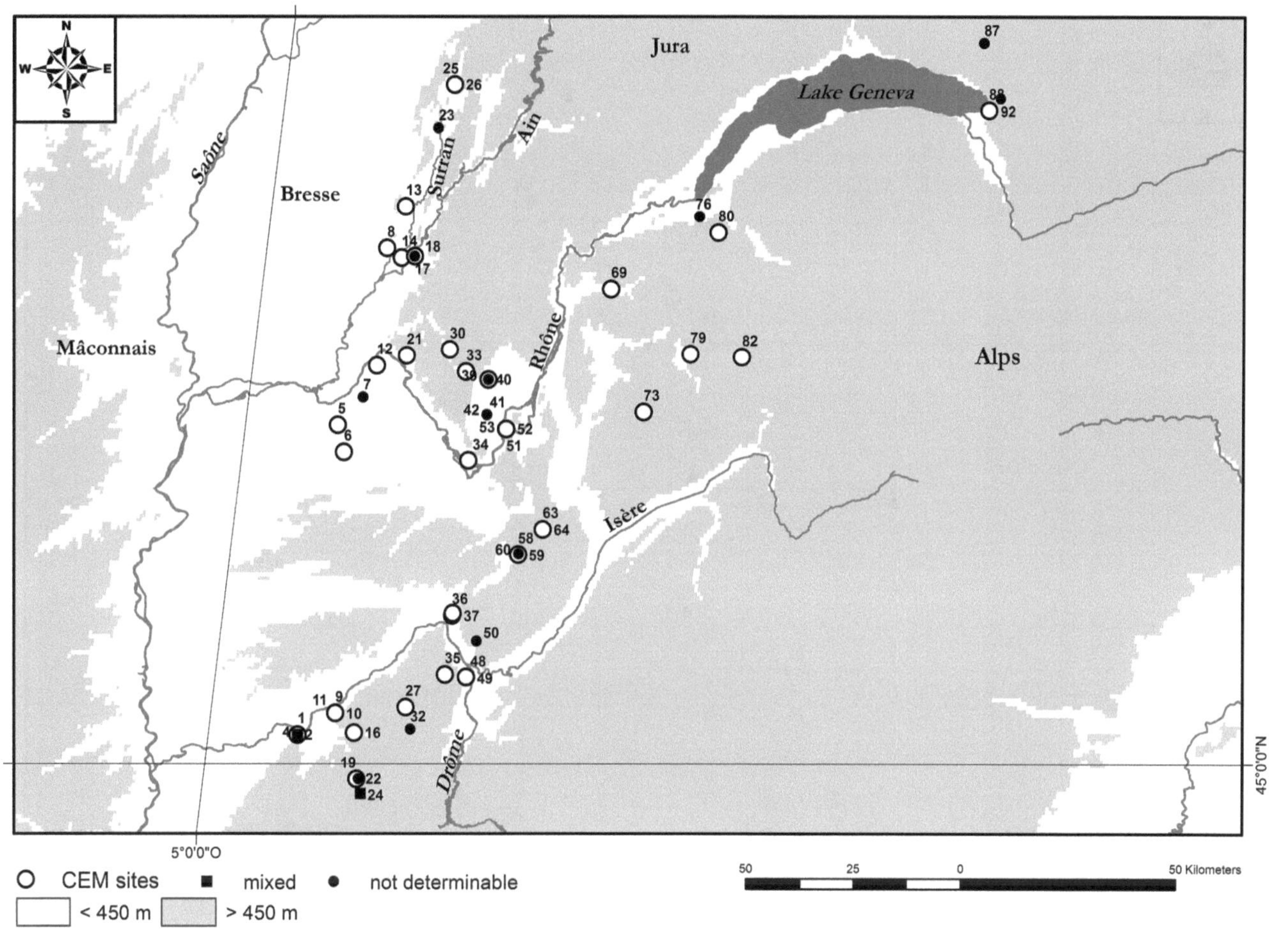

Fig. 7.3 Distribution of sites in the southern part of the Circum-Jurassic Group

There are two major geographic factors which seem to have been of importance for the selection of a site's position. On the one hand, the general outlines of the site distribution follows the course of the larger rivers; on the other, almost all sites are located at middle-range altitudes mostly on the lower slopes of the mountains, whereas the valley bottoms themselves as well as the higher mountain elevations are virtually empty of CEM traces. This observation accords well with the evidence that during the Magdalenian period, most of the depressions in the landscape were filled with meltwater from the retreating glaciers (see above). Hence, the known sites are generally situated around the Jura Massif itself as well as at the foot of the Alps, and it seems that most of the sites were located near the shores of prehistoric glacial lakes. With regard to the two major river systems of the region, the site cluster of the Circum-Jurassic Group can be divided into two sub-clusters. The first is situated to the south and west of the Jura Mountains along the drainage system of the Rhône, whereas the second can be found along the currents of Aar-Rhine river system.

(I) The sites of the first sub-cluster are situated along the Isère, Rhône, Ain, Doubs, and Saône rivers; together they form a north-south oriented, strung-out pattern along the western slopes of the Jura up to the Upper Saône valley and the southwestern border of the Vosges Mountains. The maximum distance between two sites in north-south direction is about 350 km, whereas the average east-west distance amounts to only some 10–20 km.

In the southern part of this sub-cluster, the southern end of the Jura together with the western slopes of the pre-Alps forms a small, semicircular bay that opens to the west. Coming from the Alps, the Isère and Rhône rivers enter this open bay via narrow valleys. The position of sites in this area closely traces this semicircular structure (Fig. 7.3).

Several factors likely contributed to the emergence of this region as a particularly suitable settlement area for CEM hunter-gatherers, among them the view over the wide Rhône valley, the advantageous hunting situations around the shores of the glacial paleolakes, and

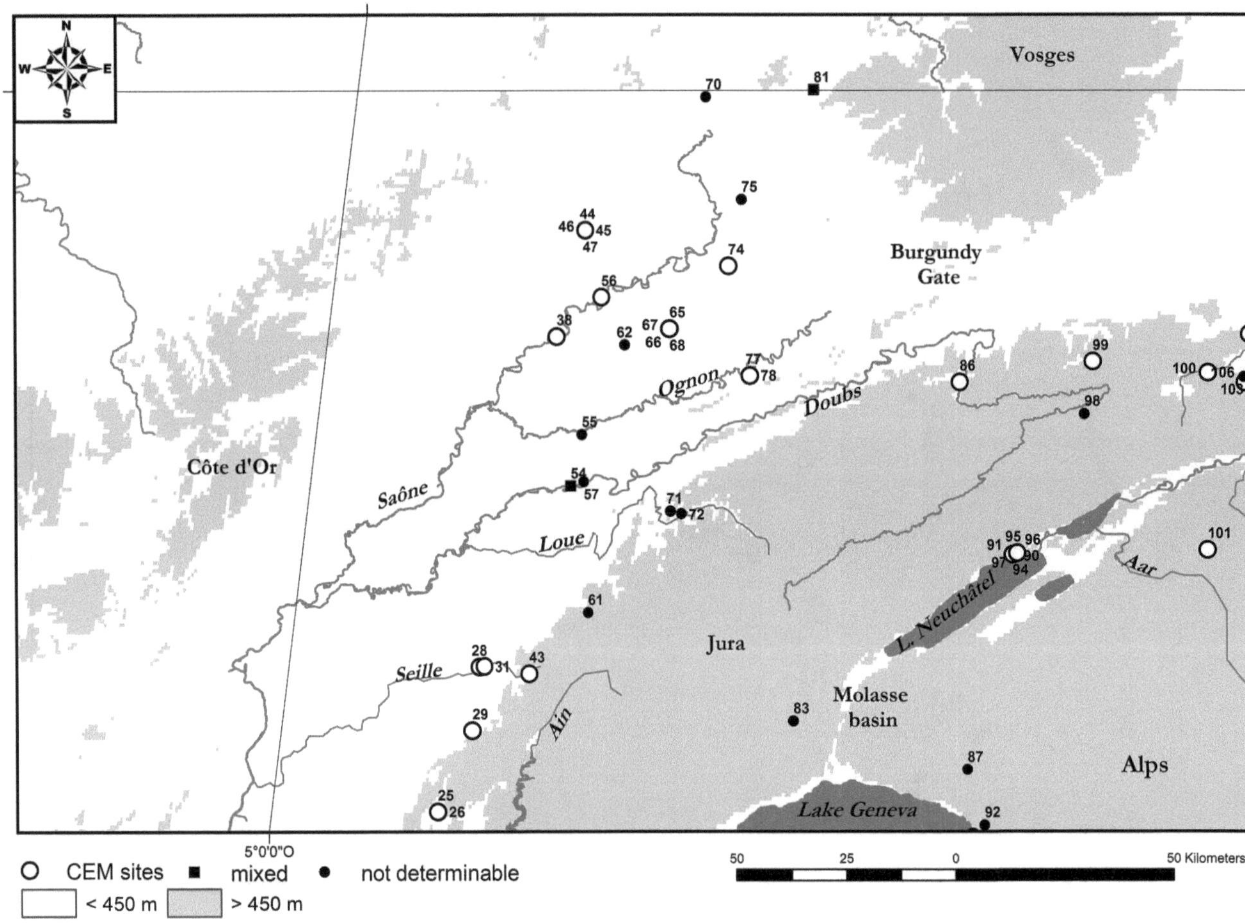

Fig. 7.4 Distribution of sites in the northwestern part of the Circum-Jurassic Group

the climatically favorable bay situation. In the central part of this sub-cluster, the sites are positioned along the Jura's western slopes. The elevated positions allowed for a good overview over the valleys adjacent to the west and the bottleneck situations shortly before the confluence of the Surran and Ain rivers presumably provided highly suitable hunting stands. However, the lower parts of the Bresse valley appear to have been less favorable; indeed, no CEM site has been found so far in this area, which at that time was probably characterized by the braided current of the Saône and Seille as well as numerous other melt-water rivers coming from the Jura Massif.

The almost strictly linear alignment of the sites in the southern and central part of the first sub-cluster appears to give way in the north to a more dispersed pattern. However, no evidence for CEM settlement has thus far been found in the lower part of the Saône and Doubs valleys so far, even though several sites such as Cabônes (ID 54), Rigney (ID 77, 78), and Seveux (ID 56) are situated in the upper valleys of these currents (Fig. 7.4).

Other sites exhibit the same spatial selection criteria as were observed in the central and southern parts of this sub-cluster. The sites of Baume Noire (ID 65–68) and Farincourt (ID 45–47), for instance, lie at an elevated position with a good view over the upper Saône valley, and the bottleneck situations in the vicinity of Rochdane (ID 86) and Bains-les-Bains (ID 81) were probably attractive factors for the selection of these locations from a hunting-strategy point of view.

(II) The second sub-cluster of the Circum-Jurassic Group is mainly concentrated around the northeastern tip of the Jura. In contrast to the first sub-cluster, this one occupies very little space, and its sites exhibit a strong tendency toward spatial aggregation. It extends in a north-south direction for approximately 100 km and is about 170 km from east to west. Two river systems seem to be of primary importance for the site distribution: the Birs river that joins the High Rhine near Basel and the Aar river

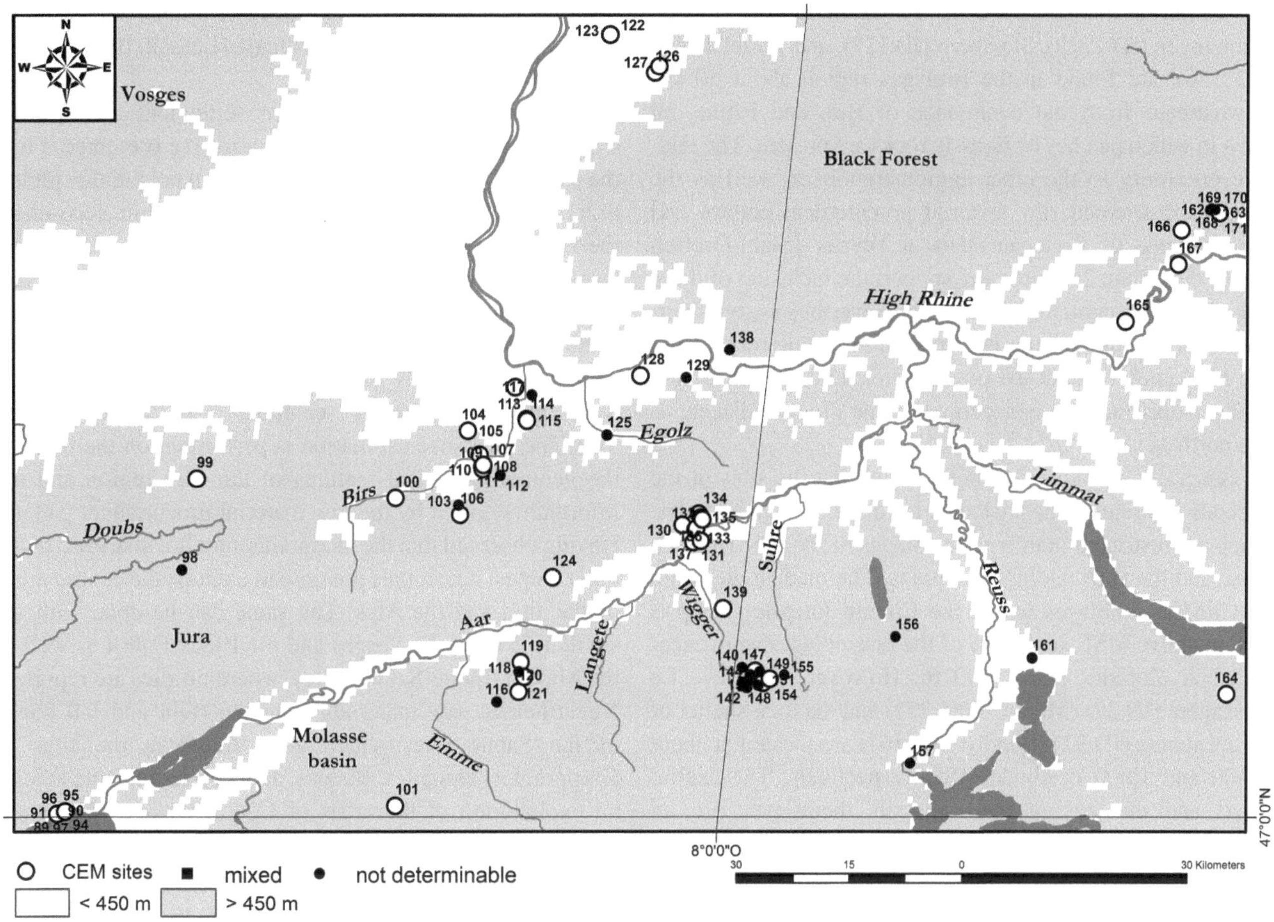

Fig. 7.5 Distribution of sites in the northeastern part of the Circum-Jurassic Group

with some of its tributaries. The Aar's source is located close to that of the Rhône. Whereas the latter flows in a southeasterly direction, however, the former traverses the Molasse basin toward the northwest and joins the High Rhine at Koblenz (CH), approximately 50 km upstream from Basel (Fig. 7.5).

In the drainage area of the Aar, there are three major agglomerations of sites. The southwestern agglomeration between the Emme and Langete rivers and the southeastern one between the upper part of the Wigger and Suhre rivers comprise mainly surface collections; in addition to some possible Magdalenian finds, they yield large quantities of Late Paleolithic and Mesolithic artifacts (Nielsen 2009). Hence, the attribution to the Magdalenian is somewhat uncertain for the majority of these sites. At the third aggregation, however, the situation is different. Here in the region of Olten, there are eight sites located in close proximity to each other which very probably belong to the CEM. The entire region was once characterized by glacial melt-water lakes (Nielsen 2009: 186). The patchwork of water bodies and broader as well as narrower passages of dry land probably provided a strategically favorable situation when it came to the animal groups which crossed this area. Together with the outcrops of high-quality raw material in the vicinity (Affolter 1997: 69f.; 2006), these factors probably made this region an attractive settlement area.

The valleys of the Birs and Ergolz rivers shape the site distribution pattern on the northern slope of the Jura. Six of ten sites are located on the slopes around the Laufen Basin, which is traversed by the Birs river and probably filled to form a melt-water lake during the Magdalenian. These sites, namely, Liesberg (ID 100), Chesselgraben (ID 106), Brügglihöhle (ID 107), Heidenküche (ID 111), Kohlerhöhle (ID 108), and Kastelhöhle Nord (ID 109, 110) are all restricted to the bottleneck situations at the narrow entries of the basin, however. The remaining four sites, Büttenloch (ID 104, 105), Birseck-Ermitage (ID 114), Hollenberg-Höhle 3 (ID 115), and Bruderholz (ID 113), are found in elevated positions overlooking the valley of the Birs river.

The northernmost sites of the second sub-cluster, namely, Munzingen (ID 1229), Steinberg (ID 127), and Teufelsküche (ID 126), are found in the Breisgau region about 50 km downstream from the confluence of Birs and Rhine, an area in which patches of Jurassic rocks are present. The relative proximity to the other agglomerations as well as the southward-oriented raw material procurement pattern and the presence of Breisgau chert at Veyrier (Stahl-Gretsch 2006) more than 250 km away support the inclusion of these sites in the Circum-Jurassic Group. Nevertheless, typological observations and the presence of siliceous rocks from the Markgräflerland at the sites of Dreieich-Götzenhain and Gönnersdorf indicate connections to the groups adjacent to the north as well.

Generally speaking, it can be stated that the sites of the Circum-Jurassic Group are located in middle-range altitudes. They are restricted mainly to the slopes of the pre-Alps, the Jura, and the surrounding mountains. The median elevation of CEM assemblages within the Circum-Jurassic Group is 408 m above MSL, and 90 % of the assemblages were excavated at altitudes below 700 m. However, the Cave La Passagère (ID 27) (Monin 2000: 272) and the rock shelter of Le Denteaux (ID 92) (Bandi 1947: 167) are located at about 1,100 and 1,600 m above MSL, respectively. The central areas and plateaus of these massifs themselves are, in contrast, virtually empty of remains of CEM occupation. An absence of sites can be observed in the lower parts of the Rhône-Saône rift valley itself. Nevertheless, it is likely that the uplands and lowlands were also exploited by CEM hunter-gatherers, an assumption that is strongly supported by the raw material procurement pattern (see Fig. 7.8). However, it seems that the higher elevations of the mountains and the wet and unstable environments of large braided rivers were not considered suitable settings for medium- or long-term camps.

The extensive investigations of raw material source tracing undertaken in recent decades allow for instructive insights into the catchment areas of the Circum-Jurassic Group's sites (e.g., Affolter 1997, 2006; Stahl-Gretsch 2006). The mapping of information obtained from the archaeological literature results in a dense and complex acquisition pattern (see Figs. 6.9 and 7.8). Looking at the contact zone between the two sub-clusters, a significant overlap in the site catchments of both units becomes apparent. This speaks in favor of strong interaction and intensive contacts between them. Scattered sites along the route through the Molasse basin from Lake Geneva via Lake Neuchâtel to the Aar valley indicate that this axis was an important corridor for exchange and communication. This hypothesis is supported by observations of several assemblages at Hauterive-Champréveyres and Neuchâtel-Monruz, where raw material imports from both ends of this corridor are reported, i.e., from the region of Bellegarde-Seyssel in the southwest and the region of Olten in the northeast (Leesch 1997; Leesch et al. 2004).

In estimating the subsistence settlement area of the Circum-Jurassic Group, three scenarios are presented. First, the estimation is calculated according to a potential exploitation range around each site. If a radius of 25 km is assumed, the resulting area amounts to 61,000 km² (Fig. 7.6). For a larger exploitation range with a radius of 40 km around each site, the calculation results in a subsistence settlement area of 83,000 km² (Fig. 7.7). In both cases, the area's calculated extent is considerably larger than the respective median values of all estimated areas (see Tables 7.3, 7.4, 7.5, 7.6, and 7.7). The third size estimation is performed on the basis of the geomorphological features of the Jura region and the information given by the raw material procurement pattern. Having observed that the mountains sites are restricted to the outer slopes, it becomes possible to exclude the central areas of the Jura and the Alps. The same can be done with the southern part of the Vosges and the Black Forest as well as the Middle Rhône-Saône valley, where no sites are reported. Nevertheless, raw materials from the right and left banks of the Saône river which were found at the sites of Grappin, Les Douattes, Banges, and Veyrier indicate at least an exploitation of the areas of Côte d'Or and Mâconnais (Fig. 7.8). This area is therefore also considered in the estimation. In the northeast, there is a noticeable gap between the Circum-Jurassic cluster and the one in the Danube valley, indicating a border of the exploitation range in this direction (see Chap. 6, "Procurement Pattern of Lithic Raw Materials"). Taking into account all these observations, a subsistence settlement area of about 47,000 km² can be assumed.

Land-Use Pattern

In the following, the observations presented thus far are interpreted with regard to the possible land-use pattern of the Circum-Jurassic Group. The size of the group's territory is estimated at approximately 47,000 km² (see above). With regard to the location of sites and the structure of the raw material procurement pattern, three exploitation zones can be distinguished—a northwestern, southern, and northeastern one—that might correspond to areas occupied multi-annually or to palimpsests of such areas (Fig. 7.9).

The northwestern exploitation zone (Fig. 7.10) is delimited by the Doubs river in the southeast, the Vosges and Faucilles Mountains in the north, and the chain of the Côte d'Or in the west. In this interpretation, the latter area is seen as a part of the northwestern zone for several reasons. First, the region between the Saône and the Côte d'Or Mountains

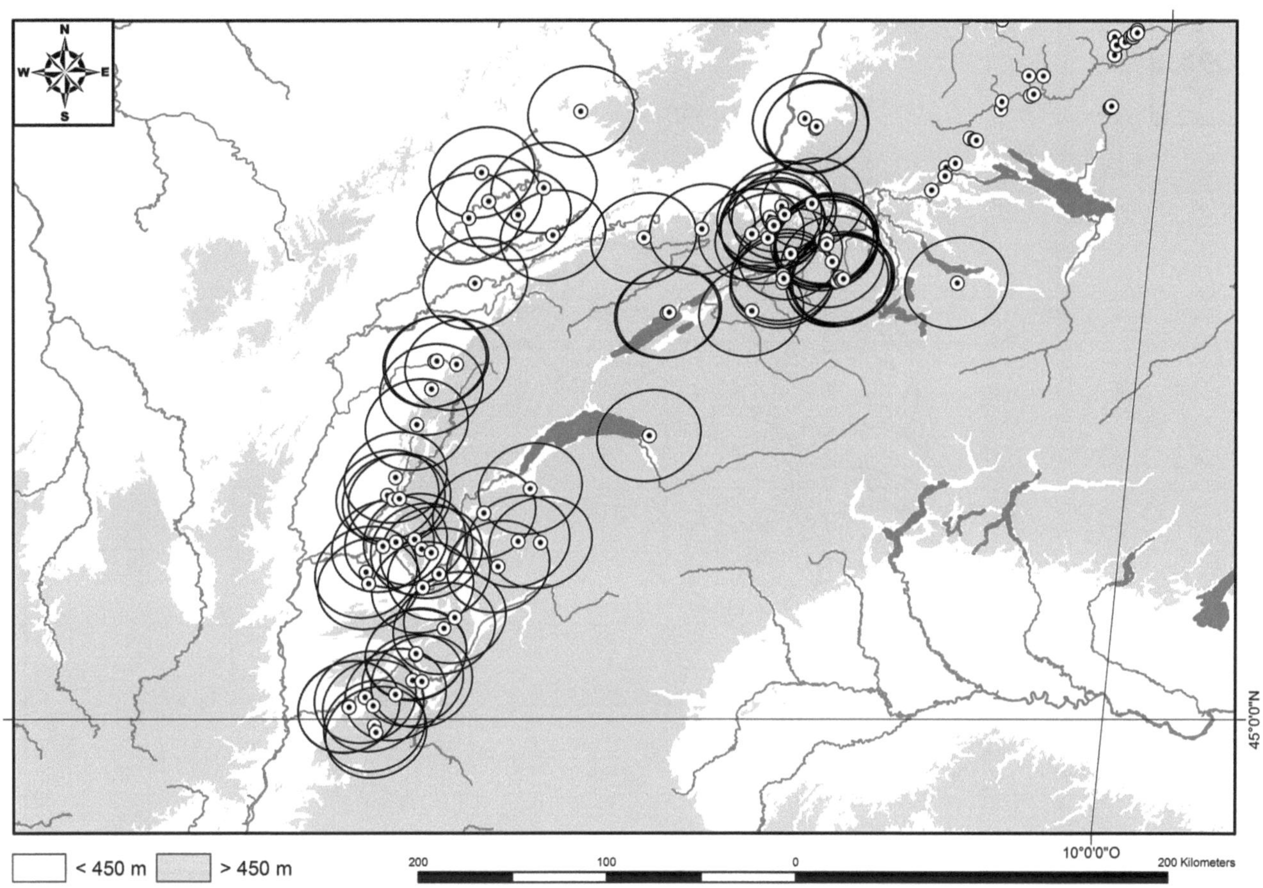

Fig. 7.6 Circum-Jurassic Group, territory estimation with 25 km radius around each site

appears too small to be interpreted as an independent exploitation zone. Its allocation to the sites along the western slope of the Jura is contradicted by the fact that no Magdalenian sites are yet known from the valley between the Jura and the Côte d'Or. Additionally, this water rich valley was probably characterized throughout the Magdalenian by swampy conditions caused by the braided currents of the Saône, Doubs, and Loue as well as other numerous smaller melt-water creeks coming from the Jura Massif. Braided rivers certainly did not represent serious obstacles to hunter-gatherers, and yet this region seems not to have been used intensively. From the northwestern zone, in contrast, the Côte d'Or region is easy to access and is not separated by any kind of natural obstacle. Thus, it appears most plausible to think of the two areas as one coherent exploitation zone.

The northwestern zone comprises the sites located between Rigney (ID 77, 78), Abri Cabônes (ID 54), Farincourt (ID 45–47), and Bains-les-Bains (ID 81) and covers an area of about 13,000 km^2. Its southernmost site, Abri Cabônes, is situated about 90 km north of the closest adjacent camp ground to the south, which includes the sites of Grappin (ID 28), Chaze 2 (ID 31), and Grotte de Chaumois-Boivin (ID 43). Because this distance is more than 4 times greater than the assumed average distance traveled by this regional group during a residential move (ARMD = 20 km, see Chap. 7, "Territories and Land-Use Patterns of the Five Regional Groups"), the border between the northeastern and the southern exploitation zones is set between these two seasonal camp grounds.

The southern exploitation zone extends over an area of about 18,000 km^2 (Fig. 7.11). In its northern part, it seems to be restricted to the western slopes of the Jura Mountains. Further to the south, its eastern border is formed by the onset of the Alps as well as the Isère and Drac rivers, and its southern border is marked by the Drôme river. To the southwest, the delimitation follows approximately the courses of the Isère and the Upper Rhône, and in the northwest, it extends as far as the Mâconnais Mountains.

At the site of Grappin (ID 28), raw materials of Oligocene and probably Cretaceous origin were found which presumably were acquired in the region of the Côte d'Or, about 70 km away as the crow flies (Cupillard and Welte 2006).

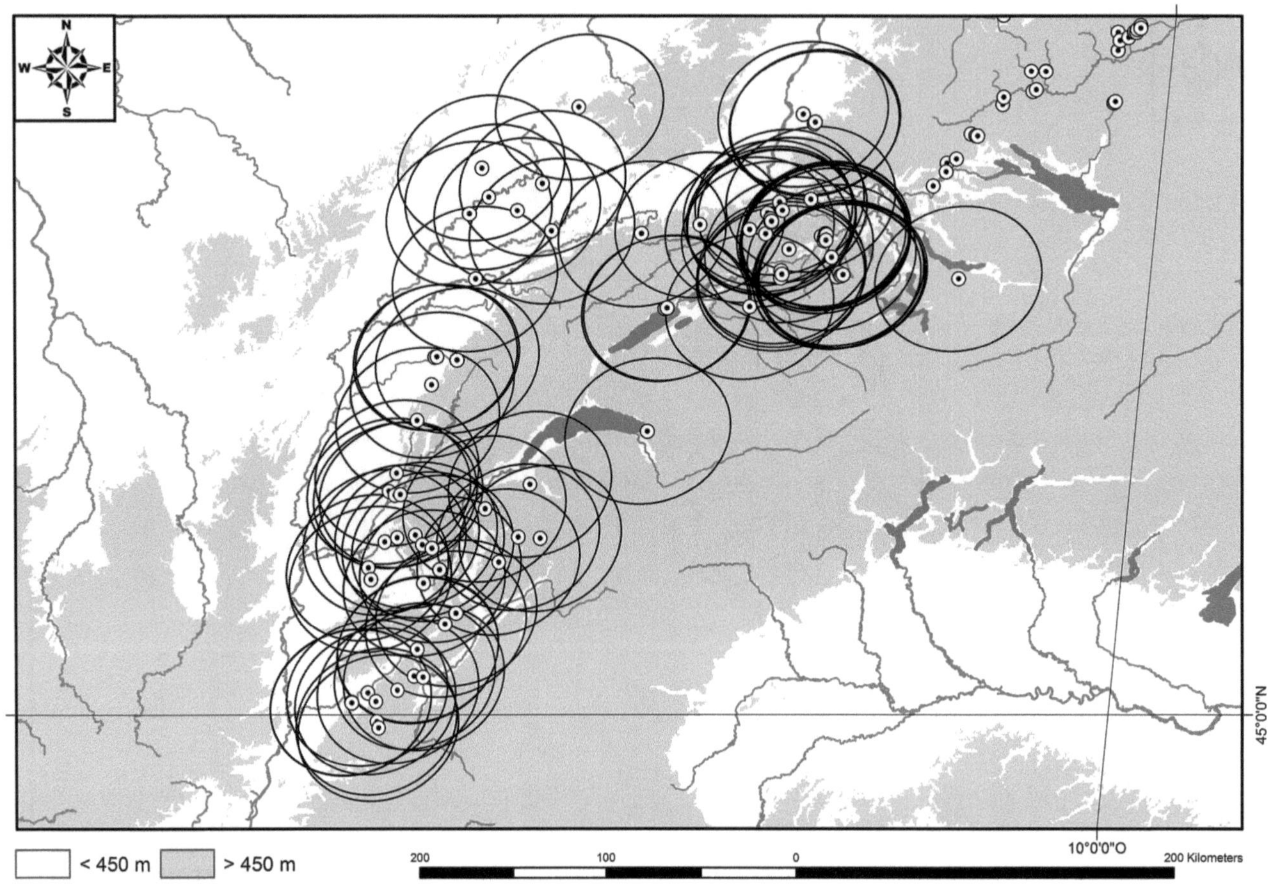

Fig. 7.7 Circum-Jurassic Group, territory estimation with 40 km radius around each site

Table 7.4 Statistical data on the Danube Group

Archaeological record	n	%	Land-use	
Sites total	97		Area r = 25 km	30,000 km²
Assemblages total	143	100	Area r = 40 km	47,000 km²
Probably Magdalenian	93	65	Area geographic	23,000 km²
Probably mixed	5	3	ARMD	24 km
Probably younger	15	11	Altitude (Median)	546 m
Not determinable	30	21		
Sites in this study	84		Environment	
Assemblages in this study	128	100	Vegetation	Tundra/steppe-tundra
Cave	68	53	Main game	Reindeer/horse
Rock shelter	22	17		
Open-air/Surface coll.	38	30	Earliest Magdalenian settlement	
>100 tools	24	19	Between 17,500–17,200 cal BP	
>1000 tools	2	2		

For the reasons given above (see Chap. 7, "Territories and Land-Use Patterns of the Five Regional Groups"), it seems most plausible to interpret these finds as evidence of a macro-move from the northwestern to the southern exploitation zone. In this view, the raw materials were procured during the course of seasonal migrations and were taken to the site of Grappin several months later, when the group changed the exploitation zone. Two artifacts from the site of Veyrier (ID 79) that were made using raw material from the northwestern slope of the Jura (Stahl Gretsch 2006: 151) are interpreted in a similar fashion: they were likely acquired during an annual cycle in the northwestern zone. After the group changed

Table 7.5 Statistical data on the Meuse-Rhine Group

Archaeological record	n	%	Land-use	
Sites	58		Area r=25 km	41,000 km²
Assemblages	73	100	Area r=40 km	60,000 km²
Probably Magdalenian	50	68	Area geographically	35,000 km²
Probably mixed	0	0	ARMD	18 km
Probably younger	8	11	Altitude (Median)	130
Not determinable	15	21		
Sites in this study	50		Environment	
Assemblages in this study	65	100	Vegetation	Tundra-steppe
Cave	25	38	Main game	Horse/reindeer
Rock shelter	0	0		
Open-air/Surface coll.	40	62	Earliest Magdalenian settlement	
>100 tools	27	42	Between 15,500–15,400 cal BP	
>1000 tools	3	5		

Table 7.6 Statistical data on the Vltava-Saale Group

Archaeological record	n	%	Land-use	
Sites	124		Area r=25 km	47,000 km²
Assemblages	137	100	Area r=40 km	75,000 km²
Probably Magdalenian	76	56	Area geographically	40,000 km²
Probably mixed	4	3	ARMD km	22 km
Probably younger	6	4	Altitude (Median)	204 m
Not determinable	51	37		
Sites in this study	118		Environment	
Assemblages in this study	131	100	Vegetation	Park-steppe
Cave	22	17	Main game	Horse
Rock shelter	1	1		
Open-air/Surface	108	82	Earliest Magdalenian settlement	
>100 tools	22	17	Between 16,600 and 16,000 cal BP	
>1000 tools	6	5		

to the southern zone, these artifacts were carried along for several seasonal stays until they were finally discarded at Veyrier. Raw materials from the region of the Mâconnais Mountains and the Drôme region found at Veyrier, Banges (ID 73), and Les Douattes (ID 69), in contrast, are seen as having been brought to these sites embedded in the context of seasonal cyclic migrations. In all cases, the pieces probably came to the sites as "curated" items of a sort (Binford 1979), because for one reason or another, they—or the core from which they were obtained—were considered worth carrying a comparatively long way.

The third and last identifiable exploitation zone (Fig. 7.12) corresponds to the second sub-cluster (see above) and encloses an area of roughly 16,000 km². It is separated from the northwestern zone by a gap approximately 50 km wide between the sites of Rigney (ID 77, 78) and Rochedane (ID 86).

At about 120 km, the distance between the northeastern and the southern zone is more than twice as long. It is measured between the sites of Hauterive-Champréveyres (ID 94–97) and Neuchâtel-Monruz (ID 89–91) on the one side and the Grotte du Étrembières (ID 80) on the other side. This separation likely reflects the environmental conditions at the time of the CEM, when wide stretches of the Molasse basin were submerged under melt-water lakes (see above). However, an intensive transport of raw materials between the two groups can be observed. Within the third exploitation zone, a close-meshed pattern of raw material transports illustrates the annual migrations of hunter-gatherers in this region. The great majority of the transportation distances do not exceed 80 km. Assuming an ARMD and an exploitation radius of 20 km around each site, the rocks in question could have been acquired within the range of two consecutive stays.

The territory of the Circum-Jurassic Group can thus be divided in at least three exploitation zones, which were probably used successively in a multi-annual rotation. Within each zone, seasonal and annual migrations took place until the resources became degraded and the hunter-gatherer group moved on to the next zone (cf. Binford 1983a: 117ff.). Raw material was often transported from one seasonally

Table 7.7 Statistical data on the Polish-Moravian Group

Archaeological record	n	%	Land-use	
Sites	103		Area r=25 km	61,000 km²
Assemblages	109	100	Area r=40 km	92,000 km²
Probably Magdalenian	50	46	Area geographically	84,000 km²
Probably mixed	0	0	ARMD	30 km
Probably younger	6	6	Altitude (median)	360 m
Not determinable	53	48		
Sites in this study	96		Environment	
Assemblages in this study	103	100	Vegetation	Park-steppe
Cave	58	56	Main game	Horse
Rock shelter	5	5		
Open-air/surface Coll	40	39	Earliest Magdalenian settlement	
>100 tools	15	14	Between 20,000 and 18,700 cal BP	
>1000 tools	2	2		

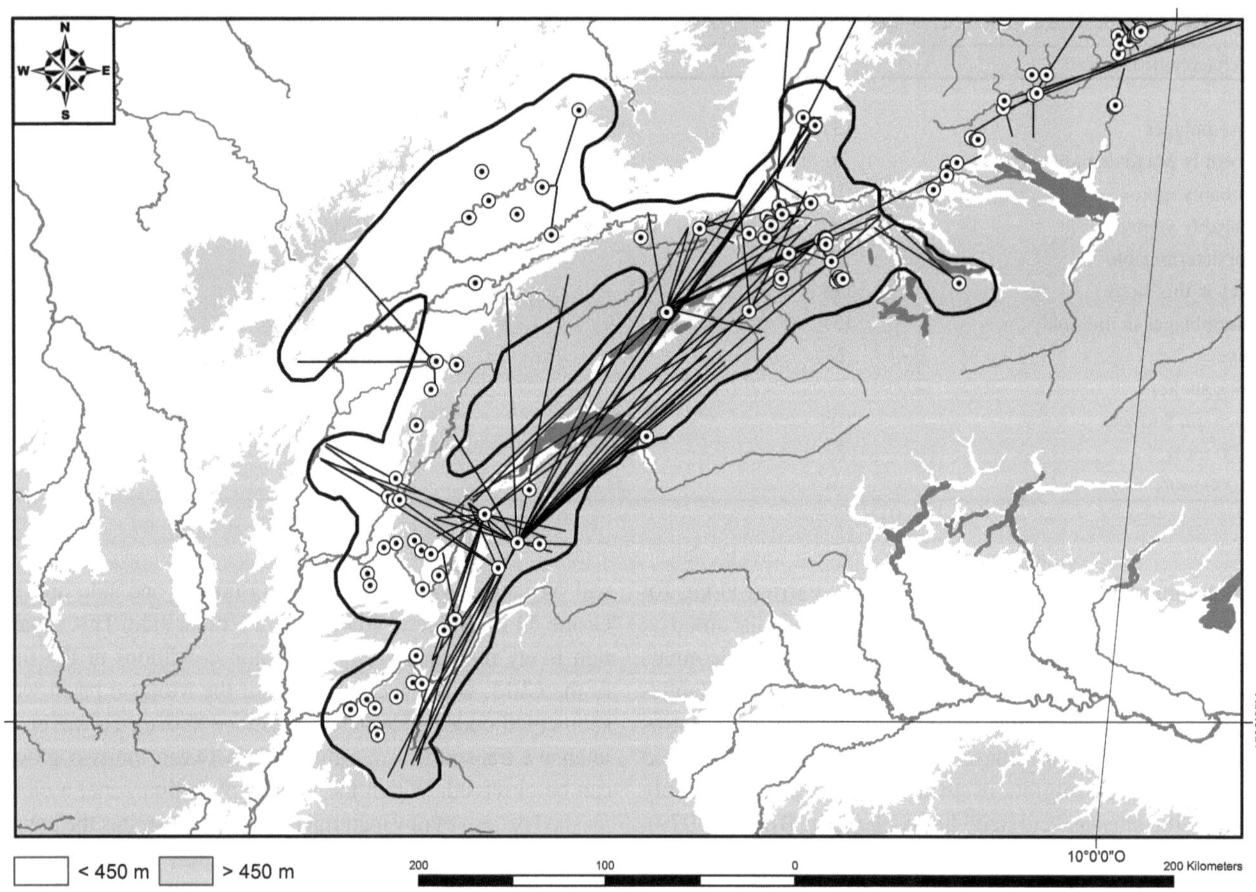

Fig. 7.8 Circum-Jurassic Group, territory estimation based on raw material procurement and geographic features

exploited area to the next. In rarer cases, artifacts were carried along for the duration of several stays, and sometimes lithic pieces were taken from one exploitation zone to the next. An important migration route was probably the Molasse basin, a lake landscape during the time of the CEM. Taking into account the diversity of the tool compositions, it seems that the archaeological record of the Circum-Jurassic Group consists of assemblages that evidence pronounced differences in the usage and function of the respective sites. A clear focus on hunting activities, for instance, can be postulated for the sites of Auzary-Thônes (ID 82) (Perpère 1981) and Abri de la Fru I (ID 58) and II (ID 59) (Pion et al. 1990), which have yielded 70 %, 81 %, and 83 % backed bladelets, respectively. From the site of Farincourt, in contrast, an

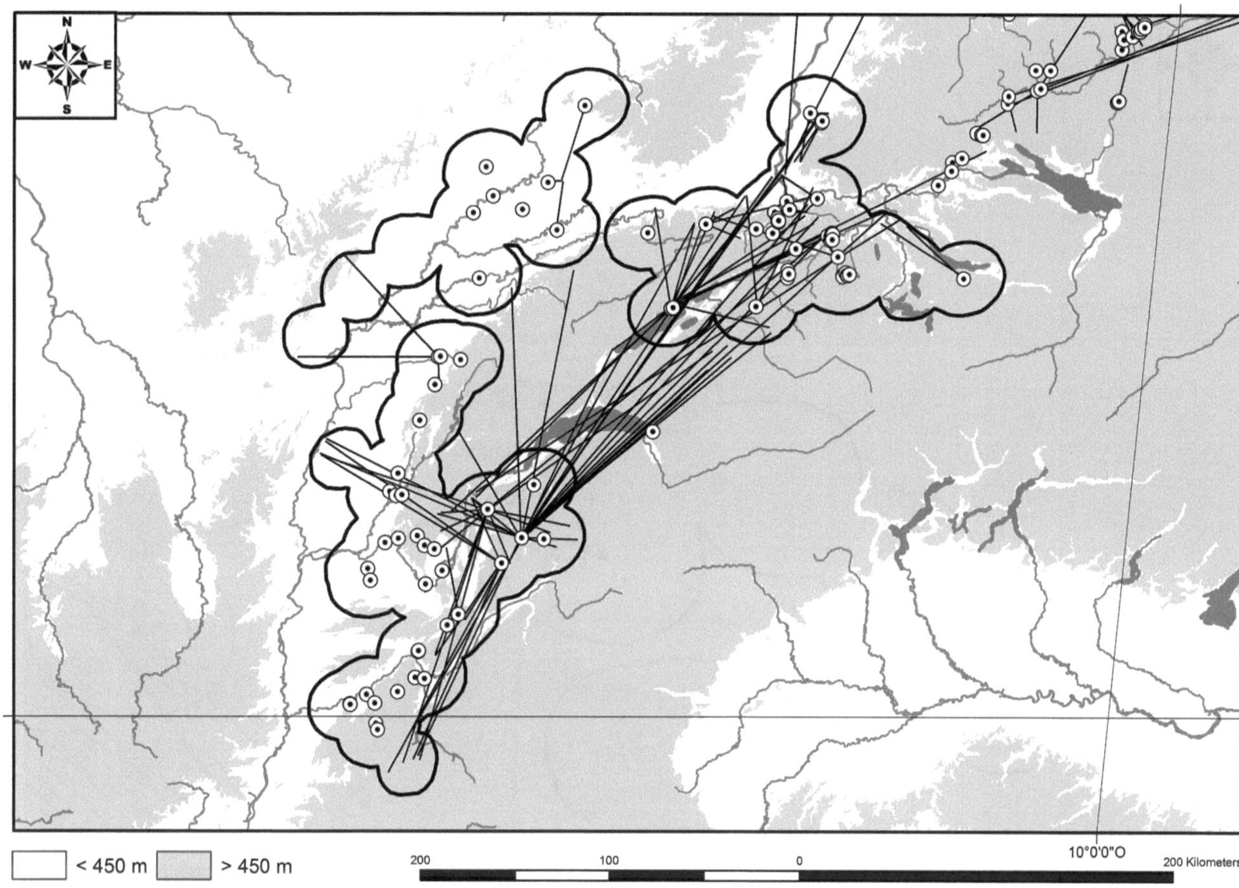

Fig. 7.9 The three exploitation zones of the Circum-Jurassic Group

assemblage with 61 % burins is reported (David 1996). At the same time, several sites show a rather diverse tool spectrum. Thus, it can be assumed that the hunter-gatherers of the Circum-Jurassic Group were likely organized logistically (cf. Binford 1980). Besides residential camps, which yield numerous finds and exhibit a more diverse tool spectrum, hunting camps and other specialized sites seem to have existed in the subsistence system.

The Danube Group

The second regional group to be discussed is the Danube Group. In the west, its territory shares a border with that of the Circum-Jurassic Group. The vast majority of the Danube Group's sites are located north of the Danube river. The earliest reliable evidence for Magdalenian settlement in this region comes from the sites of Kesslerloch and Schweizersbild, where human-modified bone fragments have been dated between 17,500 and 17,180 cal BP and 17,160 and 17,020 cal BP, respectively (Housley et al. 1997).

Group Statistics

The archaeological literature concerning the Magdalenian in the area of the Danube Group records 143 assemblages from 97 sites. For 93 inventories, an allocation to the CEM seems certain or at least highly probable. An admixture of CEM artifacts and younger finds is assumed for 5 collections. In the case of 30 inventories, however, the available information does not allow a clear decision for or against an assignment to this archaeological entity. In 15 cases, the evidence speaks more in favor of assignment to archaeological units of post-Magdalenian age. These respective assemblages are therefore excluded from further analyses (Table 7.4).

Thus, in total 128 assemblages are considered in this study. This represents the second highest number of inventories among the five observed regional groups, topped only by the Circum-Jurassic Group. Looking at the number of sites, however, the Danube Group drops to fourth place among the regional groups, with a total of 84. This contrast can be explained by the large quantity of cave sites and rock shelters, which often represent a complex of several assemblages. In the Danube Group, these comprise fully 70 % of

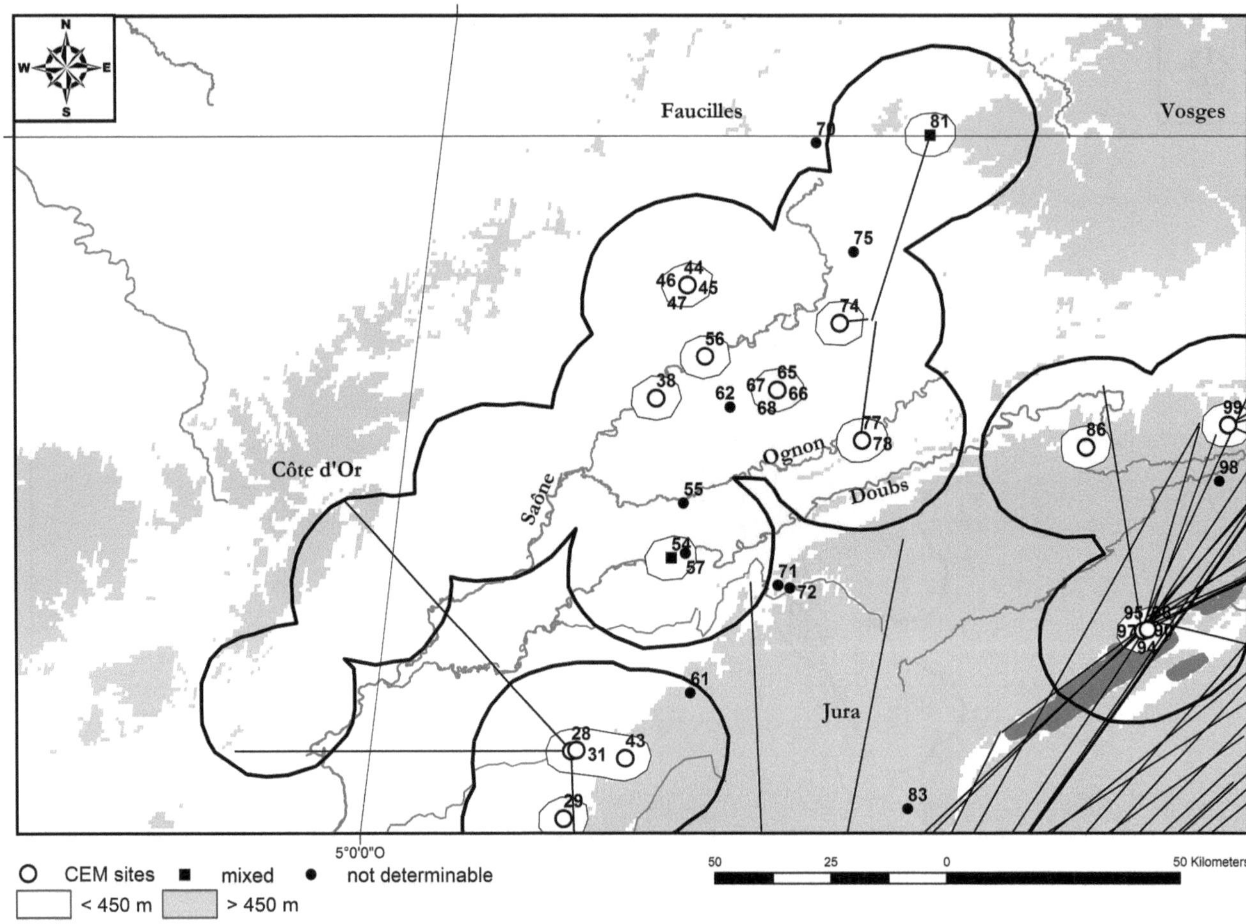

Fig. 7.10 The northwestern exploitation zone of the Circum-Jurassic Group. *Lines* indicate raw material transports between the source areas and the respective sites of discard

the total sites, the highest ratio in the entire investigated area. Open-air camps, on the other hand, play only a minor role, comprising just with 30 % of the total. A pronounced difference to the Circum-Jurassic Group can be observed in the inventories with more than 100 lithic tools: at just 19 %, these are about 1/3 less abundant than they are in the territory adjacent to the west, and only the Vltava-Saale and the Polish-Moravian Groups yield fewer such inventories. Similar relationships can be found with regard to sites with more than 1,000 lithic tools.

Landscape and Environmental Setting

The sites of the Danube Group are located in an area about 350 km long but only 30 km wide that stretches from the High Rhine in the southwest up to the region where the Regen river joins the Danube in the northeast (Fig. 7.13). Except for the sites in the Federsee region, almost all evidence for the presence of CEM hunter-gatherers is restricted to the Jurassic limestone formations west of Lake Constance and north of the Danube river, which belong to the same geological unit as the French and Swiss Jura. From west to east, the area of the Danube Group can be subdivided into the following geographic units: Randen and Hegau, Swabian Alb, Nördlinger Ries, and Franconian Alb, South of the Swabian Alb, the Federsee region is also of interest.

The southwesternmost part of the Late Jurassic cuesta which constitutes the main part of the Swabian Alb is often referred to as the Randen region (cf. Geyer and Gwinner 1962; Eriksen 1991). It is characterized by deeply incised valleys which were cut into the Jurassic sediments by erosion. The landscape of the Hegau to the east, in contrast, features about a dozen volcanic stubs. The eruptions took place during the middle and late Miocene, but did not break through the Molasse sediments which covered the area at the time. It was only during the Riß Glacial that the glaciers removed the comparatively soft Molasse sediments and tuff layers, leaving the harder and more resistant basaltic and phonolitic cores free of their surrounding matrix (Walter 1995: 374).

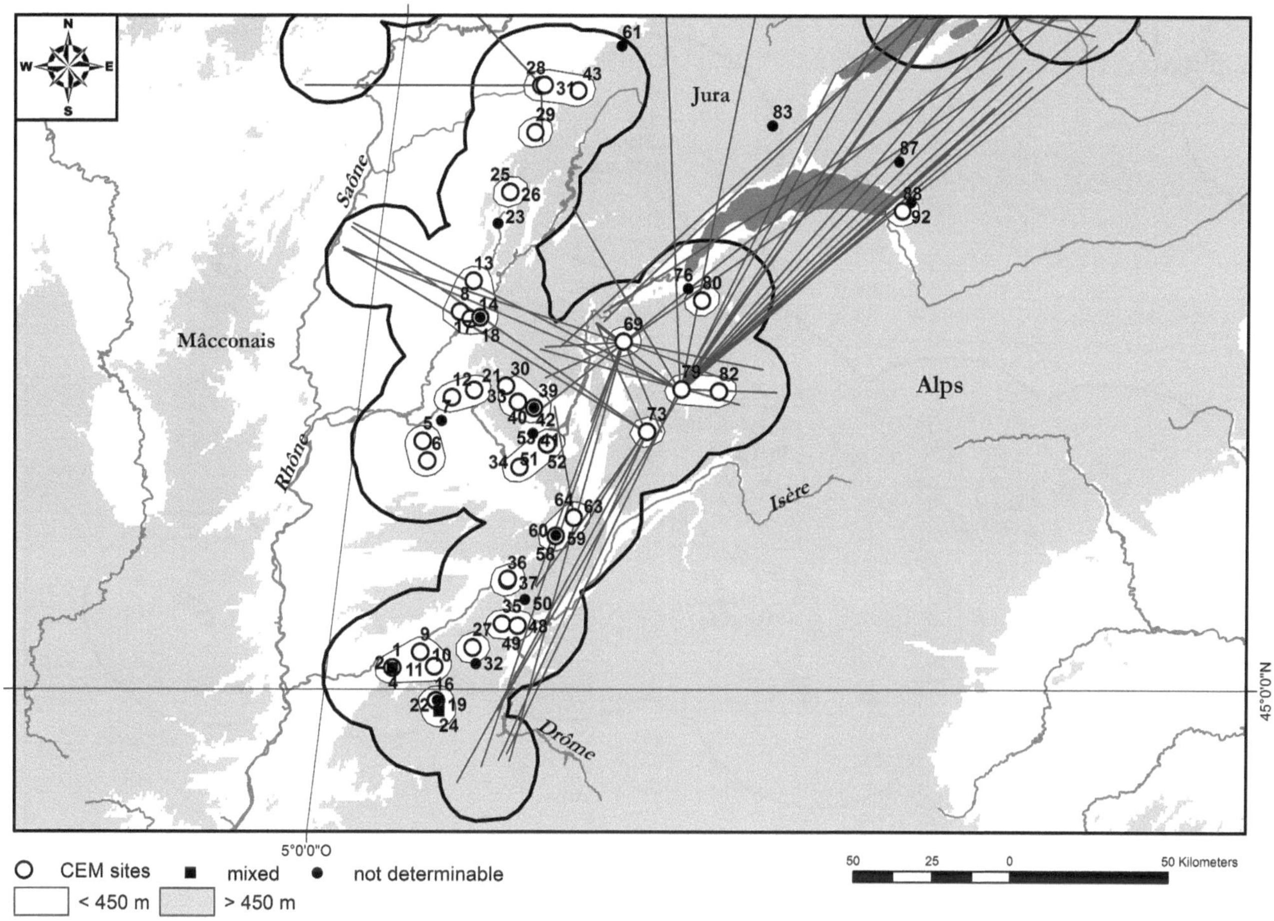

Fig. 7.11 The southern exploitation zone of the Circum-Jurassic Group. *Lines* indicate raw material transports between the source areas and the respective sites of discard

In the northeast, the Randen-Hegau region borders on the Swabian Alb. This mountain range is a Late Jurassic cuesta and extends around 40 km from southeast to northwest and approximately 220 km from southwest to northeast, where it is separated from the Franconian Alb by a large meteoric crater, the so-called Nördlinger Ries. Its southeastern border follows the Danube river for most of its length. In the northwest, the mountain range is delimited by a steep escarpment up to 400 m in height, the so-called Albtrauf (Dongus 1977, 2000: 127ff.; Walter 1995: 373). The course of the Albtrauf coincides more or less with the northern border of the Danube Group's site distribution. The Alb itself is bisected longitudinally by a 50–70 m high escarpment. This escarpment represents the cliff line of the Miocene Molasse Sea and divides the Alb into a southeastern and a northwestern bank. The southeastern bank is called "Flächenalb" and is generally flat but is traversed by deep river valleys. The northwestern "Kuppenalb" is characterized by large, rounded hills which overlook widely ramified dry valleys (Geyer and Gwinner 1962; Dongus 2000: 127ff.; Walter 1995: 373). From west to east, the Swabian Alb is usually divided into a western, central, and eastern Alb. Its highest elevations are found in the western Alb, which rises slightly more than 1000 m above MSL. Toward the northeast, there is a decrease in elevation to 700–900 m in the central Alb and 600–700 m in the eastern Alb (Geyer and Gwinner 1962: 3; Grees 1993: 332).

The Swabian Alb is separated from the Franconian Alb by the Nördlinger Ries, an almost circular meteoric crater approximately 26 km in diameter. The impact took place during the middle Miocene, about 15 Ma ago (Walter 1995: 374). The Franconian Alb is the continuation of the Late Jurassic limestone cuesta of the Swabian Alb with elevations up to 660 AMSL. Most relevant for this study is its southern part, which stretches from the Nördlinger Ries in the west up to the Altmühl valley and the basin of Regensburg in the east (Walter 1995: 375f.). In the central and northern Franconian Albs, evidence of CEM settlement activities is rather scarce (see below).

During the Magdalenian, the Federsee was a much larger melt-water lake which was fed by the retreating Rhine glacier. Many other lakes and ponds of the same origin formed

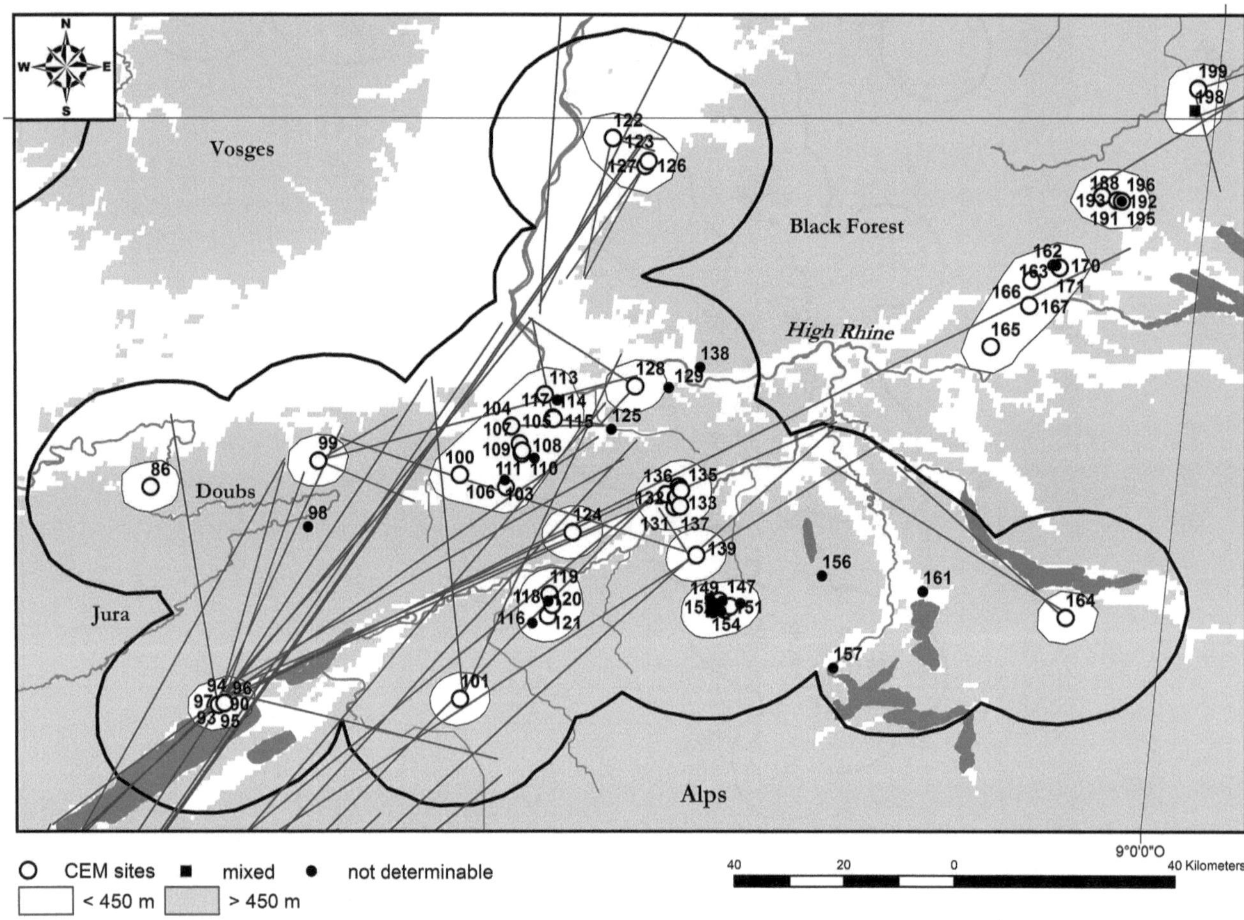

Fig. 7.12 The northeastern exploitation zone of the Circum-Jurassic Group. *Lines* indicate raw material transports between the source areas and the respective sites of discard

a patchwork of water bodies and dry land areas in the same area (Grünvogel 1958; Keller and Krayss 1987: Fig. 2). Thus, the situation was probably comparable to that in the area of the Circum-Jurassic Group (see above). The remains left in this area by CEM hunter-gatherers are remarkable in that they represent the only known assemblages of this period south of the Danube outside the Hegau region.

Overall, the area of the Danube Group is mainly characterized by two features: the Late Jurassic limestone mountains and the Danube river. Erosional and karstic processes formed a highly diversified landscape with deep valleys, steep escarpments, plain and hilly plateaus, caves, and rock shelters. Furthermore, the relatively dry karstic landscape, in which the calcareous bedrock allowed water to seep away easily, contrasted markedly with the wet conditions in the pre-alpine Molasse basin. There, braided rivers and glacial melt-water lakes dominated the landscape (Walter 1995: 407f.; Keller and Krayss 1987). The palynological evidence for the environmental conditions in the area of the Danube Group throughout the Late Pleniglacial and the Oldest Dryas is very scarce. Nevertheless, pollen analyses from sites in the region of the Black Forest and Lake Constance as well as the Isar and Inn regions suggest that the vegetation primarily consisted of treeless steppe-tundra. Here, herbaceous plants prevailed, and Artemisia was abundant (see Chap. 5, "The Oldest Dryas"). Thus, the vegetation in the Hegau and Federsee region as well as in the Danube valley and the areas adjacent to the south was likely very similar to the vegetation in comparable areas of the Circum-Jurassic Group (see above). Taking into account the presumably much dryer conditions in the karstic area of the Swabian and Franconian Alb, it seems reasonable to assume that steppic elements such as grasses prevailed there. During the Bølling Interstadial (GI-1e), juniper shrubs and dendriform birches spread over the landscape. Whereas the latter dominated in the valleys and areas with more humid soil conditions, juniper was probably most abundant in dryer places. With regard to the faunal assemblages, the sites of the Danube Group yield the most diverse spectrum of remains in the entire investigated area. At the same time, there is a clear spatial trend indicating a pronounced decline in faunal diversity from southwest to northeast. This seems only to reflect the scarcity of

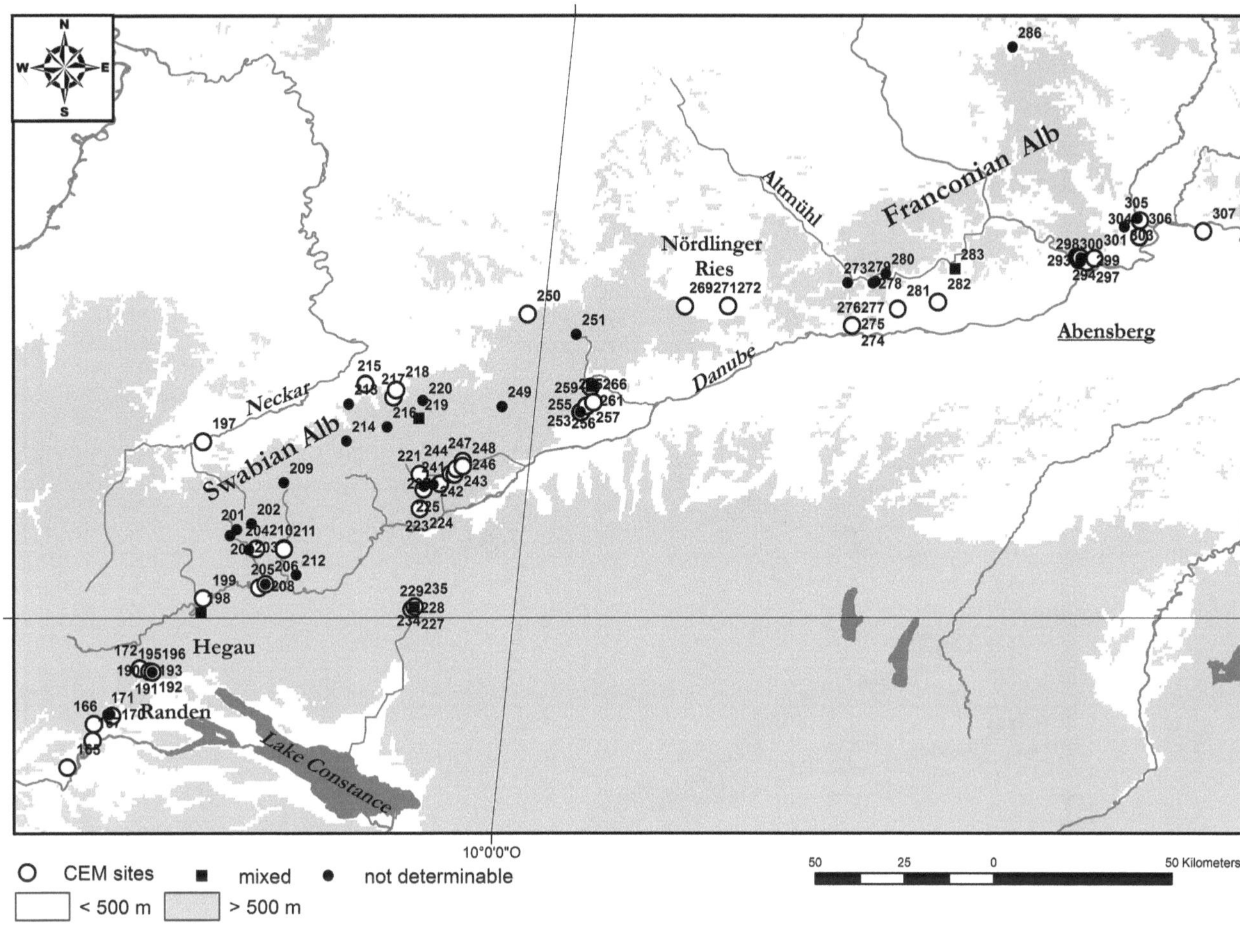

Fig. 7.13 Distribution of sites in the Danube Group

evidence from the Franconian Alb, however, rather than a real shift in animal abundances during the CEM. Animals which appear in assemblages of both the Swabian and the Franconian Alb include mammoth, ibex, chamois, and red deer; in contrast, moose, row deer, rhinoceros, and bovids are only reported in the Swabian Alb. Skeletal remains of saiga antelopes are entirely absent from the whole area. So far, the faunal spectrum is very similar to that of the Circum-Jurassic Group, a circumstance which underlines the similarities in the environmental conditions that were already noted with regard to the pollen spectra. A remarkable difference, however, can be observed in the species that were hunted the most: reindeer and horse. Here, reindeer is clearly dominant in assemblages of the Randen, Hegau, and western Alb, while horse gains significantly in importance in the areas adjacent to the east. A peculiarity of the faunal record in the assemblages of the Danube Group is the presence of musk-ox bones. The only evidence of this species in a CEM context comes from the sites of Kesslerloch, Gnirshöhle, and Burghöhle Dietfurt in the region of the Hegau and the western Alb (see Chap. 5, "Other Animal Remains in the Faunal Record").

Location of Sites and Estimation of the Group's Territory

The distribution pattern of the sites assigned to the Danube Group extends roughly 350 km from west to east and averages 30 km from north to south; it forms a strung-out pattern which at first appears reminiscent of the one described for the western sub-cluster of the Circum-Jurassic Group (Fig. 7.13). However, a closer look shows that the sites of the Danube Group are not equally distributed but rather aggregated into eight smaller sub-clusters, which are on average 30 km apart and coincide spatially with the geographic subsections of this area, namely, the Randen and Hegau region; the western, central, and eastern Albs; the Federsee area; the Nördlinger Ries; the southern Franconian Alb; and the Altmühl valley and Regensburg Basin.

The Randen/Hegau sub-cluster (Fig. 7.14) is situated between the High Rhine and the Upper Danube and consists of 12 sites; the literature mentions 32 archaeological levels, the majority of them is reported from the site of Petersfels (ID 173–193) (n=21). From the Randen region near

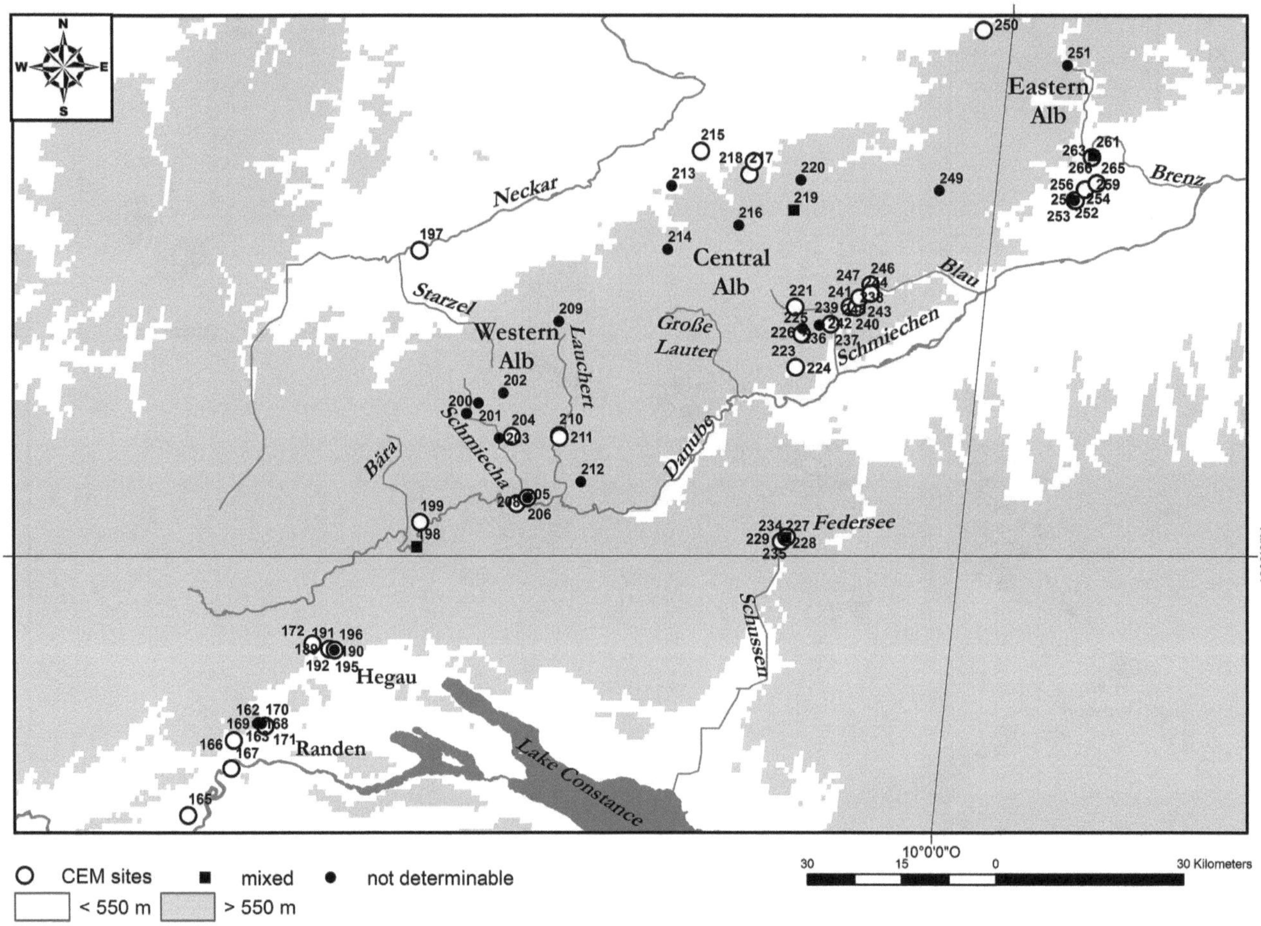

Fig. 7.14 Distribution of sites in the western part of the area of the Danube Group

Schaffhausen, where the landscape is characterized by marked relief, 8 sites are reported. In 4 of these, namely, Solgen im Grauen (ID 172), Freudenthalhöhle (ID 174), Schweizersbild (ID 173), and Kesslerloch (ID 177), a Magdalenian occupation is very probable. For the remaining 4 sites of Untere Bsetzi (ID 162), Kerzenstübli (ID 168), Neue Höhle (ID 169), and Vorderer Eichen (ID 163), an allocation to the CEM must remain questionable because of the scarcity of the archaeological evidence (see Bandi 1947). Here, several valleys open up toward the Rhine river plain forming bottleneck situations and elevated locations that simultaneously allow for a wide view. Both features make the area favorable from a hunting-strategy point of view. A similar situation, but with less pronounced relief, can be found in the region of Engen, where the sites of Bildstockfels (ID 178), Petersfels (ID 173–193), and Gnirshöhle I and II (ID 201, 202) are located. Furthermore, both areas lie at the outer margins of a melt-water lake landscape that was formed by the retreating Rhine glacier (Keller and Krayss 1987: Fig. 2).

On the northern side of the Danube, the sub-cluster of the western Alb comprises 14 sites, 6 of which are classified as not determinable (see Table A.1). The majority (13 sites) are located in valleys of rivers that flow into the Danube. Thus, the sites of Probstfels (ID 199) and Buttentalhöhle (ID 198) are located where the Bära river joins the Danube, while Annakapellenhöhe (ID 210), Nikolaushöhle (ID 211), Kuhstallhöhle (ID 204), Zigeunerfels (ID 206–208), and Burghöhle Dietfurt (ID 205), as well as the 6 not determinable sites, are located in the neighboring valleys of Schmiecha (also called Schmeie) and Lauchert. Where these tributary valleys open toward the broad Danube valley, they form bottleneck situations and thus provide favorable locations for hunting. The 14th site, Napoleonskopf (ID 197), is the only one in the area which is located in the Neckar valley. It lies at the mouth of the Starzel river, which has its source in the Swabian Alb in the vicinity of Albstadt. Hence, the landscape situation here is comparable to that of the Danube valley. The Alb plateaus at higher elevations thus far remain mostly empty of Magdalenian remains.

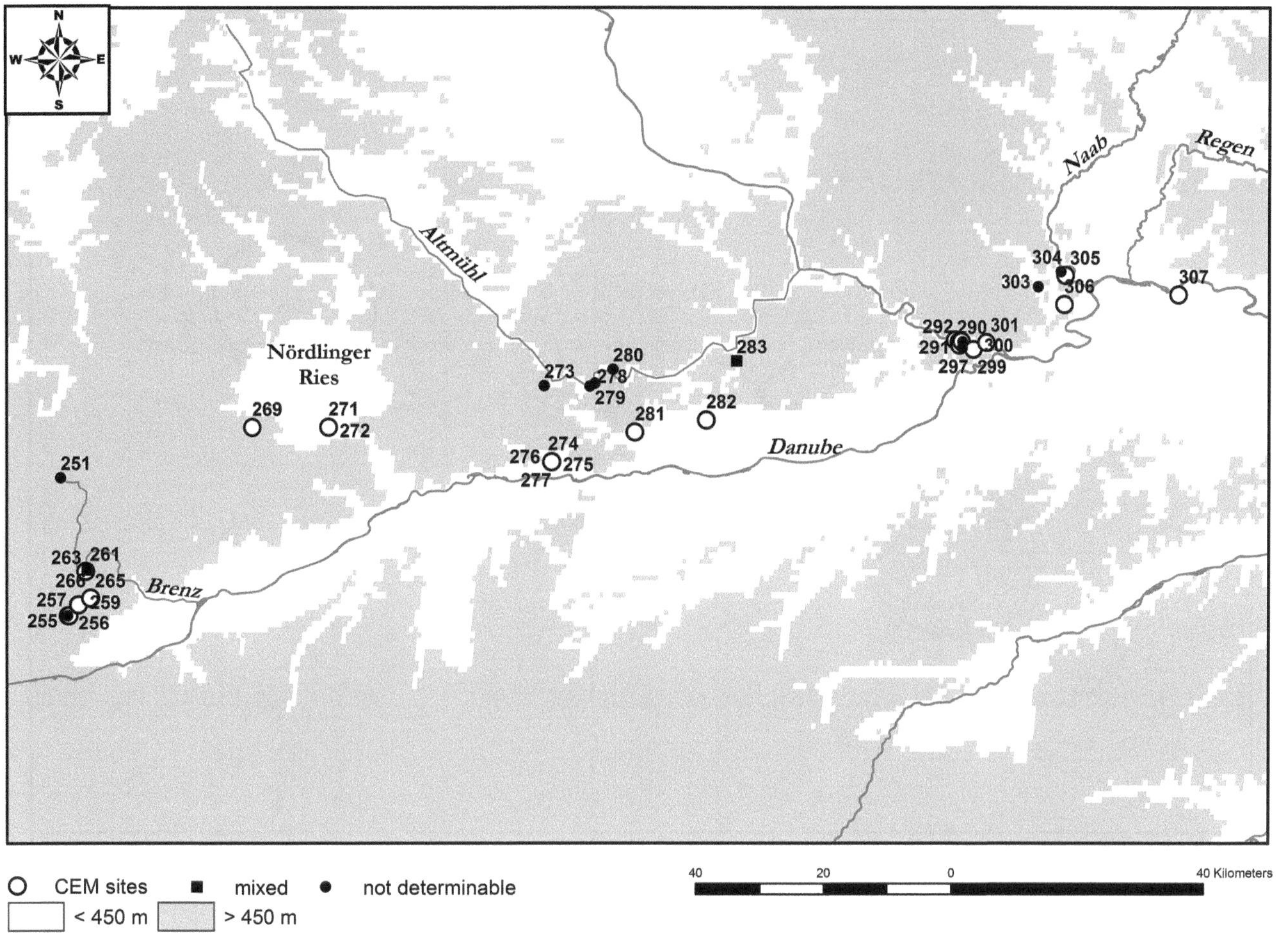

Fig. 7.15 Distribution of sites in the eastern part of the area of the Danube Group

Further to the east is the central Alb sub-cluster, which can be divided into a northern and a southern part. The northern part includes 8 assemblages: Käppele (ID 215), Braunfirst (ID 218), Randecker Maar (ID 217), and Burkhardtshöhle (ID 219), which are very probably Magdalenian, and Barnberger Höhle (ID 213), Klopfjörgelshütte (ID 214), Rappenfels (ID 216), and Papierfels (ID 220), which are classified as not determinable. All of these sites are located relatively close to the Albtrauf escarpment, which in this area is comparatively deeply indented by the steep valleys of the Erms, Lauter, and Lindach rivers. In the southern part of the sub-cluster, 18 concentrations in 11 sites can be found. The sites are located along the slopes of the pre-Weichselian Danube valley, which today is traversed by the Schmiech and Blau rivers, overlooking the wide Danube river plain. Both parts of the central Alb thus provide favorable conditions for hunting and for observation of the landscape.

The 8 sites of the eastern Alb sub-cluster contain 12 Magdalenian levels and are restricted almost exclusively to the Lone valley, where they are located close together on the southeast-facing slope with a good view over the wide Danube river plain. The only exceptions are the Kleine Scheuer (ID 250) which is situated at the Albtrauf and the site of Herwartstein (ID 251) in the Brenz valley; the attribution of the latter to the CEM is unclear.

On the southern bank of the Danube, opposite to the central Alb and south of the Federsee, is a small sub-cluster containing the two sites of Schussenquelle (ID 227–229) and Aichbühl (ID 234, 235). The latter consists of 6 concentrations, of which 4 are probably of post-Magdalenian age. The two concentrations B and C, however, may well belong to the Magdalenian. The sites are located close to a pass marking the transition from the broader Schussen river valley to the Federsee region.

Two sites, the Hohlenstein bei Edernheim (ID 269) and the Kaufertsberg with its two Magdalenian layers (ID 271, 272), are located in the southern part of the impact crater Nördlinger Ries (Fig. 7.15). East of the Ries, 8 sites with 11 Magdalenian levels constitute the sub-cluster of the southern Franconian Alb. Most important for the spatial distribution is the west-east-oriented course of the Altmühl river valley; five sites are located along it, although only the assemblage of Hohlenstein bei Schambach (ID 283) can be classified as probably of Magdalenian age. In the south, the remaining

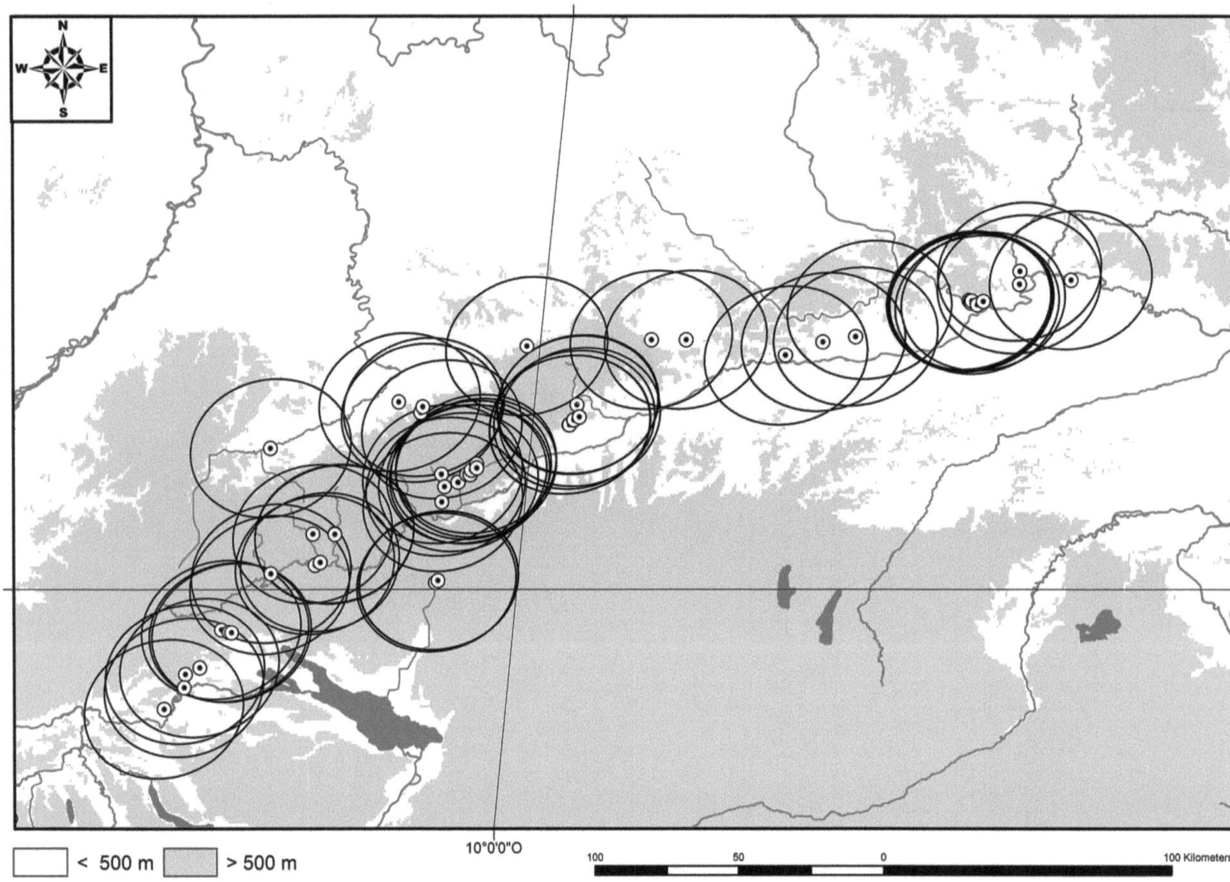

Fig. 7.16 Danube Group, territory estimation with 25 km radius around each site

three sites of Weinberghöhlen (ID 274–277), Speckberg (ID 281), and Eitensheim (ID 282) likewise form a west-east-oriented line that follows more or less closely the course of the Schutter river.

The 11 sites and 17 Magdalenian levels of the eastern-most sub-cluster are situated between Kelheim and Regensburg. The majority (8 sites) are located just before the confluence of the Altmühl and Danube rivers, where the narrow Altmühl valley opens toward the larger Danube river plain and forms a bottleneck situation. Of the remaining 4 sites, 2 are located in the narrow valley of the Schwarze Laber and 2 on the left bank of the Naab river.

In the northern part of the Franconian Alb, only 5 sites potentially belonging to the CEM are reported, namely, Rennerfels (ID 284), Fuchsenloch (ID 285), Hohlenstein bei Happurg (ID 286), Steinbergwand (ID 302), and perhaps Mäanderhöhle (Bosinski 2011). However, the scarcity of the available information makes a clear archaeological allocation impossible.

The median altitude for the assemblages in the Danube Group is about 546 m above MSL, with a standard deviation of 119 m. This is the highest elevation value among all interaction units. The estimations for the settlement area of the Danube Group are 30,000 km^2 for an assumed single site-catchment radius of 25 km and 47,000 km^2 for a radius of 40 km (Figs. 7.16 and 7.17). For the geography-based estimation (Fig. 7.18), the following parameters are considered: east of the Iller and Blau rivers, no Magdalenian sites are reported south of the Danube or north of the Albtrauf. Thus, in the eastern part of the Danube Group, the approximate course of these margins is taken as a limit for the estimation of the settlement area. In the western part, the limits are extended to the Neckar valley in the north and the region of Lake Constance in the south. The resulting estimated area extends over about 23,000 km^2.

Land-Use Pattern

The observations and results obtained so far offer clues to the nature of the land-use pattern of the Danube Group. The following section presents a possible interpretation of the findings using ethnographic analogies. At 24 km, the estimated

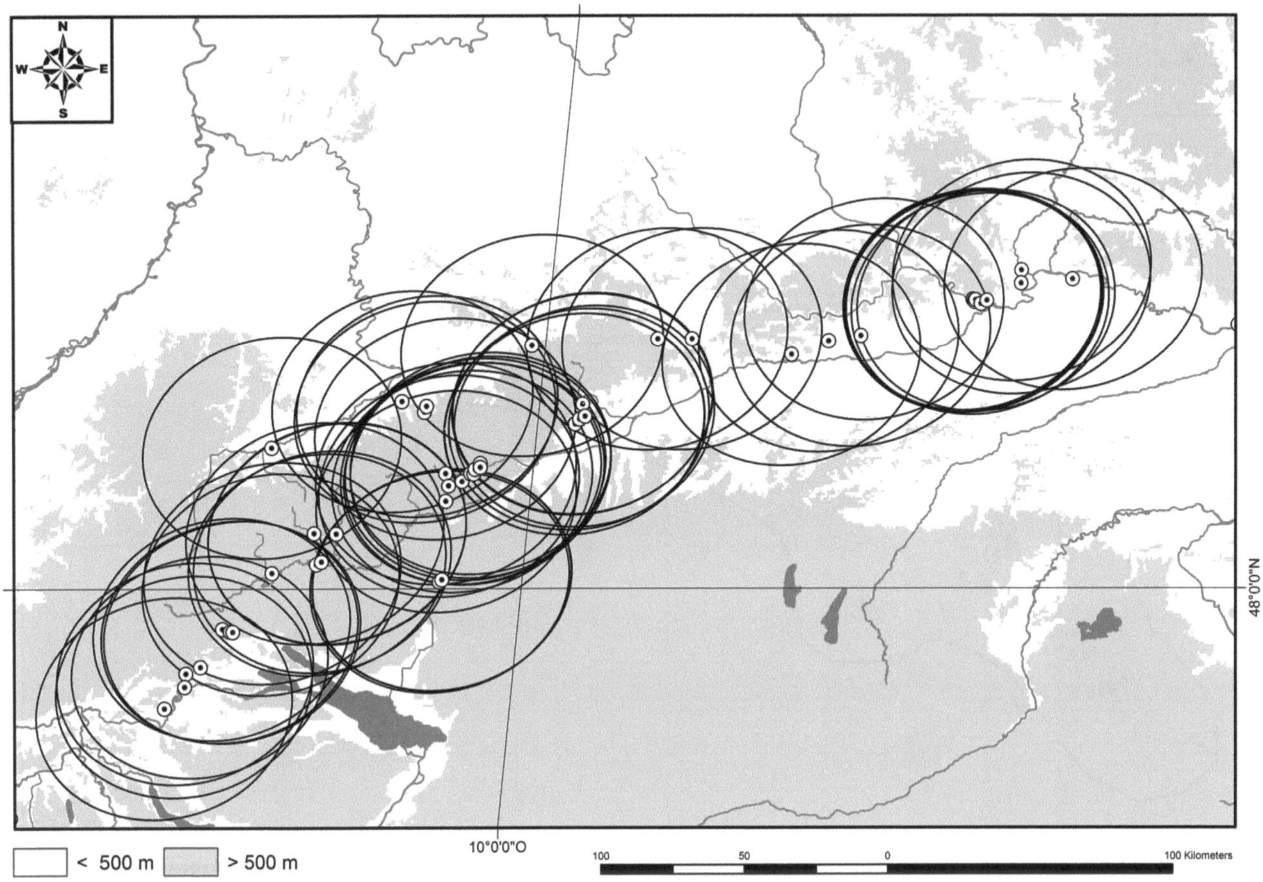

Fig. 7.17 Danube Group, territory estimation with 40 km radius around each site

average distance traveled during a residential move (ARMD) of the Danube Group is slightly larger than that of the Circum-Jurassic Group (20 km). Taking into account both site location and ARMD, the circa 23,000 km^2 territory of the Danube Group can be subdivided into three exploitation zones: a western, a central, and an eastern zone (Fig. 7.19).

The western zone covers an area of about 8,000 km^2 and extends from the Limmat and Aar rivers in the west up to the Lauter and Echaz rivers in the east and from the Neckar in the north to the area of Lake Constance in the south. It thus comprises the regions of Randen, Hegau, and the Western Alb. The central zone is the largest, at approximately 10,000 km^2. It consists of the central and eastern Alb and also encloses the Federsee region and the Nördlinger Ries. In the east, the southern Franconian Alb corresponds to the third exploitation zone, which extends from the Ries and the Wornitz river in the west up to the area where the Regen river joins the Danube in the east. It covers an area of about 7,000 km^2.

Generally, the raw material procurement distances do not exceed 48 km and thus are within the margins of the double ARMD. Thus, they could have been acquired within the range of two consecutively used residential camps. An exception to this rule is the chert from the Abensberg region, which can be found in all three exploitation zones and up to 250 km away from the outcrops where it originates. It is striking that without exception, the long-distance transport of items procurable within the area of the Danube Group took place from east to west. This is the case not only for the Abensberg chert, but applies as well to the fossil mollusks from the Steinheimer Basin and the Tertiary sediments in the region of Ulm. Although the likely palimpsest situations in the numerous caves and rock shelters that comprise 70 % of the total sites might tend to bias the tool spectrum toward homogeneity, the diversity in site function is nevertheless pronounced. Assemblages which appear to be the remains of specialized hunting camps, such as Schussenquelle or Braunfirst, alternate with inventories that exhibit a more diverse tool spectrum suggestive of residential camps, for instance Kesslerloch or Felsställe. It therefore seems that the subsistence strategies of the hunter-gatherers in the Danube Group were organized logistically (sensu Binford 1980), just as they seem to be in the Circum-Jurassic Group.

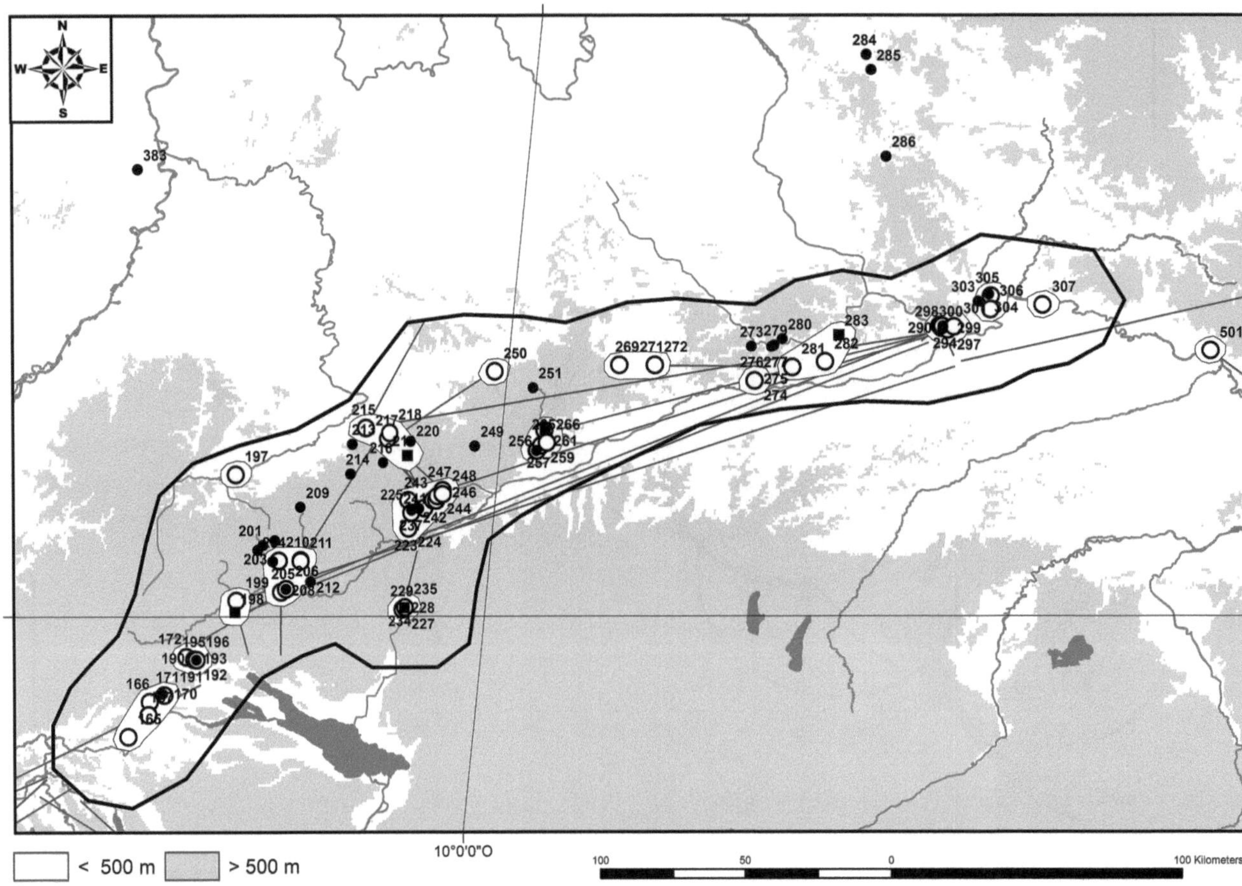

Fig. 7.18 Danube Group, territory estimation based on raw material procurement and geographic features

The Meuse-Rhine Group

In the northwestern part of the investigated area, the sites of the Meuse-Rhine Group are clustered. The earliest Magdalenian settlement in this area probably took place around 16,000 cal BP. The oldest finds are reported at the sites of Andernach-Martinsberg and Gönnersdorf in the Neuwied Basin and at Trou Walou, Trou de Chaleux, and Trou du Frontal in the Ardennes (see Table A.6). Bones from the cave site Trou des Blaireaux have also been dated to about 16,000 cal BP; however, it is not entirely clear whether these relate to human or to animal activities (Charles 1996; Housley et al. 1997).

Group Statistics

According to the archaeological literature, the Magdalenian record in the area of the Meuse-Rhine Group comprises a total of 73 assemblages from 58 sites. After a critical review, 50 inventories appear to be assignable to the CEM with certainty or high probability. In 15 cases, a clear allocation to an archaeological entity was not possible, and in the case of 8 collections, the typology rather indicates a post-Magdalenian age (Table 7.5). These 8 collections are therefore excluded from the analysis.

Thus, the Meuse-Rhine Group consists of 65 assemblages from 50 sites. Both numbers represent the lowest values of the five analyzed interaction units. In contrast, 42 % of the assemblages (27 inventories) yield more than 100 lithic tools. This is the highest percentage among the five regional groups. Three collections (4.5 %) contain more than 1000 tools. With 25 assemblages from cave sites (38 %) and 40 from open-air camps or surface collections (62 %), inventories from locations without any natural shelter are almost twice as numerous as those from naturally sheltered places.

Landscape and Environmental Setting

The sites of the Meuse-Rhine Group are located along the river valleys which traverse the Rhenish Massif and the Lower Rhine Bay (Fig. 7.20). The Rhenish Massif is a low mountain range approximately 400 km long and 150 km

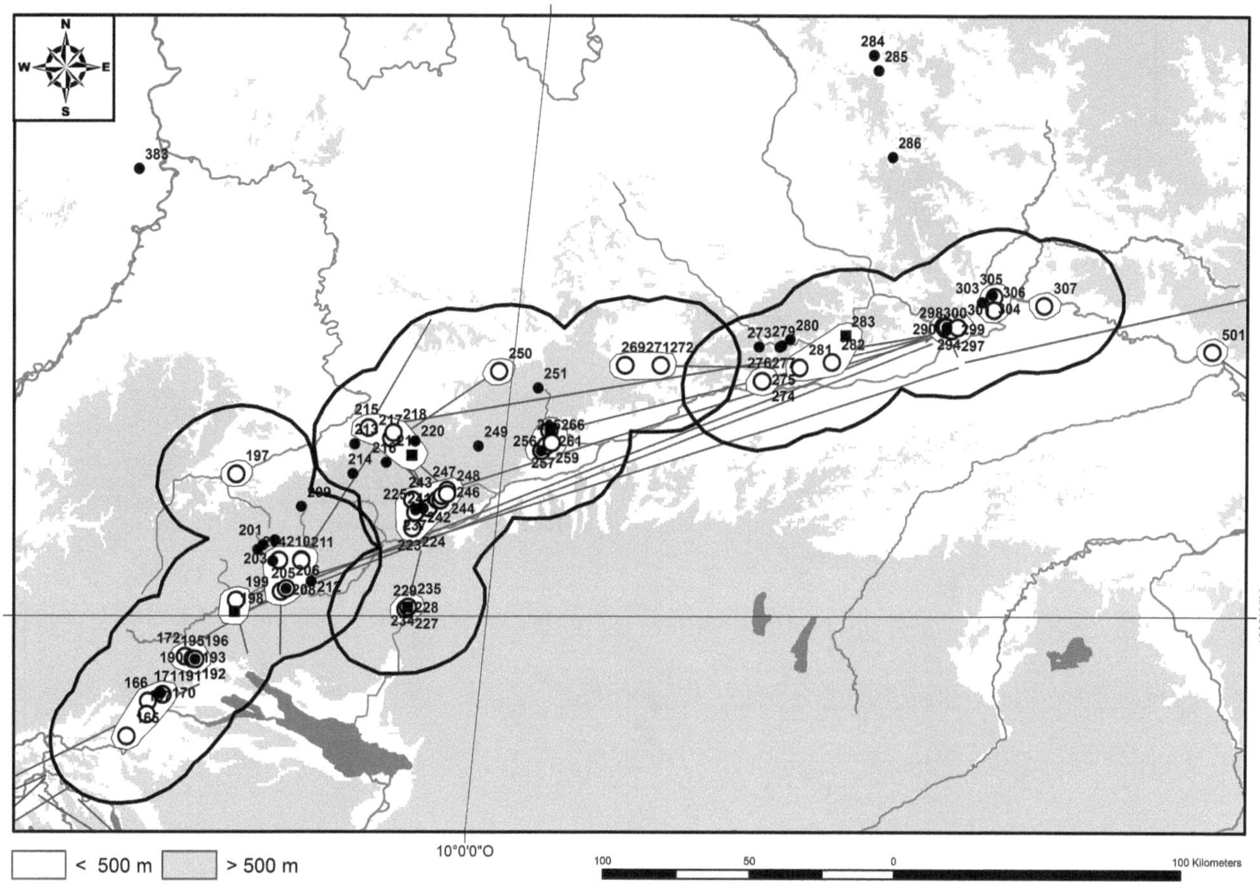

Fig. 7.19 The three exploitation zones of the Danube Group. Lines indicate raw material transports between the source areas and the sites of discard

wide, which consists primarily of Devonian rock (Walter 1995: 160ff.) and extends roughly from the valleys of the Sambre, Meuse, and Ruhr rivers in the north to the Nahe and the Main in the south. Its elevations are between 200 m and 500 m above MSL on average, but locally they sometimes exceed 800 m.

The Rhenish Massif is divided in two by the Middle Rhine Valley. Its eastern part comprises the Ardennes and Eifel, whereas the western part can be subdivided from north to south into the Sauerland, the Westerwald, and Taunus. At the northern margins of the Ardennes, patches of Cambrian, Carboniferous and Cretaceous sediments as well as loess can be found close to one another; the Westerwald and the Neuwied Basin, a broader part of the Rhine valley between Westerwald and Eifel, is characterized geologically by the occurrence of volcanic rocks (Walter 1995: 182ff, Abb. 74; Zahn 1996: 74–75). North of the Ardennes and in the Lower Rhine Bay, Meuse gravels and Meuse eggs occur in a large area (Floss 1994: Fig. 29). Tertiary quartzite can be found in many parts of the Rhenish Massif, whereas chalcedony occurs mainly in the region of Bonn (Ibid.: Figs. 8 and 11).

The resettlement of the Meuse-Rhine area took place during the Oldest Dryas. During this period, the landscape was characterized by treeless tundra-steppe. In the dryer elevated parts, the vegetation was dominated by grasses (Poaceae), whereas *Artemisia* and other herbaceous plants were much less frequent. With the onset of the Bølling Interstadial (GI-1e), temperatures rose markedly, and precipitation probably also increased significantly. As a result, the proportion of *Artemisia* and herbaceous plants increased at the expanse of the grasses, which, nevertheless, remained dominant. Juniper shrubs began to grow at favorable stands, and the first dendriform birches also likely arrived in the region during this period (see Chap. 5, "Climate and Vegetation"). The faunal record suggests that reindeer was an abundant type of game at this time, although horse seems to have been an even more successful competitor in this steppic environment. Besides these two species, a rich faunal community is indicated by finds of mammoth, woolly rhinoceros, red deer, row deer, ibex, chamois, wisent, aurochs, and moose (see Chap. 5, "Fauna"). The presence of saiga antelope in the assemblages of Gönnersdorf and Trou de Chaleux

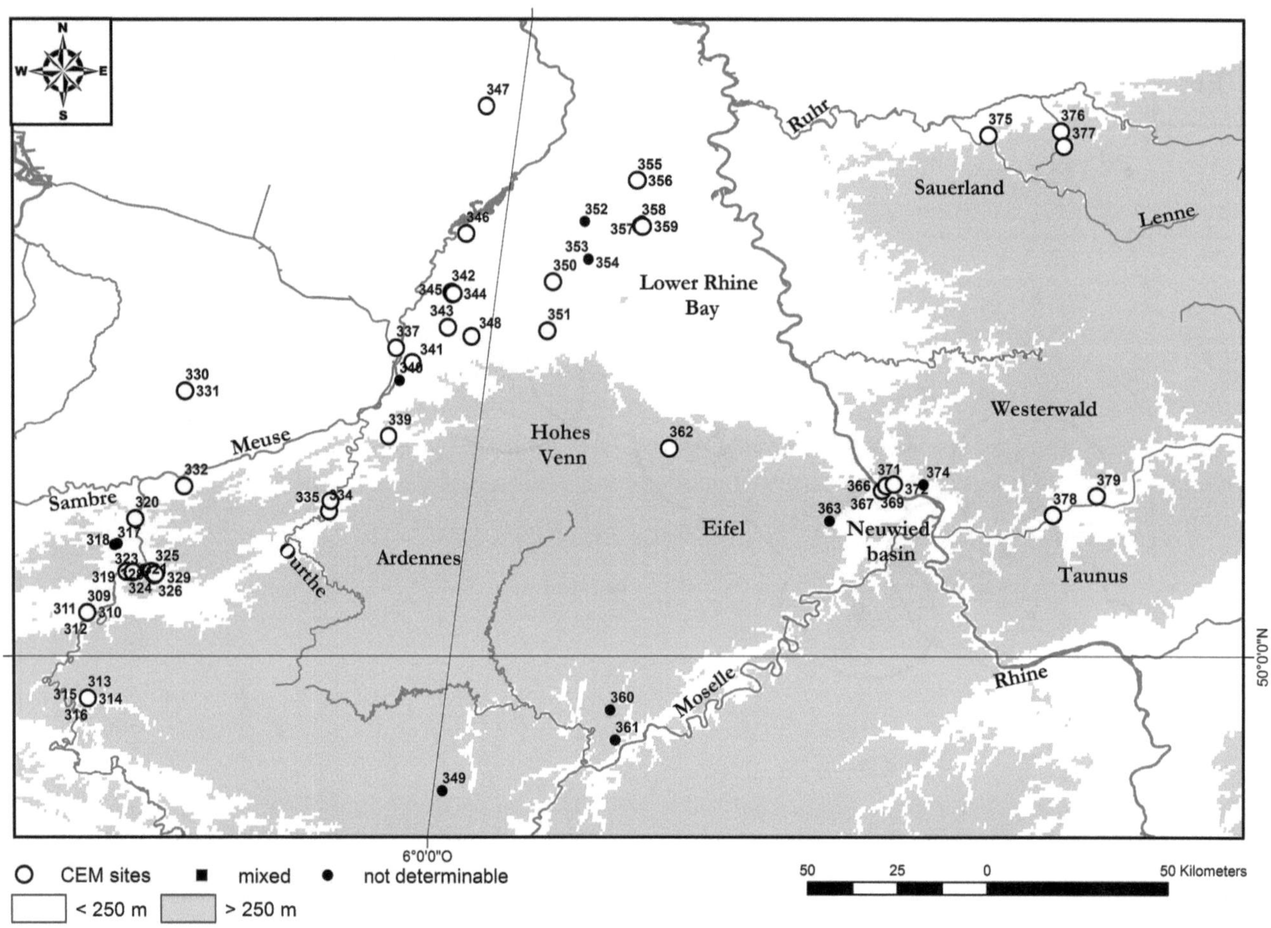

Fig. 7.20 Distribution of sites in the Meuse-Rhine Group

as well as in the more southerly located site of Roche Plate deserves special attention. Because this species is particularly well adapted to cold and arid environments, its presence highlights the rather dry and steppic character of the landscape at least during the Oldest Dryas. In this context, the remains of wild boar at Trou de Chaleux, Trou des Nutons, Trou du Frontal, and Grotte de Sy are also of interest. On the one hand, the presence of this more temperate-zone species in CEM assemblages can be seen as reflecting hunting activities during the already warmer Bølling Interstadial; if, on the other hand, saiga antelope and wild boar are assumed to be present more or less contemporaneously, this finding cloud point to the existence of climatically favorable micro-regions, for instance, in the Meuse valley, or to "cryptic refugia" in the sense of Stewart and Lister (2001).

Location of Sites and Estimation of the Group's Territory

The sites in the northwestern part of the investigated area are located mainly along the currents of the Meuse and the Rhine and their tributaries (cf. Rensink 1993: 84ff.). Here, the drainage system of the Meuse contains the majority of sites, while the site density in the western and southern part of the Rhenish Massif is considerably lower. The distribution pattern spans about 300 km from west to east and 150 km from north to south. Generally, the sites are restricted to the lower outer slopes of the Rhenish Massif and its river valleys and basins and are particularly numerous at the northern foothills of the Ardennes (Fig. 7.20). The median site elevation in the Meuse-Rhine Group is 130 m AMSL. Only two sites, Roc-la-Tour and Kartstein, are situated over 350 m: both are at about 400 m AMSL. The higher elevations of the region apparently were not attractive for prolonged stays. This is probably the reason for the conspicuous absence of assemblages from the central part of the Rhenish Massif, which is virtually empty of any evidence of Magdalenian occupation. Here, the eastern part of the Ardennes together with the Eifel and the Hohes Venn forms a continuous mountain range with altitudes generally above 350 m MSL.

With regard to their geographic location, the sites of the Meuse-Rhine Group can be divided into two major groups: (A) sites located in or close to relatively deeply incised river valleys forming natural bottleneck situations and (B) sites located at more exposed positions.

(A) In the western part of the Ardennes, all of the sites found between Charleville-Mézières and Namur, including Roc-la-Tour (ID 313–316), Trou des Blaireaux (ID 309–312), Trou des Nutons (ID 326), Trou de Chaleux (ID 324, 325), Trou du Frontal (ID 328), Trou da Somme (ID 319), Bois Laiterie (ID 320), and Goyet 3 (ID 332), are located in the immediate vicinity of the confluence of two rivers, in most cases the Meuse and one of its tributaries. Most sites are reported from the region around Dinant, at the confluence of the Lesse and Meuse rivers. In the valley of the river Ourthe, again a tributary of the Meuse, the sites Grotte de Coléoptère (ID 334), Grotte de Sy (ID 335), and Trou Walou (ID 339) are situated at comparable geographic locations. The bottleneck effects that are generated by the narrowing tributary valleys likely provided favorable conditions from the point of view of hunting strategy. At the transition from the Ardennes to the plain adjacent to the north, the Meuse valley widens and forms a funnel-shaped valley which was probably equally well suited for hunting. Here the sites of Mesch-Steenberg (ID 341) and Kanne (ID 337) are found. The same presumably applies to the sites of Martinshöhle (ID 375), Balverhöhle (ID 377), and Feldhofhöhle (ID 376) in the northern Sauerland, the northeasternmost part of the Meuse-Rhine Group, where the Lenne and Hönne rivers flow toward the Ruhr. Unfortunately, the early date of excavation and the poor condition of the collections do not allow for extensive analyses. However, the numerous caves in the vicinity of these sites and the geographic conditions suggest that the area was suitable for a Magdalenian occupation. Further to the south, at the so-called Andernacher Pforte or "Gate of Andernach," the broad valley of the Neuwied Basin narrows considerably, and the Rhine, which traverses the basin, flows through a steep valley. The open-air camps of Gönnersdorf (ID 368–371) and Andernach-Martinsberg (ID 364–367) lie on the eastern and western river banks of this natural bottleneck. Upstream from this area at a distance of only about 15 and 20 km, respectively, the Moselle and Lahn rivers join the Rhine. In the basin of Limburg, the otherwise narrow Lahn valley broadens. On both sides of the valley, where the sites of Wildweiberlei (ID 378) and Wildscheuer (ID 379) are located, the geographic situation provides suitable hunting conditions.

(B) North of the Ardennes and in the Lower Rhine Bay, the geographic setting of the CEM sites differs from that described above. Here, in the transitional area between the uplands of the Ardennes and the Eifel and the lowlands adjacent to the north, the relief is less pronounced, and the landscape is comparatively flat. Nevertheless, the sites are restricted to the more elevated parts which allow for a good view. This also applies to Griendtsveen (ID 347), the northernmost site, which is situated at the northern end of an elevated spur that rises slightly above the surrounding area. Here—as described above—the course of rivers seems to be likewise important for the location of sites. The open-air camp of Orp le Grand (ID 330, 331), for instance, is situated at the confluence of two smaller rivers. From the southwestern part of the Rhenish Massif, in the area of Luxembourg and Trier, three surface collections are mentioned. However, the typological composition is not very clear, and in all three cases, an allocation to the Magdalenian is therefore questionable, though it cannot be ruled out. A saurian vertebra found at the site of Gönnersdorf may also come from this region. It may indicate an occasional or sporadic exploitation of this area, if any took place at all.

The estimation of the territory of the Meuse-Rhine Group is roughly 41,000 km^2 if a range of 25 km radius around each site is assumed (Fig. 7.21). By increasing this range to a radius of 40 km, the estimated area grows to about 60,000 km^2 (Fig. 7.22). Taking into account the observed geographic characteristics of the sites, it is necessary to consider that the higher elevations of the Rhenish Massif as well as the northern lowlands are virtually empty of Magdalenian remains (Fig. 7.23). The raw material procurement pattern, on the other hand, points to the exploitation of outcrops in the region of Charleroi, where no CEM site has been reported so far. If those features and the ARMD of 18 km are considered in the calculation, the estimated minimum territory for this group comprises approximately 35,000 km^2.

Land-Use Pattern

In the following, the observations obtained so far are interpreted with regard to the possible land-use pattern of the Meuse-Rhine Group. At just 18 km only, the group's estimated average distance traveled during a residential move

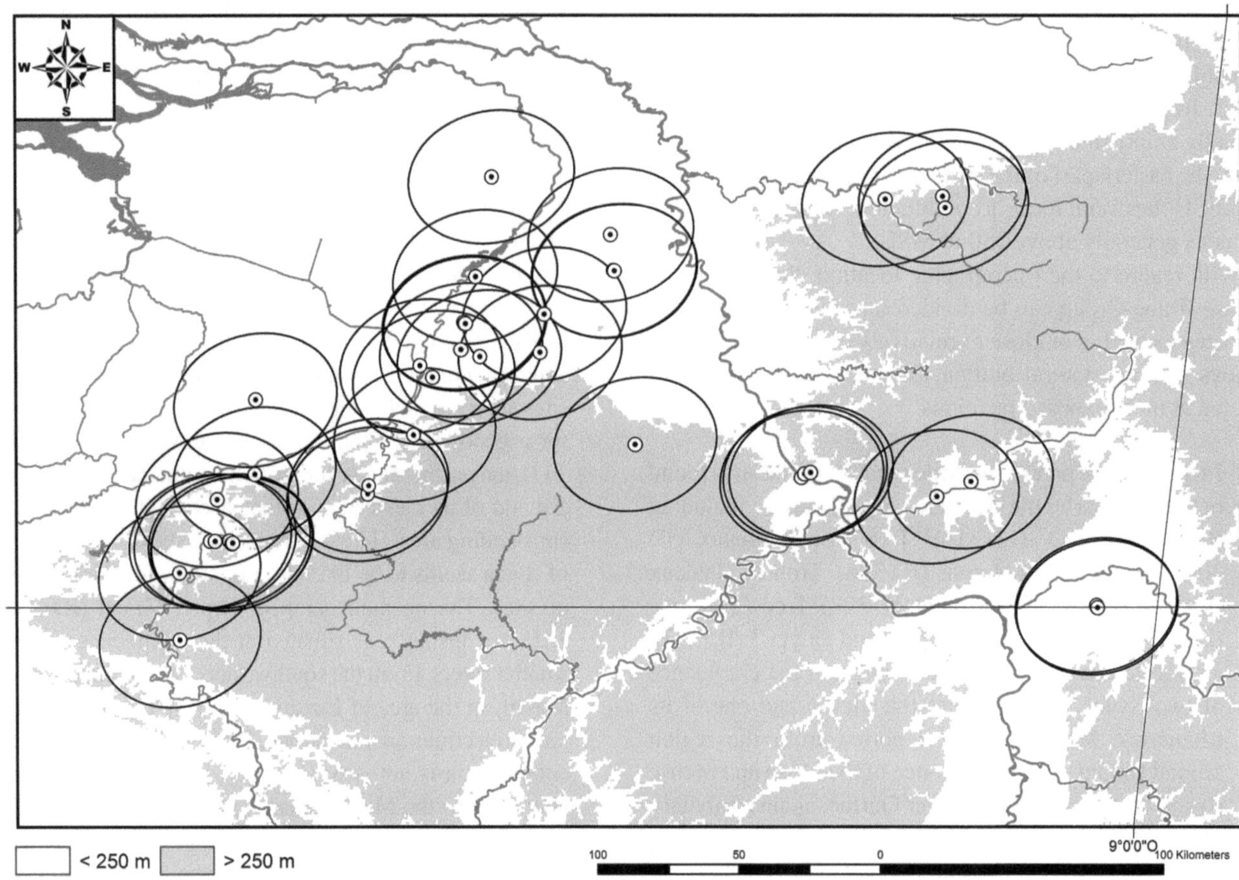

Fig. 7.21 Meuse-Rhine Group, territory estimation with 25 km radius around each site

(ARMD) is the smallest in the entire investigated area. Together with the information on site location, the territory of approximately 35,000 km^2 can be subdivided into 4 exploitation zones, which are thought to correspond to areas occupied multi-annually or to palimpsests of such areas (Fig. 7.24).

The first and westernmost zone expands from Roc-la-Tour (ID 313–316) in the south to Orp (ID 330, 331) in the north and from the region of Charleroi in the west to the site of Goyet 3 (ID 332) in the east. Thus, it covers an area of about 6,000 km^2. The sites in this zone are located along a more or less straight axis that follows the course of the Meuse in the southern and central sections and has its northern terminus at Orp. The east-west extent is delimited primarily by the raw material procurement pattern. Siliceous rocks from the region of Charleroi are reported at Bois Laiterie (ID 320) and were probably collected during a seasonal stay in this area. The pieces from Bois Laiterie may have reached their place of discard within two or three consecutive residential moves.

The second exploitation zone borders on the first in its eastern part. Here, the sites of Grotte de Sy (ID 335) and Grotte de Coléoptère (ID 334) are located more than 40 km away from Goyet 3, the easternmost site of the first zone. The second zone encompasses an area of about 9,000 km^2. It stretches from the Ourthe river in the southwest to the western bank of the Rhine in the east and the site of Griendtsveen (ID 347) in the north. This zone is rich in raw material: Meuse gravels and Meuse eggs can be found at many outcrops and are easily accessible. Consequently, raw material acquisition took place mostly on a local scale. Larger transport distances, as may be assumed for the pieces of Baltic flint found at Beeck (ID 350) or Alsdorf (ID 351), could all be traveled within the range of two or three consecutive residential moves.

The third exploitation zone is located in the northeastern part of the Rhenish Massif. Ideas on the extent of this zone are mainly speculative, since evidence of sites is largely absent. However, the cave sites in the valleys of Lenne and Hönne have yielded archaeological layers with presumably Magdalenian assemblages. The suggestion that Magdalenian people may have traveled nearly 100 km for a short-term stay in this area not only contradicts the observation of economic efficiency in hunter-gatherer land-use patterns which is reported from ethnographic examples, but it also conflicts with the observations obtained so far for the CEM.

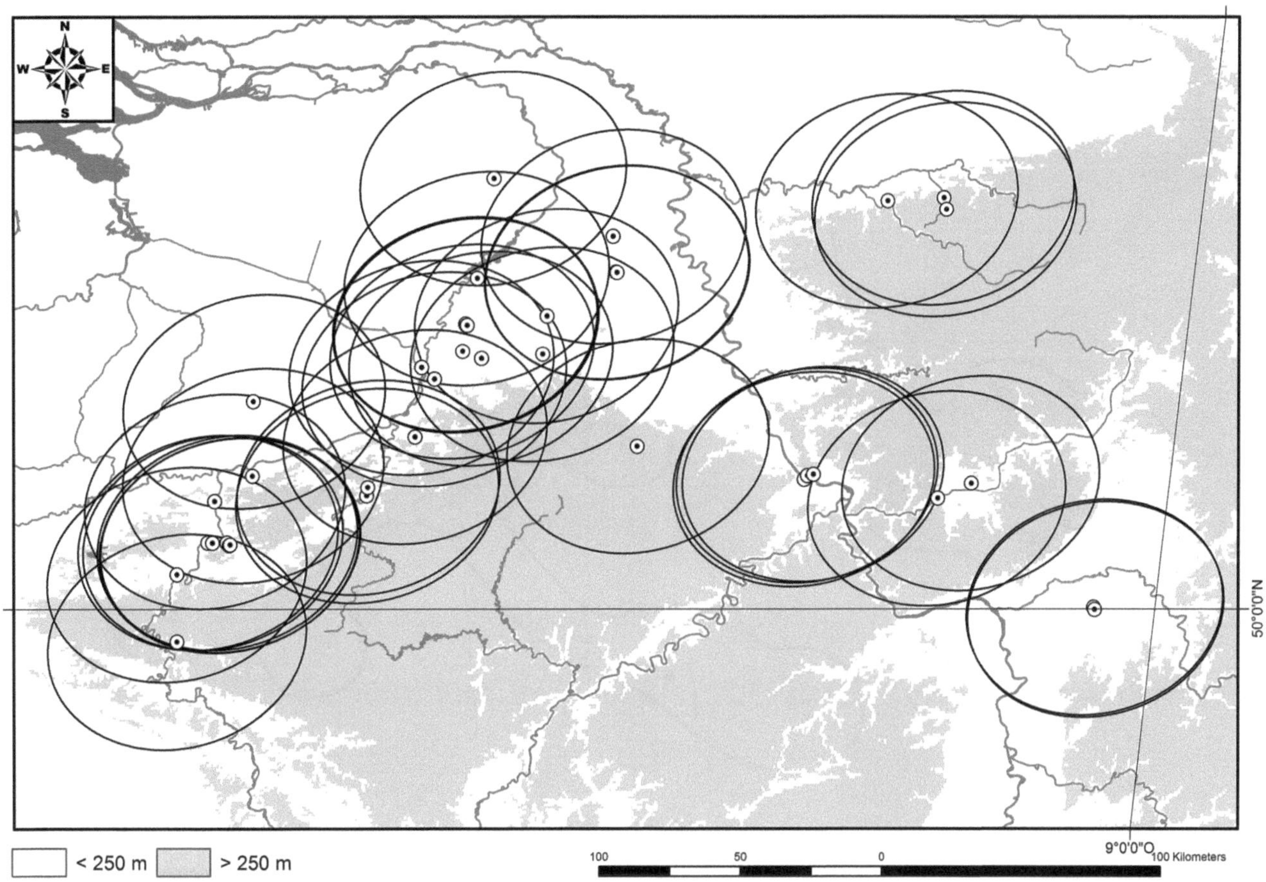

Fig. 7.22 Meuse-Rhine Group, territory estimation with 40 km radius around each site

Thus, it appears more reasonable to interpret these assemblages as evidence of regular settlement activities, such as have already been observed in many other instances. Based on the location of the surrounding sites and the geographic features such as the course of the Rhine and the northern border of the Rhenish Massif, the minimum estimated size of the third exploitation zone is about 5,000 km^2.

The fourth and southernmost zone presents similar problems, since the density of known sites is comparatively low. However, information about raw material and mollusk-shell procurement are useful in reconstructing the minimum exploitation zone. This zone comprises the site of Kartstein (ID 362) and the area of Ratingen, where the Tertiary quartzite used at Gönnersdorf may originate (Floss 1994: 221). Further to the south, it comprises the Neuwied Basin, as well as the lower Lahn valley up to the site of Wildscheuer. In its southern part, it includes the confluence area of Main and Rhine, the area around Dreieich-Götzenhain, and probably the outcrops of fossil mollusks in the Mainz Basin. In all, it thus covers an area of some 9,000 km^2. In the interpretation scheme applied in this study, the presence of Meuse gravels and Baltic flint at the sites of Gönnersdorf, Andernach-Martinsberg, Wildscheuer, Wildweiberlei, and Dreieich-Götzenhain is seen as the result of shifts from the northern exploitation zones to the southern one. In the course of these shifts, the raw material was transported southward in consecutive seasonal residential moves to the Gate of Andernach and further down to the Rhine-Main area.

The hunter-gatherers of the Meuse-Rhine Group were probably organized logistically. This is indicated by the differences in the tool composition of the assemblages, which point to a coexistence of highly specialized sites together with camps that yield a highly diverse tool spectrum. Sites such as Roc-la-Tour or Bois Laiterie, where many lithic projectile implements have been found and which therefore probably served as hunting stations, alternate with residential camps such as Andernach or Gönnersdorf, which are characterized by a more diverse spectrum of lithic tools. Notably, assemblages with high numbers of backed bladelets are numerous in the Ardennes south of the Meuse river, whereas inventories further to the north generally exhibit lower ratios of this tool class. This accords with the observation that the sites in the Ardennes are located at places that appear especially well suited for hunting purposes.

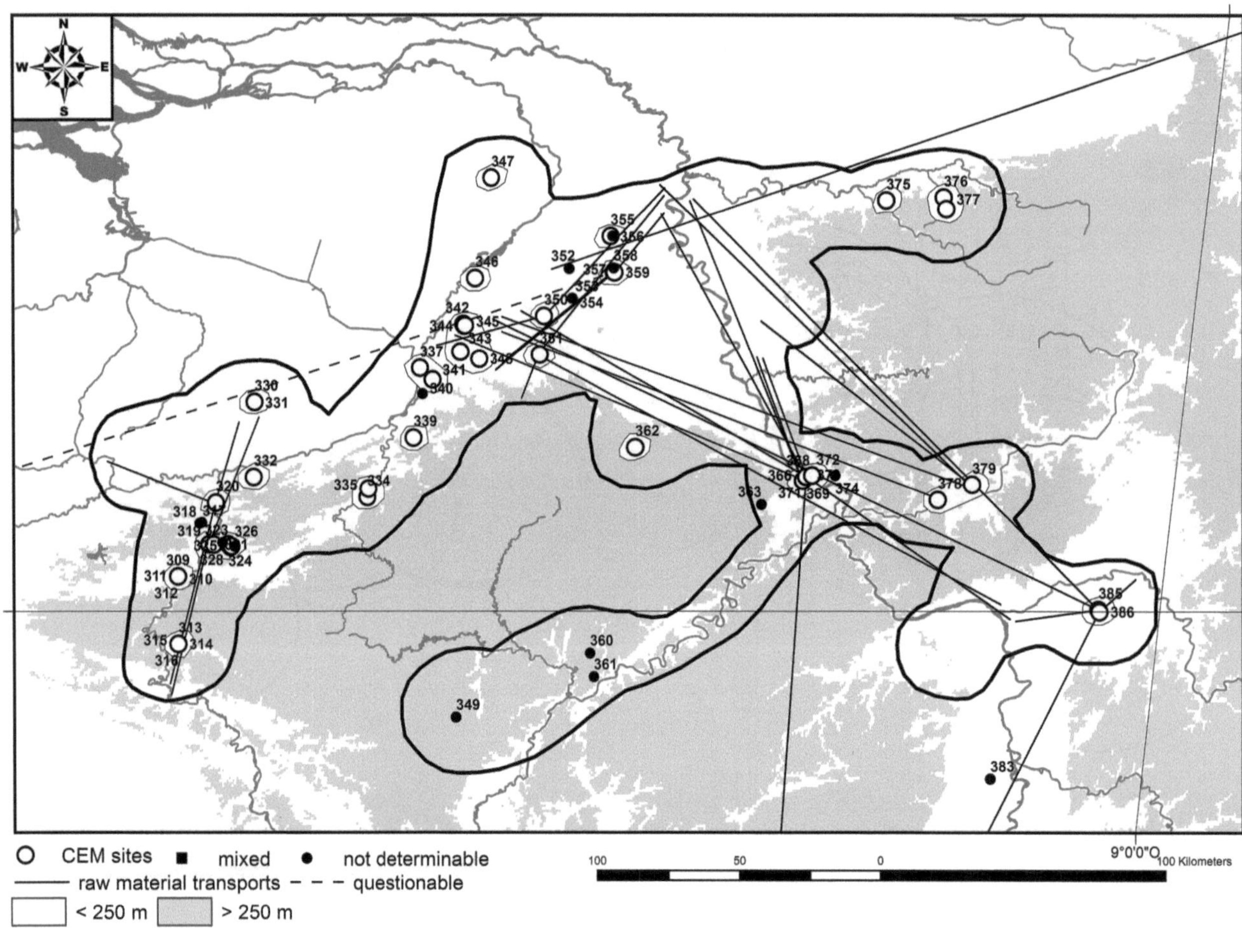

Fig. 7.23 Meuse-Rhine Group, territory estimation based on raw material procurement and geographic features. Lines indicate raw material transports between the source areas and the sites of discard

The Vltava-Saale Group

The forth regional group is situated in the eastern part of the investigated area. The sites of this group are located within the drainage system of the Elbe river, mainly along its most important tributaries, the Vltava and Saale. The earliest reliable dates for a Magdalenian occupation of this region are reported from the site of Kniegrotte in Thuringia. A set of 8 consistent ^{14}C-dates, 5 of which have been obtained on human-modified bones, give a weighted average date between 16,370 and 15,830 cal BP (13,213 ± 45 ^{14}C BP).

Group Statistics

The archaeological literature records 137 Magdalenian assemblages from 124 sites in the area of Vltava-Saale Group. During a critical review, this classification could be confirmed for only 76 assemblages (56 %). For a large part of the inventories (n = 51, 37 %), the state of publication or the scarcity of finds does not allow for a more precise determination. In the case of 4 collections, Magdalenian finds are probably mixed with artifacts of other archaeological periods, and 6 sites appear to be of post-Magdalenian age and are therefore excluded from further analysis (Table 7.6).

The following descriptions are therefore based on a total of 118 sites with 131 assemblages. These represent the second highest number of assemblages and sites of all the five interaction units under discussion. However, 39 % of these assemblages lack detailed information. The number of collections with more than 100 tools is nevertheless the third highest of all regional groups at 22 (17 %). When assemblages with more than 1000 tools are considered, the Vltava-Saale Group together with the Circum-Jurassic Group has the highest number (6 inventories), and together with the Meuse-Rhine Group, the highest percentage value (5 %) of these assemblages. With regard to the topographic position of the sites, the Vltava-Saale Group shows a pronounced difference in the distribution of sheltered sites and open-air camps: with 108 assemblages (82 %), this group has by far the highest

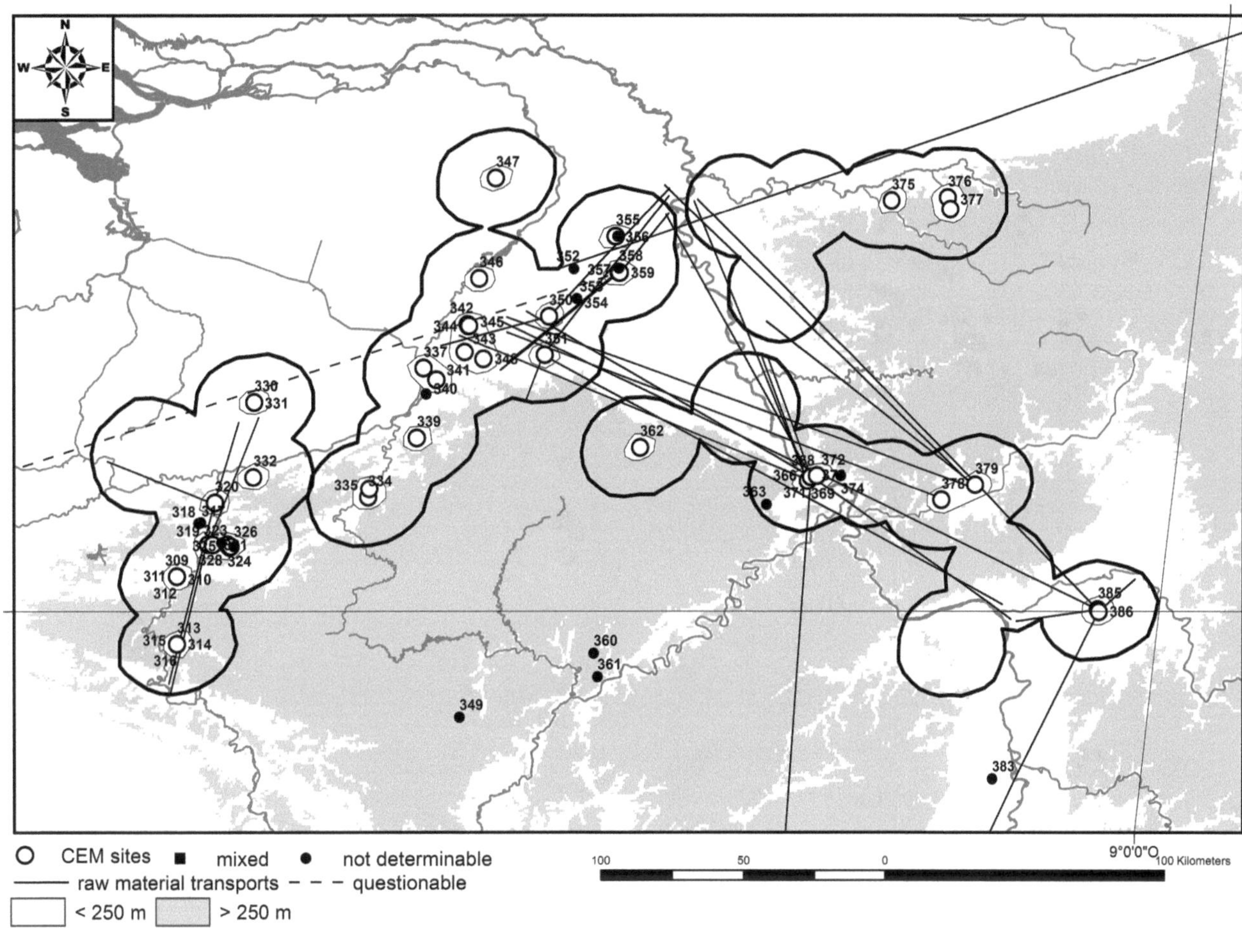

Fig. 7.24 The four exploitation zones of the Meuse-Rhine Group

number of inventories stemming from open-air situations (excavations and surface collections) and the lowest number of assemblages from cave sites (n = 22, 17 %) and rock shelters (n = 1, 1 %). Hence, open-air camps are about five times more numerous than camps at locations with a natural shelter.

Landscape and Environmental Setting

The area of the Vltava-Saale Group is characterized geographically by several low mountain chains which delimit it and subdivide it into different parts (Fig. 7.25).

To the west, the site distribution is delimited by the Thuringian-Franconian Highlands (including the Thuringian Forest, the Thuringian Slate Mountains, the Franconian Forest, and the Fichtelgebirge), the Upper Palatinate Forest, and the Bohemian Forest. The Erzgebirge or Ore Mountains, which run more or less perpendicular to the Thuringian-Franconian Highlands, divide the group into a northwestern and a southeastern part.

The Northwestern Part

The northwestern part of the territory covered by the Vltava-Saale Group is characterized geographically by the transition from the uplands of the Thuringian-Franconian Highlands, the Ore Mountains, and the Hercynian Forest, to the lowlands adjacent to the north (Fig. 7.26).

The Franconian Highlands, a mountain chain approximately 150 km long and 35 km wide with elevations of up to 1000 m, merges into the Ore Mountains in the east. The latter range is around 150 km long and 50 km wide and has a gently rising northwestern slope, whereas the southeastern part is marked by a steep scarp. The maximum elevation is 1200 m. The Hercynian Forest is a low mountain range about 110 m long and 35 km wide. In its western part, it rises to an elevation of over 1100 m; its eastern part is over 400 m high on average. The Magdalenian occupation mainly focused on the middle elevations between these uplands and the lowlands to the north. The drainage system of this area includes the Saale and Mulde rivers and their tributaries. Both currents are themselves tributaries of the Elbe river. Most important

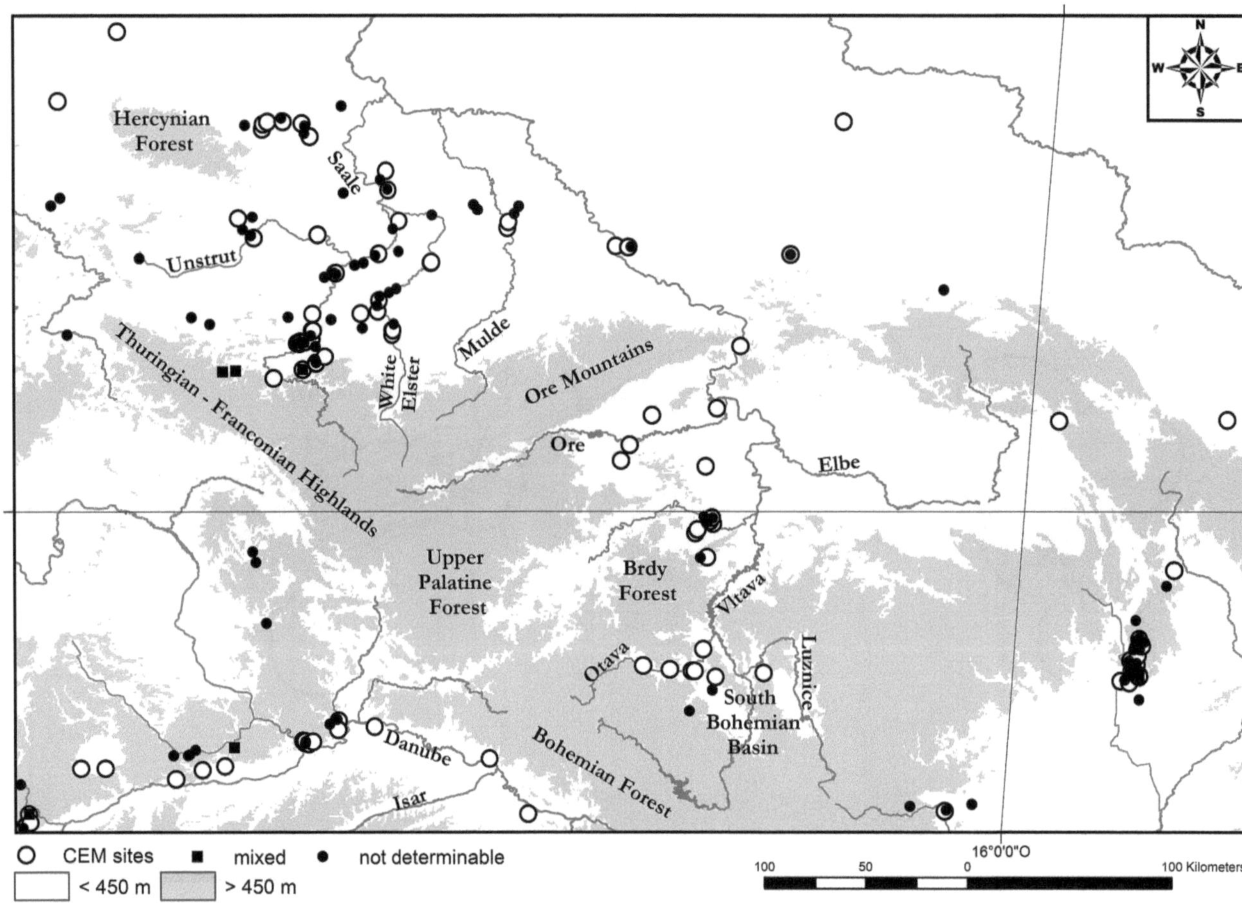

Fig. 7.25 Distribution of sites in the Vltava-Saale Group

for the Magdalenian occupation were the Saale river and one of its tributaries, the White Elster. Both rivers rise in the borderland between the Thuringian Forest and the Ore Mountains and cross the Triassic hill land, where they have cut narrow and sometimes steep valleys into the Lower Triassic (*Buntsandstein*) and Middle Triassic (Muschelkalk) sediments. Further north, they flow through the Leipzig Lowland Bay, which is characterized geologically by loess deposits in its southern part and by moraine deposits in the north (Walter 1995: Abb. 114; Zahn 1996: 74–75). Baltic flint can also be found here. West of the Saale river lies the Thuringian Basin. This region is dominated geologically by Late Triassic sediments (*Keuper*) and is surrounded by Middle Triassic formations. Lower Triassic sediments are characteristic of the outer part of the basin (Walter 1995: Abb. 114). Chert from Late Triassic sediments is often marked by clefts, rendering this raw material rather unsuitable for Magdalenian tool production (Floss 1994: 108). The occupied area is thus characterized by a hilly landscape with some deeply incised river valleys.

The Southeastern Part

South of the Ore Mountains, the geographic situation is characterized by the mountain chains of the Bohemian Massif and the river valleys of the Vltava and its tributaries. Both contribute to a landscape of mountainous regions alternating with smaller basins and river valleys (Fig. 7.27).

The Vltava sources in the Bohemian Forest, a mountain range about 120 km in length and, together with the Bavarian Forest, about 50 km in width. North of the Bohemian Forest, the Vltava crosses the South Bohemian Basin (*Jihočeské pánve*), where it is joined by the Otava and Lužnice rivers. The South Bohemian Basin is characterized by granite and metamorphic rocks, but in the vicinity of Budweis (*České Budějovice*), Cretaceous sediments which potentially yield raw material of good quality can also be found (Zahn 1996: 74–75). The hilly landscape with elevations averaging 400 m above sea level is characterized by numerous lakes. To the north, the basin is limited by the Brdy Forest, a mountainous region which is mainly

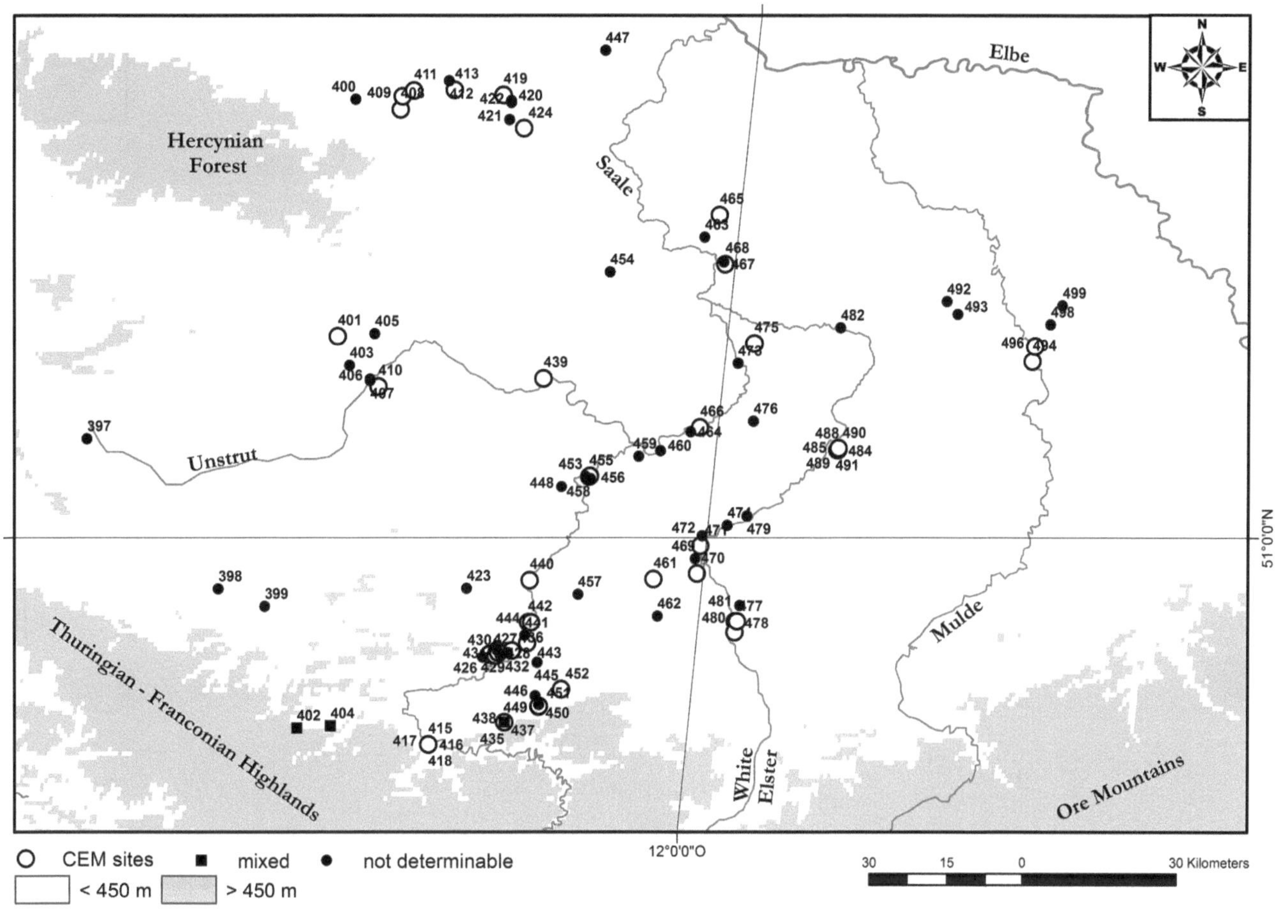

Fig. 7.26 Distribution of sites in the northwestern part of the Vltava-Saale Group

composed of Cambrian and Precambrian rocks. Here the Tok Mountain, the highest point in central Bohemia, rises to an elevation of 865 m. The North Bohemian tableland between the Berounka river, the Ore Mountains, and the Elbe (Labe) river is dominated in its eastern part by Cretaceous sediments, whereas the western part is marked by Tertiary sediments and volcanic rocks (Walter 1995: Abb. 127). The region is traversed by the Ohře (Eger) river, a tributary of the Vltava.

The palynological evidence for the Late Pleniglacial and the Oldest Dryas in the area of the Vltava-Saale Group is rather thin. Nevertheless, on the basis of information from the remaining parts of the investigated area, the following assumptions can be made. So far, the environmental settings of all of the regions discussed have been characterized by treeless steppe-tundra or tundra-steppe, with variation only in the relative abundances of grasses and herbaceous plants. East of the line marked by the Hercynian Forest, the Thuringian-Franconian Highlands, the Upper Palatinate Forest, and the Bohemian Forest, however, the character of the landscape changes. High proportions of *Pinus* pollen which make up more than 60 % of the pollen spectra indicate that stands of pines were probably quite abundant in the landscape during the Late Pleniglacial. The decline in temperatures during the Oldest Dryas, probably accompanied by a reduction in precipitation, led to a decrease in pine growth, as is indicated by the lower ratios of *Pinus* pollen in the respective spectra. At the same time, the amount of *Betula* pollen increased considerably. Together, the ratio of arboreous pollens in the spectra of the Oldest Dryas amounts to more than 50 %. Thus, it can be assumed that despite a deterioration in the conditions for tree growth, scattered stands of pines and perhaps birch trees grew at sheltered stands having a favorable microclimate. With the amelioration of the Bølling Interstadial, pines once again spread and were presumably fairly abundant in the area of the Vltava-Saale Group. In the end it can be said that the Magdalenian occupation of the area of the Vltava-Saale Group took place in an environment where trees, especially pines, were probably a common feature of the landscape (see Chap. 5, "Climate and Vegetation"). With regard to the faunal assemblages, it is clear that horse was the most important game in this area. Reindeer, on the other hand, was noticeably less important. The skeletal remains of bovids, ibex, woolly rhinoceros, and

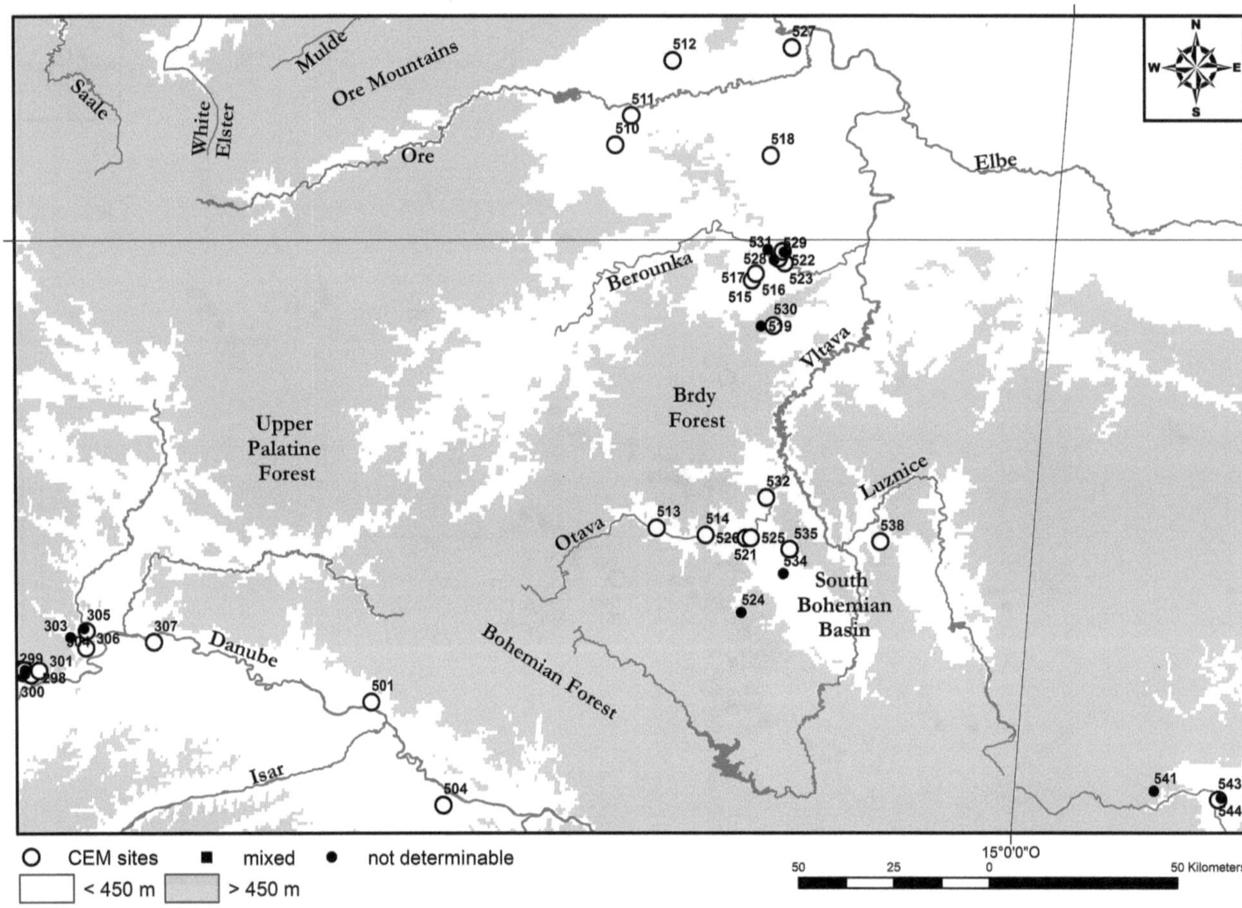

Fig. 7.27 Distribution of sites in the southeastern part of the Vltava-Saale Group

mammoth in CEM assemblages also show the presence of these species in the area during the Magdalenian. Remains of red deer, row deer, wild boar, and saiga antelope have only rarely been reported, and to date, no finds of moose or chamois have yet been mentioned (see Chap. 5, "Fauna").

Location of Sites and Estimation of the Group's Territory

On the basis of the spatial distribution and the geographic situation discussed above, the sites in the Vltava-Saale Group can be subdivided into two major sub-clusters. The first is located in the northwest, mainly along the rivers of the Saale and White Elster, while the second is situated along the Vltava river and its tributaries.

The northwestern cluster (Fig. 7.26) extends approximately 200 km from east to west and about 150 km from north to south. Its southern border is marked by the Thuringian Forest and the Ore Mountains, while its northern limit coincides with the northern slopes of the Hercynian Forest. In the east, the cluster is delimited by the Hainich hill chain, and the Mulde river can be seen as the western border. As in the case of the groups discussed previously, the inner structure of the site distribution in the northwestern cluster appears to be influenced mainly by the course of the large rivers of this region. Most of the sites are located along the currents of the Saale river and its tributaries, most notably the White Elster. Here or more specifically at the site of Groitzsch (ID 483–491), a high-quality raw material outcrop has been reported (Hanitzsch 1972). Further to the north, the site density decreases. Notably, no sites have been reported so far from the lowlands between the Elbe river or the onset of the low mountain ranges a few kilometers north of Halle and Leipzig. In the Thuringian Basin, the site distribution is rather thin. The few sites located in this region are mainly situated in a small area with special geographic characteristics. At about 11°10'00"E and 51°18'00"N, the Unstrut river has cut through the hill chain of Hainleite and Finne, forming a narrow bottleneck situation, suitable for hunting purposes. The well-known horse hunting camp of Bad Frankenhausen (ID 401) (Feustel 1977; Küßner 1998) is located here, as are other sites whose assignment to the CEM remains in question. The only group of sites east of the White Elster is situated on both sides of the Mulde river in the area around the city of Wurzen. However, for most of the sites, the allocation to the CEM is likewise undeterminable. The northernmost sub-cluster is situated along the currents of the Bode and the Holtemme, which have cut narrow valleys into the northeastern slope of the Hercynian Forest; these valleys open toward the northern lowlands, forming bottleneck situations. In addition to offering a strategically favorable hunting situation at the transition from the uplands to the lowlands, the Baltic flint occurring in the moraine deposits of the lowlands represented a source of high-quality raw material. Most of the sites in the northwestern cluster are located in the hilly, slightly elevated parts of the landscape. The highlands of the Thuringian Forest, Ore Mountains, and Hercynian Forest as well as the lowlands in the northern part of this area appear virtually empty of Magdalenian occupation. With a median elevation of 173 m and a standard deviation of 83 m, the sites are located at comparatively low altitudes.

The sites of the southeastern cluster are mainly located in the Bohemian part of the Czech Republic along the Vltava river and its tributaries (Fig. 7.27). Following the course of the Vltava, there is a strong north-south trend in the spatial distribution of this cluster. It spans about 270 km from the southern slopes of the Bavarian Forest to the northern slopes of the Ore Mountains, while its east-west extent is only about 60 km. The distribution of sites does not form a continuous pattern, but is subdivided into several smaller clusters. This separation may reflect surveying and research activities to a certain extent. However, it is probably also a result of the local geography and the structure of the landscape, in which less elevated areas alternate with mountainous regions.

The two southernmost sites assigned to the Vltava-Saale Group, namely, Metten (ID 501) and Vilshofen-Kuffing (ID 504), are situated south of the Bohemian Forest and close to the Danube valley, where the Isar river joins the Danube. Here, it was possible to gather the high-quality chert of Ortenburg and Flintsbach. North of the Bohemian Forest, another agglomeration of sites is situated in the western part of the South Bohemian Basin, the so-called Budweis Basin (*Českobudějovická pánev*). With the exception of Breznice, all sites of this sub-cluster lie comparatively evenly spaced on the western side of the Vltava, along the valleys of the Otava river and its tributary the Blanice. This region, in which elevated areas alternate with lakes and river valleys, was likely a habitat for various plants and animals during the Magdalenian and probably offered many well-suited hunting stands.

Further to the north, the next group of sites is located in the area around the city of Beroun near the confluence of the Berounka and Litavka rivers, about 20 km before the former joins the Vltava to the south of Prague. Here, in the vicinity of the northeastern slope of the Brdy Forest are patches of Ordovician, Silurian, Devonian, Carboniferous, and Cretaceous sediments (Walter 1995: Abb.102; Zahn 1996: 74–75). The narrow valleys at the transition from the Brdy Forest to the North Bohemian lowlands were probably attractive hunting areas for Magdalenian hunter-gatherers. Rather exceptional in this context is the position of Hostim (ID 530), the largest open-air camp thus far excavated in Bohemia. It is located on the southeast-facing slope of the Brdy, overlooking large parts of the Vltava valley not far from where the Litavka river cuts a broad pass through the Brdy Mountain chain. In contrast to the abovementioned sites, which all have the character of hunting camps, the site of Hostim has yielded more than 1000 lithic tools with a diverse typological spectrum and may thus represent a base camp.

In the North Bohemian Tableland, there is a fourth group of sites located at comparatively large distances from each other. Most of the sites are located along the Valley of the Ore (Ohře or Eger) river and its tributaries. The sites of Keblice (ID 527) and Rytiřrská (ID 533) lie at similar geographic positions. Both are situated at the southern entrances to narrow valleys, where the Elbe (Labe) breaks through mountain ranges and thereby forms pronounced bottleneck situations. In the case of Keblice, the range is the Central Bohemian Uplands, and in that of Rytiřrská, it is the Elbe Sandstone Mountains. North of the Ore Mountains, the three sites of Niederlommatzsch (ID 505), Wantewitz (ID 509), and Baselitz (ID 506–509) are situated between the northwestern and the southeastern cluster. In this region, Baltic flint occurs in moraine deposits.

The median value of the site elevation in the southeastern cluster is 339 m with a standard deviation of 97 m and is thus about twice as high as the median elevation in the northwestern cluster. The estimations for the territory's area are 47,000 km^2 if an exploitation radius of 25 km is assumed for each site (Fig. 7.28) and 75,000 km^2 if a radius of 40 km is assumed (Fig. 7.29). Taking into account the virtual absence of sites in the elevated parts of Bohemia and the information obtained from the raw material procurement pattern, a territory estimation based on these parameters results in an area of approximately 40,000 km^2 (Fig. 7.30).

Land-Use Pattern

The spatial position of the sites, the estimated territory of 40,000 km^2, and the estimated average distance traveled during a residential move (ARMD) of 22 km result in a subdivision of the area into three exploitation zones.

The first exploitation zone coincides with the northwestern site cluster discussed above (Fig. 7.31). It covers an area of about 18,000 km^2 and is thus comparatively large. Given the uniformity of the occupied landscape and the

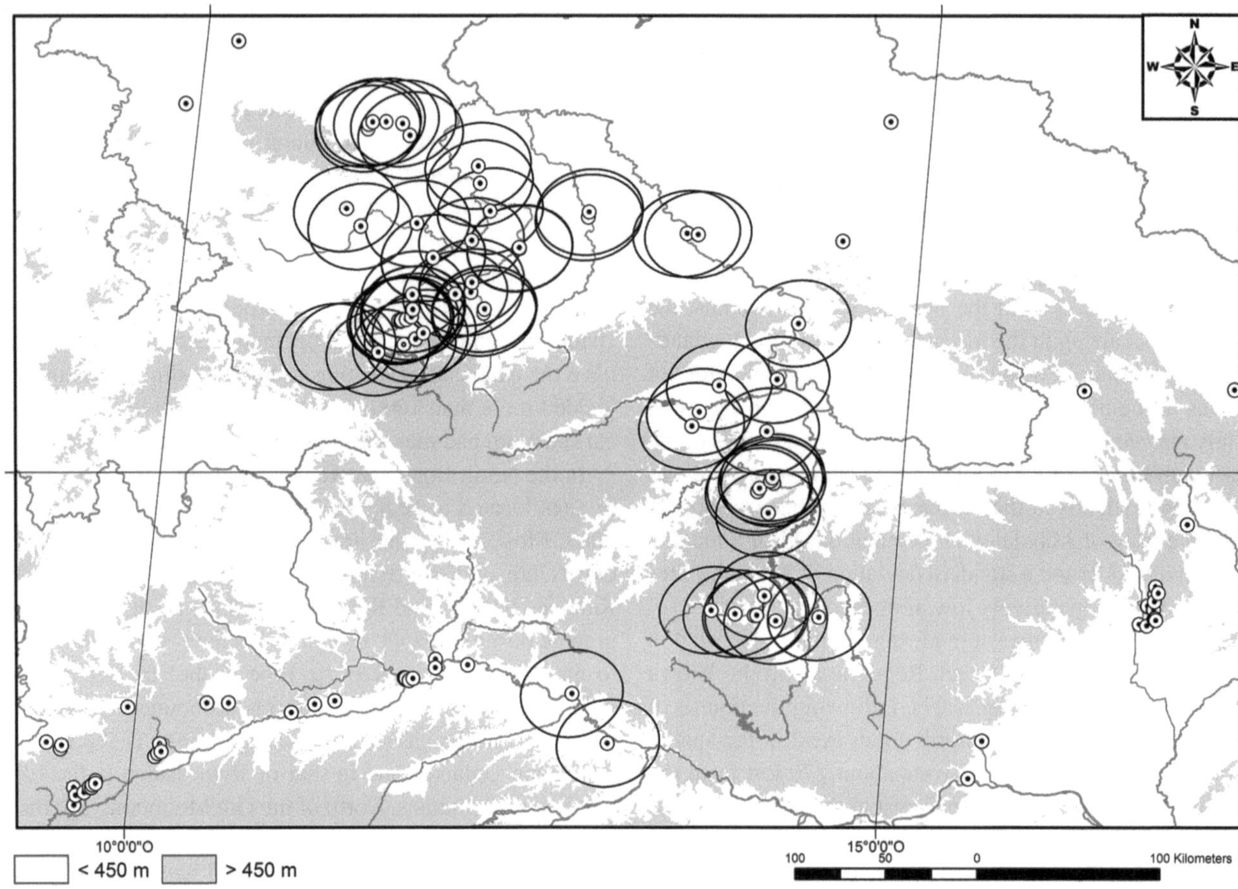

Fig. 7.28 Vltava-Saale Group, territory estimation with 25 km radius around each site

rather unstructured pattern of the site distribution, it is highly likely that this zone represents a palimpsest of several areas that were occupied multi-annually and are no longer distinguishable from one another.

Here, raw material procurement was mainly carried out on a local scale. The hunter-gatherers exploited chert, which could be gathered in the vicinity of some sites and in the gravel beds of the Saale and White Elster rivers. They also frequently used Baltic flint from the moraine deposits of the Elster as well as Saalian Glaciation in the northeastern part of the exploitation zone. The archaeological findings suggest that these raw materials were of sufficiently high quality for the purposes of CEM hunter-gatherers, and long-distance imports were therefore unnecessary. The same seems to be true in the central exploitation zone of the Meuse-Rhine Group, where Meuse gravels from local outcrops were used for tool production (see Chap. 7, "The Meuse-Rhine Group").

The second exploitation zone is located further to the southeast (Fig. 7.32). In its northern part, it comprises the Dresden Basin, where the Elbe valley widens again after the river has passed the Elbe Sandstone Mountains. To the south, it continues up to the northeastern part of the Brdy Forest and includes the eastern part of the Ore (Ohře or Eger) valley. In total, it covers an area of around 14,000 km^2.

As is the case for the northwestern zone, many assemblages yield artifacts made on Baltic flint. The closest source of this raw material can be found north of the Elbe Sandstone Mountains. Because it is ubiquitously in the northern part of the zone, it is difficult to determine whether its presence at the southern sites is the result of seasonal migration within one exploitation zone or whether it reached the place of discard in the context of a change of multi-annually occupied areas.

The southernmost exploitation zone covers an area of approximately 10,000 km^2. It comprises the Budweis Basin, where the rivers of Otava and Lužnice join the Vltava, the Bohemian Forest, and the area of the so-called Gäuboden, where the Isar joins the Danube at the southern slope of the Bavarian forest (Fig. 7.33). Here, in the vicinity of the sites Vilshofen-Kuffing and Metten, raw material from the outcrops at Ortenburg and Flintsbach can be gathered. Artifacts made of these materials are reported at the sites of Hostim, Putim, and Žd'ár 1 (Přichystal 2006: 357).

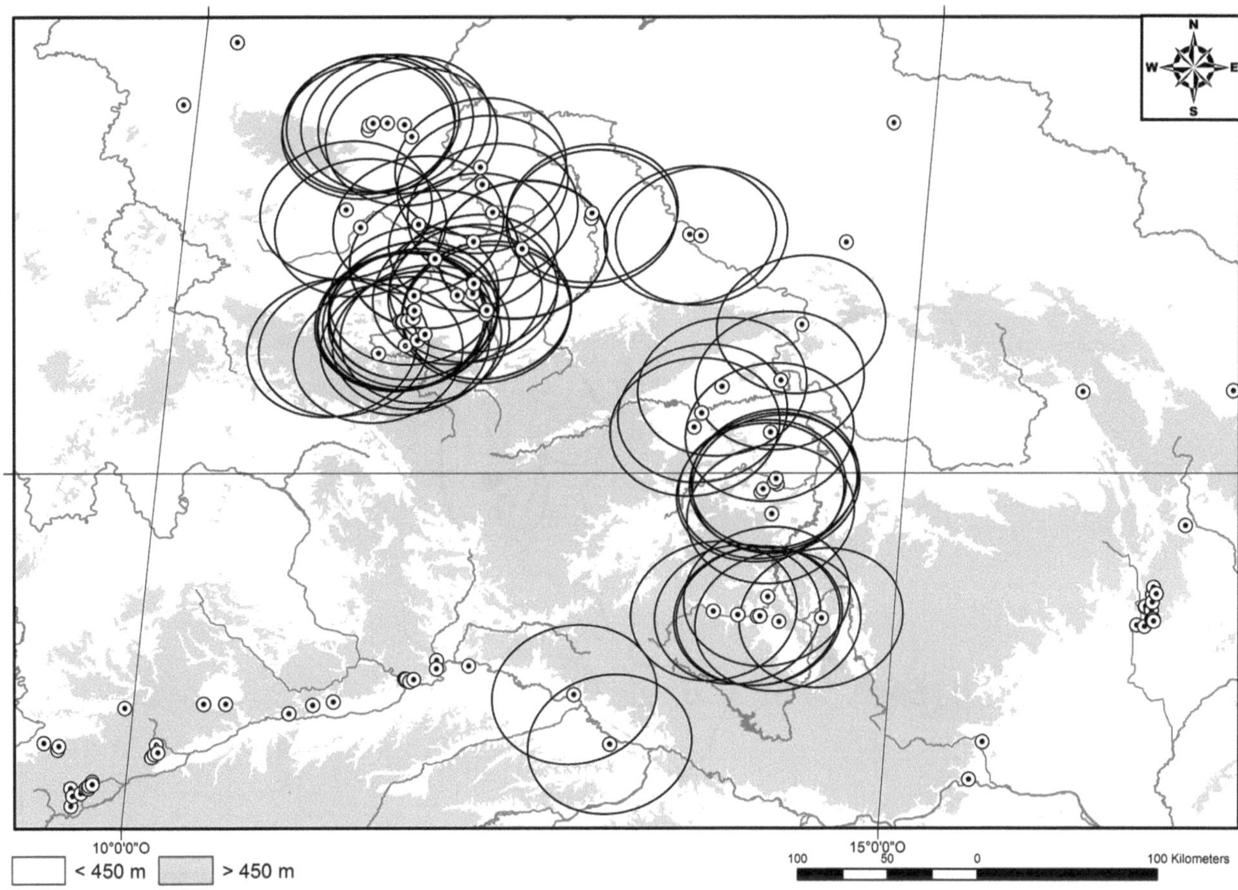

Fig. 7.29 Vltava-Saale Group, territory estimation with 40 km radius around each site

From the site cluster in the Franconian Alb, which is located only about 60 km to the west, no artifacts of these raw materials have been mentioned. For this reason, the sites of Vilshofen-Kuffing and Metten are assigned to the Vltava-Saale Group, although they are separated from the site in the Otava valley by the approximately 50 km wide massif of the Bohemian and Bavarian Forest. From Metten, there is immediate access to the mountain pass that leads through both the Bavarian and Bohemian Forest via Regen, Zwiesel, and Eisenstein, a route that is traced today by the railway line connecting Munich and Plzen. Finds of Baltic flint in assemblages of the southern exploitation zone may reflect the curation of certain tools (Binford 1979) during at least one seasonal cycle and a subsequent change of exploitation zones. The diversity in the lithic tool spectra of the Vltava-Saale Group, which indicates sites ranging from short-term camps of rather specialized character such as Dobříčany to long-term residential camps such as Hostim, Nebra, or Oelknitz, point to a logistically organized subsistence system.

The Polish-Moravian Group

The fifth and largest regional group included in this study is located in the eastern part of the investigated area. The famous *navettes* from Maszycka Cave, which have been dated to about 18,500 cal BP (15,490 ± 310 ^{14}C BP; Lv-2454) and 17,600 ± 250 cal BP (14,520 ± 240 ^{14}C BP; Ly-2453), are usually considered to be the earliest evidence for a Magdalenian occupation of this area. However, upon reflection, it seems more plausible to assume an earlier date for the first occupation of this area after the LGM (see Chap. 8,

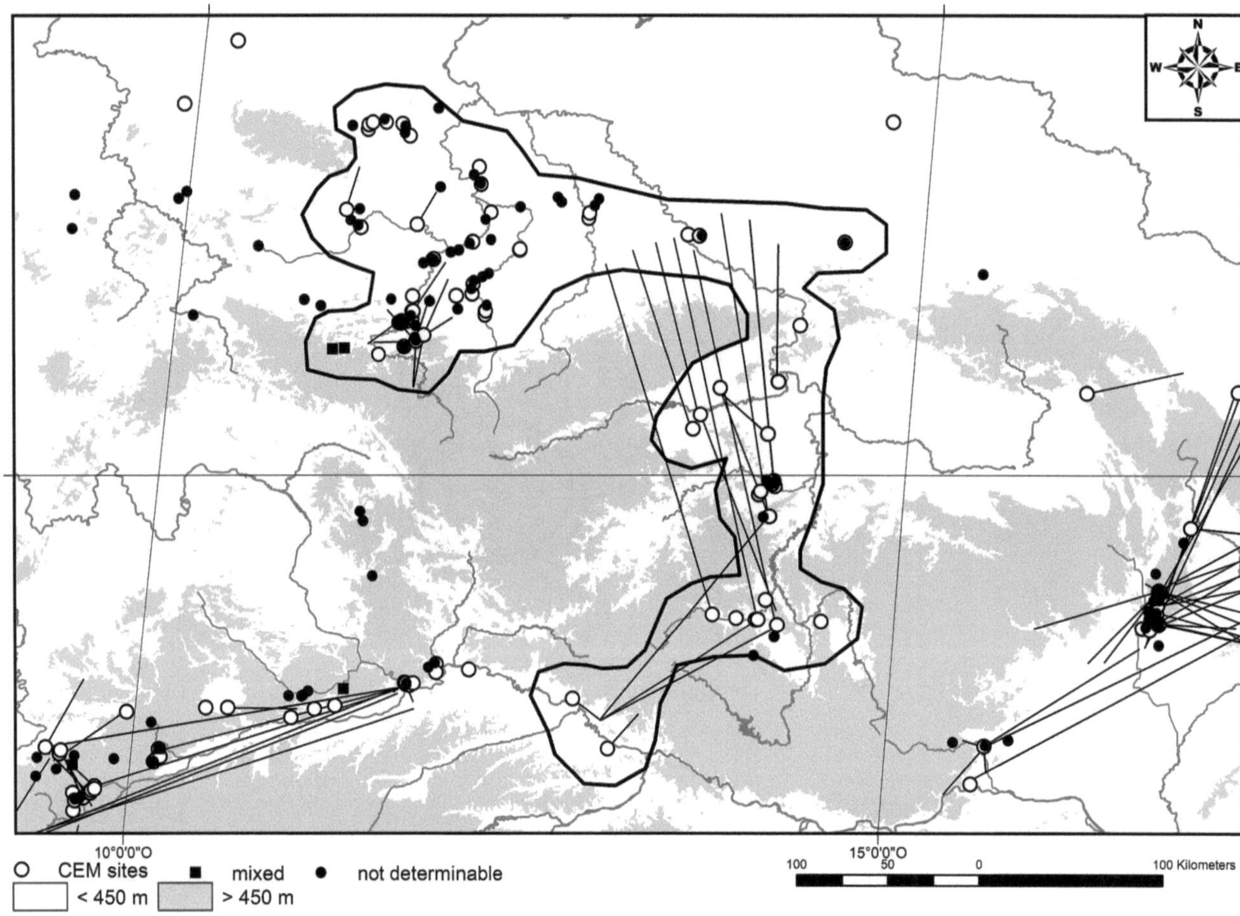

Fig. 7.30 Vltava-Saale Group, territory estimation based on raw material procurement and geographic features

"The Westward Expansion"). Nevertheless, the following reflections refer to the set of sites which are traditionally considered to be Magdalenian in the archaeological literature.

Group Statistics

Despite the large area covered by the Polish-Moravian Group, the number of sites and assemblages assigned to it in the archaeological literature is comparatively low. With 112 assemblages at 103 sites, this regional group places fourth and third, respectively, relative to the other groups (Table 7.7).

After a critical review, 6 assemblages (6 %) have been classified as likely of post-Magdalenian age. In 53 cases (48 %), an evaluation is not possible due to the state of publication or the scarcity of finds (see Table A.1). This study therefore considers 103 inventories from 96 sites. The majority (61 %) of the sites are located in caves or rock shelters, while 39 % are open-air sites and surface collections. This ratio is strongly biased by the geographic conditions of the Moravian Karst, where the site density is highest and almost all known sites are situated at naturally sheltered positions. With only 15 assemblages (14 %) yielding more than 100 lithic tools and just 2 assemblages (2 %) yielding more than 1000, the Polish-Moravian Group presents the lowest values of all regional groups in both categories.

Landscape and Environmental Setting

The settlement area of the Polish-Moravian Group comprises many geographically diverse units, such as the Waldviertel and Vienna Basin, the Moravian Karst, the eastern Sudetes and Carpathians, The Polish Jura and Holy Cross Mountains, and the valleys of the lower Vistula river and its tributaries. The uneven distribution of sites in this area may be a result of surveying patterns and research activities, but it probably also reflects the structure of the landscape with its more and less favorable settlement regions (Fig. 7.34).

The Moravian Karst consists primarily of Devonian limestone (Walter 1995: 287) and measures about 100 km^2 in area. Typical karstic formations such as dolines and caves

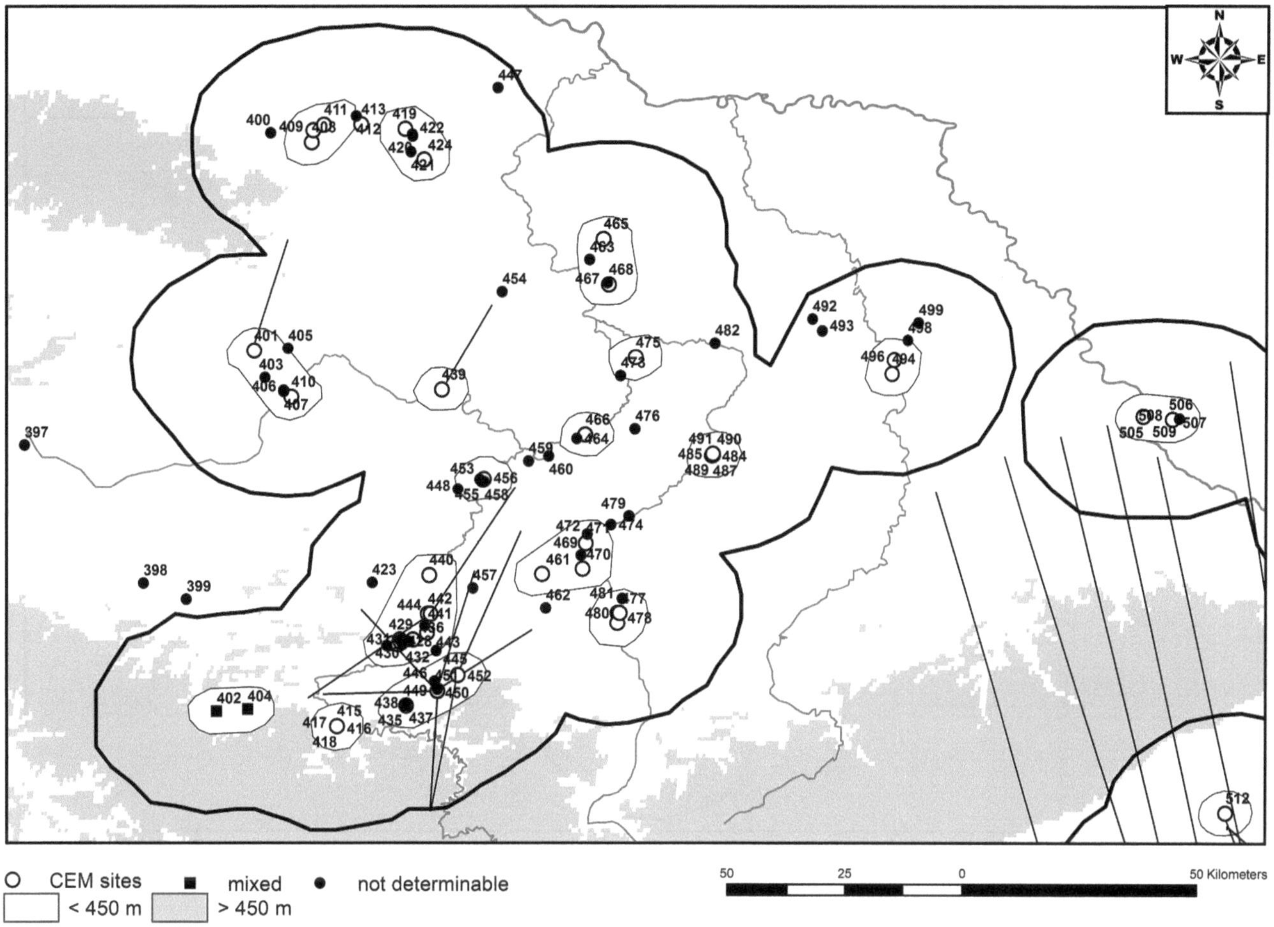

Fig. 7.31 The northwestern exploitation zone of the Vltava-Saale Group

characterize the region. Its most important river is the Svitava. A similar geographic situation can be found within the area of the Kraków-Częstochowa Upland, the so-called Polish Jura. It stretches from Kraków in the south over Częstochowa and up to Wieluń in the north. This Upper Jurassic formation is characterized by limestone escarpments that are more than 500 m thick in some places (Walter 1995: 115), and karstic processes have formed cliffs, valleys, and caves. In the northeast, the Jurassic chain merges into the Miechów depression, separating it from the Holy Cross Mountains. The depression itself is covered with Cretaceous sediments in the south and with moraine sediments in the north (Walter 1995: 117; Zahn 1996: 74–75). In their southwestern part, the Holy Cross Mountains (*Góry Świętokrzyskie*) consist mainly of formations of Cambrian, Ordovician/Silurian, Devonian, and Carboniferous rock (Walter 1995: 110f.). Łyscia and Łysa Góra are the highest elevations in this region, at 612 m and 593 m, respectively. South of the Polish Jura lie the mountains of the High Tatras. This massif is a part of the Beskids, which for their part coincide with the Outer Western Carpathians. Tertiary rocks are the dominant geological formation of the Beskids. However, patches of Jurassic to Cretaceous sediments also appear frequently, especially in the northern part (Zahn 1996: 74–75). The highest elevations rise more than 2,500 m. The largest river in the territory of the Polish-Moravian Group is the Vistula. It sources in the Beskids and along its course collects the water of several rivers coming from the Polish Jura and the Holy Cross Mountains in the north and the Carpathians in the south. One of its major tributaries is the San river. It sources in the western part of the Outer Eastern Carpathians and joins the Vistula close to Sandomierz, east of the Holy Cross Mountains. In its upper part, the San flows through the Tertiary formations of the Carpathians, where patches of Cretaceous sediments appear frequently; Loess and moraine deposits are characteristic of its middle and lower reaches (Zahn 1996: 74–75).

During the LGM, the Carpathian basin was a refuge for a number of boreal trees such as pine, birch, and larch (see Chap. 5, "The Last Glacial Maximum"). Due to climatic amelioration, these species had already spread during the late Pleniglacial and had formed fairly abundant populations

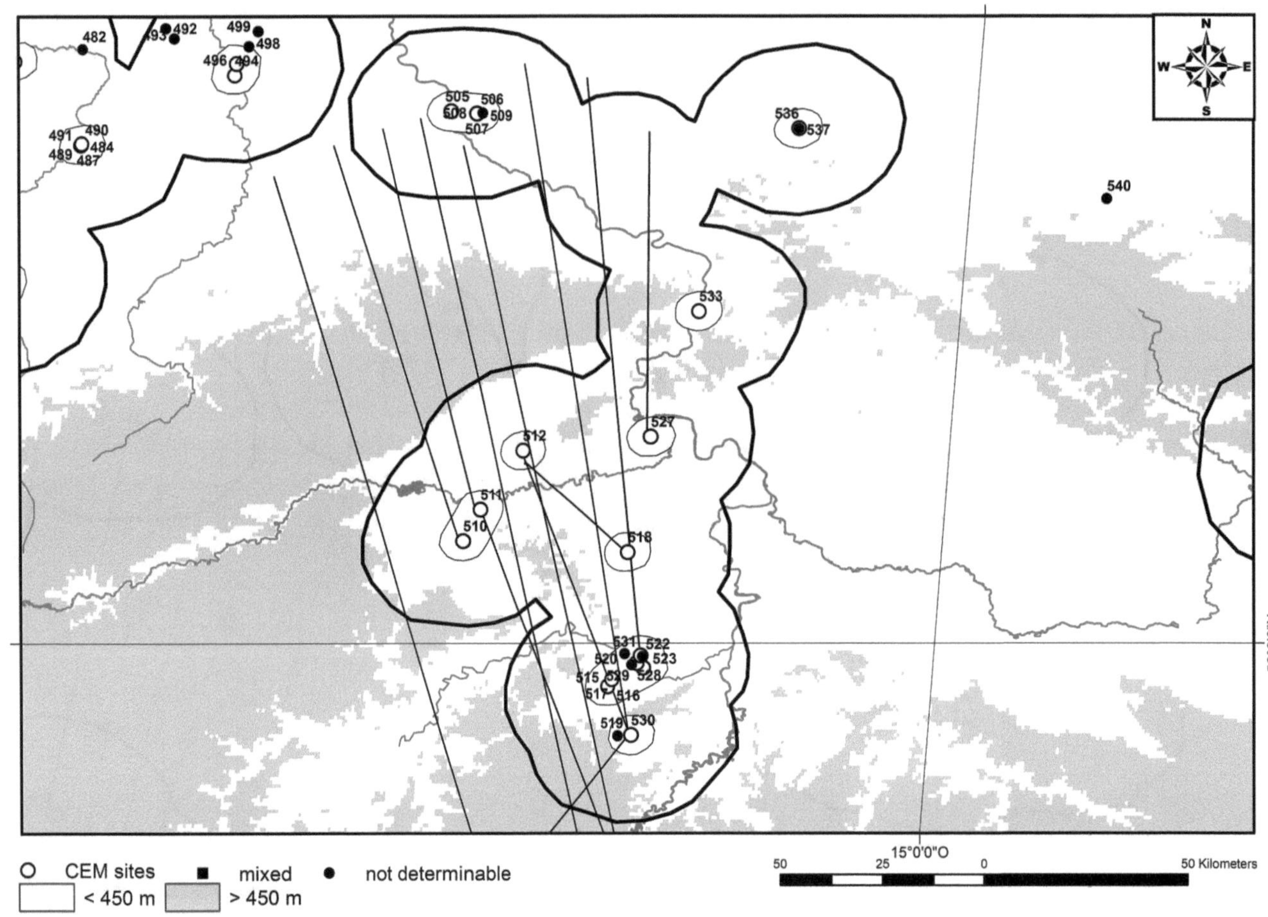

Fig. 7.32 The central exploitation zone of the Vltava-Saale Group

in the area of the Carpathian Mountains. In the Moravian Karst, for example, the pollen profile from Hrabetice shows a ratio of 70 % *Pinus* pollen, indicating extensive occurrences of pine trees in the landscape. North of the Carpathians, pines were still present, but the landscape was probably more of a park-steppe with wide grasslands and scattered pine stands. Although the climatic cooling of the Oldest Dryas had its strongest effects in the western part of Europe, it also affected the vegetation in Central Europe. The decrease in temperature, in all probability paired with a lower precipitation rate, led to a decrease in tree growth and favored the spread of tundra elements in the area of the Waldviertel and Moravia as well as the increase of grasslands north of the Carpathians. However, in contrast to the western part of the investigated area, trees persisted in the eastern regions. The ratios of *Pinus* pollen lie between 30 % and 50 % on average, whereas the percentage of *Betula* pollen ranges between 10 % and 70 % depending on the local habitat situation (see Fig. 5.5). Thus, more or less abundant pine and birch stands probably grew along the courses of braided rivers in protected valleys and other favorable stands. During the Bølling Interstadial, the landscape was characterized by pines and birch trees which grew in a landscape with varying ratios of grasses and herbaceous plants. The most hunted game in the Polish-Moravian Group was horse, the remains of which dominate the faunal assemblages of the group especially in the northeastern part of its territory. Reindeer remains mainly occur south of the Odra river, where the number of individuals hunted frequently equals that of horse (see Chap. 5, "Horse and Reindeer Remains"). This border seems to coincide approximately with the transition from the southwestern part of the territory, which was characterized by tundra elements and a comparatively moderate tree growth during the Oldest Dryas, to the northeastern part, which was probably more of a park-steppe at that time. Apparently, horse was a more successful competitor in the latter environment and as a consequence was also more abundant. The other species in the faunal assemblages are highly diverse: after horse and reindeer, the most frequently occurring species is woolly rhinoceros. Red deer, ibex,

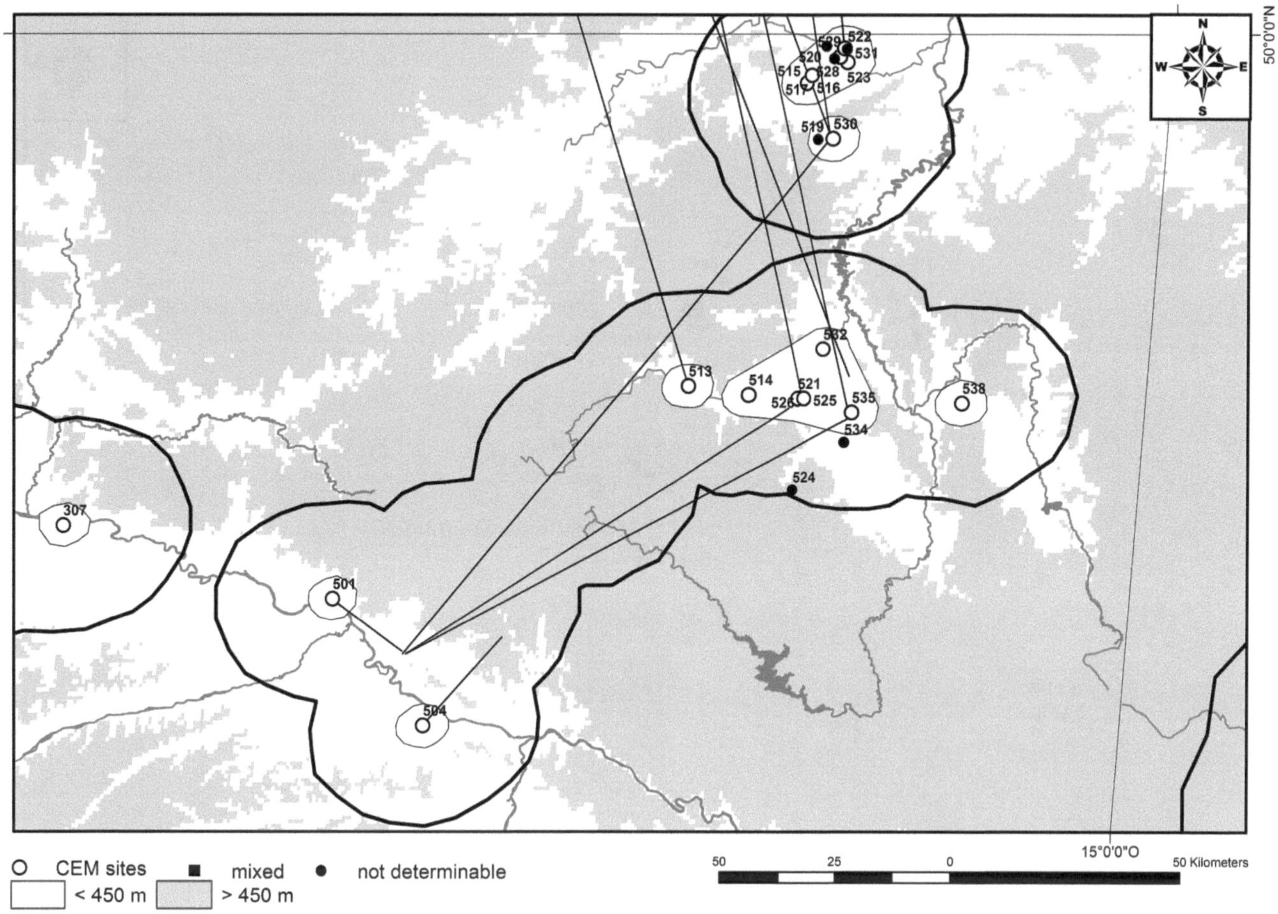

Fig. 7.33 The southern exploitation zone of the Vltava-Saale Group

chamois, mammoth, and finds of bovids have also been reported. Interestingly, finds of saiga antelope are only reported from the Deszczowa Cave and the Maszycka Cave north of the Odra river. This highlights the gradient from the rather tundra-like conditions in the southwest to the dryer, more steppe-like condition in the northeast.

Location of Sites and Estimation of the Group's Territory

The sites of the Polish-Moravian Group show a very inhomogeneous distribution (Fig. 7.34). The southwesternmost sites are situated in the Waldviertel. Besides the assemblages of Kamegg (ID 543) and Gudenushöhle (ID 542), which can very probably be attributed to the Magdalenian, finds from the sites of Frauenlucke (ID 541), Teufelslucke (ID 545), and Graselhöhle (ID 544) are sometimes also discussed in this context. All five sites are located in the area of the Kamp and Krems rivers, both of which are tributaries of the Danube. The relatively narrow valleys widen toward the Danube valley, thus forming natural bottleneck situations suitable for hunting purposes. The region with the highest density of sites in the entire investigated area is the Moravian Karst. The marked relief of the landscape and the high number of natural shelters apparently offered a favorable environment for Magdalenian hunter-gatherers.

Further east, the sites of Hranice (ID 599), Čertova dira (ID 604), and Šipka dira (ID 603) lie within the Moravian Gate, a natural pass approximately 65 km long and 15 km wide between the Eastern Sudetes and the Western Carpathians. Its northeastern part features outcrops of Cretaceous sediments can be found (Zahn 1996: 74–75). Further to the north, a group of sites clusters in the vicinity of the Troja and Zinna rivers, left bank tributaries to the Odra river, which in turn sources north of the Moravian Gate. Loessic sediments predominate south of the Zinna river, while moraine deposits become more frequent on its northern side. The sites of Cyprzanów 3 (ID 602) and Dzierżysław 35 (ID 601), among others, are located here.

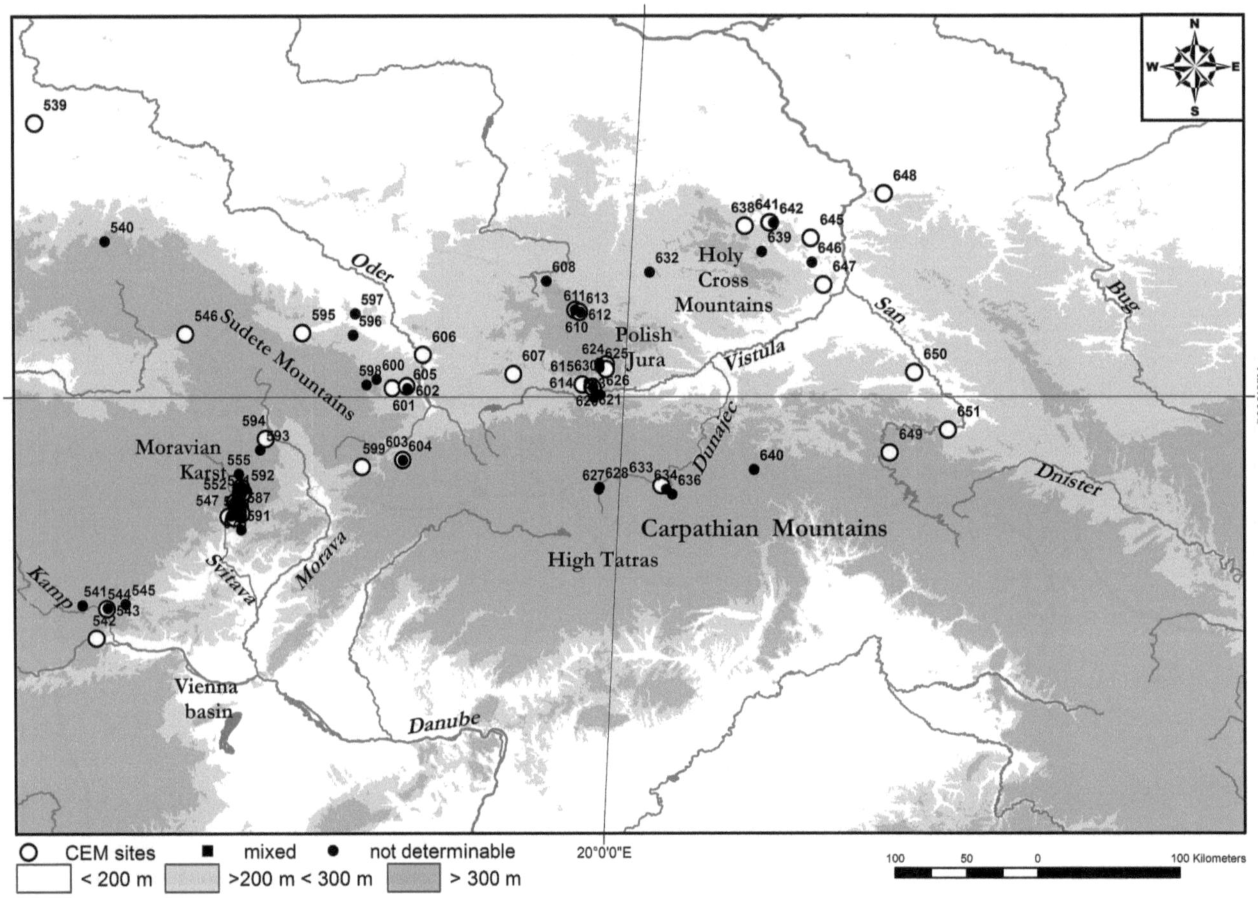

Fig. 7.34 Distribution of sites in the Polish-Moravian Group

The diversified landscape of the Polish Jura, where elevated parts alternate with deeply incised valleys that cut through the limestone massif and open toward the relatively wide plain of the Vistula valley, provided suitable conditions for CEM hunter-gatherers. Sites such as Maszycka Cave (ID 629), Puchaczowa Skála (ID 630), Zalas (ID 615), or Zawalona (ID 619) are located here. Between the Polish Jura and the Holy Cross Mountains lies the Miechów depression, in which no Magdalenian finds have been reported to date. The only exception may be the site of Kozłow 4 (ID 632), where three lithic tools have been found and attributed to the Magdalenian. Within the area of the Holy Cross Mountains, Magdalenian occupation is restricted to the lower, northeastern slopes, where Jurassic limestone frequently occurs (Walter 1995: Fig. 52; Zahn 1996: 74–75). Most of the sites are located along and around the Kamienna river, while Wilczyce (ID 647) is situated further to the south at the Patowka valley. Both rivers are left bank tributaries to the Vistula.

The presence of Volhynian flint in many assemblages of the Polish-Moravian Group shows that its territory extended at least as far as the Bug river in the east (Połtowicz 2006). Artifacts made of Volhynian flint have been found, for example, in Klementowice-Kolonia (ID 648), which is the most northeastern Magdalenian site known so far (Wiśniewski 2003); others have also been found at Grodzisko Dolne (ID 650) and Hłomcza (ID 649), two sites which, together with Przemysl 3 (ID 651), form the southeasternmost site group of the Polish-Moravian Group. They are located along the San river on the northern slope of the Carpathian Mountains. Hłomcza is located at a valley crossing, where a small tributary joins the San. At Przemysl, the San broadens its previously narrow valley and forms a bottleneck situation. The situation at Grodzisko Dolne is again similar to that at Hłomcza: located at the confluence of the Wislok and the San, the site overlooks the comparatively wide valley. Thus, all of the sites lie at positions which are favorable for hunting purposes.

About 70 km south of the Polish Jura, the sites of High Tatras sub-cluster are situated along the headwaters of the Dunajec river. The site of Katy 1 (ID 633) is situated close to the river outlet of the broad Upper Dunajec valley. The sites of Sromowce Niżne (ID 634) and Aksamitka Cave (ID 636) are also located in the vicinity. A notable feature of the Polish-Moravian Group's site distribution is the fact that the broad Vistula valley between Kraków and Sandomierz

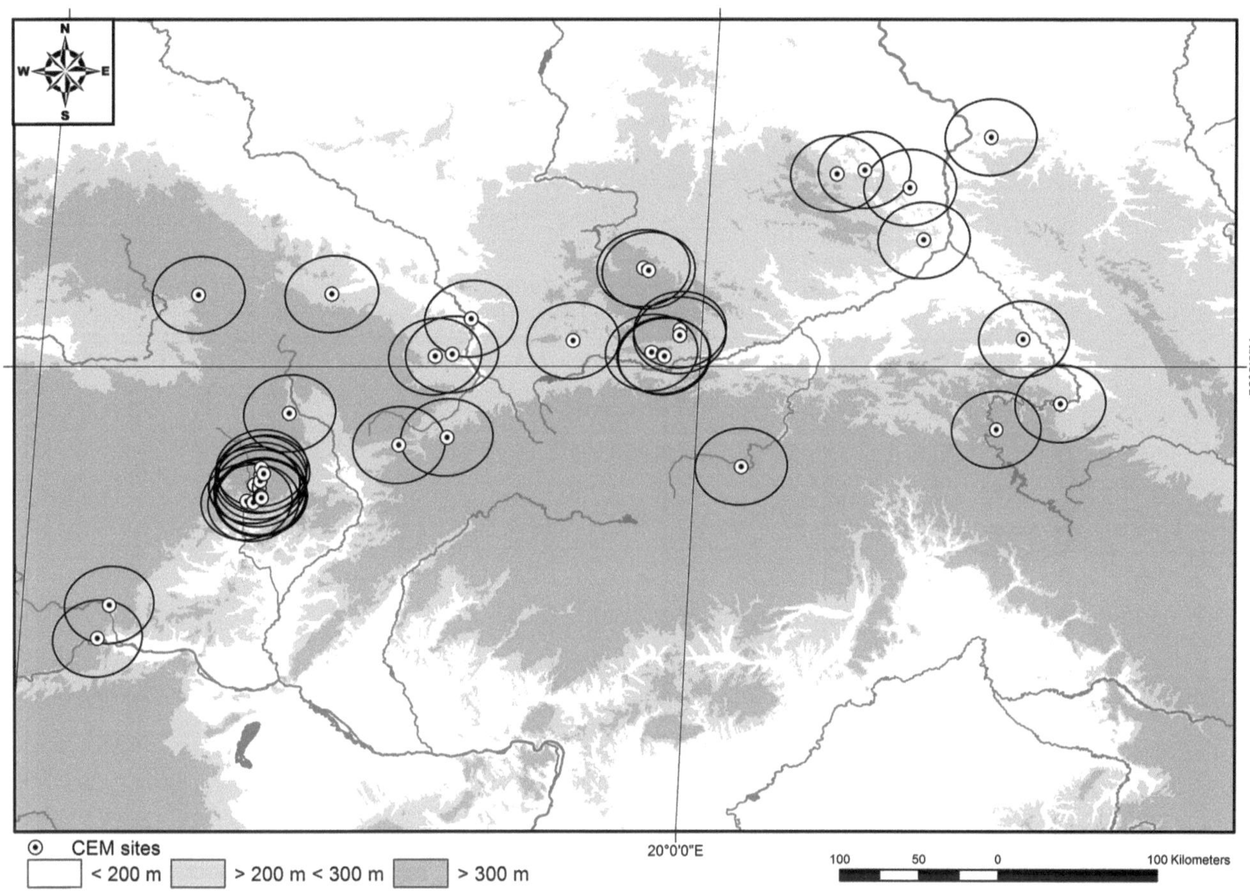

Fig. 7.35 Polish-Moravian Group, territory estimation with 25 km radius around each site

together with the lower reaches of the San is virtually empty of Magdalenian occupation. Similar observations can be made for the Rhône-Saône valley and the Rhine valley north of Bonn.

The median elevation of the sites in the Polish-Moravian group is 360 m AMSL with a standard deviation of 120 m. If an exploitation radius of 25 km around each site is assumed (Fig. 7.35), the total exploited territory for the Polish-Moravian Group is about 61,000 km^2. Increasing the radius to 40 km (Fig. 7.36) results in an estimation of 92,000 km^2. Taking into account the locations of the sites, geographic features such as the course of the major rivers and mountain massifs as well as the information obtained from the analysis of the raw material procurement (Fig. 7.37), an extent of 84,000 km^2 is estimated for the territory.

Land-Use Pattern

With an estimated extent of 84,000 km^2, the territory of the Polish-Moravian Group is the largest in the whole investigated area. At the same time, the raw material procurement pattern reveals its sites to be intensely interconnected, and the typical transportation distances for raw materials are comparatively large. Assuming that raw materials were not procured directly or exchanged in a trade system, but were acquired embedded in other daily activities, it can be concluded that the procurement pattern reflects the displacement of hunter-gatherer groups (Binford 1979: 259; Floss 1994: 325; see also Chap. 6, "Procurement Pattern of Lithic Raw Materials"). The fact that all areas of the territory are connected with each other by means of raw material transports is the main argument for conceiving of the Polish-Moravian Group as a unit in this study.

The group's ARMD is the largest in the entire investigated area at 30 km. The current state of knowledge does not allow a determination of whether this reflects prehistoric reality or is a product of research activities. Nevertheless, a larger distance traveled between one residential camp and another correlates well with the observation that raw materials were transported over larger distances. On the basis of the raw material procurement pattern and the geographic position of the sites, the Polish-Moravian Group is subdivided into six exploitation zones (Fig. 7.38). Noting that the north-south procurement distances are generally shorter than those

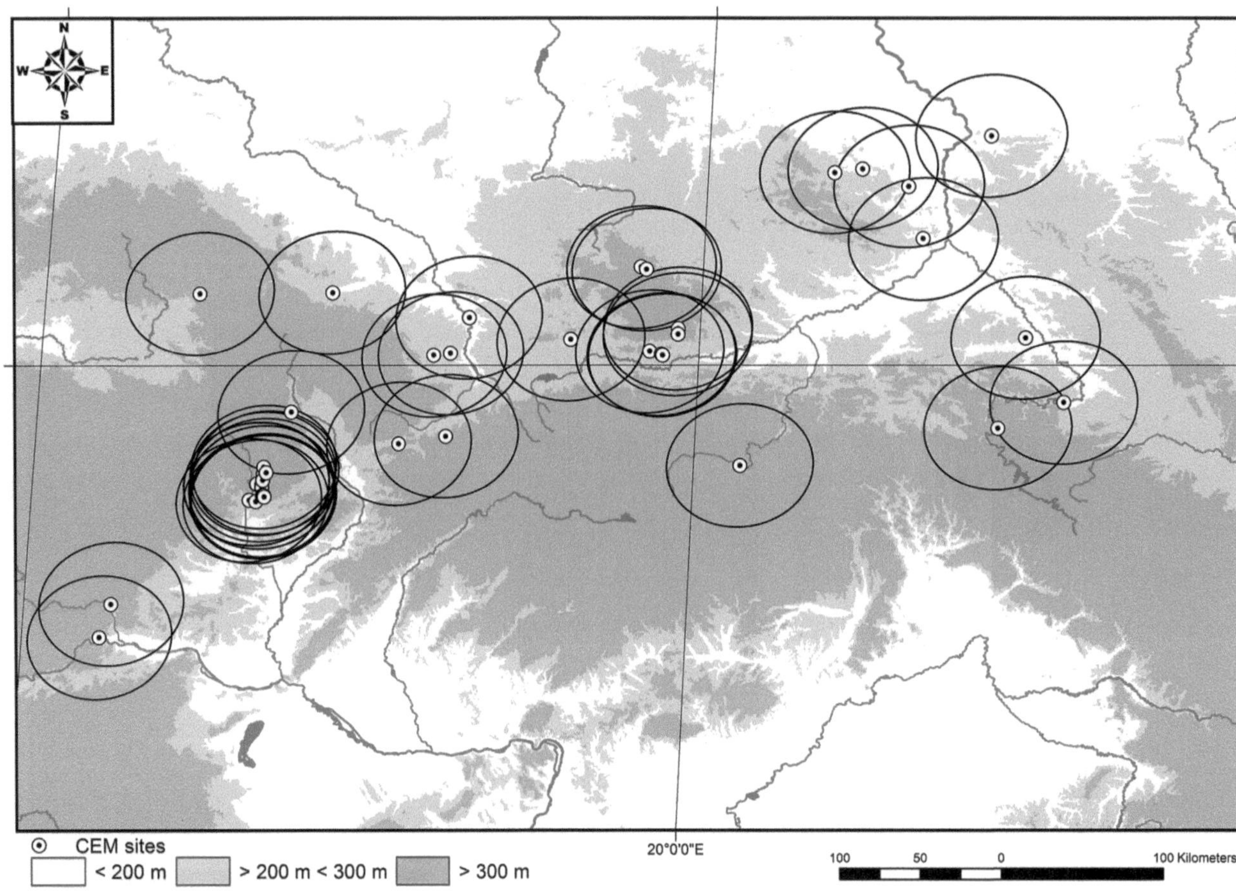

Fig. 7.36 Polish-Moravian Group, territory estimation with 40 km radius around each site

oriented east-west, it is assumed that the north-south moves mainly reflect seasonal migrations within one exploitation zone, while long-distance east-west migration mirrors the move from one exploitation zone to another.

The southwesternmost exploitation zone mainly comprises the area of the Waldviertel and the Vienna Basin (Fig. 7.39). With an extent of around 8,000 km^2, it is relatively small. At the sites of Kamegg (ID 543) and Gudenushöhle (ID 542), raw materials from the Moravian Karst, the area of the Moravian Gate, and the Polish Jura have been reported (Brandtner 1955; Połtowicz 2006; see Table A.9). Depending on the assumed migration routes, artifacts of material from the latter may have been curated for the duration of two or three multi-annual cycles.

The closest sites of the next exploitation zone are located about 110 km to the northeast, in the Moravian Karst. This zone also comprises the eastern part of the Sudetes and the lowlands adjacent to the north. It covers thus an area of about 15,000 km^2. Baltic flint, which can be found in many assemblages from the Moravian Karst, probably comes from the moraine deposits of the northern lowlands. Thus, it could have reached its place of discard within two or three consecutive residential moves. Additionally, finds of radiolarite from the western Carpathians and chert from the Moravian Gate, the Polish Jura, and the region of the Holy Cross Mountains have been excavated here. To the east, the third exploitation zone comprises an area of roughly 11,000 km^2 and includes the western Carpathians, the Moravian Gate, and the area of the Odra valley to the north. The fourth exploitation zone spans about 14,000 km^2 from the Polish Jura in the north to the Dunajec valley in the south (Fig. 7.40). In addition to raw materials which could be found within this region, Świeciechów and Chocolate flint from the Holy Cross Mountains was also used occasionally, and this is the furthest west that artifacts of Volhynian flint have been excavated.

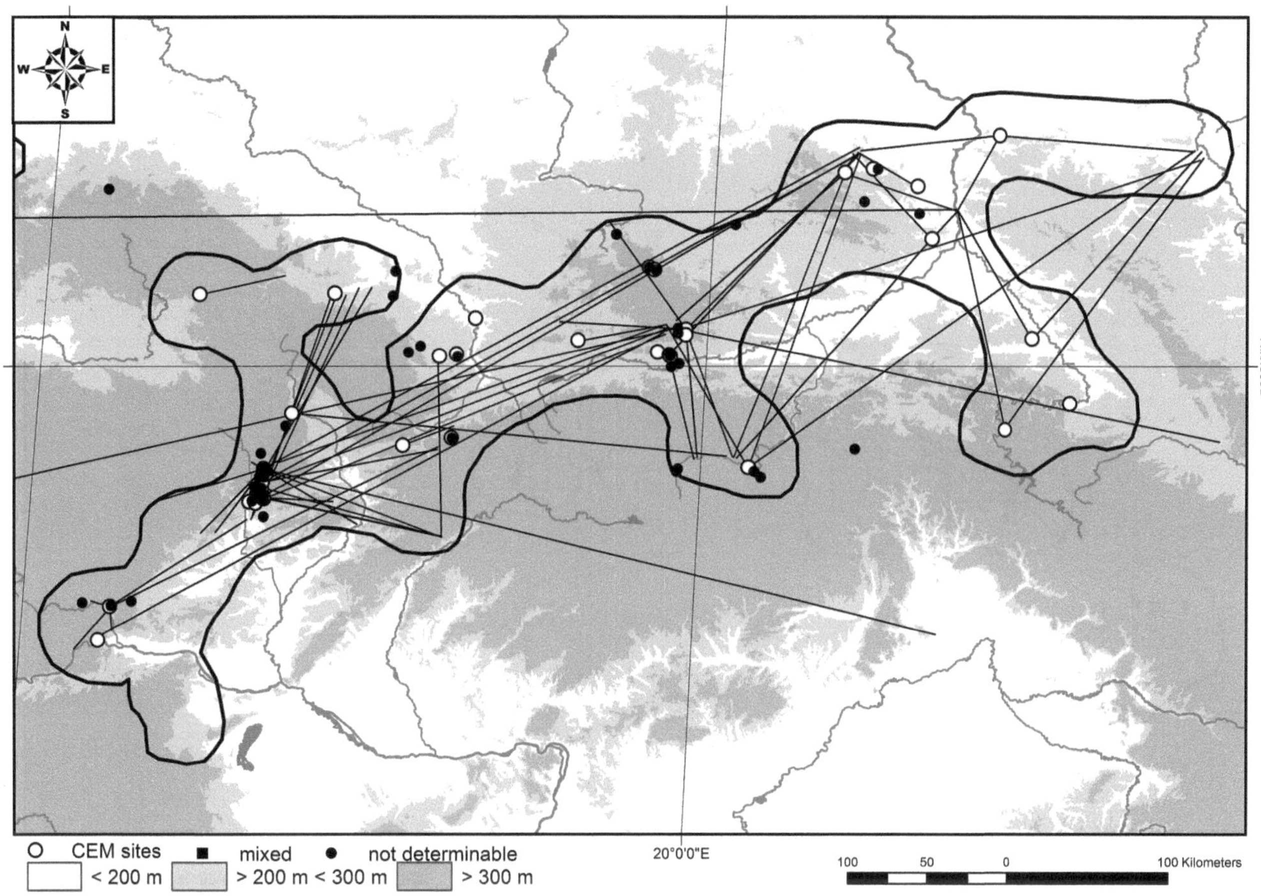

Fig. 7.37 Polish-Moravian Group, territory estimation based on raw material procurement and geographic features

In the area of the Holy Cross Mountains lie the sites of the fifth and northeasternmost exploitation zone, which also includes the outcrops of Volhynian flint in the area of the Bug river due to their spatial proximity and the absence of sites in the area of the outcrops itself. The estimated area covered by this zone is thus approximately 17,000 km^2. Apart from the local raw materials, siliceous rocks from the northern Carpathians were also used. The sixth and last exploitation zone is located along the course of the San river and extends about 8,000 km^2. The finds of Volhynian as well as Świeciechów flint (Kostruhn 2004) were probably brought to this area in the context of a change of exploitation zones.

The subsistence system of the Polish-Moravian Group appears to have been logistically organized. Pronounced differences in the artifact spectra speak in favor of this hypothesis. Sites such as Žitného or Balcarka Cave, where 44 % and 49 % of the lithic tools are backed bladelets, seem to reflect hunting activities. Klementowice-Kolonia, in contrast, is an open-air site which probably represents the remnants of a residential camp.

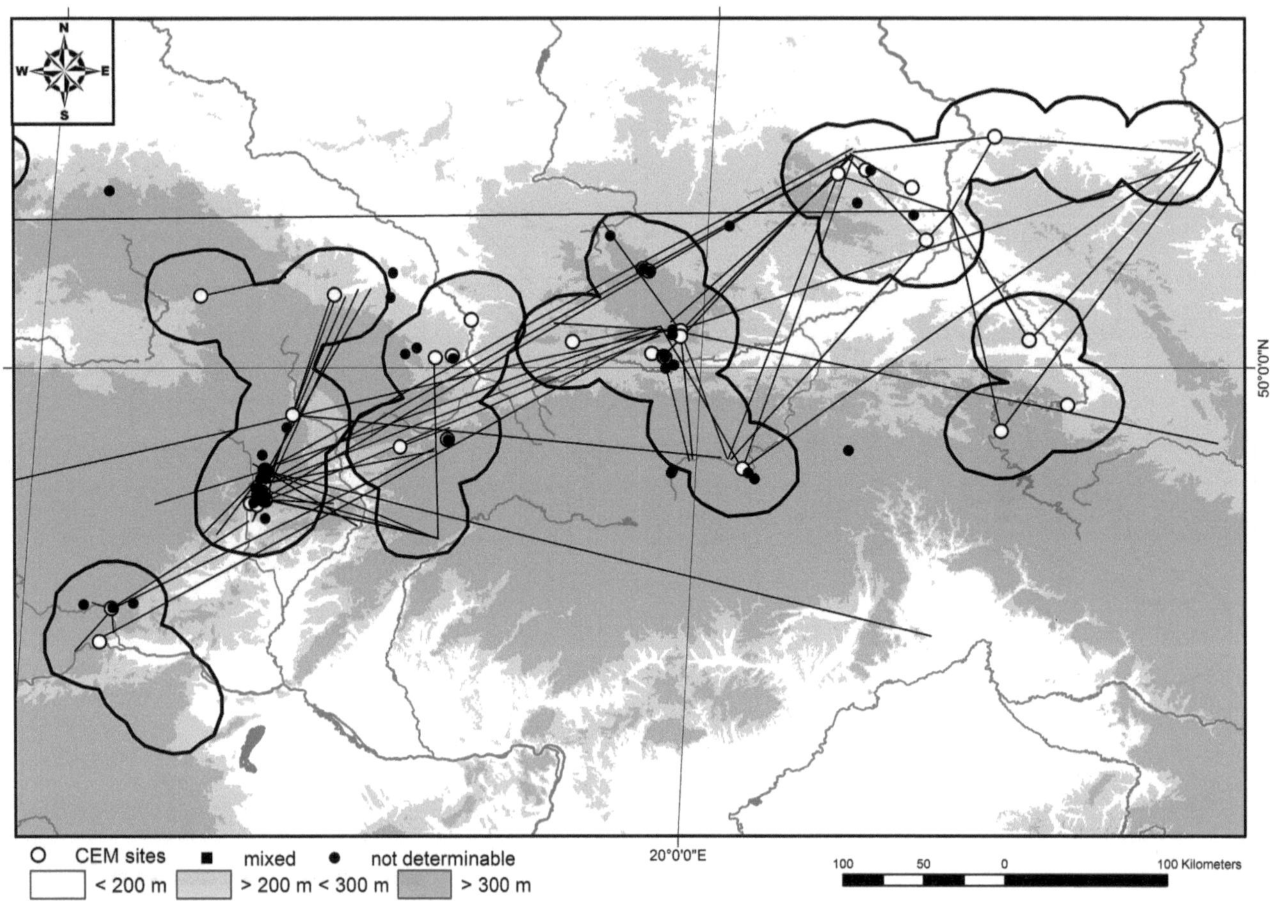

Fig. 7.38 The exploitation zones of the Polish-Moravian Group

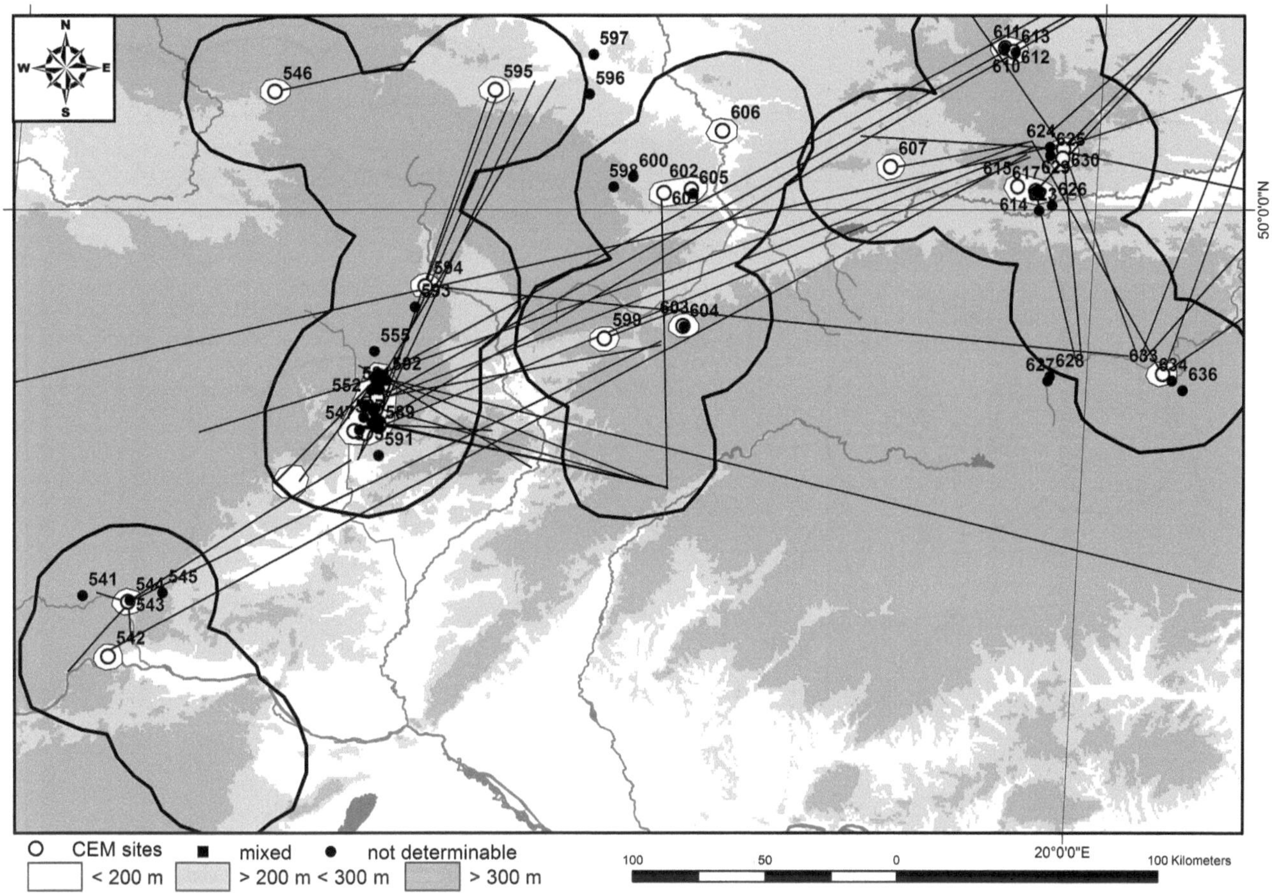

Fig. 7.39 The three western exploitation zones of the Polish-Moravian Group

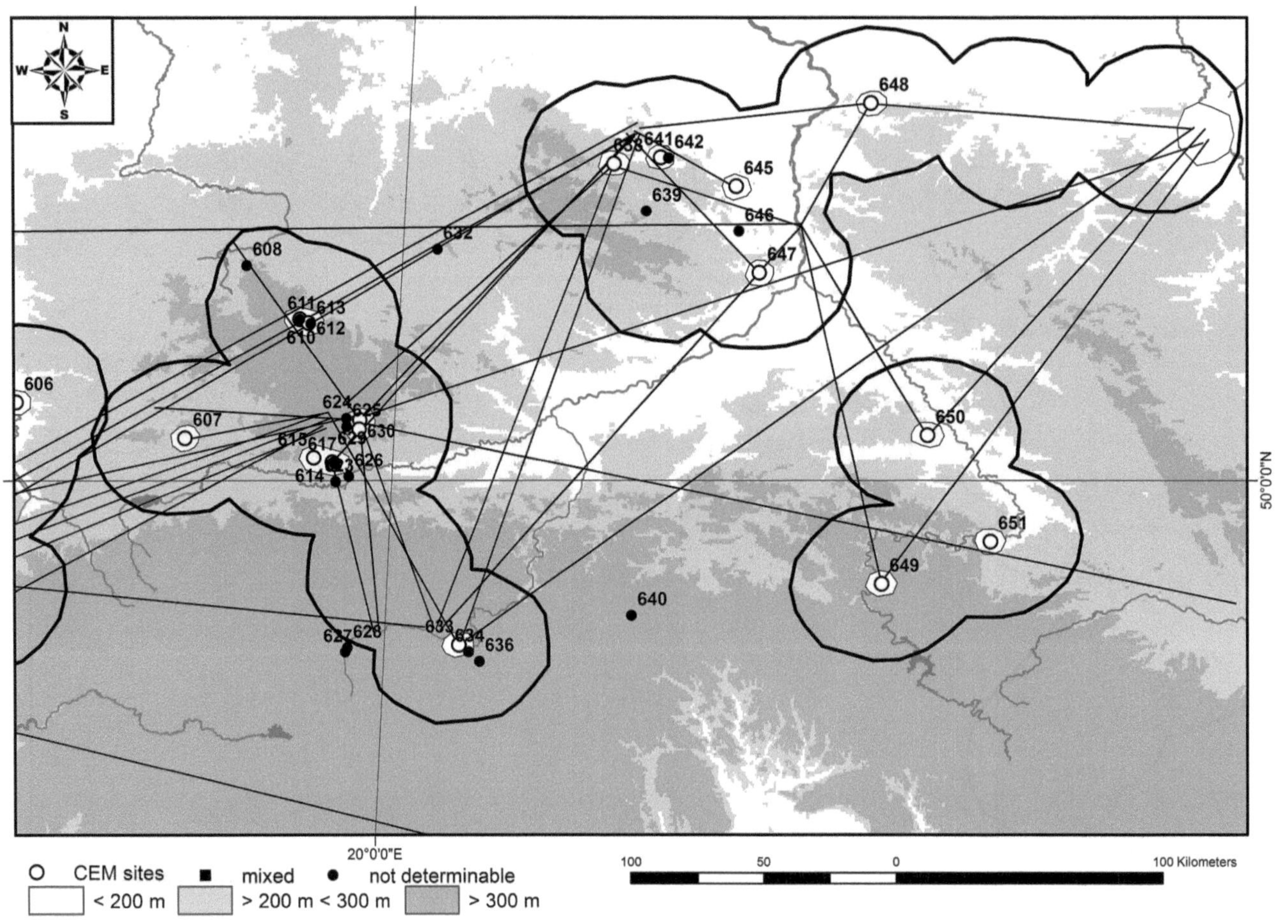

Fig. 7.40 The three eastern exploitation zones of the Polish-Moravian Group

Chapter 8
The Recolonization of Central Europe

One of the most interesting features of the Late Upper Paleolithic is the spatiotemporal dynamic of the recolonization of Central Europe following the Last Glacial Maximum (LGM). The current theory on this process, though widely accepted, is nevertheless plagued by inconsistencies and contradictions. Not all of these contradictions are obvious, but they become particularly evident against the background of the extensive spatial and temporal analyses conducted in the course of this study. Thus, this chapter uses the information obtained during the preceding investigation of the variability in the CEM record to discuss the resettlement process in Central Europe. To do so, the current hypothesis on recolonization after the LGM is presented, and four major objections are raised to it. After identifying the most important contradictions, more consistent arguments will be presented that support a new reading of the archaeological evidence and contribute to a recolonization hypothesis that better accords with the archaeological record than does the current one. Briefly stated, this new hypothesis argues that contact between western and eastern Central Europe already during the LGM triggered a parallel development of the lithic and organic industry in the two areas. In the course of the climatic amelioration process, the previously unsettled parts of Central Europe were recolonized in a bidirectional expansion from the west *and* the east.

The Current Theory and the Major Critiques to Which It Is Subject

Today, the question of how the process of recolonization in Central Europe took place is largely regarded as definitively and conclusively settled in terms of the following hypothesis (Bosinski 1990; Otte 1992; Svoboda et al. 1996; 171ff.; Valoch 1996; Floss and Terberger 2002; Küßner 2009):

A climatic deterioration during the younger half of the Greenland ice-core Stadial 3 (NGRIP GS 3) led to unfavorable environmental conditions for hunter-gatherer groups in Central Europe. Besides low temperatures, there were also strong winds which blew cold and dry air from the northeast, as is indicated by high proportions of K^+-ions in the NGRIP ice cores (Mayewski et al. 1997; Figs. 5.1 and 5.2). The cooling climate, the advancing glaciers, and the concomitant periglacial permafrost soil led to a deterioration of living conditions for humans and thus caused a dramatic drop in population. This demographic decrease in turn led to a rupture in communication between the populations of Eastern and Western Europe during the late Gravettian. According to the current theory, Central Europe remained virtually empty of human occupation for a period of about 10,000 years, and no contact existed between the people east and west of this zone. During this period, these two populations developed independently of one another: in Western Europe there emerged the Solutrean followed by the Badegoulian and eventually the Magdalenian, while Eastern Europe saw the development of the Epigravettian. Eventually, this state of separation ended when expanding Magdalenian hunter-gatherer groups from Western Europe arrived in the Carpathian region and thus reestablished the interrupted contact between east and west. First indications of this postulated eastern expansion are thought to become apparent with the Magdalenian "*à navettes,*" which is dated to the period around 18,000 cal BP. However, it is only with the beginning of the Upper Magdalenian, the Magdalenian IV *sensu* Breuil, that the extensive recolonization of Central Europe is said to have taken place. From the southwestern heartland, hunter-gatherers spread in a northeast direction, moving along the Rhône-Saône corridor arriving at the Upper Rhine valley about 17,400 cal BP (Housley et al. 1997). After this, expansion continued along the Rhine, the Danube, and the Saale, crossing over Bohemia and Moravia as far as the Kraków Jura and the Vistula basin, where the colonization process finally stopped (Fig. 8.1).

This hypothesis of a unidirectional recolonization during which western hunter-gatherers expanded from the

Andreas Maier, *The Central European Magdalenian*, Vertebrate Paleobiology and Paleoanthropology,
DOI 10.1007/978-94-017-7206-8_8

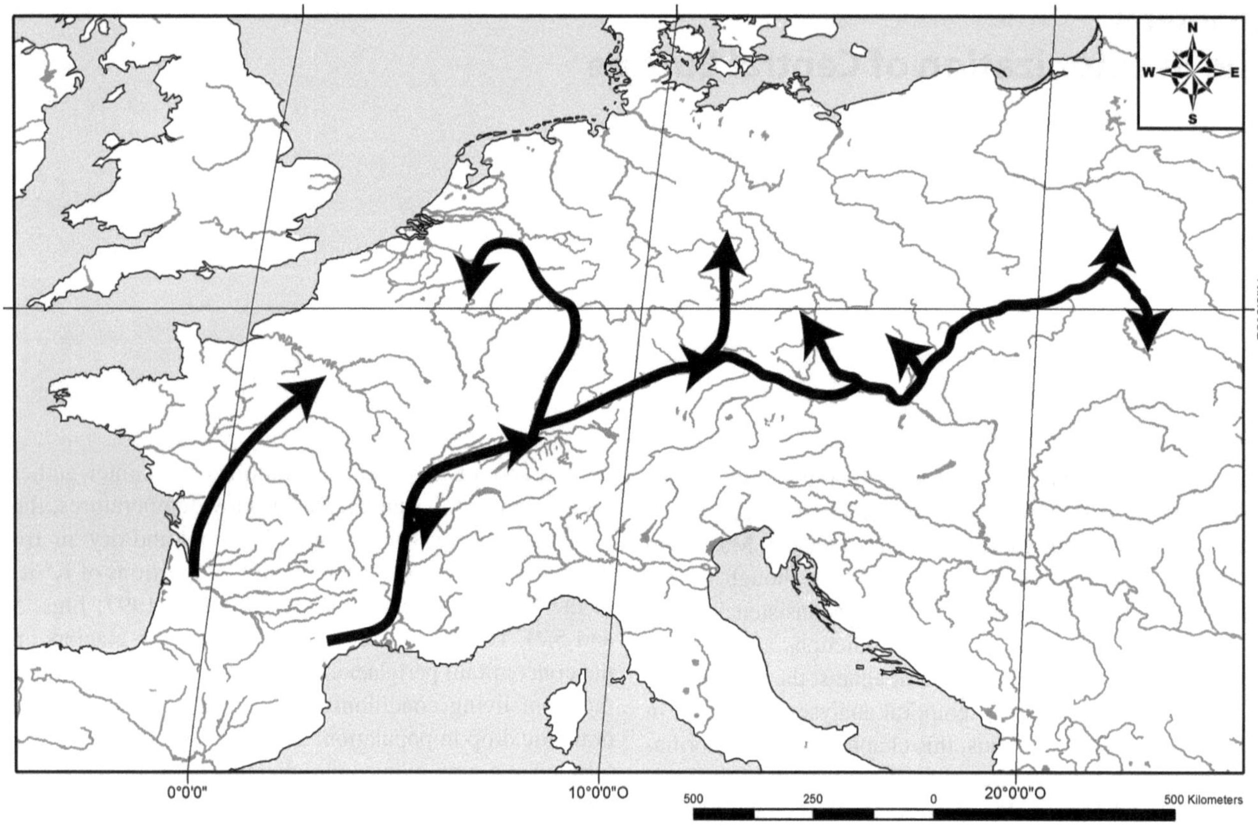

Fig. 8.1 The unidirectional expansion model of the Central European Magdalenian

Franco-Cantabrian region up to the Vistula and San rivers in the East is so widely acknowledged that it virtually became an established fact in Paleolithic research. Karel Valoch notes, for example: "Le Magdalénien représente la seule grande culture paléolithique dont l'origine et l'expansion ne font aucun doute" (Valoch 1996: 128). In other words, the hypothesis of a unidirectional expansion after the LGM became a paradigm in Paleolithic research, and the paradigmatic efficacy of this hypothesis is so strong that even conspicuous inconsistencies are disregarded or neglected in favor of its maintenance.

Paradigms understood as defined by Th. S. Kuhn (1962) are vital to scientific research. They are universal in the sense that they guide and structure the entire scientific process by direct modeling and abstract rules (Kuhn 1996: 47). They determine which problems are addressed and which questions are asked (Ibid.: 27), and they ensure that there are solutions for these problems and questions (Ibid.: 37). Paradigms are a prerequisite for perception itself (Ibid.: 113) as well as for the discovery of laws (Ibid.: 28), and it is only against their background that anomalies become visible (Ibid.: 65). In the case of competing paradigms, their role becomes circular (Kuhn 1996: 94), as each party in a debate argues its position within its own paradigm. Therefore, and because the assumptions that constitute a paradigm are often nonempirical in character, a conflict of paradigms cannot be resolved by proof (Ibid.: 148), unless it can be shown that one of them contains logical inconsistencies. The problem of paradigmatic assumptions was already addressed polemically by D.L. Clarke, who stated that paradigms "die very slowly as their substance is reincorporated in fresh patterns of research" and so "some archaeologists are still content to work in terms of nineteenth-century paradigms where only the inevitable freshness of new material prevents them from actually writing nineteenth-century textbooks" (Clarke 1972a: 8). In the same context, B.G. Trigger remarked critically: "Indeed, archaeologists often seem to build more on what their predecessors concluded about the past than on the evidence on which these conclusions were based" (Trigger 1989: 16). In this regard, the current view on the CEM seems to be indeed in some parts a view through the glasses of nineteenth- and twentieth-century archaeologists and carries some burdening heirlooms from the paradigmatic history of those times. In respect of these reflections, the hypothesis of a long-term rupture between Eastern and Western Europe and a subsequent unidirectional expansion of western hunter-gatherers seems to mirror rather the geopolitical situation of the last century, than it appears to be a stringent conclusion from the archaeological evidence. The idea of a separation of Europe into east and west, with little or no contact between

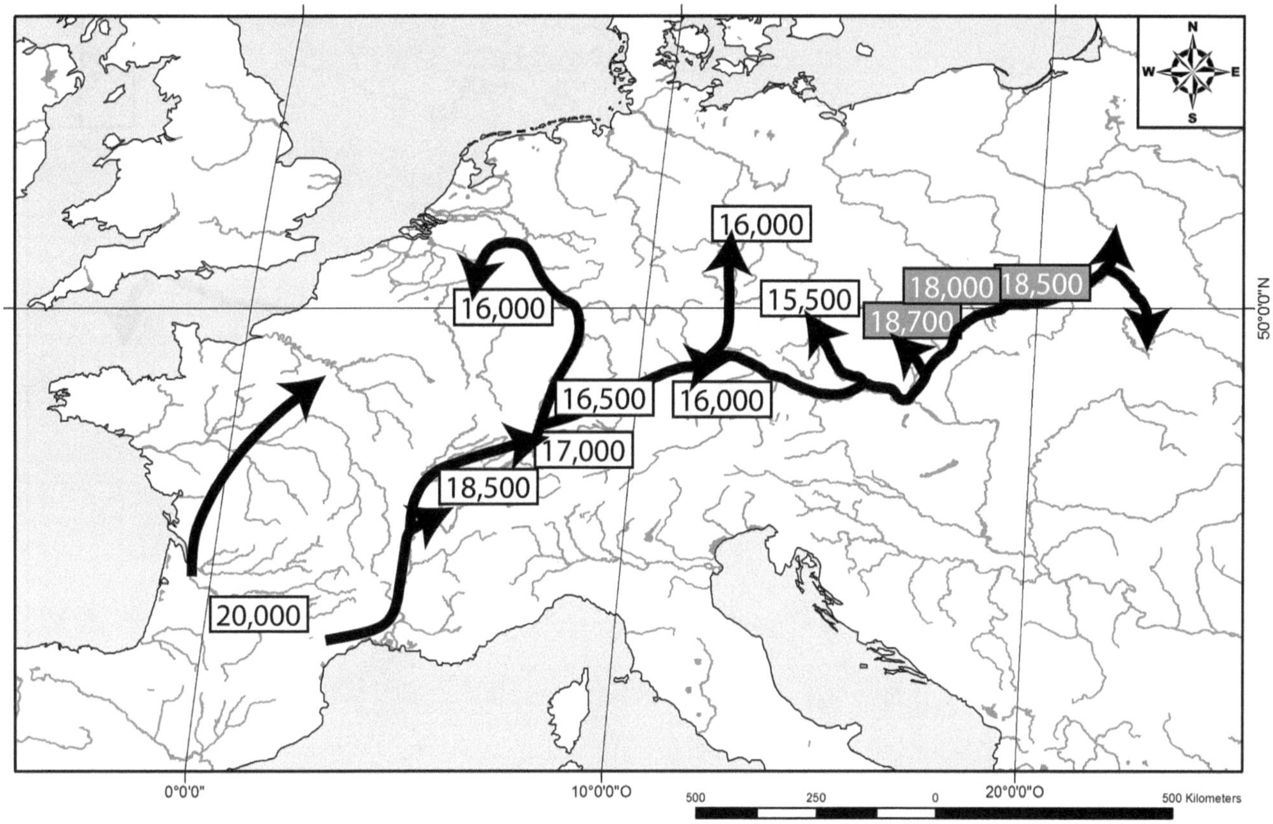

Fig. 8.2 The unidirectional expansion model and contradicting radiocarbon dates. *Numbers* give the calibrated radiocarbon dates before present (cal BP)

the two parts—a daily reality for many, as well as an ideological conception encouraged by the state—probably found its way into scientific paradigms as well (cf. Clarke 1979: 85; Trigger 1989: 12f.). Once the idea of a cultural rupture is accepted, it is only consistent to assume that the development of eastern and western hunter-gatherer groups proceeded separately. However, it need not to follow that the rupture was of long duration or that recolonization was unidirectional; what may seem to be in accord with the archaeological record at first sight turns out to be rather at odds with it on closer consideration. Generally, there are four major objections which can be raised to the current theory:

1. If a unidirectional expansion pattern is assumed, sites should gradually become younger the farther away they are located from the source area in the direction of the assumed expansion.

   Guided by the paradigm's norms, most researchers working on the Magdalenian expect the respective oldest known sites to become increasingly younger the further they are located to the east (e.g., Bosinski 1990). As a consequence, radiometric dates that contradict this view are usually treated as invalid or at least questionable and are therefore neglected in argumentation and theory building (cf. Housley et al. 1997). However, there is a growing body of consistent radiometric measurements of Magdalenian sites from Poland, Moravia, and Thuringia which provide ^{14}C-dates that are older than the oldest dates from Magdalenian sites in the Rhineland and Danube region (Fig. 8.2; see Table A.6)—a fact that is in urgent need of explanation. For instance, the oldest assemblages of the Middle Late Upper Paleolithic in Eastern Central Europe, such as Grubgraben AL1 (ca. 20,000 cal BP) or Brno-Vídeňska (ca. 18,700 cal BP)—two assemblages with a typology equivalent to that of early Magdalenian assemblages—show radiocarbon ages that are similar to those at analogous sites in southwestern Europe. Other examples are the *navette*-bearing site of Maszycka Cave (ca. 18,500 cal BP) and the triangle-bearing site of Dzierżysław 35 (ca. 18.000 cal BP). At the same time, they are older than the oldest assemblages in western the Danube area, namely, Kesslerloch and Schweizersbild (both ca. 17,000 cal BP).
2. When a group of hunter-gatherers migrate into an unsettled area, they can be expected to maintain strong relations with their origin group for various reasons, among them maintenance of subsistence security, the need for marriage partners, the existence of family ties, and the

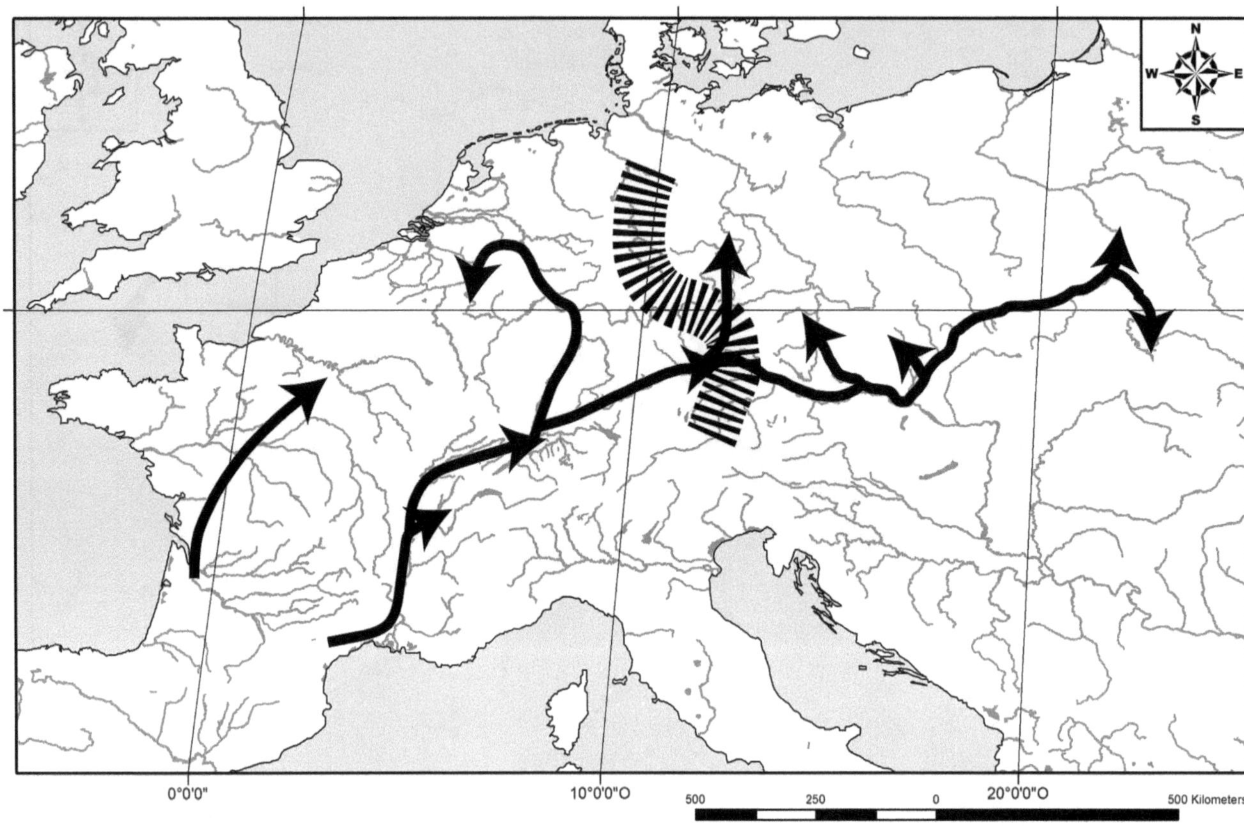

Fig. 8.3 The unidirectional expansion model. The *dashed line* indicates the course of the dividing line which could be observed with regard to interaction patterns and distributions of many typological concepts

desire to exchange information. Consequently intensive interaction between groups in adjacent areas as mirrored in the exchange of concepts and ideas should be observable.

Since "a colonizing hunter-gatherer population would in all likelihood be small" and "[a] group of 25 is fine for foraging, but not for reproduction" (Kelly 2003: 51), it can be expected that an expanding group will maintain strong relations to its origin group. If one accepts a west-east-oriented, unidirectional expansion model, it should therefore be assumed that the easternmost group interacted very strongly with its neighboring groups to the west, and similar concepts should occur most frequently in neighboring groups and should be relatively homogeneous in the whole area. However, the reconstruction of CEM interaction patterns based on long-distant transports of raw materials and mollusk shells reveals some evidence to the contrary. It shows a clear separation of the investigated area into two parts, with only minor interaction between neighboring groups on each side of the dividing line. Sites located around the Jura Mountains, along the Danube, and in the Meuse-Rhine area show a close-meshed interconnection. In contrast, there are almost no traces of such interaction between hunter-gatherers in the Danube and Meuse-Rhine areas on the one hand and in Eastern Germany and Bohemia on the other (see Chap. 6, "Procurement Pattern of Lithic Raw Materials" and "Procurement Pattern of Mollusk Shells and Amber"). The same pattern repeats itself in the distribution of typological and technological concepts (see Chap. 6, "Typological and Technological Variability"). Moreover, except for representations of female figures, which are realized in a similar and standardized way from the Atlantic Ocean to the Dnieper valley and the occurrence of phalliform-sculpted objects, all other typological concepts so far analyzed do not meet this expectation. On the contrary, the distributions of these concepts mirror the same separation that could be observed for the raw materials and mollusk shells or are restricted to single regions or even individual sites (see Chap. 6, "Some Aspects of Central European Magdalenian Sculptures and Engravings"). The results of the analyses thus speak strongly in favor of a western and an eastern community with pronounced internal interaction but comparatively little contact with each other (Fig. 8.3). The western community comprises the sites in France as well as Southern and Western Germany, and the eastern community en_ compasses assemblages in Eastern Germany, Bohemia,

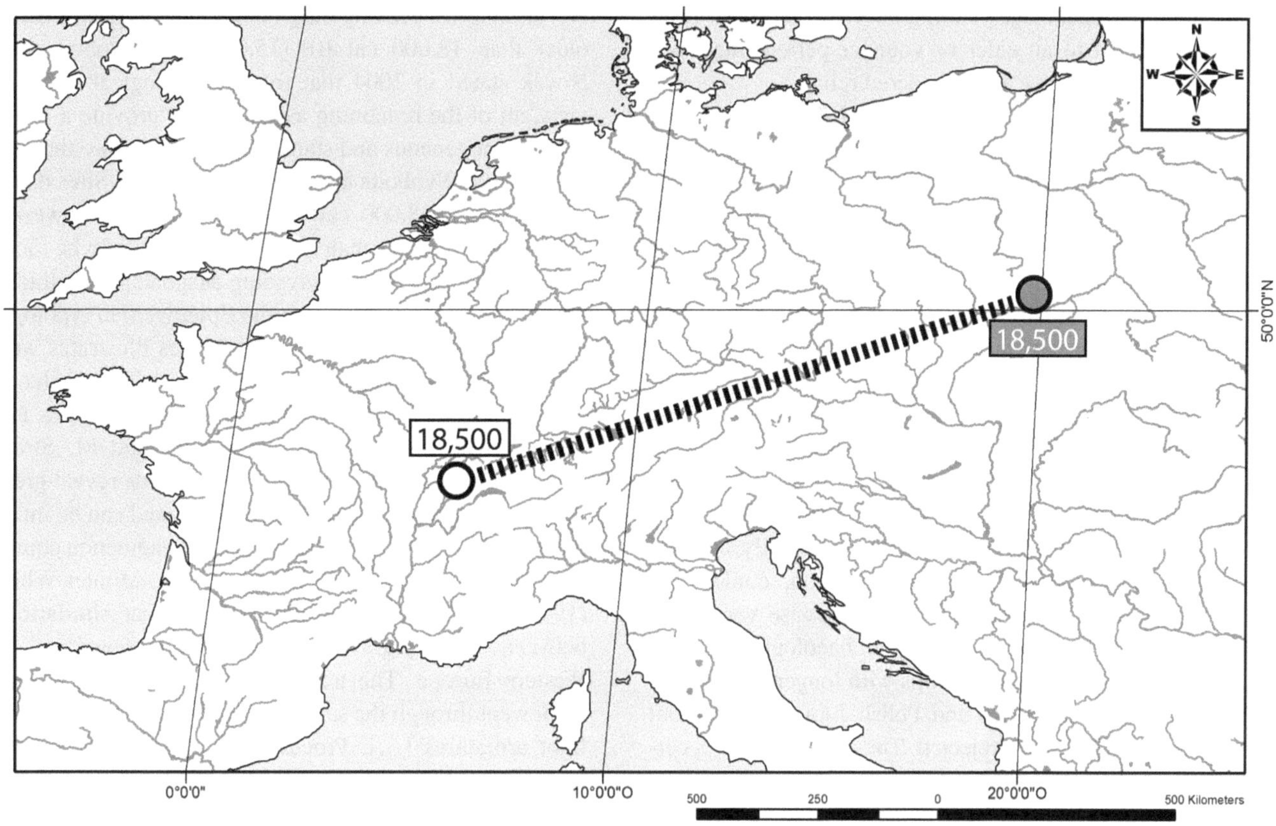

Fig. 8.4 The two contemporaneous *navettes*-bearing sites of Grappin (*west*) and Maszycka (*east*). *Numbers* give the calibrated radiocarbon dates before present (cal BP)

Moravia, and Poland. In the case of a unidirectional expansion, this pattern should be continuous; the observed pattern, however, meets the expectations associated with bidirectional expansion.

3. In the case of a unidirectional expansion into unsettled areas, innovations which are unique to the expanding group should be present only within this group and should not be found outside of the colonized territory.

The emergence of Solutrean points was restricted to southwestern Europe. These tools are a good example of an innovation which spread in an isolated community. With regard to the Magdalenian, however, the situation is different. In this context, Magdalenian lithic triangles are of importance. As demonstrated above, this tool class is highly indicative of the older part of the Magdalenian (stages II and III *sensu* Breuil), or more precisely, for the time between about 20,000 and 16,000 cal BP. According to the current theory, however, even at the end of this period, the unidirectional expansion had hardly reached the Meuse-Rhine area and the Franconian Alb (Housley et al. 1997). Indeed, ^{14}C-dates from these regions show no traces of occupation prior to about 16,000 cal BP (see below). Nonetheless, Magdalenian lithic triangles are also reported from sites further to the east and can be found in Thuringia (Kniegrotte), Moravia (Hranice), and even in Poland (Dzierżysław).

Another distinctive tool class which is widely recognized as being exclusively associated with an earlier stage of the Magdalenian is the so-called *navette* (Allain et al. 1988; see Chap. 4, "The Subdivision of the Magdalenian into Six Phases and the Critique to Which It Is Subject"). The easternmost site where *navettes* are reported—and the only one east of 6° eastern longitude—is the Maszycka Cave in the Polish Jura. Although the reliability of the obtained dates of 18,520±400 cal BP (15,490±310 ^{14}C BP; Ly-2454) and 17,600±250 cal BP (14,520±240 ^{14}C BP; Ly-2453) has been questioned (see the discussion in Housley et al. 1991), they are in perfect accord with the dates from the closest comparable collection, an assemblages with *navettes* from Arlay (Grotte Grappin) in the French Jura, the closest *navette*-yielding site to the west. Between the two sites, which are approximately 1300 km apart as the crow flies, stretches a vast area in which no other Magdalenian site of this age is yet known (Fig. 8.4). At Arlay, the radiocarbon dates (see Table A.6) give two weighted averages of 18,550±380 cal BP (15,533±268 ^{14}C BP, 40.3 % probability) and 17,780±150 cal BP (14,647±182 ^{14}C BP; 74.1 % probability). Therefore, and because no assemblage yielding *navettes* has so far been reliably dated to an older or younger period, the dates from Maszycka Cave are considered reliable.

For reasons discussed under points 1 and 2 (see above), it seems not reasonable to explain the occurrence of lithic triangles and *navettes* in eastern central Europe with the migration of a small number of people that—in a first wave of expansion (cf. Otte 2012: 358ff.)—left their group of origin, moved about 1000 km to the east to find themselves eventually isolated and to suffer subsequent "local extinction." A gradual expansion of several contact-keeping groups appears likewise improbable, since such an expansion should have taken place at a comparatively low pace and with longer—if not permanent—stays in the areas between the French and the Polish Jura to assure stable exchange networks. It is quite unlikely that these contact-keeping groups, which must necessarily be assumed in a scenario of gradual expansion, could have been flying below the radar in an otherwise very well-investigated area, which is why an archaeologically invisible expansion of several groups with longer stays in the area between the French and Polish Jura prior to about 16,000 cal BP must be rejected. The simultaneous occurrence of the same concept (here lithic triangles and *navettes*) in two distant areas rather seems to be the result of interaction and exchange of ideas between two separated but communicating hunter-gatherer communities. Single individuals traveling fast between both groups probably left only marginal and not necessarily recognizable traces of their presence. However, the ideas traveling with them could spread in both communities.

4. If a cultural separation of Eastern and Western Europe is assumed, it must be expected that the assemblages of the archaeological entities in the two regions—namely, the Badegoulian and Magdalenian on the one hand and the Epigravettian on the other—will be more similar to assemblages of the same entity than they are to assemblages from the other respective entities.

   Addressing the question of the relation between the Badegoulian, Magdalenian, and Epigravettian in all necessary detail clearly exceeds the scope of this dissertation and is not its objective. Since it is however relevant for the resettlement of Central Europe, some important aspects shall be mentioned in brief.

   In 1996, Svoboda et al. wrote that "the term *Epigravettian* covers typologically variable assemblages, and the meaning is still more chronological than cultural" (Svoboda et al. 1996: 143 [italics in original]). In the course of the following years, several Central European assemblages previously labeled "Epigravettian" were redated or reclassified. Hranice and Brno-Vídeňska in Moravia, for instance, were initially labeled Epigravettian, but later reclassified as Magdalenian (Svoboda et al. 1996: 145; Neruda and Kostruhn 2002; Svoboda and Novák 2004). Focusing only on those assemblages dated older than 18,000 cal BP (15,000 BP), Svoboda and Novák stated in 2004 that from a typological point of view, all of the remaining assemblages "provide a much more homogeneous and standard picture than we thought previously" (Svoboda and Novák 2004: 463). Sites dated younger than 18,000 cal BP were excluded, however, since "the sites later than 15 ky BP may already be more or less influenced by the invading Magdalenian" (Ibid.). This tendency to homogenize the Epigravettian typologically by excluding younger assemblages illustrates well the diachronic heterogeneity of the record. The similarity between Badegoulian and early Epigravettian sites has already been pointed out. T. Terberger and M. Street (2002: 696) state that "the Grubgraben finds reveal great similarities to Badegoulian assemblages and can be interpreted as an eastern central European phenomenon equivalent to the latter." In the same sense, A. Montet-White (1994: 505f.) writes that "[t]here are clear similarities between the Epigravettian and the Badegoulian of Western Europe. The transformation of hunting equipment went through the same stages as the development of light armatures [...]. Processing tools in both regions include chunky scrapers, carinates, and transverse burins on tabular flakes." The fact that typologically similar assemblages are sometimes labeled Badegoulian while others are termed early Epigravettian seems to confirm the growing sense that these two labels are in fact interchangeable and that their analytical value is questionable.

   Where Magdalenian and younger Epigravettian sites occur in the same region, it is almost impossible to decide typologically whether an assemblage belongs to the one or the other entity. Therefore, allocation is often carried out according to less distinctive features. An argument sometimes used to support the distinction between the two entities is that mammoth bones appear to be much more abundant in Epigravettian sites than in those of the Magdalenian. Whether this is due to different hunting preferences, however, or whether mammoth bones and tusks were simply more abundant and hence easier to collect on the eastern plains, remains an open question. Moreover, Montet-White states that while the mammoth was important to Gravettian subsistence, the proportion of mammoth bones in Epigravettian assemblages declined considerably in favor of those of reindeer and horse, which became the most important game for hunters (Montet-White 1994: 492, 505). Leaving that aside, the often very poor preservation of faunal remains at many sites does not allow for a clear distinction between the two. Another indicator often called on to distinguish between Epigravettian and Magdalenian is the usage of different raw materials, particularly obsidian (Kozłowski 1992), which is thought to be virtually exclusive to the

Epigravettian people. In contrast to typological concepts, which exist always in the context of a specific culture and are usually evaluated as a whole, the mere occurrence of a specific raw material is at best a weak argument for the demarcation of archaeological entities, if it can be considered diagnostic at all. Here it is important to point out the danger of circular reasoning: if all sites with obsidian are per se allocated to the Epigravettian, then of course only Epigravettian sites yield obsidian artifacts.

The argument of this section is that since populations will typo-technologically diverge from each other in the case of non-interaction (see Chap. 3, "Types as Indicators of Interaction"), overall similarities in technological and typological concepts speak in favor of contact and communication. With the above said in mind, it seems that synchronous similarities between early Epigravettian and Badegoulian assemblages as well as between later Epigravettian and Magdalenian inventories are much stronger than the diachronic similarities within the Epigravettian assemblages in the early and later phase. This observation is neither in accordance with the assumption of a long-term separation of hunter-gatherer groups in the East and West nor with the existence of two self-contained and separated communities, the Epigravettian on the one hand and the Badegoulian on the other. Rather, it points toward a parallel development in Eastern and Western Europe throughout the Late Upper Paleolithic as well as already during the LGM. On the assemblage scale, differences can be observed between several Late LGM sites, which are therefore given a number of different labels such as the Badegoulian, early Epigravettian, Ságvárian, Kašovian, or Grubgrabian. By taking a step back and observing these assemblages on a larger scale, however, a number of common features become visible, such as a pronounced production of flakes, the modification of many flakes to tools, or the occurrence of carinated pieces. These similarities suggest that the labels mentioned above are perhaps better seen as regional facies of a common typo-technological background that was shared during the Late LGM than as independent industries which developed and existed apart from each other. In this context, it is interesting to note that the Late LGM assemblages in eastern and western Central Europe provided older dates than those of the known Badegoulian sites in Western Europe; this is why it is thought quite possible that eastern influences were a significant factor during the emergence of the French Badegoulian (Terberger and Street 2002; Terberger 2003: 61).

Toward a New Theory of the Recolonization of Central Europe

Given the numerous and at times severe inconsistencies in the current view on the recolonization of Central Europe after the LGM as outlined above, it appears necessary to propose a new theory to explain this large-scale process. The theory presented below not only differs from the current one in the way it visualizes recolonization as having taken place, but also—and this is most important—draws on large-scale observations of recurrent patterns which emerged from the multi-site analyses of independent variables conducted in the course of this study. These patterns indicate very intensive interaction and a high typological similarity between assemblages located in the Jura and Danube region as well as in the Meuse-Rhine area on the one hand and inventories in Eastern Germany, Bohemia, Moravia, and Poland on the other (see Chap. 6). Between these two supra-regional groups, the typological similarities and intensity of interaction are less strongly pronounced, although the lithic tool types in both groups bear a strong likeness to each other, as becomes apparent when a comparison is made to Hamburgian assemblages (see Chap. 6, "Linear Discriminate Function Analysis: Multivariate Investigation of Spatial and Temporal Trends Among Concepts of Lithic Tools").

New Light on the Recolonization of Central Europe

In the course of the climatic cooling during Greenland Stadial 3, people withdrew to Southwestern Europe as well as eastern Central and Eastern Europe. In Central Europe north of the Alps, only few traces of ephemeral presence of hunter-gatherers are attested (e.g., Terberger and Street 2002; Terberger 2003; Küßner and Terberger 2006). For a certain period of time, contact between eastern and western hunter-gatherer groups appears to have been at a very low level or even to have ceased completely. This is indicated by the emergence of the Solutrean in southwestern Europe at a time when the early Epigravettian occurred east of the Alps and the pronounced technological and typological differences between these entities. However, already around 23,500 cal BP, the interrupted contact seems to have been reestablished, as is shown by typological similarities between sites of the western Badegoulian and the eastern early Epigravettian. The role of northern Italy in this process is today poorly understood and requires additional research. Be that as it may, with the restarting diffusion of concepts, both regions once again developed homogeneously to a large extent, from a typological and technological point of view. Regional variation within this space of interaction naturally occurred. However, the synchronous similarities between assemblages in Western Europe and eastern Central Europe are far more striking than the diachronic similarities within each of the respective regions (see above; cf. Montet-White 1994; Terberger and Street 2002; Terberger 2003).

While the Solutrean and Badegoulian population in southwestern Europe is documented by numerous sites (e.g., Ducasse 2010; Schmidt 2013), the situation in eastern

Central Europe and Eastern Europe is different. Here, the documented sites clearly attributable to the LGM period are less numerous and rather scattered in their distribution (Duwe 2013). However, in comparison to Solutrean and Badegoulian sites in southwestern Europe, early Epigravettian assemblages are more difficult to distinguish from later Epigravettian assemblages in terms of their typo-technological composition. Their resulting broad-meshed pattern may also reflect this problem. However, radiometrically dated and stratified sites such as Grubgraben (Haesaerts 1990; Montet-White (Ed.) 1990b), Kašov (Bánesz et al. 1992), Ságvár (Gábori 1959, 1965; Krolopp and Sümegi 2002; Lengyel 2010, 2011) or Stránská Skála (Table 8.1) prove the presence of hunter-gatherers in eastern Central Europe between 25,000 and 20,000 cal BP. An ephemeral presence in western Central Europe is attested by ^{14}C-dated assemblages from Mittlere Klause, Bockstein-Törle, Kastelhöhle-Nord, and Wiesbaden-Igstadt (Table 8.1).

The reestablished interaction did not lead immediately to an extensive recolonization process of the previously abandoned northern regions. Apparently, the climatic, environmental, and demographic conditions in Central Europe did not yet meet the requirements for such a development. Thus, the first forays by Late Upper Paleolithic hunter-gatherers into and through these regions probably had the character of temporary stopovers rather than prolonged periods of residence. Consequently, the respective traces they left behind are likely to take the form of very small and rather undiagnostic assemblages, which are difficult to detect and to classify. Moreover, in the case of rock shelters and caves, where occupations by later groups often led to the formation of palimpsest situations, these kinds of small assemblages likely disappeared into the extensive set of younger finds.

Only in the course of the following climatic amelioration after the LGM, both communities gradually expanded their settlement areas and in the process reoccupied the previously abandoned regions. Here, it is interesting to note that the dividing line between the two expanding groups, which can be traced from the Hercynian Mountain along the Thuringian and Franconian Highlands up to the Upper Palatinate Forest and Bohemian Forest, coincides with the line observed for the larger vegetation units in the investigated area. To the east of this line, a park-steppe landscape with changing proportions of trees prevailed throughout the whole period of the Late Upper Paleolithic; on the western side, a treeless steppe-tundra (southwest) and tundra-steppe (northwest) characterized the landscape. In order to account for the observed division of the investigated area into a western and an eastern group, it is most plausible to assume a bidirectional expansion from two different centers: one branch expanding from the Franco-Cantabrian region toward the east and the other one heading from the Carpathian region toward the west (Fig. 8.5).

The Eastward Expansion

During the period of climatic amelioration that followed the LGM, hunter-gatherer groups from the Franco-Cantabrian region spread continuously toward the north and east. In the western part of the investigated area, the majority of sites that are dated prior to about 17,700 cal BP (ca. 14,500 ^{14}C BP) are located to the west of the Rhône-Saône valley (Évin et al. 1994: 352). An exception is the cave site of Arlay (Grappin) in the Jura region, where occupation has been proven for the period 18,930 to 18,170 cal BP (weighted average 15,533±268 ^{14}C BP; see Table A.6 for details and references). Two dates from the lower horizon of the site of Schussenquelle, H 860-970 (15,900±360 ^{14}C BP; 19,200±550 cal BP) and GRO 468 (14,470±385 ^{14}C BP; 17,750±550 cal BP), conflict with a set of four younger dates from the same horizon which date the site to about 15,200 cal BP (see Table A.6). Since the dated chemical fraction of H 860-970 was calcium carbonate and both measurements were obtained during early years of ^{14}C-dating, they appear to be unreliable, and thus cannot be considered for this discussion (Housley et al. 1997: 28). The next step in a process of eastward expansion can be traced already around 17,000 cal BP (14,000 ^{14}C BP), when hunter-gatherers settled in the Hegau region. Here, human presence has been proven on the basis of modified bone and antler pieces found at the sites of Kesslerloch and Schweizersbild (see Table A.6). An occupation of the open-air site of Munzingen possibly took place between 18,000 and 17,000 cal BP, which would correspond with its location between Arlay and the Hegau region. Four out of eight AMS dates from this assemblage fall within this time range. However, since all dates were obtained from unmodified bones, and since there is a high variance within the collection of dates, Munzingen has to be considered "a Magdalenian site of unknown age" (Housley et al. 1997: 32). The oldest dates from the Swabian Alb come from the cave sites of Geißenklösterle, Spitzbubenhöhle, and Hohlefels at Schelklingen. From Geißenklösterle, only unmodified bones could be sampled, which produced dates of 16,940±180 ^{14}C BP (OxA-5156), 13,230±130 ^{14}C BP (OxA-4854), and 12,450±120 ^{14}C BP (OxA-5158) (see Table A.6). The same applies for the three samples from the Spitzbubenhöhle, dated to 15,230±100 ^{14}C BP (H 4149-3348), 13,840±120 ^{14}C BP (H 4314-3715), and 12,747±110 ^{14}C BP (H 4052-3212), respectively. Because it cannot be determined which of these dates relate to human activities in the caves, their strength as a basis for attribution is quite low (see the discussion in Housley at al. 1997). From the site of Hohlefels, seven dates on unmodified bones are available for the period after the LGM. The two oldest dates of 15,670±140 ^{14}C BP (H 5312-4907) and 17,100±150 ^{14}C BP (H 5120-4569) are bulk samples which are regarded as an admixture of bones from the Magdalenian and the

Table 8.1 ^{14}C-dates of selected assemblages from the LGM and early post-LGM period

Site	M	Sp	Lab.-Nr.	^{14}C BP	Std	Cal BP	Std	References
Bockstein-Törle VI	b	Bulk	H-4058-3355	20,400	220	24,300	360	INQUA
Bockstein-Törle VI	b	In	KIA-8956	20,990	120	24,930	130	
Bockstein-Törle VI	b	Bulk	H-4058-3526	23,440	290	28,280	310	
Brno-Vídeňska	bb	Bulk	GrN-9350	14,450	90	17,660	110	Verpoorte (2004) and Nerudova and Neruda (2013)
Brno-Vídeňska	b*	R	GrA-20002	14,820	120	18,130	280	
Brno-Vídeňska	t	Ma	OxA-26961	15,625	75	18,902	226	
Deszczowa; 2	n	In	Gd-9464	16,150	280	19,430	490	Kozłowski (2001)
Deszczowa; 3	h	In	Gd-10212	17,480	150	20,970	260	
Grubgraben; AL1	b	E,R	Lv-1825	16,800	280	20,130	490	Terberger and Street (2002) and INQUA
Grubgraben; AL2b	b	E,R	Lv-1821	17,350	190	20,850	320	
Grubgraben; AL3	b	E,R	Lv-1810	18,030	270	21,720	510	
Grubgraben; AL2a	b	E,R	Lv-1823	18,070	270	21,760	500	
Grubgraben; AL2-4	b	E,R	Lv-1660	18,170	300	21,850	510	
Grubgraben; AL4	n	In	Lv-1680	18,400	330	22,120	550	
Grubgraben; AL2b	n	In	Lv-1822	18,620	220	22,280	450	
Grubgraben; AL4	n	In	GrN-21893	18,820	160	22,750	230	
Grubgraben; AL4	b	In	GrN-21530	18,920	90	22,800	170	
Grubgraben; AL4	n	In	AA-1746	18,960	290	22,860	380	
Grubgraben; AL4	n	In	GrN-21790	19,270	80	23,150	150	
Kašov, upper	ch	In	Gd-6569	18,600	390	22,340	660	Svoboda and Nowák (2004) and INQUA
Kašov, lower	ch	IN	?	20,700	350	24,680	590	
Kastelhöhle-Nord	b	R	OxA-9737	18,530	150	22,170	340	Terberger and Street (2002)
Kastelhöhle-Nord	b	R	OxA-9739	19,200	150	23,040	240	
Kastelhöhle-Nord	b	R	OxA-9738	19,620	140	23,510	170	
Langmannersdorf	bb	In	GrN-6585	19,340	100	23,210	150	Verpoorte (2004)
Langmannersdorf	bb	In	GrN-6586	19,520	120	23,420	130	
Langmannersdorf	bb	In	GrN-6660	20,260	200	24,190	360	
Langmannersdorf	bb	In	GrN-6659	20,580	170	24,550	230	
Langmannersdorf	b*	R	GrA-16567	20,590	110	24,610	140	
Mittlere Klause	n	In	UCLA-1869	18,200	200	21,920	390	Terberger and Street (2002)
Mittlere Klause	n	In	OxA-9856	18,590	260	22,280	500	
Ságvár, upper	ch	In	GrN-1959	17,760	150	21,217	361	Vogel and Waterbolk (1964) and Krolopp and Sümegi (2002)
Ságvár	Mo	In	Deb-8822	18,510	160	22,110	343	
Ságvár, lower	ch	In	GrN-1783	18,900	100	22,783	286	
Ságvár	ch	In	Deb-8821	19,770	150	23,680	316	
Stránská Skála IV	b	E	GrN-13945	18,220	120	21,940	320	Svoboda (1991b)
Stránská Skála IV	b	E	GrN-14351	17,740	90	21,260	140	
Wiesbaden-Igstadt	b	In	UZ-3768	17,210	135	20,670	220	Terberger and Street (2002)
Wiesbaden-Igstadt	b	R	OxA-7500	17,820	240	21,400	400	
Wiesbaden-Igstadt	b	R	OxA-7501	18,220	180	21,940	370	
Wiesbaden-Igstadt	b	E	OxA-7502	19,320	240	23,140	340	
Wiesbaden-Igstadt	b	R	OxA-6809	18,670	160	22,540	210	
Wiesbaden-Igstadt	b	E	OxA-6808	19,080	160	22,960	260	
Wiesbaden-Igstadt	b	E	OxA-7406	19,200	160	23,040	250	
Zawalona E	b	In	Ly-2271	15,380	340	18,510	460	Alexandrowicz et al. (1992)
Zawalona E	b	In	Ly-2270	14,060	340	21,940	370	
Zytnia skála	b	In	OxA.6563	20,080	320	23,140	340	Verpoorte (2004)

Gravettian occupation of the site (Housley et al. 1997: 28). The next oldest measurement produced an age of 13,350 ± 140 ^{14}C BP (OxA-4977), which is quite comparable with the date OxA-4854 from the Geißenklösterle and only slightly younger than the second date from the Spitzbubenhöhle. In the end, it is most plausible to assume a resettlement of the Swabian Alb around or slightly before 16,500 cal BP (13,300 ^{14}C BP). Further to the east, the available dates are of younger age. Measurements from the sites of Mittlere and Obere Klause, Sesselfelsgrotte, Kaufertsberg, Hohlenstein bei

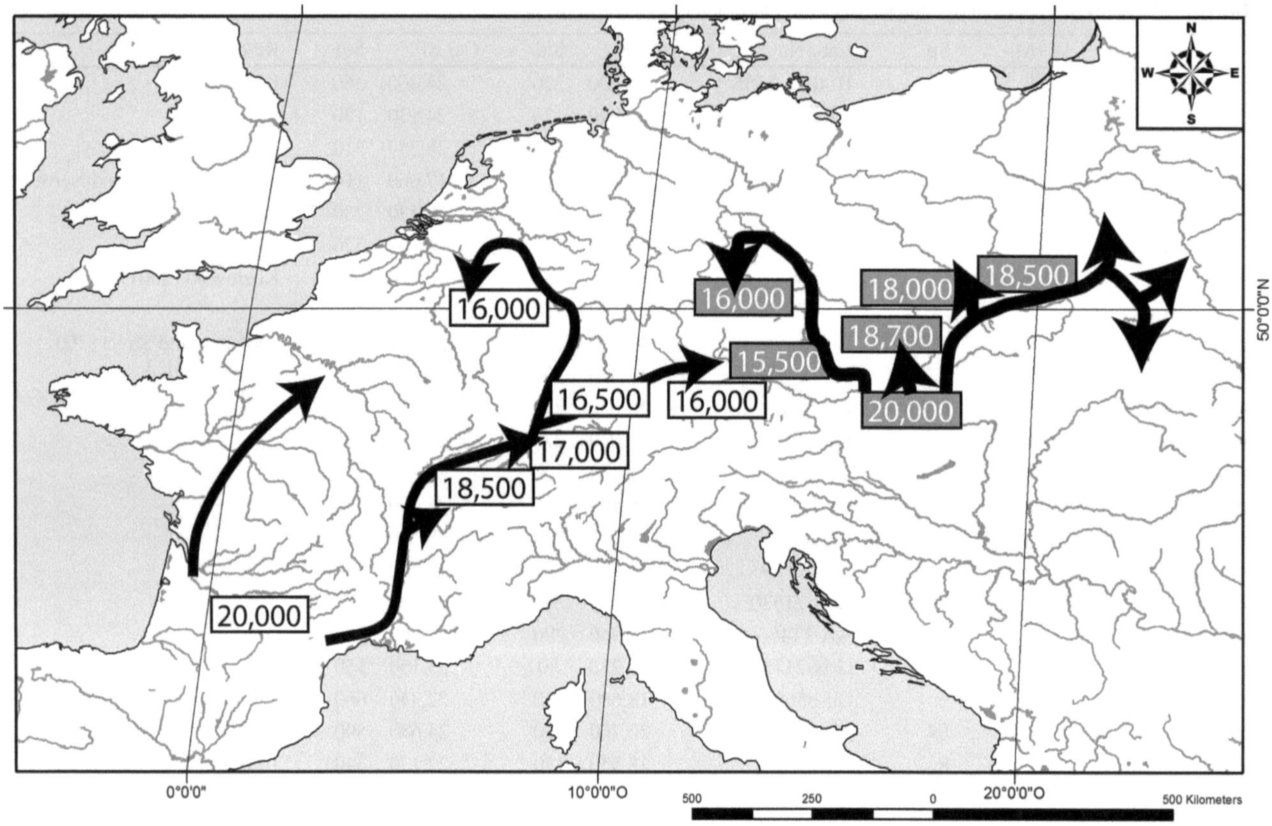

Fig. 8.5 The bidirectional expansion model of the Central European Magdalenian. *Numbers* give the calibrated radiocarbon dates before present (cal BP)

Schambach, and Kastelhänghöhle suggest a recolonization of the Altmühl valley and Regensburg Basin not before 16,000 cal BP (13,200 ^{14}C BP) (cf. Housley et al. 1997: 41; see also Table A.6). In the northern part of western Central Europe, the oldest dates are reported from the cave site of Trou des Blaireaux. Here, four dates were measured on bones from the middle layer of level III, namely, 13,930 ± 120 ^{14}C BP (LV-1433), 13,850 ± 335 ^{14}C BP (LV-1309D), 13,730 ± 400 ^{14}C BP (LV-1434D), and 13,300 ± 160 ^{14}C BP (OxA-4201). The latter date was measured on a horse ulna, which was initially thought to be cut-marked; however, later examinations raised doubts about this interpretation.

It is conspicuous that all of the other early dates obtained from assemblages in the Middle Rhine and Meuse area (see Table A.6), sampled at the sites of Andernach-Martinsberg, Gönnersdorf, Trou Walou, and Trou Frontal, form a homogeneous cluster around 16,000 cal BP (13,200 ^{14}C BP).

The Westward Expansion

The earliest dated assemblage in eastern Central Europe which can be argued as showing features attributable to the Magdalenian is the archaeological Layer 1 from Grubgraben; it differs clearly in its typological spectrum from the underlying levels (Montet-White (Ed.) 1990b) and is dated to about 20,000 cal BP (16,800 ± 280 ^{14}C BP; Lv-1821) (see Table 8.1 for details and references). Since an interpretation of the collection from Grubgraben, Layer 1 as Magdalenian is not yet generally accepted and a current review of the material not yet completed, it must be stressed that such an allocation is preliminary and requires further research. The first piece of unequivocal evidence for a Magdalenian settlement east of the Alps (and north of the Carpathian Mountains) was found during excavations at the site of Maszycka Cave in the Polish Jura, where an inventory of Magdalenian *à navettes* was brought to light (Kozłowski et al. 1995). A date of about 18,500 cal BP (15,490 ± 310 ^{14}C BP; Lv-2454), which has been questioned by some authors, is in perfect accordance with respective dates of the *navettes*-bearing site of Arlay (Allain et al. 1985) and is therefore considered as reliable (see above). Slightly further to the south lies the site of Zawalona, where only a small and fairly undiagnostic collection is reported (Alexandrowicz et al. 1992). However, two dates of 18,510 ± 460 cal BP (15,380 ± 340 ^{14}C BP; Ly-2271) and 17,260 ± 410 cal BP (14,060 ± 340 ^{14}C BP; Ly-2270) suggest an early date for this site. The collection of Brno-Vídeňska (former Koněvova) (Valoch 1975) can probably be attributed to the Magdalenian as well (Svboboda and Nowák 2004: 463) and may therefore be seen as the oldest known assemblage of this entity in Moravia. The two oldest radiocarbon dates measured on a mammoth molar (15,650±70 ^{14}C BP; GdA-495) and a cut-marked reindeer bone (14,820±120 ^{14}C BP; GrA-20002) place the occupation in the period between 18,800 and 17,900 cal BP (see Table A.6). With dates around 17,000 cal BP (14,000 ^{14}C BP), the sites of Balcarka and Nová Drátenická in the Moravian Karst and Kamegg in the vicinity of Grubgraben prove the continuous presence of hunter-gatherers in these regions. North of the Sudetes Mountains, the site of Dzierżysław 35 yielded a Magdalenian assemblage with 114 lithic triangles (Ginter and Połtowicz 2002a, b, 2004; Ginter et al. 2002, 2005) and marks with a radiocarbon date between 18,440 and 17,800 cal BP (14,850±280 ^{14}C BP; Ki-8997), the oldest occupation of this entity in the area. The earliest Magdalenian occupation of Thuringia (Eastern Germany) known so far is reported at the site of Kniegrotte. A weighted average of seven dates places the triangle-bearing assemblage around 16,100 cal BP (13,213±45 ^{14}C BP; 27.9 % probability; see Table A.6). For Bohemia, the number of dated Magdalenian sites is very low: in fact there are only two dates from two sites. The oldest measurement comes from a human cranium found at the site of Koněprusy which results in an age estimation between 15,520 and 15,360 cal BP (12,870±70 ^{14}C BP; GrA-13696). The second date comes from the site of Hostim. Due to its high standard deviation, the calendric time span covers the period between 15,540 and 14,020 cal BP (12,420±470 ^{14}C BP; Ly-1108). With the dated cranium, human presence in Bohemia is proven at about 15,500 cal BP. However, taking into account the geographic position, the low number of dated sites, and the fact that the oldest occupation in Thuringia took place around 16,100 cal BP, this might be regarded as a minimum estimation.

Chapter 9
Small-World Networks: Backbone of the Magdalenian Society?

The traces of the past that survive today reflect only a restricted set of aspects of a Paleolithic culture or society. With very few exceptions, all that remains are stone, bone, antler, and ivory finds and perhaps some spatial structures at scattered sites. However, the fact that many aspects of Magdalenian hunter-gatherer lifeways are inaccessible by archaeological means does not mean that Paleolithic research is restricted to the description of the preserved material objects and the strategies to produce them. While the reconstruction of a "Paleolithic culture" (for a discussion of this term see, e.g., Tylor 1871; Clarke 1968; Hill 1972: 73; Hahn 1977e; Stiles et al. 1979; Wotzka 1993; Gamble 1999: 366) seems too far-reaching for the actual amount of information accessible and the spatial as well as temporal resolution of the available data, the narrow reduction to the material remains seems to exclude unnecessarily the opportunity for further interpretations. This is because it is indeed possible to say something about the degree of relatedness of excavated find assemblages and consequently about the nature of the relation between the people who produced these assemblages. Here, abduction (also called retroduction), seems to be an appropriate way of reasoning. Following Peirce, "Deduction proves that something must be; Induction shows that something actually is operative; Abduction merely suggests that something may be" (Peirce 1931: VI.§171). Abduction is the construction of a hypothesis; it is a "mode of inference in which events are explained by postulating (and identifying) mechanisms which are capable of producing them" (Sayer 1992: 107). In this context, analogies are a valuable tool (see Chap. 3 "The use of ethnographic analogies in this study"), since the reconstruction of past social systems can only approximate past reality and it is not possible to prove the validity of hypotheses about the social meaning assigned to archaeological finds using the archaeological record itself (cf. Binford and Sabloff 1982: 138; Binford 1983b: 12).

From the analyses and observations presented above, it can be concluded that the CEM must, on a social level, represent a "unit of interaction" (Jochim et al. 1999: 131). The overall typo-technological similarities of the assemblages or the widespread distribution of certain schematizations, such as the stylized feminine figurines and engravings (e.g., Bosinski et al. 2001), can be explained meaningfully only against the background of interaction and exchange of ideas (cf. Binford 1972: 116). Otherwise, these phenomena would be only explicable by coincidence. Since coincidence appears to be a rather poor explanation for the observed patterns, social interaction and thus communication are assumed to be the driving forces behind their formation. Analyzing the spatial diversity of the CEM—that is, the regional differences in the results of communication and decision-making processes—brings about evidence for the existence of five regional interaction units, conscious of their territory and with marked differences in the intensity of their interregional communication. These differences make it possible to distinguish two further, supra-regional groups: a western one comprising three regional groups and an eastern one with two regional groups. On the largest scale, however, the ubiquitous distribution of tool types and abstract representational concepts suggests the existence of a wide-reaching network that encompassed all five regional groups as well all other hunter-gatherers of the Magdalenian entity.

These observations raise an interesting question: How must a social system be organized that is able to generate a structure with comparatively strong regionalization on the one hand and large-scale distribution of certain concepts on the other? A system capable of generating such a structure would probably resemble a so-called small-world network (Watts and Strogatz 1998; Bentley and Maschner 2008). The term small-world network was chosen by analogy with the small-world phenomenon described by S. Milgram in 1967 (Watts and Strogatz 1998). In his classic experiment, Milgram was able to show that almost every person in the United States was connected with every other person in the United States via just six intermediaries on average (Milgram 1967). In order to understand the special character

Andreas Maier, *The Central European Magdalenian*, Vertebrate Paleobiology and Paleoanthropology,
DOI 10.1007/978-94-017-7206-8_9

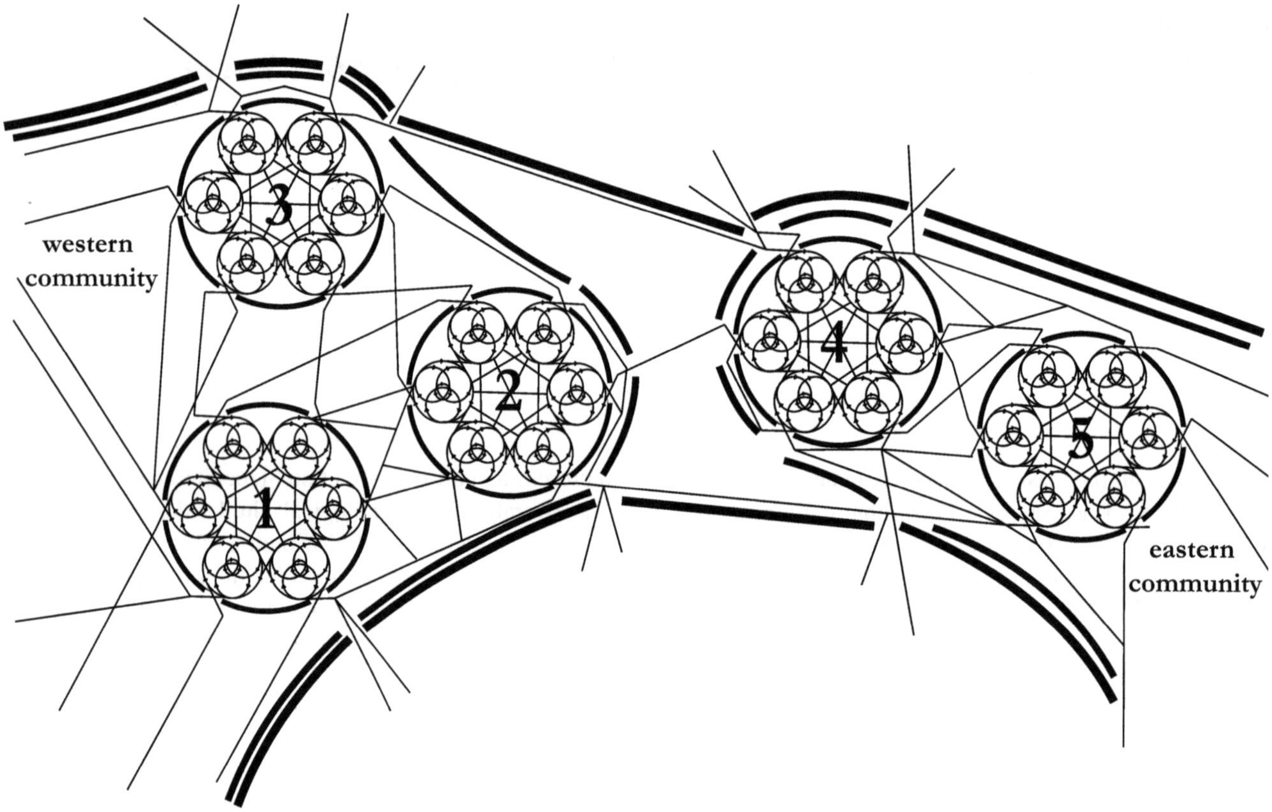

Fig. 9.1 A small-world interpretation of the social system of the Central European Magdalenian. Five regional groups constitute a kind of small-world network, where several shortcut links between members of different regional groups reduce the characteristic path length in an otherwise highly clustered network. (1) Circum-Jurassic Group, (2) Danube Group, (3) Meuse-Rhine Group, (4) Vltava-Saale Group, (5) Polish-Moravian Group

of small-world networks, it is necessary to elaborate briefly on this subject.

In 1998, D. Watts and S. Strogatz discussed the collective dynamics of small-world networks by comparing them to regular and random networks. For this purpose, they used three ring networks consisting of nodes (or agents) and connections. These networks differed only in terms of their "degree of clustering" and their "characteristic path length." Clustering refers to "the degree to which the connections of a typical node are also connected to one another," while the characteristic path length of a network is "the typical number of network links between one agent and another" (Bentley and Maschner 2008: 252). In a regular network, each agent is directly connected only to its immediate neighbors. It is thus "highly clustered because all connections are local, and it also has a large characteristic path length because crossing the network requires many small jumps between neighboring agents" (Bentley and Maschner 2008: 252). In a random network, in contrast, each agent is connected to randomly chosen others. Because of the large increase in the number of distant connections, a random network is not clustered; and since the number of shortcutting links is high, its characteristic path length is short. Small-world networks, however, are almost as clustered as regular networks. At the same time, they exhibit a characteristic path length which is almost as short as that of random networks, despite the fact that the number of shortcutting links is comparatively low (Watts and Strogatz 1998). Thus, "in a small world, an agent perceives itself to be in a clustered neighborhood, and yet the communication distance to any other agents is much shorter than if all agents were equally well connected" (Bentley and Maschner 2008: 252).

Because cultural transmission essentially depends on communication, it can be assumed that the structure of communication networks has a strong influence on the transmission of cultural traits as well as on the introduction of innovations into society and their diffusion within it. With regard to the CEM, it appears likely that the communication network was highly clustered, since it can be assumed that small- to medium-sized hunter-gatherer groups, closely connected by kinship bonds, were the most important social unit at the time. A number of such social units—which aggregated to a larger group and split up again according to annual or multi-annual cycles—formed a community, which corresponds to one of the five regional groups identified in this study. Thus, it can be suggested that most CEM hunter-gatherers spent much of their lifetime in the social context of their regional group and consequently that most connections in their social network were local or regional. However, regular annual and multi-annual meetings or intergroup marriages probably also established shortcut connections between the regional groups. These shortcuts in an otherwise highly clustered network would have decreased the characteristic path length significantly. As a result, the communication structure of the Late Upper Paleolithic social system most likely resembled a small-world network, i.e., a network which is highly clustered but has a short characteristic path length because of some shortcut connections (Fig. 9.1).

The distribution pattern of subrecent shells of sea mollusks may be indicative of these shortcut links, since the transport of items over a distance of more than 800 km is easier to conduct within the structures of a small-world network than within those of a regular network with exclusively local connections. This is because in a small-world network, a CEM individual in the Meuse-Rhine area might obtain an Atlantic sea shell that had passed through the hands of only a small number of intermediate individuals, whereas the passage of such an item from hand to hand in a regular, local network would necessarily have involved many more people. Taking mollusk shell distribution as indicative of the intensity of shortcut connections, the three western regional groups appear to have been closely connected to each other as well as to other groups in the Paris Basin and in southwestern France. Between the eastern and western communities, in contrast, these shortcuts were very rare. This pattern of an eastern and a western community exhibiting very tightly meshed internal interaction networks and at the same time relatively weak connections to the other community mirrors the bidirectional recolonization which emanated from two communicating but spatially separated populations. However, by 16,000 cal BP at the latest, communication network must have been close-knit enough to ensure the diffusion of innovations throughout the whole investigated area at a speed which is below the resolution of ^{14}C-dating, as it is shown by the spread of organic barbed points around 16,000 cal BP (see Chap. 4, section "The Long Debate about Lithic Triangles). The comparatively strong interconnectedness of the different hunter-gather groups and their presumable organization in a small-world network was probably also the reason for the high conformity of the lithic inventories at that time (see Chap. 6, section "Temporal Aspects").

Chapter 10
Summary

The general objective of this study has been to better understand the regional diversity of Central European Magdalenian (CEM) hunter-gatherer groups and thereby to gain a more profound comprehension of Late Upper Paleolithic interaction networks and of the spatiotemporal processes during the recolonization of Central Europe after the Last Glacial Maximum. In addition, this study aimed at a comprehensive overview of the current state of knowledge about the CEM and the provision of a detailed compilation of relevant data. To this end, information on 651 assemblages from 540 sites in the Netherlands, Belgium, eastern France, Switzerland, Germany, Austria, Czech Republic, and Poland was gathered from the archaeological literature. The recorded data comprises the geographic context (including location and setting), the raw material and mollusk shell spectrum, tool typology, blade production technology, ^{14}C-dates, and faunal remains, as well as sculpted and engraved objects. Additionally, a sample of 26 assemblages from all over the investigated area was analyzed directly by the author with regard to similarities and differences in blade production technology. As a reference sample, 171 Hamburgian inventories from 161 sites were collected, and data from 59 palynological sites was analyzed for an analysis of the spatial and temporal variability of the vegetation in Central Europe during the period of the Late Upper Paleolithic.

Spatial Diversity

The analyses show that the hunter-gatherers of the CEM were not a uniform social community living in an undiversified environment with homogeneous plant cover and a uniform fauna nor was it a cultural system within which all groups were equally strongly interconnected and used the same technological, typological, and representational concepts in similar proportions. On the contrary, the evaluation of the available data shows that the cultural and natural space of the Late Upper Paleolithic was heterogeneous in composition and was divided into several smaller units, which become visible at different scales of observation.

Already on the largest scale, the overall distribution of the CEM sites reveals a highly differentiated spatial pattern, where densely clustered areas alternate with regions where virtually no signs of a Late Upper Paleolithic occupation have been reported so far. Indeed, the occurrence of natural shelters such as rock shelters and caves, or the presence of places with a distinctive and strategically favorable topography, fosters the possibility of recurrent usage of a site; consequently, they favor the accumulation of anthropogenic material and hence the archaeological visibility of a hunter-gatherer population. On closer examination, the bias introduced by the presence or absence of natural shelters, erosion, and research intensity seems however negligible on this scale of observation, and the spatial pattern is therefore thought to mirror prehistoric decision-making processes and thus the actual settlement behavior of CEM hunter-gatherers to a great extent (Chap. 6, section "Possible Factors for the Formation of the Large-Scale Site Pattern").

There are social and environmental factors which seem to be relevant for the choice of camp locations and the extent of their respective catchment areas. The main environmental factors, which account together for the location of 98 % of all CEM sites, are proximity to larger rivers and the occurrence of high-quality raw material sources. These nonsocial factors are particularly relevant for the large-scale distribution of sites and the general structure of the spatial pattern, as the receptiveness of the landscape for CEM hunter-gatherer camps varies with the environment across space and is thus strongly influenced by first-order effects. The social factors, in contrast, are most relevant on the local and regional scale as they result in suppression or stimulation of the occurrence of nearby camps. Therefore, second-order effects appear to be most relevant here. For example, zones without Magdalenian finds between two neighboring site clusters can be interpreted as a kind of "no man's land" in the border area between two territory-conscious communities, where

Andreas Maier, *The Central European Magdalenian*, Vertebrate Paleobiology and Paleoanthropology,
DOI 10.1007/978-94-017-7206-8_10

social conventions suppressed settlement activities (see Chap. 6, section "Possible Factors for the Formation of the Large-Scale Site Pattern").

Regional Diversity

Taking raw material procurement as a starting point for estimating the minimum catchment area of a site, the joint consideration of different catchment areas can give information on the intensity of interaction between the hunter-gatherer groups that settled at these sites. Extensive overlapping of catchment areas and collective exploitation of raw material sources point to common usage of the same area, whereas mutually exclusive catchment areas suggest a comparatively low level of interaction. Common exploitation of the same area indicates a strong coherence of the people involved, whereas weak or no interaction rather points to independence and segregation. On this assumption, the investigated area can be divided into five regional groups located (1) around the French and Swiss Jura, (2) in the Swabian and Franconian Alb, (3) in the Meuse-Rhine region, (4) in Eastern Germany and Bohemia, and (5) in Moravia and Poland (Chap. 6, "Procurement Pattern of Lithic Raw Materials").

The remarkably clear distinction between these groups gives reason to suggest that Magdalenian hunter-gatherers were conscious of the limits of their exploitation territories and recognized their own and their neighbors' respective boundaries. These boundaries thus delimited a territory, which should not be understood as a tightly circumscribed and impermeable zone, but instead as an area exploited by a specific regional hunter-gatherer group during several annual cycles, an area whose boundaries are permeable—in accordance with social mechanisms and conventions—to members of other groups (see Chap. 3, "Territories, Territoriality and Land Tenure"). Taking into account the fact that the observed pattern was generated over a period of about 2000 years, its surprising temporal and spatial integrity emphasizes the necessary existence of social conventions that were passed on from one generation to the next; without these conventions, the clarity of the pattern would be very difficult to explain.

Supra-Regional Diversity

The five regional groups can be assigned to two supra-regional clusters. They are divided by the largest and likely most important border area which could be identified during this study and which is constituted by both social and environmental factors. It is oriented along an approximately northwest-southeast line and follows a course running from the Hercynian Mountains along the Thuringian and Franconian Highlands up to the Upper Palatinate Forest and the Bohemian Forest.

In the eastern part, the landscape was characterized by a park-steppe environment with a continuous presence of pine and birch trees, the abundance of which changed according to the climatic development of the Late Weichselian Glacial. The western part, on the other hand, remained treeless. In the southwestern part of the investigated area, a herb-dominated steppe-tundra prevailed, whereas the northwestern part was characterized by a dryer and grass-dominated tundra-steppe (see Chap. 5). These observations correspond well with those of the faunal record. On the assumption that the proportion of horse and reindeer among Late Upper Paleolithic hunting remains reflects the abundance of these animals in the vicinity of a site; it can be inferred that horses, which feed predominantly on grasses, were more abundant in the park-steppe and tundra-steppe environments, whereas reindeer were probably better adapted to the steppe-tundra environment and thus had competitive advantages in the southwestern part of the investigated area. Other species likewise show identifiable trends in their spatial distribution. Whereas remains of ibex, chamois, red deer, elk, and wild boar are mainly found in the southwestern part of the investigated area, wisent, aurochs, roe deer, and mammoths show a rather even distribution. Saiga bones only occur north of the Saône-Danube-Vistula line, and woolly rhinoceros remains are predominantly reported in eastern Central Europe (see Chap. 5, "Fauna").

A conspicuous match between the spatial distribution of natural features and cultural traits can be observed with regard to the number of representations of a certain animal class, such as red deer, woolly rhinoceros, musk ox, and wisent/aurochs, which seem to reflect the abundance of each of these animals in the area surrounding a site as indicated in the faunal remains. Again, this correlation can be observed most clearly with regard to sculptures and engravings of horses and cervids. Whereas representations of cervids are most numerous in the southwestern part of the investigated area, images of horses prevail in the northwestern and eastern part. Thus, it seems that those animals which were preferentially hunted and hence presumably more common in the environment also served as inspiration for representations more often than did less common ones (see Chap. 6, "Some Aspects of Central European Magdalenian Sculptures and Engravings").

The border between the tree-bearing and treeless biomes coincides precisely with the border between two CEM hunter-gatherer populations, which can be interpreted as two distinctive but at the same time interdependent and communicating communities. The western part comprises the sites around the French and Swiss Jura, the Swabian and Franconian Alb, and the in the Meuse-Rhine region, whereas the eastern part consists of the sites in Eastern Germany,

Bohemia, Moravia, and Poland. On closer inspection, the regional groups of each part could be shown to have had stronger relations to each other than to any regional group of the other part (see Chap. 6, "Typological and Technological Variability").

Certain representational concepts, such as the female figures of the Gönnersdorf type or phalliform sculpted objects, occur ubiquitously and thus show an exchange of ideas over the entire investigated area (see Chap. 6, "Figural Representations"). More complex and composed geometrical signs, however, are mostly restricted to single sites. Other features are restricted to one of the two supra-regional groups. The distribution of *baguettes demi-rondes* with small raised humps, and sculpted animal faces in frontal view, for instance, is restricted exclusively to the western part of the investigated area.

The analysis of the individual spatial distribution of lithic tools shows that the most common tool classes, such as scrapers, burins, borers, backed bladelets, truncated pieces, and laterally retouched pieces, do not exhibit any kind of directed spatial trends or clustering whatsoever. They are abundant in all regional groups, and variations in their relative frequencies can be best explained in terms of differences in site function. In contrast, a group of nine tool classes or types shows a clear spatial patterning. These nine types, namely, backed points, notched and denticulated pieces, long (Lacan) burins, long borers, *zinken*, short scrapers, denticulated backed bladelets, shouldered points, and splintered pieces, show clear spatial trends. These trends reveal for the most part a southwest-northeast-oriented opposition in terms of increasing and decreasing proportional frequencies. Interestingly, the abovementioned five regional groups appear to reflect preferences for certain types as a whole, and the boundary lines of the spatial trends often respect the boundaries of the regional groups to a large extent (Chap. 6, "Spatial Distribution of Concepts of Lithic Tools"). To investigate this pattern more closely, the assemblages were evaluated collectively and in comparison with a sample of Hamburgian sites. As a result, the most marked differences can be observed between Hamburgian and CEM assemblages, indicating that both industries represent distinguishable archaeological entities. Within the CEM, a clear separation can again be observed, with assemblages in the Jura, Danube, and Meuse-Rhine region on the one hand and assemblages in Eastern Germany, Bohemia, Moravia, and Poland on the other. The analysis indicates that the typological composition of an assemblage corresponds strongly with its geographical position and with the structure of the interaction pattern between the regional groups, as is indicated by the procurement patterns of raw materials and mollusk shells (see Chap. 6, "Analysis of Spatial Grouping"). Here, in the west a spatially long-span network of intensive interaction connects the three western regional groups with each other, as well as with the Paris Basin and the areas adjacent to the west, up to the coasts of the Mediterranean Sea and Atlantic Ocean. The eastern group, in contrast, exhibits a less close-meshed exchange pattern. Since most of the excavated shells can be gathered locally in the Moravian Karst, the Carpathian Foredeep, and the Vienna Basin, long-distance transports may simply not be visible. Nevertheless, certain specimens show interaction between hunter-gatherers in Bohemia, Moravia, and the Vienna Basin. Evidence for an exchange of mollusk shells between the western and eastern groups, however, is virtually absent (see Chap. 6, "Procurement Pattern of Mollusk Shells and Amber").

These observations indicate that the way in which cultural transmission took place during the CEM must have been capable of ensuring an overall homogeneity in the distribution of concepts, but at the same time allowing for regional differences in the popularity and hence the frequency of certain traits, which in turn impart an individual character to each regional group. Although the precise identification of modes of cultural transmission remains beyond the resolution of the data, it seems that within a regional group, the bias toward conformity was quite strong and affected in particular payoff-correlated tool types. Interaction, then, counteracted the forces of drift, reduced typological divergence among the regional groups and therefore led to more or less strong similarities in the application of concepts depending on the intensity of interaction (see Chap. 3, "Types as Indicators of Interaction"). This interregional interaction was probably the force that kept the CEM in a state of relative overall similarity, which in turn allows today's archaeologists to subsume these groups under the term Magdalenian.

Temporal Diversity

With regard to typological and technological similarities between Badegoulian and early Epigravettian sites (see Chap. 8, "The Current Theory and the Major Critiques to Which It Is Subject"), it can be argued that the joint development of Late Upper Paleolithic hunter-gatherers in Central Europe started already between 23,500 cal BP to 20,000 cal BP. During this period, the assemblages were characterized by the frequent usage of flakes for tool production and the occurrence of nosed and carinated scrapers as well as side scrapers. Dihedral burins and backed bladelets were rather scarce and lithic triangles virtually absent.

The period between 20,000 and 16,000 cal BP contrasts markedly with the previous one in technological and typological terms: the focus of the blank production was now clearly on blades, which were the preferred blanks for tool production at that time, and typically produced with a soft stone hammer (see Chap. 6, "Spatial and Temporal Trends Among Technological Concepts"). The typological characteristics are dihedral burins, backed bladelets, and in particular lithic triangles and *navettes* (see Chap. 4, "The Subdivision of the Magdalenian into Six Phases and the Critique to Which It Is Subject" and "The Long Debate About Lithic Triangles"). From the occurrence of lithic triangles, it can be inferred that composite barbed points were a characteristic feature of this period.

Lithic triangles disappeared—probably together with composite barbed points—around 16,000 cal BP. At the same time, the dispersal of massive organic barbed points took place (see Chap. 4, "The Long Debate About Lithic Triangles"). This shift in the production of hunting gear seems to have happened simultaneously in the entire investigated area, indicating that it took place at a rate which is below the resolution of ^{14}C-dating. Aside from this change, the typological spectrum remained largely the same, although the typological conformity increases with regard to the previous period (see Chap. 6, "Analysis of Temporal Grouping").

The rapid diffusion of massive barbed points and the increasing typological similarities between the assemblages highlights in exemplar fashion the effectiveness of the communication network at that time (see Chap. 9, "Small-World Networks: Backbone of the Magdalenian Society?"). Organic hammers were the main tool for the production of blades, and the characteristic *en éperon* preparation was applied extensively (see Chap. 6, "Spatial and Temporal Trends Among Technological Concepts").

After 14,000 cal BP, soft stone hammers once again gained in significance for blade production, and the *en éperon* preparation method was no longer in use. The typological conformity of the lithic assemblages decreases again, and backed points became more frequent, indicating the transition to the more dispersed hunter-gatherer groups of the Late Paleolithic (see Chap. 6, "Spatial and Temporal Trends Among Technological Concepts").

Subsistence

With regard to the subsistence system, the archaeological evidence in all regional groups speaks in favor of the hypothesis that CEM hunter-gatherers were organized logistically, rather than following a forager strategy (*sensu* Binford 1980). Differences in the artifact compositions of the assemblages suggest the existence of residential camps as well as rather specialized places such as hunting camps. Based on the analysis of the raw material procurement pattern, the five regional groups exploited territories with an estimated minimum range of between 23,000 and 84,000 km^2. In analogy to recent hunter-gatherer groups, it appears reasonable to assume that only parts of the territory were exploited in the course of an annual or multi-annual cycle (see also Chap. 7). The attempt to identify these multi-annual exploitation zones indicates a number between 3 and 6 zones per regional group with sizes between 5000 and 18,000 km^2. While the land-use pattern seems fairly similar for almost all regional groups, the Polish-Moravian Group shows exceptional values: its estimated territory is more than twice as large as the median value of the other territories; the number of reconstructed exploitation zones is twice as high as that of most other groups; the estimated average distance of a residential move is the largest in the entire investigated area; and the transport distances of raw materials are in many cases particularly large. Two possible explanations can be offered. On the one hand, it is possible that the Polish-Moravian Group itself represents an amalgam of two regional groups. This is contradicted, however, by the close-meshed raw material procurement pattern, which makes the division of this group into two separate units impossible. Furthermore, such an interpretation ignores the correspondence between the increased size of the territory and the increased estimated distance of a residential move. On the other hand, the different environmental conditions, especially the abundance of trees in the landscape, and the probably corresponding lower densities of large herbivorous could have led to an increase in the exploitation range and, in turn, an increase in the average distance of residential moves. In such a subsistence system, where large distances are traveled at comparatively short intervals, curated artifacts have a higher chance of being carried along over large distances, producing a raw material procurement pattern similar to the one observed for the Polish-Moravian Group.

The Bidirectional Recolonization of Central Europe

Since sites do not become gradually younger from west to east and the distribution of concepts and objects in the CEM record is discontinuous and mirrors two populations, the current hypothesis of a unidirectional expansion of Magdalenian hunter-gatherers from the Franco-Cantabrian region up the Vistula river appears largely untenable. It rather seems that the recolonization of Central Europe was the result of a bidirectional process. The reinvigorated interaction, communication, and intellectual exchange between east and west from about 23,500 cal BP onward did, however, not lead immediately to an extensive recolonization process. Apparently, the climatic, environmental, and demographic conditions prevailing in Central Europe were not yet favorable.

Nevertheless, in the course of the post-LGM climatic amelioration, communities in both east and west gradually expanded the range of their activity. The western hunter-gatherers reached the French Jura in about 17,500 cal BP at the latest and settled in the Hegau region already around 17,000 cal BP. At circa 16,500 cal BP, they settled in the Swabian Alb and reached the Regensburg area not before 16,000 cal BP. At approximately the same time, the Meuse-Rhine area was recolonized (see Chap. 8, "The Eastward Expansion"). The expansion of the eastern community reached Moravia already around 18,700 cal BP and can be traced north of the Carpathian Mountains by about 18,500 cal BP. The area north of the Sudetes Mountains was resettled by roughly 17,800 cal BP. The expanding hunter-gatherers eventually reached Thuringia around 16,100 cal BP and the oldest finds from Bohemia point toward a recolonization in approximately 15,500 cal BP (see Chap. 8, "The Westward Expansion").

The Central European Magdalenian: A Small-World Network?

With regard to the observable distribution pattern of typological, technological, and representational concepts, it appears that the sociocultural system of the Late Upper Paleolithic hunter-gatherers strongly resembled a small-world network. Since it can be assumed that small- to medium-sized, regional hunter-gatherer groups were the most important social unit at that time, the corresponding communication network was probably highly clustered. However, regular meetings and cross marriages between these regional groups—indicated by the distribution pattern of subrecent and fossil mollusk shells—likely established several shortcutting connections which decreased the characteristic path length of the social network significantly (see Chap. 9, "Small-World Networks: Backbone of the Magdalenian Society?").

Chapter 11
Zusammenfassung

(German Translation of Chapter 10 Summary)

Das Hauptanliegen dieser Dissertation war es, die regionale Diversität der Jäger-Sammler Gruppen des Mitteleuropäischen Magdalénien (MEM) besser zu verstehen um darüber ein tieferes Verständnis spätjungpaläolithischer Interaktionsnetzwerke sowie der räumlich zeitlichen Prozesse während der Wiederbesiedlung Mitteleuropas nach dem letzten Kältemaximum zu erhalten. Weitere Ziele dieser Arbeit waren einen umfassenden Überblick über den aktuellen Wissensstand zum MEM zu präsentieren und eine detaillierte Zusammenstellung relevanter Daten zu geben. Zu diesem Zweck wurden entsprechende Informationen zu 651 Inventaren aus 540 Fundstellen in den Niederlanden, Belgien, Ostfrankreich, der Schweiz, Deutschland, Österreich, Tschechien und Polen aus der archäologischen Literatur zusammengestellt. Die aufgenommen Daten beinhalten Angaben zur geographischen Position und landschaftlichen Lage, zum Spektrum der Rohmaterialien und Molluskenschalen, zur Werkzeugtypologie und Klingentechnologie, zu ^{14}C-Daten und Faunenresten sowie zu skulptierten und gravierten Objekten. Zusätzlich wurde eine Stichprobe von 26 Inventaren aus dem gesamten Untersuchungsgebiet in Hinblick auf technologische Gemeinsamkeiten und Unterschiede bei der Klingenproduktion vom Autor analysiert. Als Vergleichsdatensatz wurden 171 Inventare aus 161 Fundstellen der Hamburger Kultur aufgenommen sowie pollenanalytische Daten aus 59 palynologischen Fundstellen gesammelt um die räumliche und zeitliche Variabilität der Vegetation in Mitteleuropa während des Späten Jungpaläolithikums zu analysieren (siehe Kap. 2).

Räumliche Diversität

Die Analysen zeigen, dass die Jäger-Sammler des MEM keine einheitliche soziale Gemeinschaft waren, die in einer undiversifizierten Umwelt mit homogener Vegetation und uniformer Faunenzusammensetzung lebten. Genausowenig konstituierten sie ein kulturelles System, innerhalb dessen alle Gruppen gleich stark miteinander vernetzt waren und dieselben typologischen, technologischen und darstellerischen Konzepte in gleichen Proportionen verwendeten. Im Gegenteil, die Auswertung der zur Verfügung stehenden Daten zeigt, dass der Kultur- und Naturraum des Späten Jungpaläolithikums heterogen zusammengesetzt und in mehrere kleinere Einheiten unterteilt war, die auf verschiedenen Beobachtungsniveaus sichtbar werden.

Bereits auf der größten Skale zeigt die Gesamtverteilung der MEM-Fundstellen ein hochgradig differenziertes, räumliches Muster, bei dem Gebiete hoher Funddichte sich mit solchen abwechseln, von denen bisher nahezu keinerlei Hinweise auf eine spätjungpaläolithische Besiedlung vorliegen. Das Vorkommen natürlicher Schutzdächer wie Abris oder Höhlen fördern sicherlich die Möglichkeit der wiederholten Nutzung einer Fundstelle, daher die Akkumulation anthropogenen Materials und somit die archäologische Sichtbarkeit einer Jäger-Sammler Population. Bei näherer Betrachtung scheint die Verzerrung durch Höhlen, Abris, Erosion und Forschungsintensität auf dieser Beobachtungs-Skale aber vernachlässigbar zu sein. Das räumliche Muster scheint daher prähistorische Entscheidungsprozesse und somit weitestgehend das tatsächliche Siedlungsverhalten der MEM Jäger-Sammler widerzuspiegeln (Kap. 6, section "Possible Factors for the Formation of the Large-Scale Site Pattern").

Sowohl Sozial- als auch Umweltfaktoren scheinen für die Wahl eines Siedlungsplatzes und die Größe seines Ressourcen-Einzugsgebiets zu sein. Die wichtigsten Umweltfaktoren, die gemeinsam 98 % der Siedlungsplatzauswahl erklären, ist die Nähe zu größeren Flüssen und das Vorkommen geologischer Formationen mit Aufschlüssen qualitativ hochwertiger Rohmaterialien. Diese nicht-sozialen Faktoren sind insbesondere für die groß-skalige Verteilung der Fundstellen und die generelle Struktur des räumlichen Musters relevant, da die naturräumlichen Voraussetzungen für Lagerplätze magdalénien-zeitlicher Jäger-Sammler je nach Landschaft variieren und ihre Lage deshalb stark von

Andreas Maier, *The Central European Magdalenian*, Vertebrate Paleobiology and Paleoanthropology,
DOI 10.1007/978-94-017-7206-8_11

Effekten erster Ordnung beeinflusst wird. Die sozialen Faktoren sind im Gegensatz dazu auf der lokalen und regionalen Skale am wichtigsten, da sie in der Suppression oder Stimulation von Lagerplätzen in der näheren Umgebung resultieren. Daher scheinen Effekte zweiter Ordnung hier die größte Relevanz zu besitzen. Ein fundfreies Gebiet zwischen zwei benachbarten Fundstellenagglomerationen können als „Niemandsland" im Grenzgebiet zweier territoriumsbewusster Gemeinschaften gedeutet werden, in dem soziale Konventionen Siedlungsaktivitäten verhindern (siehe Kap. 6, "Some Aspects of Central European Magdalenian Sculptures and Engravings").

Regionale Diversität

Wenn man das räumliche Muster der Rohmaterialbeschaffung als Ausgangspunkt für eine Minimum-Schätzung des Einzugsgebiets einer Fundstelle heranzieht, so kann die gemeinsame Bewertung verschiedener Einzugsgebiete Aufschluss über die Interaktionsintensität zwischen den Jäger-Sammler Gruppen bieten, die an den entsprechenden Fundstellen siedelten. Ein starkes Überlappen mehrerer Einzugsgebiete und eine damit verbundene Ausbeutung gleicher Rohmaterialquellen deuten auf eine gemeinschaftliche Nutzung des gleichen Gebiets und somit auf einen starken Zusammenhang zwischen den beteiligten Personen hin. Demgegenüber weisen sich wechselseitig ausschließende Einzugsgebiete auf ein geringeres Interaktionsniveau und eine größere Unabhängigkeit der entsprechenden Jäger-Sammler Gruppen hin. Vor dem Hintergrund dieser Annahme weist die Analyse auf die Existenz von fünf Regionalgruppen hin. Diese fünf Gruppen befinden sich (1) im Gebiet des Französischen und Schweizerischen Juras, (2) im Gebiet der Schwäbischen und Fränkischen Alb, (3) im Maas-Rhein Gebiet, (4) in Ostdeutschland und Böhmen und (5) in Mähren und Polen (siehe Kap. 6, "Procurement Pattern of Lithic Raw Materials").

Die bemerkenswert klare Unterscheidbarkeit dieser Gruppen gibt Anlass zu der Annahme, dass magdalénienzeitliche Jäger-Sammler sich der Grenzen Ihrer Nutzungsgebiete bewusst waren und sowohl die eigenen Grenzen als auch die Grenzen Ihrer Nachbarn respektierten. Diese Grenzen schlossen also Territorien ein, die jedoch nicht als klar abgegrenzte und undurchdringliche Zonen verstanden werden sollten, sondern als Gebiete, die von bestimmten Regionalgruppen während mehrerer Jahreszyklen genutzt wurden und deren Grenzen—in Übereinstimmung mit sozialen Mechanismen und Konventionen—für Mitglieder andere Gruppen durchlässig waren (sieh Kap. 3, "Territories, Territoriality and Land Tenure"). Wenn man bedenkt, dass das beobachtete Muster über einen Zeitraum von 2000 Jahren entstand, unterstreichen seine erstaunliche zeitliche und räumliche Integrität die notwendige Existenz sozialer Konventionen, die von einer Generation zur nächsten Weitergegeben wurden. Ohne solche Konventionen wäre ein derart klares Muster nur sehr schwer erklärbar.

Überregionale Diversität

Die fünf Regionalgruppen lassen sich in zwei überregionale Gruppen aufteilen. Sie sind durch die größte und wahrscheinlich wichtigste Grenze getrennt, die in dieser Arbeit identifiziert werden konnte und die sowohl durch Sozial- als auch durch Umweltfaktoren konstituiert wird. Sie folgt einer annährend nordwest-südost ausgerichteten Linie, die vom Harz über das Thüringisch-Fränkische Mittelgebirge bis zum Oberpfälzer Wald und dem Böhmischen Wald verläuft.

Im östlichen Teil war die Landschaft durch eine Park-Steppe mit zeitlich schwankenden Anteilen von Kiefern und Birken geprägt, wohingegen auf der westlichen Seite eine baumlose Steppen-Tundra (im Südwesten) beziehungsweise Tundren-Steppe (im Nordwesten) vorherrschte. Diese Dreiteilung des Untersuchungsgebietes bestand über den gesamten Zeitraum des Späten Jungpaläolithikums hinweg, auch wenn während des Bølling Interstadials Wachholderbüsche im Südwesten deutlich an Bedeutung gewannen (siehe Kap. 5, "Climate and Vegetation").

Die Verbreitung der Jagdreste von Pferd und Rentier in den Fundstellen korrespondiert mit dieser landschaftlichen Differenzierung. Während Skelettteile des Rentiers in den Fundstellen des durch eine Steppen-Tundra geprägten Südwestens überwiegen, dominieren Überreste vom Pferd die Fundstellen im Bereich der Tundren-Steppe und Park-Steppe. Auch die Verteilung anderer Arten zeigen räumliche Trends. Während Skeletteile von Steinbock, Gämse, Rothirsch, Elch und Wildschwein vorwiegend im Südwesten zu finden sind zeigen Wisent, Ur, Reh und Mammut eine eher gleichmäßige Verteilung über das gesamte Arbeitsgebiet. Knochen der Saiga Antilope finden sich nur nördlich einer gedachten Linie von der Saône über die Donau bis zur Weichsel und Reste des Wollnashorns scheinen im östlichen Mitteleuropa besonders zahlreich zu sein (siehe Kap. 5, „Fauna").

Die räumliche Verbreitung von Pferd- und Rentierdarstellungen ähnelt in erstaunlicher Art und Weise der Verbreitung der entsprechenden Faunenreste. In beiden Fällen ist das Ren im Südwesten über- und im Nordosten unterrepräsentiert, während das gegenteilige Bild auf das Pferd zutrifft. Der einzige Unterschied zwischen der Verteilung der Faunenreste und der Darstellungen besteht in einer leichten Verschiebung der Grenze zwischen den von Rentieren einerseits und Pferden andererseits dominierten Gebieten in Richtung Südwesten. Ein ähnliches Phänomen kann ebenfalls für Steinbock, Rothirsch, Wollnashorn, Mammut, Moschusochse und Wisent/Ur beobachtet werden. Auch hier deckt sich die Verteilung ihrer Skelettreste mit der Verteilung ihrer Darstellungen. Für die Saiga Antilope, deren Darstellung bisher nur von zwei Fundstellen bekannt ist, erlaubt die geringe Stichprobengröße keine Aussage. Generell scheint, dass solche Tiere, die vorzugsweise gejagt wurden und somit auch vermutlich häufiger in der Umgebung der jeweiligen Fundstellen vorkamen, auch häufiger als Inspiration für Darstellungen dienten (siehe Kap. 6, "Some Aspects of Central European Magdalenian Sculptures and Engravings").

Die Grenze zwischen den baumlosen und baumführenden Biomen fällt genau mit der Grenze zwischen zwei MEM Jäger-Sammler Populationen zusammen, die als klar unterscheidbare aber von einander abhängige und kommunizierende Gemeinschaften interpretiert werden können. Der westliche Teil umfasst die Fundstellen im Französischen und Schweizerischen Jura, der Schwäbischen und Fränkischen Alb und im Maas-Rhein Gebiet, wohingegen der östliche Teil aus den Fundstellen in Ostdeutschland, Böhmen, Mähren und Polen besteht. Eine genauere Untersuchung konnte zeigen, dass die Regionalgruppen innerhalb der zwei Teile jeweils stärkerer Verbindungen untereinander zeigen als zu irgendeiner Regionalgruppe des jeweils anderen Teils (siehe Kap. 6, "Typological and Technological Variability").

Gewisse Darstellungskonzepte, wie beispielsweise Frauendarstellungen vom Typ Gönnersdorf oder phalliforme Objekte kommen im gesamten Arbeitsgebiet vor und zeigen einen Austausch von Ideen. Komplexere und zusammengesetzte geometrischer Zeichen sind jedoch meist auf einzelne Fundstellen beschränkt. Die Verteilung von *baguettes demi-rondes* mit Höckerverzierung und skulptierte Tiergesichter in Frontalansicht sind auf den westlichen Teil des Arbeitsgebiets beschränkt.

Die Analyse der räumlichen Verteilung der lithischen Werkzeuge zeigt, dass die häufigsten Werkzeugklassen wie Kratzer, Bohrer, Rückenmesser, Endretuschen und lateral retuschierte Stücke keine wie auch immer gearteten räumlichen Tendenzen oder Clusterbildungen erkennen lassen. Sie sind in allen Regionalgruppen zahlreich, und Variationen in ihren relativen Häufigkeiten sind wohl am besten durch funktionale Unterschiede der entsprechenden Fundstellen zu erklären. Im Gegensatz dazu zeigt eine Gruppe von neun morphologisch distinkten Werkzeugklassen bzw. Werkzeugtypen eine klare räumliche Musterbildung, nämlich Rückenspitzen, gekerbte und gezähnte Stücke, Langstichel (Lacan Stichel), Langbohrer, Zinken, kurze Kratzer, gezähnte Rückenmesser, Kerbspitzen und ausgesplitterte Stücke. Die beobachtbaren räumlichen Trends zeigen im Hinblick auf zu- oder abnehmende relative Häufigkeiten größtenteils eine in Richtung Südwest-Nordost orientierte Opposition. Interessanterweise scheinen die erwähnten fünf Regionalgruppen Präferenzen für bestimmte Typen zumeist als Ganzes widerzuspiegeln, so dass die Grenzen der räumlichen Verteilungen dieser Formen sich weitestgehend mit den Grenzen der Regionalgruppen decken (siehe Kap. 6, "Spatial Distribution of Concepts of Lithic Tools").

Um die typologische Zusammensetzung der Inventare in ihre Gesamtheit bewerten zu können und um zu testen, ob sich diese Zusammensetzung zwischen den Regionalgruppen signifikant unterscheidet und welche Typen in diesem Fall für eine Unterscheidung besonders relevant sind, wurde eine Diskriminanzanalyse durchgeführt. Die Analyseklassen wurden entsprechend der fünf Regionalgruppen zuzüglich des Vergleichsdatensatzes von Inventaren der Hamburger Kultur gebildet, während die (Hellinger-transformierten) Anteile der Werkzeugtypen pro Inventar als unabhängige Variablen dienten. Im Ergebnis lassen sich die stärksten Unterschiede zwischen Inventaren der Hamburger Kultur und solchen des Magdalénien feststellen, wobei einfachen Rückenmessern und Kerbspitzen die größte diskriminatorische Kraft zukommt. Innerhalb des MEM lässt sich eine deutliche Teilung in zwei Gruppen beobachten. Die erste Gruppe umfasst die Inventare der Jura, Donau und Maas-Rhein Region wohingegen die zweite Gruppe die Inventare in Ostdeutschland, Böhmen, Mähren und Polen einschließt. Bei dieser Unterscheidung zeigen Langstichel und Langbohrer die größte Trennkraft für die erste Diskriminanzfunktion, wohingegen Langbohrer und einfache Rückenmesser für die zweite Diskriminanzfunktion am wichtigsten sind. In solchen Fällen, in denen die ursprünglich definierte Gruppenzugehörigkeit eines Inventars vom Ergebnis der Diskriminanzanalyse abweicht, ist die Neuzuordnung weitestgehend auf räumlich unmittelbar angrenzende Gruppen beschränkt. Die Analyse zeigt also, dass die typologische Zusammensetzung eines Inventars sowohl von seiner geographischen Lage als auch von dem Interaktionsmuster der Regionalgruppen—wie es vor dem Hintergrund der Transportmuster der Rohmaterialien und Molluskenschalen erscheint—deutlich beeinflusst wird (siehe Kap. 6, "Analysis of Spatial Grouping").

Die westliche Gruppe ist in ein weit gespanntes Netzwerk intensiver Interaktion eingebunden, das die drei Regionalgruppen sowohl untereinander als auch mit Fundstellen im westlich anschließenden Pariser Becken, und den Küsten des Mittelmeeres und des Atlantiks verbindet. Die östliche Gruppe zeigt im Gegensatz dazu ein weniger engmaschiges Austauschmuster. Da hier die meisten Molluskenschalen vor Ort im Mährischen Karst, der Karpatensenke und dem Wiener Becken gesammelt werden können, sind großräumige Transportdistanzen möglicherweise im archäologischen Befund nicht sichtbar. Trotzdem belegen manche Exemplare Interaktionen zwischen Jäger-Sammler Gruppen in Böhmen, Mähren und im Wiener Becken. Belege für einen Austausch von Molluskenschalen zwischen der westlichen und der östlichen überregionalen Gruppe fehlen hingegen quasi völlig (siehe Kap. 6, "Procurement Pattern of Mollusk Shells and Amber").

Mit Hinblick auf diese Beobachtungen kann gesagt werden, dass die Art und Weise wie Kulturgut im MEM weitergegeben wurde einerseits eine allseitige Verbreitung gleicher typologischer, technologischer und darstellungsbezogener Konzepte leistete, andererseits aber auch die Ausprägung regionaler Unterschiede in der Häufigkeit (und somit Popularität) gewisser Konzepte erlaubte, die jeder Regionalgruppe einen individuellen Charakter verlieh. Auch wenn die genaue Art der Weitergabe immaterieller Kulturgüter jenseits des Auflösungsvermögens der vorhandenen Daten liegt, so scheint es, dass innerhalb einer Regionalgruppe die Neigung zur Konformität vergleichsweise stark war. Interaktion zwischen den Gruppen wirkte gleichzeitig einem konzeptuellen Auseinanderdriften entgegen, reduzierte die typologische Divergenz und führte somit zu weitreichenden Ähnlichkeiten, die abhängig vom Grad der Interaktion stärker oder schwächer ausgeprägt waren (siehe Kap. 3, "Types as Indicators of Interaction"). Diese inter-regionale Interaktion war wohl die Kraft, die das MEM in einem Zustand relativer genereller Ähnlichkeit hielt, die es heutigen Archäologen wiederum erlaubt diese Gruppen unter dem Begriff Magdalénien zusammenzufassen.

Zeitliche Diversität

Angesichts typologischer und technologischer Ähnlichkeiten zwischen Badegoulien- und älteren Epigravettien-Inventaren kann angenommen werden, dass die gemeinsame Entwicklung spätjungpaläolithischer Jäger-Sammler in Mitteleuropa bereits zwischen 23.500 und 20.000 cal BP wieder einsetzte. Während dieser Zeit sind die Inventare durch die häufige Verwendung von Abschlägen zur Geräteproduktion sowie dem Vorkommen von Nasenkratzern, Kielkratzern, Schabern charakterisiert. Mehrschlagstichel und Rückenmesser sind hingegen selten und Dreieck fehlen quasi völlig.

Der Zeitraum von etwa 20.000 bis 16.000 cal BP bildet typologisch einen deutlichen Kontrast zur vorangehenden Periode. Der Fokus der Grundformenproduktion lag nun deutlich auf der Herstellung von Klingen, die typischerweise mit einem weichen Schlagstein gewonnen und vorzugsweise zu Geräten modifiziert wurden. Die typologischen Charakteristika dieser Zeit sind Mehrschlagstichel, Rückenmesser und insbesondere Dreiecke sowie *Navettes* (Kap. 4, "The Subdivision of the Magdalenian into Six Phases and the Critique to Which It Is Subject" und "The Long Debate About Lithic Triangles"). Durch das Auftreten lithischer Dreiecke kann geschlossen werden, dass Komposit-Widerhakenspitzen ein Charakteristikum dieser Periode waren.

Lithische Dreiecke verschwinden um circa 16.000 cal BP wahrscheinlich zusammen mit Komposit-Widerhakenspitzen. Gleichzeitig breiten sich massive organische Widerhakenspitzen aus (Kap. 4, "The Long Debate About Lithic Triangles"). Dieser Wechsel des Jagdgeräts scheint gleichzeitig im gesamten Untersuchungsgebiet stattgefunden zu haben, was darauf hinweist, dass er mit einer Geschwindigkeit vonstatten ging, die unterhalb des Auflösungsvermögens der ^{14}C-datierung liegt. Abgesehen von diesem Wechsel bleibt das typologische Spektrum zwischen 16.000 und 14.000 cal BP weitestgehend gleich, allerdings sind die Inventare nun typologisch untereinander noch ähnlicher als in der vorangegangenen Periode der Fall war (Kap. 6, "Analysis of Temporal Grouping"). Die rasante Ausbreitung massiver Widerhakenspitzen und die zunehmende Ähnlichkeit der Inventare illustriert exemplarisch die Effektivität der Kommunikationsnetzwerke dieser Zeit (Kap. 9, "Small-World Networks: Backbone of the Magdalenian Society?"). Die Klingenproduktion fand vorwiegend mit einem organischen Schlägel und unter Verwendung der charakteristische *en éperon* Präparation statt (Kap. 6, "Spatial Distribution of Concepts of Lithic Tools").

Nach 14.000 cal BP gewannen mineralische Schlägel wieder an Wichtigkeit und die *en éperon* Präparation war nicht länger in Gebrauch. Die typologische Ähnlichkeit der Inventare nimmt wieder ab und rückengestumpfte Spitzen werden häufiger, was wohl den Übergang zu den verstreuter lebenden Gruppen des Spätpaläolithikums anzeigt (Kap. 6, "Spatial Distribution of Concepts of Lithic Tools").

Subsistenz

Im Hinblick auf das Subsistenzsystem spricht die archäologische Befundlage für die Hypothese, dass MEM Jäger-Sammler eher logistisch organisiert waren als einem Forager-System zu folgen (*sensu* Binford 1980). Unterschiede in der Artefaktzusammensetzung der Inventare deuten auf die Existenz von Basislagern und spezialisierten Plätzen wie Jagdlagern hin. Auf Basis der Rohmaterialbeschaffungsmuster können Territoriengrößen der fünf Regionalgruppen geschätzt werden, die zwischen 23.000 und 84.000 km^2 liegen. Analog zu rezenten Jäger-Sammler Gruppen erscheint die Annahme vernünftig, dass immer nur ein Teil dieser Territorien während jährlicher oder Mehrjähriger Zyklen genutzt wurde (siehe Kap. 7). Der Versuch diese jährlichen oder mehrjährigen Schweifgebiete zu identifizieren weist auf 3 bis 6 Schweifgebiete pro

Regionalgruppe hin mit Ausmaßen zwischen 5.000 und 18.000 km². Während die Landnutzungsmuster der meisten Regionalgruppen sehr ähnlich sind weist die Polnisch-Mährische Gruppe exzeptionelle Werte auf. Das geschätzte Territorium ist mehr als doppelt so groß wie der median der der übrigen Territorien, die Anzahl der Schweifgebiete ist doppelt so hoch wie in den meisten anderen Gruppen, die mittlere Distanz eines Residential Moves ist die größte im ganzen Untersuchungsgebiet und die Transportstrecken der Rohmaterialien sind oft sehr lang. Zwei mögliche Erklärungen sind hier denkbar. Einerseits ist es möglich, dass die Polnisch-Mährische Gruppe einen Palimpsest aus zwei regionalgruppen darstellt. Dies wird allerdings von dem engmaschigen Rohmaterialbeschaffungsmuster konterkariert, welches eine Teilung der Gruppe in zwei Teile unmöglich macht. Darüber hinaus übersieht eine solche Deutung den Zusammenhang zwischen gestiegener Territoriumsgröße und gestiegener Distanz eines Residential Moves. Andererseits könnten die unterschiedlichen Umweltbedingungen, insbesondere das Vorkommen von Bäumen, und die damit einhergehende mögliche geringere Dichte großer Herbivoren zu einer Zunahme der Schweifgebiete und darüber zu einer größeren Distanz der Residential Moves geführt haben. In einem solchen Subsistenzsystem, in dem große Distanzen in vergleichsweise kurzer Zeit zurückgelegt wurden, haben Artefakte mit einer längeren Biographie eine größere Chance über weite Entfernungen transportiert zu werden. Dies könnte im Resultat ein Muster ergeben, das sehr ähnlich zu dem für die Polnisch-Mährische Gruppe beobachteten Muster ist.

Die bidirektionale Wiederbesiedlung Mitteleuropas

Da die jeweils ältesten MEM Fundstellen entlang eines West-Ost-Gradienten nicht graduell jünger werden und die Verteilung von Konzepten und Objekten im archäologischen Befund des MEM diskontinuierlich verläuft und zwei Populationen widerspiegelt, erscheint die aktuelle Hypothese der unidirektionalen Ausbreitung von Jägern und Sammlern des Magdalénien aus Franko-Kantabrien bis zur Weichsel zunehmend unwahrscheinlich. Es scheint vielmehr, dass die Wiesebesiedlung Mitteleuropas ein bidirektionaler Prozess war.

Mit der wiedererstarkenden Interaktion zwischen Ost und West ab etwa 23.500 cal BP entwickelten sich beide Gemeinschaften parallel. Jedoch waren die klimatischen und demographischen Verhältnisse in Mitteleuropa anscheinend noch nicht günstig für eine flächendeckende Wiederbesiedlung. Erst während des Verlaufs der Klimaverbesserung nach dem LGM erweiterten beide Gemeinschaften sukzessive ihre Schweif- und Aktivitätsgebiete. Westliche Jäger-Sammler-Gruppen erreichten den Französischen Jura spätestens um 17.500 cal BP und siedelten bereits um 17.000 cal BP im Hegau. Um etwa 16.500 cal BP wurde die Schwäbische Alb und nicht vor 16.000 cal BP die Fränkische Alb wiederbesiedelt (siehe Kap. 8, "The Eastward Expansion"). Die Ausbreitung der östlichen Gemeinschaft erreichte Mähren bereits gegen 18.700 cal BP und lässt sich nördlich der Karpaten um 18.500 cal BP fassen. Das Gebiet nördlich der Sudeten wurde um 17.800 wiederbesiedelt und die sich ausbreitenden Jäger-Sammler erreichten Thüringen um 16.100 cal BP. Die ältesten Daten aus Böhmen weisen auf eine Besiedlung dieses Gebiets spätestens um 15.500 cal BP hin (siehe Kap. 8, "The Westward Expansion").

Das Mitteleuropäische Magdalénien: ein „Small-World" Netzwerk?

Im Hinblick auf die Verteilungsmuster typologischer, technologischer und darstellerischer Konzepte scheint es, dass das soziokulturelle System der spätjungpaläolithischen Jäger-Sammler stark einem „small-world" Netzwerk ähnelt. Da anzunehmen ist, dass kleine bis mittelgroße Jäger-Sammler Gruppen mit starken Verwandtschaftlichen Bindungen die wichtigsten sozialen Einheiten dieser Zeit darstellten, erscheint es wahrscheinlich, dass das Kommunikationsnetzwerk stark geclustert und die meisten Verbindungen innerhalb des Netzwerks lokaler beziehungsweise regionaler Natur waren. Regelmäßige überregionale Zusammenkünfte und Heiraten zwischen Regionalgruppen führten daneben wahrscheinlich zur Ausbildung einiger weniger überregionaler Verbindungen zwischen den Regionalgruppen, die die charakteristische Pfadlänge innerhalb der Netzwerke beträchtlich verkürzten. (siehe Kap. 9, "Small-World Networks: Backbone of the Magdalenian Society?").

References

Absolon, K. (1945). *Praehistorický výzkum jeskyně Býčí skály na Moravě na srovnávacím základě, III. kritický příspěvek k poznání praaurignacien*. Brno: Nákladem Akciové Moravské knihtiskárny polygrafie.

Absolon, K., & Czižek, R. (1926). Palaeolithický výzkum jeskyně Pekárny na Moravě. *Časopis Moravského musea, scientiae sociales, 24*, 1–59.

Absolon, K., & Czižek, R. (1927). Palaeolithický výzkum jeskyně Pekárny na Moravě. *Časopis Moravského musea, scientiae sociales, 25*, 112–201.

Absolon, K., & Czižek, R. (1932). Palaeolithický výzkum jeskyně Pekárny na Moravě. *Časopis Moravského musea, scientiae sociales*, 26–27.

Adaileh, A. (2011). *Die Oberflächenfunde der Magdalénien-Fundstation Bad Kösen-Lengefeld aus dem Landesmuseum für Vorgeschichte in Halle*. Master thesis, University of Cologne.

Adler, D., & Murdoch, D. (2011). *rgl: 3D Visualization Device System (OpenGL). R package version 0.92.798*. http://CRAN.R-project.org/package=rgl

Adrian, W. (1982). *Die Altsteinzeit in Oswestfalen und Lippe*. Köln: Böhlau.

Affolter, J. (1997). Description pétrographique. In D. Leesch (Ed.), *Un campement magdalénien au bord du lac de Neuchâtel. Cadre chronologique et culturel, mobilier et structures, analyse spatiale (secteur 1)* (pp. 68–87). Neuchâtel: Musée Cantonal d'Archéologie.

Affolter, J. (2006). Origine de matières premières. In J. Bullinger, D. Leesch, & N. Plumettaz (Eds.), *Le site magdalénien de Monruz, 1. Premiers éléments pour l'analyse d'un habitat de plein air* (pp. 75–76). Neuchâtel: Musée Cantonal d'Archéologie.

Affolter, J. (2009a). Les Douattes (Musièges, Haute-Savoie): les séries magdaléniennes et aziliennes des fouilles Jayet et Pion (1999–2002). In G. Pion (Ed.), *La fin du Paléolithique supérieur dans les Alpes du nord françaises et le Jura méridional. Approches culturelles et environmentales* (pp. 161–171). Paris: Société Préhistorique Française.

Affolter, J. (2009b). La grotte de Bange (Allèves, Haute-Savoie). In G. Pion (Ed.), *La fin du Paléolithique supérieur dans les Alpes du nord françaises et le Jura méridional. Approches culturelles et environmentales* (pp. 173–181). Paris: Société Préhistorique Française.

Affolter, J., & Bressy, C. (2009). Les matières premières siliceuse: méthodes d'études et ressources. In G. Pion (Ed.), *La fin du Paléolithique supérieur dans les Alpes du nord françaises et le Jura méridional. Approches culturelles et environmntales* (pp. 143–159). Paris: Société Préhistorique Française.

Affolter, J., Cattin, M.-I., Leesch, D., Morel, P., Plumettaz, N., Thew, N., et al. (1994). Monruz – Une nouvelle station magdalénienne au bord du lac de Neuchâtel. *Archäologie der Schweiz, 17*(3), 94–104.

Ahlheim, K.-H., & Preuß, G. (Eds.). (1983). *Meyers Großes Universallexikon* (Vol. 9). Mannheim: Bibliographisches Institut.

Albert, T. (2014). *Die Silexartefakte aus dem Nordschnitt der Magdalénien-Freilandstation Bad Kösen-Lengefeld*. Master thesis, University of Erlangen-Nürnberg.

Albrecht, G. (1974). Neue Untersuchungen am Petersfels bei Engen im Hegau. *Archäologisches Korrespondenzblatt, 4*, 285–287.

Albrecht, G. (1979). *Magdalénien-Inventare vom Petersfels. Siedlungsarchäologische Ergebnisse der Ausgrabungen 1974–1976*. Tübingen: Archaeologica Venatoria.

Albrecht, G. (1981a). Magdalénienfunde aus der Gnirshöhle bei Engen im Hegau. *Mitteilungsblatt der Archaeologica Venatoria e.V., 1*, 16–18.

Albrecht, G. (1981b). Die neuen Ausgrabungen in Munzingen 1976/77. *Mitteilungsblatt der Archaeologica Venatoria e.V., 2*, 21–23.

Albrecht, G. (1983). Die Jäger der späten Eiszeit. In H. Müller-Beck (Ed.), *Urgeschichte in Baden-Württemberg* (pp. 331–353). Stuttgart: Theiss.

Albrecht, G. (1984). Zwei Sammlungen von Silexartefakten vom Abri "Klemmer" im Eselsburger Tal. In J. Hahn (Ed.), *Die steinzeitliche Besiedlung des Eselsburger Tales bei Heidenheim (Schwäbische Alb)* (p. 148). Stuttgart: Theiss.

Albrecht, G., Berke, H., & Poplin, F. (Eds.). (1983). *Naturwissenschaftliche Untersuchungen an Magdalénien-Inventaren vom Petersfels, Grabungen 1974–1976*. Tübingen: Archaeologica Venatoria.

Albrecht, G., Campen, I., Hahn, J., Kind, J., Lindenborn, J., Torke, W. G., et al. (1977a). Eine Nachgrabung in der Burkhardtshöhle, Gemeinde Westerheim, Alb-Donau-Kreis. *Fundberichte aus Baden-Württemberg, 3*, 6–13.

Albrecht, G., Drautz, D., & Kind, C.-J. (1977b). Eine Station des Magdalénien in der Gnirshöhle bei Engen-Bittelbrunn im Hegau. *Archäologisches Korrespondenzblatt, 7*, 161–179.

Albrecht, G., & Hahn, A. (1991). *Rentierjäger im Brudertal am Ende der Eiszeit: die jungpaläolithischen Fundstellen um den Petersfels und die urgeschichtliche Abteilung des Städtischen Museums Engen*. Stuttgart: Theiss.

Alexandrowicz, S. W., Drobniewicz, B., Ginter, B., Kozłowski, J. K., Madeyska, T., Nadachowski, A., et al. (1992). Excavations in the Zawalona Cave at Mników (Cracow upland, southern Poland). *Folia Quaternaria, 63*, 43–76.

Allain, J. (1968). A propos du Badegoulien: méthode et typologie. *Bulletin de la Société Préhistorique Française, 65*(2), 36–38.

Allain, J. (1979). L'industrie lithique et osseuse de Lascaux. In A. Leroi-Gourhan & J. Allain (Eds.), *Lascaux inconnu* (pp. 87–145). Paris: Éditions du CNRS.

Allain, J. (1989). La fin du Paléolithique supérieur en région Centre. In J.-P. Rigaud (Ed.), *Le Magdalénien en Europe. "La structuration du Magdalénien". Actes du Colloque de Mayence 1987 dans le cadre du XIème Congrès U.I.S.P.P* (ERAUL, Vol. 38, pp. 193–217). Liège: Université de Liège.

Andreas Maier, *The Central European Magdalenian*, Vertebrate Paleobiology and Paleoanthropology,
DOI 10.1007/978-94-017-7206-8

Allain, J., Desbrosse, R., Kozłowski, J. K., Rigaud, A., Jeannet, M., & Leroi-Gourhan, A. (1985). Le Magdalénien à Navettes. *Gallia Préhistoire, 28*, 37–124.

Allain, J., & Fritsch, R. (1967). Le Badegoulien de l'abri Fritsch aux Roches de Pouligny-Saint-Pierre (Indre). *Bulletin de la Société Préhistorique Française, 64*, 83–93.

Allain, J., & Rigaud, A. (1986). Décor et Fonction, quelques exemples tirés du Magdalénien. *L'Anthropologie, 90*, 713–738.

Álvarez Alonso, D. (2007). The Cantabrian Magdalenian. Lateglacial chronology in Northern Iberian Peninsula. *Portugalia, N.S., 27–28*, 5–18.

Álverez Fernández, E. (1999). Las Perlas de madera fósil del terciario y los objetos de adorno-colgantes sobre dientes de zorro y ciervo del magdaleniense de Gönnersdorf y de Andernach-Martinsberg-2, (Neuwied, Rheinland Pfalz, Alemania). *Zephyrus, 52*, 79–106.

Álvarez Fernández, E. (2001). L'axe Rhin-Rhône au Paléolithique supérieur récent: l'example des mollusques utilisés comme objets de parure. *L'Anthropologie, 105*, 547–564.

Álvarez Fernández, E. (2005). Éloignés mais pas isolés": la parure hors de la "frontière française" pendant le Magdalénien. In V. Dujardin (Ed.), *Industrie osseuse et parures du Solutréen au Magdalénien en Europe. Table ronde sur le paléolithique supérieur récent. Angoulême (Charente) 28–30 Mars 2003*. Paris: Société Préhistorique Française.

Álvarez Fernández, E. (2009). Magdalenian personal ornaments on the move: A review of the current evidence in Central Europe. *Zephyrus, LXIII*, 45–59.

Ammann, B. (1985). Lobsigensee—Late-glacial and Holocene environments of a lake on the Central Swiss Plateau. 1 Introduction and Palynology: Vegetational history and core correlation at Lobsigensee (Swiss Plateau). In G. Lang (Ed.), *Swiss lake and mire environments during the last 15000 years* (pp. 127–134). Vaduz: Lubrecht and Cramer.

Ammann, B., Bandi, H.-G., & Buser, M. (1988). *Neue Untersuchungen am Kesslerloch bei Thayngen/SH. Sondierbohrungen im östlichen Vorplatzbereich und ihre naturwissenschaftlich-archäologische Auswertung*. Basel: Schweizerische Gesellschaft für Ur- u. Frühgeschichte.

Andersen, K. K., Bigler, M., Clausen, H. B., Dahl-Jensen, D., Johnsen, S. J., Rasmussen, S. O., et al. (2007). A 60,000 year Greenland stratigraphic ice core chronology. *Climate of the Past Discussions, 3*, 1235–1260.

Andersen, K. K., Svensson, A., Johnsen, S. J., Rasmussen, S. O., Bigler, M., Röthlisberger, R., et al. (2006). The Greenland ice core chronology 2005, 15-42 kyr. Part 1: Constructing the time scale. *Quaternary Science Reviews, 25*, 3246–3257.

Anderson, M. J. (2001). A new method for non-parametric multivariate analysis of variance. *Austral Ecology, 2001*(26), 32–46.

Anderson, M. J. (2006). Distance-based tests for homogeneity of multivariate dispersions. *Biometrics, 62*, 245–253.

Andrée, J. (1927). Altsteinzeitliche Funde aus Westfalen. Die Bilsteinhöhle bei Warstein i.W. *Mannus, 19*, 87–91.

Andrée, J. (1928). *Das Paläolithikum der Höhlen des Hönnetales in Westfalen*. Leipzig: Kabitzsch.

Andrée, J. (1939). *Der eiszeitliche Mensch in Deutschland und seine Kulturen*. Stuttgart: Enke.

Andrée, J., & Grimm, P. (1929). Das Herdloch (Herthahöhle) bei Ranis, Kr. Ziegenrück. *Jahresschrift für die Vorgeschichte der sächsisch-thüringischen Länder, 17*, 1–15.

Arambourou, R. (1978). *Le gisement préhistorique de Duruthy à Sorde-l'Abbaye (Landes). Bilan des recherches de 1958 à 1975*. Mémoires de la Société Préhistorique Française (Vol. 13). Paris: Ed. SPF.

Arts, N., & Deeben, J. (1987). On the northwestern border of Late Magdalénian territory: Ecology and archaeology of early late glacial band societies in Northwestern Europe. In J. M. Burdukiewicz & M. Kobusiewicz (Eds.), *Late glacial in central Europe. Culture and environment* (pp. 25–66). Wrocław: Zakład Narodowy.

Ascher, R. (1961). Analogy in archaeological interpretation. *Southwestern Journal of Anthropology, 17*, 317–325.

Aubry, D., Guélat, M., Detrey, J., & Blaise, O.-G. (2000). *Dernier cycle glaciaire et occupations paléolithiques à Alle, Noir Bois (Jura, Suisse)*. Porrentruy: Cercle d'archéologie de la société jurassienne d'émulation.

Audouze, F., Karlin, C., Cahen, D., De Croisset, E., Coudret, P., Larrière, M., et al. (1988). Taille du silex et finalité du débitage dans le Magdalénien du bassin Parisien. In M. Otte (Ed.), *De la Loire à l'Oder. Les civilisations du paléolithique final dans le nord-ouest europeen: actes du colloque de Liège, decembre 1985* (BAR International Series, Vol. 444, pp. 55–84). Oxford: Archaeopress.

Auerbach, A. (1929). Die Lindentaler Hyänenhöhle bei Gera. *Die Thüringer Höhlen, 1*(3), 46–56.

Auerbach, A. (1930). *Die vor- und frühgeschichtlichen Altertümer Ostthüringens*. Jena: Fischer.

Auffermann, B. (1991). Die Bruckersbergstationen, Gemeinde Giengen, Kreis Heidenheim. In J. Hahn & C.-J. Kind (Eds.), *Urgeschichte in Oberschwaben und der mittleren Schwäbischen Alb. Zum Stand neuerer Untersuchungen der Steinzeit-Archäologie* (pp. 122–123). Stuttgart: Röck.

Auffermann, B. (1998). *Rohmaterialnutzung im Magdalénien. Fundstellen am Nordrand der Schwäbischen Alb*. Bad Bellingen: Folio-Verlag.

Auffermann, B., Burkert, W., Hahn, J., Pasda, C., & Simon, U. (1990). Ein Merkmalsystem zur Auswertung von Steinartefaktinventaren. *Archäologisches Korrespondenzblatt, 20*, 259–268.

Aura Tortosa, J. E. (1989). Solutrenses y Magdalenienses al sur del Ebro. Primera aproximación a un proceso de cambio tecnoindustrial: el ejemplo de Parpalló. *Saguntum, 22*, 35–65.

Baales, M. (1996). *Umwelt und Jagdökonomie der Ahrensburger Rentierjäger im Mittelgebirge*. Bonn: Habelt.

Baales, M. (2002). *Der spätpaläolithische Fundplatz Kettig. Untersuchungen zur Siedlungsarchäologie der Federmessergruppen am Mittelrhein*. Monographien des Römisch-Germanischen Zentralmuseums 51, Mainz.

Baales, M. (2005a). Ein kurzer Gang durch die älteste Geschichte Westfalens. *Archäologie in Ostwestfalen, 9*, 10–37.

Baales, M. (2005b). Die steinzeitlichen Funde aus der Bilsteinhöhle im südwestfälischen Bergland. In H. Obermaier-Gesellschaft (Ed.), *Hugo Obermaier-Gesellschaft für Erforschung des Eiszeitalters und der Steinzeit e.V. 47. Tagung der Gesellschaft in Neuchâtel, Schweiz, 29. März – 02. April 2005* (pp. 10–11). Erlangen: PrintCom.

Baales, M. (2006a). Die Feldhofhöhle bei Volkringhausen, Stadt Balve, Märkischer Kreis. In H. G. Horn (Ed.), *Neandertaler + Co* (pp. 122–125). Mainz: Zabern.

Baales, M. (2006b). Der Kartstein, Weyer, Stadt Mechernich, Kreis Euskirchen. In H. G. Horn (Ed.), *Neandertaler und Co* (pp. 176–179). Mainz: Zabern.

Backhaus, K., Erichson, B., Plinke, W., & Weiber, R. (2008). *Multivariate Analysemethoden. Eine anwendungsorientierte Einführung*. Berlin: Springer.

Bahn, P. G. (1985). *Pyrenean prehistory. A palaeoeconomic survey of the French sites*. Warminster: Aris and Phillips.

Bahn, P. G., & Vertut, J. (1988). *Images of the ice age*. Leicester: Windward.

Bally, E., Heierli, J., Schwerz, F., & Hescheler, K. (1908). Höhlenfunde im sog. Käsloch bei Winznau (Kt. Solothurn). *Anzeiger für Schweizerische Altertumskunde, N.F. X*(1), 1–12.

BANADORA Banque Nationale de Données Radocarbone pour l'Europe et le Proche Orient. (2012). http://www.archeometrie.mom.fr/banadora

Bandi, H.-G. (1947). *Die Schweiz zur Rentierzeit*. Frauenfeld: Huber & Co.

Bandi, H.-G. (1960). Das Jungpaläolithikum. In W. Drack (Ed.), *Die aeltere und mittlere Steinzeit der Schweiz* (pp. 31–36). Basel: Schweizerische Gesellschaft für Ur- und Frühgeschichte.
Bandi, H.-G. (1968). Das Jungpläolithikum. In W. Drack (Ed.), *Ur- und frühgeschichtliche Archäologie der Schweiz. Band I, Die Ältere und Mittlere Steinzeit* (pp. 107–122). Basel: Schweizerische Gesellschaft für Ur- und Frühgeschichte.
Bandi, H.-G., Lüdin, C., Mamber, W., Schaub, S., Schmid, E., & Welten, M. (1954). Die Brügglihöhle an der Kohlholzhalde bei Nenzlingen (Kt. Bern), eine neue Fundstelle des Spätmagdalénien im unteren Birstal. *Jahrbuch des Bernischen Historischen Museums, 32*(33), 45–76.
Bando, C., Dagnan-Ginter, A., Kozlowski, J. K., Montet-White, A., Pawlikowski, M., & Sobczyk, K. (1993). Fosses d'extraction et ateliers de taille à Wolowice, près de Cracovie, Pologne. *L'Anthropologie, 97*(2/3), 271–290.
Bánesz, L., Hromada, J., Desbrosse, R., Margerand, I., Kozlowski, J. K., Sobczyk, K., et al. (1992). Le site de plein air du Paléolithique supérieur de Kašov 1 en Slovaquie orientale (Etude préliminaire d'une structure spatiale des outillages épigravettiens en obsidienne). *Slovenská Archeológia, XL*(1), 5–28.
Barandiarán, I. (1967). *El Paleomesolitico del Pireneo Occidental: Bases para una sistematización typológica del instrumental óseo paleolítico.* Monografias Arqueológicas III, Seminario de Prehistoria y Protohistoria, Facultad de Filosofía y Letras, Zaragoza.
Barner, W. (1962). Die würmeiszeitliche Rentierjägerstation vom Aschenstein bei Freden, Kreis Alfeld. *Nachrichten aus Niedersachsens Urgeschichte, 31*, 115–120.
Barr, J. H. (1977). Die Rislisberghöhle, ein neuer Magdalénien-Fundplatz im Schweizer Jura. *Archäologisches Korrespondenzblatt, 7*, 85–87.
Barth, F. (Ed.). (1969). *Ethnic groups and boundaries. The social organization of culture difference.* Boston: Little, Brown & Co.
Bartlett, M. S. (1947). Multivariate analysis. *Journal of the Royal Statistical Society, 9*, 176–197.
Barton, R. N. E., Jacobi, R. M., Stapert, D., & Street, M. J. (2003). The late-glacial reoccupation of the British Isles and the Creswellian. *Journal of Quaternary Science, 18*, 631–643.
Baumann, W. (1968). Jungpaläolithische Funde aus den Fluren von Baselitz und Wantewitz, Kr. Großenhain. *Ausgrabungen und Funde, 13*(2), 58–62.
Bay, R. (1953). Die Magdalénienstation am Hollenberg bei Arlesheim (Kanton Baselland). *Tätigkeitsberichte der Naturforschenden Gesellschaft Baselland, 19*, 164–178.
Bayer, J. (1925). Die ältere Steinzeit in den Sudetenländern. *Sudeta, 1*, 19–120.
Bayer, J. (1927). Die Teufelslucken bei Eggenburg in Nieder-Österreich, eine Station de Eiszeitmenschen. *Die Eiszeit, 4*, 104–107.
Bayer, J. (1928). Das zeitliche und kulturelle Verhältnis zwischen den Kulturen des Schmalklingenkulturkreises während des Diluviums in Europa. *Eiszeit und Urgeschichte, 5*, 9–23.
Bayer, J. (1929). Die Olschewakultur. *Eiszeit und Urgeschichte, 6*, 83–100.
Beck, C. W. (1997). Détermination de la provenance des résines fossiles par l'analyse spectrale en infrarouge. In D. Leesch (Ed.), *Un campment magdalénien au bord du lac de Neuchâtel. Cadre chronologique et culturel, mobilier et structures, analyse spatiale (secteur 1)* (pp. 105–107). Neuchâtel: Musée Cantonal d'Archéologie.
Becker, C. J. (1970). Eine Kerbspitze der Hamburger Stufe aus Jütland. In K. Gripp, R. Schütrumpf, & H. Schwabedissen (Eds.), *Frühe Menschheit und Umwelt, Teil 1* (pp. 362–364). Köln: Böhlau.
Becker, C. J. (1971). Late Palaeolithic finds from Denmark. *Proceedings of the Prehistoric Society, 37*(2), 131–139.
Bednarik, R. G. (2008). Pedogenetic dating of loess strata. *Journal of Archaeological Sciences, 35*(12), 3124–3129.
Behlen, H. (1905). Eine neue Nachgrabung vor der Steedener Höhle Wildscheuer, nebst einem Exkurs über die diluvialen Höhlenablagerungen im allgemeinen. *Nassauische Annalen, 35*, 290–307.
Behlen, H. (1909). Die Steedener Höhle Wildscheuer. *Nassauische Annalen, 39*, 219–351.
Behm-Blancke, G. (1956). Magdalénienzeitliche Zeltplätze eines Wildpferdjägerlagers im Kyffhäusergebiet bei Bad Frankenhausen. *Ausgrabungen und Funde, 1*, 263–266.
Behm-Blancke, G. (1960). Zur Typologie der jungpaläolithischen Zelt- und Hüttenanlagen Europas. *Ausgrabungen und Funde, 5*, 203–209.
Behm-Blancke, G. (1961). Das Paläolithikum in Thüringen. *Geologie, 10*(4/5), 550–569.
Behm-Blancke, G. (1970). Eine Kultstätte magdalénienzeitlicher Wildpferdjäger in Thüringen. In J. Filip (Ed.), *Actes du VIIe Congrès International des Sciences Préhistoriques et Protohistoriques* (pp. 360–363). Prague: Academia.
Behm-Blancke, G. (1976). Das jungpaläolithische Zeltlager von Oelknitz bei Jena. *Ausgrabungen und Funde, 21*, 30–32.
Behre, K.-E. (1967). The late glacial and early postglacial history of vegetation and climate in northwestern Germany. *Review of Palaeobotany and Palynology, 4*, 149–161.
Bellier, C., & Cattelain, P. (1986). Le Trou de Blaireaux à Vaucelles. *Helinium, 26*, 46–57.
Benecke, N., Küßner, M., & Al, E. (2006). Zur Datierung und Fauna des spätglazialen Schichtkomplexes am Abri Fuchskirche I bei Allendorf, Lkr. Saalfeld-Rudolstadt. *Alt-Thüringen, 39*, 121–130.
Bennett, K. D., Tzedakis, P. C., & Willis, K. J. (1991). Quaternary refugia of north European trees. *Journal of Biogeography, 18*, 103–115.
Bennike, O., Sarmaja-Korjonen, K., & Seppänen, A. (2004). Reinvestigation of the classic late-glacial Bølling Sø sequence, Denmark: Chronology, macrofossils, Cladocera and chydorid ephippia. *Journal of Quaternary Science, 19*(5), 465–478.
Bentley, R. A., & Maschner, H. D. G. (2008). Complexity theory. In R. A. Bentley, H. D. G. Maschner, & C. Chippindale (Eds.), *Handbook of archaeological theories* (pp. 245–270). Lanham, MD: Rowman and Littlefield.
Bentley, A., & Shennan, S. J. (2003). Cultural transmission and stochastic network growth. *American Antiquity, 68*(3), 459–485.
Benzécri, J.-P. (1973). *L'analyse des données. Tome 1: La taxinomie. Tome 2: L'analyse des correspondance.* Paris: Dunod.
Berg, F. (1966). Die prähistorischen Funde. In K. Ehrenberg (Ed.), *Die Teufels- oder Fuchsenlucke bei Eggenburg (N.Ö.)* (pp. 123 136). Wien: Springer.
Berglund, B. E., Björck, S., Lemdahl, G., Bergsten, H., Nordberg, K., & Kolstrup, E. (1994). Late Weichselian environmental change in southern Sweden and Denmark. *Journal of Quaternary Science, 9*, 127–132.
Bergmann, J. (1964). *Urgeschichte des Wolfhagener Landes.* Kassel: Schanze.
Berke, H. (1987). *Archäozoologische Detailuntersuchungen an Knochen aus südwestdeutschen Magdalénien-Inventaren.* Tübingen: Archaeologica Venatoria.
Berke, H. (1989). Archaeozoology and site catchment in the Magdalenian: Solutré, Petersfels, Pekarna Cave, Kniegrotte. A preliminary report. *Early Man News, 14*, 15–32.
Bernbeck, R. (1997). *Theorien in der Archäologie.* Tübingen: Francke.
Bernhardt, W., & Rudolph, A. (1997). Artefaktfunde der Altsteinzeit aus der Elster-Luppe-Aue bei Schkeuditz. *Veröffentlichungen des Naturkundemuseum Leipzig, 15*, 1–32.
Bettinger, R. L., Boyd, R., & Richerson, P. J. (1996). Style, function and cultural evolutionary processes. In H. D. G. Maschner (Ed.), *Darwinian archaeologies* (pp. 133–164). New York: Plenum.

Bettinger, R. L., & Eerkens, J. W. (1999). Point typologies, cultural transmission, and the spread of bow-and-arrow technology in the prehistoric Great Basin. *American Antiquity, 64*, 231–242.

Beug, H.-J. (1976). Die spätglaziale und frühpostglaziale Vegetationsgeschichte im Gebiet des ehemaligen Rosenheimer Sees (Oberbayern). *Botanische Jahrbücher für Systematik, 95*(3), 373–400.

Bicker, F. K. (1939). Ein mitteldeutsches Spät-Aurignacien bei Aschersleben. In J. Andrée (Ed.), *Der eiszeitliche Mensch und seine Kulturen* (pp. 72–105). Stuttgart: Enke.

Binder, H. (1977). *Höhlenführer Schwäbische Alb*. Stuttgart: Theiss Verlag.

Binder, H., & Jantschke, H. (2003). *Höhlenführer Schwäbische Alb*. Leinfelden-Echterdingen: DRW-Verlag.

Binford, L. R. (1968). Archaeological perspectives. In S. R. Binford & L. R. Binford (Eds.), *New perspectives in archaeology* (pp. 5–32). Chicago: Aldine.

Binford, L. R. (1972). Contemporary model building: Paradigms and the current state of Palaeolithic research. In D. L. Clarke (Ed.), *Models in archaeology* (pp. 109–166). London: Methuen.

Binford, L. R. (1973). Interassemblage variability – the Mousterian and the "functional" argument. In C. Renfrew (Ed.), *The explanation of culture change* (pp. 227–254). London: Duckworth.

Binford, L. R. (1978). *Nunamiut ethnoarchaeology*. New York: Academic Press.

Binford, L. R. (1979). Organization and formation processes: Looking at curated technologies. *Journal of Anthropological Research, 35*(3), 255–273.

Binford, L. R. (1980). Willow smoke and dogs' tails: Hunter-gatherer settlement systems and archaeological site formation. *American Antiquity, 45*(1), 4–20.

Binford, L. R. (1982). The archaeology of place. *Journal of Anthropological Archaeology, 1*, 5–31.

Binford, L. R. (1983a). *In pursuit of the past*. London: Thames and Hudson.

Binford, L. R. (1983b). *Working at archaeology*. New York: Academic Press.

Binford, L. R. (2001). *Constructing frames of reference. An analytical method for archaeological theory building using hunter-gatherer and environmental data sets*. Berkeley: University of California Press.

Binford, L. R., & Binford, S. R. (1966). A preliminary analysis of functional variability in the Mousterian of Levallois facies. *American Anthropologist, 68*, 238–295.

Binford, L. R., & Sabloff, J. A. (1982). Paradigms, systematics and archaeology. *Journal of Anthropological Research, 38*, 137–153.

Bintz, P. (1985). Choranche: Balme-Rousse. *Gallia Préhistoire, 28*(2), 404–405.

Bintz, P. (1995). Le site magdalénien et azilien de Bobache, La Chapelle-en-Vercors (Drôme). In P. Bintz & R. Picavet (Eds.), *Epipaléolithique et mésolithique en Europe: prehistoire et quaternaire en Vercors: livret – guide de l'excursion: Universités Joseph Fourier et Pierre Mendès France, Grenoble 18–23 septembre 1995* (pp. 118–123). Grenoble: Université de Grenoble.

Bintz, P., & Grunwald, C. (1995). Les grottes Jean-Pierre 1 et 2 à Saint-Thibaud-de-Couz (Savoie). Paléoenvironnement et cultures du Tardiglaciaire à l'Holocène dans les Alpes du Nord. I. Analyse de l'industrie lithique. *Gallia Préhistoire, 37*, 155–272.

Bintz, P., & Roche, J.-M. (1995). Un atelier de taille de la fin du Paléolithique supérieur au col de la Charmette (Proveysieux, Isère). In P. Bintz & R. Picavet (Eds.), *Epipaléolithique et mésolithique en Europe: prehistoire et quaternaire en Vercors: livret – guide de l'excursion: Universités Joseph Fourier et Pierre Mendès France, Grenoble 18–23 septembre 1995*. Grenoble: Université de Grenoble.

Birkner, F. (1926). Das Jungpaläolithikum in Bayern. *Der Bayerische Vorgeschichtsfreund, 6*, 13–16.

Birkner, F. (1936). *Ur- und Vorzeit Bayerns*. München: Knorr and Hirth.

Björck, S., Walker, M. J., Cwynar, L. C., Johnsen, S., Knudsen, K.-L., Lowe, J. J., et al. (1998). An event stratigraphy for the last termination in the North Atlantic region based on the Greenland ice-core record: A proposal by the INTIMATE group. *Journal of Quaternary Science, 13*(4), 283–292.

Bleich, K. E. (1965). Die ur- und frühgeschichtlichen Funde der Rosensteinhöhlen. *Jahreshefte für Karst- und Höhlenkunde, 6*, 159–160.

Blockley, S. P. E., Donahue, R. E., & Pollard, A. M. (2000a). Radiocarbon calibration and the late glacial occupation of northwest Europe. *Antiquity, 74*, 112–119.

Blockley, S. P. E., Donahue, R. E., & Pollard, A. M. (2000b). Rapid human response to late glacial climate change: A reply to Housley et al. (2000). *Antiquity, 74*, 427–428.

Bocheński, Z., Ginter, B., Kozłowski, J. K., Mook, W. G., Muszyński, M., Nadachowski, A., et al. (1985). Badania osadów schronisk podskalnych w Zalasie koło Krakowa. *Folia Quaternaria, 56*, 3–56.

Bocquet, A. (1965). Une nouvelle station magdalénienne dans le Vercors. *Bulletin de la Société Dauphinoise d'Ethnologie et d'Archéologie, 41*, 66–68.

Bocquet, A. (1969). L'Isère préhistorique et protohostorique. *Gallia Préhistoire, 12*(1–2), 121–258, 273–400.

Bocquet, A., & Lequatre, P. (1973). La grotte des Freydières à Saint-Agnan-en-Vercors (Drôme). Gisement du Magdalénien final. *Bulletin de la Société Préhistorique Française, 70*, 324–329.

Bocquet-Appel, J.-P., Demars, P.-Y., Noiret, L., & Dobrowsky, D. (2005). Estimates of upper Palaeolithic meta-population size in Europe from archaeological data. *Journal of Archaeological Science, 32*, 1656–1668.

Bodu, P. (1993). *Analyse typo-technologique du matériel lithique de quelques unités du site magdalénien de Pincevent (Seine-et-Marne). Applications spatiale, économiques et sociales*. Ph.D. Dissertation, University of Paris-I.

Bodu, P., Julien, M., Valentin, B., & Debout, G. (2006). Un dernier hiver à Pincevent: les Magdaléniens du niveau IV0 (Pincevent, La Grande Paroisse, Seine-et-Marne). *Gallia Préhistore, 48*, 1–180.

Bodu, P., & Valentin, B. (1997). Groups à Federmesser ou Aziliens dans le sud et l'ouest du Bassin parisien. Propositions pour un nouveau modèle d'évolution. *Bulletin de la Société Préhistorique Française, 94*(3), 341–348.

Boecking, H. (1992). Jung- und Endplaäolithische Feuersteinwerkzeuge vom Mittellauf der Mosel. *Bulletin de la Sociétße Préhistorique Luxembourgeoise, 14*, 39–73.

Boëda, E. (2013). *Techno-logique et technologie: une paléo-histoire des objets tranchnants*. Paris: Archeo Editions.

Bohmers, A. (1947). Jong-Palaeolithicum en Vroeg-Mesolithikum. In H. E. V. Gelder, P. Glazema, G. A. Bontekoe, H. Halbertsma, & W. Glasbergen (Eds.), *Oudheidkundig bodemderzoek in Nederland, Gedenkboek A.E. van Giffen* (pp. 129–201). Meppel: Boom & Zoon.

Bohmers, A. (1951). *Die Höhlen von Mauern*. Groningen: Wolters.

Bohmers, A. (1956). Statistics and graphs in the study of flint assemblages. II. A preliminary report on the statistical analysis of the Younger Palaeolithic in Northwestern Europe. *Palaeohistoria, 5*, 7–26.

Bohmers, A. (1960). Statistiques et Graphiques dans l'Étude des industries lithiques préhistoriques. *Palaeohistoria, 8*, 15–37.

Bohnsack, D. (1961). Ausgrabungen und Funde in Hamburg 1958–1960. *Hammaburg, XIII*, 143–168.

Bolus, M. (2012). Rückenmesser. In H. Floss (Ed.), *Steinartefakte vom Altpaläolithikum bis in die Neuzeit* (pp. 429–434). Tübingen: Kerns.

Boone, J. L., & Smith, E. A. (1998). Is it evolution yet? A critique of evolutionary archaeology. *Current Anthropology, 39*, 141–173.

Bordes, F. (1958). Nouvelles fouilles à Laugerie-Haute Est. Premiers résultats. *L'Anthropologie, 62*, 205–244.

Bordes, F. (1967). Considérations sur la typologie et les techniques dans le Paléolithique. *Quatär, 18*, 25–55.

Bordes, F. (1972). Du paléolithique moyen au paléolithique supérieur – continuité ou discontinuité? In F. Bordes (Ed.), *The origin of Homo sapiens* (pp. 211–218). Paris: UNESCO.

Bordes, F. (1973). On the chronology and contemporaneity of different paleolithic cultures in France. In C. Renfrew (Ed.), *The explanation of culture change: Models in prehistory* (pp. 217–266). London: Duckworth.

Bordes, F., & Sonneville-Bordes, D. D. (1970). The significance of variability in palaeolithic assemblages. *World Archaeology, 2*, 61–73.

Bortenschlager, S. (1967). Pollenanalytische Untersuchungen des Seemooses im Lungau (Salzburg). *Verhandlungen der Zoologisch-Botanischen Gesellschaft in Wien, 107*, 57–74.

Bortenschlager, S. (1984). Die Vegetationsentwicklung im Spätglazial: Das Moor beim Lanser See III, ein Typprofil für die Ostalpen. In G. Lang (Ed.), *Festschrift Max Welten* (pp. 71–79). Vaduz: Lubrecht and Cramer.

Bos, J. A. A. (1998). *Aspects of the lateglacial-early Holocene vegetation development in Western Europe. Palynological and palaeobotanical investigations in Brabant (The Netherlands) and Hessen (Germany).* Utrecht: LPP Foundation.

Bosinski, H. (1977). Die Rondelle des Magdalénien-Fundplatzes Gönnersdorf. Ein Arbeitsbericht. *Quartär, 27*, 153–163.

Bosinski, G. (1978). Eine zusammengesetzte Magdalénien-Geschossspitze aus der Höhle im Freudenthal, Kanton Schaffhausen. *Archäologisches Korrespondenzblatt, 8*(2), 87–89.

Bosinski, G. (1979). *Die Ausgrabungen in Gönnersdorf 1968–1976 und die Siedlungsbefunde der Grabung 1968.* Wiesbaden: Steiner.

Bosinski, G. (1982). *Die Kunst der Eiszeit in Deutschland und in der Schweiz.* Bonn: Habelt Verlag.

Bosinski, G. (1990). *Homo sapiens. L'histoire des chasseurs du Paléolithique supérieur en Europe (40000–10000 av. J.-C.).* Paris: Édition Errance.

Bosinski, G. (1994). Die Gravierungen des Magdalénien-Fundplatzes Andernach-Martinsberg. *Jahrbuch des Römisch-Germanischen Zentralmuseums, 41*(1), 19–58.

Bosinski, G. (2008a). *Tierdarstellungen von Gönnersdorf.* Mainz: Verlag des Römisch-Germanischen Zentralmuseums.

Bosinski, G. (2008b). *Urgeschichte am Rhein.* Kerns: Tübingen.

Bosinski, G. (2011). Les figurations féminines de la fin des temps glaciaires. In N. Aujoulat (Ed.), *Mille et une femmes de la fin de temps glaciaires. Musée National de Préhistoire – Les Eyzies-de-Tayac, 17 juin–19 septembre* (pp. 49–67). Paris: Réunion des musées nationaux.

Bosinski, G., D'errico, F., & Schiller, P. (2001). *Die gravierten Frauendarstellungen von Gönnersdorf.* Stuttgart: Steiner.

Bosinski, G., & Fischer, G. (1980). *Mammut- und Pferdedarstellungen von Gönnersdorf.* Wiesbanden: Steiner.

Bosselin, B., & Djindjian, F. (1988). Un essai de structuration du Magdalénien français à partir de l'outillage lithique. *Bulletin de la Société Préhistorique Française, 85*(10–12), 304–331.

Bouchud, J., & Desbrosse, R. (1973). La faune de la grotte des Freydières à Saint-Agnan-en-Vercors (Drôme). *Bulletin de la Société Préhistorique Française, 70*, 330–336.

Bourdier, F., & De Lumley, H. (1954). Existence d'une industrie proto-Azilienne contemporaine du Renne en Dauphiné. *Bulletin de la Société Préhistorique Française, 51*, 307–309.

Bourdier, F., & De Lumley, H. (1956). Magdalénien et romanello-azilien en Dauphiné. *Bulletin du Musée d'Anthropologie et Préhistoire Monaco, 3*, 123–176.

Bouvier, J.-M. (1977). *Un gisement préhistorique: La Madeleine.* Les Eyzies: Fanlac.

Bouyssonie, J., Lejeune, L., & Pérol, J.-F. (1935). La station de Lacan (Corrèze) et son outillage. In *Compte Rendu de la XIe Session du Congrès Préhistorique de France, Périgueux, 1934* (pp. 318–323). Paris: Bureaux de la Société Préhistorique Française.

Boyd, R., & Richerson, P. J. (1985). *Culture and the evolutionary process.* Chicago: University of Chicago Press.

Boyle, E. A. (2000). Is ocean thermohaline circulation linked to abrupt stadial/interstadial transitions? *Quaternary Science Reviews, 19*, 255–272.

Brandt, K. (1960). Eine Werkgrube des Magdalénien in Burk bei Bautzen. *Praehistorische Zeitschrift, 38*, 119–125.

Brandtner, F. (1955). Kamegg, eine Freilandstation des späten Paläolithikums in Niederösterreich. *Mitteilungen der Prähistorischen Kommission der Österreichischen Akademie der Wissenschaften, 7*, 1–93.

Brantingham, P. J. (2007). A unified evolutionary model of archaeological style and function based on the price equation. *American Antiquity, 72*(3), 395–416.

Bratlund, B. (2002). The faunal remains from Wilczyce. In V. Eriksen & B. Bratlund (Eds.), *Recent studies in the final Palaeolithic of the European plain. Proceedings of a U.I.S.P.P. Symposium, Stockholm, 14–17 October 1999* (pp. 101–108). Aarhus: Aarhus University Press.

Breest, K. (2000). Spätglaziale Fundstellen im Landkreis Celle. Erkenntnisse, die auf Grund ehrenamtlicher Sammeltätigkeit gewonnen werden konnten. *Die Kunde N.F., 51*, 151–178.

Breest, K., & Oldenburg, H. (2000). Der spätjungpaläolithiche Fundplatz Wittlohe FStNr.8, Gde. Kirchlinteln, Ldkr. Verden (Aller). *Die Kunde N.F., 51*, 1–40.

Breest, K., & Veil, S. (1991). The late upper Palaeolithic site of Schweskau, Ldkr Lüchow-Dannenberg (Germany), and some comments on the relationship between the Magdalenian and Hamburgian. In N. Barton, A. J. Roberts, & D. A. Roe (Eds.), *The late glacial in north-west Europe. Human adaptation and environmental change at the end of the Pleistocene* (pp. 82–99). Oxford: Council for British Archaeology.

Breuil, H. (1937). *Les subdivisions du Paléolithique supérieur et leur signification. Congrès International d'Anthropologie et d'Archéologie Préhistoriques; compte rendu de la XIVe session, Genève, 1912.* Lagny: Grevin et fils.

Breuil, H., & Obermaier, H. (1908). Die Gudenushöhle in Niederösterreich. *Mitteilungen der Anthropologischen Gesellschaft in Wien, 38*, 277–294.

Brézillon, M. (1983). *La dénomination des objets de pierre taillée.* Paris: CNRS.

Bridault, A., & Chaix, L. (2009). Réflexions sur la recomposition des spectres fauniques dans le massif jurassien et les Alpes françaises du nord durant le Tardiglaciaire. In G. Pion (Ed.), *La fin du Paléolithique supérieur dans les Alpes du nord françaises et le Jura Méridional. Approches culturelles et environmentales* (pp. 59–71). Paris: Société Préhistorique Française.

Bridault, A., Chaix, L., Pion, G., Oberlin, C., Thiébault, S., & Argant, J. (2000). Position chronologique du renne (Rangifer tarandus L.) à la fin du Tardiglaciaire dans les Alpes du Nord française et le Jura méridional. In G. Pion (Ed.), *Le Paléolithique supérieur récent: nouvelles données sur le peuplement et l'environnement. Table ronde de Chambéry 12–13 mars 1999* (pp. 47–57). Paris: Société Préhistorique Française.

Brochier, J. É., & Livache, M. (1997). *La Grotte du Taï et l'abri de Campalou à St Nazaire en Royans. Éléments typologiques préliminaires.* Rapport de synthèse 1997 du PCR "La fin du Paléolithique supérieur dans les Alpes française du Nord et le Jura méridional" 2p, 3fig.

Broecker, W. S., & Denton, G. H. (1989). The role of ocean-atmosphere reorganization in glacial cycles. *Geochimica et Cosmocimica Acta, 53*(10), 2465–2501.

Bronk Ramsey, C., Higham, T. F. G., Brock, F., Baker, D., & Ditchfield, P. (2009). Radiocarbon dates from the Oxford AMS System: Archaeometry datelist 33. *Archaeometry, 51*(2), 323–349.

Bronk Ramsey, C., Higham, T. F. G., Owen, D. C., Pike, A. W. G., & Hedges, R. E. M. (2002). Radiocarbon dates from the Oxford AMS system: Archaeometry datelist 31. *Archaeometry, 44*(3), 1–149.

Brüchmann, W. (1952). Die Fundplätze der Hamburger Stufe im Gebiet der Hamburger Walddörfer. *Hammaburg, 8*, 147–156.

Brückner, G. H. (1954). Artefakte der jüngeren Hamburger Stufe in einer Grundmoräne bei Ostseebad Grömitz. *Meyniana, 2*, 191–192.

Brückner, G. H. (1970). Grömitz und Bismarkstein, zwei Fundplätze des Jungpaläolithikums. In K. Gripp, R. Schütrumpf, & H. Schwabedissen (Eds.), *Frühe Menschheit und Umwelt 1* (pp. 365–368). Köln: Böhlau.

Brumann, C. (2007). Stamm, Volk, Ethnizität, Kultur: die gegenwärtige Diskussion. In S. Rieckhoff & U. Sommer (Eds.), *Auf der Suche nach Identitäten: Volk – Stamm – Kultur – Ethnos* (BAR International Series, Vol. 1705, pp. 31–53). Oxford: Archaeopress.

Brunnacker, K. (1978). Stratigraphie im Bereich der Gönnersdorfer Siedlung: C14-Datierung. In K. Brunnacker (Ed.), *Geowissenschaftliche Untersuchungen in Gönnersdorf* (pp. 44–45). Wiesbaden: Steiner.

Brunnacker, K., Von Koenigswald, W., Rähle, W., Schweingruber, F. H., Taute, W., & Wille, W. (1981). Der Übergang vom Pleistozän zum Holozän in der Burghöhle von Dietfurt bei Sigmaringen. *Kölner Jahrbuch für Vor- und Frühgeschichte, 15*, 86–160.

Buchanan, B., & Hamilton, M. J. (2009). A formal test of the origin of variation in North American early Palaeoindian projectile points. *American Antiquity, 74*(2), 279–298.

Bullinger, J. (1996). *Le secteur XIV du site Magdalénien de Moosbühl (Canton de Berne, Suisse)*. Master thesis, Université of Genève, Geneva.

Bullinger, J. (2006). Le jais. In J. Bullinger, D. Leesch, & N. Plumettaz (Eds.), *Le site Magdalénien de Monruz, 1. Premiers éléments pour l'analyse d'un habitat de plain air* (pp. 158–165). Neuchâtel: Musée Cantonal d'Archéologie.

Bullinger, J., Lämmli, M., & Catherine, L.-P. (1997). Le site magdalénien de plein air de Moosbühl: nouveaux éléments de datation et essai d'interpretation des données spatiales. *Jahrbuch der Schweizerischen Gesellschaft für Ur- und Frühgeschichte, 80*, 7–26.

Bullinger, J., Leesch, D., & Plumettaz, N. (Eds.). (2006). *Le site magdalénien de Monruz, 1. Premiers éléments pour l'analyse d'un habitat de plein air*. Neuchâtel: Musée Cantonal d'Archéologie.

Bullinger, J., & Thew, N. (2006). Les coquillages fossiles. In J. Bullinger, D. Leesch, & N. Plumettaz (Eds.), *Le site Magdalénien de Monruz, 1. Premiers éléments pour l'analyse d'un habitat de plain air* (pp. 154–158). Neuchâtel: Musée Cantonal d'Archéologie.

Burdukiewicz, J. M. (1980). Sprawozdanie z badań stanowiska schyłkowopaleolitycznego Siedlnica 17, gmina Wschowa /III sezon badań wykopaliskowych. *Śląskie Sprawozdania Archeologiczne, 21*, 5–6.

Burdukiewicz, J. M. (1986a). *The late Pleistocene shouldered point assemblages in Western Europe*. Leiden: Brill.

Burdukiewicz, J. M. (1986b). Siedlinca 17a—Eine neue Fundstelle der Hamburger Kultur im Odergebiet. *Archäologisches Korrespondenzblatt, 16*, 399–406.

Burdukiewicz, J. M. (1997). Concerning late Palaeolithic settlement of the Sudetes Mountains and neighbouring areas. *Studia Archeologiczne, 29*, 7–24.

Burdukiewicz, J. M., Herman, C. F., Haesaerts, P., Damblon, F., Langohr, R., Mikkelsen, J. H., et al. (1996). A new Hamburgian concentration at Siedlnica 17 in the Kopanica Valley (SW Poland). *Anthropologie et Préhistoire, 107*, 13–36.

Burdukiewicz, J. M., & Szynkiewicz, A. (2002). Badania archeologiczne i paleogeograficzne stanowiska Łęgoń, pow. Wschowa. *Śląskie Sprawozdania Archeologiczne, 44*, 57–78.

Burdukiewicz, J. M., Szynkiewicz, A., & Malkiewicz, M. (2003). Dalsze badania osadnictwa schyłkowopaleolitycznego na tle warunków paleoekologicznych w Łęgoniu, pow. Wschowa. *Śląskie Sprawozdania Archeologiczne, 45*, 17–34.

Burg, A. V. (1994). *Die Silexkomplexe des Oberaargaus (Kt. Bern)*. Licence thesis, University of Bern.

Büttner, P. (1936). Ein eiszeitlicher Wohnplatz am Pennenmoor bei Cuxhaven. *Mannus, 28*, 501–510.

Büttner, P. (1940a). Paläolithische und mesolithische Funde am Pennenmoor bei Sahlenburg. *Nachrichten aus Niedersachsens Urgeschichte, 14*, 1–27.

Büttner, P. (1940b). Der eiszeitliche Fundplatz am Pennenmoor bei Cuxhaven. *Mannus, 32*, 448–454.

Cacho, C. (1989). Structuration du Magdalénien dans l'Espagne méditerranéenne. In J.-P. Rigaud (Ed.), *Le Magdalénien en Europe. "La structuration du Magdalénien". Actes du Colloque de Mayence 1987 dans le cadre du XIème Congrès U.I.S.P.P* (ERAUL, Vol. 38, pp. 459–478). Liège: Université de Liège.

Campy, M. (1985). Circonscription de Franche-Comté. *Gallia Préhistoire, 28*(2), 235–258.

Campy, M., Chaline, J., & Vuillemey, M. (1989). *La Baume de Gigny (Jura)*. Paris: CNRS.

Capitan, L., Breuil, H., & Peyrony, D. (1924). *Les Combarelles aux Eyzies (Dordogne). Peintures et gravures murales des cavernes paléolithiques*. Paris: Masson.

Capitan, L., & Peyrony, D. (1928). *La Madeleine. Son gisement – son industrie, ses oevres d'art*. Paris: Nourry.

Cartailhac, É. (1903). Les stations de Bruniquel sur les bords de l'Aveyron. *L'Anthropologie, 14*(129–150), 295–315.

Cartonnet, M., & Naton, H.-G. (2000). Le Magdalénien de la grotte de Chênelaz à Hostias (Ain). In G. Pion (Ed.), *Le Paléolithique Supérieur récent: nouvelles données sur le peuplement et l'environnement. Actes de la Table-Ronde de Chambéry, 12–13 mars 1999* (pp. 235–243). Paris: Société Préhistorique Française.

Casti, J. (1989). *Paradigms lost: Images of man in the mirror of science*. New York: Morrow.

Cattin, M.-I. (2009). Exploitation des ressources lithiques du Magdalénien à l'Azilien: la situation du Jura oriental. In G. Pion (Ed.), *La fin du Paléolithique supérieur dans les Alpes du nord françaises et le Jura méridional. Approches culturelles et environmntales* (pp. 139–142). Paris: Société Préhistorique Française.

Chaix, L. (2003). L'exploitation de la faune au Magdalénien en Suisse et dans les régions limitrophes. In S. Costamagno & V. Laroulandie (Eds.), *Mode de vie au Magdalénien: Apports de l'Archéozoologie. Zooarchaeological insights into Magdalenian Lifeways* (BAR International Series, Vol. 1144, pp. 65–72). Oxford: Archaeopress.

Chantre, E. (1866). Note sur des cavernes à ossements et à silex tailles du Daphiné. *Bulletin de la Société Geologique de France, 23*, 532–536.

Chantre, E. (1901). *L'homme quaternaire dans le bassin du Rhône. Etude géologique et anthropologique*. Lyon: Rey.

Charles, R. (1993). Towards a new chronolgy for the Belgian Lategacial: Recent radiocarbon dates from the Oxford AMS system. *Notae Praehistoricae, 12*, 59–62.

Charles, R. (1994). Towards a new chronology for the lateglacial archaeology of Belgium. II: Recent radiocarbon dates from the Oxford AMS system. *Notae Praehistoricae, 13*, 31–39.

Charles, R. (1996). Back into the north: The radiocarbon evidence for the human recolonisation of the north-western Ardennes after the last glacial maximum. *Proceedings of the Prehistoric Society, 62*, 1–17.

Charles, R. (1998). *Food for thought. Late Magdalenian chronology and faunal exploitation in the North-Western Ardennes*. Ph.D. Dissertation, University of Oxford.

Chen, S.-H. (1982). *Neue Untersuchungen über die spät- und postglaziale Vegetationsgeschichte im Gebiet zwischen Harz und Leine*. Ph.D. Dissertation, University of Göttingen.

Cheynier, A. (1939). Le Magdalénien primitif de Badegoul et les niveaux à raclettes. *Bulletin de la Société Préhistorique Française, 36*, 354–396.

Cheynier, A. (1951). Les industries protomagdaléniennes. *Bulletin de la Société Préhistorique Française, 48*(3–4), 189–192.

Cheynier, A. (1956). Les Gros Monts à Nemours (Seine-et-Marne). In *Congrès préhistorique de France. 15e session, Poitiers-Angoulême* (pp. 344–365). Paris: Société Préhistorique Française.

Cheynier, A. (1965). *Comment vivait l'homme des cavernes à l'âge du renne*. Paris: Les éditions du Scorpion.

Childe, V. G. (1951). *Social evolution*. New York: Schumann.

Childe, V. G. (1956). *Piecing together the past. The interpretation of archaeological data*. London: Routledge and Kegan Paul.

Chłodnicki, M., & Kabaciński, J. (1997). Mirkowice – another settlement of the Hamburgian culture at the Polish Plain. *Przeglad Archeologiczny, 45*, 5–23.

Chmielewski, W. (1961). *Civilisation de Jerzmanowice*. Warszawa: Polskiej Akademii Nauk.

Chmielewski, W., & Kozłowski, S. K. (1969). Schronisko nad Jaskinią Niedostępną. *Informator Archeologiczny, 1968*, 23–24.

Chollet, A., & Dujardin, V. (Eds.). (2005). *La grotte du Bois-Ragot à Gouex (Vienne), Magdalénien et Azilien, essais sur les hommes et leur environnement*. Paris: Société Préhistorique Française.

Christiansen, H. H., & Svensson, H. (1998). Wind-polished boulders as indicators of a late Weichselian wind regime in Denmark in relation to neighbouring areas. *Permafrost and Periglacial Processes, 9*, 1–21.

Clark, G. A. (1994). Migration as an explanatory concept in Paleolithic archaeology. *Journal of Archaeological Method and Theory, 1*(4), 305–343.

Clarke, D. L. (1968). *Analytical archaeology*. London: Methuen.

Clarke, D. L. (1972a). Models and paradigms in contemporary archaeology. In D. L. Clarke (Ed.), *Models in archaeology* (pp. 1–60). London: Methuen.

Clarke, D. L. (Ed.). (1972b). *Models in archaeology*. London: Methuen.

Clarke, D. L. (1979). *Analytical archaeologist. Collected papers of David L. Clarke*. London: Academic Press.

Claude, J. (2008). *Morphometrics with R*. New York: Springer.

Cleyet-Merle, J.-J. (1989). Nouvelles données sur le Magdalénien de la vallée de la Couze. *Paléo, 1*, 107–116.

Cleyet-Merle, J.-J., & Madelaine, S. (1991). La pendeloque magdalénienne gravée d'un "Equus hydrontinus" de la grotte du Putois II, commune de Montmaurin (Haute-Garonne). *Paléo, 3*, 119–129.

Cohausen, A. V. (1874). Die Nachgrabungen an der alten Wallburg und den Höhlen bei Steeten and der Lahn. *Nassauische Annalen, 13*, 379–389.

Cohausen, A. V. (1879). Die Höhlen und die Wallburg bei Steeten an der Lahn. *Nassauische Annalen, 15*, 323–338.

Cohën, A.-S., Lambert, R., & Vuillemey, M. (1991). Préhistoire dans le Jura. Les grottes de Gigny et d'Arlay. In Musée d'Archéologie du Jura (Ed.), *Catalogue d'exposition* (pp. 63). Lons-le-Saunier: Cercle Girardot.

Collins, D. (1973). Epistemology and cultur tradition theory. In C. Renfrew (Ed.), *The explanation of culture change: Models in prehistory* (pp. 53–58). London: Duckworth.

Collot, J. (1947). L'Homme préhistorique en Haute-Saône: quelques notes techniques sur la Baume Noire de Frétigney. *Bulletin de l'Association Spéléologique de l'Est, 5*, 29–33.

Combier, J. (1977). Circonscription Rhône-Alpes. *Gallia Préhistoire, 20*(2), 561–672.

Combier, J. (1980). Informations Archéologiques. Circonscription Rhône-Alpes, Ain. *Gallia Préhistoire, 23*, 473–479.

Combier, J., & Desbrosse, R. (1964). Magdalénien final à pointe de Teyjat dans le Jura méridional. *L'Anthropologie, 68*, 190–194.

Cordy, J. M. (1974). Biostratigraphie et chronostratigraphie des dépôts tardiglaciaires et holocènes de la grotte du Coléoptère à Bomal-sur-Ourthe. *Annales de la Société Géologique de Belgique, 98*(2), 291–296.

Cosack, E. (1970). Eine Fundstelle des Hamburgium bei Uetze, Kreis Burgdorf. *Neue Ausgrabungen und Forschungen in Niedersachsen, 6*, 1–5.

Couchard, J. (1957). Essai d'application des méthodes typologie et de statistiques au gisment remanié de Bellet près Brive. *Bulletin de la Société Archéologique de la Corrèze*, 63–72.

Couchard, J. (1960). Gisment de Bellet, près Brive, Corrèze; Note complémentaire. *Bulletin de la Société Préhistorique Française, 57*, 282–286.

Crémilleux, A. (1972). L'abri de la Baume-Loire No° 1, commune de Solignac-sur-Loire (Haute-Loire). Premiers résultat. In SPF (Ed.), *Congrès Préhistorique de France. Compte rondu de la XIXe session, Auvergne, 6–14 juillet1969* (pp. 172–176). Paris: Société Préhistorique Française.

Crémilleux, A. (1977). Liste des datations ^{14}C en haute vallée de la Loire. *Bulletin Historique, Scientifique, Literaise, Artistique et Agricole, 53*, 41–43.

Crotti, P. (1993). Spätpaläolithikum und Mesolithikum in der Schweiz. Die letzten Jäger. In J.-M. Le Tensorer, U. Niffeler, & E. Gross (Eds.), *SPM I. Die Schweiz vom Paläolithikum bis zum frühen Mittelalter* (pp. 203–244). Basel: Verl. Schweizer. Gesellschaft für Ur- u. Frühgeschichte.

Cupillard, C. (1998). L'abri des Cabônes à Ranchot (Jura). In C. Cupillard & A. Richard (Eds.), *Les derniers chasseurs-cuilleurs du massif jurassien et de ses marges (13000-5500 avant Jésus-Christ)* (pp. 131–132). Lons-le-Saunier: Centre Jurassien du Patrimoine.

Cupillard, C. (Ed.) (2008). *Le Tardiglaciaire et le debut de l'Holocene dans le massif du Jura et ses marges. 20000 – 5000 av. J.-C.: des derniers chasseurs de rennes aux premiers agriculteurs. Projet Collectif de Recherche (PCR) 2005–2008*. Rapport 2007. Besançon: SRA de Franche-Comté et UMR 6249 du CNRS, Laboratoire de Chrono-environnement.

Cupillard, C., & Welte, A.-C. (2006). Le Magdalénien de la grotte "Grappin" à Arlay (Jura, France): nouveaux regards. *L'Anthropologie, 110*(4), 624–683.

Cyrek, K. (1986a). Magdaleńskie obozowisko w Górach Świętokrzyskich (Mosty, Stanowisko 13). *Acta Archaeologica Carpathica, 25*, 11–55.

Cyrek, K. (1986b). Die Technologie des magdalenischen Stein-Komplexes aus der Gudenushöhle. *Mitteilungen der Österreichischen Arbeitsgemeinschaft für Ur- und Frühgeschichte, 36*, 7–24.

Cyrek, K. (1994). Der jungpaläolithische Höhlenfundplatz Krucza Skala in Mittelpolen. *Archäologisches Korrespondenzblatt, 24*, 367–374.

Cyrek, K. (1999). Menschliche Penetration der Höhlen im Mittleren Teil der Krakowsko-Częstochowska-Hochebene zwischen dem 18. und dem 11. Jahrtausend (vom Interstadial Lascaux bis zum Interstadial Alleröd). *Folia Quaternaria, 70*, 269–288.

Cyrek, K. (2001). *The Deszczowa Cave at Kostkowice, Kroczyce district. Comprehensive sediment research at the Deszczowa Cave. Archaeology*.

Cyrek, K., Nadachowski, A., Madeyska, T., Bocheński, Z., Tomek, T., Wojtal, P., et al. (2000). Excavation in the Deszczowa Cave (Kroczyckie Rocks, Częstochowa upland, central Poland). *Folia Quaternaria, 71*, 5–84.

Czarnowski, S. J. (1904). Jaskinie i schroniska w Górze Smardzewskiej. *Pamiętnik fizjograficzny, 18*, 3–24.

Cziesla, E. (1998). Die Mittlere Steinzeit im südlichen Rheinland-Pfalz. In N. J. Conard & K.-J. Kind (Eds.), *Aktuelle Forschungen zum Mesolithikum. Current mesolithic reserach* (pp. 111–120). Mo Vince: Tübingen.

Cziesla, E. (2008). Als man über das Wetter zu reden begann. Die Altsteinzeit in der Niederlausitz. *Arbeitsberichte zur Bodendenkmalpflege in Brandenburg, 20*, 29–54.

Cziesla, E., & Masojć, M. (2007). Die einreihige Widerhakenspitze aus Węgliny (polnische Niederlausitz) und ihre kutlrurgeschichtliche Deutung. *Archäologisches Korrespondenzblatt, 37*, 457–469.

D'Errico, F., & David, S. (1993). Analyse technologique de l'art mobilier. La cas de l'abri des Cabônes à Ranchot (Jura). *Gallia Préhistoire, 35*, 139–176.

D'Errico, F., & Laroulandie, V. (1998). Authentification et analyse technologique de répresentations aviaire de galets de la Croze. In Musée de Brou (Ed.), *Pré, proto, gallo, méro, histoire de l'Ain en achéo. Catalogue du Musée de Brou* (pp. 24–26). Bourg-en-Bresse: Musée de Brou.

Dämmer, H.-W., Reim, H., & Taute, W. (1974). Probegrabungen in der Burghöhle Dietfurt im oberen Donautal. *Fundberichte aus Baden-Württemberg, 1*, 1–25.
Darasse, P., & Guffroy, S. (1960). La Magdalénien supérieur de l'abri de Fontalès, près Saint Antonin (Tarn-et-Garonne). *L'Anthropologie, 64*, 1–35.
Darpeix, A. (1934). Station préhistorique de la Forge, commune de Plazac. *Bulletin de la Société Historique et Archéologique du Périgord*, 328–350.
Dauber, A., & Birkner, F. (1934). Neue Beobachtungen in der kleinen Ofnet bei Hohlheim, B.-W. Nördlingen. *Bayerische Vorgeschblätter, 12*, 62–66.
Daum, C. (1999). *Die Grabung von K. Brandt 1932 auf dem Magdalénienfundplatz Burk bei Bautzen (Oberlausitz)*. Master thesis, University of Cologne.
David, P. (1947). La Chair à Calvin (Commune de Mouthiers, Charente). *Bulletin de la Société Préhistorique Française, 44*, 31–32.
David, S. (1982). Les chasseurs magdaléniens de l'abri des Cabônes à Ranchot (Jura). In Service régional de l'archéologie (Ed.), *Cinq années de fouilles archéologiques en Franche-Compté* (pp. 51–53). Besancon: Directions des Antiquités Préhistoriques et Historiques de Franche-Comté.
David, S. (1984). Découverte d'un galet gravé magdalénien dans l'abri de Cabônes à Ranchot (Jura). *Bulletin de la Société Préhistorique Française, 81*(4), 111–112.
David, S. (1992). L'art mobilier magdalénien de l'abri des Cabônes (Ranchot, Jura). *Revue Archéologique de l'Est et du Center-Est, 43*(1), 3–28.
David, S. (1993). *Magdalénien et Azilien en Franche-Comté: contribution à l'étude des cultures du Tardiglaciaire dans l'Est de la France*. Ph.D. Dissertation, University of Franche-Comté.
David, S. (1996). La fin du Paléolithique supérieur en Franche-Comté: Environnement, cultures, chronologie. *Gallia Préhistoire, 38*, 111–248.
David, S., & Richard, H. (1989). Les cultures du Tardiglaciaire dans le Nord-Est de la France. In J.-P. Rigaud (Ed.), *Le Magdalénien en Europe. "La structuration du Magdalénien". Actes du Colloque de Mayence 1987. XIe Congrès U.I.S.P.P* (ERAUL, Vol. 38, pp. 101–156). Liège: Université de Liège.
De Bayle Des Hermens, R. (1971). Le Magdalénien final de la couche E de la grotte du Rand du Barry. *L'Anthropologie, 75*, 585–604.
De Bayle Des Hermens, R. (1974). Note préliminaire sur le Magdalénien ancien de la couche F2 de la grotte du Rond du Barry. *L'Anthropologie, 78*, 17–36.
De Bayle Des Hermens, R. (1979). Les niveaux supérieurs du Magdalénien de la grotte du Rond du Barry, Polignac (Haute-Loire). In D. de Sonneville-Bordes (Ed.), *La fin de temps glaciaires en Europe. Chronostratigraphie et écologie des cultures du Paléolithique final. Colloques internationales du CNRS N° 271, Talence 24-28 mai 1977* (pp. 601–611). Paris: Éditions CNRS.
De Beaulieu, J.-L., & Reille, M. (1983). Paléoenvironment tardiglaciaire et holocène des lacs de Pelléautier t Siguret (Hautes-Alpes, France). Histoire de la végétation d'après les analyses polliniques. *Ecologia Mediterranea, 9*, 19–36.
De Bie, M., & Caspar, J.-P. (2000). *Rekem, a Federmesser camp on the Meuse river bank*. Asse-Zellik: Leuven University Press.
De Klerk, P. (2004). Confusing concepts in Lateglacial stratigraphy and geochronology: Origin, consequences, conclusions (with special emphasis on the type locality Bøllingsø). *Review of Palaeobotany and Palynology, 129*, 265–298.
de Sonneville-Bordes, D. (1960). *Le Paléolithique supérieur en Périgord*. Bordeaux: Delmas.
de Sonneville-Bordes, D. (1961). Le Paléolithique supérieur en Belgique. *L'Anthropologie, 65*, 421–443.
de Sonneville-Bordes, D. (1963). Le Paléolithique supérieur en Suisse. *L'Anthropologie, 67*(3–4), 205–268.
de Sonneville-Bordes, D. (1967). Observations au sujet de la communication du Docteur Allain (23 février 1967): le Badegoulien de l'abri Fritsch, aux Roches de Pouligny-Saint-Pierre. *Bulletin de la Société Préhistorique Française, 64*, 227–229.
de Sonneville-Bordes, D., & Perrot, J. (1954). Lexique typologique du Paléolithique supérieur. Outillage lithique: I Grattoirs – II Outils solutréens. *Bulletin de la Société Préhistorique Française, 51*, 327–355.
de Sonneville-Bordes, D., & Perrot, J. (1955). Lexique typologique du Paléolithique supérieur. Outillage lithique – III Outils composites – Percoirs. *Bulletin de la Société Préhistorique Française, 52*, 76–79.
de Sonneville-Bordes, D., & Perrot, J. (1956a). Lexique typologique du Paléolithique supérieur. Outillage lithique – IV Burins. *Bulletin de la Société Préhistorique Française, 53*, 408–412.
de Sonneville-Bordes, D., & Perrot, J. (1956b). Lexique typologique du Paléolithique supérieur. Outillage lithique (suite et fin) – V Outillage à bord abattu – VI Pièce tronquées – VII Lames retouchées – VIII Pièces variées – IX Outillage lamellaire. Pointe azilienne. *Bulletin de la Société Préhistorique Française, 53*, 547–559.
Decker, H. (2011). *Die Oberflächenfunde von Bad-Kösen-Lengefeld aus der Sammlung Dietrich Mania*. Master thesis, University of Cologne.
Decker, R., Rašković, S., & Brunsdiek, K. (2010). Diskriminanzanalyse. In C. Wolf & H. Best (Eds.), *Handbuch der sozialwissenschaftlichen Datenanalyse* (pp. 495–523). Wiesbaden: VS Verlag.
Deeben, J. (1992). Eine Siedlung aus dem Magdalénien in Sweikhuizen. In L. Rheinland (Ed.), *Spurensicherung. Archäologische Bodendenkmalpflege in der Euregio Maas-Rhein* (pp. 497–502). Zabern: Mainz.
Deecke, W. (1933). *Die mitteleuropäischen Silices nach Vorkommen, Eigenschaften und Verwendung in der Prähistorie*. Jena: Fischer.
Deetz, J. (1967). *Invitation to archaeology*. New York: Natural History Press.
Deetz, J. (1968). Cultural patterning of behaviour as reflected by archaeological materials. In K. C. Chang (Ed.), *Settlement archaeology* (pp. 31–42). Palo Alto: Ed. National Press.
Dehn, W., & Sangmeister, E. (1954). *Die Steinzeit im Ries*. Kallmünz/Opf.: Lassleben.
Delage, F. (1935). Les Roches de Sergeac (Dordogne). *L'Anthropologie, 45*, 281–317.
Delannoy, J.-J., Bintz, P., Naton, H.-G., Cartonnet, M., & Tillet, T. (1997). Environnements karstiques dans les Alpes du nord et le Jura méridional: spéléogénèse, sédimentation, climat et archéologie. *Quarternaire, 8*(2), 197–212.
Délibras, G., & Évin, J. (1974). Sommaire des datations ^{14}C concernant le Préhistoire en France. I. Dates parues de 1955 à 1974. *Bulletin de la Société Préhistorique Française, 71*, 149–156.
Délibras, G., & Évin, J. (1976). Datations absolues des dépôts quaternaieres et des sites préhistoriques par la methode du Carbone 14. In H. De Lumley (Ed.), *La Préhistoire Française I* (pp. 1499–1514). Paris: CNRS.
Delibrias, G., & Évin, J. (1980). Sommaires des datations ^{14}C concernant la préhistoire en France II. Dates parues de 1974 à 1978. *Bulletin de la Société Préhistorique Française, 77*, 215–224.
Delpech, F. (1989). Le temps de l'Antilope saiga. In J. P. Mohen (Ed.), *Le temps de la Préhistoire* (pp. 48–49). Paris: Éditions SPF.
Delpech, F. (1992). Le monde Magdalénien d'après le milieu animal. In J.-P. Rigaud (Ed.), *Le peuplement Magdalénien. Paléogéographie physique et humaine. Colloque de Chancelade, 10–15 octobre 1988* (pp. 127–135). Paris: Eds. du CTHS.
Demars, P.-Y., & Laurent, P. (1992). *Types d'outils lithiques du Paléolithique supérieur en Europe*. Paris: Ed. CNRS.
Demetrykiewicz, W., & Kuźniar, W. (1914). Najstarszy paleolit na ziemiach polskich oraz inne wykopaliska odkryte w jaskini „Okiennik". *MAAiE, 13*, 10–43.
Desbrosse, R. (1976a). Les civilisation du Paléolithique supérieur dans le Jura méridional et dans les Alpes du Nord. In H. Lumley (Ed.), *La*

Préhistoire Française. Les Civilisations Paléolithique et Mésolithique de la France (pp. 1196–1213). Paris: CNRS.

Desbrosse, R. (1976b). Les civilisations du Paléolithique supérieur dans le Jura et en Franche-Comté. In H. Lumley (Ed.), *La Préhistoire Française. Les Civilisations Paléolithique et Mésolithique de la France* (pp. 1348–1358). Paris: CNRS.

Desbrosse, R. (1980). Le Paléolithique du Jura Méridional. *Bulletin de l'Association française pour l'etude du quaternaire, 17*(3), 135–142.

Desbrosse, R., & Girard, M. (1974). Azilien et Magdalénien des Douattes (Haute-Savoie). *L'Anthropologie, 78*, 481–498.

Desbrosse, R., & Mourer-Chauviré, C. (1973). Les oiseaux magdaléniens de Pierre-Châtel (Ain). *Quartär, 23–24*, 149–164.

Desbrosse, R., & Prat, F. (1974). L'Elan magdalénien de Pierre-Châtel (Ain). *Quartär, 25*, 143–158.

Destinez, P., & Moreels, L. (1888). Exploration de la caverne de Verlaine-sur-Ourthe (Luxembourg). *Bulletin de la société belge de Géologie, de Paléontologie et d'Hydrologie, 2*, 202.

Dewez, M. (1975). Nouvelles recherches à la grotte du Coléoptère à Bomal-sur-Ourthe (Province du Luxembourg). Rapport provisoire de la première campagne de fouille. *Helinium, 15*, 105–133.

Dewez, M. (1976). Remarques à propos d'un fragment de concrétion excentrique. *Bulletin de la Socieété royale belge d'Etudes géologiques et anthropologie et de préhistoire, 23*, 193–207.

Dewez, M. (1987). *Le Paléolithique Supérieur Récent dans les Grottes de Belgique*. Louvain-la-Neuve: Insitit Supérieur d'Archéologie et d'Histore de l'Art.

Dewez, M. (1989). Données nouvelles sur le Gravettien de Belgique. *Bulletin de la Société Préhistorique Française, 86*, 138–142.

Dewez, M. (1992). La grotte Walou à Trooz (province Liège, Belgique), présentation du site. In M. Toussaint (Ed.), *Cinq millions d'années, l'aventure humaine* (ERAUL, Vol. 56, pp. 311–318). Liège: Université de Liège.

Dietzel, A., & Geupel, V. (1989). Jungpaläolithische Funde aus Niederlommatzsch. *Kr. Meißen. Ausgrabungen und Funde, 34*(1), 5–12.

Dirian, A. (2003). *Das späte Jungpaläolithikum und das Spätpaläolithikum der oberen Schichten der Sesselfelsgrotte*. Saarbrücken: Saarbrücker Druckerei und Verlag.

Domański, G., & Burdukiewicz, J. M. (1994). Harpun jednorzędowy z Węglin, gm. Gubin, woj. Zielnogórskie. *Śląskie Sprawozdania Archeologiczne, 35*, 529–532.

Dongus, H. (1977). *Die Oberflächenformen der Schwäbischen Alb und ihres Vorlandes*. Marburg: Selbstverlag des Geographischen Institiuts der Universität Marburg.

Dongus, H. (2000). *Die Oberflächenformen Südwestdeutschlands*. Berlin: Borntraeger.

Doudou, E. (1903). *Exploration scientifique dans les cavernes, les abîmes et les tous qui fument de la Province de Liège*. Liège: Thone.

Drucker, D. G., Bocherens, H., & Billiou, D. (2009). Quelle valence écologique pour les rennes et autres Cervidés au Tardiglaciaire dans les Alpes du nord et le Jura? Résultat de l'analyse des isotopes stables (^{13}C, ^{15}N) du collagène. In G. Pion (Ed.), *La fin du Paléolithique supérieur dans les Alpes du nord françaises et le Jura méridional. Approches culturelles et environnementales* (pp. 73–86). Joué-lès-Tours: Société Préhistorique Française.

Ducasse, S. (2010). *La « parenthèse » badegoulienne: Fondements et statut d'une discordance industrielle au travers de l'analyse techno-économique de plusieurs ensembles lithiques méridionaux du Dernier Maximum Glaciaires*. Ph.D. Dissertation, University of Toulouse.

Dunell, R. C. (1978). Style and function – fundamental dichotomy. *American Antiquity, 43*, 192–202.

Dupont, É. (1867). Découverte d'objects gravés et sculptés dans le Trou Magrite à Pont-à-Lesse. *Bulletin de l'Académie royale de Belgique, séries 2, 24*, 129–132.

Dupont, É. (1869). Note sur la découverte d'une caverne à Goyet. *Bulletins de l'Académie royale des Sciences, des Lettres et des Beaux-Arts de Belgique, 27*, 193.

Dupont, É. (1873). *L'Homme pendant les âges de la Pierre dans les environs de Dinant-sur-Meuse*. Bruxelles: Muqurdt.

Dupont, É. (1969). Dur deux fragments d'objets appelés "baton de commandement" découvert dans la caverne de Goyet (province de Namour). *Bulletins de l'Académie royale des Sciences, des Lettres et des Beaux-Arts de Belgique, 27*, 274–275.

Duwe, A. (2013). *Die Besiedlung Mittel- und Osteuropas während des Hochglazials 26 bis 18 Tausend Jahre vor Heute*. Master thesis, University of Cologne.

Eberhardt, H., Keefer, E., Kind, C.-J., Rensch, H., & Ziegler, H. (1987). Jungpaläolithische und mesolithische Fundstellen aus der Aichbühler Bucht. Auswertung von Oberflächenfunden aus dem südlichen Federseegebiet. *Fundberichte aus Baden-Württemberg, 12*, 1–51.

Eerkens, J. W., & Lipo, C. P. (2005). Cultural transmission, copying errors, and the generation of variation in material culture and the archaeological record. *Journal of Anthropological Archaeology, 24*, 316–334.

Eggert, M. K. H. (1995). Anthropologie, Ethnologie und Urgeschichte: Zur Relativierung eines forschungsgeschichtlichen Mythologems. *Mitteilungen der Berliner Gesellschaft für Anthropologie, Ethnologie und Urgeschichte, 16*, 33–38.

Eicher, U., & Siegenthaler, U. (1976). Palynological and oxygen isotope investigations on LateGlacial sediment cores from Swiss lakes. *Boreas, 5*, 109–117.

Eichhorn, G. (Ed.). (1929). *Die Entdeckung der Wallburg auf dem Jenzig bei Jena durch Kloppfleisch, sein Ausgrabungen auf dem Bergpalteau und am Fuße des Jenzig*. Jena: Fischer.

Eisel, R. (1886). Höhlenausgrabung bei Döbritz unweit Oppurg. *Zeitschrift für Ethnologie, 18*, 50–52.

Ellenberg, H. (1978). *Vegetation Mitteleuropas mit den Alpen in ökologischer Sicht*. Stuttgart: Ulmer.

Enloe, J. G. (2003). Acquisition and processing of reindeer in the Paris Basin. In S. Costamagno & V. Laroulandie (Eds.), *Mode de Vie au Magdalénien: Apports de l'Archéozoologie* (BAR International Series, Vol. 1144, pp. 23–31). Oxford: Archaeopress.

Erdmann, W. (1980). Zur archaeologischen Arbeitsweise in natürlichen Schichten. In K. Frerichs (Ed.), *Archaologie in Lubeck: Erkenntnisse von Archäologie und Bauforschung zur Geschichte und Vorgeschichte der Hansestadt* (pp. 138–140). Lübeck: Schmidt-Römhild.

Eriksen, B. V. (1991). *Change and continuety in a prehistoric hunter-gatherer society: A study of cultural adaptation in late glacial – early postglacial southwestern Germany*. Tübingen: Archaeologica Venatoria.

Eriksen, B. V. (2002). Fossil mollusks and exotic raw materials in late glacial and early postglacial find contexts: A complement to lithic studies. In L. E. Fisher & B. V. Eriksen (Eds.), *Lithic raw material economies in late glacial and early postglacial Europe* (BAR International Series, Vol. 1093, pp. 27–52). Oxford: Archaeopress.

Escalon de Fonton, M. (1962). Fouilles dans le Gard (campagne 1961). Grotte de Salpétrière (Remoulins). *Cahirs ligures de Préhistoire et d'Archéologie, 11*, 218–220.

Escola, M. (1989). La plaquette osseuse gravée de la Baume Noire de Frétigney (Haute-Saône). *Bulletin de la Société Préhistorique Arriège-Pyrénées, XLIV*, 181–192.

Évin, J. (1994). Les grottes Jean Pierre 1 et 2 à Saint-Thibaud-de-Couz (Savoie). IV. Les datations radiométriques. *Gallia Préhistoire, 36*, 183–188.

Évin, J., Bintz, P., & Monjuvent, G. (1994). Human settlement and the last deglaciation in the French Alps. *Radiocarbon, 36*(3), 345–357.

Évin, J., Marechal, J., & Marien, G. (1985). Lyon natural radiocarbon measurements X. *Radiocarbon, 27*, 386–454.

Évin, J., Marien, G., & Pachiaudi, C. (1973). Lyon natural radiocarbon measurements III. *Radiocarbon, 15*(1), 134–155.

Évin, J., Marien, G., & Pachiaudi, C. (1975). Lyon natural radiocarbon measurements V. *Radiocarbon, 17*, 4–34.

Évin, J., Marien, G., & Pachiaudi, C. (1979). Lyon natural radiocarbon measurements VIII. *Radiocarbon, 21*(3), 405–452.

Fahlke, J. M. (2009). *Der Austausch der terrestrischen Säugetierfauna and der Pleistozän/Holozän-Grenze in Mitteleuropa.* Ph.D. Dissertation, University of Bonn.

Feustel, R. (1957). Vier jungpaläolithische Freilandstationen in Ostthüringen. *Alt-Thüringen, 2*, 1–26.

Feustel, R. (1959). Bemerkungen zur statistischen Methode in der Paläolithforschung. *Ausgrabungen und Funde, 4*, 225–229.

Feustel, R. (1961a). *Jungpaläolithische Wildbeuter in Thüringen.* Weimar: Museum für Ur- und Frühgeschichte Thüringens.

Feustel, R. (1961b). Remarques sur le Magdalénien suisse. *Archives suisses d'anthropologie générale, 26*(1–2), 29–40.

Feustel, R. (1974). *Die Kniegrotte. Eine Magdalénien-Station in Thüringen.* Weimar: Böhlau.

Feustel, R. (1977). Das Fundmaterial der Wildpferdjägerstation Bad Frankenhausen. *Alt-Thüringen, 14*, 25–59.

Feustel, R. (1979). Le Magdalénien final en Thuringe (R.D.A.). In D. de Sonneville-Bordes (Ed.), *La Fin des Temps Glacieres en Europe* (pp. 877–886). Paris: Editions CNRS.

Feustel, R. (1980). *Magdalénienstation Teufelsbrücke.* Weimar: Museum für Ur- und Frühgeschichte Thüringens.

Feustel, R. (1983). Ein jungpaläolithischer Stichel aus dem südwestlichen Thüringen. *Ausgrabungen und Funde, 28*(5), 205–206.

Feustel, R. (1989). Der Homo sapiens und das Jungpaläolithikum. In J. Herrmann (Ed.), *Archäologie in der Deutschen Demokratischen Republik* (pp. 41–54). Leipzig: Theiss.

Feustel, R., Kerkmann, K., Schmid, E., Musil, R., & Jacob, H. (1971a). Der Bärenkeller bei Königsee-Garsitz, eine jungpaläolithische Kulthöhle (I). *Alt-Thüringen, 11*, 81–130.

Feustel, R., Kerkmann, K., Schmid, E., Musil, R., Mania, D., Knorre, V. D., et al. (1971b). Die Urdhöhle bei Döbritz. *Alt-Thüringen, 11*, 131–226.

Feustel, R., & Musil, R. (1977a). Der Bärenkeller bei Königsee-Garsitz, eine jungpaläolithische Kulthöhle (II). *Alt-Thüringen, 14*, 60–81.

Feustel, R., & Musil, R. (1977b). Der epipaläolithische Rastplatz am Abri Fuchskirche bei Allendorf. *Alt-Thüringen, 14*, 82–107.

Feustel, R., Teichert, M., & Unger, K. P. (1963). Die Magdalénienstation Lausnitz in der Orlasenke. *Alt-Thüringen, VI*, 57–103.

Fiedler, L. (1990). Die Alt- und Mittelsteinzeit. In F.-R. Herrmann & A. Jockenhövel (Eds.), *Die Vorgeschichte Hessens* (pp. 70–120). Stuttgart: Theiss.

Fiedler, L. (1994). *Alt- und mittelsteinzeiltliche Funde in Hessen.* Stuttgart: Theiss.

Fiedorczuk, J., Bratlund, B., Kolstrup, E., & Schild, R. (2007). Late Magdalenian feminine flint plaquettes from Poland. *Antiquity, 81*, 97–105.

Fiedorczuk, J., & Schild, R. (2002). Wilczyce – a new late Magdalenian site in Poland. In V. Eriksen & B. Bratlund (Eds.), *Recent studies in the Final Palaeolithic of the European plain. Proceedings of a U.I.S.P.P. Symposium, Stockholm, 14–17 Oktober 1999* (pp. 91–100). Århus: Aarhus University Press.

Fischer, P. (2004). *Philosophie der Technik.* München: Wilhelm Fink.

Fisher, R. A. (1936). The use of multiple measurements in taxonomic problems. *Anals of Eugenics, 7*, 179–188.

Fisher, L. E. (2002). Retooling and raw material economies: Technological change in late glacial and early postglacial Southern Germany. In L. E. Fisher & B. V. Eriksen (Eds.), *Lithic raw material economies in late glacial and early postglacial Europe* (BAR International Series, Vol. 1093, pp. 53–78). Oxford: Archaeopress.

Fitte, P., & Sonneville-Bordes, D. D. (1962). Le Magdalénien VI de la Gare de Couze, commune de Lalinde (Dordogne). *L'Anthropologie, 66*, 217–246.

Florineth, D., & Schlüchter, C. (2000). Alpine evidence for atmospheric circulation patterns in Europe during the last glacial maximum. *Quaternary Research, 54*, 295–308.

Floss, H. (1994). *Rohmaterialversorgung im Paläolithikum des Mittelrheingebietes.* Bonn: Habelt.

Floss, H. (2000). Le couloir Rhin-Saône-Rhône – axe de communication au tardiglaciaire? In *Les derniers chasseurs-cueilleurs d'Europe occidentale (13000-5500 av J.-C). Actes du Colloque de Besancon (Doubs, France), 23–25 octobre 1998* (pp. 313–321). Paris: Les Belles Lettres.

Floss, H. (2006). Le Paléolithique supérieur en Allemagne méridionale (2001–2005). In P. Noiret (Ed.), *Le Paléolithique supérieur européen. Bilan quinquuennal 2001–2006* (ERAUL, Vol. 115, pp. 71–80). Liège: Université de Liège.

Floss, H. (2012a). Bohrer. In H. Floss (Ed.), *Steinartefakte vom Altpaläolithikum bis in die Neuzeit* (pp. 477–480). Tübingen: Kerns.

Floss, H. (2012b). Lithische Spitzen des Jungpaläolithikums. In H. Floss (Ed.), *Steinartefakte vom Altpaläolithikum bis in die Neuzeit* (pp. 399–413). Tübingen: Kerns.

Floss, H. (2012c). Grundbegriffe der Artefaktmorphologie und der bruchmechanik. In H. Floss (Ed.), *Steinartefakte vom Altpaläolithikum bis in die Neuzeit* (pp. 117–132). Tübingen: Kerns.

Floss, H. (Ed.). (2012d). *Steinartefakte vom Altpaläolithikum bis in die Neuzeit.* Tübingen: Kerns.

Floss, H., & Siegeris, M. (2012). Bedeutende Silices in Europa – Historie, Bestimmungsmethodik und archäologische Bedeutung. In H. Floss (Ed.), *Steinartefakte vom Altpaläolithikum bis in die Neuzeit* (pp. 15–30). Tübingen: Kerns.

Floss, H., & Terberger, T. (2002). *Die Steinartefakte des Magdalénien von Andernach (Mittelrhein). Die Grabungen 1979–1983.* Rahden/Westf.: Leidorf.

Floss, H., & Weber, M.-J. (2012a). Schlagtechniken. In H. Floss (Ed.), *Steinartefakte vom Altpaläolithikum bis in die Neuzeit* (pp. 133–136). Tübingen: Kerns.

Floss, H., & Weber, M.-J. (2012b). Lithische Projektilspitzen im Spätglazial. In H. Floss (Ed.), *Steinartefakte vom Altpaläolithikum bis in die Neuzeit* (pp. 509–516). Tübingen: Kerns.

Födisch, H. (1961). *Das Saazer Land in ur- und frühgeschichtlicher Zeit.* München: Lerche.

Foltyn, E., & Foltyn, E. (2007). Magdalenian flint workshop at Samborowice, site 2a (Upper Silesia, Poland). In M. Masojć, T. Płonka, B. Ginter, & S. K. Kozłowski (Eds.), *Contributions to the Central European Stone Age. Papers dedicated to the late Professor Zbigniew Bagniewski* (pp. 53–75). Wrocław: Institute of Archaeology, University of Worcław.

Fraas, O. (1867). Die neuesten Erfunde an der Schussenquelle bei Schussenried. *Jahreshefte des Vereins für vaterländische Naturkunde in Württemberg, 23*, 48–74.

Fraas, O. (1872). Resultate der Ausgrabung im Hohlefels bei Schelklingen. *Württembergische naturwissenschaftliche Jahreshefte, 28*, 21–36.

Fraipont, J. (1896). *Les cavernes et leurs habitants.* Paris: Baillière.

Franken, E., & Veil, S. (1983). *Die Steinartefakte von Gönnersdorf.* Wiesbaden: Steiner.

Frauenholz, J., Obermaier, H., & Schlosser, M. (1911). Die Kastlhäng-Höhle, eine Renntierjägerstation im bayrischen Altmühlthale. *Beiträge zur Anthropologie und Urgeschichte Bayerns, 18*, 119–164.

Frenzel, W. (1934). Burk. In W. Frenzel, W. Radig, & O. Reche (Eds.), *Grundriß der Vorgeschichte Sachsens* (pp. 310–313). Leipzig: Richter.

Freund, G. (1963). Die ältere und mittlere Steinzeit in Bayern. *Jahresbericht der Bayerischen Bodendenkmalpfege, 4*, 9–167.

Freund, G. (1987). *Das Paläolithikum der Oberneder-Höhle (Landkreis Kelheim/Donau).* Bonn: Röhrscheid.

Frey, W., & Lösch, R. (2010). *Geobotanik: Pflanzen und Vegetation in Raum und Zeit.* Heidelberg: Spektrum.

Frickhinger, E. (1937). Grabungen in der kleinen Ofnethöhle bei Hohlheim, B.-W. Nördlingen. *Bayerische Vorgeschblätter, 14*, 35–38.

Fridrich, J. (1964). Vyobrazeníryby v paleolitickém umění českych zemí. *Archeologické rozhledy, 16*, 716–741.

Fridrich, J. (1972). Paleolitické osídlení v Bečově, okr. Most. *Archeologické rozhledy, 24*, 249–259.

Fridrich, J., & Sklenář, K. (1976). *Die paläolithische und mesolithische Höhlenbesiedlung des Böhmischen Karstes*. Praha: Museum Nationale Pragae – Sectio Praehistorica.

Friendly, M., & Fox, J. (2011). *candisc: Generalized canonical discriminant function analysis. R package version 0.5-21*. http://CRAN.R-project.org/package=candisc

Fritz, J. M., & Plog, F. T. (1970). The nature of archaeological explanation. *American Antiquity, 35*(4), 405–412.

Furmanek, M., Molenda, G., & Rapiński, A. (2001). Wstępne wyniki badań na stanowisku schyłkowopaleolitycznym w Sowinie, pow. Niemodlin w 2000 roku. *Śląskie Sprawozdania Archeologiczne, 43*, 485–490.

Gábori, M. (1959). A ságvári paleolitikus telep újabb ásatásának eredményei. *Archaeológiai Értesítő, 86*, 3–19.

Gábori, M. (1965). Der zweite Paläolithische Hausgrundriss von Ságvár. *Acta Archaeologica Academiae Scientiarum Hungaricae, 17*, 111–127.

Gadamer, H.-G. (2006). *Truth and method*. London: Continuum.

Gaffrey, J. (1997). Borken-Gemenkrückling. Fundchronik Archäologische Bodendenkmalpflege 1989–1990. Regierungsbezirk Münster. *Ausgrabungen und Funde in Westfalen-Lippe, 9*(A), 313.

Gaillard, M.-J. (1985). Late-Glacial and Holocene environments of some ancient lakes in the western Swiss Plateau. In G. Lang (Ed.), *Swiss lake and mire environments during the last 15000 years* (pp. 273–336). Vaduz: Lubrecht and Cramer.

Gailli, R. (2003). Description de la cavité. In J. Clottes & H. Delporte (Eds.), *La Grotte de La vache (Arriège). Fouilles Romain Robert. I – Les occupations du Magdalénien* (pp. 43–50). Paris: Éditions de la Réunion des musées nationaux.

Galiński, T. (1987). An investigation into Palaeolithic settlement of the Pomeranian Territories. In J. M. Burdukiewicz & M. Kobusiewicz (Eds.), *Late glacial in Central Europe. Culture and environment* (pp. 140–163). Warsaw: Zakad Narodowy im. Ossolinskich, Wydawn. Polskiej Akademii Nauk.

Gallay, A. (1986). *L'Archéologie demain*. Paris: Belfond.

Gamble, C. S. (1980). Information exchange in the Palaeolithic. *Nature, 283*, 522–523.

Gamble, C. S. (1986). *The Palaeolithic settlement of Europe*. Cambridge: Cambridge University Press.

Gamble, C. S. (1999). *The Palaeolithic societies of Europe*. Cambridge: Cambridge University Press.

Ganzelewski, M., & Slotta, R. (1996). *Bernstein, Tränen der Götter*. Bochum: Glückauf.

Garczarek, U. M. (2002). *Classification rules in standardized partition spaces*. Ph.D. Dissertation, University of Dortmund.

Gaudzinski, S., & Street, M. (2003). Reconsidering hunting specialisation in the German Magdalenian faunal record. In S. Costamagno & V. Laroulandie (Eds.), *Mode de Vie au Magdalénien: Apports de l'Archéozoologie* (BAR International Series, Vol. 1144, pp. 11–21). Oxford: Archaeopress.

Gaussen, J. (1994). Le Plateau Parrain (l'Organisation). *L'Anthropologie, 98*(2–3), 418–426.

Gautier, A., Cordy, J.-M., Straus, L. G., & Otte, M. (1997). Taphonomic, chronostratigraphic, paleoenvironmental and anthropogenetic implications of the Upper Pleistocene faunas from Le Trou Magrite, Belgium. *Anthropozoologica, 25–26*(1997), 343–354.

Gehlen, A. (1986). *Urmensch und Spätkultur. Philosophische Ergebnisse und Aussagen*. Wiesbaden: AULA.

Gehlen, B. (2010). *Innovationen und Netzwerke. Das Spätmesolithikum vom Forggensee (Südbayern) im Kontext des ausgehenden Mesolithikums und des Altneolithikums in der Südhälfte Europas*. Kerpen-Loogh: Welt und Erde.

Gerhardt, K., & Naber, F. B. (1983). Die mesolithische Doppelbestattung bei Altessing, Gem. Essing, Ldkr. Kelheim/Ndb. *Bayerische Vorgeschblätter, 48*, 1–30.

Gerlach, T., & Geupel, V. (1980). Ein Spätmagdalénien-Fundplatz auf dem "Burzelberg" bei Hohburg. *Kr. Wurzen. Ausgrabungen und Funde, 25*(1), 1–5.

Germonpré, M. (1997). AMS dating of the Upper Magdalenian horizon 1 at Goyet Cave. *Bulletin de L'Institut Royal des Sciences Naturelles de Belgique, 67*, 167–182.

Gersbach, E. (1925). Der Röthekopf bei Säckingen in Baden. *Berichte der Naturforschenden Gesellschaft zu Freiburg, 24*, 353–368.

Gersbach, E. (1989). *Ausgrabung heute: Methoden und Techniken der Fedlgrabung. Mit einem Beitrag von Joachim Hahn*. Darmstadt: Wissenschaftliche Buchgesellschaft.

Geupel, V. (1967). Zwei spätpaläolithische Fundplätze aus dem Wurzener Land. *Arbeits- und Fortschrittsberichte zur Sächsischen Bodendenkmalpflege, 16*(17), 7–25.

Geupel, V. (1969). Die Steinzeitlichen Funde von der "Wüsten Kirche" in Flur Wurzen-Dehnitz. *Jahresschrift für mitteldeutsche Vorgeschichte, 53*, 199–214.

Geupel, V. (1972). Altsteinzeitliche Feuersteingeräte vom "Spitzberg" bei Lüptitz. *Kr. Wurzen. Ausgrabungen und Funde, 17*(2), 51–58.

Geupel, V. (1976). Die spätaltsteinzeitlichen und mittelsteinzeitlichen Feuersteingeräte von der "Schanze" in Westewitz, Kr. Döbeln. *Arbeits- und Fortschrittsberichte zur Sächsischen Bodendenkmalpflege, 20*(21), 15–26.

Geupel, V. (1988). Spätpaläolithische Fundstelle Wurzen-Dehnitz. In H.-J. Vogt (Ed.), *Archäologische Feldforschung in Sachsen. Fünfzig Jahre Landesmuseum für Vorgeschichte in Dresden. (Arbeits- und Forschungsbericht zur sächsischen Bodendenkmalpflege, Beihesft 18)* (pp. 40–43). Berlin: Deutscher Verlag der Wissenschaften.

Geyer, O. F., & Gwinner, M. P. (1962). *Der Schwäbische Jura*. Berlin: Borntraeger.

Gietz, F. J. (2001). *Spätes Jungpaläolithikum und Mesolithikum in der Burghöhle Dietfurt an der oberen Donau*. Stuttgart: Theiss.

Gilot, É. (1997). *Index général des dates Lv, Laboratoire du Carbone 14 de Louvain, Louvain-la-Neuve*. Liège: Leuven University Press.

Ginter, B. (1972). Badania schyłkowopaleolitycznych i mmezolitycznych osad wydmowych w miejscowościach Dąbrówka, pow. Włoszczowa, i Kozłów, pow. Jędrzejów. *Sprawozdania archeologiczne, 24*, 11–20.

Ginter, B. (1974). *Spätpaläolithikum in Oberschlesien und im Flussgebiet der oberen Warta*. Kraków: Universytetu Jagiellońskiego.

Ginter, B., & Połtowicz, M. (2002a). Badania archeologiczne w Dzierżysławiu pow. Głubczyce w 2001 roku. *Śląskie Sprawozdania Archeologiczne, 44*, 47–56.

Ginter, B., & Połtowicz, M. (2002b). Badania stanowiska kultury magdaleńskiej w Dzierżysławiu, województwo opolskie. In *Badania Archeologiczne Na Górnym Śląsku I Ziemiach Pogranicznych W Latach, 1999–2000* (pp. 64–75).

Ginter, B., & Połtowicz, M. (2004). Badania w Dzierżysławiu pow. Głubczyce w latach 2002–2003. *Śląskie Sprawozdania Archeologiczne, 46*, 5–16.

Ginter, B., Połtowicz, M., Pawlikowski, M., Skiba, S., Trąbska, J., Wacnik, A., et al. (2002). Dzierżysław 35 – stanowisko magdaleńskie na przedpolu Bramy Morawskiej. In J. Gancarski (Ed.), *Starsza i środkowa epoka kamienia w Karpatach polskich* (pp. 111–146). Krosno: Muzeum Podkarpackie.

Ginter, B., Połtowicz, M., Pawlikowski, M., Skiba, S., Trąbska, J., Wacnik, A., et al. (2005). Dzierżysław 35 – Ein neuer Fundplatz des Magdalénien in Oberschlesien. *Archäologisches Korrespondenzblatt, 35*(4), 431–446.

Glory, A. (1961). La grotte de Rigney (Doubs). Anciennes fouilles de M. Jaques Collot. *Bulletin de la Société Préhistorique Française, 58*, 389–400.

Goetze, A. (1930). Die vorgeschichtliche Besiedlung der Döbritzer Grotte. *Die Thüringer Höhlen, 1*, 83–92.

Göttlich, K. (1961). Neue Beiträge zur Stratigraphie und Entwicklungsgeschichte des Federseemoores. In W. Zimmermann (Ed.), *Der Federsee* (pp. 368–386). Stuttgart: Verl. des Schwäbischen Albvereins.

Gramsch, A. (Ed.). (2000). *Vergleichen als archäologische Methode. Analogien in den Archäologien* (BAR International Series, Vol. 825). Oxford: Archaeopress.

Greenacre, M. J. (2007). *Correspondence analysis in practice*. Boca Raton: CRC.

Greenacre, M. J., & Nenadic, O. (2010). *ca: Simple, multiple and joint correspondence analysis. R package version 0.33*. http://CRAN.R--project.org/package=ca

Grees, H. (1993). Die Schwäbische Alb. In C. Borcherdt (Ed.), *Geographische Landeskunde von Baden-Württemberg* (pp. 332–362). Stuttgart: Kohlhammer.

Greppin, J. B. (1875). Drei neue Stationen des Steinzeitalters in der Umgebung von Basel. *Archiv für Anthropologie, 8*, 139–141.

Grimm, S., & Weber, M.-J. (2008). The chronological framework of the Hamburgian in the light of old and new ^{14}C dates. *Quartär, 55*, 17–40.

Grootes, P. M., Stuiver, M., White, J. W. C., Johnsen, S., & Jouzel, J. (1993). Comparison of oxygen isotope records from the GISP2 and GRIP Greenland ice core. *Nature, 366*, 552–554.

Grote, K. (1993). *Die Abris im südlichen Leinebergland bei Göttingen. Archäologische Befunde zum Leben unter Felsschutzdächern in urgeschichtlicher Zeit I*. Isensee: Archäologischer Teil. Oldenburg.

Grote, K. (1998). Eiszeitliche Jäger in Südniedersachsen. *Archäologie in Niedersachsen, 1*, 50–53.

Grünberg, J. M. (2004). Das Leben des modernen Menschen zur Zeit des Magdalénien. Jagd- und Sammelstrategien und der Fundplatz bei Saaleck. In H. Meller (Ed.), *Paläolithikum und Mesolithikum. Kataloge zur Dauerausstellung im Landesmuseum für Vorgeschichte in Halle I* (pp. 251–260). Landesamt für Denkmalpflege und Archäologie Sachsen-Anhalt, Landesmuseum für Vorgeschichte: Halle.

Grünvogel, E. (1958). Zur Talgeschichte der Bodenseezuflüsse Rotach. *Schussen und Argen auf Grund ihres Gefälls. Schriften des Vereins für Geschichte des Bodensees und seiner Umgebung, 76*, 103–151.

Guenther, E. W. (1968). Ist die Rentierjägerstation von Munzingen ein Lößmagdalénien? *Quartär, 19*, 93–124.

Gulder, A. (1952). Die Paläolithstation von Kamegg im Kamptal, N.-Ö. *Archaeologia Austriaca, 10*, 16–27.

Gumpert, K. (1931). Der madeleinezeitliche "Rennerfels" in der Fränkischen Schweiz. *Prähistorische Zeitschrift, 22*, 56–77.

Gumpert, K. (1932). Der Rennerfels. Eine neue Madeleine-Station in der Fränkischen Schweiz. *Bayerische Vorgeschblätter, 10*, 60–66.

Gumpert, K. (1933a). Die Abrisiedlung Steinbergwand bei Ensdorf, B.-A. Amberg, Oberpfalz. *Bayerische Vorgeschblätter, 11*, 57–69.

Gumpert, K. (1933b). Eine paläolithische und mesolithische Abri-Siedlung an der Steinbergwand bei Ensdorf in der Oberpfalz. *Mannus, 25*, 176–206.

Gumpert, K. (1954). Die alt- und mittelsteinzeitliche Grottensiedlung am Mühlberg bei Dollnstein (Landkreis Eichstätt). *Jahresbericht des Historischen Vereins für Mittelfranken, 74*, 1–12.

Gumpert, K. (1956). Der altsteinzeitliche "Hohle Stein" bei Schambach (Ldkr. Eichstätt). *Bayrische Vorgeschichtsblätter, 21*, 13–21.

Günther, K. (1988). *Alt- und mittelsteinszeitliche Fundplätze in Westfalen*. Münster: Westfälisches Museum für Archäologie.

Guthrie, R. D. (1990). *Frozen fauna of the mammoth steppe: The story of Blue Baby*. Chicago: University of Chicago Press.

Hacker, P. L. (1884). Die Gudenus-Höhle, eine Renthierstation im niederösterreichischen Kremsthale. *Mitteilungen der Anthropologischen Gesellschaft in Wien, 14*, 145–153.

Haesaerts, P. (1990). Stratigraphy of the Grubgraben Loess Sequence. In A. Montet-White (Ed.), *The Epigravettian site of Grubgraben, Lower Austria: The 1986 and 1987 excavations* (ERAUL, Vol. 40, pp. 15–35). Université de Liège: Liège.

Haesaerts, P., Borziak, I., Chirica, V., Bamblon, F., & Koulakovska, L. (2004). Cadre stratigraphique et chronologique du Gravettien en Europe Centrale. *Paleo, 19*, 31–51.

Hahn, J. (1974). Ausgrabungen eines spätpaläolithischen Rastplatzes in der Spitzbubenhöhle, Gem. Herbrechtingen-Eselburg, Kreis Heidenheim (Baden-Württemberg). *Archäologische Informationen, 2–3*, 55–58.

Hahn, J. (1975). Neuere urgeschichtliche Ausgrabungen auf der Ostalb. *Mitteilungen Verband Deutscher Höhlen- und Karstforscher, 21*, 27–33.

Hahn, J. (1977a). Die Spitzbubenhöhle, eine Magdalénien-Fundstelle bei Eselsburg, Gemeinde Herbrechtingen, Kreis Heidenheim. *Fundberichte aus Baden-Württemberg, 3*, 38–50.

Hahn, J. (1977b). Nachgrabungen im "Hohlen Fels" bei Schelklingen, Alb-Donau-Kreis. *Archäologische Ausgrabungen, 1977*, 10–14.

Hahn, J. (1977c). Fossilvergesellschaftungen NR. 72. Nachgrabungen im Hohlen Felsen bei Schelklingen, Alb-Donau-Kreis. *Archäologisches Korrespondenzblatt, 7*, 241–248.

Hahn, J. (1977d). Die altsteinzeitliche Schichtenfolge des "Geißenklösterle" bei Blaubeuren nach den Grabungen 1977. *Archäologische Ausgrabungen, 1977*, 20–22.

Hahn, J. (1977e). *Aurignacian. Das ältere Jungpaläolithikum in Mittel- und Osteuropa*. Köln: Böhlau.

Hahn, J. (1978). Ausgrabungen in urgeschichtlichen Fundstellen des Achtales zwischen Blaubeuren und Schelklingen, Alb-Donau-Kreis. *Archäologische Ausgrabungen, 1978*, 11–15.

Hahn, J. (1980). Die Ausgrabungen des Jahres 1980 im Geißenklösterle, Gemeinde Blaubeuren-Weiler, Alb-Donau-Kreis. *Archäologische Ausgrabungen, 1980*, 14–19.

Hahn, J. (1982). *Der Speckberg bei Meilenhofen*. Kallmünz: Lassleben.

Hahn, J. (Ed.). (1984a). *Die steinzeitliche Besiedlung des Eselsburger Tales bei Heidenheim (Schwäbische Alb)*. Stuttgart: Theiss.

Hahn, J. (1984b). Zum Abschluß der Ausgrabungen im Helga-Abri, Gemeinde Schelklingen, Alb-Donau-Kreis. *Archäologische Ausgrabungen in Baden-Württemberg, 1984*, 15–17.

Hahn, J. (1989). Das Jungpläolithikum vom Speckberg bei Meilenhofen. In Stadtmuseum Ingolstadt (Ed.), *Steinzeitliche Kulturen an Donau und Altmühl. Begleitheft zur Ausstellung im Stadtmuseum Ingolstadt* (pp. 87–91). Ingolstadt: Courier Druckhaus.

Hahn, J. (1991). *Erkennen und Bestimmen von Stein- und Knochenartefakten. Einführung in die Artefaktmorphologie*. Tübingen: Archaeologica Venatoria.

Hahn, J. (1995). Die Buttentalhöhle. Eine spät-jungpaläolithische Abristation im Oberen Donautal. *Fundberichte aus Baden-Württemberg, 20*, 13–158.

Hahn, J. (1996). Le Paléolithique supérieur en Allemagne méridionale (1991–1995). In M. Otte (Ed.), *Le Paléolithique Supérieur européen, UISPP, Congrès de Forli 1996, Commission Paléolithique Supérieur, Bilan 1991–1996* (ERAUL, Vol. 76, pp. 181–186). Liège: Université de Liège.

Hahn, J. (2002). Mobility and lithic economy at the Buttental site: A case stude. In L. E. Fisher & B. V. Eriksen (Eds.), *Lithic raw material economies in late glacial and early postglacial Europe* (BAR International Series, Vol. 1093, pp. 19–25). Oxford: Archaeopress.

Hahn, J., & Koenigswald, W. V. (1977). Die steinzeitlichen Funde und die spätglaziale Nagerschicht aus der Kleinen Scheuer am Hohlenstein im Lonetal. *Fundberichte aus Baden-Württemberg, 3*, 51–75.

Hahn, J., Koenigswald, W. V., Wagner, E., & Wille, W. (1977). Das Geißenklösterle bei Blaubeuren, Alb-Donau-Kreis. Eine altsteinzeitliche Höhlenstation der mittleren Alb. *Fundberichte aus Baden-Württemberg, 3*, 14–37.

Hahn, J., Müller-Beck, H., & Taute, W. (1985). *Eiszeithöhlen im Lonetal. Archäologie einer Landschaft auf der Schwäbischen Alb*. Stuttgart: Theiss.

Hahn, J., & Rottländer, R. C. A. (1981). Ein Magdalénien-Siedlungshorizont aus dem Helga-Abri, Stadt Schelklingen, Alb-Donau-Kreis. *Archäologische Ausgrabungen in Baden-Württemberg, 1981*, 21–26.

Hahn, J., & Scheer, A. (1983a). Das Helga-Abri am Hohlenfelsen bei Schelklingen: Eine mesolithische und jungpaläolithische Schichtenfolge. *Archäologisches Korrespondenzblatt, 13*, 19–28.

Hahn, J., & Scheer, A. (1983b). Jungpaläolithische Feuerstellen in der Höhle Geißenklösterle bei Blaubeuren, Alb-Donau-Kreis. *Archäologische Ausgrabungen in Baden-Württemberg*, 29–32.

Hahn, J., & Scheer, A. (1984). Eine magdalénienzeitliche Siedlungsstruktur im Helga-Abri am Hohlefels, Gemeinde Schelklingen, Alb-Donau-Kreis. *Archäologische Ausgrabungen in Baden-Württemberg, 1983*, 35–37.

Haïd, N., & Margerand, I. (1996). Les Lamelles à bord abattu magdaléniennes de la Grotte des Romains à Pierre-Châtel (Virignin, Ain, France). *L'Anthropologie, 100*(1), 42–54.

Hamal-Nandrin, J., & Servais, J. (1925). La grotte dite "du Coléoptère", Rapport sur les fouilles, 1923–1924. *Revue Anthropologique, 4–5*, 120–144.

Hamal-Nandrin, J., Servais, J., Louis, M., Formarier, P., Fraipont, C., & Leclercq, S. (1934). Fouilles dans la terrasse des deux grottes de Fond-de-Forêt (province de Liège) 1931–1933. *Bulletin de la Société Préhistorique Française, 31*(11), 484–499.

Handl, A. (2002). *Multivariate Analysemethoden. Theorie und Praxis unter besonderer Brücksichtigung von S-Plus*. Berlin: Springer.

Hanitzsch, H. (1956a). Die spätpaläolithischen Silexgeräte von Friedensdorf (früher Kriegsdorf), Kreis Merseburg. *Jahresschrift für mitteldeutsche Vorgeschichte, 40*, 5–8.

Hanitzsch, H. (1956b). Ein jungpaläolithischer Fundplatz auf dem Zinkenberg in den Hohburger Bergen. *Arbeits- und Fortschrittsberichte zur Sächsischen Bodendenkmalpflege, 5*, 9–16.

Hanitzsch, H. (1957). Die Spätmagdalénien-Station Groitzsch bei Eilenburg (Fundplatz A). *Forschungen zur Vor- und Frühgeschichte (Varia Praehistorica), 2*, 5–40.

Hanitzsch, H. (1969). Zur Gliederung des mitteldeutschen Magdalénien. *Jahresschrift für mitteldeutsche Vorgeschichte, 53*, 179–192.

Hanitzsch, H. (1972). *Groitzsch bei Eilenburg. Schlag- und Siedlungsplätze der späten Altsteinzeit*. Berlin: Deutscher Verlag der Wissenschaften.

Hanitzsch, H. (1978). Gravierte Schieferplatten aus dem Spätmagdalénien von Saaleck, Kreis Naumburg. *Archäologisches Korrespondenzblatt, 8*, 265–267.

Hans, L.-M., & Thévenin, A. (1993). Magdalénien Final en Lorraine et peuplement du nord-est de la France er régions limitrophes entre Dryas I et alleröd. *Revue Archéologique de l'Est et du Center-Est, 44*(1), 3–29.

Hanson, N. R. (1967). Observation and interpretation. In S. Morgenbesser (Ed.), *Philosophy of science today* (pp. 89–99). New York: Basic Books.

Harris, E. C. (1989). *Principles of archaeological stratigraphy*. London: Academic Press.

Hauser, O. (1925). *Die große zentraleuropäische Urrasse*. Langensalza: Beltz.

Häusler, R. (1914a). Die ältesten Spuren vom Menschen am Schweizersbild. *Mannus, VI*, 181–183.

Häusler, R. (1914b). Die Ausgrabungen beim Schweizersbild. *Mannus, VI*, 245–260.

Heck, H. (1920). Eine neue paläolithische Station im Lahntal. Die Wildweiberlei bei Diez an der Lahn. *Korrespondenzblatt der Deutschen Gesellschaft für Anthropologie Ethnologie und Urgeschichte, 51*, 56–59.

Hedges, R. E. M., Housley, R. A., Bronk, C. R., & Klinken, G. J. V. (1992). Radiocarbon dates from the Oxford AMS system: Archaeometry datelist 15. *Archaeometry, 34*, 337–357.

Hedges, R. E. M., Housley, R. A., Law, L. A., Perry, C., & Gowlett, J. A. J. (1987). Radiocarbon dates from the Oxford AMS system: Archaeometry datelist 6. *Archaeometry, 29*(2), 289–306.

Hedges, R. E. M., Pettitt, P. B., Bronk Ramsey, C., & Van Klinken, G. J. (1997). Radiocarbon dates from the Oxford AMS system: *Archaeometry* datelist 24. *Archaeometry, 39*(2), 445–471.

Hedges, R. E. M., Pettitt, P. B., Bronk Ramsey, C., & Van Klinken, G. J. (1998). Radiocarbon dates from the Oxford AMS system: *Archaeometry* datelist 25. *Archaeometry, 40*(1), 227–239.

Hegmon, M. (1995). *The social dynamics of pottery style in the early Puebloan Southwest*. Cortez: Crow Canyon Archaeological Center.

Heinen, M. (1993). *Archäologische Fundstellen und Funde im Stadtgebiet Viersen*. Viersen: Stadt Viersen - Stadtarchiv.

Heinen, M., & Arora, S.-K. (1995). *Archäologie im Kreis Heinsberg II*. Geilenkirchen: Kreis Heinsberg.

Heinrich, W. (1973). *Das Jungpaläolithikum in Niederösterreich*. Ph.D. Dissertation, University of Salzburg.

Heinz, H. J. (1972). Territoriality among the Bushmen in general and the !ko in particular. *Anthropos, 67*, 405–416.

Heller, F. (1937). Die fossile Mikrofauna der Magdalénien-Schicht in der Nikolaushöhle bei Veringenstadt (Hohenzollern). In *Sitzungsberichte der Heidelberger Akademie der Wissenschaften, Mathematisch-naturwissenschaftliche Klasse, 1936* (Vol. 11, pp. 3–23). Heidelberg: Weiss.

Heller, F. (1956). *Die Fauna der Breitenfurter Höhle im Landkreis Eichstätt*. Erlangen: Junge and Sohn.

Heller, F. (1957). Zur fossilen Fauna der jungpaläolithischen Stationen am Bruckersberg in Giengen an der Brenz. In G. Riek (Ed.), *Drei jungpaläolithischen Stationen am Bruckersberg in Giengen an der Brenz* (pp. 53–68). Stuttgart: Silberburg.

Hemmann, L. (2005). *Drei Sammelfunde aus Rothenstein, Saale Holzland-Kreis*. Bachelor thesis, University of Jena.

Hemmann, L., Pasda, C., & Völker, D. (2008). Ölknitz, Jena, Rothenstein. Drei Fundstellen des Magdalénien im Saale-Tal in Thüringen. *Archäologisches Korrespondenzblatt, 38*(1), 1–12.

Henneböle, E. (1963). *Die Vor- und Frühgeschichte des Warsteiner Raumes*. Warstein: Hennecke.

Henrich, J. (2004). Demography and cultural evolution: How adaptive cultural processes can produce maladaptive losses – the Tasmanian case. *American Antiquity, 69*(2), 197–214.

Heptner, V. G., Nasimovic, A. A., & Bannikov, A. G. (1966). *Die Säugetiere der Sowjetunion. Band I. Paarhufer und Unpaarhufer*. Jena: Fischer.

Herrmann, F. (1955). Die diluvialgeologischen, faunistischen und urgeschichtlichen Verhältnisse in der Burghöhle Loch. *Acta Albertina Ratisbonensia, 21*, 74–85.

Higgs, E. S., & Vita-Finzi, C. (1972). Prehistoric economies: A territorial approach. In E. S. Higgs (Ed.), *Papers in economic prehistory. Studies by members and associates of the British Academy major research project in the early history of agriculture* (pp. 27–36). Cambridge: Cambridge University Press.

Higham, T. F. G., Bronk Ramsey, C., Brock, F., Baker, D., & Ditchfield, P. (2007). Radiocarbon dates from the Oxford AMS system: *Archaeometry* datelist 32. *Archaeometry, 49*, 1–60.

Higham, T. F. G., Bronk Ramsey, C., Brock, F., Baker, D., & Ditchfield, P. (2011). Radiocarbon dates from the Oxford AMS system: *Archaeometry* datelist 34. *Archaeometry, 53*(5), 1067–1084.

Hill, J. N. (1972). The methodological debate in contemporary archaeology: A model. In D. L. Clarke (Ed.), *Models in archaeology* (pp. 61–107). London: Methuen.

Hiller, R., & Veil, S. (1997). Gadenstedt FStNr. 121, Gde Lahstedt, Ldkr. Peine, Reg. Bez. BS. *Nachrichten aus Niedersachsens Urgeschichte, 65*, 247–249.

Hingst, H. (1959). *Vorgeschuchte des Kreises Stormann*. Neumünster: Wachholtz.

Hinz, H. (1954). *Vorgeschichte des nordfriesischen Festlandes*. Neumünster: Karl Wachholtz.

Hladilová, Š. (1999). Miocenní fosilie z magdalénienských lokalit Moravském krasu. *Acta Musei Moraviae, Scientiae Sociales, 84*, 117–132.

Höck, C. (1993). Die Frauenstatuetten des Magdalénien von Gönnersdorf und Andernach. *Jahrbuch des Römisch-Germanischen Zentralmuseums, 40*(1), 253–316.

Höck, C. (2000). *Das Magdalénien der Kniegrotte – Ein Höhlenfundplatz bei Döbritz, Saale-Orla-Kreis*. Stuttgart: Theiss.

Höck, C. (2012). Dreiecke des Magdalénien. In H. Floss (Ed.), *Steinartefakte vom Altpaläolithikum bis in die Neuzeit* (pp. 497–508). Tübingen: Kerns.

Hodder, I. (1977). The distribution of material culture items in the Baringo District, western Kenya. *Man, 12*, 239–296.

Hodder, I. (1982a). *The present past: An introduction to anthropology for archaeologists*. London: Batsford.

Hodder, I. (1982b). *Symbols in action: Ethnoarchaeological studies of material culture*. Cambridge: Cambridge University Press.

Hodder, I. (1990). Style as historical. In M. W. Conkey & C. Hastorf (Eds.), *The uses of style in archaeology* (pp. 44–51). Cambridge: Cambridge University Press.

Hoek, W. (1997). *Palaeogeography of lateglacial vegetations. Aspects of lateglacial and early Holocene vegetation, abiotic landscape, and climate in The Netherlands*. Utrecht: Koninklijk Nederlands Aardrijkskundig Genootschap.

Hoffmann, W. (1968). Ausgewählte Neufunde aus dem Jahre 1966. *Jahresschrift für mitteldeutsche Vorgeschichte, 52*, 337–355.

Hoffmann, W. (1972). Ausgewählte Neufunde aus dem Jahre 1969. *Jahresschrift für mitteldeutsche Vorgeschichte, 56*, 237–248.

Holen, S. R. (1990). Bison hunting territories and lithic acquisition among the Pawnee: An ethnohistoric and archaeological study. In A. Montet-White & S. R. Holen (Eds.), *Raw material economies among prehistoric hunter-gatherers: Conference Kansas, 22–24 March 1990*. Lawrence: University of Kansa.

Holm, J., & Rieck, F. (1983). Jels I: The first Danish site of the Hamburgian culture. A preliminary report. *Journal of Danish Archaeology, 2*, 7–11.

Holtorf, C. (2000). Making sense of the past beyond analogies. In A. Gramsch (Ed.), *Vergleichen als archäologische Methode. Analogien in den Archäologien* (BAR International Series, Vol. 825, pp. 165–175). Oxford: Archaeopress.

Holzkämper, J. (2006). *Die Konzentration IV des Magdalénien von Andernach-Martinsberg, Grabung 1994–1996*. Ph.D. Dissertation, University of Cologne. http://kups.ub.uni-koeln.de/volltexte/2007/2151/index.html

Holzkämper, J., Maier, A., & Koch, I. (2014). Ergebnisse einer Suche nach Aurignacien- und Magdalénien-Fundstellen in Nordrhein-Westfalen. *Archäologisches Korrespondenzblatt, 44*(3), 293–306.

Homola, V. (1943). Jeskyně zvaná "Krápníková" na Šanovém koutě u Berouna. *Příroda, 36*, 19–23.

Höneisen, M. (1986). Kesslerloch und Schweizersbild: Zwei Rentierjäger-Stationen in der Nordschweiz. *Archäologie der Schweiz, 9*(2), 28–33.

Höneisen, M., Leesch, D., & Le Tensorer, J.-M. (1993). Das späte Jungpaläolithikum. In J.-M. Le Tensorer (Ed.), *SPM I, La Suisse du Paléolithique à l'aube du Moyen-Âge, I: Paléolithique et Mésolithique* (pp. 153–202). Bâle: Verl. Schweizer. Gesellschaft für Ur- u. Frühgeschichte.

Höneisen, M., & Peyer, S. (1994). *Schweizersbild – ein Jagdlager der Späteiszeit. Beiträge und Dokumente zur Ausgrabung vor 100 Jahren*. Schaffhausen: Kantonsarchäologie Schaffhausen.

Höpken, C. (1995). Die Steinartefakte von Kamphausen und Galgenberg bei Mönchengladbach und deren Stellung im späten Jungpaläolithikum Nordwesteuropas. *Archäologische Informationen, 18*(1), 99–103.

Horáček, I., Ložek, V., Svoboda, J., & Šajnerová, A. (2002). Přírodní prostředí a osídlení krasu v Pozdním Paleolitu a Mezolitu. In J. Svoboda (Ed.), *Prehistorické jeskyně* (pp. 313–343). Brno: AV ČR.

Hörmann, K. (1913). Der hohle Fels bei Happurg. In N. G. z. Nürnberg (Ed.), *Festschrift zum XLIV Anthropologen-Kongreß in Nürnberg, 3. bis 10. August 1913* (pp. 21–64). Nürnberg: Nister.

Housley, R. A., Gamble, C. S., Street, M., & Pettitt, P. B. (1997). Radiocarbon evidence for the lateglacial human recolonisation of northern Europe. *Proceedings of the Prehistoric Society, 63*, 25–54.

Housley, R., Gamble, C. S., & Pettit, P. (2000). Reply to Blockley, Donahue and Pollard. *Antiquity, 74*, 119–121.

Hülle, W. M. (1932). Ein Fundplatz der Spätmagdalénien-Zeit bei Saaleck, Kr. Naumburg. *Nachrichten-Blatt für Deutsche Vorzeit, 8*, 85–88.

Hülle, W. M. (1933). Eine neue paläolithische Fundstelle in Mitteldeutschland: Die Ilsenhöhle bei Ranis, Kreis Ziegenrück. *Nachrichten-Blatt für Deutsche Vorzeit, 9*, 91–93.

Hülle, W. M. (1977). *Die Ilsenhöhle unter Burg Ranis/Thüringen. Eine paläolithische Jägerstation*. Stuttgart: Fischer.

Huntley, B., & Birks, H. J. B. (1983). *An atlas of past and present pollen maps for Europe: 0-13000 years ago*. Cambridge: Cambridge University Press.

Husserl, E. (1992). *Cartesianische Meditationen. Die Krisis der europäischen Wissenschaften und die transzendentale Phänomenologie; eine Einleitung in die phänomenologische Philosophie; Text nach Husserliana VI*. Hamburg: Meiner.

Ikinger, E.-M. (1998). *Der endeiszeitliche Rückenspitzen-Kreis Mitteleuropas*. Münster: Lit.

INQUA International Union for Quaternary Research, Radiocarbon Palaeolithic Database Europe (v13). http://ees.kuleuven.be/geography/projects/14c-palaeolithic/

Ioviţă, R. P. (2009). Ontogenetic scaling and lithic systematics: Method and application. *Journal of Archaeological Science, 36*, 1447–1457.

Isaac, G. (1977). *Olorgesailie: Archaeological studies of a middle Pleistocene lake basin in Kenya*. Chicago: University of Chicago Press.

Iversen, J. (1942). En pollenanalytisk Tidfaestelse af Ferskvandslagene ved Nørre-Lyngby. *Meddelelser Dansk Geologisk Forening, 10*, 130–151.

Iversen, J. (1954). The late-glacial flora of Denmark and its relation to climate and soil. *Danmarks Geologiske Undersøgelse, 2*(80), 87–119.

Iversen, J. (1973). *The development of Denmark's nature since the last glacial*. Copenhagen: Reitzels.

Jacob-Friesen, K. H. (1959). *Einführung in Niedersachsens Urgeschichte*. Hildesheim: Lax.

Jacobi, R. M., & Higham, T. (2009). The early lateglacial re-colonization of Britain: New radiocarbon evidences from Gough's Cave, southwest England. *Quaternary Science Reviews, 28*, 1895–1913.

Jacot-Guillarmod, J. (1921). Les Grottes des Dentaux. *Bulletin Société Vaudoise des Sciences Naturelles, 203*, 193–204.

Jacquement, E. (1895). Contribution à l'étude géologique de l'Ile Crémieu. *Annales de la Société Linnéenne de Lyon, 42*, 206–218.

Jaguttis-Emden, M. (1983). Die Radiokarbondatierung der Ausgrabung Petersfels. In G. Albrecht, H. Berke, & F. Poplin (Eds.), *Naturwissenschaftliche Untersuchungen an Magdalénien-Inventaren vom Petersfels, Grabungen 1974-76* (pp. 47–57). Tübingen: Archaeologica Venatoria.

Jaguttis-Emden, M. (1987). Die Radiokarbondatierung des archäologischen Horizontes IIIb innerhalb der geologischen Schicht 3b der Magdalénienstation Felsställe bei Ehingen, Ehingen-Mühlen, Alb-Donau-Kreis. In C.-J. Kind (Ed.), *Das Felsställe. Eine jungpaläolithische-frühmesolithische Abri-Station bei Ehingen-Mühlen, Alb-Donau-Kreis* (pp. 375–382). Stuttgart: Theiss.

Jahns, S. (2007). Palynological investigations ino the later Pleistocene and Holocene history of vegetation and settlement at the Löddigsee, Mecklenburg, Germany. *Vegetation History and Archaeobotany, 16*(2–3), 157–169.

Jankovská, V., & Pokorný, P. (2008). Forest vegetation of the last full-glacial period in the Western Carpathians (Slovakia and Czech Republic). *Preslia, 80*, 307–324.

Jarošová, L. (2002). Researches by Josef Skutil in the northern part of the Moravian Karst. In J. Svoboda (Ed.), *Prehistorické Jeskyně* (pp. 255–287). Brno: Archeologický ústav AV ČR.

Jastrzębski, S., & Libera, J. (1984). A Magdalenian camp in Klementowic-Kolonia, Site 20 (Province of Lublin). *Archaeologia Interregionalis, 5*, 95–104.

Jastrzębski, S., & Libera, J. (1987). Stanowisko Późnomagdaleńskie w Klementowicach-Kolonii w Świetle badań 1981–1982. *Sprawozdania archeologiczne, 39*, 9–52.

Jayet, A. (1937). Les stations magdaléniennes de Veyrier. *Genava, 15*, 36–45.

Jayet, A. (1943). Le Paléolithique de la région Genève. *Le Globe, 42*, 49–120.

Jayet, A., & Constantin, E. (1945). Résultats préliminaires des nouvelles fouilles dans la caverne à Ursus spelaeus des Dentaux aux Rocher de Naye. *Jb. SGU, 36*, 94–100.

Jochim, M. A. (1998). *A hunter-gatherer landscape. Southwest Germany in the late Paleolithic and Mesolithic*. New York: Plenum Press.

Jochim, M., Herhahn, C., & Starr, H. (1999). The Magdalenian colonization of Southern Germany. *American Anthropologist, 101*(1), 129–142.

Johansen, L., & Stapert, D. (2004). *Oldeholtwolde. A Hamburgian family encampment around a hearth*. Lisse: A.A. Balkema Publishers.

Jones, S. (2008). Ethnicity: Theoretical approaches, methodological implications. In R. A. Bentley, H. D. G. Maschner, & C. Chippindale (Eds.), *Handbook of archaeological theories* (pp. 321–333). Lanham: Altamira Press.

Jöris, O. (1997). Ein magdalénienzeitlicher Einzelfund in der Rheinniederung bei Heimbach-Weis. *Berichte zur Archäologie an Mittelrhein und Mosel, 5*, 29–39.

Jöris, O., Schmitz, R.-W., & Thissen, J. (1993). Beek: Ein Special-Task-Camp des Magdalénien. Neue Aspekte zum späten Jungpaläolithikum im Rheinland. *Archäologisches Korrespondenzblatt, 23*, 259–273.

Jöris, O., Schmitz, R.-W., & Thissen, J. (1995). Beek. In W. Schirmer (Ed.), *Quarternary field trips in central Europe* (pp. 954–957). Pfeil: München.

Julien, M. (1982). *Les Harpons Magdaléniens*. Paris: Ed. CNRS.

Kabaciński, J., & Kobusiewicz, M. (2007). Kragola near Koło (Central Poland) – the easternmost settlement of Hamburgian culture. In M. Kobusiewicz & J. Kabaciński (Eds.), *Studies in the final Palaeolithic settlement of the Great European Plain* (pp. 21–51). Poznań: Institute of Archaeology and Ethnology Polish Academy of Sciences.

Kabaciński, J., & Schild, R. (2005). The Hamburgian site at Mircowoce: A chronological framework. *Fontes Archaeologici Posnanienses, 41*, 15–18.

Kapff, R. (1935). Steinzeitrastplatz Rappenfels. *Blätter das Schwäbischen Albvereins, 47*(2), 29–31.

Kaplan, A. (1964). *The conduct of inquiry*. San Francisco: Chandler.

Karlin, C. (1972). Le débitage. In A. Leroi-Gourhan & M. Brézillon (Eds.), *Fouilles de Pincevent. Essai d'analyse ethnographiqie d'un habitat Magdalénien* (pp. 263–278). Paris: Ed. CNRS.

Karsten, H. (1874). Studie der Urgeschichte des Menschen in einer Höhle des Schaffhauser Jura. *Mittheilungen der Antiquarischen Gesellschaft in Zürich, 18*(6), 137–162.

Kaulich, B. (1983). Das Paläolithikum des Kaufertsberg bei Lierheim. *Quartär, 33*(34), 29–97.

Kaulich, B. (1994). Die Klausenhöhle. In R. K. F. Meyer & H. Schmidt-Kaler (Eds.), *Wanderungen in die Erdgeschichte 6: Unteres Altmühltal und Weltenburger Enge* (pp. 81–86). München: Pfeil.

Kaulich, B. (1995). Eine Harpune des späten Jungpaläolithikums aus der Fürst-Albrecht-Höhle bei Alling, Gde Sinzig; Lkv. Regensberg. *Jahresbericht der Bayrischen Bodendenkmalpflege, 34/35, 1993/94*, 17–29.

Kaulich, B. (1996). Hohler Fels, Petershöhle und Hunas – drei altsteinzeitliche Höhlenfundstellen im Herzen der Hersbrucker Alb. In K. Heinlein (Ed.), *Schichten einer Landschaft: die Hersbrucker Alb; Entstehung einer Kulturlandschaft* (pp. 13–23). Nürnbeg: Bund Naturschutz in Bayern.

Kegler, J. (2002). Die federmesserzeitliche Fundschicht des paläolithischen Siedlungsplatzes Andernach-Martinsberg, Grabung 1994–1996. *Archäologisches Korrespondenzblatt, 32*(4), 501–516.

Keller, W. A. (1987). Einfluß von Störfaktoren auf die ^{14}C-Datierungen pleistozäner und holozäner Materialien. *Geographica Helvetica, 42*(2), 105–107.

Keller, O., & Krayss, E. (1987). Die hochwürmzeitlichen Rückzugsphasen des Rhein-Vorlandgletschers und der erste alpine Eisrandkomplex im Spätglazial. *Geographica Helvetica, 42*(2), 169–178.

Kelly, R. L. (1995). *The foraging spectrum. Diversity in hunter-gatherer lifeways*. Washington: Smithsonian Institution Press.

Kelly, R. L. (2003). Colonization of new land by hunter-gatherers: Expectations and implications based on ethnographic data. In M. Rockman & J. Steele (Eds.), *Colonization of unfamiliar landscapes. The archaeology of adaptation* (pp. 44–58). London: Routledge.

Kersten, K. (1951). *Vorgeschichte im Herzogtum Lauenburg*. Neumünster: Wachholtz.

Kersten, K. (1961). Kreis Herzogtum Lauenburg (Nachträge zur archäologischen Landesaufnahme). *Offa, 17*(18), 205–224.

Kieselbach, P., & Maurer, U. (2002). Neue archäologische Untersuchungen in der Bernlochhöhle bei Albstadt-Truchtelfingen, Zollernalbkreis. *Archäologische Ausgrabungen in Baden-Württemberg, 2001*, 27–29.

Kind, C.-J. (1982). Das Magdalénien aus der Station "Felsställe" bei Ehingen-Mühlen, Alb-Donau-Kreis. *Archaeologica Venatoria, Mitteilungsblatt, 3*, 12 19.

Kind, C.-J. (1984). The Felsställe, a rockshelter near Ehingen-Mühlen, Alb-Donau-Kreis (Southwest Germany) with particular emphasis on the occupational structures from the Magdalenian horizon IIIb. In H. Berke, J. Hahn, & C.-J. Kind (Eds.), *Jungpaläolithische Siedlungsstrukturen in Europa* (pp. 32–39). Tübingen: Mo-Vince.

Kind, C.-J. (1985). Das Hornstein-Rohmaterial-Spektrum aus dem Magdalénien-Horizont IIIb vom Felsställe bei Ehingen-Mühlen, Alb-Donau-Kreis. *Archaeologica Venatoria, Mitteilungsblatt, 9*, 32–39.

Kind, C.-J. (1987). *Das Felsställe. Eine jungpaläolithisch-frühmesolithische Abri-Station bei Ehingen-Mühlen, Alb-Donau-Kreis. Die Grabungen 1975–1980*. Stuttgart: Theiss.

Kind, C.-J. (2012). Kratzer. In H. Floss (Ed.), *Steinartefakte vom Altpaläolithikum bis in die Neuzeit* (pp. 415–420). Tübingen: Kerns.

Kind, C.-J., & Torke, W. (1977). Die Ausgrabungen in dem Abri "Felsställe" in Mühlen, Gemeinde Ehingen-Kirchen, Alb-Donau-Kreis. *Archäologische Ausgrabungen, 1977*, 14–20.

Kind, C.-J., & Torke, W. (1978). Die steinzeitliche Schichtenfolge in dem Abri "Felsställe" in Mühlen, Gemeinde Ehingen-Kirchen, Alb-Donau-Kreis. *Archäologische Ausgrabungen, 1978*, 16–18.

Klaatsch, H. (1903). Anthropologische und paläolithische Ergebnisse einer Studienreise durch Deutschland, Belgien und Frankreich. *Zeitschrift für Ethnologie, 1*, 93–132.

Klecka, W. R. (1980). *Discriminant function analysis*. Beverly Hills: SAGE.

Klejn, L. S. (1962). *Archaeological typology* (BAR International Series, Vol. 153). Oxford: Archeopress.

Kleman, J., Hättestrand, C., Borgström, I., & Stroven, A. (1997). Fennoscandian paleoglaciology reconstructed using a glacial geological inversion model. *Journal of Glaciology, 43*, 283–299.

Klíma, B. (1949). Výzkum jeskyně "Nové Drátenické" u Křtin. *Časopis Moravského musea, scientiae sociales, 34*, 1–15.

Klíma, B. (1951). Križova jeskyně v Moravském krasu. *Archeologické rozhledy, 3*, 109–130.

Klíma, B. (1953). Archeologický výzkum jeskyně Adlerovy. *Československý kras, 6*, 94–102.

Klíma, B. (1960). Zjištovací výzkum v jeskyni "Liščí díra" u Hoštěnic. *Přehled Výzkumů, 1959*, 11–12.

Klíma, B. (1961). Archeologický výzkum jeskyně Hadí (Mokrá u Brna). Die archäologische Erforschung der Hadí-Höhle (Mokrá bei Brno). *Anthropozoikum, 9*, 277–289.

Klíma, B. (1962). *Die archäologische Erforschung der Höhle "Švédův stůl" in Mähren*. Brno: Anthropos.

Klíma, B. (1970). Eine jungpaläolithische Behausung im Mährischen Karst. *Anthropologie, 8*, 31–34.

Klíma, B. (1985). Gravierte Tierbilder aus der Děravá-Höhle in Böhmen. In R. Fellmann, G. Germann, & K. Zimmermann (Eds.), *Jagen und Sammeln. Festschrift für Hans-Georg Bandi zum 65. Geburtstag (3.September 1985)* (pp. 109–209). Stampfli: Bern.

Klíma, B. (1988). A triple burial from the Upper Palaleolithic of Dolní Věstonice. *Czechoslovakia Journal of Human Evolution, 16*(7-8), 831–835.

Klíma, B. (2002). The caves in the Říčka valley. Kůlnička, Liší a Klímova. In J. Svoboda (Ed.), *Prehistorické Jeskyně* (pp. 158–172). Brno: Archeologický ústav AV ČR.

Kobusiewicz, M. (1996). Final Palaeolithic in Poland and Sweden. In M. Otte (Ed.), *Le Paléolithique Supérieur européen, UISPP, Congrès de Forli 1996, Commission Paléolithique Supérieur, Bilan 1991–1996* (ERAUL, Vol. 76, pp. 145–150). Liège: Université de Liège.

Kobusiewicz, M., & Kabaciński, J. (1992). Late Palaeolithic site at Wojnowo, Zielona Góra voivodeship. *Fontes Archaeologici Posnanienses, 37*, 23–46.

Koenigswald, W. V. (1972). Der Faunenwandel an der Pleistozän-Holozän-Grenze in der steinzeitlichen Schichtenfolge vom Zigeunerfels bei Sigmaringen (Vorbereicht). *Archäologische Informationen, 1*, 41–45.

Koenigswald, W. V. (2002). *Lebendige Eiszeit. Klima und Tierwelt im Wandel*. Darmstadt: Wiss. Buchges.

Koenigswald, W. V., Müller-Beck, H., & Pressmar, E. (1974). *Die Archäologie und Paläontologie in den Weinberghöhlen bei Mauern (Bayern). Grabungen 1937–1967*. Tübingen: Institut für Urgeschichte Tübingen.

Kohler, T. A., Vanbuskirk, S., & Ruscavage-Barz, S. (2004). Vessels and villages: Evidence for conformist transmission in early village aggregation on the Pajarito Plateau, New Mexico. *Journal of Anthropological Archaeology, 23*, 100–118.

Koken, E. (1912). Die Geologie und Tierwelt der Paläolithischen Kulturstätten Deutschlands. In R. R. Schmidt (Ed.), *Die diluviale Vorzeit Deutschlands* (pp. 159–226). Stuttgart: Schweizerbart.

Kopacz, J., & Skalski, A. W. (1976). Excavations of the cave system in the Towarne Mountains near Częstochowa. *Archaeologia Polona, 27*, 163–175.

Kostruhn, P. (2004). Přehled lokalit polského Magdalénienu. *Acta Musei Moraviae, Scientiae Sociale, 89*, 91–128.

Kostruhn, P. (2005). Štípana industrie Magdalénienu z jeskyně Kůlny. *Acta Musei Moraviae, Scientiae Sociale, 90*, 79–128.

Kowalski, K. (1951). *Jaskinie Polski*. Warszawa: Nakładem Państwowego Muzeum Archeologicznego.

Kozłowski, L. (1922). *Starsza epoka kamienia w Polsce*. Poznan: Ksiegarnia Gebethnera i Wolffa.

Kozłowski, J. K. (1960). Paleolit. In Z. N. U. Jagiellońskiego (Ed.), *Prace Archeologiczne* (pp. 17–65). Kraków: Nakładem Universitetu Jagiellońskiego.

Kozłowski, J. K. (1964a). Quelques problèmes de la subdivision chronologique du Magdalénien en Europe centrale. *Archaeologia Polona, 7*, 180–192.

Kozłowski, J. K. (1964b). *Paleolit na Górnym Śląsku*. Wrocław: Polska Akademia Nauk.

Kozłowski, J. K. (1971). Les problèmes du Magdalénien en Europe centre-est. In *Actes du VIIIe Congrès International des Sciences Préhistorique et Protohistorique, Beogard 9–15 septembre 1971*, Beogard (pp. 53–70).

Kozłowski, J. K. (1987). Le Magdalénien en Pologne. In J.-P. Rigaud (Ed.), *Le Magdalénien en Europe. "La structuration du Magdalénien". Actes du Colloque de Mayence 1987 dans le cadre du XIème Congrès U.I.S.P.P* (ERAUL, Vol. 38, pp. 31–41). Liège: Université de Liège.

Kozłowski, J. K. (1992a). The Balkans in the middle and upper Palaeolithic: The gate to Europe or a cul-de-sac? *Proceedings of the Prehistoric Society, 58*, 1–20.

Kozłowski, J. K. (1992b). *L'art de la préhistoire en Europe orientale*. Paris: Ed. CNRS.

Kozłowski, J. K. (2001). Nouvelles découvertes du Paléolithique supérieur en Pologne. In P. Noirret (Ed.), *Le Paléolithique supérieur européen. Bilan quinquennal 1996–2001. XIVe congrès, Liège, 2–8 septembre 2001* (ERAUL, Vol. 97, pp. 89–92). Liège: Université de Liège.

Kozłowski, J. K. (2006). Les recherches sur le Paeolithique supérieur en Pologne entre 2002 et 2005. In P. Noiret (Ed.), *Le Paléolithique supérieur européen. Bilan quinquennal 2001–2006, Comission VIII. XVe Congrès UISPP, Lisbonne (4–9 septembre 2006)* (ERAUL, Vol. 115, pp. 67–69). Liège: Université de Liège.

Kozłowski, J. K., & Kozłowski, S. K. (1977). *Epoka kamienia na ziemiach polskich*. Warschau: Państ. Wydaw. Naukowe.

Kozłowski, J. K., & Kozłowski, S. K. (1996). *Le Paléolithique en Pologne*. Grenoble: Millon.

Kozłowski, J. K., & Pettitt, P. B. (2001). Absolute dating of the Polish Magdalénien. *Fontes Archaeologici Posnanienses, 39*, 31–35.

Kozłowski, S. K. (1992). The West carpathians and Sudeten at the end of the upper Palaeolithic. *Preistoria Alpina, 28*, 127–137.

Kozłowski, S. K., Sachse-Kozłowski, E., Marshak, A., Madeyska, T., Kierdorf, H., Lasota-Moskalewska, A., et al. (1995). Maszycka Cave, a Magdalenian Site in Southern Poland. *Jahrbuch des Römisch-Germanischen Zentralmuseums, 40*(1), 1993, 115–252.

Krawczyk, M., Płonka, T., & Wiśniewski, A. (2004). Nowe stanowisko Magdaleńskie w Broniszowiach na Górnym Śląsku. *Śląskie Sprawozdania Archeologiczne, 46*, 235–240.

Kretschmer, I. (2014). *Dempgraphie der Jäger und Sammler im späten Jungpaläolithikum*. Ph.D. Dissertation, University of Cologne.

Krolopp, E., & Sümegi, P. (2002). A ságvári lösz rétegsor csigafaunája. *Malakológiai Tájékoztató, 20*, 7–14.

Krönneck, P. (2008). Der Bockstein – neue Erkenntnisse zur Paläoumwelt. *Mitteilungen der Gesellschaft für Urgeschichte, 17*, 39–53.

Krukowski, S. (1921). Badania jaskiń pasma Krakowsko-Wieluńskiego w roku 1914. *Archiwum Warszawskiego Towarzystwa Nauk Antropologicznych, 1*, 1–8.

Krukowski, S. (1927). Nowe materiałi paleolityczne w dawnych nabytkach Muzeum Archeologicznego PAU. *Sprawozdania z czynności i posiedzeń, 31*(5), 8–9.

Kuhn, T. S. (1996). *The structure of scientific revolutions*. Chicago: University of Chicago Press.

Kurt, F. (2002). *Das Reh in der Kulturlandschaft. Ökologie, Sozialverhalten, Jagd und Hege*. Stuttgart: Franckh-Kosmos.

Küßner, M. (1998). Die Oberflächenfunde von der jungpaläolithischen Freilandstation Bad Frankenhausen, Kosackenberg. *Alt-Thüringen, 32*, 7–97.

Küßner, M. (2003). Gravierungen von Fundstellen des späten Jungpaläolithikums im mittleren Elstertal um Gera, Ostthüringen. In J. M. Burdukiewicz, L. Fiedler, W.-D. Heinrich, A. Justus, & E. Brühl (Eds.), *Erkenntnisjäger. Kultur und Umwelt des frühen Menschen. Festschrift für Dietrich Mania* (pp. 337–348). Halle: Landesamtes für Archäologie Sachsen-Anhalt.

Küßner, M. (2007). Mitteldeutsche Fundstellen der Zeit zwischen dem Weichsel-Pleniglazial und dem Ende des Alleröd im Kartenbild. In A. G. I. Thüringen (Ed.), *Terra Praehistorica. Festschrift für Klaus-Dieter Jäger zum 70. Geburtstag* (pp. 211–223). Langenweißbach: Beier and Beran.

Küßner, M. (2009). *Die späte Altsteinzeit im Einzugsgebiet der Saale. Untersuchungen an ausgewählten Fundstellen.* Weimar: Beier and Beran.

Küßner, M., & Terberger, T. (2006). Die Fundstelle Gera-Zoitzberg und die Zeit zwischen Gravettien und Magdalénien in Mitteldeutschland. *Alt-Thüringen, 39*, 69–120.

Łanczont, M., Madeyska, T., Muzyczuk, A., & Valde-Nowak, P. (2002). Hłomcza – stanowisko kultury magdaleńskiej w Karpatach polskich. In J. Gancarski (Ed.), *Starsza i środkowa epoka kamienia w Karpatach polskich* (pp. 147–188). Krosno: Muzeum Podkarpackie.

Lang, G. (1952). Zur späteiszeitlichen Vegetations- und Florengeschichte Südwestdeutschlands. *Flora, 139*, 243–294.

Lang, G. (1954). Neue Untersuchungen über die spät- und nacheiszeitliche Vegetationsgeschichte des Schwarzwaldes I. Der Hotzenwald im Südschwarzwald. *Beiträge zur naturkundlichen Forschung in Südwestdeutschland, 13*(1), 3–42.

Lanting, J. N., & Van Der Pflicht, J. (1996). Die ^{14}C-Chronologie van de Nederlandse Pre- en Protohistorie I: Laat-Paleolithicum. *Palaeohistoria, 37*(38), 71–126.

Laplace, G. (1972). La Typologie analytique et structurale: Base rationelle d'étude des industries lithiques et osseuses. In CNRS (Ed.), *Banques de données archéologiques. Colloques nationaux CNRS 932* (pp. 92–143). Paris: CNRS.

Lartet, E. (1861). Nouvelle recherches sur la coexistence de l'homme et des grandes mammifères fossiles réputés caractèristiques de la dernière période géologique. *Annales des Sciences Naturelles, Partie Zoologique, 15*, 177–261.

Lartet, É., & Christy, H. (1864). *Sur des figures d'animaux gravées ou sculptées et autres produits d'art et d'industries rapportables aux temps primordiaux de la période humaine* (pp. 233–267). IX: Revue Archéologique.

Lartet, É., Christy, H., & Jones, T. R. (1875). *Reliquiae Aquitanicae; being contributions to the archaeology and palaeonology of Périgord and the adjoining provinces of southern France.* London, Paris, Leipzig: Williams and Norgate, Baillière and Fils, Brockhaus.

Larue, M., Combier, J., & Roche, J. (1955). Les gisments périgordien et magdalénien du Saut-du-Perron (Loire). *L'Anthropologie, 59*, 401–428.

Latałowa, M., Tobolski, K., & Nalepka, D. (2004). *Pinus* L. subgenus *Pinus* (subgen. *Diploxylon* (Koehne) Pilger) – Pine. In M. Ralska-Jasiewiczowa, K. Wasylikowa, K. Tobolski, E. Madeyska, H. E. Wright, & C. Turner (Eds.), *Late Glacial and Holocene history of vegetation in Poland based on isopollen maps* (pp. 165–177). Kraków: W. Szafer Institute of Botany.

Laville, H., Rigaud, J.-P., & Sackett, J. (1980). *Rock shelters of the Perigord: Geological stratigraphy and archaeological succession.* New York: Academic Press.

Lázničková-Galetová, M. (2010). Le travail des matières d'origine dure animale dans le Magdalénien Morave: l'exemple des aiguilles à chas. *L'Anthropologie, 114*(1), 68–96.

Le Brun-Ricalens, F. (2012). Ausgesplitterte Stücke. Kenntnisstand nach einem Jahrhundert Forschung. In H. Floss (Ed.), *Steinartefakte vom Altpaläolithikum bis in die Neuzeit* (pp. 439–456). Kerns: Tübingen.

Le Mignot, Y. (2005). Approche technotypologique de l'outillage du niveau 6 du Bois-Ragot. In A. Chollet & V. Dujardin (Eds.), *La grotte du Bois-Ragot à Gouex (Vienne) Magdalénien et Azilien. Essais sur les hommes et leur environnement* (pp. 43–66). Joué-Lès-Tours: La Simarre.

Le Tensorer, J.-M. (1981). *Le Paléolithique de l'Agenais.* Paris: Editions CNRS.

Le Tensorer, J.-M. (Ed.). (1993). *SPM I, La Suisse du Paléolithique à l'aube du Moyen-Âge, tome 1: Paléolithique et Mésolithique.* Bâle: Editions de la Société Suisse de Préhistoire et d'Archéologie.

Le Tensorer, J.-M. (1998). *Le Paléolithique en Suisse.* Grenoble: Millon.

Le Tensorer, J.-M. (1999). The middle and upper Palaeolithic of Switzerland (isotopic stages 2&3): Chronology, cultures and environment. In P. M. Vermeersch & J. Renault-Miskovsky (Eds.), *European Late Pleistocene, isotope stages 2 and 3: Humans, their ecology and cultural adaptations. Inqua Congress in Durban South Africa, 3–11 August 1999* (ERAUL, Vol. 90, pp. 57–80). Liège: Université de Liège.

Lech, J., & Mateiciucová, I. (1995). Bečov, Most district. *Archaeologia Polona, 33*, 276–278.

Leesch, D. (Ed.). (1997). *Hauterive-Champréveyres, 10. Un Campement magdalénien au bord du lac de Neuchâtel: cadre chronologique et culturel, mobilier et structures, analyse spatial (secteur 1).* Hauterive: Musée cantonal dàrchéologie de Neuchâtel.

Leesch, D., Marie-Isabelle, C., & Müller, W. (Eds.). (2004). *Hauterive-Champréveyres et Neuchâtel-Monruz. Témoins d'implantations magdaléniennes et aziliennes sur la rive nord du lac de Neuchâtel.* Hauterive: Service et musée cantonal d'archéologie de Neuchâtel.

Legendre, L., & Gallagher, E. D. (2001). Ecologically meaningful transformations for ordination of species data. *Oecologia, 129*, 271–280.

Legendre, P., & Legendre, L. (2003). *Numerical ecology.* Amsterdam: Elsevier.

Lehmann, U. (1954). Die Fauna des Vogelherds bei Stetten ob Lonetal (Württemberg). *Neues Jahrbuch für Geologie und Paläontologie, Abhandlungen, 99*(1), 33–146.

Lenardi, P. (1990). *Per una cultura faunistico venatoria.* Udine: Arti Grafiche Friulane.

Lengeyel, G. (2009). Radiocarbondates of the "Gravettian Entity" in Hungary. *Praehistoria, 9–10*, 241–263.

Lengyel, G. (2010). An aspect to the reevaluation of Ságvár (Lyukas-domb) upper Palaeolithic site. *Folia Archaeologica, 54*, 26–37.

Lengyel, G. (2011). The pebble, the block and the tabular. Lithic raw material use at Ságvár Lyukas-domb upper Palaeolithic site. In K. Biró & A. Markó (Eds.), *Emlékkönyv Violának. Tanulmányok T. Dobosi Viola tiszteletére. Papers in Honour of Viola T. Dobosi* (pp. 93–102). Budapest: Magyar Nemzeti Múzeum.

Lenoir, M. (1983). *Le Paléolithique de Basses Vallées de la Dordogne et de la Garonne.* Ph.D. Dissertation, University of Bordeaux I.

Lenoir, M., & Terraza, S. (1971). Le Magdalenien supérieur de l'abri Faustin commune de Cessac (Gironde). *Bulletin de la Société Préhistorique Française, 68*, 311–327.

Leroi-Gourhan, A. (1980). Interstades Würmiens: Laugerie et Lascaux. *Bulletin de l'Association française pour l'etude du quaternaire, 17*(3), 95–100.

Leroi-Gourhan, A. (1983). Une tête de sagaie à armature de lamelles de silex à Pincevent (Seine-et-Marne). *Bulletin de la Société Préhistorique Française, 80*(5), 154–156.

Leroi-Gourhan, A. (1988). *Dictionnaire de la Préhistoire.* Paris: Presses universitaires de France.

Leroi-Gourhan, A., & Evin, J. (1979). Les datations de Lascaux. In A. Leroi-Gourhan (Ed.), *Lascaux inconnu* (pp. 81–84). Paris: Éditions du CNRS.

Leroi-Gourhan, A., & Renault-Miskovsky, J. (1977). La palynologie appliquée à l'archéologie: méthodes, limites et résultats. In H. Laville & J. Renault-Miskovsky (Eds.), *Approche écologique de l'Homme fossile* (pp. 36–48). Paris: Universite Pierre Marie Curie.

Leuthardt, F. (1911). Eine neu entdeckte Station des Steinzeitmenschen in Lausen (Baselland). *Tätigkeitsberichte der Naturforschenden Gesellschaft Baselland, 4*, 95–108.

Leuthardt, F. (1916). Zur Altersfrage der steinzeitlichen Station von Lausen (Kanton Baselland). *Tätigkeitsberichte der Naturforschenden Gesellschaft Baselland, 5*, 169–174.

Leuzinger-Piccand, C. (1996). Einsiedeln SZ-Langrüti: eine spätmagdalénienzeitliche und mesolithische Freilandstation in den Voralpen. *Jahrbuch der Schweizerischen Gesellschaft für Ur- und Frühgeschichte, 79*, 7–26.

Leyer, I., & Wesche, K. (2007). *Multivariate Statistik in der Ökologie. Eine Einführung*. Berlin: Springer.

Libera, J., Zakościelna, A., Schild, R., & Bluszcz, A. (2005). Późnoplejstoceńskie obozowisko zespołów technokompleksu z tylczakami w Pawłowie koło Zawichostu w świetle wstępnej analizy stratygraficznej (badania 2001–2002). *Archeologia Polski Środkowowschodniej, 7*, 9–19.

Liebe, K. T. (1876). Die Lindenthaler Hyänenhöhle und andere diluviale Knochenfunde Ostthüringens. *Archiv der deutschen Gesellschaft für Anthropologie und Ethnologie, 9*, 155–172.

Lindner, H. (1961). *Die altsteinzeitlichen Kulturen der Räuberhöhle am Schelmengraben bei Sinzig*. Kallmünz: Lassleben.

Litt, T., Behre, K.-E., Meyer, K.-D., Stephan, H.-J., & Wansa, S. (2007). Stratigraphische Begriffe für das Quartär des norddeutschen Vereisungsgebietes. *Eiszeitalter und Gegenwart, 56*(1–2), 7–65.

Litt, T., & Stebich, M. (1999). Bio- and chronostratigraphy of the lateglacial in the Eifel region, Germany. *Quaternary International, 61*, 5–16.

Loë, A. D., & Rahir, E. (1905). Fouilles au Trou de Blaireaux à Vaucelles. *Bulletin Société royale belge d'Anthropologie et de Préhistoire, 24*(mém. 1), 1–17.

Loew, S. (2006). *Rüsselsheim 122 und die Federmessergruppen am Unteren Main*. Ph.D. Dissertation, University of Cologne.

Löhr, H. (1978). Vom Altpaläolithikum bis zum Mittelalter: Die Grabungen des Jahres 1977 am Kartstein, Gemeinde Mechernich, Kreis Euskirchen. *Ausgrabungen im Rheinland, 1977*, 40–46.

Löhr, H. (1979). *Der Magdalénien-Fundplatz Alsdorf, Kreis Aachen-Land. Ein Beitrag zur Kenntnis der funktionalen Variabilität jungpaläolithischer Stationen*. Ph.D Dissertation, University of Tübingen.

López Bayón, I. (2000). La recolonisation tardiglaciaire de la Belgique. In B. Valentin, P. Bodu, & M. Christensen (Eds.), *L'Europe centrale et septentrionale au Tardiglaciaire. Confrontation des modèles régionaux de peuplement. Actes de la Table-ronde international de Nemours 14-16 mai 1997* (pp. 139–149). Nemours: A.P.R.A.I.F.

López Bayón, I., Jean-Marc, L., Otte, M., Quinif, Y., Ancion, V., Lacroix, P., et al. (1997). Nouvelles recherches dans le site magdalénien du Trou da Somme (Hastière). *Notae Praehistoricae, 17*, 63–75.

Lorblanchet, M. (1997). *Höhlenmalerei: Ein Handbuch*. Sigmaringen: Thorbecke.

Lotter, A. (1985). Amsoldingersee – Late-Glacial and Holocene environments of a lake at the southern edge of the Swiss Plateau. In G. Lang (Ed.), *Swiss lake and mire environments during the last 15000 years* (pp. 185–208). Vaduz: Lubrecht and Cramer.

Lowe, J. J., Hoek, W. Z., & INTIMATE Group. (2001). Inter-regional correlation of palaeoclimatic records for the last glacial-interglacial transition: A protocol for improved precision recommended by the INTIMATE project group. *Quaternary Science Reviews, 20*, 1175–1187.

Lowe, J. J., Rasmussen, S. O., Björck, S., Hoek, W. Z., Steffensen, J. P., Walker, M. J. C., et al. (2008). Synchronisation of palaeoenvironmental events in the North Atlantic region during the last termination: A revised protocol recommended by the INTIMATE group. *Quaternary Science Reviews, 27*, 6–17.

Lüdin, C. (1963). Die Silexartefakte aus dem Spätmagdalénien der Kohlerhöhle. *Jahrbuch der Schweizerischen Gesellschaft für Ur- und Frühgeschichte, 1963*(50), 33–42.

Luhmann, N. (2009). *Einführung in die Systemtheorie*. Heidelberg: Auer.

Madeyska, T., & Kozłowski, S. K. (1995). Human settlement and paleoenvironmental changes in Poland 20,000–8,000 years ago. *Biuletyn Peryglacjalny, 34*, 103–123.

Madsen, B. (1983). New evidence of late Paleolithic settlement in East Jutland. *Journal of Danish Archaeology, 2*, 12–31.

Maier, H. (1936). Die altsteinzeitliche Wohnhöhle "Kleine Scheuer" im Rosenstein (Schwäb. Alb). *Mannus, 28*, 235–252.

Malina, J. (1974). Petrographische Charakteristik der Steinindustrie. In R. Feustel (Ed.), *Die Kniegrotte* (pp. 97–98). Weimar: Böhlau.

Mandera, H.-E. (1954). Bericht über die Nachuntersuchung der Höhle "Wildscheuer" und ihres Vorplatzes 1953. *Nassauische Annalen, 65*, 35–45.

Mania, D. (1999). *Nebra – eine jungpaläolithische Freilandstation im Saale-Unstrut-Gebiet*. Halle: Landesamt für Archäologie.

Mania, D. (2008). *Orlamünde und Umgebung. Ein Abriss der Besiedlungsgeschichte von den Anfängen bis zum Mittelalater*. Langenweissbach: Beier and Beran.

Mania, D., & Mania, W. (1994). *Orlamünde vor und nach 1194*. Jena: Eigenverlag.

Mania, D., & Toepfer, V. (1973). *Königsaue. Gliederung, Ökologie und mittelpaläolithische Funde der letzten Eiszeit. Zum IX. Kongreß der Internationalen Quatärvereinigung (INQUA), Christchurch, Neu-Seeland 1973*. Berlin: VEB Deutscher Verlag der Wissenschaften.

Maška, K. J. (1884). Pravěké nálezy ve Štramberku. *Casopis Vlasteneckého musejního spolku v Olomouci, I*, 15–24, 64–69, 152–159.

Maška, K. J. (1886). Pravěké nálezy ve Štramberku. *Casopis Vlasteneckého musejního spolku v Olomouci, III*, 57–65, 119–123, 163–174.

Maška, K. J. (1888). Nové výzkumy v jeskyních štramberských. *Casopis Vlasteneckého musejního spolku v Olomouci, V*, 121–124.

Maturana, H. R. (1982). *Erkennen: Die Organisation und Verkörperung von Wirklichkeit: Ausgewählte Arbeiten zur biologischen Epistemologie*. Braunschweig: Vieweg.

Matušek, V. (1987). Předběžná zpráva o výzkumu jeskyně č. 1119 u Koněprus. *Český kras, 13*, 47–50.

Mauser, P. F. (1963). Die paläolithische Besiedlung der Brillenhöhle bei Blaubeuren. *Jahreshefte für Karst- und Höhlenkunde, 4*, 317–334.

Mauser, P. F. (1970). Die jungpaläolithische Höhlenstation Petersfels im Hegau. *Badische Fundberichte, Sonderheft, 13*. 127 pp.

Mauser, P. F. (1974). Eine späteiszeitliche Freilandstation auf dem Steinberg bei Bolschweil, Lkr. Breisgau-Hochschwarzwald. *Fundberichte aus Baden-Württemberg, 1*, 37–44.

Mauser, P. F. (1976). Das eiszeitliche Jagdlager im Katzenbachtal. *Der Sülchgau, 20*, 35–44.

Mayet, L., & Pissot, J. (1915). *Abri-sous-roche préhistorique de La Colombière près Poncin (Ain)*. Lyon: Rey.

Mayewski, P., Meeker, L. D., Twickler, M. S., Whitlow, S., Yang, Q., & Prentice, M. (1997). Major features and forcing of high latitude northern hemisphere circulation using a 110000-year-long glaciochemical series. *Journal of Geophysical Research, 102*, 26345–26366.

Mayewski, P. A., Rohling, E. E., Stager, J. C., Karlén, W., Maasch, K. A., Meeker, L. D., et al. (2004). Holocene climate variability. *Quaternary Research, 62*, 243–255.

Mazálek, M. (1952). Druhý rok výzkumů na sídlišti z doby kamenné u Ražic. *Archeologické rozhledy, 4*, 198–206.

Mazálek, M. (1953). Třetí rok výzkumů paleo-mesolitické oblasti u Ražic. *Archeologické rozhledy, 5*, 577–589.

Mcardle, B. H., & Anderson, M. J. (2001). Fitting multivariate models to community data: A comment on distance-based redundancy analysis. *Ecology, 82*, 290–297.

McCabe, A. M., & Clark, P. U. (1998). Ice-sheet variability around the North Atlantic Ocean during the last deglaciation. *Nature, 392*, 373–377.

Meier-Riva, K. (1987). *Die Steinartefakte vom Hardberg bei Istein (Efringen-Kirchen, Kr. Lörrach)*. Stuttgart: Theiss.

Mellars, P. (1992). Archaeology and the population-dispersal hypothesis of modern human origins. *Philosophical Transactions of the Royal Society of London, Series B, 337*, 225–234.

Menke, B. (1968). Das Spätglazial von Glüsing. Ein Beitrag zur Kenntnis der spätglazialen Vegetationsgeschichte in Westholstein. *Eiszeitalter und Gegenwart, 19*, 73–84.

Merkt, J., & Müller, H. (1999). Varve chronology and palynology of the Lateglacial in Northwest Germany from lacustrine sediments of Hämelsee in Lower Saxony. *Quaternary International, 61*, 41–59.

Mevel, L. (2013). Magdalenian pioneers in the northern French Alps, 17000 cal BP. *Antiquity, 87*(336), 384–404.

Mey, W. (1960). Junpaläolithikum und Mesolithikum in Brandenburg. *Quartär, 12*, 1–52.

Meyer, H.-H., & Kottmeier, C. (1989). Die atmosphärische Zirkulation in Europa im Hochglazial der Weichsel-Eiszeit – abgeleitet von Paläowind-Indikatoren und Modellsimulationen. *Eiszeitalter und Gegenwart, 39*, 10–18.

Mieg, M. (1901). Mote sure une station de l'époque paléolithique des Grottes d'Istein (Grand-duché de Bade). *Bulletin des Séances de la Société des Sciences de Nancy, 3*(2), 3–6.

Mieg, M. (1910). Mote sur l'age d'industrie paléolithique des Grottes d'Istein (Gand-duché de Bade). *Bulletin des Séances de la Société des Sciences de Nancy, 3*(11), 3–10.

Mieg, M., & Stehlin, H. G. (1903). Sur l'âge et la faune de la station préhistorique d'Istein (Grand-duché de Bade). *Bulletin des Séances de la Société des Sciences de Nancy, 3*(4), 1–19.

Milgram, S. (1967). The small world problem. *Psychology Today, 1*(1), 61–67.

Miller, R., & Noiret, P. (2009). Recent results for the Belgian Magdalenian. In M. Street, N. Barton, & T. Terberger (Eds.), *Humans, environment and chronology of the late glacial of the North European Plain: Proceedings of Workshop 14 of the 15th U.I.S.P.P Congress, Lisbon, September 2006* (pp. 39–53). Mainz: Verlag des Römisch-Germanischen Zentralmuseums.

Mix, A. C., Bard, E., & Schneider, R. (2001). Environmental processes of the ice age: Land, oceans, glaciers (EPILOG). *Quaternary Science Reviews, 20*, 627–657.

Monin, G. (2000). Apport de la technologie lithique à l'étude des séries anciennes. Les assemblages tardiglaciaires des chasseurs de marmottes des grottes Colomb et de la Passagère à Méaude (Vercors, Isère). In G. Pion (Ed.), *Le Paléolithique supérieur récent: nouvelles données sur le peuplement et l'environnement. Table ronde de Chambéry 12-13 mars 1999* (pp. 271–278). Joué-lès-Tours: La Simarre.

Monjuvent, G. (1994). Les grottes Jean Pierre 1 et 2 à Saint-Thibaud-de-Couz (Savoie). II. L'environnement géomorphologique. *Gallia Préhistoire, 36*, 150–163.

Montet-White, A. (1990a). The artifact assemblage. In A. Montet-White (Ed.), *The Epigravettian site of Grubgraben, Lower Austria: The 1986 and 1987 excavations* (ERAUL, Vol. 40, pp. 133–157). Liège: Université de Liège.

Montet-White, A. (Ed.). (1990b). *The Epigravettian site of Grubgraben, Lower Austria: The 1986 and 1987 excavations* (ERAUL, Vol. 40). Liège: Université de Liège.

Montet-White, A. (1994). Alternative interpretations of the late upper Paleolithic in central Europe. *Annual Review of Anthropology, 23*, 483–508.

Morrow, J. E., & Morrow, T. A. (1999). Geographic variation in fluted projectile points: A hemispheric perspective. *American Antiquity, 64*, 215–231.

Mortillet, G. D. (1869). Essai d'une classification des cavernes et des stations sous abri fondée sur les produits de l'industrie humaine. *Bulletin de la Société Géologique de France, 2*, 583–587.

Moutel, P. (1989). Le niveau M3 de l'abri Sud de la baume Noire à Frétigney (Haute-Saône). A. L'industrie lithique. In J.-P. Rigaud (Ed.), *Le Magdalénien en Europe. "La structuration du Magdalénien". Actes du Colloque de Mayence 1987 dans le cadre du XIème Congrès U.I.S.P.P* (ERAUL, Vol. 38, pp. 159–165). Liège: Université de Liège.

Mouton, P., & Joffroy, R. (1956). Précisions nouvelles sur les stations de Farincourt (Haute-Marne). *Revue archéologique de l'Est et du Centre-Est, 7*, 193–223.

Movius, H. L., & Judson, S. (1956). *The rock-shelter of La Colombière*. Cambridge: Peabody Museum.

Müller, H. (1894). Fouilles pratiquées dans une grotte située aux Balmes de Buisse, près Voreppe (Isère). *Bulletin Société dauphinoise d'Ethnologie et d'Archéologie, 1*, 187–191.

Müller, H. (1912). Un harpon magdalénien dans le Vercors (station de Bobache). *Bulletin de la Société Préhistorique Française, 9*, 658.

Müller, H. (1953). Zur spät- und nacheiszeitlichen Vegetationsgeschichte des mitteldeutschen Trockengebietes. *Nova Acta Leopoldina, N.F., 16*, 1–67.

Müller, D. W. (1980). Die ur- und frühgeschichtliche Besiedlung des Gothaer Landes. *Alt-Thüringen, 17*, 19–180.

Müller, D. W., & Rommeis, W. (1970). Ur- und frühgeschichtliche Bodenaltertümer in einer Privatsammlung aus Wandersleben. *Ausgrabungen und Funde, 15*(5), 279–283.

Müller, J., & Zimmermann, A. (Eds.). (1997). *Archäologie und Korrespondenzanalyse. Beispiele, Fragen, Perspektiven*. Espelkamp: Leidorf.

Müller, W. (2004). Les vestiges osseux. In D. Leesch, M.-I. Cattin, & W. Müller (Eds.), *Hauterive-Champréveyres et Neuchâtel-Monruz. Témoins d'implantations magdaléniennes et aziliennes sur la rive nord du lac de Neuchâtel* (pp. 198–202). Neuchâtel: Service et musée cantonal d'archéologie de Neuchatel.

Müller, W. (2006). Les témoins animaux. In J. Bullinger, D. Leesch, & N. Plumettaz (Eds.), *Le site magdalénien de Monruz, 1. Premiers éléments pour l'analyse d'un habitat de plein air* (pp. 123–137). Neuchâtel: Service et musée cantonal d'archéologie.

Müller, W., & Leesch, D. (2011). Einige Neubestimmungen aus der Magdalénien-Fundstelle Hollenberg-Höhle 3 bei Arlesheim (Basel-Landschaft) und daraus folgende Überlegungen zur Nutzung der Höhle. *Jahrbuch der Archäologie Schweiz, 94*, 7–20.

Murray, T., & Walker, M. J. (1988). Like WHAT? A practical question of analogical inference and archaeological meaningfulness. *Journal of Anthropological Archaeology, 7*, 248–287.

Musil, R. (1957). Magdalénska fauna Žitného jeskyně. In J. Dvořák (Ed.), *Komplexni výzkum Žitného jeskyně v Moravskem krasu* (pp. 558–572). Praha: CSVA.

Musil, R. (1958). Fauna moravských magdalénských stanic. *Anthropozoikum, 7*, 7–26.

Musil, R. (1985). *Die Fauna der Magdalénien-Siedlung Oelknitz. Weimarer Monographien zu Ur- und Frühgeschichte 17*. Weimar: Museum für Ur- und Frühgeschichte Thüringens.

Musil, R. (2002). Osteologický materiál z jeskyně Kůlničky (výzkum 1959). In J. Svoboda (Ed.), *Prehistorické jeskyně* (p. 179). Brno: Archeologický ústav AV ČR.

Musil, R. (2003). The middle and upper Palaeolithic game suite in central and southeastern Europe. In T. H. van Andel & W. Davies (Eds.), *Neanderthals and modern humans in the European landscape during the last glaciation* (pp. 167–190). Cambridge: McDonald Institute for Archaeological Research.

Naber, F. B. (1977). Schellnecker Wänd—Abri II, Gemeinde Essing, Lkr. Kelheim, Niederbayern. *Archäologisches Korrespondenzblatt, 7*, 185–194.

Nagel, E. (1967). The nature and aim of science. In S. Morgenbesser (Ed.), *Philosophy of science today* (pp. 3–13). New York: Basic Books.

Nagy, P., & Spörri, D. (1998). Archäologische Forschungen im Rafzer Feld. *Berichte der Kantonsarchäologie Zürich, 14*, 285–297.

Narr, K. J. (1963). *Kultur, Umwelt und Leiblichkeit des Eiszeitmenschen. Studien zu ihrem geistigen Verhältnis*. Stuttgart: Fischer.

Narr, K. J. (1965). Die Altsteinzeitfunde aus dem Hohlenstein bei Nördlingen. *Bayerische Vorgeschichtsblätter, 30*, 1–9.

Narr, K. J. (1968). *Studien zur älteren und mittleren Steinzeit der niederen Lande*. Bonn: Habelt.

Neiman, F. (1995). Stylistic variation in evolutionary perspective: Inferences from decorative diversity and interassembalge distance in Illinois woodland ceramic assemblages. *American Antiquity, 60*, 7–36.

Neruda, P., & Kostruhn, P. (2002). Hranice-Velká-Kobylanka. Mladopaleolitická stanice v Moravské branĕ. *Acta Musei Moraviae, Scientiae Sociale, 87*, 105–156.

Neruda, P., & Nerudová, Z. (2008). Loštice I – výzkum nové magdalénienské stanice na střední Moravĕ. *Archeologické rozhledy, 60*, 509–528.

Neruda, P., Nerudová, Z., & Čulíková, V. (2009). Loštice I – Kozí Vrch. A Magdalenian site in northern Moravia. *Acta Musei Moraviae, Scientiae Sociale, XCIV*, 39–64.

Neugebauer-Maresch, C. (1999). *Le Paléolithique en Autriche*. Grenoble: Millon.

Neumann, G. (1933a). Leben und Treiben in Thüringen vor 20.000 Jahren. *Das Thüringer Fähnlein, 2*, 321–331.

Neumann, G. (1933b). Eine Freilandstation das Hochmagdalénien. *Beiträge zur Geologie von Thüringen, 3*, 362–364.

Neumann, G. (1958). Magdalénien in Ostthüringen. *Ausgrabungen und Funde, 3*, 159–161.

Neustupný, J. (1938). Nové paleolitické nálezy z Čech v Národním museu. *Časopis Národního musea, 112*, 164.

Nielsen, E. H. (1992). Paläolithische und mesolithische Fundstellen im zentralschweizerischen Wauwilermoos. *Archäologisches Korrespondenzblatt, 22*, 27–40.

Nielsen, E. H. (1994). Rentierjäger in Reiden – Eine neue Fundstelle der Altsteinzeit im Kanton Luzern. *Archäologie der Schweiz, 17*(4), 126–133.

Nielsen, E. H. (1996). Steinzeitliche Jäger im Kanton Zug. *Archäologie der Schweiz, 19*(2), 36–42.

Nielsen, E. H. (1999). Das Magdalénien im Schweizer Mittelland. *Archäologisches Korrespondenzblatt, 29*, 11–20.

Nielsen, E. H. (2002). The lateglacial settlement of the central Swiss Plateau. In B. V. Eriksen & B. Bratlund (Eds.), *Recent studies in the final Palaeolithic of the European plain. Proceedings of a U.I.S.P.P. Symposium, Stockholm, 14.-17. October 1999* (pp. 189–202). Århus: Aarhus University Press.

Nielsen, E. H. (2009). *Paläolithikum und Mesolithikum in der Zentralschweiz. Mesch und Umwelt zwischen 17'000 und 5500 v. Chr.* Luzern: Kantoner Lehrmittelverlag.

Nieves Zedeño, M., & Stoffle, R. W. (2003). Tracking the role of pathways in the evolution of a human landscape. In M. Rockman & J. Steele (Eds.), *Colonisation of unfamiliar landscapes. The archaeology of adaptation* (pp. 59–80). London: Routledge.

Nitzschke, W., & Stahlhofen, H. (1976). Ausgewählte Neufunde aus den Jahren 1972/73. *Jahresschrift für mitteldeutsche Vorgeschichte, 59*, 343–361.

Nitzschke, W., & Stahlhofen, H. (1978). Ausgewählte Neufunde aus den Jahren 1975/76. *Jahresschrift für mitteldeutsche Vorgeschichte, 62*, 221–233.

Noble, G. K. (1939). The role of dominance in the social life of birds. *The Auk, 56*(3), 263–273.

Nüesch, J. (1896). *Die praehistorische Niederlassung am Schweizersbild bei Schaffhausen. Die Schichten und ihre Einschlüsse*. Zürich: Zürchner and Furrer.

Nüesch, J. (1902). *Das Schweizersbild, eine Niederlassung aus Paläolithischer und neolithischer Zeit*. Zürich: Zürchner and Furrer.

Nývltová-Fišáková, M. (2002a). Pleistocene and Holocene faunal remains from excavations of Josef Skutil in the Moravian Karst. In J. Svoboda (Ed.), *Prehistorické jeskynĕ* (pp. 293–304). Brno: Archeologický ústav AV ČR.

Nývltová-Fišáková, M. (2002b). Osteologický materiál z Liščí jeskynĕ. In J. Svoboda (Ed.), *Prehistorické jeskynĕ* (p. 179). Brno: Archeologický ústav AV ČR.

O'Sullivan, D., & Unwin, D. J. (2003). *Geographic information analysis*. New Jersey: Wiley.

Oberlin, C., & Pion, G. (2009). Le corpus des datations radiocarbone et la disparition du Renne. In G. Pion (Ed.), *La fin du Paléolithique supérieur dans les Alpes du nord française et le Jura méridional. Approches culturelles et environnementales* (pp. 51–57). Joué-les-Tours: Société Préhistorique Française.

Obermaier, H. (1914). III: Fouilles en Bavière. *L'Anthropologie, 25*, 254–262.

Obermaier, H., & Frauenholz, J. (1926). Eine Mammutdarstellung aus Süddeutschland. *Jahrbuch für Prähistorische und Ethnographische Kunst, 3*, 29–32.

Obermaier, H., & Wernert, P. (1914). Paläolithbeiträge aus Nordbayern. *Mitteilungen der Anthropologischen Gesellschaft in Wien, 14*, 44–62.

Obermaier, H., & Frauenholz, J. (1927). Der skulptierte Rengeweihstab aus der Mittleren Klausenhöhle bei Essing (Niederbayern). *Jahrbuch für Prähistorische und Ethnographische Kunst, 4*, 1–10.

Oksanen, J., Blanchet, F. G., Kindt, R., Legendre, P., Minchin, P. R., O'Hara, R. B., et al. (2012). *Vegan: Community ecology package. R package version 2.0-3*. http://CRAN.R-project.org/package=vegan

Oliva, M. (1995). Das Paläolithikum der Býči-skála-Höhle. *Pravek, 5*, 25–38.

Oliva, M. (2003). Géographie du Magdalénien Morave sous l'aspect économique et social. In S. A. Vasil'ev, O. Soffer, & J. K. Kozlowski (Eds.), *Perceived landscapes and built environments. The cultural geography of late Paleolithic Eurasia. Actes du XIVéme Congrès UISPP, Université de Liège, 2–8 Sept. 2001* (BAR International Series, Vol. 1122, pp. 131–137). Oxford: Archaeopress.

Ondroušková, S. (2008). *Využívání jeskyní Moravského krasu v období popelnicových polí a v dobĕ železné*. Bachelor thesis, University of Prague.

Onoratini, G. (1996). Le Paléolithique Supérieur dans le bassin du Rhône, dans les Alpes et en Provence (1991–1996). In M. Otte (Ed.), *Le Paléolithique Supérieur européen, UISPP, Congrès de Forli 1996, Commission Paléolithique Supérieur, Bilan 1991–1996* (ERAUL, Vol. 76, pp. 257–268). Liège: Université de Liège.

Onoratini, G., Defleur, A., & Joris, C. (1996). Mise en évidence du Magdalénien ancien II dans les Gorges de l'Ardeche. *Bulletin de la Société Préhistorique Française, 93*, 25–32.

Onoratini, G., & Joris, C. (1995). Une nouvelle séquence du Paléolithique supérieur dans la grotte de Huguenots. *Ardèche Archéologie, 12*, 23–27.

ORAU Oxford Radiocarbon Accelerator Unit, online database. http://c14.arch.ox.ac.uk/embed.php?File=

Otte, M. (1981). *Le Gravettien en Europe Centrale*. Bruges: De Tempel.

Otte, M. (1992). Processus de diffusion à long terme au Magdalénien. In J.-P. Rigaud (Ed.), *Le peuplement magdalénien. Actes du colloque de Chancelade 1988* (pp. 399–416). Paris: Eds. du CTHS.

Otte, M. (Ed.). (1994). *Le Magdalénien du Trou de Chaleux (Hulsonniaux-Belgique)* (ERAUL, Vol. 69). Liège: Université de Liège.

Otte, M. (2009). *La Préhistoire*. Bruxelles: De Boeck.

Otte, M. (2012). Appearance, expansion and dilution of the Magdalenian civilization. *Quaternary International, 272–273*, 354–361.

Otte, M., & Keeley, L. H. (1990). The impact of regionalism on Palaeolithic studies. *Current Anthropology, 31*(5), 577–582.

Otte, M., & Straus, L. G. (Eds.). (1995). *Le Trou Magrite: Fouilles 1991–1992* (ERAUL, Vol. 69). Liège: Université de Liège.

Otte, M., & Miller, R. (1999). Chronologie paléolithique de Benelux: Phase récente (40.-10.000 BP). In P. M. Vermeersch & J. Renault-Miskovsky (Eds.), *European late Pleistocene, isotope stages 2 and 3: Humans, their ecology and cultural adaptations. INQUA congress in Durban South Africa, 3–11 August 1999* (ERAUL, Vol. 90, pp. 81–96). Liège: Université de Liège.

Otte, M., & Straus, L. G. (Eds.). (1997). *La Grotte du Bois Laiterie* (ERAUL, Vol. 80). Liège: Université de Liège.

Owen, L. R. (1988). *Blade and microblade technology: Selected assemblages from the North American Arctic and the upper Paleolithic of southwest Germany* (BAR International Series, Vol. 441). Oxford: Archaeopress.

Padtberg, A. (1925). *Das altsteinzeitliche Lösslager bei Munzingen*. Augsburg: Filser.

Pailhaugue, N. (1998). Faune et saisons d'occupation de la salle Monique au Magdalénien Pyrénéen, Grotte de la Vache (Alliat, Arriège, France). *Quarternaire, 9*(4), 385–400.

Pajot, B. (1969). *Les Civilisations du Paléolithique supérieur du bassin de l'Aveyron*. Ph.D. Dissertation, University of Toulouse.

Papp, A. (1952). Die Schmuckschnecken aus Kamegg, N.-Ö. *Archaeologia Austriaca, 10*, 28–33.

Paret, O. (1951). Hornstein. *Fundbreichte aus Schwaben, NF XI*, 17.

Paret, O. (1961). *Württemberg in vor- und frühgeschichtlicher Zeit*. Stuttgart: Kohlhammer Verlag.

Pasda, C. (1990). Der Probstfels bei Beuron – Probleme bei der Auswertung eines alt gegrabenen Inventars. *Archäologisches Korrespondenzblatt, 20*, 1–9.

Pasda, C. (1994). *Das Magdalénien in der Freiburger Bucht*. Stuttgart: Theiss.

Pasda, C. (1998). Der Beginn des Magdaléniens in Mitteleuropa. *Archäologisches Korrespondenzblatt, 28*, 175–190.

Pasda, C. (2012a). Stichel. In H. Floss (Ed.), *Steinartefakte vom Altpaläolithikum bis in die Neuzeit* (pp. 421–428). Tübingen: Kerns.

Pasda, C. (2012b). Spitzklingen. In H. Floss (Ed.), *Steinartefakte vom Altpaläolithikum bis in die Neuzeit* (pp. 457–458). Tübingen: Kerns.

Pasda, C. (2012c). Lateralretuschen. In H. Floss (Ed.), *Steinartefakte vom Altpaläolithikum bis in die Neuzeit* (pp. 467–476). Tübingen: Kerns.

Pasda, C. (2012d). Endretuschen. In H. Floss (Ed.), *Steinartefakte vom Altpaläolithikum bis in die Neuzeit* (pp. 435–437). Tübingen: Kerns.

Paus, A. (1992). *Late Weichselian vegetation, climate and floral migration in Rogaland, southwestern Norway; pollen analytical evidence from four Late-Glacial basins*. Ph.D. Dissertation, University of Bergen.

Peirce, C. S. (1931). *Collected papers of Charles Sanders Peirce*. Cambridge, MA: Harvard University Press.

Pelegrin, J. (2000). Les techniques de débitage laminaire au Tardiglaciaire: critères de diagnose et quelques réflexions. In B. Valentin, P. Bodu, & M. Christensen (Eds.), *L'Europe Centrale et Septentrionale au Tardiglaciaire. Confrontation des modèles régionaux de peuplement. Actes de la Table-ronde internationale de Nemours 14, 15, 16 mai 1997. Mémoires du Musée de Préhistoire d'Ile-de-France n°7* (pp. 73–86). Nemours: Ed. A.P.R.A.I.F.

Pericot, L. G. (1942). *La Cueva del Parpallo (Gandia)*. Madrid: Consejo Superior de Investigaciones Cientificas, Inst. Diego Velázquez.

Pernot, P. (1996). Les occupations du Magdalénien moyen à scalènes des grottes I et II de Farincourt (Haute-Marne). Approche typo-technologique. In Y. Pautrat (Ed.), *Paléolithique supérieur et Epipaléolithique dans le Nord-Est de la France. Actes de la Table ronde de Dijon, 7. et 8.Octobre 1995* (pp. 27–35). Dijon: Service Régional de l'Archéologie de Bourgogne.

Perpère, M. (1981). Le Magdalénien final de la grotte Auzary (Puy-de-Dôme). *Bulletin de la Société Préhistorique Française, 78*(10–12), 342–353.

Peschke, P. (1972). Pollenanalytische Untersuchungen im Waldviertel Niederösterreichs. *Flora, 161*, 256–284.

Peters, E. (1930). *Die altsteinzeitliche Kulturstätte Petersfels*. Augsburg: Filser.

Peters, E. (1935a). Die mesolithische Silex- und Knochenindustrie vom Rappenfels auf der Schwäbischen Alb. *Germania, 19*, 281–286.

Peters, E. (1935b). Upfingen. *Fundbreichte aus Schwaben, NF, VIII*, 16.

Peters, E. (1936a). Die altsteinzeitlichen Kulturen von Veringenstadt (Hohenzollern). *Prähistorische Zeitschrift, 27*, 173–195.

Peters, E. (1936b). Die Buttentalhöhle an der Donau, eine neue Magdalénienstation. *Badische Fundberichte, 3*, 13–19.

Peters, E. (1946). *Meine Tätigkeit im Dienste der Vorgeschichte Südwestdeutschlands*. Veringenstadt: Privatdruck.

Peters, E., & Rieth, A. (1936). Die Höhlen von Veringenstadt und ihre Bedeutung für die Vor- und Frühgeschichte Hohenzollerns. *Jahresschrift des Vereins für Geschichte, Kultur und Landeskunde Hohenzollern, 3*, 1–26.

Peters, E., & Toepfer, V. (1932). Der Abschluß der Grabungen am Petersfels bei Engen im badischen Hegau. *Prähistorische Zeitschrift, 23*, 155–199.

Peterson, N. (1975). Hunter-gatherer territoriality: The perspective from Australia. *American Anthropologist, 77*(1), 53–68.

Pétillon, J.-M. (2008). What are these barbes for? Preliminary study on the function of the Upper Magdalenian barbed weapon tips. In J.-M. Pétillon, M.-H. Dias-Meirinho, P. Cattelain, M. Honegger, C. Normand, & N. Valdeyron (Eds.), *Projectile Weapon elements from the Upper Palaeolithic to the Neolithic. Proceedings of session C83, XVth World Congress UISPP, Lisbon, September 4–9, 2006* (pp. 66–97).

Petrbok, J. (1956). Český Kras ve výzkumu do roku 1950. *Anthropozoikum, 5*(1955), 9–46.

Peyrony, D. (1936). L'abri de Villepin, commune de Tursac (Dordogne), Magdalénien supérieur et Azilien. *Bulletin Société Préhistorique Française, 33*, 253–272.

Peyrony, D. (1941). Gisment préhistoique de Crabillat. Ses rapports avec les dépôts à formes géometriques du Paléolithique supérieur et du Mésolithique. *Bulletin de la Société Préhistorique Française, XXXVIII*, 245–262.

Peyrony, D. (1947). La grotte d'Abzac à Gorge-d'Enfer. *Bulletin de la Société Historique et Archéologique du Périgord*, 167–171.

Peyrony, D., & Peyrony, E. (1934). Station Préhistorique de Jean Blancs. *Bulletin de la Société Historique et Archéologique du Périgord, 61*, 179–202.

Peyrony, D., & Peyrony, É. (1938). *Laugerie-Haute près Des Eyzies (Dordogne)*. Paris: Masson.

Philibert, M. (1982). *Le Peuplement Préhistorique du bassin supérieur de la Loire*. Saint-Etienne: Centre d'études foréziennes.

Pigeot, M. (1987). *Magdaléniens d'Étiolles. Économie du débitage et organisation sociale (l'unité d'habitation U5)*. Paris: Editions CNRS.

Pignat, G., & Winiger, A. (1998). *Les occupations mésolithique de l'abri du Mollendruz. Abri Freymond, commune de Mont-la-Ville (Vd, Suisse)*. Lausanne: Cahiers d'Archéologie Romande.

Pion, G. (2000a). Le Magdalénien de deux Savoies et du Jura méridional: synthèse préliminaire des données accessibles; proposition d'une chronologie pour les principaux gisments. In G. Pion (Ed.), *Le Paléolithique Supérieur récent: nouvelles données sur le peuplement et l'environnement. Actes de la Table Ronde de Chambéry 12–13 mars 1999* (pp. 147–164). Paris: Société Préhistorique Française.

Pion, G. (2000b). Les pointes à dos courbe dans les industries de la fin du Dryas ancien et du Bølling dans quelques sites des Alpes du Nord française et du Jura méridional. Analyse préliminaire. In C. Cupillard & A. Richard (Eds.), *Les derniers chasseurs-cuilleurs d'Europe occidentale, Actes du colloque internatinal de Besançon, octobre 1998* (pp. 103–111). Besançon: Presses Universitaires Franc-Comtoises.

Pion, G., Billard, M., Caillat, B., Cataliotti-Valdina, J., Durand, J.-M., Girard, M., et al. (1990). L'abri de la Fru á Saint-Christophe (Savoie). *Gallia Préhistoire, 32*, 65–123.

Pion, G., & Thévenin, A. (2007). Le Mésolithique de l'abri de la Fru à Saint-Christophe-la-Grotte (Savoie). *Bulletin de la Société Préhistorique Française, 104*, 483–515.

Połtowicz, M. (2006). The Magdalenian period in Poland and neighbouring areas. *Archaeologia Baltica, 7*(2), 21–28.

Poplin, F. (1976). *Les Grands Vertébrés de Gönnersdorf. Fouilles 1968*. Wiesbaden: Steiner.

Popping, H. J. (1931). Een magdalénien-station op de Veluwe. *De levende Natuur, 35*, 340–349.

Popping, H. J. (1934). De jong-palaeolithische Kuinderkulturen. Station Makkinga. *Mensch en Maatschappij, 10*, 378–400.

Pozzi, E. (2004). *Les magdaléniens. Art, civilisations, modes de vie, environnements*. Grenoble: Millon.

Prentice, I. C., Guiot, J., & Harrison, S. P. (1992). Mediterranean vegetation, lake levels and paleoclimate at the last glacial maximum. *Nature, 360*, 658–660.

Přichystal, A. (2002a). Research of the raw materials of chipped stone artifacts from the Magdalenian in the Ochozská Cave. In J. Svoboda (Ed.), *Prehistorické Jeskyně* (pp. 226–228). Brno: Archeologický ústav AV ČR.

Přichystal, A. (2002b). Suroviny paleolitických artefaktů z jeskyni Kůnička a Liščí díra. In J. Svoboda (Ed.), *Prehistorické jeskyně* (pp. 179–180). Brno: Archeologický ústav AV ČR.

Přichystal, A. (2006). Kamenné suroviny předneolitického osídlení v jižních Čechách. In S. Vencel (Ed.), *Nejstarší osídlení jižních Čech. Paleolit a Mesolit* (pp. 345–365). Praha: Archeologický ústav Akademie věd ČR.

Prošek, F. (1947). Paleolitické osídlení Čech ve světle novějších výzkumů. *Časopis Národního musea, 116*, 1–13.

Prošek, F. (1952). Výzkum Děravé jeskyně na Katýzu u Koněprus. *Archeologické rozhledy, 4*, 97–100.

Prošek, F. (1959). Novyj tip paleolitičeskogo oaga iz stojanki Šanovkout II u s. Gostim. *Sovetskaja archeologija, 1959*(1), 243–245.

Prošek, F. (1961). Mladopaleolitická obydlí v Československu. *Památky archeologické, 52*, 57–75.

R_Development_Core_Team. (2011). *R: A language and environment for statistical computing. R foundation for statistical computing*, Vienna.

Rademacher, C. (1911). Der Kartstein bei Eiserfey in der Eifel. *Prähistorische Zeitschrift, 3*(4), 201–232.

Rähle, W. (1981). Schmuckschnecken aus jungpaläolithischen Fundschichten vom Hohlen Fels bei Schelklingen (Alb-Donau-Kreis). *Archäologisches Korrespondenzblatt, 11*, 179–180.

Rähle, W. (1983a). Die Mollsuken der Grabung Helga Abri bei Schelklingen mit einer Anmerkung zum Fund einiger mesolithischer Schmuckschnecken. *Archäologisches Korrespondenzblatt, 13*, 29–36.

Rähle, W. (1983b). Schmuck aus Molluskenschalen von der Magdalénien-Station Petersfels. In G. Albrecht, H. Berke, & F. Poplin (Eds.), *Naturwissenschaftliche Untersuchungen an Magdalénien-Inventaren vom Petersfels, Grabungen 1974–1976* (pp. 154–158). Tübingen: Archaeologica Venatoria.

Ralska-Jasiewiczowa et al. (Eds.). (2004a). *Late galcial and Holocene history of vegetation in Poland based on isopollen maps*. Kraków: W. Szafer Institute of Botany.

Ralska-Jasiewiczowa, M., Wacnik, A., Mamakova, K., & Nalepka, D. (2004b). Betula L. – Birch. In M. Ralska-Jasiewiczowa, K. Wasylikowa, K. Tobolski, E. Madeyska, H. E. Wright, & C. Turner (Eds.), *Late glacial and Holocene history of vegetation in Poland based on isopollen maps* (pp. 57–68). Kraków: W. Szafer Institute of Botany.

Rasmussen, S. O., Andersen, K. K., Svensson, A. M., Steffensen, J. P., Vinther, B. M., Clausen, H. B., et al. (2006). A new Greenland ice core chronology for the last glacial terminantion. *Journal of Geophysical Research, 111*, 1–15.

Rasquin, L. D. (1911). La caverne funéraire d'Engis. *Bulletin des Chercheurs de la Walloni, 4*, 186–195.

Rausch, K.-A. (1975). Untersuchungen zur spät- und nacheiszeitlichen Vegetationsgeschichte im Gebiet des ehemaligen Inn-Chiemseegletschers. *Flora, 164*, 235–282.

Reisch, L. (1974). Eine spätjungpaläolithische Freilandstation bei Barbing, Ldkr. Regensburg. *Quartär, 25*, 53–72.

Renault-Miskovsky, J., & Leroi-Gourhan, A. (1981). Palynologie et archéologie: nouveaux résultats, du Paléolithique superieur au Mésolithique. *Bulletin de l'Association française pour l'étude du quaternaire, 18*(3–4), 121–128.

Rensink, E. (1986). Eeen jongpaleolithische vindplaats te Meesch-Steenberg (gemeente Eijsden). *Archeologie in Limburg, 29*, 173–176.

Rensink, E. (1991). L'observation du gibier et le débitage des nucléus: un poste de guet du Magdalénien à Mesch (Limbourg, Pays-Bas). *Helinium, XXXI*(1), 5–59.

Rensink, E. (1992). Eyserheide: A late Magdalenian site on the fringe of the northern loessbelt (Limburg, the Netherlands). *Archäologisches Korrespondenzblatt, 22*, 315–327.

Rensink, E. (1993). *Moving into the north: Magdalenian occupation and exploitation of the loess landscapes of northwestern Europe*, Leiden, s.n.

Renssen, H., Kasse, E., Vandenberghe, J., & Lorenz, S. J. (2007). Weichselian late pleniglacial surface winds over northwest and central Europe: A model-data comparison. *Journal of Quaternary Science, 22*(3), 281–293.

Reuter, A. (1942). Eine jungpaläolithische Siedlung auf dem Binsenacker von Liebschwitz. *Mannus, 34*, 148–164.

Reuter, U. (1993). Wichtige Neufunde der Jahre 1980–1990 aus den Regierungsbezirken Chemnitz, Dresden und Leipzig. *Arbeits- und Fortschrittsberichte zur Sächsischen Bodendenkmalpflege, 36*, 257–407.

Reuter, U. (1995). Wichtige Neufunde der Jahre 1991–1993 aus den Regierungsbezirken Dresden, Chemnitz und Leipzig. *Arbeits- und Fortschrittsberichte zur Sächsischen Bodendenkmalpflege, 37*, 243–278.

Reverdin, L. (1924). *La station préhistorique du "Sälihöhle Oben" près d'Olten (Soleure, Suisse)* (pp. 1–19). XXVI: Anzeiger für Schweizerische Altertumskunde.

Reynolds, V. (1966). Open groups in hominid evolution. *Man, 1*(4), 441–452.

Rhezak, A. (2000). Taphonomical, paleoeconomical and paleoecological investigation of the animal remains from the Abri Chesselgraben, canton of Solothurn, Swiss Jura (Late Magdalénien-Mesolithic). In P. Crotti (Ed.), *Meso '97. Actes de la Table ronde "Epipaléolithique et Mesolithique", Lausanne, 21–23 Novembre 1997* (pp. 109–114). Lausanne: Cahiers Archeologie Romande.

Rice, P. C., & Paterson, A. L. (1985). Cave art and bones: Exploring the interrelationships. *American Anthropologist, 87*(1), 94–100.

Rice, P. C., & Paterson, A. L. (1986). Validating the cave art – archaeofaunal relationship in Cantabrian Spain. *American Anthropologist, 88*(3), 658–667.

Richter, M. (1955). Die jügere Altsteinzeit im Ostthüringer Orlagau. *Alt-Thüringen, 1*, 11–42.

Ricoeur, P. (1984). *The reality of the historical past*. Milwaukee: Marquette University Press.

Rieder, K.-H. (1988). Ein Jagdplatz eiszeitlicher Jäger in der Frankenalp, Gemeinde Eitensheim, Lkr. Eichstätt, Obb. *Das Archäologische Jahr in Bayern, 1987*, 30–31.

Rieder, K.-H. (1989). Eine eiszeitliche Jagdstation auf der Windhöhe nördlich von Eitensheim Lkr. Eichstätt. In K.-H. Rieder, A. Tillmann, & J. Weinig (Eds.), *Steinzeitliche Kulturen an Donau und Altmühl. Begleitheft zur Ausstellung im Stadtmuseum Ingolstadt* (pp. 123–124). Ingolstadt: Stadt Ingolstadt.

Rieder, K. H. (1992). *Kritische Analyse alter Grabungsergebnisse aus dem Hohlen Stein bei Schambach aus der Sicht der Profiluntersuchungen von 1977–1982. Aspekte zur Geschichte der Höhlenverfüllung, ihrer Paläontologie und Archäologie*. Ph.D. Dissertation, University of Tübingen.

Riek, G. (1931). Eine mittel- und jungpaläolithische Freilandstation am Randecker Maar. *Nachrichtenblätter für Deutsche Vorzeit, 7*, 175–176.

Riek, G. (1932). Eine Mischkultur am Randecker Maar und deren Datierungsfrage. *Germania, 16*, 257–264.

Riek, G. (1933). Der Stand der Paläolithischen Fporschung in Württemberg. *Korrespondenzblatt des Gesamtvereins der deutschen Geschichts- und Altertumsvereine, 1933*, 18–30.

Riek, G. (1934). *Die Eiszeitjägerstation am Vogelherd im Lonetal*. Tübingen: Heine.

Riek, G. (1935). *Kulturbilder aus der Altsteinzeit Württembergs*. Tübingen: Heine.

Riek, G. (1938). Ein Beitrag zur Kenntnis des süddeutschen Solutréen. *Germania, 22*, 147–150.

Riek, G. (1957a). Neuffen. *Fundberichte aus Schwaben, NF, 14*, 158.

Riek, G. (1957b). *Drei jungpaläolithische Stationen am Bruckersberg in Giengen an der Brenz*. Stuttgart: Silberburg.

Riek, G. (1958). Das Paläolithikum der Brillenhöhle im Achtal bei Blaubeuren (Schwäb. Jura). In R.-G.K.d.D.A. Instituts (Ed.), *Neue Ausgrabungen in Deutschland* (pp. 6–22). Berlin: Gebr. Mann.

Riek, G. (1959a). Das federmesserführende Magdalénien der Burkhardtshöle bei Westerheim im Kreis Münsingen (Schwäbische Alb). *Fundberichte aus Schwaben, NF, 15*, 9–29.

Riek, G. (1959b). Ein magdalénienzeitlicher Rastplatz unter der Südwand des Sirgensteinfelsens. *Fundberichte aus Schwaben, NF, 15*, 30–42.

Riek, G. (1959c). Die Brillenhöhle im Achtal bei Blaubeuren, eine neue Magdalénienstation. *Fundberichte aus Schwaben, NF, 15*, 97–101.

Riek, G. (1960). Das Paläolithikum der Höhlen des Lone- und Brenztales. *Jahrbuch für Karst- und Höhlenkunde, 1*, 57–104.

Riek, G. (1973). *Das Paläolithikum der Brillenhöhle bei Blaubeuren (Schwäbische Alb)*. Stuttgrat: Müller and Gräff.

Rieth, A. (1938). *Vorgeschichte der Schwäbischen Alb*. Leipzig: Kabitzsch.

Rigaud, J.-P., & Simek, J. F. (1987). "Arms too short to box with God" problems and prospects for Paleolithic prehistory in Dordogne, France. In O. Soffer (Ed.), *The pleistocene old world. Regional perspectives* (pp. 47–61). New York: Plenum Press.

Ripotot, P., & Seibel, R. (1958). Coup d'oeil sur la préhistoire comptoise d'après les collections de Dôle. *Annales littéraires de l'Université de Besancon, 20*, 1–9.

Robert, R., & Allard, M. (2003). Historique des recherches et fouilles de Romain Robert. In J. Clottes & H. Delporte (Eds.), *La Grotte de La vache (Arriège). Fouilles Romain Robert. I – Les occupations du Magdalénien* (pp. 29–42). Paris: Éditions de la Réunion de Musées Nationaux.

Roche, J. (1971). Stratigraphie de la grotte du Placard (fouilles 1958–1968). *Mémoires Société Archéologique et Historique de la Charente, 1971*, 253–259.

Rockman, M., & Steele, J. (Eds.). (2003). *Colonization of unfamiliar landscapes. The archaology of adaptation*. London: Routledge.

Rogačev, A. N., & Anikovič, M. V. (1984). Pozdnij paleolit Russkoj ravniny i Kryma. In P. I. Boriskovskij (Ed.), *Paleolit SSSR* (pp. 162–271). Moskva: Nauka.

Rögel, F. (1998). Palaeogeographic condiserations for Mediterranean and Paratethys seaways (Oligocene to Miocene). *Annalen des Naturhistorischen Musemus Wien, 99*(A), 279–310.

Rögel, F., & Steininger, F. F. (1983). Vom Zerfall der Tethys zu Mediterran und Paratethys. Die neogene Paläogeographie und Palinspastik des zirkum-mediterranen Raumes. *Annalen des Naturhistorischen Musemus Wien, 85*(A), 135–163.

Roller, O. (1954). Ein Beitrag zur Kenntnis des Magdalénien in Südwestdeutschland. In K. Bittel (Ed.), *Festschrift für Peter Goessler* (pp. 142–149). Stuttgart: Kohlhammer.

Rösch, M. (1985). Nussbaumer Seen – Spät- und Postglaziale Umweltsveränderungen einer Seengruppe im östlichen Schweizer Mittelland. In G. Lang (Ed.), *Swiss lake and mire environments during the last 1000 years* (pp. 337–379). Vaduz: Lubrecht and Cramer.

Rosenstein, A. (1998). Vor mehr als 12000 Jahren – Jäger und Sammler des Magdalénien in Dreieich. *Landschaft Dreieich, Blätter für Heimatforschung, 1998*, 17–32.

Roth, G. (2011). http://web763.lin-www.df-webhosting.de/rchaeology/wordpress/wp-content/uploads/2011/01/Korrespondenzanalyse_version_1_ROTH_2011.pdf

Rothert, L. (1936). *Die Mittlere Steinzeit in Schlesien. Die Feuersteingeräte und ihre Einordnung*. Leipzig: Rabitzsch.

Rozoy, D. J.-G. (1978). *Les derniers chasseurs. L'Epipaléolithique en France et en Belgique*. Charleville: Société Archéologique Champenoise.

Rozoy, D. J.-G. (1988a). Le Magdalénien supérieur de Roc-la-Tour I. *Helinium, 28*(2), 157–191.

Rozoy, D. J.-G. (1988b). Le Magdalénien en Europe: démographie, groups régionaux. *Bulletin Société Préhistorique du Luxembourg, 10*, 139–158.

Rozoy, D. J.-G. (1997). Séjours d'été en Ardenne des Magdaléniens du Bassin Parisien. In A. Thévenin (Ed.), *Le Paléolithique supérieur de l'Est de la France: de l'Aurignacien à l'Ahrensbourgien, Actes du colloque de Chaumont 1994* (pp. 139–156). Reims: Société Archéologique Champenoise.

Rozoy, C., & Rozoy, D. J.-G. (2004). Roc-la-Tour, le site des Esprits. Le Magdalénien VI à Monthermé (Ardennes): topographie, chronologie, art. *Bulletin Société Préhistorique Luxembourgeoise, 26*, 9–226.

Ruddiman, W. F., & Mcintyre, A. (1973). Time-transgressive deglacial retreat of polar waters from the North Atlantic. *Quaternary Research, 3*, 117–130.

Ruddiman, W. F., & Mcintyre, A. (1981). The North Atlantic Ocean during the last deglaciation. *Palaeogeography, Palaeoclimatology, Plaeoecology, 35*, 145–214.

Rust, A. (1937). *Das altsteinzeitliche Rentierjägerlager Meiendorf*. Neumünster: Wachholtz.

Rust, A. (1942). Eine notwendige Stellungnahme. *Quartär, 4*, 197–227.

Rust, A. (1943). *Die alt- und mittelsteinzeitlichen Funde von Stellmoor*. Neumünster: Wachholtz.

Rust, A. (1958). *Die jungpaläolithischen Zeltanlagen von Ahrensburg*. Neumünster: Wachholtz.

Rydlewski, J. (1989). Pienińskie złoża radiolarytu i ich eksploatacja w epoce kamienia i wczesnej epoce brązu na Podchalu. *Acta Archaeologica Carpathica, 28*, 25–80.

Rydlewski, J., & Valde-Nowak, P. (1981a). Koniówka, gm. Czarny Dunajec, stan.1. *Informator Archeologiczny, 1980*, 12–13.

Rydlewski, J., & Valde-Nowak, P. (1981b). Sromowce Wyżne-Kąty, gm. Czorsztyn. *Informator Archeologiczny, 1980*, 21–22.

Rydlewski, J., & Valde-Nowak, P. (1981c). Podczerwone, gm. Czarny Dunajec, stan. 1. *Informator Archeologiczny, 1980*, 19.

Rydlewski, J., & Valde-Nowak, P. (1982). Sromowce Wyżne-Kąty, gm. Czorsztyn. *Informator Archeologiczny, 1981*, 21–22.

Sacchi, D. (1986). *Le Paléolithique supérieur du Languedoc occidental et du Roussillon*. Paris: Editions CNRS.

Sachse-Kozłowska, E., & Kozłowski, S. K. (2004). *Piekary ptès de Cracovie (Pologne). Complexe des sites Paléolithiques*. Kraków: Polska Akademia Umiejętności.

Sackett, J. R. (1977). The meaning of style in archaeology: A general model. *American Antiquity, 42*, 369–380.

Sano, K. (2012). *Functional variability in the late upper Palaeolithic of north-western Europe. A traceological approach. Universitätsforschungen zur prähistorischen Archäologie 219*. Bonn: Habelt.

Sano, K., Maier, A., & Heidenreich, S. (2011). Bois Laiterie revisited: Functional, morphological and technological analysis of a late glacial hunting camp in north-western Europe. *Journal of Archaeological Science, 38*(7), 1468–1484.

Sarasin, F. (1918). *Die steinzeitlichen Stationen des Birstales zwischen Basel und Delsberg*. Basel: Georg & Co.

Sarasin, F., & Stehlin, H. G. (1924). *Die Magdalénien-Station bei Ettingen (Baselland). Denkschrift der Schweizerischen Naturforschenden Gesellschaft* (vol. 61, pp. 1–26).

Saussure, H. D. (1870). La Grotte du Scé, près Villeneuve, Station Suisse du renne. *Archives des sciences physiques et naturelles, XXXVIII*, 105.

Sawicki, L. (1960). Stanowisko otwarte madleńskie Antoniów Mały. *Biuletyn Instytuto Geologicznego, 9*, 171–208.

Sayer, A. (1992). *Method in social science: A realist aproach*. London: Routledge.

Scharl, S. (2010). *Versorgungsstrategien und Tauschnetzwerke im Alt- und Mittelneolithikum – Die Silexversorgung im westlichen Franken*. Rahden/Westf.: Marie Leidorf.

Schibler, J., & Sedlmeier, J. (1993). Die Schneehuhn- und Schneehasenknochen aus dem Abri Büttenloch (Ettingen Bl, Schweiz). Ein Beitrag zur Kenntnis der Jagdbeutenutzung im Spätmagdalénien. *Archäologisches Korrespondenzblatt, 23*, 15–35.

Schiffer, M. B. (1987). *Formation processes of the archaeological record*. Albuquerque: University of New Mexico Press.

Schild, R. (1975). Późny paleolit. In W. Chmielewski, R. Schild, & H. Więckowska (Eds.), *Paleolit i Mezolit* (pp. 159–338). Wrocław: Gdańsk Polska Akademia Nauk.

Schild, R., & Królik, H. (1981). Rydno. A final Paleolithic ochre mining complex. *Przegląd Archeologiczny, 29*, 53–100.

Schindler, R. (1960). *Die Bodenaltertümer der Freien und Hansestadt Hamburg*. Hamburg: Christians.

Schirmeisen, K. (1925). Neue eiszeitliche Funde in der Stierfels- und in der Backofenhöhle. *Zeitschrift des Deutschen Vereines für die Geschichte Mährens und Schlesiens, 27*, 1–11.

Schirwitz, K. (1926). Ein Lößfund aus dem Harzvorlande. *Jahresschrift für die Vorgeschichte der sächs.- thüring. Länder, 14*, 3–6.

Schirwitz, K. (1936). Eine spätaltsteinzeitliche Uferrandsiedlung aus dem Harzvorland. *Jahresschrift für Vorgeschichte der sächsisch-thüringischen Länder, 24*, 34–43.

Schirwitz, K. (1940). Der Grabhügel auf der Bockshornschanze bei Quedlinburg. *Mitteldeutsche Volkheit, 7*, 52–56.

Schirwitz, K. (1961). Zu den Ausgrabungen von 1937 vor der Steinkirche bei Scharzfeld am Harz. *Harz-Zeitschrift, 13*, 1–8.

Schlummer, M., Hoffmann, T., Dikau, R., Eickmeier, M., Fischer, P., Gerlach, R., et al. (2014). From point to area: Upscaling approaches for late quaternary archaeological and environmental data. *Earth-Science Reviews, 131*, 22–48.

Schmeidl, H. (1971). Ein Beitrag zur spätglazialen Vegetations- und Waldentwicklung im westlichen Salzachgletschergebiet. *Eiszeitalter und Gegenwart, 22*, 110–126.

Schmider, B. (1992). *Marsangy, un campement des derniers chasseurs magdaléniens sur le bords de l'Yonne* (ERAUL, Vol. 55). Liège: Université de Liège.

Schmidt, R. R. (1910). *Der Sirgenstein und die diluvialen Kulturstätten Württembergs*. Stuttgart: Schweizerbart.

Schmidt, R. R. (1912). *Die diluviale Vorzeit Deutschlands*. Stuttgart: Schweizerbart.

Schmidt, I. (2013). *Solutrean Points of the Iberian Peninsula – Tool making and using behaviour of hunter-gatherers during the Last Glacial Maximum*. Ph.D. Dissertation, University of Cologne.

Schmidt, I., & Mom, V. (2012). Computer-aided classification of archaeological objects – methodological issues in comparing shapes of Solutrean points. In J. Cascalheira & C. Goncalves (Eds.), *Proceedings of the IV Conference of Jovens em Investigação Arqueológica, Universidade do Algarve, Faro (Portugal) from 11th to 14th May 2011. Núcleo de Arqueologia e Paleoecologia (NAP)* (pp. 39–46). Faro: NAP.

Schmidt, B., & Schneider, J. (1974). Ausgewählte Neufunde aus den Jahren 1971/1972. *Jahresschrift für mitteldeutsche Vorgeschichte, 58*, 321–342.

Schmitz, R.-W., Thissen, J., & Weiner, J. (1998). "Latente" und "evidente" Strukturen auf dem Magdalénien-Fundplatz Beeck. *Archäologie im Rheinland, 1997*, 22–26.

Schönweiß, W. (1963). Paläolithische Funde vom Hahnenberg im schwäbisch-bayerischen Ries. *Quartär, 14*, 95–104.

Schotensack, O. (1903). Über die Gleichzeitigkeit der menschlichen Niederlassung aus der Renntierzeit im Löß bei Munzingen unweit Freiburg i. B. und der paläolithischen Schicht von Thaingen und Schweizersbild bei Schaffhausen. *Archiv Anthr. N.F., I*(2), 69–77.

Schotensack, O. (1907). Über die Gleichzeitigkeit der menschlichen Niederlassung aus der Renntierzeit im Löß bei Munzingen unweit Freiburg i. B. und der paläolithischen Schicht von Thaingen und Schweizersbild bei Schaffhausen. *Archiv Anthr. N.F., VI*(2–3), 169–179.

Schránil, J. (1928). *Die Vorgeschichte Böhmens und Mährens*. Berlin: de Gruyter.

Schroeder, A. (2000). *Arlay – Grotte Grappin. Un habitat du Magdalénien moyen entre montagnes Jurassiens et plaine de la Bresse – étude archéologique et archéozoologique*. Master thesis, University of Bâle.

Schuler, A. (1989). Das Magdalénien der Schussenquelle. Die Steinartefakte der Grabung von Oskar Fraas (1866). *Archäologisches Korrespondenzblatt, 19*, 11–22.

Schuler, A. (1994). *Die Schussenquelle: eine Freilandstation des Magdalénien in Oberschwaben*. Stuttgart: Theiss.

Schütrumpf, R. (1958). Die pollenanalytischen Untersuchungen an den altsteinzeitlichen Moorfundplätzen Borneck und Poggenwisch. In A. Rust (Ed.), *Die junpaläolithischen Zeltanlagen von Ahrensburg* (pp. 11–22). Neumünster: Wachholtz.

Schwabedissen, H. (1937). Die Hamburger Stufe im nordwestlichen Deutschland. *Offa, 2*(1-30).

Schwabedissen, H. (1944). *Die mittlere Steinzeit im westlichen Norddeutschland*. Neumünster: Wachholtz.

Schwabedissen, H. (1950). Die Bedeutung der Moorarchäologie für die Urgeschichtsforschung. *Offa, 8*, 46–74.

Schwabedissen, H. (1954). *Die Federmesser-Gruppen des nordwesteuropäischen Flachlandes*. Neumünster: Wachholtz.

Schwantes, G. (1928). Nördliches Paläolithikum und Mesolithikum. *Mitteilungen aus dem Museum für Völkerkunde Hamburg, 13*, 158–253.

Schwarz-Mackensien, G. (1978). *Jägerkulturen zwischen Harz und Aller*. Hildesheim: August Lax.

Schweizer, T. (1937). *Urgeschichtliche Funde in Olten und Umgebung*. Olten: Genossenschafts-Druckerei Olten.

Schweizer, T., Schmid, E., Bay, R., Stampfli, H. R., Forcart, L., & Fey, L. (1959). Die "Kastelhöhle" im Kaltbrunnental, Gemeinde Himmelried (Solothurn). *Jahrbuch für Solothurnische Geschichte, 32*, 1–88.

Seager, R., Battistit, D. S., Yin, J., Gordon, N., Naik, N., Clement, A. C., et al. (2002). Is the Gulf Stream responsible for Europe's mild winters? *Quaterly Journal of the Royal Meterological Society, 128*, 2563–2586.

Séara, F. (1997). Le gisement magdalénien de plein air de Seveux (Haute-Saône). In A. Thévenin & A. Villes (Eds.), *Le Paléolithique supérieur de l'est de la France: De l'Aurignacien à l'Ahrensbourgien. Actes du Colloques interrégional sur le Paléolithique, Chaumont, 17–18 octobre 1994* (pp. 91–103). Joué-les-Tours: La Simarre.

Sedlmeier, J. (1982). *Die Hollenberg-Höhle 3. Eine Magdalénienfundstelle bei Arlesheim, Kanton Basel-Landschaft*. Derendingen: Habegger.

Sedlmeier, J. (1988). Jungpaläolithischer Molluskenschalen-Schmuck aus nordwestschweizerischen Fundstellen als Nachweis für Fernverbindungen. *Archäologisches Korrespondenzblatt, 18*, 1–6.

Sedlmeier, J. (1989). *Jungpaläolithikum und Spätpaläolithikum in der Nordwestschweiz. Ein Beitrag zur regionalen Erforschung des Paläolithikums auf Grund ausgewählter Fundinventare aus Grabungen der Jahre zwischen 1910 und 1956*. Ph.D. Dissertation, University of Bern.

Sedlmeier, J. (1990). Die "Kleine Ganghöhle" im Kaltbrunnental, Gemeinde Himmelried (Kanton Solothurn). Ein bisher unbekannter Fundstellentyp im Schweizerischen Jungpaläolithikum. In J. Schibler, J. Sedlmeier, & H. Spycher (Eds.), *Festschrift für Hans R. Stampfli. Beiträge zur Archäozoologie, Archäologie, Anthropologie, Geologie und Paläontologie* (pp. 241–248). Basel: Helbing and Lichtenhahn.

Sedlmeier, J. (1993). Altsteinzeitliche Funde aus der Kohlerhöhle im Laufental. *Archäologie der Schweiz, 16*, 40–45.

Seibert, P. (1958). *Die Pflanzengesellschaften im Naturschutzgebiet "Pupplinger Au"*. München: Eigenverlag.

Seiffert, H. (1991). *Einführung in die Wissenschaftstheorie 2. Phänomenologie, Hermeneutik und historische Methode, Dialektik*. München: Beck.

Seiler, M. (1996). Auf den Spuren der Wildpferdjäger. Spätpaläolithische Funde im sächsischen Muldegebiet. *Archäologie in Sachsen-Anhalt, 6*, 15–17.

Seitl, L., Svoboda, J., Ložek, V., Přichystal, A., & Svobodová, H. (1986). Das Spätgalzial in der Barová-Höhle im Mährischen Karst. *Archäologisches Korrespondenzblatt, 16*, 393–398.

Sensburg, M. (2007). *Die räumliche Organisation de Konzentration IIa von Gönnersdorf. Struktur und Dynamik eines magdalénienzeitlichen Siedlungsbefundes am Mittelrhein*. Mainz: Verlag des Römisch-Germanischen Zentralmuseums.

Serangeli, J. (2006). *Verbreitung der großen Jagdfauna in Mittel- und Westeuropa im oberen Jungpleistozän. Ein kritischer Beitrag*. Rahden/Westf.: Leidorf.

Serangeli, J., & Terberger, T. (2006). Sondierungen an der Magdalénien-Fundstelle von Götzenhain-Ost, Kreis Offenbach. *Mitteilungen der Gesellschaft für Urgeschichte, 15*, 49–61.

Séronie-Vivien, M., Bensch, C., Capdegelle, R., Chevet, B., Hemingway, M., & Leroi-Gourhan, A. (1981). La grotte de Pégourié à Canniac-du-Causse (Lot). Le gisement azilien et magdalénien initial. In *Congrès Préhistorique de France, 21e session (Montauban-Cahors 2–9 Septembre 1979)*, Paris (pp. 250–265).

Shanks, M., & Tilley, C. (1987). *Re-constructing archaeology*. Cambridge: Cambridge University Press.

Shennan, S. J. (2000). Population, culture history and the dynamics of culture change. *Current Anthropology, 41*, 811–835.

Shennan, S. J. (2001). Demography and cultural innovation: a model and its implications for the emergence of modern human culture. *Cambridge Archaeological Journal, 11*(1), 5–16.

Shennan, S. J. (2004). *Quantifying archaeology*. Edinburgh: University Press.

Shennan, S. J. (2008). Evolution in archaeology. *Annual Review of Anthropology, 37*, 75–91.

Siekmann, W. (1976). Ein Vorkommen der Hamburger Kultur bei Beckedorf, Gemeinde Hermannsburg, Kr. Celle. *Nachrichten aus Niedersachsens Urgeschichte, 45*, 313–315.

Sieveking, A. (1979). *The cave artists*. London: Thames and Hudson.

Simon, U. (1993). *Die Burkhardtshöhle – eine Magdalénienstation am Nordrand der Schwäbischen Alb*. Master thesis, University of Tübingen.

Simoneit, B., & Veil, S. (1999). Gadenstedt FStNr. 121, Gde Lahstedt, Ldkr. Peine, Reg. Bez. BS. *Nachrichten aus Niedersachsens Urgeschichte, Beiheft, 2*, 10–12.

Sisk, M. L. (2011). *Settlement and site location in the middle and upper Paleolithic of the Vézère valley, France*. Ph.D. Dissertation, Stony Brook University.

Škrdla, P. (1990). Dalši paleolitické nálezy z lokality Líšeň (okr. Brno-město). *Přehled Výzkumů, 1987*, 15.

Škrdla, P. (2002). Magdalenian settlement structure in the southern part of the Moravian Karst. The problems of open-air settlements. In J. Svoboda (Ed.), *Prehistorické Jeskyně* (pp. 229–254). Brno: Archeologický ústav AV ČR.

Škrdla, P., Kos, P., & Přichystal, A. (1999). Nová magdalénská stanice v jižní části moravského krasu. *Přehled Výzkumů, 40*, 51–64.

Skutil, J. (1927). Palaeolithická stanice ve Výpustku. *Obzor prehistoricky, 5–6*, 47–58.

Skutil, J. (1940). K paleolitické stanici v jeskyni "Jáchymce" v Josefovském údolí u Adamova. *Časopis Vlastivědného spolku musejního Olomouc, 53*, 1–4.

Skutil, J. (1962). Nález figurální plastiky na volutové keramice z jeskyně Koňské jámy v Moravském. *Přehled Výzkumů, 1961*, 33–37.

Skutil, J. (1963). Výsledky výzkumu jeskyně Koňské jámy (Suchdol, okr. Blansko) v Mor. *Krase Přehled Výzkumů, 1962*, 10–11.

Sobczyk, K. (1993). *The Late Palaeolithic flint workshops at Brzoskwinia-Krzemionki near Kraków*. Kraków: Nakł. Uniwersytet Jagiellónskiego.

Sollas, W. J. (1924). *Ancient hunters and their modern representatives*. London: Macmillan.

Spaulding, A. C. (1953). Statistical techniques for the discovery of artifact types. *American Antiquity, 18*, 305–313.

Spaulding, A. C. (1968). Explanation in archaeology. In S. R. Binford & L. R. Binford (Eds.), *New perspectives in archaeology* (pp. 33–39). Chicago: Aldine.

Springensguth, F., & Buttler, W. (1935). Unbekannte Magdalénien-Funde aus der Karsteinhöhle. *Germania, 19*, 1–4.

Spycher, H., & Sedlmeier, J. (1985). Steinzeitfunde bei Erschwil im Schwarzbubenland. *Helvetia Archaeologica, 63–64*(16), 78–80.

Staesche, U. (1999). Die Säugetierfunde aus der jungpaläolithischen Jägerstation vom Aschenstein bei Freden, Landkreis Hildesheim, Niedersachsen. In C. Becker, H. Manhart, J. Peters, & J. Schiebler (Eds.), *Beiträge zur Paläoanatomie, Archäolgie, Ägyptologie, Ethnologie und Geschichte der Tiermedizin. Festschrift für Angela von Driesch zum 65. Geburtstag* (pp. 437–446). Rahden/Westf.: Leidorf.

Staesche, U. (2005). A peculiar composition of animl remains under a Magdalénian rock shelter south of Göttingen, Northwest Germany. *Munibe, 57*(1), 265–277.

Stahl Gretsch, L.-I. (2006). *Les occupations Magdaléniennes de Veyrier. Histoire et préhistore des abris-sous-blocs*. Lausanne: Bibl. Histor. Vaudoise.

Stahlhofen, H. (1981). *Halle-Neustadt, junge sozialistische Stadt auf historischem Boden*. Naumburg: Fachbuchdruck Naumburg.

Stampfli, H. R. (Ed.). (1983). *Rislisberghöhle. Archäologie und Ökologie einer Fundstelle aus dem Spätmagdalénien bei Oensingen im Solothurner Jura*. Bern: Haupt.

Stanner, W. (1965). Aboriginal territorial organitation: Estate, range, domain and regime. *Oceania, 36*, 1–26.

Stapert, D. (1981). A site of the Hamburgian tradition on the wadden island of Texel (province of North-Holland, Netherlands). *Palaeohistoria, 23*, 1–27.

Stapert, D. (1982). A site of Hamburgian tradition with a constructed hearth near Oldeholtwolde (Province of Friesland, the Netherlande); first report. *Palaeohistoria, 24*, 53–89.

Steffensen, J. P., Andersen, K. K., Bigler, M., Clausen, H. B., Dahl-Jensen, D., Fischer, H., et al. (2008). High-resolution Greenland ice core data show abrupt climate change happens in few years. *Science, 321*, 680–684.

Stevens, R. E., Germonpré, M., Petrie, C. A., & O'connell, T. C. (2009). Palaeoenvironmental and chronological investigations of the Magdalenian site of Goyet Cave and Trou de Chaleux (Belgium), via stable isotope and radiocarbon analyses of horse skeletal remains. *Journal of Archaeological Sciences, 36*(3), 653–662.

Stewart, J. R., & Lister, A. M. (2001). Cryptic northern refugia and the origins of the modern biota. *Trends in Ecology and Evolution, 16*, 608–613.

Stiles, D., Ammerman, A. J., Benzécri, J.-P., Bhattacharya, D. K., Bordes, F., Bricker, H. M., et al. (1979). Paleolithic culture and culture change: Experiment in theory and method [and comments and reply]. *Current Anthropology, 20*(1), 1–21.

Stocker, C., Jean-Marie, C., Patou-Mathis, M., & Thévenin, A. (2003). Le gisment magdalénien de la Roche Plate à Saint Mihiel (Meuse, France). *Bulletin Société Préhistorique Luxembourgeoise, 25*, 23–41.

Stodiek, U. (1987). Fussgöhnheim - Zwei Spätjungplaäolithische Fundplätze in der Vorderpfalz. *Archäologisches Korrespondenzblatt, 17*, 31–41.

Straus, L. G., & Otte, M. (1998). Bois Laiterie Cave and the Magdalenian of Belgium. *Antiquity, 72*, 253–268.

Street, M. (2000). Aspects of late upper Palaeolithic settlement and chronology in northern central Europe. In B. Valentin, P. Bodu, & M. Christensen (Eds.), *L'Europe Centrale et Septentrionale au Tardiglaciaire. Table ronde de Nemours, 13–16 mai 1997* (pp. 55–71). Paris: Musée de Préhistoire d'Ile-de-France.

Street, M. (2007). Comment (on samples of bone from Gönnersdorf). In T. F. G. Higham, C. Bronk Ramsey, F. Brock, D. Baker, & P. Ditchfield (Eds.), *Radiocabon dates from the Oxford AMS system: Archaeometry datelist 32. Archaeometry, 49*(Suppl 1), 16–17.

Street, M., & Terberger, T. (2002). Comment (on the sample series "Absolute Chronology of the German upper Palaeolithic series"). In C. Bronk Ramsey, T. F. G. Higham, D. C. Owen, A. W. G. Pike, & R. E. M. Hedges (Eds.), *Radiocarbon dates from the Oxford AMS system: Archaeometry datelist 31. Archaeometry, 44*(3; Suppl 1), 28–30.

Street, M., & Terberger, T. (2004). The radiocarbon chronology of the German upper Palaeolithic: Fifteen years of cooperation with ORAU. In T. F. G. Higham, C. Bronk Ramsey, & D. C. Owen (Eds.), *Radiocarbon and archaeology. Fourth International Symposium, St. Catherine's College, Oxford, 9–14 April 2002: Conference Proceedings* (pp. 281–302). Oxford: Oxford University Press.

Stringer, C., & Gamble, C. (1993). *In search of the Neanderthals. Solving the puzzle of human origins*. New York: Thames and Hudson.

Stuhlfrauth, A. (1953). *Die Steinzeit*. Bayreuth: Lorenz Ellwanger.

Sulzberger, K. (1921). *Die urgeschichtlichen Fundstellen des Reyath*, Thayngen.

Surmely, F., & Alix, P. (2005). Note sur les talons en éperon du Protomagdalénien. *Paléo, 17*, 157–176.

Svenning, J.-C., Normand, S., & Kageyama, M. (2008). Glacial refugia of temperate trees in Europe: Insights from species distribution modelling. *Journal of Ecology, 96*, 1117–1127.

Svensson, A., Andersen, K. K., Bigler, M., Clausen, H. B., Dahl-Jensen, D., Davies, S. M., et al. (2006). The Greenland ice core chronology 2005, 15–42 kyr. Part 2: Comparison to other records. *Quaternary Science Reviews, 25*, 3258–3267.

Svoboda, J. (1976). Zur Problematik der magdalénienzeitlichen Kunst Mitteleuropas. *Anthropologie, 14*(3), 163–193.

Svoboda, J. (1987). Výzkum Stránské skály a okolí v roce 1985 (k. o. Slatina a Líšeň, okr. Brno-město). *Přehled Výzkumů, 1985*, 13.

Svoboda, J. (1991a). Neue Erkenntnisse zur Pekárna-Höhle im Mährischen Karst. *Archäologisches Korrespondenzblatt, 21*, 39–43.

Svoboda, J. (1991b). Stránská skála. Výsledky výzkumu v letech 1985–1987. *Památky Archeologické, 82*(1), 5–47.

Svoboda, J., Ložek, V., & Vlček, E. (1996). *Hunters between east and west. The Paleolithic of Moravia*. New York: Plenum Press.

Svoboda, J., & Novák, M. (2004). Eastern central Europe after the upper Pleniglacial: Changing points of observation. *Archäologisches Korrespondenzblatt, 34*, 463–477.

Svoboda, J., Přichystal, A., Ložek, V., Svobodová, H., & Toul, J. (1995). Kolíbky, a Magdalenian site in the Moravian Karst. *Quartär, 45*(46), 135–159.

Svoboda, J., Seitl, L., & Štrof, A. (1983). Výzkumy jeskynních výplní v severní části Moravského krasu (zpráva za rok 1981). *Přehled výzkumů*, 9–13.

Svoboda, J., Van der Plicht, H., & Balák, I. (2005). Bycí Skála Cave, Czech Republic: Radiocarbon dates of rock paintings. *International Newsletter on Rock Art, 4*, 7–9.

Svoboda, J., Van der Plicht, H., & Kuželka, V. (2002). Upper Palaeolithic and Mesolithic human fossils from Moravia and Bohemia (Czech Republic): Some new ^{14}C dates. *Antiquity, 76*, 957–962.

Szafer, W. (1964). *Ogólna geografia roślin*. Warszawa: Państwowe Wydawnictwo Naukowe.

Szyja, S. (2012). *Die modifizierten Steinartefakte aus dem Befund 12 der Magdalénien-Freilandstation Bad Kösen-Lengefeld in Sachsen-Anhalt*. Bachelor thesis, University of Cologne.

Taberlet, P., & Cheddadi, R. (2002). Quaternary refugia and persistence of biodiversity. *Science, 297*, 2009–2010.

Taborin, Y. (1993). *La parure en coquillage au Paléolithique*. Paris: Éditions CNRS.

Taute, W. (1959). Neu entdeckte Lagerplätze der Hamburger und Ahrensburger Kultur bei Deimern, Kr. Soltau, in der Lüneburger Heide. *Die Kunde, N.F., 10*(3–4), 182–191.

Taute, W. (1967). Grabungen zur mittleren Steinzeit in Höhlen und unter Felsdächern der Schwäbischen Alb 1961 bis 1965. *Fundberichte aus Schwaben, N.F., 18*, 14–21.

Taute, W. (1971). *Untersuchungen zum Mesolithikum und zum Spätpaläolithikum im südlichen Mitteleuropa. Band 1: Chronologie Süddeutschlands*. Habilitation Dissertation, University of Tübingen.

Taute, W. (1972). Die spätpaläolithisch-frühmesolithische Schichtenfolge im Zigeunerfels bei Sigmaringen (Vorbericht). *Archäologische Informationen, 1*, 29–40.

Taute, W. (1975). Ausgrabungen zum Spätpaläolithikum und Mesolithikum in Süddeutschland. In K. Böhner (Ed.), *Ausgrabungen in Deutschland, Teil 1* (pp. 64–73). Mainz: Deutsche Forschungsgemeinschaft.

Tavoso, A. (1987). Le remplissage de la grotte Tournal à Bize-Minervois (Aude). *Cypsela, 6*, 23–25.

Taylor, W. W. (1967). *A study of archaeology*. Carbondale: Southern Illinois University Press.

Terberger, K. (1987). Funde der Magdalénien-Station Saaleck. *Jahresschrift für mitteldeutsche Vorgeschichte, 70*, 95–134.

Terberger, K. (1993). *Das Lahntal-Paläolithikum*. Wiesbaden: Landesamt für Denkmalpflege Hessen.

Terberger, T. (1997). *Die Siedlungsbefunde des Magdalénien-Fundplatzes Gönnersdorf. Konzentrationen III und IV*. Stuttgart: Steiner.

Terberger, T. (2002). Das Andernacher Inventar und die erweiterte Nebraer Gruppe. In H. Floss & T. Terberger (Eds.), *Die Steinartefakte des Magdalénien von Andernach (Mittelrhein). Die Grabungen 1979–1983* (pp. 135–138). Rahden/Westf.: Leidorf.

Terberger, T. (2003). Vom Gravettien zum Magdalénien in Mitteleuropa. Aspekte der meschlichen Besiedlungsgeschichte in der Zeit um das zweite Kältemaximum der letzten Eiszeit. *Archäologisches Nachrichtenblatt, 8*, 55–62.

Terberger, T. (2006). From the first humans to the Mesolithic hunters in the northern German lowlands. In K. Møller Hansen & K. Buck

Pedersen (Eds.), *Across the western Baltic. Proceedings from an archaeological conference in Vordingborg* (pp. 23–56). Vordingborg: Udgivet af Sydsjællands Museum.
Terberger, T., Küßner, M., Schüler, T., & Street, M. (2003). Mesolithische Menschenreste aus der Urdhöhle bei Döbritz, Saale-Orla-Kreis. *Alt-Thüringen, 36*, 4–20.
Terberger, T., Serangeli, J., & Woertz, P. (2009a). Eiszeitjäger aus dem Süden – ein 15.000 Jahre alter Lagerplatz bei Dreieich-Götzenhain. *Hessen Archäologie, 2008*, 20–24.
Terberger, T., Tromnau, G., Street, M., & Weniger, G.-C. (2009b). Die jungpaläolithische Fundstelle Aschenstein bei Freden an der Leine, Kr. Hildesheim (Niedersachsen). *Quartär, 56*, 87–103.
Terberger, T., & Street, M. (2002). Hiatus or continuity? New results for the question of plenigalcial settlement in central Europe. *Antiquity, 76*(291), 691–698.
Thévenin, A. (1956). Une station du Magdalénien III à Chariez (Haute-Saône). *Revue Archéologique de l'Est et du Centre-Est, 7*, 360–361.
Thévenin, A. (1965). L'outillagepaléolithique et mésolithique du bassin supérieur de la Saône. *Annales scientifiques de l'Université de Besancon. 3e série, Géologie, 1*, 13–61.
Thévenin, A. (1968). L'abri Rochedane à Villars-sous-Dampjoux, près de Pont-de-Roide. *Bulletin de la Société Préhistorique Française, 65*(9), 235–236.
Thévenin, A. (1982). *Rochedane: l'Azilien, l'Epipaléolithique de l'Est de la France et les civilisations épipaléolithiques de l'Europe occidentale*. Ph.D. Dissertation, University of Strasbourg.
Thévenin, A. (1997). Le Magdalénien à scalènes du gisementr de la Guillotine à Chariez (Haute-Saône). *Bulletin Société Préhistorique Luxembourgeoise, 19*, 67–84.
Thieme, H. (1991). Alt- und Mittelsteinzeit. In H.-J. Häßler (Ed.), *Ur- und Frühgeschichte in Niedersachsen* (pp. 77–108). Stuttgart: Theiss.
Thiessing, J. B. (1886). Neuer Höhlenfund im Jura. *Mitteilungen der Naturforschenden Gesellschaft in Bern, 1885*, 128.
Thissen, J. (1989). Ein Fundplatz des Magdalénien am linken Niederrhein bei Kamphausen, Gem. Jüchen, Kreis Neuss. *Archäologisches Korrespondenzblatt, 19*, 315–323.
Thissen, J. (1997). *Jäger und Sammler. Paläolithikum und Mesolithikum im Gebiet des Linken Niederrhein*. Ph.D. Dissertation, University of Cologne.
Thomas, J. (1991). *Rethinking the Neolithic*. Cambridge: Cambridge University Press.
Tilley, C. (1996). *An ethnography of the Neolithic. Early prehistoric societies in southern Scandinavia*. Cambridge: Cambridge University Press.
Tillmann, A. (1989a). Die Lochschlaghöhle bei Obereichstätt. In *Steinzeitliche Kulturen an Donau und Altmühl. Begleitheft zur Ausstellung im Stadtmuseum Ingolstadt* (pp. 108–113). Ingolstadt: Courier Druckhaus.
Tillmann, A. (1989b). Die Mühlberggrotte bei Dollnstein. In *Steinzeitliche Kulturen an Donau und Altmühl. Begleitheft zur Ausstellung im Stadtmuseum Ingolstadt* (pp. 108–113). Ingolstadt: Courier Druckhaus.
Tinnes, J. (1995). *Die Geweih-, Elfenbein- und Knochenartefakte der Magdalénienfundplätze Gönnersdorf und Andernach*. Ph.D. Dissertation, University of Cologne.
Tixier, J. (1963). *Typologie de l'Épipaléolithique du Maghreb*. Paris: Arts et Métiers Graphiques.
Tixier, J. (1974). Microburins du Magdalénien V a la Faurélie II (Dordogne). *L'Anthropologie, 78*, 189–196.
Tixier, J. (1976). L'abri sous roche de la Faurélie. In J.-P. Rigaud & B. Vandermeersch (Eds.), *Livret-guide de l'excursion A 4 Sud-Ouest (Aquitaine et Charente) du IX Congrès U.I.S.P.P. 13.-18. Septembre 1976*, Nice (pp. 114–116).
Toepfer, V. (1955). Die spätpaläolithischen und mesolithischen Silexgeräte vom Galgenberg bei Halle/Saale. *Jahresschrift für mitteldeutsche Vorgeschichte, 39*, 15–35.
Toepfer, V. (1958). Altsteinzeit. *Ausgrabungen und Funde, 3*, 149–157.
Toepfer, V. (1965). Paläolithische Fundstätten am ehemaligen Aschersleben-Gaterslebener See. *Ausgrabungen und Funde, 10*(1), 3–10.
Toepfer, V. (1970). Saaleck, Kr. Naumburg. In H. Richter (Ed.), *Periglazial-Löß-Paläolithikum im Jungpleistozän der Deutschen Demokratischen Republik. Exkursionsmaterial* (pp. 393–395). Leipzig: Haack.
Toepfer, V., & Mania, D. (1973). Paläolithische Funde vom Petersberg nördlich von Halle (Saale). *Jahresschrift für mitteldeutsche Vorgeschichte, 57*, 49–69.
Torke, W. (1975). Ausgrabungen im "Felsställe" bei Mühlen, Alb-Donau-Kreis. *Archäologische Ausgrabungen, 1975*, 10–13.
Torke, W. (1976). Ausgrabungen in dem Abri "Felsställe" bei Mühlen, Alb-Donau-Kreis. *Archäologische Ausgrabungen, 1976*, 8–10.
Torke, W. (1981). *Fischreste als Quellen der Ökologie und Ökonomie in der Steinzeit Südwest-Deutschlands*. Tübingen: Archaeologica Venatoria.
Tournier, J. (1911). Les premiers habitants du Bugey, époque paléolithique, la grotte abri de la Bonne-Femme, Brégnier-Cordon (Ain). *Bulletin de la Société scientifique, historique et littéraire Le Bugey, 3*, 273–286.
Tournier, J. (1924). *La grotte des Hoteaux (Ain)*. Belley: Chaduc.
Tournier, J., & Costa De Beauregard, T. (1922). Deux stations préhistoriques du Jura occidental de l'Ain dans la vallée du Suran. *L'Anthropologie, 32*, 383–408.
Tournier, J., & Guillon, C. (1895). *Les hommes préhistoriques dans l'Ain*. Bourg: Villefranche.
Trigger, B. G. (1989). *A history of archaeological thought*. Cambridge: Cambridge University Press.
Tromnau, G. (1974a). Zur jungpaläolithischen Fundstelle der Hamburger Kultur im Dörgener Moor, Kr. Meppen. *Die Kunde, 25*, 9–14.
Tromnau, G. (1974b). Der jungpaläolithische Fundplatz Schalkholz, Kreis Dithmarschen. *Hammaburg, 1*, 9–22.
Tromnau, G. (1974c). Jungpaläolithische Funde der Hamburger Kultur vom Itzstedter See, Kreis Segeberg. *Hammaburg, 1*, 99–102.
Tromnau, G. (1975a). *Die Fundplätze der Hamburger Kultur von Heber und Deimern, Kreis Soltau*. Hildesheim: Lax.
Tromnau, G. (1975b). *Neue Ausgrabungen im Ahrensburger Tunneltal. Ein Beitrag zur Erforschung des Jungpaläolithikums im nordwesteuropäischen Flachland*. Neumünster: Wachholz.
Tromnau, G. (1976). *Rentierjäger der Späteiszeit in Norddeutschland*. Hildesheim: Lax.
Trusheim, F. (1937). Die diluvialen Tierreste aus der kleinen Ofnethöhle. *Bayerische Vorgeschblätter, 14*, 39–43.
Tylor, E. B. (1871). *Primitive culture: Researches into the development of mythology, philosophy, religion, language, art and custom*. London: Murray.
Ucko, P. J. (Ed.). (1995). *Theory in archaeology. A world perspective*. London: Routledge.
Usinger, H. (1975). *Pollenanalytische und stratigraphische Untersuchungen an zwei Spätglazial-Vorkommen in Schleswig-Holstein*. Kiel: Arbeitsgemeinschaft Geobotanik in Schleswig-Holstein u. Hamburg.
Utrilla, M. P. (1981). *El Magdaleniense inferior y medio en la Costa Cantábrica*. Santander: Min. de Cultura, Dir. General de Bellas Artes, Archivos y Bibliotecas.
Utrilla, M. P. (1982). El yacimiento de la cueva de Abauntz. *Trabajos de Arqueologia Navarra, 3*, 203–346.
Utrilla, M. P. (1995). El valle del Ebro durante el Tardiglaciar y comienzos del Holoceno. Relaciones con el Magdaleniense Cantábrico. In A. Moure Romanillo & C. González Sáinz (Eds.), *El final del Paleolítico Cantábrico. Transformaciones ambientales y culturales durante el Tardiglaciar y comienzos del Holoceno en la Región Cantábrica* (pp. 281–312). Santander: Servicio de Publ. de la Univ. de Cantabria.
Valde-Nowak, P. (1985). Uście Gorlickie, gm. Loco, stan. 2. *Informator Archeologiczny 1984*, 12.

Valde-Nowak, P. (1991). Studies in pleistocene settlement in the Polish Carpathians. *Antiquity, 65*, 593–606.

Valde-Nowak, P., & Muzyczuk, A. (2000). Magdalenian settlement at Hłomcza (Polish Carpathians). *Acta Archaeologica Carpathica, 25*, 5–32.

Valde-Nowak, P., Soják, M., & Wąs, M. (2007). On the problems of late paleolithic settlement in northern Slovakia. Example of Stará L'ubovňa. *Slovenská Archeológia, 55*(1), 1–22.

Valentin, B. (1995). *Les groupes humains et leurs traditions au Tardiglaciaire dans le Bassin Parisien. Apports de la technologie lithique comparée*. Ph.D. Dissertation, University of Paris-I.

Valentin, B. (2000). L'usage des percuteurs en pierre tendre pour le débitage des lames. Circonstances de quelques innovations au cours du Tardiglaciaire dans le Bassin parisien. In G. Pion (Ed.), *Le Paléolithique Supérieur récent: nouvelles données sur le peuplement et l'environnement. Table Ronde de Chambéry 12–13 mars 1999. Mémoires de la Société Préhistorique Française 28* (pp. 253–260). Chambery: Société Préhistorique Française.

Valentin, B. (2009). Typologie vs Typologie (*SIC*!). Comment la technologie contribue à raffiner la typologie des armatures lithiques. In T. Aubry, F. Almeida, A. C. Araújo, & M. Tiffagom (Eds.), *Space and time: Which diachronies, which synchronies, which scales? Typology vs. Technology. X.V. Congress of the U.I.S.P.P., Lisboa, Session C64 and C65* (BAR British Series, Vol. 1831, pp. 189–196). Oxford: Archaeopress.

Valentin, B., Fosse, G., & Billard, C. (2004). Aspects et rythmes de l'azilianisation dans le Bassin parisien. Caractérisation de l'industrie lithique recueillie au Cornet (locus 33) à Ambenay, Eure. *Gallia Préhistore, 46*, 171–209.

Vallois, H. V. (1961). The social life of early man: The evidence of skeletons. In A. L. Washburn (Ed.), *Social life of early man* (pp. 214–235). Chicago: Aldine.

Valoch, K. (1953). Paleolitické sidlište u Ochozke jeskyně v Moravském kras. *Čas. Moravského musea, sc. soc., 38*, 11–26.

Valoch, K. (1957). Paleolitické osidleni Žitného jeskyně. In J. Dvořák (Ed.), *Komplexni výzkum Žitného jeskyně v Moravskem krasu* (pp. 573–600). Praha: CSVA.

Valoch, K. (1960). *Magdalénien na Moravě. Das Magdalénien in Mähren*. Brno: Krajské Nakladatelství.

Valoch, K. (1963). Borky I, eine Freilandsiedlung des Magdalénien in Brno Maloměřice. *Čas. Moravského musea sc. soc., 48*, 5–30.

Valoch, K. (1965). Paleolitické nálezy z Rytířské jeskyně v Moravském krasu. *Anthropozoikum N.S., 3*, 141–155.

Valoch, K. (1974). Eine datierte Feuerstelle des Magdalénien in der Kůlna-Höhle bei Sloup im Mährischen Karst. *Anthropozoikum, 10*, 111–130.

Valoch, K. (1975). Paleolitická stanice v Koněvově ulici c Brně. *Archeologické rozhledy, 27*, 3–17.

Valoch, K. (1980). La fin de temps glaciaires en Moravie (Tchécoslovaquie). *L'Anthropologie, 84*(3), 380–390.

Valoch, K. (1988). *Die Erforschung der Kůlna-Höhle 1961–1976*. Brno: Moravské Muzeum – Anthropos Institut.

Valoch, K. (1996). *Le Paléolithique en Tchéquie et en Slovaquie*. Grenoble: Jérome Millon.

Valoch, K. (2001). Das Magdalénien in Mähren. 130 Jahre Forschung. *Jahrbuch des Römisch-Germanischen Zentralmuseums, 48*(1), 103–159.

Valoch, K. (2009). Magdalénien na Moravě – Po padesáti letech. *Acta Musei Moraviae, Scientiae Sociale, 94*, 3–37.

Valoch, K., Musil, R., & Jelínek, J. (1965). *Jeskyně Šipka a Čertova Díra u Štramberku – Die Höhlen Šipka und Čertova Díra bei Štramberk*. Brno: Moravské Museum.

Valoch, K., & Neruda, P. (2005). *On the chronology of the Moravian Magdalenian* (pp. 459–476). LVII: Archeologické rozhledy.

van Benden, P. J. (1864). Sur les fouilles faites dans le Trou des Nutons près de Furfooz. *Bulletin de l'Académie royale de Belgique, séries 2, 18*, 30–32, 228.

van Der Hammen, T. (1949a). De Allerød-oscillatie in Nederland. Pollenanalytisch onderzoek van een laatglaciale meerafzetting in Drente I. *Koninklijke Nederlandsche Akademie van Wetenschappen Proceedings, 52*, 69–75.

van Der Hammen, T. (1949b). De Allerød-oscillatie in Nederland. Pollenanalytisch onderzoek van een laatglaciale meerafzetting in Drente II. *Koninklijke Nederlandsche Akademie van Wetenschappen Proceedings, 52*, 169–176.

van Geel, B., Coope, G. R., & Van Der Hammen, T. (1989). Palaeoecology and stratigraphy of the lateglacial type section at Usselo (the Netherlands). *Review of Palaeobotany and Palynology, 60*(1–2), 25–129.

van Noort, G. (2007). De migratie van jagers/verzamelaars van de Hamburgcultuur in de Noord-Europese laagvlakte (13.000–11.000BP). *APAN/Extern, 12*, 50–70.

van Reybrouck, D. (2000). Beyond ethnoarchaeology? A critical history on the role of ethnographic analogy in contextual and post-processual archaeology. In A. Gramsch (Ed.), *Vergleich als archäologische Methode. Analogien in den Archäologien* (BAR International Series, Vol. 825, pp. 39–51). Oxford: Archaeopress.

Veil, S. (1996). Le Paléloithique supérieur et final en Allemagne au nord du Main. In M. Otte (Ed.), *Le Paléolithique supérieur européen. UISPP, Congrès de Forli, Commission Paléolithique Supérieuere, Bilan 1991–1996* (ERAUL, Vol. 76, pp. 151–180). Liège: Université de Liège.

Veil, S., Böhmer, M., & Breest, K. (1998). s.v. Bonstorf. *Nachrichten aus Niedersachsens Urgeschichte, Beiheft, 1*, 5–6.

Veil, S., & Hiller, R. (1998). Gadenstedt FStNr. 121, Gde. Lahnstedt, Landkreis Peine, Reg. Bez. BS. *Nachrichten aus Niedersachsens Urgeschichte, Beiheft, 1*, 6–8.

Venables, W. N., & Ripley, B. D. (2002). *Modern applied statistics with S*. New York: Springer.

Vencl, S. (1962). Příspěvek k poznání magdalénského osídlení Čech. *Archeologické rozhledy, 14*, 517–541.

Vencl, S. (1964a). Magdalénské nálezy z Náchoda. *Archeologické rozhledy, 16*, 161–164.

Vencl, S. (1964b). Pokus o klasifikaci pozdně glaciálních a staroholocenních osídlení okolí Řežabince. *Památky archeologické, 55*, 233–245.

Vencl, S. (1970). Das Spätpaläolithikum in Böhmen. *Anthropologie, 8*(1), 3–68.

Vencl, S. (1979). Le Paléolithique tardif en Tchécoslovaquie. In D. de Sonneville-Bordes (Ed.), *La fin de temps glaciaires en Europe. Chronostratigraphie et écologie des cultures du Paléolithique final* (pp. 847–857). Paris: Éditions CNRS.

Vencl, S. (1991). Bemerkungen zum Magdalénien in Böhmen. *Anthropologie, 29*(1), 85–93.

Vencl, S. (1992). Magdalenian finds from Keblice. *District of Litoměřice Archeologické rozhledy, 44*, 294–297.

Vencl, S. (1995). *Hostim. Magdalenian in Bohemia*. Prague: Institute of Archaeology.

Vencl, S., Michálek, J., & Fröhlich, J. (2006). Soupis pramenů předneolitického osídlení jižních Čech. In S. Vencel (Ed.), *Nejstarší osídlení jižních Čech. Paleolit a Mesolit* (pp. 17–324). Praha: Archeologický ústav Akademie věd ČR.

Vermeersch, P. M., Lauwers, R., & Van Peer, P. (1985). Un site magdalénien à Kanne (Limbourg). *Archaeologia Belgica, 1*, 17–54.

Vermeersch, P. M., & Symens, N. (1988). Le magdalénien de plein air en Belgique. In M. Otte (Ed.), *De la Loire à l'Oder. Les civilisations du Paléolithique final dans le nord-ouest européen. Actes du Colloque de Liège décembre 1985* (BAR International Series, Vol. 444, pp. 243–258). Oxford: Archaeopress.

Vermeersch, P. M., Symens, N., Vynckier, P., Gijselings, G., & Lauwers, R. (1987). Orp, site Magdalénien de plain air (comm. de Orp-Jauche). *Archaeologia Belgica, III*, 7–56.

Verpoorte, A. (2004). Eastern central Europe during the Pleniglacial. *Antiquity, 300*, 257–266.

Verpoorte, A., & Šída, P. (2009). The Magdalenian colonisation of Bohemia (Czech Republic). *Archäologisches Korrespondenzblatt, 39*, 325–332.

Vértes, L. (1965). Das Jungpaläolithikum von Arka in Nord-Ungarn. *Quartär, 15*(16), 79–132.

Vignard, É. (1965). Le Badegoulien. *Bulletin Société Préhistorique Française, 62*(8), CCLXII–CCLXIII.

Villaverde Bonilla, V. (1981). El magdaleniense de la Cova de les Cendres (Teulada, Alicante) y su magdaleniensemediterráno peninsular. *Saguntum, 16*, 9–35.

Virchow, R. (1874). Einleitendes Vorwort. In J. Lubbock (Ed.), *Die vorgeschichtliche Zeit erläutert durch die Ueberreste des Alterthums und die Sitten und Gebräuche der jetzigen Wilden* (pp. V–VIII). Jena: Herman Gostenoble.

Vita-Finzi, C., & Higgs, E. S. (1970). Prehistoric economy in the Mount Carmel area of Palestine: Site catchment analysis. *Proceedings of the Prehistoric Society, 36*, 1–37.

Vogel, J. C., & Waterbolk, H. T. (1964). Groningen radiocarbon dates V. *Radiocarbon, 6*, 349–369.

Vogel, J. C., & Waterbolk, H. T. (1967). Groningen radiocarbon dates VII. *Radiocarbon, 9*, 107–155.

Völker, D. (2005). *Jungpaläolithische Funde aus Jena von der "Hohen Saale" am Fuße des Jenzig*. Bachelor thesis, University of Jena.

Völzing, O. (1938). Die Grabung 1937 am Hohlenstein im Lonetal, Markung Asselfingen, Kr. Ulm. *Fundberichte aus Schwaben, N.F., 9*, 1–7.

Vuillemey, M. (1963). *Grotte de La Balme (Cuiseaux, 71): années 1962–1963*. Rapport de fouilles.

Vuillemey, M. (1964). *La grotte périgordienne de La Balme, Cuiseaux (71)*. Rapport de fouilles.

Vuillemey, M. (1989). La grotte de l'Abbé Dumont à Chaumois-Boivin. In M.-J. Roulière-Lambert (Ed.), *Aux origines de Lons et sa région de la Préhistore aux Mérovingiens, Catalogue d'exposition* (pp. 33–34). Lons-le-Saunier: Cercle Giradot.

Wagner, E. (1979). *Eiszeitjäger im Blaubeurener Tal*. Stuttgart: Theiss.

Wagner, E. (1983). Das Eiszeitjägerlager am Napoleonskopf im Katzenbachtal (Rottenburg-Weiler). In West-, Nordwest-, und Süddeutscher Verband für Altertumsforschung (Ed.), *Führer zu archäologischen Denkmälern in Deutschland. Tübingen und das Obere Gäu* (pp. 210–216). Stuttgart: Theiss.

Walker, M. J. C., Bohncke, S.-J. P., Coope, C. R., O'Connell, M., Usinger, H., & Verbruggen, C. (1994). The Devensian/Weichselian late-glacial in northwest Europe (Ireland, Britain, north Belgium, The Netherlands, northwest Germany). *Journal of Quaternary Science, 9*, 109–118.

Wall, E. (1961). Der Federsee von der Eiszeit bis zur Gegenwart. In W. Zimmermann (Ed.), *Der Federsee* (pp. 228–315). Stuttgart: Schwäbischer Albverein.

Walter, H. (1943). *Die Vegetation Osteuropas unter Berücksichtigung von Klima, Boden und wirtschaftlicher Nutzung*. Berlin: Parey.

Walter, R. (1995). *Geologie von Mitteleuropa*. Stuttgart: Schweizerbart'sche Verlagsbuchhandlung.

Wankel, H. (1881). Prähistorische Funde in der Pekárna-Höhle in Mähren. *Mitteilungen der Anthropologischen Gesellschaft in Wien, 10*, 347–348.

Wanner, G. (1900). Frühgeschichtliche Altertümer des Kantons Schaffhausen. *Beiträge zur vaterländischen Geschichte, 7*, 18.

Washburn, A. L. (1979). *Geocryology: A survey of periglacial processes and environments*. London: Arnold.

Watts, D. J., & Strogatz, S. H. (1998). Collective dynamics of "small-world" networks. *Nature, 393*, 440–442.

Weaver, A. J., Saenko, O. A., Clark, P. U., & Mitrovica, J. X. (2003). Meltwater pulse 1A from Antarctica as a trigger of the Bølling-Allerød warm phase. *Science, 299*, 1709–1713.

Weber, M.-J. (2006). Typologische und technologische Aspekte des Fundplatzes Le Tureau des Gardes 7 (Seine-et-Marne, Frankreich). Ein Beitrag zur Erforschung des Magdaléniens im Pariser Becken. *Archäologisches Korrespondenzblatt, 36*(2), 159–178.

Wegewitz, W. (1941). Ein Fundplatz der Hamburger Stufe in Ketzendorf, Kr. Harburg. *Harburger Jahrbuch, 2*, 227–250.

Wegewitz, W. (1950). *Harburger Heimat. Die Landschaft um Hamburg-Haburg*. Hamburg: Verlag der Gesellschaft der Freunde des vaterländischen Schul- und Erziehungswesens e.V.

Weihs, C., Ligges, U., Luebke, K., & Raabe, N. (2005). klaR, Analyzing german business cycles. In D. Baier, R. Decker, & L. Schmidt-Thieme (Eds.), *Data analysis and decision support* (pp. 335–343). Berlin: Springer.

Weiner, J. (1989). Zur Steingerätetechnologie bei Jäger- und Sammlerkulturen. In Stadtmuseum Ingolstadt (Ed.), *Steinzeitliche Kulturen an Donau und Altmühl. Begleitheft zur Ausstellung im Stadtmuseum Ingolstadt* (pp. 199–217). Ingolstadt: Courier Druckhaus.

Weißmüller, W. (1995). Eine ungewöhnlich reiche Fundstelle des Jungpaläolithikums bei Vilshofen, Lkr. Passau, Ndb. In *Ausgrabungen und Funde in Altbayern 1992–1994*, Straubing.

Weißmüller, W. (2002). Alt- und Mittelsteinzeit am bayerischen Donaulauf zwischen Lech und Inn – Ein Überblick. In K. Schmotz (Ed.), *Vorträge des 20. Niederbayerischen Archäologentages* (pp. 165–201). Rahden/Westf.: Leidorf.

Weniger, G.-C. (1982). *Wildbeuter und ihre Umwelt. Ein Beitrag zum Magdalénien Südwestdeutschlands aus ökologischer und ethnoarchäologischer Sicht*. Tübingen: Archaeologica Venatoria.

Weniger, G.-C. (1989). The Magdalénien in western central Europe: Settlement pattern and regionality. *Journal of World Prehistory, 3*(3), 323–372.

Weniger, G. C. (1995). *Widerhakenspitzen des Magdalénien Westeuropas. Ein Vergleich mit ethnohistorischen Jägergruppen Nordamerikas*. Mainz: von Zabern.

Weninger, B., & Jöris, O. (2008). A ^{14}C age calibration curve for the last 60 ka: The Greenland-Hulu U/Th timescale and its impact on understanding the middle to upper Paleolithic transition in Western Eurasia. *Journal of Human Evolution, 55*(5), 772–781.

Weninger, B., Jöris, O., & Danzeglocke, U. (2012). *CalPal-2007*. Cologne Radiocarbon Calibration and Palaeclimate Research Package. http://www.calpal.de/

Wetzel, O. (1933). Die in organischer Substanz erhaltenen Mikrofossilien des baltischen Kreidefeuersteins. *Palaeontographica, 77*, 141–186.

Wetzel, R. (1935). Die Bocksteinschmiede im Lonetal, Markung Rammingen. *Fundberichte aus Schwaben, N.F, 8*, 16–20.

Wetzel, R. (1954a). Das Törle an der alten Bocksteinhöhle. *Mitteilungen des Vereins für Naturwissenschaft und Mathematik in Ulm (Donau), 24*, 3–19.

Wetzel, R. (1954b). Quatärforschung im Lonetal. *Eiszeitalter und Gegenwart, 4*(5), 106–141.

Wetzel, R. (1958). *Die Bocksteinschmiede mit dem Bocksteinloch, der Brandplatte und dem Abhang sowie der Bocksteingrotte*. Stuttgart: Kohlhammer.

Wetzel, R., & Bosinski, G. (1969). *Die Bocksteinschmiede im Lonetal (Markung Rammingen, Kreis Ulm)*. Stuttgart: Müller and Gräf.

White, R. (1985). *Upper Paleolithic land use in the Périgord. A topographic approach to subsistence and settlement* (BAR International Series, Vol. 253). Oxford: Archaeopress.

Wichmann, H. E., & Bayer, J. (1924). Die "Frauenlucken" bei Schmerbach im oberen Kamptale, eine Höhlenstation des Magdalénien in Nieder-Österreich. *Die Eiszeit*, 65–67.

Wiermann, R. R. (2000). Hinter'm Horizont geht's weiter… Vom Nutzen der ethnographischen Analogie für die Urgeschichtsforschung. In A. Gramsch (Ed.), *Vergleichen als archäologische Methode Analogien in den Archäologien* (BAR International Series, Vol. 825, pp. 113–118). Oxford: Archaeopress.

Wiessner, P. (1983). Style and social information in Kalahari-San projectile points. *American Antiquity, 48*, 253–276.

Wiessner, P. (1984). Reconsidering the behavioural basis for style: A case study among the Kalahari San. *Journal of Anthropological Archaeology, 3*, 190–234.

Williams, C. C. (2009). *Meltwater and abrupt climate change during the last deglaciation: A Gulf of Mexico perspective*. Master thesis, University of South Florida.

Willis, K. J., Rudner, E., & Sümegi, P. (2000). The full-glacial forests of central and southeastern Europe. *Quaternary Research, 53*, 203–213.

Willis, K. J., & Van Andel, T. H. (2004). Trees or no trees? The environments of central and eastern Europe during the last glaciation. *Quaternary Science Reviews, 23*, 2369–2387.

Wiśniewski, T. (2003). *Kultura magdaleńska na ziemiach polskich. Stan badań nad inwentarzami tzw. "magdaleńskimi"*. Master thesis, University of Toruń.

Wobst, H. M. (1976). Locational relationships in Paleolithic society. *Journal of Human Evolution, 5*, 49–58.

Wobst, H. M. (1978). The archaeo-ethnology of hunter-gatherers or the tyranny of the ethnographic record in archaeology. *American Antiquity, 43*(2), 303–309.

Worm, H. (1980). Zwei Anhänger aus der höhle im Freudenthal, Kt. Schaffhausen. *Archäologisches Korrespondenzblatt, 10*(1), 9–11.

Wotzka, H. P. (1993). Zum traditionellen Kulturbegriff in der prähistorischen Archäologie. *Paideuma, 39*, 25–43.

Wouters, A. M. (1982). Magdalénien in Echt (L)? Site M 3. *Archaeologische Berichten, 11*(12), 114–121.

Wouters, A. M. (1983). Uit de oude doos. Magdalénien uit het Peelgebied. *Archaeologische Berichten, 14*, 99–108.

Wüst, K. (1998). Die gravierten Gerölle und Plättchen des Magdalénien-Fundplatzes "Teufelsbrücke bei Saalfeld". *Alt-Thüringen, 32*, 98–142.

Wylie, A. (1985). The reaction against analogy. In M. B. Schiffer (Ed.), *Advances in archaeological method and theorie* (pp. 63–111). New York: Academic Press.

Wyss, R. (1953). *Beiträge zur Typologie der paläolithisch-mesolithischen Übergangsformen im Schweizerischen Mittelland. Schriftenreihe des Institutes für Ur- und Frühgeschichte der Schweiz, Basel* (Vol. 9). Basel: Institutes für Ur- und Frühgeschichte.

Zahn, U. (1996). Mitteleuropa Geologie. In U. Zahn (Ed.), *Diercke Weltatlas* (pp. 74–75). Braunschweig: Westermann Schulbuchverlag.

Zarzycki, K. (1979). Outline of the ecology. In S. Białobok (Ed.), *Brzozy. Betula L. Nasze drzewa leśne* (pp. 265–291). Warszawa: Państwowe Wydawnictwo Naukowe.

Žebera, K. (1955). Die Magdaléniensiedlung mit Wohngruben unter freiem Himmel und mit einer reichen Feuersteinindustrie in Kvíc bei Salný. *Anthropozoikum, 4*(1954), 289–290.

Žebera, K. (1958). *Československo ve starší době kamenné*. Praha: Naklad. Československé Akad. Věd.

Žebera, K. (1966). Exploitation préhistorique du quarzite près de Bečov dan le "České stčředohoří" (Bohême). In J. Filip (Ed.), *Investigations archéologiques en Tchécoslovaquie* (pp. 27–28). Prague: Academia.

Zeitz, B. (1969). *Paläolithische und mesolithische Fundplätze aus dem Kreis Gifhorn*. Hildesheim: Lax.

Ziesaire, P., & Thill, G. (1985). Junpaläolithische Funde von Kehlen-Juckelsboesch. *Bulletin Société Préhistorique Luxembourgeoise, 7*, 45–56.

Zoller, D. (1962). Die Ergebnisse der Grabung auf der Querenstedener Düne bis Juni 1962. *Nachrichten aus Niedersachsens Urgeschichte, 31*, 189–192.

Zoller, D. (1963). Vorläufiger Bericht über eine Rentierstation der hamburger Stufe bei Querenstede, Kreis Ammerland. *Die Kunde, 14*, 17–25.

Zotz, L. F. (1925). Eine neuentdeckte Station des Menschen der älteren Steinzeit in Baden. *Mitteilungen des Badischen Landesvereins für Naturkunde und Naturschutz in Freiburg im Breisgau, 1*, 474–478.

Zotz, L. F. (1955). *Das Paläolithikum in den Weinberghöhlen bei Mauern*. Bonn: Ludwig Röhrscheid.

Zotz, L. F. (1961). Kleine Mitteilungen. Die Forschungen des Instituts für Urgeschichte der Universität Erlangen im Altmühltal. *Prähistorische Zeitschrift, 39*, 266–293.

Zotz, L. F., Stehlin, H. G., & Neuweiler, E. (1928). Die paläolithische Besiedlung der Teufelsküchen am Oelberg beim Kuckusbad. *Prähistorische Zeitschrift, 19*(1–2), 1–53.

Zuberbühler Koch, K. (2002). Die magdalénienzeitliche Höhlenstation Winznau/Käsloch im Kanton Solothurn. *Archäologie und Denkmalpflege im Kanton Solothurn, 7*, 7–49.

Züchner, C. (1979). Alt- und Mittelsteinzeit im Ries. In D. Baatz (Ed.), *Führer zu Vor- und Frühgeschichtlichen Denkmälern 40: Nördlingen – Bopfingen – Oettingen – Harburg* (pp. 20–57). Mainz: Philipp von Zabern.

Züchner, C. (2008). Die Klausenhöhlen bei Neuessing, Ldkr. Kelheim. In H. Obermaier-Gesellschaft (Ed.), *Jahrestagung in Erlangen* (Vol. 50, pp. 48–49). Erlangen: PrintCom.

Zwahlen, R. (1985). Lörmoos – Late-glacial and Holocene environments of an ancient lake on the central Swiss Plateau. In G. Lang (Ed.), *Swiss lake and mire environments during the last 15000 years* (pp. 171–184). Vaduz: Lubrecht and Cramer.

Index

Andreas Maier, *The Central European Magdalenian*, Vertebrate Paleobiology and Paleoanthropology,
DOI 10.1007/978-94-017-7206-8

Appendix

Table A.1 Recorded sites ordered by ID

ID	Site; Assemblage	Long	Lat	Co	G	C	References
1	Campalou; CA 3a	5.2414	45.0625	F	1	1	Crémilleux (1977), Monjuvent (1994), Brochier and Livache (1997), Oberlin and Pion (2009)
2	Campalou; CA 2	5.2414	45.0625	F	1	3	
3	Tai, G. du; C′1	5.2422	45.0603	F	1	4	Monjuvent (1994), Brochier and Livache (1997), Oberlin and Pion (2009)
4	Tai, G. du; C″1	5.2422	45.0603	F	1	3	
5	Beptenaz, G. de	5.2506	45.7242	F	1	2	Chantre (1901), Desbrosse (1976a), Monjuvent (1994)
6	La Garenne, Vénérieu	5.2761	45.6667	F	1	2	Monjuvent (1994)
7	Brotel, G. de	5.3108	45.7833	F	1	5	Chantre (1901), Bocquet (1969), Desbrosse (1976a), Monjuvent (1994)
8	Croze-sur-Suran; R7	5.3281	46.1019	F	1	1	Tournier and Costa De Beauregard (1922), Desbrosse (1976a, 1980), Monjuvent (1994), D'Errico and Laroulandie (1998), Pion (2000a)
9	Calvaire; IV inf.	5.3367	45.1086	F	1	1	Bourdier and De Lumley (1956), Bocquet (1969), Desbrosse (1976a), Monjuvent (1994), Oberlin and Pion (2009)
10	Calvaire; III moy.	5.3367	45.1086	F	1	1	
11	Calvaire; II sup.	5.3367	45.1086	F	1	1	
12	Balme d'Isère; hearth A	5.3381	45.8511	F	1	1	Chantre (1866), Jacquement (1895), Bourdier and De Lumley (1956), Desbrosse (1976a), Monjuvent (1994)
13	Tessonière, G. de la	5.3658	46.1914	F	1	2	Desbrosse (1976a, 1980), Monjuvent (1994)
14	Colombière, La; B	5.3711	46.0814	F	1	1	Mayet and Pissot (1915), Movius and Judson (1956), Desbrosse (1976a), Sieveking (1979), Monjuvent (1994), Pion (2000a), Pozzi (2004)
15	Balme de Cuiseaux	5.3892	46.4939	F	1	4	Vuillemey (1963, 1964), David (1993, 1996), Monjuvent (1994), Cupillard (2008)
16	Balme-Rousse D2a	5.3936	45.0667	F	1	2	Bintz (1985), Monjuvent (1994), Delannoy et al. (1997)
17	Grand'Baile, G. de la	5.4067	46.0847	F	1	5	Desbrosse (1976a), Monjuvent (1994), Pion (2000a)
18	Gay, A.	5.4069	46.0847	F	1	2	Desbrosse (1976a), Combier (1980), Monjuvent (1994), Pion (2000a, b)
19	Bobache, A. 4+3	5.4153	44.9672	F	1	2	Müller (1912), Bourdier and de Lumley (1956), Bocquet (1969), Monjuvent (1994), Bintz (1995)
20	Bobache, A. 1+2	5.4153	44.9672	F	1	4	
21	Raillarde, La	5.4169	45.8725	F	1	1	Desbrosse and Girard (1974), Desbrosse (1976a), Monjuvent (1994), Pion (2000a)
22	Le Cimetière	5.4203	44.9678	F	1	5	Monjuvent (1994)
23	Sous-la-Roche	5.4300	46.3594	F	1	5	Desbrosse (1976b), David and Richard (1989)
24	Freydières, G. des	5.4308	44.9361	F	1	3	Bocquet (1965, 1969), Bocquet and Lequatre (1973), Bouchud and Desbrosse (1973), Monjuvent (1994)
25	Gigny, G.; IV sup.	5.4603	46.4519	F	1	2	Campy et al. (1989), Cohën et al. (1991), David (1996), Cupillard (2008)
26	Gigny, G.; IV inf.	5.4603	46.4519	F	1	2	

(continued)

Andreas Maier, *The Central European Magdalenian*, Vertebrate Paleobiology and Paleoanthropology,
DOI 10.1007/978-94-017-7206-8

Table A.1 (continued)

ID	Site; Assemblage	Long	Lat	Co	G	C	References
27	Passagère	5.5264	45.1211	F	1	1	Monin (2000)
28	Grappin, Arlay	5.5269	46.7628	F	1	1	Allain et al. (1985), David (1993, 1996), Schroeder (2000), Cupillard and Welte (2006), Cupillard (2008)
29	Colomb	5.5275	46.6267	F	1	1	Bocquet (1969), Desbrosse (1976a)
30	Chênelaz 2C/2	5.5333	45.8850	F	1	1	Onoratini (1996), Cartonnet and Naton (2000)
31	Chaze 2	5.5400	46.7644	F	1	2	Cupillard and Welte (2006), Cupillard (2008)
32	Val Molière	5.5464	45.0744	F	1	5	Monjuvent (1994)
33	Hoteaux, Les	5.5850	45.8378	F	1	2	Tournier and Guillon (1895), Tournier (1924), Desbrosse (1976a), Monjuvent (1994), David (1996), Bridault et al. (2000), Pion (2000a)
34	Bonne Femme	5.6197	45.6486	F	1	2	Tournier (1911), Combier and Desbrosse (1964), Desbrosse (1976a, 1980), Monjuvent (1994)
35	Olette, Gotte d l'	5.6244	45.1908	F	1	2	Bourdier and de Lumley (1956), Monjuvent (1994)
36	Fontabert	5.6247	45.3172	F	1	2	Müller (1894), Bourdier and de Lumley (1956), Monjuvent (1994)
37	Bibi, G. (Eremitage)	5.6258	45.3219	F	1	1	Bourdier and de Lumley (1956), Bocquet (1969), Desbrosse (1976a), Monjuvent (1994)
38	Rigny	5.6372	47.4717	F	1	2	Campy (1985), David (1996)
39	Pugieu, A. de	5.6483	45.8206	F	1	2	Desbrosse (1976a)
40	Henri Martin, A.	5.6483	45.8206	F	1	5	Monjuvent (1994)
41	Thoys, A. de; TH 2	5.6556	45.7461	F	1	5	Desbrosse (1976a, 1980), Monjuvent (1994)
42	Thoys, A. de; TH1	5.6556	45.7461	F	1	5	
43	Chaumois-Boivin, G. de; 2	5.6692	46.7481	F	1	2	Vuillemey (1989), David (1996), Cupillard (2008)
44	Farincourt; surf.	5.6819	47.6994	F	1	1	Mouton and Joffroy (1956), Desbrosse (1976b), David (1996)
45	Farincourt; I+II	5.6819	47.6994	F	1	1	
46	Farincourt; III-B	5.6819	47.6994	F	1	1	
47	Farincourt; III-E	5.6819	47.6994	F	1	1	
48	Balme d.Glos 10-11	5.6828	45.1853	F	1	1	Bourdier and de Lumley (1954), Bocquet (1969), Desbrosse (1976a), Monjuvent (1994)
49	Balme d.Glos 8-9	5.6828	45.1853	F	1	1	
50	Col d. l. Charmotte	5.6997	45.2619	F	1	5	Monjuvent (1994), Bintz and Roche (1995)
51	Romains; 3	5.7125	45.7150	F	1	1	Desbrosse and Mourer-Chauviré (1973), Desbrosse and Prat (1974), Haïd and Margerand (1996), Pion (2000a), Stahl Gretsch (2006)
52	Romains; 2b	5.7125	45.7150	F	1	1	
53	Romains; 2a	5.7125	45.7150	F	1	1	
54	Cabônes, Abri; 4	5.7244	47.1508	F	1	3	David (1982, 1984, 1992, 1993, 1996), D'Errico and David (1993), Cupillard (1998)
55	La Roche, G. de la	5.7389	47.2611	F	1	5	Ripotot and Seibel (1958), Thévenin (1965), Desbrosse (1976b), David (1996)
56	Seveux	5.7497	47.5561	F	1	1	David (1996), Séara (1997)
57	Châteauneuf, A. de	5.7594	47.1603	F	1	5	David (1996)
58	A. de la Fru; Aire I	5.7853	45.4467	F	1	1	Pion et al. (1990), Évin (1994), Pion (2000a), Pion and Thévenin (2007), Oberlin and Pion (2009)
59	A. de la Fru; Aire II	5.7853	45.4467	F	1	1	
60	Gerbaix, A.	5.7858	45.4486	F	1	5	Monjuvent (1994)
61	Planches, G. des	5.8136	46.8797	F	1	5	David (1996)
62	Mont-les-Etrelles	5.8303	47.4539	F	1	5	Thévenin (1965), Desbrosse (1976b), David and Richard (1989)
63	Jean Pierre I; 9A+B	5.8447	45.5006	F	1	1	Évin (1994), Monjuvent (1994), Bintz and Grunwald (1995)
64	Jean Pierre II	5.8450	45.5006	F	1	1	Évin (1994), Monjuvent (1994), Bintz and Grunwald (1995), Pion (2000a)
65	Baume Noire; M2	5.9467	47.4883	F	1	1	Collot (1947), Escola (1989), Moutel (1989), David (1993, 1996), Cupillard (2008), Nielsen (2009)
66	Baume Noire; M3	5.9467	47.4883	F	1	1	
67	Baume Noire; M4	5.9467	47.4883	F	1	1	
68	Baume Noire; M5	5.9467	47.4883	F	1	1	

(continued)

Table A.1 (continued)

ID	Site; Assemblage	Long	Lat	Co	G	C	References
69	Douattes, Les; 7/6	5.9533	46.0144	F	1	1	Jayet (1943), Sonneville-Bordes (1963), Desbrosse and Girard (1974), Pion (2000a), Chaix (2003), Oberlin and Pion (2009), Affolter (2009a)
70	La Justice	5.9708	47.9864	F	1	5	David (1996)
71	La Piquette, G. de	6.0081	47.0961	F	1	5	David (1996)
72	Nahin, G. de	6.0386	47.0908	F	1	5	David (1996)
73	Banges; sup.+inf.	6.0803	45.7517	F	1	1	Pion (2000a), Affolter (2009b)
74	La Guillotine	6.0881	47.6228	F	1	1	Thévenin (1965, 1997), David (1996)
75	Faverney	6.1022	47.7661	F	1	5	David (1996)
76	Stat. d. Grenouilles	6.1706	46.1694	F	1	5	Le Tensorer (1998), Stahl Gretsch (2006)
77	Rigney 1, G. de; D	6.1814	47.3872	F	1	1	Glory (1961), Desbrosse (1976b), David (1996), Cupillard (2008), Nielsen (2009)
78	Rigney 1, G. de; M	6.1814	47.3872	F	1	1	
79	Veyrier	6.1889	45.8753	F	1	1	Jayet (1937, 1943), Sonneville-Bordes (1963), Feustel (1974), Pion (2000a), Chaix (2003), Stahl Gretsch (2006)
80	Étrembières, (G. du Four, l'Hôpital)	6.2269	46.1353	F	1	2	Jayet (1943), Monjuvent (1994), Stahl Gretsch (2006)
81	Bains-les-Bains	6.2642	48.0017	F	1	3	Hans and Thévenin (1993), David (1996)
82	Auzary-Thônes	6.3300	45.8689	F	1	1	Perpère (1981)
83	Freymond, A.; 5 inf.	6.4092	46.6467	CH	1	5	Pignat and Winiger (1998), Nielsen (2009), Oberlin and Pion (2009)
84	Cure, A. de la	6.5178	46.7925	CH	1	4	Crotti (1993), Nielsen (2009)
85	Bavans; 9	6.7294	47.4817	F	1	4	David (1996), Nielsen (2009)
86	Rochedane; D1/D2	6.7586	47.3733	F	1	2	Thévenin (1968, 1982), David (1996), Cupillard (2008), Oberlin and Pion (2009)
87	Châtel-Saint-Denis	6.8997	46.5422	CH	1	5	Nielsen (2002)
88	Scé du Châterlard, G.	6.9344	46.3972	CH	1	2	Saussure (1870), Bandi (1947), Sonneville-Bordes (1963), Le Tensorer (1998)
89	N.-Monruz; 2M/A	6.9589	47.0036	CH	1	1	Affolter et al. (1994), Leesch et al. (2004), Bullinger et al. (2006)
90	N.-Monruz, 1M	6.9589	47.0036	CH	1	1	
91	N.-Monruz; 1A	6.9589	47.0036	CH	1	2	
92	Dentaux, A. le	6.9639	46.4231	CH	1	5	Jacot-Guillarmod (1921), Jayet and Constantin (1945), Bandi (1947), Sonneville-Bordes (1963)
93	H.-Champré.; 1 pri.	6.9711	47.0067	CH	1	1	Leesch (1997), Leesch et al. (2004)
94	H.-Champré.; 1 inf.	6.9711	47.0067	CH	1	1	
95	H.-Champré.; 2	6.9711	47.0067	CH	1	1	
96	H.-Champré.; 3	6.9711	47.0067	CH	1	1	
97	H.-Champré; 1A	6.9711	47.0067	CH	1	2	
98	Saint-Brais II	7.1125	47.3053	CH	1	5	Bandi (1947), Le Tensorer (1998)
99	Alle, Noir Bois	7.1192	47.4178	CH	1	1	Aubry et al. (2000)
100	Liesberg	7.4364	47.3936	CH	1	1	Sarasin (1918), Bandi (1947), Sonneville-Bordes (1963), Le Tensorer (1998)
101	Moosseedorf-Moosbühl	7.4903	47.0136	CH	1	1	Feustel (1974), Bullinger (1996), Bullinger et al. (1997), Nielsen (2009)
102	Hardberggrotten	7.5303	47.6622	DE	1	4	Mieg (1901, 1910), Mieg and Stehlin (1903), Meier-Riva (1987)
103	Thierstein	7.5375	47.3847	CH	1	5	Bandi (1947), Sonneville-Bordes (1963), Le Tensorer (1998)
104	Büttenloch; A	7.5389	47.4767	CH	1	2	Sarasin and Stehlin (1924), Sedlmeier (1989), Schibler and Sedlmeier (1993), Le Tensorer (1998)
105	Büttenloch; B	7.5389	47.4767	CH	1	2	
106	Chesselgraben	7.5417	47.3728	CH	1	2	Spycher and Sedlmeier (1985), Sedlmeier (1988), Le Tensorer (1998), Rhezak (2000), Nielsen (2009)
107	Brügglihöhle	7.5617	47.4472	CH	1	1	Bandi et al. (1954), Sonneville-Bordes (1963), Rozoy (1978), Le Tensorer (1998, 1999), Nielsen (2009)

(continued)

Table A.1 (continued)

ID	Site; Assemblage	Long	Lat	Co	G	C	References
108	Kohlerhöhle	7.5678	47.4328	CH	1	2	Bandi (1947), Lüdin (1963), Sonneville-Bordes (1963), Sedlmeier (1988), Sedlmeier (1993), Le Tensorer (1998)
109	Kastelhöhle-Nord; middle	7.5692	47.4294	CH	1	1	Schweizer et al. (1959), Le Tensorer (1993, 1998), Eriksen (2002), Terberger and Street (2002), Chaix (2003), Nielsen (2009)
110	Kastelh.-N.; upper	7.5692	47.4294	CH	1	1	
111	Heidenküche	7.5694	47.4347	CH	1	1	Thiessing (1886), Sarasin (1918), Bandi (1947), Sonneville-Bordes (1963)
112	Kleine Ganghöhle	7.5978	47.4219	CH	1	5	Sedlmeier (1990)
113	Bruderholz	7.6056	47.5303	CH	1	1	Greppin (1875), Bandi (1947), Sonneville-Bordes (1963), Le Tensorer (1998)
114	Birseck-Ermitage	7.6292	47.4908	CH	1	1	Sarasin (1918), Bandi (1947), Feustel (1974), Sedlmeier (1988), Le Tensorer (1998), Höck (2000), Nielsen (2009)
115	Hollenberg-Höhle 3; D	7.6303	47.4892	CH	1	1	Bay (1953), Sonneville-Bordes (1963), Sedlmeier (1982), Müller and Leesch (2011)
116	Rüteliacher	7.6322	47.1422	CH	1	5	Burg (1994)
117	Rütihard	7.6336	47.5208	CH	1	5	Nielsen (2009)
118	Aeschi, Moosmatt.	7.6617	47.1792	CH	1	5	Nielsen (2009)
119	Bolken, Inkwilersee	7.6631	47.1911	CH	1	1	Le Tensorer (1999), Nielsen (2002)
120	Fürsteiner	7.6647	47.1547	CH	1	2	Wyss (1953), Burg (1994), Nielsen (2009)
121	Burgäschi, H. Burg	7.6714	47.1731	CH	1	2	Wyss (1953), Burg (1994), Nielsen (1999, 2009)
122	Munzingen	7.6922	47.9664	DE	1	1	Schotensack (1903, 1907), Schmidt (1912), Padtberg (1925), Andrée (1939), Guenther (1968), Weniger (1982), Pasda (1994, 1998), Eriksen (2002)
123	Munzingen; GH4 1976–1977	7.6922	47.9664	DE	1	1	
124	Rislisberghöhle	7.6981	47.2958	CH	1	2	Barr (1977), Stampfli (1983), Sedlmeier (1988), Le Tensorer (1998), Chaix (2003), Nielsen (2009)
125	Lausen	7.7592	47.4714	CH	1	5	Leuthardt (1911, 1916), Bandi (1947), Le Tensorer (1998)
126	Teufelsküche	7.7692	47.9194	DE	1	1	Zotz (1925), Zotz et al. (1928), Weniger (1982), Pasda (1994)
127	Steinberg	7.7744	47.9272	DE	1	2	Mauser (1974), Pasda (1994)
128	Rheinfelden-Eremi.	7.8003	47.5444	CH	1	1	Sedlmeier (1989), Le Tensorer (1993)
129	Uff Wigg	7.8728	47.5419	CH	1	5	Le Tensorer (1998)
130	Trimbach	7.8928	47.3603	CH	1	2	Bandi (1947), Sonneville-Bordes (1963), Le Tensorer (1998)
131	Sälihöhle-Oben	7.9133	47.3392	CH	1	1	Reverdin (1924), Schweizer (1937), Bandi (1947), Sonneville-Bordes (1963), Le Tensorer (1998), Zuberbühler Koch (2002)
132	Hard I	7.9142	47.3544	CH	1	1	Schweizer (1937), Bandi (1947), Sonneville-Bordes (1963), Zuberbühler Koch (2002), Nielsen (2009)
133	Hard II	7.9156	47.3558	CH	1	1	Bandi (1947), Sonneville-Bordes (1963), Zuberbühler Koch (2002)
134	Wilmatt	7.9172	47.3739	CH	1	1	Bandi (1947), Sonneville-Bordes (1963), Le Tensorer (1998), Zuberbühler Koch (2002), Nielsen (2009)
135	Winznau-Köpfli	7.9211	47.3694	CH	1	2	Schweizer (1937), Bandi (1947), Sonneville-Bordes (1963), Le Tensorer (1998), Nielsen (1999)
136	Käsloch	7.9236	47.3669	CH	1	1	Bally et al. (1908), Bandi (1947), Sonneville-Bordes (1963), Zuberbühler Koch (2002), Nielsen (2009)
137	Mühleloch, A.	7.9250	47.3406	CH	1	1	Reverdin (1924), Schweizer (1937), Bandi (1947), Sonneville-Bordes (1963), Le Tensorer (1998), Zuberbühler Koch (2002), Nielsen (2009)
138	Röthekopf	7.9364	47.5764	DE	1	5	Gersbach (1925), Andrée (1939), Eriksen (2002)
139	Reiden-Stumpen	7.9719	47.2575	CH	1	1	Nielsen (1994, 1999, 2002, 2009)
140	Egolzwil 3	8.0117	47.1844	CH	1	5	Nielsen (2009)
141	Houenbüel, 13	8.0167	47.1664	CH	1	3	Nielsen (1992, 2009)
142	Houenbüel, 12	8.0175	47.1617	CH	1	3	Nielsen (1992, 1994, 2009)
143	Houenbüel(L2-105)	8.0228	47.1597	CH	1	5	Nielsen (1992, 1994, 2009)

(continued)

Table A.1 (continued)

ID	Site; Assemblage	Long	Lat	Co	G	C	References
144	Obermoos, 28	8.0256	47.1756	CH	1	5	Nielsen (2009)
145	Moosmatt, 11	8.0286	47.1692	CH	1	4	Nielsen (2002, 2009)
146	Hubelweid L2-52	8.0286	47.1689	CH	1	5	Nielsen (1994, 1999, 2009)
147	Sandmatt, 25	8.0325	47.1803	CH	1	2	Nielsen (2009)
148	Göldimoos 2	8.0397	47.1625	CH	1	5	Nielsen (2009)
149	Obermoos, 206	8.0397	47.1792	CH	1	5	Nielsen (2009)
150	Fischerhäusern, 1	8.0428	47.1706	CH	1	4	Nielsen (1999, 2009)
151	Moos, 14	8.0436	47.1669	CH	1	2	Nielsen (2002, 2009)
152	Seewagermoos, 17	8.0467	47.1656	CH	1	2	Nielsen (2009)
153	Göldimoos 1	8.0503	47.1736	CH	1	4	Nielsen (2009)
154	Seewagen, 24	8.0572	47.1706	CH	1	2	Nielsen (1992, 1999, 2002)
155	Moos (L2.55)	8.0794	47.1756	CH	1	5	Nielsen (2009)
156	Hitzkirch-Bald. See	8.2519	47.2225	CH	1	5	Nielsen (1999, 2002, 2009)
157	Sedel 2	8.2975	47.0661	CH	1	5	Nielsen (2009)
158	Cham-Grindel II	8.4694	47.1972	CH	1	4	Nielsen (1999, 2009)
159	Cham-Grindel I	8.4722	47.1964	CH	1	4	Nielsen (1996, 1999, 2002, 2009)
160	Cham-Grindel III	8.4728	47.1975	CH	1	4	Nielsen (1996, 2002, 2009)
161	Hinterberg I (1117)	8.4747	47.1961	CH	1	5	Nielsen (2009)
162	Untere Bsetzi	8.6822	47.7514	CH	1	5	Bandi (1947), Sonneville-Bordes (1963), Le Tensorer (1998)
163	Vorderer Eichen	8.6875	47.7503	CH	1	5	Sulzberger (1921), Bandi (1947), Sonneville-Bordes (1963), Le Tensorer (1998)
164	Einsiedeln-Langrüti	8.7892	47.1519	CH	1	1	Leuzinger-Piccand (1996), Nielsen (2002, 2009)
165	Solgen Im Grauen	8.5642	47.6119	CH	2	2	Nagy and Spörri (1998)
166	Schweizersbild	8.6375	47.7247	CH	2	1	Nüesch (1896, 1902), Häusler (1914a, b), Feustel (1974), Bosinski (1982), Eriksen (1991, 2002), Le Tensorer (1993), Höneisen and Peyer (1994), Street (2000)
167	Freudenthal-Höhle, Rosenhalde	8.6386	47.6822	CH	2	1	Karsten (1874), Wanner (1900), Bandi (1947), Bosinski (1978, 1982), Worm (1980), Eriksen (1991, 2002)
168	Kerzenstübli	8.6800	47.7500	CH	2	5	Wanner (1900), Sulzberger (1921), Bandi (1947), Sonneville-Bordes (1963)
169	Neue Höhle	8.6800	47.7500	CH	2	5	Bandi (1947)
170	Kesslerloch; II	8.6956	47.7461	CH	2	1	Bandi (1947), Sonneville-Bordes (1963), Feustel (1974), Weniger (1982), Höneisen (1986), Ammann et al. (1988), Le Tensorer (1993, 1998), Floss (1994), Eriksen (2002)
171	Kesslerloch; III	8.6956	47.7461	CH	2	1	
172	Bildstockfels	8.7706	47.8686	DE	2	2	Peters (1930), Andrée (1939), Weniger (1982)
173	Petersfels; P1AH2	8.8042	47.8608	DE	2	1	Peters (1930), Peters and Toepfer (1932), Mauser (1970), Albrecht (1974, 1979), Albrecht et al. (1983), Berke (1987, 1989), Eriksen (1991), Street (2000)
174	Peters.; P1AH3A	8.8042	47.8608	DE	2	1	
175	Peters.; P1AH3B	8.8042	47.8608	DE	2	1	
176	Peters.; P1AH4	8.8042	47.8608	DE	2	1	
177	Peters.; P2	8.8042	47.8608	DE	2	1	
178	Peters.;P3AH1	8.8042	47.8608	DE	2	1	
179	Peters.; P3AH1/2	8.8042	47.8608	DE	2	1	
180	Peters.; P3AH2	8.8042	47.8608	DE	2	1	
181	Peters.; P3AH2/3	8.8042	47.8608	DE	2	1	
182	Peters.; P3AH3	8.8042	47.8608	DE	2	1	
183	Peters.; P3AH3/4	8.8042	47.8608	DE	2	1	
184	Peters.; P3AH4	8.8042	47.8608	DE	2	1	
185	Peters.; P3AH5	8.8042	47.8608	DE	2	1	
186	Peters.; P5AH2	8.8042	47.8608	DE	2	1	
187	Peters.; P5AH4	8.8042	47.8608	DE	2	1	
188	Peters.; P5AH5	8.8042	47.8608	DE	2	1	
189	Peters.; P6AH2	8.8042	47.8608	DE	2	1	

(continued)

Table A.1 (continued)

ID	Site; Assemblage	Long	Lat	Co	G	C	References
190	Peters.; P6AH3/4	8.8042	47.8608	DE	2	1	
191	Peters.; P6AH4	8.8042	47.8608	DE	2	1	
192	Peters.; P6AH5	8.8042	47.8608	DE	2	2	
193	Peters.; P6AH6	8.8042	47.8608	DE	2	2	
194	Drexlerloch	8.8142	47.8594	DE	2	5	Albrecht and Hahn (1991)
195	Gnirshöhle 1	8.8150	47.8594	DE	2	1	Albrecht et al. (1977), Albrecht (1981a, 1983), Weniger (1982), Eriksen (1991), Floss (2006)
196	Gnirshöhle 2	8.8150	47.8594	DE	2	1	Weniger (1982), Fisher (2002), Floss (2006)
197	Napoleonskopf	8.9003	48.4575	DE	2	1	Koken (1912), Schmidt (1912), Riek (1935), Andrée (1939), Mauser (1976), Wagner (1983), Eriksen (1991, 2002)
198	Buttentalhöhle	8.9547	48.0139	DE	2	3	Peters (1936b), Andrée (1939), Roller (1954), Eriksen (1991), Hahn (1995, 2002)
199	Probstfels; I-IV	8.9564	48.0506	DE	2	2	Koken (1912), Schmidt (1912), Andrée (1939), Pasda (1990), Eriksen (1991)
200	Heidenstein	9.0239	48.2133	DE	2	5	Paret (1961), Eriksen (1991)
201	Bernlochhöhle	9.0444	48.2292	DE	2	5	Rieth (1938), Eriksen (1991), Kieselbach and Maurer (2002)
202	Hüttenkirchle	9.0914	48.2439	DE	2	5	Rieth (1938), Roller (1954), Paret (1961), Eriksen (1991)
203	Straßberger G.	9.0917	48.1767	DE	2	5	Schmidt (1912), Rieth (1938), Roller (1954), Paret (1961), Eriksen (1991), Binder and Jantschke (2003)
204	Kuhstallhöhle	9.1164	48.1792	DE	2	2	Koken (1912), Schmidt (1912), Rieth (1938), Andrée (1939), Paret (1961), Eriksen (1991)
205	Burghöhle Dietfurt	9.1392	48.0783	DE	2	2	Dämmer et al. (1974), Brunnacker et al. (1981), Weniger (1982), Eriksen (1991), Gietz (2001)
206	Zigeunerfels; H/I	9.1597	48.0872	DE	2	2	Koenigswald (1972), Taute (1972, 1975), Weniger (1982), Fahlke (2009)
207	Zigeunerfels; G	9.1597	48.0872	DE	2	4	
208	Zigeunerfels; H	9.1597	48.0872	DE	2	5	
209	Guppenlochfels	9.1856	48.3514	DE	2	5	Riek (1933), Andrée (1939), Binder (1977), Eriksen (1991)
210	Annakapellenhöhe	9.2089	48.1803	DE	2	2	Peters and Rieth (1936), Peters (1936a), Andrée (1939), Roller (1954), Eriksen (1991)
211	Nikolaushöhle; 2	9.2094	48.1778	DE	2	2	Peters (1936a), Heller (1937), Andrée (1939), Roller (1954), Eriksen (1991)
212	Kohltalhöhle	9.2597	48.1114	DE	2	5	Peters (1946), Paret (1951), Eriksen (1991)
213	Barnberger Höhle	9.3767	48.5547	DE	2	5	Riek (1957a), Eriksen (1991)
214	Klopfjörgelshütte	9.3817	48.4589	DE	2	5	Peters (1935b), Andrée (1939), Roller (1954), Paret (1961), Eriksen (1991)
215	Käppele	9.4261	48.6067	DE	2	1	Auffermann (1998)
216	Rappenfels	9.5142	48.4953	DE	2	5	Kapff (1935), Peters (1935a, 1946), Eriksen (1991)
217	Randecker Maar	9.5244	48.5725	DE	2	2	Riek (1931, 1932), Andrée (1939), Eriksen (1991), Auffermann (1998)
218	Braunfirst	9.5311	48.5911	DE	2	1	Eriksen (1991), Auffermann (1998)
219	Burkhardtshöhle; V	9.6167	48.5181	DE	2	3	Riek (1959a), Albrecht et al. (1977), Simon (1993), Hahn (1996), Auffermann (1998), Floss (2006)
220	Papierfels	9.6239	48.5636	DE	2	5	Riek (1935), Andrée (1939), Roller (1954), Eriksen (1991)
221	Hohlefels H.; yellow	9.6383	48.3731	DE	2	2	Koken (1912), Schmidt (1912), Wagner (1979), Weniger (1982), Eriksen (1991)
222	Hohlefels H.; grey	9.6383	48.3731	DE	2	4	
223	Felsställe; IIIb	9.6514	48.2825	DE	2	1	Torke (1975, 1976), Kind and Torke (1977, 1978), Kind (1982, 1984, 1985, 1987), Eriksen (1991), Floss (2006)
224	Felsställe; IIIa	9.6514	48.2825	DE	2	1	
225	Schuntershöhle	9.6564	48.3325	DE	2	1	Taute (1967, 1971, 1975), Eriksen 1991
226	Rabensteinhöhle	9.6567	48.3411	DE	2	5	Taute (1971), Eriksen (1991)
227	Schussenquelle main	9.6581	48.0219	DE	2	1	Fraas (1867), Koken (1912), Schmidt (1912), Wall (1961), Bosinski (1982), Schuler (1989, 1994)
228	Schussenquelle; A	9.6581	48.0219	DE	2	1	
229	Schussenquelle; C	9.6581	48.0219	DE	2	1	

(continued)

Table A.1 (continued)

ID	Site; Assemblage	Long	Lat	Co	G	C	References
230	Aichbühl A	9.6622	48.0286	DE	2	4	Eberhardt et al. (1987), Nielsen (2009)
231	Aichbühl B	9.6625	48.0306	DE	2	4	
232	Aichbühl C	9.6625	48.0292	DE	2	4	
233	Aichbühl D	9.6656	48.0311	DE	2	4	
234	Aichbühl E	9.6661	48.0275	DE	2	3	
235	Aichbühl F	9.6675	48.0286	DE	2	2	
236	Ganserfels	9.6878	48.3458	DE	2	5	Schmidt (1912), Riek (1935), Rieth (1938), Wagner (1979), Eriksen (1991)
237	Schmiechenfels; 2	9.7094	48.3472	DE	2	2	Koken (1912), Schmidt (1912), Riek (1935), Andrée (1939), Wagner (1979), Weniger (1982)
238	Helga-Abri; IIIa+b	9.7425	48.3733	DE	2	1	Hahn and Rottländer (1981), Hahn and Scheer (1983a, 1984), Rähle (1983a), Hahn (1984b), Owen (1988), Eriksen (1991), Auffermann (1998), Fisher (2002)
239	Helga-Abri; IIId	9.7425	48.3733	DE	2	1	
240	Hohle Fels, Slk.; Ia	9.7556	48.3722	DE	2	1	Fraas (1872), Hahn (1977b, c, 1978), Rähle (1981), Weniger (1982), Owen (1988), Eriksen (1991, 2002), Floss (2006)
241	Hohle Fels, Slk.; Ib	9.7556	48.3722	DE	2	1	
242	Hohle Fels, Slk.; Ic	9.7556	48.3722	DE	2	1	
243	Hohle Fels, Slk.; IIa	9.7556	48.3722	DE	2	1	
244	Sirgenstein Südwand	9.7594	48.3867	DE	2	2	Riek (1959b), Wagner (1979), Eriksen (1991)
245	Sirgenstein	9.7619	48.3881	DE	2	4	Schmidt (1910, 1912), Koken (1912), Riek (1935), Andrée (1939), Wagner (1979), Weniger (1982), Eriksen (1991)
246	Brillenhöhle; IV lw.	9.7783	48.4061	DE	2	1	Riek (1958, 1959c, 1973), Mauser (1963), Weniger (1982), Eriksen (1991), Auffermann (1998)
247	Brillenhöhle; IV up.	9.7783	48.4061	DE	2	1	
248	Geißenklösterle; AH Io	9.7803	48.3933	DE	2	2	Hahn et al. (1977), Hahn (1977d, 1980), Torke (1981), Weniger (1982), Hahn and Scheer (1983b), Eriksen (1991), Floss (2006)
249	Haldensteinhöhle; VI	9.8908	48.5478	DE	2	5	Riek (1938, 1960), Hahn et al. (1985), Eriksen (1991)
250	Kleine Scheuer im Rosenstein	9.9442	48.7886	DE	2	1	Maier (1936), Bleich (1965), Weniger (1982), Eriksen (1991), Auffermann (1998)
251	Herwartstein	10.1142	48.7356	DE	2	5	Andrée (1939), Roller (1954), Eriksen (1991)
252	Bockstein Westloch; c	10.1542	48.5333	DE	2	5	Wetzel (1954b), Hahn et al. (1985), Eriksen (1991), Krönneck (2008)
253	Bocksteinhöhle; III	10.1547	48.5331	DE	2	2	Schmidt (1912), Riek (1935, 1960), Wetzel (1954b), Hahn et al. (1985), Krönneck (2008)
254	Bocksteinschmiedeand Bocksteinloch; c	10.1558	48.5331	DE	2	2	Wetzel (1935, 1958), Wetzel and Bosinski (1969), Hahn et al. (1985)
255	Bockstein-Törle; AH II, III	10.1567	48.5331	DE	2	2	Wetzel (1954a), Hahn et al. (1985), Floss (2006), Krönneck (2008)
256	Hohlenstein-Std.; II	10.1722	48.5492	DE	2	2	Schmidt (1912), Völzing (1938), Riek (1960), Weniger (1982), Hahn et al. (1985), Eriksen (1991)
257	Hohlenstein-Std.; III	10.1722	48.5492	DE	2	2	
258	Bärenhöhle; Ib	10.1722	48.5492	DE	2	4	Riek (1960), Hahn et al. (1985)
259	Kl. Scheuer, Hohlenstein-St.; III	10.1725	48.5492	DE	2	2	Hahn and Koenigswald (1977), Floss (2006)
260	Klemmer-Abri	10.1750	48.6128	DE	2	4	Albrecht (1984), Hahn (1984a), Eriksen (1991)
261	Spitzbubenhöhle GH8+9, AH2	10.1794	48.5978	DE	2	1	Hahn (1974, 1975, 1977a, 1984a), Weniger (1982), Eriksen (1991), Fisher (2002)
262	Klingenfels-Abri.	10.1819	48.6006	DE	2	4	Heller (1957), Riek (1957a), Auffermann (1991)
263	Bärenfelsgrotte; III	10.1822	48.6003	DE	2	3	Heller (1957), Riek (1957a), Weniger (1982), Auffermann (1991)
264	Spitalfels; VIIIa-VIIIb	10.1825	48.6000	DE	2	4	Heller (1957), Riek (1957a), Weniger (1982), Auffermann (1991), Eriksen (1991)
265	Vogelherd; III	10.1939	48.5586	DE	2	2	Riek (1934), Andrée (1939), Lehmann (1954), Weniger (1982), Hahn et al. (1985), Eriksen (1991)
266	Vogelherd; II	10.1939	48.5586	DE	2	2	
267	Ofnet 1, Kleine	10.4500	48.8333	DE	2	4	Schmidt (1912), Dauber and Birkner (1934), Frickhinger (1937), Trusheim (1937), Dehn and Sangmeister (1954), Weniger (1982), Eriksen (1991)

(continued)

Table A.1 (continued)

ID	Site; Assemblage	Long	Lat	Co	G	C	References
268	Ofnet, Große; 6	10.4503	48.8183	DE	2	4	Schmidt (1912), Birkner (1936), Andrée (1939), Dehn and Sangmeister (1954), Eriksen (1991)
269	Hohlenstein bei Ederheim	10.4653	48.8089	DE	2	1	Birkner (1936), Freund (1963), Narr (1965), Eriksen (1991)
270	Hahnenberg	10.6036	48.8167	DE	2	4	Schönweiß (1963), Züchner (1979), Eriksen (1991)
271	Kaufertsberg; lw.	10.6108	48.8092	DE	2	1	Kaulich (1983), Fisher (2002)
272	Kaufertsberg; up.	10.6108	48.8092	DE	2	1	
273	Stein. Rosenkranz	11.0081	48.8694	DE	2	5	Freund (1963)
274	Weinberghöhlen; Vorplatz 1+2	11.0361	48.7583	DE	2	2	Bohmers (1951), Zotz (1955), Koenigswald et al. (1974)
275	Wein.; Saazer Loch	11.0361	48.7583	DE	2	2	
276	Wein.; Grab. außen	11.0361	48.7583	DE	2	2	
277	Wein.; Vorplatz 3	11.0361	48.7583	DE	2	2	
278	Mühlberg-Grotten	11.0939	48.8689	DE	2	5	Gumpert (1954), Tillmann (1989b)
279	Breitenfurter H.; C	11.1028	48.8733	DE	2	5	Heller (1956), Freund (1963)
280	Lochschlaghöhle	11.1344	48.8936	DE	2	5	Tillmann (1989a)
281	Speckberg; A+B	11.1869	48.8019	DE	2	1	Hahn (1982, 1989)
282	Eitensheim	11.3203	48.8194	DE	2	2	Rieder (1988, 1989)
283	Hohlenstein Schambach; AHF	11.3669	48.9058	DE	2	3	Gumpert (1956), Rieder (1992), Floss (2006)
284	Rennerfels	11.3678	49.8136	DE	2	5	Gumpert (1931, 1932), Birkner (1936), Andrée (1939)
285	Fuchsenloch	11.3939	49.7644	DE	2	5	Stuhlfrauth (1953), Freund (1963)
286	Hohlenstein bei Happurg	11.4908	49.4836	DE	2	5	Hörmann (1913), Obermaier and Wernert (1914), Kaulich (1996)
287	Obere Klause; lw.	11.7772	48.9369	DE	2	1	Andrée (1939), Freund (1963), Kaulich (1994), Züchner (2008)
288	Obere Klause; up.	11.7772	48.9369	DE	2	1	
289	Mittlere Klause; AH1	11.7844	48.9344	DE	2	2	Obermaier (1914), Obermaier and Frauenholz (1927), Birkner (1936), Andrée (1939), Freund (1963)
290	Kastelhänghöhle; 5	11.7864	48.9292	DE	2	1	Frauenholz et al. (1911), Birkner (1936), Andrée (1939)
291	Sesselfelsgrotte; C2	11.7894	48.9350	DE	2	2	Dirian (2003)
292	Sesselfelsgrotte; C1	11.7894	48.9350	DE	2	2	
293	Sesselfelsgrotte; B3	11.7894	48.9350	DE	2	2	
294	Sesselfelsgrotte; B2	11.7894	48.9350	DE	2	2	
295	Heidenstein	11.7922	48.9256	DE	2	5	Andrée (1939), Zotz (1961)
296	Siegfriedfelsen; C 3	11.7922	48.9344	DE	2	5	Naber (1977), Gerhardt and Naber (1983)
297	Siegfriedfelsen; D1	11.8136	48.9231	DE	2	2	
298	Siegfriedfelsen; D2	11.8136	48.9231	DE	2	2	
299	Siegfriedfelsen; D3	11.8136	48.9231	DE	2	2	
300	Siegfriedfelsen; D4	11.8136	48.9231	DE	2	2	
301	Obernederhöhle	11.8375	48.9333	DE	2	2	Freund (1987)
302	Steinbergwand; VIII	11.9258	49.3442	DE	2	4	Gumpert (1933a, b), Birkner (1936), Andrée (1939), Freund (1963)
303	Burgh. Loch; 2 up.	11.9289	49.0150	DE	2	5	Herrmann (1955)
304	Räuberhöhle	11.9697	49.0375	DE	2	5	Lindner (1961)
305	Tunnelhöhle	11.9789	49.0317	DE	2	2	Birkner (1926, 1936), Lindner (1961)
306	Fürst-Albrecht-H.	11.9817	48.9892	DE	2	2	Kaulich (1995)
307	Barbing	12.1978	49.0033	DE	2	1	Reisch (1974)
308	Presle, Grotte de	4.5867	50.3831	BE	3	4	Dewez (1987)
309	Trou de Blaireaux; III 1	4.7433	50.1150	BE	3	2	Loë and Rahir (1905), Bellier and Cattelain (1986), Dewez (1987), López Bayón (2000), Pozzi (2004)
310	Trou de Blair.; III 2	4.7433	50.1150	BE	3	2	
311	Trou de Blair.; III 3	4.7433	50.1150	BE	3	2	
312	Trou de Blair.; II1	4.7433	50.1150	BE	3	2	

(continued)

Table A.1 (continued)

ID	Site; Assemblage	Long	Lat	Co	G	C	References
313	Roc-la-Tour; 1	4.7831	49.8906	F	3	1	Rozoy (1988a, 1997), Rozoy and Rozoy (2004)
314	Roc-la-Tour; 2	4.7831	49.8906	F	3	1	
315	Roc-la-Tour; 3	4.7831	49.8906	F	3	1	
316	Roc-la-Tour; 4	4.7831	49.8906	F	3	1	
317	Trou du Sureau	4.8061	50.2917	BE	3	5	Dupont (1873), Sonneville-Bordes (1961)
318	Trou de Chêne	4.8150	50.2944	BE	3	5	Dupont (1873), Sonneville-Bordes (1961), Rozoy (1997)
319	Trou da Somme	4.8586	50.2214	BE	3	1	López Bayón et al. (1997), López Bayón (2000), Miller and Noiret (2009), Sano (2012)
320	Bois Laiterie	4.8636	50.3597	BE	3	1	Otte and Straus (1997), Straus and Otte (1998), López Bayón (2000), Sano et al. (2011), Sano (2012)
321	Roche-al-Rue	4.8797	50.2203	BE	3	1	Dewez (1987)
322	Chauveau	4.8797	50.3569	BE	3	4	Straus and Otte (1998)
323	Trou Magrite	4.9164	50.2269	BE	3	5	Dupont (1867), Klaatsch (1903), Dewez (1987), Otte and Straus (1995), Rozoy (1997), López Bayón (2000), Pozzi (2004)
324	Trou de Chaleux; 5	4.9417	50.2217	BE	3	1	Dupont (1873), Sonneville-Bordes (1961), Dewez (1987), Otte (1994), Charles (1998), López Bayón (2000)
325	Trou de Chaleux; ex.Otte	4.9417	50.2217	BE	3	1	
326	Trou des Nutons (Furfooz)	4.9531	50.2122	BE	3	1	Van Benden (1864), Dupont (1873), Sonneville-Bordes (1961), Dewez (1987), Charles (1998), López Bayón (2000), Pozzi (2004)
327	Trou d. l. Mâchoire	4.9564	50.2139	BE	3	4	Dewez (1987)
328	Trou du Frontal	4.9567	50.2144	BE	3	2	Van Benden (1864), Dupont (1873), Sonneville-Bordes (1961), Dewez (1987), Charles (1998), López Bayón (2000)
329	Trou Reuviau	4.9708	50.2144	BE	3	5	Sonneville-Bordes (1961)
330	Orp; east	4.9731	50.6944	BE	3	1	Vermeersch et al. (1987), Pozzi (2004)
331	Orp; west	4.9731	50.6944	BE	3	1	
332	Goyet 3; Chamber A, 1	5.0133	50.4456	BE	3	1	Dupont (1869a, 1869b, 1873), Sonneville-Bordes (1961), Germonpré (1997), Pozzi (2004), Stevens et al. (2009)
333	Engis	5.4136	50.5919	BE	3	4	Fraipont (1896), Doudou (1903), Rasquin (1911), Sonneville-Bordes (1961), Dewez (1987)
334	Coléoptère, Grotte de; B	5.5186	50.3786	BE	3	1	Hamal-Nandrin and Servais (1925), Cordy (1974), Dewez (1975, 1987), Charles (1998), Otte and Miller (1999), Pozzi (2004)
335	Verlaine, G. de Sy	5.5189	50.4061	BE	3	1	Destinez and Moreels (1888), Dewez (1987), Charles (1998), Otte and Miller (1999), Pozzi (2004)
336	Roche Plate; 12	5.5433	48.8994		-	2	Stocker et al. (2003)
337	Kanne; sect. central	5.6794	50.8081	BE	3	1	Vermeersch et al. (1985), Arts and Deeben (1987)
338	Bay-Bonnet, G de	5.6825	50.5872	BE	3	4	Hamal-Nandrin et al. (1934), Sonneville-Bordes (1961), Dewez (1987), Rozoy (1997)
339	Trou Walou	5.6919	50.5761	BE	3	1	Dewez (1987, 1989, 1992), Lanting and Van der Pflicht (1996), Rozoy (1997)
340	Lixhe - Sur le Bois	5.7058	50.7217	BE	3	5	Vermeersch and Symens (1988), Boecking (1992)
341	Mesch-Steenberg	5.7417	50.7692	NL	3	1	Rensink (1986, 1991)
342	Schaelsberg, Valkenburg	5.8467	50.8622	NL	3	2	Arts and Deeben (1987)
343	Sweikhuizen- Oude Stort	5.8433	50.9542	NL	3	2	Arts and Deeben (1987), Deeben (1992)
344	Sweik. Groene Paal	5.8478	50.9494	NL	3	1	
345	Sweik.-Koolweg	5.8503	50.9503	NL	3	2	
346	Echt	5.8669	51.1075	NL	3	2	Wouters (1982), Arts and Deeben (1987)
347	Griendtsveen	5.8792	51.4433	NL	3	1	Wouters (1983), Arts and Deeben (1987)
348	Eyserheide	5.9319	50.8383	NL	3	1	Rensink (1992), Sano (2012)
349	Kehlen-Juckelsboesch	6.0303	49.6483	LU	3	5	Ziesaire and Thill (1985)
350	Beeck	6.1917	50.9811	DE	3	1	Jöris et al. (1993, 1995), Schmitz et al. (1998)
351	Alsdorf	6.1950	50.8528	DE	3	1	Löhr (1979), Floss (1994)
352	Wegberg-Berg	6.2736	51.1394	DE	3	5	Heinen and Arora (1995)

(continued)

Table A.1 (continued)

ID	Site; Assemblage	Long	Lat	Co	G	C	References
353	Erkelenz 2/Haberg; I/VII	6.3039	51.0403	DE	3	5	Heinen and Arora (1995)
354	Erkelenz 2/Hab.; VII	6.3039	51.0403	DE	3	5	
355	Viersener Donk I	6.4350	51.2478	DE	3	2	Heinen (1993), Thissen (1997)
356	Viersener Donk II	6.4447	51.2478	DE	3	5	Heinen (1993), Thissen (1997)
357	Galgenberg	6.4678	51.1411	DE	3	5	Höpken (1995), Thissen (1997)
358	Eierberg	6.4689	51.1283	DE	3	2	Holzkämper et al. (2014)
359	Kamphausen	6.4728	51.1264	DE	3	2	Thissen (1989, 1997), Höpken (1995)
360	Winterdorf	6.5758	49.8597	DE	3	5	Boecking (1992)
361	Trier Gläsgesberg	6.6050	49.7817	DE	3	5	Boecking (1992)
362	Kartstein; 5	6.6594	50.5444	DE	3	1	Rademacher (1911), Schmidt (1912), Springensguth and Buttler (1935), Andrée (1939), Löhr (1978), Arts and Deeben (1987), Lanting and Van der Pflicht (1996), Baales (2006b)
363	Ettringer Bellerberg	7.2344	50.3539	DE	3	5	Bosinski (2008b)
364	Andernach-Martinsberg; I	7.4003	50.4336	DE	3	1	Höck (1993), Bosinski (1994), Tinnes (1995), Álverez Fernández (1999), Floss and Terberger (2002), Holzkämper (2006)
365	Andernach-M.; II	7.4003	50.4336	DE	3	1	
366	Andernach-M.; III	7.4003	50.4336	DE	3	1	
367	Andernach-M.; IV	7.4003	50.4336	DE	3	1	
368	Gönnersdorf I	7.4125	50.4461	DE	3	1	Poplin (1976), Bosinski (1977, 1979, 2008a), Bosinski and Fischer (1980), Franken and Veil (1983), Höck (1993), Tinnes (1995), Terberger (1997), Álverez Fernández (1999), Bosinski et al. (2001), Sensburg (2007)
369	Gönnersdorf II	7.4125	50.4461	DE	3	1	
370	Gönnersdorf III	7.4125	50.4461	DE	3	1	
371	Gönnersdorf IV	7.4125	50.4461	DE	3	1	
372	Irlich, Sandgrube	7.4378	50.4506	DE	3	2	Baales (2002), Bosinski (2008b)
373	Irlich, Geiersberg	7.4456	50.4469	DE	3	4	Bosinski (2008b)
374	Heimbach-Weis	7.5386	50.4503	DE	3	5	Jöris (1997), Bosinski (2008b)
375	Martinshöhle, Letmathe	7.6103	51.3675	DE	3	1	Schmidt (1912), Andrée (1939), Günther (1988)
376	Feldhofhöhle	7.8547	51.3778	DE	3	2	Andrée (1928, 1939), Günther (1988), Baales (2006a), Bosinski (2008b)
377	Balverhöhle	7.8719	51.3392	DE	3	2	Schmidt (1912), Andrée (1928, 1939), Günther (1988), Baales (2005a), Bosinski (2008b)
378	Wildweiberlei	7.9908	50.3694	DE	3	1	Heck (1920), Terberger (1993)
379	Wildscheuer; V	8.1297	50.4200	DE	3	1	Cohausen (1874, 1879), Behlen (1905, 1909), Andrée (1939), Mandera (1954), Terberger (1993), Floss (1994)
380	Fußgönheim 2	8.2850	49.4503	DE	–	2	Stodiek (1987), Bosinski (2008b)
381	Fußgönheim 1	8.2883	49.4567	DE	–	2	Stodiek (1987), Bosinski (2008b)
382	Bilsteinhöhle	8.2911	51.3975	DE	3	4	Andrée (1927, 1939), Henneböle (1963), Adrian (1982), Baales (2005b)
383	Mutterstadt	8.3542	49.4439	DE	–	5	van Noort (2007)
384	Trebur	8.4092	49.9242	DE	3	4	Fiedler (1990, 1994)
385	Dreieich-Götzenhain N. (II)	8.7342	50.0056	DE	–	2	Rosenstein (1998), Serangeli and Terberger (2006), Terberger et al. (2009a)
386	Dreieich-Götzenhain E. (I)	8.7397	49.9986	DE	–	1	
387	Frille	8.9883	52.3411	DE	–	4	Günther (1988), Baales (2005a)
388	Oberlistingen	9.2283	51.4494	DE	–	5	Bergmann (1964), Adrian (1982)
389	Balhorn	9.2358	51.2725	DE	–	5	Bergmann (1964)
390	Aschenstein	9.9008	51.9192	DE	–	2	Barner (1962), Tromnau (1976), Thieme (1991), Grote (1993), Staesche (1999), Terberger et al. (2009b)
391	Stendel, A.	9.9328	51.4300	DE	–	5	Grote (1993, 1998), Staesche (2005), Terberger et al. (2009b)
392	Allerberg	9.9831	51.4656	DE	–	5	Grote (1993)
393	Bettenroder Berg IX; 18a	10.0217	51.4711	DE	–	4	Grote (1993), Veil (1996)
394	Merkers	10.1167	50.8264	DE	–	5	Feustel (1983), Küßner (2009)

(continued)

Table A.1 (continued)

ID	Site; Assemblage	Long	Lat	Co	G	C	References
395	Gadenstedt	10.2150	52.2467	DE	–	1	Hiller and Veil (1997), Veil and Hiller (1998), Simoneit and Veil (1999)
396	Steinkirche bei Scharzfeld	10.3764	51.6311	DE	–	4	Andrée (1939), Jacob-Friesen (1959), Schirwitz (1961)
397	Kleingrabe, Kalkkopf	10.5036	51.1819	DE	–	5	Küßner (2007, 2009)
398	Wandersleben (Am Bahnhof)	10.8531	50.9064	DE	4	5	Müller and Rommeis (1970), Müller (1980), Küßner (2007, 2009)
399	Ichtershausen	10.9694	50.8744	DE	4	5	Küßner (2007, 2009)
400	Westerhausen	11.0547	51.8064	DE	4	5	Schirwitz (1926), Feustel (1979), Küßner (2007, 2009)
401	Bad Frankenhausen	11.0739	51.3700	DE	4	1	Behm-Blancke (1956), Feustel (1977), Küßner (1998, 2009)
402	Bärenkeller, Königsee-Garsitz	11.0772	50.6511	DE	4	3	Feustel et al. (1971a), Feustel and Musil (1977a), Bosinski (1982)
403	Seehausen, Mutzenbrunnen	11.1097	51.3167	DE	4	5	Küßner (1998, 2003, 2009)
404	Fuchskirche I; 3	11.1561	50.6553	DE	4	3	Feustel and Musil (1977b), Benecke et al. (2006), Küßner (2009)
405	Udersleben	11.1614	51.3753	DE	4	5	Küßner (1998, 2003, 2009)
406	Oldisleben, Sachsenburg	11.1619	51.2911	DE	4	5	Küßner (1998, 2007, 2009)
407	Sachsenburg	11.1625	51.2886	DE	4	5	Küßner (2003)
408	Quedlinburg, Bockshornschanze	11.1639	51.7875	DE	4	2	Schirwitz (1940), Küßner (2007, 2009)
409	Quedlinburg, Lehof	11.1650	51.8114	DE	4	2	Schirwitz (1936), Andrée (1939)
410	Gorsleben	11.1833	51.2775	DE	4	2	Nitzschke and Stahlhofen (1976), Feustel (1979), Küßner (2003, 2007, 2009)
411	Quedlinburg, Mühlgraben	11.1906	51.8222	DE	4	1	Schirwitz (1936), Andrée (1939)
412	Hausneindorf	11.2717	51.8403	DE	4	5	Schirwitz (1936)
413	Gatersleben	11.2869	51.8233	DE	4	2	Toepfer (1965), Hoffmann (1972), Küßner (2009)
414	Schweskau	11.2883	52.9206	DE	–	2	Breest and Veil (1991), Veil (1996)
415	Teufelsbrücke; 1	11.3911	50.6208	DE	4	1	Feustel (1980), Wüst (1998)
416	Teufelsbrücke; 2	11.3911	50.6208	DE	4	1	
417	Teufelsbrücke; 3	11.3911	50.6208	DE	4	1	
418	Teufelsbrücke; 4	11.3911	50.6208	DE	4	1	
419	Königsaue, Bruchberg	11.4033	51.8142	DE	4	2	Schirwitz (1936), Andrée (1939), Toepfer (1965), Mania and Toepfer (1973), Weniger (1982), Küßner (2009)
420	Wilsleben, Ortsausgang	11.4233	51.8036	DE	4	5	Toepfer (1965)
421	Aschersleben-Bruchfeld	11.4244	51.7686	DE	4	5	Schirwitz (1936), Toepfer (1965)
422	Wilsleben, Klintberg	11.4247	51.8003	DE	4	5	Toepfer (1965), Küßner (2009)
423	Magdala	11.4431	50.9072	DE	4	5	Küßner (2007, 2009)
424	Aschersleben	11.4611	51.7531	DE	4	2	Bicker (1939), Rust (1942), Toepfer (1965), Hanitzsch (1972), Feustel (1979)
425	Brückenfeld, Windhügel	11.4947	50.7750	DE	4	4	Mania (2008)
426	Himmelreich	11.4978	50.7811	DE	4	5	Mania (2008)
427	Rennerweg, Anhöhe	11.5125	50.7858	DE	4	2	Feustel (1979), Mania and Mania (1994), Mania (2008)
428	Forstbergplateau über dem Hain	11.5272	50.7833	DE	4	2	Mania (2008)
429	Dehnamühle, üb. d.	11.5300	50.7931	DE	4	2	Mania (2008)
430	Eichenberg	11.5300	50.7931	DE	4	5	Behm-Blancke (1961), Feustel (1979), Küßner (2003, 2009)
431	Forstebrg	11.5331	50.7886	DE	4	2	Mania (2008)
432	Nähe Forstberg	11.5386	50.7836	DE	4	5	Mania (2008)
433	Vogelherd,u. Rittergutsfeld	11.5453	50.7967	DE	4	4	Mania (2008)
434	Königswinkel (Großeutersdorf)	11.5553	50.7892	DE	4	5	Mania and Mania (1994), Küßner (2007, 2009), Mania (2008)

(continued)

Table A.1 (continued)

ID	Site; Assemblage	Long	Lat	Co	G	C	References
435	Herdloch bei Ranis; 1+2	11.5642	50.6619	DE	4	2	Andrée and Grimm (1929), Andrée (1939), Weniger (1989)
436	Kirchberg	11.5642	50.7928	DE	4	2	Mania (2008), Küßner (2009)
437	Ilsenhöhle; 4	11.5642	50.6622	DE	4	3	Hülle (1933, 1977), Andrée (1939), Küßner (2009)
438	Ilsenhöhle; 5	11.5642	50.6622	DE	4	3	
439	Nebra	11.5711	51.2922	DE	4	1	Hanitzsch (1972), Küßner (1998), Mania (1999)
440	Jena "an der hohen Saale"	11.5894	50.9219	DE	4	1	Eichhorn (1929), Neumann (1933a), Völker (2005), Hemmann et al. (2008), Küßner (2009)
441	Großpürschütz	11.5917	50.8208	DE	4	5	Behm-Blancke (1961), Feustel (1979), Küßner (2003, 2009)
442	Rothestein "Kuhberg"	11.5958	50.8447	DE	4	1	Neumann (1933a), Hemmann (2005), Hemmann et al. (2008), Küßner (2009)
443	Kahla-Löbschütz	11.5989	50.8067	DE	4	1	Andrée (1939), Hanitzsch (1957), Feustel (1974), Küßner (2009)
444	Oelknitz	11.6031	50.8447	DE	4	1	Neumann (1933a, b, 1958), Behm-Blancke (1960, 1970, 1976), Feustel (1974, 1989), Bosinski (1982), Musil (1985), Gaudzinski and Street (2003), Hemmann et al. (2008)
445	Hummelshain, Bremsstall	11.6275	50.7714	DE	4	5	Andrée (1939), Hanitzsch (1972), Weniger (1982), Küßner (2009)
446	Rehmen	11.6311	50.7100	DE	4	5	Behm-Blancke (1961), Feustel (1979), Küßner (2003, 2009)
447	Förderstedt	11.6342	51.8967	DE	4	5	Küßner (2007, 2009)
448	Bad Sulza, Krähenhütte	11.6406	51.0939	DE	4	5	Behm-Blancke (1961), Feustel (1979), Küßner (2003, 2009)
449	Urdhöhle	11.6411	50.6906	DE	4	2	Behm-Blancke (1961), Feustel et al. (1971b), Terberger et al. (2003), Küßner (2009)
450	Wüste Scheuer	11.6411	50.6942	DE	4	5	Eisel (1886), Goetze (1930), Richter (1955), Behm-Blancke (1961), Küßner (2003, 2007, 2009)
451	Kniegrotte	11.6414	50.6914	DE	4	1	Feustel (1974), Berke (1989), Höck (2000)
452	Lausnitz, A. Theure	11.6900	50.7222	DE	4	1	Feustel et al. (1963), Feustel (1979), Küßner (2009)
453	Stendorf	11.6944	51.1133	DE	4	5	Schmidt and Schneider (1974), Feustel (1979), Küßner (2003)
454	Seeburg	11.7019	51.4883	DE	4	5	Hanitzsch (1972), Feustel (1979), Küßner (2009)
455	Saaleck	11.7033	51.1125	DE	4	1	Hülle (1932), Toepfer (1970), Feustel (1974), Hanitzsch (1978), Terberger (1987), Küßner (2003, 2009), Grünberg (2004)
456	Lengefeld	11.7053	51.1136	DE	4	1	Hanitzsch (1957), Hoffmann (1968), Küßner (2009), Adaileh (2011), Decker (2011), Szyja (2012), Albert (2014)
457	Trockhausen	11.7064	50.8964	DE	4	5	Hanitzsch (1957), Feustel (1979), Küßner (2003)
458	Bad Kösen, Rudelsburg	11.7072	51.1086	DE	4	5	Feustel (1979), Küßner (2003, 2009)
459	Prinz-Heinrich-Stein	11.8167	51.1500	DE	4	5	Hanitzsch (1972), Schmidt and Schneider (1974), Feustel (1979), Küßner (2003, 2009)
460	Schönburg	11.8672	51.1600	DE	4	5	Hanitzsch (1972), Feustel (1979), Küßner (2003)
461	Weißenborn, Speckberg	11.8819	50.9242	DE	4	2	Hanitzsch (1957), Feustel (1979), Küßner (2003, 2007, 2009)
462	Sankt Gangloff	11.9006	50.8567	DE	4	5	Behm-Blancke (1961), Feustel (1979), Küßner (2003, 2009)
463	Morl	11.9200	51.5522	DE	4	5	Hanitzsch (1972), Feustel (1979), Küßner (2009)
464	Gröbitz	11.9350	51.1947	DE	4	5	Hanitzsch (1972), Feustel (1979), Küßner (2003)
465	Petersberg	11.9500	51.5933	DE	4	2	Toepfer and Mania (1973), Feustel (1979), Küßner (2009)
466	Weißenfels	11.9561	51.2025	DE	4	2	Nitzschke and Stahlhofen (1976), Küßner (2007, 2009)
467	Halle Neustadt, Galgenberg	11.9719	51.5064	DE	4	5	Stahlhofen (1981), Küßner (2009)
468	Galgenberg, Halle	11.9758	51.5025	DE	4	1	Toepfer (1955)
469	Hartmannsdorf	11.9772	50.9617	DE	4	5	Hanitzsch (1957), Feustel (1979), Küßner (2003, 2009)
470	Gleina	11.9850	50.9344	DE	4	1	Feustel (1957), Küßner (2003)
471	Ahlendorf	11.9861	50.9853	DE	4	1	Feustel (1957, 1974)
472	Pötewitz	11.9881	51.0036	DE	4	5	Hanitzsch (1957), Küßner (2003, 2009)
473	Leuna-Göhlitzsch	12.0308	51.3203	DE	4	5	Schmidt and Schneider (1974), Feustel (1979), Küßner (2003, 2009)

(continued)

Table A.1 (continued)

ID	Site; Assemblage	Long	Lat	Co	G	C	References
474	Schkauditz	12.0458	51.0228	DE	4	5	Hanitzsch (1957), Feustel (1979), Küßner (2003, 2009)
475	Wallendorf (Frie-densdorf/Kriegsd.)	12.0642	51.3569	DE	4	2	Andrée (1939), Hanitzsch (1956a), Feustel (1979), Küßner (2009)
476	Poserna, Burgstättl	12.0817	51.2142	DE	4	5	Nitzschke and Stahlhofen (1978), Küßner (2007, 2009)
477	Pfortener Berg	12.0864	50.8469	DE	4	2	Feustel (1957), Küßner (2009)
478	Gera-Liebschwitz, Binsenacker	12.0886	50.8264	DE	4	1	Reuter (1942), Feustel (1974), Küßner (2003, 2009)
479	Salsitz	12.0894	51.0397	DE	4	5	Hanitzsch (1957), Feustel (1979), Küßner (2003, 2009)
480	Schafgraben, Zwötzen	12.0917	50.8467	DE	4	2	Andrée (1939), Küßner (2003, 2009)
481	Lindentaler Hyänenhöhle	12.0947	50.8758	DE	4	5	Liebe (1876), Schmidt (1912), Auerbach (1929, 1930), Küßner (2003, 2009)
482	Hänichen	12.2658	51.3856	DE	4	5	Bernhardt and Rudolph (1997), Küßner (2007, 2009)
483	Groitzsch A1	12.2881	51.1611	DE	4	1	Hanitzsch (1957, 1972)
484	Groitzsch A2	12.2881	51.1611	DE	4	1	
485	Groitzsch B	12.2889	51.1614	DE	4	1	
486	Groitzsch B-Nord	12.2889	51.1614	DE	4	1	
487	Groitzsch C1	12.2900	51.1619	DE	4	1	
488	Groitzsch C2	12.2900	51.1619	DE	4	1	
489	Groitzsch C3+DNord	12.2900	51.1619	DE	4	1	
490	Groitzsch D1	12.2900	51.1644	DE	4	1	
491	Groitzsch D2	12.2900	51.1644	DE	4	1	
492	Liemehna	12.5128	51.4339	DE	4	5	Seiler (1996), Küßner (2007, 2009)
493	Gordemitz, Fuchsberg	12.5417	51.4103	DE	4	5	Seiler (1996), Küßner (2007, 2009)
494	Kuppelberg	12.7361	51.3239	DE	4	2	Geupel (1967)
495	Wurzen-Dehnitz, Wüste Kirche	12.7369	51.3525	DE	4	4	Geupel (1967, 1969, 1988), Feustel (1979), Küßner (2009)
496	Wachtelberg A	12.7386	51.3514	DE	4	2	Geupel (1967)
497	Wachtelberg	12.7386	51.3517	DE	4	4	Geupel (1967, 1988), Küßner (2009)
498	Lüptitz (Spitzberg)	12.7708	51.3917	DE	4	5	Geupel (1972), Feustel (1979)
499	Zinkenberg	12.7947	51.4269	DE	4	5	Hanitzsch (1956b, 1957), Feustel (1979), Küßner (2003, 2007, 2009)
500	Burzelberg	12.7967	51.4203	DE	4	4	Gerlach and Geupel (1980)
501	Metten; I+II	12.9156	48.8550	DE	4	2	Weißmüller (1995, 2002)
502	Kremmen	13.0242	52.7606	DE	–	5	Ginter (1974)
503	Westewitz	13.0356	51.1381	DE	4	4	Geupel (1976), Küßner (2009)
504	Vilshofen-Kuffing	13.1806	48.5989	DE	4	2	Weißmüller (2002)
505	Niederlommatzsch	13.4069	51.2394	DE	4	2	Dietzel and Geupel (1989)
506	Baselitz; I	13.4836	51.2339	DE	4	2	Baumann (1968), Feustel (1979), Reuter (1993, 1995)
507	Baselitz; III	13.4836	51.2339	DE	4	2	
508	Baselitz; IV	13.4836	51.2339	DE	4	2	
509	Wantewitz; II	13.5011	51.2356	DE	4	5	Baumann (1968), Reuter (1993), Küßner (2009)
510	Želeč	13.5575	50.2381	CZ	4	2	Födisch (1961), Vencl (1962, 1995)
511	Dobříčany	13.6003	50.3111	CZ	4	1	Vencl (1962, 1995)
512	Bečov	13.7175	50.4483	CZ	4	2	Žebera (1966), Fridrich (1972), Lech and Mateiciucová (1995), Vencl (1995), Valoch (1996)
513	Dolni Poříčí 8	13.7922	49.2858	CZ	4	2	Vencl et al. (2006)
514	Slaník I	13.9522	49.2681	CZ	4	2	Vencl et al. (2006)
515	Děravá cave; low. I	14.0358	49.9000	CZ	4	1	Prošek (1952, 1961), Fridrich and Sklenář (1976), Klíma (1985), Vencl (1995)
516	Děravá cave; upp. I	14.0358	49.9000	CZ	4	1	
517	Koněprusy	14.0472	49.9164	CZ	4	2	Fridrich and Sklenář (1976), Matušek (1987), Vencl (1995), Svoboda et al. (2002)
518	Kvíc	14.0639	50.2117	CZ	4	1	Žebera (1955, 1958), Vencl (1979, 1995)
519	Turské maštale cave	14.0783	49.7864	CZ	4	5	Fridrich and Sklenář (1976), Verpoorte and Šída (2009)

(continued)

Table A.1 (continued)

ID	Site; Assemblage	Long	Lat	Co	G	C	References
520	Beroun-Závodí	14.0803	49.9761	CZ	4	5	Vencl (1995)
521	Lhota	14.0822	49.2606	CZ	4	2	Vencl (1970, 1979, 1995)
522	Kačák cave	14.1342	49.9697	CZ	4	5	Fridrich and Sklenář (1976), Verpoorte and Šída (2009)
523	Barrandová cave	14.1383	49.9431	CZ	4	2	Fridrich and Sklenář (1976), Verpoorte and Šída (2009)
524	Protivec 3	14.0875	49.0767	CZ	4	5	Vencl et al. (2006)
525	Putim; 773	14.0986	49.2611	CZ	4	1	Mazálek (1952, 1953), Vencl (1964b, 1995), Vencl et al. (2006)
526	Putim; 162, 767	14.0986	49.2611	CZ	4	1	
527	Keblice	14.1022	50.4803	CZ	4	1	Neustupný (1938), Fridrich (1964), Vencl (1992, 1995)
528	Tetín	14.1036	49.9503	CZ	4	5	Vencl (1991, 1995)
529	Krápníková	14.1178	49.9542	CZ	4	2	Homola (1943), Prošek (1947, 1959), Fridrich and Sklenář (1976), Vencl (1995)
530	Hostim	14.1183	49.7878	CZ	4	1	Prošek (1947), Vencl (1995), Svoboda et al. (1996, (2002)
531	Na Průchodě	14.1281	49.9711	CZ	4	2	Petrbok (1956), Fridrich and Sklenář (1976), Vencl (1995)
532	Borečnice 2	14.1392	49.3611	CZ	4	2	Vencl et al. (2006)
533	Rytiřská	14.2128	50.7725	CZ	4	2	Valoch (1960, 1965, 2001), Nývltová-Fišáková (2002a)
534	Milenovice 1	14.2142	49.1728	CZ	4	5	Vencl et al. (2006)
535	Žd'ár 1	14.2275	49.2328	CZ	4	2	Vencl et al. (2006)
536	Burk 2, Bautzen; 2	14.4658	51.1992	DE	–	2	Frenzel (1934), Andrée (1939), Brandt (1960), Daum (1999)
537	Burk 2, Bautzen; 1	14.4658	51.1992	DE	–	5	
538	Březnice	14.5147	49.2514	CZ	4	2	Vencl et al. (2006)
539	Węgliny	14.7189	51.8233	PL	–	1	Domański and Burdukiewicz (1994), Burdukiewicz (1997), Cziesla and Masojć (2007), Cziesla (2008)
540	Gryfów Śląski 40	15.4122	51.0350	PL	–	5	Cziesla and Masojć (2007)
541	Frauenlucke	15.4500	48.6333	AT	5	5	Wichmann and Bayer (1924), Heinrich (1973), Neugebauer-Maresch (1999)
542	Gudenushöhle	15.5858	48.4167	AT	5	1	Hacker (1884), Breuil and Obermaier (1908), Cyrek (1986b), Bednarik (2008), Álvarez Fernández (2009)
543	Kamegg	15.6575	48.6114	AT	5	1	Gulder (1952), Papp (1952), Otte (1981), Neugebauer-Maresch (1999), Haesaerts et al. (2004), Verpoorte (2004)
544	Graselhöhle	15.6667	48.6167	AT	5	5	Bayer (1928), Heinrich (1973), Neugebauer-Maresch (1999)
545	Teufelslucke	15.8161	48.6431	AT	5	5	Bayer (1927, 1929), Berg (1966), Heinrich (1973), Neugebauer-Maresch (1999)
546	Náchod	16.1658	50.4214	CZ	–	2	Vencl (1964a, 1991, 1995)
547	Maloměřice-BorkyI	16.6425	49.2150	CZ	5	1	Valoch (1963), Feustel (1974), Oliva (2003)
548	Občiny	16.6672	49.2200	CZ	5	5	Škrdla (2002), Oliva (2003)
549	Rozcestí	16.6828	49.2644	CZ	5	5	Oliva (2003)
550	Jáchymka	16.6861	49.3078	CZ	5	5	Wankel (1881), Skutil (1940), Valoch (2001), Oliva (2003)
551	Býči skála	16.6908	49.3089	CZ	5	1	Wankel (1881), Bayer (1925), Hauser (1925), Schirmeisen (1925), Absolon (1945), Valoch (1960, 2001), Oliva (1995, 2003), Hladilová (1999)
552	Barová Cave; 11–12	16.6936	49.3114	CZ	5	2	Seitl et al. (1986), Valoch (2001), Horáček et al. (2002), Oliva (2003)
553	Líšň-Čtvrtě	16.6947	49.2069	CZ	5	2	Svoboda (1987), Škrdla (1990, 2002)
554	Koňská jáma	16.7083	49.3603	CZ	5	5	Skutil (1962, 1963), Valoch (2001), Jarošová (2002)
555	Skleb	16.7103	49.4972	CZ	5	5	Oliva (2003)
556	Kateřinská	16.7158	49.3628	CZ	5	5	Svoboda et al. (1983), Valoch (2001), Oliva (2003)
557	Osmnáctka	16.7203	49.4025	CZ	5	5	Oliva (2003) Ondroušková (2008), Valoch (2009)
558	Puklinová; 6	16.7222	49.2439	CZ	5	5	Oliva (2003), Valoch (2009)
559	Výpustek	16.7239	49.2914	CZ	5	5	Skutil (1927), Svoboda et al. (1996), Valoch (2001), Oliva (2003)
560	Macucha, Horákov	16.7297	49.3731	CZ	5	5	Svoboda et al. (1996), Škrdla (2002), Oliva (2003)
561	Pod koňským spádem 9&10	16.7306	49.3792	CZ	5	5	Svoboda et al. (1996) Valoch (2001)

(continued)

Table A.1 (continued)

ID	Site; Assemblage	Long	Lat	Co	G	C	References
562	Sedmnáctka	16.7308	49.3933	CZ	5	5	Oliva (2003)
563	Nová Drátenická Cave; 15	16.7319	49.2914	CZ	5	2	Klíma (1949), Valoch (1980, 2001), Svoboda et al. (1996), Oliva (2003), Valoch and Neruda (2005)
564	Smrtní	16.7319	49.3636	CZ	5	5	Jarošová (2002)
565	Horní mlýn, Říčky	16.7328	49.2425	CZ	5	5	Valoch (1960), Škrdla (2002), Oliva (2003)
566	Vinckova	16.7331	49.2964	CZ	5	5	Svoboda et al. (1996), Valoch (2001), Oliva (2003), Lázničková-Galetová (2010)
567	Žitného	16.7342	49.2911	CZ	5	1	Schránil (1928), Musil (1957), Valoch (1957, 1960, 2001), Svoboda et al. (1996), Hladilová (1999), Oliva (2003), Valoch and Neruda (2005)
568	Verunčina	16.7356	49.3697	CZ	5	2	Valoch (1960, 2001), Jarošová (2002), Nývltová-Fišáková (2002a), Oliva (2003)
569	Kolíbky; 5	16.7367	49.3331	CZ	5	1	Valoch (1960, 2001), Svoboda et al. (1995, 1996), Hladilová (1999), Oliva (2003)
570	Kolíbky; 6	16.7367	49.3331	CZ	5	1	
571	Kůlna Cave; 5	16.7378	49.4092	CZ	5	1	Valoch (1974, 1988, 2001), Kostruhn (2005), Valoch and Neruda (2005)
572	Kůlna Cave; 6	16.7378	49.4092	CZ	5	1	
573	Šosůvská	16.7381	49.4081	CZ	5	5	Valoch (2001), Oliva (2003)
574	Poustevna	16.7394	49.4153	CZ	5	5	Svoboda et al. (1996), Valoch (2001, 2009)
575	Pod vyhlídkou; Klímova	16.7397	49.2614	CZ	5	5	Klíma (2002), Oliva (2003), Valoch (2009)
576	Kulnička	16.7406	49.2431	CZ	5	2	Valoch (2001), Klíma (2002), Musil (2002), Přichystal (2002b), Oliva (2003)
577	Srnči	16.7414	49.3711	CZ	5	5	Valoch (2001), Jarošová (2002), Nývltová-Fišáková (2002a)
578	Hadí	16.7419	49.2400	CZ	5	1	Klíma (1961), Svoboda et al. (1996), Valoch (2001), Oliva (2003)
579	Pekárna; g/h	16.7425	49.2431	CZ	5	1	Wankel (1881), Absolon and Czižek (1926, 1927, 1932), Schránil (1928), Feustel (1974), Berke (1989), Svoboda (1991a), Svoboda et al. (1996), Valoch (2001), Oliva (2003)
580	Pekárna; i	16.7425	49.2431	CZ	5	1	
581	Kravská	16.7458	49.3617	CZ	5	5	Oliva (2003)
582	Adlerova	16.7483	49.2406	CZ	5	1	Klíma (1953), Valoch (1960, 2001), Hladilová (1999), Oliva (2003)
583	Křižova	16.7486	49.2431	CZ	5	1	Klíma (1951), Valoch (1960, 2001), Svoboda et al. (1996), Oliva (2003)
584	Švédův stůl; 50	16.7486	49.2447	CZ	5	2	Klíma (1962), Valoch (2001), Oliva (2003)
585	Ochozska Cave	16.7508	49.2439	CZ	5	1	Valoch (1953, 2001, 2002), Klíma (1970), Feustel (1974), Přichystal (2002a), Valoch and Neruda (2005)
586	Mokrá, Lom I	16.7508	49.2342	CZ	5	2	Valoch (2001), Škrdla (2002), Oliva (2003)
587	Lišči	16.7539	49.2461	CZ	5	5	Klíma (1960, 2002), Valoch (2001), Nývltová-Fišáková (2002b), Přichystal (2002b), Oliva (2003)
588	Mokrá, Lom V	16.7544	49.2347	CZ	5	1	Škrdla et al. (1999), Valoch (2001), Škrdla (2002), Oliva (2003)
589	A. III	16.7572	49.2239	CZ	5	5	Oliva (2003)
590	Balcarka Cave	16.7583	49.3764	CZ	5	1	Musil (1958), Valoch (1960, 2001), Oliva (2003), Valoch and Neruda (2005)
591	Pracký kopec	16.7622	49.1289	CZ	5	5	Škrdla (2002)
592	Michalova	16.7728	49.3950	CZ	5	5	Valoch (1960, 2001), Oliva (2003)
593	Průchodnice	16.8817	49.6550	CZ	5	5	Svoboda et al. (1996), Horáček et al. (2002), Oliva (2003)
594	Loštice I	16.9208	49.7281	CZ	5	1	Neruda and Nerudová (2008), Neruda et al. (2009), Valoch (2009)
595	Broniszowice 2	17.1775	50.4269	PL	5	1	Krawczyk et al. (2004)
596	Śmicz B	17.6153	50.4117	PL	5	5	Kozłowski (1964b), Ginter (1974), Wiśniewski (2003), Kostruhn (2004)
597	Sowin 7	17.6214	50.5542	PL	5	5	Burdukiewicz (1997), Furmanek et al. (2001), Połtowicz (2006)
598	Bliszczyce, Głubczyce; 16	17.7567	50.0817	PL	5	5	Kozłowski (1964b), Ginter (1974), Wiśniewski (2003), Kostruhn (2004)
599	Hranice; III	17.7583	49.5433	CZ	5	1	Feustel (1974), Neruda and Kostruhn (2002), Oliva (2003)

(continued)

Table A.1 (continued)

ID	Site; Assemblage	Long	Lat	Co	G	C	References
600	Włodzienin 3	17.8431	50.1181	PL	5	5	Foltyn and Foltyn (2007)
601	Dzierzyslaw 35;	17.9886	50.0597	PL	5	1	Ginter et al. (2002), Ginter and Połtowicz (2002a, b, 2004), Kostruhn (2004), Ginter et al. (2005), Kozłowski (2006)
602	Cyprzanów 3	18.1153	50.0719	PL	5	2	Kozłowski (1964a, b, 1987), Ginter (1974), Wiśniewski (2003), Kostruhn (2004)
603	Šipka dira	18.1164	49.5872	CZ	5	2	Maška (1884, 1886, 1888), Schránil (1928), Valoch et al. (1965), Svoboda et al. (1996)
604	Čertova dira	18.1211	49.5844	CZ	5	5	Maška (1884, 1886, 1888), Schránil (1928) Valoch et al. (1965), Svoboda et al. (1996)
605	Samborowice 2a	18.1256	50.0567	PL	5	5	Cziesla and Masojć (2007), Foltyn and Foltyn (2007)
606	Bierawa	18.2414	50.2808	PL	5	2	Ginter (1974)
607	Dra Majera	19.0272	50.1525	PL	5	1	Kowalski (1951), Kozłowski (1960, 1987), Kozłowski and Kozłowski (1996), Kostruhn (2004)
608	Towarna Góra	19.2700	50.7686	PL	5	5	Kopacz and Skalski (1976), Kozłowski and Kozłowski (1977, 1996), Wiśniewski (2003), Kostruhn (2004)
609	Okiennik	19.5222	50.5242	PL	5	4	Demetrykiewicz and Kuźniar (1914), Krukowski (1921), Kozłowski and Kozłowski (1977, 1996), Wiśniewski (2003)
610	Deszczowa; 3	19.5300	50.5772	PL	5	2	Cyrek et al. (2000), Cyrek (2001), Kozłowski (2001), Kozłowski and Pettitt (2001)
611	Deszczowa; 2	19.5300	50.5772	PL	5	5	
612	Krusza Skála	19.5706	50.5636	PL	5	2	Cyrek (1994, 1999), Kobusiewicz (1996), Kozłowski and Kozłowski (1996), Wiśniewski (2003)
613	Kroczycka Cave	19.5819	50.5611	PL	5	5	Cyrek (1999)
614	Zalas; 7–9	19.6281	50.0828	PL	5	2	Bocheński et al. (1985), Kozłowski and Kozłowski (1996), Kozłowski and Pettitt (2001), Wiśniewski (2003), Kostruhn (2004)
615	Zalas; 10–11	19.6281	50.0828	PL	5	2	
616	Brzoskwinia-Krzemionki 7; 1	19.7167	50.0667	PL	5	1	Kozłowski (1987), Sobczyk (1993), Kozłowski and Kozłowski (1996), Wiśniewski (2003), Kostruhn (2004)
617	Brzoskwinia-K..7; 2	19.7167	50.0667	PL	5	1	
618	Łopiankach II	19.7264	50.0608	PL	5	5	Krukowski (1927), Kozłowski (1960), Kostruhn (2004)
619	Zawalona; E	19.7267	50.0600	PL	5	2	Alexandrowicz et al. (1992), Kostruhn (2004)
620	Łopiankach I	19.7267	50.0611	PL	5	5	Krukowski (1927), Kozłowski (1960), Kozłowski and Kozłowski (1996), Kostruhn (2004)
621	Gaik 1	19.7272	50.0664	PL	5	5	Kozłowski (1960), Kozłowski and Kozłowski (1996), Wiśniewski (2003), Kostruhn (2004)
622	Gaik 2	19.7278	50.0667	PL	5	4	Kozłowski (1960), Kozłowski and Kozłowski (1996), Wiśniewski (2003), Kostruhn (2004)
623	Wołowice 1	19.7367	49.9969	PL	5	5	Bando et al. (1993), Sobczyk (1993), Kozłowski and Kozłowski (1996), Wiśniewski (2003), Kostruhn (2004)
624	Sąspów, Niedostępną	19.7719	50.2231	PL	5	5	Chmielewski and Kozłowski (1969), Kozłowski and Kozłowski (1977), Madeyska and Kozłowski (1995), Wiśniewski (2003)
625	Nietoperzowa	19.7742	50.1939	PL	5	5	Kozłowski (1922), Chmielewski (1961), Kozłowski and Kozłowski (1996), Wiśniewski (2003)
626	Okrążek	19.7967	50.0158	PL	5	5	Kozłowski and Kozłowski (1996), Sachse-Kozłowska and Kozłowski (2004)
627	Koniówka	19.8194	49.3922	PL	5	5	Rydlewski and Valde-Nowak (1981a), Valde-Nowak (1991), Wiśniewski (2003), Kostruhn (2004)
628	Podczerwone 1	19.8289	49.4072	PL	5	5	Rydlewski and Valde-Nowak (1981c), Valde-Nowak (1991), Wiśniewski (2003), Kostruhn (2004)
629	Maszycka Cave	19.8333	50.2167	PL	5	1	Schild (1975), Kozłowski et al. (1993), Kozłowski and Kozłowski (1996), Wiśniewski (2003), Kostruhn (2004)
630	Puchaczowa Skála	19.8333	50.1836	PL	5	2	Czarnowski (1904), Kozłowski (1922), Kozłowski and Kozłowski (1977, 1996), Wiśniewski (2003), Kostruhn (2004)
631	Dział 1	19.9136	49.4967	PL	5	4	Rydlewski (1989), Valde-Nowak (1991), Wiśniewski (2003) Kostruhn (2004)

(continued)

Table A.1 (continued)

ID	Site; Assemblage	Long	Lat	Co	G	C	References
632	Kozłowskilów 4	20.1608	50.8264	PL	5	5	Ginter (1972), Kozłowski and Kozłowski (1977), Wiśniewski (2003), Kostruhn (2004)
633	Katy 1	20.3503	49.4167	PL	5	1	Rydlewski and Valde-Nowak (1981b, 1982), Kozłowski (1987), Valde-Nowak (1991), Kozłowski and Kozłowski (1996)
634	Sromowce Niżne	20.3950	49.3925	PL	5	5	Rydlewski (1989), Kozłowski and Kozłowski (1996), Kostruhn (2004)
635	Mosty 13	20.4089	50.7817	PL	5	4	Cyrek (1986a), Kozłowski and Pettitt (2001) Wiśniewski (2003), Kostruhn (2004)
636	Aksamitka Cave	20.4469	49.3589	SK	5	5	Valde-Nowak et al. (2007)
637	Stará L'ubovňa	20.7044	49.3053	SK	5	4	Valde-Nowak et al. (2007)
638	Rydno, II/59	20.9617	51.1328	PL	5	2	Kozłowski (1987), Sobczyk (1993), Kostruhn (2004)
639	Pawlow 4	21.1206	50.9622	PL	5	5	Libera et al. (2005), Kozłowski (2006)
640	Uście Gorlickie, Oreszów	21.1400	49.5219	PL	5	5	Valde-Nowak (1985), Kozłowski and Kozłowski (1996)
641	Seredzice	21.1736	51.1550	PL	5	2	Kozłowski and Kozłowski (1996), Kostruhn (2004)
642	Kolonia Seredzice 7	21.2103	51.1533	PL	5	5	Wiśniewski (2003), Kostruhn (2004)
643	Całowanie; II	21.3192	52.0044	PL	5	4	Wiśniewski (2003), Kostruhn (2004)
644	Całowanie; I	21.3192	52.0044	PL	–	4	
645	Maly Antoniów	21.5303	51.0517	PL	5	1	Sawicki (1960), Kozłowski and Kozłowski (1996), Kostruhn (2004)
646	Ćmielów	21.5528	50.8911	PL	5	5	www.archeo.uw.edu.pl/szablon.php?id=333
647	Wilczyce; 10	21.6567	50.7428	PL	5	1	Bratlund (2002), Fiedorczuk and Schild (2002), Kostruhn (2004), Fiedorczuk et al. (2007)
648	Klementowice-Kolonia 20	22.1397	51.3489	PL	5	1	Jastrzębski and Libera (1984, 1987), Kozłowski (1987), Kostruhn (2004)
649	Hłomcza	22.2806	49.6336	PL	5	1	Valde-Nowak and Muzyczuk (2000), Łanczont et al. (2002), Wiśniewski (2003), Kostruhn (2004), Kozłowski (2006)
650	Grodzisko Dolne, Leżajsk	22.4625	50.1625	PL	5	1	Kozłowski and Kozłowski (1977, 1996), Valde-Nowak and Muzyczuk (2000)
651	Przemysl 3	22.7678	49.7836	PL	5	1	Kozłowski and Kozłowski (1977, 1996), Wiśniewski (2003), Kostruhn (2004)

Long Longitude in decimal degrees (WGS 1984), *Lat* Latitude in decimal degrees (WGS 1984), *Co* Country
G Regional Group. 1: Circum-Jurassic Group, 2: Danube Group, 3: Meuse-Rhine Group, 4: Vltava-Saale Group, 5: Polish-Moravian Group, –: No group allocation possible
C Classification. 1: Surely or very probably Magdalenian, 2: Probably Magdalenian, 3: Probably mixed, 4: Probably post-Magdalenian, 5: Not determinable
Abbreviations see list of abbreviations

Table A.2 Recorded sites sorted in alphabetical order

Site; Assemblage	ID	Site; Assemblage	ID	Site; Assemblage	ID
Abri III	589	Bavans, G. Nord de; 9	85	Calvaire, A. du; II sup.	9
Adlerova	582	Bay-Bonnet, G de	338	Calvaire, A. du; III moy.	10
Aeschi, Moosmatten	118	Bečov	512	Calvaire, A. du; IV inf.	11
Ahlendorf	471	Beeck	350	Campalou; CA 2	1
Aichbühl A	230	Beptenaz, G. de	5	Campalou; CA 3a	2
Aichbühl B	231	Bernlochhöhle	201	Čertova dira	604
Aichbühl C	232	Beroun-Závodí	520	Cham-Grindel; I	159
Aichbühl D	233	Bettenroder Berg IX; 18a	393	Cham-Grindel; II West	158
Aichbühl E	234	Bibi, G. (Eremitage)	37	Cham-Grindel; III	160
Aichbühl F	235	Bierawa	606	Châteauneuf, A. de	57
Aksamitka Cave	636	Bildstockfels	172	Châtel-Saint-Denis	87
Alle, Noir Bois	99	Bilsteinhöhle	382	Chaumois-Boivin, G. de; 2	43
Allerberg, Reinhausen	392	Birseck-Ermitage	114	Chauveau	322
Alsdorf	351	Bliszczyce, Głubczyce; 16	598	Chaze 2	31
Andernach-Martinsberg; I	364	Bobache, A. de; 1+2	19	Chênelaz 2C/2	30
Andernach-Martinsberg; II	365	Bobache, A. de; 4+3	20	Chesselgraben	106
Andernach-Martinsberg; III	366	Bockstein Westloch; c	252	Ćmielów	646
Andernach-Martinsberg; IV	367	Bocksteinhöhle; III	253	Col de la Charmette	50
Annakapellenhöhe	210	Bocksteinshmiede and B.-loch	254	Coléoptère, G. de; B	334
Aschenstein	390	Bockstein-Törle; AH II, III	255	Colomb	29
Aschersleben	424	Bois Laiterie	320	Colombière, La; B	14
Aschersleben-Bruchfeld	421	Bolken, Inkwilersee	119	Croze-sur-Suran; R7	8
Auzary-Thônes	82	Bonne Femme	34	Cure, A. de la	84
Bad Frankenhausen	401	Borečnice 2	532	Cyprzanów 3	602
Bad Kösen, Rudelsburg	458	Braunfirst	218	Dehnamühle, üb. d.	429
Bad Sulza, Krähenhütte	448	Breitenfurter Höhle; C	279	Dentaux, A. le	92
Bains-les-Bains	81	Březnice	538	Deszcrowa; 2	610
Balcarka Cave	590	Brillenhöhle; IV lower	246	Deszczowa; 3	611
Balhorn	389	Brillenhöhle; IV upper	247	Dobříčany	511
Balme de Cuiseaux	15	Broniszowice 2	595	Dolni Poříčí 8	513
Balme de Glos	48	Brotel, G. de	7	Douattes, Les; 7/6	69
Balme de Glos 10–11	49	Brückenfeld, Windhügel	425	Dra Majera	607
Balme d'Isère; foyer A	12	Bruderholz	113	Dreieich-Götzenhain Ost (I)	386
Balme-Rousse D2a	16	Brügglihöhle	107	D.-Götzenhain Nord (II)	385
Balverhöhle	377	Brzoskwinia-Krzemionki 7; 1	616	Drexlerloch	194
Banges, G. du; G sup.+inf.	73	Brzoskwinia-Krzemionki 7; 2	617	Dział 1	631
Barbing	307	Burgäschi Hintere Burg	121	Dzierżysław 35	601
Bärenfelsgrotte; III	263	Burghöhle Dietfurt	205	Echt	346
Bärenhöhle; Ib	258	Burghöhle Loch; 2 upper	303	Egolzwil 3	140
Bärenkeller	402	Burk 2, Bautzen; 1	536	Eichenberg	430
Barnberger Höhle	213	Burk 2, Bautzen; 2	537	Eierberg	358
Barová Cave; 11–12	552	Burkhardtshöhle; V	219	Einsiedeln-Langrüti	164
Barrandová cave	523	Burzelberg	500	Eitensheim	282
Baselitz; I	506	Büttenloch; A	104	Engis	333
Baselitz; III	507	Büttenloch; B	105	Erkelenz 2/Haberg; I/VII	353
Baselitz; IV	508	Buttentalhöhle	198	Erkelenz 2/Haberg; VII	354
Baume Noire; M2	65	Býči skála	551	Étrembières	80
Baume Noire; M3	66	Cabônes, Abri; 4	54	Ettringer Bellerberg	363
Baume Noire; M4	67	Całowanie, Otwock; I	643	Eyserheide	348
Baume Noire; M5	68	Całowanie, Otwock; II	644	Farincourt, G. de, surf.	44

(continued)

Table A.2 (continued)

Site; Assemblage	ID	Site; Assemblage	ID	Site; Assemblage	ID
Farincourt, G. de; I+II	45	Graselhöhle	544	Hohlenstein bei Ederheim	269
Farincourt, G. de; III-B	46	Griendtsveen	347	Hohlenstein bei Happurg	286
Farincourt, G. de; III-E	47	Gröbitz	464	Hohlenstein Smb; AHF	283
Faverney	75	Grodzisko Dolne, Leżajsk	650	Hohlenstein-Stadel; II	256
Feldhofhöhle	376	Groitzsch A1	483	Hohlenstein-Stadel; III	257
Felsställe; IIIa	223	Groitzsch A2	484	Hollenberg-Höhle 3; D	115
Felsställe; IIIb	224	Groitzsch B	485	Horní mlýn, Říčky	565
Fischerhäusern, 1	150	Groitzsch B-Nord	486	Hostim	530
Fontabert	36	Groitzsch C1	487	Hoteaux, Les	33
Förderstedt	447	Groitzsch C2	488	Houenbüel 4, 12 (L2-105)	143
Forstbergplateau ü.d. Hain	428	Groitzsch C3+D Nord	489	Houenbüel, 12	142
Forstebrg	431	Groitzsch D1	490	Houenbüel, 13	141
Frauenlucke	541	Groitzsch D2	491	Hranice; III	599
Freudenthal-Höhle	167	Großpürschütz	441	Hubelweid L2-52	146
Freydières, G. des	24	Gryfów Śląski 40	540	Hummelshain, Bremsstall	445
Freymond, A.; 5 inf.	83	Gudenushöhle	542	Hüttenkirchle	202
Frille	387	Guppenlochfels	209	Ichtershausen	399
Fru, A. de la; Aire I	58	Hadí	578	Ilsenhöhle bei Ranis; 4	437
Fru, A. de la; Aire II	59	Hahnenberg	270	Ilsenhöhle bei Ranis; 5	438
Fuchsenloch	285	Haldensteinhöhle; VI	249	Irlich, Geiersberg	373
Fuchskirche I; 3	404	Halle Neustadt, Galgenberg	467	Irlich, Sandgrube	372
Fürst-Albrecht-Höhle	306	Hänichen	482	Jáchymka	550
Fürsteiner, Grab: 43/44	120	Hard I	132	Jean Pierre I; 9A+B	63
Fußgönheim 1	381	Hard II	133	Jean Pierre II	64
Fußgönheim 2	380	Hardberggrotten	102	Jena "an der hohen Saale"	440
Gadenstedt	395	Hartmannsdorf	469	Kačák cave	522
Gaik 1	621	Hausneindorf	412	Kahla-Löbschütz	443
Gaik 2	622	Hauterive-Champréveyres; 1 pri.	93	Kamegg	543
Galgenberg	357	H.-Champréveyres; 1 inf.	94	Kamphausen	359
Galgenberg, Halle	468	H.-Champréveyres; 2	95	Kanne; secteur central	337
Ganserfels	236	H.-Champréveyres; 3	96	Käppele	215
Gatersleben	413	H.-Champréveyres; 1A	97	Kartstein; 5	362
Gay, A.	18	Heidenküche	111	Käsloch	136
Geißenklösterle; AH Io	248	Heidenstein	200	Kastelhänghöhle; 5	290
Gera-Liebschwitz	478	Heidenstein	295	Kastelhöhle-Nord; middle	109
Gerbaix, A.	60	Heimbach-Weis	374	Kastelhohle-Nord; upper	110
Gigny, G. de; IV inf.	25	Helga-Abri; IIIa+b	238	Kateřinská	556
Gigny, G. de; IV sup.	26	Helga-Abri; IIId	239	Katy 1	633
Gleina	470	Henri Martin, A.	40	Kaufertsberg; lower	271
Gnirshöhle 1	195	Herdloch bei Ranis; 1+2	435	Kaufertsberg; upper	272
Gnirshöhle 2	196	Herwartstein	251	Keblice	527
Göldimoos 2 (L2.125)	148	Himmelreich	426	Kehlen-Juckelsboesch	349
Gönnersdorf I	368	Hinterberg I (1117)	161	Kerzenstübli	168
Gönnersdorf II	369	Hitzkirch-Baldegger See	156	Kesslerloch; II	170
Gönnersdorf III	370	Hłomcza	649	Kesslerloch; III	171
Gönnersdorf IV	371	Hohle Fels, Schelk. Ia	240	Kirchberg	436
Gordemitz, Fuchsberg	493	Hohle Fels, Schelk. Ib	241	Kleine Ganghöhle	112
Gorsleben	410	Hohle Fels, Schelk. Ic	242	Kleine Scheuer im Hohlenstein-Stadel III	259
Goyet 3; chambre A, 1	332	Hohle Fels, Schelk. IIa	243	Kl. Scheuer im Rosenstein	250
Grand'Baile, G. de la	17	Hohlefels Hütten; grey	221	Kleingrabe, Kalkkopf	397
Grappin (G. St.-Vincent)	28	Hohlefels Hütten; yellow	222	Klementowice-Kolonia 20	648

(continued)

Table A.2 (continued)

Site; Assemblage	ID	Site; Assemblage	ID	Site; Assemblage	ID
Klemmer-A.	260	Maloměřice-BorkyI	547	Okiennik	609
Klingenfels-Abri.	262	Maly Antoniów	645	Okrążek	626
Klopfjörgelshütte	214	Martinshöhle, Letmathe	375	Oldisleben, Sachsenburg	406
Kniegrotte	451	Maszycka Cave	629	Olette, Gortte d l'	35
Kohlerhöhle	108	Merkers	394	Orp; est	330
Kohltalhöhle	212	Mesch-Steenberg	341	Orp; ouest	331
Kolíbky; 5	569	Metten; I+II	501	Osmnáctka	557
Kolíbky; 6	570	Michalova	592	Papierfels	220
Kolonia Seredzice 7	642	Milenovice 1	534	Passagère	27
Koněprusy	517	Mittlere Klause; AH1	289	Pawlow 4	639
Königsaue, Bruchberg	419	Mokrá, Lom I	586	Pekárna; g/h	579
Königswinkel	434	Mokrá, Lom V	588	Pekárna; i	580
Koniówka	627	Mont-les-Etrelles	62	Petersberg	465
Koňská jáma	554	Moos (L2.55)	155	Petersfels; P1AH2	173
Kozlów 4	632	Moos, 14	151	Petersfels; P1AH3A	174
Krápníková	529	Göldimoos 1 (L2.124)	153	Petersfels; P1AH3B	175
Kravská	581	Moosmatt, 11	145	Petersfels; P1AH4	176
Kremmen	502	Moosseedorf-Moosbühl	101	Petersfels; P2	177
Křižova	583	Morl	463	Petersfels; P3AH1/2	178
Kroczycka Cave	613	Mosty 13	635	Petersfels; P3AH2	179
Krusza Skála	612	Mühlberg-Grotten	278	Petersfels; P3AH2/3	180
Kuhstallhöhle	204	Mühleloch, A.	137	Petersfels; P3AH3	181
Kůlna Cave; 5	571	Munzingen	122	Petersfels; P3AH3/4	182
Kůlna Cave; 6	572	Munzingen; GH4 1976–1977	123	Petersfels; P3AH4	183
Kulnička	576	Mutterstadt	383	Petersfels; P3AH5	184
Kuppelberg	494	Na Průchodě	531	Petersfels; P5AH2	185
Kvíc	518	Náchod	546	Petersfels; P5AH4	186
La Garenne; Vénérieu	6	Nähe Forstberg	432	Petersfels; P5AH5	187
La Guillotine, Chariez	74	Nahin, G. de	72	Petersfels; P6AH2	188
La Justice	70	Napoleonskopf	197	Petersfels; P6AH3/4	189
La Piquette, G. de	71	Prinz-Heinrich-Stein	459	Petersfels; P6AH4	190
La Roche, G. de la	55	Nebra	439	Petersfels; P6AH5	191
Lausen	125	Neuchâtel-Monruz; 1A	89	Petersfels; P6AH6	192
Lausnitz, A. Theure	452	Neuchâtel-Monruz; 1M	90	Petersfels;P3AH1	193
Le Cimetière	22	Neuchâtel-Monruz; 2M/A	91	Pfortener Berg, Gera	477
Lengefeld, surface	456	Neue Höhle, Thayngen	169	Planches, G. des	61
Leuna-Göhlitzsch	473	Niederlommatzsch	505	Pod koňským spádem 9 & 10	561
Lhota	521	Nietoperzowa	625	Pod vyhlídkou = Klímova	575
Liemehna	492	Nikolaushöhle; 2	211	Podczerwone 1	628
Liesberg	100	Nová Drátenická Cave; 15	563	Poserna, Burgstättl	476
Lindentaler Hyänenhöhle	481	Občiny	548	Pötewitz	472
Lišči	587	Obere Klause; lower	287	Poustevna	574
Líšň-Čtvrtě	553	Obere Klause; upper	288	Pracký kopec	591
Lixhe – Sur le Bois	340	Oberlistingen	388	Presle, G de	308
Lochschlaghöhle	280	Obermoos, 206	149	Probstfels; I–IV	199
Łopiankach I	620	Obermoos, 28	144	Protivec 3	524
Łopiankach II	618	Obernederhöhle 2up. + 1	301	Průchodnice	593
Loštice I	594	Ochozska Cave	585	Przemysl 3	651
Lüptitz (Spitzberg)	498	Oelknitz	444	Puchaczowa Skála	630
Macucha, Horákov	560	Ofnet 1, Kleine	267	Pugieu, A. de	39
Magdala	423	Ofnet, Große; 6	268	Puklinová; 6	558

(continued)

Table A.2 (continued)

Site; Assemblage	ID	Site; Assemblage	ID	Site; Assemblage	ID
Putim; 162, 767	525	Schönburg	460	Sweikhuizen-Oude Stort	343
Putim; 773	526	Schuntershöhle	225	Tai, G. du; C′1	3
Qued., Bockshornschanze	408	Schussenquelle; Com.A	227	Tai, G. du; C″1	4
Quedlinburg, Lehof	409	Schussenquelle; Com.C	228	Tessonière, G. de la	13
Quedlinburg, Mühlgraben	411	Schussenquelle; main	229	Tetín	528
Rabensteinhöhle	226	Schweizersbild	166	Teufelsbrücke; 1	415
Raillarde, La	21	Schweskau	414	Teufelsbrücke; 2	416
Randecker Maar	217	Sedel 2	157	Teufelsbrücke; 3	417
Rappenfels	216	Sedmnáctka	562	Teufelsbrücke; 4	418
Räuberhöhle	304	Seeburg	454	Teufelsküche	126
Rehmen	446	Seehausen, Mutzenbrunnen	403	Teufelslucke	545
Reiden-Stumpen	139	Seewagen, 24 (Bolt VIa)	154	Thierstein	103
Rennerfels	284	Seewagermoos, 17	152	Thoys, A. de; TH 2	41
Rennerweg, Anhöhe	427	Seredzice	641	Thoys, A. de; TH1	42
Rheinfelden-Eremitage; 1	128	Sesselfelsgrotte; B2	291	Tmaň, Děravá cave; lower I	515
Rigney 1, G. de; D	77	Sesselfelsgrotte; B3	292	Tmaň, Děravá cave; upper I	516
Rigney 1, G. de; M	78	Sesselfelsgrotte; C1	293	Towarna Góra	608
Rigny "En Terredey"	38	Sesselfelsgrotte; C2	294	Trebur	384
Rislisberghöhle	124	Seveux	56	Trier Gläsgesberg	361
Roche Plate; 12	336	Siegfriedfelsen; D1	297	Trimbach	130
Roche-al-Rue	321	Siegfriedfelsen; D2	298	Trockhausen	457
Rochedane; D1/D2	86	Siegfriedfelsen; D3	299	Trou da Somme	319
Roc-la-Tour; 1	313	Siegfriedfelsen; D4	300	Trou de Blaireaux; II1	309
Roc-la-Tour; 2	314	Šipka dira	603	Trou de Blaireaux; III 1	310
Roc-la-Tour; 3	315	Sirgenstein	245	Trou de Blaireaux; III 2	311
Roc-la-Tour; 4	316	Sirgenstein Südwand	244	Trou de Blaireaux; III 3	312
Romains; 2a	51	Skleb	555	Trou de Chaleux; 5	324
Romains; 2b	52	Slaník 1	514	Trou de Chaleux; ex.Otte	325
Romains; 3	53	Śmicz B	596	Trou de Chêne	318
Röthekopf	138	Smrtní	564	Trou de la Mâchoire	327
Rothestein "Kuhberg"	442	Solgen Im Grauen	165	Trou des Nutons (Furfooz)	326
Rozcestí	549	Šosůvská	573	Trou du Frontal	328
Rüteliacher	116	Sous-la-Roche	23	Trou du Sureau	317
Rütihard	117	Sowin 7	597	Trou Magrite	323
Rydno, II/59	638	Speckberg; A+B	281	Trou Reuviau	329
Rytířská	533	Spitalfels; VIIIa-VIIIb	264	Trou Walou	339
Saaleck	455	Spitzbubenhöhle; AH2	261	Tunnelhöhle	305
Sachsenburg	407	Srnči	577	Turské maštale cave	519
Saint-Brais II	98	Sromowce Niżne	634	Udersleben	405
Sälihöhle-Oben	131	Stará L'ubovňa	637	Uff Wigg	129
Salsitz	479	Station d. Grenouilles	76	Untere Bsetzi	162
Samborowice 2a	605	Steinberg	127	Urdhöhle	449
Sandmatt, 25	147	Steinbergwand; VIII	302	Uście Gorlickie, Oreszów	640
Sankt Gangloff	462	Steinerner Rosenkranz	273	Val Molière	32
Sąspów, Niedostępną	624	Steinkirche bei Scharzfeld	396	Verlaine, G. de Sy	335
Scé du Châterlard, G.	88	Stendel, A.	391	Verunčina	568
Schaelsberg, Valkenburg	342	Stendorf	453	Veyrier	79
Schafgraben, Zwötzen	480	Straßberger G.	203	Viersener Donk I	355
Schellneck	296	Švédův stůl; 50	584	Viersener Donk II	356
Schkauditz	474	Sweikhuizen-Groene Paal	344	Vilshofen-Kuffing	504
Schmiechenfels; 2	237	Sweikhuizen-Koolweg	345	Vinckova	566

(continued)

Table A.2 (continued)

Site; Assemblage	ID	Site; Assemblage	ID	Site; Assemblage	ID
Vogelherd u. Rittergutsfeld	433	Weinbergh.; Graben außen	276	Włodzienin 3	600
Vogelherd; II	266	Weinbergh.; Vorplatz 3	277	Wołowice 1	623
Vogelherd; III	265	Weißenborn, Speckberg	461	Wurzen-Dehnitz, Wüste Kirche	495
Vorderer Eichen	163	Weißenfels	466	Wüste Scheuer	450
Výpustek	559	Westerhausen	400	Zalas; 10–11	614
Wachtelberg	497	Westewitz	503	Zalas; 7–9	615
Wachtelberg; A	496	Wilczyce; 10	647	Zawalona; E	619
Wallendorf	475	Wildscheuer; V	379	Žd'ár 1	535
Wandersleben, a. Bahnhof	398	Wildweiberlei	378	Želeč	510
Wantewitz; II	509	Wilmatt	134	Zigeunerfels; G	206
Wegberg-Berg	352	Wilsleben, Klintberg	422	Zigeunerfels; H	207
Węgliny	539	Wilsleben, Ortsausgang	420	Zigeunerfels; H/I	208
Weinberghöhlen; Vorplatz 1+2	274	Winterdorf	360	Zinkenberg	499
Weinbergh.; Saazer Loch	275	Winznau-Köpfli	135	Žitného	567

Table A.3 Reference sample of Hamburgian sites

ID	Site	Long	Lat	Co	References
H1	Hooge Berg B	4.4906	53.0356	NL	van Noort (2007)
H2	Sassenheim	4.5142	52.2292	NL	van Noort (2007)
H3	Paal 9	4.7119	53.0214	NL	van Noort (2007)
H4	Den Hoorn	4.7533	53.0228	NL	van Noort (2007)
H5	Den Burg	4.7958	53.0536	NL	Stapert (1981), van Noort (2007)
H6	Hooge Berg A	4.8153	53.0364	NL	van Noort (2007)
H7	Aardjesberg	5.2022	52.2547	NL	van Noort (2007)
H8	Zeist	5.2383	52.0819	NL	van Noort (2007)
H9	Soesterduinen	5.2914	52.1528	NL	Bohmers (1956), Narr (1968), Tromnau (1975b), Burdukiewicz (1986a)
H10	Kolderwolde	5.5100	52.8972	NL	van Noort (2007)
H11	Rhenen, Laarscherberg	5.5594	51.9758	NL	van Noort (2007)
H12	Ermelo	5.6278	52.3006	NL	Popping (1931), van Noort (2007)
H13	Harderwijk	5.6469	52.3317	NL	Bohmers (1947), Narr (1968), Burdukiewicz (1986a)
H14	Stroe	5.6928	52.1858	NL	Bohmers (1947), Burdukiewicz (1986a)
H15	Ginkelse Heide II	5.7111	52.0456	NL	van Noort (2007)
H16	Garderen	5.7144	52.2328	NL	van Noort (2007)
H17	Ginkelse Heide II	5.7289	52.0314	NL	van Noort (2007)
H18	Hulshorster Zand	5.7333	52.3428	NL	van Noort (2007)
H19	Elspeet	5.7889	52.2911	NL	Popping (1931), Bohmers (1947)
H20	Schaveren	5.9572	52.3225	NL	van Noort (2007)
H21	Neer II	5.9889	51.2603	NL	Bohmers (1956), Burdukiewicz (1986a)
H22	Suameer	6.0006	53.1778	NL	Bohmers (1947), Burdukiewicz (1986a)
H23	Oldeholtwolde	6.0092	52.9142	NL	Johansen and Stapert (2004), van Noort (2007)
H24	Oostermeer	6.0611	53.1742	NL	Bohmers (1947), Burdukiewicz (1986a)
H25	Houtigehage I	6.1444	53.1539	NL	Bohmers (1947), Tromnau (1975b), Burdukiewicz (1986a)
H26	Kallenkote	6.1636	52.7911	NL	van Noort (2007)
H27	Ureterp I	6.1683	53.0958	NL	Bohmers (1947), Burdukiewicz (1986a)
H28	Hoornsterzwaag	6.1725	52.9983	NL	van Noort (2007)
H29	Ureterp II	6.1731	53.0906	NL	Bohmers (1947), Burdukiewicz (1986a)
H30	Wijnjeterp	6.1942	53.0461	NL	Bohmers (1947), Burdukiewicz (1986a)
H31	Vledder	6.2103	52.8572	NL	van Noort (2007)
H32	Makkinga	6.2172	52.9681	NL	Popping (1934), Narr (1968), Tromnau (1975b), Burdukiewicz (1986a)

(continued)

Table A.3 (continued)

ID	Site	Long	Lat	Co	References
H33	Duurswoude II	6.2364	53.0811	NL	Burdukiewicz (1986a)
H34	Duurswoude IV	6.2367	53.0803	NL	Burdukiewicz (1986a)
H35	Havelte Holtingerzand	6.2397	52.7697	NL	Bohmers (1947), Narr (1968), Tromnau (1975b), Burdukiewicz (1986a)
H36	Havelte Westerzand	6.2419	52.7703	NL	Bohmers (1947), Narr (1968), Tromnau (1975b), Burdukiewicz (1986a)
H37	Duurswoude I	6.2428	53.0856	NL	van Noort (2007)
H38	Bakkeveen	6.2500	53.0833	NL	Bohmers (1947), Narr (1968), Burdukiewicz (1986a)
H39	Marum	6.2681	53.1442	NL	Bohmers (1956), Narr (1968), Tromnau (1975b), Burdukiewicz (1986a)
H40	Weper	6.3042	53.0078	NL	Bohmers (1947), Schwabedissen (1950), Tromnau (1975b), Burdukiewicz (1986a)
H41	Haule I	6.3144	53.0519	NL	van Noort (2007)
H42	Diever	6.3158	52.8550	NL	Stapert (1982), van Noort (2007)
H43	Appelscha	6.3503	52.9553	NL	Bohmers (1947), Narr (1968), Burdukiewicz (1986a)
H44	Luttenberg	6.3683	52.4039	NL	van Noort (2007)
H45	Zuidwolde II	6.4103	52.6725	NL	van Noort (2007)
H46	Zuidwolde I	6.4156	52.6767	NL	van Noort (2007)
H47	Norg	6.4611	53.0664	NL	Stapert (1982), van Noort (2007)
H48	Zwarte Water	6.4864	52.7789	NL	Bohmers (1947), Burdukiewicz (1986a), van Noort (2007)
H49	Gijzelte	6.5025	52.7744	NL	Bohmers (1960), Burdukiewicz (1986a)
H50	Wijster	6.5156	52.8150	NL	van Noort (2007)
H51	Zeijen	6.5464	53.0481	NL	Bohmers (1956), Burdukiewicz (1986a)
H52	Eelde	6.5633	53.1347	NL	Bohmers (1947), Narr (1968), Tromnau (1975b), Burdukiewicz (1986a)
H53	Yde	6.5933	53.1144	NL	Bohmers (1947), van Noort (2007)
H54	Rolde	6.6456	52.9878	NL	Bohmers (1947), Narr (1968), Tromnau (1975b), Burdukiewicz (1986a)
H55	Gees	6.6889	52.7475	NL	Bohmers (1947), Schwabedissen (1954), Narr (1963), Tromnau (1975b), Burdukiewicz (1986a)
H56	Anloo	6.6972	53.0428	NL	van Noort (2007)
H57	Aalden	6.7333	52.7833	NL	Bohmers (1947), Narr (1968), Burdukiewicz (1986a)
H58	Gasselte	6.7911	52.9722	NL	Bohmers (1947), Narr (1968), Tromnau (1975b), Burdukiewicz (1986a)
H59	Buinen	6.8336	52.9311	NL	van Noort (2007)
H60	Reutum	6.8450	52.3919	NL	van Noort (2007)
H61	Weerselo II	6.8536	52.3456	NL	van Noort (2007)
H62	Weerselo I	6.8553	52.3497	NL	van Noort (2007)
H63	Borken-Gemenkrückling	6.8908	51.8728	DE	Gaffrey (1997)
H64	Emmen	6.8947	52.7875	NL	van Noort (2007)
H65	Agelerbroek	6.9178	52.3856	NL	van Noort (2007)
H66	Dörgener Moor	7.2939	52.7250	DE	Schwabedissen (1937), Tromnau (1974a)
H67	Peheim	7.8358	52.8828	DE	Schwabedissen (1937), Tromnau (1975b), Burdukiewicz (1986a)
H68	Querenstede I+II	7.9647	53.1497	DE	Zoller (1962, 1963), Tromnau (1975b), Burdukiewicz (1986a)
H69	Querenstede III	7.9647	53.1497	DE	Zoller (1962, 1963), Tromnau (1975b), Burdukiewicz (1986a)
H70	Glaner Heide	8.4508	52.8758	DE	Schwabedissen (1937), Bohmers (1947), Tromnau (1975b), Burdukiewicz (1986a)
H71	Hasenworth	8.6178	53.8508	DE	Tromnau (1975b), Burdukiewicz (1986a)
H72	Prangenbusch	8.6281	53.8644	DE	Büttner (1940a) Tromnau (1975b), Burdukiewicz (1986a)

(continued)

Table A.3 (continued)

ID	Site	Long	Lat	Co	References
H73	Pennenmoor, Sahlenburg	8.6300	53.8639	DE	Büttner (1936, 1940b), Schwabedissen (1937), Tromnau (1975b), Burdukiewicz (1986a)
H74	Twellbergsmoor	8.6469	53.8792	DE	Wegewitz (1941), Tromnau (1975b), Burdukiewicz (1986a)
H75	Ahrenshöft 61	9.0667	54.5589	DE	Hinz (1954), Tromnau (1975b) Burdukiewicz (1986a)
H76	Oldersbek	9.1656	54.4425	DE	Hinz (1954), Tromnau (1975b), Burdukiewicz (1986a)
H77	Reederstall	9.2064	54.2283	DE	Tromnau (1975b), Burdukiewicz (1986a)
H78	Jels 1	9.2083	55.3606	DK	Holm and Rieck (1983)
H79	Schalkholz	9.2569	54.2461	DE	Tromnau (1974b, 1975b), Burdukiewicz (1986a)
H80	Bunsoh	9.3242	54.1644	DE	Schwabedissen (1944), Tromnau (1975b), Burdukiewicz (1986a)
H81	Wittlohe	9.3497	52.8564	DE	Breest and Oldenburg (2000)
H82	Hvejsel, Vejle	9.4319	55.8122	DK	Becker (1970)
H83	Bjerlev	9.4333	55.8197	DK	Becker (1970, 1971), Burdukiewicz (1986a)
H84	Stade-Campe	9.4847	53.5919	DE	Schwabedissen (1937), Tromnau (1975b), Burdukiewicz (1986a)
H85	Drage	9.5200	54.0042	DE	Schwabedissen (1937), Burdukiewicz (1986a)
H86	Agathenburg	9.5333	53.5667	DE	Wegewitz (1950), Tromnau (1975b)
H87	Dollern	9.5458	53.5383	DE	Wegewitz (1941), Schwabedissen (1944), Tromnau (1975b), Burdukiewicz (1986a)
H88	Neuenkirchen	9.7067	53.0325	DE	Taute (1959), Tromnau (1975b), Burdukiewicz (1986a)
H89	Ketzendorf 3	9.7764	53.4431	DE	Wegewitz (1941), Bohmers (1947), Tromnau (1975b), Burdukiewicz (1986a)
H90	Ketzendorf 2	9.7764	53.4431	DE	Wegewitz (1941), Bohmers (1947), Tromnau (1975b), Burdukiewicz (1986a)
H91	Ketzendorf surface	9.7764	53.4431	DE	Wegewitz (1941), Bohmers (1947), Burdukiewicz (1986a)
H92	Hitzhusen	9.8514	53.9228	DE	Tromnau (1975b), Burdukiewicz (1986a)
H93	Heber 118	9.8781	53.0786	DE	Tromnau (1975a), Burdukiewicz (1986a)
H94	Buchholz-Bünsen	9.8781	53.3275	DE	Tromnau (1975b), Burdukiewicz (1986a)
H95	Heber 127	9.8850	53.0767	DE	Tromnau (1975a), Burdukiewicz (1986a)
H96	Heber 128	9.8864	53.0783	DE	Tromnau (1975a), Burdukiewicz (1986a)
H97	Deimern 42	9.9000	53.0331	DE	Tromnau (1975a), Burdukiewicz (1986a)
H98	Deimern 41a	9.9000	53.0319	DE	Tromnau (1975a), Burdukiewicz (1986a)
H99	Deimern 41	9.9003	53.0325	DE	Tromnau (1975a), Burdukiewicz (1986a)
H100	Deimern 41b	9.9003	53.0303	DE	Tromnau (1975a), Burdukiewicz (1986a)
H101	Deimern 44	9.9006	53.0300	DE	Tromnau (1975a), Burdukiewicz (1986a)
H102	Deimern 1002	9.9006	53.0325	DE	Tromnau (1975a), Burdukiewicz (1986a)
H103	Deimern 1007	9.9006	53.0319	DE	Tromnau (1975a), Burdukiewicz (1986a)
H104	Deimern 1003	9.9008	53.0328	DE	Tromnau (1975a), Burdukiewicz (1986a)
H105	Deimern 1004	9.9011	53.0322	DE	Tromnau (1975a), Burdukiewicz (1986a)
H106	Deimern 46	9.9014	53.0325	DE	Tromnau (1975a), Burdukiewicz (1986a)
H107	Dageförde	9.9472	52.8431	DE	Breest (2000)
H108	Marmstorf	9.9683	53.4358	DE	Wegewitz (1941), Tromnau (1975b), Burdukiewicz (1986a)
H109	Rönneburg	10.0072	53.4367	DE	Tromnau (1974b)
H110	Glüsingen	10.0178	53.4097	DE	Wegewitz (1941), Tromnau (1975b), Burdukiewicz (1986a)
H111	Meckelfeld	10.0233	53.4208	DE	Tromnau (1974b)
H112	Bonstorf	10.0486	52.8642	DE	Veil et al. (1998), Breest (2000)
H113	Beckedorf	10.0583	52.8156	DE	Siekmann (1976), Breest (2000)

(continued)

Table A.3 (continued)

ID	Site	Long	Lat	Co	References
H114	Wellingsbüttel	10.0797	53.6408	DE	Schindler (1960), Tromnau (1975b), Burdukiewicz (1986a)
H115	Sasel	10.1067	53.6536	DE	Brüchmann (1952), Schindler (1960), Tromnau (1975b), Burdukiewicz (1986a)
H116	Naherfurth	10.1250	53.7678	DE	Tromnau (1974c)
H117	Bergstedt	10.1269	53.6711	DE	Brüchmann (1952), Schindler (1960), Tromnau (1975b), Burdukiewicz (1986a)
H118	Itzstedt	10.1439	53.8086	DE	Tromnau (1974c, 1975b), Burdukiewicz (1986a)
H119	Volksdorf	10.1578	53.6486	DE	Schwabedissen (1937), Tromnau (1975b), Burdukiewicz (1986a)
H120	Meiendorf D	10.1667	53.6333	DE	Tromnau (1975b), Burdukiewicz (1986a)
H121	Volksdorf	10.1706	53.6417	DE	Schindler (1960), Tromnau (1975b), Burdukiewicz (1986a)
H122	Meiendorf 2	10.1944	53.6386	DE	Rust (1937), Burdukiewicz (1986a)
H123	Meiendorf 6	10.1978	53.6317	DE	Schindler (1960), Bohnsack (1961), Tromnau (1975b), Burdukiewicz (1986a)
H124	Hasewisch	10.2008	53.6383	DE	Rust (1958), Tromnau (1975b), Burdukiewicz (1986a)
H125	Poggenwisch	10.2017	53.6389	DE	Rust (1958), Tromnau (1975b), Burdukiewicz (1986a)
H126	Meiendorf 7a	10.2017	53.6361	DE	Tromnau (1975b), Burdukiewicz (1986a)
H127	Teltwisch 1	10.2025	53.6425	DE	Tromnau (1975b), Burdukiewicz (1986a)
H128	Meiendorf-Pond	10.2036	53.6419	DE	Rust (1937), Burdukiewicz (1986a)
H129	Teltwisch 3	10.2036	53.6419	DE	Tromnau (1975b)
H130	Meiendorf 9	10.2042	53.6411	DE	Tromnau (1975b), Burdukiewicz (1986a)
H131	Uetze	10.2047	52.4656	DE	Cosack (1970), Schwarz-Mackensien (1978), Burdukiewicz (1986a)
H132	Stellmoor D	10.2069	53.6450	DE	Tromnau (1975b), Burdukiewicz (1986a)
H133	Stellmoor 1	10.2072	53.6447	DE	Tromnau (1975b), Burdukiewicz (1986a)
H134	Stellmoor 2	10.2072	53.6447	DE	Tromnau (1975b), Burdukiewicz (1986a)
H135	Stellmoor 3	10.2072	53.6447	DE	Tromnau (1975b), Burdukiewicz (1986a)
H136	Stellmoor 4	10.2072	53.6447	DE	Tromnau (1975b), Burdukiewicz (1986a)
H137	Stellmoor 5	10.2072	53.6447	DE	Tromnau (1975b), Burdukiewicz (1986a)
H138	Stellmoor 6	10.2072	53.6447	DE	Tromnau (1975b), Burdukiewicz (1986a)
H139	Stellmoor 7	10.2072	53.6447	DE	Tromnau (1975b), Burdukiewicz (1986a)
H140	Stellmoor-Pond	10.2081	53.6456	DE	Rust (1943), Burdukiewicz (1986a)
H141	Krattwisch	10.2150	53.6553	DE	Tromnau (1975b), Van Noort (2007)
H142	Meiendorf 7	10.2183	53.6358	DE	Tromnau (1975b), Burdukiewicz (1986a)
H143	Borneck	10.2417	53.6592	DE	Rust (1958), Burdukiewicz (1986a)
H144	Lachendorf	10.2467	52.6219	DE	Breest (2000)
H145	Ohof Bossel	10.2911	52.4528	DE	Schwarz-Mackensien (1978), Burdukiewicz (1986a), Valoch (2001)
H146	Klein Niendorf	10.2994	53.9494	DE	Tromnau (1975b), Burdukiewicz (1986a)
H147	Kröppelshagen-Fahrendorf	10.3156	53.4869	DE	Kersten (1951, 1961), Tromnau (1975b), Burdukiewicz (1986a)
H148	Bimöhlen	10.3167	53.9319	DE	Tromnau (1975b), Burdukiewicz (1986a)
H149	Grande	10.3908	53.5814	DE	Rust (1943), Hingst (1959), Tromnau (1975b), Burdukiewicz (1986a)
H150	Kudderwörde	10.4000	53.5869	DE	Kersten (1951, 1961), Tromnau (1975b), Burdukiewicz (1986a)
H151	Grünhof-Tesperhude	10.4317	53.4067	DE	Tromnau (1975b), Burdukiewicz (1986a)
H152	Klein-Vollbüttel	10.4911	52.4158	DE	Schwabedissen (1937), Zeitz (1969), Schwarz-Mackensien (1978), Burdukiewicz (1986a)
H153	Gifhorn-Kästorf II	10.5381	52.5200	DE	Schwarz-Mackensien (1978), Burdukiewicz (1986a)

(continued)

Table A.3 (continued)

ID	Site	Long	Lat	Co	References
H154	Rieseberg	10.7869	52.2956	DE	Taute (1959), Tromnau (1975b), Burdukiewicz (1986a)
H155	Grömitz	10.9561	54.1511	DE	Brückner (1954, 1970), Tromnau (1975b), Burdukiewicz (1986a)
H156	Großwusterwitz	12.3789	52.3781	DE	Schwantes (1928), Mey (1960), Bosinski (1982)
H157	Tanowo 2	14.4661	53.5392	PL	Galiński (1987)
H158	Wojnowo I/75	15.7861	52.0989	PL	Kobusiewicz and Kabaciński (1992)
H159	Siedlnica 17a	16.3317	51.7647	PL	Burdukiewicz (1980), Burdukiewicz (1986a, b)
H160	Siedlnica 17-I/95	16.3319	51.7650	PL	Burdukiewicz et al. (1996)
H161	Olbrachcice 8	16.3644	51.7867	PL	Burdukiewicz (1986a)
H162	Łęgon 5	16.3903	51.7792	PL	Burdukiewicz and Szynkiewicz (2002), Burdukiewicz et al. (2003)
H163	Liny 1	16.6578	51.3433	PL	Mey (1960), Tromnau (1975b), Burdukiewicz (1986a)
H164	Mircowice 33	17.3947	52.7911	PL	Chłodnicki and Kabaciński (1997), Kabaciński and Schild (2005)
H165	Olbrachice 14	17.7178	50.3614	PL	Burdukiewicz (1986a)
H166	Rogow Opolski 9	17.9347	50.5189	PL	Rothert (1936), Mey (1960), Burdukiewicz (1986a)
H167	Krągola 25	18.2425	52.1603	PL	Kabaciński and Kobusiewicz (2007)
H168	Nowy Młyn I/77	20.9572	50.1172	PL	Schild and Królik (1981), Burdukiewicz (1986a)
H169	Nowy Mlyn 3	20.9614	50.1319	PL	Kozłowski and Kozłowski (1977), Burdukiewicz (1986a)
H170	Nowy Mlyn 3St.Ia/23	20.9614	50.1319	PL	Kozłowski and Kozłowski (1977), Burdukiewicz (1986a)

Abbreviations see list of abbreviations

Long Longitude in decimal degrees (WGS 1984), *Lat* Latitude in decimal degrees (WGS 1984), *Co* Country

Table A.4 Geographical position of the recorded palynological sites

ID	Site	Long.	Lat.	References	ID	Site	Long.	Lat.	References
1	Milheeze	5.7969	51.5097	Bos (1998)	29	Schwarzer See	12.2481	47.7683	Rausch (1975)
2	Gulickshof I	5.8972	51.0603	Hoek (1997)	30	Lauter	12.7133	47.8736	Schmeidl (1971)
3	Pelléautier Vb	6.1833	44.5172	de Beaulieu and Reille (1983)	31	Moosham	12.9108	47.9197	Schmeidl (1971)
4	Marais du Rosey	6.4756	46.6050	Gaillard (1985)	32	Seemoos I	13.7831	47.0833	Bortenschlager (1967)
5	Marais de Rances	6.5128	46.7725	Gaillard (1985)	33	Bełchatów	14.8711	53.7772	Ralska-Jasiewiczowa et al. (Eds.) (2004a)
6	Siguret	6.5500	44.7917	de Beaulieu and Reille (1983)	34	Kiensass	15.0283	48.8147	Peschke (1972)
7	Le Tronchet	6.7292	46.5144	Gaillard (1985)	35	Grossweissenbach	15.1606	48.5506	Peschke (1977)
8	Meerfelder Maar	6.7564	50.1006	Litt and Stebich (1999)	36	Hrabetice	16.4133	48.7981	Peschke (1977)
9	Usselo	6.8325	52.2033	van Geel et al. (1989)	37	Hrabetice	16.4133	48.7981	Peschke (1977)
10	Holzmaar	6.8792	50.1189	Litt and Stebich (1999)	38	Węglewice	18.2167	51.4167	Ralska-Jasiewiczowa et al. (Eds.) (2004a)
11	La Joux des Pont	6.9564	46.6042	Gaillard (1985)	39	Lake Gopło III	18.3167	52.6667	Ralska-Jasiewiczowa et al. (Eds.) (2004a)
12	Lobsigensee	7.2981	47.0306	Ammann (1985)	40	Rośle	18.7667	52.1167	Ralska-Jasiewiczowa et al. (Eds.) (2004a)
13	Lörmoos	7.4122	46.9828	Zwahlen (1985)	41	Bronów	18.9000	49.7500	Ralska-Jasiewiczowa et al. (Eds.) (2004a)
14	Westrhauderfehn	7.5539	53.1336	Behre (1967)	42	Niechorze	19.1667	51.2000	Ralska-Jasiewiczowa et al. (Eds.) (2004a)
15	Amsoldingersee	7.5750	46.7250	Lotter (1985)	43	Aleksandrów	19.2000	51.7333	Ralska-Jasiewiczowa et al. (Eds.) (2004a)
16	Horbacher Moor	8.1061	47.7344	Lang (1954)	44	Wolbrom 1	19.5833	50.5833	Ralska-Jasiewiczowa et al. (Eds.) (2004a)
17	Erlenbruckmoor	8.1103	47.8914	Lang (1952)	45	Wolbrom 3	19.6667	50.6667	Ralska-Jasiewiczowa et al. (Eds.) (2004a)
18	Nussbaumer See 1	8.8156	47.6161	Rösch (1985)	46	Wolbrom 2	19.7500	50.7500	Ralska-Jasiewiczowa et al. (Eds.) (2004a)
19	Hämelsee	9.3106	52.7594	Merkt and Müller (1999)	47	Stary Cieszyn St.C2	19.8667	54.0500	Ralska-Jasiewiczowa et al. (Eds.) (2004a)
20	Federsee, Schienen	9.6631	48.0453	Göttlich (1961)	48	Stary Cieszyn St.C9	19.9000	54.1667	Ralska-Jasiewiczowa et al. (Eds.) (2004a)
21	Schleinsee, 2	9.6675	47.5933	Lang (1952)	49	Błoto	20.3500	50.0667	Ralska-Jasiewiczowa et al. (Eds.) (2004a)
22	Kubitzbergmoor	10.1333	54.4000	Usinger (1975)	50	Jasło	21.4500	49.5333	Ralska-Jasiewiczowa et al. (Eds.) (2004a)
23	Luttersee	10.1644	51.5650	Chen (1982)	51	Lake Mikołajki	21.6000	53.5667	Ralska-Jasiewiczowa et al. (Eds.) (2004a)
24	Gaterslebener See	11.3528	51.8222	Müller (1953)	52	Lake Miłkowskie	21.8833	53.6833	Ralska-Jasiewiczowa et al. (Eds.) (2004a)
25	Moor bei Lanser See	11.4192	47.2422	Bortenschlager (1984)	53	Podbukowina I	22.3167	49.8000	Ralska-Jasiewiczowa et al. (Eds.) (2004a)
26	Kirchseeon I	11.9042	48.0575	Rausch (1975)	54	Lake Perespilno	22.4500	51.5667	Ralska-Jasiewiczowa et al. (Eds.) (2004a)
27	Löddigsee	11.9092	53.4114	Jahns (2007)	55	Łukcze II	22.7500	51.4833	Ralska-Jasiewiczowa et al. (Eds.) (2004a)
28	Kolbermoor,b, Prof	12.0608	47.8503	Beug (1976)					

Table A.5 Recorded values for the palynological analysis

Pleniglacial

ID	Site	Pin.	Bet.	Jun.	NAP	Art.	Poa.
40	Bronów	42	13	1	2	6	36
24	Gaterslebener See	68	8	0	9	3	12
2	Gulickshof I	47	11	1	16	2	23
19	Hämelsee	24	2	1	2	1	70
10	Holzmaar	56	4	0	4	8	28
36	Hrabetice	71	7	0	15	3	4
11	La Joux des Ponts	25	2	1	27	20	25
50	Lake Mikołajki	32	36	0	9	5	18
51	Lake Miłkowskie	48	24	1	6	7	14
7	Le Tronchet	20	3	1	31	20	25
13	Lörmoos	13	4	0	28	18	37
23	Luttersee	56	3	0	13	2	26
5	Marais de Rances	25	3	1	31	10	30
1	Milheeze	18	23	1	11	0	47
3	Pelléautier Vb	13	1	17	33	17	19
29	Schwarzer See	27	4	1	27	14	27
6	Siguret	18	4	20	22	24	12
9	Usselo	8	16	0	16	5	55

Dryas I

ID	Site	Pin.	Bet.	Jun.	NAP	Art.	Poa.
42	Aleksandrów	48	38	1	2	2	9
15	Amsoldingersee	3	17	2	21	31	23
33	Bełchatów	19	44	2	6	6	23
48	Błoto	32	20	3	10	6	29
17	Erlenbruckmoor	19	4	0	30	37	10
20	Federsee, Schienen	6	9	0	47	19	19
24	Gaterslebener See	25	38	0	8	6	23
2	Gulickshof I	5	37	2	11	3	42
19	Hämelsee	7	8	1	20	11	53
10	Holzmaar	14	7	0	14	10	55
16	Horbacher Moor	8	5	0	30	40	17
36	Hrabetice	35	22	0	20	13	10
49	Jasło	60	6	4	8	10	12
34	Kiensass	33	25	3	15	12	12
26	Kirchseeon I	17	5	2	34	17	25
28	Kolbermoor	20	9	2	12	7	50
11	La Joux des Ponts	8	6	1	25	35	25
38	Lake Gopło III	57	22	1	3	4	13
50	Lake Mikołajki	36	40	0	4	6	14
51	Lake Miłkowskie	31	33	3	5	10	18
53	Lake Perespilno	27	58	1	2	4	8
30	Lauter	12	6	1	31	25	25
7	Le Tronchet	8	10	7	23	30	22
12	Lobsigensee	5	25	1	30	22	17
13	Lörmoos	6	24	1	18	21	30
54	Łukcze II	16	66	0	3	6	9
23	Luttersee	22	10	3	8	3	54
5	Marais de Rances	9	10	5	27	22	27

(continued)

Table A.5 (continued)

Dryas I							
ID	Site	Pin.	Bet.	Jun.	NAP	Art.	Poa.
4	Marais du Rosey	5	20	2	18	35	20
8	Meerfelder Maar	17	6	0	19	6	52
1	Milheeze	6	40	3	5	6	40
25	Moor bei Lanser Se	20	5	7	22	38	13
31	Moosham	4	28	7	14	10	37
41	Niechorze	32	40	1	8	3	16
18	Nussbaumer See 1	6	12	4	36	18	24
3	Pelléautier Vb	18	2	30	14	18	18
52	Podbukowina I	62	10	0	7	2	19
39	Rośle	39	48	1	2	3	7
21	Schleinsee, 2	9	29	0	29	24	9
29	Schwarzer See	9	3	1	27	38	22
32	Seemoos I	32	3	3	14	32	16
6	Siguret	18	4	20	22	24	12
46	Stary Cieszyn St.C2	55	30	1	3	3	8
47	Stary Cieszyn St.C9	48	30	1	4	5	11
9	Usselo	8	24	1	9	8	50
37	Węglewice	41	40	0	4	5	10
45	Wolbrom 2	49	17	5	6	2	21
43	Wolbrom 1	46	8	7	3	3	33
44	Wolbrom 3	41	8	5	10	3	33

Bølling							
ID	Site	Pin.	Bet.	Jun.	NAP	Art.	Poa.
42	Aleksandrów	41	41	1	1	2	14
15	Amsoldingersee	2	36	35	15	3	9
33	Bełchatów	24	49	9	1	1	16
48	Błoto	42	15	5	21	3	14
17	Erlenbruckmoor	3	16	0	28	20	33
20	Federsee, Schien	4	46	0	11	7	33
24	Gaterslebener Se	50	36	0	4	2	8
35	Grossweissenbach	42	5	5	28	7	13
2	Gulickshof I	6	31	5	18	3	37
19	Hämelsee	3	23	7	17	6	44
10	Holzmaar	3	16	5	19	13	44
49	Jasło	49	16	4	12	4	15
34	Kiensass	45	18	6	7	6	18
26	Kirchseeon I	6	43	11	7	4	29
28	Kolbermoor,b, Pr	9	13	26	11	3	38
22	Kubitzbergmoor	14	37	4	23	5	16
11	La Joux des Pont	2	30	50	9	3	6
38	Lake Gopło III	41	40	2	2	1	14
50	Lake Mikołajki	30	50	2	8	2	7
51	Lake Miłkowskie	31	49	2	8	2	7
53	Lake Perespilno	37	48	1	4	3	7
30	Lauter	7	22	22	14	5	30
7	Le Tronchet	3	20	60	5	5	7
12	Lobsigensee	3	50	17	15	3	12
27	Löddigsee	50	19	4	13	1	13

(continued)

Table A.5 (continued)

Bølling							
ID	Site	Pin.	Bet.	Jun.	NAP	Art.	Poa.
13	Lörmoos	6	63	11	6	3	11
54	Łukcze II	37	47	1	3	3	9
23	Luttersee	20	12	8	5	3	52
5	Marais de Rances	2	25	50	13	5	5
4	Marais du Rosey	3	50	20	9	2	8
8	Meerfelder Maar	3	13	9	15	13	38
1	Milheeze	4	39	10	6	9	32
25	Moor bei Lanser	10	30	30	19	4	7
31	Moosham	4	20	20	11	4	41
41	Niechorze	43	29	2	11	2	13
18	Nussbaumer See 1	4	25	52	5	7	7
3	Pelléautier Vb	18	8	59	2	5	8
52	Podbukowina I	58	19	1	6	3	13
39	Rośle	41	41	1	1	2	14
21	Schleinsee, 2	6	70	0	8	7	9
29	Schwarzer See	12	14	18	9	6	41
32	Seemoos I	33	10	8	20	7	22
6	Siguret	20	13	52	4	5	6
46	Stary Cieszyn St.C2	46	37	2	7	2	6
47	Stary Cieszyn St.C9	47	36	2	7	2	6
9	Usselo	2	37	12	9	11	29
37	Węglewice	35	27	1	11	2	12
14	Westerhauderfehn	3	40	8	24	6	19
45	Wolbrom 2	49	16	4	2	1	27
43	Wolbrom 1	49	16	4	2	1	27
44	Wolbrom 3	49	16	4	2	1	27

Pin= *Pinus*, Bet=*Betula*, Jun= *Juniperus*, NAP=non-arboreal pollen, Art=*Artemisia*, Poa=Poaceae

Table A.6 Radiocarbon dates of Central European Magdalenian assemblages

ID	Origin	M	Sp	Lab.-Nr.	BP	Std	cal BP	Std	References
	Andernach-Martinsberg 1883	b	E	OxA-18409	13025	50	15590	70	Hedges et al. (1987), Street (2000), Kegler (2002), Street and Terberger (2002), Stevens et al. (2009), Higham et al. (2011)
		b	E	OxA-10493	13185	80	16110	360	
		b	E	OxA-10651	13270	180	16200	480	
		b	E	OxA-10492[a]	13500	90	16740	140	
364	I; Pit 20	b	E	OxA-1126	12890	140	15520	300	
		b	E	OxA-1125	12930	180	15770	520	
	I	b	E	OxA-V-2218-38	13015	50	15580	70	
		b	In	GrA-16985	13110	80	16010	340	
		b	E	OxA-V-2216-43	13135	55	16020	300	
		b	In	GrA-16986	13180	70	16100	350	
365	II; Pit 12	b	E	OxA-1127	12820	130	15310	230	
	II; Pit 35	b	E	OxA-1129	13090	130	16020	420	
		b	E	OxA-1128	13200	140	16140	440	
	II	b	E	OxA-V-2218-40	13110	50	15970	280	
366	III; Pit 26-279	b	E	OxA-V-2223-37	12675	55	15180	150	
	III; Pit 28	b	E	OxA-1130	12950	140	15790	410	
390	*Aschenstein*	a	R	KIA-33772	12366	61	14560	250	Terberger et al. (2009b)
82	*Auzary-Thônes*	b	In	?	13950	70	17090	60	INQUA
		b	E	?	15100	70	18250	220	
590	*Balcarka Cave*	n	In	GrN-28448	13930	100	17080	90	Valoch and Neruda (2005)

(continued)

Table A.6 (continued)

ID	Origin	M	Sp	Lab.-Nr.	BP	Std	cal BP	Std	References
73	*Banges; G.*	n	In	Ly-3640	11680	190	13590	250	Bridault et al. (2000), Oberlin and Pion (2009)
		b	In	OxA-538	12080	180	14210	410	
		b	R	OxA-540	12200	120	14310	340	
		b	Ce	GrA-13409	12300	60	14420	260	
		b	E	GrA-13969	12900	60	15470	90	
523	*Barrandová*	b	In	GrA-36106	12790	60	15350	120	Verpoorte and Šida (2009)
402	*Bärenkeller*	c	In	B 980	11190	180	13100	240	Küßner (2009)
		c	In	Bln 220	13700	380	16630	750	
66	*Baume Noire*; M3	n	R	Ly-3292 (Poz)	13045	75	15880	280	Cupillard (2008)
		b	In	Ly-4716	13950	135	17190	180	
67	M4	b	In	Ly-4715	14100	180	17350	270	
85	*Bavans*; 9	n	Ce	Inédit	12170	60	14290	230	Cupillard (2008)
	10	b	In	Gif-6061	12200	170	14360	420	
393	*Bettenroder Berg IX*; *18a*	b	In	Hv-17372	10390	260	12010	520	Veil (1996)
		b	In	Hv-17371	11770	350	13920	640	
114	*Birseck-Erm.*	b	In	B-4260	11860	100	13770	130	Le Tensorer (1998)
		b	In	B-4261	12040	80	14020	180	
255	*Bockst-Törle*	b	In	H-4057-3354	12980	103	15760	290	Hahn et al. (1985)
320	*Bois Laiterie*	b	In	GX-20433	12625	117	15090	250	Charles (1996)
		a	R	OxA-4198	12660	140	15120	280	
		b	In	GX-20434	12665	96	15150	220	
205	*Burghöhle Dietfurt*	b	Ho	KIA 3837	12210	60	14320	240	Gietz (2001)
		b	Ho	KIA 3838	12420	60	14690	230	
219	*Burkhardtsh. V*	b	Ho	ETH-7613	12450	110	14720	350	Hahn (1996)
104	*Büttenloch*; *A*	b	Le	B-4185	12330	60	14500	240	Le Tensorer (1998)
	A/B	b	R; Le	B-4186	12170	50	14270	220	
105	B	b	La	B-4187	12090	70	14110	220	
198	*Buttental-höhle*	b	E	OxA-4981	12040	120	14100	320	Housley et al. (1997)
		b	R	OxA-4602	13020	130	15920	410	
		b	Ur	OxA-4982	13100	140	16040	440	
54	*Cabônes*; *4*	b	In	Ly-3079	11520	190	13430	250	Cupillard (2008)
		b	In	Gif-8396	12150	80	14250	280	
		b	In	Ly-2296	12620	250	14830	570	
		n	R	GrA-?	13965	101	17110	100	
644	*Całowanie, Otwock*; *II*	n	In	Grn-5967	11390	95	13300	150	Kozłowski and Kozłowski (1996)
10	*Calvaire*; III	b	In	Ly-431	12970	300	15780	730	Évin et al. (1973), Oberlin and Pion (2009)
11	IV	b	In	Ly-432	13450	300	16290	580	
		a	R	Ly-710/OxA-8031	14370	90	17520	190	
1	*Campalou* 2	b	R	Lyon-2301	12300	60	14420	260	Évin et al. (1973, 1985), Drucker et al. (2009), Oberlin and Pion (2009)
		b	Ib	Ly-2214/ GrA-23731	12470	140	14730	410	
		b	In	Ly-436	12800	300	15440	860	
2	3a	b	In	Ly-1958	13400	350	16210	650	
		b	E	Ly-2212/ GrA23940	13940	70	17080	60	
		b	R	Ly-2211/ GrA23730	14100	70	17300	150	
		w	In	Ly-2215/ GrA-23734	13390	70	16600	160	
43	*Chaumois-Boivin 2*	b	In	Ly-440	12040	270	14210	570	Cupillard (2008)
		n	R	Ly-3280 (Poz)	12530	120	14900	310	

(continued)

Table A.6 (continued)

ID	Origin	M	Sp	Lab.-Nr.	BP	Std	cal BP	Std	References
322	*Chauveau*	b	In	Lv-1961	12000	130	14030	290	De Bie and Caspar (2000)
31	*Chaze 2*	n	Cl	Ly-3291 (Poz)	13890	80	17050	70	Cupillard (2008)
30	*Chênelaz*; *c2*	t	In	Ly-383	12190	80	14300	270	Cupillard (2008), Oberlin and Pion (2009), INQUA
		b	R	OxA-8027/Ly-703	12460	65	14850	170	
		b	In	Ly-4790	12610	200	14880	480	
		t	In	OxA-8068/Ly-742	12780	75	15320	150	
334	*Coléoptère Niv. B*	b	In	Lv-686	12150	150	14290	370	Charles (1993, 1996)
		b	In	Lv-717	12400	110	14620	360	
		b	E	OxA-3635	12870	95	15420	140	
29	*Colomb*	b	In	Ly-598	12480	80	14890	210	Oberlin and Pion (2009)
14	*Colombière*	c	In	W-150	11750	600	14080	990	Cupillard (2008), Oberlin and Pion (2009)
	D	b	Ma	Ly-433	13390	300	16230	590	
	D	n	In	L.	13400	400	16190	760	
	Hearth D2	c	In	L-177	14150	450	17270	590	
	E7	a	R	GrA-9713/Ly-644	14390	70	17610	120	
	Hearth D	c	In	L.	14700	300	17880	490	
	Hearth D1	c	In	L.	15500	700	18750	1080	
8	*Croze-sur-Suran*	b	R	Ly-638/GrA9704	14260	70	17440	180	Oberlin and Pion (2009)
		i	Ma	Ly-357	14330	260	17450	310	
		c	In	Ly-434	14850	350	17960	510	
15	*Balme de Cuiseaux*	n	Ho	Ly- en cours					Cupillard (2008)
		n	Vu	Ly-3727 (GrA)	11990	50	13870	80	
516	*Děravá Cave*	b	In	GrA-37870	11560	40	13450	90	Verpoorte (2009)
610	*Deszczowa*; 2	n	In	Gd-9464	16150	280	19430	490	Kozłowski (2001)
611	3	h	In	Gd-10212	17480	150	20970	260	
69	*Douattes, Les* c2	b	Al	Ly-2199/ OxA-12465	12255	55	14370	240	Évin et al. (1973), Drucker et al. (2009), Oberlin and Pion (2009)
		b	Ce	Ly-2210/ GrA-23536	12860	100	15400	150	
	c5	b	Ce	Ly-1453	12100	60	14110	210	
		t	E	Ly-1850/ GrA-20630	12240	70	14350	260	
		b	Cv	Ly-1416	12255	95	14400	300	
		b	Al	Ly-1851/ GrA-20632	12340	70	14520	260	
		b	R	Ly-1852/ GrA-20633	12870	80	15430	120	
	-	b	Ce	GrA-22132	12570	80	15030	200	
	c7	c	In	Ly-435	12480	260	14700	600	
		b	R	Ly-647/GrA-9725	12680	60	15180	150	
		b	In	Ly-435	12980	240	15830	640	
	Ouest	b	E	Ly-1853	12660	80	15150	190	
601	*Dzierżysław 35*; 9a	b	Ma	GdA-70	13220	70	16140	360	Ginter et al. (2002), Ginter and Połtowicz (2004)
		n	In	GdA-69	13500	80	16750	120	
		o	In	Ki-8997	14850	280	18100	370	
		n	In	?	13370	80	16420	290	
	6	o	In	Ki-8951	13700	350	16630	720	
80	*Étrembières*	so	In	ETH-3937	12300	130	14480	370	Evin et al. (1994), Bridault et al. (2000)
		n	In	B-3787	12310	140	14500	390	
		n	In	GrA-9703/Ly-637	12590	60	15060	170	

(continued)

Table A.6 (continued)

ID	Origin	M	Sp	Lab.-Nr.	BP	Std	cal BP	Std	References
224	*Felsställe* IIIb	b	E	H-7442-7491	11285	246	13200	300	Jaguttis-Emden (1987) ORAU
		b	R	H7394-7429	11800	200	13750	320	
		b	R	H7391-7426	12050	115	14130	300	
		b	R	H7383-7419	12100	60	14110	210	
		b	R	H7382-7419	12320	70	14460	270	
		b	R	H7392-7428	12400	145	14610	440	
		n	In	OxA-6426	12475	80	14870	210	
		n	In	OxA-6428	12505	80	14940	200	
223	IIIa	b	R	H7393-7428	12680	120	15150	250	
		n	In	OxA-6427	13390	90	16440	290	
24	*Freydières*	b	In	Ly-451	11380	180	13270	240	Évin et al. (1973)
83	*Freymond, Abri*; *5 inf.*	b	In	ETH-5600	11760	145	13640	220	Oberlin and Pion (2009), INQUA
		b	In	ETH-5601	12020	120	14040	290	
		b	R	Ly-1135/ OxA-9460	12780	80	15310	150	
404	*Fuchskirche I* 3a+b	b	BB	KIA-12927	12030	52	13930	90	Küßner (2009)
		b	Al	KIA-12925	12158	50	14260	220	
		b	Al	KIA-12926	12232	50	14330	240	
		b	E	KIA-12928	12721	65	15240	150	
18	*Gay, Abri*	b	R	Ly-640/GrA-9705	12160	60	14260	230	Oberlin and Pion (2009)
		b	Al	Lyon-1454	12505	65	14940	170	
		b	R	Ly-639/GrA-9720	12980	70	15550	90	
		b	Gu	Lyon-1542	13795	100	16990	90	
248	*Geißen-klösterle*; AH I0	b	E	OxA-5158	12450	120	14710	370	Housley et al. (1997), Street (2000)
		b	E	OxA-6253	12750	130	15230	250	
		b	R	OxA-6254	13130	100	16060	380	
		b	Ur	OxA-4854	13230	130	16170	430	
		b	Ib	OxA-5156	16940	180	20360	290	
26	*Gigny*; IV sup.	b	In	Ly-1798	12370	460	14920	1090	Cupillard (2008)
25	IV inf.	b	In	Ly-1702	13540	480	16400	930	
195	*Gnirshöhle 1*	n	In	H-6273-5833	12830	100	15360	170	Weniger (1982)
196	*Gnirshöhle 2*	n	In	H-6270-5832	12360	200	14580	520	Weniger (1982)
		n	In	H-62715831	12600	320	15110	880	
		n	In	H-6272-5831	13050	300	15860	710	
368	*Gönnersdorf I*	b	In	Ly-1173	11100	650	12980	1200	Évin et al. (1975), Brunnacker (1978), Housley et al. (1997), Street and Terberger (2004), Street (2007), Stevens et al. (2009)
		b	In	Ly-768	12380	230	14620	560	
		b	In	Ly-1172	12660	370	15250	990	
		b	E	OxA-5728	12730	130	15200	250	
		b	E	OxA-5730	12790	120	15280	220	
		sh	Mo	Kn-1980	12910	105	15460	150	
		b	E	OxA-5729	12910	130	15560	270	
		b	R	OxA-V-2223-42	12990	55	15560	80	
		b	E	OxA-V-2223-39	13270	55	16190	350	
369	II	i	Ma	OxA-2069	11830	110	13760	140	
		b	R	OxA-V-2222-31	13010	55	15580	80	
		b	BB	OxA-V-2223-41	13095	55	15950	280	
		b	E	OxA-V-2223-40	13165	55	16060	320	
		t	Cl	OxA-10201	13610	80	16860	90	
370	III	b	R	OxA-15295	13060	60	15860	240	
		b	R	OxA-V-2223-43	13075	55	15900	250	
	Southwest	b	Al	OxA-15296	12385	65	14600	250	

(continued)

Table A.6 (continued)

ID	Origin	M	Sp	Lab.-Nr.	BP	Std	cal BP	Std	References
	Goyet 3; chamber A, H2	b	Ax	KIA-22275	12380	60	14590	240	Germonpré (1997), Stevens et al. (2009)
332	H1	b	E	UtC-8957	12560	50	15020	150	
		b	Ov	GrA-3238	12620	90	15090	210	
		b	E	GrA-3237	12770	90	15290	180	
		b	Ov	OxA-12121	12775	50	15330	110	
		b	E	OxA-V-2223-48	12775	55	15330	110	
		b	Cl	OxA-6592	16320	140	19550	300	
28	*Grappin, Arlay*	b	Cr	Ly-3162 (Poz)	15335	115	18360	280	BANADORA, ORAU, Cupillard (2008), Bridault and Chaix (2009)
		b	R	OxA-19632	15540	55	18700	40	
	C	b	R	Lyon-4865 (GrA)	13450	50	16720	100	
		b	R	Ly-3878 (GrA)	13640	60	16890	70	
		b	R	Lyon-4868	14180	50	17360	150	
		b	E	Lyon-4866 (GrA)	14220	50	17380	160	
		a	R	Ly-1509	14220	560	17130	1010	
		b	R	Lyon-4871 (GrA)	14520	50	17750	50	
		a	R	Ly-1535	14530	290	17790	520	
		a	R	Ly-1510	14820	370	17930	530	
		ab	In	Ly-1536	14840	360	17950	520	
		b	R	Lyon-4869 (GrA)	14850	50	17900	40	
		n	R	Lyon-4867(GrA)	14940	50	18180	250	
		n	R	Ly-3877 (GrA)	15260	70	18320	260	
		b	E	Ly-3161 (Poz)	15335	100	18360	280	
		b	R	Ly-3160 (Poz)	16840	110	20240	180	
	C/D	b	In	Ly-457	15320	370	18490	470	
		b	In	Ly-559	15770	390	18930	670	
97	*Hauterive-Champréveyres* Azilian	c	In	UZ-2284	12120	170	14240	420	Leesch (1997), Leesch et al. (2004)
		o	In	B-4768	12330	70	14490	260	
		o	In	B-4769	12330	100	14490	330	
		c	In	UZ-2170	12550	130	14930	320	
93	niveau principale	c	In	UZ-2174	12510	130	14830	370	
		c	In	UZ-2173	12540	140	14880	370	
		c	In	UZ-2177	12600	145	15030	320	
		c	In	UZ-2172	12620	145	15070	300	
		c	In	UZ-2175	12630	130	15090	270	
		c	In	UZ-2171	12730	135	15200	260	
		c	In	UZ-2286	12870	135	15380	220	
		c	In	UZ-2283	12950	155	15800	450	
		c	In	UZ-2285	13050	155	15980	460	
	foyer D11	c	In	UZ-2282	12825	155	15300	270	
95	Secteur 2	c	In	UZ-2287	12500	145	14780	420	
241	*Hohle Fels*; AH Ib	n	In	H-5312-4907	12770	220	15480	570	Housley et al. (1997)
		n	In	H-5119-4601	13085	95	15990	350	
241/243	*Hohle Fels*; AH Ib/IIa	n	In	OxA-4975	12520	130	14850	360	Street (2000)
		b	In	OxA-4956	13240	110	16180	410	
		b	In	OxA-4977	13350	140	16260	440	
243	AH IIa	n	In	H-5313-4898	15760	140	18910	170	Weniger (1982)
		n	In	H-5120-4569	17100	150	20560	230	
283	*Hohlenstein, Schambach*	b	In	OxA-5752	12410	90	14650	310	Housley et al. (1997)
257	*Hohlenstein-Stadel*; III	b	In	H-3799-3045	13110	160	16050	470	Hahn et al. (1985)
		b	In	H-3779-3044	13550	130	16750	180	

(continued)

Table A.6 (continued)

ID	Origin	M	Sp	Lab.-Nr.	BP	Std	cal BP	Std	References
530	*Hostim*	b	E	Ly-1108	12420	470	14990	1120	Valoch (1996)
33	*Hoteaux, Les*	b	R	OxA-9457/ Ly-1132	12830	75	15390	130	Cupillard (2008)
438	*Ilsenhöhle bei Ranis*; 5	b	La	OxA-10251	10145	50	11790	170	Higham et al. (2007), Küßner (2009)
		b	E	OxA-12052	12615	60	15090	160	
	2	b	Ca	OxA-12047	13450	60	16710	110	
437	4	b	In	Gd-4537	14400	450	17690	610	
		b	BB	OxA-12049	14780	60	17870	40	
373	*Irlich, Geiersberg*	b	Ho	OxA-9848	11965	65	13850	100	Bronk Ramsey et al. (2002)
		b	In	UtC-9221	12110	90	14190	270	
372	*Irlich, Sandgrube*	b	Ho	OxA-9847	11910	70	13780	120	Bronk Ramsey et al. (2002)
		b	Ho	OxA-9736	12310	120	14480	350	
63	*Jean Pierre I*	c	In	Ly-625	10470	200	12200	380	Évin (1994)
		c	In	Ly-692	11590	330	13580	530	
		c	In	Ly-693	11630	240	13530	320	
		c	In	Ly-829	12720	230	15160	550	
		c	In	Ly-830	13070	210	15980	550	
64	*Jean Pierre II*	b	In	Ly-925	12400	240	14640	570	Oberlin and Pion (2009)
		c	In	Ly-828	12470	200	14720	510	
		b	R	Ly-711/Oxa 8032	13105	75	16000	320	
		c	In	Ly-926	13280	290	16140	610	
		c	In	Ly-390	13300	280	16170	590	
543	*Kamegg*	n	In	GrN-22883	13840	120	17020	100	Verpoorte (2004)
		n	In	GrN-23182	14130	110	17360	210	
290	*Kastelhäng-höhle*	a	R	OxA-5756	11590	90	13470	120	Housley et al. (1997)
		a	R	OxA-5755	12060	90	14090	240	
109	*Kastelhöhle-Nord* middle level	b	R	B-4636	13900	150	17090	150	Street (2000), Bronk Ramsey et al. (2002)
		b	R	OxA-9737	18530	150	22170	340	
		b	R	OxA-9739	19200	150	23040	240	
		b	R	OxA-9738	19620	140	23510	170	
110	Upper level	b	BB	B-4254	11320	130	13230	190	Le Tensorer (1998)
		b	BB	B-4637	11380	80	13270	130	
		b	BB	B-4253	11380	150	13290	200	
		b	R	B-4255	11680	50	13570	100	
		b	R	B-4256	12110	60	14130	210	
271	*Kaufertsberg*	a	R	OxA-5751	12610	90	15080	210	Housley et al. (1997)
527	*Keblice*	t	E	GrA-37168	12680	50	15190	140	Verpoorte and Šída (2009)
		t	E	GrA-37169	12730	60	15260	140	
170	*Kesslerloch* II	b	In	B-3327	11220	120	13100	160	Weniger (1982), Street (2000)
171	III	b	In	Hv-10652	12890	90	15440	130	
		b	In	B-3329	12970	180	15840	520	
		a	R	OxA-5746[b]	13120	90	16040	360	
	III Bc	b	In	OxA-5748	12770	90	15290	180	
		a	R	OxA-5747[b]	13430	100	16610	200	
	III Ac	a	R	OxA-5750[b]	13670	100	16900	100	
		a	R	OxA-5749[b]	14150	100	17370	200	
259	*Kl. Scheuer Hohlenst.-Sta.*	b	sm	H-4183-3416	13252	98	16180	400	Hahn et al. (1985)

(continued)

Table A.6 (continued)

ID	Origin	M	Sp	Lab.-Nr.	BP	Std	cal BP	Std	References
451	*Kniegrotte*	b	Sg	OxA-4853	13090	130	16020	420	Street (2000)
		b	R	OxA-4845	13120	130	16060	420	
		b	Sg	OxA-4849	13130	120	16070	410	
		b	E	OxA-4848	13150	130	16090	430	
		b	Ax	OxA-4850	13160	140	16110	440	
		b	E	OxA-4846	13190	130	16130	430	
		b	R	OxA-4832	13310	110	16230	410	
		b	E	OxA-4852	13520	130	16700	200	
		b	In	Bln-1564	13585	165	16740	230	
		b	Ma	OxA-4851	14470	140	17560	210	
108	*Kohlerhöhle*	b	In	B-4971	11640	150	13520	200	Le Tensorer (1998)
		b	In	B-4969	11820	160	13710	240	
569	*Kolíbky*	b	In	OxA-5973	12680	110	15160	230	Valoch and Neruda (2005)
517	*Koněprusy*	b	Ho	GrA-13696	12870	70	15430	110	Svoboda et al. (2002)
419	*Königsaue, Bruchberg*	n	In	H-89	12700	320	15290	900	Mania and Toepfer (1973), Weniger (1982)
		n	In	Fr-24	12890	190	15690	540	
		n	In	H-106/89	13250	280	16120	600	
612	*Krusza Skála*	i	In	Lod-407	11400	200	13290	270	Kostruhn (2004)
572	*Kůlna Cave*; 6	c	In	GrN-11053	11450	90	13360	140	Valoch and Neruda (2005)
		c	In	GrN-5097	11590	80	13470	100	
58	*Fru, A.; Aire I* 3 inf.	b	Ce	Ly-134/OxA-5264	12110	110	14220	310	Hedges et al. (1997), Oberlin and Pion (2009)
		b	Ce	GrA-34354	12200	50	14310	230	
	4A	b	E	Ly-29/OxA-4408	12600	120	15060	260	
		b	E	Ly-133/Oxa 5263	12740	110	15230	220	
		b	E	Ly-132/Oxa 5262	12770	110	15270	210	
		b	E	Ly-131/Oxa 5261	12810	110	15320	190	
		b	In	Gif A 92351	13040	140	15960	430	
	4B	b	R	Ly-708/Oxa 8029	13455	75	16700	130	
		b	R	Ly-89/Oxa 4937	13810	110	17000	90	
		b	R	Ly-130/Oxa 5260	14060	130	17300	210	
452	*Lausnitz, Abri Theure*	b	E	Hd-22168	9107	39	10250	40	Küßner (2009)
		b	E	Hd 22140	9533	24	10890	140	
		b	E	Hd 21899	10477	24	12490	100	
629	*Maszycka Cave*	b	E	Ly-2453	14520	240	17540	280	Kozłowski et al. (1993)
		a	R	Ly-2454[b]	15490	310	18580	460	
289	*Mittlere Klause* AH1	a	R	OxA-5718	13160	130	16100	430	Housley et al. (1997), Terberger and Street (2002)
		n	In	UCLA-1869	18200	200	21920	390	
		n	In	OxA-9856	18590	260	22280	500	
101	*Moosseedorf-Moosbühl*	c	In	B-2306	10140	120	11760	340	Le Tensorer (1998)
		c	In	B-359	10300	180	12000	470	
		c	In	B-2313	11180	120	13060	180	
		c	In	B-2310	11900	130	13830	170	
		c	In	B-2316	12060	150	14160	370	
635	*Mosty 13*	c	In	Lod-107	11290	280	13230	350	Kozłowski and Kozłowski (1996)
122	*Munzingen* I	b	In	OxA-4783	15400	130	18390	300	Pasda (1998), Street (2000)
		b	In	OxA-4786	15670	140	18820	130	
		b	R	ETH-7499	15700	135	18840	130	
		b	In	H4156-3373	15870	135	19010	200	
		b	In	OxA-4785	16060	140	19220	280	

(continued)

Table A.6 (continued)

ID	Origin	M	Sp	Lab.-Nr.	BP	Std	cal BP	Std	References
122	II	b	In	OxA-4788	14270	120	17440	220	Pasda (1998), Street (2000)
		b	In	OxA-4784	14510	110	17700	110	
	III	b	In	OxA-4820	13230	110	16170	410	
		b	R	ETH-7500	13560	120	16780	160	
	IV	n	In	H4738-4660	12130	95	14230	290	
		b	In	OxA-4787	12370	100	14550	330	
439	*Nebra*; Pit 1	b	E	OxA-11892	13070	60	15900	270	Küßner (2009)
	Pit 50	b	E	OxA-11893	13160	60	16060	320	
90	*Neuchâtel-Monruz* Magdalenian	c	Sa	ETH-20727	12800	85	15340	160	Leesch (1997), Bullinger et al. (2006)
		c	Sa	ETH-6414	12840	120	15350	200	
		c	Sa	ETH-6419	12880	120	15410	180	
		c	Sa	ETH-6415	12900	120	15440	180	
		c	Sa	ETH-6412	12970	110	15760	310	
		c	Sa	ETH-6417	13030	120	15920	390	
		c	Sa	ETH-6416	13070	130	16000	420	
		c	In	ETH-6418	13110	120	16040	410	
		c	Sa	ETH-6420	13120	120	16060	410	
		c	In	ETH-6421	13140	120	16080	410	
		c	Sa	ETH-6413	13330	110	16250	410	
91	Magdalenian/Azilian	c	Ju	ETH-23272	12355	85	14530	300	
		c	Ju	ETH-23271	12570	90	15030	210	
		c	Sa	ETH-6421	13140	120	16080	410	
89	Azilian	c	In	ETH-17973	12165	130	14290	350	
		c	In	ETH-17974	12370	110	14560	360	
563	*Nová Drátenická*	a	R	OxA-1952	11670	150	13550	200	Valoch and Neruda (2005)
		b	In	OxA-1954	12900	140	15540	300	
		b	In	OxA-1953	13870	140	17040	130	
287	*Obere Klause*; lower level	a	R	OxA-5719	12350	130	14540	390	Housley et al. (1997)
		a	R	OxA-5720	12440	140	14680	420	
585	*Ochozska Cave*	n	In	GrN-28452	12440	170	14670	480	Valoch and Neruda (2005)
444	*Oelknitz*	so	In	H-830/840	11750	250	13740	420	Feustel (1980), Hedges et al. (1998), Street (2000)
	Pit 26/60	b	E	OxA-5715	11810	110	13720	160	
	Pit 136/60	b	E	OxA-5711[b]	12050	110	14130	300	
	Pit 162/60	b	E	OxA-5710[b]	12080	110	14180	300	
	Pit 136/60	b	R	OxA-5712[b]	12270	110	14430	320	
	Pit 119/60	b	E	OxA-5709[b]	12270	120	14420	350	
	Trench VII, H4	b	In	Bln-1726	12350	85	14520	300	
	AH 4a	b	In	Bln-1565	12545	80	15000	200	
	Pit 146/60	b	R	OxA-5714[b]	12620	120	15080	260	
	Pit 113/60	b	R	OxA-5717	12670	110	15150	240	
	Pit 146/63/b	b	E	OxA-5713[b]	12740	120	15220	230	
	Pit 19/60	b	E	OxA-5716	12790	110	15290	200	
27	*Passagère*	b	R	GrA-9708/Ly-641	12440	60	14800	160	Oberlin and Pion (2009)
579	*Pekárna* g/h	b	In	OxA-5972	12500	110	14840	310	Évin et al. (1985), Valoch and Neruda (2005)
		b	In	GrN 14828	12670	80	15160	190	
		b	E	Ly-2553	12940	250	15770	660	

(continued)

Table A.6 (continued)

ID	Origin	M	Sp	Lab.-Nr.	BP	Std	cal BP	Std	References
173	*Petersfels* P1AH2	b	E;R	H-5209-4865	11975	100	13930	220	Albrecht (1979), Jaguttis-Emden (1983)
		n	In	H7132-6984	12120	100	14230	300	
174	P1AH3A	b	In	H-5210-4866	12160	95	14260	300	
		b	In	H-7139-7300	12180	100	14280	310	
		b	In	H-7136-6890	12320	90	14470	310	
		b	In	H-7137-7067	12470	100	14790	310	
		b	In	H-7134-6876	12500	120	14820	340	
		b	In	H-7140-7058	12530	90	14980	210	
		b	In	H-7215-7349	12570	95	15030	220	
		b	In	H-7133-6877	12580	130	15020	280	
		b	In	H-4277-3531	12650	100	15130	220	
		b	In	H-7214-7350	12660	100	15140	220	
		b	In	H-7217-7364	12670	90	15160	200	
		b	In	H-7135-6879	12670	100	15150	220	
		b	In	H-7141-6985	12680	110	15160	230	
		b	In	H-7138-7057	12685	75	15180	180	
176	P1AH4	b	In	H-7144-7302	12630	95	15100	220	
		b	In	H-7145-7303	12700	100	15190	210	
		b	R	H-5211-4891	12940	125	15710	330	
		b	In	H-7142-7348	12980	90	15650	190	
		b	In	H-7143-7301	13030	100	15900	340	
		b	In	H-7216-7363	13110	90	16020	350	
179	P3AH2	b	In	H-4276-3535	11700	90	13580	130	
180	P3AH2/3	b	E	H-4741-4145	11300	85	13200	100	
		b	In	H-6650-6779	11700	100	13570	150	
183	P3AH4	b	In	H-4743-4137	11890	130	13820	170	
188	P6AH2	b	In	H-6651-6780	12400	100	14630	330	
189	P6AH3/4	b	In	H-6652-6783	12230	240	14450	570	
		b	In	H-6653-6786	12600	100	15060	230	
190	P6AH4	b	In	H-6654-6787	12100	280	14310	610	
191	P6AH5	b	In	KN-2884	12400	95	14630	320	
		b	In	KN-2883	12400	100	14630	330	
		b	In	H-6655-6788	12500	220	14750	530	
192	P6AH6	b	In	H-6656-6793	12900	90	15450	130	
525	*Putim*	c	Pi	GrA-36010	13010	60	15580	80	Verpoorte and Šída (2009)
21	*Raillarde, La*	b	R	OxA-8162	12180	80	14290	270	Oberlin and Pion (2009)
128	*Rheinfelden-Eremitage*	b	E	B-4183/C137	11600	120	13490	170	Le Tensorer (1998)
		b	E	B-4184	11950	50	13840	80	
		b	R	B-4262	12570	170	14870	430	
77	*Rigney 1*	b	Cl	Ly-1191	14940	500	18030	650	Bridault and Chaix (2009)
		b	Cl	Gra-21036	16680	80	20070	160	
78	*Rigney 2*	b	In	Ly-4872 (GrA)	11805	50	13690	70	BANADORA
		b	In	Ly-4873(GrA)	11810	50	13690	70	
124	*Rislisbergh.*	b	In	Ly-1099	11860	230	13960	460	Le Tensorer (1998)
336	*Roche plate*	a	R	Lv 2096	13160	110	16100	400	Stocker et al. (2003)
86	*Rochedane*	b	In	Ly-1193	11060	470	12820	720	Cupillard (2008), Drucker et al. (2009), Oberlin and Pion (2009)
		b	Ce	GrA-21512	12250	70	14370	270	
		b	R	Ly-709/OxA-8030	12420	75	14670	270	

(continued)

Table A.6 (continued)

ID	Origin	M	Sp	Lab.-Nr.	BP	Std	cal BP	Std	References
51	*Romains*; 2a	c	In	Ly-1549	10100	350	11680	730	Évin et al. (1973, 1979), Delibrias and Évin (1980), Bridault et al. (2000), Cupillard (2008), Oberlin and Pion (2009)
52	2b	c	In	Ly-1307	10280	630	11810	1090	
		sh	Mo	MC-1276	12540	230	14770	550	
		sh	Mo	MC-1215	12540	400	15110	1020	
		sh	Mo	MC-1275	12540	400	15110	1020	
		b	R	GrA-9710/Ly-648	12830	60	15400	110	
		b	In	Ly-356	12980	240	15830	640	
		b	R	GrA-9710/Ly-643	13380	60	16600	150	
53	3	c	In	Ly-1308	10770	410	12420	750	
		b	R	GrA-9709/Ly-642	12690	60	15200	150	
		c	In	LYON-16	14380	380	17700	590	
	4?	i	Ma	Ly-1772	13140	80	16060	350	
455	*Saaleck*	b	E	OxA-11890	12780	60	15330	120	Küßner (2009), INQUA
		b	E	OxA-11891	12945	60	15510	90	
		t	In	KN-2562	13200	400	15920	810	
88	*Scé du Chât.*	b	R	Ly-1133/ OxA-9458	12695	70	15200	170	Oberlin and Pion (2009)
229	*Schussen-quelle*	?	R	ETH-6155	12510	130	14830	370	Schuler (1994)
		c	In	ETH-6154	12630	120	15090	260	
		?	R	KN-4250	12860	120	15380	190	
		?	R	KN-4251	13050	120	15960	390	
		p	In	GRO-468	14470	385	17750	550	
		p	In	H-680-970	15900	360	19200	550	
166	*Schweizers-bild*	b	In	OxA-5744[b]	11780	90	13670	120	Street (2000)
		b	In	OxA-5745[b]	13940	100	17090	90	
294	*Sesselfels-grotte* C2	b	In	OxA-5754	12680	100	15160	220	Housley et al. (1997)
		b	In	OxA-5733	12740	90	15250	190	
261	*Spitzbuben-höhle* AH2	b	In	H-4052-3212	12747	110	15240	220	Hahn (1984a)
		b	In	H-4314-3715	13840	120	17020	100	
		b	In	H 4149-3348	15230	100	18310	250	
391	*Stendel, Abri*	b	E	OxA-16878	12668	59	15170	150	Bronk Ramsey et al. (2002, 2009)
		a	R	OxA-10471	12860	75	15420	120	
		b	E	OxA-10494	12970	70	15530	100	
		b	E	OxA-10470	13105	70	15990	310	
3	*Tai, Grotte du*; C′1	b	R	GrA-9715/Ly-646	11220	60	13130	80	Oberlin and Pion (2009), BANADORA
		?	R	Ly-1381/ OxA-10302	12090	150	14190	370	
4	C″1	b	R	GrA-9714/Ly-645	12400	60	14630	240	
	II; Azilian	b	In	Ly-4216	11270	190	13170	280	
		b	R	OxA-9459/ Ly-1134	12580	90	15040	210	
415	*Teufelsbr.* 1	b	In	OxA-5725[b]	12900	130	15520	260	Street (2000), Housley et al. (1997)
416	2	b	E	OxA-5722[b]	12860	130	15370	210	
		b	In	OxA-5724	12940	140	15770	400	
		b	In	OxA-5723[b]	13080	140	16010	440	
417	3	b	E	OxA5727	10040	120	11640	290	
		b	In	Bln-1821	12300	85	14440	300	
		b	In	Bln-1924	12315	100	14460	330	
		b	R	OxA-5726[b]	12640	130	15100	270	
		b	In	Bln-1573	13025	85	15860	290	
418	4	b	In	Bln-1727	12480	90	14850	250	

(continued)

Table A.6 (continued)

ID	Origin	M	Sp	Lab.-Nr.	BP	Std	cal BP	Std	References
126	*Teufelsküche*	b	R	ETH 7502	11960	120	13930	240	Pasda (1994)
		b	R	ETH-7503	12040	120	14100	320	
		b	R	ETH-7501	13080	120	16000	400	
319	*Trou da Somme*	a	R	OxA-4199[b]	12240	130	14380	370	Charles (1994), Bronk Ramsey et al. (2002)
		b	Ov	OxA-8308	12815	75	15370	130	
324	*Trou de Chaleux*	b	In	Lv-1568	12370	170	14610	460	Otte (1994), Stevens et al. (2009)
		n	E	OxA-V-2216-44	12375	50	14590	220	
		n	E	OxA-V-2216-45	12630	55	15120	150	
		b	In	Lv-1136	12710	150	15170	290	
		b	E	OxA-3632[b]	12790	100	15300	190	
		b	Ov	Oxa-4192[b]	12860	140	15360	230	
		b	E	OxA-3633[b]	12880	100	15430	150	
		b	In	Lv-1569	12990	140	15870	420	
		n	In	MC-919	13000	200	15880	560	
326	*Trou des Nutons*	b	E	OxA-4195[b]	12630	140	15090	280	Charles (1994)
328	*Trou du Frontal*	b	E	OxA-4117[b]	12800	130	15290	230	Charles (1994), Gilot (1997), Street (2000)
		b	In	Lv-1749	12950	170	15810	490	
		b	Cn	Lv-1750	13130	170	16070	480	
339	*Trou Walou*	n	In	Lv-1582	13030	140	15940	430	Charles (1996)
		n	In	Lv-1593	13120	190	16050	510	
309	*Vaucelles*; II 1	ab	In	Lv-1386	12440	180	14680	490	Charles (1996)
310	III 1	ab	In	Lv-1558	16130	250	19390	440	
		ab	In	Lv-1385	16270	230	19500	460	
311	III 2	b	E	OxA-4200	13330	160	16240	460	
		b	In	Lv-1434 D	13730	400	16640	770	
		a	R	Lv-1309 D	13850	335	16850	630	
		n	In	Lv-1433	13930	120	17090	110	
312	III 3	a	R	Lv-1314	13790	150	16980	130	
335	*Verlaine, Grotte de Sy*	b	E	OxA-4041[b]	12870	110	15400	170	Charles (1994, 1996), Street (2000)
		b	In	Lv-690	13780	220	17010	290	
79	*Veyrier*	bb	In	Eth-3937	12300	130	14480	370	Oberlin and Pion (2009)
		?	R	Ly-637/GrA-9703	12590	60	15060	170	
475	*Wallendorf*	b	E	OxA-13849	12685	55	15190	140	Küßner (2009)
539	*Węgliny*	b	In	Poz.-10674	12120	60	14140	210	Cziesla and Masojć (2007)
647	*Wilczyce*	t	Cl	Ua-15720	11400	135	13320	180	Fiedorczuk and Schild (2002), Fiedorczuk et al. (2007)
		t	Cs	Ua-15722	11665	135	13540	180	
		t	Ax	Ua-15723	11890	105	13780	140	
		c	Pi	Poz-14892	12770	120	15260	230	
		c	In	Poz-3914	12960	60	15530	80	
		c	In	Poz-14891	13020	60	15590	90	
378	*Wildweiberlei*	a	R	OxA-18410	12835	50	15410	90	Higham et al. (2011)
623	*Wołowice 1*	n	In	Gd-4654	10920	200	12910	220	Sobczyk (1993)
614	*Zalas*; 11	b	Vu	OxA-6591	12530	110	14940	270	Kozłowski and Pettitt (2001)
615	7	b	Le	OxA-6625	12820	80	15370	140	
619	*Zawalona*	b	In	Ly-2270	14060	340	17260	410	Alexandrowicz et al. (1992)
		b	In	Ly-2271	15380	340	18510	460	

(continued)

Table A.6 (continued)

ID	Origin	M	Sp	Lab.-Nr.	BP	Std	cal BP	Std	References
	Zigeunerfels F	t	Pa	OxQ-17268	12375	50	14590	220	Fahlke (2009), INQUA
		b	In	Poz-20369	12430	60	14750	190	
206	G	b	Ar	Poz-20321	12150	60	14230	240	
		b	In	KN-3472	12690	280	15220	820	
208	H	b	Ce	KIA-26997	12200	45	14300	230	
	H/I	b	In	KN-3473	12100	100	14200	290	
		b	Ar	Poz-20368	12300	60	14420	260	
567	*Žitného*	n	In	GrN-28453	13220	90	16150	380	Valoch and Neruda (2005)

M Material. *A* Antler, *b* Bone, *bb* Burnt bone, *c* Charcoal, *h* Horn, *i* Ivory, *n* Not determined, *o* Organic, *sh* Shell, *so* Soil, *t* Tooth, *p* Peat, *w* Wood
SP Species. *Al* Alces, *Ar* Artiodactyla, *Ax* Alopex, *B/B* Bos/Bison, *Ca* Capreolus, *Ce* Cervus, *Cn* Canis, *Cl* Coelodonta, *Cr* Crocuta, *Cs* Castor, *Cv* Cervid, *E* Equus, *Gu* Gulo, *Ho* Homo, *Ib* Ibex, *In* Indeterminate, *Ju* Juniperus, *La* Lagopus, *Le* Lepus, *Ma* Mammuthus, *Mi* Microfauna, *Mo* Mollusc, *Ov* Ovibos, *Pa* Panthera, *Pi* Pinus, *R* Rangifer, *Sa* Salix, *Sg* Saiga, *sm* Small mammals, *Ur* Ursus, *Vu* Vulpes
Lab.-Nr. Laboratory Number. [a]Redated=OxA-18409, [b]Humanly modified
Other abbreviations see list of abbreviations

Table A.7 Proportion of horse and reindeer remains

ID	Site	F	ID	Site	F	ID	Site	F
15	Croze-sur-Suran; R7	3	179	Petersfels; P1AH2	4	333	Trou des Nutons	3
21	Colombière, La; B	4	180	Petersfels; P1AH3A	4	335	Trou du Frontal	2
25	Gay, A.	4	182	Petersfels; P1AH4	4	341	Coléoptère, G. de; B	2
35	Grappin; Arlay	3	185	Petersfels; P3AH1/2	4	342	Verlaine, G. de Sy	3
36	Colomb	5	186	Petersfels; P3AH2	3	343	Roche Plate; 12	4
41	Bonne Femme	5	188	Petersfels; P3AH3	4	374	Andernach-M.; IV	2
51	Farincourt, I+II	5	189	Petersfels; P3AH3/4	4	375	Gönnersdorf I	2
53	Farincourt, III-E	4	190	Petersfels; P3AH4	4	378	Gönnersdorf IV	3
54	Farincourt, III-B	3	191	Petersfels; P3AH5	4	383	Feldhofhöhle	1
55	Balme de Glos 10–11	5	201	Gnirshöhle 1	3	385	Wildweiberlei	3
56	Balme de Glos	5	203	Napoleonskopf	3	386	Wildscheuer; V	3
61	Cabônes, A.; 4	5	204	Buttentalhöhle	3	408	Bad Frankenhausen	1
65	Fru, A. de la; I	3	205	Probstfels; I IV	3	422 423	Teufelsbrücke; 1–2	3
66	Fru, A. de la; II	3	211	Burghöhle Dietfurt	5	424	Teufelsbrücke; 3	2
70	Jean Pierre I; 9A+B	5	216	Annakapellenhöhe	3	425	Teufelsbrücke; 4	2
72	Baume Noire; M2	4	217	Nikolaushöhle; 2	3	442	Herdloch; 1+2	3
73	Baume Noire; M3	4	227	Hohlefels Hütten; yellow	2	446	Nebra	3
74	Baume Noire; M4	5	229	Felsställe; IIIb	4	451	Ölknitz	2
76	Douattes, Les; 7/6	4	230	Felsställe; IIIa	3	458	Kniegrotte	2
80	Banges; sup.+inf.	3	233	Schussenquelle; main	4	459	Lausnitz, A. Theure	2
84	Rigney 1, G. de; D	4	243	Schmiechenfels; 2	3	463	Lengefeld	2
86	Veyrier	4	244	Helga-Abri IIIa+b	3	519	Bečov	1
89	Auzary-Thônes	3	246	Hohle Fels, Schelk. Ia	3	534	Keblice	1
93	Rochedane; D1/D2	5	247	Hohle Fels, Schelk. Ib	3	536	Krápníková	1
95	Scé du Châterlard, G.	5	249	Hohle Fels, Schelk. IIa	3	537	Hostim	2
97	Neuchâtel-Monruz; 1M	2	251	Sirgenstein	1	550	Kamegg	3
100	H.-Champréveyres; 1pri.	2	252253	Brillenhöhle	4	558	Býči skála	2
102	H.-Champréveyres; 2	3	254	Geißenklösterle AH Io	3	559	Barová Cave; 11–12	4
108	Moosseedorf-Moosbühl	3	256	Kl. Sch. im Rosenstein	2	570	Nová Drátenická; 15	4
111	Büttenloch; A	5	259	Bocksteinhöhle; III	1	574	Žitného	4
114	Brügglihöhle	3	261	Bockstein-Törle; AH II, III	2	575	Verunčina	3
115	Kohlerhöhle	4	262	Hohlenstein-Stadel; II	3	576	Kolíbky; 1	3

(continued)

Table A.7 (continued)

ID	Site	F	ID	Site	F	ID	Site	F
122	Hollenberg-Höhle 3; D	5	265	Kl. Sch. H.-Stadel III	4	580	Kůlna Cave; 6	3
129	Munzingen	4	267	Spitzbubenhöhle; AH2	2	588	Hadí	3
131	Rislisberghöhle	4	269	Bärenfelsgrotte; III	3	592	Adlerova	4
133	Teufelsküche	4	271	Vogelherd; III	3	595	Ochozska Cave	2
135	Rheinf.-Eremitage; 1	4	277	Kaufertsberg; lower	3	600	Balcarka Cave	4
143	Käsloch	3	297	Kastelhänghöhle; 5	3	611	Dzierzyslaw 35	3
144	Mühleloch, A.	5	316	Trou de Blaireaux; III 1	3	613	Šipka dira	3
173	Schweizersbild	4	326	Trou da Somme	2	617	Dra Majera	1
174	Freudenthal-Höhle	4	327	Bois Laiterie	2	639	Maszycka Cave	2
177	Kesslerloch	4	328	Roche-al-Rue	2	657	Wilczyce; 10	2
178	Bildstockfels	3	331	Trou de Chaleux; 5	2	658	Klementowice-K. 20	1

F Faunal composition, horse/reindeer ratio
1: Exclusively horse; 2: Horse dominated; 3: Equal proportion of horse and reindeer; 4: Reindeer dominated; 5: Exclusively reindeer. For references see IDs in Table A.1

Table A.8 Presence of other species in the faunal record

ID	Site	Ibex	Chamois	Red deer	Roe deer	Wild boar	Moose	Bison and Wisent	Musk ox	Saiga	Rhino-ceros	Mammoth
8	Croze-sur-Suran	–	–	–	–	–	–	X	–	–	–	X
14	Colombière	X	–	X	–	X	–	–	–	–	–	–
28	Grappin	–	–	X	–	–	–	X	–	–	X	X
43	Chaumois-Boivin	–	X	–	–	–	–	X	–	–	–	
45–47	Farincourt	–	X	X	–	–	–	X	–	–	–	X
48–49	Balme de Glos	X	–	–	–	X	–	–	–	–	–	–
51–53	Romains	–	–	–	–	–	X	–	–	–	–	–
54	Cabônes	–	–	X	X	X	–	–	–	–	–	–
63	Jean Pierre I	X	–	X	–	–	X	X	–	–	–	–
65–68	Baume Noire	–	–	–	–	X	–	–	–	–	–	–
73	Banges	X	–	X	–	–	–	–	–	–	–	–
77–78	Rigny	–	–	X	–	X	–	X	–	–	X	X
79	Veyrier	X	X	X	–	X	X	X	–	–	–	
82	Auzary-Thônes	X	–	–	–	–	–	–	–	–	–	–
86	Rochedane	X	–	–	–	–	–	–	–	–	–	–
88	Scé du Châterlard	X	–	–	–	–	–	–	–	–	–	–
90–91	Neuchâtel-Monruz	X	–	X	–	–	–	X	–	–	–	–
93–96	H.-Champréveyres	X	–	X	–	–	–	X	–	–	–	–
107	Brügglihöhle	X	X	–	–	X	–	X	–	–	–	–
108	Kohlerhöhle	X	X	–	–	–	–	X	–	–	–	–
122	Munzingen	–	–	–	–	–	–	–	–	–	X	–
124	Rislisberghöhle	X	–	–	–	–	–	–	–	–	–	–
126	Teufelsküche	X	X	X	–	X	–	X	–	–	–	–
128	Rheinfelden-Erem.	–	–	X	–	–	–	–	–	–	–	–
136	Käsloch	X	–	–	–	–	–	–	–	–	–	–
166	Schweizersbild	X	–	–	X	–	–	X	–	–	–	–
167	Freudenthalh.	X	–	X	–	–	X	–	–	–	–	X

(continued)

Table A.8 (continued)

ID	Site	Ibex	Chamois	Red deer	Roe deer	Wild boar	Moose	Bison and Wisent	Musk ox	Saiga	Rhino-ceros	Mammoth
170–171	Kesslerloch	X	–	–	–	–	–	X	X	–	X	X
172	Bildstockfels	–	X	X	X	–	–	–	–	–	–	–
173–193	Petersfels	X	X	–	–	X	X	X	–	–	–	X
195	Gnirshöhle	–	–	–	–	–	–	–	X	–	–	–
197	Napoleonskopf	–	–	X	–	–	–	–	–	–	X	X
198	Buttentalhöhle	X	–	–	–	–	–	X	–	–	–	–
199	Probstfels	X	–	–	–	–	–	X	–	–	–	–
204	Kuhstallhöhle	–	–		–	–	–	–	–	–	X	–
205	Burghöhle Dietfurt	–	–	X	X	–	–	–	X	–	–	–
221–222	Hohlefels Hütten	–	–	X	–	–	–	X	–	–	–	–
223–224	Felsställe	X	–	–	–	–	–	–	–	–	–	–
229	Schussenquelle	–	–	–	–	–	X	–	–	–	–	–
237	Schmiechenfels	–	–	X	–	–	–	–	–	–	–	–
240–243	Hohle Fels, Schelk.	–	–	X	–	–	–	X	–	–	–	X
246–247	Brillenhöhle	–	X	X	–	–	–	X	–	–	–	X
250	Kl. Scheuer, Rosen.	X	X	–	–	–	–	–	–	–	–	–
256–257	Hohlenstein-Stadel	–	–	X	–	–	X	–	–	–	–	–
259	Kl. Scheuer, Hohl.	–	X	–	–	–	–	–	–	–	–	–
261	Spitzbubenhöhle	–	–	–	–	–	–	–	–	–	X	–
263	Bärenfelsgrotte	–	–	X	–	–	X	–	–	–	–	–
265–266	Vogelherd		–	X	–	–	–	X	–	–	X	–
290	Kastelhänghöhle	X	X	X	–	–	–	–	–	–	–	X
319	Trou da Somme	X	X	–	–	–	–	–	–	–	–	–
321	Roche-al-rue	–	–	–	–	–	–	–	–	–	X	–
324–325	Trou de Chaleux	X	X	X	X	X	–	X	–	X	–	–
326	Trou des Nutons	–	–	X	X	X	–	X	–	–	–	–
328	Trou du Frontal	–	–	–	X	X	–	X	–	–	–	–
334	Coléoptère, G. d.	–	–	X	–	–	–	–	–	–	–	X
335	Verlaine	–	–	–	–	X	–	–	–	–	–	X
336	Roche Plate	X	–	–	–	–	–	X	–	X	–	–
368–371	Gönnersdorf	–	X	X	–	–	X	X	–	X	X	X
376	Feldhofhöhle	–	–	X	–	–	–	–	–	–	–	X
378	Wildweiberlei	X	–	–	–	–	–	–		–	X	–
379	Wildscheuer	–	–	X	–	–	–	–	–	–	X	X
390	Aschenstein	–	–	–	–	–	–	X	–	–	–	X
415–418	Teufelsbrücke	–	–	–	X	–	–	X	–	X	–	X
435	Herdloch Ranis	–	–	–	–	–	–	X	–	–	X	–
444	Oelknitz	X	–	–	–	X	–	X	–	–	X	X
451	Kniegrotte	–	–	–	–	–	–	–	–	X	–	X
452	Lausnitz	–	–	–	X	–	–	–	–	–	X	X
480	Schafgraben	–	–	–	–	–	–	–	–	–	X	X
530	Hostim	X	–	–	–	–	–	X	–	–	X	–
531	Na Průchodě	?	–	–	–	–	–	–	–	–	X	–
533	Rytiřská	X	–	X	–	–	–	–	–	–	–	–
543	Kamegg	–	–	–	–	–	–	X	–	–	X	–
551	Býci skála	–	–	X	–	–	–	X	–	–	X	–

(continued)

Table A.8 (continued)

ID	Site	Ibex	Chamois	Red deer	Roe deer	Wild boar	Moose	Bison and Wisent	Musk ox	Saiga	Rhino-ceros	Mammoth
552	Barová Cave	–	–	–	–	–	–	–	–	–	X	–
563	Nová Drátenická	X	–	X	–	–	–	X	–	–	X	–
566	Vinckova	–	–	–	–	–	–	X	–	–	X	–
567	Žitného	–	–	X	–	–	–	X	–	–	–	–
576	Kulnička	–	–	X	–	–	–	X	–	–	–	–
578	Hadí	–	–	–	–	–	–	X	–	–	–	–
585	Ochozska Cave	–	–	X	X	–	–	–	–	–	–	–
590	Balcarka Cave	–	–	–	–	–	–	–	–	–	X	X
601	Dzierżysław 35	–	–	–	–	–	–	–	–	–	–	–
603	Šipka dira	–	–	–	–	–	–	–	–	–	X	X
604	Čertova dira	–	–	–	–	–	–	–	–	–	X	
610	Deszczowa	–	–	–	–	–	–	–	–	X	X	X
611	Douattes, Les	X	–	X	X	X	X	X	–	–	–	–
619	Zawalona	–	X	–	–	–	–	–	–	–	X	–
629	Maszycka Cave	–	–	X	–	–	–	X	–	X	X	–
647	Wilczyce	–	–	–	–	–	–	–	–	–	X	–

For references see IDs in Table A.1

Table A.9 Raw material spectrum of Central European Magdalenian assemblages

ID	Site	Raw material/Source	km	%
28	Grappin	Cretaceous flint, S	17	67
		Oligocene flint, NW	70	9
		Jurassic chert, L	–	9
		Flint, W	70	–
		Flint, L	–	–
		Quarzite	?	1
		Radiolarite	?	1
54	Cabônes	Tertiary flint, Haute-Saone	–	1
		Cretaceous Silex	?	17
		Forêt d'Arne, flint	–	–
56	Seveux	Chert, L	–	–
58	Abri Fru	Sénonien/Valanginiens, L	0.5	–
59	Abri Fru	Sénonien/Valanginiens, L	0.5	–
63	Jean Pierre I	Valanginien, L	–	–
		Sénonien, L	–	–
		Bugey, NW	25	–
		Unidentified	–	–
64	Jean Pierre II	Valanginien, L	–	46
		Sénonien, L	–	43
		Bugey, NW	25	2
		Unidentified	–	2
65	Baume Noire	Local	–	–

(continued)

Table A.9 (continued)

ID	Site	Raw material/Source	km	%
69	Douattes, Les	Aleyrac	186	–
		Allèves-Bourchigny	32	–
		Andert et Condon	34	–
		Balme-de-Thuy	28	–
		Bellegarde	11	–
		Bugey?	25	–
		Chambotte	33	–
		Champagne-en-Valromey	25	–
		Chartreuse	60	–
		Châtel-Saint-Denis?	95	–
		Mâconnais	100	–
		Manigod	36	–
		Olten-Wangen	210	–
		Préalpes?	100	–
		Regional	30	–
		Rhône?	10	–
		SE of Isère	70	–
		Thuy	26	–
		Vassieux-en-Vercors	133	–
		Vauche, S	1	–
73	Banges, Grotte du	Mâconnais, WNW	135	–
		Sennecé-lès-Mâcon, WNW	120	–
		Treschenu-Creyers/Pellebit SSW	120	–
		Aleyrac? SW	165	–
		Bellegarde NNW	40	–
		Chartreuse SW	35	–
		SW	60	–
		L	–	–
74	La Guillotine	Mont-les-Étrelles, L	–	–
77	Rigney 1	Mont-les-Étrelles, L	–	–
79	Veyrier	Olten-Wangen, NE	220	3
		Lausen, NE	220	2P
		Mâconnais, W	110	1
		Pleigne, NE	190	1
		Lägern, NE	250	1
		Auggen, NE	240	4P
		Event, SE	130	1
		Susdorf, NE	250	1P
		Bellegarde, W	30	17
		Monruz?, NE	140	5P
		Friborgoise, NE	100	2
		Simmenthal, NE	160	4P
		Lains, NW	70	3P
		Vasieux-en-Vercours, S	120	2
		Balme de Thuy, E	20	6
		Kleinmontbach, NE	150	9P
		Valanginien, SW	40	13
		Dinkelberg, NE	260	10P
		Balme de thuy, E	20	15
		Yverdon, NE	110	0.5
		Champréveyres?, NE	140	2P
		Lampenberg, NE	220	1P
		Saxel Mont Pèrlin, NE	110	5P
		Charmey, NE	130	1P

(continued)

Table A.9 (continued)

ID	Site	Raw material/Source	km	%
81	Bains-les-Bains	Quarzite, L	–	–
		Mont-les-Etrelles, S	40	–
82	Auzary-Thônes	Local	–	–
83	Freymond	Flint	?	–
90	Neuchâtel-Monruz 1M	Lausen, NE	80	2
		Bellegarde	120	10
		Haurtrivien, L	0	60
		Olten, NE	80	22
		Himmelried, NE	80	2
91	Neuchâtel-Monruz 2M/A	Hauterivien, L	–	25
		Flint	?	24
		Olten/Wangen/Aarau, NE	80	21
		Malm, Lausen, NE	80	4
		Lampenberg	80	3
		Bellegarde/Seyssel, SW	120	2
		Silicic Sandstone	–	1
		Chasseral?	–	1
		Längern	120	1
		Alle	50	1
		Pâquier	–	1
		Oberbuchsiten/Egerkingen, NE	80	1
		Locle, N	10	1
		Bugey, SW	180	1
		Yverdon	25	1
93	Hauterive-Champré-veyres	Hauterivien flint, L	–	43
		Bellegarde/Seyssel, SW	130	21
		Olten, NE	80	19
95	Hauterive-Champré-veyres 2	Olten, NE	85	63
		Hauterivien flint, L	–	23
		Bellegarde/Seyssel, SW	130	5
		Creatceous flint?	–	4
		Valanginien du Bugey, SW	160	3
		Radiolarite, Moraines	–	1
		Malm, Lausen?	75	1
		Quarzite, Moraines	–	1
		Séquanien, Lampenberg, NE	80	1
99	Alle, Noir Bois	L	–	61
		Bendorf, NE	15	36
		Dossenbach, NE	50	1P
		Ajoie, E	5	–
		Delémont, SE	17	–
104	Büttenloch A	Chert	–	76
		Dotted chert, Lausen	17	2
		Banded chert, Lausen	17	7
105	Büttenloch B	Chert, L	–	71
		Dotted chert, Lausen	17	5
		Jaspis?, L?	–	1
		Banded chert, Lausen	17	7
108	Kohlerhöhle	Flint	–	–
		Chert	–	–
109	Kastelhäng-höhle	Chert, L	–	–
		Flint	–	–
		Radiolarite	–	–

(continued)

Table A.9 (continued)

ID	Site	Raw material/Source	km	%
115	Hollenberg-Höhle 3	Chert, lausen, L	10	34
		Chert	?	59
		Jaspis, Kleinkems, N	23	3
		Chalzedony	0	2
122	Munzingen	Rock crystal, Rhine terrace	–	1
		Markgräflerland	24	–
		Chert, Villingen	8	–
		Chalzedony, L	0	–
		Chert, Bollschweil	10	–
		Radiolarite, S	18	–
		Chert, L	–	–
		Quarzite	–	–
		Flint	–	–
		Silicic rock, Black Forest	20	–
124	Rislisberg-höhle	Chert, L	–	–
		Jaspis, L	–	–
		Cretaceous flint	–	–
125	Lausen	Chert, L	–	–
		Jaspis, L	–	–
126	Teufels-küche	Markgräflerland	20	68
		Chert, L	–	27
		Radiolarite, Rhine terrace	10	6
		Banded chert	–	–
		Chert	–	–
		Chalzedony, L	–	–
128	Rheinfelden-Eremitage	Chert, L	–	32
		Chert, Lausen, L	–	1
		Chert, L	–	13
		Quartzite, L	–	3
		Jaspis, Kleinkems, NW	25	1
136	Käsloch	Flint, L	–	–
		Quarzite	–	–
		Radiolarite	–	–
139	Reiden-Stumpen	Olten, L	0	52
		Otelfingen, NE	30	–
		Pleigne, NW	70	–
		Bellegrad/Seyssel, SW	230	–
164	Einsiedeln-Langrüti	Jura, Wettingen/Dielsdorf	40	56
		Malm, Olten/Wettingen/Dielsdorf	–	2
		Jura, Baden AG/Burgg	–	2
		rock crystal, L	–	1
		Radiolarite, L	–	8
		Quarzite, L	–	2
		Silicic rock, L	–	20
		Unidentified	–	9
166	Schweizers-bild	Local	10	–
		Regional	50	–
170	Kesslerloch	Local	–	100
173–193	Petersfels	Flint, L	–	–
		Radiolarit	–	–

(continued)

Table A.9 (continued)

ID	Site	Raw material/Source	km	%
195	Gnirshöhle 1	Chert, L	–	83
		Rdiolarite, L	–	17
		Baltic?	–	–
		Regional	50	–
	Probstfels	Chert, L	0	98
		Banded chert, Abensberg, E	240	1
		Chert, moraines	25	1
		Radiolarite, moraines	25	1
		Quartz, L	0	1
205	Burghöhle Dietfurt	Chert, L	–	–
		Chert, Kelheim E	220	–
		Chert, Hegau, SW	40	–
		Radiolarite, S	20	–
		Randecker Maar chert, N	60	–
215	Käppele	Randecker Maar, L	8	–
		Mainhardter Wald, NE	60	–
		Chert, Alb	–	–
		Quarzite	–	–
		Chert, Kelheim E	180	2P
217	Randecker Maar	Silicic Maar sediment, L	0	95
		Chert, L	0	4
		Radiolarite, Danube terrace	40	1
218	Braunfirst	Randecker Maar chert, L	2	99
		Radiolarite, Danube gravels	40	1
		Chert, grey	–	3P
		Chert, Alb	–	1
		Quarzite	–	2P
223	Felsställe IIIa	Chert, L?	?	99
		Flint	?	–
		Chert	?	–
		Radiolarite	?	–
		Quarzite	?	–
224	Felsställe IIIb	Chert, L	–	99
		Flint	?	–
		Chert	?	–
		Radiolarite	?	–
		Quarzite	?	–
229	Schussen-quelle	Chert, L	–	85
		Radiolarite, L	–	6
		Flint, L	–	1
		Quartzite, L	–	1
238	Helga-Abri	Randecker Maar, NW	30	3
240	Hohle Fels, Schelklingen	Local	–	100
244	Sirgenstein Südwand	Local	10	98
		Regional	50	2
246	Brillenhöhle	Randecker Maar	–	1
		Chert, Kelheim E	200	–
250	Kleine Scheuer im Rosenstein	Flint East-Württemberg	25	33
		Chert, L	–	49
		Randecker Maar chert	40	10
		Radiolarit	–	4

(continued)

Table A.9 (continued)

ID	Site	Raw material/Source	km	%
256	Hohlenstein-Stadel	Local	10	95
		Regional	50	5
257	Hohlenstein-Stadel	Local	10	95
		Regional	50	5
260	Klemmer-Abri	Local	10	90
		Regional	50	10
261	Spitzbuben-höhle	Jaspis, L	–	10
		Chert, L	–	90
263	Bärenfels-grotte	Chert, L	–	–
		Chert, Kelheim E	120	–
271	Kaufertsberg	Local	10	70
		Regional	50	30
278	Mühlberg-Grotten	Chert, L	–	–
281	Speckberg	Chert, L	–	–
282	Eitensheim	Quarzite, Neuburg	20	–
		Chert, L	–	–
283	Hohlenstein Schambach	Local	–	–
291	Sesselfels-grotte B2	Chert, L	–	91
		Quarzite, L	–	6
		Lydite	–	2P
292	Sesselfels-grotte B3	Chert, L	–	92
		Banded Chert, L	–	–
		Rock crystal	–	–
		Radiolarite	–	–
		Quarzite	–	–
293	Sesselfels-grotte C1	Chert, L	–	80
		Banded chert, L	–	–
		Quarzite, L	–	–
		Radiolarite	20	–
294	Sesselfels-grotte C2	Chert, L	–	37
		Quarzite, L	–	37
		Radiolarite	–	–
		Banded chert, L	–	–
319	Trou da Somme	Flint, L	–	–
		Charleville, S	50	–
		Quartzite, N?	30	–
320	Bois Laiterie	Doisch Agimont/Charleville	–	4
		Hesbaye/Spiennes	–	79
		Meuse gravel	–	14
		Cretaceous flint	–	1
		Tertiary flint	–	1
330	Orp	Flint, L	–	100
337	Kanne	Flint, L	–	100
341	Mesch-Steenberg	Rullen, L	0	86
		Ryckholt, L	0	14
		Simpelveld	10	1P
342	Sweikhuizen-Oude Stort	Quartzite	–	–
344	Sweikhuizen-Groene Paal	Feuerstein	–	–
		Quartzite	–	–
346	Echt	Flint	–	–
		Quarzite	–	–

(continued)

Table A.9 (continued)

ID	Site	Raw material/Source	km	%
347	Griendtsveen	Flint, L	–	–
		Quarzite	–	10P
350	Beeck	Meuse terrace, L	–	99
		Baltic	80	–
		Cretaceous	–	–
351	Alsdorf	Meuse terrace	–	99
		Vetschau, SW	15	1
		Baltic	70	1
		Main terrace (chert), L	200	1
355	Viersener Donk I	Tertiary quartzite	–	–
		Meuse terrace	–	–
356	Viersener Donk II	Baltic	–	–
		Meuse terrace	–	–
357	Galgenberg	Meuse terrace	–	–
359	Kamp-hausen	Meuse terrace	–	58
		Simpelveld?	–	–
		Baltic	30	–
		Chalzedony	–	–
		Obourg?	–	–
		Orsbach, SW	50	36
364	Andernach-Martinsberg I	Tertiary quarzite, local	–	90
		Baltic	100	8
		Chalcedon, L	–	2
		Tertiary quarzite, Ratingen	45	10P
365	Andernach-Martinsberg II	Tertiary quarzite, L	–	5
		Baltic	100	1
		Chalcedon, L	–	1
		Meuse terrace	100	87
		Palaeozoic quarzite	–	8
366	Andernach-Martinsberg III	Tertiary quarzite, L	–	84
		Baltic	100	7
		Chalcedon, L	–	9
368	Gönnersdorf I	Tertiary quarzite, NE	12	45
		Tertiary quarzite, Ratingen	45	1
		Baltic, N	100	40
		Meuse terrace, NW	100	10
		Radiolarite, L	–	2
		Chalzedony, L	–	1
		Quarzite, NW	–	–
		Rock crystal, L	–	–
		Jaspis, L	–	–
		Oolite, Mainz Basin, S	70	–
369	Gönnersdorf II	Meuse terrace, NW	100	82
		Tertiary quarzite, L	–	3
		Chalzedony, L	–	2
		Radiolarite, L	–	5
		Oolite, Mainz Basin, S	70	1
		Rock crystal	–	–
		Markgräflerland, S	290	–

(continued)

Table A.9 (continued)

ID	Site	Raw material/Source	km	%
370	Gönnersdorf III	Quarzite, NW	–	2
		Oolite, Mainz Basin, S	70	8
		Chalzedony, N?	35?	7
		Opal, N?	35?	–
		Quarzite, L	–	15
		Baltic, N	100	5
		Radiolarite, L	–	41
		Meuse terrace, NW	100	11
		Rock crystal	–	–
		Markgräflerland, S	290	–
371	Gönnersdorf IV	Chalzedony, NE?	35?	21
		Tertiary quarzite, L	–	18
		Baltic, N	100	11
		Oolite, Mainz Basin, S	70	8
		Quarzite, NW	–	2
		Radiolarite, L	–	32
		Meuse terrace, NW	100	6
		Markgräflerland, S	290	1
		Rock crystal	–	1
376	Feldhofhöhle	Flint	–	–
		Radiolarite	–	–
378	Wild-weiberlei	Radiolarite, L	0	88
		Tertiary quartzite, L	0	4
		Basalt, L	0	1
		Meuse terrace	150	7
379	Wildscheuer	Radiolarite, L	–	37
		Chalcedony, L	–	33
		Chalzedony, NW	75	–
		Baltic, N	130	23
		Meuse terrace, NW	170	3
		Tertiary quartzite, L	10	3
380	Fußgönheim 2	Createous flint, N	200	84
		Chert, L	–	15
		Unidentified	–	1
381	Fußgönheim 1	Certaceous flint, N	200	51
		Chert, L	–	25
		Chalzedony, L	–	9
		Porphyrit, L	–	12
385	Dreieich-Götzenhain Nord (II)	Baltic?	–	–
		Chalcedony	–	–
		Radiolarite	–	–
		Meuse terrace	–	–
		Chert	–	–
		Quarzite	–	–
386	Dreieich-Götzenhain Ost (I)	Markgräflerland	–	–
		Banded chert	?	1P
		Mainz basin	–	3P
		Chalcedony, Gailenberg, NE	14	6P
		Radiolarite	–	–
395	Gadenstedt	Baltic, L	–	–
		Meuse terrace, W	300	–

(continued)

Table A.9 (continued)

ID	Site	Raw material/Source	km	%
401	Bad Franken-hausen	Baltic, L	–	99
		Radiolarit, Hercynian Forest?	20?	1P
		Quarzite, L	–	–
414	Schweskau	Baltic, L	–	–
433	Vogelherd	Local	10	95
		Regional	50	5
435	Herdloch bei Ranis	Flint	–	–
		Radiolarite	–	2P
439	Nebra	Chert, Querfurter Platte?	15	99
		Quarzite	0	–
		Silicic rock	0	–
		Radiolarite	0	–
444	Oelknitz	Baltic	0	–
		Quartzite, Jena/Naumburg	–	–
		Silicic rock	–	–
		Chert, L	–	–
		Świeciechów flint?	–	–
449	Urdhöhle	Cretaceous flint	–	–
		Quartzite, N	–	–
451	Kniegrotte	Baltic, N	25	98
		Chert, N	20	1
		Radiolarite, Gamsbach	32	1
		Quarzit3, Lobensten, E	25	P
		Quartz, Gamsbach	32	P
		Carnelian, Rudolstadt	18	P
452	Lausnitz, Abri Theure	Baltic, L	–	–
		Radiolarite, L	–	–
		Rock crystal	–	–
		Jaspis, L	–	–
468	Galgenberg, Halle	Baltic, L	–	–
475	Wallendorf	Baltic, L	–	100
480	Schafgraben	Baltic, L	–	–
483–491	Groitzsch	Baltic, L	–	99
		Quarzite, E-Thuringia	–	1
494	Kuppelberg	Chert	–	–
501	Metten	Chert, L	–	–
504	Vilshofen-Kuffing	Keratophyr, Saldenburg, NE	–	–
		Quarzite, L	–	–
		Chert, L	–	–
515	Tmaň, Děrava cave	Slate, Liteň, L	–	–
		Flint	–	–
		Chert	–	–
		Opal	–	–
		Jaspis	–	–
521	Lhota	Chert	–	43
		Flint	–	23
		Quarzite	–	11
		Opal	–	6
		Radiolarite	–	1
		Rock crystal	–	1P
		Porzelanite	–	2P
		Banded chert	?	–

(continued)

Table A.9 (continued)

ID	Site	Raw material/Source	km	%
525	Putim	Baltic, N	180	–
		Vilshofen, SW	100	–
		Quartzite	–	–
530	Hostim	Rock crystal	0	1
		Radiolarite	0	1
		Quarzite, Bečov, N	75	1
		Chert, Vilshofen, SW	140	–
		Radiolarite, L	0.5	5
		Baltic, N	120	94
531	Na Průchodě	Chert	–	–
535	Žd'ár 1	Flintsbach	100	20
		Baltic	180	6
		Opal	–	15
		Rock crystal	–	3
		Serpentinite	–	5
542	Gudenus-höhle	Chalcedony	–	41
		Radiolarite, L	–	35
		Chert, L	–	20
		Polish Jura chert	400	2P
		Quarzite, L	–	–
		Serpentinite, L	–	–
		Rock crystal	–	–
543	Kamegg	Moravian chert	–	60
		Chert, SW of Brno	22	–
		Jaspis, L	–	–
		Flint, L	–	–
		Chalzedony, Röhrenbach, NW	12	–
		Serpentinite, L	–	–
		Hämatit, Langenlois, S	16	–
		Others, Spitz	–	–
546	Náchod	Baltic, NE	50	–
549	Rozcestí	Local	–	–
551	Býči skála	Chert, L	–	–
		Baltic	–	–
		Moravian chert	–	–
		Polish Jura chert	250	–
		Chocolate Flint	400	–
552	Barová Cave	Rudice Formation, L	–	31
		Olomučany, L	–	25
		Býčí skála, L	–	9
		Not local	–	5
		Radiolarite, Carpathians?	?	1
		Rock crystal, western Moravia	–	1
567	Žitného	Rock crystal	–	24
569	Kolíbky	Quarzite, L	–	3P
		Baltic, N?E?	90	7P
570	Kolíbky	Baltic, N?E?	90	59
		Olomučany, L	–	8
		Rudice Formation	–	19
		Chert, NW	15	11

(continued)

Table A.9 (continued)

ID	Site	Raw material/Source	km	%
571	Kůlna Cave	Polish Jura, E	260	–
		Olomučany, L	–	–
		Chert, W	11	–
		Radiolarite, Carpathians	100	3P
		Rock crystal	–	2P
		Krumlov chert, L	–	4P
		Obsidian, SE	420	1P
572	Kůlna Cave	Krumlov chert, L	–	1
		Rock crystal	–	1
		Nemcce	–	1
		Olomučany, L	–	5
		Quartz	–	2
		Radiolarite	–	1
		Moravian chert, L	–	3
		Baltic, N?E?	100	18
		Chokolate flint	380	1
		Polish Jura	260	11
		Cretaceous chert, L	–	52
576	Kulnička	Rudice Formation, L	–	–
		Olomučany, L	–	–
		Cretaceous chert, L	–	–
		Baltic, N?E?	–	–
578	Hadí	Baltic, N	100	45
		Moravian chert, L	–	20
		Západomoravske kridy	–	20
		Radiolarite, Carpathians	100	10
		Jemnozrnných kremencu	–	5
579	Pekárna g/h	Baltic, NÈ?	100	82
		Polish Jura	260	4
		Chocolate flint	390	1
		Radiolarite	0	6
		Cretaceous flint, L	0	5
		Troubky-Zdislavice chert	0	1
		Olomučany, L	0	4
		Stránská skála chert, L	0	1
		Býčí skála chert, L	0	3
		Rock crystal	0	1
585	Ochozska Cave	Baltic, N?E?	100	63
		Radiolarite	0	22
		Chert, L	0	15
587	Lišči	Rudice Formation, L	10	–
588	Mokrá, Lom V	Baltic, N?E?	100	32
		Olomučany, L	0	16
		Cretaceous chert, L	0	14
		Atypical Cretaceous chert	0	20
594	Loštice I	Baltic	–	–
		Radiolarite	–	1

(continued)

Table A.9 (continued)

ID	Site	Raw material/Source	km	%
599	Hranice	Chert	–	1
		Menilite	–	1
		Quarz	–	1
		Radiolarite	–	4
		Baltic, L	–	85
		Chalcedony	–	13
		Quarzite	–	1
		Silicis sandstone	–	1
601	Dzierzyslaw 35	Moraine deposits, L	–	–
		Radiolarite	?	–
607	Dra Majera	Polish Jura chert	–	–
616	Brzoskwinia-Krzemionki7	Radiolarite, Carpathians	200	–
		Chocolate Flint, N	140	–
619	Zawalona	Jurassic chert, L	–	–
621	Gaik 1	Polish Jura chert	–	–
		Świeciechów flint	–	–
622	Gaik 2	Świeciechów flint	–	–
		Polish Jura chert	–	–
629	Maszycka Cave	Polish Jura, L	–	95
		Upper Silesian	80	–
		Dniester	340	–
		Chocolate flint, NE	130	–
		Volhynian flint	330	–
		Banded chert, Franconian Alb	660	–
		Banded chert	?	–
633	Katy 1	Radiolarite, L	–	97
		Volhynian flint, NE	340	1
		Chocolate flint, N	200	1
		Polish Jura chert, N	90	1
638	Rydno, II/59	Radiolarite, Carpathians	170	–
		Polish Jura	140	–
		Chocolate flint, L	–	–
		Świeciechów flint, N	70	–
645	Maly Antoniów	Chokolate flint	60	–
		Baltic, L	0	–
	Wilczyce	Turonian flint, N	20	–
		Flint, L	–	–
		Chocolate flint, NW	70	–
		Świeciechów flint, NE	20	–
		Quarzite	–	–
		Radiolarite?	–	–
648	Klemento-wice Kolonia 20	Chocolate flint, W	100	22
		Świeciechów flint, SW	70	37
		Volhynian flint, E	120	–
		Baltic, L	–	15
649	Hłomcza	Świeciechów flint, N	140	2P
		Bircza flint, L	–	–
		Volhynian flint, NE	200	1?
650	Grodzisko Dolne	Baltic, L	–	66
		Volhynian flint, NE	150	16
		Świeciechów flint, N	90	9
		Chocolate flint, NW	170	9

?=data questionable; P= pieces. In case of very few items, the number of pieces is given instead of percentages.

Table A.10 Fossil and sub-recent mollusk shells and other fossils from Central European Magdalenian assemblages

Source		Atlantic			Atlantic/Mediterranean			Mediterranean						Mainz basin	
Species	ID	*Astarte montagui* (Dillwyn)	*Littorina obtusata* (L.)	*Nucella (Purpura) lapillus* (L.)	*Dentalium vulgare* da Costa	*Homalopoma sanguineum* (L.)	*Turritella communis* (Risso)	*Cyclope nerita* (L.)	*Dentalium inaequicostatum* (Dautzenberg)	*Euthria cornia* (L.) [Syn. *Buccinulum corneum* (L.)]	*Glycymeris insubrica* (Brocchi)	*Glycymeris pilosa* (L.)?	*Spisula subtruncata* (da Costa)	*Cerithium margaritaceum* (Brocchi)	*Cerithium plicatum* (Lamarck)
Cabônes	54	–	–	–	–	–	–	–	–	–	–	–	–	–	–
Douattes, Les	69	–	–	–	–	–	–	–	–	–	–	–	–	–	–
Veyrier	79	–	–	–	–	–	–	–	–	1	25	–	1	–	–
Neuchâtel-Monruz	90	–	–	–	–	–	–	–	–	–	–	–	–	–	–
Chesselgraben	106	–	–	–	–	–	–	–	–	–	–	–	–	–	–
Kohlerhöhle	108	–	–	–	–	1	–	–	–	–	–	–	–	–	–
Kastelhöhle	110	–	–	–	–	–	–	–	–	–	–	–	–	–	–
Birseck-Ermiit.	114	–	–	–	–	–	–	–	–	–	–	1	–	–	–
Hollenberg-H. 3	115	–	–	–	–	–	–	–	–	–	–	–	–	–	–
Munzingen	122	–	–	1	–	1	–	–	–	–	–	–	–	–	–
Rislisberghöhle	124	–	–	–	–	–	–	–	–	–	–	–	–	–	–
Teufelsküche	126	–	–	–	–	–	–	–	–	–	–	–	–	–	–
Käsloch	136	–	–	–	–	–	–	–	–	–	–	–	–	–	–
Schweizersbild	166	–	–	–	–	–	–	–	–	–	–	–	–	–	–
Freudenthal-H.	167	–	–	–	–	–	–	–	–	–	–	–	–	–	–
Kesslerloch	170	–	–	–	–	–	–	–	–	–	–	–	–	2	2
Petersfels	173ff.	–	–	–	–	1	–	2	–	–	–	–	–	–	–
Gnirshöhle 1	195	1	–	2	–	–	–	–	–	–	–	–	–	–	–
Napoleonskopf	197	–	–	–	–	–	–	–	–	–	–	–	–	–	–
Hohle Fels, Ib	241	–	–	–	–	–	–	–	–	–	–	–	–	–	–
Hohle Fels, Ic	242	–	–	–	–	–	–	2	–	–	–	–	–	–	–
Hohle Fels, IIa	243	–	1	–	–	–	–	–	–	–	–	–	–	–	–
Kaufertsberg	271	–	–	–	–	–	–	–	–	–	–	–	–	–	–
Bois Laiterie	320	–	–	–	–	–	–	–	–	–	–	–	–	–	–
Goyet 3	332	–	–	–	–	–	–	–	–	–	–	–	–	–	–
Bay-Bonnet,	338	–	–	–	–	–	–	–	–	–	–	–	–	–	–
Andernach KI	364	–	–	–	–	–	–	–	–	–	–	–	–	–	–
Andernach KII	365	–	–	–	–	46	–	1	–	–	–	–	–	–	–
Gönnersdorf KI	368	–	–	–	–	5	–	–	–	–	–	–	–	–	–
Gönnersdorf KII	369	–	–	–	–	–	–	–	–	–	–	–	–	–	–
Gönnersdorf KIII	370	–	–	–	–	–	–	–	–	–	–	–	–	–	–
Gönnersdorf KI/III	368/70	–	–	–	17	–	–	–	5	–	–	–	–	–	–
Nebra, Altenburg	439	–	–	–	–	–	–	–	–	–	–	–	–	–	–
Kniegrotte	451	–	–	–	–	–	–	–	–	–	–	–	–	–	–
Schafgraben	480	–	–	–	–	–	1	–	–	–	–	–	–	–	–

				Mainz, Paris or Belgian Tertiary basin										
Cyrene spec.	Glycynaeus spec.	*Pectunculus angiocostatus*	Shark tooth	*Cyrena convexa* (Brongniart) [Syn. *Cyrena semistriata* (Desh.)]	*Dentalium fissura* (Lamarck)? *sandbergeri* (Bosqu.)?	*Dentalium geminatum* (Goldfuss)	Glycymeris spec.	*Glycymeris obovata* (Lamarck)/*pilosa lunulata* (Nyst.)	*Pirenella* cf. *plicata* (Bruguiere)	Polynices achatensis (Koninck) (?)	*Potamides lamarcki* (Brongniart)	*Potamides plicatus galeottii* (Nyst)	*Proadusta meyeri* (Boettger)	*Tympanotonos margaritaceaus* (Broc.)
–	–	–	–	–	–	–	–	–	2	–	–	–	–	–
–	–	–	–	–	–	–	1	–	–	–	–	–	–	–
–	–	–	–	–	–	–	–	–	–	–	–	–	–	–
–	–	–	–	–	–	–	7	–	–	–	–	–	–	–
–	–	–	–	–	–	–	1	–	–	–	–	–	–	–
–	–	–	–	–	–	–	18	–	19	–	–	–	–	1
–	–	–	–	–	–	–	10	–	2	–	–	–	–	1
–	–	–	–	–	–	–	1	–	–	–	–	–	–	–
–	–	–	–	–	–	–	20	–	–	–	–	–	–	14
–	–	–	–	2	–	–	–	–	–	–	–	–	–	–
–	–	–	–	–	–	–	1	–	1	–	–	–	–	6
–	–	–	–	–	–	–	–	3	3	–	2	–	–	–
–	–	–	–	–	–	–	2	–	–	–	–	–	–	–
1	–	–	–	–	–	–	5	–	–	–	–	–	–	–
–	–	–	–	–	–	–	1	–	–	–	–	–	–	–
–	–	3	1	–	–	–	–	–	x	–	–	–	–	–
–	–	–	–	–	14	2	–	21	6	–	1	–	1	1
–	–	–	–	–	–	–	–	–	–	–	–	–	–	–
–	–	–	–	–	–	–	1	–	–	–	–	–	–	–
–	–	–	–	–	–	–	–	–	2	–	–	–	–	–
–	–	–	–	–	–	–	–	–	–	–	–	–	–	–
–	–	–	–	–	–	–	–	–	2	1	–	–	–	–
–	–	–	–	x	–	–	–	–	–	–	–	–	–	–
–	–	–	–	–	–	–	1	–	–	–	–	–	–	–
–	–	–	–	–	–	–	1	–	–	–	–	–	–	–
–	–	–	–	–	–	–	1	–	–	–	–	–	–	–
–	–	–	–	–	–	–	–	–	1	–	–	–	–	1
–	–	–	–	–	–	–	–	–	–	–	–	–	–	–
–	–	–	–	–	–	–	–	–	–	–	–	–	–	–
–	1	–	–	–	–	–	–	–	–	–	–	–	–	–
–	–	–	1	–	–	–	–	–	–	–	–	–	–	–
–	–	–	–	–	–	–	–	–	–	–	–	–	–	–
–	–	–	–	–	–	–	x	–	–	–	–	–	–	–
–	–	–	–	64	–	–	3	–	–	–	–	1	–	–
–	–	–	–	–	–	–	–	–	–	–	–	–	–	–

Table A.10 (continued)

Source		Paris Basin														
Species	ID	*Ancillaria dubia* (Deshayes)	*Ancillaria buccinoides* (Lamarck)	*Bayana lactea* (Lamarck)	Bayana spec.	Campanile sp.?	*Campanile giganteum* (Lamarck)	*Cerithium crenatulatum* (Deshayes)	*Cerithium echinoides* (Lamarck)	*Glycymeris symetricus* (Deshayes)	*Glycymeris pulvinata* (Lamarck)	*Gravesicerithium gravesi* (Deshayes)	*Cryptochorda stromboides* (Hermann)	*Cyprea media* (Deshayes)	*Hipponyx cornucopiae* (Lamarck)	*Lyra harpula*
Cabônes	54	–	–	1	–	–	–	–	–	–	–	–	–	–	–	–
Jean Pierre I, A	63	–	–	–	–	–	–	–	–	–	–	–	–	–	–	–
Douattes, Les	69	–	–	–	–	–	–	–	–	–	–	–	–	–	–	–
Kastelhöhle	110	–	–	–	–	–	–	–	–	–	–	–	–	–	–	–
Birseck-Ermiit.	114	–	–	–	–	–	–	–	–	–	–	–	–	–	–	–
Rislisberghöhle	124	–	–	–	–	–	–	–	–	–	–	–	–	–	–	–
Schweizersbild	166	–	–	–	–	–	–	–	–	–	–	–	–	–	–	–
Petersfels	173ff.	–	–	1	–	–	–	–	–	–	–	–	–	–	–	–
Gnirshöhle 1	195	–	–	–	–	–	–	–	–	–	–	–	–	–	–	–
Trou da Somme	319	–	–	1	–	–	–	–	–	–	–	–	–	–	–	–
Bois Laiterie	320	–	–	4	–	–	1	–	–	–	–	–	–	–	–	–
Trou Magrite	323	–	–	1	–	–	–	–	–	–	–	–	–	–	–	–
Trou de Chaleux	324	x	x	x	x	–	–	x	–	–	x	–	–	x	x	–
Trou du Frontal	328	–	–	–	–	–	–	–	1	–	3	–	–	–	–	–
Goyet 3	332	–	–	–	–	–	–	–	–	–	–	–	–	–	–	–
Coléoptère	334	–	–	x	–	–	–	–	–	x	–	–	–	–	–	x
Verlaine	335	–	–	–	–	–	1	–	–	–	–	2	1	–	1	–

Table A.10 (continued)

Source		Paris basin/Northern Alps			Steinheimer basin			Upper Danube, Kirchberg layers				Middle Danube, Vienna basin		
Species	ID	*Keepingia uniserialis* (Sanberger)	*Turritella granulosa* (Deshayes)	*Turritella imbricataria* (Solander)	*Gyraulus trochiformis* (Stahl)	*Gyraulus sulcatus* (Hilgendorf)	*Radix socialis* (Zieten)	*Brotia escheri* (Brongniart)	Congeria spec.	*Melanopsis kleini* (Kurr)	*Viviparus suevicus* (Wenz)	*Melanopsis inermis* (Handmann)	*Melanopsis varicosa* (Handmann)	*Melanopsis vindobonensis* (Fuchs)
Jean Pierre I, B	63	1	–	–	–	–	–	–	–	–	–	–	–	–
Neuchâtel-Monruz	90	–	–	–	19	–	–	1	–	–	11	–	–	–
Kohlerhöhle	108	–	–	–	–	–	–	–	–	–	1	–	–	–
Schweizersbild	166	–	–	–	–	–	–	–	–	–	1	–	–	–
Petersfels	173ff.	–	–	–	1	–	–	4	–	–	19	–	–	–
Gnirshöhle 1	195	–	–	–	–	–	–	–	–	–	1	–	–	–
Napoleonskopf	197	–	–	–	–	–	–	–	–	–	–	–	–	–
Nikolaushöhle	211	–	–	–	–	–	–	–	–	–	–	–	–	–
Felsställe	223	–	–	–	1	–	–	–	5	–	43	–	–	–
Hohle Fels, Ia	240	–	–	–	–	–	–	–	1	–	3	–	–	–
Hohle Fels, Ib	241	–	–	–	1	–	1	–	–	–	4	–	–	–
Hohle Fels, Ic	242	–	–	–	1	–	–	2	–	–	14	–	–	–
Hohle Fels, IIa	243	–	–	–	4	1	2	–	–	3	–	–	–	–
Trou des Nutons	326	–	–	1	–	–	–	–	–	–	–	–	–	–
Kamegg,	543	–	–	–	–	–	–	–	–	–	–	3	4	79
Žitného	567	–	–	–	–	–	–	–	–	–	–	–	–	–
Verunčina	568	–	–	–	–	–	–	–	–	–	–	–	–	–
Kolíbky	570	–	–	–	–	–	–	–	–	–	–	–	–	–
Kůlna-Höhle, 6	572	–	–	–	–	–	–	–	–	–	–	–	–	–
Hadí	578	–	–	–	–	–	–	–	–	–	–	–	–	–
Pekárna g/h	579	–	–	–	–	–	–	–	–	–	–	–	–	–
Adlerova	582	–	–	–	–	–	–	–	–	–	–	–	–	–
Ochozska-Höhle	585	–	–	–	–	–	–	–	–	–	–	–	–	–
Maszycka Cave	629	–	–	–	–	–	–	–	–	–	–	–	–	–

Melania lactea (Lamarck)	*Melanopsis buccinoides* (Ferussac)	Natica spec	*Natica willemeti* (Deshayes)	*Pleurotoma filosa* (Lamarck)	*Sigmesalia regularis* (Deshayes)	*Sycostoma bulbus* (Solander)	*Sycostoma pirus* (Solander)	Sycum spec	*Sygmesalia intermedia* (Deshayes)	*Terebralia bidentata* (Defrance)	*Trivia bouryi* (Coseman)	Turritella spec.?	*Turritella terebellata* (Lam)	*Turritella intermedia*	*Turritella adulterata* (Deshayes)	*Turritella* cf. *solanderi* (Mayer)	*Turritella* cf. *vandini* (Desh.)	*Tympanotonos conoidens* (Lamarck)	*Tympanotonos labyrinthum* (Nyst)
–	–	–	–	–	–	–	–	–	–	–	–	–	–	–	–	–	–	–	1
–	–	–	–	–	–	–	–	–	–	–	–	9	–	–	–	–	–	–	–
–	–	–	–	–	–	–	–	–	–	–	–	x	–	–	–	–	–	–	–
–	–	–	–	–	–	–	–	–	–	–	–	3	–	–	–	–	–	–	–
–	–	–	–	–	–	–	–	–	–	–	–	3	–	–	–	–	–	–	–
–	–	–	–	–	–	–	–	–	–	–	–	1	–	–	–	–	–	–	–
–	–	1	–	–	–	–	–	–	–	–	–	–	–	–	–	–	–	–	–
–	–	–	–	1	–	–	–	–	–	–	–	–	–	–	–	–	–	–	–
–	–	–	–	–	–	–	–	2	–	–	–	–	–	–	–	–	–	–	–
–	–	–	–	–	–	–	–	–	–	–	–	–	–	–	–	–	–	–	–
–	–	–	–	–	–	–	–	–	1	1	–	–	–	–	–	–	–	–	1
–	–	–	–	–	–	2	4	–	–	–	–	–	–	–	–	–	–	–	–
–	–	x	–	–	–	–	–	–	–	–	–	–	–	–	x	x	x	–	–
2	–	–	1	1	–	–	–	–	–	–	–	–	–	2	–	–	–	–	–
–	–	–	–	–	–	–	–	–	–	–	–	18	–	–	–	–	–	–	–
–	–	–	x	–	–	–	–	–	x	–	–	x	–	–	–	–	–	–	–
–	3	–	–	–	4	1	–	–	–	–	–	–	1	–	–	–	–	1	–

Vienna basin										Moravian Karst and Carpathian Foredeep				
Arca [Anadara] spec.	*Arca [Anadara] turonensis* (Dujardin)	*Ancillaria obsoleta* (Brocchi)	*Cyprea fabagina* (Lamarck)	Melongena cornuta (Agassiz) cf. *pseudobasilica* (Strausz)	*Mitra substriatula* (d'Orbigny)	*Mitra fusiformis* (Brocchi)	*Nassa vindobonense* (Mayer)	*Nassa hoernesi* (Mayer)	*Natica catena helicina* (Brocchi)	*Ancilla glandiformis* (Lamarck)	*Dentalium badense* (Partsch)	*Protoma cathedralis paucicincta* (Sacco)	*Terebralia bidentata lignitarum* (Eichwald)	*Turitella turris* (Basterot)
–	–	–	–	–	–	–	–	–	–	–	–	–	–	–
–	–	–	–	–	–	–	–	–	–	–	–	–	–	–
–	–	–	–	–	–	–	–			–	–	–	–	–
–	–	–	–	–	–	–	–	–	–	–	–	–	–	–
–	–	–	–	–	–	–	–	–	–	–	–	–	–	–
–	–	–	–	–	–	–	–	–	–	–	–	–	–	–
–	–	–	–	–	–	–	–	–	–	–	–	–	–	–
–	–	–	–	–	–	–	–	–	–	–	–	–	–	–
–	–	–	–	–	–	–	–	–	–	–	–	–	–	–
–	–	–	–	–	–	–	–	–	–	–	–	–	–	–
–	–	–	–	–	–	–	–	–	–	–	–	–	–	–
–	–	–	–	–	–	–	–	–	–	–	–	–	–	–
–	–	–	–	–	–	–	–	–	–	–	–	–	–	–
–	–	–	–	–	–	–	–	–	–	–	–	–	–	–
21	4	6	3	–	–	1	5	14	10	–	–	–	–	–
–	–	–	–	–	–	–	–	–	–	–	–	–	1	–
–	–	–	–	–	–	–	–	–	–	–	2	–	–	–
–	–	–	–	–	–	–	–	–	–	–	–	1	–	–
–	–	–	–	–	–	–	–	–	–	–	–	–	–	1
–	–	–	–	–	–	–	–	–	–	–	2	–	–	–
–	–	–	–	1	1	–	–	–	–	–	3	–	–	–
–	–	–	–	–	–	–	–	–	–	1	–	–	–	–
–	–	–	–	–	–	–	–	–	–	–	1	–	–	–
–	–	–	–	–	–	–	–	–	–	–	2	–	–	–

Table A.10 (continued)

Source		Paris basin?	Luxembourg/Lorraine	Touraine, Gers, Basses–Pýrénées	Jurassic sediments	Swabian Alb	Vienna basin and Carpathian Foredeep	Vienna basin, Korytnica Bay
Species	ID	Rhinoclavis spec.	Ichthyosaurus, vertebra	*Potamides papaveraceus* (Basterot)	Amonites	*Hyalotraros patella*	*Pirenella* cf. *nodosoplicata* (Hörnes)	*Arca [Anadara] dilluvii* (Lamarck)
Colomb	29	–	–	–	–	–	–	–
Jean Pierre I, A	63	2	–	–	–	–	–	–
Hauterive–Ch.	95	–	–	–	–	–	–	–
Moosseedorf	101	–	–	–	–	–	–	–
Hollenberg–H. 3	115	–	–	–	x	–	–	–
Schweizersbild	166	–	–	1	–	–	–	–
Kesslerloch	170	–	–	–	x	–	–	–
Petersfels	173ff.	–	–	–	–	–	–	–
Nikolaushöhle	211	–	–	–	x	–	–	–
Felsställe	223	–	–	–	9	–	–	–
Schussenquelle	229	–	–	–	–	1	–	–
Hohlenstein, Eder.	269	–	–	–	–	–	–	–
Trou Magrite	323	–	–	–	–	–	–	–
Goyet 3	332	–	–	1	–	–	–	–
Gönnersdorf KIII	370	–	1	–	–	–	–	–
Tmaň, Děravá C.	515	–	–	–	–	–	–	1
Kamegg,	543	–	–	–	–	–	–	9
Býči skála	551	–	–	–	–	–	–	1
Pekárna g/h	579	–	–	–	–	–	1	1
Adlerova	582	–	–	–	–	–	–	–
Maszycka Cave	629	–	–	–	–	–	–	3

Ödenburger bay, Hartberg	Badenian Sea	Samartian Sea	Baltic region, Moraine depositis	Switzerland	Marin Miozän? Mediterranean? Atlantic?	?	?	?
Calliostoma podolicoformis nudostriata (Papp)	*Melanopsis impressa* (Krauss) ssp.	*Cerithium pictum* (Defrance)	Fossil resin	Fossil resin	Trivia spec.	*Glycymeris pectunculus* (L.)	*Nassa neritea* (L.)	*Turitella bicarinata* (Eichwald)
–	–	–	–	–	–		x	–
–	–	–	–	–	–	–	–	–
–	–	–	1	5	–	–	–	–
–	–	–	2	–	–	–	–	–
–	–	–	–	–	–	–	–	–
–	–	–	–	–	–	–	–	–
–	–	–	–	–	–	–	–	–
–	–	–	–	–	1	–	–	–
–	–	–	–	–	–	–	–	–
–	–	–	–	–	–	–	–	–
–	–	–	–	–	–	–	–	–
–	–	–	–	–	–	2	–	–
–	–	–	–	–	1	–	–	–
–	–	–	–	–	–	–	–	–
–	–	–	–	–	–	–	–	–
–	–	–	–	–	–	–	–	1
11	–	9	–	–	–	–	–	–
–	–	–	–	–	–	–	–	–
–	5	–	–	–	–	–	–	–
–	1	–	–	–	–	–	–	–
–	–	–	–	–	–	–	–	–

In case no exact quantaties were available, x indicates the presence of a species.

Table A.11 Typological composition of lithic assemblages with ≥10 tools

ID	Site	nTool	Sc	%	SSc	%	RSc	%	Bu	%	Lbu	%	Fbu	%	Zi	%	Bo	%	Lbo	%
2	Campalou	125	12	10	0	0	0	0	22	18	0	0	0	0	0	0	0	0	0	0
6	La Garenne	298	36	12	0	0	3	1	75	25	0	0	1	0	0	0	4	1	0	0
8	Croze-sur-S.	400	56	14	0	0	0	0	144	36	0	0	0	0	0	0	108	27	0	0
9	Calvaire II	35	7	20	0	0	0	0	4	11	0	0	0	0	0	0	2	6	0	0
10	Calvaire III	70	7	10	0	0	1	1	14	20	0	0	0	0	0	0	1	1	0	0
11	Calvaire IV	36	4	11	0	0	0	0	13	36	0	0	0	0	0	0	2	6	0	0
12	B. d'Isère	119	44	37	0	0	3	3	8	7	0	0	0	0	0	0	0	0	0	0
16	B.-Rouss	34	0	0	0	0	0	0	8	24	0	0	0	0	0	0	0	0	0	0
18	Gay, A.	153	51	33	2	1	0	0	32	21	0	0	0	0	0	0	8	5	0	0
20	Bobache, A.	34	6	18	0	0	2	6	6	18	0	0	0	0	0	0	1	3	0	0
24	Freydières	35	2	6	0	0	0	0	2	6	0	0	0	0	0	0	2	6	0	0
26	Grappin	175	20	11	0	0	2	1	47	27	0	0	0	0	0	0	0	0	0	0
29	Colomb	263	91	35	0	0	0	0	24	9	0	0	0	0	0	0	9	3	0	0
30	Chênelaz	235	20	9	0	0	0	0	21	9	0	0	0	0	0	0	31	13	0	0
34	Bonne Fem.	38	11	29	0	0	1	3	9	24	0	0	0	0	0	0	5	13	0	0
35	Olette, Got	62	15	24	0	0	2	3	5	8	0	0	0	0	0	0	1	2	0	0
37	Bibi	189	82	43	0	0	3	2	23	12	0	0	0	0	0	0	4	2	0	0
44	Farincourt sur.	127	11	9	0	0	2	2	93	73	0	0	0	0	0	0	4	3	0	0
45	Farincourt I+II	179	24	13	1	1	5	3	110	61	0	0	0	0	0	0	6	3	0	0
46	Farincourt IIIE	106	3	3	0	0	1	1	69	65	0	0	0	0	0	0	3	3	0	0
47	Farincourt IIIB	46	4	9	0	0	0	0	32	70	0	0	0	0	0	0	0	0	0	0
48	B. de Glos	22	5	23	0	0	0	0	6	27	0	0	0	0	0	0	5	23	0	0
49	B. de Glos	65	20	31	0	0	0	0	5	8	0	0	0	0	0	0	0	0	0	0
53	Romains	1000	0	0	0	0	0	0	0	0	0	0	0	0	0	0	33	3	3	0
56	Seveux	692	80	12	0	0	0	0	273	39	0	0	0	0	5	1	169	24	3	0
58	Fru, Abri I	65	4	6	0	0	0	0	6	9	0	0	0	0	0	0	2	3	0	0
59	Fru, Abri II	118	7	6	0	0	0	0	9	8	0	0	0	0	0	0	4	3	0	0
63	Jean Pierre B	174	6	3	0	0	0	0	23	13	0	0	0	0	4	2	10	6	1	1
64	Jean Pierre	50	4	8	0	0	1	2	6	12	0	0	0	0	0	0	5	10	0	0
66	Baume Noire	215	18	8	0	0	0	0	118	55	0	0	0	0	0	0	19	9	0	0
69	Douattes	637	103	16	23	4	4	1	168	26	0	0	0	0	1	0	44	7	0	0
73	Banges	82	7	9	0	0	0	0	11	13	0	0	0	0	0	0	6	7	0	0
74	La Guillotine	19	1	5	0	0	0	0	1	5	0	0	0	0	0	0	0	0	0	0
79	Veyrier	593	94	16	9	2	8	1	93	16	0	0	0	0	0	0	78	13	0	0
82	Auzary-Th.	177	9	5	0	0	1	1	16	9	0	0	0	0	0	0	3	2	0	0
89	Neuchâtel A	14	9	64	7	50	1	7	2	14	0	0	0	0	0	0	1	7	0	0
90	Neuchâtel 1	1413	66	5	5	0	2	0	251	18	1	0	0	0	7	1	155	11	0	0
91	Neuchâtel 2	59	4	7	1	2	0	0	5	8	0	0	0	0	0	0	2	3	0	0
93	Hauterive 1	594	54	9	0	0	18	3	90	15	0	0	0	0	4	1	33	6	0	0
95	Hauterive 2	169	9	5	1	1	3	2	48	28	0	0	0	0	1	1	5	3	0	0
97	Hauterive 1A	64	8	13	1	2	3	5	16	25	0	0	0	0	0	0	4	6	0	0
99	Alle, Noir	26	0	0	0	0	0	0	8	31	0	0	0	0	0	0	1	4	0	0
100	Liesberg	50	15	30	2	4	0	0	13	26	0	0	0	0	0	0	6	12	1	2
101	Moosseedorf	5196	443	9	0	0	34	1	839	16	0	0	0	0	0	0	716	14	17	0
104	Büttenloch	60	1	2	0	0	0	0	1	2	0	0	0	0	0	0	5	8	0	0
107	Brügglihöhle	192	24	13	4	2	5	3	20	10	0	0	0	0	0	0	11	6	0	0
108	Kohlerhöhle	254	140	55	0	0	9	4	22	9	0	0	0	0	0	0	38	15	1	0
110	Kastelh. u.	604	109	18	2	0	2	0	116	19	0	0	3	1	0	0	91	1	0	0
111	Heidenküche	35	5	14	1	3	0	0	2	6	0	0	0	0	0	0	2	6	0	0
114	Birseck-Erm.	197	26	13	0	0	1	1	48	24	0	0	0	0	0	0	4	2	0	0

(continued)

Table A.11 (continued)

ID	Site	nTool	Sc	%	SSc	%	RSc	%	Bu	%	Lbu	%	Fbu	%	Zi	%	Bo	%	Lbo	%
115	Hollenberg-H.	87	17	20	0	0	9	10	14	16	0	0	0	0	0	0	16	18	0	0
119	Bolken	13	0	0	0	0	0	0	4	31	0	0	0	0	0	0	0	0	0	0
120	Fürsteiner	209	39	19	0	0	0	0	133	64	0	0	0	0	0	0	3	1	0	0
121	Burgäschi	15	2	13	0	0	0	0	4	27	0	0	0	0	0	0	2	13	0	0
122	Munzingen	687	92	13	0	0	6	1	160	23	0	0	0	0	2	0	35	5	1	0
126	Teufelsküche	445	68	15	0	0	4	1	90	20	2	0	0	0	0	0	48	11	1	0
127	Steinberg	21	2	10	0	0	1	5	3	14	0	0	0	0	1	5	3	14	0	0
128	Rheinfelden	79	5	6	0	0	1	1	22	28	1	1	0	0	0	0	11	14	0	0
131	Sälihöhle	267	38	14	0	0	0	0	63	24	1	0	0	0	0	0	17	6	0	0
132	Hard I	259	20	8	1	0	0	0	40	15	0	0	0	0	0	0	16	6	0	0
133	Hard II	62	3	5	0	0	0	0	9	15	0	0	0	0	0	0	1	2	0	0
134	Wilmatt	107	10	9	0	0	1	1	21	20	0	0	0	0	0	0	4	4	0	0
135	Winznau-Kö.	1876	147	8	0	0	0	0	242	13	0	0	0	0	0	0	129	7	0	0
136	Käsloch	1015	129	13	4	0	38	4	205	20	3	0	0	0	0	0	94	9	5	0
137	Mühleloch	302	6	2	0	0	0	0	36	12	0	0	0	0	0	0	3	1	0	0
139	Reiden-Stum.	245	21	9	0	0	1	0	50	20	0	0	0	0	0	0	14	6	1	0
152	Seewager.	14	1	7	0	0	0	0	1	7	0	0	0	0	0	0	0	0	0	0
164	Einsiedeln-L.	216	36	17	10	5	2	1	49	23	0	0	0	0	0	0	7	3	0	0
165	Solgen Im G.	45	6	13	0	0	0	0	3	7	0	0	0	0	0	0	17	38	0	0
166	Schweizersb.	560	81	14	0	0	11	2	320	57	0	0	0	0	7	1	87	16	0	0
170	Kesslerloch	1775	784	44	0	0	4	0	741	42	12	1	0	0	13	1	87	5	10	1
173	P1AH2	17	4	24	0	0	1	6	1	6	0	0	0	0	0	0	4	24	0	0
174	P1AH3A	40	5	13	0	0	0	0	1	3	0	0	0	0	0	0	7	18	0	0
175	P1AH3B	57	8	14	0	0	2	4	14	25	0	0	0	0	0	0	1	2	1	2
176	P1AH4	14	0	0	0	0	0	0	0	0	0	0	0	0	0	0	4	29	0	0
178	P3AH1/2	26	5	19	0	0	0	0	4	15	0	0	0	0	0	0	0	0	0	0
179	P3AH2	259	26	10	1	0	5	2	25	10	0	0	0	0	1	0	28	11	1	0
183	P3AH4	12	0	0	0	0	0	0	0	0	0	0	0	0	0	0	3	25	0	0
185	P5AH2	25	5	20	0	0	0	0	0	0	0	0	0	0	0	0	5	20	0	0
190	P6AH4	12	1	8	0	0	0	0	4	33	0	0	0	0	0	0	2	17	0	0
191	P6AH5	20	1	5	0	0	0	0	3	15	0	0	0	0	0	0	2	10	0	0
195	Gnirshöhle 1	34	8	24	0	0	2	6	5	15	0	0	0	0	0	0	5	15	0	0
197	Napoleonsk.	13	0	0	0	0	0	0	5	38	0	0	1	8	0	0	4	30	0	0
198	Buttentalh.	46	2	4	0	0	1	2	10	22	0	0	0	0	0	0	4	9	0	0
199	Probstfels	148	15	10	0	0	1	1	49	33	0	0	0	0	0	0	3	2	0	0
205	Burghöhle D.	127	11	9	3	2	1	1	14	11	0	0	0	0	1	1	5	4	0	0
210	Annakapell.	18	0	0	0	0	0	0	1	6	0	0	0	0	0	0	0	0	0	0
215	Käppele	294	25	9	0	0	0	0	35	12	0	0	0	0	0	0	108	37	2	1
217	Randecker M.	23	5	22	0	0	0	0	2	9	0	0	0	0	1	4	1	4	0	0
218	Braunfirst	240	28	12	0	0	1	0	56	23	2	1	0	0	1	0	16	7	0	0
219	Burkhardtsh.	78	5	6	0	0	0	0	14	18	0	0	0	0	0	0	4	5	0	0
221	Hohlefels H.	30	1	3	0	0	0	0	7	23	1	3	0	0	0	0	3	10	0	0
223	Felsställe IIIa	229	12	5	0	0	2	1	30	13	0	0	0	0	0	0	33	14	1	0
224	Felsställe IIIb	607	54	9	0	0	4	1	77	13	0	0	0	0	0	0	87	14	2	0
228	Schussenq.	16	2	13	0	0	0	0	5	31	0	0	0	0	0	0	3	19	0	0
229	Schussenq.	33	4	12	0	0	2	6	7	21	0	0	0	0	0	0	2	6	0	0
231	Aichbühl B	134	27	20	5	4	3	2	43	32	0	0	0	0	1	1	5	4	0	0
238	Helga-Abri	39	7	18	0	0	5	13	4	10	0	0	0	0	0	0	1	3	0	0
239	Helga-Abri	11	7	64	0	0	5	45	1	9	0	0	0	0	0	0	0	0	0	0
243	Hohle Fels	113	4	4	0	0	0	0	8	7	0	0	0	0	0	0	2	2	0	0
244	Sirgenstein	37	6	16	0	0	5	14	12	32	0	0	0	0	0	0	0	0	0	0

(continued)

Table A.11 (continued)

ID	Site	nTool	Sc	%	SSc	%	RSc	%	Bu	%	Lbu	%	Fbu	%	Zi	%	Bo	%	Lbo	%
246	Brillenhöhle	282	22	8	0	0	1	0	86	31	2	1	0	0	0	0	24	9	1	0
247	Brillenhöhle	300	20	7	0	0	1	0	90	30	2	1	0	0	0	0	14	5	0	0
250	Kl. Sch. Ros.	22	9	41	0	0	3	14	7	32	0	0	0	0	0	0	2	9	0	0
256	H. Stadel II	21	3	14	3	14	0	0	1	5	0	0	0	0	0	0	1	5	0	0
257	H. Stadel III	91	10	11	1	1	0	0	14	15	0	0	0	0	0	0	11	12	0	0
261	Spitzbubenh.	21	2	10	0	0	1	5	5	24	0	0	0	0	0	0	5	24	0	0
263	Bärenfelsg.	60	7	12	2	3	1	2	9	15	0	0	0	0	4	7	6	10	0	0
265	Vogelherd III	69	4	6	0	0	1	1	13	19	0	0	0	0	0	0	4	6	0	0
266	Vogelherd II	72	4	6	0	0	2	3	13	18	0	0	0	0	0	0	3	4	0	0
269	Hohlenst. E.	33	3	9	0	0	0	0	4	12	0	0	0	0	0	0	9	27	6	18
271	Kaufertsb. l.	229	26	11	0	0	7	3	117	51	7	3	2	1	1	0	12	5	5	2
272	Kaufertsb. u.	135	16	12	1	1	2	1	31	23	4	3	2	1	0	0	12	9	1	1
281	Speckberg	1777	70	4	0	0	17	1	282	16	0	0	0	0	19	1	142	8	14	1
282	Eitensheim	16	4	25	0	0	0	0	7	44	2	13	0	0	0	0	1	6	0	0
283	Hohlenst. S	28	3	11	0	0	0	0	3	11	0	0	0	0	0	0	0	0	0	0
290	Kastelhängh.	131	14	11	0	0	3	2	16	12	1	1	0	0	0	0	7	5	1	1
291	Sesselfelsg.	30	1	3	0	0	0	0	5	17	0	0	0	0	0	0	0	0	0	0
292	Sesselfelsg.	61	10	16	0	0	0	0	15	25	0	0	0	0	0	0	1	2	0	0
293	Sesselfelsg.	99	5	5	0	0	0	0	21	21	1	1	0	0	2	2	2	2	1	1
294	Sesselfelsg.	117	7	6	0	0	2	2	19	16	0	0	1	1	0	0	6	5	0	0
301	Obernederh.	11	3	27	0	0	1	9	4	36	0	0	1	9	1	9	0	0	0	0
307	Barbing	120	16	13	0	0	1	1	32	27	0	0	0	0	1	1	2	2	0	0
313	Roc-la-Tour	424	14	3	0	0	0	0	32	8	0	0	0	0	0	0	49	12	0	0
320	Bois Laiterie	276	23	8	0	0	10	4	36	13	1	0	2	1	0	0	22	8	0	0
321	Roche-al-Rue	80	8	10	0	0	0	0	12	15	0	0	0	0	2	3	8	10	1	1
324	T. Chaleux 5	3013	369	12	2	0	40	1	739	25	3	0	3	0	50	2	408	14	0	0
325	T. Chaleux O.	308	26	8	1	0	0	0	32	10	0	0	0	0	3	1	40	13	0	0
326	Trou Nutons	60	8	13	0	0	0	0	10	17	0	0	0	0	3	5	15	25	0	0
328	Trou Frontal	178	26	15	0	0	0	0	42	24	0	0	0	0	2	1	26	15	0	0
330	Orp east	465	60	13	0	0	3	1	245	53	71	15	0	0	0	0	14	3	4	1
331	Orp west	159	25	16	0	0	6	4	78	49	21	13	0	0	0	0	10	6	0	0
332	Goyet 3	15	0	0	0	0	0	0	2	13	0	0	0	0	0	0	4	27	0	0
334	Coléoptère	125	18	14	0	0	0	0	17	14	0	0	0	0	0	0	12	10	0	0
335	Verlaine	143	24	17	0	0	0	0	25	17	0	0	2	1	1	1	22	15	0	0
336	Roche Plate	17	5	29	0	0	1	6	3	18	0	0	0	0	0	0	2	12	0	0
337	Kanne	86	11	13	0	0	2	2	38	44	11	13	0	0	0	0	3	3	0	0
341	Mesch	67	10	15	0	0	3	4	9	13	0	0	0	0	3	4	12	18	1	1
344	Sweikhuizen	381	47	12	0	0	0	0	94	25	2	1	0	0	0	0	3	1	0	0
346	Echt	122	27	22	0	0	0	0	40	33	0	0	0	0	4	3	14	11	0	0
347	Griendtsveen	137	16	12	0	0	0	0	53	39	3	2	0	0	0	0	20	15	3	2
350	Beeck	45	3	7	0	0	0	0	18	40	6	13	0	0	0	0	11	24	4	9
351	Alsdorf	385	17	4	0	0	0	0	180	47	1	0	0	0	0	0	15	4	0	0
355	Viersener D.	10	5	50	0	0	1	10	3	30	1	10	0	0	0	0	0	0	0	0
359	Kamphausen	33	1	3	0	0	0	0	5	15	0	0	0	0	1	3	6	18	0	0
362	Kartstein	13	11	85	0	0	7	54	2	15	0	0	0	0	0	0	0	0	0	0
364	Andernach I	503	172	34	0	0	0	0	92	18	3	1	0	0	0	0	42	8	0	0
365	Andernach II	459	18	4	0	0	0	0	154	34	31	7	0	0	0	0	47	10	0	0
366	Andernach III	172	41	24	0	0	0	0	44	26	9	5	0	0	0	0	8	5	0	0
367	Andernach IV	310	27	9	0	0	19	6	129	42	4	1	0	0	0	0	11	4	0	0
368	Gönnersdorf	1809	100	6	0	0	0	0	657	36	115	6	32	2	0	0	238	13	7	0

(continued)

Table A.11 (continued)

ID	Site	nTool	Sc	%	SSc	%	RSc	%	Bu	%	Lbu	%	Fbu	%	Zi	%	Bo	%	Lbo	%
369	Gönnersdorf	2198	136	6	0	0	0	0	535	24	42	2	14	1	0	0	211	10	7	0
370	Gönnersdorf	893	54	6	0	0	0	0	232	26	35	4	8	1	0	0	151	17	6	1
371	Gönnersdorf	174	10	6	0	0	0	0	39	22	6	3	0	0	0	0	14	8	1	1
375	Martinshöhle	15	1	7	0	0	0	0	7	47	0	0	0	0	0	0	2	13	0	0
376	Feldhofhöhle	13	1	8	0	0	0	0	0	0	0	0	0	0	0	0	1	8	0	0
378	Wildweiberlei	131	9	7	0	0	3	2	15	11	0	0	0	0	2	2	9	7	0	0
379	Wildscheuer	134	8	6	0	0	6	4	18	13	0	0	0	0	6	4	1	1	0	0
381	Fußgönheim	66	7	11	1	2	0	0	8	12	0	0	0	0	0	0	7	11	0	0
385	Dreieich-G. II	10	2	20	0	0	0	0	4	40	2	20	0	0	0	0	0	0	0	0
386	Dreieich-G. I	60	0	0	0	0	0	0	25	42	2	3	0	0	0	0	0	0	0	0
395	Gadenstedt	109	0	0	0	0	0	0	12	11	0	0	0	0	0	0	6	6	1	1
401	Bad Franken.	681	189	28	0	0	0	0	116	17	0	0	0	0	54	8	60	9	0	0
402	Bärenkeller	20	3	15	0	0	0	0	0	0	0	0	0	0	0	0	0	0	0	0
414	Schweskau	77	1	1	0	0	0	0	7	9	0	0	0	0	3	4	29	38	9	12
416	Teufelsbr. 2	44	4	9	4	9	1	2	6	14	0	0	0	0	1	2	3	7	0	0
417	Teufelsbr. 3	1170	162	14	5	0	14	1	342	29	22	2	0	0	14	1	109	9	11	1
418	Teufelsbr. 4	423	41	10	1	0	0	0	39	9	0	0	0	0	19	4	57	13	4	1
424	Aschersleben	60	6	10	0	0	0	0	10	17	0	0	0	0	3	5	16	27	1	2
435	Herdl. Ranis	41	3	7	0	0	1	2	3	7	0	0	0	0	0	0	4	10	0	0
437	Ilsenhöhle 4	31	2	6	1	3	0	0	1	3	0	0	0	0	1	3	4	13	0	0
438	Ilsenhöhle 5	19	0	0	0	0	0	0	0	0	0	0	0	0	0	0	2	11	0	0
439	Nebra	2320	202	9	6	0	3	0	1026	44	56	2	24	1	10	0	293	13	13	1
443	Kahla-Löbs.	848	93	11	0	0	0	0	193	23	11	1	0	0	1	0	95	11	34	4
444	Oelknitz	1584	82	5	0	0	0	0	252	16	1	0	0	0	0	0	139	9	3	0
449	Urdhöhle	21	2	10	0	0	1	5	0	0	0	0	0	0	0	0	1	5	0	0
451	Kniegrotte	2315	234	10	6	0	13	1	514	22	14	1	0	0	0	0	205	9	2	0
452	Lausnitz, A	349	44	13	0	0	16	5	76	22	0	0	0	0	34	10	42	12	0	0
455	Saaleck	2021	161	8	5	0	44	2	213	11	0	0	0	0	22	1	387	19	120	6
456	Lengefeld	258	27	10	0	0	4	2	69	27	10	4	0	0	2	1	32	12	5	2
468	Galgenberg	119	8	7	0	0	0	0	30	25	1	1	0	0	1	1	6	5	1	1
470	Gleina	13	0	0	0	0	0	0	3	23	0	0	0	0	0	0	0	0	0	0
471	Ahlendorf	301	19	6	0	0	2	1	48	16	1	0	0	0	0	0	32	11	16	5
475	Wallendorf	79	15	19	0	0	4	5	14	18	0	0	0	0	3	4	1	1	0	0
477	Pfortener B.	38	2	5	0	0	0	0	4	11	1	3	0	0	0	0	2	5	1	3
478	G.-Liebschw.	601	69	11	0	0	0	0	201	33	17	3	0	0	5	1	39	6	16	3
480	Schafgraben	162	31	19	0	0	0	0	86	53	0	0	0	0	1	1	10	6	1	1
483	Groitzsch A1	229	31	14	2	1	1	0	32	14	0	0	0	0	7	3	28	12	18	8
484	Groitzsch A2	37	7	19	0	0	1	3	5	14	0	0	0	0	2	5	3	8	1	3
485	Groitzsch B	406	87	21	7	2	9	2	37	9	0	0	0	0	29	7	40	10	6	1
486	Groitzsch BN	55	6	11	1	2	1	2	10	18	1	2	0	0	1	2	3	5	0	0
487	Groitzsch C1	536	94	18	2	0	6	1	160	30	0	0	0	0	23	4	21	4	3	1
488	Groitzsch C2	25	9	36	0	0	1	4	6	24	0	0	0	0	4	16	0	0	0	0
489	Groitzsch C3	655	117	18	5	1	2	0	119	18	0	0	0	0	38	6	61	9	19	3
490	Groitzsch D1	196	36	18	4	2	7	4	21	11	0	0	0	0	6	3	26	13	4	2
491	Groitzsch D2	559	55	10	3	1	7	1	57	10	0	0	0	0	25	4	59	11	22	4
494	Kuppelberg	18	3	17	0	0	0	0	9	50	2	11	0	0	0	0	2	11	0	0
496	Wachtelberg	37	12	32	0	0	2	5	10	27	0	0	0	0	0	0	6	16	0	0
505	Niederlomm.	25	6	24	0	0	0	0	6	24	0	0	0	0	0	0	2	8	0	0
511	Dobříčany	19	1	5	0	0	0	0	0	0	0	0	0	0	0	0	1	5	0	0
518	Kvíc	188	20	11	0	0	2	1	34	18	0	0	0	0	1	1	28	15	0	0

(continued)

Table A.11 (continued)

ID	Site	nTool	Sc	%	SSc	%	RSc	%	Bu	%	Lbu	%	Fbu	%	Zi	%	Bo	%	Lbo	%
521	Lhota	95	24	25	10	11	0	0	12	13	0	0	0	0	0	0	0	0	0	0
526	Putim	58	5	9	2	3	0	0	13	22	0	0	0	0	1	2	24	41	1	2
530	Hostim	1871	209	11	0	0	31	2	221	12	3	0	15	1	19	1	431	23	14	1
533	Rytířská	53	11	21	1	2	4	8	13	25	0	0	0	0	2	4	4	8	0	0
535	Žd'ár 1	18	2	11	0	0	1	6	6	33	2	11	0	0	0	0	0	0	0	0
537	Burk 2	61	0	0	0	0	0	0	1	2	0	0	0	0	0	0	4	7	1	2
542	Gudenush.	70	13	19	2	3	0	0	14	20	0	0	0	0	2	3	15	21	0	0
543	Kamegg	282	48	17	0	0	0	0	135	48	0	0	0	0	3	1	17	6	0	0
546	Náchod	11	0	0	0	0	0	0	1	9	0	0	0	0	2	18	0	0	0	0
547	Maloměřice	653	57	9	3	0	6	1	252	39	13	2	0	0	1	0	117	18	13	2
551	Býči skála	1000	59	6	0	0	0	0	84	8	0	0	0	0	1	0	44	4	1	0
563	Nová Dráten.	12	0	0	0	0	0	0	0	0	0	0	0	0	0	0	0	0	0	0
567	Žitného	533	75	14	0	0	10	2	89	17	0	0	1	0	8	2	54	10	0	0
568	Verunčina	55	12	22	0	0	2	4	10	18	0	0	0	0	1	2	6	11	1	2
570	Kolíbky 6	20	2	10	0	0	0	0	5	25	0	0	0	0	1	5	0	0	0	0
571	Kůlna 5	105	16	15	8	8	0	0	15	14	0	0	1	1	2	2	8	8	0	0
572	Kůlna 6	162	36	22	10	6	0	0	17	10	0	0	1	1	8	5	8	5	0	0
578	Hadí	95	20	21	0	0	8	8	22	23	0	0	0	0	1	1	18	19	4	4
579	Pekárna g/h	918	185	20	0	0	0	0	205	22	4	0	0	0	36	4	175	19	42	5
580	Pekárna i	49	9	18	0	0	6	12	10	20	0	0	0	0	2	4	11	22	1	2
582	Adlerova	53	9	17	0	0	0	0	22	42	0	0	0	0	0	0	7	13	2	4
583	Křižova	66	7	11	0	0	1	2	11	17	2	3	0	0	0	0	11	17	0	0
584	Švédův stůl	21	4	19	1	5	0	0	7	33	0	0	2	10	1	5	2	10	0	0
585	Ochozska	355	40	11	0	0	3	1	56	16	1	0	3	1	9	3	34	10	3	1
586	Mokrá, I	13	1	8	0	0	0	0	3	23	0	0	0	0	0	0	2	15	0	0
588	Mokrá, V	72	5	7	0	0	0	0	11	15	0	0	0	0	0	0	2	3	0	0
590	Balcarka Cave	123	20	16	0	0	4	3	13	11	0	0	0	0	3	2	7	6	0	0
594	Loštice I	221	14	6	0	0	2	1	38	17	10	5	0	0	0	0	8	4	0	0
595	Broniszowice	20	2	10	0	0	1	5	4	20	0	0	0	0	0	0	0	0	0	0
599	Hranice	558	35	6	0	0	1	0	288	52	0	0	3	1	11	2	74	13	1	0
601	Dzierżysław	2000	0	0	0	0	0	0	491	25	1	0	0	0	2	0	176	9	1	0
602	Cyprzanów 3	43	10	23	0	0	1	2	1	2	0	0	0	0	1	2	25	58	0	0
616	Brzoskwinia 1	30	6	20	0	0	0	0	3	10	0	0	0	0	0	0	1	3	0	0
617	Brzoskwinia 2	33	6	18	0	0	0	0	8	24	0	0	0	0	0	0	2	6	0	0
629	Maszycka C.	66	20	30	0	0	6	9	7	11	0	0	0	0	0	0	1	2	0	0
647	Wilczyce	375	41	11	0	0	0	0	166	44	21	6	0	0	0	0	5	1	0	0
648	Klementowice	752	32	4	0	0	3	0	134	18	2	0	0	0	3	0	157	21	25	3
649	Hłomcza	31	3	10	0	0	1	3	22	71	3	10	0	0	1	3	1	3	0	0

Table A.11 (continued)

ID	Site	nTool	Bb	%	Bbd	%	Bbt	%	Bbdt	%	BBtt	%	Bbc	%	Bbp	%	Bbd	%
2	Campalou	125	40	32	0	0	7	6	0	0	0	0	0	0	0	0	0	0
6	La Garenne	298	175	59	0	0	40	13	0	0	0	0	0	0	0	0	4	1
8	Croze-sur-S.	400	5	1	0	0	3	1	0	0	0	0	0	0	0	0	0	0
9	Calvaire II	35	9	26	0	0	0	0	0	0	0	0	0	0	0	0	0	0
10	Calvaire III	70	29	41	0	0	2	3	0	0	0	0	0	0	0	0	0	0
11	Calvaire IV	36	4	11	0	0	0	0	0	0	0	0	0	0	0	0	0	0
12	B. d'Isère	119	13	11	0	0	0	0	0	0	0	0	0	0	0	0	0	0
16	B.-Rouss	34	0	0	0	0	0	0	0	0	0	0	0	0	0	0	0	0
18	Gay, A.	153	42	27	0	0	0	0	0	0	0	0	0	0	0	0	0	0
20	Bobache, A.	34	16	47	0	0	0	0	0	0	0	0	0	0	0	0	0	0
24	Freydières	35	3	9	1	3	0	0	0	0	0	0	0	0	0	0	0	0
26	Grappin	175	51	29	0	0	18	10	0	0	0	0	0	0	0	0	2	1
29	Colomb	263	20	8	0	0	0	0	0	0	0	0	0	0	0	0	0	0
30	Chênelaz	235	109	46	0	0	13	6	0	0	2	1	0	0	0	0	4	2
34	Bonne Fem.	38	4	11	0	0	0	0	0	0	1	3	0	0	0	0	1	3
35	Olette, Got	62	5	8	0	0	0	0	0	0	0	0	0	0	0	0	0	0
37	Bibi	189	39	21	0	0	3	2	0	0	0	0	0	0	0	0	0	0
44	Farincourt sur.	127	2	2	0	0	0	0	0	0	0	0	0	0	0	0	0	0
45	Farincourt I, II	179	4	2	0	0	1	1	0	0	0	0	0	0	0	0	0	0
46	Farincourt IIIE	106	2	2	0	0	0	0	0	0	0	0	0	0	0	0	0	0
47	Farincourt IIIB	46	7	15	0	0	0	0	0	0	0	0	0	0	0	0	0	0
48	B. de Glos	22	1	5	0	0	0	0	0	0	0	0	0	0	0	0	0	0
49	B. de Glos	65	25	38	0	0	0	0	0	0	0	0	0	0	0	0	0	0
53	Romains	1000	977	98	0	0	104	10	0	0	12	1	0	0	0	0	0	0
56	Seveux	692	41	6	0	0	4	1	0	0	0	0	0	0	0	0	0	0
58	Fru, Abri I	65	53	82	4	6	0	0	0	0	0	0	0	0	0	0	3	5
59	Fru, Abri II	118	98	83	2	2	0	0	0	0	0	0	0	0	0	0	0	0
63	Jean Pierre B	174	94	54	6	3	8	5	7	4	2	1	3	2	1	1	5	3
64	Jean Pierre	50	24	48	3	6	2	4	4	8	0	0	1	2	3	6	0	0
66	Baume Noire	215	36	17	0	0	0	0	0	0	0	0	0	0	0	0	0	0
69	Douattes	637	282	44	0	0	20	3	0	0	1	0	0	0	0	0	11	2
73	Banges	82	52	63	0	0	0	0	0	0	0	0	0	0	0	0	0	0
74	La Guillotine	19	4	21	0	0	0	0	0	0	0	0	0	0	0	0	0	0
79	Veyrier	593	207	35	0	0	50	8	0	0	17	3	0	0	0	0	9	2
82	Auzary-Th.	177	123	69	7	4	1	1	0	0	0	0	0	0	0	0	2	1
89	Neuchâtel A	14	0	0	0	0	0	0	0	0	0	0	0	0	0	0	0	0
90	Neuchâtel 1	1413	883	62	36	3	33	2	5	0	0	0	0	0	0	0	5	0
91	Neuchâtel 2	59	35	59	1	2	2	3	0	0	0	0	1	2	0	0	0	0
93	Hauterive 1	594	369	62	0	0	13	2	0	0	0	0	1	0	0	0	1	0
95	Hauterive 2	169	99	59	1	1	3	2	0	0	0	0	0	0	0	0	1	1
97	Hauterive 1A	64	19	30	0	0	0	0	0	0	0	0	0	0	0	0	0	0
99	Alle, Noir	26	7	27	4	15	0	0	0	0	0	0	0	0	0	0	0	0
100	Liesberg	50	8	16	0	0	0	0	0	0	0	0	0	0	0	0	1	2
101	Moosseedorf	5196	2669	51	15	0	460	9	15	0	15	0	15	0	0	0	20	0
104	Büttenloch	60	3	5	0	0	0	0	3	5	0	0	0	0	0	0	0	0
107	Brügglihöhle	192	65	34	0	0	0	0	0	0	0	0	0	0	0	0	0	0
108	Kohlerhöhle	254	20	8	0	0	3	1	0	0	0	0	0	0	0	0	0	0
110	Kastelh. u.	604	213	35	0	0	13	2	0	0	6	1	0	0	0	0	2	0
111	Heidenküche	35	19	54	0	0	2	6	0	0	0	0	0	0	0	0	0	0
114	Birseck-Erm.	197	105	53	16	8	6	3	0	0	0	0	0	0	0	0	0	0
115	Hollenberg-H.	87	19	22	2	2	2	2	1	1	2	2	0	0	0	0	0	0
119	Bolken	13	5	38	3	23	2	15	0	0	0	0	0	0	0	0	0	0

(continued)

Table A.11 (continued)

ID	Site	nTool	Bb	%	Bbd	%	Bbt	%	Bbdt	%	BBtt	%	Bbc	%	Bbp	%	Bbd	%
120	Fürsteiner	209	21	10	0	0	4	2	0	0	0	0	0	0	0	0	0	0
121	Burgäschi	15	6	40	0	0	1	7	0	0	0	0	0	0	0	0	0	0
122	Munzingen	687	68	10	6	1	10	1	2	0	0	0	0	0	0	0	3	0
126	Teufelsküche	445	98	22	6	1	13	3	0	0	0	0	0	0	0	0	0	0
127	Steinberg	21	6	29	1	5	1	5	0	0	0	0	0	0	0	0	0	0
128	Rheinfelden	79	21	27	3	4	0	0	0	0	0	0	0	0	0	0	0	0
131	Sälihöhle	267	90	34	15	6	35	13	0	0	4	1	0	0	0	0	0	0
132	Hard I	259	15	6	0	0	0	0	0	0	0	0	0	0	0	0	0	0
133	Hard II	62	3	5	0	0	0	0	0	0	0	0	0	0	0	0	0	0
134	Wilmatt	107	5	5	0	0	1	1	0	0	0	0	0	0	0	0	0	0
135	Winznau-Kö.	1876	736	39	0	0	0	0	0	0	0	0	0	0	0	0	0	0
136	Käsloch	1015	159	16	19	2	2	0	3	0	0	0	1	0	0	0	0	0
137	Mühleloch	302	48	16	0	0	1	0	1	0	5	2	0	0	0	0	0	0
139	Reiden-Stum.	245	28	11	0	0	1	0	0	0	0	0	0	0	0	0	0	0
152	Seewager.	14	9	64	1	7	0	0	0	0	0	0	0	0	0	0	1	7
164	Einsiedeln-L.	216	37	17	0	0	5	2	0	0	0	0	0	0	0	0	0	0
165	Solgen Im G.	45	13	29	3	7	2	4	0	0	1	2	1	2	0	0	0	0
166	Schweizersb.	560	127	23	12	2	5	1	5	1	0	0	0	0	0	0	0	0
170	Kesslerloch	1775	143	8	16	1	17	1	6	0	0	0	0	0	0	0	0	0
173	P1AH2	17	3	18	0	0	0	0	0	0	0	0	0	0	0	0	0	0
174	P1AH3A	40	17	43	6	15	0	0	0	0	0	0	0	0	0	0	0	0
175	P1AH3B	57	29	51	0	0	0	0	0	0	0	0	0	0	0	0	0	0
176	P1AH4	14	2	14	0	0	0	0	0	0	0	0	0	0	0	0	0	0
178	P3AH1/2	26	8	31	0	0	0	0	0	0	0	0	0	0	0	0	0	0
179	P3AH2	259	115	44	12	5	1	0	0	0	0	0	0	0	0	0	0	0
183	P3AH4	12	8	67	0	0	0	0	0	0	0	0	0	0	0	0	0	0
185	P5AH2	25	8	32	0	0	1	4	0	0	0	0	0	0	0	0	0	0
190	P6AH4	12	3	25	0	0	0	0	0	0	0	0	0	0	0	0	0	0
191	P6AH5	20	7	35	0	0	0	0	0	0	0	0	0	0	0	0	0	0
195	Gnirshöhle 1	34	7	21	2	6	0	0	1	3	0	0	0	0	0	0	0	0
197	Napoleonsk.	13	0	0	0	0	0	0	0	0	0	0	0	0	0	0	0	0
198	Buttentalh.	46	4	9	0	0	0	0	0	0	0	0	0	0	0	0	0	0
199	Probstfels	148	29	20	0	0	0	0	0	0	0	0	0	0	0	0	0	0
205	Burghöhle D.	127	33	26	0	0	2	2	2	2	0	0	0	0	0	0	1	1
210	Annakapell.	18	5	28	0	0	0	0	0	0	0	0	0	0	0	0	1	6
215	Käppele	294	70	24	10	3	7	2	0	0	2	1	0	0	0	0	0	0
217	Randecker M.	23	0	0	0	0	0	0	0	0	0	0	0	0	0	0	0	0
218	Braunfirst	240	92	38	17	7	11	5	2	1	0	0	0	0	0	0	0	0
219	Burkhardtsh.	78	5	6	0	0	0	0	0	0	0	0	0	0	0	0	0	0
221	Hohlefels H.	30	1	3	0	0	0	0	0	0	0	0	0	0	0	0	0	0
223	Felsställe IIIa	229	65	28	17	7	4	2	2	1	0	0	0	0	0	0	8	3
224	Felsställe IIIb	607	114	19	26	4	3	0	2	0	0	0	0	0	1	0	20	3
228	Schussenq.	16	5	31	0	0	1	6	0	0	0	0	0	0	0	0	0	0
229	Schussenq.	33	11	33	4	12	0	0	0	0	0	0	0	0	0	0	0	0
231	Aichbühl B	134	13	10	0	0	2	1	0	0	0	0	0	0	0	0	0	0
238	Helga-Abri	39	10	26	0	0	1	3	1	3	0	0	0	0	0	0	0	0
239	Helga-Abri	11	3	27	0	0	1	9	1	9	0	0	0	0	0	0	0	0
243	Hohle Fels	113	20	18	4	4	0	0	0	0	0	0	0	0	0	0	0	0
244	Sirgenstein	37	0	0	0	0	0	0	0	0	0	0	0	0	0	0	0	0
246	Brillenhöhle	282	11	4	0	0	0	0	0	0	0	0	0	0	0	0	0	0
247	Brillenhöhle	300	11	4	0	0	0	0	1	0	0	0	0	0	1	0	0	0
250	Kl. Sch. Ros.	22	0	0	0	0	0	0	0	0	0	0	0	0	0	0	0	0

(continued)

Table A.11 (continued)

ID	Site	nTool	Bb	%	Bbd	%	Bbt	%	Bbdt	%	BBtt	%	Bbc	%	Bbp	%	Bbd	%
256	H. Stadel II	21	7	33	0	0	0	0	0	0	0	0	0	0	0	0	0	0
257	H. Stadel III	91	6	7	0	0	0	0	0	0	0	0	0	0	0	0	0	0
261	Spitzbubenh.	21	1	5	0	0	0	0	0	0	0	0	0	0	0	0	0	0
263	Bärenfelsg.	60	4	7	0	0	0	0	0	0	0	0	0	0	0	0	0	0
265	Vogelherd III	69	0	0	0	0	0	0	0	0	0	0	0	0	0	0	0	0
266	Vogelherd II	72	0	0	0	0	0	0	0	0	0	0	0	0	0	0	0	0
269	Hohlenst. E.	33	5	15	1	3	2	6	0	0	0	0	0	0	0	0	0	0
271	Kaufertsb. l.	229	33	14	5	2	2	1	0	0	1	0	0	0	0	0	0	0
272	Kaufertsb. u.	135	44	33	2	1	0	0	0	0	0	0	0	0	0	0	0	0
281	Speckberg	1777	8	0	0	0	0	0	0	0	0	0	0	0	0	0	0	0
282	Eitensheim	16	2	13	0	0	0	0	0	0	0	0	0	0	0	0	0	0
283	Hohlenst. S	28	0	0	0	0	0	0	0	0	0	0	0	0	0	0	0	0
290	Kastelhängh.	131	25	19	0	0	0	0	0	0	0	0	0	0	0	0	1	1
291	Sesselfelsg.	30	5	17	0	0	1	3	0	0	0	0	0	0	0	0	0	0
292	Sesselfelsg.	61	14	23	0	0	1	2	0	0	0	0	0	0	0	0	0	0
293	Sesselfelsg.	99	50	51	12	12	3	3	0	0	0	0	0	0	0	0	0	0
294	Sesselfelsg.	117	71	61	17	15	7	6	1	1	0	0	0	0	0	0	0	0
301	Obernederh.	11	1	9	0	0	0	0	0	0	0	0	0	0	0	0	0	0
307	Barbing	120	52	43	10	8	8	7	0	0	2	2	0	0	0	0	0	0
313	Roc-la-Tour	424	215	51	0	0	0	0	0	0	0	0	0	0	0	0	0	0
320	Bois Laiterie	276	55	20	0	0	1	0	0	0	0	0	0	0	0	0	0	0
321	Roche-al-Rue	80	29	36	3	4	0	0	0	0	0	0	0	0	0	0	0	0
324	T. Chaleux 5	3013	802	27	91	3	29	1	0	0	0	0	0	0	0	0	9	0
325	T. Chaleux O.	308	135	44	27	9	3	1	0	0	0	0	0	0	0	0	2	1
326	Trou Nutons	60	14	23	0	0	5	8	0	0	0	0	0	0	0	0	0	0
328	Trou Frontal	178	39	22	3	2	0	0	0	0	1	1	0	0	0	0	0	0
330	Orp east	465	43	9	0	0	0	0	0	0	0	0	0	0	0	0	0	0
331	Orp west	159	3	2	0	0	0	0	0	0	0	0	0	0	0	0	0	0
332	Goyet 3	15	13	87	0	0	2	13	0	0	9	60	1	7	1	7	0	0
334	Coléoptère	125	59	47	1	1	1	1	0	0	0	0	0	0	0	0	0	0
335	Verlaine	143	15	10	2	1	0	0	1	1	0	0	0	0	3	2	0	0
336	Roche Plate	17	1	6	0	0	0	0	0	0	0	0	0	0	0	0	0	0
337	Kanne	86	2	2	0	0	0	0	0	0	0	0	0	0	0	0	0	0
341	Mesch	67	9	13	0	0	0	0	0	0	0	0	0	0	0	0	0	0
344	Sweikhuizen	381	110	29	2	1	0	0	0	0	0	0	0	0	0	0	0	0
346	Echt	122	10	8	0	0	3	2	0	0	2	2	1	1	1	1	0	0
347	Griendtsveen	137	40	29	0	0	3	2	0	0	1	1	0	0	0	0	0	0
350	Beeck	45	2	4	0	0	0	0	0	0	0	0	0	0	0	0	0	0
351	Alsdorf	385	103	27	11	3	0	0	0	0	0	0	0	0	0	0	1	0
355	Viersener D.	10	0	0	0	0	0	0	0	0	0	0	0	0	0	0	0	0
359	Kamphausen	33	6	18	1	3	0	0	0	0	0	0	0	0	0	0	0	0
362	Kartstein	13	0	0	0	0	0	0	0	0	0	0	0	0	0	0	0	0
364	Andernach I	503	122	24	0	0	0	0	0	0	0	0	0	0	0	0	0	0
365	Andernach II	459	149	32	0	0	0	0	0	0	0	0	0	0	0	0	0	0
366	Andernach III	172	56	33	0	0	0	0	0	0	0	0	0	0	0	0	0	0
367	Andernach IV	310	37	12	6	2	0	0	0	0	0	0	0	0	0	0	0	0
368	Gönnersdorf	1809	571	32	87	5	2	0	0	0	0	0	0	0	0	0	1	0
369	Gönnersdorf	2198	991	45	107	5	2	0	0	0	0	0	0	0	0	0	4	0
370	Gönnersdorf	893	302	34	44	5	7	1	0	0	0	0	0	0	0	0	3	0
371	Gönnersdorf	174	98	56	20	11	3	2	0	0	0	0	0	0	0	0	0	0
375	Martinshöhle	15	0	0	0	0	0	0	0	0	0	0	0	0	0	0	0	0
376	Feldhofhöhle	13	2	15	0	0	0	0	0	0	0	0	0	0	0	0	0	0

(continued)

Table A.11 (continued)

ID	Site	nTool	Bb	%	Bbd	%	Bbt	%	Bbdt	%	BBtt	%	Bbc	%	Bbp	%	Bbd	%
378	Wildweiberlei	131	20	15	1	1	5	4	0	0	0	0	0	0	0	0	0	0
379	Wildscheuer	134	5	4	1	1	1	1	0	0	0	0	0	0	0	0	0	0
381	Fußgönheim	66	1	2	0	0	0	0	0	0	0	0	0	0	0	0	0	0
385	Dreieich-G. II	10	3	30	0	0	0	0	0	0	0	0	0	0	0	0	0	0
386	Dreieich-G. I	60	30	50	0	0	0	0	0	0	0	0	0	0	0	0	0	0
395	Gadenstedt	109	46	42	0	0	1	1	0	0	1	1	0	0	0	0	0	0
401	Bad Franken.	681	69	10	0	0	0	0	0	0	0	0	0	0	0	0	0	0
402	Bärenkeller	20	5	25	0	0	1	5	0	0	0	0	0	0	0	0	0	0
414	Schweskau	77	0	0	0	0	0	0	0	0	0	0	0	0	0	0	0	0
416	Teufelsbr. 2	44	24	55	1	2	0	0	0	0	0	0	0	0	0	0	0	0
417	Teufelsbr. 3	1170	487	42	15	1	17	1	0	0	0	0	1	0	0	0	12	1
418	Teufelsbr. 4	423	254	60	17	4	10	2	5	1	0	0	1	0	0	0	4	1
424	Aschersleben	60	7	12	0	0	0	0	0	0	1	2	0	0	0	0	0	0
435	Herdl. Ranis	41	3	7	0	0	0	0	0	0	0	0	1	2	0	0	1	2
437	Ilsenhöhle 4	31	4	13	1	3	0	0	0	0	1	3	0	0	1	3	0	0
438	Ilsenhöhle 5	19	3	16	0	0	0	0	1	5	0	0	0	0	0	0	0	0
439	Nebra	2320	587	25	91	4	1	0	0	0	0	0	0	0	2	0	0	0
443	Kahla-Löbs.	848	458	54	85	10	5	1	5	1	0	0	0	0	0	0	0	0
444	Oelknitz	1584	857	54	0	0	0	0	0	0	0	0	0	0	0	0	0	0
449	Urdhöhle	21	3	14	0	0	1	5	0	0	0	0	0	0	0	0	0	0
451	Kniegrotte	2315	789	34	73	3	78	3	18	1	9	0	4	0	0	0	4	0
452	Lausnitz, A	349	160	46	8	2	1	0	0	0	0	0	0	0	0	0	0	0
455	Saaleck	2021	1165	58	105	5	188	9	48	2	21	1	14	1	8	0	11	1
456	Lengefeld	258	59	23	4	2	13	5	1	0	0	0	0	0	2	1	0	0
468	Galgenberg	119	47	40	4	3	3	3	0	0	0	0	0	0	0	0	0	0
470	Gleina	13	8	62	0	0	2	15	0	0	0	0	0	0	0	0	0	0
471	Ahlendorf	301	195	65	20	7	30	10	6	2	0	0	1	0	0	0	0	0
475	Wallendorf	79	11	14	2	3	2	3	0	0	0	0	1	1	0	0	0	0
477	Pfortener B.	38	20	53	0	0	0	0	0	0	0	0	0	0	0	0	0	0
478	G.-Liebschw.	601	175	29	32	5	3	1	0	0	0	0	0	0	0	0	0	0
480	Schafgraben	162	2	1	0	0	1	1	0	0	0	0	0	0	0	0	0	0
483	Groitzsch A1	229	111	48	9	4	5	2	3	1	0	0	1	0	1	0	2	1
484	Groitzsch A2	37	7	19	1	3	0	0	0	0	0	0	0	0	0	0	0	0
485	Groitzsch B	406	155	38	23	6	20	5	8	2	5	1	2	0	2	0	2	0
486	Groitzsch BN	55	29	53	2	4	4	7	2	4	0	0	0	0	0	0	0	0
487	Groitzsch C1	536	179	33	14	3	14	3	2	0	3	1	1	0	1	0	0	0
488	Groitzsch C2	25	0	0	0	0	0	0	0	0	0	0	0	0	0	0	0	0
489	Groitzsch C3	655	210	32	21	3	39	6	9	1	10	2	0	0	2	0	4	1
490	Groitzsch D1	196	94	48	7	4	11	6	7	4	2	1	0	0	2	1	5	3
491	Groitzsch D2	559	303	54	34	6	37	7	10	2	8	1	2	0	8	1	7	1
494	Kuppelberg	18	2	11	0	0	0	0	0	0	0	0	0	0	0	0	0	0
496	Wachtelberg	37	2	5	0	0	0	0	0	0	0	0	0	0	0	0	0	0
505	Niederlomm.	25	11	44	2	8	0	0	0	0	0	0	0	0	0	0	0	0
511	Dobřičany	19	10	53	3	16	1	5	0	0	0	0	2	11	0	0	0	0
518	Kvíc	188	44	23	2	1	19	10	4	2	1	1	0	0	0	0	3	2
521	Lhota	95	31	33	13	14	4	4	2	2	0	0	0	0	0	0	0	0
526	Putim	58	3	5	0	0	0	0	0	0	0	0	0	0	0	0	0	0
530	Hostim	1871	525	28	0	0	12	1	0	0	2	0	0	0	0	0	5	0
533	Rytířská	53	2	4	1	2	0	0	0	0	0	0	0	0	0	0	0	0

(continued)

Table A.11 (continued)

ID	Site	nTool	Bb	%	Bbd	%	Bbt	%	Bbdt	%	BBtt	%	Bbc	%	Bbp	%	Bbd	%
535	Žd'ár 1	18	5	28	0	0	0	0	0	0	0	0	0	0	0	0	0	0
537	Burk 2	61	3	5	1	2	0	0	0	0	0	0	0	0	0	0	0	0
542	Gudenush.	70	18	26	0	0	1	1	0	0	0	0	0	0	0	0	0	0
543	Kamegg	282	77	27	17	6	4	1	0	0	2	1	0	0	0	0	0	0
546	Náchod	11	1	9	0	0	0	0	0	0	1	9	0	0	0	0	0	0
547	Maloměřice	653	135	21	4	1	8	1	30	5	0	0	0	0	0	0	0	0
551	Býči skála	1000	50	5	0	0	0	0	2	0	2	0	0	0	0	0	0	0
563	Nová Dráten.	12	8	67	0	0	3	25	0	0	1	8	0	0	0	0	0	0
567	Žitného	533	234	44	0	0	2	0	0	0	0	0	0	0	0	0	0	0
568	Verunčina	55	11	20	0	0	1	2	1	2	0	0	0	0	0	0	0	0
570	Kolíbky 6	20	2	10	0	0	0	0	0	0	0	0	0	0	0	0	0	0
571	Kůlna 5	105	23	22	0	0	3	3	0	0	0	0	0	0	0	0	0	0
572	Kůlna 6	162	33	20	0	0	1	1	0	0	0	0	0	0	0	0	0	0
578	Hadí	95	6	6	0	0	2	2	0	0	0	0	0	0	0	0	0	0
579	Pekárna g/h	918	234	25	21	2	32	3	5	1	4	0	2	0	0	0	22	2
580	Pekárna i	49	10	20	1	2	3	6	0	0	0	0	0	0	0	0	1	2
582	Adlerova	53	10	19	0	0	0	0	0	0	0	0	0	0	0	0	0	0
583	Křižova	66	19	29	0	0	1	2	0	0	0	0	0	0	0	0	0	0
584	Švédův stůl	21	5	24	0	0	2	10	0	0	0	0	1	5	0	0	0	0
585	Ochozska	355	82	23	4	1	17	5	3	1	0	0	0	0	0	0	6	2
586	Mokrá, I	13	4	31	0	0	0	0	0	0	0	0	0	0	0	0	0	0
588	Mokrá, V	72	39	54	7	10	2	3	3	4	0	0	0	0	1	1	0	0
590	Balcarka Cave	123	60	49	3	2	2	2	1	1	0	0	0	0	0	0	4	3
594	Loštice I	221	110	50	8	4	0	0	0	0	0	0	1	0	0	0	0	0
595	Broniszowice	20	2	10	0	0	0	0	0	0	0	0	0	0	0	0	0	0
599	Hranice	558	23	4	0	0	8	1	0	0	0	0	0	0	0	0	0	0
601	Dzierżysław	2000	14	1	0	0	0	0	0	0	0	0	0	0	0	0	0	0
602	Cyprzanów 3	43	0	0	0	0	0	0	0	0	0	0	0	0	0	0	0	0
616	Brzoskwinia 1	30	1	3	0	0	0	0	0	0	0	0	0	0	0	0	0	0
617	Brzoskwinia 2	33	1	3	0	0	0	0	0	0	0	0	0	0	0	0	0	0
629	Maszycka C.	66	10	15	0	0	5	8	0	0	0	0	0	0	0	0	0	0
647	Wilczyce	375	85	23	0	0	1	0	0	0	0	0	0	0	0	0	0	0
648	Klementowice	752	120	16	26	3	11	1	12	2	1	0	0	0	0	0	0	0
649	Hłomcza	31	3	10	1	3	0	0	0	0	0	0	0	0	0	0	0	0

(continued)

Table A.11 (continued)

ID	Site	nTool	Tru	%	SpP	%	PB	%	ND	%	LR	%	BP	%	ShP	%	Tri	%
2	Campalou	125	8	6	0	0	0	0	15	12	0	0	3	2	0	0	0	0
6	La Garenne	298	6	2	0	0	0	0	2	1	0	0	0	0	0	0	0	0
8	Croze-sur-S.	400	0	0	0	0	0	0	0	0	0	0	0	0	1	0	0	0
9	Calvaire II	35	1	3	0	0	0	0	1	3	11	31	0	0	0	0	0	0
10	Calvaire III	70	1	1	0	0	0	0	3	4	15	21	0	0	0	0	0	0
11	Calvaire IV	36	0	0	0	0	0	0	2	6	11	31	0	0	0	0	0	0
12	B. d'Isère	119	13	11	0	0	0	0	6	5	32	27	3	3	0	0	0	0
16	B.-Rouss	34	0	0	0	0	0	0	0	0	0	0	0	0	0	0	0	0
18	Gay, A.	153	0	0	0	0	0	0	0	0	5	3	15	10	0	0	0	0
20	Bobache, A.	34	0	0	0	0	0	0	2	6	1	3	2	6	0	0	0	0
24	Freydières	35	0	0	0	0	1	3	0	0	23	66	2	6	0	0	0	0
26	Grappin	175	12	7	4	2	0	0	11	6	30	17	0	0	0	0	0	0
29	Colomb	263	53	20	0	0	0	0	17	6	46	17	4	2	0	0	0	0
30	Chênelaz	235	8	3	0	0	0	0	27	11	12	5	7	3	0	0	0	0
34	Bonne Fem.	38	3	8	0	0	0	0	0	0	1	3	4	11	0	0	0	0
35	Olette, Got	62	8	13	0	0	0	0	6	10	20	32	2	3	0	0	0	0
37	Bibi	189	13	7	0	0	0	0	3	2	24	13	0	0	0	0	0	0
44	Farincourt sur.	127	12	9	0	0	0	0	3	2	6	5	0	0	0	0	9	7
45	Farincourt I, II	179	19	11	1	1	0	0	1	1	20	11	0	0	0	0	25	14
46	Farinc. IIIE	106	13	12	1	1	0	0	1	1	12	11	0	0	0	0	0	0
47	Farinc. IIIB	46	5	11	0	0	0	0	0	0	3	7	2	4	0	0	0	0
48	B. de Glos	22	4	18	0	0	0	0	1	5	0	0	0	0	0	0	0	0
49	B. de Glos	65	0	0	0	0	0	0	2	3	1	2	2	3	0	0	0	0
53	Romains	1000	0	0	0	0	0	0	0	0	0	0	0	0	0	0	0	0
56	Seveux	692	21	3	0	0	22	3	30	4	51	7	0	0	0	0	0	0
58	Fru, Abri I	65	0	0	0	0	0	0	0	0	0	0	0	0	0	0	0	0
59	Fru, Abri II	118	0	0	0	0	0	0	0	0	0	0	0	0	0	0	0	0
63	Jean Pierre B	174	3	2	0	0	0	0	17	10	15	9	1	1	0	0	1	1
64	Jean Pierre	50	1	2	0	0	0	0	6	12	4	8	0	0	0	0	0	0
66	Baume Noire	215	12	6	0	0	0	0	36	17	3	1	5	2	0	0	0	0
69	Douattes	637	17	3	0	0	0	0	5	1	13	2	4	1	0	0	0	0
73	Banges	82	0	0	0	0	0	0	2	2	3	4	1	1	0	0	0	0
74	La Guillotine	19	0	0	0	0	0	0	1	5	5	26	0	0	0	0	7	37
79	Veyrier	593	53	9	0	0	0	0	39	7	24	4	5	1	0	0	0	0
82	Auzary-Th.	177	6	3	6	3	0	0	3	2	11	6	0	0	0	0	0	0
89	Neuchâtel A	14	0	0	1	7	0	0	0	0	1	7	0	0	0	0	0	0
90	Neuchâtel 1	1413	13	1	20	1	0	0	6	0	12	1	0	0	0	0	0	0
91	Neuchâtel 2	59	3	5	1	2	0	0	0	0	6	10	3	5	0	0	0	0
93	Hauterive 1	594	7	1	13	2	0	0	8	1	16	3	0	0	0	0	0	0
95	Hauterive 2	169	1	1	1	1	0	0	0	0	4	2	1	1	0	0	0	0
97	Hauterive 1A	64	2	3	2	3	0	0	0	0	3	5	10	16	0	0	0	0
99	Alle, Noir	26	3	12	0	0	0	0	0	0	5	19	1	4	0	0	0	0
100	Liesberg	50	3	6	0	0	0	0	2	4	3	6	0	0	0	0	0	0
101	Moosseedorf	5196	266	5	15	0	0	0	112	2	132	3	3	0	0	0	1	0
104	Büttenloch	60	2	3	0	0	0	0	12	20	35	58	0	0	0	0	0	0
107	Brügglihöhle	192	0	0	4	2	0	0	0	0	50	26	7	4	3	2	0	0
108	Kohlerhöhle	254	5	2	0	0	0	0	12	5	6	2	8	3	3	1	0	0
110	Kastelh. u.	604	10	2	0	0	0	0	0	0	0	0	12	2	16	3	0	0
111	Heidenküche	35	3	9	0	0	0	0	2	6	2	6	0	0	0	0	0	0
114	Birseck-Erm.	197	0	0	0	0	0	0	0	0	0	0	1	1	0	0	13	7
115	Hollenberg-H.	87	3	3	0	0	0	0	13	15	4	5	1	1	0	0	0	0
119	Bolken	13	2	15	1	8	0	0	0	0	1	8	0	0	0	0	0	0

(continued)

Table A.11 (continued)

ID	Site	nTool	Tru	%	SpP	%	PB	%	ND	%	LR	%	BP	%	ShP	%	Tri	%
120	Fürsteiner	209	0	0	5	2	0	0	0	0	0	0	9	4	0	0	0	0
121	Burgäschi	15	0	0	0	0	0	0	0	0	0	0	1	7	0	0	0	0
122	Munzingen	687	112	16	52	8	6	1	0	0	155	23	0	0	0	0	0	0
126	Teufelsküche	445	25	6	26	6	0	0	9	2	77	17	4	1	0	0	0	0
127	Steinberg	21	2	10	2	10	0	0	0	0	2	10	0	0	0	0	0	0
128	Rheinfelden	79	2	3	0	0	0	0	11	14	4	5	1	1	0	0	0	0
131	Sälihöhle	267	19	7	0	0	1	0	35	13	3	1	1	0	0	0	0	0
132	Hard I	259	3	1	0	0	0	0	5	2	160	62	0	0	0	0	0	0
133	Hard II	62	4	6	0	0	0	0	2	3	40	65	0	0	0	0	0	0
134	Wilmatt	107	1	1	0	0	0	0	0	0	65	61	0	0	1	1	0	0
135	Winznau-Kö.	1876	0	0	0	0	0	0	0	0	460	25	30	2	132	7	0	0
136	Käsloch	1015	6	1	1	0	0	0	0	0	383	38	21	2	17	2	0	0
137	Mühleloch	302	0	0	0	0	0	0	4	1	203	67	2	1	0	0	0	0
139	Reiden-Stum.	245	26	11	9	4	0	0	47	19	49	20	1	0	0	0	0	0
152	Seewager.	14	0	0	0	0	0	0	0	0	0	0	2	14	1	7	0	0
164	Einsiedeln-L.	216	11	5	7	3	0	0	3	1	33	15	7	3	0	0	5	2
165	Solgen Im G.	45	1	2	0	0	0	0	0	0	0	0	0	0	8	18	0	0
166	Schweizersb.	560	8	1	12	2	1	0	0	0	0	0	4	1	8	1	2	0
170	Kesslerloch	1775	2	0	0	0	0	0	0	0	0	0	5	0	0	0	0	0
173	P1AH2	17	0	0	0	0	0	0	0	0	5	29	0	0	0	0	0	0
174	P1AH3A	40	1	3	0	0	0	0	0	0	9	23	0	0	0	0	0	0
175	P1AH3B	57	0	0	0	0	0	0	0	0	5	9	0	0	0	0	0	0
176	P1AH4	14	0	0	1	7	0	0	0	0	5	36	1	7	0	0	0	0
178	P3AH1/2	26	0	0	0	0	0	0	0	0	9	35	0	0	0	0	0	0
179	P3AH2	259	0	0	2	1	0	0	0	0	61	24	1	0	0	0	0	0
183	P3AH4	12	0	0	0	0	0	0	0	0	1	8	0	0	0	0	0	0
185	P5AH2	25	0	0	0	0	0	0	0	0	7	28	0	0	0	0	0	0
190	P6AH4	12	0	0	0	0	0	0	0	0	2	17	0	0	0	0	0	0
191	P6AH5	20	0	0	0	0	0	0	0	0	7	35	0	0	0	0	0	0
195	Gnirshöhle 1	34	3	9	0	0	2	6	1	3	3	9	0	0	0	0	0	0
197	Napoleonsk.	13	2	15	0	0	0	0	1	8	1	8	0	0	0	0	0	0
198	Buttentalh.	46	3	7	1	2	1	2	0	0	7	15	18	39	0	0	0	0
199	Probstfels	148	5	3	7	5	0	0	0	0	16	11	24	16	0	0	0	0
205	Burghöhle D.	127	15	12	0	0	0	0	7	6	18	14	15	12	0	0	0	0
210	Annakapell.	18	0	0	0	0	0	0	0	0	0	0	0	0	0	0	0	0
215	Käppele	294	5	2	1	0	0	0	0	0	48	16	2	1	0	0	0	0
217	Randecker M.	23	2	9	0	0	0	0	0	0	12	52	0	0	0	0	0	0
218	Braunfirst	240	17	7	2	1	2	1	0	0	23	10	3	1	0	0	0	0
219	Burkhardtsh.	78	11	14	6	8	0	0	2	3	19	24	12	15	0	0	0	0
221	Hohlefels H.	30	7	23	5	17	0	0	0	0	5	17	1	3	0	0	0	0
223	Felsställe IIIa	229	29	13	3	1	0	0	0	0	53	23	4	2	0	0	0	0
224	Felsställe IIIb	607	55	9	9	1	0	0	0	0	198	33	4	1	0	0	0	0
228	Schussenq.	16	0	0	0	0	0	0	0	0	1	6	0	0	0	0	0	0
229	Schussenq.	33	2	6	0	0	0	0	1	3	6	18	0	0	0	0	0	0
231	Aichbühl B	134	4	3	2	1	0	0	0	0	36	27	1	1	0	0	0	0
238	Helga-Abri	39	3	8	1	3	0	0	0	0	7	18	6	15	0	0	0	0
239	Helga-Abri	11	0	0	0	0	0	0	0	0	0	0	0	0	0	0	0	0
243	Hohle Fels	113	8	7	5	4	0	0	50	44	14	12	0	0	0	0	0	0
244	Sirgenstein	37	5	14	0	0	0	0	1	3	13	35	0	0	0	0	0	0
246	Brillenhöhle	282	26	9	13	5	0	0	73	26	21	7	6	2	0	0	0	0
247	Brillenhöhle	300	33	11	19	6	0	0	73	24	37	12	3	1	0	0	0	0
250	Kl. Sch. Ros.	22	1	5	0	0	0	0	0	0	1	5	0	0	1	5	0	0

(continued)

Table A.11 (continued)

ID	Site	nTool	Tru	%	SpP	%	PB	%	ND	%	LR	%	BP	%	ShP	%	Tri	%
256	H. Stadel II	21	5	24	0	0	0	0	0	0	4	19	0	0	0	0	0	0
257	H. Stadel III	91	5	5	1	1	0	0	6	7	22	24	16	18	0	0	0	0
261	Spitzbubenh.	21	0	0	0	0	0	0	1	5	6	29	0	0	0	0	0	0
263	Bärenfelsg.	60	9	15	1	2	0	0	7	12	10	17	6	10	2	3	0	0
265	Vogelherd III	69	6	9	6	9	0	0	12	17	24	35	0	0	0	0	0	0
266	Vogelherd II	72	5	7	3	4	0	0	10	14	28	39	2	3	0	0	0	0
269	Hohlenst. E.	33	1	3	0	0	1	3	1	3	9	27	0	0	0	0	0	0
271	Kaufertsb. l.	229	13	6	0	0	2	1	4	2	17	7	0	0	0	0	0	0
272	Kaufertsb. u.	135	8	6	0	0	0	0	3	2	9	7	12	9	0	0	1	1
281	Speckberg	1777	109	6	35	2	441	25	142	8	313	18	2	0	0	0	0	0
282	Eitensheim	16	0	0	0	0	0	0	0	0	0	0	0	0	1	6	1	6
283	Hohlenst. S	28	3	11	0	0	0	0	2	7	10	36	5	18	0	0	0	0
290	Kastelhängh.	131	2	2	0	0	1	1	25	19	40	31	1	1	0	0	0	0
291	Sesselfelsg.	30	4	13	1	3	0	0	4	13	8	27	2	7	0	0	0	0
292	Sesselfelsg.	61	6	10	2	3	0	0	5	8	5	8	3	5	0	0	0	0
293	Sesselfelsg.	99	1	1	0	0	0	0	1	1	4	4	13	13	0	0	0	0
294	Sesselfelsg.	117	0	0	0	0	2	2	5	4	4	3	3	3	0	0	0	0
301	Obernederh.	11	0	0	0	0	0	0	0	0	2	18	0	0	0	0	0	0
307	Barbing	120	2	2	0	0	0	0	4	3	7	6	4	3	0	0	0	0
313	Roc-la-Tour	424	15	4	10	2	0	0	19	4	55	13	1	0	0	0	0	0
320	Bois Laiterie	276	25	9	2	1	0	0	20	7	91	33	23	8	0	0	0	0
321	Roche-al-Rue	80	12	15	6	8	0	0	0	0	3	4	0	0	0	0	0	0
324	T. Chaleux 5	3013	277	9	194	6	0	0	106	4	94	3	3	0	0	0	0	0
325	T. Chaleux O.	308	5	2	14	5	0	0	17	6	36	12	0	0	0	0	0	0
326	Trou Nutons	60	5	8	0	0	0	0	2	3	3	5	0	0	0	0	0	0
328	Trou Frontal	178	22	12	1	1	0	0	11	6	6	3	3	2	0	0	0	0
330	Orp east	465	17	4	0	0	0	0	26	6	60	13	0	0	0	0	0	0
331	Orp west	159	11	7	0	0	0	0	8	5	24	15	0	0	0	0	0	0
332	Goyet 3	15	0	0	0	0	0	0	0	0	0	0	4	27	0	0	0	0
334	Coléoptère	125	9	7	0	0	0	0	0	0	8	6	2	2	0	0	0	0
335	Verlaine	143	14	10	18	13	0	0	9	6	13	9	1	1	1	1	0	0
336	Roche Plate	17	0	0	0	0	0	0	0	0	2	12	2	12	1	6	0	0
337	Kanne	86	3	3	0	0	0	0	2	2	21	24	2	2	0	0	0	0
341	Mesch	67	4	6	0	0	0	0	0	0	20	30	0	0	0	0	0	0
344	Sweikhuizen	381	6	2	0	0	0	0	13	3	108	28	0	0	0	0	0	0
346	Echt	122	10	8	1	1	0	0	0	0	4	3	11	9	1	1	0	0
347	Griendtsveen	137	3	2	3	2	0	0	0	0	0	0	2	1	0	0	0	0
350	Beeck	45	7	16	0	0	2	4	1	2	4	9	0	0	0	0	0	0
351	Alsdorf	385	35	9	0	0	0	0	0	0	34	9	1	0	0	0	0	0
355	Viersener D.	10	0	0	2	20	0	0	0	0	0	0	0	0	0	0	0	0
359	Kamphausen	33	4	12	0	0	0	0	0	0	9	27	1	3	0	0	0	0
362	Kartstein	13	0	0	0	0	0	0	0	0	0	0	0	0	0	0	0	0
364	Andernach I	503	32	6	43	9	0	0	0	0	0	0	0	0	0	0	0	0
365	Andernach II	459	19	4	72	16	0	0	0	0	0	0	0	0	0	0	0	0
366	Andernach III	172	10	6	13	8	0	0	0	0	0	0	0	0	0	0	0	0
367	Andernach IV	310	11	4	95	31	0	0	0	0	0	0	0	0	0	0	0	0
368	Gönnersdorf	1809	51	3	192	11	0	0	0	0	0	0	0	0	0	0	0	0
369	Gönnersdorf	2198	51	2	456	21	0	0	0	0	0	0	0	0	0	0	0	0
370	Gönnersdorf	893	17	2	137	15	0	0	0	0	0	0	0	0	0	0	0	0
371	Gönnersdorf	174	3	2	10	6	0	0	0	0	0	0	0	0	0	0	0	0
375	Martinshöhle	15	2	13	0	0	0	0	0	0	0	0	3	20	0	0	0	0
376	Feldhofhöhle	13	1	8	0	0	0	0	0	0	7	54	1	8	0	0	0	0

(continued)

Table A.11 (continued)

ID	Site	nTool	Tru	%	SpP	%	PB	%	ND	%	LR	%	BP	%	ShP	%	Tri	%
378	Wildweiberlei	131	4	3	0	0	1	1	0	0	68	52	0	0	0	0	0	0
379	Wildscheuer	134	9	7	5	4	2	1	1	1	74	55	0	0	0	0	0	0
381	Fußgönheim	66	1	2	9	14	0	0	1	2	19	29	2	3	8	12	0	0
385	Dreieich-G. II	10	0	0	1	10	0	0	0	0	2	20	0	0	0	0	0	0
386	Dreieich-G. I	60	2	3	1	0	0	0	0	0	0	0	0	0	0	0	0	0
395	Gadenstedt	109	0	0	0	0	0	0	0	0	19	17	0	0	0	0	0	0
401	Bad Franken.	681	35	5	34	5	0	0	0	0	115	17	3	0	0	0	0	0
402	Bärenkeller	20	2	10	0	0	0	0	0	0	0	0	8	40	2	10	0	0
414	Schweskau	77	8	10	0	0	0	0	1	1	13	17	7	9	5	6	0	0
416	Teufelsbr. 2	44	2	5	4	9	0	0	0	0	0	0	0	0	0	0	0	0
417	Teufelsbr. 3	1170	31	3	23	2	0	0	0	0	0	0	2	0	0	0	0	0
418	Teufelsbr. 4	423	5	1	7	2	0	0	0	0	0	0	1	0	0	0	0	0
424	Aschersleben	60	5	8	1	2	0	0	3	5	7	12	0	0	1	2	0	0
435	Herdl. Ranis	41	1	2	0	0	0	0	1	2	1	2	2	5	0	0	0	0
437	Ilsenhöhle 4	31	11	35	0	0	3	10	1	3	5	16	8	26	0	0	0	0
438	Ilsenhöhle 5	19	2	11	0	0	1	5	0	0	6	32	5	26	0	0	0	0
439	Nebra	2320	89	4	70	3	0	0	0	0	36	2	7	0	0	0	0	0
443	Kahla-Löbs.	848	8	1	0	0	0	0	0	0	0	0	0	0	0	0	0	0
444	Oelknitz	1584	47	3	134	8	0	0	0	0	66	4	0	0	0	0	0	0
449	Urdhöhle	21	1	5	1	5	0	0	1	5	7	33	1	5	3	14	0	0
451	Kniegrotte	2315	97	4	128	6	1	0	0	0	347	15	8	0	3	0	187	8
452	Lausnitz, A	349	12	3	4	1	0	0	3	1	40	11	14	4	0	0	0	0
455	Saaleck	2021	60	3	6	0	0	0	3	0	2	0	2	0	0	0	0	0
456	Lengefeld	258	19	7	5	2	0	0	0	0	44	17	0	0	1	0	0	0
468	Galgenberg	119	2	2	0	0	1	1	0	0	4	3	4	3	0	0	0	0
470	Gleina	13	1	8	0	0	0	0	0	0	1	8	0	0	0	0	0	0
471	Ahlendorf	301	7	2	0	0	0	0	0	0	0	0	0	0	0	0	0	0
475	Wallendorf	79	10	13	2	3	0	0	2	3	18	23	3	4	0	0	0	0
477	Pfortener B.	38	1	3	1	3	0	0	0	0	3	8	5	13	0	0	0	0
478	G.-Liebschw.	601	17	3	32	5	0	0	2	0	66	11	6	1	0	0	0	0
480	Schafgraben	162	5	3	7	4	2	1	2	1	16	10	0	0	0	0	0	0
483	Groitzsch A1	229	17	7	0	0	0	0	0	0	2	1	0	0	0	0	0	0
484	Groitzsch A2	37	4	11	0	0	0	0	0	0	5	14	2	5	0	0	0	0
485	Groitzsch B	406	23	6	0	0	0	0	12	3	9	2	14	3	0	0	0	0
486	Groitzsch BN	55	3	5	0	0	0	0	0	0	2	4	1	2	0	0	0	0
487	Groitzsch C1	536	40	7	0	0	0	0	7	1	11	2	1	0	0	0	0	0
488	Groitzsch C2	25	3	12	0	0	0	0	0	0	3	12	0	0	0	0	0	0
489	Groitzsch C3	655	88	13	1	0	0	0	9	1	9	1	2	0	0	0	1	0
490	Groitzsch D1	196	11	6	0	0	0	0	0	0	2	1	0	0	0	0	0	0
491	Groitzsch D2	559	41	7	1	0	0	0	8	1	9	2	1	0	0	0	0	0
494	Kuppelberg	18	1	6	0	0	0	0	0	0	1	6	0	0	0	0	0	0
496	Wachtelberg	37	4	11	0	0	0	0	0	0	2	5	1	3	0	0	0	0
505	Niederlomm.	25	0	0	0	0	0	0	0	0	0	0	0	0	0	0	0	0
511	Dobříčany	19	1	5	0	0	0	0	2	11	3	16	1	5	0	0	0	0
518	Kvíc	188	16	9	19	10	1	1	4	2	19	10	1	1	1	1	0	0
521	Lhota	95	0	0	0	0	0	0	5	5	0	0	1	1	0	0	0	0
526	Putim	58	4	7	0	0	1	2	2	3	0	0	4	7	1	2	0	0
530	Hostim	1871	47	3	343	18	0	0	70	4	4	0	0	0	2	0	0	0
533	Rytířská	53	7	13	2	4	0	0	0	0	12	23	0	0	0	0	0	0
535	Žd'ár 1	18	0	0	0	0	0	0	0	0	4	22	0	0	0	0	0	0
537	Burk 2	61	1	2	0	0	0	0	0	0	1	2	1	2	2	3	0	0
542	Gudenush.	70	8	11	0	0	0	0	0	0	0	0	0	0	0	0	0	0

(continued)

Table A.11 (continued)

ID	Site	nTool	Tru	%	SpP	%	PB	%	ND	%	LR	%	BP	%	ShP	%	Tri	%
543	Kamegg	282	4	1	0	0	1	0	1	0	14	5	5	2	4	1	0	0
546	Náchod	11	0	0	0	0	0	0	0	0	7	64	0	0	0	0	0	0
547	Maloměřice	653	29	4	18	3	5	1	9	1	27	4	2	0	1	0	0	0
551	Býči skála	1000	0	0	0	0	0	0	0	0	0	0	0	0	0	0	0	0
563	Nová Dráten.	12	0	0	0	0	0	0	0	0	1	8	3	25	0	0	0	0
567	Žitného	533	39	7	7	1	0	0	16	3	10	2	1	0	0	0	0	0
568	Verunčina	55	2	4	0	0	0	0	1	2	9	16	1	2	0	0	0	0
570	Kolíbky 6	20	2	10	1	5	0	0	1	5	6	30	0	0	0	0	0	0
571	Kůlna 5	105	3	3	2	2	0	0	6	6	27	26	2	2	0	0	0	0
572	Kůlna 6	162	17	10	0	0	0	0	11	7	12	7	2	1	0	0	0	0
578	Hadí	95	4	4	11	12	0	0	3	3	13	14	1	1	0	0	0	0
579	Pekárna g/h	918	22	2	50	5	0	0	0	0	1	0	3	0	0	0	0	0
580	Pekárna i	49	2	4	0	0	0	0	0	0	2	4	3	6	0	0	0	0
582	Adlerova	53	0	0	0	0	0	0	2	4	3	6	0	0	0	0	0	0
583	Křižova	66	11	17	0	0	0	0	5	8	1	2	1	2	0	0	0	0
584	Švédův stůl	21	0	0	0	0	0	0	1	5	1	5	1	5	0	0	0	0
585	Ochozska	355	28	8	20	6	0	0	47	13	35	10	2	1	0	0	0	0
586	Mokrá, I	13	1	8	0	0	0	0	0	0	2	15	0	0	0	0	0	0
588	Mokrá, V	72	3	4	1	1	0	0	0	0	10	14	0	0	0	0	0	0
590	Balcarka Cave	123	3	2	1	1	0	0	6	5	7	6	3	2	0	0	0	0
594	Loštice I	221	10	5	9	4	0	0	4	2	17	8	0	0	0	0	2	1
595	Broniszowice	20	0	0	0	0	0	0	0	0	1	5	0	0	0	0	0	0
599	Hranice	558	21	4	8	1	1	0	14	3	48	9	0	0	0	0	34	6
601	Dzierżysław	2000	0	0	0	0	0	0	0	0	115	6	0	0	0	0	114	6
602	Cyprzanów 3	43	6	14	0	0	0	0	0	0	0	0	0	0	0	0	0	0
616	Brzoskwinia 1	30	2	7	0	0	0	0	1	3	16	53	0	0	0	0	0	0
617	Brzoskwinia 2	33	3	9	0	0	0	0	0	0	12	36	1	3	0	0	0	0
629	Maszycka C.	66	3	5	0	0	3	5	2	3	19	29	1	2	0	0	0	0
647	Wilczyce	375	14	4	0	0	0	0	0	0	64	17	0	0	0	0	0	0
648	Klementowice	752	87	12	0	0	0	0	2	0	169	22	17	2	0	0	0	0
649	Hłomcza	31	0	0	0	0	0	0	0	0	1	3	0	0	0	0	0	0

Abbreviations see list of abbreviations

Table A.12 Typological composition of organic asemblages

ID	Site	n mentioned	n recorded	Points	Needles	Barbed points	Awls	B. d.-rondes	B. percé	ID	Site	n mentioned	n recorded	Points	Needles	Barbed points	Awls	B. d.-rondes	B. percé
8	Croze-s-Suran	3	3	3	0	0	0	0	0	174	Peter. 1AH3A	4	1	0	1	0	0	0	0
11	Calvaire, A.	3	2	2	0	0	0	0	0	178	Peter. 3AH1/2	3	1	1	0	0	0	0	0
14	Colombière	3	3	3	0	0	0	0	0	182	Peter. 3AH3/4	2	1	0	0	0	1	0	0
20	Bobache, A.	1	1	0	0	1	0	0	0	183	Petersf. 3AH4	5	3	1	0	0	2	0	0
24	Freydières, G.	2	2	0	0	1	1	0	0	184	Petersf. 3AH5	2	1	1	0	0	0	0	0
28	Grappin	66	60	60	0	0	0	0	0	186	Petersf. 5AH4	1	0	0	0	0	0	0	0
29	Colomb	4	4	1	0	0	3	0	0	190	Petersf. 6AH4	2	1	0	1	0	0	0	0
30	Chênelaz	1	1	1	0	0	0	0	0	191	Petersf. 6AH5	2	1	0	1	0	0	0	0
33	Hoteaux, Les	1	1	0	0	0	0	0	1	–	Pet. backdirt	289	289	0	288	0	0	0	1
37	Bibi, Grotte	1	1	0	1	0	0	0	0	195	Gnirshöhle 1	19	8	1	7	0	0	0	0
39	Pugieu, Abri	1	1	1	0	0	0	0	0	196	Gnirshöhle 2	4	1	0	1	0	0	0	0
45	Farinc. I+II	13	13	2	0	0	0	10	1	197	Napoleonsk.	5	5	4	0	0	1	0	0
46	Farinc. III-B	2	2	2	0	0	0	0	0	198	Buttentalh.	1	1	1	0	0	0	0	0
47	Farinc. III-E	2	2	2	0	0	0	0	0	199	Probstfels	15	6	3	2	0	0	0	1
52	Romains 2a	2	2	0	0	1	1	0	0	204	B. Dietfurt	3	2	0	2	0	0	0	0
53	Romains 2b	7	7	2	2	1	0	0	2	207	Zigeunerfels	1	1	0	0	0	1	0	0
69	Douattes, Les	8	8	7	0	0	0	0	1	211	Nikolaushöhle	1	1	0	0	0	0	0	1
73	Banges, G.	3	3	2	0	1	0	0	0	223	Felsställe	7	7	0	7	0	0	0	0
77	Rigney 1, D	27	27	10	3	1	3	9	1	224	Felsställe	41	25	4	21	0	0	0	0
78	Rigney 1, M	2	2	0	0	0	0	2	0	229	Schussenqu.	10	10	5	0	1	0	0	4
79	Veyrier	99	88	70	9	1	0	0	8	239	Helga-Abri	3	3	0	3	0	0	0	0
82	Auzary-Thôn.	1	1	1	0	0	0	0	0	240	Hohle Fels, Ia	5	1	0	0	1	0	0	0
86	Rochedane	1	0	0	0	0	0	0	0	241	Hohle Fels, Ib	3	0	0	0	0	0	0	0
88	Scé Châterl.	1	1	0	0	0	1	0	0	242	Hohle Fels, Ic	1	1	1	0	0	0	0	0
90	N.-Monruz	31	31	9	20	1	0	0	1	243	Hohle Fels, IIa	7	0	0	0	0	0	0	0
92	Dentaux, A.	1	1	1	0	0	0	0	0	245	Sirgenstein	4	1	1	0	0	0	0	0
93	H.-Champrév.	30	30	4	26	0	0	0	0	246	Brillenhöhle	64	31	19	2	9	0	0	1
94	H.-Champrév.	1	1	1	0	0	0	0	0	247	Brillenhöhle	17	5	5	0	0	0	0	0
95	H.-Champrév.	7	7	1	5	0	1	0	0	250	K. Scheuer R.	5	5	3	1	1	0	0	0
103	Thierstein	4	4	4	0	0	0	0	0	253	Bocksteinh.	7	7	6	1	0	0	0	0
108	Kohlerhöhle	47	47	6	40	1	0	0	0	257	Hohl.-Stadel	32	6	6	0	0	0	0	0
110	Kastelh.-Nord	25	6	4	0	0	2	0	0	261	Spitzbubenh.	1	1	1	0	0	0	0	0
111	Heidenküche	1	1	0	0	1	0	0	0	265	Vogelherd	2	0	0	0	0	0	0	0
114	Birsek-Ermit.	6	6	2	1	0	0	3	0	266	Vogelherd	12	11	10	0	1	0	0	0
115	Hollenberg-H3	3	3	3	0	0	0	0	0	269	Hohlenstein, E.	4	4	4	0	0	0	0	0
122	Munzingen	200	69	67	0	0	0	0	2	271	Kaufertsberg	6	6	4	0	0	0	2	0
124	Rislisbergh.	81	81	6	73	2	0	0	0	272	Kaufertsberg	8	7	7	0	0	0	0	0
126	Teufelsküche	19	10	7	1	0	0	0	2	287	O. Klause lw.	6	6	2	2	0	2	0	0
128	Rheinfelden-E.	1	1	0	0	0	0	0	1	288	O. Klause up	10	10	3	2	2	0	0	3

(continued)

Table A.12 (continued)

ID	Site	n mentioned	n recorded	Points	Needles	Barbed points	Awls	B. d.-rondes	B. percé	ID	Site	n mentioned	n recorded	Points	Needles	Barbed points	Awls	B. d.-rondes	B. percé
136	Käsloch	3	2	2	0	0	0	0	0	289	M. Klause	1	1	0	0	0	0	0	1
166	Schweizersb.	1255	420	160	240	3	0	2	15	290	Kastelhängh.	20	20	15	1	1	3	0	0
167	Freudenthalh.	4	3	0	0	0	0	3	0	294	Sesselfelsg.	1	1	0	0	0	1	0	0
170 171	Kesslerloch	752	357	300	35	10	0	5	7	296	Schellneck A.	1	1	1	0	0	0	0	0
172	Bildstockfels	1	1	0	0	0	1	0	0	296	Schellneck A.	1	1	1	0	0	0	0	0
173	Petersf. 1AH2	2	2	1	1	0	0	0	0	301	Obernederh.	1	0	0	0	0	0	0	0
306	F.-Albrecht-H.	1	1	0	0	1	0	0	0	451	Kniegrotte	354	199	142	37	1	12	2	5
319	T. da Somme	1	1	1	0	0	0	0	0	455	Saaleck	1	1	0	0	0	1	0	0
320	Bois Laiterie	4	4	4	0	0	0	0	0	528	Tetín	1	1	1	0	0	0	0	0
323	Trou Magrite	3	3	2	1	0	0	0	0	531	Na Průchodě	2	2	2	0	0	0	0	0
324 325	T. de Chaleux	117	113	65	35	0	13	0	0	533	Rytířská	3	3	3	0	0	0	0	0
326	T. des Nutons	4	4	4	0	0	0	0	0	539	Węgliny	1	1	0	0	1	0	0	0
328	T. du Frontal	18	18	7	7	0	4	0	0	543	Kamegg	11	9	6	3	0	0	0	0
332	Goyet 3	41	41	20	17	1	0	1	2	551	Býčí skála	13	13	3	6	0	4	0	0
334	Coléoptère	14	14	1	7	5	1	0	0	559	Výpustek	1	1	0	0	0	0	0	1
335	Verlaine	22	22	7	8	1	6	0	0	563	Nová Drát.	3	3	3	0	0	0	0	0
336	Roche Plate	1	1	1	0	0	0	0	0	566	Vinckova	1	1	1	0	0	0	0	0
362	Kartstein	5	5	2	0	1	2	0	0	567	Žitného	9	9	1	7	0	0	1	0
–	And. 79–83	81	80	45	17	10	1	2	5	568	Verunčina	1	1	0	1	0	0	0	0
364	Andernach I	41	40	20	7	6	1	1	5	569	Kolíbky	1	1	1	0	0	0	0	0
365	Andernach II	16	16	12	4	0	0	0	0	571	Kůlna Cave	4	4	2	1	0	1	0	0
366	Andernach III	12	12	9	2	0	0	1	0	572	Kůlna Cave	14	13	7	4	0	1	1	0
368	Gönnersd. I	59	59	22	33	1	1	1	1	578	Hadí	4	4	4	0	0	0	0	0
369	Gönnersd. II	41	40	16	22	0	1	1	0	579	Pekárna g/h	323	323	300	0	3	0	0	20
370	Gönnersd. III	24	23	8	15	0	0	0	0	582	Adlerova	3	3	3	0	0	0	0	0
372	Irlich, Sandg.	1	1	1	0	0	0	0	0	583	Křížova	1	1	0	0	0	0	0	1
375	Martinshöhle	2	2	0	0	1	1	0	0	584	Švédův stůl	1	1	1	0	0	0	0	0
376	Feldhofhöhle	1	1	1	0	0	0	0	0	585	Ochozska Cave	4	4	4	0	0	0	0	0
378	Wildweiberlei	3	1	1	0	0	0	0	0	590	Balcarka Cave	4	4	4	0	0	0	0	0
379	Wildscheuer	1	0	0	0	0	0	0	0	610	Deszczowa	5	0	0	0	0	0	0	0
401	B. Frankenh.	3	3	3	0	0	0	0	0	611	Deszczowa	5	0	0	0	0	0	0	0
402	Bärenkeller	56	24	23	0	0	1	0	0	612	Krusza Skála	2	2	1	0	0	0	1	0
411	Quedlinburg, Mühlgraben	1	1	0	0	1	0	0	0	624	Sąspów, Niedostępną	1	0	0	0	0	0	0	0
415 416	Teufelsbrücke 1+2	2	2	2	0	0	0	0	0	629	Maszycka Cave	62	53	45	0	0	7	0	1
417 418	Teufelsbrücke 3+4	94	94	73	11	0	9	0	1	630	Puchaczowa Skála	1	1	1	0	0	0	0	0
438	Ilsenhöhle	1	1	0	0	0	1	0	0	647	Wilczyce	10	10	5	5	0	0	0	0
439	Nebra	21	11	11	0	0	0	0	0	651	Przemysl 3	1	1	0	0	1	0	0	0
444	Oelknitz	3	3	0	0	2	0	0	1										

B. d-rondes Baguettes demi-rondes

Table A.13 Blade technology analysis—Balcarka Cave

Nr.	Blank	Length	Width	Thickness	Pres.: L	Pres.: W/T	Cortex	SPR: kind	SPR: form	SPR: length	SPR: width	DR	Lip	Bulb	Bulbar scar	Impact point	Esq. du bulbe	en éperon	Flaking angle	Profile	Distal	Hammer	n Negatives	Direc. Neg.	Reg Neg.	Tool
5	1	22	8	2	2	1	1	5	2	4.5	1.5	4	4	4	5	5	5	5	70	2	1	0	2	1	1	1
12	1	25	8	3	2	1	1	4	2	3	1	4	5	4	4	5	5	5	65	2	1	0	2	1	1	1
18	1	28	7.5	2.5	1	1	1	4	5	3	2	4	4	7	5	5	5	4	65	2	5	0	2	1	1	1
25	1	33	9.5	2.5	2	1	2	5	5	4	2	4	4	4	4	5	5	4	80	2	1	0	4	1	3	1
26	1	26	12	3	1	1	1	5	4	5	2	3	6	4	5	4	5	5	45	2	4	2	2	1	1	1
29	1	26	13.5	1.5	2	1	1	4	4	3.4	4	4	4	4	5	5	5	5	75	2	2	0	3	1	1	17
43	1	36.5	13	4	1	1	1	5	4	5	1.5	8	5	4	5	5	5	5	75	2	4	0	2	1	1	1
45	1	35	14.5	2.5	2	1	1	4	2	2	1	8	5	4	4	5	5	5	80	2	1	0	2	1	1	1
47	1	36.5	13	5.5	1	1	1	4	5	6	3	2	4	7	4	4	5	5	70	1	4	2	4	1	3	1
48	1	35	14.5	3.5	2	1	1	5	2	11.5	4.5	4	4	7	5	5	5	4	50	2	1	1	2	1	1	1
49	1	42.5	14	4	1	1	1	4	5	2.5	1.5	8	4	4	5	5	5	5	80	1	7	0	2	1	2	1
50	1	40	13.5	3	2	1	1	4	4	3	1	4	5	4	4	5	5	5	80	2	1	0	4	1	3	1
51	1	42	15	3	1	1	1	5	5	5	1	4	6	4	5	4	5	5	75	2	5	2	6	?	3	1
52	1	45	16.5	5	1	1	1	5	1	11	0.5	8	5	4	4	4	5	5	80	1	1	2	4	1	3	1
54	1	42	13.5	4	2	1	1	4	2	4	1	8	5	4	4	4	5	5	65	2	1	2	3	1	1	1
55	1	45	15.5	3	1	1	1	4	4	5.5	1.5	8	4	7	5	4	5	5	75	2	4	2	3	1	1	1
56	4	45	16	4	1	1	1	4	4	3.5	1.5	4	4	4	5	4	5	5	50	2	4	2	?	?	4	1
57	1	42.5	18	3	1	1	1	5	2	8	3.5	8	4	4	4	5	5	4	80	1	6	0	4	3	2	1
58	1	45.5	15	4	2	2	3	4	7	6.5	2	4	4	7	5	4	5	5	65	1	1	2	1	1	1	1
60	4	47.5	16	4.5	2	1	1	5	2	3	1.5	4	5	7	5	5	5	5	75	2	1	0	?	?	4	1
61	1	53	16	5	1	1	1	5	4	5	1	8	5	5	5	5	4	5	80	1	5	2	2	1	1	1
62	1	50	19	3.5	1	1	1	5	2	6	3	8	4	7	4	?	5	5	70	1	4	0	2	1	3	1
64	1	52	17.5	7	1	1	1	4	2	2	1	4	4	4	5	5	4	5	55	1	6	2	4	1	3	1
85	1	31	5	2	1	1	1	4	2	4	2	8	5	4	5	5	5	5	70	2	4	0	2	1	1	1
88	1	34.5	9.5	2	1	1	1	5	2	6	2	4	4	4	4	5	5	5	70	1	4	0	3	1	2	1
89	1	35	10	3	1	1	1	5	7	6.5	1.5	4	4	7	5	5	5	5	45	1	6	0	3	1	1	1
91	1	35	10	5	2	1	1	5	5	2.5	2	4	4	4	5	5	5	5	75	2	1	0	4	1	2	1
92	1	41	10	4	2	1	1	5	5	6	2.5	4	4	4	5	5	5	5	75	1	1	0	3	1	1	1
94	1	42.5	9.5	2	1	1	1	5	4	4.5	1	8	5	7	5	5	5	5	80	2	4	0	2	1	1	1
95	4	43.5	11	5	2	1	1	4	2	4	2	4	4	7	5	4	5	5	80	2	1	2	?	?	1	1
97	1	45	10	4.5	2	1	1	4	6	3	2.5	3	5	4	5	5	4	5	75	2	2	2	3	1	1	2
98	1	46	12	2	2	1	1	4	3	1	1	8	5	4	5	5	5	5	75	2	1	0	5	1	3	1
99	1	47	9	2.5	1	1	1	5	9	5	1.5	8	4	4	4	4	5	5	75	2	4	2	5	1	2	1
101	1	52.5	10.5	2.5	1	1	1	4	2	5	2	8	5	4	5	4	5	5	70	2	4	2	2	1	1	1
103	1	47	20	4.5	1	1	1	5	2	4.5	2	2	5	4	4	5	5	5	75	2	4	0	5	4	3	1
104	1	57.5	24.5	7	2	1	1	5	2	9	3	8	5	4	6	4	4	5	75	3	2	2	7	4	3	4
105	4	63.5	22.5	9	2	1	1	4	2	4.5	2	4	5	4	5	5	4	5	75	2	1	2	?	?	1	1
106	1	69	27	10	1	1	1	5	2	13.5	4.5	8	4	4	4	5	5	5	80	2	5	2	4	1	4	1
110	1	49	18.5	5	2	1	1	4	2	4.5	1.5	5	4	4	4	4	5	5	75	1	1	2	3	1	1	1
111	1	58	15	6.5	2	1	1	5	2	9	4	4	4	4	4	5	5	5	65	2	1	0	3	1	1	1
112	1	60	18	4	2	1	1	5	2	6.5	2.5	8	4	7	5	5	5	5	60	2	1	0	2	1	1	1
140	1	21	11	2.5	2	1	1	5	5	5.5	2	4	5	4	5	4	5	4	45	2	1	2	3	1	1	1
143	1	22	15	3.5	2	1	1	5	2	9	3	4	4	7	5	5	5	5	45	1	1	0	2	1	1	1
144	1	27	15	4	2	1	1	4	4	5	1	8	4	4	5	5	5	5	80	2	1	0	3	1	1	1

(continued)

Table A.13 (continued)

Nr.	Blank	Length	Width	Thickness	Pres.: L	Pres.: W/T	Cortex	SPR: kind	SPR: form	SPR: length	SPR: width	DR	Lip	Bulb	Bulbar scar	Impact point	Esq. du bulbe	en éperon	Flaking angle	Profile	Distal	Hammer	n Negatives	Direc. Neg.	Reg Neg.	Tool
145	1	25.5	13	4	2	1	1	4	2	4	1.5	8	5	7	5	4	5	5	70	2	1	2	4	1	1	1
147	1	26	12.5	5	2	1	1	4	4	5	1	8	4	4	5	4	5	5	80	?	1	2	3	1	1	1
148	1	23.5	16	4.5	2	1	1	5	2	8	3	4	4	4	5	4	5	5	50	2	1	2	3	1	1	1
149	1	22	19.5	5	2	1	1	5	2	9	4	4	4	7	5	5	5	4	50	?	1	1	2	1	1	1
150	1	26	18.5	3.5	2	1	1	5	7	4	1	3	5	7	5	4	4	5	85	2	1	2	3	1	2	1
151	1	28	16	4	2	1	1	5	4	7	1	8	5	6	5	5	5	5	70	2	1	0	2	1	1	1
152	1	28	16.5	4	2	1	1	4	7	4	1.5	8	4	4	5	5	4	5	80	?	1	2	4	1	1	1
155	1	34	16	4.5	2	1	1	5	2	12	4.5	2	4	7	5	4	5	5	60	1	1	2	2	1	1	1
156	1	28	19.5	7	2	1	1	5	7	12.5	5.5	4	4	4	4	4	5	5	80	2	1	2	2	1	1	1
158	1	30	22.5	4	2	1	1	5	2	7	4	4	4	7	5	4	5	5	75	2	1	2	2	1	1	1
159	1	33	15	4	2	1	1	4	2	7	3	8	4	4	4	4	5	5	75	2	1	2	2	1	1	1
161	1	35	16	4	2	1	1	5	2	6.5	2	8	4	7	5	4	5	5	50	2	1	2	2	1	1	1
162	1	34	18	7.5	2	1	1	5	9	8.5	3	8	4	7	5	4	5	5	55	2	2	2	2	1	1	1
163	1	23.5	21	7.5	2	1	1	5	2	15	5	4	4	4	5	?	5	5	75	2	1	0	2	1	1	1
164	1	34	23.5	6.5	2	1	1	5	2	12	4	4	4	7	5	4	5	5	80	2	1	2	2	1	1	1
165	1	39	23.5	6	2	1	1	5	2	7	3.5	4	4	7	4	4	5	5	45	2	1	2	2	1	2	14
167	1	44	15	5.5	2	1	1	4	2	2.5	1.5	4	5	4	4	4	5	5	70	1	2	2	6	3	3	9
179	1	21.5	6	3	2	2	1	4	5	2	1	4	5	4	5	5	5	5	80	1	1	0	2	1	2	13
184	1	25	6.5	2.5	2	2	1	5	5	2.5	1	4	5	5	5	5	4	5	75	2	1	2	3	1	1	13
186	1	26.5	6.5	3	2	2	1	5	7	5	2	3	4	4	5	5	5	5	75	2	1	0	2	1	3	13
192	1	33.5	7	3	2	2	1	5	4	5	0.5	4	5	5	5	4	5	5	80	2	1	2	2	1	1	13
193	1	34	7	2.5	1	2	1	4	2	2	1	4	4	4	5	5	5	5	80	1	4	0	2	1	1	13
195	1	34	7.5	2.5	2	2	1	5	5	7.5	3	3	5	4	5	5	5	4	75	1	1	0	2	1	1	13
199	1	26	11.5	3.5	2	2	1	5	5	6.5	4	2	5	4	4	4	5	6	75	2	1	2	2	1	1	13
204	1	19	6	3	2	2	1	5	9	4.5	1	4	5	4	4	4	5	5	80	2	1	2	1	1	1	13
214	1	30	6	3	1	2	2	5	4	2.5	1	4	5	4	5	5	5	5	80	2	5	0	2	1	1	13
215	1	28	8	4	2	2	1	4	2	3	1	4	5	4	5	5	5	5	55	2	1	0	2	1	1	13
218	1	31	6	2	1	2	1	5	2	4	1.5	4	5	4	5	5	5	5	80	2	5	0	4	1	2	13
219	1	30	6	1.5	1	1	1	4	5	2	1.5	2	5	7	5	5	5	5	70	1	4	0	3	1	1	1
221	1	31	7	3	2	1	1	4	2	3	1	4	4	4	5	5	5	5	60	2	1	0	3	1	2	13
225	1	32.5	9.5	3	2	2	1	4	2	3	2	4	4	4	4	4	5	5	70	2	1	2	2	1	3	13
226	1	38	8	3	1	2	1	4	4	2.5	0.5	4	5	7	5	4	5	5	80	1	4	2	3	1	2	13
227	1	36	7	3	2	2	1	4	4	2	1	8	5	4	5	4	5	5	80	1	1	2	2	1	2	13
228	1	37.5	8.5	3	2	2	1	4	2	4.5	2.5	3	4	7	5	4	5	5	60	2	1	2	2	1	1	13
229	1	32	4.5	2	2	2	1	4	4	2	0.5	4	5	4	5	4	5	5	75	1	1	2	1	1	1	13
231	1	43	7	4	2	2	1	5	4	3	1	4	4	7	5	5	5	5	75	1	1	0	2	1	1	13
253	1	37.5	14.5	6.5	2	1	1	5	5	9.5	6	4	5	7	4	5	5	6	85	2	2	0	3	1	1	2
269	1	43	21.5	4.5	2	1	1	5	2	4.5	3	8	6	7	5	4	5	5	85	1	1	2	4	3	3	1
274	1	55	12.5	4	2	1	1	4	2	4	1.5	8	5	4	5	6	5	5	80	2	1	2	3	1	1	1
276	1	45.5	9.5	4	1	1	1	5	5	8.5	3.5	3	5	4	5	5	5	4	75	2	2	0	4	1	1	17

Feature explanations and keys for numbers see Chap. 2, Blade Technology, Recording of Technological Information, nd: feature expression not recordable

Table A.14 Blade technology analysis—Bois Laiterie

Nr.	Blank	Length	Width	Thickness	Pres.: L	Pres.: W/T	Cortex	SPR: kind	SPR: form	SPR: length	SPR: width	DR	Lip	Bulb	Bulbar scar	Impact point	Esq. du bulbe	en éperon	Flaking angle	Profile	Distal	Hammer	n Negatives	Direc. Neg.	Tool
U6-140	1	22	17	3.5	2	1	1	1	8	nd	nd	4	1	1	1	1	1	1	nd	1	1	2	2	1	1
V4-146	1	20	13	4	2	1	1	1	8	nd	nd	12	1	1	1	1	1	1	nd	1	1	2	2	3	14
V7-34	1	21	13.5	4	2	1	1	1	1	9	8	4	1	1	1	1	1	5	nd	nd	1	3	2	1	1
V4-32	1	21	12	5	2	1	1	1	8	nd	nd	4	1	8	5	1	4	1	nd	1	1	2	2	1	1
W9-25	1	35	15	4	2	1	1	1	8	nd	nd	1	1	1	1	1	4	1	nd	1	1	2	2	1	1
U6-28.3	1	27	13	5	2	1	3	1	8	nd	nd	4	1	7	5	1	4	1	nd	1	1	2	2	1	1
V2-24a	1	59	14	4	1	1	1	1	8	nd	nd	2	1	1	1	1	4	5	nd	1	8	3	4	3	1
W2-47	1	57	20	7	1	1	1	4	2	4	2	2	7	7	5	4	4	5	70	1	6	2	4	12	1
U7-104.1	1	49	22	8	1	1	2	5	2	5	2	10	7	7	5	4	4	5	85	2	8	2	4	11	1
W2-73	1	36	12	4	1	1	1	5	7	4	2	4	4	4	5	4	4	5	75	2	4	3	2	1	1
U4-101	1	29	13	4	1	1	1	5	7	7	2	4	5	7	5	5	4	5	75	1	4	2	?	?	1
U7-63.1	1	38	11	4	2	2	1	4	6	1	1	2	1	1	1	5	4	5	80	1	1	2	3	11	1
U4-99	1	24	16	3	2	1	1	4	7	8	2	4	7	7	5	5	4	5	65	2	1	2	4	2	1
U5-46	1	29.5	23	3	2	1	1	5	2	7	3	4	4	4	5	5	4	5	80	2	1	3	3	1	1
T5-43.7	1	22	8.5	2	1	1	1	5	7	3	0.5	4	5	5	5	5	4	5	85	2	6	3	2	1	1
W3-118	1	57	28	6	2	1	1	8	8	nd	nd	1	4	7	5	5	4	5	nd	1	2	3	2	1	5
X2c-18	1	60	14.5	5	1	1	1	4	2	5	1	3	4	7	5	5	4	5	50	1	7	3	3	1	1
V3-25	1	28	22	5	2	1	1	4	7	13	7	4	7	4	5	1	5	5	85	2	1	2	6	1	1
V6-53	1	47	25	8	1	1	1	5	7	11	4	4	5	7	5	4	5	5	85	1	6	2	3	1	1
T6-49	1	62	22	8	1	1	1	5	7	7	2	4	5	7	4	4	5	5	85	1	6	2	4	11	1
U6-41	1	24	15	5	2	1	1	5	7	13	5	4	7	7	5	4	5	5	80	1	1	2	3	1	1
U9-21	1	23	10	2	2	1	1	4	7	9	2.5	2	4	4	5	4	5	5	80	1	1	2	2	1	1
V7a-141	1	8.5	4	1	2	2	1	4	2	2	0.5	4	4	4	4	4	5	5	nd	1	1	2	2	1	13
U4-113	1	57	12	7	2	1	1	4	2	5	1.5	4	4	4	5	4	5	5	80	3	1	2	3	1	9
U4-56	1	29	17	4	2	1	1	5	2	7	2	4	5	4	4	4	5	5	60	1	1	3	3	1	1
U8-1006.1	1	68	25	7.5	1	1	1	4	2	4	2	4	4	7	5	4	5	5	85	1	8	3	6	4	1
W10B-#5	1	25.5	13.5	3	2	1	1	5	5	3	1.3	8	4	7	5	4	5	5	85	2	1	3	2	1	1
V8-3a	1	51.5	9	4.5	2	1	1	4	7	5	1.5	4	4	4	5	4	5	5	80	1	2	3	2	1	5
V7-70	1	27	12	2	2	1	1	4	7	2.5	1	3	4	4	5	4	5	5	80	1	1	3	2	1	1
V2d-8-17	1	17	17	4	2	1	1	4	2	8	3	4	4	4	4	4	5	5	85	nd	1	3	2	1	1
U7c-124	1	15	14.5	5	2	1	1	5	2	10	3	8	4	4	5	4	5	5	75	nd	1	3	1	1	1
W8c-2	1	37	16.5	3.5	1	1	1	5	7	8	3	4	4	4	5	4	5	5	80	2	4	3	3	1	1
V3-89	1	20	15	4	2	1	1	5	7	9	4	4	4	4	4	4	5	5	75	2	1	3	3	1	1
W2-48	1	26.5	18	2	2	1	1	4	2	5	2	4	4	7	5	4	5	5	80	2	1	3	3	1	1
W2-59	1	57	16	4	2	1	1	4	2	10	3	4	4	7	5	4	5	5	55	2	2	3	4	1	5
U3-56	1	52	20	5	1	1	2	4	5	7.5	3	2	4	7	5	4	5	5	80	1	4	3	3	2	1
U6-145	1	94	19	5	2	1	1	5	2	4	1.5	4	6	7	5	5	5	5	70	2	1	1	2	1	1
T6-25	1	58	19	4	1	1	1	5	2	5	2	4	7	7	5	5	5	5	65	3	1	1	2	1	1
T5-57	1	42	18	4	2	1	1	5	2	5	3	8	4	7	5	5	5	5	55	2	1	1	4	1	1
V3-44.9	1	37	20	5	2	1	1	5	2	4	1	4	7	7	5	5	5	5	60	2	1	1	4	1	1
V4b-73	1	53	15	5	1	1	1	5	5	5	3.5	4	4	7	5	5	5	6	70	1	6	1	3	11	1
V8-14	1	78	17	5	1	1	1	4	2	3	1.5	8	5	7	5	5	5	5	75	1.5	4	1	3	1	1
W3-71	1	58	14	6	1	1	1	5	7	1.5	2	4	4	7	5	5	5	5	65	2	4	1	?	?	1
W2-29	1	19	17	4	2	1	1	4	2	9	4	4	4	4	4	5	5	5	70	nd	1	1	6	1	1

(continued)

Table A.14 (continued)

Nr.	Blank	Length	Width	Thickness	Pres.: L	Pres.: W/T	Cortex	SPR: kind	SPR: form	SPR: length	SPR: width	DR	Lip	Bulb	Bulbar scar	Impact point	Esq. du bulbe	en éperon	Flaking angle	Profile	Distal	Hammer	n Negatives	Direc. Neg.	Tool
U6-27	1	48	15	5	1	1	1	5	2	4	1	2	4	7	5	5	5	5	75	1	4	1	4	12	1
U4-72	1	35	20	6	2	1	1	5	2	6	2	4	5	7	5	5	5	5	60	2	1	1	3	1	1
V7-8	1	17	12.5	4	2	1	1	5	5	11	5	2	5	5	5	5	5	4	55	2	1	1	3	1	1
U5-24	1	20	13	3	2	1	1	5	2	7.5	3	2	4	7	5	5	5	4	60	2	1	1	2	1	1
V8-1000	1	54	15	3.5	1	1	1	5	5	4	2	2	4	7	4	5	5	5	85	2	5	1	5	12	1
U5-67	1	35	16.6	4	1	1	1	5	2	20.5	8	2	5	4	5	5	5	4	70	2	6	1	3	1	1
U5-22	1	72	33	8	2	2	1	5	7	14	5	4	4	7	5	5	5	5	75	2	1	1	6	4	1
V10-13	1	53	15	4	1	1	1	4	2	6	2	4	4	7	5	5	5	5	60	1	6	1	2	1	1
W9-22	1	30	10	3	1	1	1	5	7	8.5	3	3	4	4	5	5	5	5	80	2	5	1	2	1	1
U6-28.16	1	29	14	4	1	1	1	5	7	5	1	2	5	6	5	5	5	5	80	2	5	1	2	1	1
U6-196	1	56	23	5	1	1	1	4	2	8	3	4	4	4	5	5	5	5	80	1	4	1	4	11	1
U3-5	1	30	17	4	2	1	1	4	4	8	1.5	4	4	4	4	5	5	5	60	2	1	1	4	11	1
V10-17	1	21	15	3	2	1	1	5	6	5	2	4	4	4	5	5	5	5	80	2	1	1	3	11	1
U6-163	1	20	9.5	2.5	2	1	1	5	7	4	1	3	4	5	5	5	5	5	70	2	1	1	2	1	1
U6-155	1	23	5	3	1	1	1	4	2	2	1	4	4	7	5	5	5	5	50	2	5	1	4	12	1
T4-27.1	1	20.5	12	3	2	1	1	5	2	7	3.5	3	4	7	5	5	5	4	45	2	1	1	2	1	1
V3b-57	1	25.5	14	5	2	1	1	5	2	10	4	4	4	7	5	5	5	4	60	2	1	1	3	1	1
U6-49	1	21	8.5	3	2	1	1	5	2	6	3	4	4	7	5	5	5	4	65	2	1	1	1	1	1
U4-70	1	14.5	15	2.5	2	1	1	5	2	7.5	3	11	4	4	5	5	5	5	65	nd	1	1	2	1	1
U4-94	1	27	13.5	2.5	1	1	1	4	2	8	2.5	4	4	7	5	5	5	4	60	2	4	1	1	1	1
U4-100	1	31	12	5	1	1	1	5	3	7	5	4	4	7	5	5	5	4	55	3	8	1	3	1	1
U4-1	1	12	24.5	2.5	1	1	1	4	7	7	1.5	6	4	7	5	5	5	4	55	2	4	1	2	1	1
V5-1	1	42	19.5	7	2	1	1	5	2	10	4.5	4	4	7	5	5	5	4	50	2	2	1	3	1	1
U6-212	1	21.5	8.5	2	2	2	1	4	2	3.5	1	6	4	7	5	5	5	5	50	1	1	1	?	?	13
W3-107	1	32.5	6.5	3	2	2	1	4	2	2.5	1	4	4	4	5	5	5	4	75	2	1	1	2	1	13
U6-28.26	1	28	5	1.5	2	2	1	4	5	1	1	3	4	7	5	5	5	4	80	2	1	1	2	1	13
V4-75	1	49	19	6	1	1	1	4	4	5	0.5	4	5	7	5	5	5	5	80	1	6	2	4	12	1
U5-44.1	1	26	16	4	2	1	1	5	7	7	2	4	5	7	5	5	5	5	85	1	1	2	4	?	1
U6-70	1	51	27	6	2	2	4	4	7	6	3	4	7	4	4	5	5	5	85	2	1	2	3	11	1
U6-94	1	22	12.5	2	2	1	1	5	2	5	2	4	4	4	4	5	5	5	85	2	1	2	2	1	1
U7-83.1	1	23	21	7	2	1	2	5	5	5	4	4	5	4	4	5	5	5	85	1	1	2	?	?	1
V7-35	1	21	9	2	2	1	1	5	6	2.5	1	4	7	4	5	5	5	5	85	1	1	2	5	1	1
V4-104	1	36.5	7	3	1	1	1	4	5	3	1.5	4	4	7	5	5	5	5	80	1	6	2	4	3	1
T4-27.2	1	24.5	10.5	2.5	2	1	1	4	2	1.5	1	4	4	4	5	5	5	5	85	2	1	2	3	1	1
V3-71	1	37	7.5	4	2	1	1	4	2	3	1	4	4	4	5	5	5	5	80	2	1	2	3	1	1
W3-153	1	45	18	3	2	1	1	4	4	3	1	4	4	7	5	5	5	5	85	2	1	2	3	1	1
W3-85	1	50	13	5	1	1	1	4	5	8	3	8	4	7	5	5	5	5	70	1	4	2	2	1	1
V4-125	1	36	12.5	2.5	1	1	1	5	7	4	1	4	4	4	5	5	5	5	80	1	6	2	2	1	1
V7-85	1	11.5	12.5	3	2	1	1	4	2	2.5	1	2	4	4	5	5	5	5	70	2	1	2	3	1	1
U6-204	1	29	35	3	2	1	2	4	2	11	3	4	4	4	4	5	5	5	60	2	1	2	1	1	1
U6-80	1	31	14.5	2	1	1	3	5	2	4.5	2	4	4	4	5	5	5	5	80	2	5	2	1	7	1
U6-237	1	82	22	7	1	1	5	5	7	7	3	2	4	4	6	5	5	5	80	2	4	2	0	0	1
V2c-sp.74	1	31.5	15	3	2	1	1	5	7	5.5	1.5	4	4	4	5	5	5	5	80	2	1	2	2	1	1
W3-143	1	52	18.5	4.5	2	2	1	4	2	4.5	2	2	4	7	5	5	5	5	40	1	2	2	3	1	19
T3-4.1003.1	1	39	17	4	1	1	1	4	2	4.5	2.5	4	4	4	5	5	5	5	80	1	6	2	3	1	1

T14-7	1	30.5	10	2.5	2	1	1	5	7	5	2	3	4	4	5	5	5	5	75	1	1	2	1	1	13
T3-1002.2	1	23	7.5	2	1	2	2	4	2	2	0.5	8	4	4	5	5	5	5	nd	1	1	2	2	1	1
V3-20	1	22	11.5	3	1	1	1	5	2	8	2.5	8	4	7	5	5	5	5	80	1	5	2	2	1	1
V7a-133	1	20	5	1	2	1	1	4	2	3	1	3	4	4	5	5	5	5	65	2	1	2	2	1	1
W1-70	1	36	11	2.5	1	1	1	5	2	5	1	4	4	4	4	5	5	5	85	1	4	2	2	1	1
U9a-9a	1	21	11	4	2	1	1	5	5	4	2	4	4	7	5	5	5	5	70	1	1	2	2	1	1
U13-11	1	41.5	15	5	1	1	4	5	7	6	2	4	4	7	5	5	5	5	70	2	5	2	?	?	1
W3-89	1	30	14.5	5	1	1	1	5	5	8.5	4	4	4	4	5	5	5	5	75	1	6	2	2	1	1
V4b-143.1	1	21	12	2	2	1	1	2	2	2.5	1	4	4	7	5	5	5	5	70	2	1	2	3	1	1
V4-141	1	18.5	11.5	3	1	1	1	4	7	5	1	4	4	6	5	5	5	5	75	1	6	2	2	1	1
V4-161	1	32	12.5	2.5	1	1	2	5	7	5	1.5	3	7	7	5	5	5	5	85	2	4	2	2	1	1
V4-58	1	23	10	2	1	1	1	4	4	3	0.1	4	4	7	5	5	5	5	85	2	4	2	5	4	1
V3-53	1	16	33	3	1	1	1	5	7	9	5	4	4	7	5	5	5	5	75	2	4	2	3	4	1
U7-19	1	32	9	3	2	1	1	4	2	2	1	4	7	7	5	5	5	5	65	1	1	2	4	1	1
U7-89	1	41	12	2.5	1	1	1	4	2	3	1	9	4	7	4	5	5	5	65	1	4	2	4	1	1
U7-4	1	45	19	4	2	1	1	5	7	6.5	2.5	4	5	7	5	5	5	5	80	2	1	2	3	1	1
W10-18	1	25	18	5	2	1	1	5	7	9	3.5	4	4	7	4	5	5	5	85	1	1	2	3	1	1
U4-59	1	21	11	2	2	1	1	5	4	9	2	4	4	7	4	5	5	5	65	1	1	2	3	1	1
U4-71	1	29	9	2	1	1	1	4	4	2	0.5	4	7	7	5	5	5	5	70	2	4	2	4	4	1
U4-54	1	21	10	2	2	1	1	5	7	8	2	4	4	6	5	5	5	5	70	2	1	2	3	1	1
U4-107	1	31	10	2.5	1	1	1	5	7	8	2.5	4	4	7	4	5	5	5	70	1	4	2	2	1	1
V2-26	1	35.5	16.5	3	1	1	1	5	2	6	3	4	4	7	5	5	5	5	75	1	4	2	2	1	1
V8d-45	1	79	32	7	2	1	1	5	7	15	6	5	4	4	5	5	5	5	80	1	2	2	4	4	5
V8-23	1	59	14.5	6	2	1	1	5	5	3	2.5	4	4	4	5	5	5	5	60	1	5	2	2	1	17
V4-99	1	74	26	8	2	1	1	5	2	6.5	2	4	4	4	5	5	5	5	70	2	2	2	6	1	1
W9-18.1	1	52	21	7	2	1	1	4	7	5	1	4	4	4	4	5	5	5	70	2	2	2	3	4	4
W3-153.1	1	46	22.5	5	1	1	1	5	2	7	2	4	4	7	5	5	5	5	65	1	2	2	4	1	1
W3-113	1	12.5	9	4	2	1	1	4	2	4	2	4	4	4	4	5	5	5	75	nd	1	2	2	1	1
U12-29	1	33.5	4	2.5	2	2	1	4	5	2	1	4	4	6	5	5	5	4	85	2	1	2	?	?	13
V7-28.5	1	11	6	2	2	2	1	4	2	3	1	8	4	7	5	5	5	5	85	1	1	2	1	1	13
V7-62.1	1	25	7	2	2	2	1	4	4	2	1	4	4	7	5	5	5	5	80	2	1	2	3	2	13
V4A-6	1	11	5	1.5	2	2	1	4	4	3	1	7	4	7	5	5	5	5	85	2	1	2	1	1	13
W3B-101	1	20	9	2.5	2	2	1	4	2	6	2	3	4	7	5	5	5	5	75	1	1	2	1	1	16
V3-57.14	1	13.5	5	2	2	2	1	4	2	2	1	8	4	7	5	5	5	5	55	2	1	2	?	?	13
U4-110.7	1	13.5	6	2	2	2	1	4	3	0.5	0.5	4	4	7	5	5	5	5	nd	2	1	2	1	1	13
U3-55	1	49	13	4	1	1	1	4	2	4	1	4	4	4	5	5	5	5	80	1	4	2	2	1	1
V4A-48	1	24	6	2	2	1	1	4	5	1	1	4	4	4	5	5	5	4	nd	2	1	2	2	1	1
V2-14	1	58	23	7	2	1	1	5	2	7	3	4	4	7	5	5	5	5	80	1	1	3	3	1	1
W4-25	1	38	25	6	2	1	1	5	2	12	4	4	4	7	5	5	5	5	80	2	1	3	2	4	1
V7-25	1	23	14	3	2	1	1	4	4	3	0.5	4	4	4	5	5	5	5	80	1	1	3	2	?	1
T5-24.2	1	22	7	2	2	1	1	4	2	2	0.5	4	4	5	5	5	5	5	85	nd	1	3	2	1	1
T5-43.1	1	28	21	3	2	1	1	4	4	12	1	4	4	4	5	5	5	5	80	1	1	3	1	1	1
V3-94	1	30	27	4	1	1	3	5	7	9	2	4	4	4	5	5	5	5	85	2	4	3	3	1	1
W3-38	1	36	17	6	2	1	1	4	4	4	0.1	4	4	4	4	5	5	5	80	1	1	3	1	1	1
W2-39	1	33	12	6	2	1	1	4	6	3.5	1.5	4	4	7	5	5	5	5	85	1	1	3	2	1	1
T5-95	1	47	23	7	1	1	1	4	2	5	2	4	4	4	5	5	5	5	85	1	6	3	4	4	1
U4-52	1	32	12.5	3	1	1	1	4	7	3.5	1.5	4	4	4	5	5	5	5	85	1	6	3	2	1	1

(continued)

Table A.14 (continued)

Nr.	Blank	Length	Width	Thickness	Pres.: L	Pres.: W/T	Cortex	SPR: kind	SPR: form	SPR: length	SPR: width	DR	Lip	Bulb	Bulbar scar	Impact point	Esq. du bulbe	en éperon	Flaking angle	Profile	Distal	Hammer	n Negatives	Direc. Neg.	Tool
W3-125	1	38	17	7	2	1	1	4	4	4.5	0.5	4	5	7	5	5	5	5	80	1	1	3	3	1	1
V4-38	1	67	19	5	2	1	1	4	7	5	1.5	4	4	6	4	5	5	5	85	2	1	3	5	1	10
-1005	1	51	15	6	2	1	1	5	7	5	2	4	7	7	5	6	5	5	85	2	1	2	6	12	1
U4b-93	1	53	22	7	1	1	1	5	7	8	3	4	4	7	5	6	5	5	75	2	8	3	2	4	1
W3-12	1	36	27	8	2	1	1	5	7	17	8	8	5	4	5	6	5	5	75	2	1	3	4	2	1
W1ob-71	2	25	17	3.5	1	1	1	1	8	nd	nd	4	1	1	1	1	1	1	nd	nd	4	2	2	1	1
W9-20	2	28.5	25	9	2	1	2	1	8	nd	nd	1	1	1	1	1	1	1	nd	nd	1	3	3	1	1
V3-86	2	59	32	5	2	2	1	5	5	12	6	4	5	7	5	4	4	5	89	1	1	2	3	12	1
W3-140	2	29.5	25	6	1	1	1	4	7	17	6	4	4	7	5	4	4	5	70	1	5	3	2	1	1
V4-116	2	39.5	40.5	8	1	1	1	5	7	20.5	7.5	4	4	4	4	4	4	5	60	1	4	3	2	4	1
T5-64.1	2	40	25	13	1	1	1	5	7	8	4	4	1	1	5	5	4	5	85	1	4	2	3	?	1
V7-18	2	37	28	5	1	1	2	4	7	13	5	4	4	6	5	5	4	5	85	2	6	2	2	1	1
V7-14	2	27	27	5	1	1	1	5	7	12	5	4	7	6	5	5	4	5	80	2	5	2	2	1	1
U6-182	2	42	46	8.5	1	1	1	5	2	23	8	2	4	4	5	5	4	5	80	nd	4	3	1	1	1
V4-148	2	49	25.5	5	1	1	1	5	4	12	1.5	4	5	7	5	5	4	5	85	1	6	3	4	2	1
W2-60	2	37.5	23	6	2	1	1	5	7	7.5	2.5	4	4	7	5	5	4	5	60	1	2	3	3	1	2
V7-24	2	21.5	30	8.5	2	1	2	8	9	nd	nd	1	1	4	4	1	5	1	nd	nd	1	2	1	2	1
U5-37	2	35	56	9	1	1	1	5	7	25	5	2	4	7	4	4	5	5	85	1	5	2	3	4	1
U6-24	2	32	29	5	2	1	1	4	7	8	2	4	5	7	4	4	5	5	85	2	1	2	3	1	1
U6-250.2	2	20	20	3	2	1	1	4	7	10	3	99	4	4	5	4	5	5	90	nd	1	3	1	1	1
V7-37	2	37.5	29	3.5	2	2	1	5	2	9.5	3	4	4	4	5	4	5	5	80	1	1	3	?	?	1
V75	2	24	35	7	1	1	2	4	2	5	1.5	2	5	7	5	4	5	5	85	2	5	3	2	7	1
V4-124	2	23.5	28.5	6.5	2	1	1	4	7	9	5	3	4	4	5	4	5	5	70	nd	1	3	2	1	1
W3-108	2	73	43	15	1	1	1	5	7	29	13	4	4	4	4	4	5	5	80	nd	4	3	3	1	1
U4-154	2	38.5	27.5	8.5	1	1	2	5	7	9	4	4	4	4	4	4	5	5	75	3	8	3	3	4	1
U6-28.1	2	53	37	4	1	1	2	5	7	36	4	2	4	4	5	4	5	5	80	2	4	3	2	1	1
V3c-39	2	44.5	28	7.5	1	1	2	4	7	10.5	4.5	4	4	4	4	4	5	5	75	1	5	3	3	1	1
V7-64	2	39	19.5	6	1	1	1	4	2	3.5	1	2	4	4	6	4	5	5	nd	3	4	3	7	?	1
U5-51	2	31	29	8	1	1	1	4	7	12	4	2	4	4	4	4	5	5	80	1	8	3	2	1	1
U3-65	2	35	19.5	6.5	1	1	1	4	2	11.5	5.5	8	4	4	5	4	5	5	65	1	6	3	2	1	1
V4-4	2	24.5	16	3.5	1	1	1	5	7	12	3.5	5	4	4	5	4	5	5	85	1	5	3	2	1	1
U4-51	2	22	27.5	6	2	1	1	4	7	17	4	4	4	6	4	4	5	5	85	nd	1	3	3	1	1
W2-22	2	36	25.5	13	1	1	1	4	7	13.5	4.5	4	4	7	4	4	5	5	75	2	8	3	2	1	1
V9-1a/b	2	28	18	4.5	2	1	1	4	2	7.5	2.5	4	4	4	4	4	5	5	65	2	2	3	4	2	2
V10-9	2	19	17	4	2	1	1	5	7	5	2	8	4	7	4	5	5	5	75	nd	1	1	2	1	1
W10-22	2	37	35	5	1	1	2	5	2	12	4	4	4	7	4	5	5	5	80	2	5	1	3	4	1
U6-28.8	2	18	13.5	3.5	1	1	1	5	5	5	2	2	4	4	5	5	5	5	80	1	6	2	3	8	1
V7-11	2	15	27	3	2	2	1	4	7	18.5	3	4	4	4	5	5	5	5	80	nd	1	2	2	1	1
U6-203	2	26	22	2	1	2	1	5	4	6	0.5	2	4	7	5	5	5	5	80	2	5	2	3	4	1
U5-34	2	27.5	22	5	1	1	1	4	7	11.5	3	4	4	7	5	5	5	5	80	1	5	2	2	1	1
W9-30	2	25	23	9	1	1	1	5	7	12	3	2	4	4	5	5	5	5	80	2	6	2	4	4	1
W3-25	2	38	30	6	1	1	1	4	2	6	1.5	2	4	7	5	5	5	5	85	1	6	2	2	1	1
U6-29	2	16	20	3.5	2	1	1	5	7	14.5	3.5	4	4	4	5	5	5	5	90	nd	1	2	?	?	1
U6b-167.1	2	29.5	33	8.5	2	1	1	5	2	10	3	2	4	4	5	5	5	5	90	1	1	2	1	2	1
V4C-19a	2	14	28	3.5	2	1	1	4	2	6	2	4	4	4	5	5	5	5	80	nd	1	2	1	1	1

T6-14	2	29	22	5	1	1	1	5	6	9	4.5	4	7	4	5	5	5	5	70	2	4	2	4	1	1
V4-175	2	30	33	8	2	1	1	4	2	11	4	4	4	4	4	5	5	5	75	nd	1	2	3	1	1
W3-155	2	21	16	3	1	1	1	5	2	6	2	4	4	7	5	5	5	5	75	nd	6	2	2	1	1
V3-101	2	44	30	5	1	2	2	4	2	6	2	4	4	7	5	5	5	5	65	1	5	2	1	1	7
V4d-122a	2	27	14	4	1	1	2	2	5	2	1	4	4	7	5	5	5	5	70	4	4	2	4	1	1
U7-76	2	26.5	15	5	1	1	1	5	7	12	5	4	4	7	5	5	5	5	85	1	4	2	2	1	1
W2-56	2	21	14	2.5	1	1	1	4	2	4	1	7	4	7	5	5	5	5	80	1	6	2	3	1	1
U4-94	2	23	15.5	3.5	1	1	1	4	2	7	2	4	4	7	5	5	5	5	70	1	6	2	2	1	1
U4-118	2	25	13	2.5	1	1	1	4	7	5.5	1	4	4	6	5	5	5	5	75	1	4	2	2	1	1
V3-38	2	32	26	6	1	1	1	4	5	2.5	1.5	4	4	5	5	5	5	5	80	1	6	2	2	1	1
V4-35	2	31	24	5	1	1	1	5	7	14.5	5	4	4	4	4	5	5	5	80	1	5	2	3	4	1
W3-100	2	44.5	26	3	1	1	4	5	2	6.5	2.5	5	4	7	5	5	5	5	80	1	5	2	1	1	1
U6-35	2	29	29	6	1	1	1	4	2	5	1	4	7	5	5	5	5	5	80	2	5	3	5	12	1
T5-35.1	2	49	32	12	1	1	1	5	7	9	4	2	4	7	5	5	5	5	85	2	6	3	6	2	1
U6-216	2	29	15	2.5	1	1	1	5	2	5	2	4	4	7	5	5	5	5	80	1	6	3	2	1	1
U6-247	2	38	24.5	10	1	2	1	4	7	25	4	4	4	4	5	5	5	5	75	2	4	3	?	?	1
T3-1002.3	2	21.5	17	7.5	1	1	1	5	7	13	6	11	4	4	5	5	5	5	70	nd	4	3	1	7	1
V3-32	2	37	26	6.5	1	1	3	1	1	nd	nd	4	1	4	5	5	5	5	nd	2	8	3	?	?	1
V2-18	2	37	20	4	1	1	1	5	7	9	3	2	4	4	4	5	5	5	85	1	6	3	3	1	1
T5.59	2	22	17	2	1	1	1	5	7	14	1.5	2	4	4	6	5	5	5	90	1	4	3	1	1	1
U5-36	3	56	27	18	1	1	1	5	7	25	12	4	5	7	4	4	5	5	89	1	8	2	?	?	1
U5-62	3	51	17	7	1	1	1	4	5	2	1	4	4	7	5	5	5	5	55	3	4	1	?	?	1
U6-123	3	26	12	5	2	1	1	5	2	3	1	4	4	7	5	5	5	5	55	1	1	1	?	?	1
W2-2	3	28	11	3	2	1	1	4	5	1.5	1	4	4	7	5	5	5	5	75	2	1	2	?	?	1
T6-1.1	3	74	16	8	1	1	1	4	2	4	1	4	5	7	5	5	5	5	70	1	4	3			1
W3-150	4	21	13	4	2	1	2	4	2	4	1	4	4	7	5	4	5	5	80	2	1	2	?	?	1
T8-5a	4	38	10	6	2	1	1	4	2	3	2	3	4	5	5	5	5	5	65	2	1	1	?	?	1
S5-45.1	4	47.5	15	6	1	1	1	5	2	6	2	2	4	4	5	5	5	4	80	3	5	1	?	?	1
T8-5a/b	4	59	10	6	1	1	1	4	2	3	1.5	3	4	5	5	5	5	5	45	2	4	1	?	?	1
U4-43	4	31	17	7	2	1	1	4	2	3.5	1.5	3	4	4	4	5	5	5	75	2	1	2	?	?	1
T4b-39	4	82	28	10	1	1	1	5	7	7.5	4	4	4	7	5	5	5	5	70	1	4	2	?	?	1
V7-9.1	4	34	14	6	2	1	1	5	5	6.5	4	5	4	7	5	5	5	5	60	2	5	2	?	?	1
V8-73	4	64	31	13	1	1	1	5	7	14	11	2	5	7	5	5	5	5	80	2	8	3	?	?	1
U6-135	5	53.5	37	17	1	1	2	5	2	25	10	2	4	7	5	5	5	5	85	1	5	2	?	?	1
U9-23/25	6	29	28	8	1	1	1	4	2	9	5	2	7	7	5	5	5	5	70	2	8	2	?	?	1
V4-59	6	24	23	10	2	1	2	2	5	4.5	2	5	5	7	5	5	5	5	75	1	1	99	?	?	1
U5-44.2	8	23.5	20.5	2	1	1	1	4	2	4	1	4	4	7	4	5	4	5	80	3	5	3	3	1	1
U6-180	8	12	17	2.5	1	1	1	5	2	6	2	2	4	7	5	5	4	5	90	nd	6	3	2	1	1
W2-78	8	20	23.5	5	1	1	1	5	7	12	2	4	4	7	5	5	4	5	70	1	5	3	4	4	1
U6-55	8	22.5	20	2	1	1	1	1	8	nd	nd	2	1	4	5	1	5	5	90	nd	5	2	3	4	1
V7-17	8	14	19	3.5	1	1	1	1	8	nd	nd	2	1	5	5	1	5	1	90	nd	6	3	?	?	1
U5-18	8	19	14	4	1	1	1	5	2	11	4.5	2	4	7	5	4	5	5	80	2	4	3	2	1	1
U6-206	8	31	22	4	1	1	1	4	7	3	1	4	6	4	5	4	5	5	80	3	4	3	4	4	1
V3-30	8	20	30.5	4.5	1	1	1	5	2	14	4	4	4	4	4	4	5	5	90	nd	5	3	?	?	1
U7-36	8	25	32	6	1	1	1	4	5	4.5	3.5	4	4	4	4	4	5	5	90	nd	6	3	?	?	1
U6-25	8	24	16	2	1	2	1	4	6	4	2	2	6	7	5	5	5	5	80	3	5	2	1	1	1
U6-146	8	21	11	2	1	2	1	4	2	5	2	2	6	7	5	5	5	5	90	1	4	2	1	1	1

(continued)

Table A.14 (continued)

Nr.	Blank	Length	Width	Thickness	Pres.: L	Pres.: W/T	Cortex	SPR: kind	SPR: form	SPR: length	SPR: width	DR	Lip	Bulb	Bulbar scar	Impact point	Esq. du bulbe	en éperon	Flaking angle	Profile	Distal	Hammer	n Negatives	Direc. Neg.	Tool
V7-23	8	19	17	2	1	1	1	4	2	6	2	2	4	4	4	5	5	5	85	1	6	2	1	1	1
U6-85	8	20	27	3.5	1	1	1	4	7	10	2.5	2	4	4	5	5	5	5	85	nd	6	2	2	1	1
U6-256.3	8	21	24	2.5	1	2	1	4	4	4	0.5	2	5	4	5	5	5	5	85	2	5	2	3	1	1
U6-86	8	18	21	4	1	1	1	4	7	7	2	3	4	4	5	5	5	5	85	nd	6	2	3	4	1
U6-28.7	8	12	24	6.5	1	1	1	5	2	8	3	2	4	7	5	5	5	5	90	nd	6	2	2	?	1
U6-28.4	8	19	16.5	2	2	1	1	8	9	nd	nd	1	1	4	4	5	5	5	nd	nd	1	2	?	?	1
U7-52	8	26	24	3	1	1	2	4	2	3.5	1	2	4	7	4	5	5	5	85	3	5	2	2	5	1
U6-234	8	18	17	4	1	1	1	4	5	4.5	3	2	4	4	5	5	5	5	85	3	4	2	3	4	1
V3-48	8	28	29	4	1	1	1	4	7	12	3.5	2	4	4	5	5	5	5	70	1	5	2	3	1	1
V3-57.18	8	28	17	3	1	1	1	5	5	6	4	4	5	4	5	5	5	5	85	3	4	2	5	4	1
U6-78	8	13.5	17.5	3	1	1	1	4	2	3	1	4	4	4	5	5	5	5	85	nd	6	2	2	1	1
U6-82	8	26	19	2	1	2	2	5	7	9	2	2	4	4	5	5	5	5	75	3	4	2	1	2	1
V10-8	8	20	26	2	1	1	1	4	2	4	1	4	4	4	5	5	5	5	80	1	5	2	1	1	1
W4-38	8	21	21	4	1	1	1	4	7	18	3	4	4	4	5	5	5	5	85	nd	6	2	2	1	1
U6-34	8	12	22	4.5	2	1	1	4	7	9	2	4	4	6	5	5	5	5	85	nd	1	3	?	?	1
W10-1	8	25	21	6.5	1	1	3	4	7	7	2	4	6	4	5	5	5	5	80	1	4	3	?	?	1

Feature explanations and keys for numbers see Chap. 2, Blade Technology, Recording of Technological Information, nd: feature expression not recordable

Table A.15 Blade technology analysis—Brillenhöhle, lower layer

Nr.	Blank	Length	Width	Thickness	Pres.: L	Pres.: W/T	Cortex	SPR: kind	SPR: form	SPR: length	SPR: width	DR	Lip	Bulb	Bulbar scar	Impact point	Esq. du bulbe	en éperon	Flaking angle	Profile	Distal	Hammer	n Negatives	Direc. Neg.	Reg Neg.	Tool
184	1	50	19	4	2	1	1	4	5	8	5	4	4	5	5	5	5	6	50	2	1	1	2	1	1	1
183	1	58	19	6	2	1	1	5	5	11.5	5	4	4	7	5	5	5	6	75	3	1	1	3	1	1	1
325	1	49	18	5	2	1	1	5	5	5	2	4	4	7	5	5	5	5	70	2	1	0	4	3	1	1
750	1	31	20	7.5	2	1	1	5	5	5	3	4	4	5	5	5	5	6	60	0	1	1	2	1	1	1
841	1	43	17	5	2	1	1	5	5	6	4	4	4	7	5	5	5	6	65	2	1	1	2	1	1	1
630	1	47.5	16	6	2	1	1	5	2	7	2	4	4	7	5	5	5	4	75	2	1	0	3	4	4	1
404	4	23	23	6	2	1	1	4	2	5	3.5	4	4	7	5	4	5	5	80	0	1	0	?	?	4	1
239	1	53	18.5	5	2	1	1	5	2	8.5	3	4	4	5	5	5	5	4	60	2	1	1	2	1	1	1
141	1	37	9	3	2	1	1	4	2	2	1	4	4	7	5	5	5	5	75	2	5	0	3	1	1	1
716	3	28	20	8	2	1	1	4	2	8	3	2	4	4	5	5	5	5	55	2	1	0	?	?	4	1
689	4	62	21	10	1	1	2	4	7	6	2	4	4	4	6	5	5	5	65	2	4	0	?	?	4	1
229	1	47	22	5	2	1	1	5	2	11.5	4	4	4	7	5	5	5	4	65	2	1	1	3	1	1	1
709	1	20	22	7	2	1	1	2	2	6	2.5	8	4	4	4	5	5	6	70	0	1	0	2	1	1	1
476	1	37	13	3	2	1	1	4	5	4	2	4	4	4	5	5	5	6	70	1	1	1	3	1	3	1
702	1	38.5	17.5	4	1	1	1	5	2	7.5	2	4	5	7	4	5	5	5	85	2	5	0	2	1	3	1
479	1	30	26	8	2	1	1	4	4	6	1.5	4	4	6	5	5	5	5	75	0	1	0	2	1	1	1
449	4	47	15	7	1	1	1	5	4	5	1.5	4	4	7	5	4	5	5	60	2	6	0	?	?	4	1
717	1	14	14	5	2	1	1	5	7	13	4	4	4	4	5	5	5	4	85	0	1	0	2	1	1	1
59	1	52.5	16	5	2	1	1	5	2	12	5	8	4	4	5	4	5	5	80	2	1	0	4	1	3	5
148	3	102	21	13	1	1	3	4	5	8	4	4	4	7	5	5	5	4	60	2	4	1	?	?	4	1
576	3	40	8	5	2	1	1	4	5	1.5	1	4	5	7	4	5	5	6	65	3	1	0	?	?	4	1
305	3	37	12	3	1	1	1	4	5	5	2.5	4	4	7	5	5	5	6	50	2	4	1	?	?	4	1
623	1	56	21	5	1	1	1	4	2	7.5	2.5	4	4	4	5	5	5	5	75	2	4	1	3	1	3	1
16	1	52	12	4	2	1	1	5	2	3.5	2	8	4	7	4	5	5	5	75	2	1	1	4	4	4	1
42	1	35	7	2.5	2	2	1	4	5	3	1.5	4	4	4	5	5	5	6	75	2	1	1	3	1	1	1
54	1	62	11	4	2	1	1	4	2	1	1	8	4	7	5	5	5	5	70	2	1	0	3	1	3	5
587	1	51	12.5	6.5	1	1	1	4	5	6.5	4	4	4	5	5	5	5	4	75	2	5	0	4	4	3	1
233	1	45	7.5	5	2	1	1	5	9	3.5	2	8	4	7	5	5	5	6	80	1	1	0	2	1	1	1
134	1	31	7	2	2	1	1	8	8	nd	nd	4	10	7	5	10	4	5	85	1	1	0	2	1	1	1
234	1	48.5	9	2.5	1	1	1	4	2	3	1	4	4	7	5	5	5	5	75	1	4	0	4	4	4	1
138	1	53.5	9	4	1	1	1	4	3	1	1	4	4	4	5	5	5	5	85	1	5	0	3	1	1	1
312	1	51	16	8	1	1	2	4	2	4.5	3	4	4	7	5	5	5	6	70	2	5	1	4	1	3	1
328	1	30	9	5	2	2	1	5	2	5	2.5	8	4	4	5	5	5	6	65	2	1	0	2	1	1	1
254	1	42	9.5	4	1	1	1	4	6	5	2.5	4	4	5	5	5	5	5	75	2	4	0	2	1	1	1
175	1	51	16	4	2	1	1	5	2	10	4	4	4	5	5	5	5	5	55	2	1	0	3	1	1	1
177	1	63	13	5	2	1	1	5	2	7	2	4	4	7	5	5	5	5	60	2	4	0	2	1	1	1
336	1	44	12	4	1	1	1	4	2	3	1	8	4	7	5	5	5	5	70	2	4	0	5	4	4	1
144	1	56	17	6.5	2	1	1	5	2	7	3	4	4	7	5	5	5	5	75	2	1	0	2	1	1	1
468	1	31.5	11.5	5	1	1	1	5	7	7.5	3.5	4	4	4	5	5	5	6	70	2	5	0	8	0	4	1
585	3	51	21	12	1	1	1	5	7	7.5	2.5	4	4	7	4	5	5	5	75	3	8	0	?	?	4	1
1018	4	66.5	21	6	1	1	2	5	2	9	3	4	4	4	5	5	5	5	70	3	4	0	?	?	4	1
547	1	50	13	7	2	1	3	5	5	3	3	2	10	10	10	10	10	?	80	1	1	0	3	1	3	2
132	1	66	11	5	1	1	3	4	7	5	2	4	4	7	5	5	5	5	75	2	4	0	4	1	3	1
691	1	45	16	6	1	1	2	5	7	6	3	4	4	7	5	5	5	5	80	1	5	0	2	1	1	1

(continued)

Table A.15 (continued)

Nr.	Blank	Length	Width	Thickness	Pres.: L	Pres.: W/T	Cortex	SPR: kind	SPR: form	SPR: length	SPR: width	DR	Lip	Bulb	Bulbar scar	Impact point	Esq. du bulbe	en éperon	Flaking angle	Profile	Distal	Hammer	n Negatives	Direc. Neg.	Reg Neg.	Tool
169	1	58	16	8	2	1	2	4	7	5	1	4	4	7	4	4	5	5	70	2	1	0	3	1	1	1
195	1	62	21	6.5	1	1	1	4	7	11	2	4	4	7	5	4	4	5	80	2	4	2	6	0	4	1
434	4	78	26	14	2	1	1	5	5	19	10	4	4	7	4	5	5	6	55	2	1	1	?	?	4	1
439	4	32	16	10	2	1	1	4	5	8	4	4	4	5	4	5	5	6	55	0	1	1	?	?	4	1
168	1	45	20	6	2	1	1	4	2	14	7	8	4	5	5	5	5	5	60	2	1	1	3	1	1	1
170	1	56	19	5	2	1	1	5	2	8	3	8	4	4	5	5	5	4	55	2	1	1	3	1	1	1
441	1	35	20	4	2	1	1	5	7	8	3	4	4	7	5	5	5	5	80	1	1	0	3	1	2	1
408	1	25	17.5	7.5	2	1	1	5	2	7	4	4	5	7	5	4	4	5	85	0	1	0	2	1	1	1
3	1	51	19	6	1	1	1	5	2	5	2	2	4	4	5	5	5	5	75	2	4	0	5	4	4	4
167	1	63	13	6	1	1	1	4	2	5	2	4	4	7	5	5	5	5	75	2	4	0	3	1	1	1
166	4	60	13	3	2	1	1	4	4	3.5	1.5	4	4	4	5	5	5	5	75	2	1	0	4	1	3	1
537	1	30	17.5	4	2	1	1	5	2	8	2.5	2	5	4	5	4	4	5	80	2	1	2	3	1	3	1
164	1	35	7	2.5	1	1	2	4	7	2.5	1	2	4	7	5	5	5	6	55	2	4	1	3	1	1	1
243	1	25	10	2.5	2	1	1	4	2	5.5	2.5	4	4	4	5	5	5	5	65	2	1	0	3	1	1	1
151	1	52	13	5	2	1	1	4	4	2.5	0.5	4	4	7	4	5	5	5	75	2	1	0	5	3	4	17
154	1	39	10	3	1	1	1	4	4	3	1	8	4	4	5	5	5	5	70	2	4	0	4	1	3	1
186	1	41.5	15	5	2	1	1	5	2	11	5	4	4	7	5	5	5	4	65	1	1	0	4	1	1	1
470	1	34	9	4	2	1	1	5	7	6	3	4	4	4	5	5	5	5	85	2	1	0	2	1	1	1
326	1	39	14	4	2	1	1	5	2	8	4	4	4	4	5	5	5	6	80	2	1	0	2	1	1	1
185	1	58	19.5	7	1	1	1	5	4	10	2.5	8	4	6	5	4	5	5	70	2	4	2	2	1	1	1
246	1	35	14	4	2	1	1	5	7	5	2	3	4	4	5	5	5	6	55	0	1	0	2	1	1	1
253	1	40	12	5	2	1	1	5	7	5	2	6	4	7	5	5	5	6	65	2	1	0	4	1	4	1
14	4	51	18	5	2	1	1	4	2	5	2	4	4	5	5	5	5	5	75	2	1	0	?	?	4	17
111	1	51	17	6.5	2	1	1	5	2	7	3.5	2	4	7	5	5	5	5	60	2	1	1	2	1	1	2
1049	4	50	17	4.5	1	1	1	4	2	5	2	4	4	7	5	5	5	5	80	2	4	0	?	?	4	1
416	1	44.5	9.5	5	1	1	1	5	7	6	3	6	6	7	5	5	5	6	75	3	4	1	3	1	2	1
113	4	45	19	7	2	1	1	4	7	7	4	4	4	7	4	5	5	6	75	2	1	0	?	?	4	4
633	1	29	24	5	2	1	1	5	7	11	2	4	4	7	5	10	5	5	80	0	1	0	2	1	0	1
425	1	32	16	5	2	1	1	5	2	7	2.5	4	4	7	5	5	5	6	60	2	1	1	2	1	1	1
303	4	37	15.5	5	2	1	1	5	2	6.5	2.5	4	4	7	5	5	5	5	65	2	1	0	?	?	4	1
306	1	27	23	8	2	1	1	5	2	16	5	4	4	7	5	5	5	5	75	2	1	0	2	1	1	1
288	1	47	15	4.5	2	1	1	4	9	13.5	2.5	4	4	7	5	5	5	5	75	1	1	0	4	0	3	1
582	1	35	17	6	2	1	1	4	7	9	3	4	4	4	5	4	5	5	70	2	1	0	2	1	1	1
218	1	30	7.5	3.5	2	1	1	4	2	3	1	4	4	7	5	5	5	6	70	2	1	1	2	1	1	1
172	1	55	25	5	2	1	2	5	2	13	4	8	4	4	5	5	5	6	60	2	1	1	2	1	1	1
269	1	22	25	7	2	1	2	4	2	5	2	4	4	7	5	5	5	5	75	0	1	0	1	1	1	1
270	1	38	24	6	2	1	3	5	2	9	3	2	4	7	5	5	5	5	75	2	1	0	1	1	1	1
395	1	29	13.5	5	2	1	1	5	2	9	3	3	4	7	5	5	5	6	70	2	1	1	3	1	1	1
162	1	59	13	5	2	1	1	5	2	8.5	3	3	4	5	5	5	5	4	60	2	1	1	2	1	1	1
159	1	53.5	11	3	2	1	1	4	2	2.5	2	4	4	7	4	5	5	5	70	1	1	0	3	0	1	4
452	1	38	16	6	2	1	1	5	2	6.5	3	4	4	7	5	5	5	5	60	2	1	1	2	1	1	1
163	1	47	10	5	2	1	1	5	7	5	2	8	4	7	4	5	5	5	70	2	1	0	2	1	1	1
504	1	43.5	13	4	1	1	1	4	7	6	1	4	4	4	5	4	5	5	75	2	5	2	2	1	1	1

819	1	52.5	23	4	1	1	1	5	4	15	3.5	4	5	4	4	4	5	5	90	1	4	2	4	1	3	1
653	1	46	25	7	2	1	1	5	7	14	3	4	4	5	5	5	5	6	70	2	1	0	3	1	3	1
224	1	34	14	3	2	1	1	4	4	4	0.5	8	4	4	4	5	5	5	70	2	1	0	2	1	1	1
208	1	46.5	16	3.5	2	1	1	5	5	7	3.5	4	4	7	5	5	5	6	85	2	1	0	3	1	4	1
271	1	56.5	23	8	1	1	1	5	2	15	7	2	4	4	4	4	5	5	80	2	4	2	4	1	3	1
684	1	52.5	16	7	2	1	1	5	7	10	2	4	4	4	5	5	5	5	70	2	1	0	4	1	1	1
627	1	40	17.5	4.5	2	1	1	5	2	9	3.5	4	4	7	5	5	5	6	75	1	1	0	3	1	1	1
572	1	45	12	4	2	1	1	4	4	3	1	4	4	7	5	5	5	5	60	2	1	0	3	1	1	1
432	1	49	15	7.5	2	1	4	5	5	4	2.5	4	5	5	5	5	5	6	85	2	1	0	1	1	1	1
658	4	59	12	5	1	1	1	4	5	4	2	4	4	7	4	5	5	6	80	2	4	0	?	?	4	1
340	1	70	24	12	2	1	5	5	5	8	7	2	4	7	5	5	5	5	70	2	1	0	0	0	4	1
610	4	48	14	3.5	2	1	1	4	4	4	1	6	4	7	5	5	5	5	70	2	1	0	?	?	4	1
655	4	41	17.5	5	1	1	1	5	2	5	2	4	4	5	4	5	5	5	70	2	5	0	?	?	4	1

Feature explanations and keys for numbers see Chap. 2, Blade Technology, Recording of Technological Information, nd: feature expression not recordable

Table A.16 Blade technology analysis—Brillenhöhle, upper layer

Nr.	Blank	Length	Width	Thickness	Pres.: L	Pres.: W/T	Cortex	SPR: kind	SPR: form	SPR: length	SPR: width	DR	Lip	Bulb	Bulbar scar	Impact point	Esq. du bulbe	en éperon	Flaking angle	Profile	Distal	Hammer	n Negatives	Direc. Neg.	Reg Neg.	Tool
2065	1	54	7	7	2	1	1	5	5	4	2	4	4	7	5	5	5	6	75	1	1	0	3	1	1	1
1254	1	50	22	4	2	1	1	4	2	5	4	4	4	7	4	5	5	5	85	1	1	0	2	1	1	1
1252	1	49	19	6	2	1	1	5	2	7.5	2.5	8	4	7	4	5	5	6	60	2	1	1	2	1	1	1
1256	1	19	17	3	2	1	1	5	4	9	2	4	4	7	5	5	5	5	60	0	1	0	2	1	1	1
1353	3	34	10	5	1	1	1	4	2	2	1	4	4	1	5	5	4	5	85	2	5	1	?	?	4	1
1815	1	50	18	4	1	1	1	4	2	4.5	1.5	4	4	7	5	5	4	5	80	1	4	2	2	1	1	1
2050	1	38	12	4.5	1	1	1	4	5	3	2	4	4	4	5	5	5	5	80	2	4	0	3	1	2	1
1948	1	34	6	2	1	2	1	4	6	3	1.5	4	4	7	5	5	5	5	75	1	4	0	3	1	1	13
1980	1	25	7	3	2	1	1	4	9	3	1	4	4	4	5	5	5	5	65	2	1	0	3	1	1	1
2026	1	61	24	5	1	1	1	5	2	11	5	4	4	4	4	5	5	5	75	2	4	0	3	1	3	1
1899	1	45	12	4	2	1	1	5	5	9	4	2	5	5	5	4	5	5	70	2	1	0	2	1	1	5
1922	1	34	13	5	2	1	1	5	2	6.5	2.5	2	4	5	5	5	5	6	65	2	1	0	3	1	1	5
2046	1	55	17.5	7	1	1	1	5	7	12	3	2	6	4	4	5	5	5	80	1	4	0	4	4	4	1
1219	1	45	11	4	1	1	1	5	7	7.5	2.5	8	6	7	5	5	5	5	70	1	4	0	2	1	1	1
1544	1	45	12	3	2	1	1	4	7	6	1.5	4	4	7	5	5	5	6	75	1	1	0	2	1	1	1
2016	1	17	11	5	2	1	1	5	7	6	2.5	10	4	7	4	5	5	5	75	0	1	0	1	1	1	1
1961	1	53	16	3	2	1	1	5	2	7	3	8	4	5	5	5	5	4	50	1	1	1	2	1	3	1
2063	1	39	15	5	1	1	1	4	5	4	2	4	4	4	4	4	5	5	80	2	4	2	3	0	4	1
1905	1	40	16	5	2	1	1	5	2	7.5	4.5	4	4	7	5	5	5	5	60	1	1	0	2	1	1	5
1931	1	25	15	5	2	1	1	4	5	7	3	4	6	4	5	4	5	5	80	0	1	2	2	1	3	5
1953	1	26	7.5	2	2	1	1	4	2	2	1	4	4	7	5	5	5	5	75	2	1	0	3	1	1	17
1201	1	38	11	6	2	1	1	4	7	4	2.5	4	4	7	5	5	5	5	60	2	1	0	3	4	4	1
1214	1	37	11	2.5	2	1	1	5	7	3	2	2	4	7	6	4	5	5	80	1	1	0	2	1	1	1
1204	1	33	10.5	2	2	1	1	8	8	nd	nd	4	10	7	5	10	4	5	75	2	1	0	3	1	3	1
1224	1	30	14	5	2	1	1	4	9	9	3	4	4	4	6	5	5	5	75	1	1	0	2	1	1	1
1202	1	43.5	10	3	1	1	1	4	9	3	0.5	4	4	4	5	5	5	5	75	2	4	0	4	4	4	1
1971	1	33.5	7	4	2	1	1	4	5	2.5	1.5	4	4	4	5	5	5	5	80	1	1	0	3	0	3	1
1796	1	44	14	3.5	1	1	1	4	2	4	2	4	4	7	5	5	5	5	45	2	5	0	8	0	4	1
1799	1	42.5	13.5	3	1	1	1	4	9	5	1.5	4	4	7	5	5	5	5	70	1	6	0	4	1	2	1
1797	1	32	10.5	4	1	1	1	4	5	3.5	2.5	4	4	7	5	5	4	5	75	1	6	0	3	1	1	1
1800	1	28	12	4	2	1	1	4	2	5	2.5	4	4	7	5	4	5	5	70	2	1	0	2	1	1	1
1312	1	40	18	6	1	1	1	4	2	9	3	4	5	5	5	5	5	5	85	1	4	0	2	1	3	1
1293	1	40	17	3.5	2	1	1	4	2	10	4	4	4	5	5	5	5	6	50	2	1	0	2	1	1	1
1307	1	36	14	6	2	1	1	5	2	10	4	4	4	7	4	5	5	5	75	2	1	0	2	1	1	1
1296	4	56	18	7	1	1	1	4	6	2	1	4	4	7	5	5	5	5	60	2	8	0	?	?	4	1
2024	1	79	18	5	2	1	1	5	7	7	4	4	5	7	5	5	5	5	60	2	1	0	4	4	2	1
1302	1	29	13.5	3	1	1	2	4	7	8	2	4	4	4	5	4	5	5	60	2	4	0	2	1	1	1
1812	1	26	12	4	2	1	1	4	7	3	2	3	4	4	5	5	5	6	75	2	1	0	3	1	3	1
1870	1	47	15	7	2	1	1	4	2	3	1.5	4	4	7	4	5	5	5	75	2	1	0	3	1	3	2
1947	1	50	11	4	2	2	1	5	7	9	3	4	4	7	5	5	5	6	75	2	1	0	2	1	1	4
1818	1	42	25	7	2	1	1	5	7	9	1	4	4	4	4	5	5	5	80	2	1	0	3	1	1	1
1820	1	30	26	3	2	1	1	4	2	5	3	8	4	4	5	5	5	6	70	2	1	1	3	1	1	1
1821	1	48	16	3	2	1	1	4	2	5.5	1.5	2	4	7	5	5	5	5	75	2	1	0	3	1	1	1

2023	4	99	21	8	2	1	1	4	7	7.5	3	4	4	7	6	5	5	5	75	1	1	0	?	?	4	1
2017	1	37.5	9	4	1	1	1	4	5	2.5	1.5	4	4	7	5	5	5	4	75	1	6	0	3	1	3	1
1824	1	73	22	6	1	1	3	4	7	6.5	0.5	4	4	4	5	5	5	5	75	2	4	0	1	1	1	1
1248	1	48	15	6	2	1	2	5	5	6	3	4	4	5	5	5	5	6	45	1	1	0	3	1	3	1
1242	1	46	12.5	3	1	1	3	5	5	5	3	4	4	7	5	5	5	4	65	1	5	0	3	0	1	1
1236	4	35	10	6	2	1	3	4	5	3	3	4	5	7	5	5	5	6	80	1	1	0	1	1	1	1
1299	1	49	10	6	2	1	1	5	5	5	3	6	4	7	5	5	5	6	70	2	1	1	3	1	1	1
1209	4	33	13.5	5	2	1	1	5	7	7	2	4	4	7	5	5	5	4	70	2	1	0	?	?	4	1
1247	4	49	15	4	2	1	1	4	2	3	2	4	4	7	5	5	5	6	80	2	1	0	?	?	4	1
2042	1	47	15	4	2	1	1	4	7	4	1	4	5	7	4	5	5	5	75	2	1	0	6	0	4	1
2043	1	51	16.5	3	1	1	1	4	5	5	1.5	2	10	7	10	10	10	5	70	2	5	0	3	1	3	1
1245	1	27.5	11	3.5	2	1	1	5	7	7	3	8	4	5	5	5	5	6	45	0	1	0	3	1	1	1
1240	1	31.5	10	4	2	1	1	4	5	3	2	4	4	4	5	5	5	6	70	2	1	0	2	1	1	1
2074	1	42	16	4	1	1	1	4	3	2	1	4	5	4	6	5	5	5	85	1	5	0	2	1	3	1
1219	1	41	16	6	2	1	2	5	7	5	2.5	4	4	5	4	5	5	6	60	2	1	1	3	2	2	1
327	1	31	11	6	2	1	1	5	5	8	6	4	5	5	5	5	5	6	65	1	6	0	2	1	1	1
1475	1	46	25	6	2	1	1	5	2	13	5	4	4	7	5	5	5	5	60	2	1	1	2	1	1	1
1511	1	51	14	7	1	1	1	4	5	6	3	6	4	4	6	5	5	4	70	1	5	0	3	4	4	1
1467	1	42.5	17.5	4	1	1	1	5	2	8	3	4	4	7	5	5	5	5	70	1	6	0	2	1	1	1
1459	1	49	18	6	2	1	1	5	2	6	2	4	4	4	5	5	5	5	75	2	1	0	3	4	4	1
1469	1	50	18	5	2	1	1	4	7	3	0.5	4	4	4	4	5	5	5	80	1	1	0	7	1	4	1
1465	1	46	16	5	2	1	2	4	2	7	2	8	4	7	5	5	5	5	65	1	1	0	1	1	1	1
1422	1	47	16	3	2	1	1	4	2	3	1	8	4	4	5	5	5	5	70	1	1	0	3	1	4	1
1491	1	44	12.5	7	1	1	1	5	2	6.5	3	7	4	4	5	5	5	5	70	2	5	0	5	0	4	1
1994	1	35.5	6.5	3.5	1	1	1	4	3	1	1	4	4	4	4	5	5	5	70	2	4	0	3	0	3	1
1484	1	42	18	10	1	1	1	4	7	5	1.5	4	6	6	4	5	5	5	85	2	6	0	3	0	3	1
1435	1	38	15	5	1	1	1	5	5	6.5	4	4	5	5	5	5	5	6	85	1	5	0	2	1	1	1
1454	1	28	10	4	2	2	1	5	2	6	2	4	4	4	5	5	4	5	80	1	1	0	3	1	3	1
1496	4	48	8	4	1	1	1	4	7	5.5	1.5	2	4	4	5	4	5	5	80	1	5	2	?	?	4	1
1471	1	36	14	4	2	1	1	5	7	6.5	1.5	3	5	5	5	5	4	5	80	1	2	2	2	1	1	1
1808	1	40	17	5	2	1	1	5	2	6	2.5	2	4	7	4	5	5	5	75	2	1	0	2	1	1	1
2029	1	60	15	5	2	1	1	4	6	2	1	6	4	7	5	5	5	5	70	1	1	0	2	1	1	1
1261	1	52.5	20	6	2	1	4	4	2	5	2	4	4	7	4	5	5	5	70	2	1	0	1	2	1	1
1383	1	58	17	5	1	1	1	5	5	5	3	4	4	7	5	5	5	6	55	2	8	0	3	1	1	1
2033	1	51	14	6	2	1	2	5	2	9	4	4	4	7	5	5	5	6	65	2	1	0	2	1	3	1
1330	1	49	26	4	2	1	1	5	2	10	3	4	4	4	5	5	5	4	75	2	1	0	3	1	4	1
1381	1	45	26	3	1	1	2	4	2	4	2	4	4	7	4	5	5	6	80	2	1	0	2	1	3	1
2036	1	52	15	5	2	1	1	5	2	7	4	4	4	7	5	5	5	6	60	2	1	0	3	3	1	1
2072	1	53	15	5	2	1	1	4	6	3	2	8	4	5	4	5	5	5	65	2	1	0	2	1	1	1
1400	4	42.5	15	6	2	1	1	5	5	5	3	2	4	5	5	5	5	6	70	2	1	0	?	?	4	1
1391	1	48	12	3	1	1	1	4	4	3	0.5	4	4	7	4	5	5	5	nd	2	4	0	3	1	1	1
1387	1	51	16	4.5	1	1	1	4	4	6	2	4	4	7	4	4	5	5	75	1	4	0	3	1	1	1
1259	1	46	11	5	1	1	1	4	2	4.5	2	4	4	7	4	5	5	5	50	2	4	0	3	1	1	1
1265	1	48	11	4.5	1	1	2	4	2	2	1	4	4	7	5	5	5	5	80	1	1	0	3	1	1	1
1880	1	37.5	17	5	2	1	1	5	5	8	4.5	4	4	7	4	5	5	4	55	2	1	1	2	1	1	2

(continued)

Table A.16 (continued)

Nr.	Blank	Length	Width	Thickness	Pres.: L	Pres.: W/T	Cortex	SPR: kind	SPR: form	SPR: length	SPR: width	DR	Lip	Bulb	Bulbar scar	Impact point	Esq. du bulbe	en éperon	Flaking angle	Profile	Distal	Hammer	n Negatives	Direc. Neg.	Reg Neg.	Tool
1954	1	42	9	3.5	2	1	1	5	2	2	1	4	4	7	4	5	5	5	60	2	1	0	3	1	3	17
2025	1	68	15	5	1	1	1	4	7	5	1	4	4	7	5	5	5	5	85	1	8	2	3	1	1	1
1788	4	59	15	6	1	1	3	5	5	4	3	4	5	5	5	5	5	6	80	2	4	1	?	?	4	1
1738	1	46	13	4.5	2	1	2	5	2	6.5	2.5	4	4	5	5	5	5	5	70	2	1	0	3	1	1	1
2069	1	52.5	20	4	2	1	1	4	4	6.5	2	4	4	4	5	5	5	5	65	2	1	0	3	1	3	4
2070	1	53.5	17	5	1	1	1	5	2	9.5	3.5	4	4	7	5	5	5	6	60	2	7	0	3	1	1	1
1528	4	43.5	14	6	2	1	1	5	5	3.5	2	4	4	7	5	5	5	4	70	2	1	0	?	?	4	1
1637	1	54	21	6	1	1	2	5	2	6	2.5	4	5	7	5	5	5	5	nd	2	5	0	3	1	1	1
1772	1	43.5	19	6	2	1	1	5	2	11	5	4	4	7	5	5	5	5	80	2	1	0	2	1	3	1
1611	1	42.5	12	4	1	1	1	5	7	4	1	4	4	7	4	5	5	5	75	2	4	0	2	1	1	1
1783	1	59	15	7	1	1	1	4	7	5	1.5	4	4	4	5	5	5	5	80	3	5	0	4	4	4	1
1618	1	43	11	4	2	1	1	4	3	1	1	4	4	4	5	4	5	5	75	2	1	0	2	1	1	1

Feature explanations and keys for numbers see Chap. 2, Blade Technology, Recording of Technological Information, nd: feature expression not recordable

Table A.17 Blade technology analysis—Chênelaz

Nr.	Blank	Length	Width	Thickness	Pres.: L	Pres.: W/T	Cortex	SPR: kind	SPR: form	SPR: length	SPR: width	DR	Lip	Bulb	Bulbar scar	Impact point	Esq. du bulbe	en éperon	Flaking angle	Profile	Distal	Hammer	n Negatives	Direc. Neg.	Reg Neg.	Tool
1371	1	58	14	5	1	1	1	5	2	6.5	2.5	3	5	7	5	4	4	5	75	1	4	2	2	1	1	1
2445	1	48	14	3	1	1	1	5	7	5.5	1.5	8	6	7	5	5	5	5	60	1	5	0	4	?	2	1
2717	1	60.5	16	5.5	1	1	2	5	9	7.5	1	3	6	7	5	5	5	5	85	1	4	2	3	1	1	1
3169	1	59	20	5	1	1	2	5	2	12	3.5	3	6	7	6	4	5	5	75	1	4	2	4	1	2	1
3110	1	33	16	5	2	1	1	5	2	10	4	4	5	7	5	5	5	5	65	2	1	0	3	1	1	1
3002	1	42	14	5.5	1	1	1	4	4	3.5	1	8	5	7	5	5	5	5	80	2	6	0	3	1	1	1
3101	1	60	19	3	2	1	1	5	9	13	2.5	3	5	7	5	5	5	5	75	1	1	0	3	1	1	1
2932	1	79	30	7	1	1	1	5	2	10.5	6	4	5	4	6	6	5	5	85	2	4	2	4	3	2	1
3002	1	75	25	5.5	1	1	1	4	7	4	2.5	3	5	7	4	4	5	5	60	2	5	2	2	1	1	1
3204	1	38	13.5	5	1	1	1	5	2	9.5	4	4	5	4	4	4	5	5	80	2	4	2	3	1	1	1
1662	1	25.5	15	5	2	1	1	5	9	8	2.5	8	4	7	4	4	5	5	65	1	1	2	2	1	1	1
2686	4	39	16	10.5	2	1	1	5	7	7.5	3	8	4	4	4	4	5	5	85	2	1	2	4	4	2	1
2796	1	29.5	10.5	3	2	1	1	5	2	5	1.5	8	5	7	5	5	4	5	75	2	1	2	2	1	1	1
3224	1	56	23	8	2	1	2	4	2	10.5	3.5	4	5	4	4	4	5	5	80	1	1	2	3	?	3	1
None	1	30	11.5	4	2	1	1	5	7	8.5	3	8	4	4	5	4	5	5	75	2	1	2	3	?	3	1
2450	1	33	13.5	3.5	2	1	1	4	2	5.5	2	8	5	4	4	5	5	5	75	2	1	0	2	1	1	1
2584	1	30	9.5	2	2	1	1	5	7	6.5	3	3	4	5	5	5	5	5	10	2	1	0	3	1	1	1
3015	1	33.5	12	2	2	1	1	5	2	4	1.5	4	5	4	5	5	5	5	70	2	1	0	3	1	1	1
3118	1	37	11.5	3	2	1	1	4	2	3	1.5	4	4	4	5	5	5	5	80	2	1	0	2	1	1	1
230	1	36	9	3.5	2	1	2	5	7	5	2	2	5	7	4	4	5	5	80	2	1	2	3	3	1	1
2448	1	38	11.5	4	1	1	1	4	2	3	2	8	6	4	4	4	5	5	60	2	5	2	5	1	2	1
2170	1	27.5	8	2.5	2	1	1	4	4	3	0.5	4	5	4	5	4	5	5	80	2	1	2	3	1	2	1
2198	1	24	9	3	2	1	1	5	4	4.5	0.5	4	5	7	4	4	5	5	80	2	1	2	2	1	2	1
198	1	31	12	2.5	2	1	1	4	5	5	2	4	5	4	5	5	5	6	70	2	1	0	2	1	1	1
2440	1	25	11	3	2	1	1	5	4	5.5	0.5	8	6	4	5	4	5	5	80	2	1	2	2	1	1	1
2472	1	28	13	5	1	1	1	5	7	5.5	2	4	4	4	4	4	5	5	70	2	5	2	3	1	1	1
1997	1	30.5	10.5	2	1	1	1	4	2	5	2	3	4	7	4	6	5	5	60	2	4	2	1	1	1	1
2464	1	53	15	6	1	1	1	4	2	3	1	3	5	7	5	5	5	5	45	2	4	0	6	?	3	1
2485	1	33.5	13	3	2	1	2	5	7	8	2	4	5	7	5	4	5	5	75	2	1	2	3	1	2	1
2875	1	36	12	4	1	1	1	4	2	2.5	1	3	5	7	4	4	5	5	80	2	6	2	3	1	1	1
2805	1	43	14	4	2	1	2	5	2	6.5	2	4	4	7	5	6	5	5	80	2	1	2	3	1	1	1
3217	1	57	18.5	4	2	1	2	5	5	7.5	4	3	4	5	5	5	5	4	60	2	1	1	3	1	2	1
2918	1	40	5	4	1	1	1	4	2	2	1	4	5	4	5	5	4	5	80	2	4	2	3	1	2	1
2666	1	44	15	3	1	1	1	4	9	7	1.5	8	5	7	4	4	5	5	75	1	4	2	3	1	2	1
1710	1	51	18	4	2	1	1	5	4	4	1	3	5	7	4	5	5	5	80	1	1	0	3	1	3	1
1821	1	33	22	2.5	2	1	1	4	2	4.5	1.5	4	5	4	5	5	5	5	80	2	1	0	3	1	1	1
2208	1	35	15	3	1	1	1	4	9	4.5	1	8	5	7	5	5	5	5	70	2	5	0	3	1	2	1
2469	1	49	17	4	1	1	2	4	2	4.5	2	8	4	4	5	4	5	5	60	1	5	2	3	1	1	1
2478	1	52	12.5	4.5	1	1	1	5	2	5.5	1.5	3	5	7	4	4	5	5	80	2	2	2	4	1	2	4
239	1	25	19	4.5	2	1	1	5	2	8	3	3	5	7	6	4	5	5	85	nd	1	2	3	1	1	1
230	1	40	16	4	2	1	1	5	2	5	2	8	5	7	5	5	5	5	75	1	2	0	2	1	1	5
2215	1	31	15	4	2	1	1	4	7	2	1	4	4	4	4	4	5	5	75	2	2	2	4	1	1	5
1657	1	48.5	20.5	8	2	1	1	4	7	9	4.5	3	5	4	4	4	5	5	75	2	1	2	5	1	4	1
3118	1	59	19	3.5	2	1	1	5	2	10.5	4	3	4	7	5	5	5	5	65	2	1	0	2	1	1	1

(continued)

Table A.17 (continued)

Nr.	Blank	Length	Width	Thickness	Pres.: L	Pres.: W/T	Cortex	SPR: kind	SPR: form	SPR: length	SPR: width	DR	Lip	Bulb	Bulbar scar	Impact point	Esq. du bulbe	en éperon	Flaking angle	Profile	Distal	Hammer	n Negatives	Direc. Neg.	Reg Neg.	Tool
206	1	42	19	5	1	1	1	5	9	16	4.5	4	4	7	4	4	5	5	80	1	5	2	4	1	1	1
2672	1	58.5	20	4	2	1	1	5	4	9	1.5	4	5	7	4	4	5	5	70	2	5	2	4	1	2	1
2663	1	52	16	4.5	1	1	1	5	2	10	4	3	5	4	4	4	5	5	80	2	5	2	2	1	1	1
3134	1	42	15	3.5	1	1	1	4	7	7	2	4	5	7	4	4	5	5	80	2	4	2	2	1	1	1
3235	1	47.5	17.5	9	1	1	1	4	2	14.5	7.5	4	5	7	4	4	5	5	85	1	4	2	3	1	1	1
2864	1	28	16	4	2	1	1	4	7	5	1.5	8	5	4	5	5	5	5	80	2	2	0	3	1	1	1
2038	1	30.5	7.5	3.5	2	2	1	5	5	3	1.5	3	5	7	5	5	5	5	75	1	1	0	2	1	1	13
None	1	9	6.5	3	2	2	1	4	2	5	2	4	5	4	4	5	5	5	80	nd	1	0	2	1	1	13
None	1	18	5	2	2	2	1	4	2	2	0.5	4	5	4	5	5	5	5	75	2	1	0	2	1	1	13
2185	1	8.5	9.5	2.5	2	2	1	4	2	3	1.5	3	5	4	4	4	5	5	85	nd	1	2	2	1	1	13
2217	1	15	5.5	2	2	2	1	4	2	3	1	4	6	7	5	4	5	5	65	2	1	2	2	1	1	13
2952	1	40	6.5	2	1	2	1	5	2	3.5	1.5	3	5	7	4	4	5	5	85	1	4	2	2	1	1	13
1966	1	23.5	7	1.5	2	2	1	5	2	4	1	8	5	4	5	5	5	5	80	1	1	0	3	1	1	13
2046	1	24	9	2.5	2	2	1	5	7	5.5	2	4	5	7	5	5	5	5	85	1	1	0	3	1	1	13
None	1	8	5	2	2	2	1	5	2	3.5	1	3	5	7	4	5	5	5	nd	nd	1	0	2	1	1	13
None	1	24	11	2.5	2	1	1	5	4	3.5	0.5	8	5	7	5	4	4	5	80	2	1	2	2	1	1	1
3157	1	41.5	14	5	1	1	1	5	2	5.5	1.5	4	4	4	4	4	5	5	80	1	4	2	2	1	1	1
2455	1	42	10	5	2	2	1	4	2	6	2	8	4	7	6	6	5	5	75	2	2	2	2	1	1	9
2196	1	31	14.5	4	1	1	1	4	7	5	2	2	5	4	4	5	5	5	85	1	6	0	4	1	1	1
200	1	21.5	12	3	2	1	1	5	5	4.5	2.5	8	4	7	5	5	5	5	60	1	1	1	2	1	1	1
218	1	12	12	4.5	2	1	1	5	7	11	3.5	4	6	4	4	4	5	5	80	nd	1	2	2	1	1	1
None	1	13.5	11.5	2.5	2	1	1	4	2	4	1.5	3	4	7	5	5	5	5	80	2	1	0	2	1	1	1
None	1	13	9.5	4	2	1	1	5	2	7	3	8	5	4	4	4	5	5	85	nd	1	2	2	1	1	1
None	1	20	9.5	2.5	2	1	1	5	2	4.5	1.5	4	5	7	4	6	5	5	60	2	1	2	2	1	1	1
2845	1	27	14	4	2	1	1	5	2	6	2	3	6	4	5	5	5	5	65	1	1	0	3	4	3	1
3125	1	25	17	5	2	1	1	5	7	13	4	4	4	4	5	4	4	5	80	1	1	2	3	4	3	1
3124	1	25	20	4	2	1	1	5	7	12.5	3	4	5	7	4	5	5	5	65	2	1	2	2	1	2	1
3002	1	21	21	3.5	2	1	1	4	9	14.5	2	4	4	4	4	4	5	5	80	nd	1	2	3	1	1	1
2659	1	32	16	2.5	1	1	1	5	2	6.5	2.5	4	5	7	5	4	4	5	55	1	6	2	3	1	1	1
3077	1	30	18	3	2	1	1	5	5	4	2.5	4	5	7	4	4	5	5	80	2	1	2	4	1	1	1
None	1	38	20.5	5.5	2	1	1	5	2	10	4	4	5	7	6	5	5	5	75	1	1	0	2	1	1	1
3091	1	39	20	5.5	1	1	1	5	2	6	2	4	5	7	4	4	5	5	65	2	5	2	2	1	1	1
2813	1	32	18	3.5	2	1	1	4	3	1	1	4	5	4	5	4	5	5	70	2	1	2	2	1	1	1
3051	1	29	13	4	2	1	1	5	2	8.5	2.5	4	5	7	4	4	5	5	80	2	1	2	3	1	1	1
3081	1	22	14	3	2	1	1	5	2	6	1.5	4	5	7	5	5	5	5	80	2	1	0	3	1	1	1
None	1	26	12	3.5	1	1	1	5	4	6	0.5	3	5	7	4	4	4	5	75	1	6	2	2	1	1	1
None	1	29	13	2.5	1	1	1	4	2	5.5	2.5	8	4	7	5	5	5	5	75	1	6	0	2	1	1	1
3088	1	23	7.5	2	2	1	1	4	2	5	2	3	5	7	4	5	5	5	65	2	1	0	3	1	1	1
3132	1	19.5	8	2.5	2	1	1	5	5	4	2	4	4	4	4	4	5	6	80	2	1	2	2	1	1	1
3095	1	26.6	15	4	2	1	1	4	2	4	2	4	6	4	6	6	5	5	65	1	1	2	2	1	1	1

3221	1	34	9	3	2	1	1	5	2	2.5	1	8	5	7	4	5	5	5	70	2	1	0	2	1	2	1
3185	1	24	17	5	2	1	1	4	9	7.5	2	4	5	4	6	5	5	5	75	1	1	0	2	1	1	1
None	1	17	11	2	2	1	1	5	2	9	2	4	4	7	5	5	5	5	70	1	1	0	3	1	1	1
None	1	17	14.5	3	2	1	1	4	2	3.5	2	3	4	7	5	4	5	5	80	1	1	2	2	1	1	1
3049	1	18	15.5	3	2	1	1	4	2	4	2	4	5	7	5	5	5	5	70	2	1	0	2	1	1	1
None	1	13	7.5	3	2	1	1	4	2	2	1	4	5	4	4	4	5	5	85	1	1	2	2	1	1	1
3048	1	13	10.5	3.5	2	1	1	5	2	5.5	2	8	4	7	5	5	5	5	60	2	1	0	2	1	1	1
None	1	12	11	4	2	1	1	5	2	9	3.5	3	5	7	4	4	5	5	85	nd	1	2	2	1	1	1
None	1	13.5	9.5	3	2	1	1	4	7	4.5	1.5	4	5	7	4	4	5	5	80	nd	1	2	2	1	1	1
2969	1	15	15	3	2	1	1	5	2	8	2	8	5	7	4	4	5	5	85	nd	1	2	2	1	1	1
2594	1	20	10	5	2	1	2	4	2	2	1	8	4	7	5	5	5	5	75	2	1	0	1	1	1	1
None	1	15	9	3	2	1	1	5	2	7	2	8	4	4	5	5	5	5	70	2	1	0	2	1	1	1
None	1	10	12.5	3.5	2	1	1	5	2	10	3.5	3	5	4	4	5	5	5	80	nd	1	0	2	1	1	1
217	1	11	14	2.5	2	1	1	4	9	6	2	4	5	4	6	5	5	5	70	nd	1	0	3	1	1	1
None	1	11	12.5	3	2	1	1	4	7	4.5	0.5	4	5	4	5	5	5	5	80	2	1	0	2	1	1	1
None	1	12	8	2	2	1	1	4	2	3	1	3	5	4	5	5	5	5	85	2	1	0	2	1	1	1

Feature explanations and keys for numbers see Chap. 2, Blade Technology, Recording of Technological Information, nd: feature expression not recordable

Table A.18 Blade technology analysis—Grappin, Arlay

Nr.	Blank	Length	Width	Thickness	Pres.: L	Pres.: W/T	Cortex	SPR: kind	SPR: form	SPR: length	SPR: width	DR	Lip	Bulb	Bulbar scar	Impact point	Esq. du bulbe	en éperon	Flaking angle	Profile	Distal	Hammer	n Negatives	Direc. Neg.	Reg Neg.	Tool
1655	1	59	23	6	1	1	1	5	2	9	4	2	5	7	4	5	5	5	45	1	6	0	4	1	3	14
1223	1	54	19.5	6	1	1	1	5	2	9	4.5	2	5	4	5	5	5	5	50	1	5	0	6	?	4	1
3163	1	18	13	3	2	1	1	5	9	8.5	2	8	6	4	5	5	5	5	55	1	1	0	3	1	1	1
1641	1	42	24.5	8.5	2	1	1	5	2	18.5	10	4	4	7	4	4	5	5	60	1	1	2	3	1	1	14
1370	1	56.5	18	6	2	1	1	4	2	3.5	2	4	5	5	5	4	6	5	60	1	1	2	3	1	1	1
1374	1	52	25	11.5	2	1	1	5	2	10	3.5	4	4	4	5	5	5	5	65	1	2	0	3	1	1	5
1631	1	30	19	4	2	2	1	4	7	9	3	4	5	7	5	5	5	5	65	1	1	0	3	1	1	14
1385	1	44.5	17	9	2	1	1	5	5	8	5	4	4	4	5	5	5	4	70	1	2	1	2	1	1	5
1200	1	24	11	3	1	1	1	4	2	3	2	4	4	4	4	4	5	5	70	1	6	2	3	1	1	1
1526	1	23	13	3	2	1	1	5	2	4	1	4	5	7	5	5	5	5	75	1	1	0	2	1	1	1
2732	1	31	10.5	4	2	1	1	5	7	6	2	4	5	4	5	4	5	5	80	1	1	2	2	1	1	1
2710	1	27	7.5	3.5	1	1	1	4	2	4.5	1	4	6	4	5	4	5	5	80	1	5	2	3	1	1	1
2907	1	23	10.5	1.5	2	1	1	4	4	3	1	4	5	7	5	4	5	5	80	1	1	2	3	1	1	1
1628	1	45.5	21.5	8.5	1	1	1	4	5	7	4.5	4	5	7	4	5	5	5	80	1	5	0	2	1	3	1
1704	1	43	13	3.5	2	1	1	4	2	3.5	1.5	8	4	4	5	5	5	5	80	1	1	0	2	1	1	1
2696	1	37	10	3.5	2	1	1	5	2	7	3	2	5	4	5	5	5	5	80	1	1	0	3	1	1	1
881	1	41	17.5	7	2	1	1	5	5	4.5	2.5	3	4	7	4	4	5	5	85	1	1	2	3	1	1	1
1528	1	20	15	4	2	1	1	4	9	3	1	4	6	7	6	4	5	5	85	1	1	2	3	1	1	1
1283	1	29	13	3.5	2	1	1	4	2	6.5	2.5	4	5	7	6	5	5	5	85	1	1	0	3	1	3	1
1459	1	45.5	16	4	2	1	1	4	2	7.5	4	8	4	5	5	5	5	4	45	2	1	1	5	?	3	1
1624	1	86	21.5	5	2	1	1	5	5	8	5	3	4	7	5	5	5	4	45	2	2	1	2	1	1	14
1471	1	21.5	18.5	6	2	1	1	5	2	14	5.5	8	4	7	5	5	5	4	55	2	1	1	2	1	1	1
2687	1	40.5	9.5	3.5	2	1	1	5	4	4.5	1	4	4	7	4	4	5	5	55	2	1	2	3	1	1	1
1712	1	45	26	5	2	1	1	5	2	11	5.5	2	4	4	4	4	5	4	60	2	1	0	2	1	1	14
2538	1	49	22	7	2	1	1	5	5	11	5.5	4	4	7	5	5	5	4	60	2	2	1	3	1	1	5
1464	1	18	12	3	2	1	1	4	2	4	1	8	4	4	4	4	5	5	60	2	1	2	3	1	1	1
2757	1	26	8.5	4	2	1	1	4	4	2.5	1	4	4	4	5	5	5	5	60	2	1	0	2	1	1	13
2695	1	31	12	3	1	1	1	4	5	5	2	8	4	7	5	5	5	5	60	2	5	0	3	?	2	1
1475	1	32	19	5	2	1	1	5	9	10	2	8	6	7	5	5	5	5	60	2	1	0	2	1	1	1
1625	1	83	24	7	2	1	2	5	2	6	2	8	4	7	5	5	5	4	65	2	2	1	3	1	1	14
1701	1	53	22	5.5	2	1	1	5	2	9	3	4	5	7	4	4	5	5	65	2	1	2	2	1	1	1
3052	1	17	7	2	2	1	1	4	2	2	1	3	5	4	5	5	5	5	65	2	1	0	2	1	1	1
1059	1	81.5	29	9	1	1	2	5	2	16	5.5	4	5	4	5	5	5	5	65	2	4	0	3	1	2	1
1538	1	24	13	3.5	2	1	1	4	7	6	1.5	8	5	7	5	5	5	5	65	2	1	0	3	1	1	1
1206	1	47	18.5	6	1	1	1	5	2	8	3	3	4	7	4	5	5	4	70	2	5	1	2	1	2	1
1217	1	37	12	3	1	1	1	5	2	5.5	2.5	8	5	4	4	4	5	5	70	2	5	2	4	?	3	1
1243	1	39	14	4	2	1	1	4	2	3	1	3	5	7	4	4	5	5	70	2	1	2	3	1	1	1
1310	1	31.5	22	4	2	1	1	5	2	12	4	8	4	4	5	4	5	5	70	2	1	2	2	1	2	1
1306	1	46	20.5	5	1	1	1	4	2	4	2	8	5	4	6	4	5	5	70	2	5	2	4	1	2	14
1215	1	52	14.5	8	1	1	1	4	2	4.5	1.5	8	5	4	4	5	5	5	70	2	4	0	6	?	3	1
2764	1	17	5	1.5	2	1	1	5	4	4	0.5	3	5	7	4	5	5	5	70	2	1	0	2	1	1	13
2745	1	29	6	2	1	1	2	4	2	2.5	1	3	4	4	5	5	5	5	70	2	5	0	3	4	2	1

2759	1	24	6	2.5	2	1	1	4	7	4	1	4	5	7	5	5	5	5	70	2	2	0	3	1	1	13
2884	1	29	10	4.5	2	1	1	5	7	8.5	3	8	6	7	5	5	5	5	70	2	1	0	2	1	2	1
3122	1	24	8	4	2	1	1	4	2	2.5	1	4	5	7	5	4	4	5	75	2	1	2	2	1	1	1
1188	1	57.5	15	5.5	1	1	2	5	9	10	2	4	5	7	5	4	4	5	75	2	4	2	5	4	2	1
1497	1	18.5	16	4	2	1	1	4	2	4	1.5	8	5	7	5	4	4	5	75	2	1	2	2	1	1	1
1373	1	45.5	18	8	2	1	1	5	9	9	2	8	4	4	4	4	5	5	75	2	2	2	6	?	4	2
3042	1	15	6	3	2	1	2	4	2	2.5	1	3	5	4	5	4	5	5	75	2	1	2	1	1	1	1
3212	1	12.5	6	1.5	2	1	1	4	2	2	1	8	5	4	5	4	5	5	75	2	1	2	2	1	1	1
1226	1	43	16	6	2	1	1	4	4	7	2	8	5	7	5	4	5	5	75	2	1	2	4	3	4	1
2739	1	25	8.5	4	2	1	1	4	2	4.5	2	8	5	7	4	5	5	5	75	2	1	0	2	1	1	1
1469	1	37.5	19	5.5	2	1	2	2	2	6	2	8	5	4	5	5	5	5	75	2	1	0	3	1	2	1
1227	1	43	14.5	4.5	1	1	3	4	2	5.5	2.5	3	4	7	5	5	5	5	75	2	5	0	2	1	2	1
1372	1	56.5	25	8.5	2	1	1	5	5	15	7.5	4	4	7	5	5	5	5	75	2	1	0	4	?	1	15
2909	1	21.5	9	3.5	2	1	1	5	2	6	2	4	5	7	5	5	5	5	75	2	1	0	3	?	2	1
1197	1	35	12	2.5	2	1	1	4	9	7	2	4	5	4	4	4	5	5	80	2	1	2	3	1	2	1
2702	1	37.8	8	2	2	1	1	4	7	3.5	1.5	8	5	7	4	4	5	5	80	2	1	2	2	1	1	1
3152	1	19	10.5	2.5	2	1	1	4	2	4	2	4	6	7	4	4	5	5	80	2	1	2	2	1	1	1
2722	1	26.5	12.5	4	2	1	2	4	4	5.5	1	8	5	4	5	4	5	5	80	2	1	2	1	1	1	1
1205	1	43	22.5	8.5	2	1	1	5	2	11.5	5.5	3	6	7	6	4	5	5	80	2	2	2	3	1	3	5
1504	1	33	14	3.5	1	1	2	5	2	4	1	8	5	4	4	5	5	5	80	2	6	0	2	1	2	1
3150	1	16	7	2.5	2	1	1	5	5	3	1	8	5	4	5	5	5	5	80	2	1	0	2	1	1	1
1480	1	20	15.5	4.5	2	1	1	5	9	12.5	3	2	6	4	5	5	5	5	80	2	1	0	3	1	1	1
2728	1	23.5	6.5	3	2	1	1	4	4	3.5	1	3	4	7	5	5	5	5	80	2	1	0	4	1	1	1
2660	1	20	23	5	2	1	1	5	2	9	4.5	4	5	7	6	5	5	5	80	2	1	0	2	1	1	1
1476	1	19	15	4	2	1	1	5	7	6.5	3	4	5	7	4	4	5	5	85	2	1	2	2	1	1	1
1477	1	31	15.5	5	2	1	1	5	2	10	4.5	8	5	7	5	4	5	5	85	2	1	2	2	1	2	1
3160	1	17.5	12	2.5	2	1	1	5	2	5	2	4	5	4	5	5	5	5	85	2	1	0	2	?	3	1
1456	1	14	10	3.5	2	1	2	4	2	7	3	8	4	4	5	4	5	5	85	?	1	2	2	1	2	1
3063	1	16	7	3	2	1	1	5	2	3.5	2	3	5	7	6	4	5	5	85	?	1	2	3	1	1	1

Feature explanations and keys for numbers see Chap. 2, Blade Technology, Recording of Technological Information, nd: feature expression not recordable

Table A.19 Blade technology analysis—Hostim

Nr.	Blank	Length	Width	Thickness	Pres.: L	Pres.: W/T	Cortex	SPR: kind	SPR: form	SPR: length	SPR: width	DR	Lip	Bulb	Bulbar scar	Impact point	Esq. du bulbe	en éperon	Flaking angle	Profile	Distal	Hammer	n Negatives	Direc. Neg.	Reg Neg.	Tool
H06	1	91	23	9	2	1	1	5	7	7	5.5	3	2	7	4	4	5	5	55	3	1	2	3	1	2	1
E82	1	72	19.5	5	2	1	1	5	2	8	3	4	4	7	4	5	5	4	55	2	1	1	3	1	1	1
HA15	1	94	20	6	2	1	1	5	2	8	3	8	4	4	4	5	5	4	60	3	1	1	2	1	1	1
E82	1	35.5	14.5	5.5	2	1	1	5	2	10	5	3	5	4	5	4	5	5	55	2	1	2	3	1	1	1
HB1	1	56	20	8	2	1	1	5	2	9.5	5	4	4	7	4	4	5	4	50	2	1	0	4	4	1	1
None	1	126	22.5	7.5	2	1	1	5	7	5	3	8	4	7	4	4	5	5	70	3	1	2	4	1	1	1
A22/H52-3	1	83	24	8	1	1	1	5	2	6	2	8	5	4	5	5	4	5	85	2	4	2	6	3	4	1
HB27	1	96.5	21	8	1	1	2	5	5	12	7	3	5	7	5	4	5	4	60	1	4	0	7	1	3	1
E94	1	81	16	6.5	2	1	1	5	2	7.5	3	3	5	7	4	4	5	5	60	3	2	2	3	1	1	4
None	1	46	25	7	2	2	1	5	2	17	5	4	4	4	5	4	5	5	80	1	1	2	2	1	1	14
E59	1	39	12	3	2	2	1	5	5	7	4	8	4	4	4	5	5	4	80	2	2	0	3	1	1	2
HE14	1	53	16	7	2	1	2	4	3	1	1	4	5	7	5	5	5	5	75	1	1	0	1	1	1	1
E49	1	42.5	15	6.5	2	1	2	5	2	12.5	6.5	6	5	4	4	5	5	4	60	1	2	1	3	1	1	2
E87	1	66	22	7.5	2	1	1	5	2	7.5	3	8	4	4	5	4	5	5	65	3	2	2	3	1	4	2
E78	1	51.5	16	6	2	1	2	5	2	9	3	8	4	4	4	4	5	5	70	2	2	2	3	1	2	2
HB34	1	47	27	7	2	1	1	5	2	22.6	6.5	8	5	4	5	4	5	4	60	2	1	0	3	1	1	1
HB27	1	43	12.5	4	1	1	1	4	2	3	1	3	5	4	5	5	5	5	70	3	4	0	2	1	1	1
HB6	1	31	9.5	2.5	2	1	1	4	2	2	1	8	4	4	4	5	5	5	85	2	1	0	4	?	2	1
Hs30	1	62	20	7	2	1	1	5	2	8	3	8	4	7	5	5	5	4	65	2	1	1	3	1	1	1
HC9	1	55.5	17	5	2	2	1	4	2	9	3	2	4	5	5	4	4	5	80	1	1	2	3	1	1	5
HB11	1	78	16	4	2	1	1	5	2	6.5	2	3	4	5	5	5	5	5	60	3	2	0	5	4	1	4
HB36	1	56	26	5.5	2	1	1	5	7	8.5	3	4	6	4	4	4	5	5	85	2	1	2	3	1	2	14
E51	1	37.5	25	4	2	1	1	5	7	10.5	5.5	4	5	4	5	5	5	5	70	2	1	0	3	1	2	1
E91	1	41	16	5	2	1	1	4	2	5	2	8	4	4	5	4	5	5	80	2	1	2	2	1	1	1
HB37	1	36	19	4	2	1	1	5	2	6.5	2	3	4	7	5	5	5	4	50	2	1	0	3	1	1	1
E93	1	48	13.5	5.5	2	1	1	4	2	6	4	2	5	7	5	4	5	5	75	2	1	2	3	3	1	1
E88	1	25	19.5	5.5	2	2	1	5	2	7.5	5	4	4	7	5	5	5	5	75	2	1	0	2	1	1	14
E93	1	38	21	5	2	2	2	4	4	6	1.5	8	4	6	5	4	5	5	70	2	1	2	2	1	1	14
E95	1	48	12	4	2	1	1	5	2	3	2	8	5	4	5	4	5	5	80	2	2	2	4	2	1	5
HB1	1	66.5	23	7	2	1	1	5	9	9	1	4	5	4	5	4	5	5	65	1	2	2	4	1	2	17
E73	1	59.5	18	5	2	1	1	5	2	6	3	3	4	7	5	5	5	4	60	1	2	1	3	1	1	5
HE13	1	45.5	11.5	5	2	1	1	5	5	6.5	4.5	8	4	7	5	5	5	4	60	2	1	0	2	1	1	1
Hs	1	35	17	4	2	1	1	5	7	5	2.5	8	4	4	4	5	5	5	75	2	2	0	2	1	1	17
A26H59-4	1	80	20.5	9	2	1	1	5	2	11	5.5	3	4	7	4	5	5	4	60	2	1	1	2	1	1	1
E109	1	87	26	9	2	1	1	5	2	13	5.5	4	5	4	5	5	5	5	70	2	2	0	3	1	1	17
E74	1	46	11.5	2.5	2	1	1	4	7	4.5	1.5	4	6	4	5	5	5	5	70	2	2	0	4	1	1	17
E108	1	72	14	7	2	1	2	4	2	4.5	1.5	8	5	7	4	5	5	5	60	2	2	0	4	1	1	4
HE33	1	34.5	15	4	2	2	1	5	2	6.5	3	4	6	4	4	4	5	5	60	2	1	2	3	1	1	14
E68	1	28	12	4	2	1	1	4	9	7	2	3	5	4	5	4	5	5	70	2	2	2	3	1	3	17
E82	1	42	15	4	2	1	1	5	7	5	2	3	4	4	4	5	5	5	75	2	2	0	2	1	2	17
E38	1	35	15.5	3.5	2	1	1	5	7	9	3	4	4	4	4	4	5	5	60	2	2	2	3	1	1	17
HE32	1	29	15	4.5	2	1	1	5	2	6.5	2.5	4	4	4	4	4	5	5	55	2	1	2	2	1	1	1
E100	1	30	7.5	3	2	1	1	5	2	6	2.5	8	5	4	5	4	5	5	50	2	2	2	2	1	1	17

HB2	1	15	10	3	2	1	1	5	2	4.5	2.5	3	5	7	5	4	4	5	85	?	1	2	2	1	1	1
HB22	1	17.5	13	5	2	1	1	4	3	1	1	4	4	7	5	5	5	5	75	?	1	0	2	1	1	1
Hb28	1	30	12.5	3	2	1	1	4	2	3	2	3	6	4	4	5	5	5	80	2	1	0	2	1	1	1
HB2	1	20	17	4	2	1	1	5	2	6.5	2	4	6	4	4	4	5	5	80	2	1	2	3	1	1	1
C7H43-4	1	23	11.5	5	2	1	1	5	2	5.5	3	3	4	4	5	5	5	4	55	2	1	1	3	1	1	1
Ha22H52-3	1	32	19	6.5	2	1	1	5	2	7.5	4	4	4	7	5	4	5	4	65	2	1	0	3	1	1	1
HA29	1	96.5	20	8	2	1	1	5	2	8	3.5	3	4	7	6	4	5	5	65	2	1	2	2	1	1	1
He42	1	90.5	22.5	3	1	1	1	5	2	5	2.5	8	4	7	4	4	5	5	55	3	4	2	3	1	1	1
E93	1	29	20.5	5	2	1	1	4	4	5	1	4	4	4	5	4	5	5	70	1	1	2	2	1	1	1
He7	4	80.5	19.5	7.5	2	1	2	5	2	11	5	4	4	4	5	5	5	4	45	3	1	1	?	?	4	1
E75	1	46.5	19	5	2	1	1	5	9	10.5	2	3	4	4	4	4	5	5	75	1	1	2	3	1	2	14
HA24	1	70	15.5	5.5	2	1	1	5	2	7	2.5	3	5	4	5	4	5	5	80	2	1	2	3	1	1	1
E112	1	48.5	21.5	6	2	2	1	5	2	7	3	2	5	4	4	5	5	4	60	2	2	0	2	1	1	17
E55	1	38	11.5	4	2	1	1	4	4	3	1	3	5	4	4	5	5	5	75	2	2	0	2	1	1	17
E37	1	71	27	7.5	1	1	2	4	4	4	1	8	4	4	5	5	5	5	70	3	8	0	3	1	1	1
E36	1	38	13.5	3	2	1	1	5	2	5.5	1.5	4	4	5	4	5	5	4	50	2	1	1	3	1	1	1
He27	1	18.5	22	4	2	1	1	5	2	2.5	1.5	3	5	4	4	5	5	5	85	2	1	0	2	1	1	1
He27	1	66	16	7	2	1	1	5	5	5.5	3.5	3	5	4	5	4	5	5	60	3	1	2	2	1	2	1
He13	1	45	14	4	1	1	1	4	2	3	1.5	3	5	4	4	5	5	5	80	2	4	0	2	1	2	1
He18	1	46	19	7	2	1	1	5	2	7.5	3	8	4	7	5	5	5	5	70	1	1	0	2	1	1	1
He18	1	45	14	5	2	1	1	5	2	2	2	4	4	4	5	5	5	6	40	1	1	1	2	1	1	2
He18	1	28	15	3	2	1	1	5	5	3.5	2.5	2	5	4	4	5	5	6	55	2	1	1	3	1	3	1
He18	1	37	9	2.5	1	1	1	4	2	1.5	0.5	3	5	4	5	5	5	5	70	2	4	0	2	1	1	1
HC8H42-3	1	34.5	15	5	2	1	1	5	5	10	5	2	5	4	5	5	5	4	60	2	1	1	4	1	1	1
HA 0	1	41	12	3.5	1	1	1	4	2	4	1.5	3	5	4	4	4	5	5	75	2	4	2	2	1	1	1
HA 0	1	25	10.5	2.5	2	1	1	5	2	7	2	3	4	4	5	4	5	4	60	2	1	0	3	1	1	1
HA 0	1	25	14	3	2	1	1	5	7	8	3	3	5	7	5	4	4	5	75	2	1	2	2	1	3	1
HA22..	1	47	16.5	3	1	1	1	4	3	1	1	4	5	4	4	4	5	5	80	2	5	2	3	1	1	1
HA22..	1	48	13.5	3	1	1	1	5	2	5.5	2	4	5	4	5	5	5	5	75	1	5	0	2	1	2	1
HA22..	1	43.5	13	3.5	1	1	1	4	5	4	2	3	5	4	4	4	5	4	60	2	4	0	4	?	1	1
HA22..	1	23	10.5	4	2	1	1	5	2	2.5	1	4	6	7	5	5	5	5	80	2	1	0	3	1	1	1
E94	1	51	15	5	2	1	2	4	2	2.5	1	3	6	4	4	4	5	5	85	2	2	2	1	1	1	17
Hs	1	61	22	5	2	1	1	5	2	9	3.5	2	4	4	5	5	5	4	40	2	1	1	3	?	1	1
HD8	1	84	19	6	2	1	1	5	5	6	3	3	4	7	5	5	5	4	30	2	1	1	3	1	1	1
E36	1	43.5	7	2.5	2	2	1	5	2	5	2	3	5	4	5	4	5	5	60	1	1	2	3	1	2	13
E51	1	36.5	7.5	2.5	2	2	1	4	4	3	1	3	5	4	5	5	5	5	80	2	1	0	2	1	1	13
HE20	1	21	6	1	2	2	1	4	4	2.5	1	4	5	4	5	5	5	5	70	2	1	0	2	1	1	13
E96	1	23	6	2	1	2	1	5	2	2	1	4	5	4	5	5	5	5	80	2	5	0	3	1	4	13
HA7	1	19	4.5	1	2	2	1	4	2	2	1	3	5	4	5	5	5	5	70	2	1	0	2	1	1	13
E90	4	48	7.5	2.5	1	2	1	4	2	2.5	1	3	7	4	4	4	5	5	55	2	4	2	?	?	2	13
E94	1	43	9	3	1	2	1	5	4	5.5	1.5	4	6	4	5	5	5	5	60	1	4	0	2	1	1	13
HE15	1	39	8.5	3.5	1	2	1	4	7	4.5	1.5	8	5	4	4	5	5	5	70	1	4	0	2	1	1	13
E54	1	48	7	4	1	2	1	4	7	3	1	3	5	6	5	5	5	5	80	1	4	0	1	1	1	13
E111	1	46.5	8.5	3	1	2	1	4	4	5	1	3	4	4	5	5	5	5	75	1	4	0	2	1	1	13
HE19	1	32	7	4	1	2	1	4	5	4.5	2	3	6	7	5	5	5	5	65	1	4	0	2	1	1	13

(continued)

Table A.19 (continued)

Nr.	Blank	Length	Width	Thickness	Pres.: L	Pres.: W/T	Cortex	SPR: kind	SPR: form	SPR: length	SPR: width	DR	Lip	Bulb	Bulbar scar	Impact point	Esq. du bulbe	en éperon	Flaking angle	Profile	Distal	Hammer	n Negatives	Direc. Neg.	Reg Neg.	Tool
HB4	1	20	5.5	2	2	2	1	4	5	2	1.5	8	4	7	5	5	5	5	50	2	1	0	2	1	1	13
C6H44-5	1	24	7	3	2	2	1	5	4	4	1	8	5	4	5	5	5	5	75	2	1	0	3	1	1	13
G2s	1	22	5	2.5	1	2	1	4	5	2.5	1	4	5	7	5	5	5	4	80	2	4	0	2	1	1	13
A36H50-1	1	22	7	2	2	2	1	4	4	2	1	8	5	4	4	5	5	5	85	2	1	0	2	1	1	13
Hs	1	36	18	3.5	2	2	1	5	2	4.5	2	3	5	4	6	5	5	5	85	2	1	0	1	?	1	13
Hs	1	16	9	3	2	2	1	4	6	3	2	8	4	4	5	5	5	5	60	2	1	0	2	1	1	13
Hs	1	21	6	2	2	2	1	5	4	3	0.5	8	5	4	5	5	4	5	70	2	1	2	2	1	1	13
E111	1	27.5	9	3	2	1	1	4	2	4	1.5	3	4	4	5	5	5	5	50	2	1	0	2	1	1	1
E94	1	26	6.5	1.5	2	2	1	4	3	1	1	2	5	4	4	5	5	5	75	2	1	0	4	?	4	13
HE28	1	20	8	3	2	1	1	4	4	3	1	3	5	4	5	5	5	5	75	2	1	0	2	1	1	1
HA29	1	27	6.5	3	2	2	1	4	2	2	1	4	4	4	5	4	5	5	80	2	1	2	2	1	1	13
E10	1	18.5	5	1.5	2	2	1	4	9	2.5	1	4	5	4	4	5	5	5	80	2	1	0	2	1	1	13

Feature explanations and keys for numbers see Chap. 2, Blade Technology, Recording of Technological Information, nd: feature expression not recordable

Table A.20 Blade technology analysis—Hranice

Nr.	Blank	Length	Width	Thickness	Pres.: L	Pres.: W/T	Cortex	SPR: kind	SPR: form	SPR: length	SPR: width	DR	Lip	Bulb	Bulbar scar	Impact point	Esq. du bulbe	en éperon	Flaking angle	Profile	Distal	Hammer	n Negatives	Direc. Neg.	Reg Neg.	Tool
92687	1	19	13	5	2	1	1	4	2	6.5	3	4	4	4	5	4	5	5	75	1	1	2	2	1	1	1
92626	1	22	17	4	2	1	1	5	7	5.5	3	4	5	7	5	5	4	5	70	2	1	2	2	1	1	1
92654	1	34	18	5	2	1	1	5	2	11.5	5	8	5	4	4	4	5	5	70	2	1	2	2	1	1	1
92652	1	22	10	4.5	2	1	1	5	7	4	3	8	5	4	5	4	4	5	70	2	1	2	2	1	1	1
92644	4	41	17	7	1	1	1	4	2	4	3	4	5	4	4	5	5	5	65	2	5	0	4	4	2	1
92673	1	36.5	25	7.5	2	1	1	5	4	25	3	8	5	4	4	5	5	5	85	2	1	0	2	1	1	1
93802	1	22.5	20	5	2	1	1	5	2	10	4	4	4	7	5	5	5	5	80	2	1	0	3	1	1	1
93801	1	23	17	5	2	1	1	5	2	10	3.5	4	5	4	5	4	5	5	75	2	1	2	2	1	1	1
92662	1	33	15	7.5	2	1	1	5	7	6.5	3	4	6	7	4	4	5	5	70	2	1	2	2	1	1	1
92684	1	25	12	4	2	1	1	5	2	6.5	4	4	4	5	4	4	5	5	40	2	1	2	2	1	1	1
92653	1	31	20	4	2	1	1	5	2	9	3	3	4	4	5	4	5	5	60	1	1	2	2	1	1	1
92665	1	29	14	3	2	1	1	5	9	5	2	2	7	4	4	4	5	5	75	2	1	2	2	1	2	1
92656	1	34	17	5	2	1	2	4	2	4.5	2	4	6	4	5	4	4	5	60	2	1	2	3	3	1	1
92666	1	19	15	5	2	1	1	5	2	5	1.5	4	4	4	5	5	5	5	75	2	1	0	2	1	1	1
92672	1	54	21	6	1	1	1	4	2	8	3	8	4	4	4	4	5	5	65	2	4	2	3	1	2	1
93813	1	20.5	6	2	2	1	1	4	4	3	1	4	6	4	5	6	5	5	80	2	1	2	2	1	1	1
93803	1	26	13	4	2	1	1	4	4	5	1	8	4	4	5	4	4	5	50	2	1	2	3	1	1	1
92645	1	35	13	3	1	1	1	5	4	5	0.5	8	5	7	5	4	4	5	75	2	5	2	4	?	2	1
92633	1	30	15.5	6.5	2	1	1	5	2	4.5	2	8	5	5	4	4	5	5	60	2	1	2	2	1	1	1
92657	1	17.5	9.5	2	2	1	1	5	2	6.5	2.5	8	5	7	5	4	5	5	60	1	1	2	2	1	1	1
92683	1	24	12.5	3	2	1	1	5	2	7	2.5	4	5	4	5	5	5	5	70	2	1	0	2	1	1	1
92628	1	22.5	22	4.5	2	1	1	4	2	8	3	4	5	4	6	5	5	5	80	?	1	0	3	1	2	1
92647	1	20	9.5	3.5	2	1	1	4	5	2	1	8	5	4	5	4	5	5	85	2	1	2	2	1	1	1
92704	1	24.5	18	6	2	1	1	4	2	7	3.5	8	4	7	5	4	5	5	70	2	1	2	2	1	1	1
93118	1	17.5	11	3	2	1	1	5	2	6	2.5	8	4	4	4	5	5	5	65	?	1	0	2	1	1	1
92466	1	24.5	15	4.5	2	2	1	5	2	8	4	8	4	5	5	4	5	5	65	1	1	2	2	1	3	14
92372	1	44	24	9	2	2	1	5	2	12.5	5	3	4	7	5	4	5	5	60	1	2	2	2	1	1	5
92480	4	26.4	14.5	7	2	1	1	5	5	10	5.5	2	5	4	4	4	5	4	60	1	1	2	?	?	1	14
93754	1	19.5	20	4	2	1	1	5	2	7	3	3	4	7	5	4	5	5	80	?	1	2	2	1	3	14
92478	1	25	516	5	2	1	1	5	2	8	3	4	6	4	4	4	5	5	65	2	1	2	2	1	1	14
92477	1	26.5	13.5	4	2	1	1	5	2	8	4	8	4	7	5	5	5	5	65	2	1	0	2	1	1	14
92474	1	20	12.5	4	2	1	1	5	2	6.5	2	8	4	4	5	5	5	5	50	1	1	0	2	1	1	14
92716	1	19	9	3	2	1	1	4	2	3	1	8	5	4	4	5	5	5	60	2	1	0	3	1	3	1
92737	1	19.5	7.5	2.5	2	1	1	5	4	4	1	8	5	4	5	4	4	5	80	1	1	2	2	1	1	1
92746	1	15	11	2.5	2	1	1	5	2	5	1.5	3	5	4	5	5	5	5	75	2	1	0	2	1	1	1
92727	1	19	8	2	2	1	1	5	7	4	2	8	4	4	4	4	5	5	80	1	1	2	3	1	3	1
92728	1	20	8	2	2	1	1	4	2	3	1.5	4	5	7	5	4	4	5	60	2	5	2	2	1	2	1
92723	1	18.5	7	2	2	1	1	5	4	5	0.5	4	4	7	5	4	4	5	70	2	1	2	2	1	1	1
92724	1	24	8.5	2.5	2	1	1	5	2	5	2	3	5	4	5	5	5	5	60	1	1	0	2	1	1	1
92259	1	46	21.5	10	2	2	2	5	2	13	5.5	8	4	7	4	4	5	5	70	2	2	2	3	1	3	5
92277	1	53	23.5	8.5	2	1	1	5	2	15	6	4	4	7	4	4	5	5	60	1	2	2	3	1	1	5
92244	1	55.5	23	10	2	1	1	5	2	6.5	2.5	6	4	7	5	4	5	5	75	2	2	2	1	1	1	5
92292	1	38.5	15	5	2	1	1	5	2	9	3	4	4	4	5	4	5	5	65	2	2	2	3	1	1	5
92900	1	39.5	17	5.5	2	1	1	5	2	14	5	4	5	4	5	5	5	5	70	1	2	2	2	1	1	5

(continued)

Table A.20 (continued)

Nr.	Blank	Length	Width	Thickness	Pres.: L	Pres.: W/T	Cortex	SPR: kind	SPR: form	SPR: length	SPR: width	DR	Lip	Bulb	Bulbar scar	Impact point	Esq. du bulbe	en éperon	Flaking angle	Profile	Distal	Hammer	n Negatives	Direc. Neg.	Reg Neg.	Tool
92274	1	30	21	5	2	1	1	5	7	10	3	3	6	4	4	5	5	5	85	1	2	0	2	1	2	5
92256	1	26.5	13	4.5	2	1	1	5	2	8	3	4	4	4	4	5	5	5	50	2	2	0	3	2	1	5
92275	1	25.5	16	3	2	1	1	5	4	5	1	4	5	4	5	4	4	5	85	2	2	2	2	1	1	5
92596	1	43	21.5	6	2	2	2	4	2	5	2	4	5	7	4	5	5	5	70	2	1	0	1	1	1	14
92575	1	27.5	20.5	7	2	1	1	5	2	8	3	4	5	4	4	4	5	5	85	2	1	2	3	1	1	14
92589	1	15.5	12	5.5	2	1	1	5	5	4	2	4	4	7	5	5	5	5	70	?	1	0	2	1	1	14
92322	1	36.5	21.5	7	2	1	1	5	4	14	3	4	6	7	5	4	5	5	70	1	2	2	2	1	1	5
92329	1	31.5	22	5	2	2	1	5	4	8	1.5	8	6	4	5	4	5	5	85	2	1	1	3	1	1	5
92549	1	33	14	6	2	2	1	5	5	7.5	3.5	3	4	7	5	5	5	4	65	1	2	0	2	1	1	5
92545	1	49.5	20	6	2	2	1	5	2	7	2.5	3	5	7	5	5	5	5	50	2	2	0	2	1	1	5
92405	1	31.5	17	5	2	1	1	4	2	2	1	8	4	4	5	4	5	5	70	2	2	2	2	1	1	4
92409	1	23.5	20	5	2	1	1	4	7	12	5	8	4	4	4	4	5	5	65	2	2	2	2	1	2	17
93791	1	16	13.5	3.5	2	1	1	5	4	3.5	0.5	4	5	7	5	5	4	5	80	?	1	2	?	?	?	1
92794	1	46	18	5.5	2	1	3	5	2	4.5	3	8	5	7	5	5	4	5	85	2	1	2	2	1	3	1
92788	1	27.5	15	4	2	1	3	5	2	5	2	8	4	7	5	4	5	5	70	2	1	2	1	1	1	1
92766	1	24	16	3	2	1	2	5	4	8	2	4	4	7	5	4	5	5	70	1	1	2	2	1	1	1
92771	1	23	16	4	2	1	1	5	7	11	4.5	4	5	4	5	4	5	5	80	1	1	2	3	1	1	1
92762	1	25	15	5	2	1	1	4	9	5.5	2	4	5	4	4	4	5	5	70	2	1	2	2	1	2	1
92763	1	18	18	4	2	1	2	5	9	11	4	4	5	4	4	4	5	5	80	?	1	2	1	1	1	1
92761	1	21	9	5	2	1	2	4	9	3	1	4	4	4	4	4	5	5	70	2	1	2	2	1	1	1
92823	1	22.5	10	9	2	1	2	5	2	3	1.5	4	5	4	4	5	5	5	85	2	1	0	2	1	1	1
92604	1	43	23	6	2	1	1	5	5	11	7	4	5	7	5	4	5	4	30	2	1	0	3	1	4	1
92609	1	41	15	6.5	2	1	1	5	9	8.5	3	8	4	7	5	4	5	5	70	2	1	2	2	1	1	1
93776	1	29.5	16	6	2	1	1	4	2	10	7	8	5	7	4	4	4	4	70	2	1	2	2	1	1	1
93777	1	41	18	6	2	1	2	5	2	8	4	8	4	7	5	4	5	5	60	2	2	2	2	1	1	4
92613	1	38	22.5	7	2	1	1	5	5	5	8	3	5	7	6	4	5	5	80	2	1	2	3	1	2	14
92610	1	49	22	8	1	1	2	5	2	8.5	4	4	7	4	4	5	5	5	85	2	8	0	4	1	4	1
92377	1	34	21.5	8	2	2	1	4	7	15	3	4	5	4	4	4	5	5	70	2	1	2	2	1	1	5
93010	1	18.5	17.5	3	2	1	1	5	9	15	2	4	5	4	5	5	5	5	75	2	1	0	1	1	1	1
92999	1	22	21	5	2	1	1	4	7	4	2.5	4	4	4	5	4	5	5	80	2	1	2	4	4	2	1
93006	1	15	20	4.5	2	1	1	5	2	9.5	3	4	4	4	4	4	5	5	75	2	1	2	3	1	1	1
93014	1	25.5	22	7	2	1	1	5	4	10	1.5	4	5	4	5	4	5	5	75	2	1	2	2	1	2	1
92989	1	20.5	21.5	4	2	1	1	5	4	15	4	3	5	4	4	5	5	5	75	2	1	0	2	1	1	1
93008	1	16.5	21.5	5	2	1	1	4	2	7	3	8	5	4	4	4	5	5	70	2	1	2	2	1	1	1
93004	1	21	17	4	2	1	1	4	7	6	4	4	5	4	4	5	5	5	80	?	1	0	2	1	1	1
92545	1	48	20	6	2	2	1	5	2	7	2.5	3	4	7	5	5	5	5	50	2	2	0	2	1	1	5
92549	1	34	14	6	2	2	1	5	2	7.5	3	3	4	7	5	4	5	5	70	1	2	2	2	1	1	5
92344	1	45	17	8	2	1	1	4	2	2	1	8	5	7	5	5	5	5	70	2	1	0	2	1	1	1

Feature explanations and keys for numbers see Chap. 2, Blade Technology, Recording of Technological Information, nd: feature expression not recordable

Table A.21 Blade technology analysis—Kastelhöhle Nord

Nr.	Blank	Length	Width	Thickness	Pres.: L	Pres.: W/T	Cortex	SPR: kind	SPR: form	SPR: length	SPR: width	DR	Lip	Bulb	Bulbar scar	Impact point	Esq. du bulbe	en éperon	Flaking angle	Profile	Distal	Hammer	n Negatives	Direc. Neg.	Reg Neg.	Tool
1848	1	42	11	4	1	1	3	5	5	10	6	3	4	5	5	5	5	4	40	1	4	1	1	1	1	1
1918	1	31	14	3	2	1	1	4	2	10	4	3	4	4	5	5	5	4	65	1	1	1	3	1	1	1
2097	1	41	17	6	1	1	1	5	7	17.5	4	4	5	5	5	5	5	6	80	1	5	0	7	5	4	1
1981	1	33	15	4	2	1	1	4	5	11.5	3.5	4	4	7	5	5	5	6	45	2	1	0	2	1	1	1
2059	4	46	20	7.5	2	1	1	5	5	3.5	2.5	4	4	7	4	5	5	4	55	2	1	0	?	?	4	1
1909	1	60	18	5.5	2	1	1	5	2	12	5	8	5	7	5	5	5	6	55	2	1	1	4	1	2	1
2080	1	37	11	2	2	1	1	5	5	7	3	2	4	5	5	5	5	4	60	2	1	0	3	1	2	1
2171	4	41	19	5.5	2	1	2	4	5	3.5	2	4	5	5	5	5	5	6	60	2	1	0	2	1	2	1
1926	4	38	15	5	2	1	1	5	7	4.5	2	8	4	7	5	5	5	6	60	2	1	0	?	?	4	1
1885	1	47.5	14	3	2	1	2	4	7	7	2.5	4	4	4	5	5	5	6	65	2	1	0	3	1	1	1
1919	1	51	18.5	5	2	1	1	5	2	11.5	3	8	4	4	5	5	5	4	65	2	1	0	3	1	2	1
1154	1	30.5	7.5	3	1	2	1	5	5	3	1.5	3	4	4	5	5	5	6	65	2	6	0	1	1	1	13
1891	1	33.5	12	2.5	2	1	1	4	7	3.5	2	8	4	7	5	5	5	6	65	2	1	0	2	1	1	1
2044	1	21	11.5	2	2	1	1	5	7	6	2	8	4	7	5	5	5	6	65	2	1	0	2	1	1	1
2030	1	33	16	3.5	2	1	1	4	2	2	1	8	4	7	4	5	5	6	70	2	1	0	3	1	1	1
1961	1	37.5	16.5	3	2	1	1	4	2	5	1.5	4	4	4	5	5	5	6	70	2	1	0	3	1	2	1
1974	1	30	14	5	2	1	2	4	5	3	2.5	4	4	4	5	5	5	6	70	2	1	0	2	1	1	1
1948	1	41	13	4.5	2	1	1	4	7	3.5	1.5	8	4	4	5	5	5	6	70	2	1	0	2	1	1	1
845	1	76	20.5	6	2	1	1	5	5	7	4	4	4	7	5	5	5	6	70	2	2	0	3	1	2	17
1883	1	45	21	3	2	1	1	5	7	3	1	4	4	7	5	5	5	6	70	2	1	0	3	1	1	1
1874	1	46.5	11	4	2	1	1	5	7	6	3	4	4	7	5	5	5	6	70	2	1	0	3	1	1	1
1988	1	34	15.5	4	2	1	1	5	5	5	2	8	4	7	5	5	5	6	70	2	1	0	3	1	1	1
1989	1	35	13	3	2	1	1	5	5	6	3	8	4	7	5	5	5	6	70	2	1	1	4	1	2	1
818	1	51	16	4	2	1	2	5	2	7.5	3.5	3	4	4	4	5	5	6	75	2	2	0	1	1	1	4
1939	1	43.5	21.5	9	2	1	3	5	5	7	5	4	4	7	4	5	5	6	75	2	1	1	1	1	1	1
1951	1	37	13.5	3	2	1	1	4	2	10	5	8	4	4	5	5	5	6	75	2	1	0	3	1	2	1
1935	1	35	12.5	2	2	1	1	5	5	4	2	8	4	4	5	5	5	6	75	2	1	0	2	1	1	1
627	1	71	24	6	1	1	1	4	2	5.5	2.5	4	4	7	5	5	5	6	80	2	5	0	4	1	2	1
848	1	47	10	5.5	2	1	3	5	5	7	4	4	4	7	5	5	5	6	80	2	1	0	1	1	1	1
1876	1	36.5	17.5	3	2	1	1	2	7	6	2	4	4	7	5	5	5	6	80	2	1	0	4	3	4	1
623	1	50.5	15.5	8	2	1	2	5	5	7.5	4.5	3	5	5	5	5	5	6	70	3	2	0	3	4	2	4
901	1	60	16.5	5	2	1	1	5	2	5	1.5	3	4	4	4	5	5	6	75	3	2	0	2	1	1	4
2004	1	36.5	12.5	3.5	2	1	1	5	7	5	1	4	4	7	5	5	5	5	30	1	1	0	3	1	2	1
1892	1	29	10	2.5	2	1	1	4	2	4	1.5	4	4	4	5	5	5	5	60	1	1	0	2	1	1	1
1880	1	41	12	5	2	1	1	5	5	6.5	5	8	4	7	5	5	5	5	70	1	1	0	3	1	2	1
1957	1	38	11.5	5	2	1	1	4	7	5	1	4	4	4	6	5	5	5	70	1	1	0	3	3	2	1
1889	1	38	12	4	2	1	1	5	7	9	1.5	4	4	4	5	4	5	5	75	1	1	2	2	1	1	1
2024	1	42.5	11.5	4	2	1	1	5	7	6	1	4	6	7	4	5	5	5	75	1	1	0	4	1	1	1
1987	1	33	16	5	2	1	1	5	9	12	4.5	8	4	4	5	4	5	5	80	1	1	0	3	1	1	1
2011	1	45.5	20	3	2	1	1	5	7	13	3	8	5	4	5	4	5	5	80	1	1	2	3	1	2	1
1936	1	25	10	2	2	1	1	4	7	4	1.5	4	4	4	5	5	5	5	80	1	1	0	2	1	1	1
2016	1	45	17	6	2	1	1	4	2	7	2	4	4	4	5	5	5	5	50	2	1	0	2	1	1	1
592	1	59	13.5	5	2	1	1	5	7	5	2	8	4	7	5	5	4	5	55	2	2	0	2	1	1	2
1968	1	24	13.5	5	2	1	1	5	2	6.5	2	3	4	4	5	5	5	5	55	2	1	0	3	1	1	1

(continued)

Table A.21 (continued)

Nr.	Blank	Length	Width	Thickness	Pres.: L	Pres.: W/T	Cortex	SPR: kind	SPR: form	SPR: length	SPR: width	DR	Lip	Bulb	Bulbar scar	Impact point	Esq. du bulbe	en éperon	Flaking angle	Profile	Distal	Hammer	n Negatives	Direc. Neg.	Reg Neg.	Tool
1953	1	36	19	5	2	1	2	4	2	4	1.5	4	4	7	5	5	5	5	55	2	1	0	2	1	2	1
1933	1	43	13	3	2	1	1	4	2	2	1	4	4	4	5	5	5	5	60	2	1	0	3	1	2	1
1216	1	23.5	6	1.5	2	2	1	4	2	3	1	8	4	4	5	5	5	5	60	2	1	0	1	1	1	13
823	1	53	18.5	7	2	1	2	4	2	5	2	4	4	7	5	5	5	5	60	2	2	0	2	4	1	1
1975	1	38.5	20	6	2	1	1	5	9	10	2	8	4	7	5	5	5	5	60	2	1	0	2	1	1	1
2043	1	41	24	5	2	1	1	5	2	10	3	4	4	4	5	4	5	5	65	2	1	2	3	3	2	1
1875	1	43	23	7	2	1	1	5	2	7.5	2	8	4	7	4	5	5	5	65	2	1	0	2	1	1	1
1990	1	30	20.5	3.5	2	1	1	4	2	4	2	4	4	4	5	5	5	5	65	2	1	0	3	1	1	1
1955	1	43	17.5	6	1	1	1	4	5	2	1	4	4	4	5	5	5	5	65	2	1	0	5	5	3	1
1897	1	27	13.5	3.5	2	1	1	5	7	9	3	4	4	4	5	5	5	5	65	2	1	0	3	1	1	1
751	1	43	16	8	2	1	2	4	7	5	2	4	4	7	5	5	5	5	65	2	2	0	3	1	1	5
1923	1	31	10	4	2	1	1	4	2	1.5	1	4	4	4	4	4	5	5	70	2	1	0	3	1	2	1
1941	1	42	19	5.5	2	1	2	4	7	10	2	4	4	7	4	4	5	5	70	2	1	0	3	1	1	1
2012	1	40	15	6	2	1	1	4	2	5	2	4	4	4	5	4	5	5	70	2	1	0	2	1	1	1
1924	1	35.5	16	5	2	1	1	4	2	2	1	4	4	4	4	5	5	5	70	2	1	0	2	1	3	1
729	1	49	14	4	2	1	1	5	7	7	1	4	4	7	4	5	5	5	70	2	2	0	2	1	1	17
2025	1	22.5	14.5	3	2	1	1	4	9	5	2	4	4	7	4	5	5	5	70	2	1	0	3	1	1	1
2039	1	39	21	7	2	1	1	4	2	8	4	4	5	7	4	5	5	5	70	2	1	0	3	?	4	1
1901	1	36	17.5	3.5	2	1	1	5	4	6	1.5	4	4	4	5	5	5	5	70	2	1	0	4	1	3	1
1899	1	32	22	5	2	1	2	4	2	4	2	8	4	4	5	5	5	5	70	2	1	0	2	1	2	1
2161	1	41.5	12.5	3	2	1	1	5	7	6	1	8	4	7	5	5	5	5	70	2	1	0	3	3	1	1
1908	1	40	12	4	2	1	1	5	7	7.5	1.5	8	5	7	5	5	5	5	70	2	1	2	2	1	1	1
1963	1	33	13	3.5	2	1	1	5	7	10.5	2	4	5	7	5	4	4	5	75	2	1	2	3	3	4	1
847	1	61	9	5	2	1	3	5	2	4	1.5	4	4	4	5	5	4	5	75	2	2	0	1	1	1	17
620	1	39	19	5.5	2	1	1	4	2	3.5	1.5	4	4	4	5	4	5	5	75	2	2	0	2	1	1	2
615	1	48.5	20.5	7	2	1	1	4	2	3.5	1.5	8	4	4	4	5	5	5	75	2	2	0	6	1	4	2
2035	1	55	14	4.5	2	1	1	2	7	9	2	4	4	7	4	5	5	5	75	2	1	0	3	?	4	1
2037	1	32	20	5	2	1	2	4	2	5	3	2	4	7	4	5	5	5	75	2	1	0	4	?	4	1
1878	1	59.5	18.5	4.5	2	1	1	4	7	5	1.5	3	4	7	4	5	5	5	75	2	1	0	4	3	3	1
817	1	55	14	5	2	1	1	4	2	3	1	4	4	4	5	5	5	5	75	2	2	0	4	1	1	17
2038	1	33	18	3	2	1	1	5	7	6	1	4	4	4	5	5	5	5	75	2	1	0	3	?	3	1
1157	1	20.5	7.5	2	2	1	1	4	2	5.5	2	3	4	4	5	5	5	5	75	2	1	0	2	1	1	13
2018	1	40	15	4	2	1	3	5	5	6.5	3	4	4	7	5	5	5	5	75	2	1	0	1	1	1	1
1846	1	35	13.5	4	1	1	1	4	7	6	2	4	4	7	5	5	5	5	75	2	4	0	3	1	1	1
1964	1	36	19	4	2	1	1	4	2	5	2.5	4	4	4	5	4	5	5	80	2	1	2	2	1	2	1
2021	1	33.5	11.5	3	2	1	1	4	4	2	1	4	4	4	5	4	5	5	80	2	1	2	2	2	1	1
924	1	46	20	5	2	1	1	5	2	7.5	3	4	4	4	4	5	5	5	80	2	2	0	3	1	1	4
1954	1	52	15	6.5	2	1	1	2	4	5	1	4	4	4	4	5	5	5	80	2	1	0	2	1	1	1
1894	1	30	12	2.5	2	1	1	4	2	2.5	1	8	4	4	4	5	5	5	80	2	1	0	2	1	1	1
752	4	71	20	7	2	1	1	5	7	7	2	2	4	4	4	5	5	5	80	2	2	0	?	?	4	5
1903	1	40	16.5	8	2	1	2	5	2	4.5	2.5	8	4	7	4	5	5	5	80	2	1	0	3	3	3	1

1905	1	32	12.5	3	2	1	1	4	5	1.5	1	4	4	4	5	5	5	5	80	2	1	0	2	1	1	1
749	1	55	15	7	2	1	1	4	2	4	2	8	4	4	5	5	5	5	80	2	2	0	4	1	1	5
1965	1	30	23	5	2	1	1	4	3	1.5	1.5	4	4	7	5	5	5	5	80	2	1	0	2	1	1	1
1945	1	23.5	12	4	2	1	1	5	2	8.5	4	3	4	4	6	5	5	5	80	2	1	0	3	4	4	1
1845	1	41	9.5	4.5	1	1	1	4	2	4	2	4	5	5	6	5	5	5	80	2	4	0	6	4	4	1
1931	1	23	15.5	4.5	2	1	1	4	9	3	2	4	4	4	5	4	5	5	85	2	1	0	2	1	1	1
1900	1	34	15	3.5	2	1	1	5	7	11	2.5	8	4	4	4	5	5	5	85	2	1	0	3	?	1	1
1971	1	31.5	10	3	2	1	1	4	2	3	1	2	4	4	5	5	5	5	85	2	1	0	2	1	1	1
1952	1	32	12	4	2	1	1	4	7	3	1	8	4	7	5	5	4	5	70	3	1	0	4	?	3	1
928	1	26	16.5	4.5	2	1	1	5	7	6.5	2	4	4	5	5	5	5	5	55	?	2	0	2	1	2	4
2020	1	37	22	5	2	1	1	4	2	15	5	4	4	4	5	5	5	5	80	?	1	0	3	1	2	1
1031	1	21	7	3	2	2	1	5	5	3	3.5	2	4	7	5	5	5	6	65	2	1	0	3	?	2	13
2177	1	50	12	3	2	1	1	2	5	10	4	2	5	5	5	5	5	6	80	2	1	0	2	1	1	1
2042	4	45	14	5	2	1	1	4	5	4.5	3	4	5	5	4	5	5	6	60	3	1	0	5	4	4	1
1887	1	41	13	3.5	2	1	1	5	7	8	1.5	4	4	4	5	5	5	?	85	2	1	0	4	1	2	1

Feature explanations and keys for numbers see Chap. 2, Blade Technology, Recording of Technological Information, nd: feature expression not recordable

Table A.22 Blade technology analysis—Kaufertsberg, upper level

Nr.	Blank	Length	Width	Thickness	Pres.: L	Pres.: W/T	Cortex	SPR: kind	SPR: form	SPR: length	SPR: width	DR	Lip	Bulb	Bulbar scar	Impact point	Esq. du bulbe	en éperon	Flaking angle	Profile	Distal	Hammer	n Negatives	Direc. Neg.	Reg Neg.	Tool
845	1	51	19	5	2	1	1	5	2	9	3	4	4	7	5	5	5	4	70	2	1	1	3	1	1	5
858	1	39	26	4	2	1	1	5	7	1	4	2	4	4	5	5	5	5	85	1	1	0	4	1	4	5
848	1	36	16	8	2	1	1	5	7	8	4	2	4	5	5	5	5	5	120	2	1	0	5	3	4	5
866	1	29	8	2.5	1	1	1	4	4	3	1	8	5	5	5	5	4	5	80	2	6	2	3	1	4	1
852	1	30	12	2	2	2	1	4	7	10	3	4	4	4	4	4	5	5	75	1	1	2	2	1	1	9
853	1	29	11	4	2	2	1	4	2	3	1.5	4	6	7	4	5	4	5	75	1	1	2	2	1	3	9
857	1	25	7.5	1.5	2	2	1	4	4	1	0.5	4	4	7	5	5	5	5	85	2	1	0	4	1	4	9
873	1	46	12.5	4	1	1	1	4	3	0.5	0.5	4	6	7	5	5	5	5	nd	1	4	0	2	1	1	1
872	1	59	29	10	2	1	2	4	4	13	3	8	4	4	5	4	5	5	85	1	1	2	3	1	1	1
875	1	38	16	3	1	1	1	4	4	2	1	8	4	4	5	5	5	5	80	1	4	0	3	1	4	1
780	1	50	20	5	2	1	1	5	2	9	4	4	4	7	5	5	5	5	60	1	1	0	2	1	1	1
786	1	51	12	5	2	1	1	4	6	4	2	3	4	7	5	5	5	5	70	2	1	0	4	1	1	1
782	1	63	15	5	1	1	1	4	2	6.5	2	8	4	7	5	5	5	5	70	2	4	0	3	1	2	1
804	1	41	25	5	2	1	1	5	2	7	2.5	4	4	4	4	5	5	4	65	2	1	0	3	1	3	1
788	1	46	25	6	2	1	1	5	2	10	4	2	4	7	6	5	5	5	65	2	1	0	5	1	3	1
787	1	51	16	6	2	1	1	4	2	8	3	4	4	7	5	5	5	5	45	2	1	0	3	4	4	1
789	1	42	19	5	1	1	1	5	2	9	2	4	4	7	5	5	5	5	75	1	6	0	4	1	2	1
806	1	69	19	11	1	1	1	4	2	2.5	1	4	4	7	5	4	5	5	60	2	8	2	?	?	?	1
783	1	60	30	6	2	1	1	5	9	19	4	2	4	7	5	5	5	5	65	2	1	0	3	1	1	1
801	1	49	16	6	2	1	1	4	3	1	1	4	4	4	5	5	5	5	65	2	1	0	3	1	1	1
808	1	41	28	8	2	1	1	6	8	0	0	4	1	1	1	1	1	1	85	2	1	2	3	1	4	1
4665 i	1	45	22	6	2	1	1	5	5	7	4	4	4	4	5	5	5	4	70	2	1	0	2	1	1	1
4665 i	1	25	14	3	2	1	1	5	2	7	3	4	4	4	4	5	5	5	60	2	1	0	2	1	1	1
4665 i	1	17	11	4	2	1	1	5	2	4	2	4	4	7	4	5	5	5	75	2	1	0	2	1	1	1
4665 i	1	51	14	3	2	1	1	4	9	5	1.5	4	4	7	4	5	5	5	75	2	1	0	3	1	1	1
4665 i	1	58	13	4	2	1	1	4	9	3	1	4	4	7	5	5	5	5	80	2	1	0	3	1	1	14
4665 i	1	52	19	5	1	1	1	5	2	6	3	4	4	7	5	4	5	5	85	1	8	2	2	1	1	1
4665 i	1	26	10	2.5	2	1	1	5	7	5	2	2	4	7	5	5	4	5	75	1	1	0	3	1	1	1
4665 i	1	32	25	4	2	1	1	4	4	8	2	4	4	7	5	5	5	5	45	2	1	0	2	1	1	1
4665 i	1	39	11.5	2.5	1	1	1	4	2	4	1.5	4	4	7	5	5	5	5	50	2	5	0	3	1	3	1
4665 i	1	46	13	4	1	1	1	5	2	5	2.5	3	4	7	4	5	5	5	70	2	4	0	2	1	3	1
4665 i	1	34	12	4	2	1	1	4	7	6.5	4.5	2	4	7	nd	5	nd	6	85	2	5	2	2	1	1	1
4665 i	1	33	9	2.5	1	1	1	4	4	2	1	8	6	4	5	5	5	5	75	2	5	0	2	1	1	1
4665 i	1	21	9	2	2	1	1	4	2	3	1	8	5	4	5	5	5	5	70	2	1	0	2	1	1	1
4665 i	1	22	12	3	2	1	1	4	2	4	1	8	4	7	4	5	5	5	45	2	1	0	2	1	1	1
4665 i	1	26	11	3.5	1	1	3	5	2	1	0.5	8	4	7	6	5	5	5	80	1	6	0	1	1	1	1
4665 i	1	17	12.5	2	2	1	1	5	2	5	2	4	4	7	5	5	5	5	75	?	1	0	2	1	1	1
4665 i	1	28	12	2	2	1	1	4	2	3	1	3	4	7	4	5	5	5	70	2	1	0	4	1	3	1
4665 i	1	30	10	2	2	1	1	5	2	3	1	4	4	7	5	5	5	5	65	2	1	0	3	3	4	1
4665 i	1	22.5	10	2.5	1	1	1	5	2	3.5	2	4	4	7	5	5	5	5	70	1	6	0	3	1	3	1
4666 i	1	35	17	4	1	1	1	5	4	3	1	4	4	7	5	5	5	5	75	1	6	0	4	3	4	1

Feature explanations and keys for numbers see Chap. 2, Blade Technology, Recording of Technological Information, nd: feature expression not recordable

Table A.23 Blade technology analysis—Kaufertsberg, lower level

Nr.	Blank	Length	Width	Thickness	Pres.: L	Pres.: W/T	Cortex	SPR: kind	SPR: form	SPR: length	SPR: width	DR	Lip	Bulb	Bulbar scar	Impact point	Esq. du bulbe	en éperon	Flaking angle	Profile	Distal	Hammer	n Negatives	Direc. Neg.	Reg Neg.	Tool
571	1	87	19	10	2	1	3	5	5	8	5	4	4	7	5	5	5	5	65	3	1	1	2	1	1	2
567	1	56	21	8	2	1	1	5	2	13	4	8	6	4	5	5	4	5	70	2	1	2	4	1	3	2
585	1	51	16	6	2	1	1	5	2	6	3	4	4	5	4	5	5	4	50	2	1	1	2	1	1	2
638	1	50	20	6	2	1	1	5	9	11	3	4	4	4	4	5	5	5	65	2	1	1	4	1	1	5
649	1	39	20	5	2	1	1	5	2	9	4	4	4	7	5	5	5	4	70	2	1	1	2	1	3	5
662	1	50	16	8	2	1	1	5	5	10	5	4	4	4	5	5	5	4	75	2	1	1	6	?	4	4
727	1	62	8	6	2	1	2	4	2	6	3	4	4	7	5	5	5	5	50	2	1	0	2	1	1	17
726	1	57	11	4	1	1	1	4	9	3	1	4	4	7	5	5	5	5	65	2	4	0	3	1	3	1
720	1	37	13	4	1	1	1	4	4	3	1	4	4	4	5	5	5	5	70	1	4	0	2	1	1	1
722	1	40	10	2	1	1	3	5	7	4	2	2	4	4	5	5	5	5	70	2	4	1	2	1	1	1
723	1	41	14	3	1	1	1	4	2	6	2	8	4	7	5	5	4	4	70	1	4	0	2	1	1	1
724	1	41	13	3	2	1	1	5	7	4	2	4	5	4	5	5	5	5	70	2	1	0	2	1	1	4
762	1	46	10	4	2	1	1	4	2	3	1	4	4	4	5	5	5	5	70	2	1	0	6	?	4	14
751	1	34	6	2	2	2	1	4	4	2	0.5	5	4	5	5	5	4	5	70	2	1	0	2	1	1	13
765	1	34	7	2	2	2	2	4	5	2	2	5	4	5	5	5	5	5	65	2	1	0	1	1	1	13
734	1	63	20	12	2	1	1	5	5	5	4	4	4	7	5	5	5	4	55	2	1	1	?	?	?	1
735	1	50	12	4	1	1	3	5	2	5	2	5	4	7	4	5	5	5	60	2	4	0	2	1	3	1
737	1	41	12	3	1	1	1	5	2	3	1	4	4	7	5	5	5	5	70	2	4	0	2	1	1	1
739	1	38	10	3	1	1	1	5	9	3	2	8	4	4	5	5	5	5	65	2	4	0	2	1	1	1
741	1	28	11	3	1	1	1	5	2	3	2	4	4	4	4	5	5	5	70	1	4	0	3	1	1	1
740	1	34	8	2	1	1	1	5	2	3	2	8	4	5	5	5	5	5	70	2	4	0	3	1	1	1
596	1	62	27	7	2	1	1	5	7	6	3	4	5	7	4	5	5	5	70	2	1	0	2	1	1	5
561	1	88	18	6	1	1	1	5	2	5	2.5	4	4	5	5	5	5	5	55	3	5	1	5	1	2	1
559	1	68	26	5	1	1	1	4	7	9	3	4	4	4	5	5	5	5	55	1	4	0	3	1	1	1
562	1	92	25	7	1	1	2	5	2	8	3	4	4	7	5	5	5	4	60	2	5	1	3	1	1	1
560	1	72	26	5	1	1	1	5	7	15	5	5	6	7	5	5	5	5	70	3	4	0	5	1	3	1
556	1	52	17	4	1	1	1	5	2	8	3	3	4	7	5	5	4	5	70	1	6	0	2	1	1	1
549	1	35	13	3	2	1	1	5	2	1.5	1	4	4	4	5	5	5	5	75	2	1	0	4	1	4	1
702	1	37	9	2	1	1	1	5	5	2	2	4	4	7	4	5	5	4	80	1	4	0	2	1	1	1
703	1	34	13	4	2	1	1	5	7	7	3	4	4	7	4	5	5	4	75	2	1	0	3	1	3	1
701	1	35	14	4	2	1	1	5	2	12	5	8	4	7	5	5	5	5	60	2	1	0	2	1	1	1
704	1	34	11	4	1	1	1	4	2	3	2	4	4	5	5	5	5	4	65	1	4	0	2	1	1	1
696	1	43	13	4	2	1	1	5	2	7	3	2	4	7	5	5	5	6	70	3	1	0	3	1	1	1
695	1	45	15	6	2	1	1	4	2	3	1	8	4	4	4	5	5	5	65	2	1	0	3	1	3	1
682	1	63	18	5	1	1	1	5	7	11	4	4	4	7	5	5	5	5	65	3	4	0	3	1	3	1
688	1	52	20	7	1	1	2	4	5	3	2	8	4	4	5	5	5	5	55	2	1	0	2	1	2	1
667	1	82	21	7	2	1	1	5	5	7	4	4	6	4	5	5	5	5	80	3	1	0	4	1	3	1
694	1	46	19	5	2	1	1	4	2	4	2	4	4	7	5	5	5	5	50	2	1	0	3	1	1	1
683	1	57	16	3	1	1	1	5	5	6	4	8	5	4	5	5	5	4	45	1	5	0	4	1	1	1
686	1	55	18	6	1	1	1	5	7	10	4	4	4	7	5	5	5	4	70	2	6	0	3	1	1	1
681	1	70	16	4	1	1	1	5	2	4	2	4	4	7	5	5	5	5	45	2	5	0	4	3	3	1
689	1	52	17	5	2	1	1	5	2	4	2	4	6	7	5	5	5	5	70	1	1	0	2	1	1	1
685	1	55	20	6	1	1	1	4	2	5	1.5	4	4	4	4	5	5	5	55	2	4	0	5	3	4	1

(continued)

Table A.23 (continued)

Nr.	Blank	Length	Width	Thickness	Pres.: L	Pres.: W/T	Cortex	SPR: kind	SPR: form	SPR: length	SPR: width	DR	Lip	Bulb	Bulbar scar	Impact point	Esq. du bulbe	en éperon	Flaking angle	Profile	Distal	Hammer	n Negatives	Direc. Neg.	Reg Neg.	Tool
673	1	67	26	10	1	1	1	5	2	9	3	2	4	7	5	5	4	5	70	2	5	0	3	1	3	1
677	1	52	28	6	2	1	1	4	2	5	2	8	4	7	4	5	5	5	65	2	1	0	2	1	1	1
676	1	64	35	6	2	1	1	4	3	1	1	4	5	5	5	5	5	4	80	1	1	0	2	1	4	1
668	1	78	30	8	1	1	1	5	2	15	7	4	4	5	5	5	5	4	80	2	8	0	6	3	4	1
672	1	66	37	11	2	1	1	5	7	15	11	5	nd	4	nd	5	5	6	90	2	8	0	6	3	2	1
671	1	76	25	11	1	1	2	5	5	8	4	2	4	7	5	5	5	5	80	3	8	0	3	5	4	1
675	1	64	33	8	2	1	1	4	2	11	5	2	4	7	nd	nd	nd	5	70	1	1	0	2	1	1	1
670	1	73	31	7	2	1	1	5	5	12	6	3	4	7	5	5	5	5	65	2	1	0	2	1	1	1
4665 a	1	42	20	7	2	1	1	4	2	9	4	4	4	7	4	4	5	5	80	2	1	2	2	1	1	5
4665 a	1	51	22	6	2	1	1	5	2	8	4	4	4	7	5	5	5	4	65	2	1	1	2	1	1	5
4665 a	1	50	17	5	2	1	1	4	2	3	1.5	8	4	7	5	5	5	5	60	2	1	0	2	1	1	17
4665 a	1	52	15	5	1	1	2	4	7	7	2.5	4	4	7	4	5	5	5	80	3	4	0	3	1	1	1
4665 a	1	46	24.4	6.5	2	1	1	5	2	5	3	2	4	7	5	5	5	6	70	1	1	0	2	1	1	5
4665 a	1	58	28	7	2	1	1	5	7	11	3	4	4	4	5	5	5	5	60	2	1	0	3	1	1	5
4665 a	1	72	37	10	2	1	1	4	2	15	5	4	4	4	5	5	4	5	80	1	1	2	4	1	3	5
4665 d	1	44	15	3	1	1	1	4	2	3	1	8	4	7	5	5	5	5	80	2	5	0	4	3	3	1
4665 d	1	41	12	4	1	1	1	4	9	4.5	2	4	4	7	4	5	5	5	80	1	4	0	2	1	1	1
4665 d	1	46	10	4	1	1	1	4	3	2	1	4	4	7	5	5	5	5	75	2	5	0	3	1	3	1
4665 d	1	26	12.5	4.5	2	1	1	4	5	6	3	8	4	7	4	5	5	4	65	2	1	1	2	1	1	1
4665 d	1	47	15	6.5	2	1	1	4	2	7.5	3	4	4	4	nd	5	5	5	45	2	1	0	2	1	1	1
4665 d	1	27.5	15	4	2	1	1	4	2	6.5	3	nd	4	5	4	5	5	5	65	2	1	0	2	1	1	1
4665 d	1	25	12.5	3.5	1	1	1	4	4	4	1	8	4	6	5	5	5	5	80	1	5	0	3	?	3	1
4665 d	1	31.5	15	7	2	1	1	5	5	8	5	2	4	5	5	5	5	5	55	2	1	0	2	1	1	1
4665 d	1	43	16	5	2	1	1	5	9	8	2	4	4	4	5	5	5	5	70	2	1	0	3	1	1	1
4665 d	1	25	10	2	2	1	1	4	4	2	1	4	4	7	5	5	4	5	70	2	1	0	2	1	1	1
4665 d	1	28	14	4	2	1	1	4	7	10	2.5	3	4	7	5	5	5	5	50	2	1	0	2	1	1	1
4665 d	1	30	10	2	2	1	1	4	2	4	1.5	4	4	7	5	5	5	5	75	2	1	0	2	1	1	1
4665 d	1	42	12	4	2	1	1	5	5	7	4	2	4	7	5	5	5	6	60	3	1	1	2	1	1	1
4665 d	1	55	9	6	1	1	1	5	7	6.5	3	8	4	4	5	5	5	4	65	1	4	1	2	1	1	1
4665 d	1	42.5	11	3	1	1	2	4	2	2.5	1	4	4	7	5	5	5	5	70	1	4	0	3	1	3	1
4665 d	1	58	12	5	1	1	3	4	2	4	2	3	4	4	4	5	5	5	75	2	4	0	4	1	3	1
4665 d	1	35	11	5	2	1	1	4	5	2	2	8	4	5	5	5	5	4	75	2	1	0	2	1	1	1
4665 d	1	47	15	3	1	1	1	4	4	3.5	0.5	8	4	4	5	5	5	5	80	1	4	0	4	1	3	1
4665 d	1	36	10.5	3	1	1	3	5	7	6.5	2.5	8	4	4	5	5	5	5	80	1	7	0	1	1	3	1
4665 d	1	43	10	3	1	1	1	5	7	4	2.5	4	4	4	4	5	5	5	75	2	5	0	3		1	1
4665 d	1	25	13	3.5	2	1	1	5	2	5	2	2	4	5	5	5	5	5	65	1	1	0	2	1	1	1
4665 c	1	27	23	5	2	1	1	5	2	11	5	8	4	7	5	5	5	5	65	?	1	0	3	1	1	1
4665 c	1	44	20	6	2	1	1	4	7	10	2.5	4	4	4	4	5	5	5	60	2	1	0	2	1	1	1
4665 c	1	58	19	9	2	1	1	5	2	10	5	4	4	7	5	5	5	4	70	2	1	0	5	1	1	1
4665 c	1	57	17	7	1	1	1	5	2	6	2	8	4	7	5	5	5	5	70	1	6	0	2	1	1	1
4665 c	1	53.5	17.5	5	1	1	1	4	2	6	2	2	4	7	5	5	5	5	65	1	4	0	3	1	1	1

4665 c	1	30	25	5	2	1	1	5	2	11.5	3.5	4	4	4	4	5	5	5	45	?	1	0	?	?	?	1
4665 c	1	31	22.5	5	2	1	1	5	2	10	3	4	4	5	5	5	5	5	40	?	2	0	1	1	1	1
4665 c	1	57	17	4	1	1	2	4	2	4	1	4	4	7	4	5	5	5	70	3	5	0	2	1	1	1
4665 c	1	32	14	5	2	1	1	5	2	7	3	8	4	7	5	5	5	5	60	1	1	0	2	1	1	1
4665 c	1	47	19	7	2	1	1	4	5	9	5	4	4	7	5	5	5	4	65	3	1	0	3	3	4	1
4665 c	1	32	17	5	2	1	1	5	2	9	3	4	4	7	5	5	5	5	60	2	1	0	2	1	1	1
4665 c	1	32	22	7	2	1	1	4	7	7	1.5	4	4	4	5	5	5	5	65	2	1	0	2	1	3	1
4665 c	1	31.5	15	2.5	1	1	1	5	2	7	3	2	4	7	5	5	5	5	70	1	6	0	3	1	3	1
4665 c	1	51	19	6	1	1	1	5	2	8	4	4	4	7	5	5	5	5	45	3	5	1	6	4	4	1
4665 c	1	50	21	4.5	2	1	1	5	2	7	2	4	4	4	5	5	5	5	65	2	1	0	4	3	3	1
4665 c	1	43	16	4	2	1	1	5	2	8	3.5	2	4	4	4	5	5	5	80	1	1	0	2	1	3	1
4665 c	1	46	20	3.5	2	1	1	4	2	6	2.5	8	4	4	5	5	5	5	60	2	1	0	2	1	3	1
4665 c	1	50	21	7	2	1	1	4	9	5	1	4	4	7	5	5	5	5	60	1	1	0	4	3	4	1
4665 e	1	29	9	3	2	1	1	4	2	3	1.5	3	4	4	4	5	5	5	80	3	1	0	2	1	1	1
4665 e	1	34	7	4	1	1	1	4	3	2	2	4	4	4	5	5	5	5	75	2	4	0	2	1	2	1
4665 e	1	23.5	7.5	2.5	2	1	1	4	7	3	1	4	4	7	5	5	5	5	70	2	1	0	2	1	1	1
4665 e	1	25	8	1.5	1	1	1	4	7	3	2	4	4	4	5	5	5	5	80	1	4	0	2	1	1	1

Feature explanations and keys for numbers see Chap. 2, Blade Technology, Recording of Technological Information, nd: feature expression not recordable

Table A.24 Blade technology analysis—Klementowice, earlier excavations

Nr.	Blank	Length	Width	Thickness	Pres.: L	Pres.: W/T	Cortex	SPR: kind	SPR: form	SPR: length	SPR: width	DR	Lip	Bulb	Bulbar scar	Impact point	Esq. du bulbe	en éperon	Flaking angle	Profile	Distal	Hammer	n Negatives	Direc. Neg.	Reg Neg.	Tool
01/81	1	29	10	3	2	1	1	5	5	5	2.5	4	4	7	4	5	5	4	55	2	1	1	3	1	1	1
02/82	1	49.5	15.5	5	2	2	1	5	2	8	3.5	2	4	7	5	5	5	4	50	2	1	1	2	1	1	1
05/81	1	50	20	6	2	1	1	5	2	12	4.5	3	5	4	5	5	5	4	60	2	1	1	4	1	2	1
06/82	1	58	23	3	1	1	1	5	9	11	3	2	4	7	4	5	5	5	60	1	5	0	3	1	1	1
07/81	1	36.5	14	4	2	1	1	5	2	9	4	8	5	4	5	5	5	4	75	2	2	1	3	1	1	17
08/82	1	55	16	4	1	1	2	5	2	8.5	2	3	5	7	5	5	5	4	50	2	6	1	2	1	1	1
08/82	1	34	21	4.5	2	1	1	5	2	8	2.5	4	4	4	7	5	5	4	55	2	1	1	3	1	4	1
08/82	1	34	14	4.5	2	2	1	5	2	9.5	3	3	4	7	5	5	5	4	60	2	2	1	2	1	1	5
08/82	1	58	16.5	5	2	1	1	5	9	9	1.5	4	6	4	5	5	5	5	80	2	2	0	3	1	1	17
09/82	1	46	16.5	4.5	2	1	1	5	9	10.5	4	4	4	7	4	4	5	5	80	2	1	2	3	1	1	1
11/82	1	46.5	13	4	2	1	1	5	2	4	2	3	4	7	4	5	5	5	55	2	2	0	3	3	3	19
114/82	1	46	17	5	2	2	2	5	5	9	4	8	4	5	5	5	5	4	65	3	2	1	2	1	2	5
12/81	1	32	12	2.5	2	1	1	5	9	6	1.5	3	4	4	5	4	5	5	60	2	2	2	2	1	1	17
12/82	1	20	8	2.5	2	2	1	5	2	6	2	3	4	7	5	5	5	4	60	2	1	1	2	1	1	13
13/82	1	30.5	16	4	2	1	1	5	2	8	3	4	4	7	5	5	5	4	45	2	1	1	3	1	1	1
13/82	1	58	18	7	2	1	1	5	2	11	5	3	4	4	5	5	5	4	65	2	2	1	4	1	3	5
13/82	1	36	12	3.5	2	2	1	5	5	8	4	2	7	4	5	5	5	4	70	2	2	1	3	4	1	5
13/82	1	23	9.5	2	2	1	1	5	2	5	2.5	3	4	7	6	5	5	4	75	2	2	1	3	1	2	17
15/81	1	30.5	7.5	3	1	1	1	5	4	3	1.5	5	4	7	5	5	5	5	50	1	6	0	3	1	1	1
15/81	1	37	12	4	1	1	1	5	2	6	2	3	4	7	5	5	5	4	55	1	1	1	2	1	1	1
15/81	1	29	15	4	2	1	1	5	2	7.5	2.5	4	4	7	5	5	5	4	55	?	1	1	2	1	1	1
15/81	1	55	15	4.5	1	1	2	5	2	7	2	8	4	7	5	5	5	4	55	1	6	1	1	1	1	1
15/81	1	26	13	4	2	1	1	5	2	5.5	2.5	4	4	7	5	5	5	4	65	1	1	1	2	1	1	1
15/81	1	22.5	14	3	2	1	2	5	2	3.5	1.5	4	4	7	5	5	5	4	70	2	1	1	1	1	1	1
15/81	1	35	17	4.5	2	1	1	5	2	8.5	4	8	4	7	5	5	5	4	70	2	1	1	3	1	1	1
15/81	1	20.5	12.5	2.5	2	1	1	5	9	8	1.5	4	5	7	4	4	5	5	75	2	1	2	2	1	1	1
15/81	1	61	18	6	1	1	1	5	2	9	2	4	4	7	5	5	5	4	75	2	4	1	4	?	2	1
15/81	1	27	17	5	2	1	1	5	2	8	3	2	4	7	5	5	5	4	80	2	1	1	2	1	1	1
15/82	1	47.5	15	4.5	2	1	1	5	2	6	2	4	5	4	4	5	5	5	80	2	1	0	2	1	1	1
15/82	1	71	22	13	2	1	2	5	2	9	3	4	4	7	5	5	5	5	80	2	1	0	2	1	1	1
16/81	1	45	15	5.52	2	1	1	5	2	7	2	3	4	7	5	5	5	4	35	2	2	1	3	1	1	17
16/81	1	50	16	3	2	1	1	5	2	8	3	8	4	4	4	5	5	4	60	2	2	1	3	1	1	17
16/81	1	18	10	3.5	2	2	1	5	9	7	2.5	2	4	4	5	5	5	4	75	2	1	1	3	1	1	13
16/81	1	27	13	3	2	1	1	5	2	8	3.5	4	5	4	5	5	5	4	75	2	2	1	3	1	1	17
16/82	1	16	11	3	2	1	1	5	5	7	3	3	4	7	4	5	5	4	70	2	1	1	2	1	1	1
17/81	1	33	16	4.5	2	1	1	5	7	5	2.5	8	4	7	5	4	5	5	45	2	1	2	3	1	1	1
17/82	1	21	15.5	2	2	1	1	5	9	7	1	3	6	7	5	4	5	5	85	?	1	2	2	1	1	1
18/82	1	45	13.5	4	2	2	2	5	5	4.5	2	3	4	7	5	5	5	4	65	2	2	1	2	1	1	5
19/82	1	52	11	4.5	1	1	1	5	5	3	1.5	4	5	4	6	4	5	5	70	2	4	2	3	1	1	1
19/82	1	57	27	7.5	2	1	1	5	2	9	3	4	4	7	4	5	5	5	70	1	1	0	4	1	1	1
190/82	1	62	19	4	2	1	1	5	5	5.5	2.5	3	5	7	5	5	5	4	70	1	1	1	3	1	1	1
21/82	1	49	14	4.5	1	1	1	5	9	7	2.5	8	4	7	5	5	5	5	50	1	5	0	3	1	1	1

21/82	1	25	11	3	2	1	1	5	2	7	2	2	4	7	5	5	5	4	55	1	1	1	2	1	1	1
22/82	1	35.5	18	6	2	1	1	5	2	7	3	4	4	7	5	5	5	4	65	2	1	1	2	1	1	1
22/82	1	26	16.5	4	2	1	1	5	7	11	2.5	8	5	4	6	5	5	5	75	1	1	0	2	1	1	1
23/82	1	38.5	22	5.5	2	1	1	5	2	14	4	3	4	7	5	5	5	4	55	2	1	1	2	1	1	1
23/82	1	58	14.5	4	2	2	1	5	5	3	2	8	4	7	5	5	5	4	75	2	2	1	2	1	1	2
24/82	1	31.5	10	3.5	2	2	1	4	4	3	1	8	5	4	5	5	5	5	75	2	2	0	2	1	1	17
24a/82	1	40	13	6	1	1	1	5	5	10	5	2	4	7	5	5	5	4	50	1	6	1	3	1	2	1
24a/82	1	18	8.5	3	2	2	1	5	2	7	2	4	5	4	4	5	5	5	60	2	1	0	2	1	1	13
25/82	1	60.5	19.5	8	2	1	1	5	2	6	1.5	3	4	7	5	5	5	5	60	3	1	0	3	?	2	1
25/82	1	30.5	13.5	3	2	1	2	5	2	5.5	2.5	4	5	4	4	4	5	5	75	2	1	2	2	1	1	1
27/82	1	21	11	3	2	1	1	5	6	5.5	2.5	4	4	7	5	5	5	4	55	2	1	1	2	1	1	1
27/82	1	33.5	21.5	4.5	2	1	1	5	2	8	3	4	5	7	4	5	5	5	65	2	1	0	2	1	1	1
27/82	1	52.5	17.5	7.5	2	1	1	5	2	4.5	1.5	3	4	7	5	5	5	5	75	3	1	0	2	1	3	1
27/82	1	28	16.5	6.5	2	1	1	5	2	6	2.5	3	4	7	5	5	5	4	75	2	1	1	3	1	1	1
27/82	1	12.5	10.5	2.5	2	1	1	5	4	6	1	4	5	7	4	4	5	5	75	?	1	2	3	1	1	1
28/82	1	81.5	23	9	1	1	1	5	2	5	2	4	5	7	5	5	5	5	75	1	4	0	5	3	3	1
28/82	1	35	22	4.5	2	1	1	5	2	12	4	8	4	7	5	5	5	4	80	2	1	1	3	1	1	1
29/81	1	60	25	5.5	2	1	1	5	2	13	5	2	4	7	5	5	5	4	35	2	1	1	3	1	1	14
30/81	1	12.5	12.5	4	2	1	1	5	2	6.5	2	4	4	4	5	5	5	4	75	?	1	1	2	1	1	1
32/81	1	51	16	4.5	2	1	1	5	9	8	2	4	5	4	4	4	5	5	80	2	1	2	5	?	3	1
32/82	1	57	16	4.5	2	1	1	5	2	6	2	8	4	7	5	5	5	4	50	2	1	1	2	1	1	1
34/81	1	50.5	17.5	5	2	1	2	5	5	8.5	5.5	2	5	7	4	5	5	4	45	1	1	1	3	?	1	1
34/82	1	31	14	4	2	1	1	5	2	10	4	3	4	7	5	5	5	4	45	2	1	1	2	1	1	1
34/82	1	25	16.5	3	2	1	1	5	2	7.5	3	8	4	7	4	5	5	4	50	2	1	1	3	1	2	1
34/82	1	42	11.5	4.5	2	1	2	5	5	6	3	2	4	7	5	5	5	4	55	2	1	1	2	1	1	1
34/82	1	30.5	12	3	2	1	1	5	5	4	2.5	4	4	7	5	5	5	5	60	2	2	1	3	3	3	17
34/82	1	58.5	15.5	5	1	1	1	5	7	7	4	4	5	7	4	4	5	5	65	1	7	2	2	1	1	1
34/82	4	55	8	4	2	1	1	5	5	3	2	8	4	7	5	5	5	4	70	3	1	1	6	4	2	1
34/82	1	26	10	3.5	2	1	1	5	5	4	2	8	5	4	4	5	5	4	70	2	1	1	2	1	1	1
36/81	1	22	7	3	2	2	1	5	5	4.5	3	4	4	7	5	5	5	4	55	2	1	1	2	1	1	13
36/81	1	15	9	3	2	2	1	5	2	6	2.5	3	6	4	5	5	5	4	70	?	1	1	2	1	1	13
36/81	1	34	11	4	2	2	1	5	9	5.5	0.5	4	5	4	4	5	5	5	70	1	2	0	2	1	1	17
37/81	1	50.5	16.5	5.5	2	2	2	5	2	7.5	2.5	3	4	7	5	5	5	4	55	2	1	1	2	1	1	1
37/82	1	38.5	12	3	2	2	1	5	2	3.5	1.5	3	4	7	5	5	5	4	45	2	2	1	3	1	1	17
37/82	1	43	14.5	3	2	1	1	5	5	9	6	4	5	4	5	5	5	4	50	2	1	1	2	1	1	1
37/82	1	46	20	5.5	2	2	1	5	2	12	5.5	4	5	4	5	5	5	4	70	2	2	1	3	1	2	5
38/1	1	41.5	12	4	2	1	2	5	2	6.5	2.5	4	4	7	5	5	5	4	55	2	2	1	2	1	1	17
38/81	1	41.5	13	3	2	1	1	5	2	4.5	2	4	4	7	5	5	5	4	55	2	2	1	3	1	1	17
38/81	1	33	11.5	3	2	1	1	5	2	4	1.5	2	4	4	4	4	4	4	70	2	2	0	4	1	1	17
38/82	1	61	14	3	2	1	1	5	2	6	2.5	2	4	7	5	5	5	4	40	2	1	1	3	1	3	1
38a/82	1	13.5	8.5	2	2	1	1	5	2	5	1.5	4	4	7	5	5	5	4	45	?	1	1	2	1	1	1
38a/82	1	37	10	4	2	1	1	4	2	3	1	3	5	4	4	4	5	5	50	2	1	2	2	1	1	1
38a/82	1	47	13.5	3.5	2	1	1	5	2	4	1.5	3	4	7	5	5	5	4	50	2	1	1	3	1	1	1

(continued)

Table A.24 (continued)

Nr.	Blank	Length	Width	Thickness	Pres.: L	Pres.: W/T	Cortex	SPR: kind	SPR: form	SPR: length	SPR: width	DR	Lip	Bulb	Bulbar scar	Impact point	Esq. du bulbe	en éperon	Flaking angle	profile	distal	hammer	n Negatives	Direc. Neg.	Reg Neg.	tool
38a/82	1	63	15	4	2	1	2	5	9	6	1.5	3	4	7	5	5	5	4	55	2	1	1	1	1	1	1
48/82	1	15	5.5	2	2	2	1	4	4	4	1	4	6	4	5	4	5	5	70	2	1	2	2	1	1	13
49/82	1	60.5	20	5	1	1	1	5	5	8	3.5	8	4	7	5	5	5	4	40	2	5	1	5	3	2	1
54/82	1	47.5	14	4	2	1	2	5	2	5.5	2	2	4	7	5	5	5	4	80	2	1	1	3	1	1	1
54/82	1	38.5	18	3	2	2	1	4	2	3	1.5	4	5	4	5	5	5	5	85	2	2	0	3	1	3	17
55a/82	1	65	10	5	2	1	2	5	5	4.5	2	4	4	7	5	5	5	4	45	2	1	1	2	1	2	1
55a/82	1	24.5	16	3	2	1	1	5	2	8	3.5	2	5	4	5	5	5	4	60	2	1	1	3	1	4	1
55a/82	1	46	17	3	1	1	1	5	2	7	3	4	5	4	6	4	5	5	65	2	5	2	3	1	2	1
62/82	1	27	16	4	2	1	1	4	2	4	2	4	5	7	4	5	5	5	65	2	1	0	2	1	1	1
65/82	1	66.5	16	4	1	1	1	5	2	10	2.5	8	6	4	5	5	5	4	60	2	4	1	2	1	1	1
65/82	1	33	13	3	2	1	1	5	9	11.5	2	4	5	4	4	5	5	5	75	2	1	0	3	1	1	1
99/82	1	41	11.5	5	2	1	1	5	2	3.5	1	8	5	7	4	4	5	5	80	2	1	2	2	1	1	1
None	1	39	9	4	2	1	2	5	5	7	4	2	5	4	4	5	5	4	40	2	1	1	3	?	3	1
None	1	49	16	4	2	1	1	5	5	10.5	4	8	4	7	5	5	5	4	45	2	1	1	3	1	1	1
None	1	13	6	3	2	1	1	5	7	5	2	3	5	7	4	5	5	4	85	?	1	1	3	1	1	1

Feature explanations and keys for numbers see Chap. 2, Blade Technology, Recording of Technological Information, nd: feature expression not recordable

Table A.25 Blade technology analysis—Klementowice, recent excavations

Nr.	Blank	Length	Width	Thickness	Pres.: L	Pres.: W/T	Cortex	SPR: kind	SPR: form	SPR: length	SPR: width	DR	Lip	Bulb	Bulbar scar	Impact point	Esq. du bulbe	en éperon	Flaking angle	Profile	Distal	Hammer	n Negatives	Direc. Neg.	Reg Neg.	Tool
604/08	1	32	19	3.5	2	1	1	5	5	8.5	4	2	4	7	5	5	5	4	45	1	1	1	3	1	1	1
1216/08	1	46.5	17.5	7	2	1	1	5	2	8	3.5	4	4	7	5	5	5	4	50	1	2	1	2	1	2	5
775/08	1	46	14	6	1	1	2	5	5	9	4.5	4	5	4	5	5	5	4	55	1	6	1	2	1	1	1
102/08	1	21	14	2	2	1	1	5	2	3.5	1	4	4	7	5	5	5	4	85	1	1	1	2	1	1	1
488/08	1	21	12.5	3	2	1	1	5	2	7	2	4	4	7	4	5	5	4	40	2	1	1	3	1	1	1
810/08	1	50	16	3.5	2	1	2	5	2	6	2	4	4	7	5	5	5	4	45	2	1	1	4	?	2	1
1044/08	1	15	8.5	3	2	1	1	5	2	7	2.5	4	4	7	5	5	5	4	45	2	1	1	3	1	1	1
696/08	1	20	11	3.5	2	1	1	5	2	4	2	3	4	7	4	5	5	4	50	2	1	1	2	1	1	1
968/08	1	15	10.5	3	2	1	1	5	2	6	2	3	4	7	5	5	5	4	50	2	1	1	3	1	1	1
631/08	1	27	8.5	3	2	1	1	5	5	5	2	3	4	7	5	5	5	4	50	2	2	1	2	1	1	17
750/08	1	72	17	4.5	2	1	2	5	5	6	4	3	4	7	5	5	5	4	50	2	1	1	3	1	2	1
708/08	1	32	13.5	4	2	2	1	5	2	6.5	3	8	4	7	5	5	5	4	55	2	1	1	3	1	1	1
1215/08	1	50	20.5	6	2	1	1	5	2	12.5	3	8	4	7	5	5	5	4	55	2	2	1	4	?	4	2
974/08	1	31.5	11.5	4	2	1	1	5	5	7	3	8	4	7	5	5	5	4	55	2	1	1	2	1	1	1
1195/08	1	58	15.5	7	2	1	2	5	2	7	2.5	4	4	7	5	5	5	4	60	2	1	1	3	1	1	1
555/08	1	22.5	10	3.5	2	1	1	5	2	5	2	8	4	7	5	5	5	4	60	2	1	1	2	1	1	1
656/08	1	29	13	4	2	1	1	5	2	9.5	3	8	5	7	4	5	5	4	65	2	1	1	3	1	1	1
624/08	1	27.5	20	4.5	2	1	1	5	7	14	3	4	4	4	5	5	5	4	65	2	1	1	3	1	2	1
208/08	1	10.5	7	2	2	1	1	5	7	4.5	2	3	4	7	5	5	5	4	75	2	1	1	3	1	1	1
963/08	1	40.5	11	4	2	1	1	5	2	5	2	4	4	7	5	5	5	4	80	2	1	1	3	1	1	1
617/08	1	27.5	20	3.5	2	1	1	5	2	11	4	8	4	7	5	5	5	4	80	2	1	1	2	1	2	1
951/08	1	46	16	3.5	2	1	1	5	4	10	1.5	4	5	7	5	5	4	5	85	1	2	2	3	1	3	17
300/80	1	33.5	7.5	3	2	1	1	4	2	4	1	4	5	4	5	5	5	5	55	2	1	0	3	1	1	1
406/08	1	17.5	15.5	4	2	1	1	5	7	12	2	8	4	7	4	4	5	5	65	2	1	2	2	1	1	1
742/08	1	35	11	3	2	1	1	5	2	3	1	4	4	7	5	5	5	5	65	2	1	0	3	1	1	1
791/08	1	43	16	4.5	1	1	1	5	7	9	4	3	5	4	4	5	5	5	70	2	4	0	4	?	4	1
137/08	1	30	17	6	2	1	1	5	7	9	1.5	8	5	7	4	4	5	5	75	2	1	2	2	1	1	1
28 08	1	23	12.5	3	2	1	1	5	2	6.5	2	4	4	4	5	4	5	5	75	2	1	2	?	?	?	1
982/08	1	63	14	5.5	2	1	2	5	7	12	3	3	4	4	4	5	5	5	80	2	2	2	1	1	2	4
721/08	1	30.5	7.5	3.5	2	2	1	5	2	4	1.5	8	4	4	5	5	5	5	80	2	1	0	4	?	2	13
794/08	1	27.5	17.5	3.5	2	1	1	5	2	7.5	4	4	6	7	4	4	5	5	85	2	1	2	3	1	1	1
86/08	1	20	17.5	3.5	2	1	2	4	7	2	1	4	5	4	4	5	5	5	85	2	0	1	2	1	2	1
902/08	1	42.5	14	3.5	1	1	1	5	5	5	2	3	4	4	5	5	5	6	80	2	5	0	2	1	1	1

Feature explanations and keys for numbers see Chap. 2, Blade Technology, Recording of Technological Information, nd: feature expression not recordable

Table A.26 Blade technology analysis—Lhota

Nr.	Blank	Length	Width	Thickness	Pres.: L	Pres.: W/T	Cortex	SPR: kind	SPR: form	SPR: length	SPR: width	DR	Lip	Bulb	Bulbar scar	Impact point	Esq. du bulbe	en éperon	Flaking angle	Profile	Distal	Hammer	n Negatives	Direc. Neg.	Reg Neg.	Tool
15923	1	12	9.5	2.5	2	1	1	4	4	3	1	8	4	4	5	5	5	5	45	0	1	0	2	1	1	1
15753	1	10	9.5	2	2	1	1	4	4	4	1	4	4	7	5	4	5	5	50	0	1	2	2	1	1	1
16002	1	11.5	12	4	2	1	1	4	2	2.5	1	8	5	4	4	5	5	5	75	0	1	0	2	1	1	1
15768	1	10	15	2.5	2	1	1	5	2	4.5	2	8	5	4	5	4	5	5	85	0	1	2	3	1	1	1
15926	1	11.5	10	3.5	2	1	1	5	9	9.5	2.5	3	5	4	5	5	5	5	85	0	2	0	3	1	1	1
15724	1	33	15	5	2	1	1	5	2	6	2.5	4	4	7	5	5	5	5	40	2	1	0	2	1	1	1
15935	1	45	19	6.5	1	1	1	5	2	9	4	4	4	4	5	4	5	5	45	2	8	2	4	1	1	1
15752	1	22	17	3.5	2	1	2	5	2	4	2	4	4	4	5	4	5	5	55	2	1	2	2	1	1	1
15606	1	31.5	14	4.5	2	1	1	4	2	3	1	4	4	4	4	5	5	5	60	2	1	0	4	1	2	1
15623	1	21	8	3	2	2	1	4	2	2	1	2	4	4	4	4	5	5	65	2	1	2	2	1	1	13
15649	1	19.5	11	3	2	1	1	5	4	7.5	1.5	4	4	4	4	4	5	5	65	2	1	2	3	1	3	1
15639	1	27	18.5	4	2	1	1	4	9	7	2	8	7	4	4	4	5	5	65	2	1	2	3	1	2	4
15936	1	37.5	20	5	2	1	2	4	2	4	1.5	3	4	4	5	4	5	5	65	2	1	2	2	1	1	1
15948	1	27	15	5	2	1	2	4	3	1	1	3	5	4	5	4	5	5	65	2	1	2	1	1	1	1
15762	1	11	5.5	2	2	1	1	4	7	1.5	1	8	4	4	5	5	5	5	65	2	1	0	2	1	1	1
15851	1	13	9	2	2	1	1	5	2	2.5	1	8	5	4	4	4	5	5	70	2	1	2	2	1	1	1
15718	1	20.5	14	4	2	1	1	5	9	8	2	4	5	4	5	4	5	5	70	2	1	2	2	1	1	1
15770	1	42	14	6.5	1	1	1	5	7	8	3	8	4	7	5	4	5	5	70	2	5	2	4	1	2	1
15775	1	13	7	2.5	2	1	1	4	2	4	1	8	4	7	5	5	5	5	70	2	1	0	2	1	1	1
15867	1	13	12	3	2	1	1	5	5	4.5	2.5	2	4	4	5	4	5	6	70	2	1	2	2	1	1	1
15630	1	8.5	10.5	3	2	1	1	5	5	7.5	3	2	5	4	4	4	5	5	75	2	1	2	2	1	1	1
15802	1	17	16.5	3	2	1	2	5	4	11.5	3	4	4	4	5	4	5	5	75	2	1	2	1	1	1	1
15610	4	23	12	5	2	1	1	5	4	4.5	1	4	5	4	5	4	4	5	80	2	1	2	3	4	2	1
15626	1	14	11	3	2	1	1	4	2	4	1.5	4	4	4	4	4	5	5	80	2	1	2	2	1	1	1
15697	1	12.5	6.5	2	2	1	1	4	2	2	1	4	5	4	4	4	5	5	80	2	1	2	2	1	1	1
15881	1	12	9	1.5	2	1	1	4	2	4.5	1	4	5	4	4	4	5	5	80	2	1	2	2	1	1	1
15819	1	13	12	2	2	1	1	5	2	4.5	1.5	3	6	4	4	4	5	5	80	2	1	2	2	1	1	1
15654	1	18	13.5	2	2	1	1	4	9	5	1	2	4	4	5	4	5	5	80	2	1	2	3	1	1	1
15611	1	29	17.5	4.5	2	1	1	5	4	3	0.5	4	5	4	5	4	5	5	80	2	1	2	4	5	1	1
15800	4	27	16	5	2	1	1	5	2	5	2.5	3	4	7	5	4	5	5	80	2	1	2	5	4	1	1
15917	1	31	9.5	3.5	1	1	2	5	4	3.5	1	3	5	4	4	5	5	5	80	2	5	0	2	3	1	1
15729	1	22	12	3.5	2	1	1	4	2	0.5	0.25	8	5	4	4	5	5	5	80	2	1	0	2	1	1	1
15941	1	16	13	2.5	2	1	1	5	9	5	1	2	5	6	4	5	5	5	80	2	2	0	3	1	1	1
15673	1	12.5	6.5	2.5	2	1	1	4	2	2	1.5	3	5	7	4	5	5	5	80	2	1	0	2	1	1	1
15906	1	15	13	3.5	2	1	1	4	2	3	1.5	4	4	4	5	5	5	5	80	2	1	0	2	1	1	1
15785	1	27	14	3.5	2	1	1	5	4	6	1.5	8	5	4	5	4	4	5	85	2	1	2	2	1	1	1
15754	1	33	25.5	11	2	1	2	4	2	4.5	3	4	5	4	4	4	5	5	85	2	1	2	2	1	1	1
15920	1	30	15.5	3	2	1	1	5	9	6	1	8	5	4	4	4	5	5	85	2	1	2	3	1	1	14
15944	1	14.5	11	2	2	1	1	5	9	6	1	4	5	6	4	4	5	5	85	2	1	2	3	1	1	1
15980	1	36	12	4.5	1	1	1	5	9	4	1	8	5	4	5	4	5	5	85	2	7	2	3	1	1	1
15869	1	15	11.5	3	2	1	2	4	2	3	1.5	4	4	4	6	4	5	5	85	2	1	2	1	1	1	1

Feature explanations and keys for numbers see Chap. 2, Blade Technology, Recording of Technological Information, nd: feature expression not recordable

Table A.27 Blade technology analysis—Mokra V

Nr.	Blank	Length	Width	Thickness	Pres.: L	Pres.: W/T	Cortex	SPR: kind	SPR: form	SPR: length	SPR: width	DR	Lip	Bulb	Bulbar scar	Impact point	Esq. du bulbe	en éperon	Flaking angle	Profile	Distal	Hammer	n Negatives	Direc. Neg.	Reg Neg.	Tool
1	1	18.5	14	3	2	1	1	4	2	7.5	3	3	4	7	5	5	5	4	50	2	1	1	?	?	?	1
2	1	26	15	4.5	2	1	1	5	2	7	3	4	4	7	5	5	5	4	75	2	1	0	3	1	1	1
3	1	30	20.5	5	2	1	1	5	2	7	3	3	4	7	4	5	5	4	70	?	1	1	2	1	1	14
4	1	21	15.5	2.5	2	1	1	4	2	6.5	1.5	8	4	4	5	5	5	5	65	1	1	0	2	1	1	1
5	1	25	13	5	2	1	1	4	9	7	2	8	4	7	4	4	5	5	80	1	1	2	2	1	1	1
6	1	28	13	5	2	1	1	5	7	9	4	4	5	6	5	5	5	5	85	1	1	2	2	1	1	1
7	1	39	17	7	2	1	1	5	7	6	3	3	5	4	4	4	5	5	55	2	1	2	3	1	1	14
8	1	37	17.5	4	2	1	1	4	2	6	3	4	4	7	5	5	5	5	55	2	1	0	4	1	2	1
9	1	20	10.5	3.5	2	1	1	4	2	5	2	8	4	7	5	5	5	5	55	2	1	1	3	1	3	1
10	1	16	8	3	2	1	1	4	3	1	1	3	5	4	5	4	5	5	60	2	1	2	2	1	1	1
11	1	33	10	3	2	1	1	5	9	6	1.5	8	5	4	4	4	5	5	75	2	1	2	2	1	1	1
12	1	25.5	25	8	2	1	1	5	2	13	6.5	4	6	4	4	4	5	5	75	2	1	2	2	1	1	1
13	1	21	10	2.5	2	1	1	5	2	3	2	4	5	4	4	4	5	5	80	2	1	2	4	1	1	1
14	1	26	13.5	3	2	1	1	5	9	9.5	1.5	8	5	4	6	4	5	5	80	2	1	2	2	1	1	1
15	1	18.5	13	3	2	1	1	5	7	10	1	4	5	6	4	6	5	5	80	2	1	2	3	1	2	1
16	1	45	22.5	4.5	2	1	2	4	2	5	3	4	4	4	5	5	5	5	85	2	1	0	2	1	2	1
17	1	25	13.5	3.5	2	1	1	5	4	3	1	4	4	7	5	5	5	5	85	2	1	0	3	1	3	1
18	1	20	11	4.5	2	1	1	4	2	4	1.5	3	4	7	5	4	5	5	45	?	1	2	2	1	1	1
19	1	18.5	16	5	2	1	1	5	2	13	5	8	4	7	5	5	5	5	45	?	1	0	2	1	1	1
20	1	14	16	3	2	1	1	4	4	4	1	8	6	4	5	5	5	5	75	?	1	0	2	1	1	1
21	1	10.5	11	1.5	2	1	1	5	7	5	2	3	5	4	4	4	5	5	80	?	1	2	?	?	?	1
22	1	16	15	3.5	2	1	1	5	2	5	2	8	6	4	6	4	5	5	80	?	1	2	2	1	2	1
23	1	15	11	3.5	2	1	1	5	4	4.5	0.5	4	5	4	5	5	4	5	85	?	1	2	2	1	1	1
24	1	16.5	10.5	2	2	1	1	4	2	4	1.5	8	5	4	5	5	5	5	85	?	1	0	2	1	1	1
25	1	13	17	4.5	2	1	1	5	7	11	3	4	5	6	5	6	5	5	85	?	1	2	2	1	1	1
26	1	16.5	17	4	2	1	1	5	4	6	2	4	5	4	5	5	5	5	nd	?	1	0	2	1	1	1
27	1	10	12	3	2	1	1	5	4	6.5	1.5	8	5	4	5	5	5	5	nd	?	1	0	2	1	1	1
28	4	81.5	30.5	12	1	1	1	4	2	11	6	4	4	7	5	5	5	6	65	2	4	1	?	?	4	1
29	1	14.5	8.5	3	2	1	1	5	5	2	2	3	5	4	5	4	4	6	85	2	1	2	3	1	2	1
30	1	19	19	5	2	1	1	5	2	3	4.5	4	6	4	5	5	5	6	75	?	1	0	3	1	2	1
31	1	11	7	2.5	2	1	1	5	5	3	2	4	5	7	5	5	5	6	85	?	1	0	2	1	1	1
32	1	25	13	5	2	1	1	4	5	3	3.5	2	4	7	5	5	5	6	55	2	1	1	2	1	1	1

Feature explanations and keys for numbers see Chap. 2, Blade Technology, Recording of Technological Information, nd: feature expression not recordable

Table A.28 Blade technology analysis—Neuchâtel-Monruz

Nr.	Blank	Length	Width	Thickness	Pres.: L	Pres.: W/T	Cortex	SPR: kind	SPR: form	SPR: length	SPR: width	DR	Lip	Bulb	Bulbar scar	Impact point	Esq. du bulbe	en éperon	Flaking angle	Profile	Distal	Hammer	n Negatives	Direc. Neg.	Reg Neg.	Tool
97	1	30	16.5	5.5	2	1	1	4	3	1	1	3	1	1	1	1	4	5	85	1	1	2	4	?	2	1
307	1	38	10	2	1	1	1	4	2	3	1	3	5	4	5	4	5	5	75	2	4	2	4	1	2	1
21	1	45	10.5	4	2	2	1	4	4	4	1	4	5	6	5	4	5	5	85	1	1	2	2	1	1	1
236	1	79.5	29	5.5	2	1	1	5	2	10.5	4	8	4	7	4	4	5	5	45	2	1	0	2	1	1	1
76	1	27.5	12.5	2.5	2	1	1	4	7	5	2.5	3	8	8	4	4	4	5	85	1	1	2	3	1	1	1
82	1	48	14	3	2	1	1	4	2	9	3	8	4	4	5	4	5	5	70	2	1	2	3	1	1	1
21	4	55	19	8	2	1	1	5	2	12	6	8	5	4	5	4	4	5	80	2	2	2	?	?	4	5
331	1	39	16.5	5	2	1	1	5	2	7	2	4	4	4	5	4	5	5	65	1	1	2	2	1	1	1
194	1	26.5	21	6	2	1	1	5	2	8	2.5	8	4	7	5	4	5	5	75	2	1	2	2	1	1	1
94	1	41.5	11	4	1	1	1	5	2	4	1	3	5	4	4	4	5	5	85	2	4	2	2	1	2	1
242	1	50.5	21	4.5	1	1	1	5	7	6.5	2	8	4	7	5	4	5	5	60	1	5	2	2	1	3	1
157	1	45	20.5	3.5	2	1	1	4	7	7	3	4	4	7	5	4	4	5	70	2	1	2	5	4	2	1
234	1	40	12	4	2	1	1	5	7	6	3	2	5	7	4	4	5	5	80	2	1	2	2	1	2	1
235	1	33.5	13.5	5	2	1	1	4	2	6.5	3	4	4	5	5	4	5	5	50	1	1	2	2	1	1	1
318	1	53	12.5	2	1	1	1	5	2	3	2	4	4	5	5	4	5	5	55	1	4	2	4	1	2	1
196	1	69.5	20	6	2	1	1	5	2	6.5	3	2	4	7	4	4	5	5	70	1	1	2	3	1	1	1
525	1	56.5	17	6	2	1	1	5	5	4.5	3	4	4	7	5	4	4	5	75	1	1	2	2	1	1	1
839	1	50	14	6.5	1	1	2	5	4	4.5	1	8	5	7	5	4	4	5	75	1	4	2	2	1	1	1
74	1	38	18	3	2	1	1	5	9	10	3	4	4	4	4	4	5	5	75	1	1	2	2	1	2	1
195	1	45	12	3.5	1	1	1	4	4	4.5	1	4	4	7	5	4	5	5	75	1	4	2	2	1	1	1
19	1	31.5	13.5	6.5	2	1	1	5	2	8	3.5	8	4	7	5	4	5	5	75	1	1	2	2	1	1	1
118	1	35	21.5	5	2	1	1	5	7	7	2	8	5	7	5	4	5	5	80	1	1	2	2	1	1	1
206	1	33	28	3.5	2	1	1	5	2	5.5	2	4	5	4	4	4	5	5	85	1	1	2	2	1	?	1
375	1	78	23	8.5	1	1	1	4	2	7.5	3	3	4	4	5	4	5	5	50	2	6	2	3	1	1	1
59	1	69	19	5	2	1	1	5	2	7	2	4	4	5	5	4	5	5	50	2	1	2	3	1	1	1
162	1	36	14	3.5	2	1	1	5	2	3	2	4	4	7	5	4	5	5	55	2	1	2	3	1	1	1
256	1	75	28	9	2	1	1	4	2	10	3	8	4	7	5	4	5	5	60	2	7	2	2	1	1	1
77	1	52.5	12	2	1	1	1	5	2	4	1.5	8	4	7	5	4	5	5	60	2	4	2	2	1	1	1
396	1	66	15	4	1	1	1	5	2	8	4	8	4	7	5	4	5	5	60	2	4	2	4	1	1	1
279	1	27.5	14	2.5	2	1	1	4	2	5	2	8	4	7	5	4	4	5	70	2	1	2	3	1	1	1
286	1	54.5	17.5	5	1	1	1	4	4	4	1	3	5	7	5	4	4	5	70	2	5	2	3	3	2	1
174	1	51	20	4.5	1	1	1	4	9	4.5	2	4	4	4	4	4	5	5	70	2	5	2	4	1	1	1
84	1	46.5	19	4	2	1	1	4	2	4	2	4	4	7	5	4	5	5	70	2	1	2	4	1	2	1
453	1	42	10	3	2	1	1	4	4	2.5	1	3	5	4	4	4	5	5	75	2	1	2	3	1	1	1
227	1	43.5	13	5	2	1	1	5	5	6	3	4	5	4	4	4	5	5	75	2	1	2	4	3	3	1
44	1	49	12	5	2	1	1	5	4	4	1	4	4	7	4	4	5	5	75	2	1	2	3	1	1	1
174	1	49.5	15	3	1	1	1	4	2	4	1.5	4	5	7	5	4	5	5	75	2	5	2	3	1	1	1
515	1	25	16	3.5	2	1	1	5	2	7	3	7	4	7	5	4	4	5	55	?	1	2	4	1	1	1
42	1	21	14	4.5	2	1	1	4	2	5	2.5	4	4	7	4	4	5	5	75	?	2	2	3	1	1	1
93	1	21.5	5	3	2	1	1	5	7	9.5	3	4	5	7	5	4	5	5	75	?	1	2	3	1	1	1
160	1	25	18	6	2	1	1	4	4	5	0.5	4	5	7	5	4	5	5	85	?	1	2	2	1	1	1
128	1	41.5	11	3	1	1	1	5	2	2.5	1.5	4	5	7	5	4	5	5	80	1	4	2	3	3	2	1

289	1	35	9.5	2.5	2	1	1	5	9	5.5	1	3	4	7	5	4	4	5	75	2	1	2	3	2	4	1
59	1	53	15	7.5	2	1	1	5	4	9	2	8	5	6	5	4	5	5	80	2	2	2	2	1	1	17
30	1	43	11	3	1	1	1	5	2	5	2	4	5	4	4	4	5	5	85	1	6	2	3	1	1	1
226	1	33	10	3	2	1	1	4	4	3.5	1.5	4	5	7	5	4	5	5	85	1	1	2	6	1	2	1
17	1	58	21.5	8	1	1	1	5	2	9	3	8	4	7	4	4	5	5	50	2	7	2	2	1	1	1
254	1	42	19	3	2	1	2	5	2	11.5	3	4	4	4	5	4	5	5	60	2	1	2	2	1	1	1
723	1	66.5	25.5	5	2	1	1	5	2	9	3	4	6	7	5	4	5	5	80	2	1	2	2	1	1	1
77	1	55	11	3.5	1	1	1	4	2	4	1	8	4	4	5	5	5	5	50	1	4	0	3	1	1	1
290	1	42.5	9.5	5	1	1	2	4	2	3.5	1	3	4	4	5	5	5	5	75	1	6	0	2	1	1	1
262	1	41	13	2.5	1	1	1	4	2	3.5	1	4	4	4	5	5	5	5	60	2	5	0	3	1	1	1
198	1	50	12.5	3	2	1	1	5	5	4	2	8	4	4	5	5	5	5	65	2	1	0	2	1	2	1
426	1	60	14	3.5	1	1	1	5	2	7	2.5	4	4	7	5	5	5	5	65	2	4	0	3	1	1	1
86	1	64	14	4	2	1	1	5	2	4	1	8	4	4	5	5	5	5	70	2	1	0	3	1	1	1
125	1	59	10	3.5	1	1	1	4	2	2	1	8	1	1	8	5	5	5	70	2	4	0	3	1	3	1
11	1	49	14.5	6	2	1	1	4	3	1	0.5	2	1	1	8	5	5	5	nd	3	2	0	4	2	2	5
296	1	28.5	15	3	2	1	1	5	7	6	1	3	1	1	5	5	4	5	70	?	1	0	2	1	?	1
7	1	34.5	15	3	2	1	1	5	2	6	2	4	4	4	5	5	5	5	60	2	1	0	2	1	2	1
104	1	41	15.5	4.5	1	1	1	4	2	2.5	1	4	5	7	8	5	5	5	85	1	8	2	5	4	3	1
121	1	17	8	3	2	1	1	5	5	6	3	3	4	4	5	5	5	4	55	2	1	1	3	1	1	1
398	1	36	8.5	2	1	1	1	4	2	2.5	1	3	5	4	5	5	5	5	55	2	4	0	3	1	2	1
400	1	54	19	3	1	1	1	5	2	4	2	8	5	4	4	5	5	5	75	2	6	0	3	1	1	1
819	1	31.5	11.5	2.5	2	1	1	4	2	5	1.5	4	5	5	5	5	5	5	80	2	1	0	3	1	1	1
79	1	34	18	4.5	2	1	1	4	2	4.5	2	4	4	7	5	5	5	5	75	2	1	0	2	1	3	1
267	1	40	13.5	2	2	1	1	5	2	3.5	1.5	8	5	4	5	5	5	5	70	1	1	0	2	1	1	1
62	1	55	16.5	5	1	1	1	4	2	3	1	8	4	4	5	5	5	5	75	2	4	0	3	1	1	1
709	1	32	20	3	2	1	1	4	4	3	0.5	4	5	7	5	5	5	5	85	2	1	2	1	1	?	1
110	1	41.5	12	3	2	1	1	4	2	2.5	1	4	4	4	5	5	5	5	75	2	1	0	3	1	1	1
19	1	30	21.5	5	2	1	1	4	4	3.5	0.5	4	8	4	5	5	5	5	80	2	1	0	2	1	1	1
25	1	40	16.5	4.5	2	1	1	5	2	4.5	2	8	4	7	4	5	5	5	50	1	1	0	2	1	1	1
150	1	37.5	20	3	2	1	1	4	2	5	1.5	8	4	4	5	5	5	5	50	2	1	0	3	1	1	1
320	1	73	20	5.5	1	1	1	5	5	9.5	4.5	4	4	7	5	5	5	4	60	1	4	1	2	1	1	1
16	1	53	13	6	1	1	1	5	2	3.5	2	3	4	7	5	5	4	5	70	1	5	2	3	3	1	1
73	1	47	13	4	2	1	1	4	9	5	1.5	8	4	4	4	5	5	5	75	1	1	0	5	3	3	1
325	1	22	13	3	2	1	1	5	9	5	1.5	8	4	7	5	5	5	5	75	1	1	0	2	1	1	1
44	1	108	26	6	1	1	1	5	2	10	3.5	8	4	5	5	5	5	5	40	2	4	0	2	1	2	1
20	1	51	16.5	4.5	2	1	1	5	5	6	4	8	4	7	5	5	5	4	50	2	1	1	3	1	1	1
45	1	42	17	3	2	1	1	4	2	9	3	8	4	7	5	5	5	5	50	2	1	0	3	1	1	1
206	1	82	19	8	2	1	1	5	5	5	2.5	4	4	7	5	5	5	4	55	2	1	1	2	1	1	1
19	1	27.5	18	6	2	1	1	5	2	7.5	3	3	4	7	5	5	5	5	55	2	1	0	2	1	3	1
144	1	29	14	4	2	1	1	5	2	5	2	4	4	7	5	5	5	4	60	2	1	1	2	1	1	1
98	1	74	26	5	2	1	1	4	2	7	2.5	3	4	7	5	5	5	5	60	2	1	0	5	3	3	1
29	4	130	23	12	1	1	1	5	2	9	3	4	4	4	5	5	5	5	65	2	5	0	5	4	2	1
814	1	35.5	16.5	5.5	2	1	1	4	2	2.5	1.5	8	4	7	5	5	5	5	70	2	1	0	2	1	1	1
258	1	61	23.5	5.5	1	1	1	5	7	6	2	8	4	7	5	5	4	5	80	2	5	2	3	1	3	1
234	1	21.5	8	3	2	1	1	4	4	3	1	4	4	7	5	5	5	5	80	2	1	0	2	1	1	1

(continued)

Table A.28 (continued)

Nr.	Blank	Length	Width	Thickness	Pres.: L	Pres.: W/T	Cortex	SPR: kind	SPR: form	SPR: length	SPR: width	DR	Lip	Bulb	Bulbar scar	Impact point	Esq. du bulbe	en éperon	Flaking angle	Profile	Distal	Hammer	n Negatives	Direc. Neg.	Reg Neg.	Tool
57	1	31.5	26.6	5	2	1	1	4	4	5	1.5	4	5	7	5	5	5	5	80	2	1	0	3	1	2	1
148	1	30	22	5	2	1	1	5	2	6.5	2.5	4	4	7	5	5	5	5	65	?	1	0	2	1	1	1
379	1	24.5	15.5	4	2	1	1	5	2	9	4.5	4	4	4	5	5	5	5	70	?	1	0	2	1	1	1
44	1	57.5	21.5	5.5	2	1	1	5	2	5.5	2	8	4	7	5	5	5	5	65	1	2	0	3	1	1	2
662	1	59	15	3	2	1	1	4	4	4	1	4	4	7	4	5	5	5	70	2	2	0	3	1	2	17
360	1	66	17	4	1	1	1	4	2	3	1.5	4	5	4	5	5	4	5	75	2	5	2	3	1	3	1
449	4	50.5	11	5	1	1	1	4	2	2.5	2	8	4	4	5	5	5	5	75	2	4	0	5	4	1	1
285	1	36	11	4	2	1	1	5	4	3	1	4	4	4	4	5	5	5	70	2	1	0	3	1	1	1
574	1	48.5	12	2.5	1	1	1	4	9	2.5	0.5	3	5	4	5	5	5	5	75	2	5	0	3	1	1	1
354	1	55	14	4.5	1	1	1	4	2	3	1.5	8	4	7	5	5	5	5	60	1	5	0	5	3	2	1
623	1	68	21.5	5.5	2	1	1	5	5	8	3	3	4	7	5	5	5	5	75	2	1	0	2	1	1	1
83	1	54	14	4	1	1	1	4	2	3	1	4	4	4	6	5	5	5	85	2	4	0	3	1	2	1
734	1	84	25	9	2	1	1	5	5	10	4.5	8	4	7	5	5	5	4	50	3	2	1	3	1	1	5

Feature explanations and keys for numbers see Chap. 2, Blade Technology, Recording of Technological Information, nd: feature expression not recordable

Table A.29 Blade technology analysis—Oelknitz

Box	Trench	Level	Pit	Blank	Length	Width	Thickness	Pres.: L	Pres.: W/T	Cortex	SPR: kind	SPR: form	SPR: length	SPR: width	DR	Lip	Bulb	Bulbar scar	Impact point	Esq. du bulbe	en éperon	Flaking angle	Profile	Distal	Hammer	n Negatives	Direc. Neg.	Reg Neg.	Tool
690	XV	3	?	1	38.5	16	4	2	1	1	5	2	7	3.5	3	4	7	5	5	5	4	60	1	1	1	3	1	1	1
690	XV	3	?	1	34	13.3	4	1	1	1	5	9	7	2	4	4	7	5	5	5	5	60	1	6	0	2	1	1	1
690	XV	3	?	1	50	20.5	5.5	2	1	1	5	5	11	4	4	4	4	5	4	5	4	65	1	1	0	5	5	4	1
690	XV	3	?	1	59	12	7	2	1	1	4	5	5	3	8	4	7	4	5	5	4	65	1	1	1	3	1	2	1
690	XIV	3	?	1	29	16.5	4.5	2	1	1	5	2	8	3	3	4	7	5	4	5	5	75	1	1	2	2	1	1	1
613	Xh	4	?	1	35	22	5	2	1	1	5	2	12	4.5	8	4	4	5	4	5	5	85	1	1	2	2	1	2	1
613	X	4	?	1	44	15	5	2	1	1	5	2	8	4	8	4	7	5	5	5	4	40	2	2	1	3	1	1	5
690	XV	3	?	1	37	18	4	2	1	1	5	2	8	3.5	4	4	5	5	5	5	4	45	2	1	1	2	1	1	1
690	XIV	3	?	1	34	10	5	2	1	1	5	2	5	2.5	8	4	7	5	5	5	4	45	2	1	1	3	1	1	1
690	XIV	3	?	1	23	13	3	2	1	1	5	2	6	2	3	4	5	4	5	5	4	50	2	1	1	2	1	1	1
613	X	4	?	1	84	26	7	2	1	1	5	2	8	3	8	4	7	5	5	5	4	55	2	1	0	2	1	1	1
690	XV	3	?	1	49	15	5	2	1	1	5	5	5.5	3	3	4	7	5	5	5	4	60	2	1	1	2	1	1	1
613	X	4a	?	1	56	17	4.5	2	2	1	4	2	2.5	1	4	4	7	6	5	5	5	60	2	2	0	2	1	2	4
613	Xh	4	?	1	49	14.5	5	2	1	1	4	4	3	1	8	4	5	5	4	5	5	65	2	2	2	3	1	2	17
690	XIV	3	?	1	40	20	5	2	1	1	5	2	7	3	3	4	7	5	5	5	4	70	2	1	1	3	4	2	1
613	Xf1	4a	?	1	25	13	4	2	1	1	5	9	8	2.5	3	4	7	5	5	5	5	70	2	1	0	2	1	1	1
690	XIV	3	?	1	27	14.5	5	2	1	1	5	7	6	2.5	4	4	7	5	5	5	5	75	2	1	1	2	1	1	1
690	XIV	3	?	1	40	11	4	1	1	1	5	9	5.5	1.5	3	4	4	4	4	5	5	80	2	4	2	3	1	2	1
690	XV	3	?	1	48	12	3.5	1	1	1	4	2	5.5	2	3	4	7	5	5	5	5	55	3	4	0	2	1	1	1
613	Xf1	4a	?	1	38.5	16	4	2	1	1	5	5	8.5	4.5	4	5	4	5	5	5	4	50	2	1	1	2	1	2	1
690	XV	3	?	1	28.5	11	3.5	2	1	1	5	5	6	3	4	5	4	4	5	5	4	55	2	1	1	3	1	1	1
690	XV	3	?	1	48	17	5	2	1	1	5	2	15	6.5	4	5	4	4	5	5	4	60	2	1	1	3	1	2	1
613	Xh	4	?	1	25.5	15	3	2	1	1	5	2	3	2.5	4	5	4	5	5	5	4	70	2	2	1	3	1	1	17
613	Xh	4	?	1	40.5	10.5	3	2	1	1	4	4	4	1	3	5	7	6	4	5	5	70	2	1	2	3	3	1	1
613	Xh	4	?	1	23	13	3	2	1	1	5	7	9	3	4	5	4	5	4	5	5	80	2	1	2	4	1	3	1
613	Xf1	4a	?	1	29.5	13	3.5	2	1	1	5	7	5	2	8	5	4	5	4	5	5	80	2	1	2	3	1	2	1
690	XV	3	?	1	55	19	5.5	2	1	1	5	2	10	3.5	4	5	7	4	5	5	5	80	2	1	0	2	1	1	1
627	XIII	?	7	1	37	15	7	2	1	1	5	2	15	5	3	4	4	5	5	5	5	70	1	1	0	3	?	2	1
627	XIII	?	8	1	35	15	5	2	1	1	5	2	7	3	3	5	4	6	5	5	5	85	1	1	0	2	1	2	1
627	XIII	?	17	1	75.5	20.5	8	1	1	1	5	2	9.5	3.5	5	4	5	5	5	5	5	50	2	5	0	2	1	1	1
612	VII	?	29	1	29	18	6	2	1	1	5	2	8	3	3	4	7	5	5	5	4	50	2	1	1	2	1	1	1
612	VII	?	29	1	51	24	5	2	1	1	4	2	12	5	3	4	7	5	5	5	5	50	2	1	0	2	1	1	1
612	VII	?	29	1	60	14.5	4	2	1	1	5	2	8	3	8	6	7	5	5	5	5	70	2	1	0	3	1	1	1
627	XIII	?	38	1	63.5	26	4	2	1	1	5	2	10	3	8	4	7	4	5	5	4	65	2	1	1	2	1	1	1
612	VII	?	46	1	46	10	4.5	2	2	1	5	2	6.5	2	4	4	7	5	5	5	4	70	2	4	1	2	1	1	14
624	XI	?	140	1	52	17	3.5	2	1	1	5	2	7	1	8	4	7	5	5	4	4	65	1	1	1	2	1	1	1
624	XI	?	140	1	75.5	22	5	2	1	1	5	2	9	3.5	4	4	7	4	4	5	5	55	2	1	2	5	4	3	1
624	XI	?	140	1	38	21	6.5	2	1	1	5	2	9.5	3.5	4	5	4	4	4	5	5	60	2	1	2	3	2	1	1
624	XI	?	147	1	61	13.5	3	1	1	1	5	2	3.5	1.5	3	4	5	5	5	5	5	60	1	6	0	2	1	1	1
624	XI	?	147	1	82	20.5	5	1	1	1	5	2	6.5	3	3	4	7	5	5	5	5	80	2	5	0	2	1	1	1
624	XI	?	147	1	69	14	5	2	1	1	5	5	3.5	2	2	5	7	5	5	5	5	50	2	1	0	2	1	1	1
624	XI	?	147	1	64	14	3	2	1	1	5	2	6	2	3	5	4	5	5	5	4	55	2	1	1	4	1	1	1
624	XI	?	147	1	115.5	22.5	4	1	1	1	5	2	5.5	2	8	5	7	5	5	5	5	60	2	4	0	3	1	1	1

(continued)

Table A.29 (continued)

Box	Trench	Level	Pit	Blank	Length	Width	Thickness	Pres.: L	Pres.: W/T	Cortex	SPR: kind	SPR: form	SPR: length	SPR: width	DR	Lip	Bulb	Bulbar scar	Impact point	Esq. du bulbe	en éperon	Flaking angle	Profile	Distal	Hammer	n Negatives	Direc. Neg.	Reg Neg.	Tool
690	XIV	4	H/I 3	1	95	24	5	1	1	1	5	2	14	5	3	4	7	5	5	5	4	55	2	5	1	3	1	1	1
627	XIII	?	12a+b	1	45	22	5	2	1	1	5	2	12	3	3	4	7	4	5	5	4	75	1	1	1	2	1	2	1
627	XIII	?	12a+b	1	66.5	15.5	5	2	1	1	5	2	8	3	3	4	7	5	5	5	4	45	2	1	1	2	1	1	1
601	V	?	173/316	1	27	8	3	2	2	1	5	5	6	3	3	6	7	5	5	5	4	60	2	1	1	2	1	1	1
601	V	?	176/318	1	48	14	5	2	1	1	5	2	6	2	2	5	7	5	4	5	5	60	2	1	2	3	1	2	1
601	V	?	181/320	1	39	12.5	2.5	1	1	1	5	9	8	2	3	4	4	5	5	5	4	80	1	5	1	3	1	1	1
601	V	?	183/322	1	51	15	7	1	1	1	5	2	9	2	8	5	7	6	5	5	4	65	1	5	0	4	1	3	1
601	V	?	187/326	1	62.5	19	7	1	1	1	5	5	15	7	8	4	4	5	5	5	4	80	1	4	0	4	?	2	1
601	V	?	187/326	1	71	19	8	1	1	1	5	4	7.5	1	4	4	7	4	4	5	5	80	1	7	2	4	1	2	1
601	V	?	187/326	1	41	19	6.5	2	1	1	5	5	10	5.5	3	4	7	5	5	5	6	70	2	1	0	3	1	2	1
601	V	?	187/326	1	70	22.5	8.5	2	1	1	5	7	8.5	4	2	5	7	4	5	5	5	80	1	1	0	2	1	2	1
601	V	?	187/326	1	33	14.5	5.5	2	1	1	5	9	10	2.5	8	5	7	4	4	5	5	85	1	1	2	3	1	1	1
601	V	?	189/331	1	44	14	3	1	1	1	5	2	6.5	2	8	6	4	4	5	5	4	70	1	5	0	2	1	1	1
612	VII	?	29/206	1	59	14	4	2	1	1	5	5	5	3	4	4	7	5	5	5	4	50	3	1	1	3	1	2	1
612	VII	?	30/207	1	35	22.5	5	2	1	1	5	2	11	4	8	4	4	5	4	5	5	80	2	1	2	3	1	3	1
613	X	4	31/44/45	1	37	10.5	4	2	1	1	5	2	8.5	3	3	4	7	4	4	5	4	55	2	1	2	2	1	1	14
613	X	4	31/44/45	1	25.5	11	2	2	1	1	4	9	5	1.5	3	4	7	4	4	5	5	60	2	1	2	3	1	2	1
612	VII	?	36/213	1	54	14	3.5	1	1	1	5	2	5	2	4	4	4	5	5	5	4	60	2	4	1	5	4	2	1
612	VII	?	36/213	1	39	16	2	2	1	1	5	2	7	2	4	4	7	5	5	5	4	60	2	1	1	3	1	2	1
612	VII	?	36/213	1	44	10	2.5	2	1	1	5	2	3	1	3	5	4	4	4	5	5	70	1	1	2	2	1	1	1
612	VII	?	36/213	1	37	7	4	2	2	1	5	5	4	2.5	3	5	4	5	5	5	5	80	2	1	0	2	1	1	13
660	VII	4	368/hearth	1	25	13.5	4	2	1	1	5	2	5.5	2	3	4	7	5	4	5	5	70	1	1	2	2	1	3	1
660	VII	4	368/hearth	1	24	24.5	4	2	1	1	5	2	7	3	2	4	7	5	5	5	4	75	1	1	1	3	1	1	1
660	VII	4	368/hearth	1	38	13	5	2	1	1	5	2	7	2.5	8	4	7	4	5	5	4	60	2	1	1	2	1	1	1
660	VII	4	368/hearth	1	51	16	7	2	1	1	5	2	9	4	3	5	4	5	5	5	4	65	1	1	1	4	4	3	1
660	VII	4	368/hearth	1	23	18	3.5	2	1	1	5	2	4.5	2	4	5	7	4	4	5	5	80	1	1	2	3	1	1	1
660	VII	4	368/hearth	1	59	18	5	2	1	1	5	2	10	3	4	6	4	5	4	5	5	80	3	1	2	5	4	2	1
612	VII	?	40/215	1	43.5	14.5	5	1	1	1	5	2	7.5	3.5	4	4	7	4	5	5	4	60	2	6	1	4	1	2	1
612	VII	?	40/215	1	60	12.5	4.5	2	1	1	5	2	12	4	2	5	5	5	5	5	4	60	2	1	1	3	1	1	1
613	X	4	41/44/45	1	30	12	3	2	1	1	5	4	9	2	3	5	4	4	4	5	5	80	2	1	2	3	?	2	1
612	VII	?	46/221	1	48	19	4	2	1	1	5	4	9	2	3	4	4	5	5	5	5	60	2	1	0	2	1	1	14
612	VII	?	48/49/223	1	43.5	12	5	2	1	2	5	5	4	2	4	4	7	5	4	5	5	70	2	1	2	2	1	1	1
612	VII	?	53/224	1	51	15	3	2	1	1	4	5	8	2	4	4	4	5	5	5	5	60	3	5	0	2	1	2	1
612	VII	?	55/225	1	29	7	3	2	2	1	5	2	3	1	4	4	5	5	5	5	4	60	2	1	1	2	1	1	13
Pits with older ^{14}C-dates																													
627	XIII	?	19	1	29	7	2.5	1	1	1	5	2	4	2	3	5	4	4	4	5	5	75	2	4	2	2	1	2	1
627	XIII	?	19	1	16	10	3	2	1	1	5	5	5	3	2	5	4	4	5	5	6	80	?	1	0	2	1	1	1
627	XIII	?	19	1	36	17	5	2	2	1	5	2	8	3	3	6	4	5	5	5	4	65	2	1	1	2	1	1	1
616	VII	?	61/229	1	50	8.5	3	1	1	1	5	2	4	2	8	4	7	4	5	5	5	75	1	5	0	2	1	2	1
616	VII	?	61/229	1	33	14	5.5	2	1	1	5	4	8	2	8	4	7	5	4	4	5	80	1	1	2	2	1	2	1
616	VII	?	61/229	1	20	10.5	3	2	1	1	5	2	3.5	1	2	4	7	5	5	5	4	40	?	1	0	2	1	1	1
616	VII	?	61/229	1	50	18	7.5	2	1	1	5	7	14	7	3	5	4	5	5	5	5	80	2	2	0	3	4	4	4
616	VII	?	61/229	1	23	13.5	3.5	2	1	2	5	2	2.5	1.5	2	5	7	5	4	5	5	85	2	1	0	2	1	1	1
585	VII	?	113/264	1	69.5	23	4	2	1	1	5	2	6	2.5	4	5	7	5	5	5	4	60	2	1	1	3	1	1	1
585	VII	?	113/264	1	39	8	3	1	1	1	5	4	3.5	1	8	5	4	4	5	5	5	80	2	4	0	3	1	2	1

585	VII	?	113/264	1	29	6	2	2	2	1	5	2	4	1.5	2	5	4	5	5	5	5	nd	2	1	0	3	1	1	13
585	VII	?	113/264	1	29	10	2.5	2	1	1	5	2	4	1.5	4	6	7	4	5	5	4	60	2	1	1	3	1	1	1
610	IV	?	146/292	1	52	20	4	2	1	1	5	2	11	3	8	4	7	5	5	5	4	45	2	2	1	3	1	2	2
610	IV	?	146/292	1	45	17	5	2	1	1	5	2	7	2	3	4	7	4	4	5	5	45	2	1	2	2	1	2	1
610	IV	?	146/292	1	39	18	4	1	1	1	5	2	4.5	2	8	5	7	5	5	5	5	75	1	6	0	2	1	1	1
610	IV	?	146/292	1	56	24	4.5	2	1	1	4	9	16	3	4	5	4	5	4	5	5	85	1	1	2	3	1	3	1
610	IV	?	146/292	1	66	19.5	5	1	1	1	5	2	7	2.5	4	5	7	4	4	5	5	70	2	5	2	3	1	1	1
610	IV	?	146/292	1	59	17.5	6	2	1	1	5	7	7	3	8	5	7	4	5	5	5	85	2	2	0	3	1	1	4
Pits with younger ^{14}C-dates																													
627	XIII	?	26	1	81	23	6	1	1	1	5	2	9	2.5	4	4	7	5	5	5	4	65	3	5	1	4	4	3	1
627	XIII	?	26	1	25	14.5	4	2	1	1	5	2	5	2	4	5	7	4	5	5	5	85	1	1	0	2	1	1	1
663	VII	?	26/203	1	86	18	4.5	1	1	1	5	2	8.5	3.5	4	5	7	5	5	5	4	55	3	5	1	5	3	3	1
585	VII	?	119/268	1	55	14	5	2	1	1	5	4	5	1	3	4	4	5	5	5	5	55	1	1	0	3	1	2	4
585	VII	?	119/268	1	76	26.5	8	2	1	1	5	5	12	6	8	4	7	4	4	5	4	55	2	2	2	3	1	2	2
585	VII	?	119/268	1	51.5	15	5.5	2	1	1	5	2	7	2.5	3	4	7	5	5	5	4	60	2	1	1	2	1	1	1
585	VII	?	119/268	1	47	14	5	2	1	1	5	2	6	3	8	5	4	5	4	5	5	85	1	1	2	2	1	1	1
585	VII	?	119/268	1	66	26.5	6	2	1	1	5	5	14.5	6.5	8	5	4	5	5	5	4	55	2	2	1	4	1	2	2
602	V	?	162/308	1	48.5	14	5	1	2	1	5	5	6	3	4	4	7	5	5	5	5	85	1	6	0	2	1	1	1
602	V	?	162/308	1	70	21.5	5	2	1	1	5	2	6	2	2	4	7	5	5	5	6	55	2	1	0	4	4	2	1
602	V	?	162/308	1	52.5	24	4.5	2	1	1	5	4	11	2	8	4	7	5	5	5	5	60	2	2	0	3	1	1	2
602	V	?	162/308	1	24.8	18.5	4	2	1	1	4	2	9	3	8	4	7	4	4	5	5	55	?	1	2	2	1	1	1
602	V	?	162/308	1	66.5	18.5	5	2	1	1	5	2	10	5	4	5	4	4	5	5	4	60	2	2	1	4	1	1	2

Feature explanations and keys for numbers see Chap. 2, Blade Technology, Recording of Technological Information, nd: feature expression not recordable

Table A.30 Blade technology analysis—Rislisberghöhle

Nr.	Blank	Length	Width	Thickness	Pres.: L	Pres.: W/T	Cortex	SPR: kind	SPR: form	SPR: length	SPR: width	DR	Lip	Bulb	Bulbar scar	Impact point	Esq. du bulbe	en éperon	Flaking angle	Profile	Distal	Hammer	n Negatives	Direc. Neg.	Reg Neg.	Tool
26b-7651	1	46	15.5	4	1	1	1	4	4	2	0.5	4	5	7	5	4	4	5	80	1	7	2	3	1	2	1
54a-16944	1	31	5	2	2	1	1	4	2	1.5	1	3	1	7	5	5	4	5	65	2	1	2	2	1	1	1
54d-17851	1	48	10	2.5	1	1	1	4	2	2	1	8	4	4	5	5	5	5	75	2	4	2	2	1	1	1
45-14871	1	29	18	3	2	1	1	4	7	2.5	1	3	5	4	5	4	5	5	75	1	1	2	2	1	1	1
56a-18200	1	44	8	2.5	1	2	1	4	2	2.5	1	8	4	4	5	4	5	5	60	2	4	2	4	1	3	9
66a-19704	4	42	18	3.5	2	1	1	4	2	2.5	1	4	4	4	5	5	5	5	75	2	1	0	5	4	2	1
52c-15772	1	44	12	3	1	1	1	4	2	2.5	1	4	4	4	5	4	5	5	80	2	4	2	2	1	1	1
54a-16946	1	46	11	4	1	1	1	4	2	3	1	8	4	4	5	4	5	5	60	2	4	2	2	1	1	1
45-14865	1	64	20	6	2	1	1	4	2	3	1	3	5	4	4	4	5	5	80	2	1	2	4	1	1	1
53a-15892	1	39	13	2.5	2	1	1	4	4	3.5	1	8	4	4	5	4	5	5	75	2	1	2	2	1	1	1
75lb-20168	1	41	11	4	1	1	1	5	7	3.5	1	8	4	4	5	4	5	5	85	2	4	2	6	?	2	1
65-19364	1	30	10	4.5	1	1	1	4	9	4	1	8	5	7	5	4	5	5	75	1	6	2	2	1	1	1
45-14874	1	42.5	13.5	4	2	1	1	4	4	4	1	8	4	4	5	4	5	5	50	2	4	2	2	1	1	1
43a-12352	1	63	11	5	1	1	1	4	3	2	2	8	4	4	4	5	5	5	75	2	4	0	2	1	1	1
66d-20063	1	28	12.5	4	2	1	1	4	4	4	1	4	4	4	4	4	5	5	80	2	1	2	4	1	1	1
66d-20069	1	42	16	3	2	1	1	4	4	5	1	8	4	7	5	4	5	5	70	2	1	2	3	1	2	1
56a-18367	1	42	17.5	3	2	1	1	4	4	5	1	4	4	4	5	4	5	5	75	2	1	2	2	1	1	1
51b-15611	4	55	10	5	2	1	1	4	2	3.5	1.5	8	4	4	5	5	5	5	55	2	1	0	7	4	2	1
55-18012	1	40.5	16.5	5	2	1	1	4	2	3.5	1.5	8	4	4	5	4	5	5	70	2	1	2	3	1	1	1
55-18015	1	47	20	6	2	1	1	4	2	3.5	1.5	8	4	7	5	4	5	5	70	2	1	2	2	1	2	1
66d-20064	1	40	10	4	2	1	1	4	2	3	2	4	4	7	5	5	5	5	80	1	1	0	2	1	1	1
51b-15607	1	51	14.5	4	1	1	1	5	2	3	2	4	4	4	5	5	5	5	65	2	5	0	3	1	2	1
24c-6358	1	24.5	11	3	2	1	1	4	5	3	2	8	4	4	5	5	5	4	70	2	1	0	2	1	1	1
51d-15707	1	58	15.5	6	2	1	1	4	2	3	2	8	4	4	5	4	5	5	70	2	2	2	3	3	2	19
24c-6465	1	33	10.5	3	1	1	1	4	4	6	1	4	5	4	5	4	5	5	80	2	4	2	2	1	1	1
43c-12876	1	33	15	3	2	1	1	4	2	4	1.5	8	6	4	5	4	5	5	80	2	1	2	2	1	1	1
66a-19742	1	19	15	2.5	2	1	1	5	2	4	1.5	4	5	4	5	5	5	5	80	2	1	0	2	1	1	1
43a-12286	1	40	14	6.5	2	1	3	4	7	4.5	1.5	8	4	7	5	5	5	5	75	2	2	0	2	1	2	2
24c-6374	1	21.5	11.5	4	2	1	1	4	2	3.5	2	4	4	7	5	5	5	5	75	2	1	0	2	1	1	1
54a-16941	1	24.5	13.5	5	2	1	1	4	5	5	1.5	4	5	7	5	4	5	5	70	2	1	2	2	1	1	1
43a-12351	1	40	13	4.5	2	1	2	4	7	4	2	4	4	7	5	5	4	5	65	1	1	0	2	1	1	1
52a-15730	1	49	14.5	3	1	1	1	5	7	4	2	3	6	4	4	4	4	5	75	1	4	2	2	1	1	1
51a-15493	1	43	12.5	3.5	2	1	1	4	2	4	2	4	4	4	4	4	5	5	60	2	1	2	2	1	1	1
26c-7931	1	22.5	11	2	2	1	1	4	7	4	2	4	4	4	5	5	5	4	75	2	1	0	2	1	2	1
24c-6395	1	44	9	4	2	1	1	4	2	4	2	8	6	4	5	5	5	5	75	2	2	0	3	1	1	17
54b-17293	1	52	14.5	3	1	1	1	4	2	4	2	8	4	4	5	4	5	5	80	2	4	2	3	1	1	1
26b-7640	1	42	14	5	2	1	1	4	7	4	2	4	4	4	5	5	5	5	80	2	2	0	2	1	2	1
24c-6411	1	38	12	4	2	1	2	5	7	5.5	1.5	8	4	6	4	4	5	5	70	2	1	2	4	1	2	1
42a-12054	4	45.5	14.5	3.5	2	1	2	5	2	4.5	2	8	4	4	5	5	5	5	50	2	1	0	4	4	4	1
56a-18371	1	42.5	19	3	2	1	1	4	2	4.5	2	3	4	7	5	4	4	5	75	2	1	2	2	1	2	1
24c-6390	1	33	15.5	4	2	1	1	5	7	6	1.5	8	5	6	5	5	5	5	75	2	1	0	3	1	2	14
57a-18961	1	56	19	3	2	1	1	4	2	6.5	1.5	8	4	7	5	4	4	5	80	2	1	2	2	1	2	1

53c-16666	1	55.5	24.5	5.5	2	1	1	4	7	5	2	4	4	7	4	4	5	5	65	2	1	2	2	1	2	1
65-19416	1	33	22	7	2	1	1	4	2	5	2	4	4	4	5	4	5	5	65	2	1	2	2	1	1	1
46b-15096	1	64	19	5	2	1	1	4	7	5	2	4	6	7	4	4	4	5	70	2	1	2	3	1	1	1
75Ib-20129	1	32	12	5	2	1	1	5	5	5	2	2	5	7	5	4	5	5	70	2	2	2	2	1	1	1
43c-12945	1	19	15	3	2	1	1	5	2	5	2	5	4	7	5	5	5	5	70	2	1	0	?	?	?	1
43a-12338	1	41	10	4	1	1	2	4	9	7	1.5	4	5	4	4	5	5	5	80	2	4	2	4	1	3	1
26c-7824	1	48	17	3.5	1	1	1	5	7	5.5	2	4	4	4	4	4	4	5	80	1	5	2	3	1	2	1
24c-6386	1	32	8	2	1	1	1	5	7	5.5	2	8	6	4	4	5	5	5	70	2	4	0	2	1	1	1
51a-15477	1	23.5	11.5	3.5	2	1	1	4	2	6	2	8	4	5	5	5	5	5	65	2	1	0	2	1	1	1
24c-6375	1	23	8.5	2	2	1	1	5	7	6	2	8	5	4	4	5	5	5	75	2	1	0	3	1	1	1
24c-6361	1	24	8.5	2.5	2	1	1	4	5	6	2	8	4	7	5	5	5	5	75	2	1	2	2	1	1	1
42a-12075	1	43.5	12	3.5	2	1	1	5	7	6	2	8	4	4	4	5	5	5	80	2	1	2	3	1	2	1
43c-12901	1	31	16	5	2	1	1	5	7	6.5	2	4	5	7	5	4	4	5	80	2	1	2	2	1	1	1
53a-15878	1	33.5	15.5	5	2	1	1	4	2	7	2	8	4	4	5	4	5	5	75	2	1	2	3	1	1	1
26b-7647	1	40	18	4.5	2	1	2	5	9	7	2	4	5	4	4	5	5	5	75	2	1	0	3	1	3	1
43a-12325	1	50	12	6	1	1	1	5	5	5	3	4	4	4	5	5	5	5	80	1	2	2	3	3	2	1
45-14870	1	51	17.5	5	2	1	1	5	2	5	3	8	4	7	5	4	4	5	60	2	1	2	2	1	1	1
24c-6469	1	27	11	13	2	1	1	5	9	7.5	2	8	4	4	5	4	5	5	65	2	1	2	2	1	2	1
75Ib-20130	1	44	17	3.5	2	1	2	5	9	7.5	2	4	5	4	4	5	5	5	80	2	1	2	2	1	1	1
64a-19149	1	33.5	16	4.5	2	1	1	5	9	8	2	4	5	4	4	5	5	5	85	2	1	2	2	1	1	1
41d-12007	1	45	15	4	2	1	1	5	4	8.5	2	4	4	7	4	5	5	5	75	2	1	2	5	3	3	1
75Ib-20145	1	24.5	12.5	3	2	1	2	5	2	7	2.5	4	4	4	5	4	5	5	70	2	1	2	2	1	1	1
56a-18201	1	41	17	4	2	1	1	5	5	5	3.5	2	4	7	5	5	5	6	70	2	1	0	3	1	2	14
24c-6373	1	38.5	12	7	1	1	2	5	7	5	3.5	4	4	4	5	5	5	4	85	2	7	0	5	?	4	1
42a-12076	1	32.5	16	3	2	1	1	5	2	6	3	8	4	4	5	5	4	5	85	1	1	2	2	?	nd	1
43a-12300	1	38	17	5	1	1	1	4	2	5	4	8	4	4	4	5	5	5	75	1	6	0	2	1	2	1
42b-12083	1	33	13.5	5	2	1	1	5	2	8	2.5	4	6	4	5	5	5	5	80	2	2	0	2	1	1	1
56a-18204	1	50	21	6	2	1	1	5	2	7	3	3	4	7	5	4	5	5	50	2	1	2	3	4	2	1
53c-16671	1	53	17.5	6	2	1	1	5	2	7	3	8	4	7	5	5	5	4	55	2	1	1	2	1	1	1
54a-17090	1	63	13	4	2	1	1	5	2	7	3	3	4	7	5	4	5	5	55	2	1	2	4	1	2	1
43c-13148	1	36	11.5	3	2	1	2	5	7	7	3	4	4	4	5	5	5	5	70	2	1	0	4	3	1	1
45-14766	1	34	11	3.3	2	1	1	4	2	8.5	2.5	8	6	6	4	4	4	5	60	2	1	2	3	1	1	1
65-19342	1	36	15	4	2	1	1	5	4	8.5	2.5	4	4	4	5	5	5	5	70	2	1	0	3	1	1	1
46c-15294	1	56.5	11.5	4.5	1	1	1	5	2	7.5	3	3	5	7	5	5	5	5	40	2	6	2	2	1	1	1
43c-13161	1	38	22.5	5	2	1	1	5	5	8	3	4	4	7	5	5	5	4	65	2	2	1	2	1	1	1
43a-12299	1	26	15	6	2	1	1	5	7	7	3.5	8	4	4	4	4	5	5	70	2	1	2	2	1	1	1
65-19347	1	25	17	5.5	2	1	1	5	2	7.5	3.5	8	5	4	5	4	5	5	75	1	1	2	2	1	1	1
43a-12355	1	22.5	16.5	4.5	2	1	1	4	2	9	3	4	4	4	4	5	5	5	75	?	1	0	2	1	1	1
41d-12010	1	64.5	25.5	6.5	1	1	1	5	4	14	2	8	4	7	5	5	5	5	80	2	4	0	3	1	2	1
46b-15091	1	78	24	8	2	1	1	5	2	9.5	3	4	4	7	5	5	5	5	70	2	1	0	2	1	1	1
26c-7916	1	106	23	6.5	1	1	1	5	2	10	3	4	4	7	5	5	5	5	55	2	5	0	3	?	3	1
43a-12364	1	48	18	5.5	1	1	1	5	2	10	3	8	4	4	5	5	5	5	75	2	4	0	3	?	4	1
24c-6355	1	25.5	18.5	4	2	1	1	5	5	7.5	4	4	4	4	5	5	5	4	80	?	1	0	2	1	1	1
26c-7890	1	32.5	14.5	3.5	2	1	1	5	2	9	3.5	4	4	4	4	4	5	5	75	1	2	2	1	1	1	2
24c-6409	1	33	16.5	6	2	1	1	4	5	8	4	4	4	4	4	4	5	4	70	1	1	2	2	1	3	1

(continued)

Table A.30 (continued)

Nr.	Blank	Length	Width	Thickness	Pres.: L	Pres.: W/T	Cortex	SPR: kind	SPR: form	SPR: length	SPR: width	DR	Lip	Bulb	Bulbar scar	Impact point	Esq. du bulbe	en éperon	Flaking angle	Profile	Distal	Hammer	n Negatives	Direc. Neg.	Reg Neg.	Tool
66d-82002	1	39	16	6	2	1	1	4	2	8	4	4	4	7	5	5	5	5	70	2	1	0	4	3	3	1
43a-12314	1	76	17.5	5	2	1	1	5	5	6.5	5	4	5	7	4	4	5	5	70	2	1	2	2	1	1	1
53a-15865	1	22	13.5	3	2	1	1	5	9	11	3	4	4	4	5	5	5	5	55	1	1	0	3	1	1	1
42a-12047	1	39	16	6.5	2	1	1	5	5	8.5	4	4	4	4	5	5	5	5	75	2	2	0	4	1	2	1
26c-7927	1	35	19.5	4.5	2	1	1	5	2	11.5	3	4	4	7	4	4	5	5	70	1	1	2	3	1	2	1
75Ib-20131	1	60.5	24	9	2	1	3	5	5	7	5	4	4	7	5	5	5	5	70	2	2	0	5	?	4	4
52d-15805	1	57.5	17.5	7.5	2	1	1	5	2	10	3.5	4	4	4	5	5	5	4	75	2	1	0	6	3	4	1
53a-15857	1	32.5	15	5	2	1	1	5	2	9.5	4	8	4	4	5	4	5	5	80	2	1	2	3	1	2	1
55-18024	1	55	15	4	1	1	1	5	2	10	4	3	4	5	5	5	5	4	50	2	6	1	3	1	1	1
43c-12939	1	42.5	16	3.5	1	1	1	5	2	11.5	4	6	4	5	5	5	5	4	45	2	5	1	4	1	3	1
24c-6335	1	30	16.5	6	2	1	1	5	5	9.5	6.5	2	4	4	5	4	5	6	75	?	1	0	2	1	2	1
43c-12842	1	57.5	17	6	1	1	2	5	7	13	5	4	4	7	5	5	5	5	75	2	5	0	2	1	2	1
51a-15485	1	38	15	3	2	1	1	5	2	12	5.5	4	4	5	5	4	5	4	50	2	1	0	3	1	1	1

Feature explanations and keys for numbers see Chap. 2, Blade Technology, Recording of Technological Information, nd: feature expression not recordable

Table A.31 Blade technology analysis—Rydno II/59

Nr.	Blank	Length	Width	Thickness	Pres.: L	Pres.: W/T	Cortex	SPR: kind	SPR: form	SPR: length	SPR: width	DR	Lip	Bulb	Bulbar scar	Impact point	Esq. du bulbe	en éperon	Flaking angle	Profile	Distal	Hammer	n Negatives	Direc. Neg.	Reg Neg.	Tool
1w	1	124	40	20	1	1	1	5	7	18.5	5	4	5	4	4	4	5	5	85	3	8	2	5	1	3	1
5w	1	82	28	9	2	1	1	5	4	7.5	1	3	5	7	5	4	4	5	75	1	2	2	3	?	2	2
13w	1	48.5	17	5	1	2	1	4	7	6.5	1.5	4	5	7	4	5	5	5	90	1	5	0	4	1	2	14
56w	1	60.5	22	5	2	1	1	4	7	12	4	3	5	4	6	4	5	5	75	1	1	2	2	1	1	1
39w	1	33.5	19.5	6	2	1	1	5	7	9	3.5	8	4	7	4	5	5	5	90	1	2	2	2	1	1	2
14w	1	45.5	21	5	2	1	1	4	2	11.5	4	8	5	4	4	4	5	5	70	1	1	2	3	1	1	1
247	1	56	24.5	8	1	1	1	5	2	14	4.5	8	4	4	5	4	5	5	70	2	5	2	4	1	4	1
9w	1	42	18	4	2	1	1	5	7	9	4	4	5	4	4	5	5	5	75	2	2	0	4	1	1	17
302	1	30.5	12	4	2	1	1	5	7	10	4	4	5	4	4	5	5	5	80	2	1	0	5	3	3	23
318	1	68.5	12	4.5	2	1	1	5	4	5.5	1	8	5	4	4	5	5	5	70	2	1	0	4	4	2	1
376	1	41	11	3	2	1	1	5	4	5	1.5	4	5	7	5	4	5	5	75	2	1	2	2	1	1	1
41	1	31	12	5	2	1	1	5	5	5.5	2.5	3	4	7	5	5	5	5	70	2	1	0	2	1	1	1
32	1	24	14.5	3	2	1	2	4	7	5.5	1.5	4	6	7	4	4	5	5	70	2	1	2	2	1	1	1
280	1	47	17	5	2	1	2	5	7	9	3	4	6	7	4	4	5	5	75	2	1	2	2	1	1	1
343	1	58	14	5.5	1	1	1	4	2	4	2	8	5	4	5	5	5	6	70	2	4	0	2	1	1	1
442	1	37	12.5	5	1	1	1	5	2	13	5	8	5	4	5	5	5	4	65	2	1	1	3	1	3	1
370	1	25	16.5	3	2	1	1	4	2	6	1.5	8	5	4	4	5	5	5	60	2	1	0	2	1	1	1
415	1	38	23.5	5	2	1	1	5	7	10	4	8	4	7	4	4	5	5	85	2	1	2	4	1	1	1
300	1	31	13	3	2	1	1	5	2	2.5	1	8	5	4	4	4	5	5	85	2	1	2	2	1	1	1
186	1	61	21	8	2	1	2	4	7	16	7	8	5	7	4	4	5	5	80	1	5	3	4	3	2	1
349	1	41.5	18.5	3	1	1	1	4	2	5	2	4	5	7	5	5	5	5	50	2	5	0	6	?	2	1
6	1	28	12	3.5	2	1	1	5	2	7	3	4	4	7	5	5	5	5	80	1	1	0	2	1	1	1
417	1	23	13.5	3.5	2	1	1	5	2	5.5	2	5	5	4	4	5	5	5	80	2	1	0	3	1	1	1
101	1	23.5	16	4	2	1	1	5	2	5	2	2	4	4	5	4	5	5	85	2	1	2	3	1	1	1
99	1	35	7	2	2	1	1	4	2	3	1	3	5	7	5	5	4	5	80	2	1	2	2	1	1	1
186	1	19	12	4	2	1	1	5	7	7	3	4	5	7	4	4	5	5	80	2	1	2	3	1	1	1
31	1	16.5	13	3	2	1	2	4	2	4.5	2	4	4	7	4	5	5	5	80	2	1	0	2	1	1	1
372	1	75	15.5	5	2	1	1	4	7	10	3	3	4	4	4	4	5	5	75	1	1	2	3	?	2	1
221	1	49.5	11.5	2	2	1	1	4	2	2.5	1	8	5	4	5	5	5	5	80	2	1	0	4	1	1	1
108	1	81	19	6	1	1	3	5	2	5.5	3	4	4	7	5	5	5	4	80	1	5	1	2	1	1	1
416	1	53	7.5	2	1	1	1	5	2	3	1	3	4	7	5	5	5	5	80	2	5	0	3	1	1	1
223	1	62.5	16	5	1	1	1	4	2	4	1	3	4	4	5	5	5	5	70	2	5	0	3	?	2	1
182	1	45	13.5	4	1	1	1	4	2	4	1	8	4	4	5	5	5	5	65	1	6	0	2	1	1	1
381	1	47.5	11	3	2	1	1	4	2	4	2.5	8	5	4	4	5	5	5	85	1	1	0	2	1	1	1
392	1	43.5	12	4	2	1	1	5	2	10.5	4	2	4	4	4	5	5	5	70	1	1	0	3	1	1	1
235	1	59	23	3	1	1	1	4	2	6	3	3	4	4	6	4	5	5	70	2	1	2	4	1	3	1
423	1	70	21	5	1	1	1	5	9	20	4.5	4	5	4	4	4	5	5	80	2	4	2	3	1	1	1
352	1	30.5	9.5	2.5	2	1	1	4	2	2	1	8	5	4	4	4	5	5	75	2	1	2	2	1	1	1
395	1	27.5	9.5	1.5	2	1	1	5	7	2	1	8	5	4	6	4	5	5	85	2	1	2	2	1	2	1
122	1	46	10.5	4	2	1	1	5	2	5	2	4	4	4	4	5	5	5	70	1	1	0	2	1	1	1

(continued)

Table A.31 (continued)

Nr.	Blank	Length	Width	Thickness	Pres.: L	Pres.: W/T	Cortex	SPR: kind	SPR: form	SPR: length	SPR: width	DR	Lip	Bulb	Bulbar scar	Impact point	Esq. du bulbe	en éperon	Flaking angle	Profile	Distal	Hammer	n Negatives	Direc. Neg.	Reg Neg.	Tool
397	1	33	19	5	2	1	1	4	2	5	2.5	8	5	4	6	4	5	5	80	2	1	2	3	1	1	1
313	1	30	21	5	2	1	1	5	2	11.5	4.5	8	4	4	4	4	5	5	75	2	1	2	3	1	2	1
199	1	69	19	5	1	1	2	4	2	4	1.5	3	4	4	5	5	5	5	60	2	4	0	3	1	2	1
370	1	41	15	5	2	1	1	4	2	9	3	4	4	4	4	5	5	5	70	2	1	0	2	1	1	1
442	1	28	16	4	2	1	1	4	4	4	1	4	5	7	4	4	5	5	80	2	1	2	2	1	1	1
316	1	24	13	6.5	2	1	1	4	2	5	2.5	4	5	7	6	5	5	5	80	2	1	0	2	1	1	1
182	1	36	15.5	5	1	1	1	4	2	4	1	4	5	7	4	4	5	5	80	1	6	2	2	4	1	1
20	1	52	22	6	2	1	1	4	4	7	1.5	8	5	7	5	5	5	5	80	2	1	0	3	?	3	1
417	1	41.5	17	3.5	2	1	1	4	2	5.5	2	3	4	7	5	5	5	5	60	2	1	0	2	1	1	1
89	1	19	8.5	2.5	2	1	1	4	7	4.5	1	3	5	4	5	5	5	5	70	2	1	0	2	1	1	1
107	1	15	10.5	3	2	1	1	4	2	4	2	4	5	7	4	4	5	5	80	2	1	2	2	1	1	1

Feature explanations and keys for numbers see Chap. 2, Blade Technology, Recording of Technological Information, nd: feature expression not recordable

Table A.32 Blade technology analysis—Teufelsbrücke, level 3

Nr.	Blank	Length	Width	Thickness	Pres.: L	Pres.: W/T	Cortex	SPR: kind	SPR: form	SPR: length	SPR: width	DR	Lip	Bulb	Bulbar scar	Impact point	Esq. du bulbe	en éperon	Flaking angle	Profile	Distal	Hammer	n Negatives	Direc. Neg.	Reg Neg.	Tool
1	4	68	20	6.5	2	1	1	5	5	3	3.5	8	4	4	5	5	5	4	45	2	1	1	5	4	3	1
2	1	35.5	15	3	2	1	1	5	2	9	3	3	4	7	5	5	5	6	45	2	1	1	2	1	1	1
3	1	62.5	15	4	2	1	1	5	2	5.5	3	2	4	7	5	5	5	4	50	2	1	1	3	1	1	1
4	1	27	17	4	2	1	1	5	2	11	3	2	4	7	5	5	5	6	50	2	1	1	2	1	1	1
5	1	13	12	2.5	2	1	1	5	2	7	2.5	3	4	7	5	5	5	6	50	2	1	1	2	1	1	1
6	1	44	15	4	2	2	1	5	2	3.5	3.5	8	4	4	5	5	5	4	55	2	2	1	2	1	1	5
7	1	37	17	5	2	1	1	5	5	5	3.5	8	4	7	4	5	5	4	60	2	2	1	4	1	2	2
8	1	38	16	5	2	1	2	5	2	10	3	8	4	4	5	5	5	4	60	2	1	1	1	1	1	1
9	1	28	14	4	2	2	1	5	2	7.5	4	4	4	4	5	5	5	6	65	2	2	1	3	1	1	5
10	1	24	16.5	4	2	1	1	5	2	10	3.5	3	4	7	5	5	5	4	65	2	1	1	4	1	1	1
11	1	30.5	21	6	2	1	1	5	2	3	3.5	4	4	7	4	5	5	4	70	2	1	1	3	1	1	1
12	1	21	24	3.5	2	1	2	5	5	7	3	2	4	7	5	5	5	4	75	2	1	1	1	1	1	1
13	1	20	6	2	2	2	1	5	2	4.5	2	4	4	7	5	5	5	4	75	2	1	1	2	1	1	13
14	1	27	11.5	5	2	1	1	5	2	11.5	5	2	5	4	5	5	5	4	70	1	2	1	4	1	2	2
15	1	41	13.5	4	2	1	1	5	2	3.5	3	3	5	4	5	5	5	4	70	1	1	1	2	1	1	1
16	1	38	12.5	4	2	1	1	5	2	2.5	1.5	3	5	4	4	5	5	6	85	1	1	0	4	1	2	1
17	1	22	5	2.5	1	2	1	5	5	4	2	3	5	4	5	5	5	4	55	2	5	1	4	1	1	13
18	1	46	27	6	2	1	1	5	2	3	6.5	8	5	4	5	5	5	6	65	2	1	1	3	?	3	1
19	1	22.5	6	1	2	2	1	5	5	2.5	1.5	4	5	4	5	5	5	6	80	2	1	0	2	1	1	14
20	1	32.5	11.5	3	2	1	1	4	2	5	2	3	4	7	4	4	5	5	45	1	1	2	2	1	1	1
21	1	24	13.5	3.5	2	1	1	5	2	9	3	4	4	7	4	5	5	5	45	1	2	0	2	1	1	2
22	1	31	13	4	2	1	2	4	2	7.5	3	5	4	7	5	5	5	5	45	1	2	0	1	1	1	2
23	1	18	12	2	2	1	1	5	2	5.5	1.5	4	4	7	4	5	5	5	50	1	1	0	3	1	1	1
24	1	56	8	4	2	1	1	4	7	3	1	8	4	7	4	4	5	5	55	1	1	2	3	1	1	1
25	1	31	10	3	2	1	1	5	2	9	2.5	8	4	7	5	5	5	5	60	1	1	0	2	1	1	1
26	1	40.5	15	3	2	2	1	5	2	8	2.5	4	4	4	4	4	5	5	65	1	2	2	3	1	1	5
27	1	31	13	3.5	2	1	1	4	7	4	1.5	3	4	7	4	4	5	5	65	1	1	2	3	1	1	1
28	1	40	14.5	4	2	2	1	5	7	6	5	4	4	7	5	4	5	5	70	1	1	2	3	1	1	1
29	1	36	15	3	2	1	1	5	7	13	4	4	4	7	5	4	5	5	70	1	1	2	2	1	1	1
30	1	29	1	2	2	1	1	4	2	2	1	2	4	7	5	5	5	5	70	1	1	0	3	1	1	1
31	1	21	10	5	2	1	1	5	5	10	4.5	8	4	7	5	4	5	5	75	1	1	2	3	1	1	1
32	1	21	11	3	2	1	1	5	7	5	1.5	8	4	4	5	5	5	5	80	1	1	0	2	1	1	1
33	1	32.5	10	3	2	1	1	4	2	3	1.5	8	4	7	5	4	5	5	45	2	1	2	3	1	1	1
34	1	19	9	2.5	2	2	1	4	2	4	1	4	4	7	4	5	5	5	45	2	1	0	2	1	1	13
35	1	28	9.5	2	2	1	1	5	2	7.5	2	4	4	4	5	5	5	5	45	2	1	0	2	1	1	1
36	1	56	10	6	1	1	2	4	2	3	1.5	4	4	7	5	5	5	5	45	2	8	0	4	?	1	1
37	1	27	6	2.5	2	2	1	4	2	4	1.4	3	4	7	5	5	5	5	55	2	1	0	2	1	1	13
38	1	21	9	2	2	1	1	4	2	2	1	8	4	4	5	5	5	5	60	2	1	0	2	1	1	1
39	1	44	15	4	2	2	1	5	2	6.5	3	4	4	7	4	4	5	5	65	2	2	2	3	1	1	5
40	1	32	16	2.5	2	1	1	4	2	6	1	8	4	7	5	4	5	5	65	2	2	2	3	1	1	2
41	1	32	17	5	2	1	1	5	9	6.5	2	3	4	4	4	5	5	5	65	2	2	0	2	1	1	2
42	1	37	13	3.5	2	1	2	5	2	5	2	8	4	4	4	5	5	5	65	2	1	0	3	1	1	1
43	1	38	7	2.5	2	2	1	5	2	3	1	4	4	4	5	5	5	5	65	2	1	0	2	1	1	13

(continued)

Table A.32 (continued)

Nr.	Blank	Length	Width	Thickness	Pres.: L	Pres.: W/T	Cortex	SPR: kind	SPR: form	SPR: length	SPR: width	DR	Lip	Bulb	Bulbar scar	Impact point	Esq. du bulbe	en éperon	Flaking angle	Profile	Distal	Hammer	n Negatives	Direc. Neg.	Reg Neg.	Tool
44	1	39	6.5	2	2	1	1	4	2	3	1.5	4	4	7	5	5	5	5	65	2	1	1	4	?	2	1
45	1	42	9	2	1	1	1	5	5	3	2	3	4	7	4	4	5	5	70	2	5	2	4	?	1	1
46	1	38	13	3	2	1	1	4	2	2.5	1	4	4	7	4	4	5	5	70	2	1	2	4	1	3	1
47	1	43	18	5	2	1	1	4	4	4.5	1	8	4	4	5	4	5	5	70	2	2	2	2	1	1	2
48	1	33	16	4.5	2	1	1	4	2	4	1.5	8	4	7	4	5	5	5	70	2	1	0	2	1	1	1
49	1	43.5	10.5	2	1	1	1	5	2	3.5	1.5	3	4	7	5	5	5	5	70	2	4	0	4	1	2	1
50	1	35	12.5	4	2	1	1	5	2	3.5	1	4	4	7	5	5	5	5	70	2	1	0	2	1	1	1
51	1	39	14	8	2	1	1	5	2	4.5	1.5	4	4	7	5	5	5	5	70	2	2	0	3	1	1	2
52	1	27	14	5	2	2	1	4	4	3	1	8	4	7	5	5	5	5	70	2	2	0	3	1	2	4
53	1	22	12.5	4	2	1	1	5	2	5	2	3	4	7	5	4	5	5	75	2	1	2	3	1	1	1
54	1	32.5	15	5	2	1	1	5	2	7	2	8	4	7	5	4	5	5	75	2	2	2	3	1	1	2
55	1	36	14	4	2	1	1	4	2	2.5	1	4	4	4	5	4	5	5	80	2	1	2	3	1	2	1
56	1	16.5	5	2	2	1	1	5	6	3	2	3	4	4	5	5	5	5	80	2	1	0	3	1	1	1
57	1	26	8	2.5	2	1	1	4	2	2	0.5	4	4	4	5	5	5	5	80	2	1	0	2	1	1	1
58	1	23.5	16	5	2	1	2	5	7	8	3	4	4	7	4	5	5	5	85	2	1	0	2	1	1	1
59	1	16.5	10	2.5	2	1	1	4	2	2.5	1.5	4	4	4	5	5	5	5	85	2	1	0	3	1	1	1
60	1	60.5	18	7	2	2	2	5	7	5	1.5	3	4	7	5	5	5	5	55	3	2	0	5	?	4	5
61	1	17	13	4.5	2	1	1	4	7	8	2	4	5	4	4	4	5	5	80	0	1	2	3	1	1	1
62	1	22.5	6	2	2	1	1	4	4	4	1	8	5	7	5	4	5	5	55	1	1	2	3	1	1	1
63	1	28	11	3	2	1	1	5	2	4	2	4	5	7	5	4	5	5	75	1	1	2	3	3	3	1
64	1	40	11	3	2	1	1	4	4	2.5	1	8	5	4	5	5	5	5	75	1	1	0	4	1	2	1
65	1	36	12.5	3	2	1	1	5	4	6	0.5	3	5	7	6	4	4	5	80	1	1	2	5	1	2	1
66	1	29	7	2	2	1	1	5	4	2.5	1	3	5	7	5	5	4	5	80	1	1	2	2	1	1	1
67	1	44	15	3	1	1	1	4	2	3.5	1.5	8	5	7	4	4	5	5	80	1	5	2	3	1	2	1
68	1	32	14	2.5	2	2	1	5	9	6.5	1	2	5	4	5	5	5	5	80	1	1	0	2	1	1	1
69	1	28	9	3	2	1	1	5	7	3	2	8	5	7	5	5	4	5	85	1	2	2	3	1	1	4
70	1	31.5	10	3	2	1	1	4	2	3	2	3	5	4	4	4	5	5	85	1	1	2	2	1	1	1
71	1	29	17	4	2	1	1	4	9	8.8	2	4	5	4	4	4	5	5	55	2	1	2	2	1	1	1
72	1	40.5	8	3	1	1	1	5	4	6	1.5	3	5	7	5	4	5	5	55	2	5	2	3	1	1	1
73	1	19	5	1.5	2	2	1	4	2	3	1	3	5	7	5	5	5	5	60	2	1	0	2	1	1	13
74	1	26	12	3	2	1	1	5	2	7	2.5	4	5	4	5	4	5	5	65	2	1	2	2	1	1	1
75	1	59	14.5	4	1	1	1	4	2	3	1.5	8	5	4	5	5	5	5	65	2	5	0	3	1	1	1
76	1	24	6	1	2	2	1	5	2	2.5	1	3	5	4	5	4	4	5	70	2	1	2	2	1	1	13
77	1	51.5	12	4	1	1	1	5	2	3	1	4	5	4	4	4	5	5	70	2	5	2	4	1	1	1
78	1	33.5	9	3	2	1	1	5	2	3	1.5	5	5	4	4	4	5	5	70	2	1	2	3	1	1	14
79	1	47.5	10	3	2	1	1	5	2	3	1.5	8	5	4	4	4	5	5	70	2	2	2	3	1	1	4
80	1	47	13	4	2	1	1	4	7	3.5	1.5	8	5	4	5	4	5	5	70	2	1	2	6	4	4	1
81	1	25	7	2.5	2	2	1	4	2	3	1	3	5	7	5	4	5	5	70	2	1	2	4	3	2	13
82	1	14	5	1.5	2	2	1	4	2	2	1	4	5	4	4	5	5	5	70	2	1	0	2	1	1	13
83	1	31	7	3	2	1	1	4	2	1.5	0.5	3	5	4	5	5	5	5	70	2	1	0	2	1	1	1
84	1	19	7	2.5	2	1	1	5	2	3	1.5	3	5	4	5	5	5	5	70	2	1	0	3	1	1	1
85	1	26	8	2.5	2	1	1	4	4	2	1	3	5	4	4	4	5	5	75	2	1	2	3	1	1	1
86	1	53	11	6	1	1	1	5	2	5	2	3	5	7	4	4	5	5	75	2	4	2	3	4	1	1
87	1	35.5	9.5	1.5	2	1	1	4	2	2	1	3	5	7	4	5	5	5	75	2	1	0	3	1	1	1

88	1	41	11	7	1	1	1	5	2	2.5	1.5	4	5	4	5	5	5	5	75	2	8	0	6	?	1	1
89	1	28	13	3	2	1	1	4	4	4	1	4	5	7	6	5	5	5	75	2	1	0	3	1	1	1
90	1	27	10	2.5	1	1	1	5	7	4	2	3	5	7	4	4	5	5	80	2	5	2	3	1	1	1
91	1	31	12	3	2	1	1	5	2	3	1.5	4	5	7	4	4	5	5	80	2	2	2	4	1	2	17
92	1	24	11.5	3	2	1	1	5	5	4	2.5	8	5	4	5	5	5	5	80	2	1	0	3	1	1	1
93	1	26	7.5	3	2	1	1	5	2	2	1	4	5	7	5	5	5	5	80	2	1	0	2	1	1	1
94	1	36	6	2	2	1	1	5	4	3.5	1	8	6	7	4	5	5	5	70	2	1	2	3	1	2	14
95	1	27	10.5	3	2	1	1	5	5	4.5	2	4	6	7	5	6	5	5	70	2	1	2	3	1	1	1
96	1	23	9.5	3	2	1	2	4	4	4.5	1	4	6	6	4	4	5	5	75	2	1	2	2	1	1	1
97	1	30	7	2	2	1	1	4	4	4	1	8	6	7	5	4	5	5	75	2	1	2	2	1	1	1
98	1	32	11	3.5	2	2	1	5	2	5	2	8	6	7	4	5	5	5	80	2	2	0	3	1	3	10
99	1	19	11.5	4	2	1	2	4	2	4.5	2	4	6	4	5	5	5	5	80	2	1	0	1	1	1	1
100	1	45	12.5	5	2	1	1	5	2	3.5	1	8	7	4	5	4	5	5	80	2	2	2	3	1	1	2

Feature explanations and keys for numbers see Chap. 2, Blade Technology, Recording of Technological Information, nd: feature expression not recordable

Table A.33 Blade technology analysis—Teufelsbrücke, level 4

Nr.	Blank	Length	Width	Thickness	Pres.: L	Pres.: W/T	Cortex	SPR: kind	SPR: form	SPR: length	SPR: width	DR	Lip	Bulb	Bulbar scar	Impact point	Esq. du bulbe	en éperon	Flaking angle	Profile	Distal	Hammer	n Negatives	Direc. Neg.	Reg Neg.	Tool
1	1	28.5	9	3	2	1	1	5	2	6	2	3	4	7	5	5	5	4	50	1	1	1	3	1	2	1
2	1	20.5	11.5	3.5	2	1	1	5	2	6	3	3	5	7	4	5	5	4	55	1	1	1	2	1	1	1
3	1	27	10	2.5	2	1	1	5	2	6	2	3	4	7	4	5	5	4	60	1	1	1	2	1	1	1
4	1	46.5	14	4.5	2	1	1	4	2	5	3	8	4	7	5	5	5	6	80	1	1	1	2	1	1	1
5	1	23	9	3	2	1	1	5	2	5	1	3	4	7	5	5	5	4	50	2	1	1	2	1	1	1
6	1	17.5	12.5	4	2	1	1	5	2	8.5	3.5	4	4	7	5	5	5	6	50	2	1	1	2	1	1	1
7	1	45	16	4	2	1	1	5	2	6.5	2.5	8	4	4	5	5	5	4	55	2	2	1	2	1	1	5
8	1	27.5	15.5	4	2	1	1	5	5	6.5	4	4	5	4	5	5	5	4	55	2	1	1	3	1	3	1
9	1	21	6.5	2.5	2	1	1	5	2	3.5	1.5	8	4	7	5	5	5	4	55	2	1	1	3	1	1	1
10	1	36	13.5	3	2	1	1	5	2	7	2.5	3	4	7	5	5	5	4	65	2	1	1	3	1	1	1
11	1	27.5	20	5	2	1	1	5	2	12	5	8	5	4	4	4	4	4	70	2	1	2	2	1	1	1
12	1	19	16	6	2	1	1	5	2	7.5	3.5	3	4	7	4	5	5	4	75	2	1	1	2	1	1	1
13	1	28	9	3	2	1	1	5	5	4	2.5	4	4	4	5	5	5	4	80	2	1	0	2	1	1	1
14	1	19.5	18.5	2.5	2	1	1	4	2	2.5	1	4	5	4	5	5	5	5	80	2	1	0	2	1	1	1
15	1	15	9	3	2	1	1	5	5	5	3	4	4	7	5	4	5	4	65	?	1	0	3	1	1	1
16	1	31	10	2.5	1	2	1	5	2	3	1	4	4	4	5	4	5	5	45	1	5	2	2	1	1	13
17	1	32	6	2.5	2	1	1	4	2	3	2	4	5	4	5	5	5	5	45	1	1	0	3	1	2	1
18	1	30.5	10	2.5	1	1	1	5	2	4	1.5	4	5	4	5	5	5	5	65	1	6	0	3	1	1	1
19	1	14	10	2	2	1	1	4	2	4	2	3	4	7	5	5	5	5	65	1	1	0	2	1	1	1
20	1	39.5	16	3	2	1	1	4	7	6.5	1.5	8	5	4	4	4	5	5	70	1	1	2	2	1	1	1
21	1	24	12.5	5	2	1	1	5	7	5	1	8	4	4	5	4	5	5	70	1	1	2	2	1	1	1
22	1	25	12.5	3.5	2	1	1	5	2	4	2	3	6	7	5	4	5	5	70	1	1	2	2	1	1	1
23	1	30	15	4	2	1	1	5	2	5	2.5	4	5	7	4	5	5	5	70	1	1	2	3	1	1	14
24	1	30.5	12	2	1	1	1	5	2	8	2	4	5	4	5	5	5	5	70	1	4	0	2	1	1	1
25	1	24	9	2.5	2	1	1	5	4	2.5	0.5	4	5	7	5	5	5	5	70	1	1	0	2	1	1	1
26	1	34.5	11.5	4.5	1	1	1	4	7	4	2	4	6	4	6	4	5	5	75	1	5	2	3	1	1	1
27	1	21	7.5	2	2	1	1	5	2	3.5	2	8	5	4	4	4	5	5	80	1	1	2	3	1	1	1
28	1	31	6.5	2	1	1	1	5	5	2.5	1.5	3	5	7	5	4	5	5	80	1	1	2	3	1	1	1
29	1	48	9.5	3.5	1	1	1	5	2	3	1	2	5	4	4	5	5	5	80	1	4	0	3	1	3	1
30	1	22	19.5	5	2	1	1	5	2	7	2.5	8	4	4	5	5	5	5	80	1	1	0	2	1	1	1
31	1	23	7	2.5	2	1	1	4	2	1.5	1	3	5	4	5	5	5	5	80	1	1	0	2	1	2	1
32	1	28	11	2.5	2	1	1	4	5	4	1.5	4	5	4	5	5	5	5	80	1	1	0	3	1	1	1
33	1	25	14	3	2	1	1	5	2	3	1.5	3	5	7	5	5	5	5	80	1	1	0	2	1	1	1
34	1	44	15	4.5	2	1	1	4	2	3	1.5	4	5	4	4	5	5	5	85	1	1	0	2	1	1	1
35	1	21.5	9	3	2	1	1	4	2	3.5	1.5	3	4	7	5	5	5	5	40	2	1	0	3	1	1	1
36	1	22	15	4	2	1	1	5	2	7.5	3	4	4	4	4	5	5	5	55	2	1	0	2	1	2	1
37	1	39	12	3.5	2	1	1	5	2	4	2	3	4	7	5	4	5	5	60	2	1	2	2	1	1	1
38	1	29	13	3	2	1	1	5	4	7	2	8	4	4	5	5	5	5	60	2	1	0	3	1	1	1
39	1	42.5	15	4	2	1	1	4	2	5.5	1.5	4	5	4	4	4	5	5	65	2	1	2	3	1	2	1
40	1	46	16	5.5	2	1	1	5	2	9	3.5	8	6	4	4	4	5	5	65	2	1	2	2	1	1	1
41	1	29	16.5	6	2	1	1	5	2	9	3	4	4	7	4	4	5	5	65	2	1	0	2	1	1	1
42	1	40	15	6	2	1	1	5	2	8.5	3	8	4	7	4	4	5	5	65	2	1	2	5	?	3	1
43	1	27	14	6	2	1	1	5	2	6	3	4	5	7	4	4	5	5	65	2	1	2	3	1	1	1

44	1	19	10	2	2	1	1	5	5	4	2	2	4	7	5	4	5	5	65	2	1	2	2	1	1	1
45	1	47	13	3	1	1	1	4	5	3	2	4	5	7	5	4	5	5	65	2	5	2	2	?	2	1
46	1	15	15	3.5	2	1	2	5	2	8	3	8	4	4	6	4	5	5	65	2	1	2	1	1	2	1
47	1	41	13.5	3	1	1	1	5	5	3	2	4	4	4	4	5	5	5	65	2	5	0	2	1	2	1
48	1	44.5	14	5	1	1	1	5	2	11	4.5	8	5	4	4	5	5	5	65	2	5	0	3	?	2	1
49	1	31	8.5	2	2	1	1	4	2	4	2	4	5	4	4	4	5	5	70	2	1	2	4	1	1	1
50	1	23	10	3	2	1	1	4	2	4	1	4	4	7	4	4	5	5	70	2	1	2	2	1	1	1
51	1	21	9.5	3	2	1	1	5	5	2.5	1.5	4	4	4	5	4	5	5	70	2	1	2	3	1	1	1
52	1	28	13	4.5	2	1	1	4	4	3	1	8	4	4	5	4	5	5	70	2	1	2	2	1	1	1
53	1	26	13	2.5	2	1	1	5	2	5.5	2	3	4	6	5	4	5	5	70	2	1	2	3	1	2	1
54	1	35	12.5	3	2	2	1	4	4	2	1	3	4	7	5	4	5	5	70	2	1	2	3	1	2	1
55	1	34	19	4	2	1	2	5	2	9	3.5	3	6	7	5	4	5	5	70	2	1	2	2	1	1	1
56	1	22	6	2	2	1	1	4	2	2.5	1	3	5	4	4	5	5	5	70	2	1	0	3	1	1	1
57	1	19	8.5	2	2	1	1	5	2	5.5	2	3	5	4	4	5	5	5	70	2	1	0	2	1	1	1
58	1	21	12	3	2	1	1	5	2	5	1.5	8	4	7	5	5	5	5	70	2	1	0	2	1	1	1
59	1	30	16	3.5	2	1	1	5	4	8	2	8	4	7	5	5	5	5	70	2	1	0	3	1	1	1
60	1	23.5	13	2	2	1	1	4	7	3.5	1	8	5	7	5	5	5	5	70	2	1	0	3	1	2	1
61	1	20	7.5	2	2	1	1	4	7	4	0.5	3	5	4	4	4	5	5	75	2	1	2	3	1	1	1
62	1	25	12.5	3	2	1	1	4	2	3	2	4	5	7	4	4	5	5	75	2	1	2	3	1	3	1
63	1	40	5.5	3	1	2	1	4	2	2.5	1	3	5	4	5	4	5	5	75	2	4	2	2	1	1	13
64	1	44.5	7.5	3	2	1	1	4	4	3	1	4	5	4	5	4	5	5	75	2	1	2	2	1	1	1
65	1	34.5	10	2.5	1	1	1	5	9	7	2	2	5	4	6	4	5	5	75	2	4	2	3	1	1	1
66	1	29	8.5	3	1	1	1	5	2	2	1	4	5	4	4	5	5	5	75	2	5	0	4	1	3	1
67	1	20.5	10	3	2	1	1	4	4	2	1	8	5	7	4	5	5	5	75	2	1	0	3	1	1	1
68	1	19	14	5	2	1	1	4	2	3	1.5	4	4	4	5	5	5	5	75	2	1	0	2	1	1	1
69	1	16	6	1.5	2	1	1	4	2	2.5	1	3	5	4	5	5	5	5	75	2	1	0	2	1	1	1
70	1	25	12	2	2	1	1	5	9	5.5	1.5	4	5	4	5	5	5	5	75	2	1	0	3	1	1	1
71	1	23	9.5	5	2	1	1	4	2	3.5	2	8	5	4	5	5	5	5	75	2	1	0	2	1	3	1
72	1	29	8.5	2	2	2	1	5	2	4	2	3	4	7	5	5	5	5	75	2	1	0	2	1	1	13
73	1	47	12	3.5	1	1	1	4	2	5	2	4	4	7	5	5	5	5	75	2	5	0	3	1	3	1
74	1	26	14	3	2	1	1	5	2	4.5	2	4	5	7	4	5	4	5	80	2	1	0	2	1	1	1
75	1	18.5	18.5	3.5	2	1	1	5	9	6	2	2	4	4	4	4	5	5	80	2	1	2	3	1	1	1
76	1	32.5	15.5	5	2	1	1	5	2	13	5	4	5	4	4	4	5	5	80	2	1	2	2	1	1	1
77	1	19	12	4	2	1	1	5	9	8	3	4	5	4	4	4	5	5	80	2	2	2	3	1	1	17
78	1	15	8	1.5	2	1	1	5	2	2.5	1	3	5	4	5	4	5	5	80	2	1	2	3	1	1	1
79	1	36	17	3	2	1	1	4	2	2.5	1	4	5	4	5	4	5	5	80	2	1	2	3	1	2	1
80	1	36	9	2.5	1	1	1	5	2	2.5	2	8	5	4	5	4	5	5	80	2	4	2	2	1	1	1
81	1	20.5	10	2	2	1	1	5	7	6	0.5	8	5	5	6	4	5	5	80	2	1	2	2	1	1	1
82	1	41.5	8	3	1	1	1	4	2	1	0.5	8	5	4	4	5	5	5	80	2	4	0	6	1	2	1
83	1	24	17.5	3.5	2	1	1	5	2	7.5	4	4	6	4	4	5	5	5	80	2	1	0	2	1	2	1
84	1	15	8	1.5	2	1	2	5	2	2	1.5	4	4	7	4	5	5	5	80	2	1	0	2	1	1	1
85	1	17.5	11	3.5	2	1	1	5	2	3.5	2	3	5	7	5	5	5	5	80	2	1	0	2	1	1	1
86	1	24	10	2	2	1	1	4	2	3.5	1.5	3	5	4	4	4	5	5	85	2	1	2	2	1	1	1

(continued)

Table A.33 (continued)

Nr.	Blank	Length	Width	Thickness	Pres.: L	Pres.: W/T	Cortex	SPR: kind	SPR: form	SPR: length	SPR: width	DR	Lip	Bulb	Bulbar scar	Impact point	Esq. du bulbe	en éperon	Flaking angle	Profile	Distal	Hammer	n Negatives	Direc. Neg.	Reg Neg.	Tool
87	1	12	11.5	2	2	1	1	5	5	4	1	4	5	4	4	4	5	5	85	2	1	2	2	1	2	1
88	1	28	13	4	2	1	1	5	2	7	2	4	6	4	4	4	5	5	85	2	1	2	2	1	1	1
89	1	20	17	2.5	2	1	1	4	9	6	1.5	4	5	4	4	5	5	5	85	2	1	0	2	1	1	1
90	1	19.5	14	4.5	2	1	1	5	2	11	4.5	4	5	4	4	4	5	5	45	?	1	2	2	1	1	1
91	1	17	12	3	2	1	1	5	2	4.5	2	8	5	4	4	4	5	5	70	?	1	2	2	1	1	1
92	1	12	12	2	2	1	2	4	9	6	2	2	5	4	4	5	5	5	75	?	1	0	2	1	1	1
93	1	16	13	4	2	1	1	5	5	8	5	8	5	4	5	4	5	5	80	?	1	2	3	1	1	1
94	1	22	12	3	2	1	1	5	2	8	2	4	5	4	4	5	5	5	80	?	1	0	2	1	1	1
95	1	15	14.5	3	2	1	1	5	2	8	3	3	6	4	4	5	5	5	80	?	1	0	3	1	1	1
96	1	16	14	3	2	1	1	5	2	7	2	3	5	7	6	5	5	5	80	?	1	0	3	1	1	1
97	1	28.5	8	1.5	1	1	1	5	5	3	1.5	3	5	4	6	5	5	6	80	1	4	0	3	1	1	1
98	1	30	6.5	2.5	2	2	1	5	5	5	2	2	5	4	5	5	5	6	80	1	1	0	2	1	1	13
99	1	26	5	3	2	2	1	5	5	2	2	2	5	7	4	5	5	6	75	2	1	0	2	1	1	13
100	1	34	8.5	3	1	1	1	4	5	3	2	2	5	4	5	5	5	6	85	2	4	0	3	1	2	1

Feature explanations and keys for numbers see Chap. 2, Blade Technology, Recording of Technological Information, nd: feature expression not recordable

Table A.34 Blade technology analysis—Wilczyce

Nr.	Blank	Length	Width	Thickness	Pres.: L	Pres.: W/T	Cortex	SPR: kind	SPR: form	SPR: length	SPR: width	DR	Lip	Bulb	Bulbar scar	Impact point	Esq. du bulbe	en éperon	Flaking angle	Profile	Distal	Hammer	n Negatives	Direc. Neg.	Reg Neg.	Tool
W10/96/186	1	55	25	5	2	1	1	5	2	9.5	3.5	4	4	7	5	4	5	4	50	1	1	0	2	1	2	1
10184	1	17	10	2	2	1	1	5	2	4.5	1.5	4	4	7	5	5	5	4	50	1	1	1	3	1	1	1
2587	1	36	20	8	2	1	1	5	2	12	4	3	4	7	5	5	5	5	50	1	1	0	3	1	1	1
OR 2006	1	32	16.5	4	2	1	1	5	2	10	3	3	4	5	5	5	5	4	55	1	1	1	3	1	1	1
OR12474	1	31	24	7	2	1	1	5	2	14	3	3	4	7	5	5	5	4	55	1	1	1	2	1	1	1
9723	1	43	17	5	2	1	1	5	2	7	3	4	4	7	5	5	5	4	55	1	1	0	3	1	1	1
3593	1	34.5	14	5	2	1	1	4	2	5	2	3	4	7	5	5	5	5	55	1	1	0	2	1	1	1
4774	1	25	12.5	5	2	1	1	5	2	9	3.5	3	4	7	5	5	5	4	60	1	1	1	2	1	1	1
OR12477	1	42	23	10	2	1	1	5	5	11.5	4.5	3	4	7	5	5	5	4	60	1	1	1	3	1	1	1
9132	1	17	8.5	4	2	1	1	5	7	6	1	3	6	7	5	4	5	5	60	1	1	2	2	1	1	1
1446	1	73	17.5	7	2	1	1	5	2	9	3	4	4	7	5	5	5	5	60	1	5	0	3	?	2	1
2842	1	53	10.5	3	1	1	1	5	2	4	1.5	3	4	7	5	5	5	5	60	1	5	0	3	?	1	1
4840	1	23	11	3.5	2	1	1	5	5	8	4	2	5	4	5	5	5	4	65	1	1	1	2	1	1	1
1736	1	51	15.5	4	2	1	1	5	7	13	4	4	5	4	4	4	5	5	65	1	1	2	3	1	1	1
1363	1	35	23.5	4.5	2	1	1	5	5	11	5	4	4	7	5	5	5	4	70	1	1	1	3	1	2	1
1079	1	44	18.5	6.5	2	1	1	5	2	6	2	4	6	7	5	4	5	5	70	1	1	2	2	1	1	1
9018	1	46	17	5	2	1	1	5	7	10	4.5	4	4	7	4	5	5	5	70	1	1	2	3	1	1	1
9908	1	66.5	25	6.5	2	1	2	5	2	16	4	3	4	7	5	5	5	4	75	1	1	1	2	1	1	1
8196	1	45	20	4.5	2	1	1	5	2	12	4	4	4	7	5	5	5	4	75	1	1	1	2	1	1	1
6040	1	52	14	5	1	1	2	5	2	9	3	4	5	4	5	5	5	5	80	1	4	2	2	1	1	1
5002	1	55	15	5	1	1	1	5	2	7	2	3	4	7	5	5	5	5	80	1	4	0	3	4	2	1
9164	1	35	10	2.5	2	1	1	4	4	3.5	1	3	5	7	5	5	5	5	80	1	6	0	2	1	1	1
5783	1	15.5	11	3	2	1	1	5	4	9	1	3	5	7	4	4	5	5	85	1	1	2	2	1	1	1
4427	1	27.5	15	6	2	1	1	4	2	6	2	4	4	7	5	5	5	4	nd	1	1	1	3	1	1	1
9822	1	53	14	5	2	1	1	5	4	5	1.5	4	4	5	5	5	5	4	40	2	1	1	3	1	1	1
10837	1	43	24	5	2	1	1	5	7	12.5	4	3	4	7	5	5	5	4	40	2	1	1	2	1	2	1
9505	1	68	24	7.5	2	2	1	5	5	13	4.5	3	4	7	4	5	5	4	45	2	2	1	4	1	1	2
4765	1	38	18	4.5	2	1	1	5	2	5	2	2	4	7	5	5	5	4	45	2	1	1	2	1	1	1
12520	1	52	19	6	2	1	1	5	2	11	3.5	2	4	7	5	5	5	4	45	2	1	1	3	1	1	1
714	1	69	28	5	2	1	1	5	2	6	2	3	4	7	5	5	5	4	45	2	1	1	4	1	3	1
7763	1	52	15.5	5	2	1	2	5	2	5.5	1.5	3	4	7	5	5	5	4	45	2	1	1	2	1	1	1
5405	1	57	26.5	7.5	2	1	1	5	2	8	3	3	4	7	5	5	5	4	45	2	1	1	4	1	2	1
11489	1	37	9	5	2	2	1	5	5	4	1.5	3	4	7	5	5	5	4	45	2	1	1	4	4	2	13
16439	1	19.5	8	3	2	2	1	5	5	4	2	4	4	5	5	5	5	4	50	2	1	1	3	1	1	13
515	1	58	12	4	2	1	1	5	5	4	2	2	4	7	5	5	5	4	50	2	1	1	2	1	1	1
1952	1	58	18	4	2	1	1	5	2	5	1.5	3	4	7	5	5	5	4	50	2	1	1	3	1	3	1
5013	1	28	13	3	2	1	1	5	2	4.5	2	3	4	7	5	5	5	4	50	2	1	1	4	1	1	1
6824	1	82	21	7	2	1	1	5	2	6	2	3	4	7	5	5	5	4	50	2	1	1	3	1	1	1
OR12477	1	28.5	20	7	2	1	1	5	2	9	2.5	3	4	7	5	5	5	4	50	2	1	1	3	1	1	1
4652	1	137	33	10.5	1	1	1	5	2	12.5	4.5	3	4	7	5	5	5	4	50	2	1	1	2	1	1	1
575	1	20.5	6.5	2.5	2	2	1	5	5	5	2	3	4	7	5	5	5	4	50	2	1	1	2	1	1	13
7767	1	34.5	19	6	2	1	1	5	5	6	3	4	4	7	5	5	5	4	50	2	1	1	3	?	1	1

(continued)

Table A.34 (continued)

Nr.	Blank	Length	Width	Thickness	Pres.: L	Pres.: W/T	Cortex	SPR: kind	SPR: form	SPR: length	SPR: width	DR	Lip	Bulb	Bulbar scar	Impact point	Esq. du bulbe	en éperon	Flaking angle	Profile	Distal	Hammer	n Negatives	Direc. Neg.	Reg Neg.	Tool
1840	1	32	14.5	4	2	1	1	5	2	4	1.5	3	4	7	4	4	5	5	50	2	1	2	3	1	1	1
W10/98/04	1	30	11.5	4	2	1	1	5	7	5	2.5	3	4	7	6	4	5	5	50	2	1	2	2	1	1	1
POW2004	1	28	11.5	3	2	1	1	5	2	4.5	2	3	4	7	4	5	5	5	50	2	1	0	3	1	1	1
9341	1	70	23	6	2	1	1	5	2	7.5	3	4	4	7	5	5	5	6	50	2	1	1	3	1	2	1
5869	1	18	8	4	2	2	1	5	2	5	2.5	3	4	7	4	5	5	4	55	2	1	1	3	1	1	13
W10/98/598	1	60	25	5	2	2	1	5	2	7	3.5	4	4	7	4	5	5	4	55	2	2	1	2	1	1	5
2086	1	39	16.5	5	2	2	1	5	5	10	5	4	5	4	5	5	5	4	55	2	1	1	2	1	1	1
1341	1	34	14	3.5	2	1	1	5	5	5.5	2.5	2	4	7	5	5	5	4	55	2	1	1	3	1	1	1
W10/98/04	1	49	17.5	5	2	1	1	5	5	7.5	4	3	4	7	5	5	5	4	55	2	1	1	4	?	3	1
8293	1	43	23	7.5	2	1	1	5	5	14	5.5	3	4	7	5	5	5	4	55	2	1	1	2	1	1	1
6218	1	29.5	17	5	2	1	1	5	2	8	3	4	4	7	5	5	5	4	55	2	1	1	3	1	1	1
5792	1	64	23	10	2	1	1	5	7	12	4	3	4	7	5	4	5	5	55	2	1	2	2	1	1	1
10640	1	27	15	4.5	2	1	1	5	5	10	4.5	4	5	4	5	5	5	4	60	2	1	1	3	1	1	1
5003	1	63.5	20	6	2	1	1	5	5	8.5	3.5	2	5	5	5	5	5	4	60	2	2	1	2	1	1	2
7042	1	38	16.5	6.5	2	1	1	5	5	9.5	5	4	4	7	5	5	5	4	60	2	1	1	3	1	1	1
W10/96/SIV	1	62	32	8	2	1	1	5	2	17	5	4	4	7	4	4	5	4	65	2	1	2	3	1	2	1
15884	1	50	17	5.5	2	2	1	5	9	9.5	2.5	3	6	4	4	5	5	4	65	2	2	0	2	1	1	5
7918	4	85.5	17	5.5	1	1	1	5	5	11	4.5	2	5	4	5	5	5	4	65	2	5	1	6	4	2	1
8299	1	53.5	27	8	2	1	1	5	2	10	3.5	3	4	7	5	5	5	4	65	2	1	1	2	1	1	1
10936	1	49	23	7	2	1	1	5	5	12.5	6	3	4	7	5	5	5	4	65	2	1	1	3	1	2	1
POW2004	1	25	12	4.5	2	1	1	5	2	8	3	4	4	7	5	5	5	4	65	2	1	1	3	1	1	1
15510	1	22	16	6	2	1	1	5	2	8	3	4	4	7	5	5	5	4	65	2	1	1	2	1	1	14
W10/96	1	28	19.5	6	2	1	1	5	2	10	4	8	4	7	5	5	5	4	65	2	1	1	2	1	1	1
8982	1	66	19	7	2	1	1	5	6	4	3	8	4	7	5	5	5	4	65	2	1	1	3	1	1	1
7352	1	31	18	6	2	1	2	5	2	13	6	4	5	4	4	5	5	4	70	2	1	1	2	1	1	1
7106	1	91	21	7	2	1	1	5	2	11	3.5	8	4	4	5	5	5	4	70	2	1	1	3	1	1	1
9799	1	37	15	7	2	1	1	5	2	16.5	5	2	4	7	5	5	5	4	70	2	1	1	2	?	2	1
OR20051b	1	70	33.5	10	2	1	1	5	2	16	5.5	3	4	7	5	5	5	4	70	2	1	1	3	1	3	1
OR12516	1	35.5	19	4	2	1	1	5	2	10	4.5	4	4	7	5	5	5	4	70	2	1	1	3	1	2	1
5112	1	57	18.5	4	2	1	1	5	5	7	3	4	4	7	5	5	5	4	70	2	1	1	5	?	3	1
2565	1	60	17	8	2	1	1	5	2	10	4	8	4	7	5	5	5	4	70	2	2	1	4	1	2	5
2569	1	79	14	8	2	1	2	5	4	5	1.5	8	4	7	5	5	5	4	70	2	2	1	3	1	1	2
9377	1	30.5	15	4	2	1	1	5	5	8.5	3	8	4	7	5	5	5	4	70	2	1	1	3	1	1	1
7875	1	30	14	4.5	2	1	1	4	2	4	2	8	4	7	5	4	5	5	70	2	1	2	4	1	1	1
8346	1	27.5	10	2.5	2	1	1	5	2	3	1	2	4	7	4	5	5	5	70	2	1	0	3	1	1	1
10745	1	28	6.5	3	2	2	1	5	9	4.5	2	4	5	4	5	5	5	5	70	2	1	0	3	1	1	13
2677	1	33	20	5.5	2	1	1	5	2	7	4	3	4	7	5	5	5	4	75	2	1	1	3	1	1	1
8404	1	61.5	24.5	9.5	2	2	1	5	7	16	4	3	4	7	5	5	5	4	75	2	2	1	2	1	1	5
OR2000	1	50	23	8	2	1	1	5	2	12	5	8	4	7	5	5	5	4	75	2	1	1	3	?	1	1
8154	1	83.5	25	9.5	2	1	1	5	7	14	7	8	4	7	5	5	5	4	75	2	1	1	4	1	2	14
1489	1	20	10	2	2	1	1	5	7	4.5	1	4	5	4	4	4	5	5	75	2	1	2	2	1	1	1
8316	1	63	12.5	6.5	2	1	1	5	7	7	2.5	8	6	5	5	5	5	5	75	2	2	0	3	1	1	2
7518	1	44	10	4	2	1	1	5	5	5.5	2.5	8	5	5	5	5	5	4	80	2	1	0	2	1	1	1
6981	1	38.5	14.5	6	2	1	1	5	2	4	2	3	4	7	5	5	5	4	80	2	1	1	2	1	1	1
6406	1	26	15	3	2	1	1	5	2	9	3	3	5	7	5	5	5	4	80	2	1	1	2	1	1	1

W10/98	1	36	19	5	2	1	1	5	7	9.5	5.5	8	4	4	4	4	5	5	80	2	1	2	4	1	4	1
X063	1	23	12	4	2	1	1	5	2	3	1.5	4	4	7	5	5	5	5	80	2	1	0	2	1	1	1
7971	1	55.5	20	5	1	1	1	5	9	7.5	1.5	8	4	4	5	4	5	5	85	2	6	2	4	?	4	1
OR=2000	1	34	22.5	6	2	1	3	5	7	14	5.5	4	5	7	6	4	5	5	85	2	1	2	2	1	1	1
16182	1	64	22	10	2	2	1	5	9	16	4.5	8	4	7	5	5	5	5	85	2	2	0	2	1	2	5
2034	1	59.5	6.5	4	2	1	1	5	5	3.5	2	8	4	7	5	5	5	4	40	3	1	1	2	1	2	1
1519	4	96.5	1	7	1	1	1	5	2	9	3	8	4	7	5	5	5	4	50	3	5	1	7	4	2	1
7808	1	42	19	7	2	1	1	5	5	13	7	2	5	4	5	5	5	4	55	3	1	1	3	1	1	1
8759	1	96	27	8	2	1	1	5	2	9.5	3	4	4	7	5	5	5	4	70	3	2	1	5	3	2	2
W10/98/621	1	62.5	13.5	6.5	2	1	2	5	5	4	2	2	5	7	5	5	5	6	75	3	1	0	3	1	1	1
7825	1	17.5	10	4	2	1	1	5	2	5	2	4	4	7	5	4	5	5	70	?	1	2	2	1	1	1
W10/96/324	1	25	16	5	2	1	1	5	5	7	3	4	4	7	5	5	5	5	80	?	1	0	2	1	1	1
5296	1	18	10	3	2	1	1	5	9	5	1	4	5	7	5	4	4	5	nd	?	1	2	2	1	1	1

Feature explanations and keys for numbers see Chap. 2, Blade Technology, Recording of Technological Information, nd: feature expression not recordable

Table A.35 Discriminant analysis. Raw data of the spatial investigation

Features SC-t – LBO

ID	Site	G	nTool	SC-t	%	SC	%	SSC	%	RSC	%	BU-t	%	BU	%	LBU	%	FBU	%	ZIN	%	BO-t	%	BO	%	LBO	%
471	Ahlendorf	4	301	19	6.3	17	5.6	0	0.0	2	0.7	48	15.9	47	15.6	1	0.3	0	0.0	0	0.0	32	10.6	5	1.7	16	5.3
231	Aichbühl B	2	132	27	20.5	19	14.4	5	3.8	3	2.3	43	32.6	43	32.6	0	0.0	0	0.0	1	0.8	5	3.8	5	3.8	0	0.0
351	Alsdorf	3	385	17	4.4	17	4.4	0	0.0	0	0.0	180	46.8	179	46.5	1	0.3	0	0.0	0	0.0	15	3.9	14	3.6	0	0.0
364	Andernach; I	3	503	172	34.2	172	34.2	0	0.0	0	0.0	92	18.3	89	17.7	3	0.6	0	0.0	0	0.0	42	8.3	42	8.3	0	0.0
365	Andernach; II	3	459	18	3.9	18	3.9	0	0.0	0	0.0	154	33.6	123	26.8	31	6.8	0	0.0	0	0.0	47	10.2	47	10.2	0	0.0
366	Andernach; III	3	172	41	23.8	41	23.8	0	0.0	0	0.0	44	25.6	35	20.3	9	5.2	0	0.0	0	0.0	8	4.7	8	4.7	0	0.0
367	Andernach; IV	3	310	27	8.7	8	2.6	0	0.0	19	6.1	129	41.6	125	40.3	4	1.3	0	0.0	0	0.0	11	3.5	11	3.5	0	0.0
82	Auzary-Thônes	1	177	9	5.1	8	4.5	0	0.0	1	0.6	16	9.0	16	9.0	0	0.0	0	0.0	0	0.0	3	1.7	3	1.7	0	0.0
401	B. Frankenh.	4	675	189	28.0	189	28.0	0	0.0	0	0.0	116	17.2	116	17.2	0	0.0	0	0.0	54	8.0	60	8.9	60	8.9	0	0.0
590	Balcarka Cave	5	123	20	16.3	16	13.0	0	0.0	4	3.3	13	10.6	13	10.6	0	0.0	0	0.0	3	2.4	7	5.7	7	5.7	0	0.0
12	Balme d'Isère	1	119	44	37.0	41	34.5	0	0.0	3	2.5	8	6.7	8	6.7	0	0.0	0	0.0	0	0.0	0	0.0	0	0.0	0	0.0
307	Barbing	2	120	16	13.3	15	12.5	0	0.0	1	0.8	32	26.7	32	26.7	0	0.0	0	0.0	1	0.8	2	1.7	2	1.7	0	0.0
67	Baume Noire	1	247	18	7.3	18	7.3	0	0.0	0	0.0	118	47.8	118	47.8	0	0.0	0	0.0	0	0.0	19	7.7	19	7.7	0	0.0
37	Bibi, G.	1	188	82	43.6	79	42.0	0	0.0	3	1.6	23	12.2	23	12.2	0	0.0	0	0.0	0	0.0	4	2.1	4	2.1	0	0.0
114	Birseck-Ermit.	1	197	26	13.2	25	12.7	0	0.0	1	0.5	48	24.4	48	24.4	0	0.0	0	0.0	0	0.0	4	2.0	4	2.0	0	0.0
320	Bois Laiterie	3	297	23	7.7	13	4.4	0	0.0	10	3.4	36	12.1	33	11.1	1	0.3	2	0.7	0	0.0	22	7.4	22	7.4	0	0.0
217	Braunfirst	2	240	28	11.7	27	11.3	0	0.0	1	0.4	56	23.3	54	22.5	2	0.8	0	0.0	1	0.4	16	6.7	16	6.7	0	0.0
246	Brillenh.; low.	2	282	22	7.8	21	7.4	0	0.0	1	0.4	86	30.5	84	29.8	2	0.7	0	0.0	0	0.0	24	8.5	23	8.2	1	0.4
247	Brillenh.; upp.	2	300	20	6.7	19	6.3	0	0.0	1	0.3	90	30.0	88	29.3	2	0.7	0	0.0	0	0.0	14	4.7	14	4.7	0	0.0
107	Brügglihöhle	1	184	24	13.0	15	8.2	4	2.2	5	2.7	20	10.9	20	10.9	0	0.0	0	0.0	0	0.0	11	6.0	11	6.0	0	0.0
205	Burgh. Dietfurt	2	119	11	9.2	7	5.9	3	2.5	1	0.8	14	11.8	14	11.8	0	0.0	0	0.0	1	0.8	5	4.2	5	4.2	0	0.0
551	Býčí skála	5	238	59	24.8	59	24.8	0	0.0	0	0.0	84	35.3	84	35.3	0	0.0	0	0.0	1	0.4	44	18.5	43	18.1	1	0.4
2	Campalou	1	100	12	12.0	12	12.0	0	0.0	0	0.0	22	22.0	22	22.0	0	0.0	0	0.0	0	0.0	0	0.0	0	0.0	0	0.0
30	Chênelaz	1	235	20	8.5	20	8.5	0	0.0	0	0.0	21	8.9	21	8.9	0	0.0	0	0.0	0	0.0	31	13.2	31	13.2	0	0.0
334	Coléoptère	3	125	18	14.4	18	14.4	0	0.0	0	0.0	17	13.6	17	13.6	0	0.0	0	0.0	0	0.0	12	9.6	12	9.6	0	0.0
29	Colomb	1	264	91	34.5	91	34.5	0	0.0	0	0.0	24	9.1	24	9.1	0	0.0	0	0.0	0	0.0	9	3.4	9	3.4	0	0.0
8	Croze-s.-Suran	1	314	56	17.8	56	17.8	0	0.0	0	0.0	144	45.9	144	45.9	0	0.0	0	0.0	0	0.0	108	34.4	108	34.4	0	0.0
69	Douattes, Les	1	637	103	16.2	76	11.9	23	3.6	4	0.6	168	26.4	168	26.4	0	0.0	0	0.0	1	0.2	44	6.9	44	6.9	0	0.0
601	Dzierzyslaw 35	5	912	0	0.0	0	0.0	0	0.0	0	0.0	491	53.8	490	53.7	1	0.1	0	0.0	2	0.2	176	19.3	175	19.2	1	0.1
346	Echt	3	122	27	22.1	27	22.1	0	0.0	0	0.0	40	32.8	40	32.8	0	0.0	0	0.0	4	3.3	14	11.5	14	11.5	0	0.0
164	Einsiedeln-Lang.	1	195	36	18.5	24	12.3	10	5.1	2	1.0	49	25.1	49	25.1	0	0.0	0	0.0	0	0.0	7	3.6	7	3.6	0	0.0
44	Farincourt; surf.	1	140	11	7.9	9	6.4	0	0.0	2	1.4	93	66.4	93	66.4	0	0.0	0	0.0	0	0.0	4	2.9	4	2.9	0	0.0
45	Farincourt; I+II	1	210	24	11.4	18	8.6	1	0.5	5	2.4	110	52.4	110	52.4	0	0.0	0	0.0	0	0.0	6	2.9	6	2.9	0	0.0
46	Farincourt; III-E	1	104	3	2.9	2	1.9	0	0.0	1	1.0	69	66.3	69	66.3	0	0.0	0	0.0	0	0.0	3	2.9	3	2.9	0	0.0
223	Felsställe; IIIa	2	229	12	5.2	10	4.4	0	0.0	2	0.9	30	13.1	30	13.1	0	0.0	0	0.0	0	0.0	33	14.4	32	14.0	1	0.4
224	Felsställe; IIIb	2	598	54	9.0	50	8.4	0	0.0	4	0.7	77	12.9	77	12.9	0	0.0	0	0.0	0	0.0	87	14.5	85	14.2	2	0.3
59	Fru, A.; II	1	118	7	5.9	7	5.9	0	0.0	0	0.0	9	7.6	9	7.6	0	0.0	0	0.0	0	0.0	4	3.4	4	3.4	0	0.0
120	Fürsteiner	1	210	39	18.6	39	18.6	0	0.0	0	0.0	133	63.3	133	63.3	0	0.0	0	0.0	0	0.0	3	1.4	3	1.4	0	0.0
468	Galgenberg	4	103	8	7.8	8	7.8	0	0.0	0	0.0	30	29.1	29	28.2	1	1.0	0	0.0	1	1.0	6	5.8	5	4.9	1	1.0
18	Gay, A.	1	153	51	33.3	49	32.0	2	1.3	0	0.0	32	20.9	32	20.9	0	0.0	0	0.0	0	0.0	8	5.2	8	5.2	0	0.0
478	Gera-Liebschw.	4	612	69	11.3	69	11.3	0	0.0	0	0.0	201	32.8	184	30.1	17	2.8	0	0.0	5	0.8	39	6.4	23	3.8	16	2.6
368	Gönnersdorf; I	3	1809	100	5.5	100	5.5	0	0.0	0	0.0	657	36.3	510	28.2	115	6.4	32	1.8	0	0.0	238	13.2	231	12.8	7	0.4
369	Gönnersdorf; II	3	2380	136	5.7	136	5.7	0	0.0	0	0.0	535	22.5	479	20.1	42	1.8	14	0.6	0	0.0	211	8.9	204	8.6	7	0.3
370	Gönnersdorf; III	3	893	54	6.0	54	6.0	0	0.0	0	0.0	232	26.0	189	21.2	35	3.9	8	0.9	0	0.0	151	16.9	145	16.2	6	0.7

371	Gönnersdorf; IV	3	174	10	5.7	10	5.7	0	0.0	0	0.0	39	22.4	33	19.0	6	3.4	0	0.0	0	0.0	14	8.0	13	7.5	1	0.6
28	Grappin	1	175	20	11.4	18	10.3	0	0.0	2	1.1	47	26.9	47	26.9	0	0.0	0	0.0	0	0.0	0	0.0	0	0.0	0	0.0
347	Griendtsveen	3	137	16	11.7	16	11.7	0	0.0	0	0.0	53	38.7	50	36.5	3	2.2	0	0.0	0	0.0	20	14.5	12	8.8	3	2.2
483	Groitzsch; A1	4	228	31	13.6	28	12.3	2	0.9	1	0.4	32	14.0	32	14.0	0	0.0	0	0.0	7	3.1	28	12.3	10	4.4	18	7.9
485	Groitzsch; B	4	406	87	21.4	71	17.5	7	1.7	9	2.2	37	9.1	37	9.1	0	0.0	0	0.0	29	7.1	40	9.9	34	8.4	6	1.5
487	Groitzsch; C1	4	536	94	17.5	86	16.0	2	0.4	6	1.1	160	29.9	160	29.9	0	0.0	0	0.0	23	4.3	21	3.9	18	3.4	3	0.6
489	Groi.; C3+DN.	4	655	117	17.9	110	16.8	5	0.8	2	0.3	119	18.2	119	18.2	0	0.0	0	0.0	38	5.8	61	9.3	42	6.4	19	2.9
490	Groitzsch; D1	4	559	55	9.8	45	8.1	3	0.5	7	1.3	57	10.2	57	10.2	0	0.0	0	0.0	25	4.5	59	10.6	37	6.6	22	3.9
491	Groitzsch; D2	4	196	36	18.4	25	12.8	4	2.0	7	3.6	21	10.7	21	10.7	0	0.0	0	0.0	6	3.1	26	13.3	22	11.2	4	2.0
95	H.-Champ.; 2	1	169	9	5.3	5	3.0	1	0.6	3	1.8	43	28.4	48	28.4	0	0.0	0	0.0	1	0.6	5	3.0	5	3.0	0	0.0
93	H.-Champ.; pri.	1	594	54	9.1	36	6.1	0	0.0	18	3.0	90	15.2	90	15.2	0	0.0	0	0.0	4	0.7	33	5.6	33	5.6	0	0.0
132	Hard I	1	259	20	7.7	19	7.3	1	0.4	0	0.0	40	15.4	40	15.4	0	0.0	0	0.0	0	0.0	16	6.2	16	6.2	0	0.0
243	Hohle Fels; IIa	2	111	4	3.6	4	3.6	0	0.0	0	0.0	8	7.2	8	7.2	0	0.0	0	0.0	0	0.0	2	1.8	2	1.8	0	0.0
530	Hostim	4	1871	209	11.2	178	9.5	0	0.0	31	1.7	221	11.8	203	10.8	3	0.2	15	0.8	19	1.0	431	23.0	417	22.3	14	0.7
599	Hranice	5	557	35	6.3	34	6.1	0	0.0	1	0.2	238	51.7	285	51.2	0	0.0	3	0.5	11	2.0	74	13.3	73	13.1	1	0.2
63	Jean P. I; 9A+B	1	174	6	3.4	6	3.4	0	0.0	0	0.0	23	13.2	23	13.2	0	0.0	0	0.0	4	2.3	10	5.7	9	5.2	1	0.6
443	Kahla-Löbschütz	4	848	93	11.0	93	11.0	0	0.0	0	0.0	193	22.8	182	21.5	11	1.3	0	0.0	1	0.1	95	11.2	61	7.2	34	4.0
543	Kamegg	5	309	48	15.5	48	15.5	0	0.0	0	0.0	135	43.7	135	43.7	0	0.0	0	0.0	3	1.0	17	5.5	17	5.5	0	0.0
214	Käppele	2	294	25	8.5	25	8.5	0	0.0	0	0.0	35	11.9	35	11.9	0	0.0	0	0.0	0	0.0	108	36.7	106	36.1	2	0.7
136	Käsloch	1	1015	129	12.7	87	8.6	4	0.4	38	3.7	205	20.2	202	19.9	3	0.3	0	0.0	0	0.0	94	9.3	89	8.8	5	0.5
290	Kastelhänghöhle	2	131	14	10.7	11	8.4	0	0.0	3	2.3	15	12.2	15	11.5	1	0.8	0	0.0	0	0.0	7	5.3	6	4.6	1	0.8
110	Kastelh.-N.; upp.	1	567	109	19.2	105	18.5	2	0.4	2	0.4	116	20.5	113	19.9	0	0.0	3	0.5	0	0.0	91	16	84	14.8	0	0.0
272	Kaufertsb.; upp.	2	136	16	11.8	13	9.6	1	0.7	2	1.5	31	22.8	25	18.4	4	2.9	2	1.5	0	0.0	12	8.8	11	8.1	1	0.7
271	Kaufertsb.; low.	2	225	26	11.6	19	8.4	0	0.0	7	3.1	117	52.0	108	48.0	7	3.1	2	0.9	1	0.4	12	5.3	7	3.1	5	2.2
170	Kesslerloch	2	1775	784	44.2	780	43.9	0	0.0	4	0.2	741	41.7	729	41.1	12	0.7	0	0.0	13	0.7	87	4.9	77	4.3	10	0.6
648	Klementowice	5	721	32	4.4	29	4.0	0	0.0	3	0.4	134	18.6	132	18.3	2	0.3	0	0.0	3	0.4	157	21.8	132	18.3	25	3.5
451	Kniegrotte	4	2513	234	9.3	215	8.6	6	0.2	13	0.5	514	20.5	500	19.9	14	0.6	0	0.0	0	0.0	205	8.2	203	8.1	2	0.1
108	Kohlerhöhle	1	254	140	55.1	131	51.6	0	0.0	9	3.5	22	8.7	22	8.7	0	0.0	0	0.0	0	0.0	38	15.0	37	14.6	1	0.4
571	Kůlna Cave; 5	5	104	16	15.4	8	7.7	8	7.7	0	0.0	15	14.4	14	13.5	0	0.0	1	1.0	2	1.9	8	7.7	8	7.7	0	0.0
572	Kůlna Cave; 6	5	144	36	25.0	26	18.1	10	6.9	0	0.0	17	11.8	16	11.1	0	0.0	1	0.7	8	5.6	8	5.6	8	5.6	0	0.0
518	Kvíc	4	188	20	10.6	18	9.6	0	0.0	2	1.1	34	18.1	34	18.1	0	0.0	0	0.0	1	0.5	28	14.9	28	14.9	0	0.0
452	Lausnitz	4	429	44	10.3	28	6.5	0	0.0	16	3.7	75	17.7	76	17.7	0	0.0	0	0.0	34	7.9	42	9.8	42	9.8	0	0.0
456	Lengefeld; surf.	4	258	27	10.5	23	8.9	0	0.0	4	1.6	69	26.7	59	22.9	10	3.9	0	0.0	2	0.8	32	12.4	27	10.5	5	1.9
594	Loštice I	5	212	14	6.6	12	5.7	0	0.0	2	0.9	33	17.9	28	13.2	10	4.7	0	0.0	0	0.0	8	3.8	8	3.8	0	0.0
547	Maloměřice-B; I	5	653	57	8.7	48	7.4	3	0.5	6	0.9	252	38.6	239	36.6	13	2.0	0	0.0	1	0.2	117	17.9	104	15.9	13	2.0
101	Moosseedorf-M.	1	5196	443	8.5	409	7.9	0	0.0	34	0.7	839	16.1	839	16.1	0	0.0	0	0.0	0	0.0	716	13.8	−17	−0.3	17	0.3
137	Mühleloch, A.	1	302	6	2.0	6	2.0	0	0.0	0	0.0	35	11.9	36	11.9	0	0.0	0	0.0	0	0.0	3	1.0	3	1.0	0	0.0
122	Munzingen	1	682	92	13.5	86	12.6	0	0.0	6	0.9	160	23.5	160	23.5	0	0.0	0	0.0	2	0.3	35	5.1	34	5.0	1	0.1
439	Nebra	4	2320	202	8.7	193	8.3	6	0.3	3	0.1	1026	44.2	946	40.8	56	2.4	24	1.0	10	0.4	293	12.6	280	12.1	13	0.6
90	N.-Monruz; 1M	1	1413	66	4.7	59	4.2	5	0.4	2	0.1	251	17.8	250	17.7	1	0.1	0	0.0	7	0.5	155	11.0	155	11.0	0	0.0
585	Ochozska-H.	5	353	40	11.3	37	10.5	0	0.0	3	0.8	55	15.9	52	14.7	1	0.3	3	0.8	9	2.5	34	9.6	31	8.8	3	0.8
444	Ölknitz	4	1577	82	5.2	82	5.2	0	0.0	0	0.0	252	16.0	251	15.9	1	0.1	0	0.0	0	0.0	139	8.8	136	8.6	3	0.2
331	Orp; ouest	3	159	25	15.7	19	11.9	0	0.0	6	3.8	73	49.1	57	35.8	21	13.2	0	0.0	0	0.0	10	6.3	10	6.3	0	0.0
330	Orp; est	3	465	60	12.9	57	12.3	0	0.0	3	0.6	245	52.7	174	37.4	71	15.3	0	0.0	0	0.0	14	3.0	10	2.2	4	0.9
579	Pekárna; gh	5	911	185	20.3	185	20.3	0	0.0	0	0.0	205	22.5	201	22.1	4	0.4	0	0.0	36	4.0	175	19.2	133	14.6	42	4.6
179	Petersfels 3AH2	2	259	26	10.0	20	7.7	1	0.4	5	1.9	25	9.7	25	9.7	0	0.0	0	0.0	1	0.4	28	10.8	27	10.4	1	0.4
199	Probstfels	2	148	15	10.1	14	9.5	0	0.0	1	0.7	49	33.1	49	33.1	0	0.0	0	0.0	0	0.0	3	2.0	3	2.0	0	0.0

(continued)

Table A.35 (continued)

Features SC-t – LBO

ID	Site	G	nTool	SC-t	%	SC	%	SSC	%	RSC	%	BU-t	%	BU	%	LBU	%	FBU	%	ZIN	%	BO-t	%	BO	%	LBO	%
139	Reiden-Stumpen	1	245	21	8.6	20	8.2	0	0.0	1	0.4	50	20.4	50	20.4	0	0.0	0	0.0	0	0.0	14	5.7	13	5.3	1	0.4
313	Roc-la-Tour	3	410	14	3.4	14	3.4	0	0.0	0	0.0	32	7.8	32	7.8	0	0.0	0	0.0	0	0.0	49	12.0	49	12.0	0	0.0
455	Saaleck	4	2021	161	8.0	112	5.5	5	0.2	44	2.2	213	10.5	213	10.5	0	0.0	0	0.0	22	1.1	387	19.1	267	13.2	120	5.9
480	Schafgraben,	4	162	31	19.1	31	19.1	0	0.0	0	0.0	86	53.1	86	53.1	0	0.0	0	0.0	1	0.6	10	6.2	9	5.6	1	0.6
166	Schweizersbild	2	657	81	12.3	70	10.7	0	0.0	11	1.7	320	48.7	320	48.7	0	0.0	0	0.0	7	1.1	87	13.2	87	13.2	0	0.0
291	Sesselfelsg.; C2	2	117	7	6.0	5	4.3	0	0.0	2	1.7	19	16.2	18	15.4	0	0.0	1	0.9	0	0.0	6	5.1	6	5.1	0	0.0
56	Seveux	1	692	80	11.6	80	11.6	0	0.0	0	0.0	273	39.5	273	39.5	0	0.0	0	0.0	5	0.7	169	24.4	166	24.0	3	0.4
281	Speckberg	2	1563	70	4.5	53	3.4	0	0.0	17	1.1	282	18.0	282	18.0	0	0.0	0	0.0	19	1.2	142	9.1	128	8.2	14	0.9
344	S.-Groene Paal	3	381	47	12.3	47	12.3	0	0.0	0	0.0	94	24.7	92	24.1	2	0.5	0	0.0	0	0.0	3	0.8	3	0.8	0	0.0
417	Teufelsbrücke; 3	4	1170	162	13.8	143	12.2	5	0.4	14	1.2	342	29.2	320	27.4	22	1.9	0	0.0	14	1.2	109	9.3	98	8.4	11	0.9
418	Teufelsbrücke; 4	4	423	41	9.7	40	9.5	1	0.2	0	0.0	39	9.2	39	9.2	0	0.0	0	0.0	19	4.5	57	13.5	53	12.5	4	0.9
126	Teufelsküche	1	445	68	15.3	64	14.4	0	0.0	4	0.9	90	20.2	88	19.8	2	0.4	0	0.0	0	0.0	48	10.8	47	10.6	1	0.2
324	T. d.Chaleux; 5	3	3042	369	12.1	327	10.7	2	0.1	40	1.3	739	24.3	733	24.1	3	0.1	3	0.1	50	1.6	408	13.4	408	13.4	0	0.0
325	T. d.Chal.; ex.O.	3	308	26	8.4	25	8.1	1	0.3	0	0.0	32	10.4	32	10.4	0	0.0	0	0.0	3	1.0	40	13.0	40	13.0	0	0.0
328	Trou du Frontal	3	178	26	14.6	26	14.6	0	0.0	0	0.0	42	23.6	42	23.6	0	0.0	0	0.0	2	1.1	26	14.6	26	14.6	0	0.0
335	Verlaine	3	143	24	16.8	24	16.8	0	0.0	0	0.0	25	17.5	23	16.1	0	0.0	2	1.4	1	0.7	22	15.4	22	15.4	0	0.0
79	Veyrier	1	593	94	15.9	77	13.0	9	1.5	8	1.3	93	15.7	93	15.7	0	0.0	0	0.0	0	0.0	78	13.2	78	13.2	0	0.0
647	Wilczyce	5	375	41	10.9	41	10.9	0	0.0	0	0.0	166	44.3	145	38.7	21	5.6	0	0.0	0	0.0	5	1.3	5	1.3	0	0.0
379	Wildscheuer	3	129	8	6.2	2	1.6	0	0.0	6	4.7	18	14.0	18	14.0	0	0.0	0	0.0	6	4.7	1	0.8	1	0.8	0	0.0
378	Wildweiberlei	3	128	9	7.0	6	4.7	0	0.0	3	2.3	15	11.7	15	11.7	0	0.0	0	0.0	2	1.6	9	7.0	9	7.0	0	0.0
134	Wilmatt	1	107	10	9.3	9	8.4	0	0.0	1	0.9	21	19.6	21	19.6	0	0.0	0	0.0	0	0.0	4	3.7	4	3.7	0	0.0
135	Winznau-Köpfli	1	1876	147	7.8	147	7.8	0	0.0	0	0.0	242	12.9	242	12.9	0	0.0	0	0.0	0	0.0	129	6.9	129	6.9	0	0.0
567	Žitného	5	533	75	14.1	65	12.2	0	0.0	10	1.9	89	16.7	88	16.5	0	0.0	1	0.2	8	1.5	54	10.1	54	10.1	0	0.0
Hamburgian																											
H143	Borneck	6	533	121	22.7	67	12.6	1	0.2	53	9.9	159	29.8	159	29.8	0	0.0	0	0.0	90	16.9	9	1.7	9	1.7	0	0.0
H99	Deimern 41	6	410	110	26.8	110	26.8	0	0.0	0	0.0	98	23.9	98	23.9	0	0.0	0	0.0	75	18.3	13	3.2	12	2.9	1	0.2
H100	Deimern 41b	6	177	55	31.1	55	31.1	0	0.0	0	0.0	14	7.9	14	7.9	0	0.0	0	0.0	38	21.5	4	2.3	4	2.3	0	0.0
H97	Deimern 42	6	859	122	14.2	122	14.2	0	0.0	0	0.0	124	14.4	124	14.4	0	0.0	0	0.0	157	18.3	4	0.5	2	0.2	2	0.2
H101	Deimern 44	6	812	124	15.3	123	15.1	1	0.1	0	0.0	172	21.2	171	21.1	1	0.1	0	0.0	126	15.5	23	2.8	19	2.3	4	0.5
H106	Deimern 46	6	109	36	33.0	10	9.2	0	0.0	26	23.9	14	12.8	13	11.9	1	0.9	0	0.0	17	15.6	17	15.6	16	14.7	1	0.9
H70	Glaner Heide	6	101	51	50.5	38	37.6	0	0.0	13	12.9	33	32.7	33	32.7	0	0.0	0	0.0	11	10.9	1	1.0	0	0.0	1	1.0
H124	Hasewisch	6	789	84	10.6	54	6.8	0	0.0	30	3.8	116	14.7	114	14.4	2	0.3	0	0.0	236	29.9	12	1.5	10	1.3	2	0.3
H35	Hav. Holtingerz.	6	259	168	64.9	168	64.9	0	0.0	0	0.0	48	18.5	48	18.5	0	0.0	0	0.0	21	8.1	0	0.0	0	0.0	0	0.0
H93	Heber 118	6	197	43	21.8	19	9.6	0	0.0	24	12.2	38	19.3	38	19.3	0	0.0	0	0.0	21	10.7	11	5.6	9	4.6	2	1.0
H95	Heber 127	6	659	212	32.2	172	26.1	2	0.3	38	5.8	104	15.8	104	15.8	0	0.0	0	0.0	130	19.7	21	3.2	18	2.7	3	0.5
H163	Liny 1	6	152	17	11.2	12	7.9	5	3.3	0	0.0	18	11.8	18	11.8	0	0.0	0	0.0	9	5.9	4	2.6	0	0.0	4	2.6
H122	Meiendorf 2	6	950	261	27.5	261	27.5	0	0.0	0	0.0	210	22.1	210	22.1	0	0.0	0	0.0	232	24.4	28	2.9	26	2.7	2	0.2
H130	Meiendorf 9	6	131	29	22.1	23	17.6	0	0.0	6	4.6	26	19.8	26	19.8	0	0.0	0	0.0	37	28.2	10	7.6	9	6.9	1	0.8
H164	Mircowice 33	6	132	7	5.3	4	3.0	0	0.0	3	2.3	22	16.7	22	16.7	0	0.0	0	0.0	15	11.4	25	18.9	25	18.9	0	0.0
H21	Neer II	6	235	84	35.7	84	35.7	0	0.0	0	0.0	65	27.7	65	27.7	0	0.0	0	0.0	24	10.2	0	0.0	0	0.0	0	0.0
H161	Olbrachcice 8	6	463	54	11.7	36	7.8	9	1.9	9	1.9	54	11.7	54	11.7	0	0.0	0	0.0	50	10.8	26	5.6	26	5.6	0	0.0
H125	Poggenwisch	6	534	210	39.3	110	20.6	0	0.0	100	18.7	52	9.7	52	9.7	0	0.0	0	0.0	98	18.4	15	2.8	14	2.6	1	0.2
H79	Schalkholz	6	106	32	30.2	15	14.2	0	0.0	17	16.0	18	17.0	18	17.0	0	0.0	0	0.0	23	21.7	8	7.5	7	6.6	1	0.9
H127	Teltwisch 1	6	1160	199	17.2	156	13.4	3	0.3	40	3.4	247	21.3	245	21.1	2	0.2	0	0.0	397	34.2	23	2.0	20	1.7	3	0.3
H27	Ureterp I	6	999	252	25.2	242	24.2	0	0.0	10	1.0	134	13.4	134	13.4	0	0.0	0	0.0	100	10.0	17	1.7	16	1.6	1	0.1
H114	Wellingsbüttel	6	179	99	55.3	80	44.7	0	0.0	19	10.6	22	12.3	22	12.3	0	0.0	0	0.0	41	22.9	4	2.2	4	2.2	0	0.0
H158	Wojnowo; I/75	6	119	55	46.2	50	42.0	0	0.0	5	4.2	61	51.3	61	51.3	0	0.0	0	0.0	0	0.0	3	2.5	3	2.5	0	0.0

Features SP – BP

ID	Site	G	nTool	SP	%	NDP	%	BB-t	%	BB	%	BBD	%	BBT	%	BBDT	%	BBTT	%	BBC	%	BBP	%	BBDE	%	BP	%
471	Ahlendorf	4	301	0	0.0	0	0.0	195	64.8	138	45.8	20	6.6	30	10.0	6	2.0	0	0.0	1	0.3	0	0.0	0	0.0	0	0.0
231	Aichbühl B	2	132	2	1.5	0	0.0	13	9.8	11	8.3	0	0.0	2	1.5	0	0.0	0	0.0	0	0.0	0	0.0	0	0.0	1	0.8
351	Alsdorf	3	385	0	0.0	0	0.0	103	26.8	91	23.6	11	2.9	0	0.0	0	0.0	0	0.0	0	0.0	0	0.0	1	0.3	1	0.3
364	Andernach; I	3	503	43	8.5	0	0.0	122	24.3	122	24.3	0	0.0	0	0.0	0	0.0	0	0.0	0	0.0	0	0.0	0	0.0	0	0.0
365	Andernach; II	3	459	72	15.7	0	0.0	149	32.5	149	32.5	0	0.0	0	0.0	0	0.0	0	0.0	0	0.0	0	0.0	0	0.0	0	0.0
366	Andernach; III	3	172	13	7.6	0	0.0	56	32.6	55	32.6	0	0.0	0	0.0	0	0.0	0	0.0	0	0.0	0	0.0	0	0.0	0	0.0
367	Andernach; IV	3	310	95	30.6	0	0.0	37	11.9	31	10.0	6	1.9	0	0.0	0	0.0	0	0.0	0	0.0	0	0.0	0	0.0	0	0.0
82	Auzary-Thô.	1	177	6	3.4	3	1.7	123	69.5	113	63.8	7	4.0	1	0.6	0	0.0	0	0.0	0	0.0	0	0.0	2	1.1	0	0.0
401	B. Frankenh.	4	675	34	5.0	0	0.0	69	10.2	69	10.2	0	0.0	0	0.0	0	0.0	0	0.0	0	0.0	0	0.0	0	0.0	3	0.4
590	Balcarka Cave	5	123	1	0.8	6	4.9	60	48.8	50	40.7	3	2.4	2	1.6	1	0.8	0	0.0	0	0.0	0	0.0	4	3.3	3	2.4
12	Balme d'Isère	1	119	0	0.0	6	5.0	13	10.9	13	10.9	0	0.0	0	0.0	0	0.0	0	0.0	0	0.0	0	0.0	0	0.0	3	2.5
307	Barbing	2	120	0	0.0	4	3.3	52	43.3	32	26.7	10	8.3	8	6.7	0	0.0	2	1.7	0	0.0	0	0.0	0	0.0	4	3.3
67	Baume Noire	1	247	0	0.0	36	14.6	36	14.6	36	14.6	0	0.0	0	0.0	0	0.0	0	0.0	0	0.0	0	0.0	0	0.0	5	2.0
37	Bibi, G.	1	188	0	0.0	3	1.6	39	20.7	36	19.1	0	0.0	3	1.6	0	0.0	0	0.0	0	0.0	0	0.0	0	0.0	0	0.0
114	Birseck-Ermit.	1	197	0	0.0	0	0.0	105	53.3	83	42.1	16	8.1	6	3.0	0	0.0	0	0.0	0	0.0	0	0.0	0	0.0	1	0.5
320	Bois Laiterie	3	297	2	0.7	20	6.7	55	18.5	54	18.2	0	0.0	1	0.3	0	0.0	0	0.0	0	0.0	0	0.0	0	0.0	23	7.7
217	Braunfirst	2	240	2	0.8	0	0.0	92	38.3	62	25.8	17	7.1	11	4.6	2	0.8	0	0.0	0	0.0	0	0.0	0	0.0	3	1.3
246	Brillenh.; low.	2	282	13	4.6	73	25.9	11	3.9	11	3.9	0	0.0	0	0.0	0	0.0	0	0.0	0	0.0	0	0.0	0	0.0	6	2.1
247	Brillenh.; upp.	2	300	19	6.3	73	24.3	11	3.7	9	3.0	0	0.0	0	0.0	1	0.3	0	0.0	0	0.0	1	0.3	0	0.0	3	1.0
107	Brügglihöhle	1	184	4	2.2	0	0.0	65	35.3	65	35.3	0	0.0	0	0.0	0	0.0	0	0.0	0	0.0	0	0.0	0	0.0	7	3.8
205	Burgh. Dietf.	2	119	0	0.0	7	5.9	33	27.7	28	23.5	0	0.0	2	1.7	2	1.7	0	0.0	0	0.0	0	0.0	1	0.8	15	12.6
551	Býčí skála	5	238	0	0.0	0	0.0	50	21.0	46	19.3	0	0.0	0	0.0	2	0.8	2	0.8	0	0.0	0	0.0	0	0.0	0	0.0
2	Campalou	1	100	0	0.0	15	15.0	40	40.0	33	33.0	0	0.0	7	7.0	0	0.0	0	0.0	0	0.0	0	0.0	0	0.0	3	3.0
30	Chênelaz	1	235	0	0.0	27	11.5	109	46.4	–2	–0.9	92	39.1	13	5.5	0	0.0	2	0.9	0	0.0	0	0.0	4	1.7	7	3.0
334	Coléoptère	3	125	0	0.0	0	0.0	59	47.2	57	45.6	1	0.8	1	0.8	0	0.0	0	0.0	0	0.0	0	0.0	0	0.0	2	1.6
29	Colomb	1	264	0	0.0	17	6.4	20	7.6	20	7.6	0	0.0	0	0.0	0	0.0	0	0.0	0	0.0	0	0.0	0	0.0	4	1.5
8	Croze-s.-Suran	1	314	0	0.0	0	0.0	5	1.6	2	0.6	0	0.0	3	1.0	0	0.0	0	0.0	0	0.0	0	0.0	0	0.0	0	0.0
69	Douattes, Les	1	637	0	0.0	5	0.8	282	44.3	250	39.2	0	0.0	20	3.1	0	0.0	1	0.2	0	0.0	0	0.0	11	1.7	4	0.6
601	Dzierzyslaw	5	912	0	0.0	0	0.0	14	1.5	14	1.5	0	0.0	0	0.0	0	0.0	0	0.0	0	0.0	0	0.0	0	0.0	0	0.0
346	Echt	3	122	1	0.8	0	0.0	10	8.2	3	2.5	0	0.0	3	2.5	0	0.0	2	1.6	1	0.8	1	0.8	0	0.0	11	9.0
164	Einsiedeln-L.	1	195	7	3.6	3	1.5	37	19.0	32	16.4	0	0.0	5	2.6	0	0.0	0	0.0	0	0.0	0	0.0	0	0.0	7	3.6
44	Farinc.; surf.	1	140	0	0.0	3	2.1	2	1.4	2	1.4	0	0.0	0	0.0	0	0.0	0	0.0	0	0.0	0	0.0	0	0.0	0	0.0
45	Farinc.; I+II	1	210	1	0.5	1	0.5	4	1.9	3	1.4	0	0.0	1	0.5	0	0.0	0	0.0	0	0.0	0	0.0	0	0.0	0	0.0
46	Farinc.; III-E	1	104	1	1.0	1	1.0	2	1.9	2	1.9	0	0.0	0	0.0	0	0.0	0	0.0	0	0.0	0	0.0	0	0.0	0	0.0
223	Felsställe; IIIa	2	229	3	1.3	0	0.0	65	28.4	34	14.8	17	7.4	4	1.7	2	0.9	0	0.0	0	0.0	0	0.0	8	3.5	4	1.7
224	Felsställe; IIIb	2	598	9	1.5	0	0.0	114	19.1	62	10.4	26	4.3	3	0.5	2	0.3	0	0.0	0	0.0	1	0.2	20	3.3	4	0.7
59	Fru, A.; II	1	118	0	0.0	0	0.0	98	83.1	96	81.4	2	1.7	0	0.0	0	0.0	0	0.0	0	0.0	0	0.0	0	0.0	0	0.0
120	Fürsteiner	1	210	5	2.4	0	0.0	21	10.0	17	8.1	0	0.0	4	1.9	0	0.0	0	0.0	0	0.0	0	0.0	0	0.0	9	4.3
468	Galgenberg	4	103	0	0.0	0	0.0	47	45.6	40	38.8	4	3.9	3	2.9	0	0.0	0	0.0	0	0.0	0	0.0	0	0.0	4	3.9
18	Gay, A.	1	153	0	0.0	0	0.0	42	27.5	42	27.5	0	0.0	0	0.0	0	0.0	0	0.0	0	0.0	0	0.0	0	0.0	15	9.8
478	Gera-Liebsch.	4	612	32	5.2	2	0.3	175	28.6	140	22.9	32	5.2	3	0.5	0	0.0	0	0.0	0	0.0	0	0.0	0	0.0	6	1.0
368	Gönnersd.; I	3	1809	192	10.6	0	0.0	571	31.6	481	26.6	87	4.8	2	0.1	0	0.0	0	0.0	0	0.0	0	0.0	1	0.1	0	0.0
369	Gönnersd.; II	3	2380	456	19.2	0	0.0	991	41.6	878	36.9	107	4.5	2	0.1	0	0.0	0	0.0	0	0.0	0	0.0	4	0.2	0	0.0
370	Gönnersd.; III	3	893	137	15.3	0	0.0	302	33.8	248	27.8	44	4.9	7	0.8	0	0.0	0	0.0	0	0.0	0	0.0	3	0.3	0	0.0

(continued)

Table A.35 (continued)

Features SP – BP

ID	Site	G	nTool	SP	%	NDP	%	BB-t	%	BB	%	BBD	%	BBT	%	BBDT	%	BBTT	%	BBC	%	BBP	%	BBDE	%	BP	%
371	Gönnersd.; IV	3	174	10	5.7	0	0.0	98	56.3	75	43.1	20	11.5	3	1.7	0	0.0	0	0.0	0	0.0	0	0.0	0	0.0	0	0.0
28	Grappin	1	175	4	2.3	11	6.3	51	29.1	31	17.7	0	0.0	18	10.3	0	0.0	0	0.0	0	0.0	0	0.0	2	1.1	0	0.0
347	Griendtsveen	3	137	3	2.2	0	0.0	40	29.2	36	26.3	0	0.0	3	2.2	0	0.0	1	0.7	0	0.0	0	0.0	0	0.0	2	1.5
483	Groitzsch; A1	4	228	0	0.0	0	0.0	111	48.7	90	39.5	9	3.9	5	2.2	3	1.3	0	0.0	1	0.4	1	0.4	2	0.9	0	0.0
485	Groitzsch; B	4	406	0	0.0	12	3.0	155	38.2	93	22.9	23	5.7	20	4.9	8	2.0	5	1.2	2	0.5	2	0.5	2	0.5	14	3.4
487	Groitzsch; C1	4	536	0	0.0	7	1.3	179	33.4	144	26.9	14	2.6	14	2.6	2	0.4	3	0.6	1	0.2	1	0.2	0	0.0	1	0.2
489	Groi.; C3+DN.	4	655	1	0.2	9	1.4	210	32.1	125	19.1	21	3.2	39	6.0	9	1.4	10	1.5	0	0.0	2	0.3	4	0.6	2	0.3
490	Groitzsch; D1	4	559	1	0.2	8	1.4	303	54.2	197	35.2	34	6.1	37	6.6	10	1.8	8	1.4	2	0.4	8	1.4	7	1.3	1	0.2
491	Groitzsch; D2	4	196	0	0.0	0	0.0	94	48.0	60	30.6	7	3.6	11	5.6	7	3.6	2	1.0	0	0.0	2	1.0	5	2.6	0	0.0
95	H.-Cham.; 2	1	169	1	0.6	0	0.0	99	58.6	94	55.6	1	0.6	3	1.8	0	0.0	0	0.0	0	0.0	0	0.0	1	0.6	1	0.6
93	H.-Cham.; pri.	1	594	13	2.2	8	1.3	369	62.1	354	59.6	0	0.0	13	2.2	0	0.0	0	0.0	1	0.2	0	0.0	1	0.2	0	0.0
132	Hard I	1	259	0	0.0	5	1.9	15	5.8	15	5.8	0	0.0	0	0.0	0	0.0	0	0.0	0	0.0	0	0.0	0	0.0	0	0.0
243	Hohle Fels; IIa	2	111	5	4.5	50	45.0	20	18.0	16	14.4	4	3.6	0	0.0	0	0.0	0	0.0	0	0.0	0	0.0	0	0.0	0	0.0
530	Hostim	4	1871	343	18.3	70	3.7	525	28.1	506	27.0	0	0.0	12	0.6	0	0.0	2	0.1	0	0.0	0	0.0	5	0.3	0	0.0
599	Hranice	5	557	8	1.4	14	2.5	23	4.1	15	2.7	0	0.0	8	1.4	0	0.0	0	0.0	0	0.0	0	0.0	0	0.0	0	0.0
63	Jean P. I	1	174	0	0.0	17	9.8	94	54.0	62	35.6	6	3.4	8	4.6	7	4.0	2	1.1	3	1.7	1	0.6	5	2.9	1	0.6
443	Kahla-Löbsch.	4	848	0	0.0	0	0.0	458	54.0	363	42.8	85	10.0	5	0.6	5	0.6	0	0.0	0	0.0	0	0.0	0	0.0	0	0.0
543	Kamegg	5	309	0	0.0	1	0.3	77	24.9	54	17.5	17	5.5	4	1.3	0	0.0	2	0.6	0	0.0	0	0.0	0	0.0	5	1.6
214	Käppele	2	294	1	0.3	0	0.0	70	23.8	51	17.3	10	3.4	7	2.4	0	0.0	2	0.7	0	0.0	0	0.0	0	0.0	2	0.7
136	Käsloch	1	1015	1	0.1	0	0.0	159	15.7	134	13.2	19	1.9	2	0.2	3	0.3	0	0.0	1	0.1	0	0.0	0	0.0	21	2.1
290	Kastelhängh.	2	131	0	0.0	25	19.1	25	19.1	24	18.3	0	0.0	0	0.0	0	0.0	0	0.0	0	0.0	0	0.0	1	0.8	1	0.8
110	Kastelh.; upp.	1	567	0	0.0	0	0.0	213	37.6	192	33.9	0	0.0	13	2.3	0	0.0	6	1.1	0	0.0	0	0.0	2	0.4	12	2.1
272	Kauferts.; upp.	2	136	0	0.0	3	2.2	44	32.4	42	30.9	2	1.5	0	0.0	0	0.0	0	0.0	0	0.0	0	0.0	0	0.0	12	8.8
271	Kauferts.; low.	2	225	0	0.0	4	1.8	33	14.7	25	11.1	5	2.2	2	0.9	0	0.0	1	0.4	0	0.0	0	0.0	0	0.0	0	0.0
170	Kesslerloch	2	1775	0	0.0	0	0.0	143	8.1	104	5.9	16	0.9	17	1.0	6	0.3	0	0.0	0	0.0	0	0.0	0	0.0	5	0.3
648	Klementowice	5	721	0	0.0	2	0.3	120	16.6	70	9.7	26	3.6	11	1.5	12	1.7	1	0.1	0	0.0	0	0.0	0	0.0	17	2.4
451	Kniegrotte	4	2513	128	5.1	0	0.0	789	31.4	603	24.0	73	2.9	78	3.1	18	0.7	9	0.4	4	0.2	0	0.0	4	0.2	8	0.3
108	Kohlerhöhle	1	254	0	0.0	12	4.7	20	7.9	17	6.7	0	0.0	3	1.2	0	0.0	0	0.0	0	0.0	0	0.0	0	0.0	8	3.1
571	Kůlna-H.; 5	5	104	2	1.9	6	5.8	23	22.1	20	19.2	0	0.0	3	2.9	0	0.0	0	0.0	0	0.0	0	0.0	0	0.0	2	1.9
572	Kůlna-H.; 6	5	144	0	0.0	11	7.6	33	22.9	32	22.2	0	0.0	1	0.7	0	0.0	0	0.0	0	0.0	0	0.0	0	0.0	2	1.4
518	Kvíc	4	188	19	10.1	4	2.1	44	23.4	15	8.0	2	1.1	19	10.1	4	2.1	1	0.5	0	0.0	0	0.0	3	1.6	1	0.5
452	Lausnitz	4	429	4	0.9	3	0.7	160	37.3	151	35.2	8	1.9	1	0.2	0	0.0	0	0.0	0	0.0	0	0.0	0	0.0	14	3.3
456	Lengef.; surf.	4	258	5	1.9	0	0.0	59	22.9	39	15.1	4	1.6	13	5.0	1	0.4	0	0.0	0	0.0	2	0.8	0	0.0	0	0.0
594	Loštice I	5	212	9	4.2	4	1.9	110	51.9	101	47.6	8	3.8	0	0.0	0	0.0	0	0.0	1	0.5	0	0.0	0	0.0	0	0.0
547	Maloměřice	5	653	18	2.8	9	1.4	135	20.7	93	14.2	4	0.6	8	1.2	30	4.6	0	0.0	0	0.0	0	0.0	0	0.0	2	0.3
101	Moosseedorf	1	5196	15	0.3	112	2.2	2669	51.4	2129	41.0	15	0.3	460	8.9	15	0.3	15	0.3	15	0.3	0	0.0	20	0.4	3	0.1
137	Mühleloch, A.	1	302	0	0.0	4	1.3	48	15.9	41	13.6	0	0.0	1	0.3	1	0.3	5	1.7	0	0.0	0	0.0	0	0.0	2	0.7
122	Munzingen	1	682	52	7.6	0	0.0	68	10.0	47	6.9	6	0.9	10	1.5	2	0.3	0	0.0	0	0.0	0	0.0	3	0.4	0	0.0
439	Nebra	4	2320	70	3.0	0	0.0	587	25.3	493	21.3	91	3.9	1	0.0	0	0.0	0	0.0	0	0.0	2	0.1	0	0.0	7	0.3
90	N.-Monruz	1	1413	20	1.4	6	0.4	883	62.5	804	56.9	36	2.5	33	2.3	5	0.4	0	0.0	0	0.0	0	0.0	5	0.4	0	0.0
585	Ochozska-H.	5	353	20	5.7	47	13.3	82	23.2	52	14.7	4	1.1	17	4.8	3	0.8	0	0.0	0	0.0	0	0.0	6	1.7	2	0.6
444	Ölknitz	4	1577	134	8.5	0	0.0	857	54.3	857	54.3	0	0.0	0	0.0	0	0.0	0	0.0	0	0.0	0	0.0	0	0.0	0	0.0
331	Orp; ouest	3	159	0	0.0	8	5.0	3	1.9	3	1.9	0	0.0	0	0.0	0	0.0	0	0.0	0	0.0	0	0.0	0	0.0	0	0.0
330	Orp; est	3	465	0	0.0	26	5.6	43	9.2	43	9.2	0	0.0	0	0.0	0	0.0	0	0.0	0	0.0	0	0.0	0	0.0	0	0.0
579	Pekárna; gh	5	911	50	5.5	0	0.0	234	25.7	148	16.2	21	2.3	32	3.5	5	0.5	4	0.4	2	0.2	0	0.0	22	2.4	3	0.3
179	Petersfels	2	259	2	0.8	0	0.0	115	44.4	102	39.4	12	4.6	1	0.4	0	0.0	0	0.0	0	0.0	0	0.0	0	0.0	1	0.4
199	Probstfels	2	148	7	4.7	0	0.0	29	19.6	29	19.6	0	0.0	0	0.0	0	0.0	0	0.0	0	0.0	0	0.0	0	0.0	24	16.2

139	Reiden-St.	1	245	9	3.7	47	19.2	28	11.4	27	11.0	0	0.0	1	0.4	0	0.0	0	0.0	0	0.0	0	0.0	0	0.0	1	0.4
313	Roc-la-Tour	3	410	10	2.4	19	4.6	215	52.4	215	52.4	0	0.0	0	0.0	0	0.0	0	0.0	0	0.0	0	0.0	0	0.0	1	0.2
455	Saaleck	4	2021	6	0.3	3	0.1	1165	57.6	770	38.1	105	5.2	188	9.3	48	2.4	21	1.0	14	0.7	8	0.4	11	0.5	2	0.1
480	Schafgraben	4	162	7	4.3	2	1.2	2	1.2	1	0.6	0	0.0	1	0.6	0	0.0	0	0.0	0	0.0	0	0.0	0	0.0	0	0.0
166	Schweizersb.	2	657	12	1.8	0	0.0	127	19.3	105	16.0	12	1.8	5	0.8	5	0.8	0	0.0	0	0.0	0	0.0	0	0.0	4	0.6
291	Sesselfels.; C2	2	117	0	0.0	5	4.3	71	60.7	45	39.3	17	14.5	7	6.0	1	0.9	0	0.0	0	0.0	0	0.0	0	0.0	3	2.6
56	Seveux	1	692	0	0.0	30	4.3	41	5.9	37	5.3	0	0.0	4	0.6	0	0.0	0	0.0	0	0.0	0	0.0	0	0.0	0	0.0
281	Speckberg	2	1563	35	2.2	142	9.1	8	0.5	8	0.5	0	0.0	0	0.0	0	0.0	0	0.0	0	0.0	0	0.0	0	0.0	2	0.1
344	S.-Groene Paal	3	381	0	0.0	13	3.4	110	28.9	108	28.3	2	0.5	0	0.0	0	0.0	0	0.0	0	0.0	0	0.0	0	0.0	0	0.0
417	Teufelsbr.e; 3	4	1170	23	2.0	0	0.0	487	41.6	442	37.8	15	1.3	17	1.5	0	0.0	0	0.0	1	0.1	0	0.0	12	1.0	2	0.2
418	Teufelsbr.; 4	4	423	7	1.7	0	0.0	254	60.0	217	51.3	17	4.0	10	2.4	5	1.2	0	0.0	1	0.2	0	0.0	4	0.9	1	0.2
126	Teufelsküche	1	445	26	5.8	9	2.0	98	22.0	79	17.8	6	1.3	13	2.9	0	0.0	0	0.0	0	0.0	0	0.0	0	0.0	4	0.9
324	T. Chaleux; 5	3	3042	194	6.4	106	3.5	802	26.4	673	22.1	91	3.0	29	1.0	0	0.0	0	0.0	0	0.0	0	0.0	9	0.3	3	0.1
325	T. Chal.; ex.O.	3	308	14	4.5	17	5.5	135	43.8	103	33.4	27	8.8	3	1.0	0	0.0	0	0.0	0	0.0	0	0.0	2	0.6	0	0.0
328	Trou Frontal	3	178	1	0.6	11	6.2	39	21.9	35	19.7	3	1.7	0	0.0	0	0.0	1	0.6	0	0.0	0	0.0	0	0.0	3	1.7
335	Verlaine	3	143	18	12.6	9	6.3	15	10.5	9	6.3	2	1.4	0	0.0	1	0.7	0	0.0	0	0.0	3	2.1	0	0.0	1	0.7
79	Veyrier	1	593	0	0.0	39	6.6	207	34.9	131	22.1	0	0.0	50	8.4	0	0.0	17	2.9	0	0.0	0	0.0	9	1.5	5	0.8
647	Wilczyce	5	375	0	0.0	0	0.0	85	22.7	84	22.4	0	0.0	1	0.3	0	0.0	0	0.0	0	0.0	0	0.0	0	0.0	0	0.0
379	Wildscheuer	3	129	5	3.9	1	0.8	5	3.9	3	2.3	1	0.8	1	0.8	0	0.0	0	0.0	0	0.0	0	0.0	0	0.0	0	0.0
378	Wildweiberlei	3	128	0	0.0	0	0.0	20	15.6	14	10.9	1	0.8	5	3.9	0	0.0	0	0.0	0	0.0	0	0.0	0	0.0	0	0.0
134	Wilmatt	1	107	0	0.0	0	0.0	5	4.7	4	3.7	0	0.0	1	0.9	0	0.0	0	0.0	0	0.0	0	0.0	0	0.0	0	0.0
135	Winznau-K.	1	1876	0	0.0	0	0.0	736	39.2	736	39.2	0	0.0	0	0.0	0	0.0	0	0.0	0	0.0	0	0.0	0	0.0	30	1.6
567	Žitného	5	533	7	1.3	16	3.0	234	43.9	232	43.5	0	0.0	2	0.4	0	0.0	0	0.0	0	0.0	0	0.0	0	0.0	1	0.2
Hamburgian																											
H143	Borneck	6	533	0	0.0	40	7.5	1	0.2	1	0.2	0	0.0	0	0.0	0	0.0	0	0.0	0	0.0	0	0.0	0	0.0	1	0.2
H99	Deimern 41	6	410	0	0.0	0	0.0	0	0.0	0	0.0	0	0.0	0	0.0	0	0.0	0	0.0	0	0.0	0	0.0	0	0.0	1	0.2
H100	Deimern 41b	6	177	0	0.0	0	0.0	0	0.0	0	0.0	0	0.0	0	0.0	0	0.0	0	0.0	0	0.0	0	0.0	0	0.0	0	0.0
H97	Deimern 42	6	859	0	0.0	0	0.0	0	0.0	0	0.0	0	0.0	0	0.0	0	0.0	0	0.0	0	0.0	0	0.0	0	0.0	0	0.0
H101	Deimern 44	6	812	0	0.0	16	2.0	2	0.2	2	0.2	0	0.0	0	0.0	0	0.0	0	0.0	0	0.0	0	0.0	0	0.0	17	2.0
H106	Deimern 46	6	109	0	0.0	5	4.6	0	0.0	0	0.0	0	0.0	0	0.0	0	0.0	0	0.0	0	0.0	0	0.0	0	0.0	1	0.1
H70	Glaner Heide	6	101	0	0.0	0	0.0	0	0.0	0	0.0	0	0.0	0	0.0	0	0.0	0	0.0	0	0.0	0	0.0	0	0.0	0	0.0
H124	Hasewisch	6	789	0	0.0	23	2.9	0	0.0	0	0.0	0	0.0	0	0.0	0	0.0	0	0.0	0	0.0	0	0.0	0	0.0	0	0.0
H35	Hav. Holtinger.	6	259	0	0.0	3	1.2	0	0.0	0	0.0	0	0.0	0	0.0	0	0.0	0	0.0	0	0.0	0	0.0	0	0.0	0	0.0
H93	Heber 118	6	197	0	0.0	15	7.6	0	0.0	0	0.0	0	0.0	0	0.0	0	0.0	0	0.0	0	0.0	0	0.0	0	0.0	2	1.0
H95	Heber 127	6	659	0	0.0	18	2.7	0	0.0	0	0.0	0	0.0	0	0.0	0	0.0	0	0.0	0	0.0	0	0.0	0	0.0	0	0.0
H163	Liny 1	6	152	0	0.0	1	0.7	0	0.0	0	0.0	0	0.0	0	0.0	0	0.0	0	0.0	0	0.0	0	0.0	0	0.0	3	2.0
H122	Meiendorf 2	6	950	0	0.0	117	12.3	0	0.0	0	0.0	0	0.0	0	0.0	0	0.0	0	0.0	0	0.0	0	0.0	0	0.0	1	0.1
H130	Meiendorf 9	6	131	0	0.0	4	3.1	0	0.0	0	0.0	0	0.0	0	0.0	0	0.0	0	0.0	0	0.0	0	0.0	0	0.0	0	0.0
H164	Mircowice 33	6	132	0	0.0	4	3.0	0	0.0	0	0.0	0	0.0	0	0.0	0	0.0	0	0.0	0	0.0	0	0.0	0	0.0	0	0.0
H21	Neer II	6	235	0	0.0	0	0.0	0	0.0	0	0.0	0	0.0	0	0.0	0	0.0	0	0.0	0	0.0	0	0.0	0	0.0	35	14.9
H161	Olbrachcice 8	6	463	0	0.0	32	6.9	0	0.0	0	0.0	0	0.0	0	0.0	0	0.0	0	0.0	0	0.0	0	0.0	0	0.0	7	1.5
H125	Poggenwisch	6	534	0	0.0	26	4.9	0	0.0	0	0.0	0	0.0	0	0.0	0	0.0	0	0.0	0	0.0	0	0.0	0	0.0	2	0.4
H79	Schalkholz	6	106	0	0.0	2	1.9	0	0.0	0	0.0	0	0.0	0	0.0	0	0.0	0	0.0	0	0.0	0	0.0	0	0.0	0	0.0
H127	Teltwisch 1	6	1160	0	0.0	28	2.4	0	0.0	0	0.0	0	0.0	0	0.0	0	0.0	0	0.0	0	0.0	0	0.0	0	0.0	11	0.9
H27	Ureterp I	6	999	0	0.0	86	8.6	0	0.0	0	0.0	0	0.0	0	0.0	0	0.0	0	0.0	0	0.0	0	0.0	0	0.0	1	0.1
H114	Wellingsbüttel	6	179	0	0.0	1	0.6	0	0.0	0	0.0	0	0.0	0	0.0	0	0.0	0	0.0	0	0.0	0	0.0	0	0.0	0	0.0
H158	Wojnowo I/75	6	119	0	0.0	0	0.0	0	0.0	0	0.0	0	0.0	0	0.0	0	0.0	0	0.0	0	0.0	0	0.0	0	0.0	0	0.0

(continued)

Table A.35 (continued)

Features TRU – SHP

ID	Site	G	nTool	TRU	%	PB	%	LR	%	TRI	%	SHP	%	ID	Site	G	nTool	TRU	%	PB	%	LR	%	TRI	%	SHP	%
471	Ahlendorf	4	301	7	2.3	0	0.0	0	0.0	0	0.0	0	0.0	346	Echt	3	122	10	8.2	0	0.0	4	3.3	0	0.0	1	0.8
231	Aichbühl B	2	132	4	3.0	0	0.0	36	27.3	0	0.0	0	0.0	164	Einsiedeln-L.	1	195	11	5.6	0	0.0	33	16.9	5	2.6	0	0.0
351	Alsdorf	3	385	35	9.1	0	0.0	34	8.8	0	0.0	0	0.0	44	Farincourt; surf.	1	140	12	8.6	0	0.0	6	4.3	9	6.4	0	0.0
364	Andernach; I	3	503	32	6.4	0	0.0	0	0.0	0	0.0	0	0.0	45	Farincourt; I+II	1	210	19	9.0	0	0.0	20	9.5	25	11.9	0	0.0
365	Andernach; II	3	459	19	4.1	0	0.0	0	0.0	0	0.0	0	0.0	46	Farincourt; III-E	1	104	13	12.5	0	0.0	12	11.5	0	0.0	0	0.0
366	Andernach; III	3	172	10	5.8	0	0.0	0	0.0	0	0.0	0	0.0	223	Felsställe; IIIa	2	229	29	12.7	0	0.0	53	23.1	0	0.0	0	0.0
367	Andernach; IV	3	310	11	3.5	0	0.0	0	0.0	0	0.0	0	0.0	224	Felsställe; IIIb	2	598	55	9.2	0	0.0	198	33.1	0	0.0	0	0.0
82	Auzary-Thônes	1	177	6	3.4	0	0.0	11	6.2	0	0.0	0	0.0	59	Fru, A.; II	1	118	0	0.0	0	0.0	0	0.0	0	0.0	0	0.0
401	B. Frankenh.	4	675	35	5.2	0	0.0	115	17.0	0	0.0	0	0.0	120	Fürsteiner	1	210	0	0.0	0	0.0	0	0.0	0	0.0	0	0.0
590	Balcarka Cave	5	123	3	2.4	0	0.0	7	5.7	0	0.0	0	0.0	468	Galgenberg	4	103	2	1.9	1	1.0	4	3.9	0	0.0	0	0.0
12	Balme d'Isère	1	119	13	10.9	0	0.0	32	26.9	0	0.0	0	0.0	18	Gay, A.	1	153	0	0.0	0	0.0	5	3.3	0	0.0	0	0.0
307	Barbing	2	120	2	1.7	0	0.0	7	5.8	0	0.0	0	0.0	478	Gera-Liebschw.	4	612	17	2.8	0	0.0	66	10.8	0	0.0	0	0.0
67	Baume Noire	1	247	12	4.9	0	0.0	3	1.2	0	0.0	0	0.0	368	Gönnersdorf; I	3	1809	51	2.8	0	0.0	0	0.0	0	0.0	0	0.0
37	Bibi, G.	1	188	13	6.9	0	0.0	24	12.8	0	0.0	0	0.0	369	Gönnersdorf; II	3	2380	51	2.1	0	0.0	0	0.0	0	0.0	0	0.0
114	Birseck-Ermit.	1	197	0	0.0	0	0.0	0	0.0	13	6.6	0	0.0	370	Gönnersdorf; III	3	893	17	1.9	0	0.0	0	0.0	0	0.0	0	0.0
320	Bois Laiterie	3	297	25	8.4	0	0.0	91	30.6	0	0.0	0	0.0	371	Gönnersdorf; IV	3	174	3	1.7	0	0.0	0	0.0	0	0.0	0	0.0
217	Braunfirst	2	240	17	7.1	2	0.8	23	9.6	0	0.0	0	0.0	28	Grappin	1	175	12	6.9	0	0.0	30	17.1	0	0.0	0	0.0
246	Brillenh.; low.	2	282	26	9.2	0	0.0	21	7.4	0	0.0	0	0.0	347	Griendtsveen	3	137	3	2.2	0	0.0	0	0.0	0	0.0	0	0.0
247	Brillenh.; upp.	2	300	33	11.0	0	0.0	37	12.3	0	0.0	0	0.0	483	Groitzsch; A1	4	228	17	7.5	0	0.0	2	0.9	0	0.0	0	0.0
107	Brügglihöhle	1	184	0	0.0	0	0.0	50	27.2	0	0.0	3	1.6	485	Groitzsch; B	4	406	23	5.7	0	0.0	9	2.2	0	0.0	0	0.0
205	Burgh. Dietfurt	2	119	15	12.6	0	0.0	18	15.1	0	0.0	0	0.0	487	Groitzsch; C1	4	536	40	7.5	0	0.0	11	2.1	0	0.0	0	0.0
551	Býčí skála	5	238	0	0.0	0	0.0	0	0.0	0	0.0	0	0.0	489	Groi.; C3+DN.	4	655	88	13.4	0	0.0	9	1.4	1	0.2	0	0.0
2	Campalou	1	100	8	8.0	0	0.0	0	0.0	0	0.0	0	0.0	490	Groitzsch; D1	4	559	41	7.3	0	0.0	9	1.6	0	0.0	0	0.0
30	Chênelaz	1	235	8	3.4	0	0.0	12	5.1	0	0.0	0	0.0	491	Groitzsch; D2	4	196	11	5.6	0	0.0	2	1.0	0	0.0	0	0.0
334	Coléoptère	3	125	9	7.2	0	0.0	8	6.4	0	0.0	0	0.0	95	H.-Champ.; 2	1	169	1	0.6	0	0.0	4	2.4	0	0.0	0	0.0
29	Colomb	1	264	53	20.1	0	0.0	46	17.4	0	0.0	0	0.0	93	H.-Champ.; pri.	1	594	7	1.2	0	0.0	16	2.7	0	0.0	0	0.0
8	Croze-s.-Suran	1	314	0	0.0	0	0.0	0	0.0	0	0.0	1	0.3	132	Hard I	1	259	3	1.2	0	0.0	160	61.8	0	0.0	0	0.0
69	Douattes, Les	1	637	17	2.7	0	0.0	13	2.0	0	0.0	0	0.0	243	Hohle Fels; IIa	2	111	8	7.2	0	0.0	14	12.6	0	0.0	0	0.0
601	Dzierzyslaw 35	5	912	0	0.0	0	0.0	115	12.6	114	12.5	0	0.0	530	Hostim	4	1871	47	2.5	0	0.0	4	0.2	0	0.0	2	0.1
599	Hranice	5	557	21	3.8	1	0.2	48	8.6	34	6.1	0	0.0	331	Orp; ouest	3	159	11	6.9	0	0.0	24	15.1	0	0.0	0	0.0
63	Jean P. I; 9A+B	1	174	3	1.7	0	0.0	15	8.6	1	0.6	0	0.0	330	Orp; est	3	465	17	3.7	0	0.0	60	12.9	0	0.0	0	0.0
443	Kahla-Löbsch.	4	848	8	0.9	0	0.0	0	0.0	0	0.0	0	0.0	579	Pekárna; gh	5	911	22	2.4	0	0.0	1	0.1	0	0.0	0	0.0
543	Kamegg	5	309	4	1.3	1	0.3	14	4.5	0	0.0	4	1.3	179	Petersfels 3AH2	2	259	0	0.0	0	0.0	61	23.6	0	0.0	0	0.0
214	Käppele	2	294	5	1.7	0	0.0	48	16.3	0	0.0	0	0.0	199	Probstfels	2	148	5	3.4	0	0.0	16	10.8	0	0.0	0	0.0
136	Käsloch	1	1015	6	0.6	0	0.0	383	37.7	0	0.0	17	1.7	139	Reiden-St.	1	245	26	10.6	0	0.0	49	20.0	0	0.0	0	0.0
290	Kastelhängh.	2	131	2	1.5	1	0.8	40	30.5	0	0.0	0	0.0	313	Roc-la-Tour	3	410	15	3.7	0	0.0	55	13.4	0	0.0	0	0.0
110	Kastelh.; upp.	1	567	10	1.8	0	0.0	0	0.0	0	0.0	16	2.8	455	Saaleck	4	2021	60	3.0	0	0.0	2	0.1	0	0.0	0	0.0
272	Kaufertsb.; upp.	2	136	8	5.9	0	0.0	9	6.6	1	0.7	0	0.0	480	Schafgraben,	4	162	5	3.1	2	1.2	16	9.9	0	0.0	0	0.0
271	Kaufertsb.; low.	2	225	13	5.8	2	0.9	17	7.6	0	0.0	0	0.0	166	Schweizersbild	2	657	8	1.2	1	0.2	0	0.0	2	0.3	8	1.2
170	Kesslerloch	2	1775	2	0.1	0	0.0	0	0.0	0	0.0	0	0.0	291	Sesselfelsg.; C2	2	117	0	0.0	2	1.7	4	3.4	0	0.0	0	0.0
648	Klementowice	5	721	87	12.1	0	0.0	169	23.4	0	0.0	0	0.0	56	Seveux	1	692	21	3.0	22	3.2	51	7.4	0	0.0	0	0.0
451	Kniegrotte	4	2513	97	3.9	1	0.0	347	13.8	187	7.4	3	0.1	281	Speckberg	2	1563	109	7.0	441	28.2	313	20.0	0	0.0	0	0.0
108	Kohlerhöhle	1	254	5	2.0	0	0.0	6	2.4	0	0.0	3	1.2	344	S.-Groene Paal	3	381	6	1.6	0	0.0	108	28.3	0	0.0	0	0.0
571	Kůlna Cave; 5	5	104	3	2.9	0	0.0	27	26.0	0	0.0	0	0.0	417	Teufelsbr.; 3	4	1170	31	2.6	0	0.0	0	0.0	0	0.0	0	0.0

572	Kůlna Cave; 6	5	144	17	11.8	0	0.0	12	8.3	0	0.0	0	0.0	418	Teufelsbr.; 4	4	423	5	1.2	0	0.0	0	0.0	0	0.0	0	0.0
518	Kvíc	4	188	16	8.5	1	0.5	19	10.1	0	0.0	1	0.5	126	Teufelsküche	1	445	25	5.6	0	0.0	77	17.3	0	0.0	0	0.0
452	Lausnitz	4	429	12	2.8	0	0.0	40	9.3	0	0.0	0	0.0	324	T. d.Chaleux; 5	3	3042	277	9.1	0	0.0	94	3.1	0	0.0	0	0.0
456	Lengefeld; surf.	4	258	19	7.4	0	0.0	44	17.1	0	0.0	1	0.4	325	T. d.Chal.; ex.O.	3	308	5	1.6	0	0.0	36	11.7	0	0.0	0	0.0
594	Loštice I	5	212	10	4.7	0	0.0	17	8.0	2	0.9	0	0.0	328	Trou du Frontal	3	178	22	12.4	0	0.0	6	3.4	0	0.0	0	0.0
547	Maloměřice-B; I	5	653	29	4.4	5	0.8	27	4.1	0	0.0	1	0.2	335	Verlaine	3	143	14	9.8	0	0.0	13	9.1	0	0.0	1	0.7
101	Moosseedorf-M.	1	5196	266	5.1	0	0.0	132	2.5	1	0.0	0	0.0	79	Veyrier	1	593	53	8.9	0	0.0	24	4.0	0	0.0	0	0.0
137	Mühleloch, A.	1	302	0	0.0	0	0.0	203	67.2	0	0.0	0	0.0	647	Wilczyce	5	375	14	3.7	0	0.0	64	17.1	0	0.0	0	0.0
122	Munzingen	1	682	112	16.4	6	0.9	155	22.7	0	0.0	0	0.0	379	Wildscheuer	3	129	9	7.0	2	1.6	74	57.4	0	0.0	0	0.0
439	Nebra	4	2320	89	3.8	0	0.0	36	1.6	0	0.0	0	0.0	378	Wildweiberlei	3	128	4	3.1	1	0.8	68	53.1	0	0.0	0	0.0
90	N.-Monruz; 1M	1	1413	13	0.9	0	0.0	12	0.8	0	0.0	0	0.0	134	Wilmatt	1	107	1	0.9	0	0.0	65	60.7	0	0.0	1	0.9
585	Ochozska-H.	5	353	28	7.9	0	0.0	35	9.9	0	0.0	0	0.0	135	Winznau-Köpfli	1	1876	0	0.0	0	0.0	460	24.5	0	0.0	132	7.0
444	Ölknitz	4	1577	47	3.0	0	0.0	66	4.2	0	0.0	0	0.0	567	Žitného	5	533	39	7.3	0	0.0	10	1.9	0	0.0	0	0.0
Hamburgian																											
H143	Borneck	6	533	13	2.4	0	0.0	60	11.3	0	0.0	39	7.3	H122	Meiendorf 2	6	950	32	3.4	0	0.0	13	1.4	0	0.0	56	5.9
H99	Deimern 41	6	410	27	6.6	0	0.0	38	9.3	0	0.0	49	12.0	H130	Meiendorf 9	6	131	20	15.3	0	0.0	1	0.8	0	0.0	4	3.1
H100	Deimern 41b	6	177	18	10.2	0	0.0	17	9.6	0	0.0	31	17.5	H164	Mircowice 33	6	132	19	14.4	1	0.8	27	20.5	0	0.0	12	9.1
H97	Deimern 42	6	859	86	10.0	0	0.0	251	29.2	0	0.0	98	11.4	H21	Neer II	6	235	15	6.4	0	0.0	0	0.0	0	0.0	12	5.1
H101	Deimern 44	6	812	82	10.1	0	0.0	188	23.2	0	0.0	78	9.6	H161	Olbrachcice 8	6	463	33	7.1	0	0.0	170	36.7	0	0.0	37	8.0
H106	Deimern 46	6	109	10	9.2	0	0.0	5	4.6	0	0.0	5	4.6	H125	Poggenwisch	6	534	28	5.2	0	0.0	58	10.9	0	0.0	45	8.4
H70	Glaner Heide	6	101	0	0.0	0	0.0	0	0.0	0	0.0	5	5.0	H79	Schalkholz	6	106	12	11.3	0	0.0	1	0.9	0	0.0	10	9.4
H124	Hasewisch	6	789	51	6.5	0	0.0	242	30.7	0	0.0	25	3.2	H127	Teltwisch 1	6	1160	53	4.6	0	0.0	147	12.7	0	0.0	55	4.7
H35	Hav. Holtingerz.	6	259	6	2.3	0	0.0	0	0.0	0	0.0	13	5.0	H27	Ureterp I	6	999	111	11.1	0	0.0	240	24.0	0	0.0	58	5.8
H93	Heber 118	6	197	31	15.7	0	0.0	0	0.0	0	0.0	36	18.3	H114	Wellingsbüttel	6	179	0	0.0	0	0.0	2	1.1	0	0.0	10	5.6
H95	Heber 127	6	659	70	10.6	0	0.0	30	4.6	0	0.0	74	11.2	H158	Wojnowo I/75	6	119	0	0.0	0	0.0	0	0.0	0	0.0	0	0.0
H163	Liny 1	6	152	19	12.5	0	0.0	69	45.4	0	0.0	12	7.9														

Abbreviations see list of abbreviations

Table A.36 Discriminant function analysis. Raw data of the temporal investigation

Features SC-t – LBO

ID	Site	T	nTool	SC-t	%	SC	%	SSC	%	RSC	%	BU-t	%	BU	%	LBU	%	FBU	%	ZIN	%	BO-t	%	BO	%	LBO	%
364	Andernach; I	2	503	172	34.2	172	34.2	0	0.0	0	0.0	92	18.3	89	17.7	3	0.6	0	0.0	0	0.0	42	8.3	42	8.3	0	0.0
365	Andernach; II	2	459	18	3.9	18	3.9	0	0.0	0	0.0	154	33.6	123	26.8	31	6.8	0	0.0	0	0.0	47	10.2	47	10.2	0	0.0
366	Andernach; III	2	172	41	23.8	41	23.8	0	0.0	0	0.0	44	25.6	35	20.3	9	5.2	0	0.0	0	0.0	8	4.7	8	4.7	0	0.0
82	Auzary-Thôn.	1	177	9	5.1	8	4.5	0	0.0	1	0.6	16	9.0	16	9.0	0	0.0	0	0.0	0	0.0	3	1.7	3	1.7	0	0.0
590	Balcarka Cave	1	123	20	16.3	16	13.0	0	0.0	4	3.3	13	10.6	13	10.6	0	0.0	0	0.0	3	2.4	7	5.7	7	5.7	0	0.0
15	B. de Cuiseaux	4	21	1	4.8	1	4.8	0	0.0	0	0.0	1	4.8	1	4.8	0	0.0	0	0.0	0	0.0	2	9.5	2	9.5	0	0.0
73	Banges	3	82	7	8.5	7	8.5	0	0.0	0	0.0	11	13.4	11	13.4	0	0.0	0	0.0	0	0.0	6	7.3	6	7.3	0	0.0
66	Baume N.; M3	2	247	18	7.3	18	7.3	0	0.0	0	0.0	118	47.8	118	47.8	0	0.0	0	0.0	0	0.0	19	7.7	19	7.7	0	0.0
114	Birseck-Ermit.	4	197	26	13.2	25	12.7	0	0.0	1	0.5	48	24.4	48	24.4	0	0.0	0	0.0	0	0.0	4	2.0	4	2.0	0	0.0
320	Bois Laiterie	2	297	23	7.7	13	4.4	0	0.0	10	3.4	36	12.1	33	11.1	1	0.3	2	0.7	0	0.0	22	7.4	22	7.4	0	0.0
205	Burgh. Dietfurt	3	119	11	9.2	7	5.9	3	2.5	1	0.8	14	11.8	14	11.8	0	0.0	0	0.0	1	0.8	5	4.2	5	4.2	0	0.0
219	Burkhardtsh.; V	3	78	5	6.4	5	6.4	0	0.0	0	0.0	14	17.9	14	17.9	0	0.0	0	0.0	0	0.0	4	5.1	4	5.1	0	0.0
104/105	Büttenloch A+B	3	59	1	1.7	1	1.7	0	0.0	0	0.0	1	1.7	1	1.7	0	0.0	0	0.0	0	0.0	5	8.5	5	8.5	0	0.0
198	Buttentalhöhle	2	50	2	4.0	1	2.0	0	0.0	1	2.0	10	20.0	10	20.0	0	0.0	0	0.0	0	0.0	4	8.0	4	8.0	0	0.0
11	Calvaire; IV inf.	1	36	4	11.1	4	11.1	0	0.0	0	0.0	13	36.1	13	36.1	0	0.0	0	0.0	0	0.0	2	5.6	2	5.6	0	0.0
10	Calv.; III moy.	2	70	7	10.0	6	8.6	0	0.0	1	1.4	14	20.0	14	20.0	0	0.0	0	0.0	0	0.0	1	1.4	1	1.4	0	0.0
2	Campalou	1	100	12	12.0	12	12.0	0	0.0	0	0.0	22	22.0	22	22.0	0	0.0	0	0.0	0	0.0	0	0.0	0	0.0	0	0.0
30	Chênelaz	3	235	20	8.5	20	8.5	0	0.0	0	0.0	21	8.9	21	8.9	0	0.0	0	0.0	0	0.0	31	13.2	31	13.2	0	0.0
334	Coléoptère	3	125	18	14.4	18	14.4	0	0.0	0	0.0	17	13.6	17	13.6	0	0.0	0	0.0	0	0.0	12	9.6	12	9.6	0	0.0
29	Colomb	3	264	91	34.5	91	34.5	0	0.0	0	0.0	24	9.1	24	9.1	0	0.0	0	0.0	0	0.0	9	3.4	9	3.4	0	0.0
8	Croze-s.-Suran	1	314	56	17.8	56	17.8	0	0.0	0	0.0	144	45.9	144	45.9	0	0.0	0	0.0	0	0.0	108	34.4	108	34.4	0	0.0
69	Douattes, Les	2	637	103	16.2	76	11.9	23	3.6	4	0.6	168	26.4	168	26.4	0	0.0	0	0.0	1	0.2	44	6.9	44	6.9	0	0.0
601	Dzierzyslaw 35	1	912	0	0.0	0	0.0	0	0.0	0	0.0	491	53.8	490	53.7	1	0.1	0	0.0	2	0.2	176	19.3	175	19.2	1	0.1
223	Felsställe; IIIa	2	229	12	5.2	10	4.4	0	0.0	2	0.9	30	13.1	30	13.1	0	0.0	0	0.0	0	0.0	33	14.4	32	14.0	1	0.4
224	Felsställe; IIIb	3	598	54	9.0	50	8.4	0	0.0	4	0.7	77	12.9	77	12.9	0	0.0	0	0.0	0	0.0	87	14.5	85	14.2	2	0.3
24	Freydières	4	35	2	5.7	2	5.7	0	0.0	0	0.0	2	5.7	2	5.7	0	0.0	0	0.0	0	0.0	2	5.7	2	5.7	0	0.0
195	Gnirshöhöle 1	2	34	8	23.5	6	17.6	0	0.0	2	5.9	5	14.7	5	14.7	0	0.0	0	0.0	0	0.0	5	14.7	5	14.7	0	0.0
368	Gönnersdorf; I	2	1809	100	5.5	100	5.5	0	0.0	0	0.0	657	36.3	510	28.2	115	6.4	32	1.8	0	0.0	238	13.2	231	12.8	7	0.4
369	Gönnersdorf; II	2	2380	136	5.7	136	5.7	0	0.0	0	0.0	535	22.5	479	20.1	42	1.8	14	0.6	0	0.0	211	8.9	204	8.6	7	0.3
370	Gönnersdorf; III	2	893	54	6.0	54	6.0	0	0.0	0	0.0	232	26.0	189	21.2	35	3.9	8	0.9	0	0.0	151	16.9	145	16.2	6	0.7
28	Grappin	1	175	20	11.4	18	10.3	0	0.0	2	1.1	47	26.9	47	26.9	0	0.0	0	0.0	0	0.0	0	0.0	0	0.0	0	0.0
93	H.-Champ.; pri.	2	594	54	9.1	36	6.1	0	0.0	18	3.0	90	15.2	90	15.2	0	0.0	0	0.0	4	0.7	33	5.6	33	5.6	0	0.0
95	H.-Champ.; 2	2	169	9	5.3	5	3.0	1	0.6	3	1.8	48	28.4	48	28.4	0	0.0	0	0.0	1	0.6	5	3.0	5	3.0	0	0.0
97	H.-Champ.; 1A	3	64	8	12.5	4	6.3	1	1.6	3	4.7	16	25.0	16	25.0	0	0.0	0	0.0	0	0.0	4	6.3	4	6.3	0	0.0
243	Hohle Fels; IIa	2	111	4	3.6	4	3.6	0	0.0	0	0.0	8	7.2	8	7.2	0	0.0	0	0.0	0	0.0	2	1.8	2	1.8	0	0.0
530	Hostim	3	1871	209	11.2	178	9.5	0	0.0	31	1.7	221	11.8	203	10.8	3	0.2	15	0.8	19	1.0	431	23.0	417	22.3	14	0.7
599	Hranice	1	557	35	6.3	34	6.1	0	0.0	1	0.2	288	51.7	285	51.2	0	0.0	3	0.5	11	2.0	74	13.3	73	13.1	1	0.2
63	Jean P. I; 9A+B	2	174	6	3.4	6	3.4	0	0.0	0	0.0	23	13.2	23	13.2	0	0.0	0	0.0	4	2.3	10	5.7	9	5.2	1	0.6
64	Jean P. II	2	50	4	8.0	3	6.0	0	0.0	1	2.0	6	12.0	6	12.0	0	0.0	0	0.0	0	0.0	5	10.0	5	10.0	0	0.0
543	Kamegg	1	309	48	15.5	48	15.5	0	0.0	0	0.0	135	43.7	135	43.7	0	0.0	0	0.0	3	1.0	17	5.5	17	5.5	0	0.0
290	Kastelhängh.	4	131	14	10.7	11	8.4	0	0.0	3	2.3	16	12.2	15	11.5	1	0.8	0	0.0	0	0.0	7	5.3	6	4.6	1	0.8
110	Kastelh.; upp.	4	567	109	19.2	105	18.5	2	0.4	2	0.4	116	20.5	113	19.9	0	0.0	3	0.5	0	0.0	91	16.0	84	14.8	0	0.0
271	Kaufertsb.; low.	2	225	26	11.6	19	8.4	0	0.0	7	3.1	117	52.0	108	48.0	7	3.1	2	0.9	1	0.4	12	5.3	7	3.1	5	2.2
170	Kesslerloch	1	1775	784	44.2	780	43.9	0	0.0	4	0.2	741	41.7	729	41.1	12	0.7	0	0.0	13	0.7	87	4.9	77	4.3	10	0.6
451	Kniegrotte	1	2513	234	9.3	215	8.6	6	0.2	13	0.5	514	20.5	500	19.9	14	0.6	0	0.0	0	0.0	205	8.2	203	8.1	2	0.1

108	Kohlerhöhle	4	254	140	55.1	131	51.6	0	0.0	9	3.5	22	8.7	22	8.7	0	0.0	0	0.0	0	0.0	38	15.0	37	14.6	1	0.4
570	Kolíbky, 6	2	20	2	10.0	2	10.0	0	0.0	0	0.0	5	25.0	5	25.0	0	0.0	0	0.0	1	5.0	0	0.0	0	0.0	0	0.0
572	Kůlna Cave; 6	4	144	36	25.0	26	18.1	10	6.9	0	0.0	17	11.8	16	11.1	0	0.0	1	0.7	8	5.6	8	5.6	8	5.6	0	0.0
101	Moosseedorf-M.	4	5196	443	8.5	409	7.9	0	0.0	34	0.7	839	16.1	839	16.1	0	0.0	0	0.0	0	0.0	716	13.8	17	0.3	17	0.3
629	Maszycka	1	66	20	30.3	14	21.2	0	0.0	6	9.1	7	10.6	7	10.6	0	0.0	0	0.0	0	0.0	1	1.5	1	1.5	0	0.0
635	Mosty	4	47	11	23.4	11	23.4	0	0.0	0	0.0	5	10.6	5	10.6	0	0.0	0	0.0	2	4.3	6	12.8	6	12.8	0	0.0
122	Munzingen	1	682	92	13.5	86	12.6	0	0.0	6	0.9	160	23.5	160	23.5	0	0.0	0	0.0	2	0.3	35	5.1	34	5.0	1	0.1
439	Nebra	2	2320	202	8.7	193	8.3	6	0.3	3	0.1	1026	44.2	946	40.8	56	2.4	24	1.0	10	0.4	293	12.6	280	12.1	13	0.6
90	N.-Monruz; 1M	2	1413	66	4.7	59	4.2	5	0.4	2	0.1	251	17.8	250	17.7	1	0.1	0	0.0	7	0.5	155	11.0	155	11.0	0	0.0
91	N.-Mon.; 2M/A	2	59	4	6.8	3	5.1	1	1.7	0	0.0	5	8.5	5	8.5	0	0.0	0	0.0	0	0.0	2	3.4	2	3.4	0	0.0
585	Ochozska-H.	3	353	40	11.3	37	10.5	0	0.0	3	0.8	56	15.9	52	14.7	1	0.3	3	0.8	9	2.5	34	9.6	31	8.8	3	0.8
444	Ölknitz	3	1577	82	5.2	82	5.2	0	0.0	0	0.0	252	16.0	251	15.9	1	0.1	0	0.0	0	0.0	139	8.8	136	8.6	3	0.2
579	Pekárna; gh	2	911	185	20.3	185	20.3	0	0.0	0	0.0	205	22.5	201	22.1	4	0.4	0	0.0	36	4.0	175	19.2	133	14.6	42	4.6
179	Petersf.; 3AH2	4	259	26	10.0	20	7.7	1	0.4	5	1.9	25	9.7	25	9.7	0	0.0	0	0.0	1	0.4	28	10.8	27	10.4	1	0.4
174	Petersf.;1AH3A	3	40	5	12.5	5	12.5	0	0.0	0	0.0	1	2.5	1	2.5	0	0.0	0	0.0	0	0.0	7	17.5	7	17.5	0	0.0
128	Rheinfelden-E.	4	77	5	6.5	4	5.2	0	0.0	1	1.3	22	28.6	21	27.3	1	1.3	0	0.0	0	0.0	11	14.3	11	14.3	0	0.0
455	Saaleck	2	2021	161	8.0	112	5.5	5	0.2	44	2.2	213	10.5	213	10.5	0	0.0	0	0.0	22	1.1	387	19.1	267	13.2	120	5.9
229	Schussen.; main	2	33	4	12.1	2	6.1	0	0.0	2	6.1	7	21.2	7	21.2	0	0.0	0	0.0	0	0.0	2	6.1	2	6.1	0	0.0
166	Schweizersbild	4	657	81	12.3	70	10.7	0	0.0	11	1.7	320	48.7	320	48.7	0	0.0	0	0.0	7	1.1	87	13.2	87	13.2	0	0.0
291	Sesselfelsg.; C2	2	117	7	6.0	5	4.3	0	0.0	2	1.7	19	16.2	18	15.4	0	0.0	1	0.9	0	0.0	6	5.1	6	5.1	0	0.0
261	Spitzbub.; AH2	2	20	2	10.0	1	5.0	0	0.0	1	5.0	5	25.0	5	25.0	0	0.0	0	0.0	0	0.0	5	25.0	5	25.0	0	0.0
416	Teufelsbr.; 2	2	44	4	9.1	0	0.0	4	9.1	1	2.3	6	13.6	6	13.6	0	0.0	0	0.0	1	2.3	3	6.8	3	6.8	0	0.0
417	Teufelsbr.; 3	3	1170	162	13.8	143	12.2	5	0.4	14	1.2	342	29.2	320	27.4	22	1.9	0	0.0	14	1.2	109	9.3	98	8.4	11	0.9
418	Teufelsbr.; 4	3	423	41	9.7	40	9.5	1	0.2	0	0.0	39	9.2	39	9.2	0	0.0	0	0.0	19	4.5	57	13.5	53	12.5	4	0.9
324	T. d.Chaleux; 5	2	3042	369	12.1	327	10.7	2	0.1	40	1.3	739	24.3	733	24.1	3	0.1	3	0.1	50	1.6	408	13.4	408	13.4	0	0.0
326	T. des Nutons	2	60	8	13.3	8	13.3	0	0.0	0	0.0	10	16.7	10	16.7	0	0.0	0	0.0	3	5.0	15	25.0	15	25.0	0	0.0
328	Trou du Frontal	2	178	26	14.6	26	14.6	0	0.0	0	0.0	42	23.6	42	23.6	0	0.0	0	0.0	2	1.1	26	14.6	26	14.6	0	0.0
335	Verlaine	2	143	24	16.8	24	16.8	0	0.0	0	0.0	25	17.5	23	16.1	0	0.0	2	1.4	1	0.7	22	15.4	22	15.4	0	0.0
79	Veyrier	3	593	94	15.9	77	13.0	9	1.5	8	1.3	93	15.7	93	15.7	0	0.0	0	0.0	0	0.0	78	13.2	78	13.2	0	0.0
475	Wallendorf	2	79	15	19.0	11	13.9	0	0.0	4	5.1	14	17.7	14	17.7	0	0.0	0	0.0	3	3.8	1	1.3	1	1.3	0	0.0
647	Wilczyce	2	375	41	10.9	41	10.9	0	0.0	0	0.0	166	44.3	145	38.7	21	5.6	0	0.0	0	0.0	5	1.3	5	1.3	0	0.0
378	Wildweiberlei	2	128	9	7.0	6	4.7	0	0.0	3	2.3	15	11.7	15	11.7	0	0.0	0	0.0	2	1.6	9	7.0	9	7.0	0	0.0
567	Žitného	1	533	75	14.1	65	12.2	0	0.0	10	1.9	89	16.7	88	16.5	0	0.0	1	0.2	8	1.5	54	10.1	54	10.1	0	0.0

(continued)

Table A.36 (continued)

Features SP – BP

ID	Site	T	nTool	SP	%	NDP	%	BB-t	%	BB	%	BBD	%	BBT	%	BBDT	%	BBTT	%	BBC	%	BBP	%	BBDE	%	BP	%
364	Andernach; I	2	503	43	8.5	0	0.0	122	24.3	122	24.3	0	0.0	0	0.0	0	0.0	0	0.0	0	0.0	0	0.0	0	0.0	0	0.0
365	Andernach; II	2	459	72	15.7	0	0.0	149	32.5	149	32.5	0	0.0	0	0.0	0	0.0	0	0.0	0	0.0	0	0.0	0	0.0	0	0.0
366	Andernach; III	2	172	13	7.6	0	0.0	56	32.6	56	32.6	0	0.0	0	0.0	0	0.0	0	0.0	0	0.0	0	0.0	0	0.0	0	0.0
82	Auzary-Thônes	1	177	6	3.4	3	1.7	123	69.5	113	63.8	7	4.0	1	0.6	0	0.0	0	0.0	0	0.0	0	0.0	2	1.1	0	0.0
590	Balcarka Cave	1	123	1	0.8	6	4.9	60	48.8	50	40.7	3	2.4	2	1.6	1	0.8	0	0.0	0	0.0	0	0.0	4	3.3	3	2.4
15	B. de Cuiseaux	4	21	0	0.0	3	14.3	6	28.6	6	28.6	0	0.0	0	0.0	0	0.0	0	0.0	0	0.0	0	0.0	0	0.0	4	19.0
73	Banges	3	82	0	0.0	2	2.4	52	63.4	52	63.4	0	0.0	0	0.0	0	0.0	0	0.0	0	0.0	0	0.0	0	0.0	1	1.2
66	Baume N.; M3	2	247	0	0.0	36	14.6	36	14.6	36	14.6	0	0.0	0	0.0	0	0.0	0	0.0	0	0.0	0	0.0	0	0.0	5	2.0
114	Birseck-Ermit.	4	197	0	0.0	0	0.0	105	53.3	83	42.1	16	8.1	6	3.0	0	0.0	0	0.0	0	0.0	0	0.0	0	0.0	1	0.5
320	Bois Laiterie	2	297	2	0.7	20	6.7	55	18.5	54	18.2	0	0.0	1	0.3	0	0.0	0	0.0	0	0.0	0	0.0	0	0.0	23	7.7
205	Burgh. Dietfurt	3	119	0	0.0	7	5.9	33	27.7	28	23.5	0	0.0	2	1.7	2	1.7	0	0.0	0	0.0	0	0.0	1	0.8	15	12.6
219	Burkhardtsh.; V	3	78	6	7.7	2	2.6	5	6.4	5	6.4	0	0.0	0	0.0	0	0.0	0	0.0	0	0.0	0	0.0	0	0.0	12	15.4
104,5	Büttenloch A+B	3	59	0	0.0	12	20.3	3	5.1	0	0.0	0	0.0	0	0.0	3	5.1	0	0.0	0	0.0	0	0.0	0	0.0	0	0.0
198	Buttentalhöhle	2	50	1	2.0	0	0.0	4	8.0	4	8.0	0	0.0	0	0.0	0	0.0	0	0.0	0	0.0	0	0.0	0	0.0	18	36.0
11	Calvaire; IV inf.	1	36	0	0.0	2	5.6	4	11.1	4	11.1	0	0.0	0	0.0	0	0.0	0	0.0	0	0.0	0	0.0	0	0.0	0	0.0
10	Calv.; III moy.	2	70	0	0.0	3	4.3	29	41.4	27	38.6	0	0.0	2	2.9	0	0.0	0	0.0	0	0.0	0	0.0	0	0.0	0	0.0
2	Campalou;	1	100	0	0.0	15	15.0	40	40.0	33	33.0	0	0.0	7	7.0	0	0.0	0	0.0	0	0.0	0	0.0	0	0.0	3	3.0
30	Chênelaz	3	235	0	0.0	27	11.5	109	46.4	–2	–0.9	92	39.1	13	5.5	0	0.0	2	0.9	0	0.0	0	0.0	4	1.7	7	3.0
334	Coléoptère	3	125	0	0.0	0	0.0	59	47.2	57	45.6	1	0.8	1	0.8	0	0.0	0	0.0	0	0.0	0	0.0	0	0.0	2	1.6
29	Colomb	3	264	0	0.0	17	6.4	20	7.6	20	7.6	0	0.0	0	0.0	0	0.0	0	0.0	0	0.0	0	0.0	0	0.0	4	1.5
8	Croze-s.-Suran	1	314	0	0.0	0	0.0	5	1.6	2	0.6	0	0.0	3	1.0	0	0.0	0	0.0	0	0.0	0	0.0	0	0.0	0	0.0
69	Douattes, Les	2	637	0	0.0	5	0.8	282	44.3	250	39.2	0	0.0	20	3.1	0	0.0	1	0.2	0	0.0	0	0.0	11	1.7	4	0.6
601	Dzierzyslaw 35	1	912	0	0.0	0	0.0	14	1.5	14	1.5	0	0.0	0	0.0	0	0.0	0	0.0	0	0.0	0	0.0	0	0.0	0	0.0
223	Felsställe; IIIa	2	229	3	1.3	0	0.0	65	28.4	34	14.8	17	7.4	4	1.7	2	0.9	0	0.0	0	0.0	0	0.0	8	3.5	4	1.7
224	Felsställe; IIIb	3	598	9	1.5	0	0.0	114	19.1	62	10.4	26	4.3	3	0.5	2	0.3	0	0.0	0	0.0	1	0.2	20	3.3	4	0.7
24	Freydières	4	35	0	0.0	0	0.0	3	8.6	2	5.7	1	2.9	0	0.0	0	0.0	0	0.0	0	0.0	0	0.0	0	0.0	2	5.7
195	Gnirshöhöle 1	2	34	0	0.0	1	2.9	7	20.6	4	11.8	2	5.9	0	0.0	1	2.9	0	0.0	0	0.0	0	0.0	0	0.0	0	0.0
368	Gönnersdorf; I	2	1809	192	10.6	0	0.0	571	31.6	481	26.6	87	4.8	2	0.1	0	0.0	0	0.0	0	0.0	0	0.0	1	0.1	0	0.0
369	Gönnersdorf; II	2	2380	456	19.2	0	0.0	991	41.6	878	36.9	107	4.5	2	0.1	0	0.0	0	0.0	0	0.0	0	0.0	4	0.2	0	0.0
370	Gönnersdorf; III	2	893	137	15.3	0	0.0	302	33.8	248	27.8	44	4.9	7	0.8	0	0.0	0	0.0	0	0.0	0	0.0	3	0.3	0	0.0
28	Grappin	1	175	4	2.3	11	6.3	51	29.1	31	17.7	0	0.0	18	10.3	0	0.0	0	0.0	0	0.0	0	0.0	2	1.1	0	0.0
93	H.-Champ.; pri.	2	594	13	2.2	8	1.3	369	62.1	354	59.6	0	0.0	13	2.2	0	0.0	0	0.0	1	0.2	0	0.0	1	0.2	0	0.0
95	H.-Champ.; 2	2	169	1	0.6	0	0.0	99	58.6	94	55.6	1	0.6	3	1.8	0	0.0	0	0.0	0	0.0	0	0.0	1	0.6	1	0.6
97	H.-Champ.; 1A	3	64	2	3.1	0	0.0	19	29.7	19	29.7	0	0.0	0	0.0	0	0.0	0	0.0	0	0.0	0	0.0	0	0.0	10	15.6
243	Hohle Fels; IIa	2	111	5	4.5	50	45.0	20	18.0	16	14.4	4	3.6	0	0.0	0	0.0	0	0.0	0	0.0	0	0.0	0	0.0	0	0.0
530	Hostim	3	1871	343	18.3	70	3.7	525	28.1	506	27.0	0	0.0	12	0.6	0	0.0	2	0.1	0	0.0	0	0.0	5	0.3	0	0.0
599	Hranice	1	557	8	1.4	14	2.5	23	4.1	15	2.7	0	0.0	8	1.4	0	0.0	0	0.0	0	0.0	0	0.0	0	0.0	0	0.0
63	Jean P. I; 9A+B	2	174	0	0.0	17	9.8	94	54.0	62	35.6	6	3.4	8	4.6	7	4.0	2	1.1	3	1.7	1	0.6	5	2.9	1	0.6
64	Jean P. II	2	50	0	0.0	6	12.0	24	48.0	11	22.0	3	6.0	2	4.0	4	8.0	0	0.0	1	2.0	3	6.0	0	0.0	0	0.0
543	Kamegg	1	309	0	0.0	1	0.3	77	24.9	54	17.5	17	5.5	4	1.3	0	0.0	2	0.6	0	0.0	0	0.0	0	0.0	5	1.6
290	Kastelhängh.	4	131	0	0.0	25	19.1	25	19.1	24	18.3	0	0.0	0	0.0	0	0.0	0	0.0	0	0.0	0	0.0	1	0.8	1	0.8
110	Kastelh.; upp.	4	567	0	0.0	0	0.0	213	37.6	192	33.9	0	0.0	13	2.3	0	0.0	6	1.1	0	0.0	0	0.0	2	0.4	12	2.1
271	Kaufertsb.; low.	2	225	0	0.0	4	1.8	33	14.7	25	11.1	5	2.2	2	0.9	0	0.0	1	0.4	0	0.0	0	0.0	0	0.0	0	0.0
170	Kesslerloch	1	1775	0	0.0	0	0.0	143	8.1	104	5.9	16	0.9	17	1.0	6	0.3	0	0.0	0	0.0	0	0.0	0	0.0	5	0.3

451	Kniegrotte	1	2513	128	5.1	0	0.0	789	31.4	603	24.0	73	2.9	78	3.1	18	0.7	9	0.4	4	0.2	0	0.0	4	0.2	8	0.3
108	Kohlerhöhle	4	254	0	0.0	12	4.7	20	7.9	17	6.7	0	0.0	3	1.2	0	0.0	0	0.0	0	0.0	0	0.0	0	0.0	8	3.1
570	Kolíbky, 6	2	20	1	5.0	1	5.0	2	10.0	2	10.0	0	0.0	0	0.0	0	0.0	0	0.0	0	0.0	0	0.0	0	0.0	0	0.0
572	Kůlna Cave; 6	4	144	0	0.0	11	7.6	33	22.9	32	22.2	0	0.0	1	0.7	0	0.0	0	0.0	0	0.0	0	0.0	0	0.0	2	1.4
101	Moosseedorf	4	5196	15	0.3	112	2.2	2659	51.4	2129	41.0	15	0.3	460	8.9	15	0.3	15	0.3	15	0.3	0	0.0	20	0.4	3	0.1
629	Maszycka	1	66	0	0.0	2	3.0	10	15.2	5	7.6	0	0.0	5	7.6	0	0.0	0	0.0	0	0.0	0	0.0	0	0.0	1	1.5
635	Mosty	4	47	0	0.0	0	0.0	3	6.4	3	6.4	0	0.0	0	0.0	0	0.0	0	0.0	0	0.0	0	0.0	0	0.0	3	6.4
122	Munzingen	1	682	52	7.6	0	0.0	68	10.0	47	6.9	6	0.9	10	1.5	2	0.3	0	0.0	0	0.0	0	0.0	3	0.4	0	0.0
439	Nebra	2	2320	70	3.0	0	0.0	587	25.3	493	21.3	91	3.9	1	0.0	0	0.0	0	0.0	0	0.0	2	0.1	0	0.0	7	0.3
90	N.-Monruz; 1M	2	1413	20	1.4	6	0.4	883	62.5	804	56.9	36	2.5	33	2.3	5	0.4	0	0.0	0	0.0	0	0.0	5	0.4	0	0.0
91	N.-Mon.; 2M/A	2	59	1	1.7	0	0.0	35	59.3	31	52.5	1	1.7	2	3.4	0	0.0	0	0.0	1	1.7	0	0.0	0	0.0	3	5.1
585	Ochozska-H.	3	353	20	5.7	47	13.3	82	23.2	52	14.7	4	1.1	17	4.8	3	0.8	0	0.0	0	0.0	0	0.0	6	1.7	2	0.6
444	Ölknitz	3	1577	134	8.5	0	0.0	857	54.3	857	54.3	0	0.0	0	0.0	0	0.0	0	0.0	0	0.0	0	0.0	0	0.0	0	0.0
579	Pekárna; gh	2	911	50	5.5	0	0.0	234	25.7	148	16.2	21	2.3	32	3.5	5	0.5	4	0.4	2	0.2	0	0.0	22	2.4	3	0.3
179	Petersf.; 3AH2	4	259	2	0.8	0	0.0	115	44.4	102	39.4	12	4.6	1	0.4	0	0.0	0	0.0	0	0.0	0	0.0	0	0.0	1	0.4
174	Petersf.;1AH3A	3	40	0	0.0	0	0.0	17	42.5	11	27.5	6	15.0	0	0.0	0	0.0	0	0.0	0	0.0	0	0.0	0	0.0	0	0.0
128	Rheinfelden-E.	4	77	0	0.0	11	14.3	21	27.3	18	23.4	3	3.9	0	0.0	0	0.0	0	0.0	0	0.0	0	0.0	0	0.0	1	1.3
455	Saaleck	2	2021	6	0.3	3	0.1	1165	57.6	770	38.1	105	5.2	188	9.3	48	2.4	21	1.0	14	0.7	8	0.4	11	0.5	2	0.1
229	Schussen.; main	2	33	0	0.0	1	3.0	11	33.3	7	21.2	4	12.1	0	0.0	0	0.0	0	0.0	0	0.0	0	0.0	0	0.0	0	0.0
166	Schweizersbild	4	657	12	1.8	0	0.0	127	19.3	105	16.0	12	1.8	5	0.8	5	0.8	0	0.0	0	0.0	0	0.0	0	0.0	4	0.6
291	Sesselfelsg.; C2	2	117	0	0.0	5	4.3	71	60.7	46	39.3	17	14.5	7	6.0	1	0.9	0	0.0	0	0.0	0	0.0	0	0.0	3	2.6
261	Spitzbub.; AH2	2	20	0	0.0	1	5.0	1	5.0	1	5.0	0	0.0	0	0.0	0	0.0	0	0.0	0	0.0	0	0.0	0	0.0	0	0.0
416	Teufelsbr.; 2	2	44	4	9.1	0	0.0	24	54.5	23	52.3	1	2.3	0	0.0	0	0.0	0	0.0	0	0.0	0	0.0	0	0.0	0	0.0
417	Teufelsbr.; 3	3	1170	23	2.0	0	0.0	487	41.6	442	37.8	15	1.3	17	1.5	0	0.0	0	0.0	1	0.1	0	0.0	12	1.0	2	0.2
418	Teufelsbr.; 4	3	423	7	1.7	0	0.0	254	60.0	217	51.3	17	4.0	10	2.4	5	1.2	0	0.0	1	0.2	0	0.0	4	0.9	1	0.2
324	T. d.Chaleux; 5	2	3042	194	6.4	106	3.5	802	26.4	673	22.1	91	3.0	29	1.0	0	0.0	0	0.0	0	0.0	0	0.0	9	0.3	3	0.1
326	T. des Nutons	2	60	0	0.0	2	3.3	14	23.3	9	15.0	0	0.0	5	8.3	0	0.0	0	0.0	0	0.0	0	0.0	0	0.0	0	0.0
328	Trou du Frontal	2	178	1	0.6	11	6.2	39	21.9	35	19.7	3	1.7	0	0.0	0	0.0	1	0.6	0	0.0	0	0.0	0	0.0	3	1.7
335	Verlaine	2	143	18	12.6	9	6.3	15	10.5	9	6.3	2	1.4	0	0.0	1	0.7	0	0.0	0	0.0	3	2.1	0	0.0	1	0.7
79	Veyrier	3	593	0	0.0	39	6.6	207	34.9	131	22.1	0	0.0	50	8.4	0	0.0	17	2.9	0	0.0	0	0.0	9	1.5	5	0.8
475	Wallendorf	2	79	2	2.5	2	2.5	11	13.9	6	7.6	2	2.5	2	2.5	0	0.0	0	0.0	1	1.3	0	0.0	0	0.0	3	3.8
647	Wilczyce	2	375	0	0.0	0	0.0	85	22.7	84	22.4	0	0.0	1	0.3	0	0.0	0	0.0	0	0.0	0	0.0	0	0.0	0	0.0
378	Wildweiberlei	2	128	0	0.0	0	0.0	20	15.6	14	10.9	1	0.8	5	3.9	0	0.0	0	0.0	0	0.0	0	0.0	0	0.0	0	0.0
567	Žitného	1	533	7	1.3	16	3.0	234	43.9	232	43.5	0	0.0	2	0.4	0	0.0	0	0.0	0	0.0	0	0.0	0	0.0	1	0.2

(continued)

Table A.36 (continued)

Features TRU – SHP

ID	Site	T	nTool	TRU	%	PB	%	LR	%	TRI	%	SHP	%	ID	Site	T	nTool	TRU	%	PB	%	LR	%	TRI	%	SHP	%
364	Andernach; I	2	503	32	6.4	0	0.0	0	0.0	0	0.0	0	0.0	8	Croze-s.-Suran	1	314	0	0.0	0	0.0	0	0.0	0	0.0	1	0.3
365	Andernach; II	2	459	19	4.1	0	0.0	0	0.0	0	0.0	0	0.0	69	Douattes, Les	2	637	17	2.7	0	0.0	13	2.0	0	0.0	0	0.0
366	Andernach; III	2	172	10	5.8	0	0.0	0	0.0	0	0.0	0	0.0	601	Dzierzyslaw 35	1	912	0	0.0	0	0.0	115	12.6	114	12.5	0	0.0
82	Auzary-Thôn.	1	177	6	3.4	0	0.0	11	6.2	0	0.0	0	0.0	223	Felsställe; IIIa	2	229	29	12.7	0	0.0	53	23.1	0	0.0	0	0.0
590	Balcarka Cave	1	123	3	2.4	0	0.0	7	5.7	0	0.0	0	0.0	224	Felsställe; IIIb	3	598	55	9.2	0	0.0	198	33.1	0	0.0	0	0.0
15	B. de Cuiseaux	4	21	0	0.0	0	0.0	4	19.0	0	0.0	0	0.0	24	Freydières	4	35	0	0.0	1	2.9	23	65.7	0	0.0	0	0.0
73	Banges	3	82	0	0.0	0	0.0	3	3.7	0	0.0	0	0.0	195	Gnirshöhöle 1	2	34	3	8.8	2	5.9	3	8.8	0	0.0	0	0.0
66	Baume N.; M3	2	247	12	4.9	0	0.0	3	1.2	0	0.0	0	0.0	368	Gönnersdorf; I	2	1809	51	2.8	0	0.0	0	0.0	0	0.0	0	0.0
114	Birseck-Ermit.	4	197	0	0.0	0	0.0	0	0.0	13	6.6	0	0.0	369	Gönnersdorf; II	2	2380	51	2.1	0	0.0	0	0.0	0	0.0	0	0.0
320	Bois Laiterie	2	297	25	8.4	0	0.0	91	30.6	0	0.0	0	0.0	370	Gönnersdorf; III	2	893	17	1.9	0	0.0	0	0.0	0	0.0	0	0.0
205	Burgh. Dietf.	3	119	15	12.6	0	0.0	18	15.1	0	0.0	0	0.0	28	Grappin	1	175	12	6.9	0	0.0	30	17.1	0	0.0	0	0.0
219	Burkhardtsh.; V	3	78	11	14.1	0	0.0	19	24.4	0	0.0	0	0.0	93	H.-Champ.; pri.	2	594	7	1.2	0	0.0	16	2.7	0	0.0	0	0.0
104,5	Büttenloch A+B	3	59	2	3.4	0	0.0	35	59.3	0	0.0	0	0.0	95	H.-Champ.; 2	2	169	1	0.6	0	0.0	4	2.4	0	0.0	0	0.0
198	Buttentalhöhle	2	50	3	6.0	1	2.0	7	14.0	0	0.0	0	0.0	97	H.-Champ.; 1A	3	64	2	3.1	0	0.0	3	4.7	0	0.0	0	0.0
11	Calvaire; IV inf.	1	36	0	0.0	0	0.0	11	30.6	0	0.0	0	0.0	243	Hohle Fels; IIa	2	111	8	7.2	0	0.0	14	12.6	0	0.0	0	0.0
10	Calv.; III moy.	2	70	1	1.4	0	0.0	15	21.4	0	0.0	0	0.0	530	Hostim	3	1871	47	2.5	0	0.0	4	0.2	0	0.0	2	0.1
2	Campalou	1	100	8	8.0	0	0.0	0	0.0	0	0.0	0	0.0	599	Hranice	1	557	21	3.8	1	0.2	48	8.6	34	6.1	0	0.0
30	Chênelaz	3	235	8	3.4	0	0.0	12	5.1	0	0.0	0	0.0	63	Jean P. I; 9A+B	2	174	3	1.7	0	0.0	15	8.6	1	0.6	0	0.0
334	Coléoptère	3	125	9	7.2	0	0.0	8	6.4	0	0.0	0	0.0	64	Jean P. II	2	50	1	2.0	0	0.0	4	8.0	0	0.0	0	0.0
29	Colomb	3	264	53	20.1	0	0.0	46	17.4	0	0.0	0	0.0	543	Kamegg	1	309	4	1.3	1	0.3	14	4.5	0	0.0	4	1.3
290	Kastelhängh.	4	131	2	1.5	1	0.8	40	30.5	0	0.0	0	0.0		Petersf.;1AH3A	3	40	1	2.5	0	0.0	9	22.5	0	0.0	0	0.0
110	Kastelh. upp.	4	567	10	1.8	0	0.0	0	0.0	0	0.0	16	2.8	128	Rheinfelden-E.	4	77	2	2.6	0	0.0	4	5.2	0	0.0	0	0.0
271	Kaufert.; low.	2	225	13	5.8	2	0.9	17	7.6	0	0.0	0	0.0	455	Saaleck	2	2021	60	3.0	0	0.0	2	0.1	0	0.0	0	0.0
170	Kesslerloch	1	1775	2	0.1	0	0.0	0	0.0	0	0.0	0	0.0	229	Schussen.; main	2	33	2	6.1	0	0.0	6	18.2	0	0.0	0	0.0
451	Kniegrotte	1	2513	97	3.9	1	0.0	347	13.8	187	7.4	3	0.1	166	Schweizersbild	4	657	8	1.2	1	0.2	0	0.0	2	0.3	8	1.2
108	Kohlerhöhle	4	254	5	2.0	0	0.0	6	2.4	0	0.0	3	1.2	291	Sesselfelsg.; C2	2	117	0	0.0	2	1.7	4	3.4	0	0.0	0	0.0
570	Kolíbky, 6	2	20	2	10.0	0	0.0	6	30.0	0	0.0	0	0.0	261	Spitzbub.; AH2	2	20	0	0.0	0	0.0	6	30.0	0	0.0	0	0.0
572	Kůlna-H.; 6	4	144	17	11.8	0	0.0	12	8.3	0	0.0	0	0.0	416	Teufelsbr.; 2	2	44	2	4.5	0	0.0	0	0.0	0	0.0	0	0.0
101	Moosseedorf	4	5196	266	5.1	0	0.0	132	2.5	1	0.0	0	0.0	417	Teufelsbr.; 3	3	1170	31	2.6	0	0.0	0	0.0	0	0.0	0	0.0
629	Maszycka	1	66	3	4.5	3	4.5	19	28.8	0	0.0	0	0.0	418	Teufelsbr.; 4	3	423	5	1.2	0	0.0	0	0.0	0	0.0	0	0.0
635	Mosty	4	47	8	17.0	0	0.0	7	14.9	2	4.3	0	0.0	324	T. d.Chaleux; 5	2	3042	277	9.1	0	0.0	94	3.1	0	0.0	0	0.0
122	Munzingen	1	682	112	16.4	6	0.9	155	22.7	0	0.0	0	0.0	326	T. des Nutons	2	60	5	8.3	0	0.0	3	5.0	0	0.0	0	0.0
439	Nebra	2	2320	89	3.8	0	0.0	36	1.6	0	0.0	0	0.0	328	Trou du Frontal	2	178	22	12.4	0	0.0	6	3.4	0	0.0	0	0.0
90	N.-Monruz; 1M	2	1413	13	0.9	0	0.0	12	0.8	0	0.0	0	0.0	335	Verlaine	2	143	14	9.8	0	0.0	13	9.1	0	0.0	1	0.7
91	N.-Mon.; 2M/A	2	59	3	5.1	0	0.0	6	10.2	0	0.0	0	0.0	79	Veyrier	3	593	53	8.9	0	0.0	24	4.0	0	0.0	0	0.0
585	Ochozska-H.	3	353	28	7.9	0	0.0	35	9.9	0	0.0	0	0.0	475	Wallendorf	2	79	10	12.7	0	0.0	18	22.8	0	0.0	0	0.0
444	Ölknitz	3	1577	47	3.0	0	0.0	66	4.2	0	0.0	0	0.0	647	Wilczyce	2	375	14	3.7	0	0.0	64	17.1	0	0.0	0	0.0
579	Pekárna; gh	2	911	22	2.4	0	0.0	1	0.1	0	0.0	0	0.0	378	Wildweiberlei	2	128	4	3.1	1	0.8	68	53.1	0	0.0	0	0.0
179	Petersf.; 3AH2	4	259	0	0.0	0	0.0	61	23.6	0	0.0	0	0.0	567	Žitného	1	533	39	7.3	0	0.0	10	1.9	0	0.0	0	0.0

Abbreviations see list of abbreviations

Table A.37 LDA of spatial grouping (Magdalenian and Hamburgian)

Standardised canonical discriminant coefficients					
	Function 1	Function 2	Function 3	Function 4	Function 5
nSC	–0.437	–0.167	–0.426	0.11	0.111
nSSC	0.024	0.041	–0.094	0.089	–0.031
nRSC	–0.096	0.133	–0.017	0.134	–0.292
nBU	–0.672	–0.154	–0.587	0.226	0.299
nLBU	0.158	–0.421	0.536	0.01	0.288
nFBU	–0.067	0.004	0.01	0.146	0.51
nZIN	0.712	–0.659	–0.17	0.049	0.294
nBO	–0.477	–0.218	–0.045	0.124	0.293
nLBO	0.118	–0.42	–0.629	0.065	–0.322
nSP	–0.214	–0.467	0.069	–0.109	–0.084
nNDP	–0.062	0.148	–0.262	0.448	0.097
nBB	–0.864	–0.482	–0.76	–0.257	0.271
nBBD	–0.02	–0.21	0.431	0.576	–0.408
nBBT	–0.099	0.099	–0.29	–0.362	–0.171
nBBDT	–0.083	–0.252	–0.398	0.438	0.515
nBBTT	–0.083	–0.139	0.149	0	0.093
nBBC	–0.115	0.173	0.011	–0.243	0.154
nBBP	–0.323	–0.192	0.225	–0.379	–0.624
nBBDE	0.05	0.325	–0.053	0.142	0.026
nBP	–0.092	–0.072	0.11	0.543	–0.252
nTRU	0.006	–0.199	0.192	–0.191	0.042
nPB	–0.271	–0.058	0.041	0.339	–0.232
nLR	–0.727	–0.248	–0.499	0.246	0.299
nTRI	0.003	–0.127	–0.43	0.016	0.261
nSHP	0.727	0.204	–0.117	–0.085	–0.066

Table A.38 LDA of spatial grouping (Magdalenian and Hamburgian)

Means of groups					
Group	Function 1	Function 2	Group	Function 1	Function 2
1	–2	1.759	4	–1.505	–1.685
2	–2.176	0.441	5	–1.45	–0.441
3	–1.979	–0.377	6	9.11	0.303

Table A.39 LDA of spatial grouping (Magdalenian and Hamburgian)

Coordinates of cases								
ID	Function 1	Function 2	ID	Function 1	Function 2	ID	Function 1	Function 2
2	−1.181	2.576	243	−0.514	1.65	483	−0.752	−2.727
8	−1.136	2.751	246	−1.479	0.68	485	−0.629	−2.607
12	−1.26	2.217	247	−2.296	0.313	487	−1.334	−1.724
18	−2.917	1.481	271	−1.775	−1.278	489	−0.136	−3.432
28	−2.166	2.819	272	−2.425	−0.771	490	−1.563	−3.231
29	−1.456	1.679	281	−1.82	−0.019	491	−1.66	−2.391
30	−2.764	2.359	290	−2.181	1.71	518	−1.541	−0.011
37	−2.498	1.873	291	−3.394	0.937	530	−1.162	−1.388
44	−1.286	2.186	307	−2.007	0.228	543	−0.67	0.406
45	−1.5	1.708	313	−3.025	0.489	547	−2.276	−2.547
46	−1.619	2.007	320	−2.616	1.172	551	−2.445	−0.374
56	−2.724	0.42	324	−1.756	−0.859	567	−1.914	−0.28
59	−1.674	1.215	325	−1.945	−0.296	571	−1.468	0.208
63	−2.565	−0.336	328	−1.869	−0.66	572	0.449	−0.412
67	−2.352	2.078	330	−0.863	−0.801	579	−0.263	−2.779
69	−2.564	2.154	331	−0.86	0.396	585	−0.781	−1.155
79	−2.84	2.373	334	−3.16	0.605	590	−1.214	0.352
82	−2.315	1.086	335	−2.701	−2.107	594	−2.542	−0.788
90	−2.245	−0.093	344	−2.654	1.088	599	−1.273	−0.344
93	−2.44	1.007	346	−0.887	−0.323	601	−0.878	0.966
95	−2.071	1.073	347	−2.565	−1.501	647	−2.225	0.1
101	−3.455	1.249	351	−2.533	0.916	648	−1.583	−2.255
107	−1.224	1.639	364	−2.354	−0.394	H21	7.835	0.835
108	−0.268	2.366	365	−1.819	−1.781	H27	7.146	0.784
110	−0.563	2.292	366	−1.862	−1.223	H35	7.565	1.64
114	−2.114	0.978	367	−1.353	−0.598	H70	7.841	1.175
120	−2.261	1.889	368	−2.014	−2.147	H79	10.705	−0.303
122	−2.352	−0.405	369	−2.155	−1.713	H93	11.663	1.704
126	−3.361	−0.43	370	−2.103	−1.964	H95	10.62	0.007
132	−2.052	2.186	371	−1.709	−1.869	H97	10.513	−0.306
134	−0.526	2.868	378	−1.444	0.705	H99	10.012	−0.335
135	1.145	2.487	379	0.744	−0.27	H100	12.461	0.155
136	−1.05	1.115	401	0.112	−2.312	H101	8.938	−0.624
137	−2.039	1.633	417	−1.252	−0.923	H106	8.212	−0.55
139	−2.211	0.909	418	−0.264	−1.865	H114	9.324	0.388
164	−2.965	0.917	439	−2.506	−2.168	H122	9.651	−0.644
166	−0.516	−0.082	443	−1.259	−2.229	H124	9.307	−1.779
170	−0.827	−0.043	444	−3.031	−0.728	H125	9.618	0.642
179	−2.583	−0.38	451	−2.956	−0.9	H127	10.078	−1.958
199	−3.01	0.563	452	−0.272	−1.932	H130	9.149	−1.883
205	−1.163	0.852	455	−2.164	−2.119	H143	8.163	0.93
214	−3.389	−0.352	456	−1.708	−3.083	H161	7.773	1.234
217	−2.839	−1.335	468	−2.001	−1.5	H163	8.247	0.581
223	−2.367	0.567	471	−1.757	−0.993	H164	7.52	0.688
224	−3.046	0.598	478	−1.452	−3.158			
231	−2.059	0.836	480	−1.908	0.074			

Abbreviations see list of abbreviations

Table A.40 LDA of spatial grouping (Magdalenian and Hamburgian)

Coordinates of arrows								
Type	Funct. 1	Funct. 2	Type	Funct. 1	Funct. 2	Type	Funct. 1	Funct. 2
nSC	0.35	0.073	nSP	–0.345	–0.336	nBBP	–0.128	–0.402
nSSC	–0.029	0.001	nNDP	0.083	0.259	nBBDE	–0.231	–0.079
nRSC	0.412	0.054	nBB	–0.717	–0.151	nBP	–0.106	0.179
nBU	–0.151	0.044	nBBD	–0.351	–0.478	nTRU	0.236	–0.148
nLBU	–0.18	–0.39	nBBT	–0.411	–0.128	nPB	–0.097	–0.021
nFBU	–0.16	–0.199	nBBDT	–0.213	–0.419	nLR	0.027	0.264
nZIN	0.894	–0.271	nBBTT	–0.195	–0.151	nTRI	–0.114	0.099
nBO	–0.359	–0.272	nBBC	–0.135	–0.26	nSHP	0.927	0.125
nLBO	0.055	–0.618						

Table A.41 LDA of spatial grouping (Magdalenian)

Standardised canonical discriminant coefficients				
	Function 1	Function 2	Function 3	Function 4
nSC	0.017	–0.332	–0.166	0.05
nSSC	0.144	–0.11	–0.12	–0.045
nRSC	0.188	0.031	–0.244	–0.328
nBU	–0.028	–0.484	–0.409	0.246
nLBU	–0.536	0.564	0.097	0.332
nFBU	0.005	0.006	–0.055	0.481
nZIN	–1.007	–0.214	0.05	0.38
nBO	–0.179	0.145	–0.194	0.253
nLBO	–0.431	–0.627	–0.196	–0.412
nSP	–0.372	0.101	0.018	–0.182
nNDP	0.166	–0.255	–0.488	0.203
nBB	–0.431	–0.626	0.034	0.117
nBBD	–0.042	0.497	–0.576	–0.294
nBBT	0.058	–0.285	0.302	–0.176
nBBDT	–0.273	–0.316	–0.413	0.537
nBBTT	–0.097	0.154	0.004	0.039
nBBC	0.252	0.009	0.242	0.099
nBBP	–0.028	0.304	0.209	–0.71
nBBDE	0.358	–0.101	–0.098	0.083
nBP	0.01	0.236	–0.585	–0.177
nTRU	–0.229	0.193	0.281	0.007
nPB	0.113	0.147	–0.372	–0.189
nLR	–0.182	–0.339	–0.356	0.313
nTRI	–0.208	–0.423	–0.01	0.243
nSHP	0.067	–0.311	0.263	0.036

Table A.42 LDA of spatial grouping (Magdalenian)

Means of groups		
Group	Function 1	Function 2
1	–1.917	0.583
2	–0.834	–0.582
3	0.227	–1.578
4	1.738	0.785
5	0.786	0.792

Table A.43 LDA of spatial grouping (Magdalenian)

Coordinates of cases								
ID	Function 1	Function 2	ID	Function 1	Function 2	ID	Function 1	Function 2
2	−2.376	0.998	179	0.06	0.023	378	−0.084	−0.403
8	−2.636	−0.399	199	−1.234	−1.129	379	1.249	−1.217
12	−2.525	−0.338	205	−0.561	0.572	401	3.7	0.299
18	−2.254	0.145	214	−0.019	−1.219	417	1.095	0.326
28	−2.801	1.755	217	0.971	−1.384	418	2.7	1.568
29	−1.751	−0.794	223	−1.022	−1.134	439	1.813	−1.845
30	−2.294	0.175	224	−1.332	−1.443	443	2.078	0.762
37	−2.161	0.88	231	−1.001	0.825	444	0.622	0.06
44	−1.928	1.204	243	−1.719	−1.431	451	0.52	1.234
45	−1.555	1.86	246	−0.807	−0.948	452	3.121	−0.306
46	−1.957	−0.72	247	−0.8	−1.85	455	1.662	1.943
56	−0.395	0.491	271	1.117	−0.201	456	3.122	−0.438
59	−1.197	−0.384	272	0.363	−1.204	468	1.545	−0.17
63	0.324	1.897	281	−0.703	−1.649	471	0.768	2.417
67	−2.156	0.035	290	−1.962	0.495	478	3.166	−0.115
69	−2.378	1.424	291	−2.012	−1.129	480	−0.22	0.961
79	−2.747	0.076	307	−0.499	−0.663	483	2.816	3.253
82	−1.426	−0.439	313	−0.4	−0.3	485	2.68	0.687
90	0.253	0.303	320	−1.302	−1.69	487	1.993	1.153
93	−0.745	1.112	324	1.011	−0.948	489	4.061	2.043
95	−0.953	0.865	325	0.222	−0.802	490	3.011	1.08
101	−1.417	1.389	328	0.824	−1.32	491	2.115	1.359
107	−1.996	0.854	330	1.16	−1.476	518	0.002	0.419
108	−2.637	0.971	331	−0.132	−3.298	530	1.723	0.195
110	−2.243	0.969	334	−0.791	−0.541	543	−0.396	0.174
114	−1.302	1.054	335	1.237	−2.363	547	2.295	1.574
120	−2.375	0.388	344	−1.171	0.315	551	0.395	1.488
122	0.091	−0.338	346	0.352	−1.714	567	0.829	0.429
126	−0.097	−0.566	347	1.337	0.189	571	0.198	1.03
132	−2.077	−0.066	351	−1.108	−1.179	572	1.527	1.077
134	−2.474	0.722	364	0.003	−1.095	579	3.362	2.03
135	−1.77	2.16	365	1.889	−3.041	585	1.893	1.443
136	−1.47	0.703	366	1.117	−2.025	590	0.019	0.844
137	−1.448	−0.008	367	−0.141	−3.249	594	0.443	−2.154
139	−1.009	0.577	368	1.849	−3.228	599	1.355	2.213
164	−1.524	1.404	369	1.202	−2.266	601	0.141	1.327
166	0.322	1.072	370	1.576	−2.584	647	0.173	−0.916
170	0.059	1.009	371	1.454	−2.462	648	2.503	0.282

Abbreviations see list of abbreviations

Table A.44 LDA of spatial grouping (Magdalenian)

Coordinates of arrows					
Type	Function 1	Function 2	Type	Function 1	Function2
nSC	0.052	–0.155	nBBT	–0.123	–0.452
nSSC	–0.019	–0.326	nBBDT	–0.403	–0.381
nRSC	0.069	–0.069	nBBTT	–0.141	–0.199
nBU	0.04	0.026	nBBC	–0.255	–0.284
nLBU	–0.347	0.515	nBBP	–0.356	–0.049
nFBU	–0.183	0.23	nBBDE	–0.073	–0.255
nZIN	–0.672	–0.312	nBP	0.212	0.011
nBO	–0.288	0.042	nTRU	–0.155	0.184
nLBO	–0.606	–0.326	nPB	0.012	0.091
nSP	–0.294	0.438	nLR	0.276	0.041
nNDP	0.259	0.041	nTRI	0.064	–0.269
nBB	–0.156	–0.112	nSHP	0.191	–0.159
nBBD	–0.423	0.122			

Table A.45 LDA of spatial grouping (Magdalenian)

Observed and predicted groups								
ID	a priori	posteriori	ID	a priori	posteriori	ID	a priori	posteriori
2	1	1	179	2	2	378	3	3
8	1	1	199	2	2	379	3	3
12	1	1	205	2	2	401	4	5
18	1	1	214	2	3	417	4	4
28	1	1	217	2	2	418	4	4
29	1	1	223	2	2	439	4	3
30	1	1	224	2	2	443	4	4
37	1	1	231	2	1	444	4	3
44	1	1	243	2	2	451	4	5
45	1	1	246	2	2	452	4	4
46	1	1	247	2	2	455	4	4
56	1	2	271	2	2	456	4	4
59	1	1	272	2	2	468	4	4
63	1	5	281	2	2	471	4	4
67	1	1	290	2	2	478	4	4
69	1	1	291	2	2	480	4	1
79	1	1	307	2	2	483	4	4
82	1	1	313	3	1	485	4	4
90	1	1	320	3	2	487	4	4
93	1	1	324	3	3	489	4	4
95	1	1	325	3	2	490	4	4
101	1	1	328	3	3	491	4	4
107	1	1	330	3	3	518	4	5
108	1	1	331	3	3	530	4	4
110	1	1	334	3	1	543	5	2
114	1	1	335	3	3	547	5	5
120	1	1	344	3	1	551	5	5
122	1	3	346	3	3	567	5	5
126	1	2	347	3	4	571	5	5

(continued)

Table A.45 (continued)

Observed and predicted groups								
ID	a priori	posteriori	ID	a priori	posteriori	ID	a priori	posteriori
132	1	1	351	3	1	572	5	5
134	1	1	364	3	3	579	5	4
135	1	1	365	3	3	585	5	5
136	1	1	366	3	3	590	5	2
137	1	1	367	3	3	594	5	3
139	1	1	368	3	3	599	5	5
164	1	1	369	3	3	601	5	5
166	2	5	370	3	3	647	5	3
170	2	1	371	3	3	648	5	4

Abbreviations see list of abbreviations

Table A.46 LDA of temporal grouping

Standardised canonical discriminant coefficients			
	Function 1	Function 2	Function 3
nSC	0.565	−0.108	0.253
nSSC	0.342	0.130	−0.148
nRSC	−0.168	−0.026	−0.206
nBU	−0.300	−0.540	0.117
nLBU	0.141	0.264	−0.808
nFBU	−0.301	0.034	0.164
nZIN	−0.037	−0.032	−0.477
nBO	0.226	0.495	−0.144
nLBO	0.526	0.325	−0.005
nSP	−0.223	−0.217	0.332
nNDP	0.532	−0.056	−0.393
nBB	0.478	0.461	−0.118
nBBD	−0.014	−0.033	−0.442
nBBT	0.114	−0.350	0.085
nBBDT	−0.220	0.239	0.691
nBBTT	−0.451	0.171	0.196
nBBC	−0.324	−0.111	−0.459
nBBP	−0.099	−0.064	−0.331
nBBDE	0.038	−0.226	0.291
nBP	0.384	0.356	−0.067
nTRU	−0.329	0.338	0.039
nPB	0.103	−0.355	−0.363
nLR	0.500	0.212	0.067
nTRI	0.754	−0.326	−0.002
nSHP	0.782	−0.052	−0.516

Table A.47 LDA of temporal grouping

Means of groups		
Group	Function 1	Function 2
1	−0.091	−1.579
2	−1.324	0.118
3	−0.261	1.135
4	1.676	0.326

Table A.48 LDA of temporal grouping

Coordinates of cases								
ID	Function 1	Function 2	ID	Function 1	Function 2	ID	Function 1	Function 2
2	0.176	−1.485	122	−1.833	−1.443	366	−0.818	0.305
8	0.105	−1.936	128	0.35	0.837	368	−2.324	0.69
10	0.365	−0.878	166	1.2	−2.167	369	−1.982	0.104
11	0.353	−1.479	170	0.177	−1.898	370	−1.488	0.558
15	3.162	2.425	174	0.921	2.228	378	−0.402	−0.747
24	1.51	−0.146	179	1.622	1.302	416	−1.533	1.153
28	−0.852	−3.178	195	−0.752	−0.415	417	0.082	0.002
29	0.865	0.916	198	−0.44	0.69	418	0.257	1.117
30	0.705	1.062	205	1.117	2.033	439	−1.567	0.446
63	−1.31	−0.016	219	−0.538	1.021	444	−0.616	0.995
64	−2.903	−0.07	223	−0.62	1.245	451	2.499	−1.214
66	−0.695	−0.903	224	−0.314	0.787	455	−0.857	2.137
69	0.466	−0.437	229	−1.162	0.09	475	−2.136	−1.203
73	1.298	1.001	243	0.276	−0.118	530	−0.179	0.9
79	−1.066	0.718	261	−0.189	−0.092	543	1.536	−1.874
82	−0.426	−0.087	271	−1.749	−0.656	567	−0.764	0.704
90	−0.406	0.244	290	2.723	0.384	570	−1.527	−1.614
91	−0.353	0.96	291	−0.268	−1.095	572	1.55	1.435
93	−0.918	−0.521	320	−0.114	1.923	579	−0.007	0.353
95	−0.819	−0.953	324	−1.317	−0.332	585	0.036	0.133
97	−0.063	1.094	326	0.066	0.119	590	0.561	0.027
101	−0.719	0.392	328	−1.194	1.217	599	0.599	−3.385
104	−0.31	1.881	334	0.399	1.416	601	1.577	−3.014
108	5.57	0.395	335	−0.065	0.15	629	0.939	−2.592
110	3.157	0.511	364	−0.736	−0.128	635	2.516	0.816
114	2.547	−2.925	365	−2.196	0.52	647	−0.754	−0.436

Abbreviations see list of abbreviations

Table A.49 LDA of temporal grouping

Coordinates of arrows					
Type	Function 1	Function 2	Type	Function 1	Function 2
nSC	0.329	–0.117	nBBT	–0.03	–0.251
nSSC	0.034	0.256	nBBDT	–0.175	0.121
nRSC	–0.045	–0.109	nBBTT	0.006	0.11
nBU	–0.199	–0.64	nBBC	–0.26	0.056
nLBU	–0.303	0.049	nBBP	–0.249	0.054
nFBU	–0.167	0.065	nBBDE	–0.067	0.123
nZIN	–0.063	–0.039	nBP	0.324	0.315
nBO	0.097	0.307	nTRU	–0.283	0.221
nLBO	–0.074	0.142	nPB	0.046	–0.267
nSSP	–0.434	0.081	nLR	0.125	0.045
nNDP	0.074	0.075	nTRI	0.339	–0.433
nBB	–0.1	0.291	nSHP	0.454	–0.142
nBBD	–0.203	0.046			

Table A.50 Correspondance analysis. Raw data

Assemblages Hostim—Klementowice new												
Site	Hostim	Grappin	Balcarka Cave	Brillenhöhle upper	Brillenhöhle lower	Chênelaz	Hranice	Kastelhöhle Nord	Kaufertsberg lower	Kaufertsberg upper	Klementowice Exc.80	Klementowice new
Abr.	Hos	Gra	Bal	Bup	Blw	Che	Hra	Kas	Klw	Kup	Kl8	Kln
n Blades	100	71	84	100	100	100	82	100	100	40	100	33
G	4	1	5	2	2	1	5	1	2	2	5	5
to5	30	18	29	23	17	25	14	18	22	16	8	4
5to15	31	23	32	43	32	44	19	49	27	9	45	13
ov15	39	30	23	34	51	31	49	33	51	15	47	16
Ssm	38	33	36	55	43	42	22	48	44	21	5	2
Sfc	62	37	48	45	56	58	60	48	56	19	95	31
Sov	57	42	40	39	47	57	47	39	53	21	59	19
Spt	4	0	1	3	1	1	0	1	3	2	0	0
Sln	12	7	17	4	10	8	13	3	5	7	4	1
Sdr	11	8	15	22	18	6	6	21	14	1	20	6
Src	1	0	1	4	1	0	0	0	0	1	1	0
Sir	11	7	6	24	21	19	10	32	19	5	4	7
Sus	4	7	3	4	2	9	6	4	6	3	12	0
DR	93	66	79	91	91	98	79	94	84	34	84	32
DRn	7	5	5	9	9	2	2	6	12	6	15	1
DRp	32	12	7	2	3	28	12	10	5	3	26	8
DRh	29	28	43	76	71	45	39	57	57	23	39	15
DRph	31	26	28	8	14	25	28	27	21	8	19	9
Lip	43	24	44	83	93	26	34	88	92	35	63	25
Lil	10	8	3	4	1	9	9	1	4	3	6	1
Bbd	27	39	27	56	57	57	31	41	52	28	67	24
Bb	70	30	54	29	30	41	48	50	36	10	32	9
Sc	38	27	23	29	18	53	31	27	22	9	31	9
Ipt	40	32	36	9	12	54	52	12	1	4	15	4
BSc	5	3	8	6	4	7	12	4	3	3	1	1
EE	23	8	8	8	9	1	4	5	19	3	66	21
A80+	25	24	27	26	22	48	21	25	17	10	12	9
A7t8	31	25	35	44	48	30	34	43	37	17	30	4
A-7	44	22	22	28	30	21	27	32	46	12	58	20
Pfs	18	19	24	35	13	31	18	12	27	13	15	5
t	3	2	2	0	0	4	2	5	3	4	0	0

Table A.50 (continued)

Assemblages Lhota—Wilczyce												
Site	Lhota	Mokra V	Monruz	Oelknitz	Oelknitz Old	Oelknitz Young	Rislisberg-Höhle	Rydno II_59	Teufelsbrücke; 3	Teufelsbrücke; 4	Bois Laiterie	Wilczyce
Abr.	Lho	Mok	Mon	Oel	OeO	OeY	Ris	Ryd	Te3	Te4	Boi	Wil
n Blades	41	32	100	108	18	13	100	51	100	100	133	100
G	4	5	1	4	4	4	1	5	4	4	3	5
to5	21	5	32	9	3	1	17	11	41	32	46	5
5to15	14	15	35	35	8	2	44	20	32	43	39	34
ov15	6	12	33	64	7	10	39	20	27	25	48	61
Ssm	19	12	45	10	1	1	47	27	37	33	64	4
Sfc	22	20	55	98	17	12	53	24	63	67	68	96
Sov	20	15	6	69	12	9	49	30	64	66	60	53
Spt	1	1	2	0	0	0	1	0	0	0	2	0
Sln	8	5	15	9	2	2	10	5	12	7	12	4
Sdr	2	3	9	17	1	3	13	1	9	15	14	25
Src	0	0	0	0	0	0	0	0	1	0	4	1
Sir	2	5	8	6	2	0	20	14	11	6	41	12
Sus	8	3	6	7	1	0	7	1	3	6	0	5
DR	36	31	97	98	14	12	97	49	92	92	120	89
DRn	5	1	3	10	4	1	3	2	6	8	13	11
DRp	7	7	17	36	4	2	9	11	28	28	10	31
DRh	17	14	45	33	4	5	47	19	35	40	93	34
DRph	12	10	34	28	6	5	39	18	29	24	10	24
Lip	19	13	66	63	5	8	74	19	54	36	102	78
Lil	2	4	3	6	2	5	7	2	7	7	14	4
Bbd	6	12	54	65	11	9	34	22	57	38	70	80
Bb	35	20	37	36	7	4	62	29	43	61	59	14
Sc	19	10	20	33	8	4	25	33	35	48	21	16
Ipt	29	13	48	31	6	3	47	23	38	43	21	16
BSc	2	2	14	2	1	0	12	2	4	2	9	1
EE	0	7	6	48	5	6	11	2	12	15	14	69
A80+	21	14	22	28	8	10	25	25	22	34	64	13
A7t8	8	6	41	20	4	0	50	20	39	37	33	26
A-7	12	10	36	60	6	3	25	6	39	29	36	59
Pfs	0	3	29	31	4	4	15	14	25	25	57	24
t	5	0	3	3/4	3	4	5	5	3	4	3	3

Abbreviations see list of abbreviations

Table A.51 Correspondance analysis. Row-data, principal coordinates

Site	CA-axis 1	CA-axis 2	CA-axis 3	CA-axis 4	CA-axis 5
Hos	−0.037353	0.218314	−0.015250	0.034733	−0.133037
Gra	−0.115778	0.149698	0.040558	−0.024010	0.051980
Bal	−0.212867	0.051124	−0.085849	−0.047199	−0.080729
Bup	−0.084462	−0.380214	0.064569	0.015168	−0.000851
Blw	0.004873	−0.328407	−0.086462	0.046299	0.042379
Che	−0.227077	0.196981	0.168975	0.004478	0.141949
Hra	−0.152992	0.208846	−0.122525	−0.073964	0.138083
Kas	−0.093023	−0.244449	−0.095843	0.213637	0.030094
Klw	0.096943	−0.270596	−0.036491	0.024181	−0.063784
Kup	−0.095274	−0.246967	0.042102	−0.183768	−0.129985
Kl8	0.528880	0.093502	−0.009052	0.053647	−0.064650
Kln	0.482330	0.006142	0.047141	0.042723	0.061145
Lho	−0.389037	0.319259	−0.046182	0.064917	−0.266762
Mok	−0.074859	0.183040	0.000665	0.067579	−0.001749
Mon	−0.169839	0.004306	−0.195033	−0.229447	0.034053
Oel	0.430596	0.157958	−0.006113	−0.041883	0.023495
OeO	0.141659	0.258867	0.165467	0.007789	0.030579
OeY	0.290892	0.110370	0.189214	−0.067133	0.138059
Ris	−0.170645	0.008460	−0.194669	0.072901	0.064711
Ryd	−0.279895	0.141919	0.137176	0.131626	0.150687
Te3	−0.059039	0.113195	0.039053	−0.056093	−0.064095
Te4	−0.086229	0.218491	0.088997	0.088912	−0.065515
Boi	−0.132414	−0.274458	0.205110	−0.091415	−0.025373
Wil	0.608560	0.007643	−0.009339	−0.048721	0.049880

Table A.52 Correspondance analysis. Row-data, standrad coordinates

Site	Mass	QLT	INR	K=1	COR	CTR	K=2	COR	CTR
Hos	54	499	32	−37	14	1	218	484	59
Gra	38	518	16	−116	194	8	150	324	20
Bal	45	499	26	−213	472	30	51	27	3
Bup	52	827	57	−84	39	6	−380	788	175
Blw	51	754	43	5	0	0	−328	754	127
Che	54	588	50	−227	335	41	197	252	49
Hra	45	458	39	−153	160	16	209	298	46
Kas	51	495	42	−93	63	7	−244	433	71
Klw	52	716	36	97	81	7	−271	635	89
Kup	21	406	22	−95	53	3	−247	353	30
Kl8	55	918	104	529	890	229	94	28	11
Kln	18	801	32	482	801	63	6	0	0
Lho	22	630	53	−389	376	49	319	254	52
Mok	17	378	10	−75	54	1	183	324	13
Mon	50	194	44	−170	194	21	4	0	0
Oel	42	896	59	431	789	117	158	106	25
OeO	10	432	12	142	100	3	259	332	15
OeY	7	246	17	291	215	9	110	31	2
Ris	54	322	30	−171	321	24	8	1	0
Ryd	28	527	31	−280	419	32	142	108	13
Te3	54	268	19	−59	57	3	113	211	16
Te4	54	599	30	−86	81	6	218	518	60
Boi	71	557	71	−132	105	19	−274	452	124
Wil	55	970	127	609	970	305	8	0	0

Abbreviations see list of abbreviations

Table A.53 Correspondance analysis. Inertia

Axis	Inertia	Inertia-%	Cumulative-%	Axis	Inertia	Inertia-%	Cumulative-%
1	0.067293	40.21	40.21	13	0.001180	0.70	97.83
2	0.043008	25.70	65.90	14	0.000958	0.57	98.41
3	0.012398	7.41	73.31	15	0.000681	0.41	98.81
4	0.008824	5.27	78.58	16	0.000610	0.36	99.18
5	0.007581	4.53	83.11	17	0.000471	0.28	99.46
6	0.005540	3.31	86.42	18	0.000376	0.22	99.68
7	0.004443	2.65	89.08	19	0.000295	0.18	99.86
8	0.004032	2.41	91.49	20	0.000122	0.07	99.93
9	0.003041	1.82	93.30	21	0.000089	0.05	99.99
10	0.002814	1.68	94.98	22	0.000017	0.01	100.00
11	0.001901	1.14	96.12	23	0.000007	0.00	100.00
12	0.001687	1.01	97.13				

Table A.54 Correspondance analysis. Column-data, principal coordinates

Feature	Mass	QLT	INR	K=1	COR	CTR	K=2	COR	CTR
to5	31	529	41	−342	529	53	10	0	0
5t15	47	7	12	−17	6	0	−1	0	0
ov15	52	443	25	185	420	26	−43	22	2
Ssm	48	923	56	−379	736	102	−191	187	41
Sfc	82	913	27	202	755	50	93	158	16
Sov	66	215	24	58	56	3	97	159	15
Spt	2	234	12	−347	99	3	−404	134	6
Sln	12	421	20	−291	319	16	164	102	8
Sdr	18	530	21	261	349	18	−188	181	15
Src	1	364	13	−33	1	0	−871	364	18
Sir	20	586	39	−180	99	10	−398	487	74
Sus	7	242	18	23	1	0	315	241	17
DRn	10	298	16	230	196	8	−166	102	6
DRp	23	767	48	247	174	21	455	593	111
DRh	62	880	34	−100	108	9	−267	772	103
DRph	33	465	25	−92	66	4	227	399	39
Lip	81	852	58	86	61	9	−308	790	179
Lil	8	233	13	−127	59	2	219	174	9
Bbd	65	534	31	180	415	32	−97	120	14
Bb	58	640	43	−238	455	49	152	185	31
Sc	40	572	34	−147	153	13	244	420	55
Ipt	40	891	85	−319	287	61	463	604	200
BSc	7	374	22	−427	374	20	5	0	0
EE	25	958	161	1006	935	375	157	23	14
A80	37	410	40	−269	396	40	50	13	2
A7t8	46	359	22	−132	214	12	−108	145	12
A7	47	765	34	304	759	64	27	6	1
Pfs	31	117	27	−56	22	1	−117	95	10

Abbreviations see list of abbreviations

Table A.55 Correspondance analysis. Column-data, standard coordinates

Feature	CA-axis 1	CA-axis 2	CA-axis 3	CA-axis 4	CA-axis 5
to5	−1.317653	0.047239	0.738043	−1.437593	−2.824770
5t15	−0.064556	−0.007031	−0.148070	1.303013	0.269173
Ov15	0.711477	−0.205407	−0.495459	−0.369827	1.353832
Ssm	−1.460884	−0.922382	−0.080317	0.188977	−0.347964
Sfc	0.780472	0.447275	−0.062451	−0.212612	0.140548
Sov	0.223480	0.469946	0.963689	1.012685	−0.237217
Spt	−1.337221	−1.948483	−0.742686	−1.624319	−5.957889
Sln	−1.120737	0.792762	−1.236322	−2.604508	−1.259408
Sdr	1.005232	−0.905584	−0.828292	1.391018	−0.159373
Src	−0.128517	−4.201215	5.129543	−3.823437	−4.266479
Sir	−0.692647	−1.918427	1.115209	1.902221	1.993922
Sus	0.087666	1.519693	−1.353138	0.573604	−1.427378
DRn	0.885922	−0.798183	1.097890	0.167908	−3.241265
DRp	0.950977	2.196230	0.967519	−0.133724	−0.133178
DRh	−0.384645	−1.285151	0.127888	−0.012414	0.154849
DRph	−0.355945	1.095858	−1.684710	0.105589	0.750233
Lip	0.330072	−1.484348	−0.935957	0.035374	−0.377509
Lil	−0.490974	1.056000	1.862540	−1.273792	−0.150328
Bbd	0.694605	−0.466381	0.624968	−1.040242	0.957054
Bh	−0.919289	0.732800	−0.558433	1.165303	−1.256775
Sc	−0.567388	1.176001	1.088155	1.473673	0.496547
Ipt	−1.229429	2.231099	−0.954447	−0.985582	0.988073
BSc	−1.644127	0.022442	−2.840971	−3.330970	2.097538
EE	3.878934	0.755456	0.176400	0.049571	−0.832180
A80	−1.037101	0.238875	2.215246	0.145741	0.559007
A7t8	−0.507586	−0.522346	−1.367284	0.352342	0.373736
A7	1.173191	0.129530	−0.653699	−0.450031	−0.906612
Pfs	−0.217415	−0.564495	1.853450	−2.364725	0.621954

Table A.56 Animal representations in Central European Magdalenian assemblages

Site	Andernach	Balverhöh.	B. Noire	Býči skála	Cabônes	Colombière	Cr.-Suran	Gönnersd.	Goyet 3	Grappin	Groitz. C	Gudenush.	Hohlenst.	Hostim	Hoteaux	Kesslerl.
ID	364–67	377	68	551	54	14	8	368–71	332	28	487	542	269	530	33	170–71
G	3	3	1	5	1	1	1	3	3	1	4	5	2	4	1	2
Horse	25	1	1	–	–	8	–	72	–	–	4	–	1	2	–	12
Reindeer	1	–	–	–	–	7	–	5	–	–	–	1	–	–	–	1
Red deer	1	–	–	–	–	–	–	2	–	–	–	–	–	–	–	–
Moose	–	–	–	–	–	1	–	–	–	–	–	–	–	–	–	–
Cervidae indet.	–	–	–	–	–	3	–	10	–	–	–	–	–	–	1	2
Cervidae total	2	–	–	–	–	11	–	17	–	–	–	1	–	–	1	3
Musk ox	–	–	–	–	–	1	–	–	–	–	–	–	–	–	–	1
Bovidae indet.	–	–	–	–	–	–	–	14	–	1	–	–	–	–	–	–
Mammoth	2	–	–	–	?	2	–	74	–	3	–	–	–	–	–	–
Rhinoceros	–	–	–	–	–	4	–	17	–	–	–	–	–	–	–	–
Ibex	–	–	–	1[a]	1	1	–	8	1	–	–	–	–	–	–	–
Chaumois	–	–	–	–	–	1	–	–	–	–	–	–	–	–	–	–
Saiga	–	–	–	–	–	–	–	4	–	–	–	–	–	–	–	–
Bear	1	–	1	–	–	3	–	5	–	–	–	–	–	–	–	–
Seal	1	–	–	–	–	–	–	12	–	–	–	–	–	–	–	–
Bird	1	–	–	?	–	–	3	21	–	–	–	–	–	–	–	–
Wolf	–	–	–	–	–	–	–	3	–	–	–	–	–	–	–	–
Felinae	–	–	–	–	–	3	–	1	1	–	–	–	–	–	–	–
Fich	–	–	–	–	–	–	–	4	5	–	–	–	–	–	–	–
Animal indet.	2	–	–	1	–	–	–	9	–	–	–	–	–	1	–	3
References	2	10	5, 8	16	5	9	6	3	7	4	11	15	1	22	9	1

References: 1: Bosinski (1982), 2: Bosinski (1994), 3: Bosinski (2008a), 4: Cupillard and Welte (2006), 5: David (1996), 6: D'Errico and Laroulandie (1998), 7: Dewez (1987), 8: Escola (1989), 9: Faure (1978), 10: Günther (1988), 11: Hanitzsch (1972), 12: Höck (2000), 13: Klíma (1985), 14: Mania (1999), 15: Breuil and Obermaier (1908), 16: Oliva (1995), 17: Otte (1994), 18: Rozoy and Rozoy (2004), 19: Stahl Gretsch (2006), 20: Stocker et al. (2003), 21: Svoboda (1976), 22: Vencl (1991)

[a]Attribution to the Magdalenian questionable

Kniegrotte	M. Klause	Nebra	O. Klause	Oelknitz	Pekárna	Petersfels	Rigney 1	Rislisbergh.	R. Plate	R.-la-Tour	Saaleck	Schweizers.	Teufelsbr.	Děravá C.	T. Chaleux	T. Nutons	Urdhöhle	Veyrier	Total min.
451	289	439	287	444	579	173–93	77–78	124	336	313–16	455	166	415–18	515–16	324–25	326	499	79	
4	2	4	2	4	5	2	1	1	5	3	4	1	4	4	3	3	4	1	
1	1	–	–	2	10	11	–	–	2	4	1	6	6	2	2	–	–	–	174
–	–	–	–	–	1	4	–	–	–	–	–	–	–	–	2	–	–	–	22
–	–	–	–	–	–	–	–	–	–	1	–	?	–	–	–	–	–	–	4
–	–	–	–	–	–	–	–	–	–	1	–	–	–	–	–	–	–	–	2
–	–	–	–	–	–	1	2	–	–	1	–	5	–	–	–	–	–	–	25
–	–	–	–	–	1	5	2	–	–	3	–	5	–	–	2	–	–	–	53
–	–	–	–	–	–	–	–	–	–	–	–	–	–	–	–	–	–	–	2
–	1	–	–	–	4	–	–	–	–	3	–	–	–	–	1	1	–	–	25
–	–	–	1[a]	–	1	–	–	–	1	1	–	–	1	–	–	–	–	–	84
1	–	–	–	–	–	–	–	–	–	–	–	–	1	2	–	–	–	–	25
–	–	–	–	–	–	–	–	1	–	–	–	–	–	1	?	–	–	1	14
–	–	–	–	–	–	–	–	–	–	–	–	–	–	–	–	–	–	–	1
–	–	–	–	–	1	–	–	–	–	–	–	–	–	–	–	–	–	–	5
–	–	–	–	–	2	–	–	–	–	1	–	–	–	–	–	–	1	–	14
–	–	–	–	–	–	–	–	–	–	–	–	–	–	–	–	–	–	–	13
–	–	–	–	–	–	–	–	–	–	–	–	–	1	–	1	–	–	–	27
–	–	–	–	–	–	–	–	–	–	1	–	–	–	–	–	–	–	–	4
–	–	–	–	–	–	–	–	–	–	1	–	–	–	–	–	–	–	–	6
–	–	–	–	–	–	4	–	–	–	6	–	–	–	–	–	–	–	–	19
1	–	1	–	–	–	2	–	–	–	–	2	1	–	–	1	–	–	–	24
12	1	14	1	1	21	1	5	1	20	18	3	1, 3	1	13	17	7	1	19	

MIX
Papier aus verantwortungsvollen Quellen
Paper from responsible sources
FSC® C105338

If you have any concerns about our products,
you can contact us on
ProductSafety@springernature.com

In case Publisher is established outside the EU,
the EU authorized representative is:
Springer Nature Customer Service Center GmbH
Europaplatz 3, 69115 Heidelberg, Germany

Printed by Libri Plureos GmbH
in Hamburg, Germany